Samantha Rowell

WATER SUPPLY AND POLLUTION CONTROL

Seventh Edition

WATER SUPPLY AND POLLUTION CONTROL

Seventh Edition

Warren Viessman, Jr.
University of Florida
Mark J. Hammer
Lincoln, Nebraska

Upper Saddle River, NJ 07458

Library of Congress Cataloging-in-Publication Data
Viessman, Warren.
Water supply and pollution control / Warren Viessman, Jr., Mark J. Hammer.—7th ed.
p. cm.
Includes bibliographical references and index.
ISBN 0-13-140970-0
1. Water-supply—Management. 2. Water—Purification. 3. Sewage—Purification.
I. Hammer, Mark J., 1931- II. Title.

TD353.V54 2004
628.1—dc22

2004045395

Vice President and Editorial Director, ECS: *Marcia J. Horton*
Editorial Assistant: *Andrea Messineo*
Vice President and Director of Production and Manufacturing, ESM: *David W. Riccardi*
Executive Managing Editor: *Vince O'Brien*
Managing Editor: *David A. George*
Production Editor: *Donna King*
Director of Creative Services: *Paul Belfanti*
Art Director: *Jayne Conte*
Cover Designer: *Bruce Kenselaar*
Art Editor: *Greg Dulles*
Manufacturing Manager: *Trudy Pisciotti*
Manufacturing Buyer: *Lynda Castillo*
Marketing Manager: *Holly Stark*

About the Cover: Cover photo courtesy of Ron Franklin, University of Florida, Gainesville.

Pearson Prentice Hall
Pearson Education, Inc.
Upper Saddle River, NJ 07458

Printed in the United States of America

10 9 8 7 6 5 4 3 2 1

ISBN 0-13-140970-0

Pearson Education Ltd., *London*
Pearson Education Australia Pty, Ltd., *Sydney*
Pearson Education Singapore, Pte. Ltd.
Pearson Education North Asia Ltd., *Hong Kong*
Pearson Education Canada, Inc., *Toronto*
Pearson Educación de Mexico, S.A. de C.V.
Pearson Education—Japan, *Tokyo*
Pearson Education Malaysia, Pte. Ltd.
Pearson Education, Inc., *Upper Saddle River, New Jersey*

To Bette and Audrey for their help and support over these many years.

Contents

Preface

This seventh edition of *Water Supply and Pollution Control* has been updated and its coverage of topics expanded to meet the contemporary needs of civil and environmental engineering students. As we embark upon the twenty-first century, engineers responsible for providing safe water supplies to the inhabitants of this planet, and for treating wastes to render them reusable, will face many challenges. These include providing needed quantities of good-quality water for drinking and other household purposes, especially in water-short areas, and dealing with wastes that sometimes contain staggering levels of harmful substances. The engineers of tomorrow must be equipped to deal with a diversity of issues, such as forecasting future levels of population; estimating the potential for technological developments to reduce water requirements; recognizing that allocating water to meet human and other traditional water needs must also compete with water requirements for sustaining natural systems; exploring the impacts of climate change on local to global water supplies; and designing water supply and wastewater management systems to take into account technical, economic, environmental, social, legal, and political elements. The notion of continually striving to provide more water is giving way to one of husbanding this precious natural resource.

Water Supply and Pollution Control has been revised to include new material on standards, water and wastewater treatment processes, water distribution system analysis and design, water quality, advanced wastewater treatment for recycling, storm water management, and urban hydrology. In particular, there are major revisions of the chapters, or sections, on water supply and use (Chapters 3 and 4), water distribution (Chapter 6), hydraulics and hydrology of sewer and storm drainage systems (Chapter 7), monitoring of drinking water for pathogens (Chapter 8), membrane filtration (Chapter 10), disinfection/disinfection by-products rule (Chapter 11), biological treatment processes (Chapter 12), and indirect reuse to augment drinking water supply (Chapter 14). New topics, such as security of potable water supplies, the use of membranes in water treatment, and the application of Geographical Information Systems (GIS) to water supply and wastewater management problems, have been introduced. There are more practical examples, and many new problems have been added. Consistent with the original intent of the book, the emphasis is on the *application* of scientific methods to problems associated with the development, movement, and treatment of water and wastewater.

The book's tradition of presenting treatment processes in the context of what they can do, rather than in the context of water or wastewater treatment, is becoming more and more appropriate as we move toward the concept of total water management, recognizing that all waters are potential sources of supply. Water reuse is increasingly becoming an important national consideration. On the water supply side, more attention is paid to the sharing of water with natural systems and the impacts that this has on the quantities of water available for traditional water-using sectors, including public water supply. Many solved examples and homework problems serve to amplify the concepts presented in the text, and appropriate Web addresses have been provided where applicable.

Numerous sources have been drawn upon to provide subject matter for the book, and the authors have endeavored to provide suitable acknowledgment for them. The authors also wish to acknowledge the advice and assistance of students, professors, and practicing engineers who have reviewed and commented on previous editions. Particular recognition is given to those who helped prepare the manuscript for the seventh edition, namely, Audrey Hammer and Bette Viessman. We are indebted to them for their perseverance and understanding and for the excellent quality of their work.

WARREN VIESSMAN, JR.
MARK J. HAMMER

WATER SUPPLY AND POLLUTION CONTROL

Seventh Edition

CHAPTER 1

Introduction

OBJECTIVES

The purpose of this chapter is to:

- Summarize the evolution of water supply and pollution control practices
- Indicate the challenges faced by environmental engineers

1.1 DRINKING WATER SYSTEMS

The human search for pure water supplies must have begun in prehistoric times. Much of that earliest activity is subject to speculation. Some individuals may have led water where they wanted it through trenches dug in the earth. Later, a hollow log was perhaps used as the first water pipe.

Thousands of years probably passed before our more recent ancestors learned to build cities and enjoy the convenience of water piped into houses and wastes carried away by water. Our earliest archeological records of central water supply and wastewater disposal date back about 5000 years, to Nippur of Sumeria. In the ruins of Nippur there is an arched drain, each stone being a wedge tapering downward into place [1]. Water was drawn from wells and cisterns. An extensive system of drainage conveyed the wastes from the palaces and residential districts of the city.

The earliest recorded knowledge of water treatment is in Sanskrit medical lore and Egyptian wall inscriptions [2]. Sanskrit writings dating about 2000 B.C. tell how to purify foul water by boiling it in copper vessels, exposing it to sunlight, filtering it through charcoal, and cooling it in an earthen vessel.

There is nothing written concerning water treatment in the biblical sanitary and hygienic code of the early Hebrews, although three incidents may be cited as examples of the importance of fresh water. At Morah, Moses is said to have sweetened bitter waters by casting into them a tree shown him by God [3]. During the wandering in the wilderness, the Lord commanded Moses to bring forth water by smiting a rock [4]. At a much later date, Elisha is said to have "healed unto this day" the spring water of Jericho by casting "salt" into it [5].

The earliest known apparatus for clarifying liquids was pictured on Egyptian walls in the fifteenth and thirteenth centuries B.C. The first picture, in a tomb of the reign of Amenhotep II (1447–1420 B.C.), represents the siphoning of either water or settled wine. A second picture, in the tomb of Rameses II (1300–1223 B.C.), shows the use of wick siphons in an Egyptian kitchen.

The first engineering report on water supply and treatment was made in A.D. 98 by Sextus Julius Frontinus, water commissioner of Rome. He produced two books on the water supply of Rome. In these, he described a settling reservoir at the head of one of the aqueducts and pebble catchers built into most of the aqueducts. His writings were first translated into English by the noted hydraulic engineer Clemens Herschel in 1899 [2].

In the eighth century A.D., an Arabian alchemist, Geber, wrote a rather specialized treatise on distillation that included various stills for water and other liquids.

The English philosopher Sir Francis Bacon wrote of his experiments on the purification of water by filtration, boiling, distillation, and clarification by coagulation. This work was published in 1627, one year after his death. Bacon also noted that clarifying water tends to improve health and increase the "pleasure of the eye."

The first known illustrated description of sand filters was published in 1685 by Luc Antonio Porzio, an Italian physician. He wrote a book on conserving the health of soldiers in camps, based on his experience in the Austro-Turkish War. This was probably the earliest published work on mass sanitation. He described and illustrated the use of sand filters and sedimentation. Porzio also stated that his filtration method was the same as that of "those who built the Wells in the Palace of the Doges in Venice and in the Palace of Cardinal Sachette, at Rome" [2].

The oldest known archeological examples of water filtration are in Venice and the colonies it occupied. The ornate heads on the cisterns bear dates, but it is not known when the filters were placed. Venice, built on a series of islands, depended on catching and storing rainwater for its principal freshwater supply for over 1300 years. Cisterns were built and many were connected with sand filters. The rainwater ran off the house tops to the streets, where it was collected in stone-grated catch basins and then filtered through sand into cisterns.

A comprehensive article on the water supply of Venice appeared in the *Practical Mechanics Journal* in 1863 [6]. The land area of Venice was 12.85 acres and the average yearly rainfall was 32 in. Nearly all of this rainfall was collected in 177 public and 1900 private cisterns. These cisterns provided a daily average supply of about 4.2 gallons per capita per day (gpcd). This low consumption was due in part to the absence of sewers, the practice of washing clothes in the lagoon, and the universal drinking of wine. The article explained in detail the construction of the cisterns. The cisterns were usually 10–12 ft deep. The earth was first excavated to the shape of a truncated inverted pyramid. Well-puddled clay was placed against the sides of the pit. A flat stone was placed in the bottom and a cylindrical wall was built from brick laid with open joints. The space between the clay walls and the central brick cylinder was filled with sand. The stone surfaces of the courtyards were sloped toward the cistern, where perforated stone blocks collected the water at the lowest point and discharged it to the filter sand. This water was always fresh and cool, with a temperature of about 52 °F. These cisterns continued to be the principal water supply of Venice until about the sixteenth century.

Many experiments were conducted in the eighteenth and nineteenth centuries in England, France, Germany, and Russia. Henry Darcy patented filters in France and England in 1856, anticipating all aspects of the American rapid-sand filter except for coagulation. He appears to be the first to apply the laws of hydraulics to filter design [7]. The first filter to supply water to a whole town was completed at Paisley, Scotland, in 1804, but this water was carted to consumers [2]. In Glasgow, Scotland, filtered water was piped to consumers in 1807 [8].

In the United States, little attention was given to water treatment until after the Civil War. Turbidity was not as urgent a problem as it was in Europe. The first filters were of the slow-sand type, similar to British design. About 1890, rapid-sand filters were developed in the United States, and coagulants were later introduced to increase their efficiency. These filters soon evolved to our present rapid-sand filters.

1.2 DRAINAGE AND SEWERAGE SYSTEMS

The drains and sewers of Nippur and Rome are among the great structures of antiquity. These drains were intended primarily to carry away runoff from storms and for the flushing of streets. There are specific instances where direct connections were made to private homes and palaces, but these were the exceptions, for most of the houses did not have such connections. The need for regular cleansing of the city and flushing of the sewers was well recognized by commissioner Frontinus of Rome, as indicated in his statement, "I desire that nobody shall conduct away any excess water without having received my permission or that of my representatives, for it is necessary that a part of the supply flowing from the water-castles shall be utilized not only for cleaning our city, but also for flushing the sewers."

It is astonishing to note that from the days of Frontinus to the middle of the nineteenth century there was no marked progress in sewerage. In 1842, after a fire destroyed the old section of the city of Hamburg, Germany, it was decided to rebuild it according to modern ideas of convenience. The work was entrusted to an English engineer, W. Lindley, who was far ahead of his time. He designed an excellent collection system that included many of the ideas now in current use. Unfortunately, the ideas of Lindley and their influence on public health were not then widely recognized.

The history of the progress of sanitation in London probably affords a more typical picture of what took place in the middle of the nineteenth century. In 1847, following an outbreak of cholera in India that had begun to work westward, a royal commission was appointed to look into the sanitary conditions of London. This royal commission found that one of the major obstacles was the political structure, especially the lack of a central authority. The city of London was only a small part of the metropolitan area, comprising approximately 9.5% of the land area and less than 6% of the total population of approximately 2.5 million. This lack of a central authority made the execution of sewerage works all but impossible. The existing sewers were at different elevations, and in some instances the wastes would have had to flow uphill. In 1848, Parliament followed the advice of this commission and created the Metropolitan Commission of Sewers. That body and its successors produced reports that clearly showed the need for extensive sewerage works and other sanitary conditions [9]. Cholera appeared in London during the summer of 1848, and 14,600 deaths were recorded during 1849. In 1854, cholera claimed the lives of 10,675 people in London. The connection was established between a contaminated water supply and spread of the disease, and it was determined that the absence of effective sewerage was a major hindrance in combating the problem.

In 1855, Parliament passed an act "for the better local management of the metropolis," thereby providing the basis for the subsequent work of the Metropolitan Commission of Sewers, which soon after undertook the development of an adequate sewerage system. It will be noted that the sewerage system of London came about as a result of the cholera epidemic, as was true of Paris.

The remedy for these foul conditions was to discharge human excrement into the existing storm sewers and install additional collection systems. This suggestion created the combined sewers of many older metropolitan areas. These storm drains had been constructed to discharge into the nearest watercourse. The addition of wastes to the small streams overtaxed the receiving capacities of the waters, and many of them were covered and converted into sewers. Much of the material was carried away from the point of entry into the drains, which in turn overtaxed the receiving waters. First the smaller and then the larger bodies of water began to ferment, creating a general health problem, especially during dry, hot weather. The solution has been the varying degrees of treatment currently practiced according to the capabilities of the receiving stream or lake to take the load.

The work on storm drainage in the United States closely paralleled that in Europe, especially England. Some difficulty was experienced because of the difference in the rainfall patterns in America and England. English rains are more frequent but less intense. In the United States, storm drains must usually be larger for the same topographical conditions.

The enormous demands being placed on water supply and wastewater disposal facilities have necessitated the development and implementation of far broader concepts in environmental engineering than those envisioned only a few years ago. The standards for water quality have significantly increased concurrently with a marked decrease in raw-water quality. Evidence of water supply contamination by toxic and hazardous materials has become common and concern about broad water-related environmental issues has heightened. As populations throughout the world multiply at an alarming rate, environmental control becomes a critical factor. Land and water management become increasingly urgent. Many European and Asian nations have reached the maximum populations that their land areas can bear comfortably. They are faced with the problem of providing for more people than their lands can conveniently support. The lesson is that populations increase, but water and land resources do not. Consequently, the use and control of these resources must be nearly perfect to maintain our way of life.

Tomorrow's environmental engineers and scientists must be able to range beyond the technicalities of design, and become players in shaping the policies that will ultimately prescribe the types of designs that will be accepted by society. Imaginative and creative engineers are needed who can perceive and respect technical, nontechnical, and combination solutions to society's problems; who can set forth and assess viable alternatives; and who can understand what is implementable and what is not. Engineers and scientists must be prepared to take leadership roles in guiding those in decision-making capacities so that they create the best possible programs and regulations for managing the world's water resources. The challenge is to produce technically qualified individuals who can relate their knowledge to the realities of the political, economic, environmental, legal, and social setting in which all water-related problems must be solved. The future of this "one world" rests upon the decisions that environmental scientists, engineers, and others will contribute to, and on the actions that will flow from these decisions. An understanding of water supply and pollution control technology must be incorporated directly into today's environmental policy making.

REFERENCES

[1] W. Durant, *Our Oriental Heritage* (New York: Simon & Schuster, 1954), 132.

[2] M. N. Baker, *The Quest for Pure Water* (New York: American Water Works Association, 1949), 1–3, 6–11.

[3] Exodus 15:22–27.

[4] Exodus 17:1–7.

[5] 2 Kings 2:19–22 (King James Version).

[6] "The Water Cistern in Venice," *J. Franklin Inst.*, 3rd ser., 70 (1860): 372–373.

[7] H. Darcy, *Les Fontaines Publiques de la Ville de Dijon; Distribution d'Eau et Filtrage des Eaux* (Paris: Victor Dalmont, 1856).

[8] D. Mackain, "On the Supply of Water to the City of Glasgow," *Proc. Inst. Civil Engrs.* (London) 2 (1842–1843): 134–136.

[9] First Report of the Metropolitan Sanitary Commission (London: 1848).

CHAPTER 2

Water Management

OBJECTIVES

The purpose of this chapter is to:

- Identify contemporary water management issues
- Summarize environmental laws and regulations
- Indicate the likely direction of future water supply and pollution control efforts in the United States

Water management is multidimensional. It embraces planning, design, construction, operation, and maintenance. It must be conducted within the constraints of technology, social goals, laws and regulations, political viewpoints, environmental concerns, and economic realities. To be effective, water management must recognize and take advantage of interconnections between surface and groundwater bodies, exploit the potential for coordinated use of existing facilities, acknowledge that water quantity and quality are a single issue, devise new ways to operate old systems, blend structural and nonstructural approaches, accept that the expansiveness of water resources systems may require regional rather than local solutions to problems, and provide equity, insofar as is possible, if not on a monetary basis, at least on a service basis, to all those affected. In concept, water management is simple; the trouble is that the boundaries of the physical systems that must be dealt with often differ markedly from the political boundaries that affect how water is used and developed. Furthermore, many historical, social, legal, and organizational factors have been narrowly focused and constrain, if not preclude, good water management.

2.1 FROM PROJECTS TO ISSUES

In the early twentieth century, the construction of dams, waterways, water treatment plants, and wastewater treatment facilities in the United States was given priority. Irrigation works helped settle the West. Waterways improvements encouraged commerce and industry in populous areas of the East, South, and Midwest. Municipal water and wastewater systems provided the basis for increasing urbanization and industrial growth in many localities. Now, however, most of the nation's rivers have been subjected to engineering controls, and many old water policies are no longer valid. Furthermore, numerous facilities constructed in the past are reaching the end of their design lives, and the question of how to rehabilitate them is becoming important.

The maturity of our water infrastructure suggests that the exercise of good management practices be the basis for correcting deficiencies and making improvements. Broad issues, rather than the local interests that historically have been satisfied on a project-by-project basis, are moving to the forefront. A transition is under way. The question is whether this new outlook can hurdle the barriers created over the years.

Of interest is the fact that the National Water Commission's 1973 recommendations still stand as a model for modernizing water management policies [1]. The NWC's seven recurring themes are relevant to the subject of water supply and pollution control. They are

1. Future water demands are not inevitable but are the result of policy decisions within the control of society. Good planning should be based on a range of plausible alternative futures.
2. National priorities are shifting from developing water resources to restoring and enhancing water quality.
3. Water resources planning must be tied more closely to land use planning.
4. Water use efficiency should be emphasized and policies to encourage wise water use and conservation practices should be promoted.
5. Sound economic principles should be incorporated into decisions on whether to build water projects. Beneficiaries should pay for the costs of the services they receive and unjustified subsidies that distort allocation of scarce resources should be eliminated.
6. Laws and legal institutions should be reexamined in the light of contemporary water problems.
7. Development, management, and protection of water resources should be controlled at that level of government nearest the problem and most capable of effectively representing the vital interests involved.

2.2 INSTITUTIONS

The need for objective water management has long been recognized, but its implementation presents a problem. Successful application requires coordinating *human, agency, government*, and *special-group* interests and resolving conflicts among them. This institutional problem underlies most water issues. It is important because our institutions—governments, their agencies, laws and ordinances, regulations, and cultures—drive water resources planning and management programs. The influences of laws and regulations, political boundaries, agency missions, financing mechanisms, social customs, and the belief that water is free for the taking have all interacted to create a "water crisis" aura. Institutions are hard to change, but they cannot be ignored. Like facilities, they must be kept current. If we are not able to do this, outcomes will likely be inefficient water management at best, and widespread water shortages at worst.

Since the late 1980s, the notion of making "watersheds" the focal point for water resources management has been rekindled. But the concept is not new. The National Water Commission's 1973 report, "Water Policies for the Future," and the many studies that preceded it and followed it, has argued for integrated watershed management to be the foundation for managing water resources on all geographic scales [2–5].

In the United States, a number of federal agencies have responsibility for one or more aspects of water resources development and management. On the water quality side, the Environmental Protection Agency (EPA, www.epa.org) is the major player. On the water supply side, the U.S. Army Corps of Engineers (USACE, www.usace.army.mil), the U.S. Bureau of Reclamation (USBR), the Natural Resources Conservation Service (NRCS, www.nrcs.usda.gov), and the Tennessee Valley Authority (TVA, www.tva.gov) are the most significant. The Water Resources Council (WRC) was, until its demise in 1982, the principal body for coordinating federal and state water programs and assessing the state of the nation's waters. Several other agencies, the Economic Development Administration (EDA), Small Business Administration (SBA), Farmers Home Administration (FmHA), and the Department of Housing and Urban Development (HUD), assist rural and economically depressed areas in building and maintaining adequate water supply and wastewater disposal facilities. A brief description of the role of each of these agencies is given in Table 2.1.

2.3 INTEGRATED WATER MANAGEMENT

Many water problems cannot be solved in the context of traditional spatial and institutional boundaries. Recognizing this, several states have taken regional or watershed management approaches. Nebraska and Florida have established statewide management districts that have broad powers to manage water resources and provide facilities. The twenty-three Nebraska Natural Resources Districts and the five Florida Water Management Districts blanket their states and have similar powers, including the authority to levy property taxes [6,7]. Furthermore, federal, state, and local government agencies are beginning to adopt and implement holistic water management practices. These watershed-oriented approaches are based on flexible frameworks that specify guidelines, define the roles and responsibilities of key players, and permit the unique attributes of the watershed to dictate appropriate actions [8]. They are analogous to ecosystems approaches in that they consider the linkages between air, water, land, and the life forms resident within the systems' boundaries. Integrated watershed management means focusing on the appropriate spatial configuration (the right problem watershed); utilizing solid science and credible data; involving the key stakeholders in decision-making processes; and applying the concepts of "sustainable development" [8].

TABLE 2.1 Principal United States Water Resources Planning and Development Agencies

Agency	Mission
Army Corps of Engineers (CE)	Planning, constructing, operating, and maintaining a wide variety of water resources facilities, including those for navigation, flood control, water supply, recreation, hydroelectric power generation, water quality control, and other purposes. Nationwide activities.
U.S. Bureau of Reclamation (USBR)	Planning, constructing, operating, and maintaining facilities for irrigation, power generation, recreation, fish and wildlife preservation, and municipal water supply. Most activities are confined to the 17 western states. Original efforts were concentrated on irrigation.
Natural Resources Conservation Service (NRCS)	Carries out a national soil and water conservation program. Provides technical and financial assistance for flood prevention, recreation, and water supply development in small watersheds (fewer than 250,000 acres). Also appraises the nation's soil, water, and related resources.
Tennessee Valley Authority (TVA)	Planning, constructing, operating, and maintaining facilities in the Tennessee River Basin for navigation, flood control, and the generation of electricity. The TVA is a unique regional organization that has worked well in the United States.
U.S. Environmental Protection Agency (EPA)	Abatement and control of pollution. Provision of financial and technical assistance to states and local governments for constructing wastewater treatment facilities and for water quality management planning. Coordination of national programs and policies relating to water quality. Its principal role is regulatory.
U.S. Water Resources Council (WRC)	Principal role was the coordination of regional and river basin plans, assessing the adequacy of the nation's water and related land resources, suggesting changes in national policy related to water matters, and assisting the states in developing water planning capability. Although terminated in 1982, the WRC exemplified the long-sought mechanism for water program coordination and water policy analysis that was recommended by many study commissions since the early 1900s. A new organization with many of the WRC's roles is almost sure to come.
Economic Development Administration (EDA), Small Business Administration (SBA), Farmers Home Administration (FmHA), and the Department of Housing and Urban Development (HUD)	In the water resources field, the principal role of these agencies is to assist rural and economically depressed areas to develop and maintain water and wastewater conveyance, processing, and other related facilities. This is accomplished mainly through grant and loan programs.

2.4 ROADBLOCKS TO BE OVERCOME

Integrated water management is conceptually sound, and should be the goal, but there are a number of barriers that must be overcome if it is to be a reality. They include complexities associated with holistic water management planning; agency, interest group, and political boundaries (boundaries of authority and space); government, agency, and professional biases and traditions; the lack of effective forums for assembling and retaining stakeholders; the narrow focus, lack of implementation capability, poor public involvement, and limited coordination attributes of many water resources planning and management processes; the separation of land and water management, water quantity and water quality management, surface water and groundwater management, and other direct linkage actions; poor coordination and collaboration among state, local, and federal water-related agencies; limited ability to value environmental systems on monetary or other scales; the public's perception of risk as opposed to the reality of risk associated with water management options; suspicion regarding the formation of partnerships; and poor communications links among planners, managers, stakeholders, and others.

2.5 ENVIRONMENTAL REGULATION AND PROTECTION

Water pollution legislation originated in 1886, when Congress passed a bill that forbade the dumping of impediments to navigation in New York Harbor. In 1899, Congress passed the Rivers and Harbors Act, which prohibited deposit of solid wastes into navigable waters. These early concerns with water pollution were strictly in the interests of navigation. The Public Health Service Act of 1912 included a section on waterborne diseases, and the Oil Pollution Act of 1924 was designed to prevent oil discharges from vessels into coastal waters; such discharges could damage aquatic life. This act gave pollution enforcement authority to the federal government if local efforts failed; and it included a provision for matching grants for waste treatment facilities. Policy was strengthened with the Water Quality Act of 1965, which set water quality standards for interstate waters.

In 1966 attention to water quality sharpened, owing to the efforts of President Johnson. It was his position that entire river basins rather than localities should be considered in pollution control efforts. He proposed a "clean rivers demonstration program" in which the federal government would provide funds to interstate and regional water pollution control authorities on a first-ready, first-served basis. Those participating in the program would be required to have permanent water quality planning organizations, water quality standards, and implementation plans in effect for all waters of the basin designated.

The Clean Rivers Restoration Act of 1966 provided for a substantial increase in the level of funding appropriated for the construction of wastewater treatment facilities. But due to the Vietnam War, the construction grant program was not funded at the levels authorized.

After the Nixon administration took office in 1969, Congress prodded it to take action in the areas of water pollution control and environmental policy. This prodding was supported by the strong, and growing, environmental movement of the late 1960s. By 1970 the Nixon administration became convinced that there was a need for a massive federal investment in sewage treatment plant construction. In his February 1970 message on environmental quality, President Nixon proposed a four-year, $10 billion program of state, federal, and local investment in wastewater treatment facilities. The federal share of this investment was to be $1 billion per year. While this amount lagged actual authorized funding levels and was less than many environmental advocates desired, it was much more than any previous presidential request [9].

In 1970 the National Environmental Policy Act was passed (NEPA, 1969). The act was praised by President Nixon, who proclaimed that the three-member Council on Environmental Quality (CEQ) would be a great asset in informing the president on important environmental issues. The Nixon administration promptly put the provisions of the NEPA into effect. On March 5, 1970, the president issued an executive order instructing all federal agencies to report on possible variances of their authorities and policies with the NEPA's purposes. Then on April 30, 1970, the CEQ issued interim guidelines for the preparation of environmental impact statements.

In December 1970, as an outgrowth of the administration's environmental interests, a new independent body, the Environmental Protection Agency (EPA), was created. This organization assumed the functions of several existing agencies relative to matters of environmental management. It brought together under one roof all of the pollution control programs related to water, air, solid wastes, pesticides, and radiation. The EPA was seen by the administration as the most effective way of recognizing that the environment should be considered as a single, interrelated system. It is noteworthy, however, that the creation of the EPA made the separation of water quality programs from other water programs even more pronounced.

Even with the enactment of the NEPA, it was clear that a comprehensive response to water pollution issues was still lacking. It became evident during Congressional hearings in 1971 that, relative to the construction grants program, the program was underfunded. To rectify this situation, Congress passed the Water Pollution Control Act Amendments of 1972 (P.L. 92-500). Responding to public demand for cleaner water, the law ended two years of intense debate, negotiation, and compromise and resulted in the most assertive step taken in the history of national water pollution control activities.

The act (P.L. 92-500) departed in several ways from previous water pollution control legislation. It expanded the federal role in water pollution control, increased the level of federal funding for construction of publicly owned waste treatment works, elevated planning to a new level of significance, opened new avenues for public participation, and created a regulatory mechanism requiring uniform technology-based effluent standards, together with a national permit system for all point-source dischargers as the means of enforcement.

In the strategy for implementation, Congress stated requirements for achievement of specific goals and objectives within specified time frames. The objective of the act was to restore and maintain the chemical, physical, and biological integrity of the nation's waters. Two goals and eight policies were articulated.

Goals

1. To reach, wherever attainable, a water quality that provides for the protection and propagation of fish, shellfish, and wildlife, and for recreation in and on the water.
2. To eliminate the discharge of pollutants into navigable waters.

Policies

1. To prohibit the discharge of toxic pollutants in toxic amounts.
2. To provide federal financial assistance for construction of publicly owned treatment works.
3. To develop and implement area-wide waste treatment management planning.
4. To mount a major research and demonstration effort in wastewater treatment technology.
5. To recognize, preserve, and protect the primary responsibilities and roles of the states to prevent, reduce, and eliminate pollution.
6. To ensure, where possible, that foreign nations act to prevent, reduce, and eliminate pollution in international waters.
7. To provide for, encourage, and assist public participation in executing the act.
8. To pursue procedures that drastically diminish paperwork and interagency decisions on procedures and prevent needless duplication and unnecessary delays at all levels of government.

The act provides for achievement of its goals and objectives in phases, with accompanying requirements and deadlines. It was intended to be more than a mandate for point-source discharge control. It embodied an entirely new approach to the traditional way Americans had used and abused their water resources. Construction grants for publicly owned treatment works were made available to encourage full waste treatment management, providing for

1. The recycling of potential sewage pollutants through the production of agriculture, silviculture, and aquaculture products, or any combination thereof.
2. The confined and contained disposal of pollutants not recycled.
3. The reclamation of wastewater.
4. The ultimate disposal of sludge in a manner that will not result in environmental hazards.

These statutory provisions outline a long-term program to reduce water use, reduce the generation of wastes, and establish financially self-sustaining, public-owned pollution control facilities.

The 1972 amendments recognized the importance and urgency of the water quality management problem. It was estimated by the National League of Cities and the U.S. Conference of Mayors, for example, that a financial commitment of from $33 billion to $37 billion would be needed for water pollution control programs during the remainder of the 1970s [6]. The 1972 act committed the federal government to 75% of the costs associated with the construction of wastewater treatment facilities and authorized $18 billion of contract authority.

After passage of Public Law 92-500, there was a transition from researching the water pollution problem to implementing the solutions [10]. For example, Section 101 of the act states goals for fishable and swimmable waters and the prohibition of toxic discharges. These goals required that programs be implemented to reverse the threats that scientists had identified. The 1972 Clean Water Act provided the framework for a concerted effort to control water pollution. Contract authority to construct treatment facilities combined with meaningful enforcement procedures set in motion a policy to reverse the water quality-degrading practices of the past.

Not long after passage of the 1972 Clean Water Act, the Safe Drinking Water Act was passed (December 16, 1974). The purpose of that legislation was to ensure that water supply systems serving the public would meet minimum standards for the protection of public health. The act was designed to achieve uniform safety and quality of drinking water in the United States by identifying contaminants and establishing maximum acceptable levels. Prior to the Safe Drinking Water Act (SDWA), it was possible to prescribe federal drinking water standards only for water supplies used by interstate carriers. After the act was passed, the EPA established federal standards to control the levels of harmful contaminants in drinking water supplied by all public water systems. It also established a joint federal-state system for ensuring compliance with these standards. The major provisions of the SDWA act were

1. Establishment of primary regulations for the protection of the public health.
2. Establishment of secondary regulations that are related to taste, odor, and appearance of drinking water.
3. Establishment of regulations to protect underground drinking water sources by the control of surface injection.
4. Initiation of research on health, economic, and technological problems related to drinking water supplies.
5. Initiation of a survey of rural water supplies.
6. Allocation of funds to states for improving their drinking water programs through technical assistance, training of personnel, and grant support.

In 1977, in response to an indicated need to address deficiencies in the 1972 act, the Clean Water Act was revised. The salient points of the 1977 act included the following:

1. States were specifically mandated primacy over water quality and water use issues.
2. Municipalities were given evidence of a federal commitment in the form of construction grants and training assistance.
3. The public received assurances of the priority of water quality in the form of effective enforcement and incentive provisions for governments and industries to achieve the goal of fishable and swimmable waters.
4. Industry received the necessary extensions of compliance deadlines under the effluent discharge limitations provision.

5. Environmental groups witnessed the incorporation of a Resource Defense Council/EPA consent decree into the law that established toxic effluent standards and set forth a comprehensible process to implement effluent limitations [11].

In 1986, the Safe Drinking Water Act of 1974 was amended [12]. The principal changes were focused on groundwater protection. A wellhead protection program was established. The program provides that states undertaking wellhead protection efforts are eligible to receive federal grants to aid them in these endeavors. The EPA guidelines for wellhead protection are somewhat unique in that they allow regional flexibility, rather than prescribe uniform national standards. The act also provides for sole-source aquifer protection. The objective is to protect from contamination recharge areas that are primary sources of drinking water. Drinking water standards apply to these areas, and underground injection of effluent is regulated. Enforcement provisions of the act are strong, and in 1987 the first criminal conviction under the act was obtained [13]. The act was again amended in 1996. The changes focused water program spending on the contaminants that pose the greatest risk to human health and that are most likely to occur in a specified water system. Rather than focusing on certain contaminants, the law gives the EPA more authority to determine which contaminants to regulate. The amendments require that the best available scientific information and objective practices be used when proposing drinking water standards and they require that the EPA and the states begin to emphasize protection of source waters.

The Clean Water Act was reauthorized in 1987 as the Water Quality Act of 1987 [14]. A major feature of the 1987 act was the addition of the goal of controlling nonpoint sources of pollution. It is the most pronounced federal excursion into this important water quality management dimension. Agricultural fields, feedlots, and urban areas, including streets, are addressed. And while mandatory controls are not authorized, Congress did direct the states to conduct planning studies for the purpose of developing strategies for abating water pollution associated with nonpoint sources. A total of $400 million of federal funds was authorized to be used by the states to implement cleanup programs. Priority is to be given regulatory programs, innovative practices, and strategies that deal with groundwater contamination. The 1987 act provides for creation, by the states, of revolving funds to facilitate low-interest loans to local governments for sewage treatment improvements. It also provides more options for state and federal sharing of programs under the National Pollution Discharge Elimination System (NPDES). The EPA and the states can now divide the categories of discharges regulated within each state.

The Pollution Prevention Act of 1990 (P.L. 101–508) established the Office of Pollution Prevention within the EPA to coordinate agency efforts at source reduction. It created a volunteer program to improve lighting efficiency, thereby reducing energy consumption, and stated that waste minimization was to be the primary means of hazardous waste management. It also promoted voluntary industry reduction of hazardous waste and mandated a source reduction and recycling report to accompany the annual toxics release inventory.

In 1996, the Safe Drinking Water Act was again reauthorized. This was the first major revision to the act in 10 years. Upon signing the bill, President Clinton said the law "replaces an inflexible approach with the authority to act on contaminants of

TABLE 2.2 A Summary of Federal Environmental Legislation: 1948–1996

Year	Act
1948	Federal Water Pollution Control Act
1968	Wild and Scenic Rivers Act
1969	National Environmental Policy Act
1972	Federal Water Pollution Control Act Amendments
1972	Marine Protection, Research and Sanctuaries Act
1973	Endangered Species Act
1974	Safe Drinking Water Act
1976	Resources Conservation and Recovery Act
1976	Toxic Substances Control Act
1977	Clean Water Act
1980	Comprehensive Environmental Response, Compensation, and Liability Act
1984	Resources Conservation and Recovery Act Amendments
1986	Superfund Amendment and Reauthorization Act
1986	Federal Safe Drinking Water Act Amendments
1987	Clean Water Act Amendments
1990	Pollution Prevention Act
1996	Safe Drinking Water Act Amendments

greatest risk and to analyze costs and benefits, while retaining public health as the paramount value. Americans do have a right to know what's in their drinking water, where it comes from, before turning on their taps. Americans have a right to trust that every precaution is being taken to protect their families from dangerous, and sometimes even deadly contaminants." The SDWA amendments focus on funding related to contaminants that pose the greatest risk to human health and that are most likely to occur in a given water system. Rather than to prescribe the contaminants that the EPA is to focus on, the law gives the EPA latitude to select which contaminants to regulate, but requires it to use the best available scientific information and objective practices when proposing drinking water standards. The act also establishes a self revolving trust fund for drinking water systems, requires that water system operators be certified, maintains requirements for setting both a maximum contaminant level and a maximum contaminant level goal for regulated contaminants based on health risk reduction and cost/benefit analyses, and requires the EPA to establish a database to monitor the presence of unregulated contaminants in water.

As a result of the water pollution control efforts since the late 1960s, the once-rising tide of pollution has diminished. But there is still much to be done, particularly in the field of nonpoint pollution control. A summary of federal statutes governing or affecting water quality protection is given in Table 2.2.

2.6 EFFECTS OF ENVIRONMENTAL REGULATIONS

The intended result of Congressional passage of pollution control laws was for the EPA and the states to issue enforceable regulations to improve the quality of the nation's waters. Pollution control programs were generated at every level of government to implement regulations, issue permits, inspect regulated facilities, and enforce established rules. In response, industries and municipalities organized internal pollution control programs to stay abreast of regulatory requirements, work with plant personnel to attain compliance with regulations, learn about environmental monitoring and sampling techniques, and work with the regulatory agencies to obtain permits. In a sense, the 1970s was a period of institutionalization of the ideals of the environmental movement prevalent in the 1960s.

The President's Council on Environmental Quality, in its 1981 report, stated that water pollution controls were showing positive results in the United States [15]. The EPA has also reported success stories of rivers and lakes slowly returning to their natural state. The point-source program is now well established and appears to be working well, at least for industrial sources. But there is still need for improvement in the operation of municipal waste treatment facilities.

By law, all waters must have designated "beneficial uses" that must be protected and met. These uses establish the water quality criteria that must be considered in pollution control efforts. Using EPA guidelines, states apply a range of chemical, biological, habitat, and other parameters to establish criteria to protect specific designated uses. The EPA must approve the water quality standards that result and the states then apply them to determine the quality of their waters, consistent with supported uses. In 1990, it was reported that of 519,000 miles of streams assessed in 1988, 30% did not meet, or partially did not meet, the standards for their designated uses [11,16,17].

2.7 A LOOK TO THE FUTURE

During the Reagan administration, a review of regulations and regulatory practices was initiated to determine whether costs of pollution control could be reduced by modifying the regulatory approach [6]. This led to a debate about the relative merits of water quality standards versus technology-based standards, which had been the keystone of the Clean Water Act since 1972. *Water quality standards* establish a designated use for a specified section of a water body, which is then balanced with the maximum amount of waste the water body can assimilate. *Technology-based standards* are effluent limitations based on the levels of pollutant removal that can be achieved by modern wastewater treatment technology.

Congress initiated the technology-based approach in 1972 because the water quality–based approach of the 1960s had failed due to difficulties of enforcement and the limited availability of data for use in water quality models. The arguments in favor of a technology-based approach are as follows:

1. Technology-based standards are easy to enforce. This is important from an institutional perspective.
2. These standards are the first step toward the ultimate goal of zero discharge of pollutants to natural waters, as opposed to merely cleaning up waters to suit human objectives (the basis for water quality standards).
3. Knowledge and resources to set water quality standards for all pollutants and locations are insufficient. Technology-based standards are an interim approach to avoid pollution.

4. Nationwide uniformity in treatment standards minimizes economic dislocations.
5. The approach promotes equity among dischargers. No one should have the right to discharge more into the environment simply because of geographic location.

The Reagan administration, however, contended that while technology-based standards were important in the past in providing impetus for local governments and industry to clean up pollution from their treatment facilities, the agency now had the ability and sophistication to regulate discharged pollutants under water quality standards and that the Clean Water Act should be amended accordingly. The Reagan administration noted the following advantages of water quality standards [6]:

1. Water quality standards and the process by which they are adopted inherently encourage an assessment of costs and benefits, which is absent in the adoption and application of technology-based standards.
2. These standards foster scientific debate, which accelerates the advancement of the state of the art in predicting the fate and effect of pollutants.
3. The debate takes place in a local and state arena and heightens awareness on the part of local government, policy makers, and the public of the importance of water pollution control in their communities.
4. The assertion of the primary right and responsibility of states to regulate pollutants is essential to establishing the appropriate balance of power between the federal establishment and state governments.
5. Water quality–based decisions can avoid requirements of treatment for treatment's sake, which can result from application of technology-based standards [9].

For the present, it appears that technology-based effluent standards will continue to be the norm, even though they may be economically and socially inefficient [18]. But someday, a shift to water quality standards may gain stronger support, particularly as holistic water management becomes reality.

The Clean Water Act was last amended in 1987, despite subsequent efforts to revise it. During the 104th Congress, the Republican majority worked with business and industry lobbyists to write a bill that would ease restrictions on the discharge of a wide variety of industrial pollutants, but environmentalists and the Clinton administration opposed this thrust. Many complex issues surround amendment of the act, a particularly significant one being the control of nonpoint pollution. About two-thirds of all pollution stems from farms, construction sites, mining, forestry, and urban runoff. Solutions to the nonpoint problem may require tougher regulations on the use of pesticides and fertilizers, as well as new land use controls to protect watersheds [19].

Another major issue that must be addressed is that of the storage, treatment, and disposal of the hazardous and toxic wastes generated by our industrial society. Even though many manufacturers have reduced their use of hazardous and toxic materials, the volume of such wastes continues to increase. According to one source, the cleanup of all civilian and military hazardous waste sites could cost in excess of $750 billion over the next 30 years [20].

For the future, it appears that the American public is strongly committed to the goal of clean water. Billions of dollars have already been invested in water quality control programs, and this trend is expected to continue. Many of the easiest problems have been solved, however, and the future agenda will pose some significant political, legal, social, and economic challenges.

2.8 CONCLUSIONS

Integrated water management is the paradigm for the twenty-first century. But if it is to become common practice, full recognition will have to be given to the true spatial, environmental, and institutional dimensions of the problems of concern. The key stakeholders must all be involved in the planning process and the approach must be holistic to the extent that it can be so. Plans for integrated water management should drive water resources decision-making processes, and serve as the basis for developing regulatory programs. And preventive rather than remedial actions should be emphasized.

PROBLEMS

2.1 Do your state's water planning agencies take an integrated approach to water resources planning? Explain.

2.2 What are the principal statutes under which the EPA controls water pollution? Which aspects of water pollution may be regulated under each act?

2.3 Define point and nonpoint water pollution sources. Give five examples of each.

2.4 List three adverse health effects that can be caused by toxic chemical pollutants.

2.5 Identify the agencies in your state responsible for managing (a) water quality and (b) water quantity.

2.6 Do you believe the primary responsibility for pollution control should rest with the states, the federal government, or some mix? Explain your viewpoint.

2.7 Why do you think that integrated water management is difficult to achieve?

REFERENCES

[1] National Water Commission, *Water Policies for the Future* (Washington, DC: U.S. Government Printing Office, June 1973.)

[2] J. A. Ballweber, "Prospects for Comprehensive, Integrated Watershed Management Under Existing Law," *Water Resources Update,* Universities Council on Water Resources (Summer 1995), no. 100 (Carbondale, IL), 19–23.

[3] J. W. Bulkley, "Integrated Water Management: Past, Present, and Future," *Water Resources Update,* Universities Council on Water Resources (Summer 1995), no. 100 (Carbondale, IL): 7–18.

[4] R. E. Deyle, "Integrated Water Management: Contending with Garbage Can Decision Making in Organized Anarchies," *Water Resources Bulletin* (American Water Resources Association), 31, no. 3 (June 1995): 387–398.

[5] M. W. Hall, "A Conceptual Model for Integrated Water Management" (paper prepared for the workshop "Total Water Environment Management for Military Installations," U.S. Army Environmental Policy Institute, Atlanta, GA, 1996).

[6] W. Viessman, Jr., and C. Welty, *Water Management: Technology and Institutions* (New York: Harper & Row, 1985).

[7] A. Kovar, "Natural Resource Districts and Groundwater Quality Protection: An Evolving Role," in *Redefining National Water Policy: New Roles and Directions*, ed. Stephen M. Born, Special Publication No. 89-1 of the American Water Resources Association (1989): 51–72.

[8] L. P. Wise and J. Pawlukiewicz, "The Watershed Approach: An Institutional Framework for Action" (paper prepared for the workshop "Total Water Environment Management for Military Installations," U.S. Army Environmental Policy Institute, Atlanta, GA, 1996).

[9] B. H. Holmes, *History of Federal Water Resources Programs and Policies, 1961–1970*, U.S. Department of Agriculture Misc. Publ. No. 1379 (Washington, DC: U.S. Government Printing Office, September 1979).

[10] R. M. Linton, "The Politics of Clean Water," *Chemtech* (July 1982).

[11] U. S. Environmental Protection Agency, "National Water Quality Inventory: 1988 Report to Congress," Report No. EPA 440-4-90-003, Office of Water (Washington, DC, April 1990).

[12] 42 U.S.C. 300(f) et seq.

[13] *United States v. Jay Woods Oil Co. Inc.* (ED Mich. No. 87 CR20012 BC), unreported opinion cited in *Environmental Reporter*, 18, no. 6 (June 5, 1987): 502.

[14] Public Law 100-4: (Feb. 4, 1987).

[15] Council on Environmental Quality, *Environmental Quality 1981: The Twelfth Annual Report of the Council on Environmental Quality* (Washington, DC: U.S. Government Printing Office, December 1981).

[16] D. H. Moreau, "The Clean Water Act in Retrospect," *Water Resources Update,* Universities Council on Water Resources (Winter 1991), no. 84 (Carbondale, IL): 5–12.

[17] R. Savage, "The Clean Water Act: Accomplishments," *Water Resources Update,* Universities Council on Water Resources (Winter 1991), no. 84 (Carbondale, IL): 30–34.

[18] L. J. MacDonnell, "Restoring and Maintaining the Integrity of the Nation's Water: An Assessment of the Clean Water Act," *Water Resources Update,* Universities Council on Water Resources (Winter 1991), no. 84 (Carbondale, IL): 19–23.

[19] N. J. Vig and M. E. Kraft, eds., "The New Environmental Agenda," Chap. 17 in *Environmental Policy in the 1990s*, 3rd ed. (Washington, DC: CQ Press, 1997).

[20] M. Russell, E. W. Colglazier, and B. E. Tonn, "The U.S. Hazardous Waste Legacy," *Environment* 34 (July–August 1992): 12–14, 34–39.

CHAPTER 3

Water Resources Development

OBJECTIVES

The purpose of this chapter is to:

- Discuss the availability of water (surface water and groundwater sources)
- Define the water budget
- Indicate how water storage capacity can be determined
- Describe groundwater sources and development methods

Not many years ago, water resources management was focused almost exclusively on water supply, flood control, and navigation. Today, protecting the environment, ensuring safe drinking water, and providing aesthetic and recreational experiences compete equally for the allocation of water resources and for funds for water management and or development. An environmentally conscious public is pressing for improved management practices with fewer structural components to solve the nation's water problems. The notion of continually striving to provide access to more water has been replaced by one of husbanding this precious resource.

Water is located in all regions of the earth. The problem is that the distribution, quality, quantity, and mode of occurrence are highly variable from one locale to another.

The most voluminous water source is the oceans. It is estimated that they contain about 1060 trillion acre · ft of water [1]. The most valuable water supply (in terms of quality or freshness) is contained within the atmosphere, on the earth's surface, or underground. This supply, however, amounts to only about 3% of that contained in the oceans.

WATER QUANTITY

Water resources vary widely in regional and local patterns of availability. The supply is dependent on topographic and meteorological conditions as they influence precipitation and evapotranspiration. Quantities of water stored are dependent to a large

TABLE 3.1 Summary Data Concerning Water Resources of the Continental United States

	Square Miles	Acre · Feet($\times 10^6$)
Gross area of continental United States	3,080,809	—
Land area, excluding inland water	2,974,726	—
Volume of average annual precipitation	—	4,750
Volume of average annual runoff (discharge to sea)	—	1,372
Estimated total usable groundwater	—	47,500
Average amount of soil moisture	—	635
Estimated total lake storage	—	13,000
Total reservoir storage (capacity of 5000 acre · ft or more)	—	365

Source: E. A. Ackerman and G. O. Löf, *Technology in American Water Development* (Baltimore, MD: The Johns Hopkins Press, 1959).

extent on the physical features of the Earth and on the Earth's geological structure. Table 3.1 shows the major components of the water resources of the continental United States.

3.1 SOIL MOISTURE

Soil moisture is the most broadly used water source on the Earth's surface. Agriculture and natural plant life are dependent on it for sustenance. The quantity of water stored as soil moisture at any specified time is small, however. Estimates indicate an equivalent layer about 4.6 in in thickness distributed over 57 million mi^2 of land surface. This in itself would be insufficient to support adequate plant growth without renewal. It is therefore important that the frequency with which this supply is renewed and the length of time it remains available for use be known. The supply of soil moisture is dependent on geographical location, climatic conditions, geologic structure, and soil type. Variations may be experienced on a seasonal, weekly, or even daily basis.

It is considered that the natural supply of soil moisture in most of the agricultural areas of this country is less than the optimum for crop growth during an average year. It is evident, then, that a greater understanding of optimal water requirements for crops is essential if we are to economically and efficiently supply water artifically to overcome natural soil moisture deficiencies.

3.2 SURFACE WATERS AND GROUNDWATER

Surface waters are nonuniformly distributed over the earth's surface. Of the U.S. land mass, only about 4% is covered by rivers, lakes, and streams. The volumes of these freshwater sources depend on geographic, landscape, and temporal variations and on the impact of human activities.

For surface waters, historic records of stream flows, lake levels, and climatic data are used to identify trends and to indicate deficiencies in databases. Since surface water supplies are always in a state of transition, hydrologic models become valuable tools for estimating future water supply scenarios based on assumed sequences of hydrologic variables, such as precipitation, temperature, and evaporation and for projected physical manipulations of the surface water containment system. The verification of hydrologic models depends heavily on adequate historic data for testing, and where data voids exist every effort should be made to fill them.

Groundwater supplies are much more widely distributed than surface waters, but local variations are found as a result of the variety of soils, rocks, and geologic structures located beneath the land surface.

The importance of groundwater to the health and well-being of humans is well documented. Groundwater is a major source of fresh water for public consumption, industrial uses, and crop irrigation. More than half of the fresh water used in Florida for all purposes, for example, comes from groundwater sources, and about 90% of that state's population depends on groundwater for its potable water supply. The need to husband this resource is clear. Quantity and quality dimensions are both very important.

The need for an adequate database to determine the stock of groundwater and to estimate its change over time is very great because groundwater systems are not as easily defined as those for surface water. Groundwater storage volumes and transmission rates are affected by soil properties and geologic conditions, and these are often highly variable in space and not amenable to simple quantification.

3.3 RUNOFF DISTRIBUTION

Approximately 30% of the average annual rainfall in the United States is estimated to appear as surface runoff. The allocation of this water is directly related to precipitation patterns and thus to meteorologic, geographic, topographic, and geologic conditions. In the West, large regions are devoid of permanent runoff, and some localities, such as Death Valley, California, receive no runoff for years at a time. In contrast, some areas in the Pacific Northwest average about 6 ft of runoff annually. Mountain regions are usually the most productive of runoff, whereas flat areas, especially those experiencing lower precipitation rates, are generally poor runoff producers.

Runoff is distributed nonuniformly over the continental United States. It is subject to seasonal and annual variations influenced by climate and weather. For example, about 75% of the runoff in the semiarid and arid regions of the United States occurs during a period of a few weeks following snowmelt in the upper portion of the watershed. Even in the well-watered East, an uneven distribution of runoff prevails, and this has an impact on the availability of water for various competing uses. The four major runoff regions in the United States are depicted in Figure 3.1.

3.4 GROUNDWATER DISTRIBUTION

The usable groundwater storage in the United States is estimated to be about 48 billion acre · ft. This vast reservoir is distributed across the nation in quantities determined primarily by precipitation, evapotranspiration, and geologic structure. There are two components to this supply: one, a part of the hydrologic cycle; the other, water trapped underground in past ages and no longer naturally circulated in the cycle.

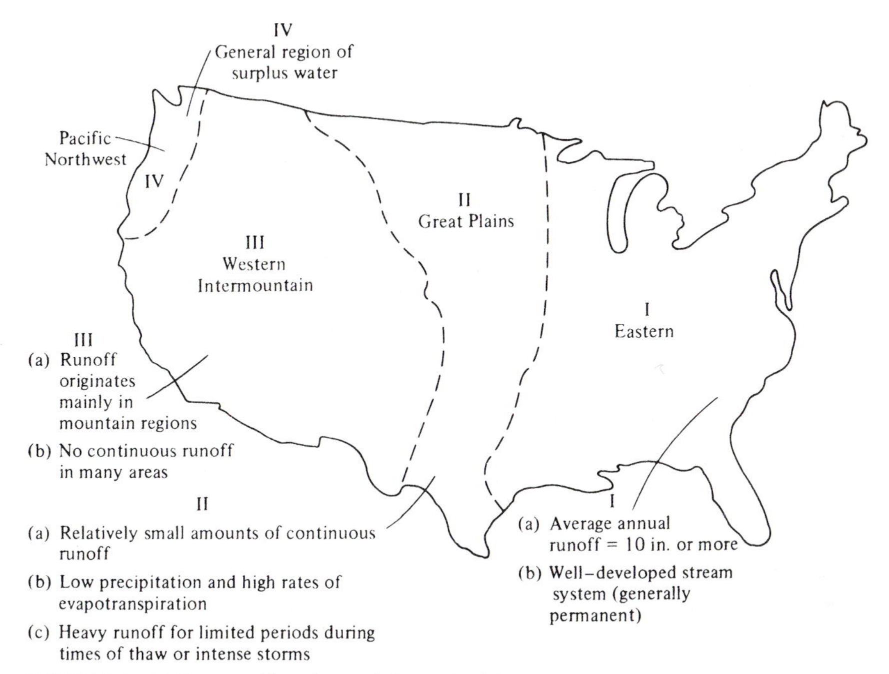

FIGURE 3.1 Major runoff regions of the United States.

Figure 3.2 shows the principal groundwater areas of the United States as depicted by Thomas [2]. Generally, it is evident that the mountain regions in the East and West, the northern Great Plains, and the granitic and metamorphic rock areas of New England and the southern Piedmont do not contain important groundwater supplies.

Aquifers may be classified into four categories:

1. Those directly connected to surface supplies that are replenished by gravitational water and that release water to surface flow. Gravels found in floodplains or river valleys are examples.
2. "Regional" aquifers occurring east of the 100th meridian. These aquifers produce some of the largest permanent groundwater yields and have moderate to high rates of recharge. Good examples are found in the Atlantic and Gulf coastal plain areas.
3. Low recharge aquifers between the 100th and 120th meridians. These aquifers have relatively little inflow compared to potential or actual drafts. Although storage volumes are often large, the low rate of replenishment indicates that the water must be considered more a minable material than a renewable resource. This possible "limited-life" category poses particular problems of development and management.
4. Aquifers subject to saline-water intrusion. These are usually found in coastal regions, but inland saline waters also exist, principally in the western states.

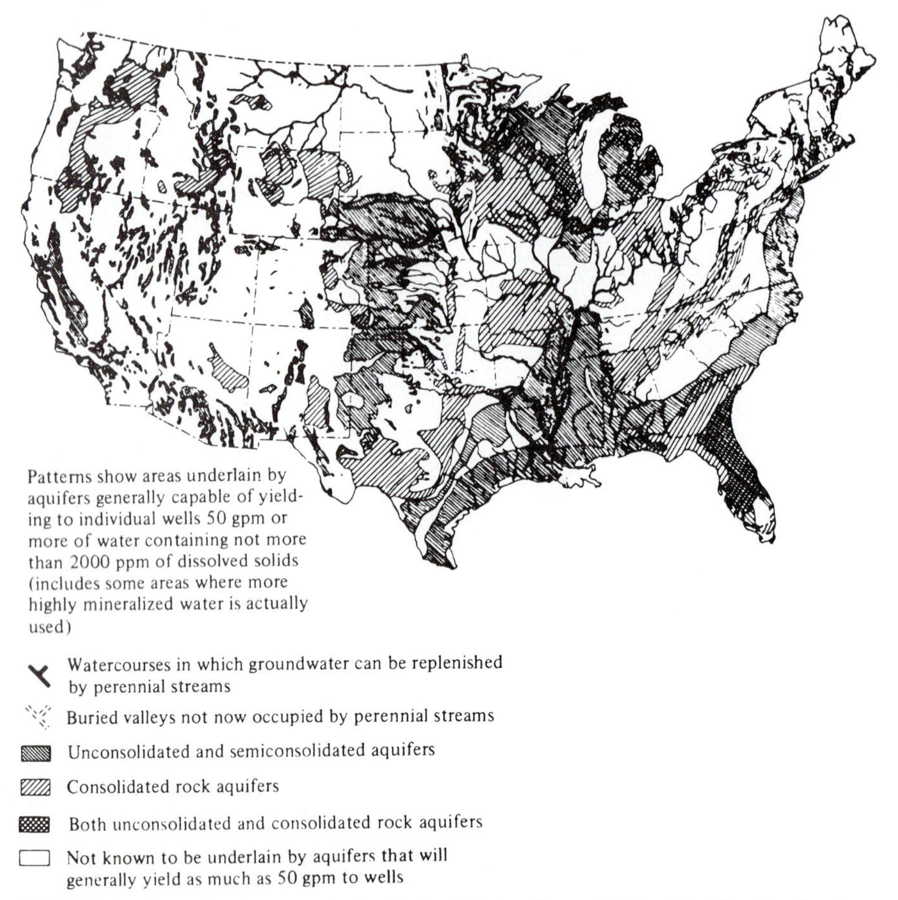

FIGURE 3.2 Groundwater areas in the United States. [From "Water," in *The Yearbook of Agriculture* (Washington, DC: U.S. Government Printing Office, 1955).]

WATER QUALITY

Although water quality and water quantity are inextricably linked, water quality deserves special attention because of its implications for affecting the public health and the quality of life. Even with the large federal investments in pollution control since 1972, the President's Council on Environmental Quality reports that the nation's waters continue to be damaged by pollution and misuse. Pollutants reach water bodies from both point and nonpoint sources. Municipal wastes, urban and agricultural runoff, and industrial wastes are principal offenders. Of special importance are the vestiges of past toxic and hazardous materials that are now being transported by surface water and groundwater systems. The impacts of polluting activities are widespread and they affect the public health, the economy, and the environment [3,4].

In 1991, the United States Geological Survey (USGS) established a National Water Quality Assessment (NAWQA) Program to develop long-term consistent and

TABLE 3.2 Some Waterborne Diseases of Concern in the United States

Disease	Microbial Agent	General Symptoms
Amebiasis	Protozoan (*Entamoeba histolytica*)	Abdominal discomfort, fatigue, diarrhea, flatulence, weight loss
Campylobacteriosis	Bacterium (*Campylobacter jejuni*)	Fever, abdominal pain, diarrhea
Cholera	Bacterium (*Vibrio cholerae*)	Watery diarrhea, vomiting, occasional muscle cramps
Cryptosporidiosis	Protozoan (*Cryptosporidium parvum*)	Diarrhea, abdominal discomfort
Giardiasis	Protozoan (*Giardia lomblia*)	Diarrhea, abdominal discomfort
Hepatitis	Virus (hepatitis A)	Fever, chills, abdominal discomfort, jaundice, dark urine
Shigellosis	Bacterium (*Shigella* species)	Fever, diarrhea, bloody stool
Typhoid fever	Bacterium (*Salmonella typhi*)	Fever, headache, constipation, appetite loss, nausea, diarrhea, vomiting, appearance of abdominal rash
Viral Gastroenteritis	Viruses (Norwalk, rotavirus, and other types)	Fever, headache, gastrointestinal discomfort, vomiting, diarrhea

Source: Courtesy of the Water Resources Research Center, University of Arizona, Tucson, AZ. Table appeared in *Arroyo* 10, no. 2 (March 1998).

comparable science-based information on national water-quality conditions. As of 2002, NAWQA assessments indicate that the waters in the United States are generally suitable for drinking water supplies, other human and recreational uses, and irrigation. Nevertheless, there are trouble spots. Protecting the nation's water resources from non-point sources emanating from pesticides, nutrients, metals, volatile organic chemicals, naturally occurring pollutants, and other contaminants remains a major problem [5].

Water-borne pathogenic microbes (Table 3.2) and pharmaceuticals are also potential threats to our water supplies. Various instances of gastrointestinal illnesses have been blamed on microbial pathogens in drinking water. In 1998, the Center for Disease Control and Prevention estimated that the annual number of deaths related to microbial illnesses associated with drinking water range from about 900 to 1000, suggesting the existence of a serious problem [6]. Microbial contaminants such as *Cryptosporidium, Giardia, Legionella*, and the Norwalk virus are among the culprits.

Pharmaceuticals have also been identified as a source of water pollution. Antibiotics, antidepressants, birth control pills, pain killers, and many other drugs have been identified in water bodies. Concern about these potential contaminants emerged in Europe in the 1980s, and during the 1990s it surfaced in the United States as well. As a result, the USGS has included the occurrence of human and veterinary pharmaceuticals in its assessment of emerging contaminants found in selected streams. The extent of threat these contaminants pose to humans and animals is not well known at present, but research is underway to better define the scale of this potential problem [7].

Finally, the threat of introduction of harmful substances into water supplies through acts of terrorism must be taken into consideration. Safeguards for water treatment plants and water supply reservoirs are called for.

3.5 GROUNDWATER

Groundwater quality is influenced by the quality of its source. Changes in source waters or degraded quality of source supplies may seriously impair the quality of the groundwater supply. Municipal and industrial wastes entering an aquifer are major sources of organic and inorganic pollution. Large-scale organic pollution of groundwaters is infrequent, however, since significant quantities of organic wastes usually cannot be easily introduced underground. The problem is quite different with inorganic solutions, since these move easily through the soil, and once introduced are removed only with great difficulty. In addition, the effects of such pollution may continue for indefinite periods since natural dilution is slow and artificial flushing or treatment is generally impractical or too expensive. The number of harmful enteric organisms is generally reduced to tolerable levels by the percolation of water through 6 or 7 ft of fine-grained soil [8]. However, as the water passes through the soil, a significant increase in the amounts of dissolved salts may occur. These salts are added by soluble products of soil weathering and of erosion by rainfall and flowing water. Locations downstream from heavily irrigated areas may find that the water they are receiving is too saline for satisfactory crop production. These saline contaminants are difficult to control because removal methods are very expensive. A possible solution is to dilute with water of lower salt concentration (wastewater treatment plant effluent, for example) so that the average water produced by mixing will be suitable for use.

3.6 SURFACE WATER

The primary causes of deterioration of surface water quality are municipal and domestic wastewater, industrial and agricultural wastes (organic, inorganic, heat), and solid and semisolid refuse. A municipality obtaining its water supply from a surface body may find its source so fouled by wastes and toxic chemicals that it is unsuitable or too costly to treat for use as a water supply. Fortunately, waste products discharged by cities and industry are being controlled at the point of initiation. This has been borne out by successes in cleaning up such watersheds as the Delaware and Susquehanna in the eastern United States.

HYDROLOGY AND WATER MANAGEMENT

In the past, water resources engineers and planners focused mainly on the developmental aspects of water resources: building dams, canals, distribution networks, and other infrastructure elements. Today, management is the byword and it centers on applying both structural and nonstructural measures to problems related to water supply, water allocation, water quality, floods, droughts, and environmental protection. The emphasis is on innovative technical approaches to problem solving, a more holistic view of water management, and improved operating efficiency of the water and wastewater systems already in place. But regardless of whether the focus is conservation, protection, or development, good water management requires an understanding of the hydrology and quality of the relevant water sources.

3.7 THE WATER BUDGET

In theory, accounting for the water resources of an area is simple. The basic procedure involves the evaluation of each component of the water budget so that a quantitative comparison of the available water resources with the known or anticipated water requirements of the area can be made. In practice, however, the evaluation of the water budget is often quite complex, and extensive and time-consuming investigations are generally required [9].

Both natural and human-induced gains and losses in water sources must be considered. The principal natural gains to surface water bodies are those resulting from direct runoff caused by precipitation and effluent seepage of groundwater. Evapotranspiration (combined losses from evaporation from water surfaces and transpiration from plants) and unrecovered infiltration are the major natural losses. Dependable dry-season supplies can be increased through diversion from other areas, through low-flow augmentation, through saline-water conversion, and perhaps in the future through induced precipitation. The major human-induced losses are from diversion of flows out of the watershed.

Once the gross dependable water supply has been estimated, the net dependable supply can be determined by subtracting the quantity of water used, detained, or lost as a result of human activities from the gross supply. When water is withdrawn from a flowing stream, a decrease in flow between the point of withdrawal and the point of return is experienced. As the water is used, part of it is lost to the atmosphere through various consumptive uses that are cumulative downstream. Decreases in dependable water supply along a watercourse are the result of withdrawals of water that occur upstream.

Although a water supply may be adequate for present needs, it may not be capable of supporting future demands for the resource. Forecasts of future water requirements are needed as are those of changes in dependable water supplies. Factors considered in making such forecasts include population, industrial development, agricultural practices, water policy, technology, and water management practices. More will be said about this in Chapter 4. Very crudely, the water budget may be represented by the following equation:

$$I - O = \Delta S \tag{3.1}$$

where the inflows (I) are all sources of water, natural and human-made, entering the region; the outflows (O) are all movements of water out of the region, including evaporation, transpiration, seepage, and stream flows; and the change in storage (ΔS) is the increase or decrease in storage over time for all natural (surface and underground) and all artificial reservoirs.

Consumption use has been defined by the American Water Works Association as water used in connection with vegetative growth, food processing, or incidental to an industrial process, which is discharged to the atmosphere or incorporated in the products of the process [10]. In short, it is water that is not returned to the watershed for potential reuse.

Withdrawal use is the use of water for any purpose that requires that it be physically removed from the source. Depending on the use to which the water is put, some of it may be returned (after use) to the original source and be available for reuse.

Nonwithdrawal use is the use of water for any purpose that does not require that it be removed from the original source. Water used for navigation and providing support for fish and wildlife are examples.

Certain water losses, although not "consumptive" by definition, may have the effect of reducing an available water supply. For example, dead storage (storage below outlet elevations) in impoundments is unavailable for downstream use. Diversion of water from one drainage basin to another represents an additional form of *nonconsumptive loss*. An example of this is the use of Delaware River basin water for the municipal supply of New York City. This decreases the total flow in the Delaware River below the point of diversion. New York is required, however, to augment low flows through compensating the downstream interests for diversion losses. Water contaminated or polluted during use to the extent that it cannot be economically treated for reuse also constitutes a real loss from the total water supply.

SURFACE WATER SOURCES

Surface water supplies may be categorized as perennial or continuous unregulated rivers, rivers or streams containing impoundments, or natural lakes. Evaluation of the capability of a region's surface water resources to sustain various uses requires assembly of data on the climate, hydrology, geology, and topography of the area. Information on industrial, agricultural, and residential development (population centers) is also needed as are forecasts of future changes in these categories. An assessment of the region's natural resources and the impact their development would have on the watershed's hydrology and economy is also of value.

3.8 BASIN CHARACTERISTICS AFFECTING RUNOFF

Important natural features of a watershed that affect stream flow are topography and geology. Topography determines the slopes and location of drainage channels and the storage capacity of the basin. Channel slope and configuration are directly related to the rate of flow in a basin and the magnitude of peak flows. A steep watershed generally indicates a rapid rate of runoff with little storage, whereas relatively flat areas are subject to considerable storage and lower rates of flow.

Prevailing soil types determine infiltration capacity and the ability of underground strata to transmit and hold groundwater. A thorough understanding of underground formations is a prerequisite for assessing groundwater storage potential.

3.9 NATURAL AND REGULATED RUNOFF

Natural runoff is defined as runoff that is unaffected by any other than natural influences. Runoff subject to withdrawals by humans or by artificial storage is defined as *regulated runoff*. If an unregulated stream is developed as a primary water source, the safe yield to be expected will be approximately the lowest dry-weather flow of the stream. Under this condition, users will always have an adequate supply, provided that their maximum requirements do not exceed the minimum flow. If demands exceed the lowest dry-weather flow during any period of time, a water shortage will occur unless supplementary water supplies can be made available.

Regulated runoff is normally the type of runoff for which information is available, although many streams for which runoff records are at hand were unregulated at the time they were first gauged. Stream-flow records are affected by artificial regulation from upstream storage works or by the diversion of flows into or out of the stream at points above a gauging station. Withdrawals from watersheds or diversions into them from outside sources affect the watershed's hydrology. Fortunately, diversions and withdrawals usually lend themselves to accurate estimates, since they are for specific purposes and are gauged or can be measured with little difficulty.

The safe yield of a stream that is regulated approaches the average annual flow as storage approaches full development. Economical yields are between the safe yields for unregulated and fully regulated flows. Safe yields of 75% to 90% of the mean annual flow can often be developed. Through regulation, the greatest benefits are normally derived from a stream since allocations of water for various uses can be made in a more nearly optimal manner.

3.10 STORAGE

Water may be stored for single or multiple purposes such as navigation, flood control, hydroelectric power, irrigation, municipal water supply, pollution abatement, recreation, and flow augmentation. Either surface or subsurface storage can be utilized, but both necessitate the use of a reservoir or reservoirs.

Reservoirs regulate stream flow for beneficial use by storing water for later release. The term *regulation* can be defined as the amount of water stored or released from storage in a period of time, usually one year. The ability of a reservoir to regulate river flow depends on the ratio of its capacity to the volume of river flow. Evaluation of the regulation provided by existing storage facilities can be made by studying the records of typical reservoirs. Information on the usable capacity, detention period, and annual regulation of a number of reservoirs having detention periods from 0.01 to 20 yr is given by Langbein [11].

About 190 million acre · ft of water, representing approximately 13% of the total river flow, has been made available through reservoir storage development in the United States [11]. The degree of storage development is variable but is generally greatest in the Colorado River basin and least in the Ohio River basin. Substantial increases in water supply can be attained through the development of additional storage, but water regulation of this type follows a law of diminishing returns. There are limitations on the amount of storage that can be used. The storage development of the Colorado River basin, for example, may be approaching (if not already in excess of) the maximum useful limit.

RESERVOIRS

Where natural storage in the form of ponds or lakes is not available, artificial impoundments or reservoirs can sometimes be built to optimize the development of surface water flows. Some interesting statistics on dams and reservoirs may be found at www2.privatei.com/~uscold_s.html.

3.11 DETERMINATION OF REQUIRED RESERVOIR CAPACITY

The amount of storage needed is a function of expected demands and the quantity of inflow to the impoundment. Mathematically this may be stated as follows:

$$\Delta S = I - O \quad (3.2)$$

where ΔS = change in storage volume during a specified time interval, I = total inflow volume during this period, and O = total outflow volume during this period. Normally, O is the draft requirement imposed by the various uses, but it may also include evaporation, flood discharges during periods of high runoff, and seepage from the bottom or sides of the reservoir.

Because the natural inflow to any impoundment area is often highly variable from year to year, season to season, or even day to day, the reservoir function must be that of redistributing inflow with respect to time so that projected demands are satisfied.

3.12 METHODS OF COMPUTATION

Several approaches may be taken to calculate reservoir capacities. Actual or synthetic records of stream flow and a knowledge of the proposed operating rules of the reservoir are fundamental to all solutions. Determination of storage may be accomplished by graphical or analytical techniques, with spread sheets being a current standard.

Commonly, storage calculations are based on comparing demands with a critical low flow period such as the most severe drought of record. Once the critical period is chosen, the required storage is usually determined using a mass-curve analysis introduced in 1883 by Rippl [12]. This method evaluates the cumulative deficiency between outflow and inflow ($O - I$) and selects the maximum cumulative value as the required storage. Examples 3.1 and 3.2 illustrate the procedure.

Example 3.1

Find the storage capacity required to provide a safe yield of 67,000 acre · ft/yr for the data given in Figure 3.3.

Solution: Construct tangents at A, B, and C having slopes equal to 67,000 acre · ft/yr. Find the maximum vertical ordinate between the inflow mass curve and the constructed draft rates. From Figure 3.3 the maximum ordinate is found to be 38,000 acre · ft, which is the required capacity.

This example shows that the magnitude of the required storage capacity depends entirely on the time period chosen. Since the period of record given covers only 5.5 yr, it is clear that a design storage of 38,000 acre · ft might be totally inadequate for the next 3 yr, for example. Unless the frequency of the flow conditions used in the design is known, little can be said regarding the long- or short-term adequacy of the design.

Example 3.1 also illustrates the fact that the period during which storage must be provided is dependent on hydrologic conditions. Since reservoir yield is defined as

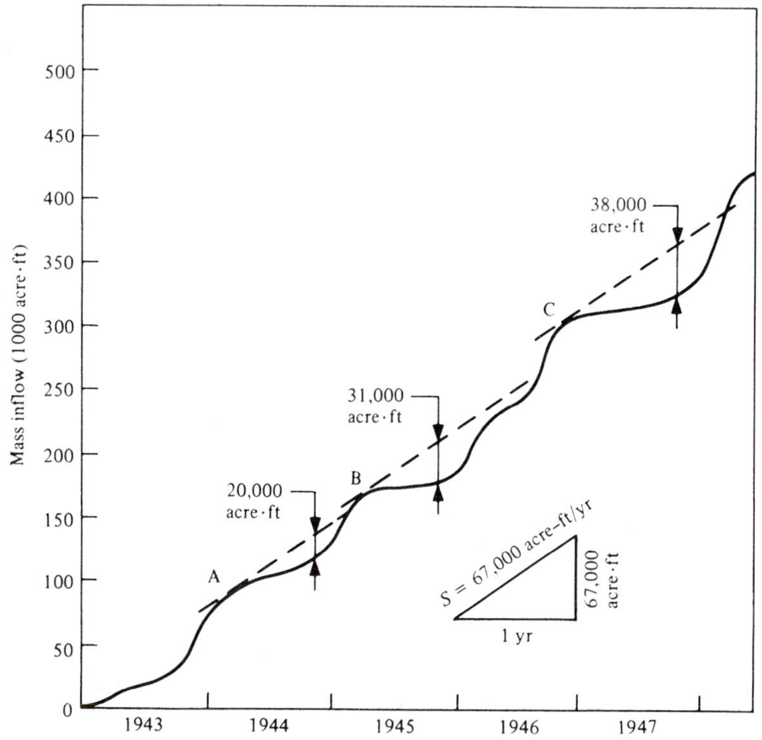

FIGURE 3.3 Reservoir capacity for a specified yield as determined by use of a mass curve.

the amount of water that can be supplied during a specific time interval, choice of the interval is critical. For distribution reservoirs, a period of 1 day is often sufficient. For large impounding reservoirs, periods of several months, a year, or several years may be required.

Example 3.2

Consider an impounding reservoir that is expected to provide for a constant draft of 637 million gallons (mil gal)/mi^2/yr. The following record of monthly mean inflow values is representative of the critical or design period. Find the storage requirement. Data on monthly inflows are given in Table 3.3 in column 2.

Solution: The calculations are shown on the spread sheet given in Table 3.3. It can be seen that the maximum cumulative deficiency is 202.2 mil gal/mi^2, which occurs in September. The number of months of draft is 202.2/53.1 = 3.8, or, stated differently, enough water must be stored to supply the region for about 3.8 months.

This example gives a numerical answer to the question posed in determining a storage design. It does not, however, give an expression for the probabilities of the shortages or excesses that may result from this design. Past practice has been to use the lowest

TABLE 3.3 Spread Sheet Storage Requirement Calculations

Month	Inflow (I)	Draft (O)	Sum Of Inflows ΣI	Deficiency $(O - I)$	Cumulative Deficiency[a] $\Sigma(I - O)$
J	37.2	53.1	37.2	15.9	15.9
F	64.8	53.1	102	−11.7	0
M	108	53.1	210	−54.9	0
A	12	53.1	222	41.1	41.1
M	8.4	53.1	230.4	44.7	85.8
J	9.6	53.1	240	43.5	129.3
J	2.4	53.1	242.4	50.7	180
A	33.6	53.1	276	19.5	199.5
S	50.4	53.1	326.4	2.7	202.2
O	129.6	53.1	456	−76.5	0
N	117.6	53.1	573.6	−64.5	0
D	26.4	53.1	600	26.7	26.7
J	60	53.1	660	−6.9	0

[a]Only positive values of cumulative deficiency are tabulated.

recorded flow of the stream as the critical period. Obviously, this approach overlooks the possibility that a more serious drought might occur, with a resultant yield less than the anticipated safe yield.

3.13 FREQUENCY OF EXTREME EVENTS

If a hydrologic event has a true recurrence interval of T_R years, the probability that this magnitude will be equaled or exceeded in any particular year is

$$P = 1/TR \tag{3.3}$$

where T_R is the recurrence interval of the event. *Recurrence interval* is defined as the average interval in years between the occurrence of an event of stated magnitude and an equal or more serious event.

Both annual series and partial duration series are used in estimating the recurrence intervals of extreme events [9]. An annual series is composed of one significant event for each year of record. The nature of the event depends on the object of the study. Usually the event will be a maximum or minimum flow. A partial duration series consists of all events exceeding, in significance, a base value. The two series compare favorably at the larger recurrence intervals, but for the smaller recurrence intervals, the partial duration series will normally indicate events of greater magnitude.

There are two possibilities regarding an event: It either will or will not occur in a specified year. The probability that at least one event of equal or greater significance than the T_R-year event will occur in any series of N years is shown in Table 3.4. For example, there exists a probability of 0.22 that the 100-yr event will occur in a design period of 25 yr.

TABLE 3.4 Probability That an Event Having a Prescribed Recurrence Interval Will Be Equaled or Exceeded During a Specified Design Period

	Design Period (yr)					
T_R (yr)	1	5	10	25	50	100
1	1.0	1.0	1.0	1.0	1.0	1.0
2	0.5	0.97	0.999	1.0[a]	1.0[a]	1.0[a]
5	0.2	0.67	0.89	0.996	1.0[a]	1.0[a]
10	0.1	0.41	0.65	0.93	0.995	1.0[a]
50	0.02	0.10	0.18	0.40	0.64	0.87
100	0.01	0.05	0.10	0.22	0.40	0.63
200	0.005	0.02	0.05	0.12	0.22	0.39

[a]Values are approximate.

Table 3.4 was derived by means of the binomial distribution, which gives the probability $p(X; N)$ that a particular event will occur X times out of N trials as

$$p(X; N) = \binom{N}{X} P^X (1 - P)^{N-X}$$
$$\binom{N}{X} = \frac{N!}{X!(N - X)!} \tag{3.4}$$

where P is the probability that an event will occur in each individual trial ($P = 1/T_R$ in this case). Now, if we let the number of occurrences equal zero ($X = 0$) in a given period of years N (number of trials) and substitute this value in Eq. (3.4) the result is

$$p(0; N) = (1 - P)^N \tag{3.5}$$

This is the probability of zero events equal to or greater than the T_R-year event. Then the probability Z that at least one event equal to or greater than the T_R-year event will occur in a sequence of N years is given by

$$Z = 1 - \left(1 - \frac{1}{T_R}\right)^N \tag{3.6}$$

Solution of Eq. (3.6) for various values of N and T_R provided the data for Table 3.4.

3.14 PROBABILISTIC MASS TYPE OF ANALYSIS

Low-flow information is the basis for reservoir design and for studies of the waste-assimilative capacity of streams. A customary critical period used by water quality personnel is the average low flow for 7 consecutive days occurring on the average of once in 10 yr. For reservoir design, the critical period is usually measured in months or years and return periods are normally on the order of 20, 50, or 100 yr.

Information on the probability of occurrence of droughts of various severities during any single year or during any specified period of years can be developed from climatological and hydrologic records. Using such information, estimates of the risks associated with various reservoir releases can be made, and the numerical odds for any specified yield tabulated.

To design a reservoir to meet a specific draft, one must determine the critical low flow period; determine the magnitude of the critical low flow; and calculate the frequency of occurrence of the critical low flow event.

The critical period can be determined by experimenting with a number of low-flow durations and then selecting, by judgment or by policy, some duration with which to work. Using existing or generated stream-flow data, a series of magnitudes of critical flows for the specified duration can be obtained. And by assigning recurrence intervals to the critical events, the frequency of events can be estimated.

3.15 LOSSES FROM STORAGE

The availability of water impounded in a reservoir is affected by losses in storage that result from natural or artificial phenomena. Natural losses occur through evaporation, seepage, and siltation, while artificial losses result from withdrawals.

After a dam has been built and the impoundment filled, the exposed water-surface area is increased significantly over that of the natural stream. The resultant effect is greatly increased opportunity for evaporation. The opportunity for generation of runoff from the flooded land is also eliminated, but this loss is countered by gains made through the catchment of direct precipitation. These *water-surface effects* tend to result in net gains in well-watered regions, but in arid lands, losses are typical, since evaporation generally exceeds precipitation.

The magnitude of seepage losses depends mainly on the geology of the region. If porous strata underlie the reservoir valley, considerable losses can occur. On the other hand, where permeability is low, seepage may be negligible. A subsurface exploration is a prerequisite to the adequate evaluation of such losses.

Since the useful life of a reservoir can be significantly affected by the deposition of sediment, a knowledge of sedimentation rates is important in reaching a decision regarding the feasibility of its construction [3,11].

The rate and characteristic of the sediment inflow can be controlled by using sedimentation basins, providing vegetative screens, and by employing various erosion control techniques [3]. Dams can also be designed so that part of the sediment load can be passed through or over them. A last resort is the physical removal of sediment deposits. Normally this is not economically feasible.

Example 3.3

Determine the expected life of the Lost Valley Reservoir. The initial capacity of the reservoir is 45,000 acre · ft. and the average annual inflow is 76,000 acre · ft. A sediment inflow of 176 acre · ft/yr is reported. Assume that the useful life of the reservoir is exceeded when 77.8% of the original capacity is lost.

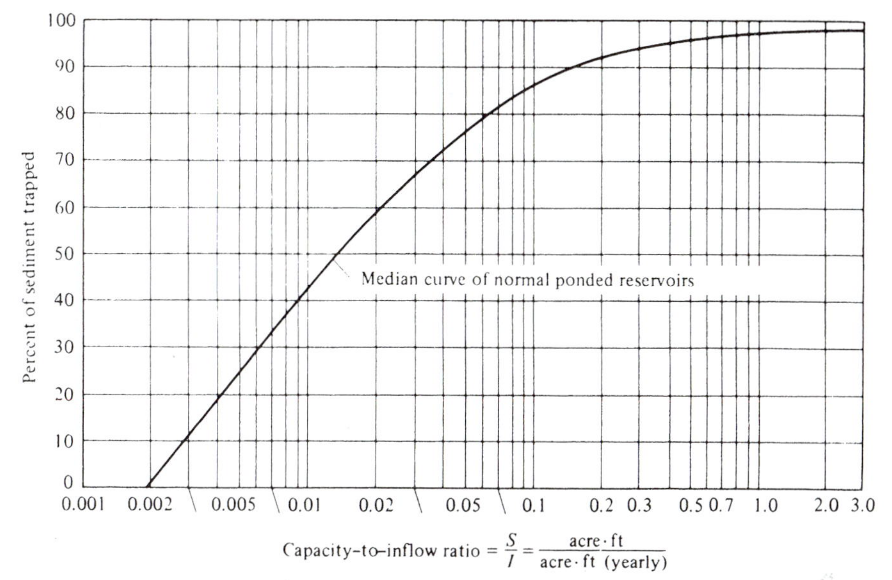

FIGURE 3.4 Relationship between reservoir sediment trap efficiency and the capacity/inflow ratio. [Developed from data by G. M. Bruce, "Trap Efficiency of Reservoirs," *Trans. Am. Geophys. Union* 34 (1953): 407–418.]

TABLE 3.5 Determination of Probable Life of the Lost Valley Reservoir

(1) Reservoir Capacity (acre · ft)	(2) Volume Increment (acre · ft)	(3) Capacity Inflow Ratio: (1) ÷ 76,000	(4) Percent Sediment Trapped, from Fig. 3.4	(5) Average Percent Sediment Trapped per Volume Increment	(6) Acre-Ft Sediment Trapped Annually: (5) × 176	(7) Number of Years Required to Fill the Volume Increment: (2) ÷ (6)
45,000	5000	0.59	96.5			
40,000	5000	0.52	96.1	96.3	169	30
35,000	5000	0.46	95.8	95.9	169	30
30,000	5000	0.39	95.0	95.4	168	30
25,000	5000	0.33	94.5	94.7	167	30
20,000	5000	0.26	93.0	93.8	165	30
15,000	5000	0.20	92.0	92.5	163	31
10,000	5000	0.13	88.0	90.0	158	32
Total number of years of useful life						213

Solution: The solution is obtained through application of the data given in Figure 3.4. The results are tabulated in Table 3.5.

Problems of the type illustrated in Examples 3.2 and 3.3 are especially suited to the use of spreadsheet analyses. These and analytic tools offer opportunity for quick adjustments to parameters and speedy recalculation of values.

GROUNDWATER

Groundwater storage is considerably in excess of all artificial and natural surface storage in the United States, including the Great Lakes [3,13]. This enormous groundwater reserve sustains the continuing outflow of streams and lakes during periods that follow those of runoff-producing rains. The relation between groundwater and surface water is one of mutual interdependence. Groundwater intercepted by a well as it moves toward a stream is the same as a diversion from the stream, for example. By developing and using surface water and groundwater sources jointly, opportunities for making water available for various uses can be optimized (see Section 3.26).

3.16 THE SUBSURFACE DISTRIBUTION OF WATER

Groundwater distribution may be generally categorized into zones of aeration and saturation. The *saturation zone* is one in which all the soil voids are filled with water under hydrostatic pressure. The *aeration zone*, in which the interstices are filled partly with air and partly with water, may be subdivided into several subzones. Todd classifies these as follows [14]:

1. The *soil-water zone* begins at the ground surface and extends downward through the major root zone. Its total depth is variable and dependent on soil type and vegetation. The zone is unsaturated except during periods of heavy infiltration. Three categories of water classification may be encountered in this region: *hygroscopic water*, which is adsorbed from the air; *capillary water*, which is held by surface tension; and *gravitational water*, which is excess soil water draining through the soil.
2. The *intermediate zone* extends from the bottom of the soil-water zone to the top of the capillary fringe and may vary from nonexistence to several hundred feet in thickness. The zone is essentially a connecting link between the near-ground surface region and the near-water table region through which infiltrating waters must pass.
3. The *capillary zone* extends from the water table to a height determined by the capillary rise that can be generated in the soil. The capillary zone thickness is a function of soil texture and may vary not only from region to region but also within a local area.
4. At the *groundwater zone,* groundwater fills the pore spaces completely, and porosity is therefore a direct measure of storage volume. Part of this water (specific retention) cannot be removed by pumping or drainage because of molecular and surface-tension forces. The specific retention is the ratio of the volume of water retained against gravity drainage to the gross volume of the soil.

The water that can be drained from a soil by gravity is known as the *specific yield.* It is expressed as the ratio of the volume of water that can be drained by gravity to the gross volume of the soil. Values of specific yield are dependent on soil particle size, shape and distribution of pores, and degree of compaction of the soil. Average values of specific yield for alluvial aquifers range from 10% to 20%. Meinzer and others have proposed numerous procedures for determining specific yield [16].

3.17 AQUIFERS

An *aquifer* is a water-bearing stratum that is capable of transmitting water in quantities sufficient to permit development. Aquifers may be classified as confined or unconfined, depending on whether or not a water table or free water surface exists under atmospheric pressure. The storage volume within an aquifer is changed whenever water is recharged to or discharged from it. In the case of an unconfined aquifer, this may be determined using the following equation:

$$\Delta S = S_y \Delta V \tag{3.7}$$

where ΔS = change in storage volume, S_y = average specific yield of the aquifer, ΔV = volume of the aquifer lying between the original water table and the water table at a later, specified time.

For saturated, confined aquifers, pressure changes produce only slight changes in storage volume. In this case, the weight of the overburden is supported partly by hydrostatic pressure and partly by the solid material in the aquifer. When the hydrostatic pressure in a confined aquifer is reduced by pumping or other means, the load on the aquifer increases, causing its compression, with the result that some water is forced from it. Decreasing the hydrostatic pressure also causes a small expansion, which in turn produces an additional release of water. For confined aquifers, the water yield is expressed in terms of a storage coefficient S_c. This storage coefficient may be defined as the volume of water that an aquifer takes in or releases per unit surface area of aquifer per unit change in head normal to the surface. Figure 3.5 illustrates the classifications of aquifers.

In addition to water-bearing strata exhibiting satisfactory rates of yield, there are also non-water-bearing and impermeable strata. An *aquiclude* is an impermeable stratum that may contain large quantities of water but whose transmission rates are not high enough to permit effective development. An *aquifuge* is a formation that is impermeable and devoid of water.

3.18 FLUCTUATIONS IN GROUNDWATER LEVEL

Any circumstance that alters the pressure imposed on underground water will also cause a variation in the groundwater level. Seasonal factors, changes in stream and river stages, evapotranspiration, atmospheric pressure changes, winds, tides, external loads, various forms of withdrawal and recharge, and earthquakes all may produce fluctuations in the level of the water table or the piezometric surface, depending on whether the aquifer is free or confined [14].

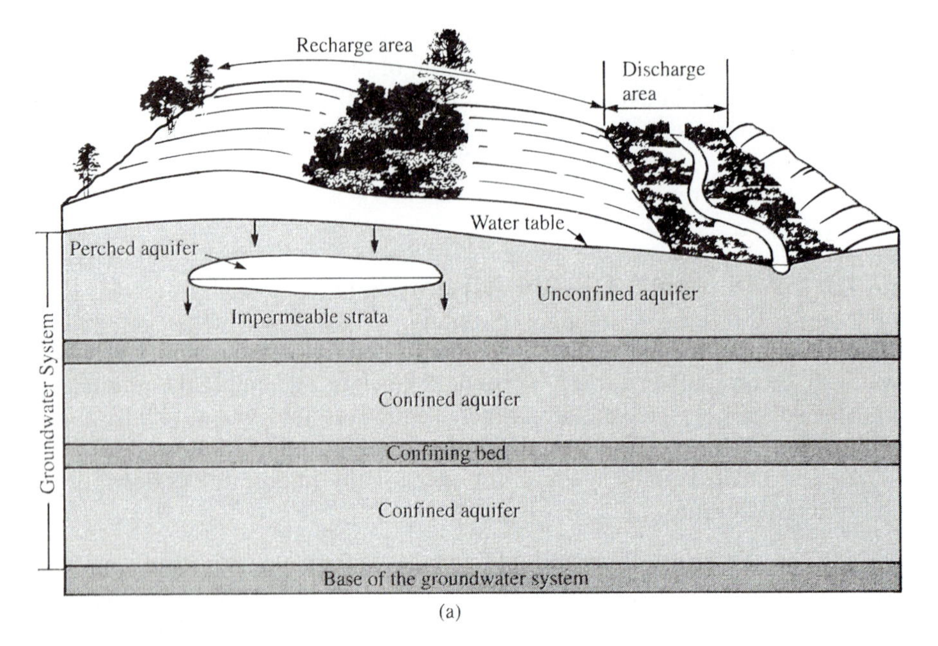

(a)

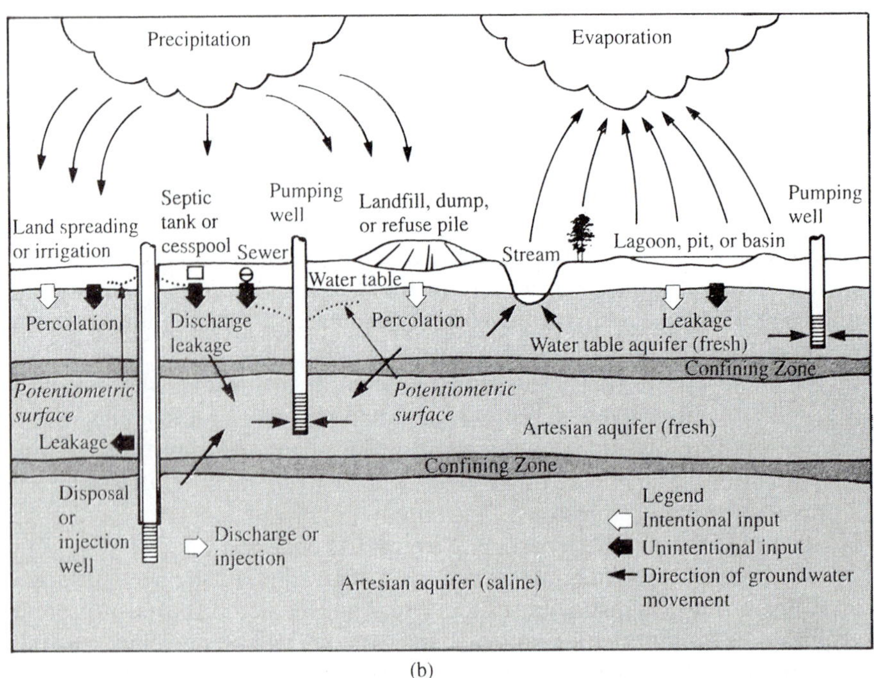

(b)

FIGURE 3.5 Definition sketches of groundwater systems and mechanisms for recharge and withdrawal. (a) Aquifer notation [17]. (b) Components of the hydrologic cycle affecting groundwater [18].

3.19 SAFE YIELD OF AN AQUIFER

Before a groundwater source is developed for use, the quantity of water that it can be expected to deliver must be estimated. This is known as the *safe yield* of the aquifer. It is the quantity of water that can be withdrawn annually without the ultimate depletion of the aquifer. Other related terms are defined as follows:

1. The *maximum sustained yield* is the maximum rate at which water can be withdrawn on a continuing basis from a given source.
2. The *permissive sustained yield* is the maximum rate at which withdrawals can be made legally and economically on a continuing basis for beneficial use without the development of undesired results.
3. The *maximum mining yield* is the total storage volume in a given source that can be withdrawn and used.
4. The *permissive mining yield* is the maximum volume of water that can be withdrawn legally and economically, to be used for beneficial purposes, without causing an undesired result.

A review of these definitions should make it clear that groundwater resources are finite and not inexhaustible. If the drafts imposed on them are such that natural and artificial recharge mechanisms will make up for these losses over a period of time, no harm will come. On the other hand, if drafts exceed recharge, groundwater storage can be mined out, or depleted to a level below which economic development is infeasible. Some areas in the United States where perennial overdrafts occur are shown on Figure 3.6 [3,14].

Methods for determining safe yield have been proposed by Hill, Harding, Simpson, and others [14]. The Hill method is based on groundwater studies in southern California and Arizona. In this method, the annual change in the elevation of the groundwater table or piezometric surface is plotted against the annual draft. The data points can be fitted by a straight line, provided that the water supply to the basin is fairly uniform.

The draft that corresponds to zero change in elevation is considered to be the safe yield. The period of record should be such that the supply during this period approximates the long-time average supply. Even though the draft during the period of record may be an overdraft, the safe yield can be determined by extending the line of best fit to an intersection with the zero change in elevation line. An example of this procedure is given in Figure 3.7.

It is important to understand that the safe yield of an aquifer can change over time if the conditions under which it was determined do not remain constant. This requires that drafts, recharge rates, and other conditions affecting the safe yield be monitored.

3.20 GROUNDWATER FLOW

The rate of movement of water through the ground is of an entirely different magnitude from that through natural or artificial channels or conduits. Typical values range from 5 ft/day to a few feet per year. Methods for determining these transmission rates

FIGURE 3.6 Groundwater reservoirs with perennial overdraft. [From H. E. Thomas, "Water," in *The Yearbook of Agriculture* (Washington, DC: U.S. Government Printing Office, 1955).]

are primarily based on the principles of fluid flow represented by *Darcy's law* [15]. This law may be stated as

$$Q = -KA\frac{dh}{dx} \tag{3.8}$$

where

A = total cross-sectional area including the space occupied by the porous material

K = hydraulic conductivity of the material

Q = flow across the control area A

In Eq. (3.8):

$$h = z + \frac{p}{\gamma} + C \tag{3.9}$$

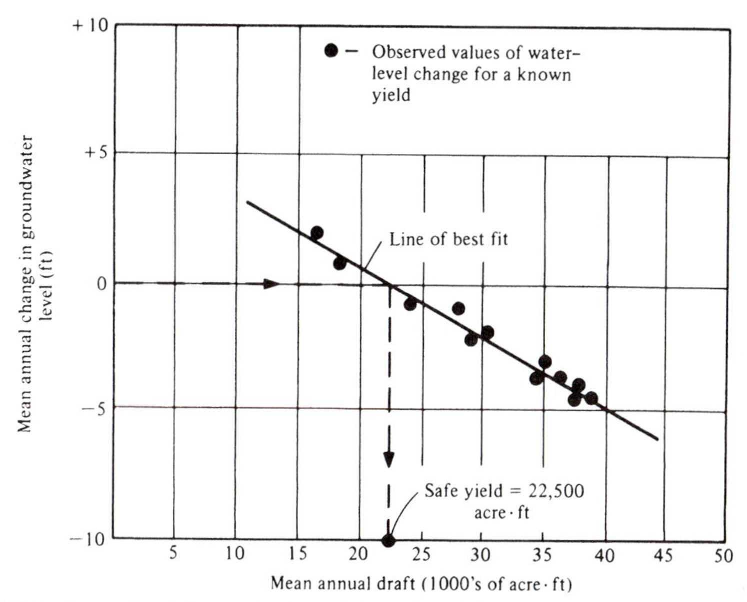

FIGURE 3.7 Example of determination of the safe yield by the Hill method.

where

h = piezometric head

z = elevation above a datum

p = hydrostatic pressure

C = an arbitrary constant

γ = specific weight of water

If the specific discharge $q = Q/A$ is substituted in Eq. (3.8):

$$q = -K\frac{d}{dx}\left(z + \frac{p}{\gamma}\right) \tag{3.10}$$

Note that q also equals the porosity n multiplied by the pore velocity V_p. Darcy's law is widely used in groundwater flow problems. Applications are illustrated later in this chapter.

Darcy's law is limited in its applicability to flows in the laminar region. The controlling criterion is the *Reynolds number*:

$$N_R = Vd/v \tag{3.11}$$

where

$$V = \text{flow velocity}$$

$$d = \text{mean grain diameter}$$

$$\nu = \text{kinematic viscosity}$$

Groundwater flow may be considered laminar for Reynolds numbers less than 1. Departure from laminar conditions normally occurs within the range of Reynolds numbers from 1 to 10, depending on grain size and shape. Under most conditions encountered, with the exception of regions in close proximity to collecting devices, the flow of groundwater is laminar and Darcy's law applies.

Example 3.4

1. Find the Reynolds number for the portion of an aquifer distant from any collection device where water temperature is 50°F ($\nu = 1.41 \times 10^{-5}$ ft^2/s), flow velocity is 1.0 ft/day, and mean grain diameter is 0.09 in.
2. Find the Reynolds number for a flow 4 ft from the centerline of a well being pumped at a rate of 3800 gpm if the well completely penetrates a confined aquifer 28 ft thick. Assume a mean grain diameter of 0.10 in, a porosity of 35%, and $\nu = 1.41 \times 10^{-5}$ ft^2/s.

Solution:

1. Using Eq. (3.11), we obtain

$$N_R = \frac{Vd}{\nu}$$

where

$$V = \frac{1.0}{86{,}400}\text{ fps} \qquad d = \frac{0.09}{12}\text{ ft}$$

$$N_R = \frac{1.0}{86{,}400} \times \frac{0.09}{12} \times \frac{1}{1.41 \times 10^{-5}}$$

$$= 0.0062 \text{ (indicating laminar flow)}$$

2. Using Eq. (3.11) yields

$$N_R = \frac{Vd}{\nu} = \frac{Q}{A}\frac{d}{\nu}$$

$$Q = 3800 \times 2.23 \times 10^{-3} = 8.46 \text{ cfs}$$

$$V = 8.46/2\pi rh \times \text{porosity}$$

$$= 8.46/8\pi \times 28 \times 0.35$$

$$= 0.0344 \text{ fps}$$

$$N_R = \frac{0.0344 \times 0.10}{1.41 \times 10^{-5} \times 12} = 20.3 \text{ (beyond Darcy's law range)}$$

To compute discharge, Eq. (3.8) can be used. Note that this equation may also be stated as

$$Q = pAkS \tag{3.12}$$

where

p = porosity or ratio of void volume to total volume of the mass

A = gross cross-sectional area

k = intrinsic permeability

S = slope of hydraulic gradient

By combining k and p into a single term, Eq. (3.12) may be written in its most common form,

$$Q = KAS \tag{3.13}$$

Several ways of expressing *hydraulic conductivity* K may be found in the literature. The U.S. Geological Survey defines standard hydraulic conductivity K_S as the number of gallons of water per day that will flow through a medium of 1-ft^2 cross-sectional area under a hydraulic gradient of unity at 60°F. The field coefficient of permeability is obtained directly from the standard coefficient by correcting for temperature:

$$K_f = K_S(\mu_{60}/\mu_f) \tag{3.14}$$

where

K_f = field coefficient

μ_{60} = dynamic viscosity at 60°F

μ_f = dynamic viscosity at field temperature

An additional term that is much used in groundwater computations is the *coefficient of transmissibility* T. It is equal to the field coefficient of permeability multiplied by the saturated thickness of the aquifer in feet. Using this terminology, Eq. (3.12) may also be written

$$Q = T \times \text{section width} \times S \tag{3.15}$$

Table 3.6 gives typical values of the standard hydraulic conductivity for a range of sedimentary materials. It should be noted that the permeabilities for specific materials vary widely. Traces of silt and clay can significantly decrease the permeability of an aquifer. Differences in particle orientation and shape can cause striking changes in permeability within aquifers composed of the same material. Careful evaluation of geologic information is essential if realistic values of permeability are to be identified for use in groundwater flow computations.

TABLE 3.6 Some Values of the Standard Hydraulic Conductivity and Intrinsic Permeability for Several Classes of Materials

Material	Approximate Range K_s (gpd/ft^2)	Approximate Range k (darcys)
Clean gravel	10^6–10^4	10^5–10^3
Clean sands; mixtures of clean gravels and sands	10^4–10	10^3–1
Very fine sands; silts; mixtures of sands, silts, clays; stratified clays	10–10^{-3}	1–10^{-4}
Unweathered clays	10^{-3}–10^{-4}	10^{-4}–10^{-5}

Example 3.5

Laboratory tests on an aquifer material indicate a standard hydraulic conductivity $K_s = 1.08 \times 10^3$ gpd/ft^2. If the field temperature is 70 °F, find the field hydraulic conductivity K_f.

Solution: Using Eq. (3.14),

$$K_f = K_s\left(\frac{\mu_{60}}{\mu_f}\right)$$

and the values of the kinematic viscosity given in Table A.8 in the Appendix for 60 and 70°F, 1.21×10^{-5} and 1.06×10^{-5}, respectively, we get

$$K_f = \frac{1.08 \times 10^3 \times 1.21 \times 10^{-5}}{1.06 \times 10^{-5}} = 1232.8 \text{ gpd/ft}^2$$

Note that the absolute viscosity and the kinematic viscosity are related as shown in the following equation,

$$\nu = \frac{\mu}{\rho}$$

and given that the density of water over the range of temperatures in this case is virtually constant, values for the kinematic viscosity may be used in place of those for the absolute velocity in Eq.(3.14).

3.21 HYDRAULICS OF WELLS

The collection of groundwater is accomplished mainly through the use of wells or infiltration galleries. Numerous factors affect the performance of these collection works and they must be taken into account when mathematical models are used to make estimates. Some cases are amenable to solution through the utilization of relatively simple mathematical expressions, others require the use of sophisticated mathematical models. Several approaches will be discussed here. The reader is cautioned not to be misled by the simplicity of some of the solutions presented, and should understand that many of these are special-case solutions and are not applicable to all groundwater flow problems. A more complete treatment of groundwater and seepage problems may be found in numerous references [13,14,19,20].

Flow to Wells

Well systems generally have three components—well structure, pump, and discharge piping. The well itself contains an open section through which water enters, and a casing through which the flow is transported to the ground surface. The open section is usually a perforated casing, or a slotted metal screen, that permits water to enter and at the same time prevents collapse of the hole. Occasionally gravel is placed at the bottom of the well casing around the screen.

When a well is pumped, water is removed from the aquifer immediately adjacent to the screen. Flow then becomes established at locations some distance from the well in order to replenish this withdrawal. Owing to the resistance to flow offered by the soil, a head loss is encountered and the piezometric surface adjacent to the well is depressed. This is known as the *cone of depression* (Fig. 3.8). The cone of depression spreads until a condition of equilibrium is reached and steady-state conditions are established.

The hydraulic characteristics of an aquifer (which are described by the storage coefficient and the aquifer permeability) can be determined by laboratory or field tests. The three most commonly used field methods are the application of tracers, use of field permeameters, and aquifer performance tests [3,13]. Aquifer performance tests are discussed here along with the development of flow equations for wells. Aquifer performance tests are classified as equilibrium or nonequilibrium. For equilibrium tests, the cone of depression must be stabilized for the flow equation to be derived. The first performance tests based on equilibrium conditions were published by Thiem in

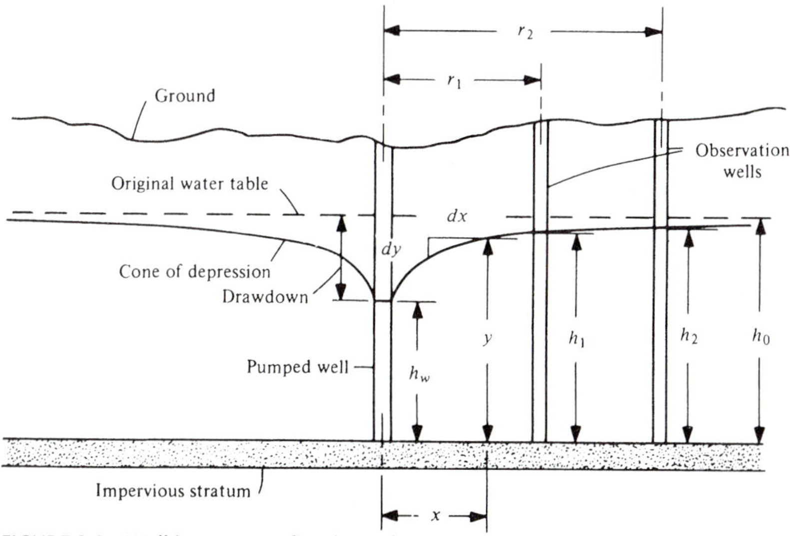

FIGURE 3.8 Well in an unconfined aquifer.

1906 [21]. In nonequilibrium tests, the derivation of the flow equation takes into consideration the condition that steady-state conditions have not been reached.

The basic equilibrium equation for an unconfined aquifer can be derived using the notation of Fig. 3.8. In this case, the flow is assumed to be radial, the original water table is considered horizontal, the well is considered to fully penetrate an aquifer of infinite areal extent, and steady-state conditions are considered to prevail. Using these assumptions, the flow toward a well at any location x from the well equals the product of the cylindrical element of area at that section and the flow velocity. Using Darcy's law, this becomes

$$Q = 2\pi x y K_f \frac{dy}{dx} \tag{3.16}$$

where

$$2\pi xy = \text{area at any section}$$

$$K_f\, dy/dx = \text{flow velocity}$$

$$Q = \text{discharge, cfs}$$

Integrating over the limits specified below yields

$$\int_{r_1}^{r_2} Q\frac{dx}{x} = 2\pi K_f \int_{h_1}^{h_2} y\, dy \tag{3.17}$$

$$Q \ln\left(\frac{r_2}{r_1}\right) = \frac{2\pi K_f(h_2^2 - h_1^2)}{2} \tag{3.18}$$

and

$$Q = \frac{\pi K_f(h_2^2 - h_1^2)}{\ln(r_2/r_1)} \tag{3.19}$$

This equation may then be solved for K_f to yield

$$K_f = \frac{1055Q \log(r_2/r_1)}{h_2^2 - h_1^2} \tag{3.20}$$

where ln has been converted to log, K_f is in gallons per day per square foot, Q is in gallons per minute, and r and h are measured in feet. If the drawdown is small compared with the total aquifer thickness, an approximate formula for the discharge of the pumped well can be obtained by inserting h_w for h_1 and the height of the aquifer for h_2 in Eq. (3.19).

The basic equilibrium equation for a confined aquifer can be obtained in a similar manner, using the notation of Figure 3.9. The same assumptions apply. Mathematically, the flow in cubic feet per second may be determined as follows:

$$Q = 2\pi x m K_f \frac{dy}{dx} \tag{3.21}$$

Integrating, we obtain

$$Q = 2\pi K_f m \frac{h_2 - h_1}{\ln(r_2/r_1)} \tag{3.22}$$

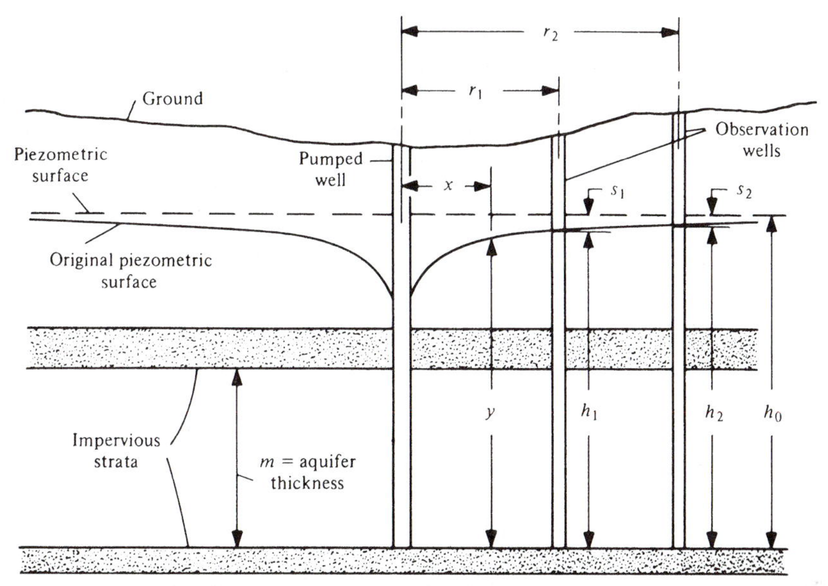

FIGURE 3.9 Radial flow to a well in a confined aquifer.

The coefficient of permeability may be determined by rearranging Eq. (3.22) to the form

$$K_f = \frac{528Q\log(r_2/r_1)}{m(h_2 - h_1)} \tag{3.23}$$

where Q is in gallons per minute, K_f is the permeability in gallons per day per square foot, and r and h are measured in feet.

Example 3.6

Find the hydraulic conductivity of an artesian aquifer being pumped by a fully penetrating well. The aquifer is 100 ft thick and composed of medium sand. The steady-state pumping rate is 1000 gpm. The drawdown at an observation well 50 ft away is 10 ft; in a second observation well 500 ft away it is 1 ft.

Solution: Using Eq. 3.23

$$\begin{aligned} K_f &= \frac{528Q\log(r_2/r_1)}{m(h_2 - h_1)} \\ &= \frac{528 \times 1000 \times 10}{100 \times (10 - 1)} \\ &= 586.7 \text{ gpd/ft}^2 \end{aligned}$$

Example 3.7

A 20 in. well fully penetrates an unconfined aquifer of 100-ft depth. Two observation wells located 90 and 240 ft from the pumped well are known to have drawdowns of 23 and 21.5 ft respectively. If the flow is steady and $K_f = 1400$ gpd/ft^2, find the discharge from the well.

Solution: Eq. (3.20) is applicable, and for the given units this is

$$Q = \frac{K(h_2^2 - h_1^2)}{1055 \log(r_2/r_1)}$$

$$\text{Log}(r_2/r_1) = \log(240/90) = 0.42651$$

$$h_2 = 100 - 21.5 = 78.5 \text{ ft}$$

$$h_1 = 100 - 23 = 77 \text{ ft}$$

$$Q = \frac{1400(78.5^2 - 77^2)}{1055 \times 0.42651}$$

$$= 725.7 \text{ gpm}$$

For a steady-state well in a uniform flow field where the original piezometric surface is not horizontal, a somewhat different situation from that previously assumed prevails. Consider the artesian aquifer shown in Figure 3.10. The heretofore assumed circular area of influence becomes distorted in this case. This problem may be solved by application of potential theory or by graphical means; or, if the slope of the piezometric surface is very slight, Eq. (3.22) may be applied without serious error.

Referring to the definition sketch of Figure 3.10, a graphical solution to this type of problem will be discussed. First, an orthogonal flow net consisting of flow lines and equipotential lines must be constructed. The construction should be performed so that the completed flow net will be composed of a number of elements that approach little squares in shape. Harr [19] is a good source of information on this subject for the interested reader. A comprehensive discussion cannot be provided here.

Once the net is complete, it may be analyzed by considering the net geometry and using Darcy's law in the manner of Todd [14]. In the definition sketch of Figure 3.10, the hydraulic gradient is

$$h_g = \frac{\Delta h}{\Delta s} \tag{3.24}$$

and the flow increment between adjacent flow lines is

$$\Delta q = K \frac{\Delta h}{\Delta s} \Delta m \tag{3.25}$$

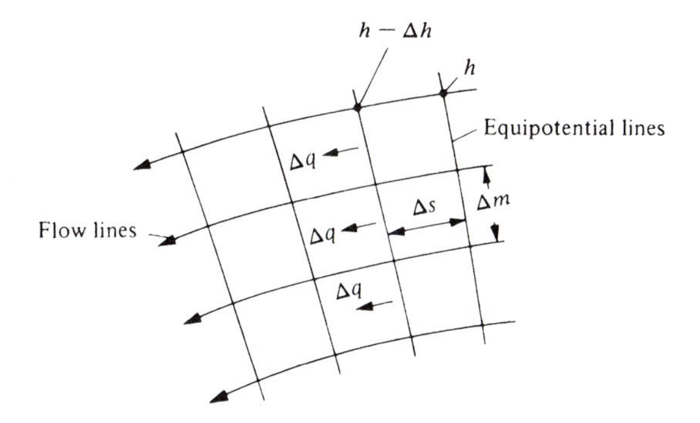

Definition Sketch of a Segment of a Flow Net

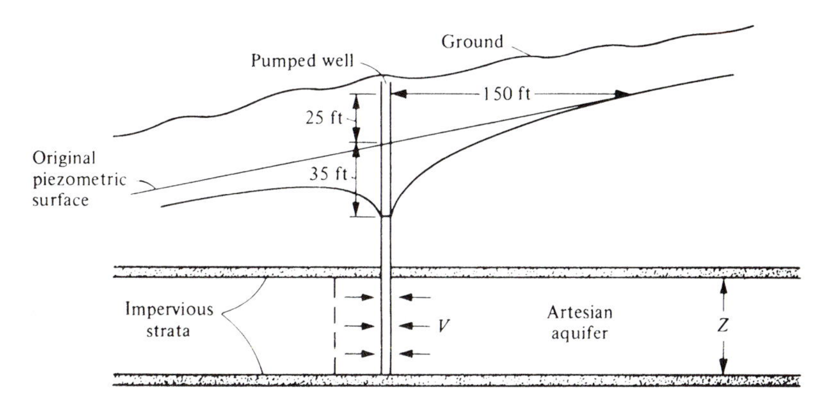

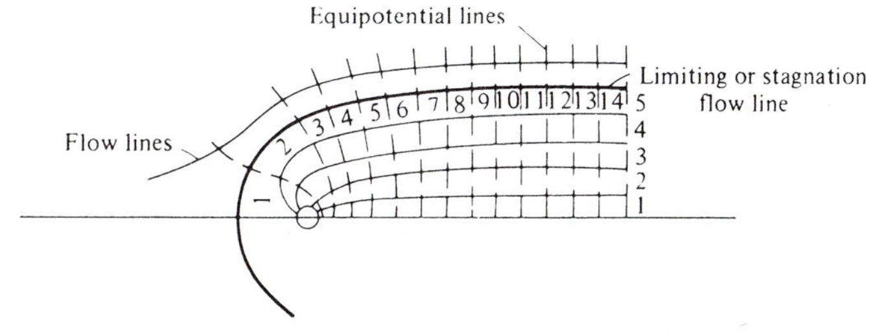

FIGURE 3.10 Well in a uniform flow field and flow-net definition.

where, for a unit thickness, Δm represents the cross-sectional area. If the flow net is properly constructed so that it is orthogonal and composed of little square elements,

$$\Delta m \approx \Delta s \tag{3.26}$$

and

$$\Delta q = K \Delta h \tag{3.27}$$

Now consider the entire flow net,

$$\Delta h = \frac{h}{n} \tag{3.28}$$

where n is the number of subdivisions between equipotential lines. If the flow is divided into m sections by the flow lines, the discharge per unit width of the aquifer will be

$$q = \frac{Kmh}{n} \tag{3.29}$$

Knowledge of the aquifer permeability and the flow-net geometry permits solution of Eq. (3.29).

Example 3.7

Find the discharge to the well of Figure 3.10 by using the applicable flow net. Consider that the aquifer is 35 ft thick, K_f is 3.65×10^{-4} fps, and the other dimensions are as shown.

Solution: Using Eq. (3.29),

$$q = \frac{Kmh}{n}$$

where

$$h = (35 + 25) = 60 \text{ ft}$$

$$m = 2 \times 5 = 10$$

$$n = 14$$

$$q = \frac{3.65 \times 10^{-4} \times 60 \times 10}{14}$$

$$= 0.0156 \text{ cfs per unit thickness of the aquifer}$$

The total discharge Q is thus

$$Q = 0.0156 \times 35 = 0.55 \text{ cfs} \qquad \text{or} \qquad 245 \text{ gpm}$$

When a new well is first pumped, a large portion of the discharge is produced directly from the storage volume released as the cone of depression develops. Under these circumstances, the equilibrium equations overestimate permeability and therefore the yield of the well. Where steady-state conditions are not encountered—as is usually the case in practice—a nonequilibrium equation must be used. Two approaches can be taken: the rather rigorous method of Theis or a simplified procedure such as that proposed by Cooper and Jacob [22,23].

In 1935, Theis published a nonequilibrium approach that takes into consideration time and the storage characteristics of an aquifer. Application of the method is appropriate for confined aquifers of constant thickness. For use under conditions of unconfined flow, vertical components of flow must be negligible, and changes in aquifer storage through water expansion and aquifer compression must also be negligible relative to the gravity drainage of pores as the water table drops as a result of pumping.

Theis states that the drawdown (s) in an observation well located at a distance r from the pumped well is given by

$$s = \frac{Q}{4\pi T}\int_{u}^{\infty}\frac{e^{-u}}{u}du \tag{3.30}$$

where Q = constant pumping rate (L^3T^{-1} units), T = aquifer transmissivity(L^2T^{-1} units), and u is a dimensionless variable defined by

$$u = r^2\frac{S_c}{4tT} \tag{3.31}$$

where r is the radial distance from the pumping well to an observation well, S_c is the aquifer storativity (dimensionless), and t is time. The integral in Eq. (3.30) is commonly called the *well function of u* and is written as $W(u)$. It can be evaluated from the infinite series

$$W(u) = -0.577216 - \ln u + u - \frac{u^2}{2 \times 2!} + \frac{u^3}{3 \times 3!}\cdots \tag{3.32}$$

Using this notation, Eq. (3.30) may be written as

$$s = \frac{QW(u)}{4\pi T} \tag{3.33}$$

The basic assumptions of the Theis equation are generally the same as those in Eq. (3.23) except for the nonsteady-state condition. Some values of the well function of u are given in Table 3.7.

In U.S. practice, Eqs. (3.30) and (3.31) commonly appear in the following form:

$$s = \frac{114.6Q}{T}\int_{u}^{\infty}\frac{e^{-u}}{u}du \tag{3.34}$$

$$u = \frac{1.87r^2S_c}{Tt} \tag{3.35}$$

where T is in units of gpd/ft, Q has units of gpm, and t is the time in days since the start of pumping.

Equations (3.30) and (3.31) can be solved by comparing a log-log plot of u versus $W(u)$, known as a "type curve," with a log-log plot of the observed data r^2/t versus s. In plotting the type curve, $W(u)$ is the ordinate and u is the abscissa. The two

TABLE 3.7 Values of $W(u)$ for Various Values of u

u	1.0	2.0	3.0	4.0	5.0	6.0	7.0	8.0	9.0
$\times 1$	0.219	0.049	0.013	0.0038	0.0011	0.00036	0.000038	0.000012	0.000012
$\times 10^{-1}$	1.82	1.22	0.91	0.70	0.56	0.45	0.37	0.31	0.26
$\times 10^{-2}$	4.04	3.35	2.96	2.68	2.47	2.30	2.15	2.03	1.92
$\times 10^{-3}$	6.33	5.64	5.23	4.95	4.73	4.54	4.39	4.26	4.14
$\times 10^{-4}$	8.63	7.94	7.53	7.25	7.02	6.84	6.69	6.55	6.44
$\times 10^{-5}$	10.94	10.24	9.84	9.55	9.33	9.14	8.99	8.86	8.74
$\times 10^{-6}$	13.24	12.55	12.14	11.85	11.63	11.45	11.29	11.16	11.04
$\times 10^{-7}$	15.54	14.85	14.44	14.15	13.93	13.75	13.60	13.46	13.34
$\times 10^{-8}$	17.84	17.15	16.74	16.46	16.23	16.05	15.90	15.76	15.65
$\times 10^{-9}$	20.15	19.45	19.05	18.76	18.54	18.35	18.20	18.07	17.95
$\times 10^{-10}$	22.45	21.76	21.35	21.06	20.84	20.66	20.50	20.37	20.25
$\times 10^{-11}$	24.75	24.06	23.65	23.36	23.14	22.96	22.81	22.67	22.55
$\times 10^{-12}$	27.05	26.36	25.96	25.67	25.44	25.26	25.11	24.97	24.86
$\times 10^{-13}$	29.36	28.66	28.26	27.97	27.75	27.56	27.41	27.28	27.16
$\times 10^{-14}$	31.66	30.97	30.56	30.27	30.05	29.87	29.71	29.58	29.46
$\times 10^{-15}$	33.96	33.27	32.86	32.58	32.35	32.17	32.02	31.88	31.76

Source: After L. K. Wenzel, "Methods for Determining Permeability of Water Bearing Materials with Special Reference to Discharging Well Methods," U.S. Geological Survey, Water-Supply Paper 887, Washington, DC, 1942.

curves are superimposed and moved about until some of their segments coincide. In doing this, the axes must be maintained parallel. A coincident point is then selected on the matched curves and both plots are marked. The type curve then yields values of u and $W(u)$ for the selected point. Corresponding values of s and r^2/t are determined from the plot of the observed data. Inserting these values in Eqs. (3.30) and (3.31) and rearranging, values for the transmissibility T and the storage coefficient S_c may be found.

Often, this procedure can be shortened and simplified. When r is small and t large, Jacob found that values of u are generally small [23]. Thus, the terms in the series of Eq. (3.32) beyond the second term become negligible and the expression for T becomes

$$T = \frac{264Q(\log t_2 - \log t_1)}{h_0 - h} \tag{3.36}$$

which can be further reduced to

$$T = \frac{264Q}{\Delta h} \tag{3.37}$$

where

Δh = drawdown per log cycle of time, $(h_0 - h)/(\log t_2 - \log t_1)$

Q = well discharge, gpm

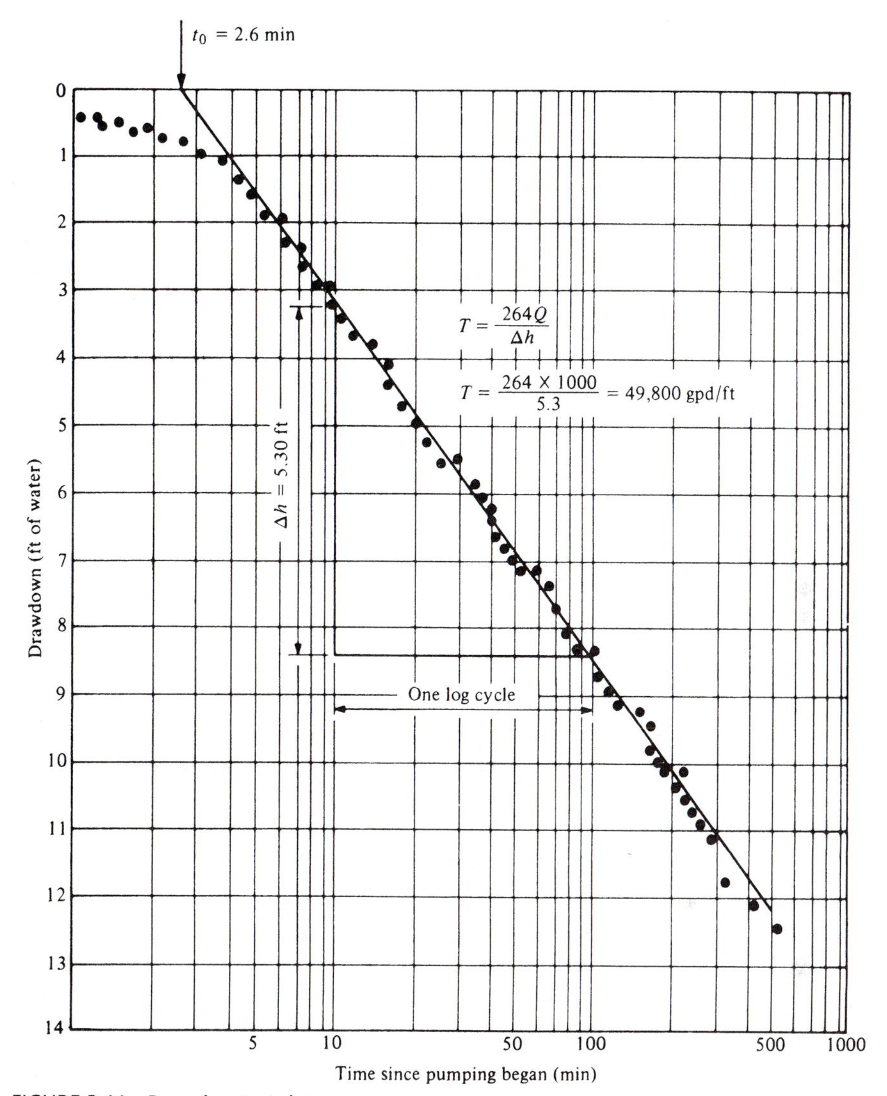

FIGURE 3.11 Pumping test data.

h_0 and h are defined as shown in Figure 3.9, and T is the transmissibility in gallons per day per foot. Field data on drawdown $h_0 - h$ versus t are plotted on semilogarithmic paper. The drawdown is plotted on the arithmetic scale as shown in Figure 3.11. This plot forms a straight line, the slope of which permits the determination of the formation constants using Eq. (3.37) and

$$S_c = \frac{0.3Tt_0}{r^2} \tag{3.38}$$

where t_0 is the time that corresponds to zero drawdown.

Example 3.9

Using the data given in Figure 3.11, find the coefficient of transmissibility T and the storage coefficient S_c for the aquifer, given $Q = 1000$ gpm and $r = 300$ ft.

Solution: Find the value of Δh from the graph. This is 5.3 ft. Then, using Eq. (3.37), we obtain

$$T = \frac{264Q}{\Delta h} = \frac{264 \times 1000}{5.3}$$

$$= 49{,}800 \text{ gpd/ft}$$

Using Eq. (3.38) yields

$$S_c = \frac{0.3Tt_0}{r^2}$$

Note from Figure 3.11 that $t_0 = 2.6$ min. Converting to days, this becomes

$$t_0 = 1.81 \times 10^{-3} \text{ days}$$

and

$$S_c = \frac{0.3 \times 49.800 \times 1.81 \times 10^{-3}}{(300)^2}$$

$$= 0.0003$$

Example 3.10

Find the drawdown at an observation point 300 ft away from a pumping well. It has been found that $T = 2.8 \times 10^4$ gpd/ft, the pumping time is 15 days, the storativity is $= 2.7 \times 10^{-4}$, and $Q = 275$ gpm.

Solution: From Eq. (3.35), u can be computed,

$$u = \frac{1.87r^2S_c}{Tt}$$

$$u = [1.87 \times (300)^2 \times 2.7 \times 10^{-4}]/[2.8 \times 10^4 \times 15] = 1.08 \times 10^{-4}$$

Referring to Table 3.7 and interpolating, we estimate $W(u)$ to be 8.62. Then using Eq. (3.34), the drawdown is found to be

$$s = \frac{114.6Q}{T}\int_u^\infty \frac{e^{-u}}{u}du$$

$$s = [114.6 \times 275 \times 8.62]/[2.8 \times 10^4] = 9.70 \text{ ft}$$

Example 3.11

A well is being pumped at a constant rate of $0.004\ m^3/s$. Given that $T = 0.0025\ m^2/s$, $r = 100$ meters, and the storage coeffcient $= 0.00087$, find the drawdown in the observation well for a time period of (a) 15 min, and (b) 20 hr.

Solution: (a) Using Eq. (3.31), u can be computed as follows:

$$u = \frac{r^2 S_c}{4tT}$$

$$u = [100 \times 100 \times 0.00087]/[4 \times 15 \times 60 \times 0.0025]$$

$$u = 0.97$$

Then from Table 3.7, W(u) is found to be 0.23.

Applying Eq. (3.33), the drawdown can be determined:

$$s = \frac{QW(u)}{4\pi t}$$

$$s = [0.004 \times 0.23]/[4 \times \pi \times 0.0025]$$

$$s = 0.029\ \text{m}$$

(b) Follow the procedure used in (a)

$$u = [100 \times 100 \times 0.00087]/[4 \times 72{,}000 \times 0.0025]$$

$$u = 0.0121$$

Then from Table 3.7, W(u) is found to be 8.49.

Applying Eq. (3.33), the drawdown can be determined,

$$s = [0.0004 \times 8.49]/[4 \times \pi \times 0.0025]$$

$$s = 1.08\ \text{m}$$

3.22 BOUNDARY EFFECTS

Only the effect of pumping a single well has been considered here. But if more than one well is pumped in a region, a composite effect (interference) due to the overlap of the cones of depression of the individual will result. In this case, the drawdown at any location is obtained by summing the individual drawdowns of the various wells involved. An additional problem is that of boundary conditions. The previous derivations have been based on the supposition of a homogeneous aquifer of infinite areal extent. A situation such as this is rarely encountered in practice. Computations based on this assumption are often sufficiently accurate, however, provided that field conditions closely approximate the basic hypotheses. Boundary effects may be evaluated by using the theory of image wells proposed by Lord Kelvin, through the use of electrical and membrane analogies, and through the use of relaxation procedures. For a detailed discussion of these topics, the reader is referred to the many references on groundwater flow [13,14,19].

3.23 REGIONAL GROUNDWATER SYSTEMS

Methods discussed so far have related mostly to the flow of water to individual wells. But regional well fields are common, and analysis of these systems is complex. An in-depth treatment of this subject is beyond the scope of this book, but some approaches used in analyzing regional groundwater systems are introduced [9,13,24–34].

Both the quantity and quality of groundwater are of importance, and must be dealt with. If groundwater sources are so contaminated that their use is impaired, extensive treatment may be required or the source may have to be abandoned.

Today, regional groundwater systems are analyzed using mathematical models. These models consist of sets of equations representing the physical, chemical, biological, and other processes that occur in an aquifer. The models may be deterministic, probabilistic, or a combination of the two. The discussions that follow are limited to deterministic mathematical models. These models describe cause–effect relationships stemming from known features of the system under study. Figure 3.12 characterizes the procedure for developing a deterministic mathematical model. Based on a study of the region of interest, and armed with an understanding of the mechanics of groundwater flow, a conceptual model is formulated. This is translated into a mathematical model of the system, which usually consists of a set of partial differential equations accompanied by appropriate boundary and initial conditions. Continuity and conservation of momentum considerations are featured in the model and are represented over the extent of the region of concern. Darcy's law, discussed earlier, is widely used to describe conservation of momentum. Other model features include artesian or water table condition designation and dimensionality, that is one-, two-, or three-dimensional. Where the modeling objective includes water quality or heat transport considerations, other equations describing conservation of mass for the chemical constituents involved and conservation of energy

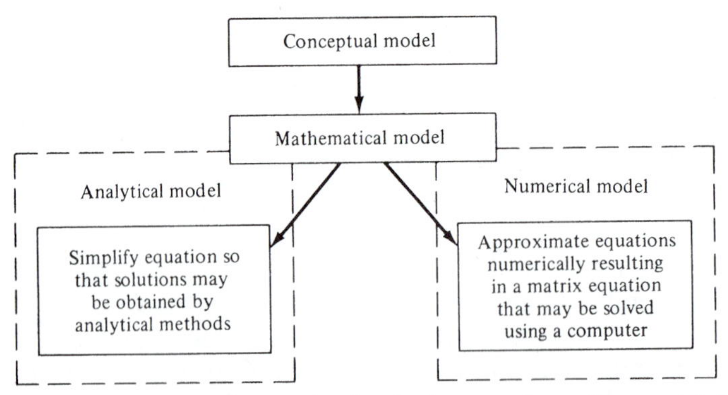

FIGURE 3.12 Logic diagram for developing a mathematical model. (Courtesy of the National Water Well Association, Worthington, OH.)

are required. Typically used relationships are Fick's law for chemical diffusion and Fourier's law for heat transport.

Once a conceptual model has been designed, the next step is to translate it into a workable model by introducing needed assumptions. It is then formulated for solution using numerical methods such as finite difference or finite elements to represent the governing partial differential equations. When using the finite difference approach, the groundwater region is divided into grid elements and the continuous variables are represented as discrete variables at nodal points in the grid. In this way, the continuous differential equation defining head or other features is replaced by a finite number of algebraic equations that define the head or other variables at nodal points. Such models find wide application in prediction of site-specific aquifer behavior. Although a choice of model should be based on the study involved, numerical models have proven to be effective where irregular boundaries, heterogeneities, or highly variable pumping or recharge rates are expected [24]. Figure 3.13 indicates four types of groundwater models and their application.

Normally, several steps are involved in the modeling process. These include data collection and preparation, matching of observed histories, and predictive simulation. The first step in data preparation involves specifying the region's boundaries. These may be physical, such as an impervious layer, or arbitrary, such as the choice of some small subregion. Once the overall boundaries have been defined, the region is divided into discrete elements by superimposing a rectangular or polygonal grid (see Fig. 3.14).

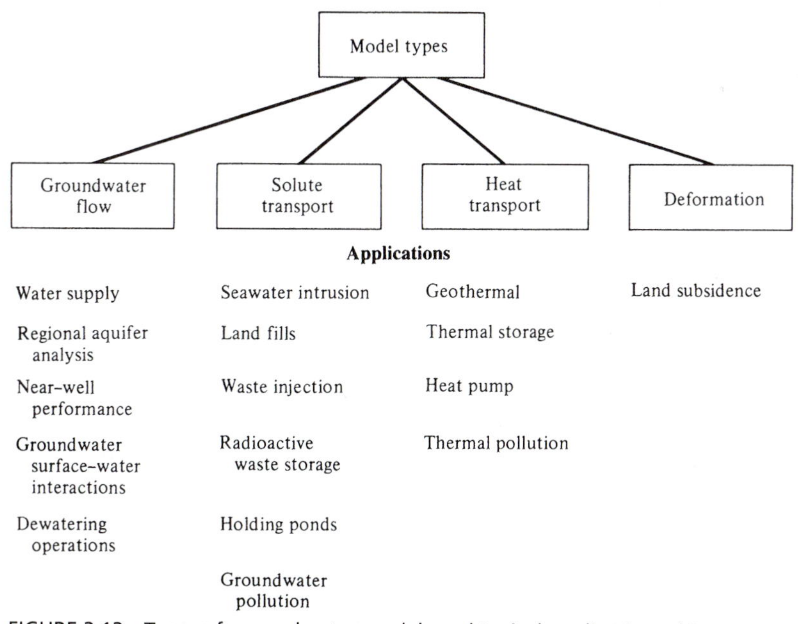

FIGURE 3.13 Types of groundwater models and typical applications. (Courtesy of the National Water Well Association, Worthington, OH.)

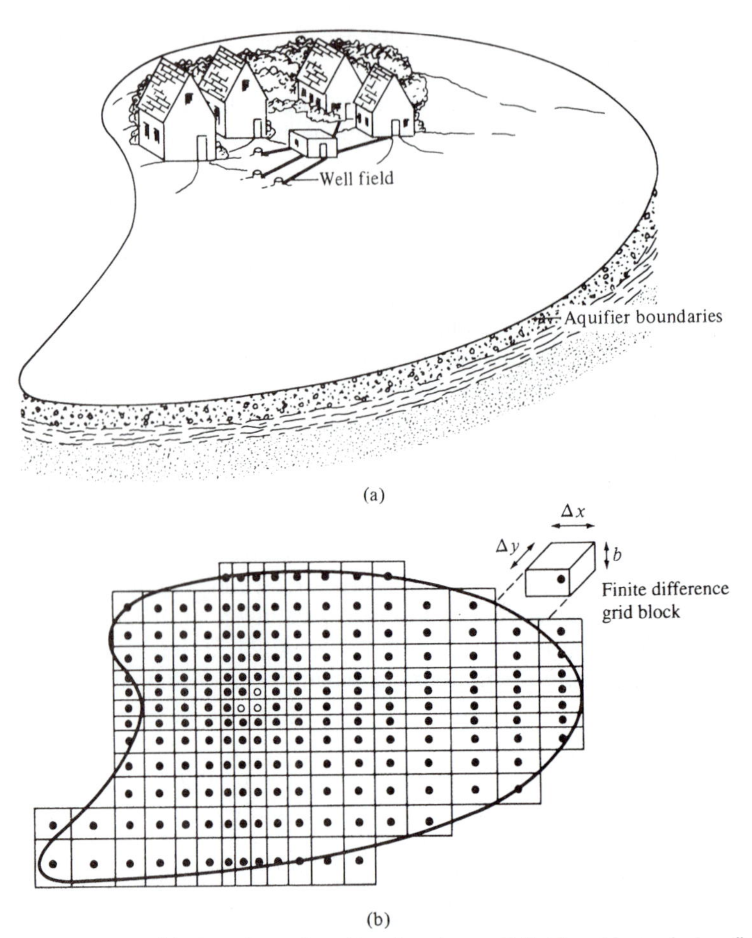

FIGURE 3.14 (a) Map view of aquifer showing well field and boundaries. (b) Finite difference grid for aquifer study, where Δx is the spacing in the x direction, Δy is the spacing in the y direction, and b is the aquifer thickness. Solid dots: block-center nodes; open circles: source-sink nodes. (Courtesy of the National Water Well Association, Worthington, OH.)

After the grid type has been selected, the modeler must specify the controlling aquifer parameters (such as storage coefficients, transmissivities, etc.) and set initial conditions. This must be accomplished for each grid element. Where solute transport models are required, additional parameters such as hydrodynamic dispersion properties must also be specified. Results of model runs include determination of hydraulic heads at node points for each time step during the period of interest. Where solute and heat transport are involved, concentrations of constituents and temperatures may also be determined at the nodes for each time interval.

After the aquifer parameters have been set, the model is operated using the initial values and the output is checked with recorded history. This process is known as history matching. This procedure is used to refine parameter values and to determine boundaries and flow conditions at the boundaries. Comparisons between historic conditions and modeled conditions are made and parameters adjusted until satisfactory fits are obtained. This is known as the *calibration process*. There is no rule that specifies when adequate matching is achieved, however. This determination must be made by the modeler based on an understanding of the problem and the use to which the model's results are to be put.

Once calibration is completed, the model can be used to analyze many different types of management and development options so that the outcomes of these courses of action can be assessed. Observations of model performance under varying conditions is an asset in determining courses of action to be prescribed for future aquifer operation or development. Groundwater models may be used to estimate natural and artificial recharge, effects of boundaries, effects of well location and spacing, effects of varying rates of drawdown and recharge, rates of movement of hazardous wastes, and saltwater intrusion [24].

Although groundwater models have much to recommend them, caution must be exercised if they are not to be misused. According to Prickett, there are three common ways to misuse models [25]: overkill, inappropriate prediction, and misinterpretation. To avoid such pitfalls, the modeler and model user must be fully aware of the limitations and sources of errors in the model used. In particular, underlying assumptions must be well understood.

Numerical models are important tools for analyzing groundwater problems. But the modeler must understand the aquifer being studied, be aware of alternative modeling approaches, and be cognizant of model limitations and restrictions imposed by simplifying assumptions. Used appropriately, models can be powerful decision-making aids. Used inappropriately, they can lead to erroneous and sometimes damaging proposals.

Several public domain computer codes for solving groundwater flow problems are referenced in Table 3.8. These codes become models when the groundwater system being studied is described to the code by inputting the system geometry and known internal operandi (aquifer and flow field parameters, initial and boundary conditions, and water use and flow stresses applied over time to all or parts of the system). Codes fall generally into four major categories: groundwater flow codes, solute transport codes, particle tracking codes, and aquifer test data analysis programs [29].

3.24 SALT WATER INTRUSION

Saltwater contamination of freshwater aquifers can be a major water quality problem in island locations; in coastal areas; and occasionally inland, as in Arizona, where some aquifers contain highly saline waters. Because freshwater is lighter than salt water (specific gravity of seawater is about 1.025), it will usually float above a layer of salt water. When an aquifer is pumped, the original equilibrium is disturbed and salt water replaces the freshwater. Under equilibrium conditions, a drawdown of 1 ft in the freshwater table will result in a rise by the salt water of approximately 40 ft. Pumping rates of wells subject to saltwater intrusion must therefore be strictly controlled. In coastal areas, recharge wells are sometimes used in an attempt to maintain a sufficient head to prevent seawater intrusion. Injection wells have been used effectively in this manner in southern California.

TABLE 3.8 Public Domain Computer Codes for Groundwater Modeling

Acronym for Code	Description	Source	Year
	Groundwater flow models		
PLASM	Two-dimensional finite difference	Ill. SWS	1971
MODFLOW	Three-dimensional finite difference	USGS	1988
AQUIFEM-1	Two- and three-dimensional finite element	MIT	1979
GWFLOW	Package of 7 analytical solutions	IGWMC	1975
GWSIM-II	Storage and movement model	TDWR	1981
GWFL3D	Three-dimensional finite difference	TDWR	1991
MODRET	Seepage from retention ponds	USGS	1992
	Solute transport models		
SUTRA	Dissolved substance transport model	USGS	1980
RANDOMWALK	Two-dimensional transient model	Ill. SWS	1981
MT3D	Three-dimensional solute transport	EPA	1990
AT123D	Analytical solution package	DOE	1981
MOC	Two-dimensional solute transport	USGS	1978
HST3D	3-D heat and solute transport model	USGS	1992
	Particle tracking models		
FLOWPATH	Two-dimensional steady state	SSG	1990
PATH3D	Three-dimensional transient solutions	Wisc. GS	1989
MODPATH	Three-dimensional transient solutions	USGS	1991
WHPA	Analytical solution package	EPA	1990
	Aquifer test analyses		
TECTYPE	Pump and slug test by curve matching	SSG	1988
PUMPTEST	Pumping and slug test	IGWMC	1980
THCVFIT	Pumping and slug test	IGWMC	1989
TGUESS	Specific capacity determination	IGWMC	1990

Note: IGWMC = International Groundwater Modeling Center; Ill. SWS = Illinois State Water Survey; SSG = Scientific Software Group; EPA = Environmental Protection Agency; USGS = U.S. Geological Survey; Wisc. GS = Wisconsin Geological Survey; MIT = Massachusetts Institute of Technology; TDWR = Texas Department of Water Resources; DOE = Department of Energy.

A prime example of freshwater contamination by seawater is noted in Long Island, New York [35]. During the first part of the 20th century, the rate of pumping far exceeded the natural recharge rate. The problem was further complicated because stormwater runoff from the highly developed land areas was transported directly to the sea. This precluded the opportunity for this water to return to the ground. As pumping continued, the water table dropped well below sea level and saline water entered the aquifer. The result was such a serious impairment of local water quality that Long Island was forced to transport its water supply from upper New York State.

3.25 GROUNDWATER RECHARGE

The volumes of groundwater replaced annually through natural mechanisms are relatively small because of the slow rates of movement of groundwater and the limited opportunity for surface water to penetrate the earth's surface. To supplement this natural recharge process, a trend toward artificial recharge has been developing since the turn of the last century. As early as the mid-1950s, over 700 million gallons of water per day were artificially recharged in the United States [36]. This water was derived from natural surface sources and returns from air-conditioning, industrial wastes, and municipal water supplies. The total recharge volume was equal, however, to only about 1.5% of the groundwater withdrawn that year. In California, for example, artificial recharge is at present a primary method of water conservation. During the period 1957–1958, a daily recharge volume of about 560 mil gal was reported for 63 projects in that state alone [36].

Numerous methods are employed in artificial recharge operations. One of the most common plans is the utilization of holding basins. The usual practice is to impound the water in a series of reservoirs arranged so that the overflow of one will enter the next, and so on. These artificial storage works are generally formed by the construction of dikes or levees. A second method is the modified streambed, which makes use of the natural water supply. The stream channel is widened, leveled, scarified, or treated by a combination of methods to increase its recharge capabilities. Ditches and furrows are also used. The basic types of arrangement are the contour type, in which the ditch follows the contour of the ground; the lateral type, in which water is diverted into a number of small furrows from the main canal or channel; and the tree-shaped or branching type, where water is diverted from the primary channel into successively smaller canals and ditches. Where slopes are relatively flat and uniform, flooding provides an economical means of recharge. Normal practice is to spread the recharge water over the ground at relatively small depths so as not to disturb the soil or native vegetation. An additional method is the use of injection wells. Recharge rates are normally less than pumping rates for the same head conditions, however, because of the clogging that is often encountered in the area adjacent to the well casing. Clogging may result from the entrapment of fine aquifer particles, from suspended material in the recharge water that is subsequently strained out and deposited in the vicinity of the well screen, from air binding, from chemical reactions between recharge and natural waters, and from bacteria. For best results the recharge water should be clear, contain little or no sodium, and be chlorinated.

3.26 CONCURRENT DEVELOPMENT OF GROUNDWATER AND SURFACE WATER SOURCES

The maximum practical conservation of our water resources is based on the coordinated development of groundwater and surface water supplies. Geologic, hydrologic, economic, and legal factors affect the process.

Concurrent utilization is primarily founded on the premise of transference of impounded surface water to groundwater storage at optimal rates [37,38]. Annual water requirements are generally met by surface storage; while groundwater storage is used to meet cyclic requirements covering periods of dry years. The operational procedure involves a lowering of groundwater levels during periods of below-average precipitation and a subsequent raising of levels during wet years. Transfer rates of surface waters to underground storage must be large enough to ensure that surface water reservoirs will be drawn down sufficiently to permit impounding significant volumes during periods of high runoff. To provide the required maximum transfer capacity, methods of artificial recharge such as spreading, ponding, injecting, or returning flows from irrigation must be used.

The coordinated use of groundwater and surface water sources results in the provision of larger quantities of water at lower costs. As an example, it has been found that the conjunctive operation of the Folsom Reservoir (California) and its groundwater basin yields a conservation and utilization efficiency of approximately 82% as compared with about 51% efficiency for the operation of the surface reservoir alone [39]. There is little doubt that the inclusion of groundwater resources should be given very careful consideration in future planning for large-scale water development projects.

In general, the analysis of a conjunctive system consisting of a dam and an aquifer requires the solution of three fundamental problems. The first is to establish the design criteria for the dam and the recharge facilities. The second is to determine the service area for the combined system. Finally, a set of operating rules that defines the reservoir drafts and pumpages to be taken from the aquifer is required. A mathematical model for an analysis such as this has been proposed by Buras [37].

3.27 AQUIFER STORAGE AND RECOVERY (ASR)

Aquifer storage and recovery (ASR) is a process involving the storage of water in an aquifer (by recharge) during wet periods and storing it for removal during dry periods when it is needed. The underground storage basin can be operated in a fashion similar to that of a surface water reservoir, but it eliminates evaporative losses, which can be significant, and it does not require inundating the large land areas that are associated with surface water reservoirs. ASR systems can also be called upon to support water supply needs during severe multiyear droughts. The concept is not new, having been used in the United States for over 30 years. The largest operational site as of 2000 is that in Las Vegas, Nevada, where there are about 30 ASR wells capable of delivering 100 mgd. In 1983, ASR systems became operational in Florida. Figure 3.15 illustrates the ASR operating system.

The Comprehensive Everglades Restoration Plan (CERP) includes ASR as a principal component [www.evergladesplan.org/docs/asr_whitepaper.pdf]. The CERP projects over 300 ASR wells in south Florida with a total capacity of 1.6 billion gallons per day by the year 2025. The ASR project is intended to provide storage for the capture of excess water in south Florida that, in 2000, was discharged to the ocean. The scale of the ASR project is without precedent, however, and potential problems have been identified that are now under study. They include source water quality; uncertainties about the regional hydrogeology of the Floridan Aquifer system; rock fracturing; regional changes in pattern of flow; water quality changes; mercury bioaccumulation; and anticipated recovery rates and the volume required for recharge.

Full-scale implementation of the ASR project is expected to proceed over at least a 20-year time period. Pilot projects and an ASR feasibility study will facilitate an informed decision about the feasibility of an ASR project of the scale originally envisioned. For more information on the CERP, see www.evergladesplan.org/.

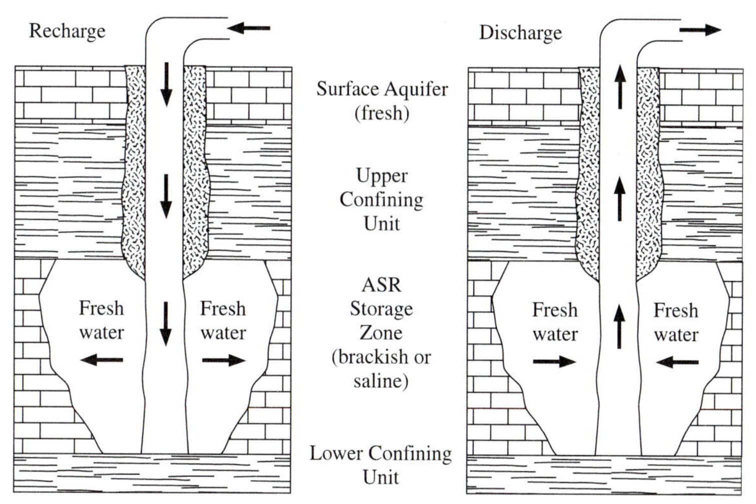

FIGURE 3.15 Schematic diagram of an aquifer storage and recovery system. (Everglades Restoration Plan, www.evergladesplan.org/docs/asr_whitepaper.pdf.)

PROBLEMS

3.1 Compare the amounts of water required by the various users in your state. What is the relative worth of water in its various uses?

3.2 A flow of 100 mgd is to be developed from a 190-mi^2 watershed. At the flow line the area's reservoir is estimated to cover 3900 acres. The annual rainfall is 40 in, the annual runoff is 14 in, and the annual evaporation is 49 in. Find the net gain or loss in storage this represents. Calculate the volume of water evaporated in acre · ft and cubic meters.

3.3 A flow of 4.8 m^3/s is to be developed from a 500-km^2 watershed. At the flow line, the area's reservoir is estimated to cover 1700 hectares. The annual rainfall is 97 cm, the annual runoff is 30 cm, and the annual evaporation is 120 cm. Find the net gain or loss in storage this represents. Calculate the volume of water evaporated in cubic meters.

3.4 Discuss how you would go about collecting data for an analysis of the water budget of a region. What agencies would you contact? What other sources of information would you seek out?

3.5 For an area of your choice, make a plot of mean monthly precipitation versus time. Explain how this fits the pattern of seasonal water uses for the area. Will the form of precipitation be an important consideration?

3.6 Given the following 10-yr record of annual precipitation, plot a rough precipitation frequency curve. Tabulate the data to be plotted and show the method of computation. The data are annual precipitation in inches: 28, 21, 33, 26, 29, 27, 19, 28, 18, 22. (*Note:* The frequency in percent of years is $1/T_R \times 100$.)

3.7 Given the 10-yr record of annual precipitation that follows, develop and plot a precipitation frequency curve. The precipitation values in cm are 70, 54, 89, 66, 75, 69, 48, 72, 46, and 56.

3.8 An impounding reservoir is expected to provide a constant draft of 448 million gallons per square mile per year. The following record of monthly mean inflow values (mg per sq. mi. per month) is representative of the critical or design period. Find the storage required.

Mo	F	M	A	M	J	J	A	S	O	N	D	J	F
In	31	54	90	10	7	8	2	28	42	108	98	22	50

3.9 Over a 100-mi^2 surface area, the average level of the water table for an unconfined aquifer has dropped 10 ft due to the removal of 128,000 acre · ft of water from the aquifer. Determine the storage coefficient. The specific yield is 0.2 and the porosity is 0.22.

3.10 Over a 100-mi^2 surface area, the average level of the piezometric surface for a confined aquifer has dropped 400 ft due to long-term pumping. Determine the volume of water in acre · ft pumped from the aquifer. The porosity is 0.3 and the coefficient of storage is 0.0002.

3.11 Find a maximum reservoir storage requirement if a uniform draft of 726,000 gpd/mi^2 from a river is to be maintained. The following record of average monthly runoff values is given (mg per sq. mi. per month):

Mo	A	M	J	J	A	S	O	N	D	J	F	M	A	M	J
R	97	136	59	14	6	5	3	7	19	13	74	96	37	63	49

3.12 Using the information given in Table 3.4, plot recurrence interval in years as the ordinate, design period in years as the abscissa, and construct a series of recurrence interval–design period probabilities that an event will not be exceeded during the design period. Use arithmetic coordinate paper. [*Note:* To conform to this, probabilities in the table must be subtracted from 1.0. Where sufficient information is not provided by the table, probabilities may be computed using $P_n = (1 - 1/T_R)^n$, *where* n is the design period in years.]

3.13 Given the following 50-month record of mean monthly discharge, find the magnitude of the 20-month low flow. The consecutive average monthly flows (cfs) were 14, 17, 19, 21, 18, 16, 18, 25, 29, 32, 34, 33, 30, 28, 20, 23, 16, 14, 12, 13, 16, 13, 12, 12, 13, 14, 16, 13, 12, 11, 10, 12, 10, 9, 8, 7, 6, 4, 6, 7, 8, 9, 11, 9, 8, 6, 7, 9, 13, 17.

3.14 Find the expected life of a reservoir having an initial capacity of 45,000 acre · ft. The average annual inflow is 63,000 acre · ft, and a sediment inflow of 180 acre · ft/yr is reported. Consider the useful life of the reservoir to be exceeded when 80% of the original capacity is lost. Use 5000-acre · ft volume increments. From Figure 3.4, obtain values of percent sediment trapped.

3.15 Given the following data relating mean annual change in groundwater level in ft to mean annual draft in thousands of acre · ft. Find the safe yield.

Change in GW level	+1	+2	−1	−3	−4	+1.5	+1.2	−2.6
Mean annual draft	23	19	31	42	44	21	19	33

3.16 Given the following data relating mean annual change in groundwater level in ft to mean annual draft in thousands of acre · ft. Find the safe yield.

GW level	−2	+2.5	−4	+0.5	−3	+2	−2.5	−0.5	+0.5	−2
Ann. draft	26	14	36	21.5	32	13	32.5	26	19	28

3.17 What would the maximum continuous constant yield be from a reservoir having a storage capacity of 750 acre · ft? Give your results in acre · ft/yr and m^3/yr.

3.18 If a constant annual yield rate of 1500 gal/min was required, what reservoir capacity would be needed to sustain it? Give the capacity in acre · ft/yr.

3.19 A mean draft of 100 mgd is to be developed from a 150 mi^2 catchment area. At the flow line, the reservoir is estimated to be 4000 acres. The annual rainfall is 38 in, the mean annual runoff is 13 in, and the mean annual evaporation is 49 in. Find the net gain or loss in storage that this represents. Compute the volume of water evaporated. State this figure in a form such as the number of years the volume could supply a given community.

3.20 For the following data, and using the well and flow-net configuration of Figure 3.10, find the discharge using a flow-net solution. The well is fully penetrating; $K = 2.87 \times 10^{-4}$ ft/s, (a) = 180 ft, (b) = 43 ft, and (c) = 50 ft.

3.21 Rework Problem 3.20 if $K = 8.4 \times 10^{-5}$ m/s, (a) = 100 m, (b) = 22 m, and (c) = 35 m.

3.22 Use the following data: $Q = 60{,}000$ ft^3/day, $T = 6200$ ft^3/day, $t = 30$ days, $r = 1$ ft, and $S_c = 6.4 \times 10^{-4}$. Consider this a nonequilibrium problem. Find the drawdown s. Note that for

$$u = 8.0 \times 10^{-10} \qquad W(u) = 20.37$$

$$u = 9.0 \times 10^{-10} \qquad W(u) = 20.25$$

3.23 Determine the permeability of an artesian aquifer being pumped by a fully penetrating well. The aquifer is composed of medium sand and is 90 ft thick. The steady-state pumping rate is 850 gpm. The drawdown of an observation well 50 ft away is 10 ft, and the drawdown in a second observation well 500 ft away is 1 ft.

3.24 An 18-in well fully penetrates an unconfined aquifer of 100-ft depth. Two observation wells located 100 and 235 ft from the pumped well have drawdowns of 22.2 and 21 ft respectively. If the flow is steady and $K_f = 1320$ gpd/ft^2, what would be the discharge?

3.25 Find the drawdown at an observation point 200 ft away from a pumping well. Given that $T = 3.0 \times 10^4$ gpd/ft, the pumping time is 12 days, $S_c = 3.0 \times 10^{-4}$, and $Q = 300$ gpm.

3.26 A well is being pumped at a constant rate of 0.0038 cubic meters per second. Given that $T = 0.0028$ m^2/s, $r = 90$ m, and the storage coefficient = 0.00098, find the drawdown in the observation well for a time period of (a) 1000 s and (b) 20 hr.

3.27 A well is being pumped at a constant rate of 0.004 m^3/s. Given that $T = 0.0028$ m^2/s, $r = 100$ m, and the storage coefficient = 0.001, find the drawdown in the observation well for a time period of (a) 1 hr and (b) 24 hr.

3.28 A well is being pumped at a constant rate of 0.003 m^3/s. Given that $T = 0.0028$ m^2/s, the storage coefficient = 0.001, and the time since pumping began is 12 hr, find the drawdown in an observation well for a radial distance of (a) 150 m and (b) 500 m.

3.29 A 12-in well fully penetrates a confined aquifer 100 ft thick. The coefficient of permeability is 600 gpd/ft^2. Two test wells located 45 and 120 ft away show a difference in drawdown between them of 8 ft. Find the rate of flow delivered by the well.

3.30 Determine the permeability of an artesian aquifer being pumped by a fully penetrating well. The aquifer is composed of medium sand and is 100 ft thick. The steady-state pumping rate is 1200 gpm. The drawdown in an observation well 75 ft away is 14 ft, and the drawdown in a second observation well 500 ft away is 1.2 ft. Find K_f in gpd/ft^2.

3.31 Consider a confined aquifer with a coefficient of transmissibility $T = 680$ ft^3/day/ft. At $t = 5$ min, the drawdown $s = 5.6$ ft; at 50 min, $s = 23.1$ ft; and at 100 min, $s = 28.2$ ft. The observation well is 75 ft away from the pumping well. Find the discharge of the well.

3.32 Assume that an aquifer is being pumped at a rate of 300 gpm. The aquifer is confined and the pumping test data are given below. Find the coefficient of transmissibility T and the storage coefficient S for $r = 60$ ft.

Time since pumping started (min)	1.3	2.5	4.2	8.0	11.0	100.0
Drawdown s (ft)	4.6	8.1	9.3	12.0	15.1	29.0

3.33 Find the drawdown at an observation point 250 ft away from a pumping well, given that $T = 3.1 \times 10^4$ gpd/ft, the pumping time is 10 days, $S_c = 3 \times 10^{-4}$, and $Q = 280$ gpm.

3.34 A 12-in. well fully penetrates a confined aquifer 100 ft thick. The coefficient of permeability is 600 gpd/ft^2. Two test wells located 40 and 120 ft away show a difference in drawdown between them of 9 ft. Find the rate of flow delivered by the well.

3.35 Find the permeability of an artesian aquifer being pumped by a fully penetrating well. The aquifer is 130 ft thick and is composed of medium sand. The steady-state pumping rate is 1300 gpm. The drawdown in an observation well 65 ft away is 12 ft, and in a second well 500 ft away it is 1.2 ft. Find K_f in gpd/ft^2.

3.36 Consider a confined aquifer with a coefficient of transmissibility of 700 ft^3/day ft. At $t = 5$ min, the drawdown is 5.1 ft; at 50 min, $s = 20.0$ ft; at 100 min, $s = 26.2$ ft. The observation well is 60 ft from the pumping well. Find the discharge from the well.

3.37 An 18-in well fully penetrates an unconfined aquifer 100 ft deep. Two observation wells located 90 and 235 ft from the pumped well are known to have drawdowns of 22.5 ft and 20.6 ft, respectively. If the flow is steady and $K_f = 1300$ gpd/ft^2, what is the discharge?

3.38 A well is pumped at the rate of 500 gpm under nonequilibrium conditions. For the data given below, find the formation constants S and T. Use the Theis method.

r^2/t	**Average Drawdown H (ft)**
1,250	3.24
5,000	2.18
11,250	1.93
20,000	1.28
45,000	0.80
80,000	0.56
125,000	0.38
180,000	0.22
245,000	0.15
320,000	0.10

3.39 A well fully penetrates the 100-ft depth of a saturated unconfined aquifer. The drawdown at the well casing is 40 ft when equilibrium conditions are established using a constant discharge of 50 gpm. What is the drawdown when equilibrium is established using a constant discharge of 66 gpm?

3.40 A confined aquifer 80 ft deep is being pumped under equilibrium conditions at a rate of 700 gpm. The well fully penetrates the aquifer. Water levels in observation wells 150 and 230 ft from the pumped well are 95 and 97 ft, respectively. Find the field coefficient of permeability.

REFERENCES

[1] E. A. Ackerman and G. O. Löf, *Technology in American Water Development* (Baltimore: The Johns Hopkins Press, 1959).

[2] H. E. Thomas, "Underground Sources of Water," *The Yearbook of Agriculture, 1955* (Washington, DC: U.S. Government Printing Office, 1956).

[3] W. Viessman, Jr., and C. Welty, *Water Management: Technology and Institutions* (New York: Harper & Row, 1985), 44–45.

[4] W. Viessman, Jr., "United States Water Resources Development," *Natl. Forum* 69, (1) (Winter 1989): 6, 7.

[5] P. A. Hamilton, "Water-Quality Patterns in Some of the Nation's Major River Basins and Aquifers," *Water Resources Impact*, 4, no. 4 (July 2002).

[6] J. Gelt, "Microbes Increasingly Viewed as Water Quality Threat," *Arroyo* 10, no. 2, Water Resources Research Center, University of Arizona, Tucson, AZ (March 1998).

[7] "Pharmaceuticals in Our Water Supplies," *Arizona Water Resource*, Water Resources Research Center, University of Arizona, Tucson, AZ (July–August 2000).

[8] J. Hirshleifer, J. DeHaven, and J. Milliman, *Water Supply—Economics, Technology, and Policy* (Chicago: University of Chicago Press, 1960).

[9] W. Viessman, Jr., and G. L. Lewis, *Introduction to Hydrology*, 5th ed. (Upper Saddle River, NJ: Prentice Hall, 2003).

[10] Task Group A4, D1, "Water Conservation in Industry," *J. Am. Water Works Assoc.* 45 (December 1958).

[11] W. B. Langbein, "Water Yield and Reservoir Storage in the United States," *U.S. Geol. Survey Circular* (1959).

[12] W. Rippl, "The Capacity of Storage Reservoirs for Water Supply," *Proc. Inst. Civil Engrs.* (London) 71 (1883): 270.

[13] R. A. Freeze and J. A. Cherry, *Groundwater* (Upper Saddle River, NJ: Prentice Hall, 1979).

[14] D. K. Todd, *Ground Water Hydrology* (New York: Wiley, 1960).

[15] Henri Darcy, *Les fontaines publiques de la ville de Dijon* (Paris: V. Dalmont, 1856).

[16] O. E. Meinzer, "Outline of Methods of Estimating Ground Water Supplies," U.S. Geol. Survey Water Supply Paper No. 638C (1932).

[17] R. C. Heath, "Groundwater Regions of the United States," Geological Survey Water Supply Paper No. 2242 (Washington, DC: U.S. Government Printing Office, 1984).

[18] U.S. Environmental Protection Agency, "The Report to Congress: Waste Disposal Practices and Their Effects on Groundwater," Executive Summary, U.S. EPA, PB 265–364 (1977).

[19] M. E. Harr, *Groundwater and Seepage* (New York: McGraw-Hill, 1962).

[20] M. Muskat, *The Flow of Homogeneous Fluids Through Porous Media* (Ann Arbor, MI: J. W. Edwards, 1946).

[21] G. Thiem, *Hydrologische Methoden* (Leipzig: Gebhart, 1906), 56.

[22] C. V. Theis, "The Relation Between the Lowering of the Piezometric Surface and the Rate and Duration of Discharge of a Well Using Ground Water Storage," *Trans. Am. Geophys. Union* 16 (1935): 519–524.

[23] H. H. Cooper, Jr., and C. E. Jacob, "A Generalized Graphical Method for Evaluating Formation Constants and Summarizing Well-Field History," *Trans. Am. Geophys. Union* 27 (1946): 526–534.

[24] J. W. Mercer and C. R. Faust, *Ground-Water Modeling* (Worthington, OH: National Water Well Association, 1981).

[25] T. A. Prickett, "Ground-water Computer Models—State of the Art," *Ground Water* 17 (2) (1979): 121–128.

[26] C. A. Appel and J. D. Bredehoeft, "Status of Groundwater Modeling in the U.S. Geological Survey," *U.S. Geol. Survey Circular* 737 (1976).

[27] Y. Bachmat, B. Andres, D. Holta, and S. Sebastian. *Utilization of Numerical Groundwater Models for Water Resource Management*, U.S. Environmental Protection Agency Report EPA-600/8-78-012.

[28] J. E. Moore, "Contribution of Ground-Water Modeling to Planning," *J. Hydrol.* 43 (October 1979).

[29] National Research Council, *Ground Water Models—Scientific and Regulatory Applications*, Water Science and Technology Board, Commission on Physical Sciences, Mathematics, and Resources, (Washington, DC: National Academy Press, 1990).

[30] R. J. Charbeneau, *Groundwater Hydraulics and Pollutant Transport* (Upper Saddle River, NJ: Prentice Hall, 2000).

[31] M. McDonald and A. Harbaugh, *A Modular Three-Dimensional Finite-Difference Ground-Water Flow Model Book 6 Modeling Techniques* (Washington, DC: Scientific Software Group, 1988).

[32] M. P. Anderson and W. W. Woessner, *Applied Groundwater Modeling: Simulation of Flow and Advective Transport*, (Orlando, FL: Academic Press, 1992).

[33] J. Istok, *Groundwater Modeling by the Finite Element Method*, Monograph 13, (Washington, DC: American Geophysical Union, 1989).

[34] H. F. Wang and M. P. Anderson, *Introduction to Groundwater Modeling: Finite Difference and Finite Element Methods*, (San Francisco: W. H. Freeman, 1982).

[35] J. F. Hoffman, "How Underground Reservoirs Provide Cool Water for Industrial Uses," *Heating, Piping, Air Conditioning* (October 1960).

[36] R. C. Richter and R. Y. D. Chun, "Artificial Recharge of Ground Water Reservoirs in California," *Proc. Am. Soc. Civil Engrs., J. Irrigation Drainage Div.* 85 (IR4) (December 1959).

[37] N. Buras, "Conjunctive Operation of Dams and Aquifers," *Proc. Am. Soc. Civil Engrs., J. Hyraulics Div.* 89 (HY6) (November 1963).

[38] F. B. Clendenen, "A Comprehensive Plan for the Conjunctive Utilization of a Surface Reservoir with Underground Storage for Basin-wide Water Supply Development: Solano Project, California" (D. Eng. thesis, University of California, Berkeley, 1959), 160.

[39] "Ground Water Basin Management," *Manual of Engineering Practice*, no. 40 (New York: American Society of Civil Engineers, 1961).

CHAPTER 4

Water Use

OBJECTIVES

The purpose of this chapter is to:

- Define the principal water-using sectors
- Discuss water use trends
- Present techniques for forecasting water requirements

The water supply problem is one of balancing supply and demand. Availability of water sources, geographically and temporally, the quality of these resources, the rates at which they are replenished and depleted, and the demands placed upon them by water users are determining factors in water management strategies. Estimates of future water uses, uncertain as they might be, are fundamental to efficient and equitable allocation of water supplies. These estimates depend on an ability to forecast changes in population, agricultural and industrial activity, economic conditions, technology, and social and other related factors.

Topics associated with water supply and use include water sources; water quality; water use permitting; estimating water requirements; inventorying water supplies; defining "reasonable beneficial uses"; infrastructure development, operation, maintenance, and replacement; policies for allocating water among competing users; strategic planning; interbasin transfers; saline water conversion; financing; education; conservation and reuse; research; drought management; coordinating water supply and water quality management programs; climate change; and assessing risks associated with proposed courses of action. Storm water runoff and sewage flows, although degraded in quality, must also be considered as source waters.

4.1 WATER SOURCES

Historically, natural water sources have been considered almost exclusively by those desiring to increase their water supplies. But now, these principal water stocks—the rivers, streams, lakes, aquifers, and oceans—must be supplemented by other nontraditional sources of water supply such as discharges from wastewater treatment plants, urban storm water flows, brackish and saline water conversion, and irrigation return flows. These flows are often significant, and at present many of them are wasted. Up to about 75% of treated potable water is returned to a discharge point after its use. Along Florida's lower east coast, for example, the quantities of water generated for potable water use and then discharged to the ocean as treated wastewater flows constitute a significant fraction of the water used daily in that area. The infrastructure needed to reuse all return flows efficiently is not in place and the economic and public health aspects of reuse are constraining factors, but these water sources should be identified in water resource assessments, and they should be considered in water supply planning strategies. In addition, many localities have access to brackish or seawater sources. And though the cost of converting saline to freshwater has constrained saline water conversion programs in the past, conversion costs are decreasing. In Florida's Tampa Bay region, the nation's largest seawater desalination plant went online in March 2003 (www.tampabaywater.org). While initially producing 4.9 million gallons per day (mgd), the plant will eventually produce 25 mgd and can be expanded to 35 mgd. The wholesale cost for desalinated water for 30 years is projected to average $2.49 per thousand gallons, competitive with traditional water supply costs ranging from about $1.00 to $8.00 per thousand gallons. An average value in the United States in 2003 of $2.05 per thousand gallons was reported by *Water World*. [1] As water shortages become more frequent and widespread, conversion of brackish and saline waters will become more common. Importation of water across political and natural boundaries may also be a viable water supply option, providing that political, economic, and environmental circumstances are favorable.

Reclaimed Water Reclamation of water that has been used for a variety of purposes offers an attractive alternative to the development or expansion of natural water sources. Municipal waste flows are widely used, especially in the more arid western and southwestern states. In California and Texas, for example, reuse projects are numerous. Wastewater reuse appears to be particularly attractive in areas where rainfall is low, evaporation is high, irrigation water use is intense, and interbasin transfers of water are being practiced or planned. Environmental regulations, water scarcity, and economic factors are expected to accelerate reuse of waste streams. This will undoubtedly affect water use trends in the future, especially in the industrial and agricultural sectors. Figure 4.1 gives a good indication of the significance of wastewater reclamation to the water supply future of Orange County, California. By 2010, it is projected that about 20% of the Orange County Water District's water needs will be met by the use of reclaimed wastewater. Benefits of wastewater reuse include improved quality of surface waters, preservation of higher-quality water for potable consumption, and added recreational opportunities [2].

Treated wastewaters may be used directly or indirectly. In direct reuse, wastewater is treated and then delivered to a user without intervening travel dilution in natural surface water or groundwater bodies [3]. Connection of a municipal wastewater plant to an irrigation site provides a direct water supply to that site. Indirect reuse involves a middle step between generation of reclaimed water and reuse. This commonly includes discharge, retention, and mixing with another water source before reuse. Recycling refers to two or more consecutive uses of water by the same

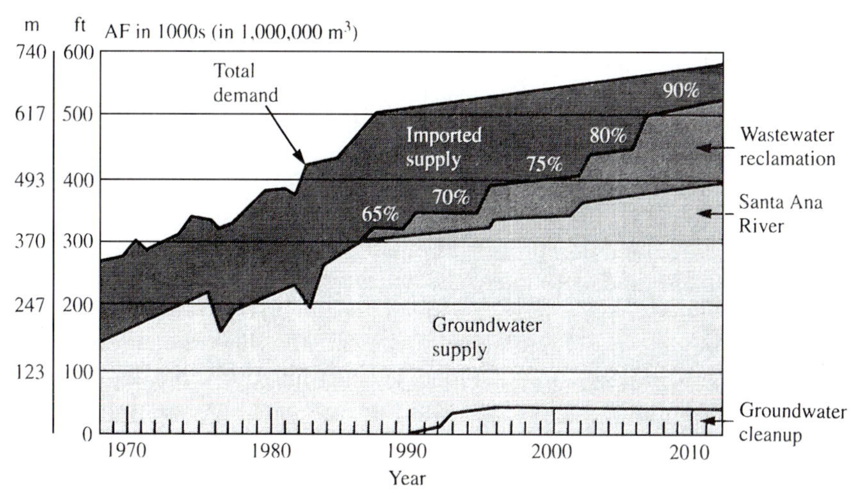

FIGURE 4.1 Projected distribution of source waters for Orange County, California, water supply to the year 2010. *Source:* From L. W. Owen and W. R. Mills, "California's Orange County Water District: A Model for Comprehensive Water Resources Management," in *Water Resources: Planning and Management and Urban Water Resources*, Proceedings of the 18th Annual Conference and Symposium of the Water Resources Planning and Management Division of ASCE, New York, NY, May 1991, p. 5.

business, industry, or person in a coordinated, planned manner, sometimes with partial treatment between uses.

The most widely available and least variable source of wastewater for reuse is municipal wastewater. It can be relied on to provide a dependable continuous flow having fairly stable physical, chemical, and biological characteristics. Reuse of municipal wastewater may be accomplished in a number of ways, with treatment ranging from none to the most advanced systems available, depending on the end use of the water. Municipal wastewater reuses are varied. In California, municipal wastewaters are reused for fiber and seed crop irrigation, landscape irrigation, orchard and vineyard irrigation, processed food crop irrigation, groundwater recharge, food crop irrigation (not processed), restricted recreational impoundments, pasture for dairy animals, and unrestricted recreational impoundments [4].

Manufacturing processes can contribute significantly to the amount of wastewater generated in an area. However, the constituents in industrial wastes may limit options for reusing them. Nevertheless, on-site wastewater reuse by industry has become more prevalent as environmental standards set by regulatory agencies have tightened.

In agriculture, return flows from irrigation projects are a potential source of reusable water. These flows are often contaminated with salts leached from the soil, however, and treatment is often required before they can be reused.

The objective of wastewater reclamation is to provide a water supply of adequate quality to meet the standards of the proposed reuse application. For municipal wastewater flows that have been subjected to secondary treatment, pollutants that may still require removal include: nitrates, phosphates, total dissolved solids, microorganisms, and refractory organics such as trace levels of pesticides [3,5]. Depending on the type of reuse, some of these constituents may enhance the value of the product water and should remain. For example, phosphates and nitrates are desirable in reclaimed water scheduled for reuse in irrigation because they are useful nutrients for the crops that are to be grown. A number of states, including California, have established water quality standards for various types of reuse [6].

To restore wastewaters to drinking water quality, *tertiary* or advanced treatment processes are usually required. These proven wastewater reclamation technologies are widely used and some are readily available as "off-the-shelf" packaged units.

4.2 WATER-USING SECTORS

Decisions on developing and allocating water resources must be based on availability, quality, type and rate of use of the resource. A complication is that source waters must usually be allocated to numerous competing uses. When supplies are limited, conflicts among users may become intense and tradeoffs will have to be made.

Agriculture

Water is critical to agriculture. In arid and semiarid regions without a dependable water supply, there is little chance of achieving success in agricultural operations. In humid areas, rainfall is often adequate to produce good crops, but even there, supplemental irrigation is being relied on more and more to prevent crop failures and to improve the quality of the products produced. In the eastern United States, the increasing use of supplemental irrigation highlights the importance of sustainable water supplies for crop production [7].

Irrigation water requirements are generally seasonal, varying with climate and type of crop. In humid regions, water withdrawals for irrigation may range from about 10% of the total annual demand in May to 30% in September, while in arid and semiarid locations, rates of withdrawal are nearly uniform during the irrigation season. The quantities and timing of water uses for irrigation conflict with many other uses. This creates special problems in arid regions. Irrigators promote storing as much water as possible during the winter, when hydropower producers are eager to release flows into their turbines to generate electricity, for example. Of considerable importance is the fact the water used for irrigation (about 40% of U.S. freshwater use in 2000) is consumptively used (evaporated or transpired) and is thus unavailable for reuse in the region.

Thermoelectric Power

The principal use of water in electric generating facilities is for cooling to dissipate rejected heat. The amount of cooling water withdrawn depends on plant size, generator thermal efficiency, cooling heat transfer efficiency, and institutionally regulated limits on effluent temperatures. In 2000, generation of electricity ranked first in total water withdrawals in the United States (fresh plus saline water). About 70% of the amount used was from freshwater sources [8]. If the demands for cooling water increase, limitations on freshwater resources will probably stimulate even greater interest in developing coastal sites with their potential for once-through cooling using saline water. *Once-through cooling* is the passage of water through cooling units followed by direct release to a receiving body of water without any recycling through water-cooling facilities. Withdrawals for once-through cooling are large, but little water is used consumptively.

Cities and Other Communities

In 2000, central water supply systems furnished water to about 285 million people residing in municipal areas of the United States [8]. Approximately 45.2 million people living outside of these service areas had their own domestic systems. The principal domestic and commercial water uses are for drinking, cooking, meeting sanitary needs, lawn watering, swimming pool maintenance, street cleaning, firefighting, and various aspects of city and park maintenance. Although the public water use sector is vital to our well-being, since it furnishes much of our drinking water, the total amount of water used by this sector is small when compared to water-using sectors such as irrigation and thermoelectric cooling. In 2000, the United States Geological Survey (USGS) reported that this sector represented about 12.5% of the nation's freshwater withdrawals. In 1975, the national average daily per capita use from public supplies was about 170 gpcd; in 1990, it had increased to about 184 gpcd [7,8]. It is noteworthy that total public water withdrawals declined to about 180 gpcd in 2000. Since 1990, public per capita water use has been declining slightly even though the population has been increasing. Average domestic water use from self-supplied systems averaged about 79 gpcd in 2000. As a result of more conservative water use, and more efficient plumbing fixtures, per capita water use may show further declines in the future. Note that the unit of measurement in the SI system is liters per capita per day (lpcd).

Residential water use rates are continually fluctuating, from hour to hour, from day to day, and from season to season. Average daily winter consumption is only about 80% of the annual daily average, whereas summer consumption averages are about 25% greater than the annual daily average. Figure 4.2 compares a typical winter day with a typical maximum summer day in Baltimore, Maryland. Note the hourly fluctuations and the tendency toward two peaks. Studies by Wolff indicate that hydrographs of systems serving predominantly residential communities generally show two peaks, the first between the hours of 7 A.M. and 1 P.M., the second in the evening between 5 and 9 P.M. [9]. During the summer, when lawn-sprinkling demands are high, the second peak is usually the greatest, while during the colder months or during periods of high rainfall, the morning peak is commonly the larger of the two.

In Figure 4.3 the effects of lawn-sprinkling demands are strikingly demonstrated. This figure shows the effects of rainfall on two different days—one a hot, dry day with no antecedent rainfall for 4 days and the other a day during which rainfall in excess of 1 in was recorded. Lawn sprinkling has been found to represent as much as 75% of total daily volumes and as much as 95% of peak hourly demands where large residential lots are involved [10]. Peak hourly demands have been found to vary from average daily demands by as much as 1500% [9], but there is no rule that can be universally applied to predetermine variations. Variations are a function of the type of development, its age,

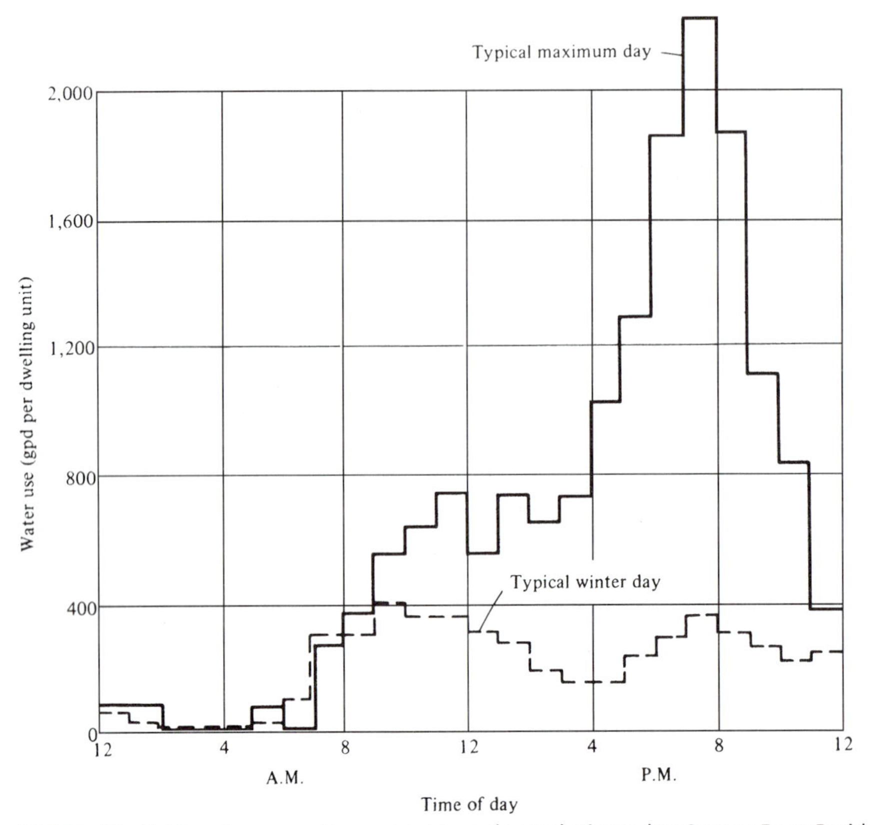

FIGURE 4.2 Daily water use patterns, maximum day and winter day. *Source:* From Residential Water-Use Research Project, Johns Hopkins University and Federal Housing Administration, 1963.

geographic location, and extent of conservation practices. Figures 4.4 and 4.5 provide some guidance on peak demand values.

Firefighting demands must also be considered in municipal water system design. The annual volumes required for firefighting purposes are small, but during periods of need the demand may be large and may govern the design of distribution systems, distribution storage, and pumping equipment. Recommendations as to the quantities of water to be used in firefighting in high-valued community districts have been published by the National Board of Fire Underwriters (American Insurance Association) [11].

Firefighting requirements for residential areas vary from 500 to 3000 gpm, the required rate being a function of population density. Hydrant pressures should generally exceed 20 psi where motor pumpers are used; otherwise, pressures in excess of 100 psi might be required. If recommended fire flows cannot be maintained for the indicated time periods, community fire insurance rates may be adjusted upward.

The coincident draft (flow to be expected at the time the fire is being fought) during firefighting is usually considered equal to the maximum daily demand, since the probability of the maximum rate of water usage for community purposes occurring simultaneously with a major conflagration is slight.

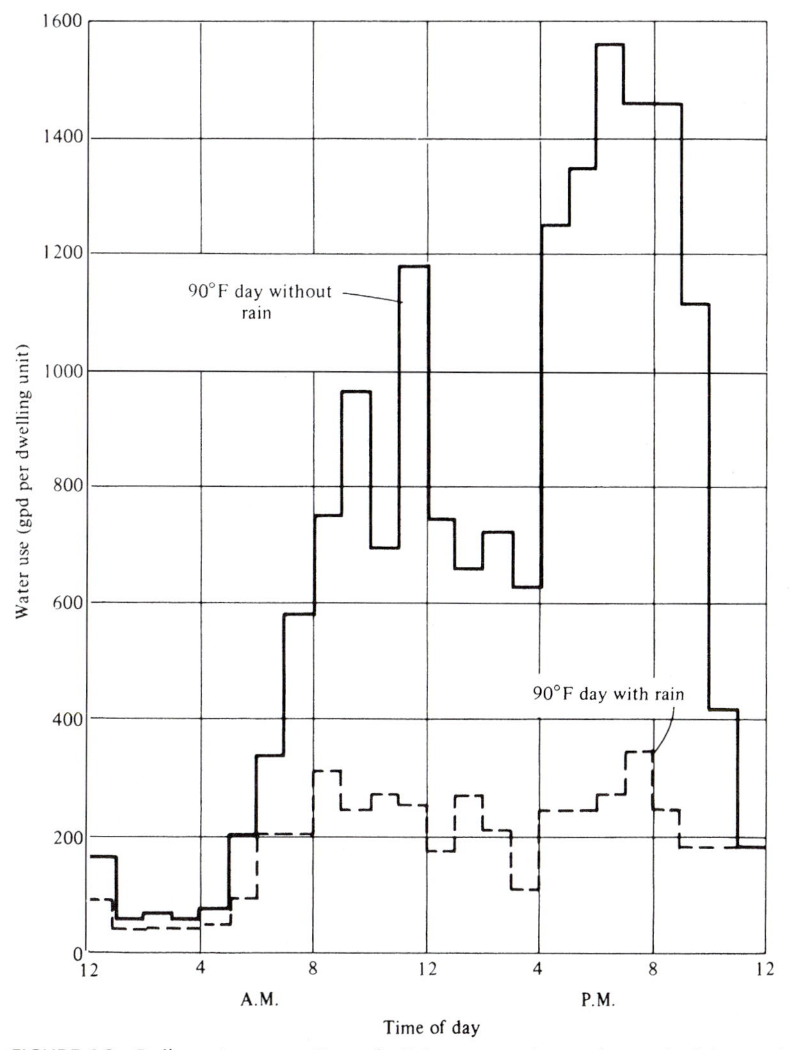

FIGURE 4.3 Daily water use patterns in R-6 area: maximum day and minimum day. *Source:* From Residential Water-Use Research Project, Johns Hopkins University and Federal Housing Administration, 1963.

The National Board of Fire Underwriters has proposed the following formula for computing fire flows of communities having less than 200,000 people:

$$Q = \sqrt{P}(1 - 0.01\sqrt{P}) \tag{4.1}$$

where

Q = demand in gpm

P = population in thousands

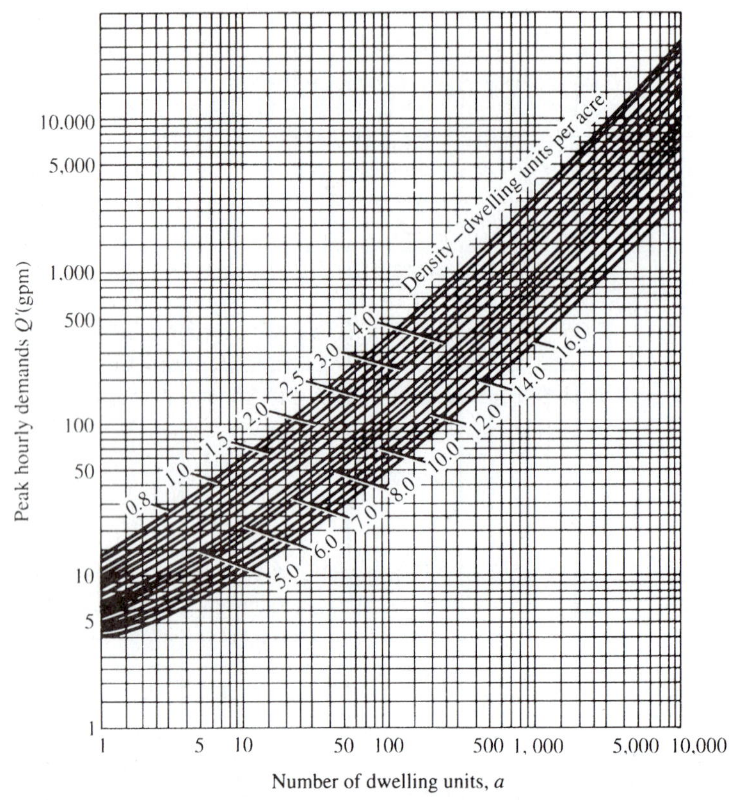

FIGURE 4.4 Relation of total peak hourly demands to number of dwelling units in terms of housing density. *Source:* Courtesy of the Residential Water-Use Research Project of The Johns Hopkins University and the Office of Technical Studies of the Architectural Standards Division of the Federal Housing Administration, 1963.

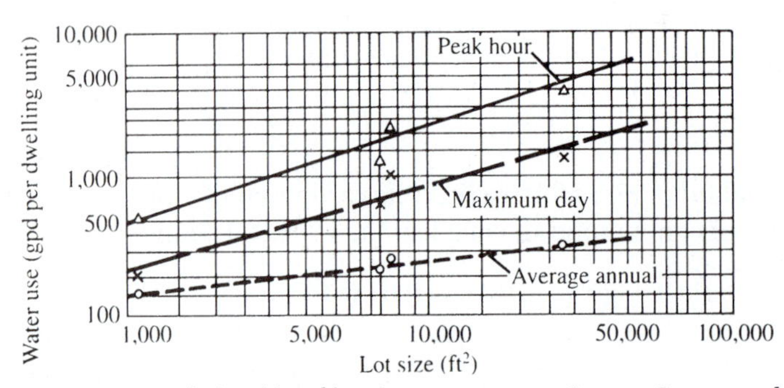

FIGURE 4.5 Relationship of lot size to water use. *Source:* Courtesy of the Residential Water-Use Research Project of The Johns Hopkins University and the Office of Technical Studies of the Architectural Standards Division of the Federal Housing Administration, 1963.

Such flows must often be provided for periods up to 10 hr or more. Gupta describes another equation for computing fire flows, based on construction type, floor area, and occupancy of a building, which was developed by the Insurance Services Office [12]. It states that

$$Q = 18C\sqrt{A} \tag{4.2}$$

where

Q = required fire flow, gpm

A = total floor area excluding the basement, ft^2

C = a coefficient having values of 1.5 for wood frames, 1.0 for ordinary construction, 0.8 for noncombustible buildings, and 0.6 for fire-resistant construction

For this equation, the flow should be more than 500 gpm, but should not exceed 6000 gpm for a single-story structure, 8000 gpm for a single building, and 12,000 gpm for a single fire.

In addition to supplying water for homes, firefighting, and perhaps some industrial purposes, most communities also must meet the needs of various commercial establishments. In considering commercial requirements, it is important to know both the magnitude and time of occurrence of peak flow. Table 4.1, although dated, provides some insight to the relative magnitudes and timing of water uses by several commercial sectors. The table provides a guide to water requirements and periods of maximum demand for apartments, motels, hotels, office buildings, shopping centers, laundries, car washes, and service stations [10]. Note that commercial uses may also include civilian and military installations and public supplies delivered to golf courses. In 1995 the USGS reported that commercial uses constituted about 3% of all freshwater uses [8]. Generally, it can be stated that commercial water users do not materially affect peak municipal demands. In fact, peak hours for many commercial establishments tend to coincide with the secondary residential peak period. The cessation of numerous commercial activities at about 6 P.M. precludes the imposition of large demands in the early evening, when sprinkling demands are often high.

The commercial peak-hour demands given in Table 4.1 are compared to peak-hour demands for a typical individual residence on a 1-acre lot. The data show that maximum commercial needs are considerably less important than peak sprinkling demands in determining peak loads on a distribution system subject to heavy sprinkling loads. An average figure of 20 gpcd is normally considered representative of commercial water consumption. The range is normally reported as 10–30 gpcd.

The many factors affecting municipal water use preclude any generalization that could apply to all areas. General trends and representative figures are useful, but it should be understood that local usage may vary considerably from reported averages. For design purposes, thorough studies must be made of past records of the type and pattern of community water use, the physical and climatic characteristics of the area, expected trends in development, projected population values, and other pertinent factors.

TABLE 4.1 Commercial Water Use

	Unit	Average Annual Demand (gpd)	Maximum Hourly Demand Rate (gpd)	Hour of Peak Occurrence	Ratio, Maximum Hourly to Average Annual	Average Annual Demand per Unit	Ratio, Maximum Hourly Demand to R-40 Demand[a]
Miscellaneous residential:							
Apartment building	22 units	3,430	11,700	5–6 p.m.	3.41	156 gpd/unit	2.2:1
Motel	166 units	11,400	21,600	7–8 a.m.	1.89	69 gpd/unit	4.0:1
Hotels:							
Belvedere	275 rooms	112,000	156,000	9–10 a.m.	1.39	407 gpd/room	29:1
Emerson	410 rooms	126,000				307 gpd/room	
Office buildings:							
Commercial Credit	490,000 ft^2	41,400	206,000	10–11 a.m.	4.89	0.084 gpd/ft^2	38:1
Internal Revenue	182,000 ft^2	14,900	74,700	11–12 a.m.	5.01	0.082 gpd/ft^2	14:1
State Office Building	389,000 ft^2	27,000	71,800	10–11 a.m.	2.58	0.070 gpd/ft^2 [b]	13:1
Shopping centers:							
Towson Plaza	240,000 ft^2	35,500	89,900	2–3 p.m.	2.50	0.15 gpd/ft^2	17:1
Hillendale	145,000 ft^2	26,000				0.18 gpd/ft^2	
Miscellaneous commercial:							
Laundries:							
Laundromat	Ten 8-lb washers	1,840	12,600	11–12 a.m.	6.85	184 gpd/washer	2.3:1
Commercial	Equivalent to ten 8-lb washers	2,510	16,200	10–11 a.m.	6.45	251 gpd/washer equivalent	3.0:1
Car wash	Capacity of 24 cars per hour	7,930	75,000	11–12 a.m.	9.46	330 gpd per car per hour of capacity	14:1
Service station	1 lift	472	12,500	6–7 p.m.	26.5	472 gpd/lift	2.3:1

[a] Lot type R-40 (1 acre) peak hourly demand for single service is 5400 gpd.

[b] Exclusive of air-conditioning.

Source: Residential Water Use Research Project of The Johns Hopkins University and the Office of Technical Studies of the Architectural Standards Division of the Federal Housing Administration, 1963.

Example 4.1

Given a residential area encompassing 500 acres with a housing density of 6 houses per acre. Assume a high-value residence with a fire flow requirement of 1000 gpm. Find (a) the combined draft and (b) the peak hourly demand.

Solution: (a) Given that 1 acre contains 43,560 ft^2, each lot will be about 7000 ft^2 in size. From Figure 4.5, this value produces a maximum day value of 700 gpd per dwelling unit. For the 3000 dwelling units, this would be $3000 \times 700 = 2{,}100{,}000$ gpd or 1458 gpm. The combined draft is thus $1458 + 1000 = 2458$ gpm. (b) From Figure 4.4, for 3000 dwelling units and a density of 6.0, find a peak hourly demand of 2500 gpm. The peak hourly flow would control since it exceeds the combined flow estimate.

Industry

From 1970 to 1980 manufacturing accounted for about 17% of total U.S. freshwater withdrawals. In 1995, industrial withdrawals represented only about 7% of the total withdrawals for all categories of water use [8]. This decline is attributed mainly to recycling and process changes. Manufacturing uses vary with the product produced, but they generally include both process waters and cooling waters. From about 1955 to 1975 the amount of fresh- and saline water withdrawn for manufacturing purposes almost doubled. This water was recycled about twice before being returned to the source and diminished somewhat less than 10% by evaporation and incorporation into products. Although manufacturing water use is expected to increase in the future, recycling is also predicted to increase substantially, with the prospect that actual water withdrawals for this purpose will continue to show a decline. Consumptive use will increase, however.

Natural Systems

Providing water for the preservation and benefit of fish and wildlife, protection of marshes and estuary areas, and for other environmentally oriented purposes is now considered a necessity. But such water uses are often in conflict with traditional uses, and resolving these conflicts is destined to become an increasingly common task. The Everglades restoration project in south Florida is an excellent example (www.evergladesplan.org). Estimation of the quantities of water needed for environmental protection and restoration is difficult. Scientific data needed to make good determinations are often lacking, and this presents special problems since the quantities of water involved can be substantial. Topics of concern include instream flow requirements, maintenance of lake levels, freshwater releases to bays and estuaries, and water requirements for protecting fish and wildlife.

The water-related aspects of restoring, protecting, and managing natural systems are abundant. Dealing with them requires special policies, a good database, and close coordination with a host of programs conducted under the auspices of various levels of government. Water management policies that encourage or result in excessive growth may trigger unwanted spillover effects on the environment. The drainage and reclamation of lands may provide additional opportunity for economic development, for example, but not without a price to be paid in disrupting prevailing ecosystems. Development and environmental protection can be partners, but only if care is exercised in modifying the landscape. Growth management policies that embrace the many dimensions of managing natural systems are needed.

Navigation

Water requirements for navigation on most river systems are seasonal. The greatest demands usually occur during the driest months of the year. Flows released for navigation limit the availability of water for irrigation and hydropower generation and for recreational uses at reservoir sites. They do, however, complement other instream uses. Where navigation depths are maintained by low dams, there is usually little effect on other water uses in a river. These structures do not impound large volumes of water; rather, they serve to provide greater uniformity of flows. Many advantages result from this type of operation, including benefits to fish and other wildlife, recreation, pollution control, and aesthetics.

Large multipurpose reservoirs, such as those on the main stem of the Missouri River, also provide storage to meet periodic navigational flows. In such cases, reservoir operating policies must be designed to accommodate the conflicting requirements of other water uses for which storage is provided.

Hydroelectric Power Generation

In the past, requirements for hydroelectric power were usually heaviest during the peak winter heating months, but with the increased use of air-conditioning, demands for electricity are less seasonal, and in some cases the summer months are the most demanding. The use of hydroelectric facilities to provide peak power, as opposed to furnishing base load, is also becoming more common. Unfortunately, this type of operation increases conflicts with recreationists and others who favor little or no short-term fluctuations in reservoir levels. In general, conflicts between water use for electric power generation and use for other purposes stem from opposing seasonal requirements. For example, heavy summertime releases for navigation dictate maximizing storage during the winter, a situation in conflict with discharging from storage during the same period to produce electricity. Hydroelectric production does not adversely affect all water uses, however; for example, water passed through turbines can also be used downstream for navigation, flow augmentation, and other purposes. It is worth noting that the largest hydroelectric project in the world is under construction in China. The Three Gorges Dam project, scheduled for completion in 2009, is projected to produce 18 gigawatts of electrical energy, approximately 10% of China's total capacity in 1993. The project has been widely criticized on technical and environmental grounds, but proponents state that the project will decrease atmospheric pollution and eliminate the burning of 40 to 50 million tons of coal each year.

Recreation

About a fourth of the nation's outdoor recreation activity depends on water. In 1975, swimming, fishing, boating, water skiing, and ice skating accounted for about 3 billion activity days. By 2000, this figure had increased to about 8 billion [7]. Water requirements for recreation are normally greatest in the summer. The sports enthusiast and vacationer desire substantial stream flows and unvarying reservoir levels during this period. Such conditions are optimal for water-based recreation activities but conflict with many withdrawal uses.

Energy Resource Development

Water can be used to produce energy via turbines driving electric generators. It can also be used to process energy-producing resources such as coal and oil shale and to aid in restoring lands despoiled during mining operations. The water requirements for extraction of coal, oil shale, uranium, and oil gas are not great, but secondary recovery operations for oil require large quantities of water. Substantial quantities of water may be used in coal slurry pipelines and for retorting oil shale. Synfuels conversion processes also require large quantities of water, and as stated earlier, withdrawal of water for cooling thermal electric plants is the largest category of total water use in the United States.

The availability of water is a factor in the location and design of energy conversion facilities, but these users can generally afford to pay high prices for water. Securing legal rights to water, rather than its availability, is often the critical issue in dry regions.

4.3 THE IMPACT OF CLIMATE CHANGE ON WATER AVAILABILITY AND USE

The most anticipated result of the buildup of carbon dioxide and other greenhouse gases in the atmosphere is global warming. But the impact on society of the changes this may bring about in the hydrologic cycle may be more serious than the warming trend [13,14]. The water cycle, although simple in concept, is complex in reality and very sensitive to local conditions. This is particularly true in water-short desert areas associated with mountainous regions such as exist in the southwestern United States. Given some stable histories of climate, these processes can be mathematically modeled to provide useful estimates of future conditions. But given the uncertainty associated with climate change, forecasting becomes much more challenging.

Issues to be dealt with include both temperature and precipitation. In a given locale, is it going to be warmer or cooler, and is precipitation going to increase or decrease? Also to be considered is the impact of temperature on the net amount of water that will be available to the region. For example, if a temperature increase were to occur with no change in precipitation, the net water availability would be reduced as a result of higher rates of evaporation and transpiration.

Changes in soil moisture and runoff resulting from increased or decreased precipitation could significantly affect water supplies for municipalities, agriculture, industry, environmental protection, and other purposes. The five major global circulation models all predict an increase in average worldwide precipitation. This would also be associated with soil moisture increases. However, the precipitation increase is not projected to be uniform across the world [15]. In fact, some regions are projected to have decreases in precipitation and lowered amounts of soil moisture. Many inconsistencies are apparent in the analyses of climate change model results, and it is clear that the prediction of changes in local and regional hydrology are far from perfect.

Global circulation models suggest that the interior of the United States is likely to be drier in the summer. If this does occur, the implications could be serious for the wheat belt of the Great Plains and the corn belt of the north central states. Whether southwestern desert regions will be drier or wetter is highly speculative, but it is certain that even slight decreases in precipitation in those regions would magnify their water supply problems.

It can be argued that the potential effects of global climate change on water resources could be more serious than warming. This is particularly true for regions where surface water supplies are generated from water stored in snowpacks in the mountains. Such a situation exists in the southwestern United States where the Colorado River is a major source of water supply for cities in Southern California and Arizona. The Colorado River Basin covers southwestern Wyoming, western Colorado, eastern Utah, all of Arizona, and small potions of southern Nevada, northwestern New Mexico, and southeastern California. In those regions, even if total precipitation does not change while temperatures rise, a decreasing water supply would be the outcome.

Colorado River water supply has its main origin in the upper basin mountains. Much of the water is derived from snowfall that is stored as snowpack on mountain slopes. The snowpack acts as a natural reservoir and holds the water until spring and summer thaws release it at a time when it is most needed for agriculture and other purposes. The natural snowpack storage is also augmented by reservoirs that have been constructed on the Colorado River. With a global warming scenario, winter temperatures would be warmer, and more of the precipitation would be in the form of rain rather than of snow. Rainfall infiltrates the ground where it falls and becomes surface runoff as well. The result of warming would be a reduction in the natural snowpack storage as well as a change in the timing of releases of water from the upper portions of the basin. This shift could significantly alter the dependable water supply of users of Colorado River water.

A report of the American Association for the Advancement of Science Panel on Climatic Variability, Climate Change and the Planning and Management of U.S. Water Resources is a good source of information on potential climate change impacts on water resources [16,17]. The report noted that stream flow is sensitive to climate change and that low flows will be more affected than high ones. It stated that dry climates will be more affected than humid ones. It also noted that atmospheric warming alone will decrease stream flow much less than warming associated with a simultaneous decrease in precipitation. The report stated that reservoir yields would be affected, and that water quality problems, which are often associated with low flow conditions, may be more widespread in arid regions [17].

Our climate change models are not perfect, but model results provide an insight to what might be expected under a variety of global change scenarios. The water policy implications of global climate change are significant and must be taken into account in planning for water resources development and management in the future. To the extent that actions can be taken now to reduce the potential impacts of climate change we should do so. Many of these actions relate to improving water use efficiency and would be wise courses of action even if potential climate change scenarios do not unfold.

4.4 WATER USE TRENDS

Analyses of water use projections made since 1970 show that a rapidly increasing rate of per capita water use is less likely than estimators of the 1950s and 1960s would have believed. Another point is that national or regional trends are not always indicative of state and local trends. Thus, planners must be equipped to deal with development and management options at several geographic levels so that special local and regional influences can be accommodated.

Every five years the U.S. Geological Survey publishes a pamphlet, "Estimated Water Use in the United States" [8]. This publication summarizes water use in each major water-using category and indicates trends over time. The data are available by state and by region. Although the overall tendency in water use to the year 2000 appears to be more conservative than that of the past, striking local variances can be expected. Table 4.2 gives USGS year 2000 water use estimates. Table 4.3 and Figure 4.6 show U.S. water use trends since 1950. Note that the total offstream water withdrawal declined from 440 bgd in 1980 to 408 bgd in 2000 (fresh and saline water). During the 20-year period from 1980 to 2000 there was an overall reduction in total withdrawals of about 7% even though the U.S. population increased almost 19% during that same period. The shift in trend that occurred in 1980 suggests that a more conservative approach to water use and water resources development is beginning to take hold. Table 4.3 and Figure 4.6 show trends for several of the water-using sectors. Note that public water use increased about 27% from 1980 to 2000. This is not surprising, since the public sector is strongly associated with population growth and as long as the population continues to increase, water use in that sector can be expected to increase, although not necessarily at the same per capita rate. Figure 4.7 indicates the sources (surface and groundwater), from which fresh water was withdrawn from 1950 to 2000. About 70% of the water withdrawn is returned to the source after use, about 5% is lost in irrigation conveyances, and about 25% is consumptively used (evaporated or transpired). Of the amount of water consumptively used, irrigated agriculture is responsible for the lion's share (about 80%).

Example 4.2

Given the population, public water withdrawal, and total freshwater withdrawal for the states shown on Table 4.4 in 1995, calculate the percent of freshwater withdrawn by the public sector and the per capita water use of the freshwater for each state.

Solution: Percentages of water use and per capita water use figures appear in columns 5 and 6 of Table 4.4. The average percent of water withdrawn by the public sector for

TABLE 4.2 Total Water Withdrawals by Water Use Category and State, 2000

	Public Supply	Domestic	Irrigation	Livestock	Aquaculture	Industrial		Mining		Thermoelectric Power		Total		
State	Fresh	Fresh	Fresh	Fresh	Fresh	Fresh	Saline	Fresh	Saline	Fresh	Saline	Fresh	Saline	Total
Alabama	834	78.9	43.1	—	10.4	833	0	—	—	8,190	0	9,990	0	9,990
Alaska	80.0	11.2	1.01	—	—	8.12	3.86	27.4	140	33.6	0	161	144	305
Arizona	1,080	28.9	5,400	—	—	19.8	0	85.7	8.17	100	0	6,720	8.17	6,730
Arkansas	421	28.5	7,910	—	198	134	0.08	2.78	0	2,180	0	10,900	0.08	10,900
California	6,120	286	30,500	409	537	188	13.6	23.7	153	352	12,600	38,400	12,800	51,200
Colorado	899	66.8	11,400	—	—	120	0	—	—	138	0	12,600	0	12,600
Connecticut	424	56.2	30.4	—	—	10.7	0	—	—	187	3,440	708	3,440	4,150
Delaware	94.9	13.3	43.5	3.92	0.07	59.4	3.25	—	—	366	738	582	741	1,320
District of Columbia	0	0	0.18	—	—	0	0	—	—	9.69	0	9.87	0	9.87
Florida	2,440	199	4,290	32.5	8.02	291	1.18	217	0	658	12,000	8,140	12,000	20,100
Georgia	1,250	110	1,140	19.4	15.4	622	30.0	9.80	0	3,250	61.7	6,410	91.7	6,500
Hawaii	250	12.0	364	—	—	14.5	0.85	—	—	0	0	640	0.85	641
Idaho	244	85.2	17,100	34.9	1,970	55.5	0	—	—	0	0	19,500	0	19,500
Illinois	1,760	135	154	37.6	—	391	0	—	—	11,300	0	13,700	0	13,700
Indiana	670	122	101	41.9	—	2,400	0	82.5	0	6,700	0	10,100	0	10,100
Iowa	383	33.2	21.5	109	—	237	0	32.8	0	2,540	0	3,360	0	3,360
Kansas	416	21.6	3,710	111	5.60	53.3	0	31.4	0	2,260	0	6,610	0	6,610
Kentucky	525	27.5	29.3	—	—	317	0	—	—	3,260	0	4,160	0	4,160
Louisiana	753	41.2	1,020	7.34	243	2,680	0	—	—	5,610	0	10,400	0	10,400
Maine	102	35.7	5.84	—	—	247	0	—	—	113	295	504	295	799
Maryland	824	77.1	42.4	10.4	19.6	65.8	227	8.31	0.02	379	6,260	1,430	6,490	7,910
Massachusetts	739	42.2	126	—	—	36.8	0	—	—	108	3,610	1,050	3,610	4,660
Michigan	1,140	239	201	11.3	—	698	0	—	—	7,710	0	10,000	0	10,000
Minnesota	500	80.8	227	52.8	—	154	0	588	0	2,270	0	3,870	0	3,870
Mississippi	359	69.3	1,410	—	371	242	0	—	—	362	148	2,810	148	2,960

Missouri	872	53.6	1,430	72.4	83.3	62.7	0	16.9	0	5,640	0	8,230	0	8,230
Montana	149	18.6	7,950	—	—	61.3	0	—	—	110	0	8,290	0	8,290
Nebraska	330	48.4	8,790	93.4	—	38.1	0	128	4.552	820	0	12,200	4.55	12,300
Nevada	629	22.4	2,110	—	—	10.3	0	—	—	36.7	0	2,810	0	2,810
New Hampshire	97.1	41.0	4.75	—	16.3	44.9	0	6.80	0	236	761	447	761	1,210
New Jersey	1,050	79.7	140	1.68	6.46	132	0	110	0	650	3,390	2,170	3,390	5,560
New Mexico	296	31.4	2,860	—	—	10.5	0	—	—	56.4	0	3,260	0	3,260
New York	2,570	270	35.5	—	—	297	0	—	—	4,040	5,010	7,210	5,010	12,200
North Carolina	945	189	287	121	7.88	293	0	36.4	0	7,850	1,620	9,730	1,620	11,400
North Dakota	63.6	11.9	145	—	—	17.6	0	—	—	902	0	1,140	0	1,140
Ohio	1,470	134	31.7	25.3	1.36	807	0	88.5	0	8,590	0	11,100	0	11,100
Oklahoma	675	25.5	718	151	16.4	25.9	0	2.48	256	146	0	1,760	256	2,020
Oregon	566	76.2	6,080	—	—	195	0	—	—	15.30	0	6,930	0	6,930
Pennsylvania	1,460	132	13.9	—	—	1,190	0	182	0	6,980	0	9,950	0	9,950
Rhode Island	119	8.99	3.45	—	—	4.28	0	—	—	2.40	290	138	290	429
South Carolina	566	63.5	267	—	—	565	0	—	—	5,710	0	7,170	0	7,170
South Dakota	93.3	9.53	373	42.0	—	5.12	0	—	—	5.24	0	528	0	528
Tennessee	890	32.6	22.4	—	—	842	0	—	—	9,040	0	10,800	0	10,800
Texas	4,230	131	8,630	308	—	1,450	907	220	504	9,820	3,440	24,800	4,850	29,600
Utah	638	16.1	3,860	—	116	42.7	5.08	26.3	198	62.20	0	4,760	203	4,970
Vermont	60.1	21.0	3.78	—	—	6.91	0	—	—	355	0	447	0	447
Virginia	720	133	26.4	—	—	470	53.3	—	—	3,850	3,580	5,200	3,640	8,830
Washington	1,020	125	3,040	—	—	577	39.9	—	—	519	0	5,270	39.9	5,310
West Virginia	190	40.4	0.04	—	—	968	0	—	—	3,950	0	5,150	0	5,150
Wisconsin	623	96.3	196	66.3	70.2	447	0	—	—	6,090	0	7,590	0	7,590
Wyoming	107	6.57	4,500	—	—	5.78	0	79.5	222	243	0	4,940	222	5,170
Puerto Rico	513	0.88	94.5	—	—	11.2	0	—	—	0	2,190	620	2,190	2,810
U.S. Virgin Islands	6.09	1.69	0.50	—	—	3.34	0	—	—	0	136	11.6	136	148
Total	43,300	3,720	137,000	1,760	3,700	18,500	1,280	2,010	1,490	136,000	59,500	346,000	62,300	408,000

[Figures may not sum to totals because of independent rounding. All values in million gallons per day. —, data not collected]
Source: U.S. Geological Survey http://water.usgs.gov/pubs/circ/2004/circ1268/htdocs/table02.html

TABLE 4.3 Trends of Estimated Water Use in the United States. 1950–1995

	Year										Percentage Change
	[a]1950	[a]1955	[b]1960	[b]1965	[c]1970	[d]1975	[d]1980	[d]1985	[d]1990	[d]1995	1990–95
Population, in millions	150.7	164.0	179.3	193.8	205.9	216.4	229.6	242.4	252.3	267.1	+6
Offstream use:											
Total withdrawals	180	240	270	310	370	420	[e]440	399	408	402	−2
Public supply	14	17	21	24	27	29	34	36.5	38.5	40.2	+4
Rural domestic and livestock	3.6	3.6	3.6	4.0	4.5	4.9	5.6	7.79	7.89	8.89	+13
Irrigation	89	110	110	120	130	140	150	137	137	134	−2
Industrial:											
Thermoelectric power use	40	72	100	130	170	200	210	187	195	190	−3
Other industrial use	37	39	38	46	47	45	45	30.5	29.9	29.1	−3
Source of water:											
Ground:											
Fresh	34	47	50	60	68	82	[e]83	73.2	79.4	76.4	−4
Saline	([f])	.6	.4	.5	1	1	.9	.652	1.22	1.11	−9
Surface:											
Fresh	140	180	190	210	250	260	290	265	259	264	+2
Saline	10	18	31	43	53	69	71	59.6	68.2	59.7	−12
Reclaimed wastewater	([f])	.2	.6	.7	.5	.5	.5	.579	.750	1.02	+36
Consumptive use	([f])	([f])	61	77	[g]87	[g]96	[g]100	[g]92.3	[g]94.0	[g]100	+6
Instream use:											
Hydroelectric power	1100	1500	2000	2300	2800	3300	3300	3050	3290	3160	−4

[a] 48 states and District of Columbia
[b] 50 states and District of Columbia
[c] 50 states and District of Columbia, and Puerto Rico
[d] 50 states and District of Columbia, Puerto Rico, and Virgin Islands
[e] Revised
[f] Data not available
[g] Freshwater only

Source: USGS 1995 [8]. [Data for 1950–1980 adapted from MacKichan (1951, 1957), MacKichan and Kammerer (1961), Murray (1968), Murray and Reeves (1972, 1977), and Solley and others (1983, 1988). The water use data are in thousands of million gallons per day and are rounded to two significant figures for 1950–1980 and to three significant figures for 1985–1990; percentage change is calculated from unrounded numbers.]

these states is calculated to be 16% (for all states in 1995, the average was 11.7%). Given that the most populated states are included in the example, it is not surprising that the average percent use for this subset of states is somewhat higher than the national average. The average per capita use for the subset is calculated to be about 173 gpcd, which is close to the national average of 179 gcpd.

The example shows that public water supplies constitute a relatively small portion of all freshwater withdrawn and that, while more efficient water use in that sector is important, major reductions in total water use can be achieved only if agricultural and thermoelectric uses are reduced.

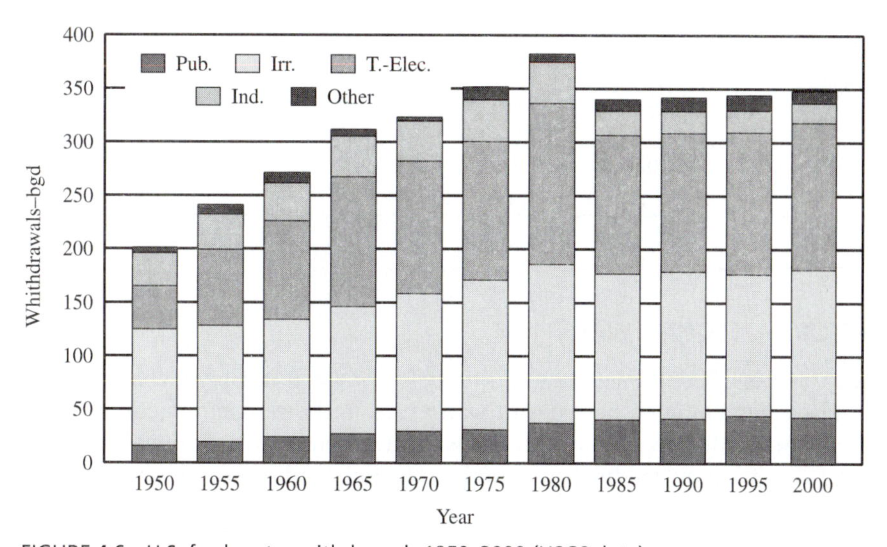

FIGURE 4.6 U.S. freshwater withdrawals 1950–2000 (USGS data).

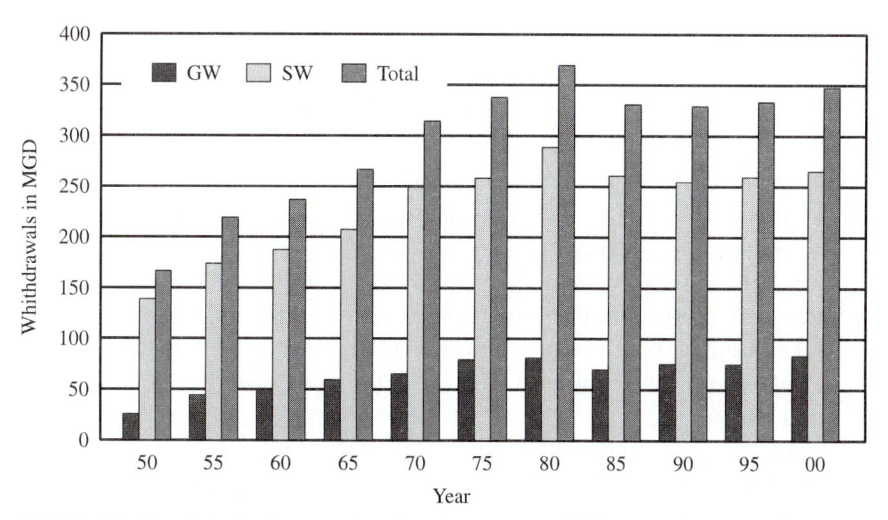

FIGURE 4.7 Trends in fresh ground and surface water withdrawals 1950–2000 (USGS data).

TABLE 4.4 Data Display and Spreadsheet Calculations for Example 4.2

State	Population, 1000s	Public Use, mgd	Total Fresh Water Use, mgd	% of Total Freshwater Used	Public Water per Capita Use
Alaska	604	81	211	38	134.1
Arizona	4,218	807	6,820	12	191.3
California	32,063	5,620	36,300	15	175.3
Florida	14,166	2,070	7,210	29	146.1
Hawaii	1,187	214	1,010	21	180.3
Illinois	11,830	1,820	19,800	9	153.8
Maine	1,241	100	221	45	80.6
Nevada	1,530	468	2,260	21	305.9
New York	16,136	3000	10300	29	185.9
Pennsylvania	12,072	1,550	9,680	16	128.4
Texas	18,724	3,290	24,300	14	175.7
Washington	5,431	1,180	8,820	13	217.3
Totals	119,202	20,200	126,932		2074.7
Averages				16	172.9

Source: Data from USGS 1995 [8].

4.5 FACTORS AFFECTING WATER USE

Many factors affect the amount and timing of water use. They include population size and character; climate; the types of water uses in the region; the cost of water; public attitude toward conservation and wastewater reuse; water management practices; public commitment to environmental protection and restoration; tourism; emerging technologies; and federal, state, and local government laws and ordinances.

Population. The amount of water used in a locality is directly related to the size, distribution, and composition of the local population. Forecasts of future water use are dependent, in part, on population forecasts as well.

Climate. The amount of water used in a locality is influenced by its climate. Lawn sprinkling, gardening, bathing, irrigation, cooling, and many other water uses are directly affected.

Types of Water Users. The type and scale of residential, commercial, industrial, and agricultural development in an area define the levels and timing of water uses.

Economic Conditions. Economic health is reflected in all aspects of resource management and development. Inflation and other economic trends influence the availability of funds for water supply, wastewater treatment, and environmental and other programs, and they affect the attitudes of individuals as well.

Environmental Protection. Social attitudes toward environmental protection and enhancement strongly affect water allocation and use. Water use forecasts must take into account the amount of water that is to be dedicated to environmental protection and restoration. This quantity can be substantial.

Conservation. Attractive alternatives to the development of new water supplies are conservation practices and the reuse of wastewaters. These approaches, although not a panacea, can at least delay the need for additional water supplies and/or the development of new facilities.

Management Practices. Water management practices influence water use trends. They include: interbasin transfers; weather modification; saline water conversion; water reclamation and reuse; and genetic engineering [18–26]. The impact of technological change on water use can be significant.

Tourism. Some states, such as Florida, have annual tourist populations that significantly exceed their resident populations. The impacts of such occurrences must be recognized when forecasting future water demands.

4.6 POPULATION

Generally, the greater the number of people residing in an area, the more water will be used. But it is not only the number of people that is important but also their ages, level of education, social background, field of employment, religious beliefs, and other factors. Factors that must be considered include the geographic distribution and growth rate of the population; measures that might be taken to influence the growth rate; and the impact of population changes on the regional economy, natural resources, labor force, energy requirements, urban infrastructure, and so on.

Historical data are basic to estimating future levels of population, but these data are not always available. Even in the United States where the Bureau of the Census (www.census.gov) maintains historic records and forecasts, errors in estimates sometime occur [27,28]. Because many uncertain factors affect population change (fertility, mortality, migration, for example), most forecasters suggest exploring at least three trends in population growth based on plausible mixes of influencing factors.

Population Trends

During the last part of the twentieth century, there have been some notable trends in population change in the United States. They include internal migration from the Northeast and other areas to the South and West, and out-migration from central cities. The U.S. Census Bureau projects that during the first quarter of the twenty-first century, net population change will be most evident in three states—California, Texas, and Florida [27]. It has also been projected that the fastest growth will be in the West. Such population trends affect water use and supply. Census Bureau population projections from 1995 to 2025 are shown on Table 4.5.

Cities confronted with declines in population often experience shifts in the character of their residents as well. In general, the more affluent residents usually move out, thus reducing the tax base and leaving a less-well-off population to shoulder tax burdens and maintain water and other municipal services. An associated problem can be that of maintaining a larger than needed infrastructure with fewer resources to do so.

Rural migration, fueled by the belief of many urban dwellers that rural locations are more attractive and in some cases less expensive, is companion to the loss of population by central cities. The shifting of population to sparsely settled areas can create problems in needed water supply and wastewater treatment services. Factors that

TABLE 4.5 Total Population and Net Change for States: 1995–2025
[Thousands, Resident population]

Region, Division, and State	Projections for July 1							July 1, 1995, to July 1, 2025				
									Components of Change			
											Net Migration	
	1995	2000	2005	2010	2015	2020	2025	Net Change	Births	Deaths	Interstate Migration	Immigration
United States	262,755	274,634	285,981	297,716	310,133	322,742	335,050	72,294	126,986	84,633	—	24,666
Northeast	51,466	52,107	52,767	53,692	54,836	56,103	57,392	5,927	21,585	16,537	(7,168)	6,830
New England	13,312	13,581	13,843	14,172	14,546	14,938	15,321	2,009	5,448	4,096	(1,041)	1,338
Maine	1,241	1,259	1,285	1,323	1,362	1,396	1,423	181	437	402	86	20
New Hampshire	1,148	1,224	1,281	1,329	1,372	1,410	1,439	291	481	344	84	31
Vermont	585	617	638	651	662	671	678	94	221	180	26	7
Massachusetts	6,074	6,199	6,310	6,431	6,574	6,734	6,902	828	2,520	1,860	(815)	831
Rhode Island	990	998	1,012	1,038	1,070	1,105	1,141	151	423	318	(94)	113
Connecticut	3,275	3,284	3,317	3,400	3,506	3,621	3,739	464	1,368	992	(329)	337
Middle Atlantic	38,153	38,526	38,923	39,520	40,289	41,164	42,071	3,918	16,136	12,441	(6,127)	5,492
New York	18,136	18,146	18,250	18,530	18,916	19,359	19,830	1,694	8,117	5,598	(5,038)	3,886
New Jersey	7,945	8,178	8,392	8,638	8,924	9,238	9,558	1,613	3,535	2,542	(747)	1,201
Pennsylvania	12,072	12,202	12,281	12,352	12,449	12,567	12,683	611	4,484	4,301	(342)	405
Midwest	61,804	63,502	64,825	65,915	67,024	68,114	69,109	7,305	26,334	19,534	(3,541)	2,365
East North Central	43,456	44,419	45,151	45,764	46,410	47,063	47,675	4,219	18,512	13,557	(3,653)	1,839
Ohio	11,151	11,319	11,428	11,505	11,588	11,671	11,744	594	4,417	3,626	(758)	247
Indiana	5,803	6,045	6,215	6,318	6,404	6,481	6,546	742	2,377	1,879	(35)	110
Illinois	11,830	12,051	12,266	12,515	12,808	13,121	13,440	1,610	5,672	3,582	(1,699)	1,037
Michigan	9,549	9,679	9,763	9,836	9,917	10,002	10,078	528	3,965	2,874	(1,122)	310
Wisconsin	5,123	5,326	5,479	5,590	5,693	5,788	5,867	744	2,081	1,596	(39)	134
West North Central	18,348	19,082	19,673	20,151	20,615	21,051	21,434	3,086	7,822	5,978	112	526
Minnesota	4,610	4,830	5,005	5,147	5,283	5,406	5,510	900	1,993	1,349	(89)	190
Iowa	2,842	2,900	2,941	2,968	2,994	3,019	3,040	198	1,073	958	(97)	83
Missouri	5,324	5,540	5,718	5,864	6,005	6,137	6,250	927	2,260	1,858	255	105
North Dakota	641	662	677	690	704	717	729	88	270	214	(6)	13
South Dakota	729	777	810	826	840	853	866	137	341	246	6	6
Nebraska	1,637	1,705	1,761	1,806	1,850	1,892	1,930	293	718	543	35	29
Kansas	2,565	2,668	2,761	2,849	2,939	3,026	3,108	543	1,167	810	7	102

South	91,890	97,613	102,788	107,597	112,384	117,060	121,448	29,558	43,142	32,054	11,067	5,273
South Atlantic	46,995	50,147	52,921	55,457	57,966	60,411	62,675	15,680	20,682	16,883	6,707	3,790
Delaware	717	768	800	817	832	847	861	144	313	249	35	24
Maryland	5,042	5,275	5,467	5,657	5,862	6,071	6,274	1,232	2,295	1,537	(251)	593
District of Columbia	554	523	529	560	594	625	655	101	334	213	(156)	135
Virginia	6,618	6,997	7,324	7,627	7,921	8,204	8,466	1,848	2,839	2,074	299	605
West Virginia	1,828	1,841	1,849	1,851	1,851	1,850	1,845	17	555	715	105	14
North Carolina	7,195	7,777	8,227	8,552	8,840	9,111	9,349	2,154	3,039	2,612	1,295	199
South Carolina	3,673	3,858	4,033	4,205	4,369	4,517	4,645	972	1,566	1,313	546	58
Georgia	7,201	7,875	8,413	8,824	9,200	9,552	9,869	2,669	3,571	2,340	953	306
Florida	14,166	15,233	16,279	17,363	18,497	19,634	20,710	6,544	6,169	5,829	3,879	1,856
East South Central	16,067	16,918	17,604	18,122	18,586	19,002	19,345	3,279	6,593	5,791	1,737	262
Kentucky	3,860	3,995	4,098	4,170	4,231	4,281	4,314	454	1,439	1,344	175	67
Tennessee	5,256	5,657	5,966	6,180	6,365	6,529	6,665	1,409	2,217	1,909	845	97
Alabama	4,253	4,451	4,631	4,798	4,956	5,100	5,224	971	1,759	1,563	577	71
Mississippi	2,697	2,816	2,908	2,974	3,035	3,093	3,142	445	1,179	975	140	27
West South Central	28,828	30,548	32,263	34,019	35,832	37,647	39,427	10,599	15,867	9,380	2,624	1,222
Arkansas	2,484	2,631	2,750	2,840	2,922	2,997	3,055	572	1,000	979	436	31
Louisiana	4,342	4,425	4,535	4,683	4,840	4,991	5,133	790	2,054	1,501	45	90
Oklahoma	3,278	3,373	3,491	3,639	3,789	3,930	4,057	779	1,411	1,224	412	92
Texas	18,724	20,119	21,487	22,857	24,280	25,729	27,183	8,459	11,403	5,676	1,730	1,008
West	57,596	61,413	65,603	70,512	75,889	81,465	87,101	29,505	35,925	16,506	(358)	10,198
Mountain	15,645	17,725	19,249	20,221	21,122	22,049	22,962	7,317	8,794	4,938	2,490	646
Montana	870	950	1,006	1,040	1,069	1,097	1,121	251	374	316	143	13
Idaho	1,163	1,347	1,480	1,557	1,622	1,683	1,739	576	627	379	257	33
Wyoming	480	525	568	607	641	670	694	214	244	160	111	2
Colorado	3,747	4,168	4,468	4,658	4,833	5,012	5,188	1,442	1,855	1,122	504	123
New Mexico	1,685	1,860	2,016	2,155	2,300	2,454	2,612	927	1,030	526	403	12
Arizona	4,218	4,798	5,230	5,522	5,808	6,111	6,412	2,195	2,542	1,434	753	276
Utah	1,951	2,207	2,411	2,551	2,670	2,781	2,883	931	1,310	486	(31)	80
Nevada	1,530	1,871	2,070	2,131	2,179	2,241	2,312	782	813	516	351	106
Pacific	41,951	43,687	46,354	50,291	54,768	59,416	64,139	22,188	27,130	11,570	(2,848)	9,553
Washington	5,431	5,858	6,258	6,658	7,058	7,446	7,808	2,377	2,600	1,708	931	394
Oregon	3,141	3,397	3,613	3,803	3,992	4,177	4,349	1,209	1,364	1,169	712	197
California	31,589	32,521	34,441	37,644	41,373	45,278	49,285	17,696	22,035	8,248	(4,429)	8,725
Alaska	604	653	700	745	791	838	885	281	422	105	(84)	28
Hawaii	1,187	1,257	1,342	1,440	1,553	1,677	1,812	625	709	339	21	209

Source: U.S. Bureau of the Census, Population Division, PPL-47 [27].

could counter this trend include escalating energy prices making transportation to jobs more costly, and urban renewal projects in downtown areas offering modern accommodations combined with convenience to city attractions.

Since the 1960s, movement to the sunbelt has brought about regional population shifts. The population in some northeastern and north central states has been declining, whereas growth in the South and West has been accelerating.

Current and emerging population trends must be recognized and carefully considered by planners and developers. The impacts of these trends on water requirements can be significant, and, if ignored, can result in costly deficit or excess capacities of proposed facilities.

It must be recognized, however, that rates of change in population are not necessarily the same as those for water use. Per capita water use can remain constant, increase, or decrease during the period of the forecast. Thus, the rate of water use could mirror the rate of population change, or be more or less than that rate.

Example 4.3

Given the population data of Table 4.5, estimate the percent increase in population from 1995 to 2025 for the following states: California, Florida, Texas, New York, North Dakota, Maine, and West Virginia.

Solution: Percentage changes in population are calculated from the data in columns 2 and 3 of Table 4.6, and are shown in column 4.

TABLE 4.6 Data Display and Calculations for Example 4.3

State	1995 Population	Projected 2025 Population	Percent Increase
California	31,589,000	49,285,000	56
Florida	14,166,000	20,710,000	46
Texas	18,724,000	27,183,000	45
New York	18,136,000	19,830,000	9
North Dakota	641,000	729,000	14
Maine	1,241,000	1,423,000	15
West Virginia	1,828,000	1,845,000	1

Note that the projected increases in population for California, Florida, and Texas are close to 50% while New York's increase is considerably less, only about 9%. For North Dakota and Maine, population increases are about 15%, and for West Virginia, only about 1%. These trends are important ingredients in forecasting the need for and allocation of water and other resources in the respective states.

Population Forecasting

Population estimates required for the operation and design of water supply and waste treatment works may be (1) short-term estimates in the range of 1 to 10 years, and (2) long-term estimates of 10 to 50 years or more.

The prediction of future population is at best complex. It should be emphasized that there is no exact solution, even though seemingly sophisticated mathematical equations are often used. War, technological developments, new scientific discoveries, government operations, and a whole host of other factors can drastically disrupt population trends. There is no sure-fire way to predict many of these occurrences; thus, their impact can only be estimated to the best of current ability. Both mathematical and graphical methods are used. Forecasts are often based on past census records for the area, or on the records of what are considered to be similar communities. But extrapolations of past trends do not take into consideration factors such as the influx of workers when new industries settle in the area, the loss of residents due to curtailment of military activities, or changes in business or transportation facilities. To optimize estimates, all possible information regarding anticipated industrial growth, local birth and death rates, government activities, and other related factors should be obtained and used. The local census bureau, the planning commission, the bureau of vital statistics, local utility companies, movers, and the chamber of commerce are all sources of information.

Trend-Based Methods

Trend-based methods assume that population growth follows natural growth patterns and can therefore be represented in mathematical or graphic form. Usually, the approach is that of extending past trends into the future. Linear, geometric, exponential, logarithmic, and other mathematical tools have been used. These methods are easy to use but they neglect to consider that past trends are not necessarily sustained. Component and other methods discussed later take this consideration into account.

Most short-term estimates (1 to 10 years) are made using trend-based methods. They often follow segments of a typical population growth curve as shown on Figure 4.8.

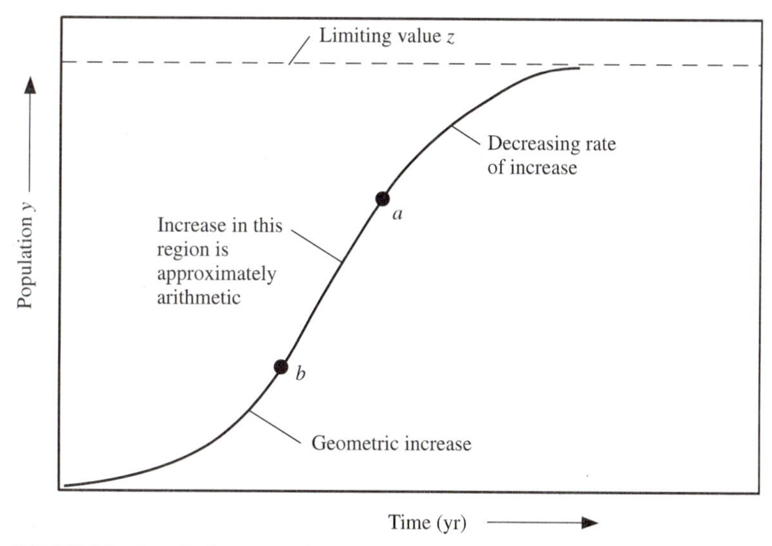

FIGURE 4.8 Population growth curve.

This S curve can be considered to consist of geometric, arithmetic, and decreasing rate of increase segments.

Arithmetical Progression This method of estimation is based on a constant increment of increase and may be stated as follows:

$$\frac{dY}{dt} = K_u \tag{4.3}$$

where

$$Y = \text{population}$$

$$t = \text{time (usually years)}$$

$$K_u = \text{uniform growth-rate constant}$$

If Y_1 represents the population at the census preceding the last census (time t_1), and Y_2 represents the population at the last census (time t_2), then

$$\int_{Y_1}^{Y_2} dY = \int_{t_1}^{t_2} dt$$

Integrating and inserting the limits, we obtain

$$Y_2 - Y_1 = K_u(t_2 - t_1)$$

Therefore,

$$K_u = \frac{Y_{2-} - Y_1}{t_2 - t_1} \tag{4.4}$$

We use Eq. (4.4) to write an expression for short-term arithmetic estimates of population growth:

$$Y = Y_2 + \frac{Y_2 - Y_1}{t_2 - t_1}(t - t_2) \tag{4.5}$$

where t represents the end of the forecast period.

Constant-Percentage Growth Rate For equal periods of time, this procedure assumes constant growth percentages. If the population increased from 90,000 to 100,000 in the past 10 years, it would be estimated that the growth in the ensuing decade would be to $100{,}000 + 0.11 \times 100{,}000$, or 111,000. Mathematically, this may be formulated as

$$\frac{dY}{dt} = K_p Y \tag{4.6}$$

where the variables are defined as before, except that K_p represents a constant percentage increase per unit time. Integrating this expression and setting the limits yields

$$K_p = \frac{\log_e Y_2 - \log_e Y_1}{t_2 - t_1} \tag{4.7}$$

A short-term geometric estimate of population growth is thus given by

$$\log_e Y = \log_e Y_2 + K_p(t - t_2) \tag{4.8}$$

Note that base 10 logs may also be used in Eqs. (4.7) and (4.8).

Decreasing Rate of Increase Estimates made on the basis of a decreasing rate of increase assume a variable rate of change. Mathematically, this may be formulated as

$$\frac{dY}{dt} = K(Z - Y) \tag{4.9}$$

where Z is the saturation or limiting value that must be estimated and the other variables are as previously defined. Then,

$$\int_{y_1}^{y_2} \frac{dY}{(Z - Y)} = K_D \int_{t_1}^{t_2} dt$$

and upon integration,

$$-\log_e \frac{Z - Y_2}{Z - Y_1} = K_D(t_2 - t_1)$$

Rearranging yields

$$Z - Y_2 = (Z - Y)e^{-K_D \Delta t}$$

Then, subtracting both sides of the equation from $(Z - Y_1)$,

$$(Z - Y_1) - (Z - Y_2) = (Z - Y_1) - (Z - Y_1)e_{-}^{K_D \Delta t}$$

and

$$Y_2 - Y_1 = (Z - Y_1)(1 - e^{-K_D \Delta t}) \tag{4.10}$$

Equation (4.10) may be used to make short-term estimates in the limiting region.

Curve Fitting Population data are also used to derive equations that fit observed trends or to select a mathematical function that appears to fit the data and then use that relationship for extrapolating into the future [29]. In general, mathematical curve fitting has its greatest utility in the study of large population centers, or nations. The Gompertz curve and the logistic curve are both used in establishing long-term population trends. Both curves are S-shaped and have upper and lower asymptotes, with the lower asymptotes being equal to zero.

The logistic curve in its simplest form is [29]

$$Y_c = \frac{K}{1 + 10^{a+bX}} \tag{4.11}$$

where

Y_c = ordinate of the curve

X = time period in years (10-year intervals are frequently chosen)

K, a, b = constants

Short-term, trend-based forecasts are more reliable than long-term, trend-based forecasts since over long periods there is considerable opportunity for unpredictable factors to affect the forecasted trend.

Methods Based on Relationship of Growth in One Area to That of Another

Long-term predictions are sometimes made by graphical comparison with growth rates of similar and larger cities. The population-time curve of a given community can be extrapolated on the basis of trends experienced by similar and larger communities. Population trends are plotted in such a manner that all the curves are coincident at the present population value of the city being studied (see Fig. 4.9). The cities selected for comparison should not have reached the reference population value too far in the past since the historical periods involved may be considerably different. It should be understood that the future growth of a city may digress significantly from the observed development of communities of similar size. In making the final projection, consideration should be given to conditions that are anticipated for the growth of the community in question. With the exercise of due caution, this method could give reasonable results.

The population growth in an area may also be estimated from the growth of a larger area of which it is a part, such as a state, geographical region, or nation. The procedure used is to compute the ratio of the population of the area of interest to the population of the larger area at the time of the most recent census [30]. Then, using a projection of the growth of the larger area, an estimate of growth for the study area may be made by applying the population ratio. This is known as the ratio method. Ratios may also be computed over time and the time series of ratios extrapolated using techniques for projecting trends discussed earlier. The ratio projected for a future date is then applied to the forecasted population of the larger area for that point in time. Ratio methods should be used with caution because historic relationships between the area of concern and the larger area may change as well as other influencing factors.

Component Methods

The rate of population change at any location at any time is determined by many factors, some of which are interactive. They include birth rate, death rate, immigration, emigration, government policies, societal attitudes, religious beliefs, education, technological

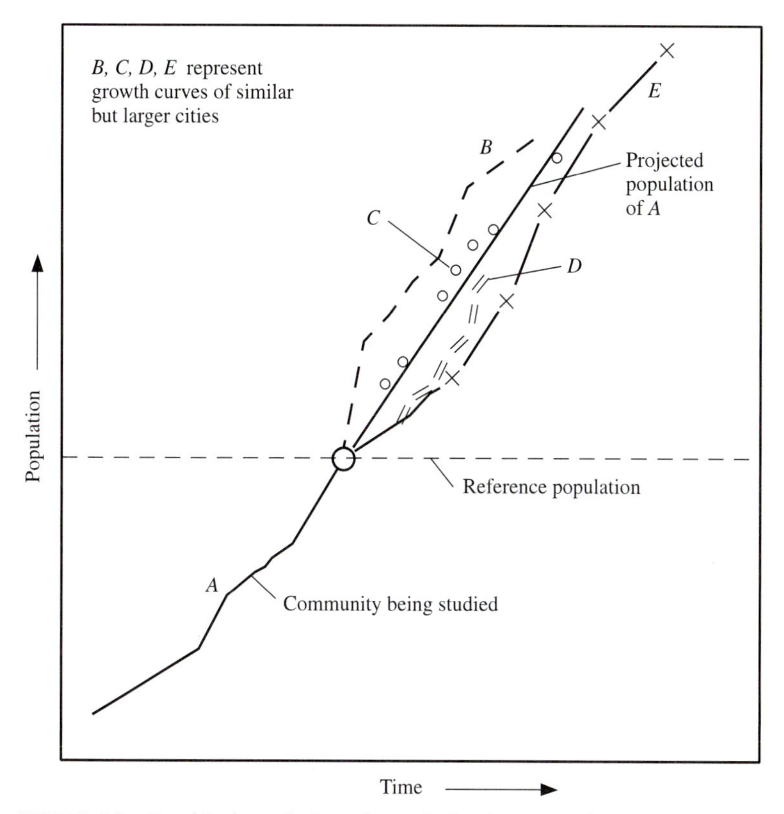

FIGURE 4.9 Graphical prediction of population by comparison.

change, and war. The components of population change (births, deaths, net migration) can be linked together to form the fundamental population equation:

$$P_2 = P_1 + B - D \pm M \tag{4.12}$$

where

P_2 = the population at the end of the time interval

P_1 = the population at the beginning of the time interval

B = number of births occurring in the population during the time interval

D = number of deaths occurring in the population during the time interval

M = net number of migrants moving to or away from the region during the time interval

The component method is widely used by population forecasters. For most areas and municipalities, it should provide better forecasts than those that would be obtained using the trend and ratio methods described earlier (Table 4.5 was developed by the U.S. Bureau of the Census using a component procedure, for example). Component

TABLE 4.7 Guide to Population Density

Area Type	Number of Persons per Acre
Residential—single-family units	5–35
Residential—multiple-family units	30–100
Apartments	100–1000
Commercial areas	15–30
Industrial areas	5–15

methods take into account the size of the study area at the start of the forecast period, and the effects of a population of that size on births, deaths, and migration. The procedure also requires the forecaster to take into account the influences on the components of Eq. (4.12) of events (industrial growth, regulatory policies, etc.) that may be expected to occur during the forecast period.

Forecasts Based on Estimates of Employment

Population growth in a locality is dependent on the ability of that location to provide jobs for the people living there. Forecasts of the labor force required to support local commercial, industrial, and institutional needs can thus be used as the basis for making population projections. Usually a labor force to population ratio is computed and used. Procedures such as this that are based on a single indicator are subject to errors that result from the failure to consider other influences on population change. They should be used with caution.

Population Density

A knowledge of the total population of a region will permit estimates of the total volume of water supply or wastewater generated to be made. In order to design conveyance systems for such flows, additional information regarding the physical distribution of the population to be served must be had. A knowledge of the population density as well as of the total population is important. Population densities may be estimated from data collected on existing areas and from zoning master plans for undeveloped areas. Table 4.7 may be used as a guide if more reliable local data are not available.

Example 4.4

A community having a population of 250,000 in 2000 estimates that its population will increase to 400,000 by the year 2020. The water treatment facilities in place can process up to 55 million gallons per day (mgd). The 2000 per capita water use rate was found to be 160 gpcd. Estimate the water requirements for the community in 2020 assuming that the per capita use rate remains unchanged. Will new treatment facilities be needed to accommodate this growth in population? If revised plumbing codes were adopted during the period of growth and if these changes resulted in an overall reduction in the community's water use by 15%, what would the water requirement be in 2020? Could expansion of treatment facilities be deferred until after the year 2020 under these conditions?

Solution:

1. The water requirement in the year 2020 for a population of 400,000 and a per capita use rate of 160 gpcd would be

$$400{,}000 \times 160 = 64 \text{ mgd}$$

2. Since 64 mgd exceeds the 2000 treatment capacity of 55 mgd, new facilities would be needed before the year 2020.
3. For a 15% reduction in water use, the per capita water requirements would be

$$160 \times 0.85 = 136 \text{ gpcd}$$

Water use in 2020 would be

$$136 \times 400{,}000 = 54.4 \text{ mgd}$$

4. Under these conditions, expansion of water treatment facilities could be deferred until after the year 2020, since the demand of 54.4 mgd is less than the treatment plant capacity of 55 mgd.

This example illustrates that an alternative to providing new facilities to meet expanding water needs could be more efficient use of the water already at hand.

Example 4.5

Consider that in 2000 a state had a total water withdrawal of 8 billion gallons per day (bgd) distributed as follows: municipal water use, 1 bgd; steam electric generation, 5 bgd; and irrigated agriculture, 2 bgd. The 2000 population was 11.7 million, and by the year 2020 it is expected that the population will increase to 13.2 million. Furthermore, it is estimated that 4000 megawatts (MW) of new electric-generating capacity will be installed by 2020 and that irrigated acreage will expand by 500,000 acres. Estimate the total water withdrawal in 2020 and the withdrawal for each of the three sectors.

Solution:

1. Estimated water use in the municipal sector can be obtained by using the projected change in population and an estimate of per capita water use in the year 2020. Assume the latter to be 140 gpcd. Change in population 2000 to 2020 $= 13.2 - 11.7 = 1.5$ million $1.5 \times 140 = 210$ mgd $= 0.21$ bgd increase in municipal water use from 2000 to 2020
2. Assume that the water requirements for the crops to be raised average 3 acre · ft per acre. 500,000 (acres) $\times$ 3 (acre · ft/per acre) $=$ 1,500,000 acre · ft irrigation water needed annually in 2020. Since 1.12 million acre · ft/yr $=$ 1 bgd, 1.5/1.12 $=$ 1.34 bgd, the added irrigation water requirement in the year 2020.
3. Assume that the plant capacity factor for the steam electric facilities to be built is 60% (this measures the percentage of the nameplate, or installed capacity, of generating facilities that is actually realized in operation). At a

60% capacity factor, one kilowatt of installed capacity would produce 14.4 kilowatt-hours (kWh) of electrical energy each day. Note also that the WRC indicates that new steam electric facilities using once-through cooling (which we shall assume here) will require about 50 gallons of water per kilowatt-hour. Then 4000 MW $\times$ 1000 (kW/MW) $\times$ 14.4 = 57,600,000 kWh per day $57.6 \times 10^6 \times 50$ (gal/kWh) = 2880 mgd or 2.88 bgd, the water requirement for steam electric cooling to be added during the interval 2000 to 2020

4. The added water requirements to the year 2020 are thus obtained by totaling the incremental increases for the three sectors: 0.21 + 1.34 + 2.88 = 4.43 bgd increase; thus, the total water withdrawal in 2020 would be 4.43 + 8.00 = 12.43 bgd. The combined withdrawal use in 2020 would thus be about 1.6 times that of 2000.

4.7 WATER USE FORECASTING

Forecasting is the art and science of looking ahead; it is the core of planning processes. Options for selecting a forecasting procedure range from the exercise of judgment to the use of complex mathematical models. Forecasts are required for periods varying from less than a day to more than 50 years. Unfortunately, the more distant the planning horizon, the more questionable the forecast. It is therefore important that great care be exercised in selecting a forecasting technique and that an understanding be had of the limitations of the method selected. Furthermore, it is recommended that an array of alternative futures, rather than a single projection, be used to guide decision-making processes. Forecasts should be recognized for what they are: approximations of what might occur, not accurate portrayals of what will be. Commonly used forecasting methods include projections based on historic data, the use of models and simulation, and qualitative and holistic techniques [31].

The Delphi technique and scenario writing are examples of qualitative/holistic approaches, while trend extrapolation and trend impact analysis are history-based procedures [31]. Simulation is the process of mimicking the dynamic behavior of a system over time. A simulation model is a surrogate of the real-world system it is designed to represent. The results of simulation computer runs are widely used to describe (forecast) the future states of water or other systems being studied.

Many types of forecasting models have been used in water supply planning. They range from rough trend extrapolations and crude correlations of variables to complex mathematical representations of the dynamics of land and water use. Historically, future water requirements were determined as the product of a projected service area population and an anticipated per capita water use rate. Such an approach employs only two of the determinants of water use. Furthermore, per capita water use rates generally vary within and among communities and over time. The bottom line, however, is that long-term planning for public water supply systems depends on reliable forecasts of water demands and on the identification and assessment of demand-side alternatives. Demand reduction options must be considered because they can increase the efficiency of water use, thereby increasing the likelihood that future water supplies and demands will be balanced at a cost below the economic, social, and environmental costs of developing new water sources.

IWR-MAIN Water Demand Analysis Software

IWR-MAIN is a computerized water use forecasting system that includes a range of forecasting models and data management techniques [32–39]. The IWR-MAIN Water Demand Management Suite (DMS, 1999©) operates in the Windows environment. It utilizes an open architecture in which the user inputs the forecasting model and specific variables within the model. The software features a high level of disaggregation of water use categories and user flexibility in selecting forecasting methods and assumptions. It is designed to deal principally with urban water uses: residential; commercial/institutional; manufacturing; and uses not accounted for in other sectors. The model also provides for forecasts of water use recognizing the influences of conservation policies. Figure 4.10 indicates the disaggregation of urban water use, Figure 4.11 summarizes estimation methods, and Figure 4.12 indicates the IWR-MAIN model inputs and outputs. The narrative presentation of the IWR-MAIN model given herein follows that of Planning and Management Consultants, Ltd. [32, 33].

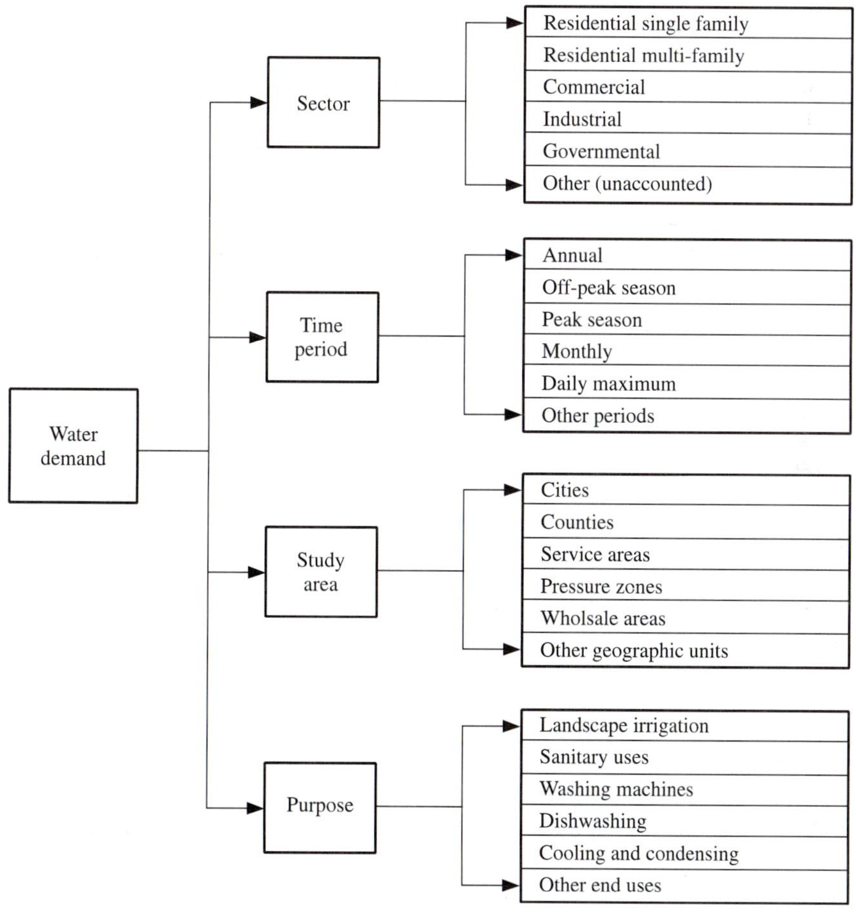

FIGURE 4.10 Disaggregation of urban water use. Courtesy of Planning and Management Consultants, Ltd., Carbondale, IL.

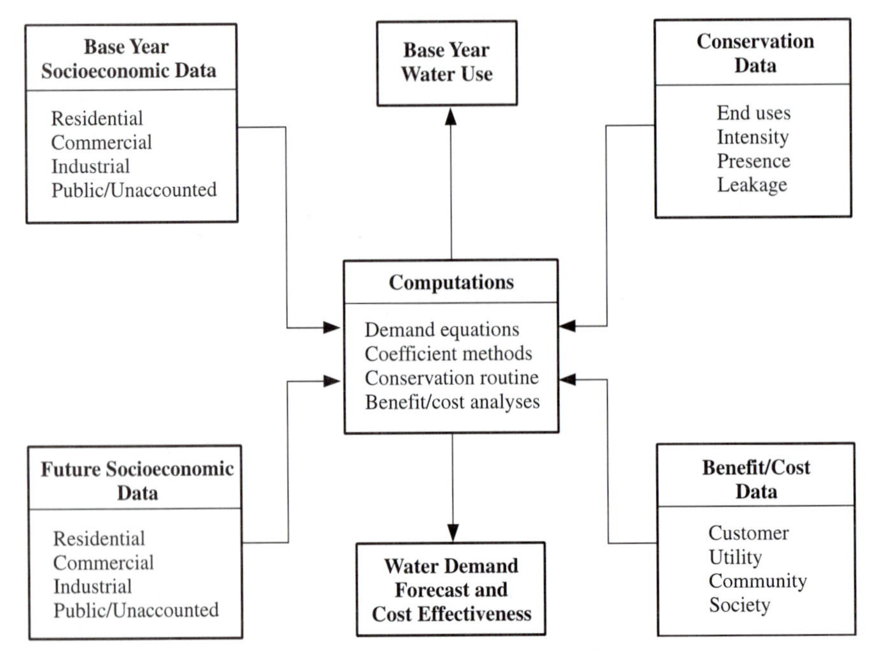

FIGURE 4.11 IWR-MAIN estimation methods. Courtesy of Planning and Management Consultants, Ltd., Carbondale, IL.

The water demand management models described here are effective predictive models developed for IWR-MAIN Version 6.1©. They have three major components: forecasting water use, estimating conservation water savings, and performing benefit/cost analyses. The DMS version of IWR-MAIN combines the conservation water saving and benefit cost calculations into one single component. IWR-MAIN uses econometric water demand models for translating the existing demographic, housing, and business statistics into estimates of existing water demands. These estimates are used to fine-tune the water use equations for translating the long-term projections of population, housing, and employment into disaggregated forecasts of water use. An extensive analysis of existing and projected demands, disaggregated by season, sector, and purpose, is conducted in order to generate estimates of water conservation savings from efficient technologies and plumbing codes as well as from utility-sponsored programs such as retrofits, water audits, financial incentives, and public education and information initiatives. Estimates of water savings can also be used in an analysis of economic effectiveness of demand-side alternatives.

The IWR-MAIN software disaggregates total urban water use by customer sectors, time periods, spatial study areas, and end use purposes. Water demands of various parts of a service area are disaggregated according to their seasonal variation and the relative needs of various customer classes and sectors (e.g., residential single-family, residential multifamily, commercial, manufacturing, and government). The water demands of each sector in a given area and time period are expressed as a product of (1) the number of users (i.e., demand drivers such as the number of residents, housing units, employees, and parkways) and (2) the average rate of water use (e.g., per household or per employee) as

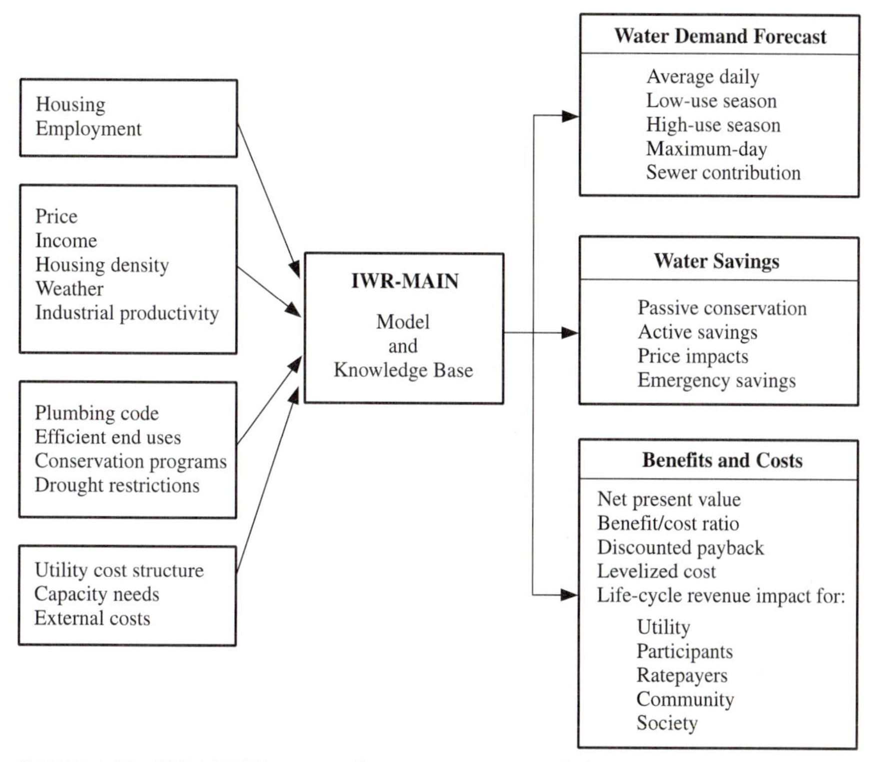

FIGURE 4.12 IWR-MAIN inputs and outputs. Courtesy of Planning and Management Consultants, Ltd., Carbondale, IL [32].

determined by a set of explanatory variables for a given sector. In the residential sector, the set of explanatory variables may include income, price of water and wastewater services, household size, housing density, air temperature, and rainfall. The seasonality of water use may be represented by including a set of binary variables for months (or seasons) of the year. In the nonresidential sector, these factors may include employment by industry type, labor productivity, weather conditions, and the price of water and wastewater services. Separate forecasts can be made for each identified sector; for specific study areas such as cities, counties, and load zones; for summer and winter daily water use; and for average daily, annual, and peak-day water use. In addition, the software provides for the inclusion of a variety of demand management or conservation practices. The conservation savings component of IWR-MAIN distinguishes among passive, active, and emergency (i.e., temporary) conservation effects. *Passive* conservation effects are represented by shifts in end use consumption from less efficient fixtures/practices to more efficient fixtures/practices brought about mainly through plumbing codes for new construction or improved water-using technologies. The conservation savings of *active* programs are estimated by noted changes in water use efficiency classes brought about by participation in a utility-sponsored program in which inefficient or standard end uses are replaced or retrofitted. These water savings can be incorporated into long-term forecasts of water demand.

IWR-MAIN can also assist the user in conducting *benefit/cost analyses of demand management alternatives*, that is, of the active conservation programs for which water savings have been estimated. A number of economic feasibility tests are used to evaluate the economic merits of conservation programs, including: net present value, discounted payback period, benefit/cost ratio, levelized cost, and life-cycle revenue impact. The results of the benefit/cost component of IWR-MAIN can be used in comparing supply augmentation alternatives with demand management alternatives using the same economic criteria.

IWR-MAIN also offers the ability to conduct *sensitivity analyses*, or what-if scenarios, regarding projected changes in the determinants of water demand and to assess the impact of these changes on long-term water demands. Thus, water planners may evaluate the impact of changes in socioeconomic conditions, weather patterns, water pricing, or alternative conservation programs.

Residential Sector Forecasting

Within the residential sector of IWR-MAIN there are seven subsectors available for forecasting water demands: (1) single family—1 attached, 1 detached units; (2) multifamily low density—2, 3, 4 units per structure; (3) multifamily high density—5 or more units per structure; (4) mobile homes; (5) nonurban; (6) user added; and (7) total residential. These categories correspond to the housing types used by the U.S. Bureau of the Census. One or more of the seven subsectors may be used, depending on the characteristics of the service area. The DMS allows the user to select the subsectors to be evaluated and allows the user to define new subsectors. The residential water demand model described here may be entered into the IWR-MAIN DMS for residential subsectors.

Average rates of water use within each residential subsector are estimated using causal water demand models, which take the following theoretical form [32,33]:

$$q_{c,s,t} = \alpha I^{\beta 1} H^{\beta 2} L^{\beta 3} T^{\beta 4} R^{\beta 5} P^{\beta 6} e^{b7B} \tag{4.13}$$

where

$q_{c,s,t}$ = predicted average water use in sector c, during season s, in year t

I = median household income

H = average household size (persons)

L = average housing density (units per acre)

T = average maximum daily temperature

R = rainfall

P = marginal price of water (including sewer charges related to water use)

B = fixed charge or rate premium

α = constant

β_i = constant elasticities of explanatory variables

b_7 = coefficient of the rate premium

e = base of the natural logarithm

The season, sectoral, and temporal indices of the explanatory variables in the above equation are suppressed herein for clarity. This water demand model conforms to economic theory and may be considered causal since the explanatory variables can be shown to cause the demand. For example, income measures the consumer's ability to pay for water and price influences the amount of water the consumer is willing to purchase.

The default elasticities of the explanatory variables of residential household water use are derived from econometric studies of water demand through a rigorous statistical analysis of empirical data. Multiple regression is used to explain the variance in the values of reported elasticities due to interstudy differences. Note that elasticity is interpreted as the percent change in quantity (e.g., water use) that is expected from a 1% change in the explanatory variable [33].

Nonresidential Sector Forecasting

The nonresidential sector of IWR-MAIN addresses water uses within the following major industry groups: (1) construction; (2) manufacturing; (3) transportation, communications, and utilities (TCU); (4) wholesale trade; (5) retail trade; (6) finance, insurance, and real estate (FIRE); (7) services; and (8) public administration. The DMS has default nonresidential subsectors for commercial, industrial, and government, yet allows the user to add additional subsectors as needed for the analysis of a given study area. The eight major industry groups are classified according to Department of Commerce Standard Industrial Classification (SIC) codes (note that in 1997, the U.S. Office of Management and Budget issued the North American Industry Classification System [NAICS] to replace the SIC classification system). The conceptual model of water use in the nonresidential (commercial/industrial) sector is [32,33]:

$$Q_i = f(GED_i, E_i, L_i, P_i, CDD, O_i) \tag{4.14}$$

where

Q_i = category-wide water use in gallons per day

GED_i = gallon per employee per day water use

E_i = category-wide employment

L_i = average productivity (of labor) in category I

P_i = marginal price of water and wastewater services in category I

CDD = cooling degree days

O_i = other variables known to affect commercial/industrial water use

Although this theoretical model is fully operational within IWR-MAIN, there are no currently available econometric (and generally applicable) models that contain model elasticities for price, productivity, cooling degree days, or the other variables for nonresidential water uses. Such models of nonresidential use may be estimated for the given study area from historic water use, economic and climatic data. The IWR-MAIN system is designed, however, to accommodate the above model specifications once data become available regarding the responsiveness of nonresidential water use to such variables. Thus, version 6.1 of IWR-MAIN calculates commercial/industrial water use based upon gallon per employee per day coefficients for SIC categories and groups [32,33]:

$$Q_i = (GED_i \cdot E_i) \tag{4.15}$$

The water use per employee coefficients contained within IWR-MAIN are the result of extensive research devoted to collecting data on employment and water use for various establishments throughout the United States. The water use coefficients within IWR-MAIN are based upon the analysis of water use and employment relationships in over 7000 establishments. Table 4.8 shows the water use per employee coefficients for the eight major industry groups.

The appropriate SIC level to be used should be based upon the structure of employment in the community, the availability of employment data at the various levels, and the need for sensitivity analyses regarding potential water use impacts.

Additional Sector Forecasting

IWR-MAIN can be used to address water uses that occur in a service area not accounted for by the other sectors. Water use in these special categories must be expressed as a function of a single explanatory variable (such as the number of acres or number of facilities). Water use coefficients (expressed in gallons per day per selected unit) must also be provided. Forecasts for these uses can be made by projecting the number of units into the future and multiplying them by the appropriate water use

TABLE 4.8 Nonresidential Water Use Coefficients Contained in TWR-MAIN Version 6 (1995)

Major Industry Group	SIC Codes	Water Use Coefficient (gallons/employee/day)[a]	
Construction	15–17	20.7	(244)
Manufacturing	20–39	132.5	(2784)
Transportation, communications, utilities (TCU)	40–49	49.3	(225)
Wholesale trade	50–51	42.8	(750)
Retail trade	52–59	93.1	(1041)
Finance, insurance, real estate (FIRE)	60–67	70.8	(233)
Services	70–89	137.5	(1870)
Public administration	91–97	105.7	(25)

[a]The numbers in parentheses represent the sample number of establishments from which the water use coefficient was calculated.

Source: Courtesy of Planning and Management Consultants, Ltd., Carbondale, IL [33].

coefficients. The following are examples of uses and defining parameters that may fall within this classification: (1) irrigation of public parks and medians (acres); (2) make-up water for public swimming pools (number of pools); and (3) irrigation of golf courses (acres) [32,33].

Other/Unaccounted-for Sector Forecasts

The difference between the total quantity of water produced (treated and delivered) and the quantity of water sold to customers is referred to as *unaccounted-for water use*. This sector may include the following types of uses/losses: (1) distribution system leakage; (2) meter slippage; (3) hydrant flushing; (4) major line breaks; (5) firefighting; (6) unmetered or nonbilled customers; (7) illegal connections; and (8) street washing/construction water [33]. Water utilities generally record unaccounted-for water use as the percentage difference between total quantity of water delivered into the distribution system and total metered sales. In the IWR-MAIN model, users specify a percentage rate for each base and forecast year to estimate the amount of unaccounted-for water.

Socioeconomic Input Data Requirements

Two types of data are required for the generation of water use estimates from the IWR-MAIN system: (1) actual values of demographic and socioeconomic determinants (or parameters) of water use for the base year and (2) projected values of selected determinants for each forecast year [32]. IWR-MAIN can accommodate a variable degree of data availability. The amount of time and effort required to prepare a forecast depends on the chosen level of detail of the input data and the chosen level of disaggregation of the user sectors. Base year input data for the IWR-MAIN system may be used to estimate current water requirements and may also be used as a reference for projections of future parameter values.

Water Conservation Savings Methods

The conservation savings module of the IWR-MAIN system further disaggregates seasonal demands of various water use sectors into a number of specific end uses such as dishwashing, toilet flushing, lawn watering, cooling, and others. This high level of disaggregation is designed to accommodate the evaluation of various demand management (conservation) measures that usually target specific end uses. The IWR-MAIN conservation savings module utilizes an end use accounting system that disaggregates the seasonal demands of various water use sectors into as many as 16 different end uses [32]. A rational representation of each end use is made using a structural end use equation. Given parameters of local end use conditions, the equation predicts the average quantity of water for each end use as a function of (1) the distribution of end uses among three classes of efficiency (i.e., nonconserving, standard, and ultraconserving), (2) average usage rate or intensity of use, (3) leakage rate and incidence of leaks, and (4) presence of end use within a given customer sector. The structure of the end use equation allows the planner to estimate the net effects of long-term conservation programs by tracking the values of each end use parameter over time. The end use relationship is expressed as [32,33]:

$$q_i = [(M_1S_1 + M_2S_2 + M_3S_3) \cdot U_N + K \cdot F_N] \cdot A_N \tag{4.16}$$

where

q_i = quantity of water used by end use i, gpd/unit

$M_{1\text{-}3}$ = mechanical parameter (e.g., volume per use, flow rate per minute)

$S_{1\text{-}3}$ = fraction of the sector for end use that is nonconserving, standard, and ultraconserving (see also Fig. 4.13)

U_N = intensity of use parameters (e.g., flushes per day/unit, minutes of use per day/unit)

K = mechanical parameter representing the rate of leakage

F_N = fraction of end uses with leakage

A_N = fraction of units in which end use i is present

$_N$ = normal use or nondrought/nonemergency

1–3 = end use or group that is nonconserving, standard, and ultraconserving 1 signifying the lowest level of efficiency

Long-term conservation savings may be achieved by increasing the fractions S_2 and S_3. This is accomplished by moving customers from one efficiency class to another. For example, for each end use, the fraction of the water users would be shifted from non-conserving to ultra-conserving or from standard to ultra-conserving. The quantifiable

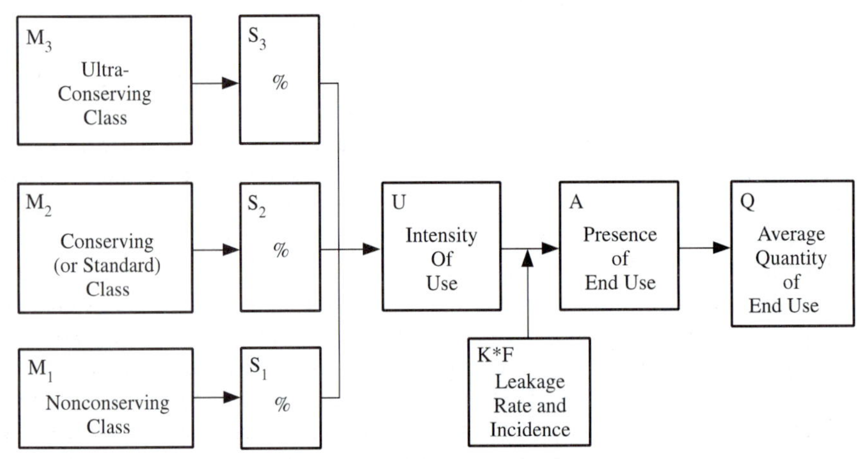

FIGURE 4.13 Structural end use relationships. Courtesy of Planning and Management Consultants, Ltd., Carbondale, IL [32].

effect of the program is accounted for directly by the numerical shift in the customer pools and the change in the fractions of customers in each efficiency class.

The conservation savings module distinguishes among passive, active, and emergency (i.e., temporary) conservation effects. Passive conservation effects are represented by shifts in end use consumption from less efficient fixtures to more efficient fixtures brought about primarily by plumbing codes for new construction (e.g., the toilet end use moves from the inefficient 5.5-gallon-per-flush toilet to the standard 3.5-gallon toilet, or the highly efficient 1.6-gallon toilet). The conservation savings of active programs are estimated by noting changes in the distribution of efficiency classes brought about by the participation in a utility-sponsored program whose inefficient or standard end uses were replaced or retrofitted.

Applications

Output from the IWR-MAIN model can be used to aid in (see also Fig. 4.12):

- Planning to meet future water supply needs
- Sizing and expanding distribution systems
- Preparing contingency plans for water shortages
- Evaluating the effectiveness of conservation practices
- Performing sensitivity analyses using varying assumptions about water prices, climatic factors, and other determinants of water use
- Assessing utility revenues with improved precision

The IWR-MAIN model has been used in Arizona, Illinois, Texas, Oklahoma, California, Nevada, Massachusetts, Oregon, and Florida [32,33]. A brief case study of an application in Oregon follows. To obtain more information on the IWR-MAIN system, or to find out how to obtain the computer program, the reader should access the IWR-MAIN Web site www.IWRMAIN.com

Eugene, Oregon, Case Study

The Eugene Water and Electric Board's (EWEB) Water Resource Management Plan called for the development of a reliable, accepted, and flexible methodology for forecasting water demands and for evaluating the potential impacts of water demand management (water conservation) strategies. In response to a contract, Planning and Management Consultants, Ltd. (PMCL) analyzed the EWEB water service area using IWR-MAIN Water Demand Analysis Software (version 6©) [33].

The entire EWEB retail and wholesale water service area comprised the study area. The year 1990 was chosen as the base year and the planning horizon included the years 1995, 2000, 2010, 2015, and 2020. Using current and projected demographic characteristics of the area, IWR-MAIN was used to forecast total urban water use by sectors (single-family residential, multifamily residential, commercial, industrial, governmental, and other) and time steps (annual average daily, summer season average daily, winter season average daily, and maximum daily). The *baseline* water use forecast accounted for projected growth of population, housing units, and employment in the study area.

TABLE 4.9 IWR-MAIN Baseline Water Use Forecast for Eugene Water and Electric Board (MGD)[a,b]

Site/Sector	1990	1992	1995	2000	2005	2010	2015	2020	1990–2020 Percent Change
Residential:									
Single-family	10.842	11.241	11.849	12.708	13.564	14.606	15.827	17.062	57
Multifamily	3.068	3.181	3.353	3.596	3.838	4.133	4.478	4.828	57
Total residential	13.910	14.422	15.202	16.304	17.402	18.739	20.305	21.891	57
Nonresidential:									
Construction	0.068	0.072	0.074	0.078	0.082	0.086	0.090	0.094	38
Manufacturing	2.619	2.363	2.459	2.653	2.806	2.956	3.105	3.250	24
Transportation	0.201	0.195	0.198	0.211	0.223	0.235	0.249	0.262	30
Wholesale	0.209	0.209	0.208	0.217	0.226	0.235	0.245	0.255	22
Retail	1.722	1.658	1.723	1.871	2.016	2.163	2.312	2.464	43
Finance	0.281	0.308	0.324	0.353	0.383	0.412	0.443	0.474	69
Services	3.848	3.923	4.136	4.528	4.936	5.346	5.760	6.177	61
Government	2.110	2.060	2.077	2.210	2.352	2.496	2.638	2.781	32
Total nonresidential	11.059	10.788	11.200	12.121	13.022	13.930	14.842	15.759	42
Other:									
Total other/unaccounted	3.213	3.867	3.910	4.210	4.506	4.838	5.205	5.576	74
Total average daily demands	28.182	29.077	30.313	32.635	34.931	37.507	40.352	43.224	53
Total maximum daily demands	60.459	63.072	65.610	70.592	75.550	81.135	87.321	93.567	55

[a]For comparison purposes, the water use estimates for 1990 and all subsequent years reflect long-term average weather conditions.
[b]Unaccounted percentages are 11.4 for 1990, 13.3 for 1992, and 12.9 for all forecast years.
Source: Courtesy of Planning and Management Consultants, Ltd., Carbondale, IL [33].

To estimate water savings from water demand management and other programs, total sector water demands were broken down into functional purposes or end uses (e.g., toilet, shower, landscape irrigation, and process). Once sector water use was allocated to these functional purposes, the IWR-MAIN system was used to determine the water savings of implementing demand management programs or efficiency improvements. A forecast of *adjusted baseline* water use was prepared that accounted for water savings from conservation programs that had already been implemented in EWEB's service area (i.e., a showerhead/faucet retrofit program) and from passive efficiency improvements that were expected to result from state and federal laws. The IWR-MAIN system was also used to conduct a benefit/cost analysis of the potential demand management programs [33].

Using appropriate socioeconomic and demographic data inputs, baseline water use projections for municipal and industrial water use were made for the years 1990, 1992, 1995, 2000, 2005, 2010, and 2020 (these forecasts did not take into consideration potential impacts of water demand management measures implemented after 1992). The baseline forecasts of average daily and maximum daily water use are given in Table 4.9. The water use estimates for 1990 and all subsequent years reflect long-term average weather conditions. The demographic data driving the water use forecasts did not take into account the closure of a food products company and the potential establishment of a new Hyundai facility. The forecast showed that average daily residential demands could be expected to increase by 57% between 1990 and 2020. This generally follows the projections of population and housing units. However, nonresidential water demands during the same time period were projected to increase by only 42%. These water demands generally follow employment statistics. The most rapid growth in nonresidential water demands was expected to occur in water use related to the finance, insurance, and real estate and services industries. Average daily demands and maximum daily demands were projected to increase by about 53% and 55%, respectively. Maximum daily demands would be expected to increase at a slightly higher rate because residential water use, which has a larger component of outdoor use, is projected to grow more rapidly than nonresidential use [33].

There are numerous models for forecasting water use. The prediction equations for many of these models are empirical and have been derived using regression analysis or other related techniques [40–44]. Most models developed since 1970 also include the cost of supplying water as a parameter [32, 34–39, 44–47]. The IWR-MAIN model, which is widely used, is typical of the types of approaches that can be taken in water use forecasting. Further information on forecasting models may be found in the references at the end of the chapter [48–58].

PROBLEMS

4.1 A reservoir has a capacity of 2.75 million acre · ft. How many years would this supply a city of 100,000 if evaporation is neglected? Assume a use rate of 180 gpcd.

4.2 If the minimum flow of a stream having a 200-mi^2 watershed is 0.10 cfs/mi^2, what population could be supplied continuously from the stream? Assume that there is only distribution storage (maximum withdrawal = stream flow) and that the water use rate is 175 gpcd.

4.3 Estimate the 2010 and 2015 population of your community by plotting the historic data and extrapolating the trend. How reliable do you think this estimate would be?

4.4 Given the population, public water withdrawal, and total freshwater withdrawal for the states shown in Table 4.2 in 1995, calculate the percent of freshwater withdrawn by the public sector and the per capita water use of the freshwater for your state and the surrounding states.

4.5 Given the population, public water withdrawal, and total freshwater withdrawal for the states shown in Table 4.2 in 1995, calculate the percent of freshwater withdrawn by the public sector and the per capita water use of the freshwater for 10 states of your choice and state why you think these percentages differ.

4.6 Given the population data of Table 4.5, estimate the percent increase in population from 1995 to 2025 for three northeastern, southeastern, north central, south central, northwestern, and southwestern states. Discuss the differences.

4.7 Make a comparison between the annual water requirements of an 1500-acre irrigated farm and a city of 130,000 population. Assume an irrigation requirement of 3 acre · ft/yr/acre and a per capita water use rate of 180 gpcd.

4.8 Obtain historic and projected water use figures for your state (consult your state water planning agency for data). Analyze the historic trends and try to explain the major influences on water use in each of the principal water-using sectors. Do you think the trends projected for the future are reasonable? Why? If not, why not? What could be done to modify these trends? How much reduction could be expected in each sector?

4.9 For the region in which you reside, make a determination of which water-using sectors are most dominant. How did you arrive at this determination? Do you think the past trends will continue into the future? If so, why? If not, why not? Are there water supply problems in the region? If so, could these be alleviated by modifying water use rates in one or more of the water-using sectors? How much of a reduction below current rates of use do you think could be achieved? What revision in facilities or systems operation would be required to bring about this reduction?

4.10 Obtain historic and future population data on the municipality in which you reside. Consult your local water department and obtain historic and projected water use trends for your city. Estimate the current per capita water use rate. Use this figure to project water requirements to the year 2005. How does your projection compare with that of the water department? By how much could the per capita water use rate be reduced, if any, by 2005?

4.11 Given a residential area encompassing 1100 acres with a housing density of four houses per acre. Assume a high-value residence with a fire flow requirement of 1000 gpm. Find (a) the combined draft and (b) the peak hourly demand.

4.12 Given that a residential community has an area of 10 mi^2, assume a population density and calculate the required fire flow. Give results in gpm and lpm.

4.13 Consider a 1000-acre residential area with a housing density of four dwellings per acre. Estimate the peak hourly water use requirement.

4.14 If 100 acres of farmland were developed for urban housing (four houses per acre), what would be the difference in average annual water requirements after the changeover? Assume that the irrigation requirement is 2.5 acre · ft of water for a growing season of six months.

4.15 The population of a state was 7 million in 2000. Consider that by the year 2015, it is expected to increase to 9 million. Consider that the amount of freshwater withdrawn in 2000 was 2.5 bgd. Estimate the amount of freshwater that might be withdrawn in 2015. State your assumptions.

4.16 A community had a population of 200,000 in 2000 and it is expected that this will increase to 260,000 by 2015. The water treatment capacity in 2000 was 43 mgd. A survey showed that the average per capita water use rate was 180 gpcd. Estimate the community's water requirements in 2015 assuming (a) no change in use rate and (b) a reduced rate of 160 gpcd. Will expanded treatment facilities be needed by 2015 for condition (a)? For condition (b)?

4.17 For the community described in Problem 4.16, assume that the treatment plant capacity in 2000 was 35 mgd. If the water use rate in 2015 is 140 gpcd, will expanded treatment facilities be needed by 2015? In about what year would they be needed if the answer above is yes? What reduction in water use rate would be needed to eliminate the need for new facilities until 2015? Would a reduction of this magnitude appear to be attainable?

4.18 Consider that a state had a total water withdrawal of 4.5 bgd in 2000. This was distributed as follows: municipal use 1.0 bgd; industrial use 1.5 bgd; and steam electric cooling 2 bgd. Assume a 2000 population of 10 million and a projection of 12 million for the year 2015. It is estimated that an additional 3000 MW of electric-generating capacity will be required by 2015 and that the cooling water requirements will be 50 gal/kWh. An industrial expansion of 10% is also expected. Estimate the total water that will be withdrawn in 2015 for each sector and for all sectors combined. State your assumptions. Assume a plant capacity factor of 0.6 (water use will relate to 3000 MW $\times$ 0.6).

4.19 Given that a residential community has an area of 26 km^2, assume a population density and calculate the required fire flow. Given results in lpm.

4.20 Consider a 450-acre residential area with a housing density of four dwellings per acre. Estimate the peak hourly water use requirement and the peak hourly sewage flow.

REFERENCES

[1] *Water World* (Northbrook, IL) 19 no. 1 (June 2003).

[2] R. Von Dohren, "Consider the Many Reasons for Reuse," *Water and Wastes Eng.* 17, no. 9 (1980): 7478.

[3] SCS Engineers, Inc., "Contaminants Associated with Direct and Indirect Reuse of Municipal Wastewater," U.S. EPA Report No. EPA600/178019, NTIS No. PB 280 482 (March 1978).

[4] J. Crook, "Reliability of Wastewater Reclamation Facilities" (Berkeley, CA: California Department of Health, Water Sanitation Section, 1976).

[5] W. E. Garrison and R. P. Miele, "Current Trends in Water Reclamation Technology," *J. Am. Water Works Assoc.* 69, no. 7 (1977): 364–369.

[6] California Department of Health, "Wastewater Reclamation Criteria," California Administrative Code, Title 22, Division 4, Water Sanitation Section (Berkeley, CA, 1975).

[7] U.S. Water Resources Council, *The Nation's Water Resources 1975 to 2000* (Washington, DC: U.S. Government Printing Office, December 1978).

[8] "Estimated Water Use in the United States, 1995," U.S. Geol. Survey Circular 1200 (1995).

[9] J. B. Wolff, "Peak Demands in Residential Areas," *J. Am. Water Works Assoc.* 87 (October 1961).

[10] F. P. Linaweaver, Jr., "Report on Phase One, Residential Water Use Research Project" (Baltimore: The Johns Hopkins University, Department of Sanitary Engineering, October 1963).

[11] *Standard Schedule for Grading Cities and Towns of the United States with Reference to Their Fire Defenses and Physical Conditions* (New York: National Board of Fire Underwriters, 1956).

[12] R. S. Gupta, *Hydrology and Hydraulic Systems* (Englewood Cliffs, NJ: Prentice-Hall, 1989).

[13] E. T. Smerdon, "Global Change and Water Resources in the Southwest," *Water Resources Update*, 111 Universities Council on Water Resources, Carbondale, IL (Spring 1998).

[14] K. P. Singh and G. S. Ramamurthy, "Climate Change and Resulting Hydrologic Response: Illinois River Basin," *Watershed Planning and Analysis in Action* (ASCE Symposium Proceedings, Durango, CO, 1990).

[15] R. M. White, "The Great Climate Debate," *Scientific American* 263 (July 1990).

[16] P. E. Waggoner, ed., *Climate Change and U.S. Water Resources* (New York: John Wiley & Sons, 1990).

[17] J. C. Schaake, "From Climate to Flow" in *Climate Change and U.S. Water Resources* (New York: John Wiley & Sons, 1990).

[18] S. W. Work, M. R. Rothberg, and K. J. Miller, "Denver's Potable Reuse Project: Pathway to Public Acceptance," *J. Am. Water Works Assoc.* 72, no. 8 (1980): 435–440.

[19] U.S. Congress, "Water Reuse Project Underway in Denver," *Congressional Record*, pp. E5191, E5192, (October 22, 1979).

[20] J. E. Matthews, "Industrial Reuse and Recycle of Wastewaters: Literature Review," EPA Report No. EPA/600/280183, U.S. EPA, Ada, OK (1980).

[21] B. A. Carnes, J. M. Eller, and J. C. Martin, "Reuse of Refinery and Petrochemical Wastewaters," *Ind. Water Eng.* 9, no. 4 (1979): 25.

[22] L. E. Streebin, L. W. Canter, and J. R. Palafox, "Water Conservation and Reuse in the Canning Industry" (proceedings of the 26th Industrial Waste Conference, Purdue University, Lafayette, IN, Part II 1971): 766.

[23] C. A. Caswell, "Water Reuse in the Steel Industry," *Complete Water Reuse: Industry's Opportunity* (New York: American Institute of Chemical Engineers, 1973); 384.

[24] R. Field and C. Fan. "Recycling Urban Stormwater for Profit," *Water/Eng. Management* 128, no. 4 (1981).

[25] J. Merrell and R. Stoyer, "Reclaimed Sewage Becomes a Community," *American City* (April 1964): 97.

[26] J. C. Merrell, Jr., W. F. Jopling, R. F. Bott, A. Katko, and H. E. Pintler, "Santee Recreational Project, Santee, California, Final Report," FWPCA Report WP-20-7 (Cincinnati: Publication Office, Ohio Basin Region, Federal Water Pollution Control Administration, 1967).

[27] P. Campbell, *Population Projections: States, 1995–2025*, Current Population Reports, Bureau of the Census, U.S. Department of Commerce, P25-1131 (Washington, DC, May 1997).

[28] F. W. Hollmann, T. J. Mulder, and J. E. Kallan, *Methodology and Assumptions for the Population Projections of the United States: 1999 to 2100*, Bureau of the Census, U.S. Department of Commerce, Washington, DC, January 2000.

[29] F. E. Croxton and D. J. Cowden, *Applied General Statistics* (Englewood Cliffs, NJ: Prentice-Hall, 1960).

[30] R. C. Schmitt, "Forecasting Population by the Ratio Method," *J. Am. Water Works Assoc.* 46 (1954): 960.

[31] W. Viessman, Jr. and C. Welty, *Water Management: Technology and Institutions* (New York: Harper & Row, 1985).

[32] Planning and Management Consultants, Ltd. "IWR-MAIN, Version 6.1©, Water Demand Analysis Software, Technical Overview" (Carbondale, IL, 1994).

[33] E. M. Opitz, B. Dziegielewski, and J. R. M. Steinbeck, "Water Use Forecasts and Conservation Evaluations for the Eugene Water and Electric Board," (Carbondale, IL; Technical Report, Planning and Management Consultants, Ltd., November 1995).

[34] J. J. Boland, "IWR-MAIN System Improvements: Phase III—Growth Models, Consultant Report," Contract DACW72-84-C-0004 (Fort Belvoir, VA: U.S. Army Engineer Institute for Water Resources, 1987).

[35] B. Dziegielewski, "The IWR-MAIN Disaggregate Water Use Model" (proceedings of the 1987 Annual UCOWR Meeting, UCOWR, Southern Illinois University, Carbondale, IL, 1987).

[36] Planning and Management Consultants, Ltd., "A Disaggregate Water Use Forecast for the Phoenix Water Service Area" (report presented to City of Phoenix, Water and Wastewater Department, March 1986).

[37] Planning and Management Consultants, Ltd., "IWR-Main Water Use Forecasting System, System Description," Carbondale, IL (January 1987).

[38] Planning and Management Consultants, Ltd., "Municipal and Industrial Water Use in the Metropolitan Water District Service Area," Carbondale, IL (November 1987).

[39] D. D. Baumann, J. J. Boland, and W. M. Hanemann, *Urban Water Demand Management and Planning*, (New York: McGraw-Hill, 1998).

[40] D. W. Berry and G. W. Bonem, "Predicting the Municipal Demand for Water," *Water Resources Res.* 10, no. 6 (1974): 1239–1242.

[41] H. S. Foster and B. R. Beattie, "Urban Residential Demand for Water in the United States," *Land Economics* 55, no. 1 (1979): 43.

[42] S. L. Franklin and D. R. Maidment, "An Evaluation of Weekly and Monthly Time Series Forecasts of Municipal Water Use," *Water Resources Bull.*, Paper No. 86038, Am. Water Resources Assoc. 2, no. 4 (August 1986).

[43] D. T. Lauria and C. H. Chiang, *Models for Municipal and Industrial Water Demand Forecasting in North Carolina* (Department of Environmental Sciences and Engineering, University of North Carolina, Chapel Hill, NC, 1975).

[44] S. T. Wong, "A Model of Municipal Water Demand: A Case Study on Northeastern Illinois," *Land Economics* 48, no. 1 (1972): 34.

[45] C. W. Howe, "The Impact of Price on Residential Water Demand: Some New Insights," *Water Resources Res.* 18, no. 4 (1982): 713–716.

[46] C. W. Howe and F. P. Linaweaver, "The Impact of Price on Residential Water Demand and Its Relation to System Design and Price Structure," *Water Resources Res.* 3, no. 1 (1967): 13–32.

[47] D. E. Agthe and R. B. Billings, "Dynamic Models of Water Demand," *Water Resources Res.* 16, no. 3 (1980): 476–480.

[48] AWWA Committee on Water Use, "Review of The Johns Hopkins University Research Project Method for Estimating Residential Water Use," *J. Am. Water Works Assoc.* 65, no. 5 (1973): 300–301.

[49] J. P. Heaney, G. D. Lynne, N. Khanal, W. C. Martin, C. L. Sova, and R. Dickinson, "Agricultural and Municipal Water Demand Projection Models" (University of Florida, Gainesville, FL, March 1981).

[50] B. Dziegielewski, J. J. Boland, and D. D. Baumann, "An Annotated Bibliography on Techniques of Forecasting Demand for Water," IWR Contract Report 81-C03 (Fort Belvoir, VA: U.S. Army Corps of Engineers, 1981.

[51] J. J. Boland, D. D. Baumann, and B. D. Ziegielewski, "An Assessment of Municipal and Industrial Water Use Forecasting Approaches," IWR Contract Report 81-C05 Fort Belvoir, VA: Army Corps of Engineers, 1981.

[52] B. T. Bower, I. Gouevsky, D. R. Maidment, and W. R. D. Sewell, *Modeling Water Demands* (New York: Academic, 1984).

[53] Comptroller General, "Water Supply for Urban Areas: Problems in Meeting Future Demand," Report to the Congress of the United States, CED-79-56, June 15, 1979.

[54] Congressional Research Service, Library of Congress, *State and National Water Use Trends to the Year 2000*, Serial No. 96-12 (Washington, DC: U.S. Government Printing Office, May 1980).

[55] J. R. Kim and R. H. McCuen, "Factors for Predicting Commercial Water Use," *Water Resources Bull.* 15, no. 4 (1979): 1073–1080.

[56] D. R. Maidment and S. Miaou, "Daily Water Use in Nine Cities," *Water Resources Res.* 22, no. 6 (1986): 845–851.

[57] C. T. Osborn, J. E. Schefter, and L. Shabman, "The Accuracy of Water Use Forecasts: Evaluation and Implications," *Water Resources Bull.*, Paper No. 84245, Am. Water Resources Assoc. 22, no. 1 (February 1986).

[58] J. J. Boland, E. M. Opitz, B. Dziegielewski, and D. D. Baumann, *IWR MAIN Modifications, Technical Consultant Report,* Contract DACW-72-84-C-0004, Task 2 (Fort Belvoir, VA: U.S. Army Engineer Institute for Water Resources, 1985).

CHAPTER 5

Wastewater Generation

OBJECTIVES

The purpose of this chapter is to:

- Define the principal wastewater generating sectors
- Discuss trends in wastewater generation
- Present techniques for forecasting wastewater volumes

5.1 QUANTITIES OF WASTEWATER

In order to transport and treat wastewater flows the amount, timing, and characteristics of these flows must be known. The quantities and timing of waste streams are discussed in this chapter. Wastewater characteristics are addressed in the chapters on water quality and treatment.

The major wastewater producers are cities, industries, and agricultural operations. Generally, their waste streams are handled independently, although some industrial wastes are transported and treated in municipal systems. The time variation of flows is an especially important consideration, since most conveyances are gravity flow and they must be able to accommodate minimum flows at sufficient velocities for self-cleansing, as well as handle peak flow events. Municipal sewer flow consists mainly of community wastes plus infiltration, although if there are illicit connections, or if the sewer is a combined one, stormwater flows must also be considered.

5.2 WASTE FLOWS FROM URBAN AREAS

The quantity of sewage generated by a community depends on its population and thus the per capita discharge to the sewer. As in the case of water use forecasting, population estimates are a requisite for estimating wastewater flows. Figure 5.1 gives a comparison of water use and wastewater flow on days without lawn sprinkling. Although the data were collected a number of years ago, the figure is still

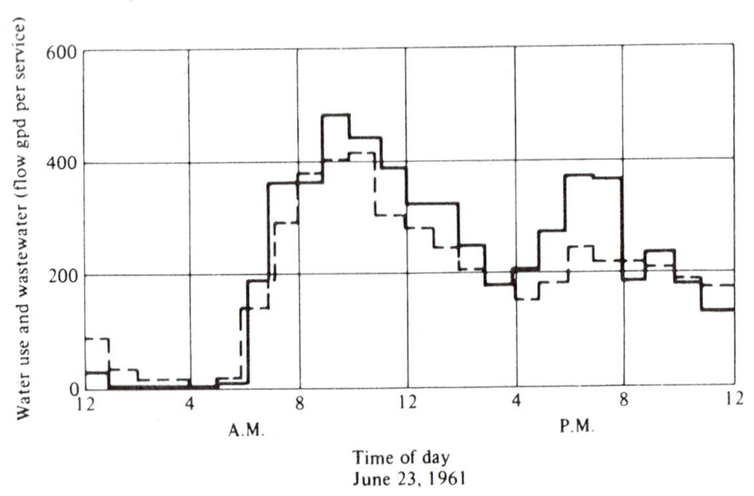

FIGURE 5.1 Comparison of water use (solid line) and wastewater flow (dashed lines) on days when little sprinkling occurred. *Source:* From Residential Water-Use Research Project, Johns Hopkins University and Federal Housing Administration, 1963.

representative of conditions today. The data are from the Pine Valley subdivision in Baltimore County, Maryland [1]. It can be seen that domestic sewage flows are highly variable throughout the day and, as in the case of the hydrograph of water use, there are two distinct peaks. The primary peak occurs early in the morning. The secondary peak occurs about dinnertime and maintains itself during the evening hours. Extraneous flows resulting from infiltration or storm runoff tend to distort the basic hydrograph shape. Infiltration rates generally tend to gradually increase the total daily volume but do not ordinarily alter the twin-peaked character of the hydrograph. Storm runoff that enters the system may impose almost instantaneous changes; if the quantity is large, the entire characteristic of the hydrograph may be changed. Estimation of the various components of the flow is essential for design purposes. A 1963 study by Lentz of wastewater flows in communities in California, Florida, Missouri, and Maryland provides considerable useful information regarding residential flows and the components of these flows [1].

Average Rates of Flow

Lentz and Linaweaver have shown that when residential water is not being used for consumptive purposes (principally lawn sprinkling) and when infiltration and exfiltration do not produce large flow components, the wastewater flow is essentially equal to the water use [1,2]. Thus, average daily water use rates that do not reflect sprinkling demands can be used to estimate annual average domestic wastewater flows. It is generally reported that about 60 to 80% of the total water supplied to a community becomes wastewater. Low ratios generally apply to semiarid regions. Ordinarily, the annual variation in the ratio of sewage to water supply in a city is not great, and thus the amount of water used by a city is a good indicator of the amount of sewage that will be generated. Sometimes, however, illicit drains and water use from privately owned sources produce quantities of sewage larger than public water withdrawals. When such external factors are not present, sewage flows from residential areas are often less than about 100 gpcd.

Variability in Sewage Flows

Sewage flow rates vary by source and with time. In most municipalities, the sources may be residences, institutions such as hospitals and schools, commercial establishments, and industries. It is thus necessary for the designer and manager to determine the mix of these elements and to estimate their contributions. Guideline figures are reported in the literature, but these may not be reliable, especially where industrial wastes are involved [3]. A case-by-case assessment should always be made before sewers are sized or treatment plant capacities set. For institutions, flows may vary from as low as 10 gpcd for schools to 175 gpcd for hospitals. Hotels may produce flows of about 100 gpcd, while small businesses may generate only about 2060 gal per day per employee.

Sewage flow patterns in residential areas closely resemble water use patterns for those areas, with the exception of a time lag, as shown in Figure 5.1. The magnitude of this lag varies with the situation, but it is usually on the order of a few hours. Where infiltration from rainfall or from other water uses enters sewers, the hydrographs produced in the sewers may vary considerably from those generated during dry periods. Otherwise, the variation in daily flow patterns for most residential areas is quite small. Where sewers receive significant quantities of wastes from industrial operations, the amounts and timing of flows are affected by prevailing industrial practices. Table 5.1 gives an indication of the variation in residential wastewater flows as ratios to the average. In the absence of site-specific data, such figures may be used to estimate high and low flows. Data on the ratio of peak flows and minimum flows to average daily flow have been summarized by Gupta [4]. They show that the peak to average daily flow ratio ranges from about 3.0 for cities of about 10,000 to about 1.5 for cities of one million. For the same size cities, the minimum flow to average daily flow ranges from about 0.3 to about 0.7, respectively.

Ordinarily, residential, industrial, and commercial sewage flows are estimated for an area and then combined to obtain the composite sewage hydrograph. In this way, detailed information regarding the nature of each flow component can be introduced into calculations.

TABLE 5.1 Residential Wastewater Flows as Ratios to the Average

Description of Flow	Ratio to the Average
Maximum daily	2.25:1
Maximum hourly	3:1
Minimum daily	0.67:1
Minimum hourly	0.33:1

Infiltration and Exfiltration

Infiltration and exfiltration are both functions of the height of the groundwater table in the vicinity of the sewer, type and tightness of sewer joints, and soil type. Exfiltration is undesirable since it may tend to pollute local groundwaters, while infiltration has the effect of reducing the capacity of the sewer for conveying the waste flows for which it was designed. If the sewer is well above the groundwater table, infiltration will occur only during or after periods of precipitation, when water is percolating downward through the soil. Where groundwater tables are high and sewers are not tight, infiltration rates in excess of 60,000 gpd/mi of sewer might be experienced. Rates of 3500 to 5000 gpd/mi/24 hr for an 8-in. pipe, 4500 to 6000 for a 12-in. pipe, and 10,000 to 12,000 for a 24-in. pipe represent the range in which most specifications fall [5]. It is often the practice to design for the peak-design rate of wastewater flow plus 30,000 gpd infiltration per mile of sewer and house connections [4–6]. The infiltration allowance should, however, be determined by the designer based on the physical characteristics of the area and the type of pipe and joint to be used.

Stormwater Runoff

Except for large combined sewers, storm runoff should be excluded from the sewerage system. Storm runoff may enter at manholes or through illicit roof drains connected to the sanitary system. Quantities of flow that enter in this manner vary with degree of enforcement of regulations and types of preventive measures taken. The American Society of Civil Engineers reports that tests on leakage through manhole covers show that 20 to 70 gpm may enter a manhole cover submerged by 1 in. of water [6]. Rates of this magnitude may be considerably in excess of average wastewater flows. Small sewers can be surcharged easily by very few roof-drain connections. For example, a rainfall of 1 in./hr on a 1000-ft^2 roof area will contribute flows in excess of 10 gpm. The average domestic sewage flow from a dwelling with this approximate roof area (consider four persons) would equal only about 1.5% of this.

Example 5.1

Estimate the maximum hourly, average daily, and minimum hourly residential sewage flows from an area occupied by 1200 people. Consider the length of sewer and house drains equal to 1.2 mi. Give results in gpd and lpd.

Solution: Based on 1995 USGS water use data and assuming a return flow of 75%, the average daily per capita flow would be 101 gpcd $\times$ 0.75 = 75.8 gpcd or 286.7 lpcd. Thus, the total average daily domestic flow would be 78.8 $\times$ 1200 = 90,960 gpd or 344,284 lpd.

$$\text{Infiltration} = 30{,}000 \text{ gpd/mi} \times 1.2 = 36{,}000 \text{ gpd or } 136{,}260 \text{ lpd}$$

Using data from Table 5.1, we have

$$\text{Maximum hourly flow} = 36{,}000 + 90{,}960 \times 3 = 308{,}880 \text{ gpd or } 1{,}169{,}111 \text{ lpd}$$

$$\text{Total average daily flow} = 36{,}000 + 90{,}960 = 126{,}960 \text{ gpd or } 480{,}544 \text{ lpd}$$

$$\text{Minimum hourly flow} = 36{,}000 + 90{,}960/3 = 66{,}320 \text{ gpd or } 251{,}021 \text{ lpd}$$

Example 5.2

Compare the maximum hourly domestic sewage flow from 10 houses (4 persons per house) with the roof drainage from these houses if the roof dimensions are 60×35 ft and the rainfall intensity is 2.0 in./hr. Approximately what size of sewer would be required to handle (a) the domestic flow alone and (b) the combined flow if the pipe is laid on a 1.1% grade?

Solution: Assume a domestic sewage flow of 100 gpcd. Referring to Table 5.1, the ratio of the maximum hour to the average is given as 3. The number of people is $4 \times 10 = 40$.

The maximum hourly domestic sewage flow is

$$100 \times 3 \times 40 = 12{,}000 \text{ gph or } 0.45 \text{ cfs}$$

The roof drainage from the houses can be computed using the rational method (see Chapter 7), $Q = cia$; where Q is the flow in cfs, c is a runoff coefficient (1.0 for impervious surfaces), I is the rainfall intensity in inches per hour, and a is the drainage area in acres. The storm flow is thus:

$$Q = cia = 1.0 \times 2.0 \times (21{,}000/43{,}560) = 0.96 \text{ cfs}$$

where the roof area is $10 \times 60 \times 35 = 12{,}000$ square ft, and $43{,}560 =$ the number of square ft in an acre.

$$\text{The combined flow is } 0.96 + 0.44 = 1.40 \text{ cfs}$$

Referring to Eq. (7.1) (Manning's equation), the pipe size can be determined. Using a value of $n = 0.015$, the pipe sizes for the given slope that would accommodate the domestic sewage flow and the combined flow are 6 in and 8 in, respectively.

This example illustrates the fact that sewage flows for a community are small when compared to those generated by significant rainfalls.

Example 5.3

Estimate the average hourly, average daily, and minimum hourly residential sewage flows from an area serving a population of 1200. Assume that the length of sewer and house drains equals 4.0 km and that infiltration occurs.

Solution: Assuming an infiltration rate 30,000 gpd/mi, and noting that 1 kilometer = 0.6214 miles, the infiltration in gph can be calculated.

$$(4 \times 0.6214 \times 30{,}000)/24 = 3107 \text{ gph}$$

Assuming an average wastewater flow of 100 gpcd, the hourly sewage flow can be calculated.

$$100 \times 1200/24 = 5000 \text{ gph}$$

Referring to Table 5.1, the maximum hourly flow is calculated as

$$3 \times 5000 + 3107 = 18{,}107 \text{ gph}$$

The average hourly flow is

$$5000 + 3107 = 8107 \text{ gph}$$

The minimum daily flow is

$$5000/3 + 3107 = 4774 \text{ gph}$$

Example 5.4

Given a 100-acre housing development with 280 houses (4 persons per house), and an average annual rainfall of 32 in., calculate the yearly volume of precipitation over the area and compare it with the annual sewage flow from the area. Repeat the calculation for an annual rainfall of 15 in. and 10 in.

Solution: The average annual sewage flow would be

$$4 \times 280 \times 100 \times 365 = 40{,}880{,}000 \text{ g/yr, or } 5{,}465{,}241 \text{ ft}^3\text{/yr.}$$

For a 32 in. rainfall over the area in one year the volume would be

$$(32/12) \times 100 \times 43{,}560 = 11{,}616{,}000 \text{ ft}^3\text{/yr.}$$

For a 15 in. rainfall over the area in one year the volume would be

$$11{,}616{,}000 \times 15/32 = 5{,}444{,}419 \text{ ft}^3\text{/yr.}$$

For a 10 in. rainfall over the area in one year the volume would be

$$11{,}616{,}000 \times 10/32 = 3{,}630{,}000 \text{ ft}^3\text{/yr.}$$

This problem shows that in the amount of water sent to waste in an urban area can exceed the amount generated in drier climates. The value of reuse is also illustrated by the example.

5.3 INDUSTRIAL WASTE VOLUMES

Industrial waste volumes are highly variable in both quantity and quality, depending principally on the product produced. Since very little water is consumed in industrial processing, large volumes are often returned as waste. These wastes may include toxic metals, chemicals, organic materials, biologic contaminants, and radioactive materials. The design of treatment processes for these wastes is a highly specialized operation. Where industrial wastes must be processed in municipal sewage treatment works, accurate estimates of the time distribution and total volume of the load are necessary, together with a complete analysis of the characteristics of the waste. Under these circumstances, metering and analyzing the industrial waste is normally required. For more complete information on volumes as well as characteristics of industrial wastes, the reader should consult appropriate references [7–27].

5.4 AGRICULTURAL WASTES

The quantities and character of wastes from agricultural lands are highly variable. The most important pollutants found in runoff from agricultural areas are sediment, animal wastes, wastes from industrial processing of raw agricultural products, plant nutrients, forest and crop residues, inorganic salts and minerals, and pesticides [28–30]. Because waste volumes are determined by numerous factors for any given area, it is impossible to indicate any general rules. Nevertheless, the importance of these wastes to any region should not be overlooked. It is clear that large quantities of agricultural drainage are discharged into streams, rivers, and lakes. Regional water quality control will not be a reality unless these wastes are considered along with those of municipalities and industries. A comprehensive treatment of this topic is beyond the scope of this book, but the student should not minimize the importance of the subject. More detailed information can be found in the references cited.

5.5 A CLOSING NOTE

Figure 5.1 and the examples given in this chapter should make it clear that the volume of water that we use and that has historically been treated as waste must be reconsidered as a resource to be reclaimed and used again. This is particularly true in arid parts of the world.

PROBLEMS

5.1 Compare the maximum hourly domestic sewage flow from 20 houses (four persons per house) with the roof drainage from these houses if the roof dimensions are 60 × 40 ft and the rainfall intensity is 1.5 in./hr. Approximately what size of sewer would be required to handle (a) the domestic flow alone and (b) the combined flow if the pipe is laid on a 2% grade?

5.2 Estimate the average hourly, average daily, and minimum hourly residential sewage flows from an area serving a population of 1500. Assume that the length of sewer and house drains equals 3.0 km and that infiltration occurs.

5.3 Estimate the maximum hourly, average daily, and minimum hourly residential sewage flows from an area occupied by 700 people. Consider the length of sewer and house drains equal to 2.0 km. Give results in lpd and gpd.

5.4 Rework Problem 5.3 for 3700 people and a sewer length of 5.2 miles.

5.5 Estimate the maximum hourly, average daily, and minimum hourly residential sewage flows from an area occupied by 650 people. Consider the length of sewer and house drains equal to 1.2 mi. Give results in gpd and lpd.

5.6 Given a 300 acre housing development with 850 houses (4 persons per house), and an average annual rainfall of 28 in., calculate the yearly volume of precipitation over the area and compare it with the annual sewage flow from the area. Repeat the calculation for an annual rainfall of 12 in.

5.7 Compare the average daily sewage flow from an apartment building of 15 floors having ten apartments on each floor with the sewage flow from a 10-acre residential area having four houses per acre.

REFERENCES

[1] J. J. Lentz, "Special Report No. 4 of the Residential Sewage Research Project to the Federal Housing Administration," The Johns Hopkins University, Department of Sanitary Engineering, Baltimore, May 1963.

[2] F. P. Linaweaver, Jr., "Report on Phase One, Residential Water Use Research Project," The Johns Hopkins University, Department of Sanitary Engineering, Baltimore, October 1963.

[3] Metcalf and Eddy, Inc., *Wastewater Engineering* (New York: McGraw-Hill, 1972).

[4] R. S. Gupta, *Hydrology and Hydraulic Systems* (Upper Saddle River, NJ: Prentice Hall, 1989).

[5] C. R. Velzy and J. M. Sprague, *Sewage Ind. Wastes* 27(3) (1955).

[6] "Design and Construction of Sanitary and Storm Sewers," Manual of Engineering Practice No. 37 (New York: American Society of Civil Engineers, 1960).

[7] "Development Document for Proposed Effluent Limitations Guidelines and New Source Performance Standards for the Iron and Steel Industry," EPA 440/1-82/024 (1982).

[8] "Development Document for Proposed Effluent Limitations Guidelines and New Source Standards of Performance for the Nonferrous Metals Industry," EPA 440/1-83/019-b (1983).

[9] "Development Document for Proposed Effluent Limitations Guidelines and New Source Performance Standards for the Bauxite Refining Subcategory of the Aluminum Segment of the Nonferrous Metals Manufacturing Point Source Category," EPA 440/1-74/091-c (1974).

[10] "Development Document for Proposed Effluent Limitations Guidelines and New Source Performance Standards for the Copper, Nickel, Chromium, and Zinc Segments of the Electroplating Point Source Category," EPA 440/1-74/003-a (1974).

[11] "Development Document for Proposed Effluent Limitations Guidelines and New Source Performance Standards for the Petroleum Refining Industry," EPA 440/1-82/014 (1982).

[12] "Development Document for Proposed Effluent Limitations Guidelines and New Source Performance Standards for Inorganic Chemicals, Alkali and Chlorine Industry," EPA Contrast No. 68-01-1513 (draft) (June 1973).

[13] "Development Document for Proposed Effluent Limitations Guidelines and New Source Performance Standards for the Organic Chemicals Industry," EPA 440/1-74/009-a (1974).

[14] "Development Document for Proposed Effluent Limitations Guidelines and New Source Performance Standards for the Fertilizer and Phosphate Manufacturing Industry," EPA 440/1-74/011-a and 74/006-a (1974).

[15] "Development Document for Proposed Effluent Limitations Guidelines and New Source Performance Standards for the Plastic and Synthetics Industry," EPA 440/1-83/009-b (1983).

[16] "Development Document for Proposed Effluent Limitations Guidelines and New Source Performance Standards for the Major Inorganic Products Segment of the Inorganic Chemicals Manufacturing Point Source Category," EPA 440/1-74/007-a (1974).

[17] "Development Document for Proposed Effluent Limitations Guidelines and New Source Performance Standards for the Phosphorus-Derived Chemicals Segment of the Phosphate Manufacturing Point Source Category," EPA 440/1-74/006-a (1974).

[18] "Development Document for Proposed Effluent Limitations Guidelines and New Source Performance Standards for the Synthetic Resins Segment of the Plastics and Synthetic Materials Manufacturing Point Source Category," EPA 440/1-74/010-a (1974).

[19] "Development Document for Proposed Effluent Limitations Guidelines and New Source Performance Standards for the Canned and Preserved Fish and Seafoods Processing Industry," EPA 440/1-80/020 (1980).

[20] "Development Document for Proposed Effluent Limitations Guidelines and New Source Performance Standards for the Meat Packing Industry," EPA 440/1-74/012-a (1974).

[21] "Development Document for Proposed Effluent Limitations Guidelines and New Source Performance Standards for the Dairy Products Processing Industry," EPA 440/1-74/021-a (1974).

[22] "Development Document for Proposed Effluent Limitations Guidelines and New Source Performance Standards for the Citrus, Apple and Potato Segment of the Canned and Preserved Fruits and Vegetables Processing Point Source Category," EPA 440/1-74/027-a (1974).

[23] "Development Document for Proposed Effluent Limitations Guidelines and New Source Performance Standards for the Unbleached Kraft and Semichemical Pulp Segment of the Pulp, Paper, and Paperboard Mills Point Source Category," EPA 440/1-74/025-a (1974).

[24] "Development Document for Proposed Effluent Limitations Guidelines and New Source Performance Standards for the Builders Paper and Roofing Felt Segment of the Builders Paper and Board Mills Point Source Category," EPA 440/1-80/025-b (1980).

[25] "Development Document for Proposed Effluent Limitations Guidelines and New Source Performance Standards for the Textile Mills," EPA 440/1-82/022 (1982).

[26] "Development Document for Proposed Effluent Limitations Guidelines and New Source Performance Standards for the Leather Tanning and Finishing Industry," EPA 440/1-82/016 (1982).

[27] "Development Document for Proposed Effluent Limitations Guidelines and New Source Performance Standards for Steam Electric Power Plants," EPA 440/1-82/029 (1982).

[28] R. C. Loehr, "Animal Wastes—A National Problem," *Proc. Am. Soc. Civil Engrs., J. San. Eng. Div.* 95 (SA2) (1969): 189–221.

[29] "Control of Agriculture-Related Pollution," A report to the President submitted by the Secretary of Agriculture and the Director of the Office of Science and Technology, Washington, DC (January 1969).

[30] "Agricultural Waste Waters," Proceedings Symposium of Agricultural Waste Waters, Report No. 10, Water Resources Center, University of California, Davis, CA (April 1966).

CHAPTER 6

Conveying and Distributing Water

OBJECTIVES

The purpose of this chapter is to:

- Acquaint the reader with the components of water supply and distribution systems
- Introduce the principles of flow in pipes and open channels
- Illustrate the application of EPANET to the analysis and design of water distribution systems
- Provide examples of pipeline and network problem solving

Water is transported over long distances through aqueducts to locations where it is to be used and/or treated. It is then conveyed to individual users or use points through distribution networks.

6.1 AQUEDUCTS

Selection of an aqueduct type rests on such factors as topography, head availability, climate, construction practices, economics, and water quality protection. Aqueducts may include or be solely composed of open channels, pipelines, or tunnels.

Open Channels

Open channels are designed to convey water under conditions of atmospheric pressure. By this definition the hydraulic gradient and free-water surface are coincident. If the channel is supported on or above the ground, it is classified as a flume. Open channels may be covered or open and may take on a variety of shapes.

The choice of an open channel as the means of conveyance is usually predicated on suitable topographic conditions that permit gravity flow with minimal excavation or fill. If the channel is unlined, the perviousness of the soil must be considered relative to seepage losses. Other important considerations are the potential for pollution and evaporative losses.

Open channels may be lined with concrete, bituminous materials, butyl rubber, vinyl, synthetic fabrics, or other products to reduce the resistance to flow, minimize seepage, and lower maintenance costs. Flumes are usually constructed of concrete, steel, or timber.

Pipelines

Pipelines are usually built where topographic conditions preclude the use of canals. Pipelines may be laid above or below ground or may be partly buried. Most modern pressure conduits are built of concrete, steel, cast iron, asbestos cement, or plastic (polyvinyl chloride, PVC).

Pipelines may require gate valves, check valves, air-release valves, drains, surge control equipment, expansion joints, insulation joints, manholes, and pumping stations. The appurtenances are provided to ensure safe and efficient operation, provide for easy inspection, and facilitate maintenance. Check valves are normally located on the upstream side of pumping equipment and at the beginning of each rise in the pipeline to prevent backflow. Gate valves are often spaced about 1200 ft apart so that the intervening section of line can be drained for inspection or repair and on either side of a check valve to permit its removal for inspection or repair. Air-release valves are needed at the high points in the line to release trapped gases and to vent the line to prevent vacuum formation. Drains are located at low points to permit removal of sediment and allow the conduit to be emptied. Surge tanks or quick-opening valves provide relief from problems of hydraulic surge.

Tunnels

Where it is not practical or economical to lay a pipeline on the surface or provide an open trench for underground installation, a tunnel is selected. Tunnels are well suited to mountain or river crossings. They may be operated under pressure or act as open channels.

6.2 HYDRAULIC CONSIDERATIONS

The analysis of flows in a water conveyance system is carried out through application of basic principles of open-channel and closed-conduit hydraulics. It is assumed that the student already has been exposed to these concepts in courses in hydraulics or fluid mechanics.

Except for sludges, most flows may be treated hydraulically in the same manner as clean water even though considerable quantities of suspended material are being carried. The Hazen–Williams and Manning formulas are used extensively in designing water conveyances. The Hazen–Williams formula is used primarily for pressure conduits, while the Manning equation has found its major application in open-channel problems. Both equations are applicable when normal temperatures prevail, a relatively high degree of turbulence is developed, and ordinary commercial materials are used [1–6]. The Hazen–Williams equation is

$$V = 1.318CR^{0.633}S^{0.54} \text{ (English units)} \tag{6.1}$$

$$V = 0.85CR^{0.63}S^{0.54} \text{ (SI units)}$$

where

V = velocity of flow (ft/s or m/s)

C = a coefficient that is a function of the construction material and age of the pipe

R = hydraulic radius (cross-sectional area divided by the wetted perimeter)(ft or m)

S = slope of energy gradient in feet per foot of length or meters per meter of length

For circular conduits flowing full, the equation may be restated as

$$Q = 0.279CD^{2.63}S^{0.54} \tag{6.2}$$

where

Q = flow, mgd

D = pipe diameter, ft

and as

$$Q = 0.278CD^{2.63}S^{0.54} \tag{6.2}$$

where

Q = flow, m³/s

D = pipe diameter, m

Some values of C for use in the Hazen–Williams formula are given in Table 6.1. A nomograph that facilitates the solution of this equation is given in Figure 6.1.

The Manning equation is stated in the form

$$V = 1.49R^{0.66}S^{0.5}/n \tag{6.3}$$

where

V = velocity of flow, fps

n = coefficient of roughness

R = hydraulic radius, ft

S = slope of energy grade line

In metric units

$$V = R^{0.66}S^{0.5}/n$$

TABLE 6.1 Roughness Coefficients

Material	Hazen–Williams, C	Manning, n
New pipes:		
Cast iron	130–140	0.012–0.015
Concrete	120–140	0.012–0.017
Concrete-lined galvanized iron	120	0.015–0.017
Plastic	140–150	0.011–0.015
Steel	140–150	0.015–0.017
Vitrified clay	110	0.013–0.015
Welded steel	120	—
Older pipes and other materials:		
5-yr-old cast iron	120	—
20-yr-old cast iron	100	—
Asbestos cement	140	—
Brick	—	0.016
Corrugated metal pipe	—	0.022
Bituminous concrete	—	0.015
Uniform, firm sodded earth	—	0.025

where

V = velocity of flow, m/s

n = coefficient of roughness

R = hydraulic radius, m

S = slope of energy grade line

The equation is applicable as long as S does not materially exceed 0.10. In channels having no uniform roughness, an average value of n is selected. Where the cross-sectional roughness changes considerably, as in a channel with a paved center section and grassed outer sections, it is common practice to compute the flow for each section independently and sum these flows to obtain the total. For most purposes, n is considered a constant. It is actually a function of pipe diameter, however, and should be adjusted for pipe diameters exceeding several feet. As in the case of the Hazen–Williams equation, nomographs are available to permit rapid computations (see Fig. 7.1, Chapter 7). Values of n for use in Manning's equation are indicated in Table 6.1.

Head loss in pipelines results from pipe friction losses and in piping auxiliaries. Minor losses include those resulting from valves, fittings, bends, changes in cross section, and changes in flow characteristics at inlets and outlets. Over long lengths of pipeline, minor losses can usually be ignored in calculations of head loss because they contribute a relatively small proportion to the total losses. On the other hand, minor losses in short water transportation systems, such as those in water and wastewater treatment plants, should not be ignored because their proportion of the total head loss is significantly larger. Minor losses are usually expressed as a function of the velocity

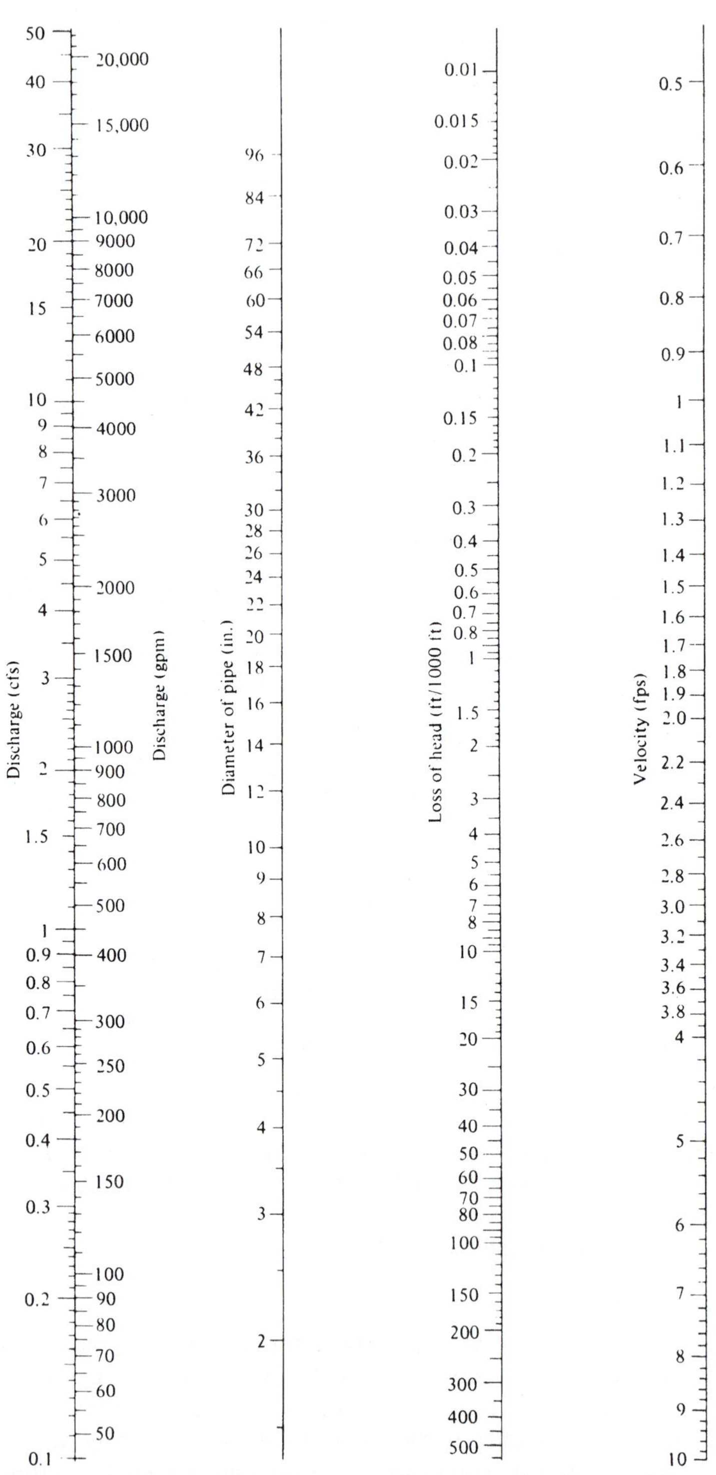

FIGURE 6.1 Nomograph for the Hazen–Williams formula with C = 100.

TABLE 6.2 Minor Head Loss Coefficients, K

Component	Loss Coefficient
Globe valve, fully open	10.0
Angle valve, fully open	5.0
Swing check valve, fully open	2.5
Gate valve, fully open	0.2
Short-radius elbow	0.9
Medium-radius elbow	0.8
Long-radius elbow	0.6
45° elbow	0.4
Closed return bend	2.2
Standard tee—flow through run	0.6
Standard tee—flow through branch	1.8
Square entrance	0.5
Exit	1.0

Source: Rossman [5].

head in performing calculations, that is, $H_L = KV^2/2g$. Some values of the minor head loss coefficient K are given in Table 6.2 (see also [1–7]).

Head loss as a result of pipe friction can be computed by solving Eq. (6.1) or (6.3) for S and multiplying by the length of the pipeline. A slightly more direct method is to use the Darcy–Weisbach equation,

$$h_L = fLV^2/2Dg \tag{6.4}$$

where

h_L = head loss, ft

L = pipe length, ft

D = pipe diameter, ft

f = friction factor

V = flow velocity, fps

The friction factor is related to the Reynolds number and the relative roughness of the pipe. For conditions of complete turbulence, Figure 6.2 relates the friction factor to pipe geometry and characteristics.

The Energy Equation

Consider flow in a straight pipe of uniform diameter. Then the energy equation for flow in a segment of length L between points 1 and 2 may be written as

$$Z_1 + \frac{P_1}{\gamma} + \frac{V_1^2}{2g} = Z_2 + \frac{P_2}{\gamma} + \frac{V_2^2}{2g} + H_L \tag{6.5}$$

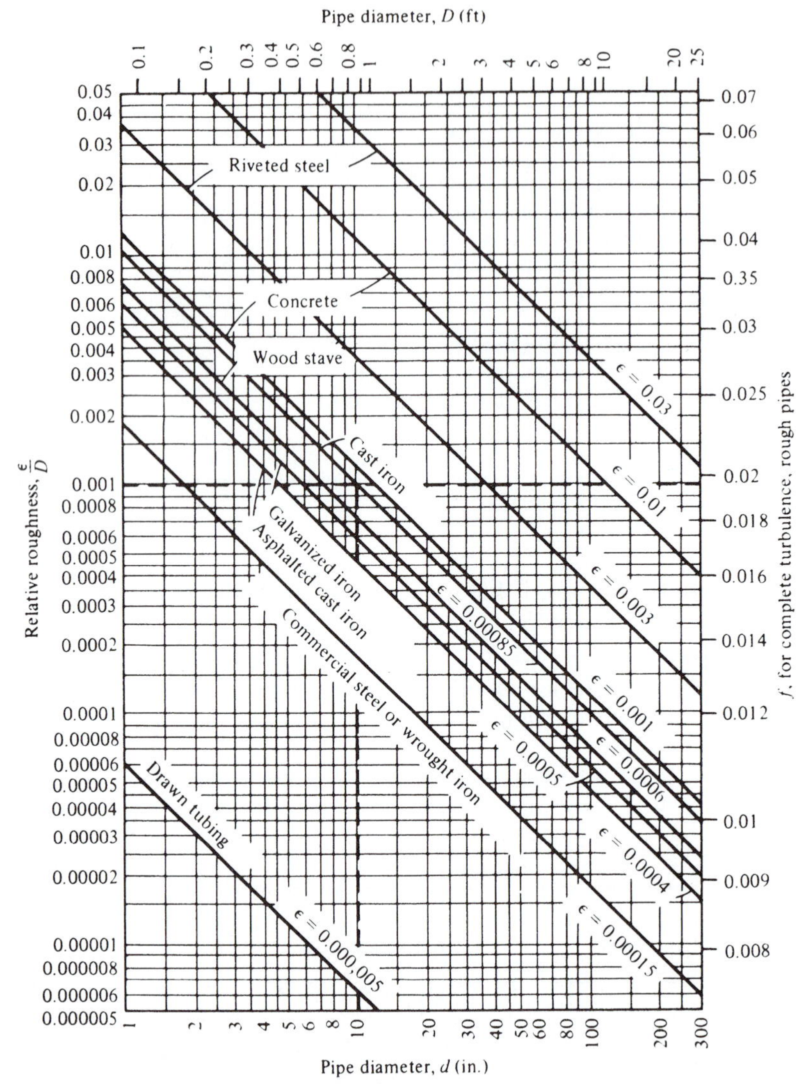

FIGURE 6.2 Relative roughness of pipe materials and friction factors for complete turbulence. *Source:* Courtesy of the Crane Company, Chicago, IL.

where Z = elevation above an arbitrary datum (ft, m); p/γ = pressure head (ft, m); V = average velocity of flow (ft/sec, m/sec); and H_L = total head loss (energy loss) between the two cross sections.

Note that the terms of the energy equation are all in units of energy per unit weight (ft-lb/lb, kg-m/kg), which reduce to units of length (ft, m). If a pump were inserted into the pipeline, the quantity Hp would be added to the left-hand side of the equation to account for the additional energy head resulting from the action of the pump. If a turbine were inserted in the pipeline in place of the pump, the positive quantity Hp would be replaced by a negative quantity H, since a turbine converts the energy of flow into mechanical work, thereby consuming energy from the pipeflow instead of imparting energy to the flow as in the case of a pump.

Example 6.1

Consider that water is pumped 12 mi from a reservoir at elevation 100 ft to a second reservoir at elevation 220 ft. The pipeline connecting the reservoirs is 48 in. in diameter and is constructed of concrete with an absolute roughness of 0.003. If the flow is 28 mgd and the efficiency of the pumping station is 80%, what will be the monthly power bill if electricity costs 15 cents/kwh?

Solution:

1. Writing the energy equation between a point on the water surface of reservoir A and a point on the water surface of reservoir B, one obtains

$$Z_A + \frac{P_A}{W} + \frac{V_A^2}{2g} + H_p = Z_B + \frac{P_B}{W} + \frac{V_B^2}{2g} + H_L$$

2. Letting $Z_A = 0$, and noting that $P_A = P_B$ is equal to the atmospheric pressure and that $V_A = V_B = 0$ for a large reservoir, one finds that the equation reduces to

$$H_p = Z_B + H_L$$

where

H_p = head developed by the pump
H_L = total head lost between A and B, including pipe friction and minor losses

The following conversion factors are used in the calculations: mgd $\times$ 1.55 = cfs; and ft-lb $\times$ (3.766×10^{-7}) = kilowatt hours (kwh)

3. Using Figure 6.2, determine the value of f as 0.0182.
4. Using Eq. (6.4), find the pipe friction head loss. Assuming that the minor losses are negligible in this problem, this is equal to H_L:

$$H_L = f\frac{L}{D}\frac{V^2}{2g}$$

The velocity V must be determined before Eq.(6.4) can be solved:

$$V = Q/A = (28 \times 1.55)/(\pi \times 4) = 3.45 \text{ ft/sec}$$
$$H_L = 0.0182 \times (5280 \times 12)/4) \times ((3.45)^2/64.4) = 53.3 \text{ ft}$$

5. $H_p = (220 - 100) + 53.3 = 130 + 53.3 = 173.3$ ft-lb/lb the energy imparted by the pump to the water.
6. The power requirement may be computed as

$$P = Q\gamma H_p$$
$$= 28 \times 1.55 \times 62.4 \times 177.3 = 469{,}324 \text{ ft-lb/s}$$

7. For 80% efficiency, the power requirement is

$$469{,}324/0.80 = 586{,}655 \text{ ft-lb/s}$$

8. $586{,}655 \times 3.766 \times 10^{-7} = 0.22$ kwh/s
The number of kilowatt-hours per 30-day month is then

$$0.22 \times 30 \times 86{,}400 = 570{,}240 \text{ kwh/month}$$

9. The monthly power cost is therefore $570{,}240 \times 0.15 = \$85{,}536$.

Flow in Branching Pipes

A common hydraulic problem is determining the direction and magnitude of flow in each pipe when several reservoirs are connected. The flow distribution will depend on the total head loss in each pipe, the diameter and length of the pipelines, and the number of connected facilities. A simple illustration is the classic *three-reservoir problem*, shown in Figure 6.3. Three reservoirs, A, B, and C, are connected by a system of pipelines that intersect at a single junction, J. Given the lengths and diameters of the pipes and the elevations of the three reservoirs, the problem is to determine the magnitude and direction of flow in each pipe.

It should be obvious that the flow will be out of reservoir A and into reservoir C, but it is not immediately evident whether the flow will be into or out of reservoir B, because it is not known whether the pressure head at J is higher or lower than the water surface elevation at reservoir B. This problem can be solved by making use of the continuity equation and the energy equation, which indicate that the flow into J equals the flow out of J, and the pressure head for all three pipes is the same at the point of intersection. Thus, by continuity,

$$Q_1 = Q_2 + Q_3 \text{ if the flow is into reservoir } B \tag{6.6a}$$

or

$$Q_1 + Q_2 = Q_3 \text{ if the flow is out of reservoir } B \tag{6.6b}$$

and, by energy equivalence,

$$p_1/\gamma = p_2/\gamma = p_3/\gamma = P \tag{6.7}$$

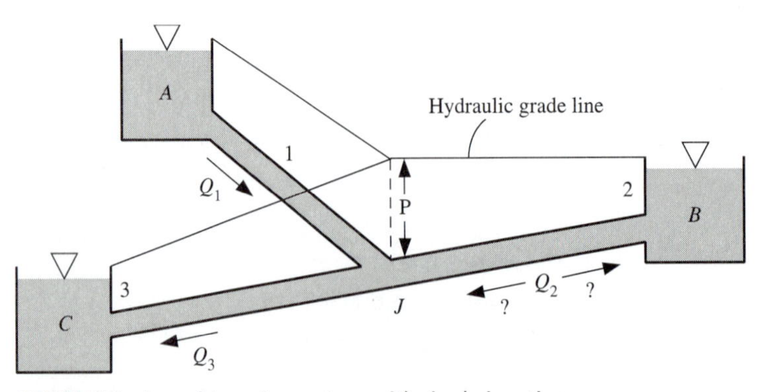

FIGURE 6.3 Branching-pipe system with single junction.

at J, where

$$Q = \text{flow in each pipe (vol/time)}$$

$$p/\gamma = \text{pressure head in each pipe (height in units of length)}$$

The solution is derived by choosing a trial height for P and solving for Q_1, Q_2, and Q_3 using the Manning equation, the Hazen–Williams equation, or the Darcy–Weisbach equation. The trial-and-error process is repeated until the continuity equation is satisfied.

An alternate, but more directly solved, branching-pipe problem is to find the elevation of one reservoir given all pipe lengths and diameters, the surface elevations of the other two reservoirs, and the flow either to or from one reservoir.

Flow in Pipes in Series

When a number of pipes of different diameters and lengths are connected in series, as depicted in Figure 6.4, the problem is either to determine the head loss given the flow, or to determine the flow given the head loss. The continuity equation allows us to state that the flow into and out of each pipe section must be the same, and the energy equation allows us to state that the head loss for the system is the sum of the head losses for each section of pipe. In other words, for the example shown in Figure 6.4,

$$Q = Q_1 = Q_2 = Q_3 \tag{6.8}$$

and

$$H_L = H_{L1} + H_{L2} + H_{L3} \tag{6.9}$$

For cases where the total head loss is given and the problem is to find the flow, the total head loss is written in terms of the dimensions of the head loss of each section, which for Figure 6.4 would be

$$\begin{aligned} H_L = {} & [f_1(L_1/D_1)(V_1^2/2g) + \Sigma K(V_1^2/2g)] \\ & + [f_2(L_2/D_2)(V_2^2/2g) + \Sigma(V_2^2/2g)] \\ & + [f_3(L_3/D_3\}(V_3^2/2g) + \Sigma(V_3^2/2g)] \end{aligned} \tag{6.10}$$

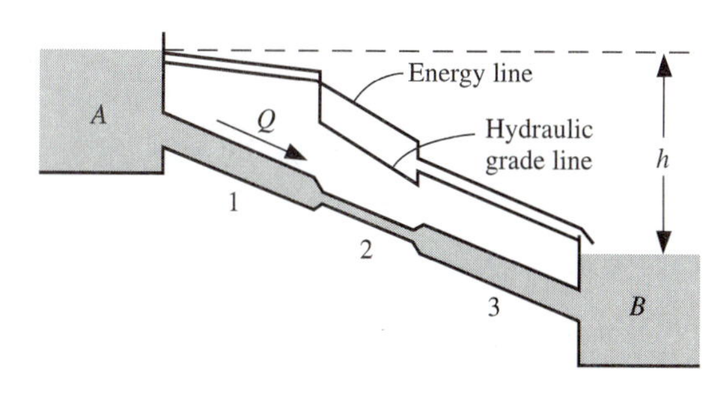

FIGURE 6.4 Flow in pipes in series.

Minor losses are designated for each section as a function of velocity head, that is, $\Sigma K(V^2/2g)$. Since the flow is equivalent for each section, by continuity the velocity head for each section can be expressed as a function of the velocity head of any one section. For example, referring to Figure 6.4:

$$V_1^2/2g = V_2^2/2g(D_2/D_1)^4 \tag{6.11}$$

and

$$V_1^2/2g = V_3^2/2g(D_3/D_1)^4 \tag{6.12}$$

If a friction factor is assumed, the velocity of one pipe section can be found and used in turn to calculate the flow, which would be the same for all pipe sections.

Example 6.2

Find the discharge from reservoir A into reservoir B in Figure 6.4 if three cast-iron pipes in series have diameters $D_1 = 15$ in., $D_2 = 10$ in., and $D_3 = 12$ in. lengths $L_1 = 1500$ ft, $L_2 = 1350$ ft, $L_3 = 2500$ ft, and the total head loss is 100 ft.

Solution:

1. Assuming $f = 0.01$ for all three pipes, and substituting the given values into the head loss equation given by Eq. (6.10), the objective is to determine V_1, V_2, and V_3.

$$\begin{aligned} 100 = {} & 0.01[1500/(15/12)](V_1^2/2g) \\ & + 0.01[1350/(10/12)](V_2^2/2g) \\ & + 0.01[2500/(12/12)](V_3^2/2g) \end{aligned}$$

2. From Eqs. (6.11) and (6.12),

$$V_1^2/2g = V_2^2/2g(10/15)^4 = 0.198(V_2^2/2g)$$
$$V_3^2/2g = V_2^2/2g(10/12)^4 = 0.482(V_2^2/2g)$$

3. Substituting back into the head loss equation:

$$100 = V_2^2/2g[12(0.198) + 16.2 + 25(0.482)]$$
$$V_2 = 14.5 \text{ ft/s}$$

4. Substituting V_2 back into the equations given in step 2,

$$V_1^2/2g = 0.198(14.5)^2/2g \qquad V_1 = 6.45 \text{ ft/s}$$
$$V_3^2/2g = 0.482(14.5)^2/2g \qquad V_3 = 10 \text{ ft/s}$$

5. Since $Q = Q_1 = Q_2 = Q_3$,

$$Q = Q_1 = V_1A_1 = 6.45(\pi)(7.5/12)^2 = 7.9 \text{ cfs}$$

As a check,

$$Q = Q_2 = V_2A_2 = 14.5(\pi)(5/12)^2 = 7.9 \text{ cfs}$$
$$Q = Q_3 = V_3A_3 = 10(\pi)(6/12)^2 = 7.9 \text{ cfs}$$

Flow in Parallel Pipes

In the case of pipes connected in parallel, the problem is again either to determine the head loss and distribution of flow for the system given the total flow, or to determine the total flow in the system given the head loss. For Figure 6.5, the continuity equation shows that the flow at the two junctions A and B is equivalent. In other words,

$$Q_A = Q_1 + Q_2 + Q_3 = Q_b \tag{6.13}$$

The head loss for the system can be shown by the energy equation to be equivalent to the head loss in each parallel pipe:

$$H_L = H_1 = H_2 = H_3 \tag{6.14}$$

Given the total flow, the head loss distribution may be determined by solving the Darcy–Weisbach head loss equation [Eq. (6.4)] for V for each pipe,

$$V = [2gH_L/f(L/D)]^{1/2}$$

then substituting the above expression for V into $Q = VA$, that is,

$$Q = A[2gH_L/f(L/D)]^{1/2} \tag{6.15}$$

and writing Q as a function of the head loss and C, where C is constant for a given pipe $(C = A[2g/f(L/D)]^{1/2}$:

$$Q = C(H_L)^{1/2} \tag{6.16}$$

The flows for each pipe can then be summed and expressed as a function of the system head loss; for Figure 6.5, which has three pipes, this would be

$$Q = C_1(H_L)^{1/2} + C_2(H_L)^{1/2} + C_3(H_L)^{1/2}$$

From Eq. (6.13), this becomes

$$Q = (H_L)^{1/2}(C_1 + C_2 + C_3)$$

An alternate method of analysis for simple systems of pipes in parallel or series is the *equivalent-pipe method*. In this method, either a series of pipes or a system of parallel pipes is replaced with a pipe of equivalent head loss, for the purpose of simplifying calculations. This method can also be used to simplify portions of complex pipe systems.

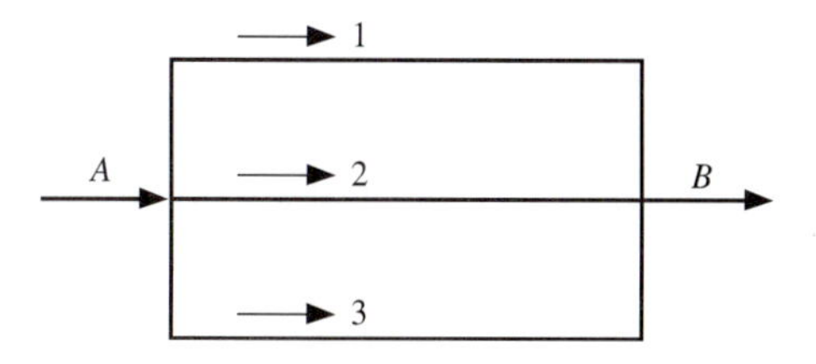

FIGURE 6.5 Flow in parallel pipes.

6.3 DESIGN CONSIDERATIONS

The design of water transportation systems involves considerations of hydraulic adequacy, structural adequacy, and economic efficiency. The waterway area needed is a function of the flow to be carried, the head available, the character of the conduit material, and limiting velocities.

Location

The routing of water conveyances is constrained by the availability of right-of-way and technical and economic considerations. An aqueduct's beginning and end points are fixed by the source of supply and the location at which the water is to be delivered. Between these two locations, the most cost-effective route must be found. The choice of location is also a determinant of the type, or types, of conveyances to be used. Aqueducts may be pipes, tunnels, flumes, canals or some combination of them. Aqueducts built to grade require topography that allows cut-and-cover operations to be closely balanced. Pressure aqueducts, on the other hand, can follow the topography. Pumping and materials and construction costs are related to topography.

Sizing

The size and configuration of an aqueduct commonly varies along the route. For a given type of aqueduct the size is usually determined on the basis of hydraulic, economic, and construction considerations. Occasionally, construction practices dictate a minimum size in excess of that required to handle the flow under the prevailing hydraulic conditions (available head). This condition is generally encountered for tunnels. Hydraulic factors that control the design are the head available and permissible velocities. Available heads are affected by reservoir drawdown and local pressure requirements. Limiting velocities are based on the character of the water to be transported and the need to protect transmission lines against excessive pressures that might develop through hydraulic surge. Where silt is transported with the water, minimum velocities of about 2.5 fps should be maintained. Maximum velocities must preclude pipe erosion or hydraulic surge problems and are ordinarily between 10 and 20 fps [8]. The usual range in velocities is from about 4 to 6 fps.

Where power generation is involved, pumping costs and/or the worth of power and conduit costs jointly determine the conduit size. For single gravity-flow pipelines, the size should be determined so that the available head is consumed by friction.

Strength

Pipelines and other conveyances must be designed to resist forces such as those resulting from water pressure within the conduit, hydraulic surge (transient internal pressure generated when the velocity of flow is rapidly reduced), external loads, forces at bends or changes in cross section, expansion and contraction, and flexural stresses. [4,7,9].

Economics

Hydraulic head has economic value. It costs money to produce the head at the upstream end of a system, but the head can then be used to increase flow, for power production, or for other purposes. A definite relationship always exists among aqueduct size, hydraulic gradient, and the value of head. In some cases, construction costs are related to the elevation of the hydraulic gradient. The elevation of the gradient also affects pumping costs and power production values, as does the slope of the hydraulic gradient. In long aqueducts composed of different types of conduits, a means of coordinating conduit types, choosing dam elevations, and selecting pump lifts or power drops is important. Dealing with this problem requires a joint application of hydraulic and economic principles [10,11].

In any conduit, there must be sufficient hydraulic slope to obtain the required flow. Steep slopes generate high velocities with smaller conduit requirements. When sufficient fall is available, steep slopes are often economical. On the other hand, if head can be generated only by pumping or through construction of a dam, flatter slopes calling for larger conduits may be needed to reduce the cost of the lift. Some combination of lift and slope will yield the optimum economy.

Usually some controlling feature establishes the elevation of the aqueduct at a specified point. Examples of possible governing features are dam heights, tunnel locations, terminal reservoirs, and hilltops.

DISTRIBUTION SYSTEMS

Water distribution systems are designed to satisfy the water requirements of domestic, commercial, industrial, and firefighting purposes. The system should be capable of meeting the demands placed on it at all times, and at satisfactory pressures. Pipe systems, pumping stations, storage facilities, fire hydrants, house service connections, meters, and other appurtenances are the main elements of the system [6].

6.4 SYSTEM CONFIGURATIONS

Water distribution systems may be classified as grid systems, branching systems, or a combination of the two. The configuration of the system is influenced by street patterns, topography, degree and type of development of the region to be served, and location of treatment and storage works. Figure 6.6 illustrates the basic types of distribution systems. Grid systems are usually preferred to branching systems, since they can supply a withdrawal point from at least two directions. Branching systems do not permit this type of circulation, because they have numerous terminals or dead ends. Both grid and combination systems can incorporate loop feeders, which can distribute water to a takeoff point from several directions. In locations where sharp changes in topography occur (hilly or mountainous regions) it is common practice to divide the distribution system into two or more service areas or zones. This precludes the difficulty of having extremely high pressure in low-lying areas in order to maintain reasonable

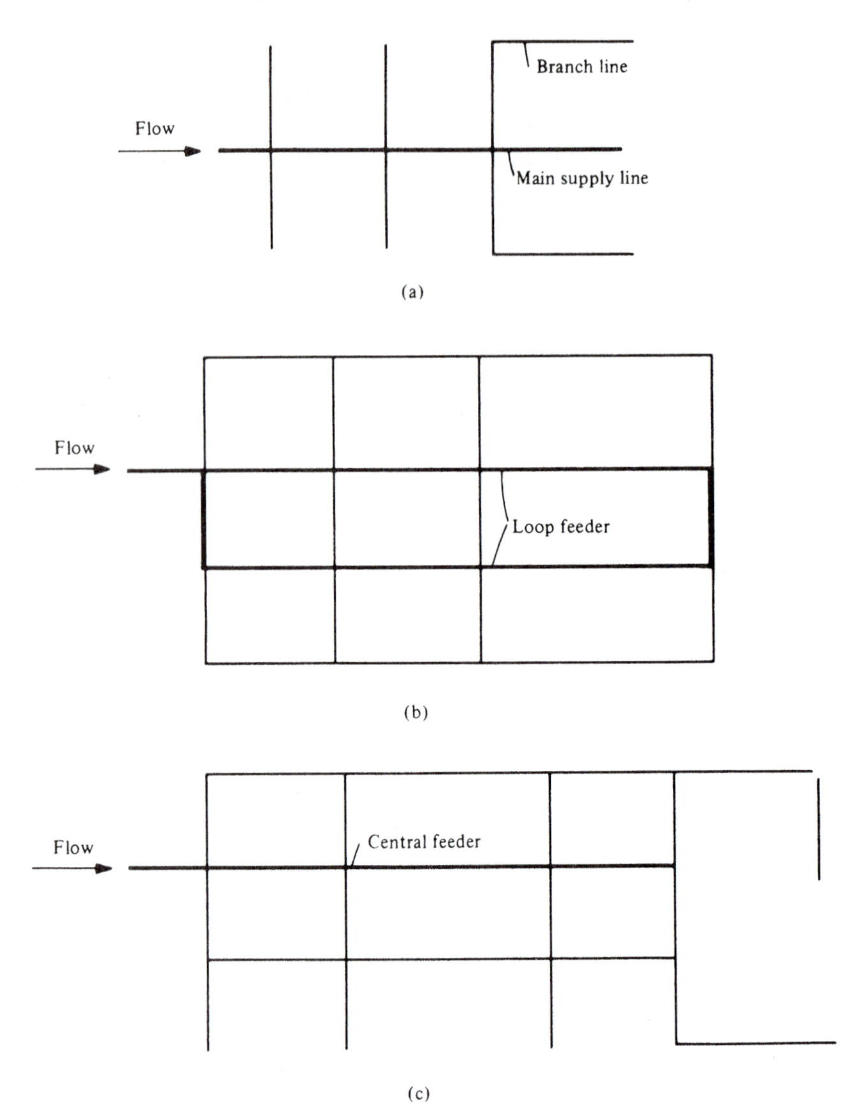

FIGURE 6.6 Types of water distribution systems. (a) Branching. (b) Grid. (c) Combination.

pressures at higher elevations. Usual practice is to interconnect the various systems, with the interconnections closed off by valves during normal operations.

6.5 DISTRIBUTION SYSTEM COMPONENTS

A water distribution network is a collection of links connected together at their end-points called nodes (refer to Fig. 6.7). Links may include pipes, pumps, and valves. Nodes may be points of water withdrawal (demand nodes), locations where water is introduced to the network (source nodes), or locations of tanks or reservoirs (storage nodes). Flows

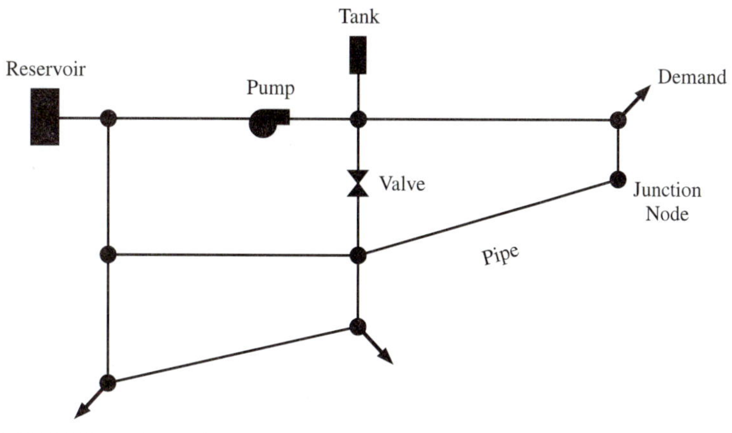

FIGURE 6.7 Network components.

may be stated as gallons per minute (gpm), cubic feet per second (cfs), million gallons per day (mgd), or liters per second (l/s).

Pipes Pipes are used to convey water. The direction of flow is from the end at higher head (potential energy per pound of water) to that at a lower head. Pipes may contain check valves that restrict flow to a specific direction. Such valves can be made to open or close at preset times, when tank levels fall below or above certain set points, or when nodal pressures fall below or above certain set points. Head lost to friction resulting from pipe flow can be expressed by the following type of equation [see also Eq. (6.4)]:

$$h_L = Aq^B \tag{6.17}$$

where h_L is the head loss (length), q is the flow in (volume/time), A is a resistance coefficient, and B is a flow exponent. The Hazen–Williams formula, the Darcy–Weisbach formula, and the Chezy–Manning formula can all be used to calculate friction head loss. The Hazen–Williams formula is the one most commonly used, but it cannot be used for liquids other than water. The Darcy–Weisbach formula is the most theoretically correct, and it is suited to all liquids. The Chezy–Manning equation is generally used for problems of open channel flow. Table 6.3 gives expressions for the resistance coefficient A, and values of the flow exponent B. The pipe roughness coefficients in these expressions must be determined empirically. Table 6.1 gives ranges of these coefficients generally encountered in practice. Note that as pipes age, these coefficients can change significantly.

Junctions Junctions (also called nodes) are points where pipes come together and where water enters or leaves the network. Storage nodes (i.e., tanks and reservoirs) are special types of nodes where a free water surface exists and the hydraulic head is the elevation of water above sea level. To determine the total hydraulic head at a node, the elevation above sea level of all nodes must be specified. The magnitude of water withdrawals (demands) or inputs at nodes that are not storage nodes must be known over the time frame that the network is being analyzed.

TABLE 6.3 Pipe Head Loss Formulas for Full Flow (head loss in ft and flow rate in cfs)

Formula	*Resistance Coefficient (A)*	*Flow Exponent (B)*
Hazen–Williams	$4.727C^{-1.852}d^{-4.871}L$	1.852
Darcy–Weisbach	$0.0252f(\varepsilon, d, q)d^{-5}L$	2
Chezy–Manning	$4.66n^2d^{-5.33}L$	2

Notes: C = Hazen–Williams roughness coefficient
ε = Darcy–Weisbach roughness coefficient (ft)
f = friction factor (dependent on ε, d, and q)
n = Manning roughness coefficient
d = pipe diameter (ft)
L = pipe length (ft)
q = flow rate (cfs)
Source: Rossman [5].

Reservoirs Reservoirs are nodes that represent lakes, rivers, and groundwater aquifers. Because of their scale, reservoirs are often considered to represent an infinite source or sink of water to the distribution system. A major feature of a reservoir is its hydraulic head, which is the water surface elevation, providing that the reservoir is not under pressure. The hydraulic head may vary over time, however, and this must be taken into account when modeling water supply systems.

Tanks Tanks are storage nodes where the volume of water can vary with time. Tank properties are the bottom elevation where the water level is zero; the diameter or shape of the tank; the initial water level; and the minimum and maximum water levels within which the tank can operate. The change in water level of a storage tank can be calculated using the following equation:

$$\Delta y = (q/A)\Delta t \tag{6.18}$$

where

Δy = change in water level, ft
q = flow rate into (+) or out of (−) tank, cfs
A = cross-sectional area of the tank, ft^2
Δt = time interval, s

Emitters Emitters are nozzles or orifices that discharge to the atmosphere. The flow through these devices is a function of the pressure available at the node:

$$q = Cp^{\gamma} \tag{6.19}$$

where

q = flow rate
C = a discharge coefficient

p = pressure

γ = pressure exponent

For nozzles and sprinkler heads, γ equals 0.5. Manufacturers of emitters generally supply the value of the discharge coefficient usually in units of gpm/psi$^{0.5}$ (stated as the flow through the device at a 1 psi pressure drop). Emitters are used to model irrigation networks and sprinkler systems. They may also be used to estimate pipe leakage if a discharge coefficient and a pressure exponent can be estimated.

Minor Losses Minor head losses are usually associated with turbulence that occurs at bends, junctions, meters, and valves. The importance of such losses depends on the nature of the pipe network and the degree of accuracy required in the analysis. Minor head losses may be accounted for by assigning a minor head loss coefficient to the appropriate fixture (see Table 6.2 for a list of such coefficients). The minor head loss is then calculated using the following formula:

$$h_L = K(v^2/2g) \tag{6.20}$$

where K is a minor head loss coefficient, v is flow velocity (length/time), and g is the acceleration of gravity (length/time^2).

Pumps Pumps are used to increase the hydraulic head of water. A pump characteristic curve (Fig. 6.8) describes the head imparted to a fluid as a function of its flow rate through the pump. Pump curves can be represented with a function of the form:

$$h_G = h_o - aq^b \tag{6.21}$$

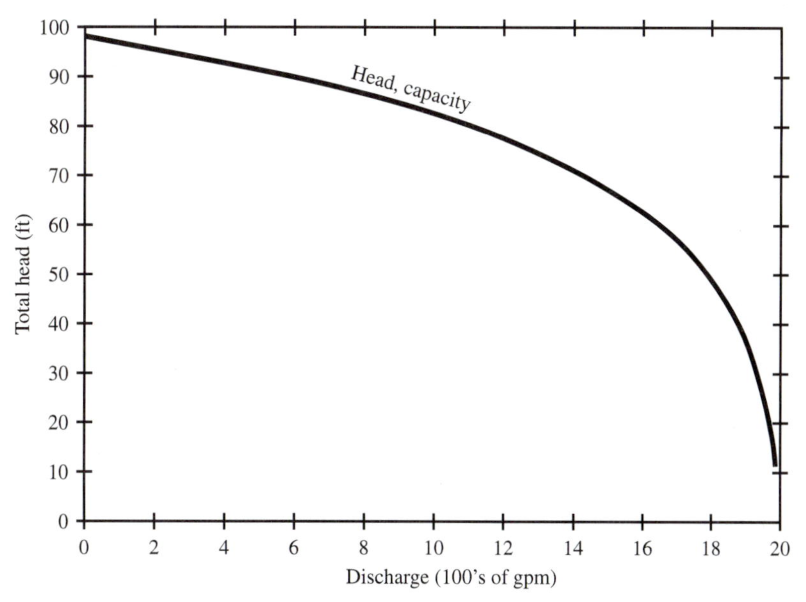

FIGURE 6.8 Typical pump curve.

where h_G is the head gain imparted by the pump in ft, q is the flow through the pump, h_o is the shutoff head (head at no flow), a is a resistance coefficient, and b is a flow exponent. Flow through a pump is unidirectional and pumps must operate within the head and flow limits imposed by their characteristic curves. Pumps may be run at constant or variable speed.

Valves Valves are links in pipelines that are used to regulate flow or pressure. There are numerous types of valves. Shutoff (gate) valves and check (nonreturn) valves completely open or close pipes. Pressure-reducing valves (PRVs) limit the pressure at a point in a pipe network. Pressure-sustaining valves (PSVs) maintain a set pressure at a specified location in a pipe network. Flow control valves (FCVs) limit the flow to a prescribed amount.

6.6 SYSTEM REQUIREMENTS

The performance of a distribution system can be based on the pressures available in the system for a specific rate of flow [12,13]. Pressures should be great enough to meet consumer and firefighting needs. At the same time, they should not be excessive, since the development of pressure head is an important cost consideration. In addition, higher pressures may cause leakage, which is associated with loss of treated water and higher costs. Costs of distribution systems are significant, and minimizing them must be a design objective.

For commercial areas, pressures in excess of 60 pounds per square inch, gauge (psig), are usually required. Adequate pressures for residential areas usually range from 40 to 50 psig. In tower buildings, it is often necessary to provide booster pumps to elevate the water to upper floors. Storage tanks are usually provided at the highest level and distribution is made directly from them.

The capacity of the distribution system is determined on the basis of local water needs plus fire demands, as outlined in Chapter 4. Pipe sizes should be selected so that high velocities are avoided. Once the flow has been determined, pipe sizes can be selected by assuming velocities of from 3 to 5 fps. Where fire-fighting requirements are to be met, a minimum diameter of 6 in. is recommended. The National Board of Fire Underwriters recommends 8 in. as a minimum but permits 6-in. pipes in grid systems, provided the length between connections does not exceed 600 ft.

6.7 DISTRIBUTION SYSTEM DESIGN AND ANALYSIS

The design of a water distribution system involves the selection of a system of pipes and other components so that design flows can be carried with head losses that do not exceed those deemed necessary for adequate operation of the system. Design flows should be based on estimated future water requirements, since distribution systems must be capable of providing service for many years in the future (sometimes as long as 100 years). A typical sequence of evaluation, design, and layout operations follows [1–3,5,6,14]:

1. Review maps, construction plans, billing records, planning studies, zoning regulations, population figures, water use studies, and any other data relevant to the system being analyzed.

2. For existing water systems, determine pipe ages, the roughness of pipe interiors, pipe lengths and diameters, and the locations of pipes and appurtenances. Water supply sources, pumping station locations and characteristics, and storage tank locations and volumes must also be determined.
3. Prepare a detailed map of the existing system or the proposed system.
4. Forecast population growth and distribution to the end of the design period (10–50 yr). Project water use patterns (spatial and temporal) for domestic, industrial, and commercial uses to be served by the system. These estimates must also extend over the design life of the network. Water use estimates are based on population projections and anticipated trends in commercial and industrial activity.
5. Develop a computer model of the existing or proposed system. In analyzing a network, it is common to simplify the system by eliminating nonessential small lines, combining pipes using the equivalent-pipe method, and assuming water use to be concentrated at takeoff points (nodes). Ordinarily, the model selected will be one of those described under network analysis.
6. Use the computer model of the existing system to evaluate historical conditions. If observed data and model runs compare favorably, the model may be considered "validated." The model can then be presumed to be adequate for evaluating proposed modifications to, or extensions of, the existing system. Applying the model to the existing system will quickly identify areas of low pressure, pipelines having high head loss characteristics, and overloaded parts of the network. Proposed replacements and extensions of the system can then be evaluated to see if they will resolve problems or meet new requirements.
7. The process for designing new or expanded water distribution systems follows:

   a. On a development plan of the area to be serviced, sketch the tentative location of all water mains that will be needed to supply the area. The completed drawing should differentiate between proposed feeder mains and smaller service mains. The various pipelines making up the system should be interconnected at intervals of 1200 ft or less. Looped feeder systems are desirable and should be used whenever possible. Two small feeder mains running parallel several blocks apart are preferable to a single large main with an equal or slightly larger capacity than the two mains combined.

   b. Using estimated values of the anticipated design flows, select appropriate pipe sizes by assuming velocities of from 3 to 5 fps.

   c. Mark the position of building service connections, fire hydrants, and valves. Service connections form the link between the distribution system and the individual consumer. Normally, the practice is one customer per service pipe. Figure 6.9 illustrates the details of a typical service connection for a private residence. Fire hydrants are located to provide complete protection to the area covered by the distribution system. Recommendations regarding average area per hydrant for various populations and required fire flow are given by the National Board of Fire Underwriters. Hydrants generally should not be farther apart than about 500 ft. Figure 6.10 illustrates a typical fire hydrant setting.

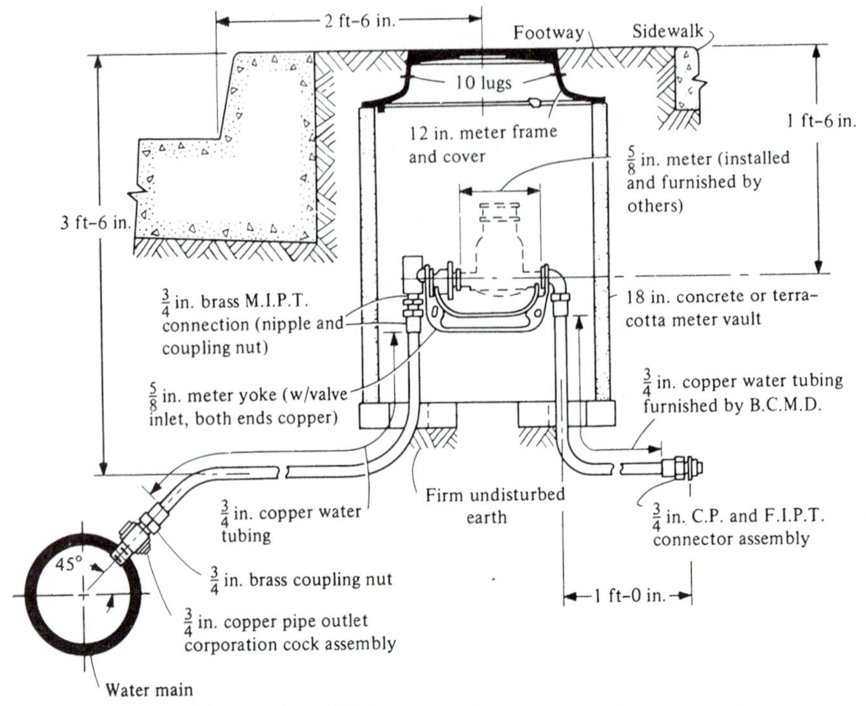

FIGURE 6.9 Typical installation of $^3/_4$-in. metered domestic service. *Source:* Courtesy of Baltimore County, MD, Department of Public Works.

d. Apply projected water demands (including fire flow) to the network and calculate residual pressures.

e. Compare calculated pressures to standards. Identify areas of projected less-than-adequate service.

f. Check head losses in individual pipes to find excesses. These will usually occur in pipes where flow velocities are greater than 1.5 m/s (5 fps).

g. Add pipes or replace high-head-loss pipes with larger pipes. Run the model again and see if residual pressures under maximum future loads are adequate. If not, try other additions or changes until the system is adequate for anticipated future loads. The system must be able to handle maximum daily loads plus fire loads without decreasing residual pressures below minimum standards. The overall objective is to design a system that will meet projected water demands at the least cost while incorporating appropriate safety measures for looping or duplicate lines so that line breaks or other disturbances will not isolate users from a water supply. Note, however, the network analysis and resulting design will be no better than the assumptions made regarding water demands, pipe roughness, and so on.

8. Estimate construction costs for the proposed improvements.

9. Prepare a construction schedule for the identified improvements or new system that is consistent with the financial capabilities of the city.

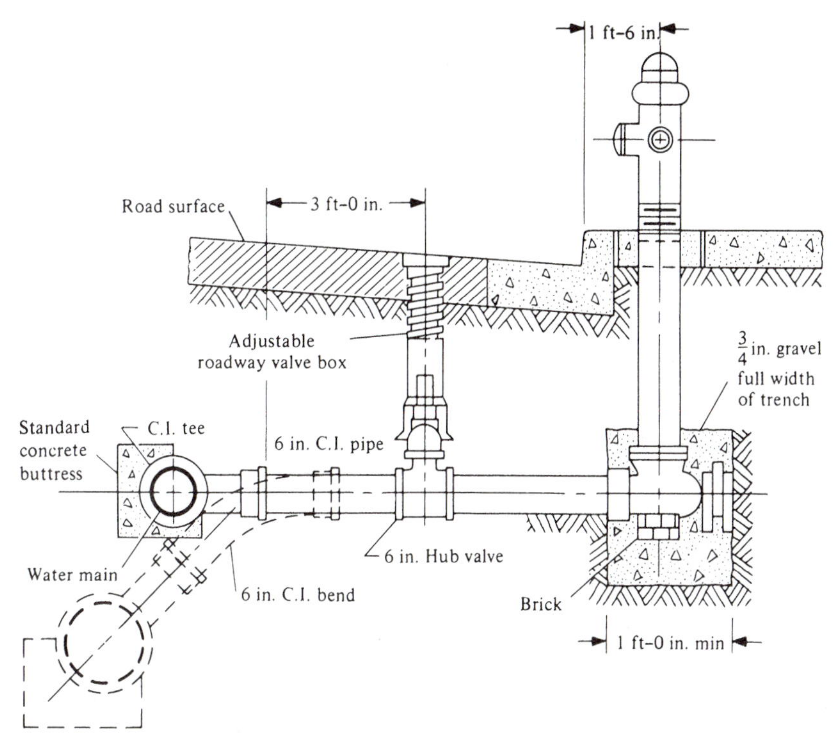

FIGURE 6.10 Typical fire hydrant setting. *Source:* Courtesy of Baltimore County, MD, Department of Public Works.

6.8 HYDRAULIC DESIGN

Hydraulic design of water distribution systems requires information on anticipated rates of water withdrawal, locations of withdrawals, and pressure gradients required for the system. Both the maximum daily rate of withdrawal plus fire protection and the maximum hourly rate of withdrawal should be investigated to determine which will govern the system design.

The spatial distribution of water use can be estimated utilizing population densities and commercial and industrial use patterns that exist or are predicted for the region. When determining the peak hour for a feeder to an area consisting of residential, commercial, and industrial users the predicted hydrograph for each type of user must be had. This will allow calculation of the specific hour in which the summation of the three component flows is greatest. Once the design demands have been determined, it is common practice to consider them to be concentrated at specified points (nodes) on the feeder–main system.

Distribution systems are generally designed so that reasonably uniform pressures prevail [1, 6, 12]. Transmission mains may carry pressures up to 250 psi, but the need for pressures exceeding 150 psi is usually limited to those transmission mains serving pressure zones at higher elevations. The working pressure for residential areas is normally in the range of 40 to 60 psi. It should be noted that few plumbing fixtures will operate well at pressures less than about 20 psi. For urban water systems, the maximum design pressure that customers should experience is considered to fall within the range of 90–110 psi, while minimum design pressures (pressures at a customer's tap) are usually in the 40–50 psi range [6]. The maximum allowable velocity for pipelines is usually 5 fps.

The analysis of a distribution system is often simplified by first skeletonizing the system. This might involve the replacement of a series of pipes of varying diameter with one equivalent pipe or replacing a system of pipes with an equivalent pipe. An equivalent pipe is one in which the loss of head for a specified flow is the same as the loss in head of the system it replaces. An example illustrates this method of analysis.

Example 6.3

Referring to the pipe system shown in Figure 6.11, replace (a) pipes *BC* and *CD* with an equivalent 12-in. pipe and (b) the system from *B* to *D* with an equivalent 20-in. pipe.

Solution:

1. Assume a discharge through *BCD* of 8 cfs. Using the Hazen–Williams formula [Eq.(6.1)], a spreadsheet is used to determine the parameters and head losses. The calculated values are shown in Table 6.4. The total head loss for pipe BC = 1.23 ft and that for pipe CD is 5.51 ft (see the right-hand column in the table). The total head loss between B and D is therefore $1.23 + 5.51 = 6.74$ ft.

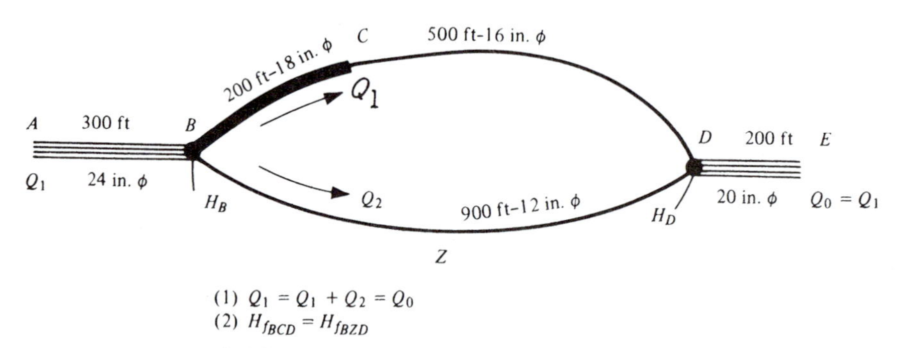

FIGURE 6.11 Example 6.3.

TABLE 6.4 Spreadsheet Calculations for Example 6.3

Pipe Size (ft)	Length (ft)	Q, cfs	Constant $1.318 \times C$	Area − A ($\pi d^{2/4}$)	Perimeter, πd	R A/P	$R^{0.63}$	$S^{0.54}$	S	Total Head Loss
1.5	200	8	131.8	1.77	4.71	0.38	0.54	0.06	0.0061	1.23
1.33	500	8	131.8	1.39	4.18	0.33	0.50	0.09	0.0110	5.51
1	900	8	131.8	0.79	3.14	0.25	0.42	0.19	0.0441	39.71
1.67		9.69	131.8	2.19	5.25	0.42	0.58	0.06	0.0052	
1		6.931	131.8	0.79		0.25	0.42	0.1585	0.033	
1		2.761	131.8	0.79		0.25	0.42	0.0631	0.006	

Note: Lower part of table refers to part 2 of the problem.

For a discharge of 8 cfs, the head loss S in ft/ft is found to be 0.0441. The equivalent length of 12-in. pipe is therefore

$$L = 6.74/0.0441 = 153.18 \text{ ft}$$

2. Assume a total head loss between B and D of 5.0 ft. For the 12-in. equivalent pipe this is 0.033 ft/ft (5/153.18). For the 900 ft of 12-in. pipe it is 0.006 ft/ft (5/900). Inserting these values in the Hazen–Williams formula, the flows are found to be 6.93 and 2.76 cfs respectively. The total flow (the sum of the two) is thus 9.69 cfs at a head loss of 5 ft. For this discharge, a 20-in. pipe will have a head loss of 0.0052 ft/ft. The equivalent 20-in. pipe to replace the whole system will be

$$5/0.0052 = 961.5 \text{ ft long}$$

The analysis of this simple hydraulic system presents little difficulty. A slightly more complex system is shown in Figure 6.12. The method of equivalent pipes will fail to yield a solution in this case because there are crossover pipes (pipes that operate in more than one circuit and a number of withdrawal points throughout the system). Solution of this type of problem requires the use of network analysis techniques.

Pipe Networks

Most municipal water distribution systems are complex mazes containing pumps, storage elements, and pipelines of a variety of sizes. As pointed out in the preceding section, the ordinary methods of hydraulic analysis must be extended to take into account the looping characteristics of networks, changing reservoir levels, pumping, etc. Special techniques of network analysis come into play in this case. In these methods, iterative solutions based on initial assumptions lead to either balancing flows in a system or balancing heads in a system. The underlying principles are those of preserving mass continuity and ensuring energy conservation.

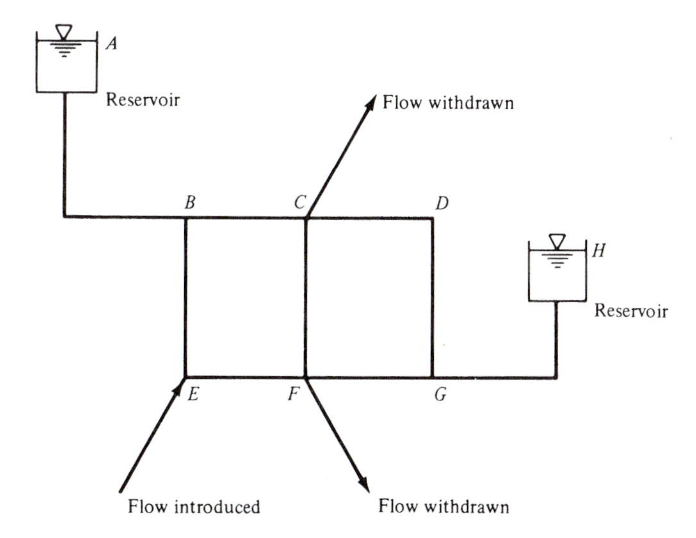

FIGURE 6.12 Pipe network showing pipe and fixed-grade nodes.

Pipe networks are composed of a number of constant-diameter pipe sections containing pumps and fittings. The ends of each pipe section are called nodes. In Figure 6.12, the lettered points are nodes which may be of either fixed grade or junction type. Junction nodes are points where pipes meet and where flow may be introduced or withdrawn. Fixed-grade nodes are points where a constant grade is maintained. Connections to storage tanks or reservoirs or constant-pressure regions are examples (Fig. 6.12). Networks are commonly divided into loops for computational purposes. Primary loops such as those shown in Figure 6.12 are closed pipe circuits in that the network has no other closed pipe circuits within them. Using the foregoing terminology, we may write

$$P = J + L + F - 1 \tag{6.22}$$

where

P = number of pipes

J = number of junction nodes

L = number of loops

F = number of fixed-grade nodes

This identity is directly related to the fundamental hydraulics equations that describe steady-state flow in a network. For Figure 6.12, $J = 6$, $L = 2$, $F = 2$, and thus the sum minus 1 is 9, the number of pipes in the complete network. Equations used to analyze steady-state pipe networks fall into two main categories—loop equations and node equations. The loop equations express mass conservation and energy conservation in terms of the discharge in a pipe section, while the node equations express mass continuity in terms of elevations or grades at junction nodes.

Many computer programs are available that can handle the analysis of flows in pipe systems of any configuration and with a variety of components such as storage tanks, pumps, check valves, pressure-regulating valves, and variable pressure water supplies [1,5,6,15–26]. Information on several of these programs is available on the Internet: KYPIPE (www.kypipe.com); WaterCAD (www.haestad.com); and EPANET (www.epa.gov/ORD/NRMRL/wswrd/epanet.html).

Loop Equations Equation (6.22) is the basis for formulation of a set of equations describing the hydraulic performance of a pipe network. In terms of unknown flows in the pipes, mass continuity and energy conservation equations can be written for the pipes and nodes. For each loop, an energy conservation equation can be written. A mass continuity equation can also be written for each node. For junction nodes, the inflow to the junction must be balanced by the outflow. This can be written

$$\Sigma Q_{in} - \Sigma Q_{out} = Q_e (\text{J equations}) \tag{6.23}$$

where Q_{in} is the inflow, Q_{out} is the outflow, and Q_e is the external flow into the system or the withdrawal from the system at the node. For the primary loops, energy conservation can be described by

$$\Sigma h_L = \Sigma E_p \text{ (L equations)} \tag{6.24}$$

where h_L is the pipe energy loss (minor losses included) and E_p is the energy introduced into the system by pumps. For loops having no pumps, the sum of the energy losses around the loop is zero. Note that a sign convention is used for loops. Clockwise flows might be considered positive and counterclockwise flows negative, for example.

Where there are F fixed-grade nodes, $F - 1$ independent energy conservation equations can be written for pipe paths between any two fixed-grade nodes. These equations take the form

$$\Delta E = \Sigma h_L - \Sigma E_p \quad (F - 1 \text{ equations}) \tag{6.25}$$

where ΔE is the difference in elevation (grade) between the two fixed-grade nodes. Any connected path between the two fixed-grade nodes can be selected. This can be done by selecting a series of pipes so that the ending node of one path is the starting node for the next, etc. This procedure will produce the needed $F - 1$ equations with no redundancy [15].

Equation (6.24) can be considered a special case of Eq. (6.25) where the difference in elevation (ΔE) is zero for a closed-loop path. It follows that the energy conservation equations for a pipe network can be expressed by $L + F - 1$ energy equations described by Eq. (6.25). The continuity and energy equations that describe the pipe network are P in number. They form a set of simultaneous nonlinear algebraic equations (loop equations) that describe steady-state flow conditions in a pipe network. To analyze an existing or proposed pipe network, the loop equations are solved to determine the flow in each pipe. In order to effect a solution, the terms in the energy equations must be expressed as functions of flow. Expressions for frictional losses in pipes, minor losses in fittings, and pump energy are needed.

Frictional losses in pipes are expressed as

$$h_{LP} = K_P Q^n \tag{6.26}$$

where K_P is a constant incorporating pipe length, diameter, and roughness and n is an exponent. The values of K_P and n are generally determined by the selection of the Darcy–Weisbach, Hazen–Williams, or Manning equations for the expression of energy losses.

The minor losses in a section of pipe result from fittings, valves, meters, or other insertions that affect the flow. They are expressed as

$$h_{LM} = K_M Q^2 \tag{6.27}$$

where K_M is the minor loss constant, a function of the sum of the minor loss coefficients for all fittings in the length of pipe (ΣM) and the pipe diameter. It is given by

$$K_M = \sum M/2gA^2 \tag{6.28}$$

where A is the cross-sectional area of the pipe.

The term in the energy equations representing pumping energy can be expressed in several ways. A constant power input can be specified or a curve can be fitted to data obtained from pump operations. In any event, the relationship between pump energy E_P (head developed by the pump) and the flow Q can be represented by

$$E_P = P(Q) \tag{6.29}$$

If the pump operates at constant power, then using the relationship for horsepower (hp $= Qwh/550$, where w is the specific weight of water in pounds per ft^3, h is the head in feet, Q is the flow in cfs, and 550 is a conversion factor), $P(Q)$ is given by 550 hp/62.4Q. Letting Z = 550 hp/62.4, $P(Q)$ can be written, for constant power, as Z/Q.

Combining Eqs. (6.26 – 6.29), the energy relationships in terms of discharge become

$$\Delta E = \sum(K_P Q^n + K_M Q^2) - P(Q) \tag{6.30}$$

Equation (6.30) and the continuity equations [Eq. (6.23)] comprise the set of P simultaneous equations that must be solved in a loop analysis. There is no direct solution to these nonlinear algebraic relationships, but several algorithms for determining an answer are available. They will be discussed in subsequent sections.

Node Equations Solution of the loop equations begins generally with an assumption of flow rates in the pipe network. Computations proceed until adjustments in flows are considered to be within tolerable levels. When using node equations, adjustments are made in initial assumptions of head.

When considering nodes, the principal relationship used is the continuity equation [Eq. (6.23)]. The discharge in a section of pipe connecting nodes such as E and B (Fig. 6.12) is expressed in terms of the grade (head) at junction node A (H_a), the grade at junction node B (H_b), and the resistance offered by the pipeline (K_{ab}). This can be expressed as

$$Q_{ab} = [(H_a - H_b)/K_{ab}]^{1/n} \tag{6.31}$$

where it is assumed that the pipe section is free of pumps and the head loss is calculated as

$$h_L = KQ^n \tag{6.32}$$

and K is determined as indicated for Eq. (6.26). Combine Eqs. (6.23) and (6.31):

$$\sum_{b=1}^{N}\left[\pm\left(\frac{H_a - H_b}{K_{ab}}\right)^{1/n}\right] = Q_e \tag{6.33}$$

which expresses continuity at a given junction node where N pipes join. The sign of the term in the summation depends on the direction of flow into or out of the junction. A total of J junction node equations result. This basic set can be expanded to include pumps where they exist. For each pump encountered, junction nodes are specified at the pump inlet and outlet, locations b and c in Figure 6.13. Two additional equations are

FIGURE 6.13 Pump notation for the node equation.

thus generated, one at the suction side and the other at the discharge side of the pump [15]. These equations involve the unknown heads (grades) on either side of the pump.

Following the notation of Figure 6.13, an equation utilizing flow continuity in the suction and discharge lines can be written as

$$H_a - H_b = \frac{K_{ab}}{K_{cd}}(H_c - H_d) \tag{6.34}$$

Another equation can be developed that relates the head change across the pump to the discharge in either the inlet or outlet pipe. Where the pump being considered is operating at constant power, the relationship in terms of the outlet line discharge, according to Eq. (6.29), is

$$H_c - H_b = P\left[\left(\frac{H_c - H_d}{K_{cd}}\right)^{1/n}\right] \tag{6.35}$$

Equations (6.33) to (6.35) constitute the complete set of pipe network node equations. All of them are expressed in terms of the unknown grades at junction nodes and in terms of the suction and discharge grades at pumps within the system. This set of equations is also nonlinear, and thus direct solution is impossible. Commonly used algorithms involving the node equations are discussed in later sections.

Algorithms for Solving Loop Equations

Several methods are widely used to solve the loop equations [15]. All use gradient methods to accommodate the nonlinear flow terms in Eq. (6.30). The gradient method is derived from the first two terms of the Taylor series expansion. Any function $f(x)$ that is continuous (differentiable) can be approximated as follows:

$$f(x) \approx f(x_0) + f'(x_0)(x - x_0) \tag{6.36}$$

Examination of the right-hand side of Eq. (6.36) reveals that the approximation has reduced $f(x)$ to a linear form. However, if f is a function of more than one variable, Eq. (6.26) can be generalized as follows:

$$\begin{aligned} f[x(1), x(2), \ldots] &= f[x(1)_0, x(2)_0, \ldots] \\ &+ \frac{\partial f}{\partial x(1)}[x(1) - x(1)_0] + \frac{\partial f}{\partial x(2)}[x(2) - x(2)_0] + \cdots \end{aligned} \tag{6.37}$$

in which the partial derivatives are evaluated at some $x(1) = x(1)_0, x(2) = x(2)_0$, etc.

The right-hand side of Eq. (6.30) represents the grade difference across a pipe carrying a discharge of Q. This can be stated as

$$f(Q) = K_P Q^n + K_M Q^2 - P(Q) \tag{6.38}$$

Substituting an estimated Q_i for Q and denoting $f(Q_i)$ by H_i, Eq. (6.38) becomes

$$H_i = f(Q_i) = K_P Q_i^n + K_M Q_i^2 - P(Q_i) \tag{6.39}$$

Differentiating Eq. (6.38) and setting $Q = Q_i$ gives the gradient of the function at $Q = Q_i$. Thus,

$$f'(Q_i) = nK_P Q_i^{n-1} + 2K_M Q_i - P'(Q_i)$$

Denoting $f'(Q_i)$ by G_i,

$$G_i = nK_P Q_i^{n-1} + 2K_M Q_i - P'(Q_i) \tag{6.40}$$

Both the function H_i and its gradient G_i evaluated at $Q = Q_i$ are used in algorithms for solving the loop equations.

Single-Path Adjustment (P) Method This solution technique is the oldest and best known of all the loop methods. It was first described by Hardy Cross [27]. Originally, the method was restricted to closed-loop networks and provided only for line losses. A generalization of the procedure is described below [15,28].

1. An initial set of flow rates that satisfy continuity at each junction node is selected.
2. A flow adjustment factor is computed for each path $(L + F - 1)$ to satisfy the energy equation for that path. Continuity is maintained in this process.
3. Step 2 is repeated, building on improved solutions until the average correction factor is within an acceptable limit.

Equation (6.30) is used to compute the adjustment factor for a path using the gradient method to linearize the energy equations. Thus,

$$f(Q) = f(Q_i) + f'(Q_i)\Delta Q \tag{6.41}$$

in which $\Delta Q = Q - Q_i$, where Q_i is the estimated discharge. Applying Eq. (6.41) to Eq. (6.30) and solving for ΔQ gives

$$\Delta Q = \frac{\Delta E - \Sigma H_i}{\Sigma G_i} \tag{6.42}$$

which is the flow adjustment factor to be applied to each pipe in the path. The numerator represents the imbalance in the energy relationship due to incorrect flow rates. The procedure reduces this to a negligible quantity. Flow adjustment is carried out for all L fundamental (closed) loops and $F - 1$ pseudoloops in the network.

The Hardy Cross method of network analysis permits the computation of rates of flow through a network and the resulting head losses in the system [27]. It is a relaxation method by which corrections are applied to assumed flows or assumed heads until an acceptable hydraulic balance of the system is achieved.

The Hardy Cross analysis is based on the principles that (1) in any system continuity must be preserved and (2) the pressure at any junction of pipes is single valued. Referring to the simple network of Figure 6.14, the elements of the procedure can be explained. First, the system must be defined in terms of pipe size, length, and roughness. Then, for any inflow Q_1, the system can be balanced hydraulically only if $H_{fBCD} = H_{fBZD}$. This restriction limits the possibilities to only one value of Q_1 and Q_2 which will satisfy the conditions.

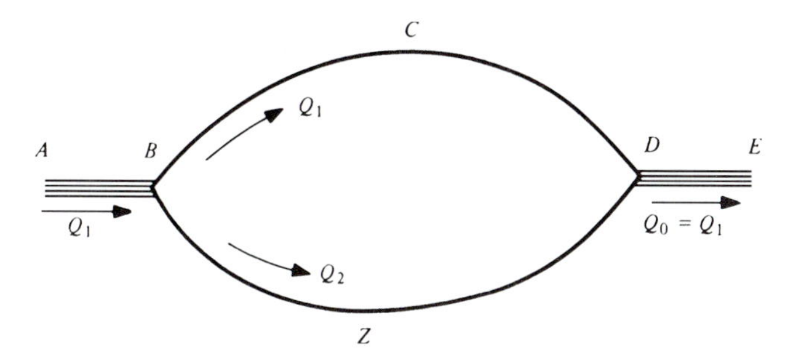

(1) $Q_1 = Q_1 + Q_2 = Q_0$
(2) $H_{fBCD} = H_{fBZD}$

FIGURE 6.14 Derivation of the Hardy Cross method.

Derivation of the basic equation for balancing heads by correcting assumed flows will now be given for the loop of Figure 6.14 [27]. First, find the required inflow Q_1. Then arbitrarily divide this flow into components Q_1 and Q_2. The only restriction on the selection of these values is that $Q_1 + Q_2 = Q_1$. An attempt should be made, however, to select realistic values. Since the procedure involves a number of trials, the amount of work involved will be dependent on the accuracy of the value originally selected. For example, in the network shown, *BCD* is considerably larger in diameter than *BZD*. A logical choice would therefore assume that Q_1 will be larger than Q_2. The final solution to the problem will be the same regardless of the original choice, but much more rapid progress results from reasonable initial assumptions.

After Q_1 and Q_2 have been chosen, H_{fBCD} and H_{fBZD} can be computed using the Hazen–Williams or some other pipe-flow formula. Remembering that the Hazen–Williams equation is of the form of Eq. (6.2),

$$Q = 0.279CD^{2.63}S^{0.54}$$

the equation may be rewritten

$$Q = K_a S^{0.54} \tag{6.43}$$

where K_a is a constant when we are dealing only with a single pipe of specified size and material. Rearranging this equation and substituting H_f/L for S yields Eq. (6.26):

$$H_f = KQ^n$$

where $n = 1.85$ in the Hazen–Williams equation. Equation (6.26) is convenient for expressing head loss as a function of flow in network analyses.

If the computed values of H_{fBCD} and H_{fBZD} are not equal (which is usually the case on the first trial), a correction must be applied to the initial values. Call this correction ΔQ. If, for example, $H_{fBCD} > H_{fBZD}$, then the new value for Q_1 will be $Q_1 - \Delta Q = Q_1'$ and the new value for Q_2 must be $Q_2 + \Delta Q = Q_2'$. The corresponding values of head loss will be H'_{fBCD} and H'_{fBZD}. If ΔQ is the true correction, then

$$H'_{fBCD} - H'_{fBZD} = 0 = K_1(Q_1 - \Delta Q)^n - K_2(Q_2 + \Delta Q)^n$$

The binomials may be expanded as follows:

$$K_1(Q_1^n - n\Delta Q Q_1^{n-1} + \cdots) - K_2(Q_2^n + n\Delta Q Q_2^{n-1} + \cdots) = 0$$

If ΔQ is small, the terms in the expansion involving ΔQ to powers greater than unity can be neglected. Therefore,

$$K_1 Q_1^n - nK_1 \Delta Q Q_1^{n-1} - K_2 Q_2^n - nK_2 \Delta Q Q_2^{n-1} = 0$$

Substituting H_{fBCD} for $K_1 Q_1^n$, H_{fBZD} for $K_2 Q_2^n$, and rewriting the terms KQ^{n-1} as $K(Q^n/Q)$, one has

$$H_{fBCD} - \Delta Q n K_1 \frac{Q_1^n}{Q_1} - H_{fBZD} - \Delta Q n K_2 \frac{Q_2^n}{Q_2} = 0$$

$$H_{fBCD} - H_{fBZD} = \Delta Q n \left(\frac{H_{fBCD}}{Q_1} + \frac{H_{fBZD}}{Q_2} \right)$$

and

$$\Delta Q = \frac{H_{fBCD} - H_{fBZD}}{n(H_{fBCD}/Q_1 + H_{fBZD}/Q_2)} \tag{6.44}$$

Expanding this expression to the more general case gives the following equation for the flow correction ΔQ:

$$\Delta Q = -\sum H/n \sum \left(\frac{H}{Q} \right) \tag{6.45}$$

Application of this equation involves an initial assumption of discharge and a sign convention for the flow. Either clockwise or counterclockwise flows may be considered positive, and the terms in the numerator will bear the appropriate sign. For example, if the counterclockwise direction is considered positive, all H values for counterclockwise flows will be positive and all H values for clockwise flows will be negative. The denominator, however, is the absolute sum without regard to sign convention. The correction ΔQ has a single direction for all pipes in the loop, and thus the sign convention must also be considered in applying the correction.

Example 6.4

Given the network, the inflow at A, and the outflows at B, C, and D in Figure 6.15, carry out a Hardy Cross analysis using a spreadsheet to find the flows in the individual pipes comprising the network. Assume that the Hazen–Williams coefficient C is 100.

Solution: The computational procedure is given in Table 6.5. The initial and final flows are also shown in Figure 6.15. In Table 6.5, columns 1–4 are self-explanatory; column 5 indicates the sign convention, the values used as multipliers for values that

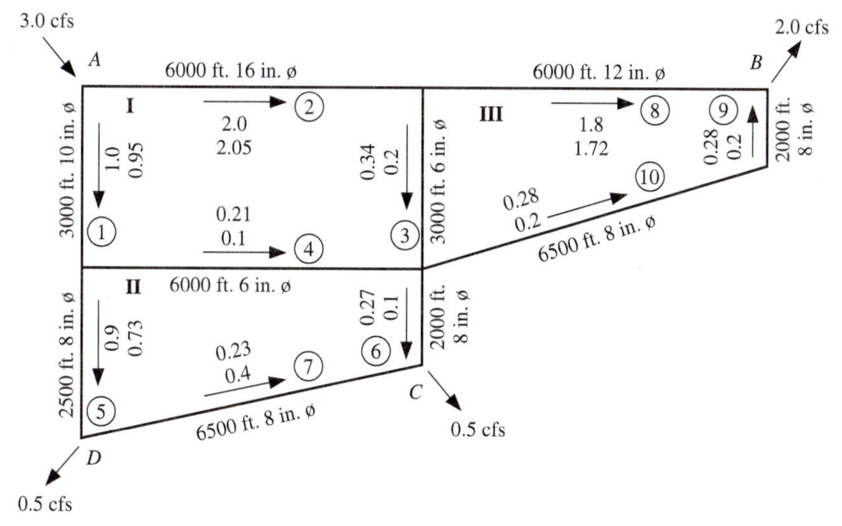

FIGURE 6.15 Pipe network analyzed by the Hardy Cross method. (The clockwise direction is considered positive. The flows are the initial assumed and final corrected values.)

are sign-dependent; column 6 gives the absolute values of Q, for Trial 1 (these are the initial assumptions); column 7 computes the values of head using Eq. (6.4) (these values are multiplied by the values in column 5 so that they carry the appropriate sign); column 8 displays the ratios of column 7 to column 6 values multiplied by the sign convention (column 5; these are absolute values); column 9 computes the denominator of Eq. (6.45) (absolute value); column 10 is the sum of the values in column 7 for each loop; column 11 computes the ΔQ values for each loop using Eq. (6.45); column 12 computes the ΔQ value for each pipe in the loop (it is the sum of all corrections that must be made if a pipe appears in more than one loop; pipes 3 and 4 in the example appear in two loops; sign convention is needed); and column 13 calculates the new value of Q for each pipe in the loop. These values become the column 6 values for the next iteration.

A similar procedure is to assume values of H and then balance the flows by correcting the assumed heads. The mechanics of the two methods are the same and the applicable relationship,

$$\Delta H = -n\sum Q \bigg/ \sum\left(\frac{Q}{H}\right) \tag{6.46}$$

can be derived in a manner similar to that for Eq. (6.45). The number of trials required for the satisfactory solution of any problem using Eq. (6.45) or (6.46) depends to a large extent on the accuracy of the initial set of assumed values and on the desired degree of accuracy of the results.

In using the Hardy Cross method to analyze large distribution systems, it is often useful to reduce the system to a skeleton network of main feeders [12]. Where the main

TABLE 6.5 Spreadsheet Solution to Example 6.4

TRIAL 1

(1) Loop No.	(2) Pipe No.	(3) Pipe Dia. (in.)	(4) Length (ft)	(5) Sign (−OR+)	(6) Q (cfs) ABS	(7) H (ft)	(8) (H/Q) ABS	(9) N(Sum(H/Q)) ABS	(10) Sum(H)	(11) Delta Q Loop	(12) Delta Q Pipe	(13) Q2	(14) Q2 ABS
1	1	10	3000	−1	1	−6.88	6.88				−0.00	−1.00	1
	2	16	6000	1	2	5.03	2.51				−0.00	2.00	2
	3	6	3000	1	0.2	4.22	21.06				−0.12	0.32	0.32
	4	6	6000	−1	0.1	−2.34	23.39	99.65	0.03	−0.00	0.17	−0.27	0.27
2	4	6	6000	1	0.1	2.34	23.39				0.17	0.27	0.27
	6	8	2000	1	0.1	0.19	1.92				0.17	0.27	0.27
	7	8	6500	−1	0.4	−8.11	20.28				0.17	−0.23	0.23
	5	8	2500	−1	0.9	−13.99	15.54	113.09	−19.57	0.17	0.17	−0.73	0.73
3	3	6	3000	−1	0.2	−4.22	21.06				−0.12	−0.32	0.32
	8	12	6000	1	1.6	16.80	9.33				−0.12	1.68	1.68
	9	8	2000	−1	0.2	−0.69	3.46				−0.12	−0.32	0.32
	10	8	6500	−1	0.2	−2.25	11.25	83.48	9.64	−0.12	−0.12	−0.32	0.32

TRIAL 2

Loop No.	Pipe No.	Pipe Dia. (in.)	Length (ft)	Sign (− OR +)	Q2 (cfs) ABS	H (ft)	(H/Q) ABS	N(Sum(H/Q)) ABS	Sum(H)	Delta Q Loop	Delta Q Pipe	Q3	Q3 ABS
1	1	10	3000	−1	1	−6.88	6.88				0.04	−0.96	0.96
	2	16	6000	1	2	5.03	2.51				0.04	2.04	2.04
	3	6	3000	1	0.32	10.06	31.43				0.06	0.34	0.34
	4	6	6000	−1	0.27	−14.69	54.41	176.19	−6.48	0.04	0.01	−0.21	0.21
2	4	6	6000	1	0.27	14.69	54.41				0.01	0.21	0.21
	6	8	2000	1	0.27	1.21	4.47				−0.02	0.25	0.25
	7	8	6500	−1	0.23	−2.91	12.67				−0.02	−0.25	0.25
	5	8	2500	−1	0.73	−9.50	13.01	156.43	3.49	−0.02	−0.02	−0.75	0.75
3	3	6	3000	−1	0.32	−10.06	31.43				0.06	−0.34	0.34
	8	12	6000	1	1.68	14.79	8.80				0.02	1.70	1.7
	9	8	2000	−1	0.32	−1.65	5.16				0.02	−0.30	0.3
	10	8	6500	−1	0.32	−5.37	16.78	115.02	−2.29	0.02	0.02	−0.30	0.3

TRIAL 3

Loop No.	Pipe No.	Pipe Dia. (in.)	Length (ft)	Sign (− OR +)	Q3 (cfs) ABS	H (ft)	(H/Q) ABS	N(Sum(H/Q)) ABS	Sum(H)	Delta Q Loop	Delta Q Pipe	Q4	Q4 ABS
1	1	10	3000	−1	0.96	−6.38	6.65				−0.01	−0.97	0.97
	2	16	6000	1	2.04	5.22	2.56				−0.01	2.03	2.03
	3	6	3000	1	0.34	11.25	33.09				0.02	0.31	0.31
	4	6	6000	−1	0.21	−9.23	43.95	159.55	0.86	−0.01	0.02	−0.24	0.24
2	4	6	6000	1	0.21	9.23	43.95				0.02	0.24	0.24
	6	8	2000	1	0.25	1.05	4.19				0.02	0.27	0.27
	7	8	6500	−1	0.25	−3.40	13.60				0.02	−0.23	0.23
	5	8	2500	−1	0.75	−9.96	13.31	138.83	−3.11	0.02	0.02	−0.73	0.73
3	3	6	3000	−1	0.34	−11.25	33.09				0.02	−0.31	0.31
	8	12	6000	1	1.7	15.11	8.89				0.02	1.72	1.72
	9	8	2000	−1	0.3	−1.47	4.89				0.02	−0.28	0.28
	10	8	6500	−1	0.3	−4.76	15.88	116.09	−2.37	0.02	0.02	−0.28	0.28

TRIAL 4

Loop No.	Pipe No.	Pipe Dia. (in.)	Length (ft)	Sign (− OR +)	Q4 (cfs) ABS	H (ft)	(H/Q) ABS	N(Sum(H/Q)) ABS	Sum(H)	Delta Q Loop	Delta Q Pipe	Q5
1	1	10	3000	−1	0.97	−6.50	6.70				0.02	−0.95
	2	16	6000	1	2.03	5.17	2.55				0.02	2.05
	3	6	3000	1	0.31	9.48	30.60				0.02	0.34
	4	6	6000	−1	0.24	−11.81	49.23	164.79	−3.66	0.02	0.02	−0.21
2	4	6	6000	1	0.24	11.81	49.23				0.02	0.21
	6	6	2000	1	0.27	1.21	4.47				−0.00	0.27
	7	8	6500	−1	0.23	−2.91	12.67				−0.00	−0.23
	5	8	2500	−1	0.73	−9.50	13.01	146.84	0.61	−0.00	−0.00	−0.73
3	3	6	3000	−1	0.31	−9.48	30.60				0.02	−0.34
	8	12	6000	1	1.72	15.44	8.96				−0.00	1.72
	9	8	2000	−1	0.28	−1.29	4.61				−0.00	−0.28
	10	8	6500	−1	0.28	−4.19	14.96	109.44	0.47	−0.00	−0.00	−0.28

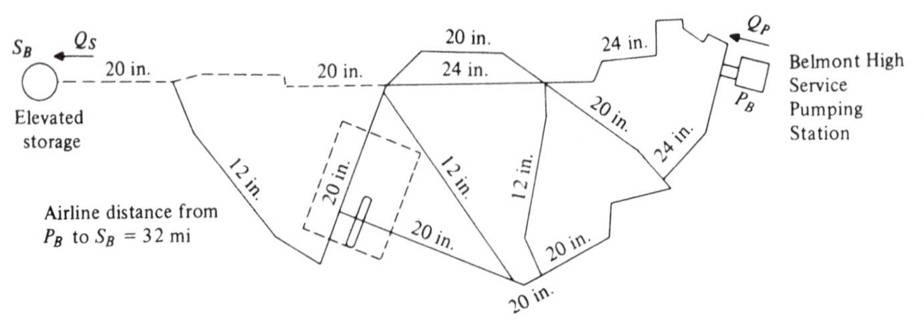

FIGURE 6.16 Arterial pipe network of the Belmont High Service District, Philadelphia.
Source: Courtesy of the Civil Engineering Department, University of Illinois, Urbana.

feeder system has a very large capacity relative to that of the smaller mains, field observations indicate that this type of skeletonizing yields reasonable results. Where no well-defined feeder system is apparent, serious errors may result from skeletonizing. Figure 6.16 illustrates a skeletonized distribution network consisting of arterial mains only. Figure 6.17 shows how a portion of the distribution system of Figure 6.16 (that part lying within the dashed rectangle) looked before skeletonizing. A more complete discussion of such procedures is given by Reh [12]. The analysis of a large network may also be expedited by balancing portions of the system successively instead of analyzing the whole network simultaneously.

Normally, minor losses are neglected in network studies, but they can easily be introduced as equivalent lengths of pipe when it is felt that they should be included. Where C values are determined from field measurements, they invariably include a component due to the various minor losses encountered. McPherson gives a good discussion of local losses in water distribution networks [28].

The construction of pressure contours helps to isolate shortcomings in the hydraulic performance of distribution systems. Contours are often drawn with intervals of 1–5 ft of head loss but may have other intervals depending on local circumstances. For a given set of operating rules applicable to a particular network, the pressure contours indicate the distribution of head loss and are helpful in showing regions where head losses are excessive. Figure 6.18 illustrates contours constructed for a distribution network.

Simultaneous Path Adjustment (SP) Method To improve convergence over the P methods, a method of solution has been developed that simultaneously adjusts the flowrate in each path of pipes represented by an energy equation. The method is as follows [15]:

1. An initial set of flowrates satisfying continuity at each junction node is determined.
2. A flow adjustment factor is simultaneously computed for each loop to satisfy the energy equations and avoid disturbance of the continuity balance.
3. Step 2 is repeated using improved solutions until the flow adjustment factor is within a specified limit.

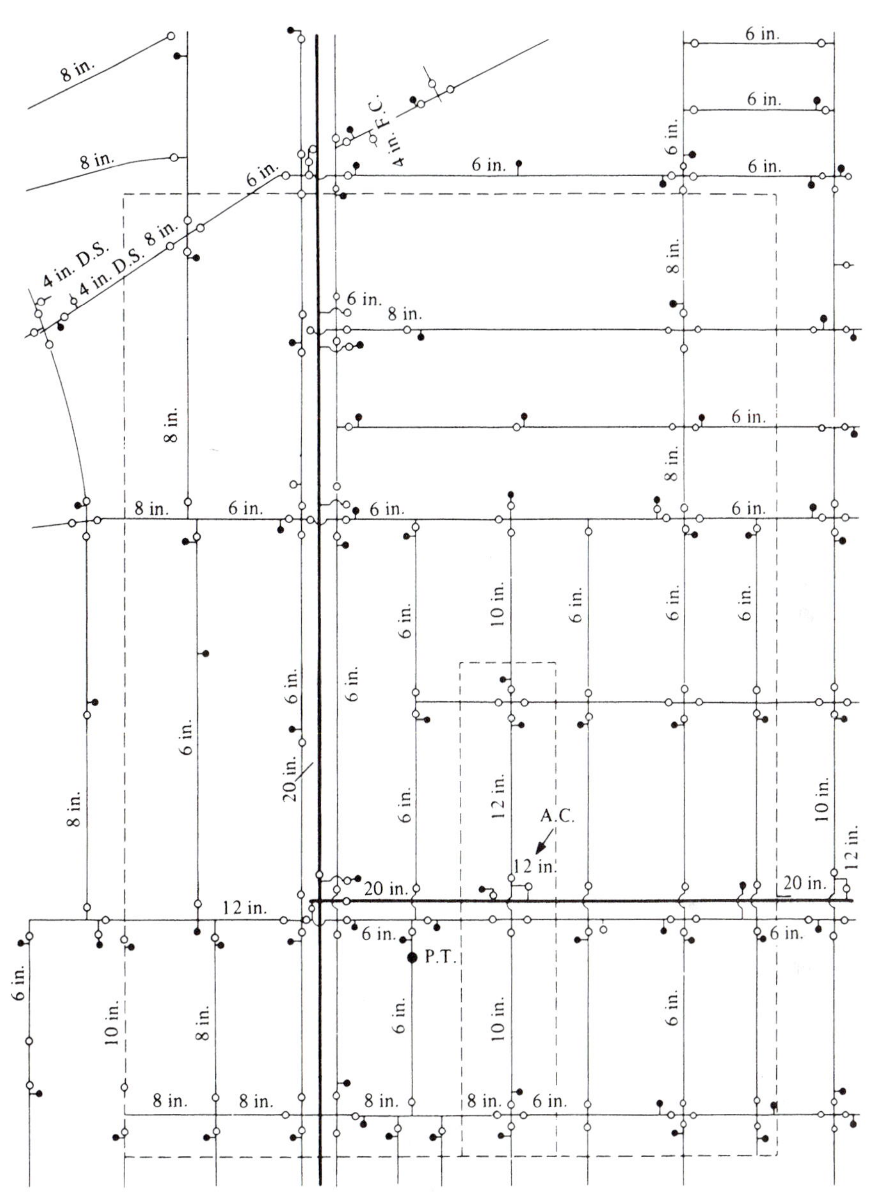

FIGURE 6.17 Intermediate grid sector, Belmont High Service District, Philadelphia. *Source:* Courtesy of the Civil Engineering Department, University of Illinois, Urbana.

The simultaneous solution of $L + F - 1$ equations is required to determine the loop flow adjustment factors. Each equation includes the contribution for a particular loop as well as contributions from all other loops that have pipes in common.

For loop j, the head change required to balance the energy equation is expressed in terms of the flow change in loop j (ΔQ_j) and the flow changes in adjacent loops (ΔQ_k) as follows:

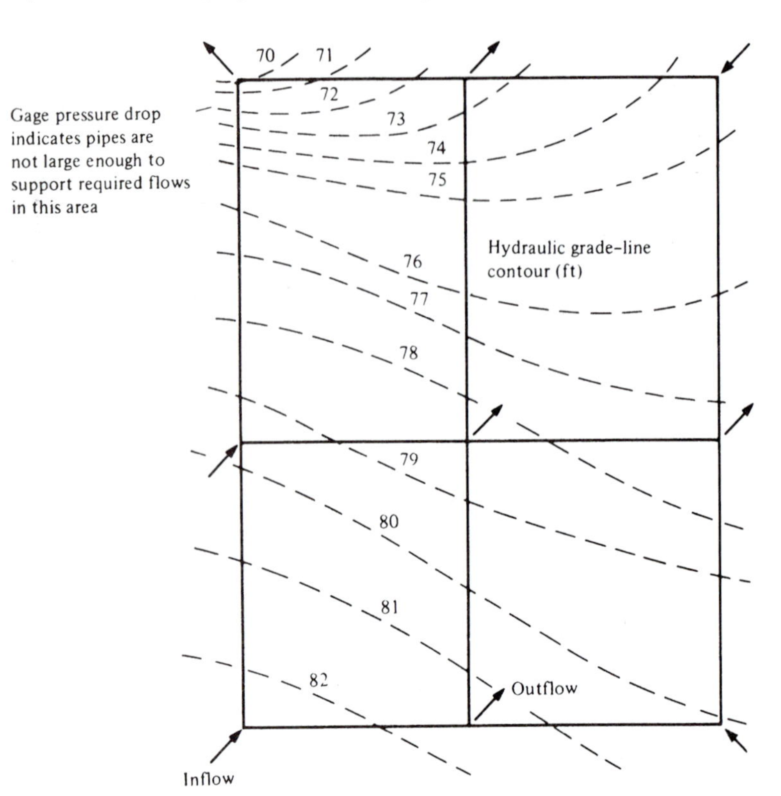

FIGURE 6.18 Pressure contours of a distribution network.

$$f(Q) = f(Q_i) + \frac{\partial f}{\partial Q}\Delta Q_j + \frac{\partial f}{\partial Q}\Delta Q_k$$

or

$$f(Q) = f(Q_i) + f'(Q_i)\Delta Q_j + f'(Q_i)\Delta Q_k \tag{6.47}$$

With the substitution $f(Q) = \Delta E, f(Q_i) = \Sigma H_i$, and $f'(Q) = \Sigma G_i$, Eq (6.47) becomes

$$\Delta E - \sum H_i = \left(\sum G_i\right)\Delta Q_j + \sum(G_i \Delta Q_k) \tag{6.48}$$

in which ΣH_i is the sum of the head changes for all pipes in loop j, $(\Sigma G_i) \times \Delta Q_j$ is the sum of all gradients for the same pipes times the flow change for loop j, and $\Sigma(G_i\Delta Q_k)$ is the sum of the gradients for pipes common to loops j and k multiplied by the flow change for loop k.

A set of simultaneous linear equations is thus formed in terms of flow adjustment factors for each loop representing an energy equation. The solution of these linear equations provides an improved solution for another trial until a specified convergence criterion is met.

Linear (L) Method This procedure involves the solution of the basic hydraulic equations for a pipe network [15]. In the method, the energy equations are linearized using a gradient approximation. This is accomplished in terms of an approximate discharge Q_i as follows:

$$f(Q) = f(Q_i) + f'(Q_i)(Q - Q_i)$$

Introducing the expressions H_i and G_i, defined as before, the foregoing equation becomes

$$\sum G_i Q = \sum (G_i Q_i - H_i) + \Delta E \tag{6.49}$$

This relationship is employed to formulate $L + F - 1$ energy equations, which, together with the J continuity equations, form a set of P simultaneous linear equations in terms of the flow rate in each pipe. A significant advantage of this scheme is that an arbitrary set of initial flow rates, which need not satisfy continuity, can be used to start the iteration. A flow rate based on a mean flow velocity of 4 fps has been used by Wood [15]. Successive trials are carried out until the change in flow rate between successive trials becomes insignificant.

The use of the linear (L) method of analysis is illustrated by an example developed by Wood [15]. The calculations are given for one trial. The system analyzed is shown in Figure 6.19, which includes the necessary data and shows the numbers assigned to the pipe sections and junction nodes. The system includes a globe valve in pipe 7 that imposes a noticeable minor loss. Other minor losses are neglected. A pump with constant-power input is included in pipe 2. The useful horsepower (hp) for this pump is given as 5. Thus, Eq. (6.29) becomes

$$E_P = P(Q) = Z/Q$$

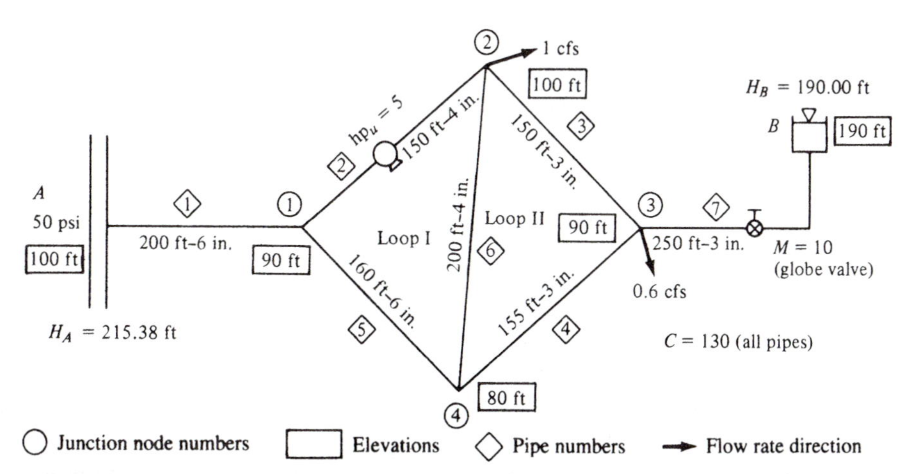

FIGURE 6.19 Wood's sample pipe system: ○, junction node numbers; □, elevations; ◇, pipe numbers; →, flow rate direction. *Source:* After D. T. Wood, "Algorithms for Pipe Network Analysis and Their Reliability." Res. Rep. No. 127, University of Kentucky, Water Resources Research Institute, Lexington, KY, 1981.

and the pump constant $Z = 550\ \text{hp}_u/62.4 = 44.07$ in the example. The pump terms used in the equations for H_i and G_i [Eqs. (6.39) and (6.40)] are

$$P(Q_i) = Z/Q_i \quad \text{and} \quad P'(Q_i) = -Z/Q_i^2$$

The Hazen–Williams equation is employed in the example for head loss calculations. Using that expression, one obtains the pipeline constant:

$$K_P = \frac{4.73L}{C^{1.852}D^{4.87}}$$

where $n = 1.852$, L is the pipe length, and D is the pipe diameter. Formulas for computing the pipeline constant in English and metric units for both the Hazen–Williams and Darcy–Weisbach equations are given in Table 6.6.

Table 6.7 summarizes values of pipeline, minor loss, and pump constants for each pipe in the example system for a C value of 130.

In the example, four mass continuity equations for the four junction nodes and three energy equations are required. The energy equations include one for each of the two loops noted ($\Delta E = 0$) and one additional energy equation for pipes connecting the two fixed-grade nodes ($\Delta E = 25.38$ ft).

An arbitrary set of initial flow rates is defined to start the procedure. A flow rate based on a mean flow velocity of 4 fps is used for this purpose. The initial flow rates and corresponding values for G and H are shown in Table 6.8. A total of four continuity equations and three energy equations are to be solved simultaneously. The continuity equations [Eq. (6.23)] are

$$-Q_1 + Q_2 + Q_5 = 0 \qquad \text{(junction 1)}$$
$$-Q_2 + Q_3 - Q_6 = -1.0 \quad \text{(junction 2)}$$

TABLE 6.6 Formulas for Determining the Pipe Constant K_P

		Units for			
Equation Type	K_P	Q	L	D	H_f
Hazen–Williams	$\frac{4.73L}{C^{1.85}D^{4.87}}$	cfs	ft	ft	ft
	$\frac{10.44L}{C^{1.85}D^{4.87}}$	gpm	ft	in.	ft
	$\frac{10.70L}{C^{1.85}D^{4.87}}$	m^3/s	m	m	m
Darcy–Weisbach	$\frac{fL}{39.70D^5}$	cfs	ft	ft	ft
	$\frac{fL}{32.15D^5}$	gpm	ft	in.	ft
	$\frac{fL}{12.10D^5}$	m^3/s	m	m	m

TABLE 6.7 Pipe System Constants

Pipe No.	K_P	K_M	Z
1	3.36	0	0
2	18.18	0	44.1
3	73.78	0	0
4	76.24	0	0
5	2.69	0	0
6	24.23	0	0
7	122.97	64.44	0

$$-Q_3 - Q_4 + Q_7 = -0.6 \quad \text{(junction 3)}$$

$$Q_4 - Q_5 + Q_6 = 0 \quad \text{(junction 4)}$$

The energy equations [Eq. (6.49)] are derived using the data in Table 6.8. Calculations on the right-hand side of these equations for the two loops and for the path AB between the two fixed-grade nodes are given in Table 6.9. The left-hand side of the equations is the sum of the products of the G_i and Q for each pipe in the loop or path. Notice that a sign convention must be used in accounting. The resulting three energy equations follow:

$$5.072Q_1 + 375.42Q_2 + 34.14Q_3 + 82.204Q_7 = 292.63 \quad \text{(path } AB\text{)}$$

$$375.42Q_2 - 4.057Q_5 - 18.308Q_6 = 250.304 \quad \text{(loop I)}$$

$$34.14Q_3 - 35.278Q_4 + 18.308Q_6 = 2.837 \quad \text{(loop II)}$$

The solution of these equations (in cfs) is $Q_1 = 1.725$, $Q_2 = 0.705$, $Q_3 = 0.262$, $Q_4 = 0.463$, $Q_5 = 1.020$, $Q_6 = 0.557$, and $Q_7 = 0.125$. These are used to formulate a second set of equations (only the energy equations change) and a second solution is obtained. The procedure continues until a specified convergence criterion is met. After five iterations the final flows were found to be $Q_1 = 1.73$, $Q_2 = 1.37$, $Q_3 = 0.37$, $Q_4 = 0.36$, $Q_5 = 0.36$, $Q_6 = 0.001$, and $Q_7 = 0.131$.

TABLE 6.8 Values of Q_i, G_i, and H_i for the Linear Method

Pipe No.	Q_i	G_i	H_i
1	0.7854	5.072	2.151
2	0.3491	375.42	−123.66
3	0.1963	34.14	3.619
4	0.1963	35.278	3.740
5	0.7854	4.057	1.721
6	0.3491	18.308	3.451
7	0.1963	82.204	8.517

TABLE 6.9 Calculations for Energy Equations for the Linear (L) Method

	Pipe No. and Sign	$G_i \times Q_i$	H_i	ΔE for Loop
Loop I	2+	375.42 × 0.3491 = +131.059	+123.66	
	6−	18.308 × 0.3491 = −6.391	+3.451	0
	5−	4.057 × 0.7854 = −3.186	+1.721	
	$\Sigma\, G_iQ = \Sigma(G_iQ_i - H_i) + \Delta E = 121.482 + 128.832 + 0.0 = 250.31$			
Loop II	6+	18.308 × 0.3491 = +6.391	−3.451	
	3+	34.14 × 0.1963 = +6.702	−3.169	0
	4−	35.278 × 0.1963 = −6.925	+3.740	
	$\Sigma\, G_iQ = \Sigma(G_iQ_i - H_i) + \Delta E = 6.168 - 3.330 + 0.0 = 2.838$			
Path *AB*	1+	5.072 × 0.7854 = +3.984	−2.151	
	2+	375.42 × 0.3491 = +131.059	+123.66	
	3+	34.14 × 0.1963 = +6.702	−3.619	25.38
	7+	82.204 × 0.1963 = +16.137	−8.517	
	$\Sigma\, G_iQ = \Sigma(G_iQ_i - H_i) + \Delta E = 157.882 + 109.373 + 25.38 = 292.635$			

Algorithms for Solving Node Equations

The two most widely used node methods are the single-node adjustment (N) method and the simultaneous node adjustment (SN) method [15]. The N method was originally described by Hardy Cross [27]. The procedure is as follows:

1. A reasonable grade is assumed for each junction node in the system. The better the initial assumptions, the fewer the required trials.
2. A grade adjustment factor for each junction node that tends to satisfy continuity is computed.
3. Step 2 is repeated using improved solutions until a specified convergence criterion is met.

The grade adjustment factor is the change in grade at a particular node (ΔH) that will result in satisfying continuity and considering the grades at adjacent nodes fixed. For convenience, the required grade correction is expressed in terms of Q_i, the flow based on the grades at adjacent nodes before adjustment. With the gradient approximation,

$$f(Q) = f(Q_i) + f'(Q_i)\Delta Q$$

and substituting terms defined previously, one derives the flow correction

$$\Delta Q = \sum\left(\frac{1}{G_i}\right)\Delta H \tag{6.50}$$

where $\Delta H = H - H_i$, and the grade adjustment factor and ΔQ represent the flow corrections required to satisfy continuity at the nodes. From Eq. (6.23),

$$\Delta Q = \sum Q_i - Q_e \tag{6.51}$$

Thus, from Eqs. (6.50) and (6.51),

$$\Delta H = \left(\sum Q_i - Q_e\right) \Big/ \sum \frac{1}{G_i} \tag{6.52}$$

In Eq. (6.52), inflow is assumed positive. The numerator represents the unbalanced flow rate at the junction node [see also Eq. (6.46)].

The flow rate Q_i in a pipe section before adjustment is computed from

$$Q_i = (\Delta H_i/K)^{1/n}$$

in which ΔH_i is the grade change based on assumed initial values of grade.

When pumps are located in a pipeline, the following expression can be used to determine Q_i:

$$\Delta H_i = KQ_i^n - P(Q_i) \tag{6.53}$$

Equation (6.53) is solved using an approximation procedure. Adjustment of the grade for each junction node is made following each trial until a selected convergence criterion is satisfied [15]. The SN method is based on simultaneous solution of the basic network node equations. These equations must be linearized in terms of approximate values of grade (head). Details of the procedure are reported in [15].

Newton–Raphson Method The Newton–Raphson method is a widely used numerical method for solving systems of nonlinear equations [1]. The method is applicable to problems that can be expressed in the form $F(H) = 0$, where the solution is that value of H that causes F to become zero. Application of the technique to a simple system where there is only one equation with one unknown illustrates the principle involved. Here, the derivative of F can be approximated as

$$\frac{dF}{dH} = \frac{F(H + \Delta H) - F(H)}{\Delta H} \tag{6.54}$$

With an initial assumption of H, the solution is obtained by determining the value of $H + \Delta H$ that forces F to zero. By setting $F(H + \Delta H)$ to zero, the solution for ΔH becomes

$$\Delta H = \frac{-F(H)}{dF/dH} \tag{6.55}$$

The value of H used in the next step of the iterative process then becomes $H + \Delta H$. Iterations continue until F closely approaches zero.

Analysis of the types of pipe networks encountered in practice usually means dealing with large numbers of equations and unknowns. The Newton–Raphson method can be applied to either the $N - 1$, ΔH equations [Eq. (6.25)] or the ΔQ equations exemplified by Eq. (6.41). For each node, a head equation of the form of Eq. (6.33) is written

$$F(H_b) = \sum_{b=1}^{N}\left[\pm\left(\frac{H_a - H_b}{K_{ab}}\right)^{1/n}\right] - Q = 0 \tag{6.56}$$

where

N = the number of pipes that join at a node

K = the pipeline constant

n = 1.85 for the Hazen–Williams equation

Q = the flow withdrawn at the node

The sign of the term in the summation depends on the direction of flow into or out of the junction. If $F(i)$ is the value of F at iteration i, then it follows that

$$dF = F(i + 1) - F(i) \tag{6.57}$$

The same change can also be approximated as

$$dF = \frac{\partial F}{\partial H_1}\Delta H_1 + \frac{\partial F}{\partial H_2}\Delta H_2 + \cdots + \frac{\partial F}{\partial H_k}\Delta H_k \tag{6.58}$$

where ΔH is the change in H in the iteration from i to $i + 1$. The problem is one of iteratively determining the values of ΔH so that the end result is that $F(i + 1)$ becomes zero. The process involves setting Eq. (6.57) equal to Eq. (6.58). A system of k linear equations with k unknowns of the form ΔH results. These equations can be solved by various linear methods [1].

The solution is obtained by selecting initial values of H, calculating the partial derivatives of each F with respect to H, and solving the set of linear equations to find a new value of H. The process is repeated until all of the calculated F values are sufficiently close to zero. Note that the derivative of the terms in Eq. (6.56) is of the form

$$\frac{dF}{dH} = \pm\frac{(H_a - H_b)^{(1/n-1)}}{n(K_{ab})^{1/n}} \tag{6.59}$$

The following example illustrates the Newton–Raphson procedure.

Example 6.5

Given the simple pipe network of Figure 6.20, find the value of H_1 if $C = 100$ and $n = 1.85$.

Solution: Equations (6.56) and (6.59) are applied along with

$$H(i + 1) = H(i) - \frac{F}{dF/dH} \tag{6.60}$$

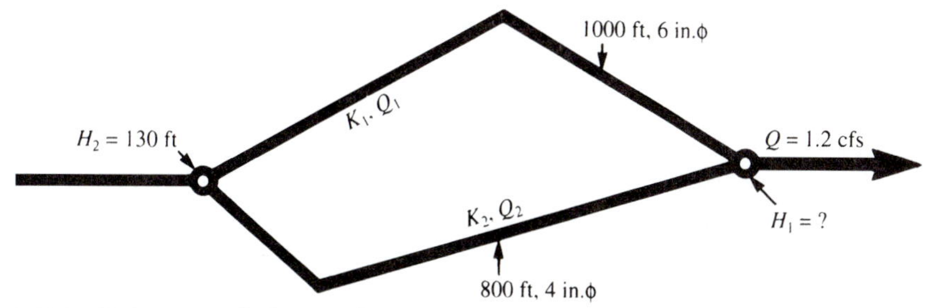

FIGURE 6.20 Network diagram for Example 6.5.

In this case, there are only two pipes, and only one equation must be solved for F, dF/dH, and $H(i + 1)$ at each iteraction. Based on an initial assumption of 100 for H_1, the calculated values of F, dF/dH, and $H(i + 1)$ are displayed in Table 6.10. The first set of calculations follows.

1. Values of K_1 and K_2 are calculated by using the formula in Table 6.6 for the Hazen–Williams equation, where Q is in cfs and L and d are in ft:

$$K_1 = (4.73 \times 1000)/[(100)^{1.85}(0.5)^{4.87}]$$
$$= 275.95$$

In like manner, by substituting the appropriate values and solving, K_2 is found to be 167.78.

2. Assuming that $H_1 = 100$ and using Eq. (6.56),

$$F(H)_1 = \left(\frac{130 - 100}{275.95}\right)^{0.54} + \left(\frac{130 - 100}{167.78}\right)^{0.54} - 1.2$$
$$= -0.504$$

3. Using Eq. (6.59),

$$\left(\frac{dF}{dH}\right)_1 = \frac{-1}{1.85(275.95)^{0.54}(130 - 100)^{0.46}} + \frac{-1}{1.85(167.78)^{0.54}(130 - 100)^{0.46}}$$
$$= -0.0125$$

TABLE 6.10 Calculated Values of H, F, and dF/dH for Example 6.5

Iteration No.	H	F	dF/dH
1.	100.00	−0.504	−0.0125
2.	59.68	−0.097	−0.0084
3.	48.13	−0.002	−0.0078
4.	47.87		

4. The new value of H is found from Eq. (6.60):

$$H_{i+1} = 100 - (-0.504/ - 0.0125)$$
$$= 59.68$$

5. The procedure is repeated, and the solutions are recorded in Table 6.7
6. Noting that the sum of the Qs must equal 1.2 cfs, the equations for Q in pipes 1 and 2 are solved by using the head loss calculated to see if their sum is correct. From Eq. (6.31),

$$Q_1 = [(130 - 47.87)/275.95]^{0.54} = 0.52 \text{ cfs}$$
$$Q_2 = [(130 - 47.87)/167.78]^{0.54} = 0.68 \text{ cfs}$$
$$Q_1 + Q_2 = 0.52 + 0.68 = 1.2 \text{ cfs},$$

which checks.

7. If the check had shown that the sum of the Qs did not equal 1.2 very closely, additional iterations would be required.

6.9 NETWORK MODELING SOFTWARE

Network modeling has come a long way, and software to accommodate it is widely available [22–26,29–32]. Graphic and interactive graphics packages support the design and analysis of water distribution systems [6]. Network models are also used to optimize pipe sizes in a network, thus supporting cost-effective designs [30]. They are also used to analyze water quality issues, fire flow requirements, and to train water systems operators.

Network modeling packages vary in how data are input to them and how results are displayed, but they all employ algorithms based on flow rates and/or head loss. Conservation of mass and conservation of energy are the governing equations. The linear or Newton–Raphson methods previously discussed are commonly used in network models to carry out the analyses. A number of proprietary and public domain network models are available and widely used. One of these will be described in the following section.

Distribution System Analysis Using EPANET

EPANET 2.0 is a Windows 95/98/NT program that performs extended period simulation of hydraulic and water quality behavior within pressurized pipe networks. EPANET is public domain software that can be freely copied and used. The program and manual are available at the Web address www.epa.gov/ORD/NRMRL/wswrd/epanet.html [5]. Components of the network include pipes, nodes (pipe junctions), pumps, valves, storage tanks, and reservoirs. The program tracks the flow of water in each pipe, the pressure at each node, the height of water in each tank or reservoir, and the concentration of a chemical constituent during a simulation period over multiple time steps. Water age and source tracing can also be simulated.

The program incorporates an extended period hydraulic analysis that can (1) handle systems of any size; (2) compute friction head loss using the Hazen–Williams,

Darcy–Weisbach, or Chezy–Manning formulas; (3) include minor head losses for bends, fittings, etc.; (4) model constant or variable speed pumps; (5) compute pumping energy and cost; (6) model various types of valves including shutoff, check, pressure regulating, and flow control valves; (7) allow storage tanks to have any shape (diameter can vary with height); (8) consider multiple demand categories at nodes, each with its own pattern of time variation; (9) model pressure-dependent flow issuing from emitters (sprinkler heads); and (10) base system operation on simple tank level or timer controls as well as on complex rule-based controls.

The EPANET water quality analyzer can (1) model the movement of nonreactive tracers through the network over time; (2) model the movement and fate of a reactive material as it grows or decays; (3) model the age of water through the network; (4) track the percent of flow from a given node reaching all other nodes over time; (5) model reactions both in the bulk flow and in the pipe wall; (6) allow growth or decay reactions to proceed up to a limiting concentration; (7) use global reaction rate coefficients that that can be modified on a pipe-by-pipe basis; (8) allow for time-varying mass or concentration inputs at any location in the network; and (9) model storage tanks as being either complete mix, plug flow, or two-compartment reactors.

Hydraulic Simulation Model

The hydraulic model used by EPANET is an extended period hydraulic simulator that solves the flow continuity and head loss equations that characterize the hydraulic state of the pipe network at a given point in time. It is a hybrid node-loop (*gradient method*) approach. The procedure developed by Todini and Pilati is used [29].

The hydraulic model solves the following equations for each storage node s (tank or reservoir) in the system:

$$\partial y_s/\partial t = q_s/\text{As} \tag{6.61}$$

$$q_s = \Sigma_i q_{is} - \Sigma_j q_{sj} \tag{6.62}$$

$$h_s = E_s + ys \tag{6.63}$$

along with the following equations for each link (between nodes I and j) and each node k:

$$h_i - h_j = f(q_{ij}) \tag{6.64}$$

$$\Sigma_i q_{ik} - \Sigma_j q_{kj} - Q_k = 0 \tag{6.65}$$

where the unknown quantities are y_s = height of water stored at node s, ft; q_s = flow into storage node s, cfs; q_{ij} = flow in link connecting nodes I and j, cfs; h_i = hydraulic grade line elevation at node I (equal to elevation head plus pressure head), ft; and the known constants are A_s = cross − sectional area of storage node s (taken as infinite for reservoirs), ft^2; E_s = elevation of node s, ft; Q_k = flow consumed (+) or supplied (−) at node k, cfs; $f(q_{ij})$ = functional relation between head loss and flow in a link.

Equation (6.61) expresses conservation of water volume at a storage node, while Eqs. (6.62) and (6.65) do the same for pipe junctions. Equation (6.64) represents the energy loss or gain due to flow within a link. For known initial storage node levels y_s at time zero, Eqs. (6.64) and (6.65) are solved for all flows qij and heads h_i using Eq. (6.63) as a boundary condition. This step is called "hydraulically balancing" the network, and is accomplished by using an iterative technique to solve the nonlinear equations involved.

The gradient algorithm method used by EPANET to solve these equations has several appealing features. The system of linear equations to be solved at each iteration is sparse, symmetric, and positive-definite. This permits very efficient sparse matrix techniques to be used for their solution [5,29]. The gradient method also maintains flow continuity at every node after the first iteration. In addition, the procedure can handle pumps and valves without having to change the structure of the equation matrix when the status of these components changes. For complete details see Appendix D in Reference 5.

Once a network hydraulic solution is obtained, flow into (or out of) each storage node q_s is found from Eq. (6.62) and used in Eq. (6.61) to find new storage node elevations after a time step Δt. This process is repeated for all subsequent time steps for the remainder of the simulation period.

The usual hydraulic time step used in EPANET is 1 hr, but it can be shortened if greater accuracy is required. Shorter time steps than normal can occur automatically whenever pipe or pump controls are activated (e.g., a tank fills to the level that causes a pump to shut off), or when a tank becomes either empty or full (causing the tank outlet/inlet line to be closed).

Water Quality Simulation Model

EPANET's dynamic water quality simulator tracks the fate of a dissolved substance flowing through the network over time. It uses the flows from the hydraulic simulation to solve a conservation of mass equation for the substance within each link. Additional information on this feature may be had from [5].

Data Assembly

Before applying EPANET, the following data assembly steps are recommended [5]:

1. Identify the network's components and their connections. The components consist of pipes, pumps, valves, storage tanks, and reservoirs. Nodes are junctions where network components connect. Tanks and reservoirs are considered nodes. The component connecting any two nodes is termed a link and may be a pipeline, pump, or valve.
2. Assign unique ID numbers to all nodes and ID numbers to each link. The same ID number may be used for both a node and a link.
3. Collect information on the following system parameters: diameter, length, roughness, and minor loss coefficient for each pipe; the characteristic curve for each pump; the diameter, minor loss coefficient, and pressure or flow setting for each control valve; diameter and lower and upper water levels for each tank; control rules that determine how pump, valve, and pipe settings change with time, tank water levels, or nodal pressures; changes in water demands for each node over

the time period being simulated; and initial water quality at all nodes and changes in water quality over time at source nodes.

Application of EPANET is demonstrated by the following examples. It is suggested that the student solve the network problems at the end of the chapter using EPANET or another water distribution network package. The software and manual can be downloaded from EPANET's Web site.

Example 6.6

Using EPANET 2.0, find the pipe flows and the pressures at the nodes of the network shown in Figure 6.21. Input data are given in Tables 6.11 and 6.12. The flows are in gpm and the heads are in ft. The tank elevation is 1100 ft, the minimum water level in the tank is zero, the initial water level is 6 ft and the maximum water level is 22 ft. The tank diameter is 55 ft. The pump can deliver a flow of 1200 gpm at a head of 280 ft. The total head at the reservoir is 950 ft. Output is to be determined at two-hour increments for a period of 72 hours. Assume that the demands for the 12 two-hour periods in each day are determined by multiplying the demands in Table 6.11 by the following multipliers— 0.2, 0.4, 0.6, 0.8, 1.1, 1.3, 1.6, 1.8, 1.5, 1.4, 1.3, and 1.2, consecutively. This cycle is assumed to be repeated for all days in the total time period. Provide output tables for hours 12 and 18. Provide a graph of demand at node 4 for the entire time period.

Solution: Using the identification numbers for the reservoir, tank, pump, pipes, and junctions displayed in Figure 6.21, and the data assembled in Tables 6.11 and 6.12, entries are made in the EPANET 2.0 property editors. Figure 6.22 shows examples for the reservoir, tank, and junction 5 and pipe 5. Once this has been accomplished, head and flow values are entered in the curve editor and a pump curve is developed for the pump, Figure 6.23. Using two-hour hydraulic time steps and a total duration of 72 hours, the multipliers provided are then entered in a pattern editor, Figure 6.24.

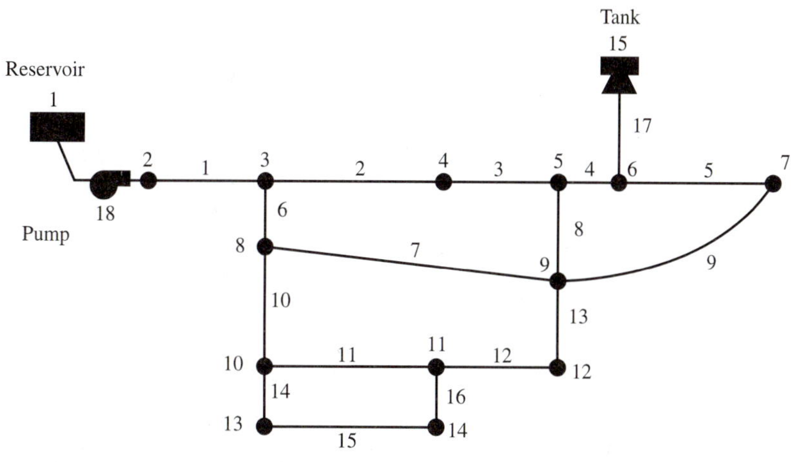

FIGURE 6.21 Pipe network for Example 6.6.

TABLE 6.11 Node Input Data for Example 6.6

Node #	Elevation ft.	Demand gpm
reservoir - 1	950	0
2	916	0
3	900	140
4	913	100
5	922	135
6	902	0
7	918	115
8	903	108
9	920	95
10	910	117
11	900	120
12	908	115
13	900	120
14	850	110
tank - 15	1100	0

This completes the data entry and the problem can be run. Figure 6.25 shows the output for one time period with arrows showing the direction of the flows. Tables 6.13 and 6.14 show the nodal values of demand, head and the pressure in psi for each node at 12:00 hours and 18:00 hours. Tables 6.15 and 6.16 show the pipe values of flow, velocity, unit head loss, and friction factor for each pipe at 12:00 hours and 18:00 hours. Figure 6.26 is a graph of the variation of demand over time for node 4.

TABLE 6.12 Pipe Input Data for Example 6.6

Pipe #	Length, ft	Diameter, in.	Roughness, C
1	2500	20	100
2	3750	20	100
3	2500	20	100
4	1000	20	100
5	3000	16	100
6	1250	14	100
7	6050	12	100
8	1500	12	100
9	6000	14	100
10	2500	12	100
11	3750	12	100
12	2550	14	100
13	1500	14	100
14	1250	14	100
15	3750	14	100
16	1250	10	100
17	1250	18	100

Reservoir 1	
Property	Value
*Reservoir ID	1
X-Coordinate	-3563.47
Y-Coordinate	17282.85
Description	
Tag	
*Total Head	950
Head Pattern	
Initial Quality	

(a)

Tank 15	
Property	Value
Tag	
*Elevation	1100
*Initial Level	6
*Minimum Level	0
*Maximum Level	22
*Diameter	55
Minimum Volume	
Volume Curve	

(b)

Junction 5	
Property	Value
Tag	
*Elevation	922
Base Demand	135
Demand Pattern	
Demand Categories	1
Emitter Coeff.	
Initial Quality	
Source Quality	

(c)

Pipe 5	
Property	Value
Tag	
*Length	3000
*Diameter	16
*Roughness	100
Loss Coeff.	0
Initial Status	Open
Bulk Coeff.	
Wall Coeff.	

(d)

FIGURE 6.22 Examples of EPANET 2.0 data entries for Example 6.6. (a) Reservoir. (b) Tank. (c) Junctions. (d) Pipes.

Example 6.7

Using EPANET 2.0, find the pipe flows and the pressures at the nodes of the network shown in Figure 6.27. This problem (6.4) was solved previously using spreadsheets. It is solved again using EPANET to show the ease of use and speed of the hydraulic network simulator.

Solution: Using the identification numbers for the pipes and junctions displayed in Figure 6.27, and the data assembled in Tables 6.17 and 6.18, entries are made in the EPANET 2.0 property editors. To facilitate solution by EPANET 2.0, a reservoir is assumed to exist at node one and a data entry under reservoir is made in the appropriate property editor. Only a snapshot solution is required in this case.

This completes the data entry and the problem can be run. Figure 6.28 shows the output for the single time period with arrows showing the direction of the flows. Table 6.19 shows the nodal values of demand, head and the pressure in psi for each

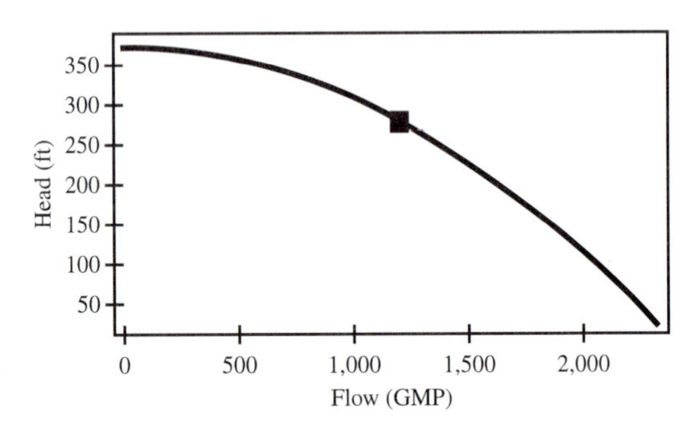

FIGURE 6.23 EPANET 2.0 generated pump curve for Example 6.6.

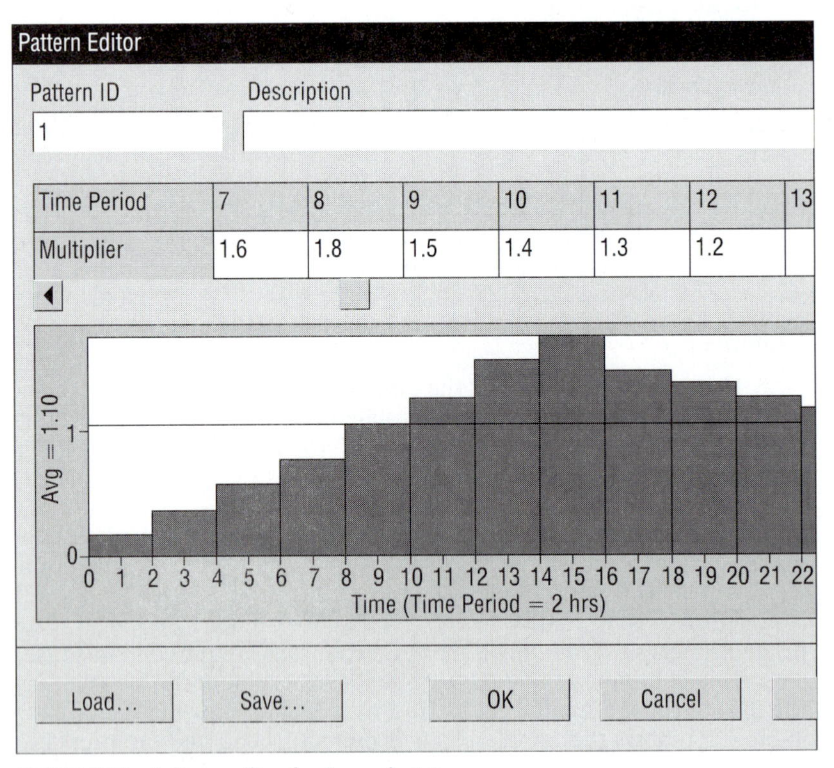

FIGURE 6.24 Pattern editor for Example 6.6.

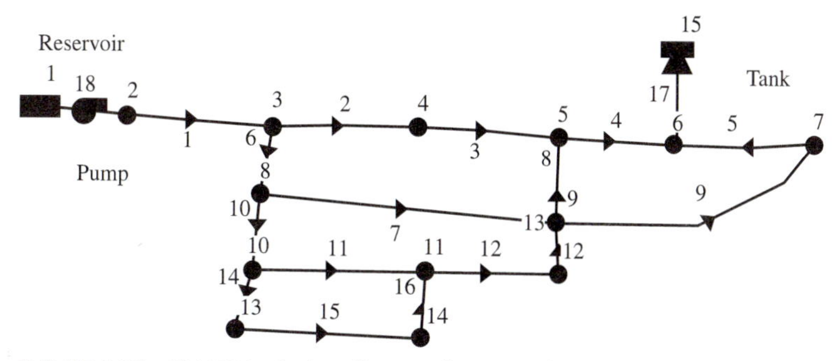

FIGURE 6.25 EPANET solution diagram for Example 6.6 showing flow directions.

TABLE 6.13 Node Values at 12:00 Hours

Node ID	Demand, gpm	Head, ft	Pressure, psi	Quality
Junc 2	0.00	1125.51	90.78	0.00
Junc 3	224.00	1123.10	96.67	0.00
Junc 4	160.00	1122.29	90.68	0.00
Junc 5	216.00	1121.93	86.63	0.00
Junc 6	0.00	1121.93	95.29	0.00
Junc 7	184.00	1121.48	88.17	0.00
Junc 8	172.80	1121.72	94.77	0.00
Junc 9	152.00	1121.05	87.12	0.00
Junc 10	187.20	1119.61	90.82	0.00
Junc 11	192.00	1119.61	95.16	0.00
Junc 12	184.00	1120.22	91.96	0.00
Junc 13	192.00	1119.42	95.07	0.00
Junc 14	176.00	1119.38	116.72	0.00
Resvr 1	−1747.04	950.00	0.00	0.00
Tank 15	−292.96	1122.00	9.53	0.00

TABLE 6.14 Node Values at 18:00 Hours

Node ID	Demand, gpm	Head, ft	Pressure, psi	Quality
Junc 2	0.00	1119.47	88.17	0.00
Junc 3	196.00	1117.00	94.03	0.00
Junc 4	140.00	1115.99	87.96	0.00
Junc 5	189.00	1115.51	83.85	0.00
Junc 6	0.00	1115.48	92.50	0.00
Junc 7	161.00	1115.18	85.44	0.00
Junc 8	151.20	1115.75	92.18	0.00
Junc 9	133.00	1114.94	84.47	0.00
Junc 10	163.80	1113.93	88.36	0.00
Junc 11	168.00	1113.93	92.70	0.00
Junc 12	161.00	1114.35	89.41	0.00
Junc 13	168.00	1113.78	92.63	0.00
Junc 14	154.00	1113.75	114.28	0.00
Resvr 1	−1773.48	950.00	0.00	0.00
Tank 15	−11.52	1115.48	6.71	0.00

TABLE 6.15 Pipe Values at 12:00 Hours

Link ID	Flow, gpm	Velocity, fps	Unit Headloss, ft/Kft	Friction Factor
Pipe 1	1747.04	1.78	0.96	0.032
Pipe 2	784.27	0.80	0.22	0.037
Pipe 3	624.27	0.64	0.14	0.038
Pipe 4	59.88	0.06	0.00	0.056
Pipe 5	352.84	0.56	0.15	0.040
Pipe 6	738.77	1.54	1.11	0.035
Pipe 7	141.32	0.40	0.11	0.044
Pipe 8	−348.39	0.99	0.58	0.039
Pipe 9	−168.84	0.35	0.07	0.044
Pipe 10	424.65	1.20	0.84	0.037
Pipe 11	−14.05	0.04	0.00	0.062
Pipe 12	−322.55	0.67	0.24	0.040
Pipe 13	−506.55	1.06	0.55	0.037
Pipe 14	251.50	0.52	0.15	0.041
Pipe 15	59.50	0.12	0.01	0.051
Pipe 16	−116.50	0.48	0.19	0.044
Pipe 17	−292.96	0.37	0.06	0.042
Pump 18	1747.04	0.00	−175.51	0.000

TABLE 6.16 Pipe Values at 18:00 Hours

Link ID	Flow, gpm	Velocity, fps	Unit Headloss, ft/Kft	Friction Factor
Pipe 1	1773.48	1.81	0.99	0.032
Pipe 2	877.60	0.90	0.27	0.036
Pipe 3	737.60	0.75	0.19	0.037
Pipe 4	272.77	0.28	0.03	0.043
Pipe 5	284.29	0.45	0.10	0.041
Pipe 6	699.89	1.46	1.00	0.035
Pipe 7	157.01	0.45	0.13	0.043
Pipe 8	−275.83	0.78	0.38	0.040
Pipe 9	−123.29	0.26	0.04	0.046
Pipe 10	391.67	1.11	0.73	0.038
Pipe 11	6.75	0.02	0.00	0.069
Pipe 12	−262.13	0.55	0.16	0.041
Pipe 13	−423.13	0.88	0.40	0.038
Pipe 14	221.12	0.46	0.12	0.042
Pipe 15	53.12	0.11	0.01	0.052
Pipe 16	−100.88	0.41	0.14	0.045
Pipe 17	−11.52	0.01	0.00	0.045
Pump 18	1773.48	0.00	−169.47	0.000

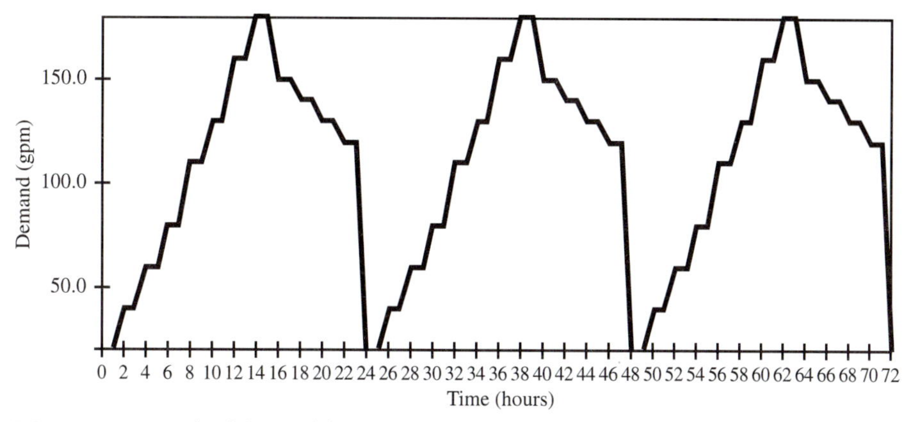

FIGURE 6.26 Graph of demand for node four over the 72 hour simulation for Example 6.6.

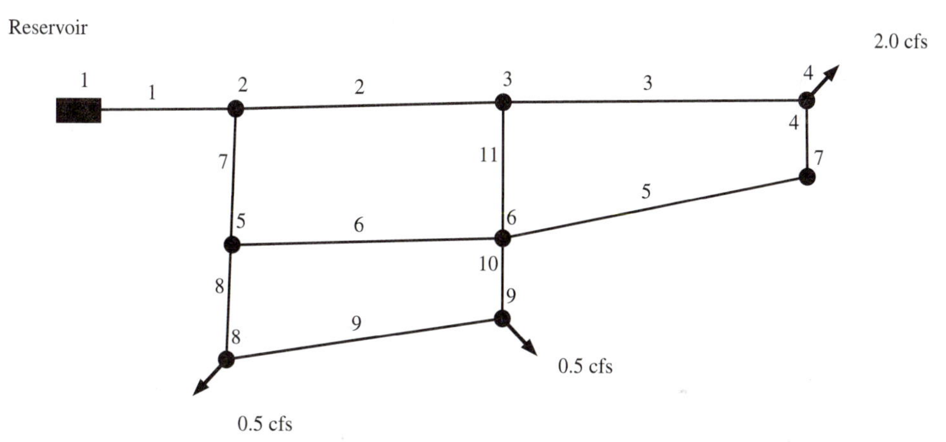

FIGURE 6.27 Pipe network for Example 6.7.

TABLE 6.17 Node Input Data for Example 6.7

Node #	Elevation, ft	Demand, cfs
reservoir - 1	250	0
2	0	0
3	0	0
4	0	0
5	0	0
6	0	0
7	0	0
8	0	0.5
9	0	0.5

TABLE 6.18 Pipe Input Data for Example 6.7

Pipe #	Length, ft	Diameter, in	Roughness, C
1	1000	20	100
2	6000	16	100
3	6000	12	100
4	2000	8	100
5	6500	8	100
6	6000	6	100
7	3000	10	100
8	2500	8	100
9	6500	8	100
10	2000	8	100
11	3000	6	100

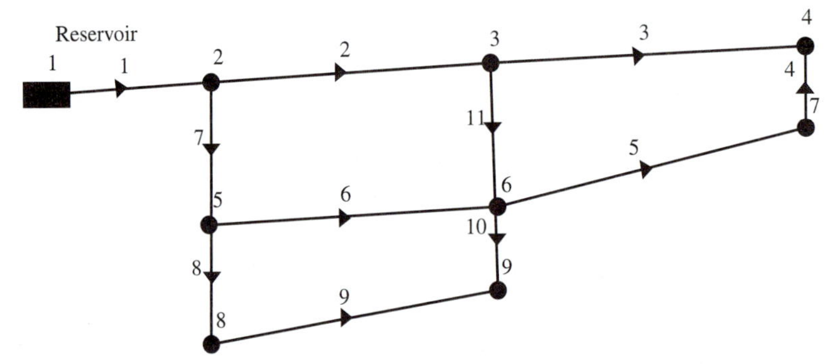

FIGURE 6.28 EPANET 2.0 solution diagram for Example 6.7 showing flow directions.

TABLE 6.19 Node Values for Example 6.7

Node ID	Demand, cfs	Head, ft	Pressure, psi	Quality
Junc 2	0.00	249.41	108.07	0.00
Junc 3	0.00	244.11	105.77	0.00
Junc 4	2.00	228.61	99.05	0.00
Junc 5	0.00	243.41	105.47	0.00
Junc 6	0.00	233.31	101.09	0.00
Junc 7	0.00	229.78	99.56	0.00
Junc 8	0.50	234.44	101.58	0.00
Junc 9	0.50	231.97	100.51	0.00
Resvr 1	−3.00	250.00	0.00	0.00

TABLE 6.20 Pipe Values for Example 6.7

Link ID	Flow, cfs	Velocity, fps	Unit Headloss, ft/Kft	Friction Factor
Pipe 1	3.00	1.38	0.59	0.034
Pipe 2	2.07	1.48	0.88	0.035
Pipe 3	1.73	2.21	2.58	0.034
Pipe 4	−0.27	0.77	0.59	0.043
Pipe 5	−0.27	0.77	0.59	0.043
Pipe 6	−0.22	1.13	1.68	0.042
Pipe 7	−0.93	1.71	2.00	0.037
Pipe 8	0.71	2.04	3.59	0.037
Pipe 9	0.21	0.61	0.38	0.044
Pipe 10	−0.29	0.83	0.67	0.042
Pipe 11	−0.33	1.70	3.60	0.040

node. Table 6.20 shows the pipe values of flow, velocity, unit head loss, and friction factor for each pipe. Note how the output compares with that of Example 6.4.

Selection of a Method of Network Analysis

In decisions regarding the selection of an appropriate analytical tool, the analyst must answer the question of what is expected of the design or analysis. The network methods described herein are all equipped to accommodate many features of pipe systems. But the methods are not all equal in their breadth, and they are not all equal in their ability to converge on a solution. Accordingly, the analyst should know something about the virtues and deficiencies of these tools [16].

Node equations are easy to formulate because they include only contributions from adjacent nodes. Loop equations require the identification of an appropriate set of energy equations, including terms for all pipes in the primary loops, and for paths between fixed-grade nodes. Formulating this set of equations is considerably more difficult than that of formulating the node equations.

The procedures described herein are iterative. Computation continues until a specified convergence criterion is met. The solutions are therefore approximate, although they can be very accurate. The ability of an algorithm to produce an acceptable solution is important, and a knowledge of the convergence problems associated with it must be understood.

A solution is considered satisfactory when all the basic equations are satisfied to a high degree of accuracy. Continuity is always exactly satisfied when loop equations are used. The loop algorithms proceed to satisfy the energy equations iteratively, and the degree to which heads are unbalanced for the energy equations is evidence of solution accuracy. For methods based on node equations, iterations are carried out to satisfy continuity at junction nodes and the unbalance in continuity is the indicator of solution accuracy.

Studies by Wood, using an extensive database, have shown that the P, N, and SN methods may exhibit significant convergence problems [15]. Thus, care should be exercised when using these tools.

Failures of the SN method are characterized by difficulty in meeting a reasonable convergence criterion. When this occurs in a limited number of trials, further trials are usually of no benefit. The failure rate for this method is high and the use of results obtained employing this method is not recommended unless accuracy is obtained in a small number of trials [15]. It has also been found that algorithms based on the node equations (N and SN methods) fail to provide reliable results where low-resistance lines are encountered. This is attributable to the fact that solution algorithms for these equations do not incorporate an exact continuity balance.

For most network methods, failure rates are reduced when assumed initial values are close to the final values. However, there is no way to assume that this condition will be met, and even when it is, an excellent set of initial conditions does not guarantee convergence.

Both the SP and L methods have been found to provide excellent convergence, and the attainment of a reasonable convergence criterion is sufficient to assure accuracy. Convergence failure for these methods is very rare, but since a gradient method is used to handle the nonlinear terms, there is always the possibility of convergence problems. Poor pump descriptions pose special problems, for example. The L method appears to have some advantages over the SP method. Assumed arbitrary flow rates need not satisfy continuity as the continuity conditions are already incorporated into the basic set of equations. This method also allows a more straightforward and reliable inclusion of hydraulic components such as check valves, closed lines, and pressure-regulating valves. Although the SP method has significantly fewer equations to solve, the use of sparse matrix techniques to handle the larger matrix generated by the L method has somewhat negated this advantage.

Wood has concluded that, if possible, either the SP or L method should be employed for pipe network analysis and that convergence is virtually assured if reasonable data are employed. Of the two methods, the L method appears to have slightly better convergence characteristics [15].

Example 6.8

Development of an 1800-acre parcel of land is underway. Figure 6.29 and Table 6.21 indicate the subdivision of the parcel into 3 sectors: (A) an area planned for commercial

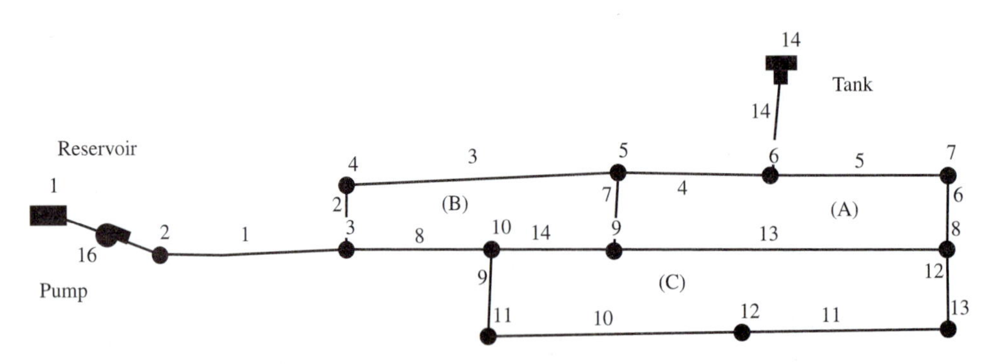

(A) Commercial Area, (B) MultiFamily Residential Area, (C) Single-Family Residential Area

FIGURE 6.29 Feeder network for Example 6.8.

TABLE 6.21 Data for Example 6.8

Land Use Type	AD Water Use	Units Per Acre	AD Water Use, gpm/Acre	Total Acreage	AD Demand, gpm	MD/AD Demand	MD Demand, gpm
Commercial	1.9 gpm/acre	n/a	1.9	300	570	1.5	855
Multifamily Residential	0.2 gpm/unit	10	2	600	1200	2.5	3000
Single-Family Residential	0.3 gpm/unit	4	1.2	900	1080	5	5400

AD = average day, MD = maximum day
Multifamily Residential = 10 units per acre
Single-Family Residential = 4 units per acre

development, (B) an area planned for multifamily residences, and (C) an area planned for single-family residences. Table 6.21 shows the estimated average day and maximum day water requirements for the three sectors. The figure shows the main feeder network that will provide flows to the internal network that will be connected to homes and businesses. Tables 6.22 and 6.23 give the node and pipe data needed to analyze the feeder network (note that the demands in Table 6.22 are for the maximum day). Using EPANET 2.0, find the pipe flows and nodal pressures.

TABLE 6.22 Node Input Data for Example 6.8

Node No.	Elevation, ft	Demand, gpm
Reservoir—1	1000	0
2	910	0
3	903	1000
4	918	1000
5	922	400
6	940	0
7	937	455
8	950	1100
9	933	1000
10	920	1000
11	938	1200
12	945	1200
13	943	1000
Tank—14	1150	0

Solution: Using the identification numbers for the pipes and junctions displayed in Figure 6.29, and the data assembled in Tables 6.22 and 6.23, entries are made in the EPANET 2.0 property editors.

Once the data entry is complete, the network model can be run. Figure 6.30 shows the direction of the flows for the maximum day demands. Table 6.24 shows the nodal values of demand, head and the pressure in psi for each node. Table 6.25 shows

TABLE 6.23 Pipe Input Data for Example 6.8

Pipe	Length, ft	Diameter, in.	Roughness, *C*
1	2000	32	120
2	3000	20	120
3	4800	18	120
4	2900	16	120
5	3100	16	120
6	3000	14	120
7	3000	12	120
8	2900	30	120
9	3100	18	120
10	3800	18	120
11	3700	16	120
12	3100	16	120
13	4500	16	120
14	3000	16	120
15	1100	20	120

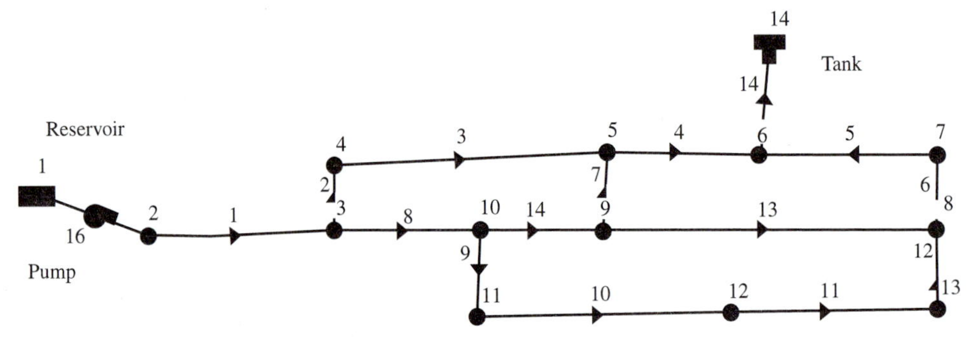

FIGURE 6.30 EPANET 2.0 solution diagram for Example 6.8 showing flow directions.

TABLE 6.24 EPANET 2.0 Node Values for Example 6.8

Node ID	Demand, gpm	Head, ft	Pressure, psi	Quality
Junc 2	0.00	1207.56	128.93	0.00
Junc 3	1000.00	1201.90	129.51	0.00
Junc 4	1000.00	1192.52	118.95	0.00
Junc 5	400.00	1177.90	110.88	0.00
Junc 6	0.00	1162.84	96.56	0.00
Junc 7	455.00	1163.94	98.33	0.00
Junc 8	1100.00	1169.19	94.98	0.00
Junc 9	1000.00	1179.07	106.62	0.00
Junc 10	1000.00	1197.31	120.16	0.00
Junc 11	1200.00	1182.22	105.82	0.00
Junc 12	1200.00	1174.72	99.54	0.00
Junc 13	1000.00	1169.63	98.20	0.00
Resvr 1	−12,931.16	1000.00	0.00	0.00
Tank 14	3576.15	1160.00	4.33	0.00

TABLE 6.25 EPANET 2.0 Pipe Values for Example 6.8

Link ID	Flow, gpm	Velocity, fps	Unit Head Loss, ft/Kft	Friction Factor
Pipe 1	12,931.16	5.16	2.83	0.018
Pipe 2	3961.52	4.05	3.13	0.020
Pipe 3	2961.52	3.73	3.05	0.021
Pipe 4	2897.15	4.62	5.19	0.021
Pipe 5	−679.01	1.08	0.35	0.026
Pipe 6	−1134.01	2.36	1.75	0.024
Pipe 7	−335.62	0.95	0.39	0.028
Pipe 8	7969.63	3.62	1.58	0.019
Pipe 9	3814.28	4.81	4.87	0.020
Pipe 10	2614.29	3.30	2.42	0.022
Pipe 11	1414.29	2.26	1.38	0.023
Pipe 12	414.29	0.66	0.14	0.028
Pipe 13	−1819.72	2.90	2.19	0.022
Pipe 14	−3155.35	5.03	6.08	0.021
Pipe 15	3576.15	3.65	2.59	0.021
Pump 16	12,931.16	0.00	−207.56	0.000

the pipe values of flow, velocity, unit head loss, and friction factor for each pipe. Note that pressures and velocities are generally within prescribed limits and are considered satisfactory for the feeder network.

6.10 DISTRIBUTION RESERVOIRS AND SERVICE STORAGE

Distribution reservoirs provide service storage to meet fluctuating demands often imposed on distribution systems, to accommodate firefighting and emergency requirements, and to equalize operating pressures. They may be elevated or below ground level.

The main categories are surface reservoirs, standpipes, and elevated tanks. Common practice is to line surface reservoirs with concrete, gunite, asphalt, or an asphaltic membrane. Surface reservoirs may be covered or uncovered. Whenever possible, a cover should be considered for the prevention of contamination of the water supply by animals or humans and to prevent the formation of algae.

Standpipes or elevated tanks are normally employed where the construction of a surface reservoir would not provide sufficient head. A standpipe is essentially a tall cylindrical tank whose storage volume includes an upper portion (the useful storage), which is above the entrance to the discharge pipe, and a lower portion (supporting storage), which acts only to support the useful storage and provide the required head. For this reason, standpipes over 50 ft high are usually uneconomical. Steel, concrete, and wood are used in the construction of standpipes and elevated tanks. When it becomes more economical to build the supporting structure for an elevated tank than to provide for the supporting storage in a standpipe, the elevated tank is used.

Distribution reservoirs should be located strategically for maximum benefit. Normally, the reservoir should be near the center of use, but in large metropolitan areas a number of distribution reservoirs may be located at key points. Reservoirs providing

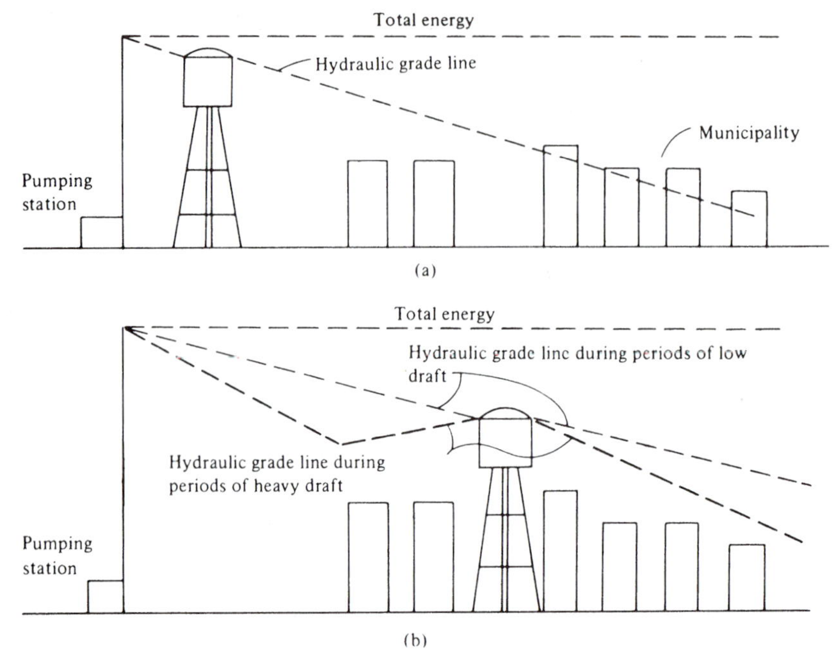

FIGURE 6.31 Pressure distribution as influenced by the location of a distribution reservoir.

service storage must be high enough to develop adequate pressures in the system they are to serve. A central location decreases friction losses by reducing the distance from supply point to the area served. Positioning the reservoir so that pressures may be approximately equalized is an additional consideration of importance. Figure 6.31 illustrates this point. The location of the tank, as shown in part (a), results in a very large loss of head by the time the far end of the municipality is reached. Thus, pressures too low will prevail at the far end or excessive pressures will be in evidence at the near end. In part (b), it is seen that pressures over the whole municipal area are more uniform for periods of both high and low demand. Note that during periods of high demand the tank is supplying flow in both directions (being emptied), whereas during periods of low demand the pump is supplying the tank and the municipality.

The amount of storage to be provided is a function of the capacity of the distribution network, the location of the service storage, and the use to which it is to be put. Water treatment plants are commonly operated at a uniform rate of flow such as the maximum daily rate. It is also desirable to operate pumping units at constant rates. Demands on the system in excess of these rates must therefore be met by operating storage. Requirements for firefighting purposes should be sufficient to provide fire flows for 10–12 hr in large communities and for 2 hr or longer in smaller ones. Emergency storage is provided to sustain the community's needs during periods when the inflow to the reservoir is shut off—for example, through a failure of the water supply works, failure of pumping equipment, or need to take a supply line out of service for maintenance or repair. The length of time the supply system is expected to be out of service dictates the amount of emergency storage to be provided. Emergency storage sufficient to last for several days is desirable.

The amount of storage required for emergency and firefighting purposes is readily computed once the time period over which these flows are to be provided has been selected [33]. An emergency storage of 3 days for a community of 8000 having an average use rate of 150 gpcd is $3 \times 150 \times 8000 = 3.6$ mil gal. Given that a fire flow of 2750 gpm must be provided for a duration of 10 hr, this means a total firefighting storage of 1.65 mil gal. To the sum of these values, an additional equalizing or operating storage requirement would be added. The determination of this volume is slightly more complex and needs further explanation.

To compute the required equalizing or operating storage, a mass diagram or hydrograph indicating the hourly rate of water use is required. The procedure used in determining the needed storage volume follows:

1. Obtain a hydrograph of hourly demands for the maximum day. This may be obtained through a study of available records, by gauging the existing system during dry periods when lawn-sprinkling demands are high, or by using available design criteria such as those presented in Chapter 4 to predict a hydrograph for a future condition of development.
2. Tabulate the hourly demand data for the maximum day as shown in Table 6.26.
3. Find the required operating storage by using mass diagrams such as in Figures 6.32 and 6.33, the hydrograph of Figure 6.34, or the values tabulated in column 6 of Table 6.26.

The required operating storage is found by using a mass diagram with the cumulative pumping curve plotted on it. Figure 6.30 illustrates this diagram for a uniform 24-hr pumping rate. Note that the total volume pumped in 24 hr must equal the total 24-hr demand, and thus the mass curve and cumulative pumping curve must be coincident at the origin and at the end of the day. Next, construct a tangent to the mass curve parallel to the pumping curve at point *A* in the figure. Then draw a second parallel tangent to the mass curve at point *C* and drop a vertical from *C* to an intersection with tangent *AB* at *B*. The required storage is equal to the magnitude of the ordinate *CB* measured on the vertical scale. In the example shown, the necessary storage volume is found to be 1.47 mil gal for a 24-hr pumping period.

Note that the reservoir is full at *A*, empty at *C*, is filling whenever the slope of the pump curve exceeds that of the cumulative demand curve, and is being drawn down when the rate of demand exceeds the rate of pumping.

It is often desirable to operate an equalizing reservoir so that pumping will take place at a uniform rate but for a period less than 24 hr. In small communities, for example, it is often advantageous to pump only during the normal working day. It may also be more economical to operate the pumping station at off-peak periods when electric power rates are low.

Figure 6.33 illustrates the operation of a storage reservoir where pumping occurs during the period between 6 A.M. and 6 P.M. only. To find the required storage in this case, construct the cumulative pumping curve *ED* so that the total volume of 6.87 mil gal is pumped uniformly from 6 A.M. to 6 P.M. Then project point *E* vertically upward to an intersection with the cumulative demand curve at *A*. Construct line *AC* parallel to

TABLE 6.26 Hourly Demand for the Maximum Day

(1) Time	(2) Average Hourly Demand Rate (gpm)	(3) Hourly Demand (gal)	(4) Cumulative Demand (gal)	(5) Hourly Demand as a Percent of Average	(6) Average Hourly Demand Minus Hourly Demand: 286,250 − (3)	
					−	+
12 A.M.	0	0	0	0	—	—
1	2170	130,000	130,000	45.4		156,250
2	2100	126,000	256,000	44.1		160,250
3	2020	121,000	377,000	42.3		165,250
4	1970	118,000	495,000	41.3		168,250
5	1980	119,000	614,000	41.6		167,250
6	2080	125,000	739,000	43.2		161,250
7	3630	218,000	957,000	76.2		68,250
8	5190	312,000	1,269,000	108.9	25,750	
9	5620	337,000	1,606,000	117.8	50,750	
10	5900	354,000	1,960,000	123.6	67,750	
11	6040	363,000	2,323,000	126.7	76,750	
12 P.M.	6320	379,000	2,702,000	132.4	92,750	
1	6440	387,000	3,089,000	135.2	100,750	
2	6370	382,000	3,471,000	133.4	95,750	
3	6320	379,000	3,850,000	132.4	92,750	
4	6340	381,000	4,231,000	133.0	94,750	
5	6640	399,000	4,630,000	139.5	112,750	
6	7320	439,000	5,069,000	153.3	152,750	
7	9333	560,000	5,629,000	195.5	273,750	
8	8320	499,000	6,128,000	174.4	212,750	
9	5050	303,000	6,431,000	105.8	16,750	
10	2570	154,000	6,585,000	53.8		132,250
11	2470	148,000	6,733,000	51.7		138,250
12 A.M.	2290	137,000	6,870,000	47.9		149,250
Total		6,870,000			1,466,500	1,466,500

$$\text{Average hourly demand} = \frac{6{,}870{,}000}{24} = 286{,}250 \text{ gal}$$

ED. Point *C* will be at the intersection of line *AC* with the vertical extended upward from 6 P.M. on the abscissa. The required storage equals the value of the ordinate *CBD*. Numerically, it is 2.55 mil gal and exceeds the storage requirement for 24-hr pumping. Another graphical solution to the storage problem may be obtained as outlined in Figure 6.34. The figure is a plot of the demand hydrograph for the maximum day. For uniform 24-hr pumping, the pumping rate will be equal to the mean hourly demand. This is shown as line *PQ*. The required storage is then obtained by planimetering or determining in some other manner the area between curve *BEC* and line *PQ*. Conversion of this area to units of volume yields the required storage of 1.47 mil gal. The required storage for 24-hr pumping may also be determined by summing either the plus or minus values of column 6 in Table 6.26.

Unless pumping follows the demand curve or demand hydrograph, storage will be required. Figure 6.32 shows that a maximum pumping rate of about 9400 gpm will

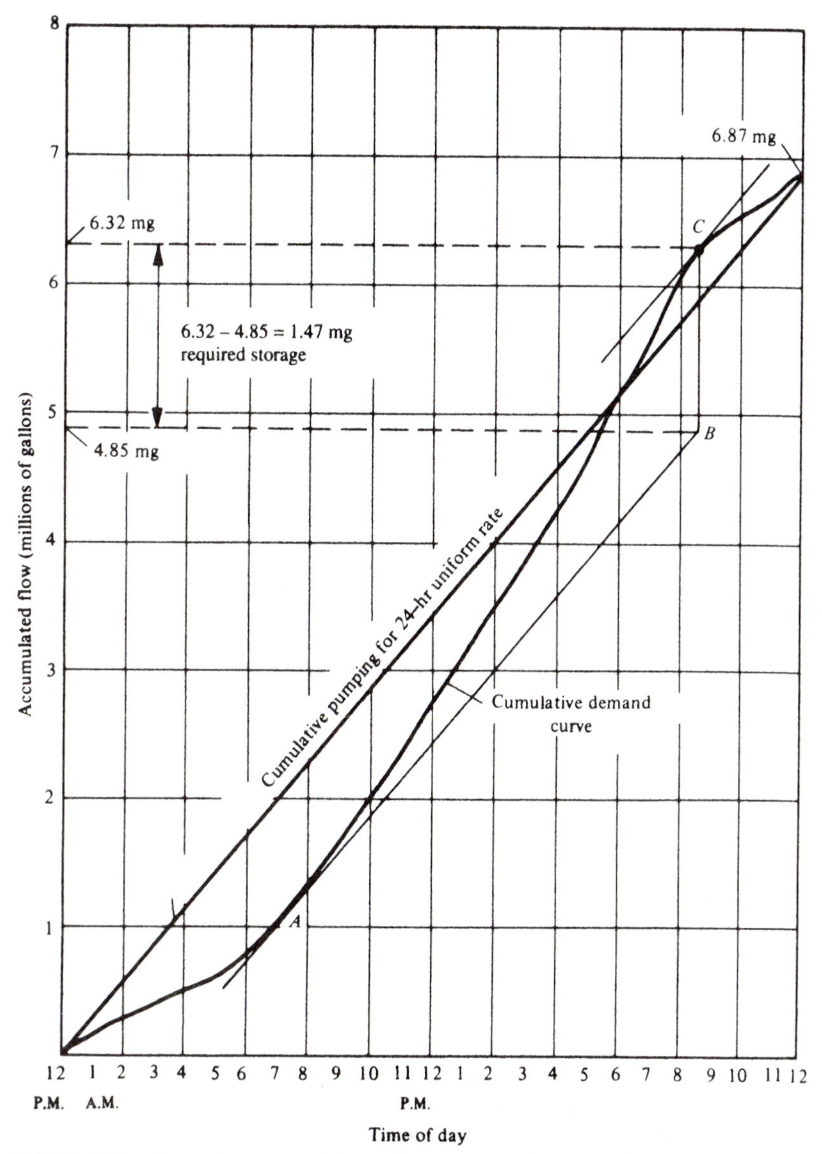

FIGURE 6.32 Operating storage for 24-hr pumping, determined by use of a mass diagram.

be required with no storage, whereas if storage is provided, a maximum pumping rate of 4775 gpm (about 50% of that required with no storage) will suffice. This example illustrates the economics of providing operating storage.

Variable-rate pumping is normally not economical. In practice, it is common to provide storage and pumping facilities so that pumping at the average rate for the maximum day can be maintained. On days of less demand, some pumping units will stand idle. Another operational procedure is to provide enough storage for pumping at the average rate for the average day, with idle reserve capacity, and then to overload all

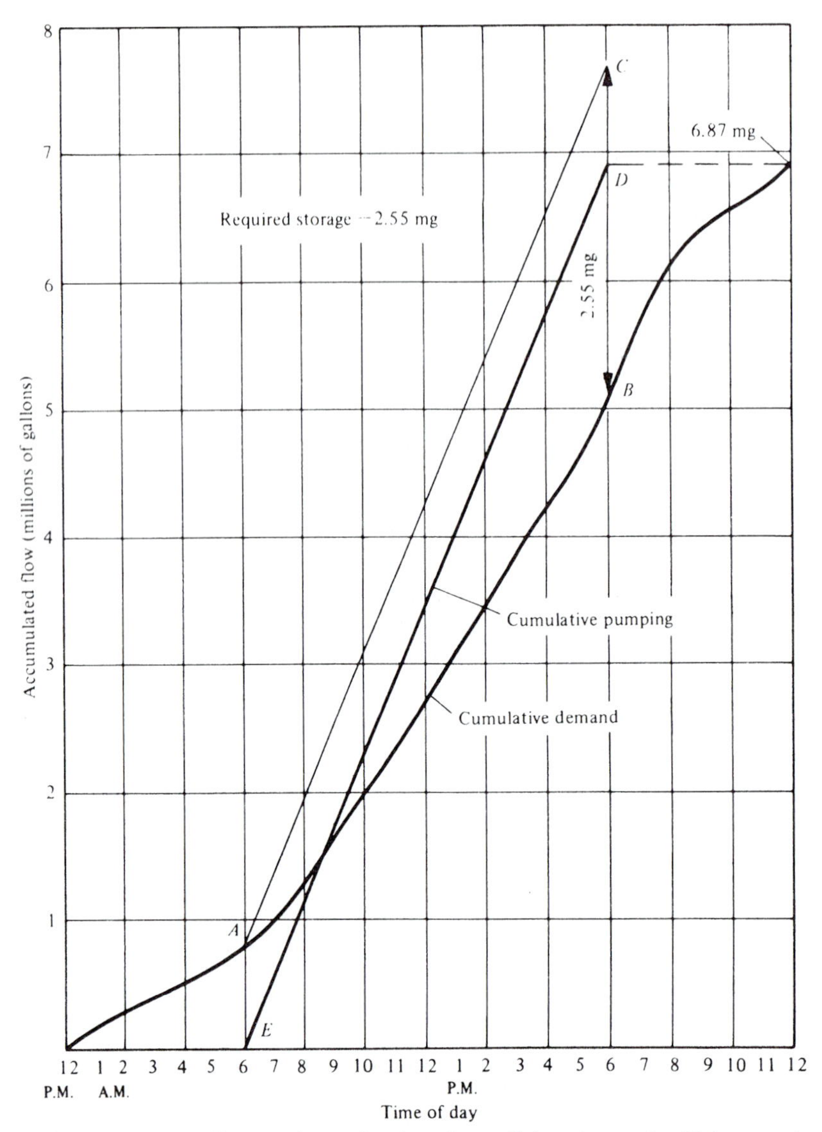

FIGURE 6.33 Mass diagram determination of equalizing storage for 12-hr pumping.

available units on the maximum day. Provision of pumping and storage capacity to meet peak demands experienced for only a few hours every few years has been found to be economically impractical.

Analyses of distribution systems are commonly concerned with the pipe network, topographic conditions, pumping station performance, and the operating characteristics of the storage system. Where multiple sources of supply operate under variable-head conditions, the hydraulic balancing of the system becomes more complex. The simple system of Figure 6.35 illustrates this point.

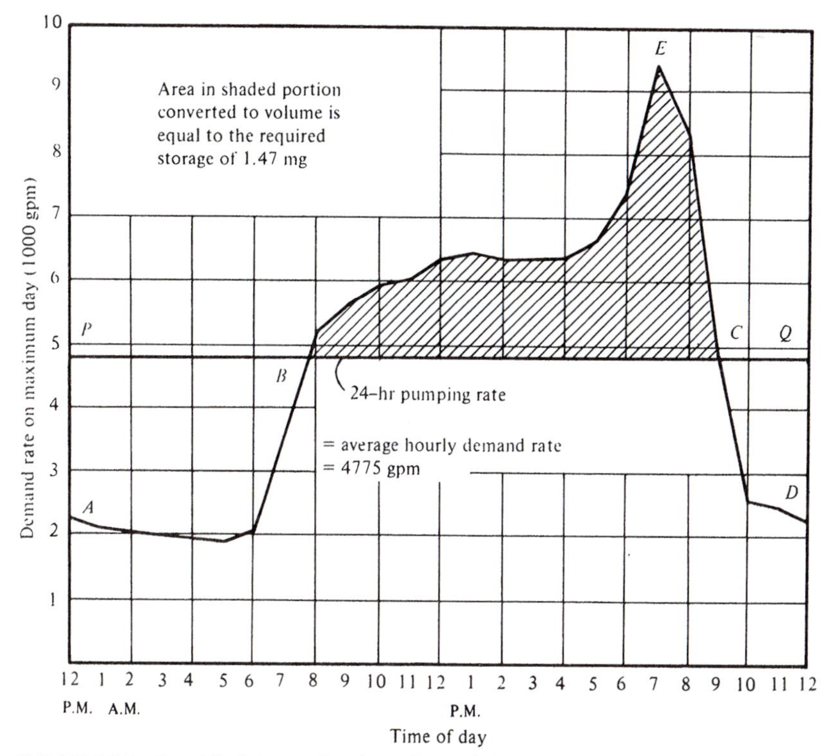

FIGURE 6.34 Graphical determination of equalizing storage.

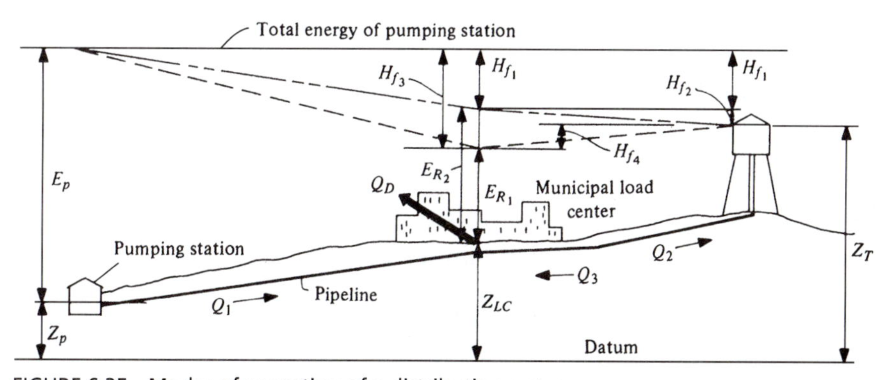

FIGURE 6.35 Modes of operation of a distribution system.

Considering that the demand for water by the municipal load center fluctuates hourly, it is evident that there are essentially two modes of operation of the given distribution system. When municipal requirements are light, such as in the early morning hours, the pumping station will meet these demands and in addition supply the reservoir. The solution of the problem may then be had by use of the following equations

$$
\begin{array}{ll}
(1) & Q_1 - Q_D = Q_2 \\
(2) & Z_P + E_P = Z_{LC} + E_{R2} + H_{f1} \\
(3) & Z_{LC} + E_{R2} = Z_T + H_{f2} \\
(2+3) & H_{f1} + H_{f2} = E_P + Z_P - Z_T
\end{array}
$$

where

Q_1 = flow from the pump

Q_D = municipal demand

Q_2 = flow to the tank

Z_P, Z_{LC}, Z_T = elevation above the arbitrary datum (Z_T = elevation of water surface in tank)

E_P = energy produced by the pump

E_{R2} = residual energy of the load center (pressure head plus velocity head)

H_{f1}, etc. = friction head losses

If Q_D, Z_P, Z_{LC}, and Z_T are specified, the equations may be solved by selecting values of E_{R2} and then solving for H_{f1} and H_{f2}. When a solution is reached so that Eq. (2 + 3) is satisfied, Q_1 and Q_2 may be computed.

When demands are high, both the tank and the pump will supply the community. The direction of flow will then be reversed in the line from the tank to the pump and the applicable equations will be

$$
\begin{array}{ll}
(1) & Q_1 + Q_3 = Q_D \\
(2) & Z_P + E_P = Z_{LC} + E_{R1} + H_{f3} \\
(3) & Z_{LC} + E_{R1} + H_{f4} = Z_T
\end{array}
$$

Again, an assumed value for E_R will be taken and trial solutions carried out until Eq. (1) is satisfied. Note that the foregoing illustration is a simple case, since Z_T has been specified. Actually, since Z_T fluctuates with time, it is necessary to have information on storage volume available versus water elevation in the tank so that at any specified condition of draft, the actual value for Z_T can be determined and used in the computations.

Water distribution systems generally are considered a composite of four basic constituents: the pipe network, the storage, the pump performance, and the pumping station and its suction source. These components must be integrated into a functioning system for various schedules of demand. A thorough analysis of each system must be made to ensure that it will operate satisfactorily under all anticipated combinations of demand and hydraulic component characteristics. The system may work well under one set of conditions but will not necessarily be operable under some other set. A comprehensive system balance requires an hourly simulation of performance for the expected operating schedule.

There is an infinite number of arrangements of the basic components in a distribution system, but the hydraulic analyses discussed herein are applicable to all of them.

PUMPING

Pumps are important components of most water conveyance systems. They are called upon to provide the energy to deliver flows ranging from a few gpm to large numbers of cfs. The primary types of pumps are centrifugal and displacement. Airlift pumps, jet pumps, and hydraulic rams are also used in special applications. In water and sewage works, centrifugal pumps are most common. *Centrifugal pumps* have a rotating element (impeller) that imparts energy to the water. *Displacement pumps* are often of the reciprocating type, in which a piston draws water into a closed chamber and then expels it under pressure. Reciprocating pumps are widely used to handle sludge in sewage treatment works.

Electric power is the primary source of energy for pumping, but gasoline, steam, and diesel power are also used. Often, a standby engine powered by one of these other forms is included in primary pumping stations to operate in emergency situations when electric power fails.

6.11 PUMPING HEAD

The first step in selecting pumps is that of determining the operating characteristics of the system in which they are to be used [34–37]. An important feature is the *total dynamic head* (TDH) against which the pump must operate. The TDH is composed of the difference in elevation between the pump centerline and the elevation to which the water is to be raised, the difference in elevation between the level of the suction pool and the pump centerline, the frictional losses encountered in the pump, pipe, valves and fittings, and the velocity head. Expressed in equation form, this becomes

$$\text{TDH} = H_L + H_F + H_V \tag{6.66}$$

where H_L is the total static head or elevation difference between the pumping source and the point of delivery, H_F is the total friction head loss, and H_V is the velocity head $V^2/2g$. Figure 6.36 illustrates the total static head.

6.12 POWER

For a known discharge and total pump lift, the theoretical horsepower (hp) required may be found by using

$$\text{hp} = Q\gamma H/550 \tag{6.67}$$

where

Q = discharge, cfs

γ = specific weight of water

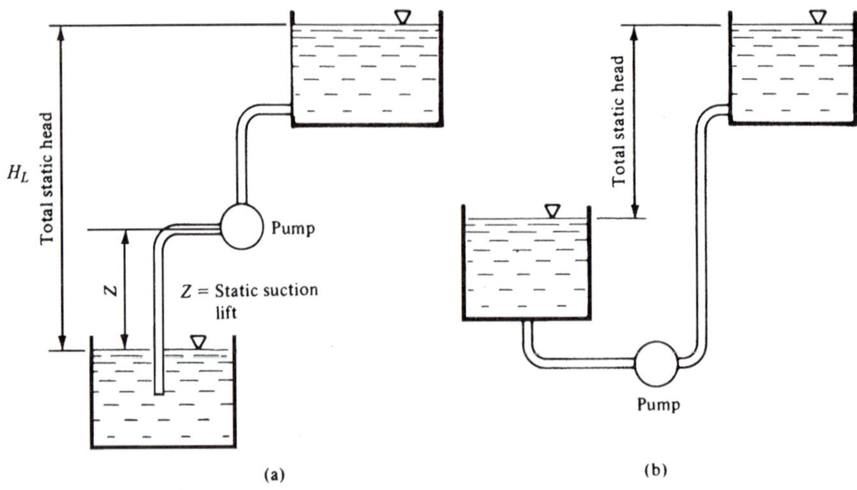

FIGURE 6.36 Total static head. (a) Intake below the pump centerline. (b) Intake above the pump center line.

H = total dynamic head

550 = conversion from foot-pounds per second to horsepower

The actual horsepower required is obtained by dividing the theoretical horsepower by the efficiency of the pump and driving unit.

6.13 CAVITATION

Cavitation is the phenomenon of cavity formation or the formation and collapse of cavities [35]. Cavities develop when the absolute pressure in a liquid reaches the vapor pressure related to the liquid temperature. Under severe conditions, cavitation can result in the breakdown of pumping equipment. As the net positive suction head (NPSH) for a pump is reduced, a point is reached where cavitation becomes detrimental. This point is usually referred to as the minimum net positive suction head ($NPSH_{min}$) and is a function of the type of pump and the discharge through the pump. NPSH is calculated as

$$\text{NPSH} = \frac{V_1^2}{2g} + \frac{p_1}{\gamma} - \frac{p_v}{\gamma} \tag{6.68}$$

where V_1 is the velocity of flow at the center line of the inlet to the pump, p_1 is the pressure at the center line of the pump inlet, and p_v is the vapor pressure of the fluid. Referring to Figure 6.36a and writing the energy equation between the intake pool and the inlet to the pump, we have

$$\frac{p_a}{\gamma} = \frac{V^2}{2g} + \frac{p_1}{\gamma} + Z + h_L$$

or

$$\frac{p_a}{\gamma} - Z - h_L = \frac{V^2}{2g} + \frac{p_1}{\gamma}$$

where p_a is atmospheric pressure and h_L is the head loss in the intake.

Subtracting p_v/γ from both sides, we have

$$\frac{p_a}{\gamma} - Z - h_L - \frac{p_v}{\gamma} = \frac{V^2}{2g} + \frac{p_1}{\gamma} - \frac{p_v}{\gamma}$$

This may be written

$$\frac{p_a}{\gamma} - Z - h_L - \frac{p_v}{\gamma} = \text{NPSH}$$

The minimum value of the static lift is then determined as

$$Z_{\text{min}} = \frac{p_a - p_v}{\gamma} - \text{NPSH}_{\text{min}} - h_L \tag{6.69}$$

The required NPSH for any pump can be obtained from the manufacturer. This value can then be checked against the proposed installation using Eqs. (6.68) and (6.69) to ensure that the available NPSH is greater than the manufacturer's requirement.

6.14 SYSTEM HEAD

The system head is represented by a plot of TDH versus discharge for the system being studied. Such plots are very useful in selecting pumping units [37]. It should be clear that the system head curve will vary with flow since H_F and H_V are both a function of discharge. In addition, the static head H_L may vary as a result of fluctuating water levels and similar factors, and it is often necessary to plot system head curves covering the range of variations in static head. Figure 6.37 illustrates typical system head curves for a fluctuating static water level.

6.15 PUMP CHARACTERISTICS

Each pump has its own characteristics relative to power requirements, efficiency, and head developed as a function of flow rate. These relationships are usually given as a set of pump characteristic curves for a specified speed. They are used in conjunction with system head curves to select correct pumping equipment for a particular installation. A set of characteristic curves is shown in Figure 6.38.

At no flow, the head is known as the *shutoff head*. The pump head may rise slightly or fall from the shutoff value as discharge increases. Ultimately, however, the head for any centrifugal pump will fall with increase in flow. At maximum efficiency, the discharge is known as the *normal* or *rated discharge* of the pump. Varying the

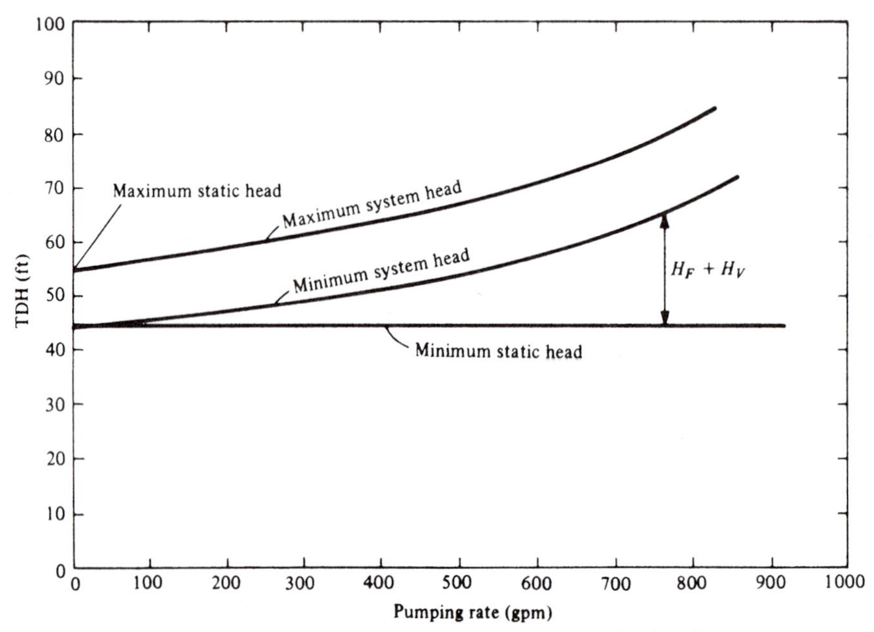

FIGURE 6.37 System head curves for a fluctuating static pumping head.

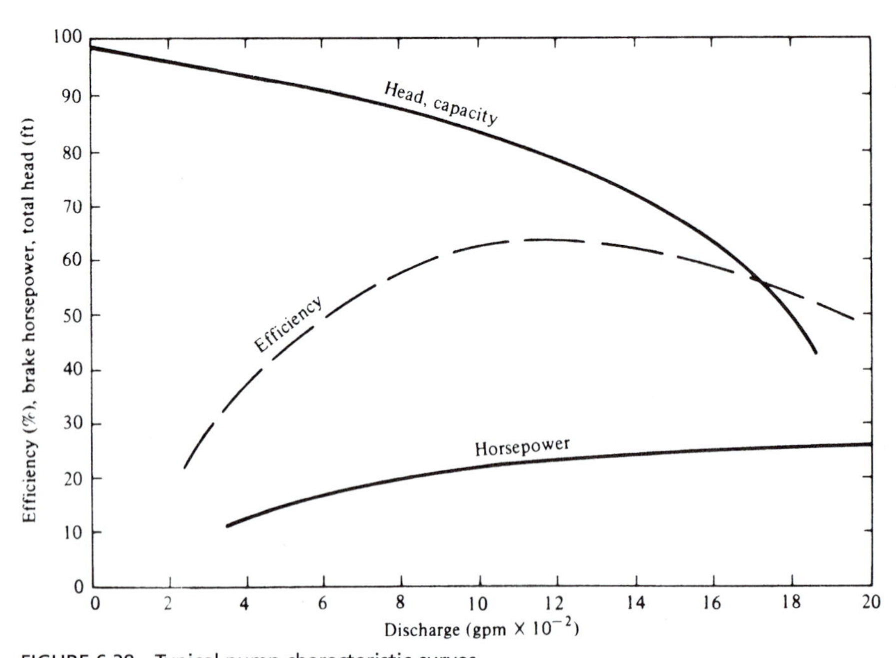

FIGURE 6.38 Typical pump characteristic curves.

pump discharge by throttling will lower the efficiency of the unit. By changing the speed of the pump, discharge can be varied within a certain range without a loss of efficiency. The most practical and efficient approach to a variable-flow problem is to provide two or more pumps in parallel so that the flow may be carried at close to the peak efficiency of the units operating.

The normal range of efficiencies for centrifugal pumps is between 50% and 85%, although efficiencies in excess of 90% have been reported. Pump efficiency usually increases with the size and capacity of the pump [36].

6.16 SELECTION OF PUMPS

Once the system head has been determined, the next step is to find a pump or pumps to deliver the required flows. This is done by plotting the system head curve on a sheet with the pump characteristic curves. The operating point is at the intersection of the system head curve and the pump head capacity curve. This gives the head and flow at which the pump will be operating. A pump should be selected so that the operating point is also as close as possible to peak efficiency. This procedure is shown in Figure 6.39.

Pumps may be connected in series or in parallel. For series operation at a given capacity, the total head equals the sum of the heads added by each pump. For parallel operation, the total discharge is multiplied by the number of pumps for a given head. It should be noted, however, that when two pumps are used in series or parallel, neither the head nor capacity for a given system head curve is doubled (Fig. 6.39).

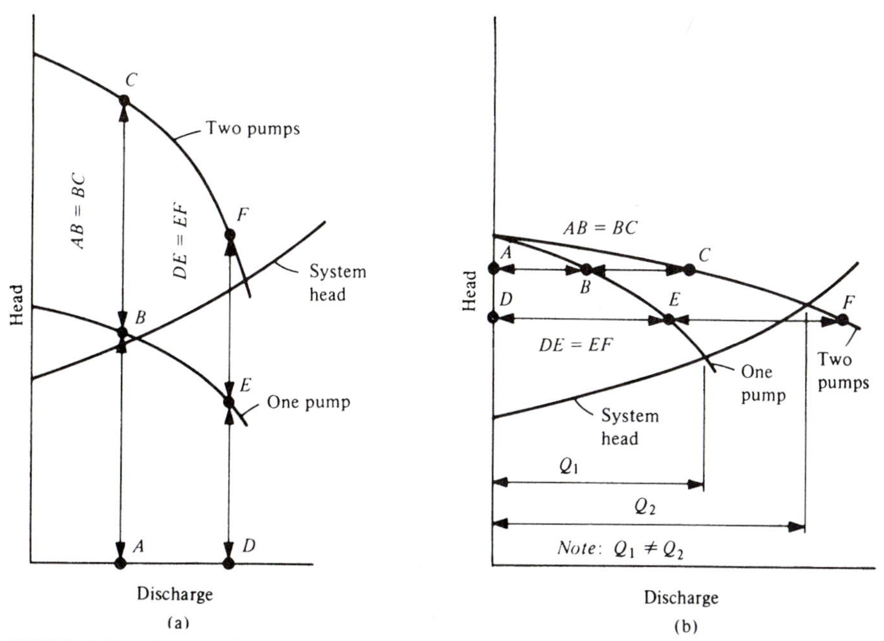

FIGURE 6.39 Characteristic curves for (a) Series. (b) Parallel pump operations of equal pumps.

Example 6.9

A pumping station is to be designed for an ultimate capacity of 1200 gpm at a total head of 80 = ft. The present requirements are that the station deliver 750 gpm at a total head of 60 = ft. One pump will be required as a standby.

Solution: The system head curve is plotted as shown in Figure 6.40. Values for the curve are obtained as indicated in Section 6.11.

Consider that three pumps will ultimately be needed (one as a standby). Determine the design flows as follows:

1. Two pumps at 1200 gpm at 80 ft of TDH
2. One pump at 1200/2 = 600 gpm at 80 ft of TDH
3. One pump must also be able to meet the requirements of 750 gpm at 60 ft of TDH.

From manufacturers' catalogs, two pumps, *A* and *B*, are found that will meet the specifications. The characteristic curves for each pump are shown in Figure 6.40. The intersection of the characteristic curves with the system head curve indicates that pump *A* can deliver 750 gpm at a TDH of 60 ft, while pump *B* can deliver 790 gpm at a TDH of 62 ft. A check of the efficiency curves for each pump indicates that pump *B* will deliver the present flow at a much greater efficiency than pump *A*. Therefore, select pump *B*.

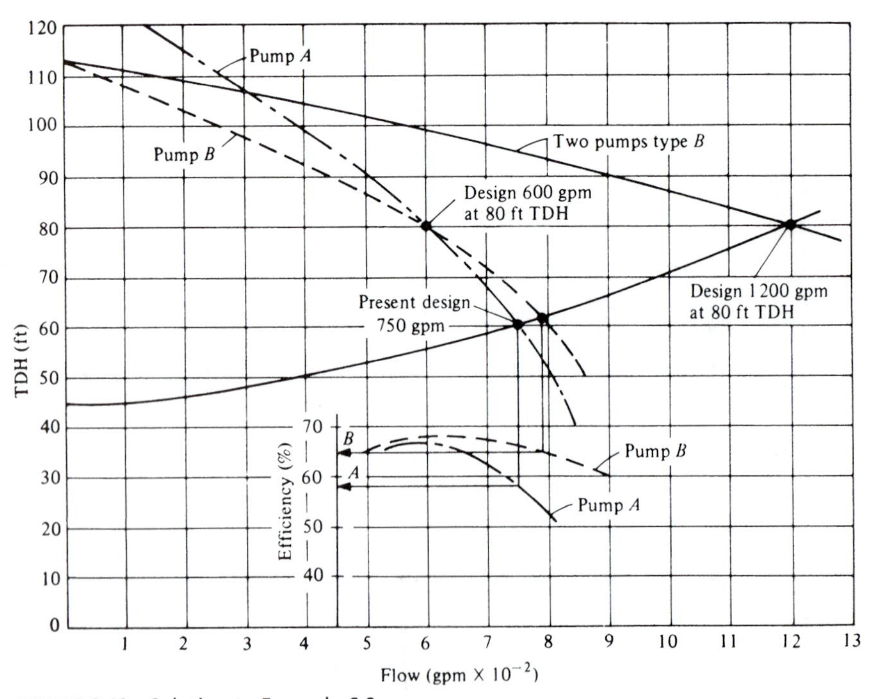

FIGURE 6.40 Solution to Example 6.9.

For the present, select two pumps of type *B* and use one as a standby. For the future, add one more pump of type *B*.

Operating characteristics for a wide range of pump sizes and speeds are available from pump manufacturers. Special equipment manufactured to satisfy a customer's prescribed requirements must customarily satisfy an acceptance test after it has been installed.

PROBLEMS

6.1 Given a V-shaped channel with a bottom slope of 0.001, a top width of 10 ft, and a depth of 5 ft, determine the velocity of flow. Find the discharge in cfs and m^3/s.

6.2 A trapezoidal channel measures 3 m across the top and 1 m across the bottom. The depth of flow is 1.5 m. For $s = 0.005$ and $n = 0.012$, determine the velocity and rate of flow.

6.3 Given an 18-in. concrete conduit with a roughness coefficient of $n = 0.013$, $s = 0.02$, and a discharge capacity of 15 cfs, what diameter pipe is required to quadruple the capacity?

6.4 Find the dimensions of a rectangular concrete channel to carry a flow of 150 m^3/s, with a bottom slope of 0.015 and a mean velocity of 10.2 m/s.

6.5 Determine the head loss in a 46-cm concrete pipe with an average velocity of flow of 1.0 m/s and a length of 30 m.

6.6 Find the discharge from a full-flowing cast-iron pipe with 24-in. diameter and a slope of 0.004.

6.7 Refer to the figure below and assume that reservoirs *A*, *B*, and *C* have water surface elevations of 150 ft, 90 ft, and 40 ft, respectively, and are connected by a system of concrete pipes of lengths $L_1 = 2400$ ft, $L_2 = 1500$ ft, and $L_3 = 5500$ ft with respective diameters of 8, 12, and 21 in. Find the discharge in each pipe and the elevation of the hydraulic gradient at *P*.

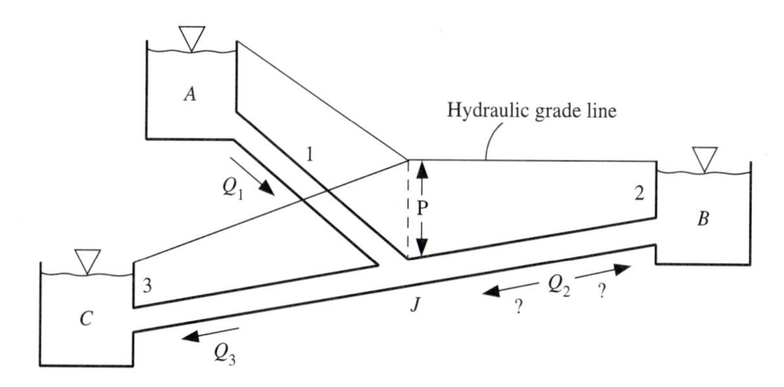

6.8 Three reservoirs *A*, *B*, and *C* are connected by a branching cast-iron pipe system. If the pipe lengths are $L_1 = 3000$ ft, $L_2 = 2200$ ft, and $L_3 = 1600$ ft, the respective pipe diameters are 15, 10, and 18 in, and the surface water elevations of two of the three reservoirs are $A = 125$ ft and $B = 55$ ft, find the surface water elevation of reservoir *C*. Assume that the flow to reservoir *C* is 15 cfs.

6.9 Solve Problem 6.8 by assuming that the flow in pipe 3 is from reservoir *C* rather than to reservoir *C*.

6.10 Three riveted steel pipes are connected in series, with the flow through the system being 1.5 m^3/s. Find the total head loss if the pipe diameters and lengths are $D_1 = 60$ cm, $D_2 = 40$ cm, and $D_3 = 54$ cm, $L_1 = 400$ m, $L_2 = 450$ m, and $L_3 = 750$ m. Assume the friction factor $f = 0.0125$.

6.11 Given the same lengths and diameters of the pipes in series as in Problem 6.10, determine the total flow if the system head loss is 60 m.

6.12 Consider the pipe system in the figure. If the flow in *BCD* is 6 cfs, find (a) the flow in *BED*, (b) the total flow, and (c) a length of 16-in pipe equivalent to the two parallel pipes.

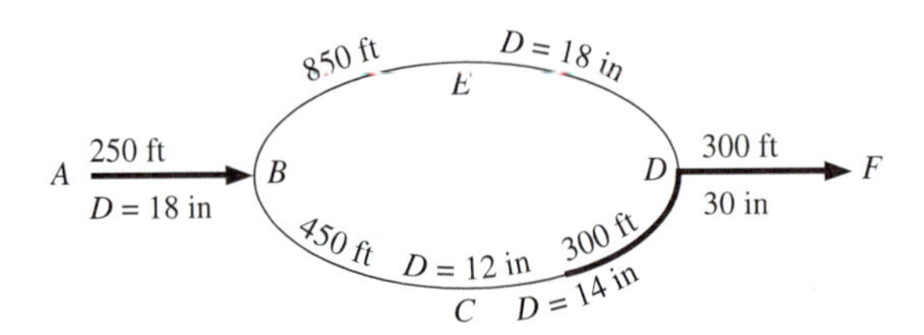

6.13 A flow of 1.3 m^3/s is divided into three parallel pipes of diameters 30, 20, and 45 cm and lengths of 30, 40, and 25 m, respectively. Find the head loss and distribution of flow. Assume $f = 0.015$.

6.14 If a system of parallel pipes has diameters of 18, 8, and 21 in and lengths of 50, 95, and 60 ft, respectively, find the total flow in the system. Assume $f = 0.024$ and the total head loss is 45 ft.

6.15 From the given layout, determine the length of an equivalent 24-cm pipe.

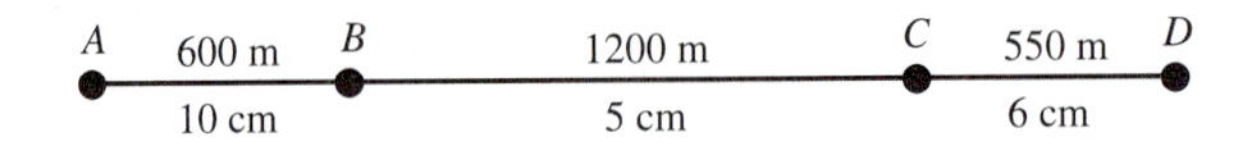

6.16 For the pipe layout in Problem 6.15, find the diameter of an equivalent 1000-m pipe.

6.17 For the pipe network shown, determine the direction and magnitude of flow in each pipe. Assume $C = 100$. Solve using a spreadsheet, EPANET, or another suitable method.

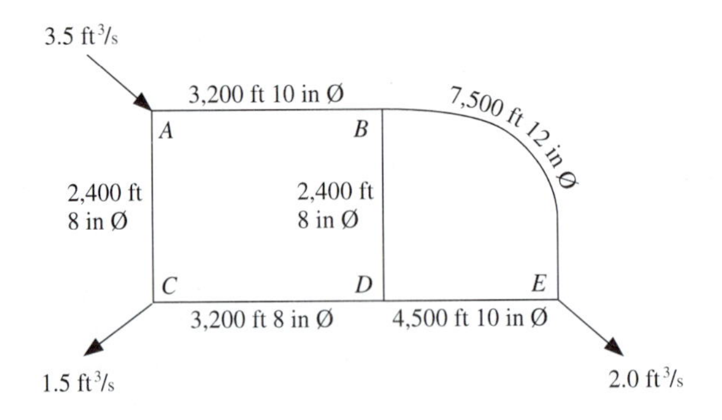

6.18 Solve Problem 6.17 if the flow at *A* is 4.0 cfs and that at *E* is 2.5 cfs.

6.19 Water is pumped 9 mi from a reservoir at elevation 100 ft to a second reservoir at elevation 210 ft. The pipeline connecting the reservoirs has a 54 in diameter. It is concrete and has an absolute roughness of 0.003. If the flow is 25 mgd and pumping station efficiency is 80%, what will be the monthly power bill if electricity costs 3 cents/kwh?

6.20 A reservoir at elevation 700 ft is to supply a second reservoir at elevation 460 ft. The reservoirs are connected by 1300 ft of 24-in cast-iron pipe and 2000 ft of 20-in cast-iron pipe in series. What will be the discharge delivered from the upper reservoir to the lower one?

6.21 Refer to the figure and tables below and analyze the network using EPANET and metric units.

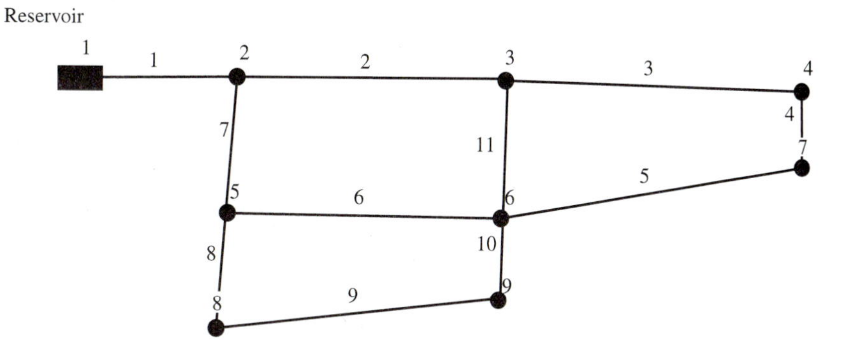

Node Input Data for Problem 6.21		
Node #	**Elevation, m**	**Demand, L/s**
reservoir - 1	70	0
2	30	0
3	30	14.2
4	30	2
5	30	0
6	30	14.2
7	30	56.6
8	30	0
9	30	0

Pipe Input Data for Problem 6.21			
Pipe #	**Length, m**	**Diameter, mm**	**Roughness, *C***
1	61	460	120
2	1829	410	120
3	1829	300	120
4	610	200	120
5	1981	200	120
6	1829	410	120
7	914	250	120
8	762	200	120
9	1981	200	120
10	610	200	120
11	914	150	120

6.22 Find the pipe flows and pressures at the nodes and pipes of the network shown below. Use EPANET for your solution. Input data are given in the tables that follow. Flows are in gpm and heads are in ft.

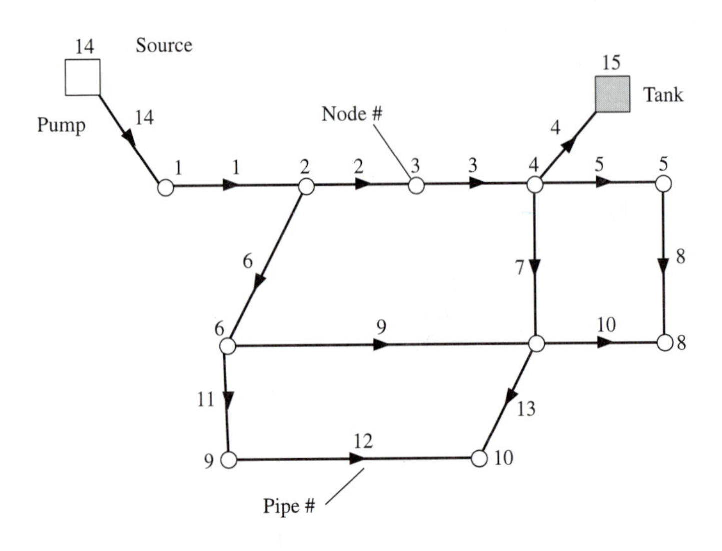

[Junctions]

ID	Elevation	Demand	(Pattern)
1	805	0	
2	800	150	
3	775	150	
4	760	165	
5	790	115	
6	800	145	
7	807	160	
8	810	155	
9	825	120	
10	815	110	

[Tanks]

ID	Elevation	Initial Level	Minimum Level	Maximum Level	Diameter	Minimum (Volume)
15	960	125	100	155	52	
14	900					

[Pipes]

ID	Head Node	Tail Node	Length	Diameter	Roughness Coefficient	(Minor Loss Coefficient)	(Check Valve)
1	1	2	11000	18	120		
2	2	3	4000	16	120		
3	3	4	3000	14	120		
4	15	4	300	18	120		
5	4	5	6000	10	120		
6	2	6	7000	10	120		
7	4	7	5500	12	120		
8	5	8	5000	10	120		
9	6	7	7700	8	120		
10	7	8	6000	8	120		
11	6	9	5700	8	120		
12	9	10	7100	8	120		
13	10	7	7200	6	120		

[Pumps]

ID	Head Node	Tail Node	Head	Flow
			Characteristics	
14	14	1	275	1600

[Controls]

Link	ID	Setting	Condition
Link 14 open if node 15 below 115			
Link 14 closed if node 15 above 140			

[Patterns]

	ID	Multipliers				
1	1.0	1.2	1.4	1.6	1.4	1.2
1	1.0	0.8	0.6	0.4	0.6	0.8

6.23 Rework solved Example 6.8 for the average day flows.

6.24 It is necessary to pump 6000 gpm of water from a reservoir at an elevation of 900 ft to a tank whose bottom is at an elevation of 1050 ft. The pumping unit is located at elevation 900 ft. The suction pipe is 24 in. in diameter and very short, so head losses may be neglected. The pipeline from the pump to the upper tank is 410 ft long and is 20 in. in diameter. Consider that minor losses in the line equal 2.5 ft of water. The maximum depth of water in the tank is 38 ft, and the supply lines are cast iron. Find the maximum lift of the pump and the horsepower required for pumping if the pump efficiency is 76%.

6.25 Rework Problem 6.24 if the water to be pumped is 7000 gpm and the elevation of the tank is 1000 ft.

6.26 If a flow of 5.0 cfs is to be carried by an 11,000-ft cast-iron pipeline without exceeding a head loss of 137 ft, what must the pipe diameter be?

6.27 Rework Problem 6.26 if the flow is 4.5 cfs and the head loss cannot exceed 120 ft.

6.28 A 48-in. water main carries 79 cfs and branches into two pipes at point A. The branching pipes are 36 and 20 in. in diameter and 2800 and 5000 ft long, respectively. These pipes rejoin at point B and again form a single 48-in. pipe. If the friction factor is 0.022 for the 36-in. pipe and 0.024 for the 20-in. pipe, what will the discharge be in each branch?

6.29 Water is pumped from a reservoir whose surface elevation is 1390 ft to a second reservoir whose surface elevation is 1475 ft. The connecting pipeline is 4500 ft long and 12 in. in diameter. If the pressure during pumping is 80 psi at a point midway on the pipe at elevation 1320 ft, find the rate of flow and the power exerted by the pumps. Also, plot the hydraulic grade line. Assume that $f = 0.022$.

6.30 A concrete channel 18 ft wide at the bottom is constructed with side slopes of 2.2 horizontal to 1 vertical. The slope of the energy gradient is 1 in 1400 and the depth of flow is 4.0 ft. Find the velocity and the discharge.

6.31 A rectangular channel is to carry 200 cfs. The mean velocity must be greater than 2.5 fps. The channel bottom width should be about twice the channel depth. Find the channel cross section and the required channel slope.

6.32 A rectangular channel carries a flow of 10 cfs/ft of width. Plot a curve of specific energy versus depth. Compute the minimum value of specific energy and the critical depth. What are the alternative depths for $Es = 5.0$?

6.33 Determine an equivalent pipe for the system shown below.

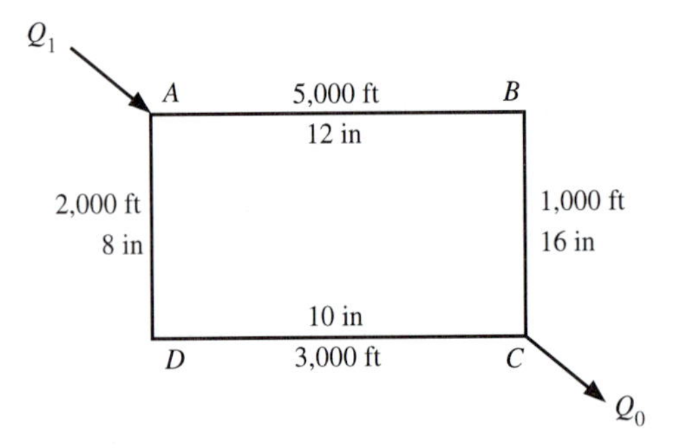

6.34 From the diagram below, compute (a) the total head loss from A to C, (b) P_a if $P_c = 25$ psi, and (c) the flow in each line.

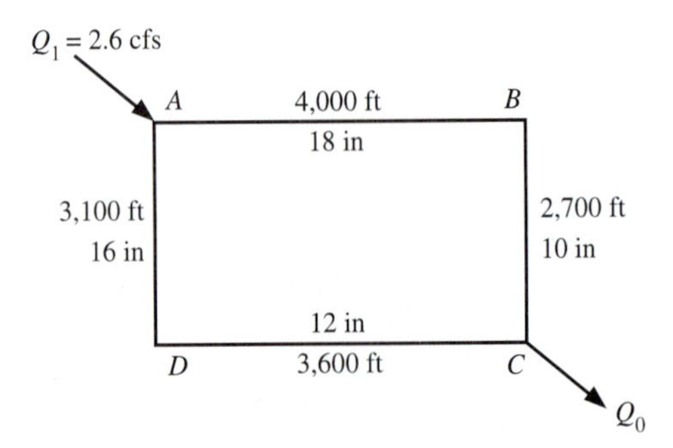

6.35 Given the pipe layout shown, determine the length of an equivalent 18-in. pipe.

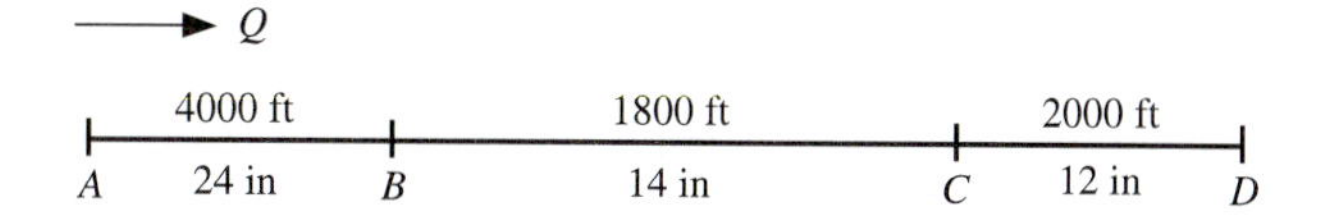

6.36 Given the following average hourly demand rates in gallons per minute, find the uniform 24-hr pumping rate and the required storage.

12 P.M.	0	12 A.M.	6300
1 A.M.	1900	1 P.M.	6500
2	1800	2	6460
3	1795	3	6430
4	1700	4	6500
5	1800	5	6700
6	1910	6	7119
7	3200	7	9000
8	5000	8	8690
9	5650	9	5220
10	6000	10	2200
11	6210	11	2100
		12 P.M.	2000

6.37 Solve Problem 6.36 if the period of pumping is from 6 A.M. to 6 P.M. only.

6.38 Solve Problem 6.36 by the method outlined in Figure 6.34, assuming 24-hr pumping.

6.39 Determine the total dynamic head of a pumping system where the total static head is 50 ft, the total friction head loss is 5 ft, and the velocity head is 10 ft.

6.40 Calculate the horsepower requirements for the system described in Problem 6.39 if the flow is 25 cfs.

6.41 Plot the system head and the pump characteristic curves for the data given. At what point should the pump operate?

Total dynamic head (ft)	55	60	65	70	75	80	85	90
Flow (gal/min)	200	500	700	850	975	1075	1200	1300

Efficiency (%)	Horsepower (hp)	Total Head (ft)	Flow (gal/min)
22	8	95	225
37	12	93	400
49	17	90	600
57	20	87	800
63	22	84	1000
64	23	78	1200
62	24	72	1400
59	25	64	1600
54	26	49	1800

REFERENCES

[1] T. M. Walski, *Analysis of Water Distribution Systems* (New York: Van Nostrand Reinhold, 1984).

[2] R. S. Gupta, *Hydrology and Hydraulic Systems* (Upper Saddle River, NJ: Prentice Hall, 1989).

[3] H. W. Gehm and J. I. Bregman, eds., *Handbook of Water Resources and Pollution Control* (New York: Van Nostrand Reinhold, 1976).

[4] H. M. Morris, *Applied Hydraulics in Engineering* (New York: The Ronald Press, 1963).

[5] L. A. Rossman, "EPANET 2 Users Manual," Water Supply and Water Resources Division, National Risk Management Research Laboratory, Office of Research and Development, U.S. Environmental Protection Agency (Cincinnati, OH, September 2000).

[6] L. Cesario, *Modeling, Analysis, and Design of Water Distribution Systems* (Denver: American Water Works Association, 1995).

[7] "Flow of Fluids Through Valves, Fittings, and Pipe," *Tech. Paper No. 410* (Chicago: Crane Co., 1957).

[8] N. Joukowsky, "Water Hammer" (translated by O. Simin), *Proc. Am. Water Works Assoc.* 24 (1904): 341–424.

[9] G. E. Russell, *Hydraulics* (New York: Holt, Rinehart and Winston, 1957).

[10] J. M. Edmonston and E. E. Jackson, "The Feather River Project and the Establishment of the Optimum Hydraulic Grade Line for the Project Aqueduct," Metropolitan Water District of Southern California (Los Angeles; 1963).

[11] J. Hinds, "Economic Water Conduit Size," *Eng. News-Record* (January 1937).

[12] C. W. Reh, "Hydraulics of Water Distribution Systems," *University of Illinois Engineering Experiment Station Circular No. 75* (February 1962).

[13] M. B. McPherson and J. V. Radziul, "Water Distribution Design and the McIlroy Network Analyzer," *Proc. Am. Soc. Civil Engrs.* 84, Paper 1588 (April 1958).

[14] E. L. Thackston, "Notes on Network Analysis," unpublished, Vanderbilt University (1982).

[15] D. J. Wood, "Algorithms for Pipe Network Analysis and Their Reliability," Research Report No. 127 (Lexington, KY: University of Kentucky, Water Resources Research Institute, 1981).

[16] T. M. Walski et al., "Battle of the Network Models: Epilogue," *J. Water Resources Planning Management*, 113, no. 2 (March 1987): 191–203.

[17] D. J. Wood and C. O. A. Charles, "Hydraulic Network Analysis Using Linear Theory," *Proc. Am. Soc. Civil Engrs., Hydraulics Div.* 98 (HY7) (July 1972).

[18] R. Epp and A. G. Fowler, "Efficient Code for Steady-State Flows in Networks," *J. Hydraulics Division* (New York: American Society of Civil Engineers, January 1970).

[19] T. M. Walski, "Sherlock Holmes Meets Hardy—Cross or Model Calibration in Austin, Texas," *J. Am. Water Works Assoc.* (March 1990): 34–38.

[20] T. M. Walski, "Case Study: Pipe Network Calibration Issues," *J. Water Resources Planning Management* 112, no. 2 (April 1986): 238–249.

[21] D. J. Wood, "Pipe Network Analysis Programs: KYPIPE," Civil Engineering Software Center, University of Kentucky (Lexington, KY, 1987).

[22] A. L. Cesario, "Computer Modeling Programs: Tools for Model Operations," *J. AWWA*, 72, no. 9 (1980).

[23] G. C. Dandy et al., "A Review of Pipe Network Optimization Techniques," in *Proc. Watercomp93, the 2nd Australian Conference on Technical Computing in the Water Industry* (Melbourne, Australia, March 30, 1993).

[24] R. W. Jeppson, *Analysis of Flow in Pipe Networks* (Ann Arbor, MI: Ann Arbor Science, 1976).

[25] L. D. Ormsbee et al., "Network Modeling for Small Distribution Systems," in Proc. *1992 Computer Conference* (Nashville, TN, April 1992).

[26] T. M. Walski et al., *Water Distribution System: Simulation and Sizing* (Chelsea, MI: Lewis Publishers, 1990).

[27] H. Cross, "Analysis of Flow in Networks of Conduits or Conductors," *University of Illinois Bull. No. 286* (November 1936).

[28] M. B. McPherson, "Generalized Distribution Network Head Loss Characteristics," *Proc. Am. Soc. Civil Engrs., J. Hydraulics Div.* 86 (HY1) (January 1960).

[29] E. Todini and S. Pilati, "A Gradient Method for the Analysis of Pipe Networks," International Conference on Computer Applications for Water Supply and Distribution (England: Leicester Polytechnic, September 8–10, 1987).

[30] B. F. Loubser and J. Gessler, "Computer-Aided Optimization of Water Distribution Networks," *The Civil Engineer in South Africa* (October 1990).

[31] L. J. Murphy et al., "Design of a Pipe Network Using Genetic Algorithms," *Water Journal*, Australian Water and Wastewater Association, 20, no. 4 (1993).

[32] A. George and J. W-H. Liu, *Computer Solution of Large Sparse Positive Definite Systems* (Upper Saddle River, NJ: Prentice Hall, 1981).

[33] J. E. Kiker, "Design Criteria for Water Distribution Storage," *Public Works* (March 1964).

[34] L. D. Benefield et al., *Treatment Plant Hydraulics for Environmental Engineers*, (Upper Saddle River, NJ: Prentice Hall, 1984).

[35] L. E. Ormsbee and T. M. Walski, "Identifying Efficient Pump Combinations," *J. Am. Water Works Assoc.* (January 1989): 30–34.

[36] R. M. Olson, *Engineering Fluid Mechanics*, 3rd ed. (New York: IEP, 1973).

[37] T. M. Walski and L. E. Ormsbee, "Developing System Head Curves for Water Distribution Planning," *J. Am. Water Works Assoc.* (July 1989): 63–66.

CHAPTER 7

Wastewater and Storm Water Systems

OBJECTIVES

The purpose of this chapter is to:

- Review the principles of open-channel flow
- Illustrate the design of sanitary sewer systems
- Define urban runoff and present methods for calculating flows to storm drainage systems
- Illustrate the design of storm drainage systems

The collection and transportation of surface drainage, sewage, and other waste flows from residential, commercial, industrial, and rural sites pose problems different from those encountered in water supply. Sewage must be delivered to treatment locations as quickly as possible to prevent development of septic conditions. And many waste flows contain solids that must be kept in suspension during transport. Most waste discharges exhibit fluctuations in time. Storm water flows are particularly characterized by rapid rises and falls. These temporal variations affect conveyance designs and must be taken into account when hydraulic calculations are made. Corrosive and other types of constituents in waste flows requires attention to the types of conveyances used and the materials of which they are constructed.

HYDRAULICS

Wastewater systems are usually designed as open channels except where lift stations are required to overcome topographic barriers. Hydraulic problems associated with wastewater flows are complicated by the quality of the fluid, the variable nature of the flows, and the fact that an unconfined or free surface usually exists. The driving force

TABLE 7.1 Values of Manning's Roughness Coefficient n

Nature of Surface	Manning's n Range
Concrete pipe	0.011–0.013
Corrugated metal pipe	0.019–0.030
Vitrified clay pipe	0.012–0.014
Steel pipe	0.009–0.011
Monolithic concrete	0.012–0.017
Cement rubble	0.017–0.025
Brick	0.014–0.017
Laminated treated wood	0.015–0.017
Open channels:	
Lined with concrete	0.013–0.022
Earth, clean, after weathering	0.018–0.020
Earth, with grass and some weeds	0.025–0.030
Excavated in rock, smooth	0.035–0.040
Excavated in rock, jagged and irregular	0.040–0.045
Natural stream channels:	
No boulders or brush	0.028–0.033
Dense growth of weeds	0.035–0.050
Bottom of cobbles with large boulders	0.050–0.070

Source: Design Charts for Open-Channel Flow, U.S. Department of Transportation, Federal Highway Administration, Hydraulic Design Series No. 3 (Washington, DC; U.S. Government Printing Office, 1961).

for open-channel flow is gravity. The forces retarding open-channel flow are derived from viscous shear along the channel bed.

7.1 UNIFORM FLOW

For the condition of uniform flow, the velocity in an open channel is usually determined by Manning's equation,

$$V = \frac{1.49}{n} R^{2/3} S^{1/2} \tag{7.1}$$

which was discussed in Chapter 6 [Eq. (6.3)]. Table 7.1 gives values of the roughness coefficient n for various materials used in open channels. Figure 7.1 is a nomograph that facilitates the solution of the equation for various values of n.

For channels with distinct changes in roughness across the width of the cross section, such as a paved channel with grassy sides, it is common to subdivide the cross section according to roughness, compute flow velocity for each subsection, and then sum the flows.

Example 7.1

Determine the discharge of a trapezoidal channel having a brick bottom and grassy sides, with the following dimensions: depth—6 ft, bottom width—12 ft, top width—18 ft. Assume $S = 0.002$.

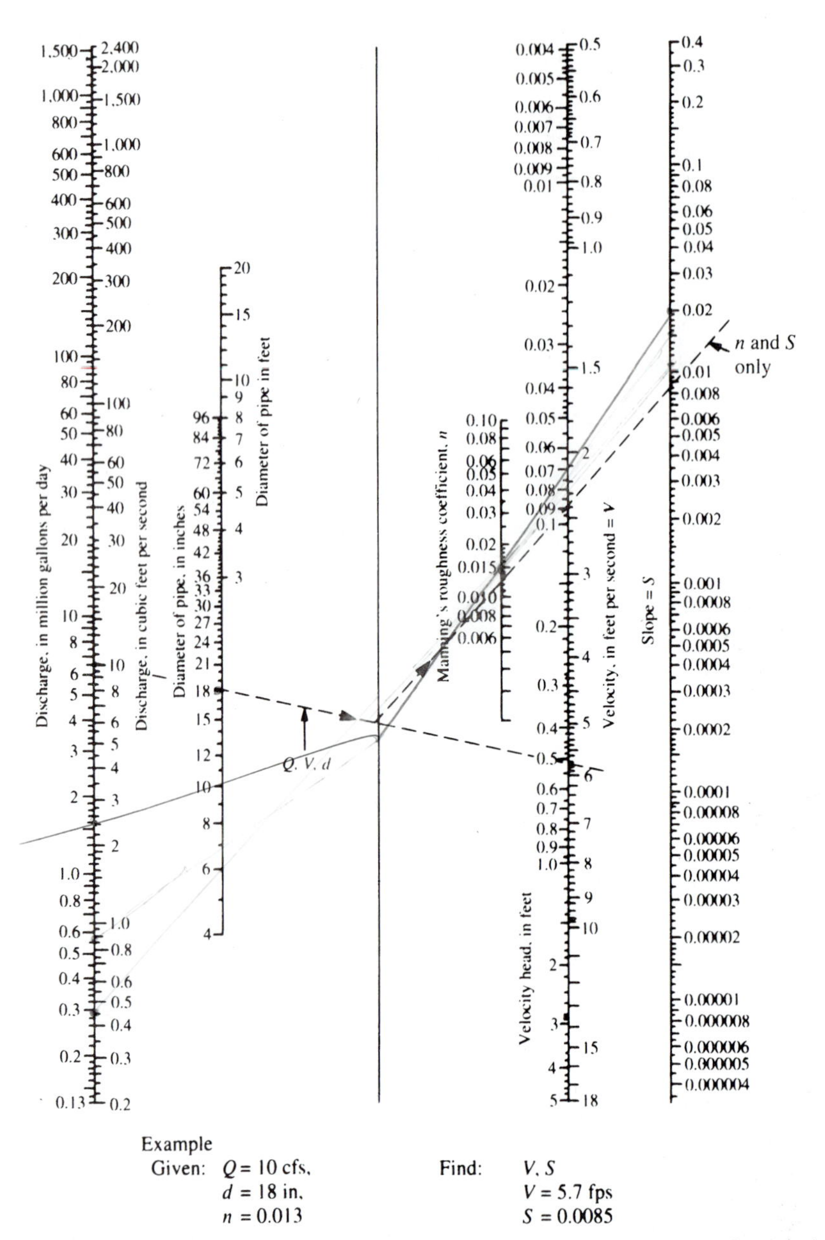

FIGURE 7.1 Nomograph based on Manning's formula for circular pipes flowing full.

Solution:

1. Using Eq. (7.1) and noting that $Q = AV$, calculate the discharge for the portion of flow in the rectangular subsection. From Table 7.1, choose $n = 0.017$.

$$R = (6)(12)/(12) = 6.0 \text{ ft}$$

$$Q = (1.49/0.017)(72)(6)^{2/3}(0.002)^{1/2} = 930.75 \text{ cfs}$$

2. Calculate the discharge for the portion of flow in the grassy subsections of the channel. From Table 7.1, choose $n = 0.025$. Then for each side

$$A = (0.5)(3)(6) = 9 \text{ ft}^2$$

$$R = 9/6.7 = 1.35 \text{ ft}$$

For both sides,

$$Q = 2[(1.49/0.025)(9)(1.35)^{2/3}(0.002)^{1/2}] = 58.59 \text{ cfs}$$

3. The total discharge for the channel is thus

$$Q = 930.75 + 58.59 = 989.34 \text{ cfs}$$

For open channels consisting of circular pipes, or tunnels flowing partly full, calculation of hydraulic radius and cross-sectional area of flow can be cumbersome. Figure 7.2 facilitates these calculations by showing the relation between the hydraulic elements of a circular pipe, which allows the conditions of a pipe flowing partly full, to be calculated from the conditions of the full-flowing pipe.

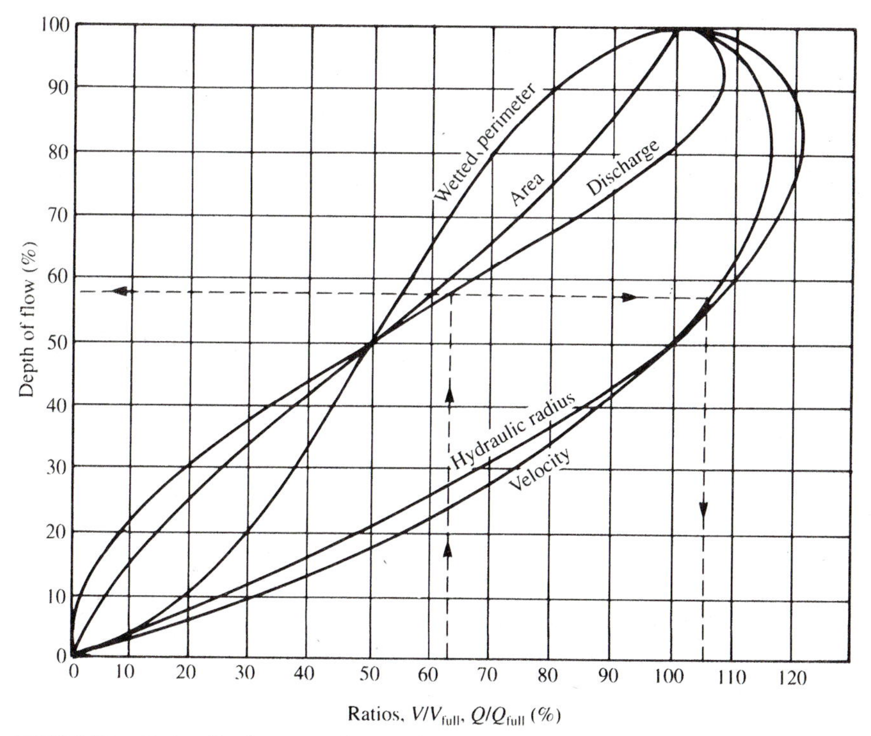

FIGURE 7.2 Hydraulic elements of a circular section for constant n.

Example 7.2

Given a pipe discharge flowing full of 16 cfs and a velocity of 8 fps, find the velocity and depth of flow when $Q = 10$ cfs.

Solution: Enter Figure 7.2 at $10/16 = 62.5\%$ of value for full section. Obtain a depth of flow of 57.5% of full-flow depth and a velocity of $1.05 \times 8 = 8.4$ fps.

The depth at which uniform flow occurs in an open channel is termed the *normal depth* d_n. This depth can be computed by using Manning's equation for discharge after the cross-sectional area A and the hydraulic radius R have been translated into functions of depth. The solution for d_n is often obtained by trial and error.

For a specified channel cross section and discharge, there are three possible values for the normal depth, all dependent on the channel slope. Uniform flow for a given discharge may occur at critical depth, at less than critical depth, or at greater than critical depth. Critical depth occurs when the specific energy is a minimum. Specific energy is defined as

$$E_s = d + V^2/2g \tag{7.2}$$

where

d = depth of flow

V = mean velocity

Flow at critical depth is highly unstable, and designs indicating flow at or near critical depth are to be avoided. For any value of E_s above the minimum, two alternative depths of flow are possible. One is greater than critical depth, the other less than critical depth. The former case indicates subcritical flow, the latter supercritical flow. The critical depth for a channel can be found by taking the derivative of Eq. (7.2) with respect to depth, setting this equal to zero, and solving for d_c. For mild slopes, the normal depth is greater than d_c and subcritical flow prevails. On steep slopes, the normal depth is less than the critical depth and flow is supercritical. Once the critical depth has been computed, critical velocity is easily obtained. The critical velocity for a channel of any cross section can be shown to be

$$V_c = \sqrt{g\frac{A}{B}} \tag{7.3}$$

where

V_c = critical velocity

A = cross-sectional area of the channel

B = width of the channel at the water surface

The critical slope can then be found by using Manning's equation.

In practice, uniform flows are encountered only in long channels after a transition from nonuniform conditions. Nevertheless, a knowledge of uniform-flow hydraulics is important, as numerous varied flow problems are solved through partial applications of uniform-flow theory. A basic assumption in gradually varied flow analyses, for example, is that energy losses are considered the same as for uniform flow at the average depth between two sections along the channel that are closely spaced.

7.2 GRADUALLY VARIED FLOW AND SURFACE PROFILES

Gradually varied flow results from gradual changes in depth that take place over relatively long reaches of a channel. Abrupt changes in the flow regime are classified as rapidly varied flow. Problems in gradually varied flow are widespread and represent the majority of flows in natural open channels and many of the flows in man-made channels. They can be caused by such factors as change in channel slope, cross-sectional area, or roughness or by obstructions to flow, such as dams, gates, culverts, and bridges. The pressure distribution in gradually varied flow is hydrostatic and the streamlines are considered approximately parallel.

The significance of gradually varied flow problems may be illustrated by considering the following case. Assume that the maximum design flow for a uniform rectangular canal will occur at a depth of 8 ft under uniform-flow conditions. For these circumstances, the requisite canal depth, including 1 ft of freeboard, will be 9 ft. Now consider that a gate is placed at the lower end of this canal. Assume that at maximum flow the depth immediately upstream from the gate will have to be 12 ft in order to produce the required flow through the gate. The depth of flow will then begin decreasing gradually in an upstream direction and approach the uniform depth of 8 ft. Obviously, unless the depth of flow is known all along the channel, the channel design depth cannot be determined.

Determination of the water-surface profile in an open channel for a given discharge is required for solving many engineering problems. There are 12 classifications of water-surface profiles, or *backwater curves* [1]. Figure 7.3 illustrates several of these and also a typical change that might take place in the transition from one type of flow regime to another. For any channel, the applicable backwater curve will be a function of the relationship between the actual depth of flow and the normal and critical depths of the channel. Often it is helpful to sketch the type curves for a given problem before attempting an actual solution. In doing this, the type curves given in Figure 6.4 are useful. Woodward and Posey provide additional type curves [1].

Numerous procedures have been proposed for computing backwater curves. The direct-step method is discussed here. Referring to Figure 7.4, the energy equation may be written as

$$Z_1' + d_1 + \frac{V_1^2}{2g} = d_2 + \frac{V_2^2}{2g} + H_f$$

A rearrangement of this equation yields

$$\left(\frac{V_2^2}{2g} + d_2\right) - \left(\frac{V_1^2}{2g} + d_1\right) = Z_1' - H_f$$

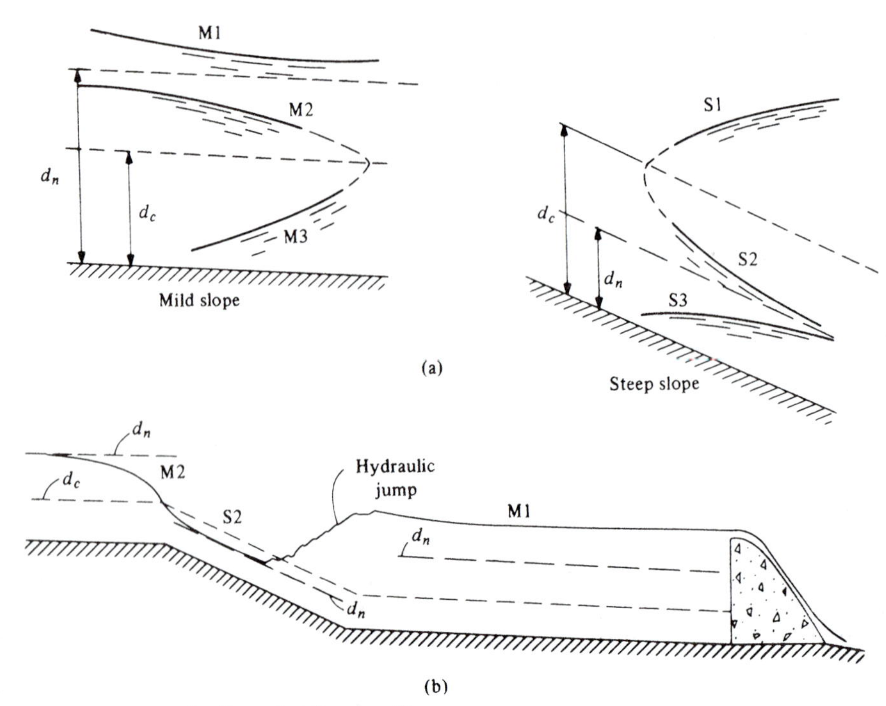

FIGURE 7.3 Gradually varied flow profiles.

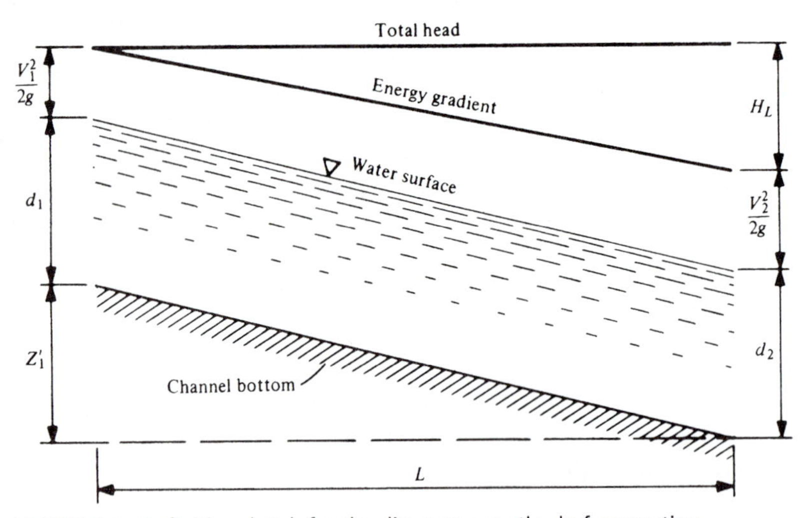

FIGURE 7.4 Definition sketch for the direct-step method of computing backwater curves.

or

$$E_2 - E_1 = S_c L - \overline{S}_e L$$

$$L = \frac{E_2 - E_1}{S_c - \overline{S}_e} \tag{7.4}$$

where

E_2, E_1 = values of specific energy at sections 1 and 2

S_c = slope of the channel bottom

$\overline{S}_e$ = slope of the energy gradient

The value of $\overline{S}_e$ is obtained by assuming that (1) the actual energy gradient is the same as that obtained for uniform flow at a velocity equal to the average of the velocities at sections 1 and 2 (this can be found using Manning's equation) or (2) the slope of the energy gradient is equal to the mean of the slopes of the energy gradients for uniform flow at the two sections. The procedure in using Eq. (7.4) is to select some starting point where the depth of flow is known. A second depth is then selected in an upstream or downstream direction, and the distance from the known depth to this point is computed. Then, using the second depth as a reference, a third is selected and another length increment calculated, and so on. The results obtained will be reasonably accurate provided the selected depth increments are small, since the energy-loss assumptions are fairly accurate under these conditions.

Example 7.3

Water flows in a rectangular concrete channel 10 ft wide, 8 ft deep, and inclined at a grade of 0.10%. The channel carries a flow of 245 cfs and has a roughness coefficient n of 0.013. At the intersection of this channel with a canal, the depth of flow is 7.5 ft. Find the distance upstream to a point where normal depth prevails, and determine the surface profile.

Solution: Using Manning's equation and the given value of discharge, the normal depth is found to be 4 ft. Critical depth y_c is determined by solving Eq. (7.3) after rearranging and substituting by_c for A and b for B, where b equals the width of the rectangular channel. Then

$$V_c = \sqrt{g\left(\frac{by_c}{b}\right)}$$

$$= \sqrt{gy_c}$$

$$y_c = \frac{V_c^2}{g} = \frac{Q^2}{A^2 g} \tag{7.5}$$

$$y_c^3 = \frac{Q^2}{gb^2}$$

TABLE 7.2 Calculations for the Surface Profile of Example 7.3

y	A	V	$\frac{V^2}{2g}$	$E = d + \frac{V^2}{2g}$	P	R	$R^{4/3}$	$S_e = \frac{V^2}{(1.49)^2 R^{4/3}/n^2}$	$\overline{S}_e$	ΔE	$S_c - \overline{S}_e$	$\Delta L = \frac{\Delta E}{S_c - S_e}$
7.5	75	3.27	0.166	7.666	25	3.0	4.33	0.000188	0.000207	0.475	0.000793	598
7.0	70	3.50	0.191	7.191	24	2.91	4.15	0.000225	0.000279	0.932	0.000721	1292
6.0	60	4.08	0.259	6.259	22	2.73	3.82	0.000333	0.000436	0.885	0.000564	1570
5.0	50	4.90	0.374	5.374	20	2.50	3.40	0.000539	0.000762	0.792	0.000238	3328
4.0	40	6.12	0.582	4.582	18	2.22	2.90	0.000984				
												$L = \Sigma \Delta L = 6788$

$$y_c = \sqrt[3]{\frac{Q^2}{gb^2}}$$

Substituting the given values for Q and b yields

$$y_c = \sqrt[3]{\frac{(245)^2}{32.2 \times (10)^2}}$$

and

$$y_c = 2.65 \text{ ft}$$

Since $y > y_n > y_c$, an M1 profile will depict the water surface. The calculations then follow the procedure indicated in Table 7.2.

The distance upstream from the junction to the point of normal depth is 6788 ft. The surface profile can be plotted by using the values in Table 7.2 for y and ΔL. In practice, greater accuracy could be obtained by using smaller depth increments, such as 0.25 ft, but the computational procedure would be the same.

7.3 VELOCITY

Both maximum and minimum velocities are prescribed for water transportation systems. Minimum velocities are set to ensure that suspended matter does not settle out in the conduit, while maximum velocities are set to prevent erosion of the channel material. Normally, velocities in excess of 20 fps are to be avoided in concrete or tile sewers, and whenever possible velocities of 10 fps or less should be used. Specially lined inverts (pipe bottoms) are sometimes employed to combat channel erosion. The average value of the channel shearing stress is related to both erosion and sediment deposition.

Using the Chézy equation for velocity, Fair and Geyer [2] have shown that the minimum velocity for self-cleansing V_m can be determined according to

$$V_m = C\sqrt{\frac{K(\gamma_s - \gamma)}{\gamma}d} \tag{7.6}$$

where γ_s is the specific weight of the particles, γ is the specific weight of water, d is the particle diameter, and C is the Chézy coefficient (equal to $1.49R^{1/6}/n$ as evaluated by Manning). The value of C is selected with consideration given to the containment of solids in the flow. The value of K must be found experimentally and appears to range from 0.04 for the initiation of scour to more than 0.8 for effective cleansing [2].

If nonflocculating particles have mean diameters between 0.05 and 0.5 mm, the minimum velocity may be determined [3] using

$$V_m = \left[25.3 \times 10^{-3} g \frac{d(\rho_s - \rho)}{\rho}\right]^{0.816} \left(\frac{D\rho_m}{\mu_m}\right)^{0.633} \tag{7.7}$$

where

V_m = minimum velocity, fps

d = mean particle diameter, ft

D = pipe diameter, ft

ρ_m = mean mass density of the suspension, pcf

ρ_s = particle density, pcf

μ_m = viscosity of the suspension, lb-sec/ft^2

ρ = liquid mass density, pcf

The value of ρ_m may be determined using

$$\rho_m = X_v \rho_s + (1 - X_v)\rho \tag{7.8}$$

where X_v is the volumetric fraction of the suspended solids. Ordinarily, minimum velocities are 2 and 3 fps for sanitary sewers and storm drains, respectively.

DESIGN OF SANITARY SEWERS

The design of sanitary sewers involves (1) the application of the principles given in Chapter 6 to the estimation of design flows and (2) the application of hydraulic engineering principles to devise appropriate conveyances for transporting these flows [2,4,5]. Most sewers are designed for a capacity life of 25–50 yr. These conveyances must be capable of handling peak and minimum flows without the deposition of suspended solids. In general, circular sanitary sewers are designed to flow at between one-half and full depth when carrying their design discharges. Ordinarily, they should not be designed to run full since some airspace is desirable for ventilation and suppression of sulfide generation. Sewers at system extremities should be designed to flow about half full at peak flow. Large interceptor and trunk sewers are less subject to flow variations than are small collecting sewers, and greater depths of flow are justified for them. Gupta notes that pipes up to 16 in. (400 mm) in diameter are usually designed to flow half full; those between 16 and 35 in. (900 mm) to flow two-thirds full; and larger pipes are usually designed to flow at three-quarters to full depth [4]. Once the flow that the sewer must carry has been determined, a pipe size and slope must be established. Since sewers are commonly located beneath roads and streets, their slopes are generally made to conform to street slopes. This practice also tends to minimize excavation costs. Where street slopes do not permit generating minimum carrying velocities (2 ft/s), sewer grades are set accordingly. Table 7.3 summarizes minimum slopes for a range of flow values. In cul-de-sacs and other short, dead-end street sections, the designer will often find it impossible to meet minimum velocity requirements. Such pipes may require periodic flushing to scour accumulated sediments.

7.4 HOUSE AND BUILDING CONNECTIONS

Connections from the main sewer to houses or other buildings are commonly constructed of vitrified clay, concrete, or asbestos cement pipe. Building connections are usually made on about a 2% grade with 6 in. or larger pipe. Grades less than 1% are to be avoided, as are extremely sharp grades.

TABLE 7.3 Slopes Required to Maintain Velocities of 2 ft/sec[a]

Q (cfs)	Slope (ft/1000)
0.1	9.2
0.2	6.1
0.3	4.8
0.4	4.1
0.6	3.22
0.8	2.73
1.0	2.39
1.5	1.89
2.0	1.59
3.0	1.26
4.0	1.06

[a]For circular pipes where the flow is not less than 0.1% and not more than 95% of capacity [6].

7.5 COLLECTING SEWERS

Collecting sewers gather flows from individual buildings and transport the material to an interceptor or main sewer. Local standards usually dictate the location of these sewers, but they are commonly located under the street paving on one side of the storm drain, which is usually centered. The collecting sewer should be capable of conveying the flow of the present and anticipated population of the area it is to serve. Design flows are the sum of the peak domestic, commercial, and industrial flows and infiltration. The collecting sewer must transport this design flow when flowing full. Grades requiring minimum excavation while meeting maximum and minimum restrictions on velocity are preferred. Manholes are normally located at changes in direction, grade, or pipe size or at intersections of collecting sewers. For 8-, 10-, or 12-in. sewers, manholes spaced no farther than about 400 ft apart permit inspection and cleaning when necessary. For larger sizes, the maximum spacing can be increased. The minimum size pipe allowed by most codes is 8 in. Note that where changes in sewer direction occur and/or where flows combine, usually at manholes, head losses occur. When flow velocities are in the usual range for sewers, a drop of about 0.1 ft or 30 mm across a manhole is usually sufficient to account for this (Fig. 7.5). For large sewers, such as interceptors and trunks, head losses may require greater adjustments (see Section 7.21). At junctions where sewer size increases, common practice is to keep the crowns of the incoming and outgoing pipes, or the 0.8 depth points of these pipes, at the same elevation. A typical plan and profile of a residential sewer are given in Figure 7.5.

7.6 INTERCEPTING SEWERS

Intercepting sewers are expected to carry flows from the collector sewers in the drainage basin to the point of treatment or disposal of the wastewater. These sewers normally follow valleys or natural streambeds of the drainage area. When the sewer is built through

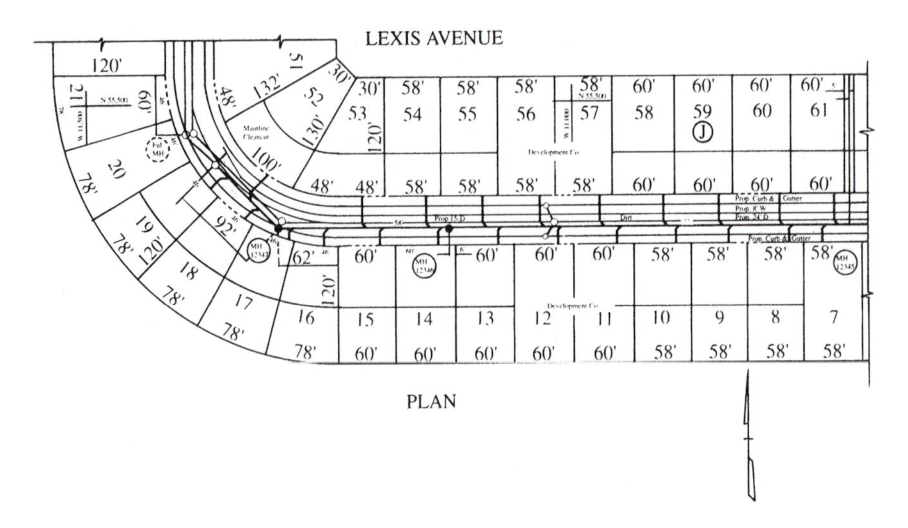

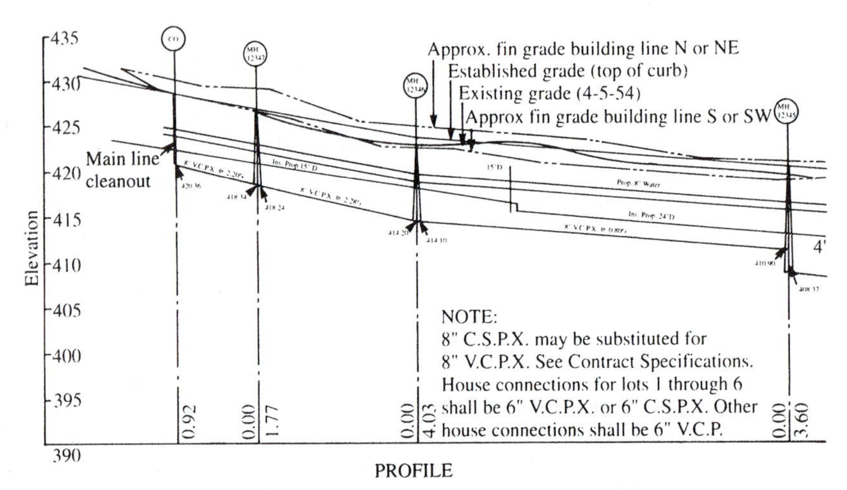

FIGURE 7.5 Typical sewer plan and profile.

underdeveloped areas, precaution should be taken to ensure the proper location of manholes for future connections. For 15- to 27-in. sewers, manholes are constructed at least every 600 ft; for larger sewers, increased spacing is common. Manholes or other transition structures are usually built at every change in pipe size, grade, or alignment. For large sewers, horizontal or vertical curves are sometimes constructed. Grades should be designed so that the criteria regarding maximum and minimum velocities are satisfied.

7.7 MATERIALS

Collecting and intercepting sewers are constructed of asbestos cement sewer pipe, cast-iron pipe, concrete pipe, vitrified clay pipe, brick, plastic or PVC, and bituminized fiber pipe. Care should be exercised in designing the system so that permissible structural loadings are not exceeded for the material selected. Information on pipe loading is readily available from the Clay Products Association, the Portland Cement Association, the Cast Iron Pipe Manufacturers, and others.

7.8 SYSTEM LAYOUT

The first step in designing a sewerage system is to establish an overall system layout that includes a plan of the area to be sewered, showing roads, streets, buildings, other utilities, topography, soil type, and the cellar or lowest floor elevation of all buildings to be drained. Where part of the drainage area to be served is undeveloped and proposed development plans are not yet available, care must be taken to provide adequate terminal manholes that can later be connected to the constructed system serving the area. If the proposed sewer connects to an existing one, an accurate location of the existing terminal manhole, giving invert elevation, size, and slope, is essential.

On the drainage area plan just described, a tentative layout of collecting sewers and the intercepting sewer or sewers should be made. When feasible, the sewer location should minimize the length required to provide service to the entire area. Length may be sacrificed if the shorter runs would require costlier excavation. Normally, the sewer slope should follow the ground surface so that waste flows can follow the approximate path of the area's surface drainage. In some instances, it may be necessary to lay the sewer slope in opposition to the surface slope, or to pump wastes across a drainage divide. This situation can occur when a developer buys land lying in two adjacent drainage basins and, for economic or other reasons, must sewer the whole tract through only one basin.

Intercepting sewers or trunk lines are located to pass through the lowest point in the drainage area (the outlet) and to extend through the entire area to the drainage divide. Normally, they follow major natural drainage ways and are located in a designated right-of-way or street. Land slopes on both sides should be toward the intercepting sewer. Lateral or collecting sewers are connected to the intercepting sewer and proceed upslope to the drainage divide. Collecting sewers should be located in all streets or rights-of-way within the area to provide service throughout the drainage basin. Figure 7.6 illustrates a typical sewer layout. Note that the interceptor sewer is located in the stream valley and collector sewers transport flows from the various tributary areas to the interceptor.

A general guideline for sewer system layout and design has been proposed by Thackston. Basic steps include the following [7]:

1. Obtain or develop a topographic map of the area to be served.
2. Locate the drainage outlet. This is usually near the lowest point in the area and is often along a stream or drainage way.
3. Sketch in a preliminary pipe system to serve all of the contributors (Fig. 7.6).
4. Pipes must be located so that users or future users can readily connect. Pipes must also be located to provide access for maintenance. This is usually accomplished by placing them in streets or other rights-of-way.

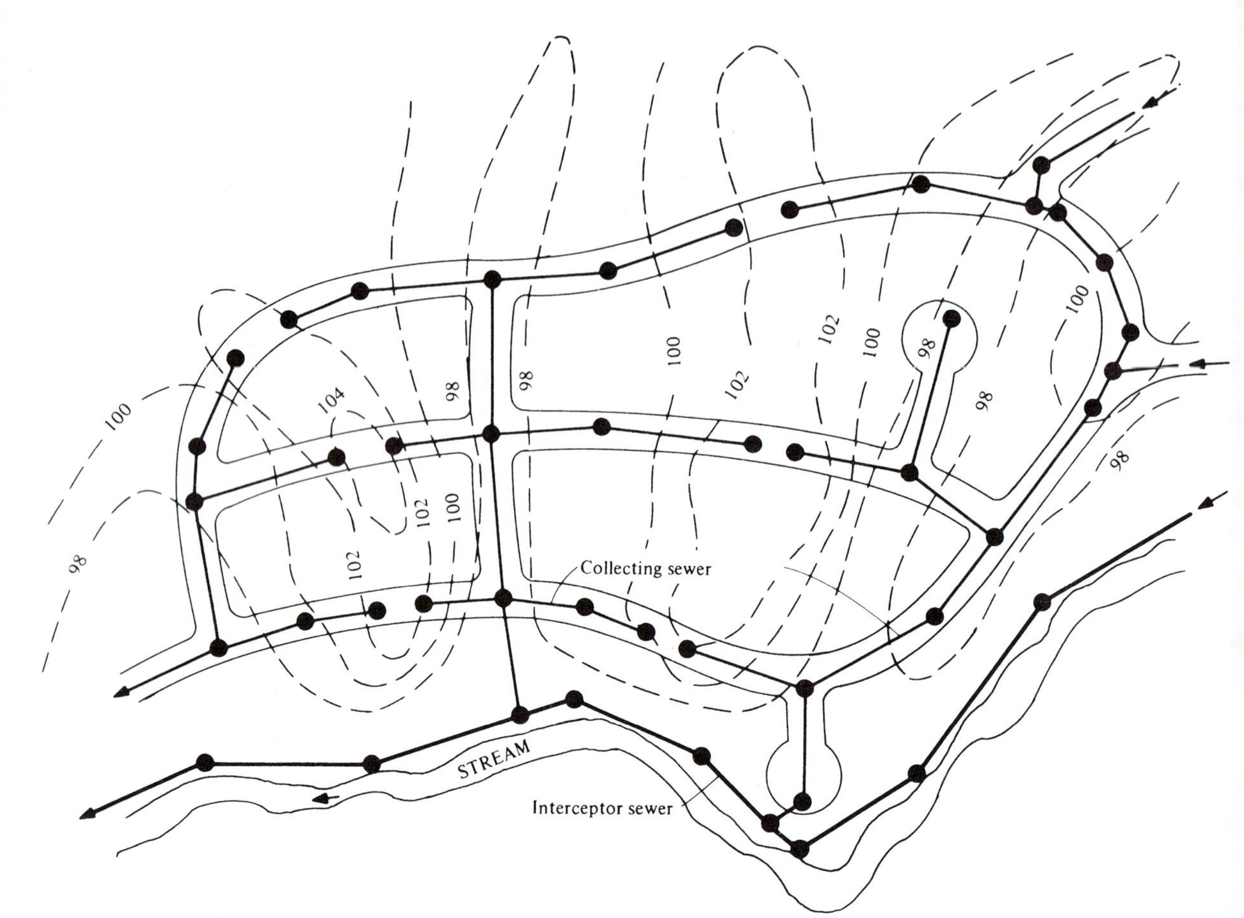

FIGURE 7.6 Typical layout for the design of a sewerage system.

5. Insofar as practical, sewers should follow natural drainage ways to minimize excavation and pumping requirements. Large trunk sewers are usually constructed in low-lying areas closely paralleling streams or channels. In general, pipes should cross contours at right angles.
6. Establish preliminary pipe sizes. Eight inches (usually the minimum allowable) can serve several hundred residences even at minimal grades.
7. Revise the layout to optimize flow-carrying capacity at minimum cost. Pipe lengths and sizes should be kept as small as possible, pipe slopes should be maximized within tolerances for velocity, excavation depth should be minimized, and the number of appurtenances should be kept as small as possible. Traditional hydraulic design procedures combined with the use of optimization techniques can be utilized in this process [8].
8. Try to avoid pumping across drainage boundaries. Pumping stations are costly and add maintenance problems. Furthermore, energy costs for pumping can be

significant. A balance must be made between excavation and pumping. While no hard-and-fast rule can be given, pumping should be considered only when excavations exceeding about 20–40 ft are contemplated.

7.9 HYDRAULIC DESIGN

Open-Channel Flow

The hydraulic design of a sanitary sewer can be carried out in a systematic manner (see solved Example 7.6). Peak design flows should be carried by pipes at velocities great enough to prevent sedimentation, yet small enough to prevent erosion. To minimize head losses at transitions and eliminate backwater effects, the hydraulic gradient must not change abruptly or slope in a direction adverse to the flow at changes in horizontal direction, pipe size, or quantity of flow. Sewers are usually designed to closely follow the grade of the ground surface or street paving under which they are laid. In addition, the depth of cover should be kept as close to the minimum (below the frost line) as possible. Sewer location must take into account the location of other subsurface utilities or structures. General practice is not to locate sewers in the same trench as water mains.

Pipe sizes are determined in the following manner. A profile of the proposed sewer route is drawn and the hydraulic gradient at the downstream end of the sewer noted (normally, the elevation of the hydraulic gradient of the sewer being met). Where discharge is to a treatment plant or an open body of water, the hydraulic gradient is the elevation of the free water surface at this point. At the beginning elevation of the hydraulic gradient, a tentative gradient approximately following the ground surface is drawn upstream to the next point of control (usually a manhole). Exceptions to this occur when the surface slope is less than adequate to provide cleansing velocities, where obstructions preclude using this slope, or where adequate cover cannot be maintained. Using the tentative gradient slope, a pipe size is then selected that comes closest to carrying the design flow at the desired depth. Usually, a standard size pipe will not be found that will carry the flow at the exact depth and gradient investigated. It is common practice then to select the next largest pipe size, modify the slope, or both. The choice depends on a comparison of pipe cost savings versus excavation cost and on local conditions, such as the placement of other utilities in the right-of-way. But regardless of velocity and pipe size calculations, pipe diameter is never decreased on downstream reaches because this can lead to sediment accumulation and blockage at the point of reduction.

Hydraulic computations using Manning's equation [Eq. (7.1), Fig. 7.1] for full pipe flow are straightforward. Commonly, however, flow at less than full is encountered or desired in practice. In such cases, relationships involving geometric properties (Table 7.4) and hydraulic elements (Fig. 7.2) of circular sections facilitate calculations. Their use is illustrated in the following examples.

Example 7.4

Find the peak hourly flow in mgd and m^3/s for an 800-acre urban area having the following features: domestic flows 90 gpcd, commercial flows 15 gpcd, infiltration 600 gpd/acre, and population density 20 persons per acre.

TABLE 7.4 Geometric Relationships for Circular Pipes

d/d_m	R/d_m	$AR^{2/3}/d_m^{8/3}$
0.01	0.0066	0.0000
0.05	0.0326	0.0015
0.10	0.0635	0.0065
0.15	0.0929	0.0152
0.20	0.1206	0.0273
0.25	0.1466	0.0427
0.30	0.1709	0.0610
0.35	0.1935	0.0820
0.40	0.2142	0.1050
0.45	0.2331	0.1298
0.50	0.2500	0.1558
0.55	0.2649	0.1825
0.60	0.2776	0.2092
0.65	0.2881	0.2358
0.70	0.2962	0.2608
0.75	0.3017	0.2840
0.80	0.3042	0.3045
0.85	0.3033	0.3212
0.90	0.2890	0.3324
0.95	0.2864	0.3349
1.00	0.2500	0.3117

d_m = full-flow depth
d = actual depth of flow
R = hydraulic radius
A = area of flow

Solution: Calculate the average daily flows for each contributor.

$$\text{Domestic: } 90 \times 20 \times 800 = 1.44 \text{ mgd}$$

$$\text{Commercial: } 15 \times 20 \times 800 = 0.24 \text{ mgd}$$

$$\text{Infiltration: } 600 \times 800 = 0.48 \text{ mgd}$$

$$\text{Total} = 1.44 + 0.24 + 0.48 = 2.16 \text{ mgd}$$

Referring to Table 5.1, select a peak hour to average day ratio of 3.0.

$$\text{Peak hourly flow} = 3.0 \times 2.16 = 6.48 \text{ mgd}$$

$$= 6.48 \times 0.044 = 0.29 \text{ m}^3/\text{s}$$

Example 7.5

A sewer with a flow of 10 cfs enters manhole 6. The distance downstream to manhole 7 is 400 ft. The finished street surface elevation at manhole 6 is 166.3 ft and at manhole 7 it is 164.3 ft. Note that the crown of the pipe must be at least 6 ft below the street surface at each manhole. For a Manning $n = 0.013$, find the required standard size pipe to carry the flow under (a) full flow and (b) one-half of full flow. The flow velocity must be at least 2 ft/s and must not exceed 10 ft/s.

Solution: The fall of the sewer in 400 ft is 2 ft if the street grade is used as the slope. The initial assumption for S is 0.005 ft/1000. If this slope is not sufficient to sustain a velocity of 2 fps, it will be modified later. In like manner, if the calculated velocity exceeds 10 fps, the slope will have to be reduced.

1. Using the nomograph in Figure 7.1 and the values $Q = 10$ and $S = 0.005$, we determine that a 21-in.-diameter pipe is required. This pipe will support a discharge somewhat higher than the design value of 10, but it is the closest standard size. The velocity associated with a 21-in. pipe flowing full at a slope of 0.005 is between 4 and 5 fps and thus meets the design criteria. For full flow, the solution is to choose a 21-in. pipe laid at a slope of 0.5 ft/100 ft.
2. For flow at $d = 0.5d_m$, Manning's equation Figure 7.2, and Table 7.4 are used. From Figure 7.2, it can be seen that for half-full depth, the velocity is the same as for full depth, and geometrically A and R are one-half their full-depth values.

The pipe size for one-half flow is calculated from the equation

$$Q = 1.49/nAR^{2/3}S^{1/2}$$

by inserting known values for Q, n, and S. From Table 7.4, and noting that $AR^{2/3}(d_m)^{8/3} = 0.156$ for $d/d_m = 0.5$, the equation can be solved for d_m:

$$10 = (1.49/0.013)(0.156(d_m)^{8/3})(0.005)^{0.5}$$

$$(d_m)^{8/3} = 7.91 \qquad \text{and} \qquad d_m = 2.17 \text{ ft}$$

Select the closest standard pipe size of 27 in. diameter.
From Figure 7.1, it can be seen that for a 27-in. pipe laid at a slope of 0.005, the velocity at full flow (also at half flow) is 5.5 fps. This satisfies the specifications.

The flow in the 27-in. pipe at half-full conditions is calculated to determine how close it comes to the design flow of 10 cfs:

$$\begin{aligned} Q &= (1.49/0.013)(0.156)(d_m)^{8/3})(0.005)^{0.5} \\ &= 10.9 \text{ cfs} \end{aligned}$$

Thus, the 27-in. pipe is flowing at slightly less than half depth under design conditions of $Q = 10$ cfs. For this pipe and slope, the full flow is $10.9 \times 2 = 21.8$ cfs. From Figure 7.2 with $Q/Q_{\text{full}} = 10/21.8$, it can be seen that the actual flow depth under design conditions is about $0.48d_m$.

Example 7.6

A sewer system is to be designed for the urban area of Figure 7.7. Street elevations at manhole locations are shown on the figure. It has been determined that the population density is 40 persons per acre and that the sewage contribution per capita is 100 gpd. In addition, there is an infiltration component of 600 gallons per acre per

Manhole	Street El.
1	101.3
2	104.18
3	105.33
4	107.25
5	109.23
6	112.19
7	116.6
8	112.04
9	115.04
10	117.46
11	113.77
12	110.29
13	115.8
14	111.92
15	108.58
16	116.37
17	112.57
18	108.89

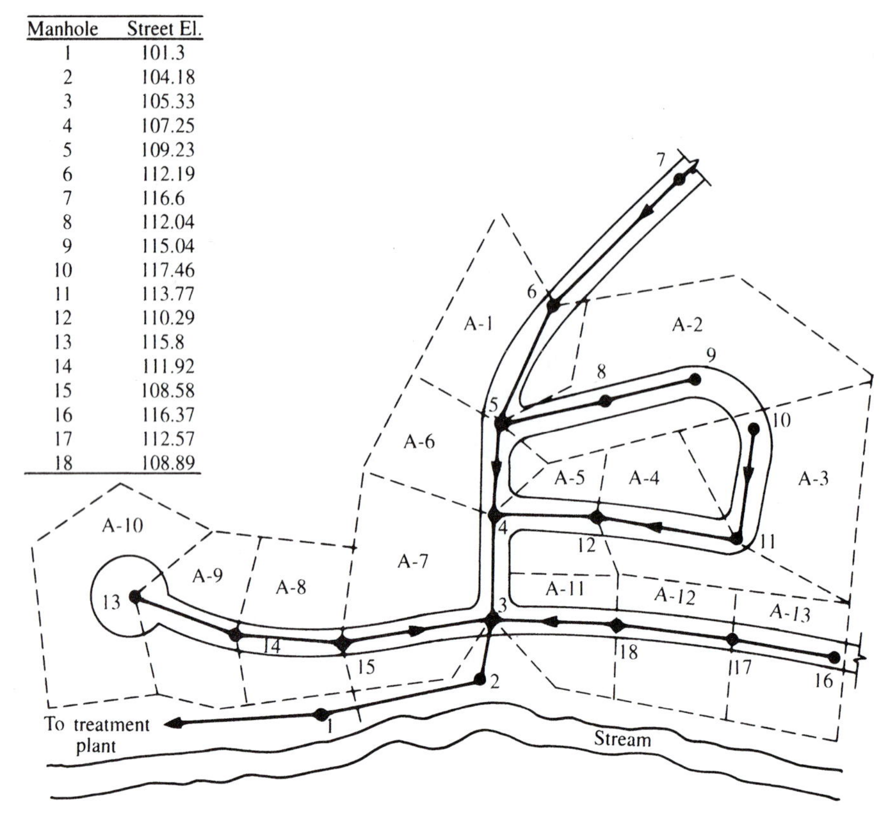

FIGURE 7.7 Sewer system for Example 7.6.

day (gpad). Local regulations require that no sewers be less than 8 in. in diameter. Assume an n value of 0.013.

Solution: The design procedure follows a series of steps and the format of Tables 7.5 and 7.6.

1. Determine manhole locations. These are already indicated for the example in Figure 7.7, but in practice they are determined on the basis of changes in direction, junctions of pipes, and maximum lengths for various pipe sizes, as discussed earlier. Pipe lengths are tabulated in column 3 of Table 7.5.
2. Determine street elevations at manhole locations. These are given in Figure 7.7, but they would be obtained from street profiles of the type shown in Figure 7.5.
3. Using the plan showing the location of manholes, determine the distance between them. For the example, the lengths are given in column 3 of Table 7.5.
4. Calculate the street slopes between manholes. This is done by taking the differences in elevations between manholes and dividing them by the distances between the manholes. For example, the slope between M-6 and M-5 is

$$S = (112.19 - 109.23)/470 = 0.0063 \text{ ft/ft}$$

Calculated slopes are shown in column 5 of Table 7.6.

TABLE 7.5 Calculation of Sewage Flows for Example 7.6

(1)	(2)		(3)	(4)	(5)	(6)	(7)	(8)	(9)	(10)
	Manhole		Pipe Lgth.	Incr. Area	Cum. for Br.	Incr. Mn.	Cum. Mn.	Sewage	Flow	Flow
Pipe No.	From	To	(ft)	(acres)	(acres)	(acres)	(acres)	(mgd/acre)	(mgd)	(cfs)
1	7	6	630	—	—	—	87	0.0138	1.20	1.86
2	6	5	470	5.1	—	5.1		0.0138	—	—
—	6	5	—	—	—	—	92.1	0.0138	1.27	1.97
3	9	8	390	12.1	—	—	—	0.0138	0.17	0.26
4	8	5	385	—	12.1	—	—	0.0138	0.17	0.26
5	5	4	330	4.8	—	4.8	—	0.0138	—	—
—	5	4	—	—	—	—	109	0.0138	1.50	2.33
6	10	11	410	8.7	8.7	—	—	0.0138	0.12	0.19
7	11	12	400	6.3	15	—	—	0.0138	0.21	0.32
8	12	4	380	4.7	19.7	—	—	0.0138	0.27	0.42
9	4	3	370	—	—	—	128.7	0.0138	1.78	2.75
10	16	17	380	5	5	—	—	0.0138	0.07	0.11
11	17	18	400	4.9	9.9	—	—	0.0138	0.14	0.21
12	18	3	405	4.3	14.2	—	—	0.0138	0.20	0.30
13	13	14	400	13.1	13.1	—	—	0.0138	0.18	0.28
14	14	15	380	5.3	18.1	—	—	0.0138	0.25	0.39
15	15	3	411	9.7	28.1	—	—	0.0138	0.39	0.60
16	3	2	230	—	—	—	171	0.0138	2.36	3.66
17	2	1	600	—	—	—	171	0.0138	2.36	3.66

Column 4 = incremental area contributing to pipe.
Column 5 = cumulative area for branches entering collecting sewer.
Column 6 = incremental area contributing directly to collecting sewer.
Column 7 = cumulative area contributing to a sewer.
Column 8 = calculated contributing sewage flow in mgd/acre.
Column 9 = pipe flow in mgd (0.0138 × area).
Column 10 = values in column 9 converted to cfs [(col.9) × 1.55].

TABLE 7.6 Calculation of Pipe Sizes and Velocities for Example 7.6

1	2		3	4	5	6	7	8	9	10	11	12	13	14	15	16
	Manhole		Pipe	Des.	Str.	Min.	Des.	Pipe	Pipe	Pipe	*Q/Q*-	*d/d*-	Flow-	*V*-	*V/V*-	
Pipe			Lgth.	Flow	Sl.	Sl.	Sl.	Size	Size	Flow	Full	Max	*d*	Full	Full	*V*
No.	From	To	(ft)	(cfs)	(ft/ft)	(ft/ft)	(ft/ft)	(cal.-in.)	(sel.-in.)	Full	(cfs)	—	(in.)	(fps)	—	(fps)
1	7	6	630	1.86	0.0070	0.0027	0.007	10.04	12	2.98	0.62	0.58	6.96	3.80	1.05	3.99
2	6	5	470	—	—	—	—	—	—	—		—	—	—	—	—
—	6	5	—	1.97	0.0063	0.0016	0.0063	10.26	12	2.83	0.70	0.62	7.44	3.60	1.08	3.89
3	9	8	390	0.26	0.0077	0.0054	0.0077	—	8	1.06	0.25	0.35	2.8	3.04	0.82	2.49
4	8	5	385	0.26	0.0073	0.0054	0.0073	—	8	1.03	0.25	0.35	2.8	2.96	0.82	2.42
5	5	4	330	—	—	—	—	—	—	—	—	—	—	—	—	—
—	5	4	—	2.33	0.006	0.0015	0.006	10.93	12	2.76	0.84	0.7	8.4	3.52	1.12	3.94
6	10	11	410	0.19	0.009	0.0064	0.009	—	8	1.14	0.17	0.29	2.32	3.28	0.73	2.40
7	11	12	400	0.32	0.0087	0.0047	0.0087	—	8	1.13	0.28	0.36	2.88	3.23	0.83	2.68
8	12	4	380	0.42	0.008	0.0037	0.008	—	8	1.08	0.39	0.44	3.52	3.09	0.92	2.85
9	4	3	370	2.75	0.0052	0.0013	0.0052	11.63	15	4.66	0.59	0.55	8.25	3.80	1.04	3.95
10	16	17	380	0.11	0.01	0.0089	0.01	—	8	1.21	0.09	0.2	1.6	3.46	0.55	1.90
11	17	18	400	0.21	0.0092	0.006	0.0092	—	8	1.16	0.18	0.3	2.4	3.32	0.74	2.46
12	18	3	405	0.3	0.0088	0.006	0.0088	—	8	1.13	0.26	0.35	2.8	3.24	0.82	2.66
13	13	14	400	0.28	0.0097	0.005	0.0097	—	8	1.19	0.24	0.34	2.7	3.41	0.8	2.73
14	14	15	380	0.39	0.0088	0.0042	0.0088	—	8	1.13	0.34	0.41	3.28	3.24	0.89	2.89
15	15	3	411	0.6	0.0079	0.0032	0.0079	—	8	1.07	0.56	0.53	4.24	3.07	1.03	3.17
16	3	2	230	3.66	0.005	0.0013	0.005	12.94	15	4.57	0.80	0.68	10.2	3.73	1.11	4.14
17	2	1	600	3.66	0.0048	0.0013	0.0048	12.94	15	4.48	0.82	0.69	10.35	3.65	1.12	4.09

Column 5 = street slope; column 6 = minimum slope from Table 7.3; column 7 = design slope, larger of values in columns 5 and 6.
Column 8 = full flow Q determined using Manning's equation; column 9 = closest standard pipe size.
Column 10 = full flow in selected pipe size (Manning's equation); column 11 = ratio of values in columns 4 and 10.
Column 12 = value from Figure 7.2 for ratio of column 11; column 13 = value in column 9 × value in column 12.
Column 14 = *Q/A* for full pipe flow; column 15 = ratio determined from Figure 7.2 for the d/d-max value of column 12.
Column 16 = value in column 14 × value in column 15.

5. Determine the size of contributing areas to the sewer segments. This is done by using finished-grade topography and knowledge of the building connection schemes for contributing elements, such as houses, to the sewer segments. For the example, the contributing area boundaries are shown in Figure 7.7, and the area sizes in acres, both incremental and cumulative, are tabulated in Table 7.5, columns 4–7.
6. Calculate the flow per contributing acre for each of the 13 areas. In this example, the per-acre flows are all the same, but they can be highly variable. They are calculated by using procedures described in Section 5.2 and in Example 7.4. For this problem, the per-acre flows are calculated as follows:

   a. residential flow = (40 per acre) × (100 gpcd) = 4000 gpad
   b. infiltration = 600 gpad
   avg. daily flow = 4600 gpad
   c. assuming the design flow = 3 × avg. daily flow, peak flow = 4600 × 3 = 13,800 gpad
   d. converting to mgd, the design flow per acre is 0.0138 mgd/acre. This value is entered in column 8 of Table 7.5

7. Calculate the sewage flows in pipes. This is done by multiplying the number of contributing acres by 0.0138 (the per-acre contribution). The flows in mgd and cfs are shown in columns 9 and 10 of Table 7.5. Note that these calculations are performed by the spreadsheet used in the example. Spreadsheets are excellent tools for solving problems of this type. In this case QUATTRO PRO 3.0 was used, but many other spreadsheet programs are available.
8. Enter street slopes and minimum slopes for pipes in Table 7.6 (column 7). Note that the minimum slopes are estimated from values in Table 7.3. The initial design slope, adjusted later if need be, is the street slope unless that slope is less than the minimum slope required or other conditions make it necessary to assign a different grade.
9. Calculate the pipe size required to handle the design flow under full-flow conditions. Calculations made in the spreadsheet use Manning's equation with $n = 0.013$. Computed diameters are given in column 8 of Table 7.6. Note that a pipe size of 8 in. (the minimum allowable) is assigned to a number of pipes without calculating a diameter. The reasoning for that assignment is as follows. A look at the design flows of column 4, Table 7.6, shows that many of them are small, less than 0.6 cfs. For these flows, the minimum design slope is 0.0073. From Manning's nomograph, Figure 7.1, it can be seen that an 8-in. pipe at a slope of 0.0073 can carry a flow of about 1 cfs at a velocity of about 3 fps. Accordingly, it is clear that an 8-in. pipe will handle all flows up to 0.6 cfs for the slopes given. An 8-in. pipe is thus selected for these flows and tabulated in column 9 of Table 7.6.

   The calculated pipe sizes (column 8, Table 7.6) are modified to the next largest standard pipe size or to a size that accommodates the design flow at a desired depth. These selected sizes are entered in column 9 of Table 7.6. Note that for the range of pipe sizes encountered in this problem, a desirable depth of flow is normally from about half to three-fourths full. Many of the 8-in. pipes, because of the low flows carried, have design flow depths that are less than half the maximum depth. This cannot be avoided and does not create a problem as long as cleansing velocities are maintained.

10. Calculate full flows for selected pipe sizes (column 10, Table 7.6). The design slope and a Manning's n of 0.013 are used.
11. Compute ratios of Q/Q_{full} (column 11, Table 7.6). Using these values and entering Figure 7.2, we determine d/dmax values and V/Vfull values and enter them in columns 12 and 15 of Table 7.6. Multiplying the depth ratios by the full-flow depths for each pipe gives the flow depths for each pipe (column 13, Table 7.6).
12. The full-flow velocity is calculated as Q/A (column 14, Table 7.6). Multiplying the full-flow velocities by the velocity ratio gives the flow velocity for each pipe (column 16, Table 7.6).
13. The calculated values of V are checked to see that they are at least 2 fps. All velocities agree except the velocity for pipe 16–17, 1.9 fps, but this value is considered close enough. Had the calculated value been much less, the design slope would have to be modified.

Example 7.7

Given an invert elevation of 105.19 for the pipe leaving M-6 in Figure 7.7, calculate the invert elevations in and out of M-5 and M-4. Use the slopes and pipe lengths given in Example 7.6.

Solution: Calculations are as follows:

$$105.19 - (0.0063)(470) = 102.23 \text{ invert at entrance to M-5}$$

$$102.23 - 0.1 = 102.13 \text{ drop across M-5}$$

$$102.13 - (0.006)(330) = 100.15 \text{ invert at entrance to M-4}$$

$$100.15 - 0.25 = 99.9 \text{ invert out of M-4}$$

A drop across a manhole of 0.1 was applied where no change in pipe size occurred. At M-4, the upstream pipe is 12 in., and the downstream pipe is 15 in. In this case, a drop of $15 - 12 = 3$ in., or 0.25 ft, was used. The profile of Figure 7.5 illustrates this process.

Pressure Flow

Where pumping of sewage is required, force mains (pressure conduits) must be designed to carry the flows. The costs of pumping and associated equipment are important considerations. The hydraulics of these systems follows the principles already discussed. As in the case of open-channel sewers, pressure sewers must be able to transport sewage at velocities sufficient to avoid deposition and yet not so high as to create pipe erosion problems. Force mains are generally 8 in. in diameter or greater, but for small pumping stations, smaller pipes may sometimes be acceptable. According to Metcalf and Eddy, the following *C* values for use in the Hazen–Williams equation for calculating friction losses in force mains are valid [9]:

100 for unlined cast-iron pipe

120 for cement-lined, cast-iron pipe, reinforced concrete pipe, asbestos-cement pipe, and plastic pipe

110 for steel pipe having bituminous or cement-mortar lining and exceeding 20 in. in diameter

Velocities encountered in force main operations are often in the range of 35 fps. In designing force mains, high points in lines should be avoided if possible. This eliminates the need for air-relief valves. Good design practice dictates that the hydraulic gradient should lie above the line at all points during periods of minimum-flow pumping.

7.10 PROTECTION AGAINST FLOODWATERS

Because the volume of sanitary wastewater is extremely small compared with flood flows, it is important that sewers be constructed to prevent admittance of large surface-runoff volumes. This will preclude overloads on treatment plants with resultant reduction in degree of treatment or complete elimination of treatment in some instances.

Where interceptor sewers are built along streambeds, manhole stacks frequently are raised above a design flood level, such as the 50-yr level. In addition, the manhole structures are waterproofed. Where stacks cannot be raised, watertight manhole covers are employed. Such measures as these keep surface drainage into the sewer at a minimum.

7.11 INVERTED SIPHONS

An inverted siphon is a section of sewer constructed below the hydraulic gradient due to some obstruction; it operates under pressure. The term *depressed sewer* is actually more appropriate, since no real siphon action is involved. Usually, two or more pipes are needed for a siphon in order to handle flow variability. Normally, the water-surface elevations at the entrance and exit to a siphon are fixed. Under these conditions, the hydraulic design consists of selecting a pipe or pipes that will carry the design flow with a head loss equivalent to the difference in entrance and exit water-surface elevations. A transition structure is generally required at the entrance and exit of the depressed sewer to properly proportion or combine the flows.

The minimum flow in a siphon must be great enough to prevent deposition of suspended solids. Normally, velocities of less than 3 fps are unsatisfactory. When the siphon is required to handle flows that vary considerably during any 24-hr period, it is customary to provide two or more pipes. A small pipe is provided to handle the low flows; intermediate flows may be carried by the smallest pipe and a larger pipe. Maximum flows may require the use of three or more pipes. By subdividing the flow in this manner, adequate cleansing velocities are ensured for all flow magnitudes. The

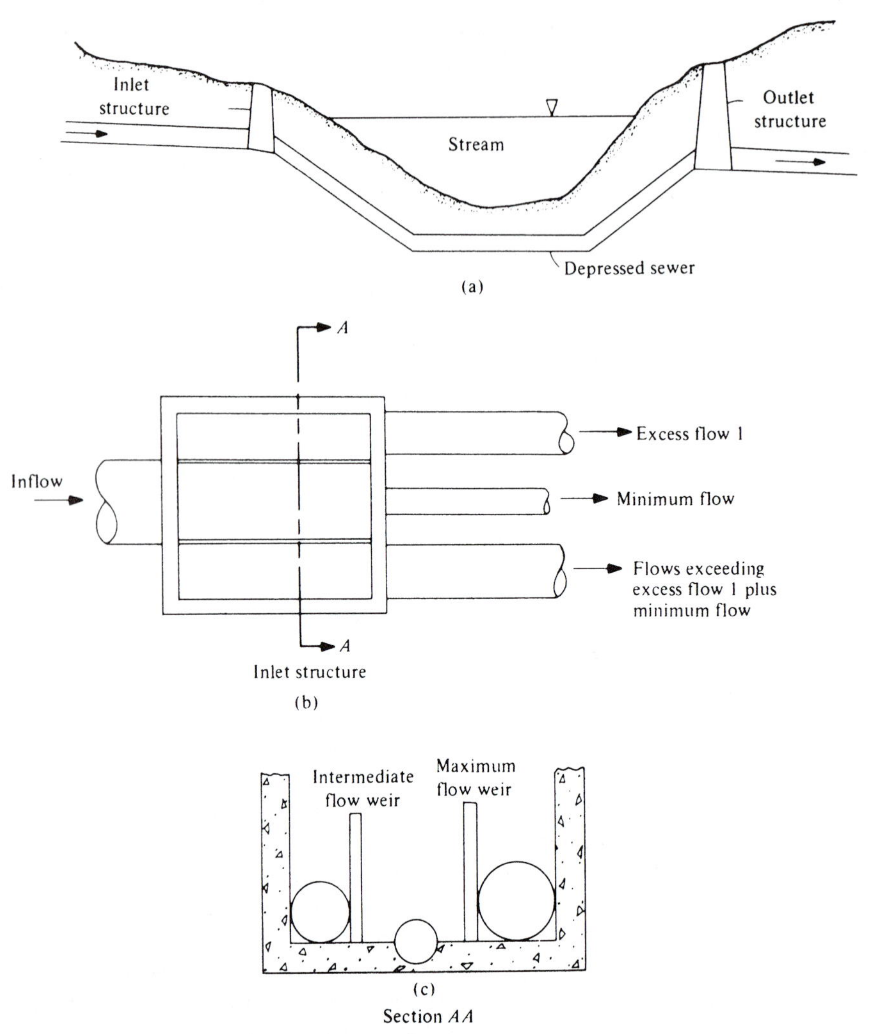

FIGURE 7.8 Inverted siphon or depressed sewer.

entrance transition structure is designed to channel the flow properly into the various pipes. Figure 7.8 illustrates a typical transition structure.

7.12 WASTEWATER PUMPING STATIONS

It is often necessary to accumulate wastewater at a low point in the collection system and pump it to treatment works or to a continuation of the system at a higher elevation. Pumping stations consist primarily of a wet well, which intercepts incoming flows and permits equalization of pump loadings, and a bank of pumps that lift the wastewater from the wet well. In most cases, centrifugal pumps are used and standby equipment is required for emergency purposes. Selection of centrifugal pumps is made according to the principles outlined in Chapter 6.

Pumping of sewage may be necessitated by a number of circumstances. The area of concern may be lower in elevation than the nearest trunk sewer. It may also be that the area to be served lies outside of the drainage area in which the sewage treatment plant to be used is located. Such circumstances require pumping sewage flows across drainage divides. Excavation costs to construct gravity-flow systems may also be such that pumping is more economical than constructing a gravity sewer. In any event, circumstances and trade-offs will dictate the course of action to be followed. Additional details of pumping station designs may be found in *Wastewater Engineering* [9].

STORM WATER MANAGEMENT

Urban runoff is the flow generated on an urban drainage area from a rainfall event. Normally, such flows travel over land until they reach a stream, lake, or ocean, a detention or retention pond, or a storm sewer [10]. Both the quality and quantity of these flows are of concern. Associated issues are flood damage, roadway safety, stream pollution, public health, and water supply (see http://stormwater.ucf.edu/).

7.13 ALTERNATIVE STRATEGIES

Up until the latter part of the twentieth century, storm water management was mainly focused on getting runoff away from the drainage area as quickly as possible. As a result, the almost universal solution to stormwater problems was the construction of a drainage network. Today, water supply and water quality issues confronting the nation have opened the door to broader alternatives in storm water management. They include building retention and detention ponds; using storm water to feed artificial wetlands; grading lawns and open spaces so that some of the drainage can be ponded to reduce irrigation requirements through infiltration; using parking lots and rooftops to pond some of the rainfall catch to reduce peak flows to drainage systems through delayed releases and/or evaporation; and constructing drainage networks. Urban storm water can also be used to recharge groundwater systems.

Drainage plans for urban areas should be designed to include as many of these options as are considered practical and suited to the scale and character of the area under consideration, the objective being to optimize the utilization of this resource, address related water quality problems, and to minimize costs.

WATER QUALITY

Second only to agriculture, urban activities that disturb the natural environment are the greatest contributors to surface water pollution in the United States. With about 2000 hectares of U.S. rural land being converted daily to urban use, accelerated urbanization has led to increased nonpoint pollution loadings from urban runoff [11]. The primary sources of urban pollution are fertilizers from lawns and gardens, animal and bird feces, oil

TABLE 7.7 Sources of Nonpoint Pollution from Urban Residential and Commercial Areas

Category	Parameters	Potential Sources
Bacteria	Total coliforms, fecal coliforms, fecal streptococci, other pathogens	Animals, birds, soil bacteria (humans)
Nutrients	Nitrogen, phosphorus	Lawn fertilizers, decomposing organic matter (leaves and grass clippings), urban street refuse, atmospheric deposition
Biodegradable chemicals	BOD, COD, TOC	Leaves, grass clippings, animals, street litter, oil and grease
Organic chemicals	Pesticides, PCBs	Pest and weed control, packaging, leaking transformers, hydraulic and lubricating fluids
Inorganic chemicals	Suspended solids, dissolved solids, toxic metals, chloride	Erosion, dust and dirt on streets, atmospheric deposition, industrial pollution, traffic, deicing salts

drippings, street litter, herbicide residues, atmospheric fallout of air pollutants, and dead animals and vegetation that have been either purposefully or incidentally placed on the land. In suburban areas, soil erosion and soil-adsorbed pollutants are also sources of non-point pollution. As runoff from precipitation travels over the land, wastes and residues are entrained by the flow and carried to receiving surface water bodies. These residues may contribute significant amounts of suspended solids, BOD, nitrates, phosphates, fecal coliforms, and toxic metals to receiving waters. A summary of the pollutant characteristics imparted to urban runoff by urban residues is given in Table 7.7 [11].

The magnitude of the pollution load transported by urban runoff to receiving water bodies is comparable to that of treated sewage or even untreated sewage in some cases. The highest concentration of pollutants in stormwater is measured during the initial stages of storm runoff, during which the storm water exhibits a so-called first flush effect by initially cleansing away the bulk of pollutants deposited during the preceding dry-weather period.

The concentration of pollutants in storm water depends on whether the storm water reaches the receiving water by overflow from combined sewers, flow from storm sewers or nonsewered overland flow, or municipal sewer overflow. A *combined sewer* is designed to receive both intercepted surface runoff and municipal sewage. *Combined sewer overflow* is the flow from a combined sewer in excess of the interceptor capacity that is discharged into a receiving body of water. A *storm sewer* carries intercepted surface runoff, street wash, and other wash waters or drainage, but excludes domestic sewage and industrial wastes. *Nonsewered urban runoff* is that part of the precipitation that runs off the surface of an urban drainage area and reaches a stream or other body of water without passing through a sewer system. It is about the same quality as water collected by a storm sewer. Storm water can enter municipal sanitary sewers and if infiltration of the sewer system is severe enough, the storm water may cause the municipal system to overflow [10].

The most significant characteristic of urban runoff is the suspended-solids content. On average, the suspended-solids concentration of combined sewer overflow is greater

TABLE 7.8 Generalized Quality Comparisons of Wastewaters[a]

Type	BOD_5 (mg/l)	SS (mg/l)	Total Coliforms (mpn/100 ml)	Total Nitrogen (mg/l as n)	Total Phosphorus (mg/l as p)
Untreated municipal	200	200	5×10^7	40	10
Treated municipal					
Primary effluent	135	80	2×10^7	35	8
Secondary effluent	25	15	1×10^3	30	5
Combined sewage	115	410	5×10^6	11	4
Surface runoff	30	630	4×10^5	3	1

[a]Values based on flow-weighted means in individual test areas.
Source: Lager and Smith [10].

than twice, and for surface runoff three times, the concentration of suspended solids in *untreated* sewage. For this reason, physical and chemical treatment of storm water to remove suspended solids are usually the principal treatment methods employed.

The BOD_5 content of combined sewer overflow and municipal sewer overflow is approximately equivalent—about half the concentration of untreated sewage. Surface runoff and storm sewer flow have about the same BOD_5 strength as secondary municipal effluent. The bacterial content of combined sewer overflow has been found to be typically one order of magnitude lower than that of untreated municipal sewage, whereas the bacterial content of surface runoff is two to four orders of magnitude lower than that of untreated sewage [10]. The bacterial concentrations in urban surface runoff, however, can be two to five orders of magnitude higher than those considered safe for water contact activities. Table 7.8 gives a comparison of the quality of untreated and treated sewage and urban runoff [10].

Water pollution from urban runoff is a problem of increasing magnitude. Researchers have devoted considerable resources to attempting to quantify the problem and determine methods for its control [10–13]. For example, Novotny and Chesters have shown that the factors for determining pollutant loads from residential areas include (1) degree of imperviousness of land surfaces, (2) street refuse accumulation and cleanliness of impervious surfaces, (3) street sweeping practices, (4) curb heights, and (5) type of stormwater drainage system [11].

7.14 BEST MANAGEMENT PRACTICES

Best management practices (BMPs) for urban runoff are nonstructural or low structurally intensive alternatives for the control of urban runoff pollution at its source [12]. BMPs are less costly and more efficient than the alternative—the use of unit processes (see Chapter 9) to reduce pollutant loads—and they include the benefits of soil erosion control, flood control, and cleaner neighborhoods. The success of such measures in controlling urban runoff pollution is very much dependent, however, on educational programs and passage of legislation or ordinances to encourage or force people to comply with the intended BMPs. The EPA has delineated the following preventive measures, construction controls, and corrective maintenance and operation practices as suggestions for BMPs [12,13]:

Preventive Measures

1. Utilization of greenways and detention ponds
2. Utilization of pervious areas for recharge
3. Avoidance of steep slopes for development
4. Maintenance of maximum land area in a natural, undisturbed state
5. Prohibiting development on floodplains
6. Utilization of porous pavements where applicable
7. Utilization of natural drainage features

Soil Erosion Controls at Construction Sites

1. Minimizing area and duration of soil exposure
2. Protecting soil with mulch and vegetative cover
3. Increasing infiltration rates
4. Construction of temporary storage basins or protective dikes to limit storm runoff

Corrective Maintenance and Operation Practices

1. Control of litter, debris, and agricultural chemicals
2. Regular street sweeping and repair
3. Improved roadway deicing and materials storage practices
4. Proper use and maintenance of catch basins and drainage collection systems
5. On-site retention or detention of stormwater runoff

7.15 TREATMENT PROCESSES

The alternative (or supplement) to BMPs for improving storm water quality is treatment by storage; physical, chemical and biological processes; or disinfection [12,13]. The unit processes (covered in later chapters) may be used separately or in sequence to achieve the desired level of treatment.

WATER QUANTITY

Urban drainage facilities have progressed from crude ditches and stepping stones to the present intricate coordinated systems of curbs, gutters, inlets, and underground conveyances.

Handling surface runoff in urban drainage areas is a complex and costly undertaking. Large volumes of water are generated during intense storms and they must be dealt with in a manner to ensure public safety, minimize economic losses, and be protective of the environment.

7.16 HYDROLOGIC CONSIDERATIONS

The hydrologic phase of urban drainage design is focused on determining the magnitude, distribution, and timing of rainfall and runoff events. Major storms are the basis for design of drainage systems. In some cases, a knowledge of peak flow will suffice, but if storage or routing considerations are important, the volume and time distribution of flows must also be known. The latter requirement is especially important for water quality studies.

The quantity of runoff generated on a drainage area is related to the character of the rainfall event and the physical features of the area. Factors involved in precipitation–runoff relationships include precipitation type; rainfall intensity, duration, and areal distribution; storm direction; antecedent precipitation; initial soil moisture conditions; soil type; evaporation; transpiration; and the size, shape, slope, elevation, directional orientation, and land use characteristics of the drainage area.

The design of effective urban drainage systems requires reliable estimates of the flows to be dealt with. Failures of drainage systems are more often the result of faulty estimates of flows than by structural inadequacies.

Runoff may be generated by rainfall, snowmelt, or snowmelt combined with rainfall. Maximum flows on urban areas usually result from high-intensity, short-duration rain storms, whereas floods on large drainage basins often result from a combination of rainfall and snowmelt.

7.17 DESIGN FLOW

Design flow is the maximum flow that can pass through a specified structure safely. An important consideration is that of selecting the probability of occurrence of the design event. Should a drainage system be designed to carry the maximum probable flow, the 5%, the 1%, or some other chance discharge? Most municipalities have standards specifying design storm frequencies for various types of development. Note, however, that there is little relationship between the expected life of the drainage works and the frequency of the design flow.

Table 7.9 shows, for example, that there is a 22% chance that the 100-yr storm will occur in any 25-yr period. This suggests that there is reason for questioning the design of a structure for a life expectancy of 25 yr, and then using design flows expected on the average of once in 25 yr or less. If such an approach is taken, the chances are good that the structure will be damaged, destroyed, or at least overloaded before it has served its useful life.

Selection of design frequency is based on the potential threat to human life, property damage, and inconvenience that would result from storm events of various frequencies. Human life cannot be judged in terms of monetary values. If it is apparent that failure of a proposed drainage system would imperil human lives, the design should be revised accordingly. Property damage is an economic consideration, and design flows can be based on the magnitude of the flow against which it is economically practical to protect. Inconvenience is an intangible factor, but it can be cast in economic terms under some circumstances.

TABLE 7.9 Probability That an Event Having a Prescribed Recurrence Interval Will Be Equaled or Exceeded During a Specified Period

T_R (yr)	Period (yr)					
	1	5	10	25	50	100
1	1.0	1.0	1.0	1.0	1.0	1.0
2	0.5	0.97	0.999	1.0[a]	1.0[a]	1.0[a]
5	0.2	0.67	0.89	0.996	1.0[a]	1.0[a]
10	0.1	0.41	0.65	0.93	0.995	1.0[a]
50	0.02	0.10	0.18	0.40	0.64	0.87
100	0.01	0.05	0.10	0.22	0.40	0.63

[a]Values are approximate.

7.18 RUNOFF ESTIMATES

Procedures used in estimating runoff magnitude and frequency can be generally categorized as (1) empirical approaches, (2) statistical and probability methods, and (3) methods relating rainfall to runoff [14]. Historically, numerous empirical equations have been developed for use in the prediction of runoff. Most of them have been based on the correlation of only two or three variables and, at best, have given only rough approximations. Many are applicable only to specific localities—a fact that should be carefully considered before they are used. In most cases, the frequency of the computed flow is unknown. Formulas of this type are useful only when a more reliable means is unavailable.

Statistical analyses provide good results if sufficient records are available and if no significant changes in stream regimen are experienced or expected in the future. Estimates are usually based on duration or probability curves. In applying these curves, the larger the sample population, the greater the validity of the estimate. For example, a determination of the 10-yr peak flow based on records of only 10 yr might be questionable, whereas the same result obtained from a 100-yr record would be expected to yield good results. About 10 independent samples normally can be expected to provide satisfactory estimates of maximum-flow magnitudes of any frequency. The need for long-term records can thus be a limiting factor in the use of probability methods. Many streams in the United States do not have reliable gauging records longer than 50 yr, and even short-term records for urban areas are scarce.

Of the methods relating rainfall to runoff, the unit hydrograph method, the rational method, and various hydrologic simulation models are widely used.

Unit Hydrograph Method

The *unit hydrograph method* (UHM) is widely used for estimating runoff magnitudes of various frequencies that may occur on a specific stream [14]. To use this approach it is necessary to have continuous records of runoff and precipitation for the drainage area under study. Infiltration capacity variations over time must also be determined. The UHM is limited to areas for which precipitation patterns do not vary markedly.

For large drainage basins, hydrographs must be developed for sub-watersheds and then synthesized into a single hydrograph at the critical location.

A unit hydrograph represents a runoff volume of 1 in. from a drainage area for a specified rainfall duration. A separate unit hydrograph is theoretically required for every possible rainfall length of interest. Ordinarily, however, variations of ±25% from any duration are considered acceptable. Unit hydrographs for short periods can be synthesized into hydrographs for longer durations, however.

Once a unit hydrograph has been derived for a drainage area and specified rainfall duration, the hydrograph for any other storm of equal duration can be obtained. The new hydrograph is developed by applying the unit hydrograph theorem, which states that the ordinates of all hydrographs resulting from equal unit time rains are proportional to the total direct runoff from that rain. The condition may be stated mathematically as

$$\frac{Q_s}{V_s} = \frac{Q_u}{1} \tag{7.9}$$

where

Q_s = magnitude of a hydrograph ordinate of direct runoff having a volume equal to V_s (in inches) at some instant of time after the start of runoff

Q_u = ordinate of the unit hydrograph having a volume of 1 in. at the same instant of time

Storms of reasonably uniform rainfall intensity, having a duration of about 25% of the drainage area lag time (the difference in time between the center of mass of the rainfall and the center of mass of the resulting runoff) and producing a total of 1 in or more of runoff, are most suitable for deriving a unit hydrograph.

Steps in the development of a unit hydrograph are

1. Analyze the stream flow hydrograph to permit separation of the surface runoff from the base flow. It can be accomplished by any one of several methods [13,15].
2. Measure the total volume of surface runoff (direct runoff) from the storm producing the original hydrograph. The volume is equal to the area under the hydrograph after the base flow has been subtracted.
3. Divide the ordinates of the direct runoff hydrograph by the total direct runoff volume in inches. The resulting plot of these values versus time is a unit hydrograph for the basin.
4. Finally, the effective duration of the runoff-producing rain for the unit hydrograph must be determined. This can be estimated from the rainfall record.

These steps are used in deriving a unit hydrograph from an isolated storm. Other procedures are required for complex storms or for developing synthetic unit hydrographs when few data are available. In some cases, unit hydrographs may also be transposed from one basin to another [13]. The following example illustrates the derivation and application of a unit hydrograph.

Example 7.8

Using the hydrograph given in Figure 7.9, derive a unit hydrograph for the 3-mi^2 drainage area. From this unit hydrograph, derive a hydrograph of direct runoff for the rainfall sequence given (see Tables 7.10 and 7.11).

Solution:

1. Subtract the base flow to obtain the total direct runoff hydrograph. A number of procedures are reported in the literature, but a common method is to construct line *AC* beginning where the hydrograph begins an appreciable rise and ending where the recession curve intersects the base-flow curve. The important point is to be consistent in methodology from storm to storm.
2. Determine the duration of the effective rainfall (rainfall that actually produces surface runoff). The effective rainfall volume must be equivalent to the

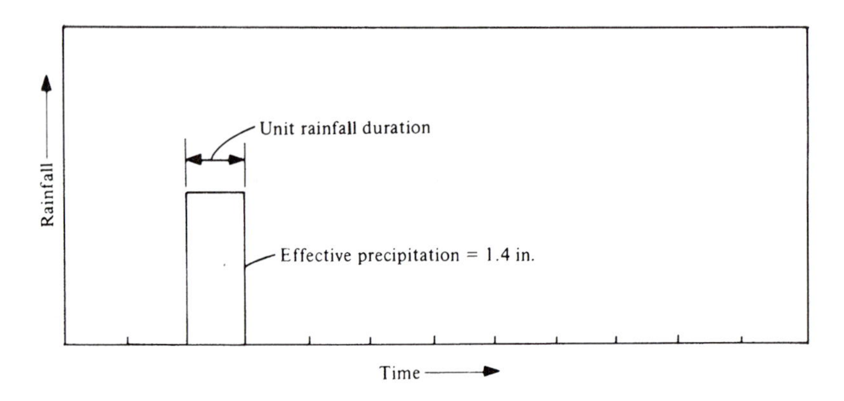

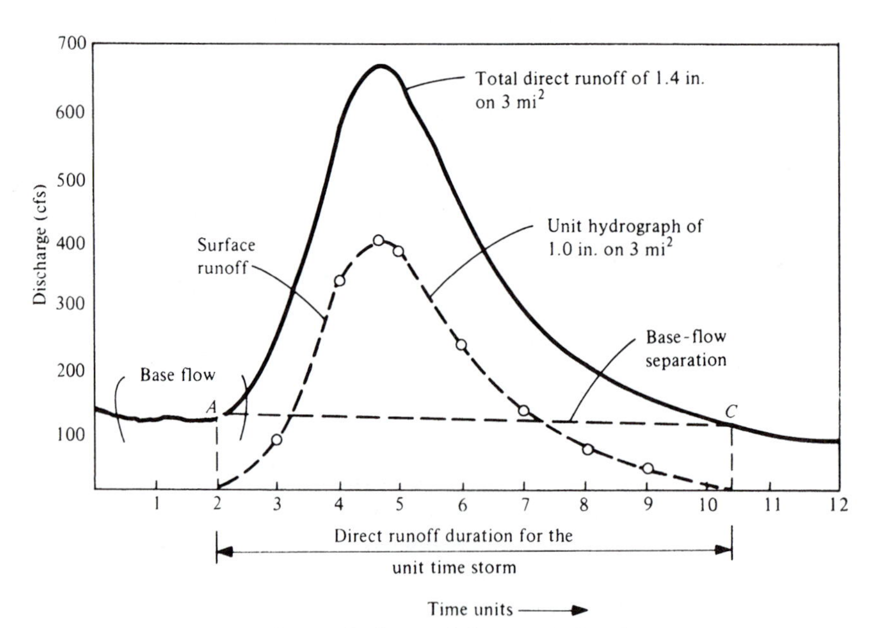

FIGURE 7.9 Derivation of a unit hydrograph from an isolated storm.

TABLE 7.10 Determination of a Unit Hydrograph from an Isolated Storm

(1) Time Unit	(2) Total Runoff (cfs)	(3) Base Flow (cfs)	(4) Total Direct Runoff: (2) − (3) (cfs)	Unit Hydrograph Ordinate: (4) ÷ 1.4 (cfs)
1	110	110	0	0
2	122	122	0	0
3	230	120	110	78.7
4	578	118	460	328
4.7	666	116	550	393
5	645	115	530	379
6	434	114	320	229
7	293	113	180	129
8	202	112	90	64.2
9	160	110	50	35.7
10	117	105	12	8.6
10.5	105	105	0	0
11	90	90	0	0
12	80	80	0	0

TABLE 7.11 Unit Hydrograph Application

Time Unit Sequence	Rain Unit Number	Effective Rainfall (in.)	Hydrograph Ordinates[a] for Rainfall Unit . . .		
			1	2	3
1	1	0.7	55.1	—	—
2	2	1.7	229	134	—
2.7	3	1.2	275	—	—
3	—	—	265	558	94.3
3.7	—	—	—	668	—
4	—	—	161	664	393
4.7	—	—	—	—	472
5	—	—	90.5	389	455
6	—	—	44.9	219	275
7	—	—	25.0	109	155
8	—	—	6.0	60.7	77
9	—	—	—	14.6	42.8
10	—	—	—	—	10.3

[a]Values are obtained by multiplying effective rainfall values by unit hydrograph ordinates.

volume of direct surface runoff. Usually the unit time of the effective rainfall will be 1 day, 1 hr, 12 hr, or some other interval appropriate for the size of drainage area being studied. The unit storm duration should not exceed about 25% of the drainage area lag time. The effective portion of the rainstorm for this example is given in Figure 7.9 together with its duration. The effective volume is 1.4 in.

3. Project the base length of the unit hydrograph down to the abscissa, giving the horizontal projection of the base-flow separation line *AC*. Unit hydrograph

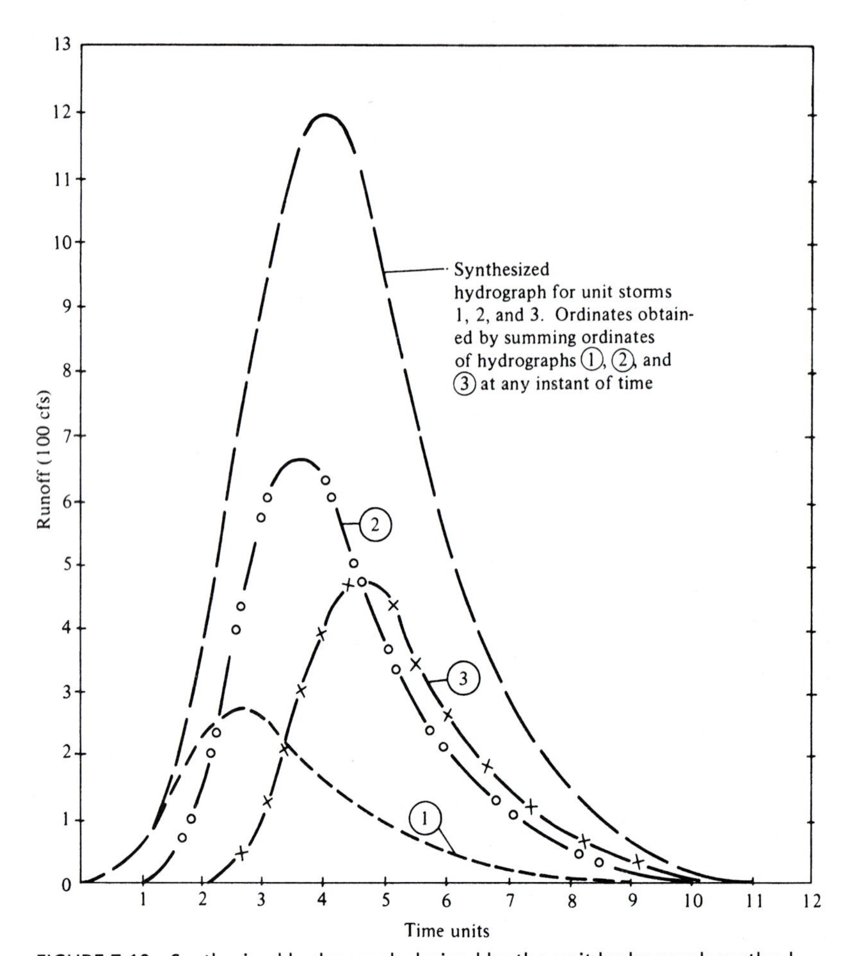

FIGURE 7.10 Synthesized hydrograph derived by the unit hydrograph method.

theory assumes that for all storms of equal duration, regardless of intensity, the period of surface runoff is the same.

4. Tabulate the ordinates of direct runoff at the peak rate of flow and at sufficient other positions to determine the hydrograph shape. Note that the direct runoff ordinate is the ordinate above the base-flow separation line. See Table 7.10.
5. Compute the ordinates of the unit hydrograph by using Eq. (7.9). In this example, the values are obtained by dividing the direct runoff ordinates by 1.4. Table 7.10 shows computation of the unit hydrograph ordinates.
6. Using the values from Table 7.10, plot the unit hydrograph as shown in Figure 7.9.
7. Using the derived ordinates of the unit hydrograph, determine the ordinates of the hydrographs for each consecutive rainfall period as given in Table 7.11.
8. Determine the synthesized hydrograph for unit storms 1–3 by plotting the three hydrographs and summing the ordinates. The procedure is indicated in Figure 7.10.

The unit hydrograph method provides the entire hydrograph resulting from a storm and offers advantages over procedures that only produce peak flows. It has the disadvantage of requiring both rainfall and runoff data for its derivation and of being limited in its application to a particular drainage basin. A number of refinements to the procedure may be found in the literature [13,15,16]. Procedures for producing synthetic unit hydrographs covering areas where adequate records are not available have also been derived [16].

Rational Method

The *rational method*, first proposed in 1889, is a widely used method for computing quantities of stormwater runoff from small areas [6]. The rational formula relates runoff to rainfall in the following manner:

$$Q = Kcia \tag{7.10}$$

where

Q = peak runoff rate, cfs

c = runoff coefficient, which is the ratio of the peak runoff rate to the average rainfall rate for a period known as the time of concentration

i = average rainfall intensity, in./hr, for a period equal to the time of concentration

a = drainage area, acres

K = a conversion factor (unity for English units and 1/360 for metric units)

It should be noted that the assignment of cfs to Q is satisfactory for all practical purposes, since 1.008 cfs equals 1 in. of rainfall in 1 hr on an area of 1 acre ($K = 1$).

Basic assumptions of the rational method are that (1) the maximum runoff rate at any location is a function of the average rainfall rate during the time of concentration for that location; and (2) the maximum rainfall rate occurs during the time of concentration. The variability of the storm pattern is not taken into consideration. The time of concentration t_c is defined as the flow time from the most remote point in the drainage area to the point in question. Usually it is considered to be composed of an overland flow time or, in most urban areas, an inlet time plus a channel flow time.

The channel flow time can be estimated with reasonable accuracy from the hydraulic characteristics of the sewer. Normally, the average full-flow velocity of the conveyance for the existing or proposed hydraulic gradient is used. The channel flow time is then determined as the flow length divided by the average velocity.

The inlet time consists of the time required for water to reach a defined channel such as a street gutter, plus the gutter flow time to the inlet. Numerous factors, such as rainfall intensity, surface slope, surface roughness, flow distance, infiltration capacity, and depression storage, affect inlet time. Because of this, accurate values are difficult to obtain. Design inlet flow times of from 5 to 30 min are used in practice. In highly developed areas with closely spaced inlets, inlet times of 5–15 min are common; for similar areas with flat slopes, periods of 10–15 min are common; and for very level areas with widely spaced inlets, inlet times of 20–30 min are frequently used [6].

Inlet times are also estimated by breaking the flow path into various components, such as grass, asphalt, and so on, then computing individual times for each surface and adding them. In theory this would seem desirable, but in practice so many variables affect the flow that the reliability of computations of this type is questionable. The standard procedure is to use inlet times that have been found through experience to be applicable to the various types of urban areas.

An example method for estimating t_c, developed by Izzard for small experimental plots without developed channels, suggests the problems with using such approaches [17]:

$$t_c = \frac{41bL_o^{1/3}}{(ki)^{2/3}} \tag{7.11}$$

where

t_c = time of concentration, min

b = coefficient

L_o = overland flow length, ft

k = rational runoff coefficient (see Table 7.12)

i = rainfall intensity, in./hr, during time t_c

The equation is valid only for laminar flow conditions where the product iL_o is less than 500. The coefficient b is found using

$$b = \frac{0.0007i + c_r}{S_o^{1/3}} \tag{7.12}$$

where

S_o = surface slope

C_r = coefficient of retardance

Values of C_r are given in Table 7.13. Note that the reliability of this equation is strongly influenced by the selection of parameters.

The runoff coefficient c is the component of the rational formula that requires the greatest exercise of judgment by the engineer. It is not amenable to exact determination, since it includes the influence of a number of variables, such as infiltration capacity, interception by vegetation, depression storage, and antecedent conditions. As used in the rational equation, the coefficient c represents a fixed ratio of runoff to rainfall, while in actuality it is not fixed and may vary for a specific drainage basin with time during a particular storm, from storm to storm, and with change in season. Fortunately, the closer the area comes to being impervious, the more reasonable

TABLE 7.12 Typical C Coefficients for 5- to 10-yr Frequency Design

Description of Area	Runoff Coefficients
Business:	
Downtown areas	0.70–0.95
Neighborhood areas	0.50–0.70
Residential:	
Single-family areas	0.30–0.50
Multiunits, detached	0.40–0.60
Multiunits, attached	0.60–0.75
Residential (suburban)	0.25–0.40
Apartment dwelling areas	0.50–0.70
Industrial:	
Light areas	0.50–0.80
Heavy areas	0.60–0.90
Parks, cemeteries	0.10–0.25
Playgrounds	0.20–0.35
Railroad yard areas	0.20–0.40
Unimproved areas	0.10–0.30
Streets:	
Asphaltic	0.70–0.95
Concrete	0.80–0.95
Brick	0.70–0.85
Drives and walks	0.75–0.85
Roofs	0.75–0.95
Lawns, sandy soil:	
Flat, 2%	0.05–0.10
Average, 2–7%	0.10–0.15
Steep, 7%	0.15–0.20
Lawns, heavy soil:	
Flat, 2%	0.13–0.17
Average, 2–7%	0.18–0.22
Steep, 7%	0.25–0.35

TABLE 7.13 Izzard's Retardance Coefficient C_r

Surface	C_r
Smooth asphalt	0.007
Concrete paving	0.012
Tar and gravel paving	0.017
Closely clipped sod	0.046
Dense bluegrass turf	0.060

the selection of c becomes. This is true, because for highly impervious areas c approaches unity, and for these areas the nature of the surface is much less variable for changing seasonal, meteorological, or antecedent conditions. The rational method is thus best suited for use in urban areas, where a high percentage of imperviousness prevails.

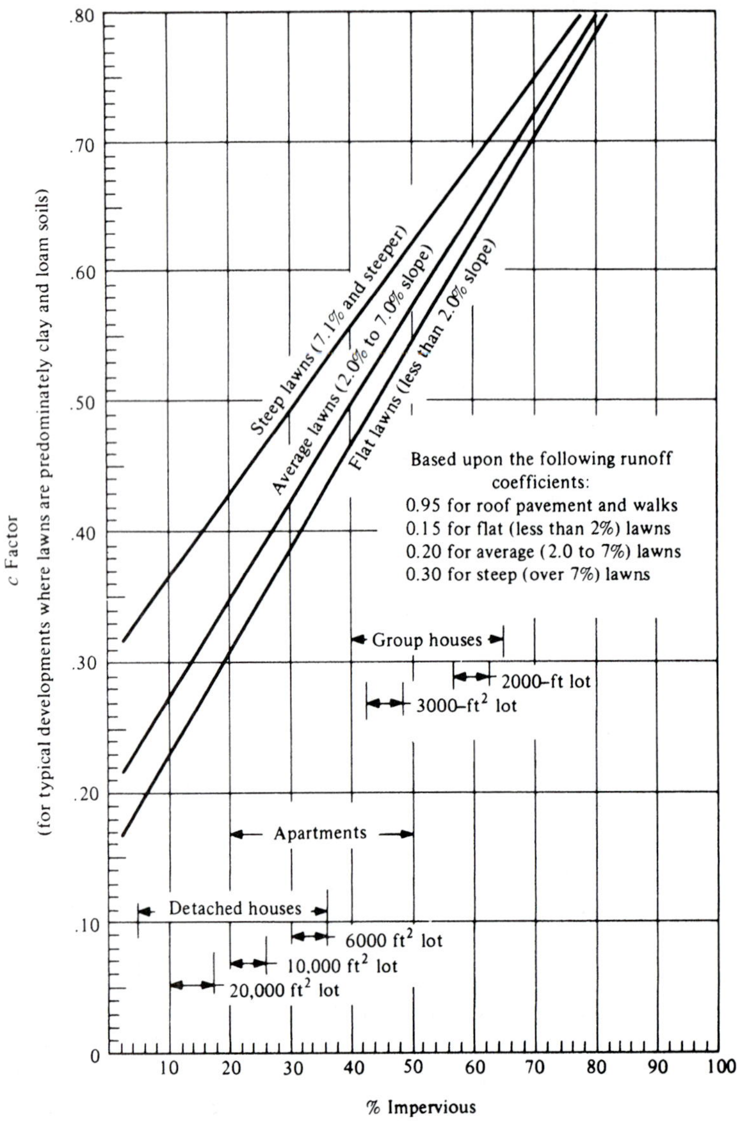

FIGURE 7.11 *c* factors for typical developments in clay soils. *Source:* Courtesy of the Baltimore County, Maryland, Department of Public Works.

There is no precise method for evaluating the runoff coefficient c, although some research has been directed toward that end [18]. Common engineering practice is to make use of average values of the coefficient for various types of land surface covers. Table 7.12 gives some values of the runoff coefficient. Figure 7.11 relates the rational c to imperviousness, soil type, and lawn slope. Most users of the rational method use information reported in similar tabular or graphical forms, using local conditions following their experience and practice.

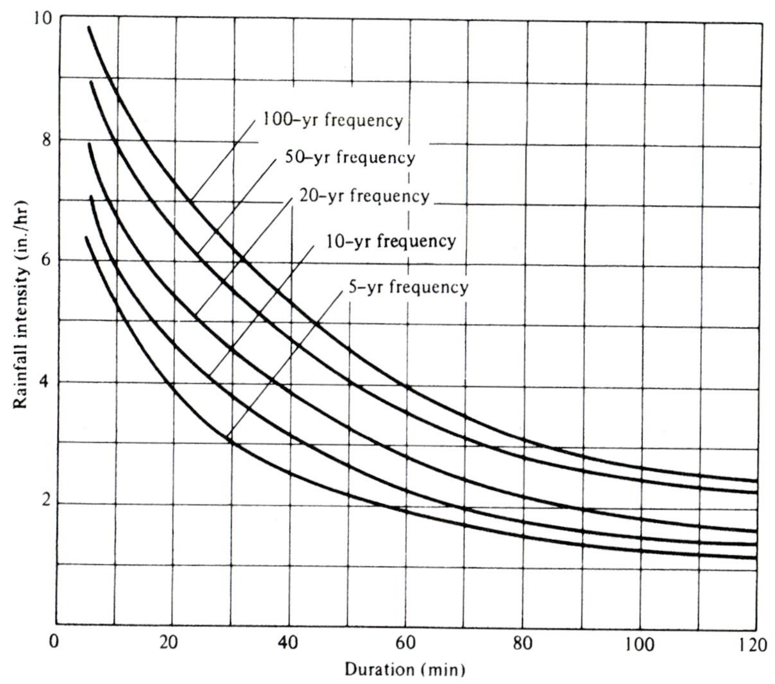

FIGURE 7.12 Typical intensity–duration–frequency curves.

To apply the rational method, an average rainfall intensity i representative of the design storm of a prescribed frequency for the time of concentration is used. The frequency chosen is largely a matter of economics. Frequencies of 1 to 10 yr are commonly used where residential areas are to be protected. For higher-value districts, 10–20 yr or higher return periods are commonly selected. Local conditions and practice normally dictate the selection of these criteria.

Once t_c and the rainfall frequency have been ascertained, the rainfall intensity i is usually obtained from a set of of rainfall intensity–duration–frequency curves such as those shown in Figure 7.12. Entering the curves on the abscissa with the appropriate value of t_c and then projecting upward to an intersection with the desired frequency curve allows i to be found by projecting this intersection point horizontally to an intersection with the ordinate. If an adequate number of years of local rainfall records is available, curves similar to Figure 7.12 may be derived. Rainfall data compiled by the Weather Bureau, the Department of Commerce, the Department of Agriculture, and other government agencies may also be retrieved and used.

Generally, the rational method should be used only on areas that are smaller than about 2 mi^2 (approximately 1280 acres) in size. For areas larger than 100 acres, due caution should be exercised. Most urban drainage areas served by storm drains become tributary to natural drainage channels or large conveyances before they reach 100 acres or more in size, however, and for these tributary areas the rational method can be put to reasonable use.

Example 7.9

Given the following data for a 25-year storm, estimate the peak rate of flow to a culvert entrance. Consider the rainfall intensity–duration–frequency curves of Figure 7.12 applicable. The contributing upstream drainage area is 20 acres. The overland flow distance is 125 ft and the average land slope is 2.5%. The land use of the overland flow portion of the drainage area is lawns, sandy soil. The land use for the drainage basin is 75% residential, multiple units, detached, and 25% lawns, sandy soil with an overall average slope of about 2.7%. A channel leading to the culvert is 1550 ft long with a slope of 0.016 ft/ft. Manning's n value for the grassed channel is 0.030. The channel is trapezoidal with a bottom width of 3 ft and side slopes of 2 ft vertical to 1 ft horizontal.

Solution:

1. Determine the rational C value

a. for the overland flow portion, using the data given and Table 7.12, C is found to be 0.15

b. for the residential portion, from Table 7.12, C is found to be 0.5
The weighted average C value for the entire area is thus

$$C = (0.75 \times 0.5 + 0.25 \times 0.15)/1.0 = 0.41$$

where 0.75 and 0.25 are the percentages of residential and lawn areas and the denominator is the sum of them = 1.0

2. Calculate the time of concentration

a. for the overland flow area
Use the Federal Aviation Administration equation [19]

$$t_c = 1.8(1.1 - C)L^{0.5}/S^{0.333}$$

where C = the rational coefficient (Table 7.12), L = overland flow length in ft, and S = surface slope in %.

$$t_c = 1.8(1.1 - 0.15)125^{0.5}/2.50^{0.333} = 14.09 \text{ min}$$

b. for the main channel
From the channel geometry and assuming a depth of 2 ft, R is calculated to be 1.72 ft. Then, using Manning's equation,

$$V = (1.49/0.03)(1.72)^{0.666}(0.016)^{0.5} = 9.01 \text{ ft/s}$$

$$t_c = L/(V \times 60) = 1550/(9.01 \times 60) = 2.86 \text{ min}$$

The total time of concentration is thus 14.09 + 2.86 = 16.95, or 17 min.

3. Calculate the peak runoff for the 25-year storm.
From Figure 7.12, using a time of concentration of 17 min and interpolating for the 25-year curve, the rainfall intensity is found to be 6.4 in/hr. Now the rational equation can be solved for Q:

$$Q = CiA = 0.41 \times 6.4 \times 20 = 52.48 \text{ cfs}$$

Modified Rational Method

The *modified rational method* provides for development of a complete hydrograph [13,20]. For this model, if the storm duration exceeds the time of concentration, the hydrograph rises to a peaking rate at the time of concentration, continues on at a peaking rate until the storm event ceases, and then decreases to zero as excess rainfall is released from the drainage area. Numerous software packages are available to accommodate this approach. Figure 7.13 illustrates the shape of the rational method hydrograph for a rainfall intensity exceeding the time of concentration t_c. In this case, it is trapezoidal. If the rainfall duration is equal to t_c, the shape would be triangular.

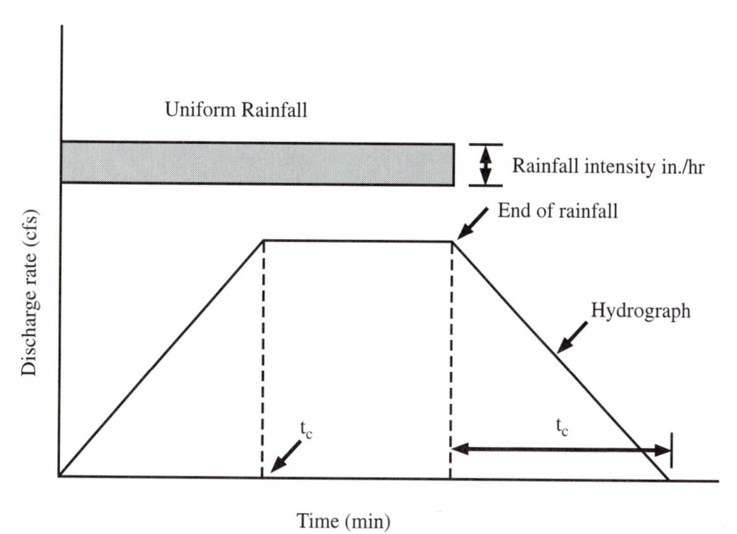

FIGURE 7.13 Rational method hydrograph for uniform rainfall intensity.

Example 7.10

A watershed of 100 acres has a t_c of 30 min and a runoff coefficient of 0.6. Consider that a rainstorm having an average intensity of 2.5 in./hr, and a duration 50 min occurs over the watershed. Use the modified rational method to develop a storm hydrograph.

Solution:

1. At the time of concentration t_c, the discharge Q will be given by

$$Q = ciA = 0.6 \times 2.5 \times 100 = 150 \text{ cfs}$$

2. Between t_c and the end of the storm, the dischage will be 150 cfs
3. At $t = 50$ minutes, rainfall ceases and the hydrograph recedes to zero over the t_c.

The resulting hydrograph is given in Figure 7.14.

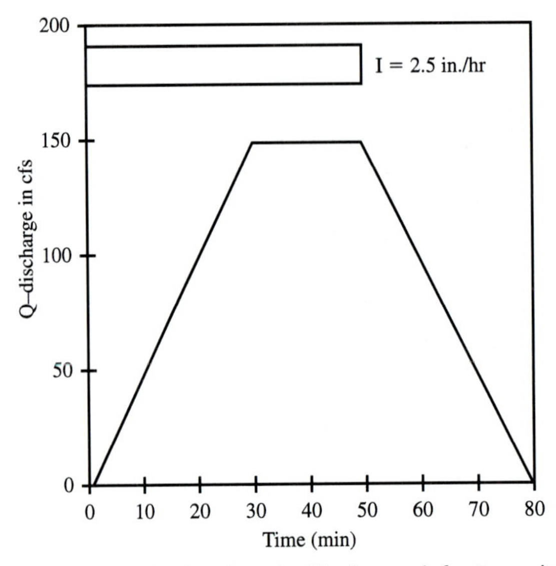

FIGURE 7.14 Rational method hydrograph for Example 7.10.

The previous discussion centered on developing a rational hydrograph for a storm having a uniform rainfall intensity. Since rainfall intensities are rarely uniform for very long time periods, the rational hydrograph method has also been modified for use with continuous nonuniform rainstorm patterns [20].

For varying rainfall intensities over time, the peaking portion of the hydrograph begins when all portions of the watershed are contributing flow to the point under consideration. Prior to that time (the time of concentration), the hydrograph is defined by a rising limb. Once the storm event ceases, there is a recession limb that lasts until all of the water being drained from the watershed reaches the point under consideration. Determination of the rational hydrograph for time-varying intensities is based on the use of moving averages of rainfall intensities. The procedure is illustrated below for each of the three hydrograph components [20].

The Rising Limb. Prior to the time of concentration when all portions of the watershed are contributing flow to the point under consideration, the rising portion of the hydrograph is characterized by increasing rates of flow. The area contributing runoff at a time T, which is less than the time of concentration T_c, may be considered proportional to the ratio of T to T_c [20]. Accordingly, the contributing portion of the drainage area A is given by

$$A_c = A\frac{T}{T_c} \tag{7.13}$$

where A_c is the contributing portion of area A. The average rainfall intensity for this segment of the hydrograph is given by

$$I_t = \frac{1}{T}\sum_{T=0}^{T} P_t \qquad \text{for } T_c \geq T \geq 0 \tag{7.14}$$

where

$$I_t = \text{the average rainfall intensity at time } T$$
$$P_t = \text{the cumulative precipitation to time } T$$

The runoff rate for the rising segment of the hydrograph is given by

$$Q_t = KCA_cI_t = KCA\frac{T}{T_c}I_t \qquad \text{for } T_c \geq T \geq 0 \tag{7.15}$$

where

Q_t = runoff at time T

K = a unit conversion that is 1 for English units and 1/360 for metric units

C = the rational runoff coefficient

The Peak Segment. The peaking portion of the hydrograph occurs during the period from T_c until rainfall ceases. During this period, all portions of the watershed are contributing to the outflow at the point of interest. The average rainfall intensity for a period of T_c before time T is determined using a moving average.

The average rainfall intensity for this period is determined by the following equation:

$$I_t = \frac{1}{T_c}\sum_{t=T-T_c}^{T} P_t \qquad \text{for } T_d \geq T \geq T_c \tag{7.16}$$

where T_d is the time that rainfall ends.

The runoff rate for the peak segment of the hydrograph is given by

$$Q_t = KCAI_t \qquad \text{for } T_d \geq T \geq T_c \tag{7.17}$$

The Recession Limb. At the time that rainfall ceases, the recession portion of the hydrograph begins. Recession curves are often represented by exponential decay functions, but for small basins a linear approximation may be made. Such an approximation is used for the rational hydrograph.

$$Q_t = Q_{T_d}\left[1 - \frac{T - T_d}{T_c}\right] \qquad \text{for } T_d \leq T \leq (T_d + T_c) \tag{7.18}$$

and

$$Q_{T_d} = KCAI_{T_d}$$

where Q_{T_d} = the rate of flow at the time that rainfall ceases T_d.

Example 7.11

A 50-min rainstorm occurs on a 9-acre urban area having a runoff coefficient of 0.65. The time of concentration has been calculated as 20 min. Incremental rainfall depths for the storm are given in Table 7.14. Using a spreadsheet, calculate the runoff from the storm at 5-minute intervals. Plot the resulting hydrograph.

Solution: The solution is obtained through the use of Eqs. (7.13) to (7.18). They are inserted in the spreadsheet shown in Table 7.14. Example computations for one point on each of the rising, peak, and falling components of the hydrograph illustrate those made by the spreadsheet.

1. Calculation for rising limb ordinate at T = 20 min. Using Eq. (7.14), I_{20} is computed as

$$I_{20} = (60/20) \times (0.1 + 0.22 + 0.33 + 0.57) = 3.6 \text{ in./hr}$$

The value − 60 min/hr converts the result to in./hr

Using Eq. (7.13), A_c is determined as

$$A_c = 9 \times (20/20) = 9 \text{ acres}$$

Using Eq. (7.15), the flow at T = 20 minutes is computed as

$$Q_{20} = 0.65 \times 9 \times 3.6 = 21.1 \text{ cfs}$$

2. Calculation for peaking ordinate at T = 40 min. Using Eq. (7.16), I_{40} is computed as

$$I_{40} = (60/20) \times (0.62 + 0.5 + 0.29 + 0.18) = 4.8 \text{ in./hr}$$

TABLE 7.14 Modified Rational Method Data and Solution to Example 7.11

Time Minutes	Incremental Rainfall Depth in inches	$\Sigma P(t)$ inches	$\Sigma P(t)$ from T-T_c to T	Average Rainfall Intensity 1 − in./hr	Runoff Rate Q − cfs
0	0.0	0.0	0.0	0.0	0.0
5	0.1	0.1		1.2	1.8
10	0.22	0.32		1.92	5.6
15	0.31	0.63		2.52	11.1
20	0.57	1.2		3.6	21.1
25	0.63		1.73	5.19	30.4
30	0.5		2.01	6.03	35.3
35	0.29		1.99	5.97	34.9
40	0.18		1.6	4.8	28.1
45	0.1		1.07	3.21	18.8
50	0.06		0.63	1.89	11.1
55	0.0			0.0	8.3
60	0.0			0.0	5.5
65	0.0			0.0	2.8
70	0.0			0.0	0.0

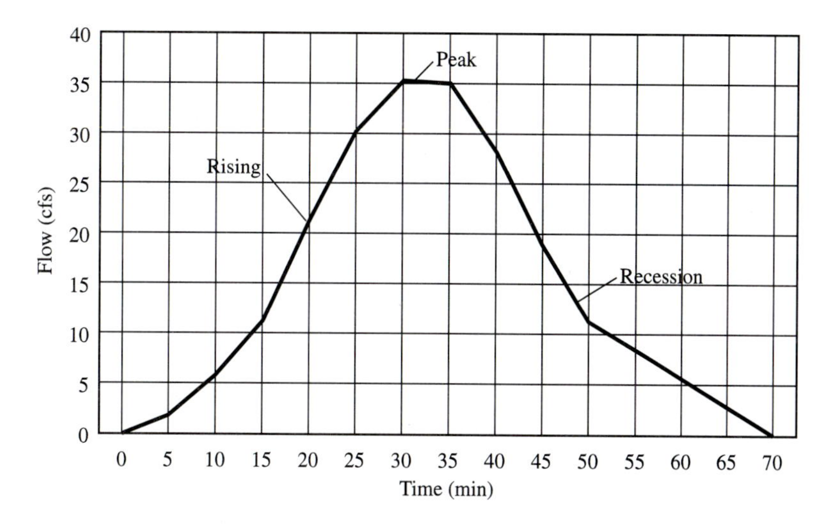

FIGURE 7.15 Rational method hydrograph for Example 7.11.

Using Eq. (7.17), the flow at $T = 40$ minutes is computed as

$$Q_{40} = 0.65 \times 9 \times 4.8 = 28.1 \text{ cfs}$$

3. Calculation for recession ordinate at $T = 60$ min. Using Eq. (7.18), Q_{60} is computed as

$$Q_{60} = Q_{50} \times (1 - (60 - 50)/20) = 5.5 \text{ cfs}$$

Figure 7.15 is a plot of the hydrograph ordinates given in the last column of the Table 7.12.

Simulation Models

Simulation models find widespread use in estimating storm water flows. Where sufficient data are available, these models can be used to test proposed designs and/or systems' operating procedures [8,13]. A few of the many event, continuous, and urban runoff computer models are listed in Table 7.15. Most of these models were developed by university researchers or federal agencies. All of them have moderate to extensive input data requirements, and all have from 1–10% judgment parameter inputs [13]. Models are usually validated through repeated trial runs. The first nine models listed in Table 7.15 under the urban runoff simulation models heading are public domain models that are periodically updated. Current versions should be requested when acquiring the computer codes. Additional information on simulation models may be found in references [21–45].

One of the earliest simulation models to find wide use was the Stanford Watershed Model, developed by Crawford and Linsley in 1966 [23]. Since then, their model has undergone many revisions by its authors and others [8,13]. The version discussed here is the *hydrocomp simulation program* (HSP) [35]. This and other versions of the

TABLE 7.15 Digital Simulation Models of Hydrologic Processes

Code Name	Model Name	Agency or Organization	Date of Original Development
	Continuous Stream Flow Simulation Models		
API	Antecedent Precipitation Index Model	Private	1969
USDAHL	1970, 1973, 1974 Revised Watershed Hydrology	ARS	1970
SWM-IV	Stanford Watershed Model IV	Stanford University	1959
HSPF	Hydrocomp Simulation Program—FORTRAN	EPA	1967
NWSRFS	National Weather Service Runoff Forecast System		1972
SSARR	Streamflow Synthesis and Reservoir Regulation	Corps	1958
PRMS	Precipitation-Runoff Modeling System	USGS	1982
SWRRB	Simulator for Water Resources in Rural Basins	USDA	1990
	Rainfall-Runoff Event-Simulation Models		
HEC-1	HEC-1 Flood Hydrograph Package	Corps	1973
TR-20	Computer Program for Project Hydrology	SCS	1965
USGS	USGS Rainfall—Runoff Model	USGS	1972
HYMO	Hydrologic Model Computer Language	ARS	1972
SWMM	Storm Water Management Model	EPA	1971
	Urban Runoff Simulation Models[a]		
CHM	Chicago Hydrograph Method	City of Chicago	1959
RRL	Road Research Laboratory Method	Road Research Lab	1962
ILLUDAS	Illinois Urban Drainage Area Simulator	Ill. Water Survey	1972
STORM	Storage, Treatment, Overflow Runoff Model	Corps of Engineers	1974
TR-55	SCS Technical Release 55	SCS	1992
DR3M	Distributed Routing Rainfall–Runoff Model	USGS	1978
HYDRA	Hydrologic Component of HYDRAIN Package	FHWA	1990
SWMM	Storm Water Management Model	EPA	1971
UCURM	U. of Cincinnati Urban Runoff Model	U. of Cincinnati	1972
MITCAT	MIT Catchment Model	MIT	1970
PSURM	Pennsylvania State Urban Runoff Model	Pennsylvania State University	1979

[a]First nine urban runoff simulation models listed are in the public domain.

model have been used extensively for synthesizing continuous hydrographs of hourly or daily stream flows for watersheds of varying size and character.

The HSP is a sequence of computational routines representing the major hydrologic processes. Input to the model includes precipitation, potential evapotranspiration, temperature, radiation, dew point, and wind. The last four of these are needed only if snowmelt is involved.

The model accounts for the initial moisture stored in the watershed and then determines the fate of a precipitation input. The hydrologic budget is balanced for selected time steps so that the precipitation input can be converted to stream flow after the appropriate allocations to interception, evapotranspiration, and upper, lower, and deep groundwater storages have been made. In simple terms, the following equation is solved at each time step for stream flow:

$$R = P - \mathrm{ET} - \Delta S \tag{7.19}$$

where

P = precipitation

R = stream flow

ET = evapotranspiration

ΔS = total change in storage

The equation is balanced at the end of each time period, and the calculations proceed from interval to interval until there is no additional input of data [35].

EPA Storm Water Management Model

A widely used simulation model is EPA's Storm Water Management Model (SWMM) (www.epa.gov/SWMMWINDOWS) [13,36–38]. SWMM is capable of simulating the movement of precipitation and pollutants from the surface of the ground through pipe and channel networks and storage treatment units to receiving waters. Single-event and continuous simulation may be performed on catchments having storm sewers and natural drainage. Flows, stages, and pollutant concentrations are outputs of the model. SWMM can be used for planning and design purposes. The planning mode is used to assess urban runoff problems and proposed abatement options. For design application, detailed catchment layouts are used, as are shorter time steps for the precipitation input. A flow diagram for the runoff portion of the model is given in Figure 7.16.

The computer program for SWMM consists of a control segment and five computational blocks: (1) executive, (2) runoff, (3) transport, (4) storage, and (5) receiving water.

The *executive* block is the control for computational blocks, with all access and transfers between the other blocks routed through its subroutine MAIN. The *runoff* block accepts rainfall data and a description of the drainage system as inputs and provides both hydrographs and time-dependent pollutional graphs (pollutographs). The *transport* block routes flows through the conveyance system, which are corrected for

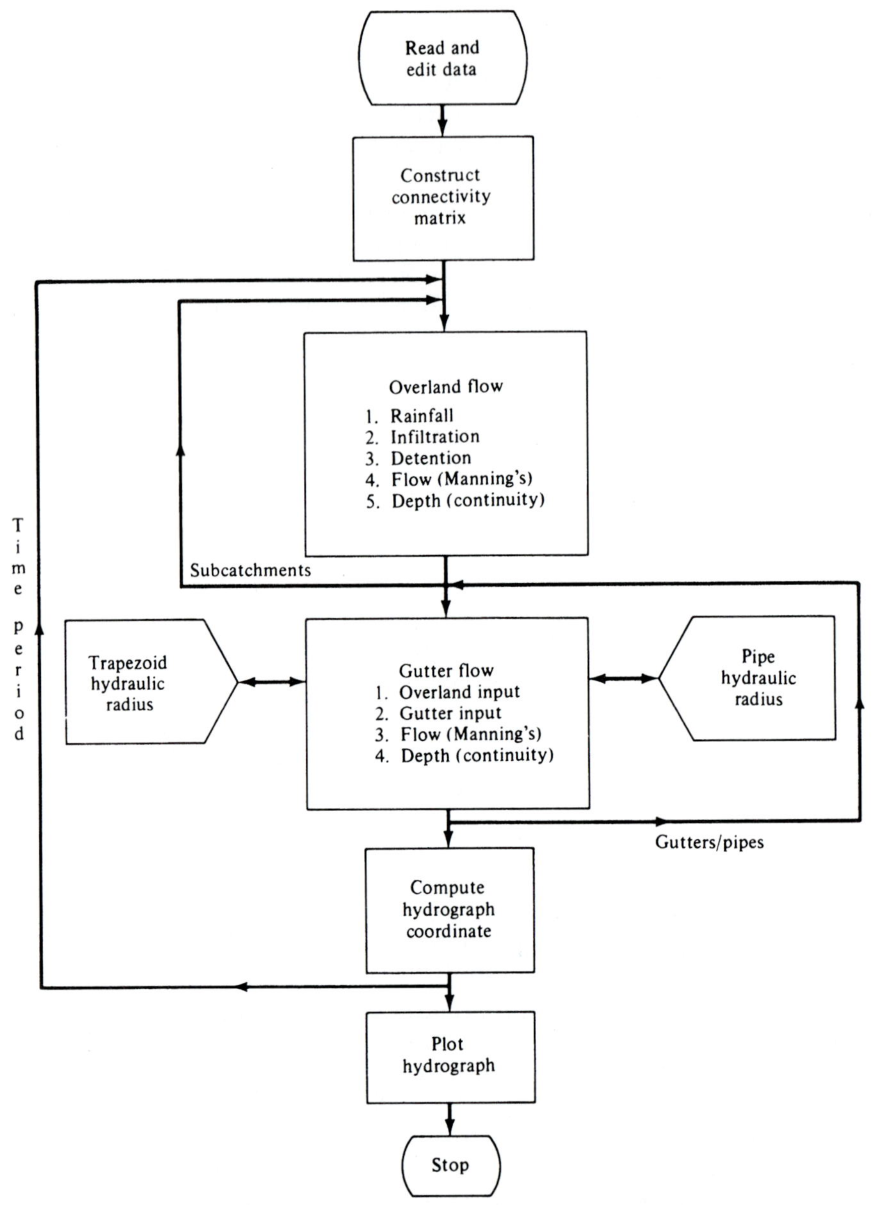

FIGURE 7.16 Flowchart for hydrographic computation. *Source*: After Metcalf and Eddy, Inc., University of Florida, and Water Resources Engineers, Inc. [36].

both dry-weather flows and infiltration. The block can also produce hydrographs and pollutographs at specified locations. In the *storage* segment, flows from the transport block are received and the effect of any treatment provided is considered. The effects of the discharge on a receiving stream are determined in the *receiving water* block. Segments may be run separately to facilitate adjustments.

The drainage area is divided into subcatchments, gutters, and pipes. Subcatchments are divided into three parts: pervious, impervious with surface detention, and impervious without surface detention. Required inputs to the model are the surface area, width of subcatchment, ground slope, Manning's roughness coefficient, infiltration rate, and detention depth. Channel descriptions are the length, Manning's roughness coefficient, invert slope, diameter for pipes, or cross-sectional dimensions for open channels. General data requirements are summarized in Table 7.16. A step-by-step process accounts for inflow, infiltration losses, and flow from upstream subcatchment areas. It also provides a discharge hydrograph at the drainage basin outlet. The logic of the model is illustrated by the following equations:

1. Rainfall is added to the subcatchment according to the specified hydrograph:

$$D_1 = D_t + R_t + \Delta t \tag{7.20}$$

where D_1 = the water depth after rainfall, D_t = the water depth of the subcatchment at time t, and R_t = the intensity of rainfall in time interval Δt.

2. Infiltration I_t is computed by Horton's exponential function, $I_t = f_c + (f_0 - f_c)e^{-kt}$, and is subtracted from the water depth existing on the subcatchment

$$D_2 = D_1 - I_t \Delta t \tag{7.21}$$

where f_c, f_0, k = coefficients in Horton's equation and D_2 = the intermediate water depth after accounting for infiltration.

3. If the resulting water depth of subcatchment D_2 is larger than the specified detention depth D_d, an outflow rate is computed using a modified Manning's equation

$$V = (1.49/n)(D_2 - D_d)^{2/3}S^{1/2} \tag{7.22}$$

and

$$Q_W = VW(D_2 - D_d) \tag{7.23}$$

where V = velocity, n = Manning's coefficient, S = ground slope, W = the width, and Q_W = the outflow rate.

TABLE 7.16 General Data Requirements for Storm Water Management Model (SWMM)

Item 1.	*Define the Study Area.* Land use, topography, population distribution, census tract data, aerial photos, and area boundaries.
Item 2.	*Define the System.* Plans of the collection system to define branching, sizes, and slopes; types and general locations of inlet structures.
Item 3.	*Define the System Specialties.* Flow diversions, regulators, and storage basins.
Item 4.	*Define the System Maintenance.* Street sweeping (description and frequency), catch-basin cleaning; trouble spots (flooding).
Item 5.	*Define the Base Flow (DWF).* Measured directly or through sewerage facility operating data; hourly variation and weekday versus weekend; the DWF characteristics (composited BOD and SS results); industrial flows (locations, average quantities, and quality).
Item 6.	*Define the Storm Flow.* Daily rainfall totals over an extended period (6 months or longer) encompassing the study events; continuous rainfall hyetographs, continuous runoff hydrographs, and combined flow quality measurements (BOD and SS) for the study events; discrete or composited samples as available (describe fully when and how taken).

4. The continuity equation is solved to determine water depth of the subcatchments resulting from rainfall, infiltration, and outflow. Thus,

$$D_{t+\Delta t} = D_2 - (Q_W/A)\,\Delta t \tag{7.24}$$

5. Steps 1–4 are repeated until computations for all subcatchments are completed.
6. Inflow (Q_{in}) to a gutter is computed as a summation of outflow from tributary subcatchments ($Q_{w,i}$) and flow rate of immediate upstream gutters ($Q_{g,i}$)

$$Q_{in} = \Sigma Q_{w,i} + \Sigma Q_{g,i} \tag{7.25}$$

7. The inflow is added to raise the existing water depth of the gutter according to its geometry. Thus,

$$Y_1 = Y_t + (Q_{in}/A_s)\,\Delta t \tag{7.26}$$

where Y_1, Y_t = the water depth of the gutter, and A_s = the mean water surface area between Y_1 and Y_t.

8. The outflow is calculated for the gutter using Manning's equation:

$$V = (1.49/n)R^{2/3}S^{1/2} \tag{7.27}$$

and

$$Q_g = VA_c \tag{7.28}$$

where R = hydraulic radius, S_i = the invert slope, and A_c = the cross-sectional area at Y_1.

9. The continuity equation is solved to determine the water depth of the gutter resulting from the inflow and outflow. Thus,

$$Y_{t+\Delta t} = Y_1 + (Q_{in} - Q_g)(\Delta t/A_s) \tag{7.29}$$

10. Steps 6–9 are repeated until all the gutters are finished.
11. The flows reaching the point concerned are added to produce a hydrograph coordinate along the time axis.
12. The processes from 1 to 11 are repeated in succeeding time periods until the complete hydrograph is computed.

Three general types of output are provided by SWMM. If waste treatment processes are simulated or proposed, the capital, land, and operation and maintenance costs are printed.

Plots of water quality constituents versus time form the second type of output. These pollutographs are produced for several locations in the system and in the receiving waters. Quality constituents handled by SWMM include suspended solids, settleable solids, BOD, nitrogen, phosphorus, and grease. The third type of output is hydrologic. Hydrographs at any point—for example, at the end of a gutter or inlet—are printed for designated time periods. A special *statistics* block provides a frequency analysis of storm events from the continuous simulation.

The input to the water quality portion of the SWMM consists of hydrographs developed in the hydrologic phase of the model. The output takes the form of pollutographs for each of the pollutants modeled. The hydrographs and pollutographs that are calculated are then introduced into the transport block, which combines them with the dry-weather and infiltrated flow components to produce the actual outfall graphs for water quality and quantity. The SWMM is capable of predicting the concentrations of suspended solids, BOD, total coliform, COD, settleable solids, nitrates, phosphates, and grease in storm water runoff [46].

The SWMM makes use of the assumption that the amount of a pollutant that can be removed from a drainage area during a storm event is a function of the storm duration and initial quantity of the pollutant. This can be represented by a first-order differential equation of the form

$$-\frac{dP}{dt} = kP \tag{7.30}$$

which integrates to

$$P_0 - P = P_0(1 - e^{-kt}) \tag{7.31}$$

where

P_0 = initial amount of pollutant per unit area

P = remaining amount of pollutant per unit area at time t

k = decay rate

The value of k is assumed to be directly proportional to the rate of runoff. In the model, this is represented as $k = br$, where b is a constant and r is the runoff rate. A value of 4.6 has been used for k based on analyses of storm event data from urban areas. This implies identical rainfall intensities and wash-off rates for all storms, a condition not met with in reality.

For the prediction of suspended solids and BOD, it has been determined that a modification of Eq. (7.31) is needed. This change incorporates an availability factor A_0, which represents the percentage of pollutant P that is available for capture by the storm [46]. Thus, Eq. (7.31) becomes

$$P_0 - P = A_0 P_0(1 - e^{-kt}) \tag{7.32}$$

Coliform densities are predicted as the product of the suspended-solids concentration and an appropriate conversion factor. The EPA provides complete information on methods used to establish the required model parameters [39].

For each time step in the modeling process, the rate of runoff is calculated using the hydrologic model. A value of P is also determined [Eq. (7.31) or (7.32)] and then becomes the value of P for the next time step. Then, during the time interval, the change in value of P can be related to the quantity of flow from the area to produce the pollutograph of the constituent of interest. Calculations proceed from one time step to the next until the storm event has ended.

Calibration of the model centers around a trial-and-error procedure to determine the ideal combination of loading rate and removal coefficient that would result in a satisfactory match of the observed and computed pollutographs. The parameters derived in this manner are valid only for the particular storm used in the calibration and should not be used for other storms unless their features are considered to be quite similar. The SWMM has the flexibility to determine pollutant loading from a variety of urban land use characterizations. It can also be used to generate input data for use in stream quality models.

7.19 SYSTEM LAYOUT

Storm drainage systems may be closed conduit, open conduit, or some combination of the two. In most urban areas, the smaller drains frequently are closed conduits, and as the system moves downstream, open channels are often employed. Because quantities of storm water runoff are usually quite large when contrasted with flows in sanitary sewers, the drainage works needed to carry them are also large and thus important from the standpoint of utilities placement. Common practice is to build storm drains under the centerline of the street, then offset water mains to one side and sanitary sewers to the other.

To produce a workable system layout, a map of the area is needed showing contours, streets, buildings, other existing utilities, natural drainage ways, and areas for future development. Storm drains will generally be located under streets or in designated drainage rights-of-way. The entire area to be served must be considered.

The first step in laying out the drainage system is to tentatively locate the inlet structures. Once this has been done, a skeleton pipe system connecting the inlets is drawn, along with the location of all proposed manholes, wye branches, special bend or junction structures, and the outfall.

Customarily, a manhole or junction structure will be required at all changes in grade, pipe size, direction of flow, and quantity of flow. Figure 7.17 illustrates the details of a typical manhole. Note that today most of these structures are precast concrete rather than brick. Storm drains are usually not built smaller than 12 in. in diameter, as smaller diameter pipes tend to clog readily with debris, thus creating malfunction problems. Maximum manhole spacing for pipes 27 in. and under should not exceed about 600 ft. For larger pipes, no maximum spacing is prescribed. The normal requirements for locating these structures should provide access to the drain for inspection, cleaning, or maintenance.

7.20 STORM WATER INLETS

Storm water inlet capacity is an important consideration. Regardless of the adequacy of the underground drainage system, proper drainage cannot result unless storm water is efficiently collected and introduced into it. A knowledge of the ability of storm water inlets to accept incoming flows has important implications: Inadequate inlet capacity can result in street flooding, inefficient use of the storm drainage system, and blocking of inlets by debris. Street grade, cross slope, and depression geometry affect the hydraulic efficiency of storm water inlets, and the choice of inlet type to be used must be made accordingly. No specific inlet type can be considered best for all conditions of

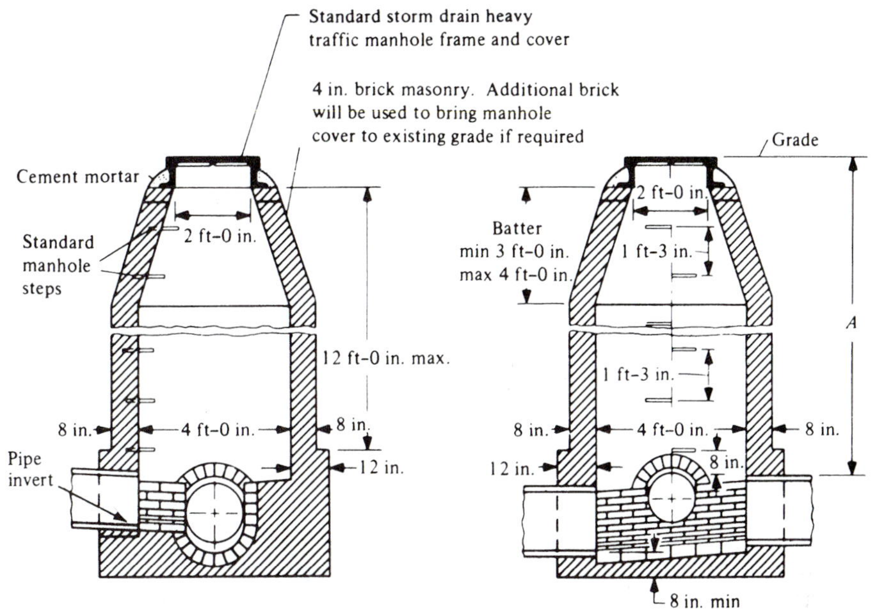

FIGURE 7.17 Typical storm drainage manhole details. *Source*: Courtesy of the Baltimore County, Maryland, Department of Public Works.

use. Ideally, a simple opening across the flow path would be the most effective type of inlet structure. But construction of this type would be impractical and unsafe. The opening must be covered with a grate or located in the curb. Unfortunately, grates obstruct the flow of water and often serve to collect debris, while curb openings are not in the direct path of flow. Thus, compromises are required.

Increasing the street cross slope will increase the depth of flow of the gutter, gutter depressions will concentrate flows at the inlet, and curb and gutter openings can be combined. These and other measures can be taken to increase inlet capacities, although some of them are not compatible with high-volume traffic.

Numerous inlet designs are in evidence, most of which have been developed from the practical experience of engineers or by rule-of-thumb procedures. The hydraulic capacity of many of these designs is unknown, and estimates of it are often in error. But laboratory studies of inlet capacities have been conducted to produce efficient designs and develop better understanding of inlet behavior [47].

Four major types of inlets are in use: curb inlets, gutter inlets, combination inlets, and multiple inlets. For each classification of design, various configurations have been made. A brief description of the basic types of inlets follows:

1. *Curb inlet*—A vertical opening in the curb through which gutter flow passes.
2. *Gutter inlet*—A depressed or undepressed grated opening in the gutter section through which the surface drainage falls.
3. *Combination inlet*—An inlet composed of both curb and gutter openings. Gutter openings may be placed directly in front of the curb opening (contiguous

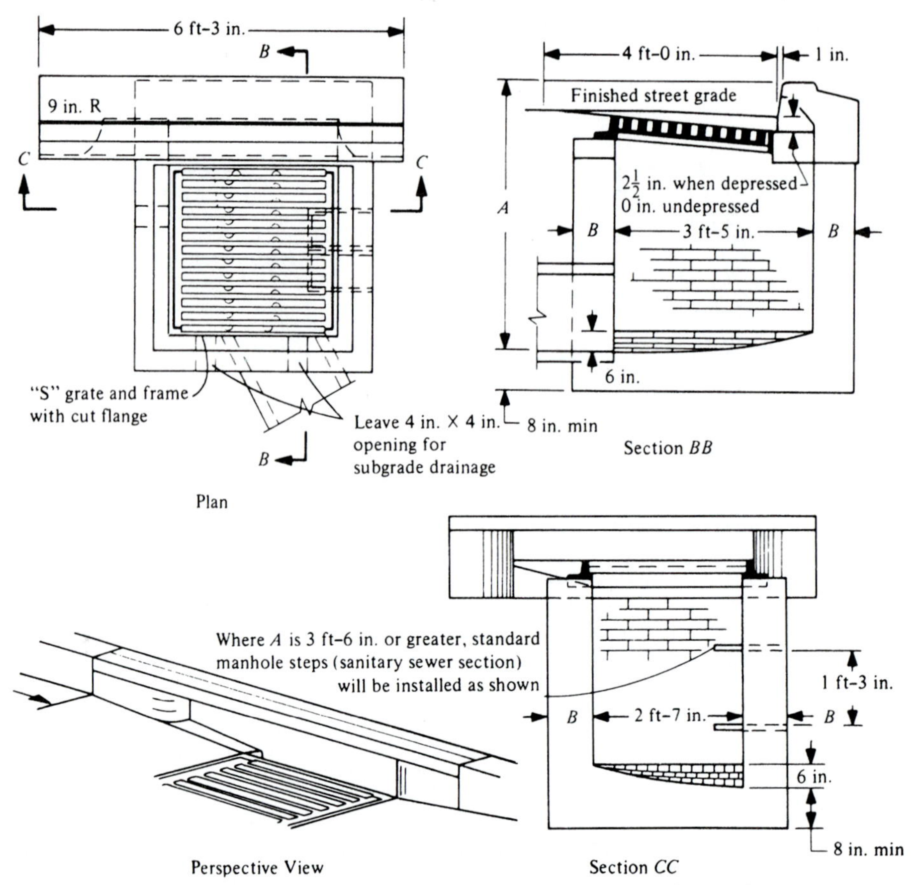

FIGURE 7.18 Baltimore type *S* combination inlet. *Source*: Courtesy of the Baltimore County, Maryland, Department of Public Works.

combination inlet) or upstream or downstream of the gutter opening (offset inlet). Combination inlets may be depressed or undepressed. Figure 7.18 depicts a typical combination inlet. Figure 7.19 relates the capacity of the inlet to the percent gutter slope.

4. *Multiple inlets*—Closely spaced interconnected inlets acting as a unit. Identical inlets end to end are called "double inlets."

The selection of an optimum inlet type for a specific location should be based on the exercise of engineering judgment relative to the importance of clogging, traffic hazard, safety, and cost. In general, consistent with vehicle safety and driver comfort, street cross slopes as steep as practical should be used. Inlets should be located and designed so that there will be a 5–10% bypass in gutter flow. This significantly increases inlet capacity. But the amount of bypass flow should not be so great as to inconvenience pedestrians or vehicular traffic. On streets where parking is permitted, or where vehicles are not expected to travel near the curb, contiguous combination curb-and-gutter

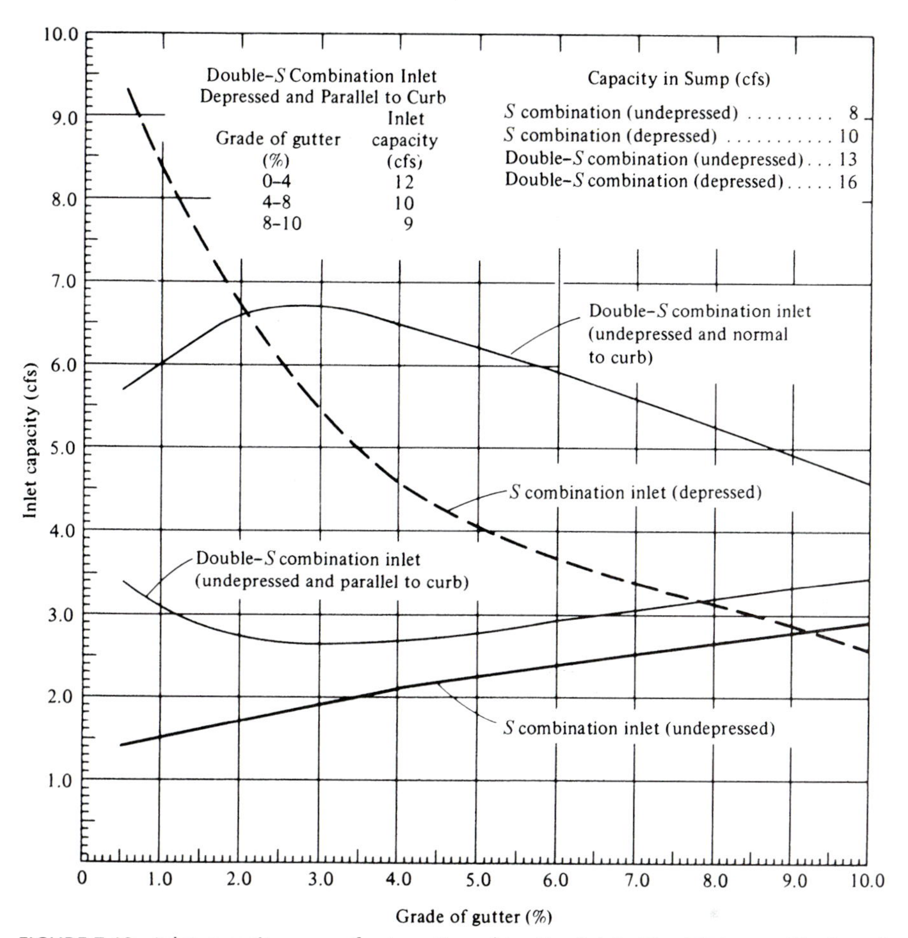

FIGURE 7.19 Inlet capacity curves for type *S* combination inlets. The inlet capacities have been reduced by 20% for the estimated effects of debris partially clogging the inlet (on slopes). The capacities are based on standard $7^3/_{16}$-in. curb and gutter and 1–18 average crown slope for a distance of about 5 ft from the face of the curve. *Source*: Courtesy of the Baltimore County, Maryland, Department of Public Works.

inlets with longitudinal grate bars could be used, or depressed gutter inlets used if clogging is not a problem. Where clogging is likely, for low gutter flows use depressed curb inlets for large flows, use depressed combination inlets with the curb openings upstream. On streets having slopes in excess of 5%, where traffic passes close to the curb, use deflector inlets if road dirt will not pack in the grooves. For flat slopes or where dirt is a problem, use undepressed gutter inlets or combination inlets with longitudinal bars only. For streets having flat grades or sumps (lows), pitch the grade toward the inlet on both ends. This will have the effect of providing a sump at the inlet. For true sump locations, use curb openings or combination inlets. The total open area, not the size and arrangement of bars, is important because the inlet will act as an orifice. Normally, sump inlets should be overdesigned because of the unique clogging problems that develop in depressed areas.

Inlets should be constructed in all sumps and at all street intersections where the quantity of flow is significant, or where nuisance conditions warrant such construction. Inlets are required at intermediate points along streets where the curb and gutter capacity would be exceeded without them. Inlet capacities should be equal to or greater than design flows. Figure 7.19 gives a relationship between inlet capacity and gutter slope. This relationship must be taken into account when selecting inlet configurations for specific locations.

Rapid and efficient removal of surface runoff from streets and highways is important for safety and nuisance minimization. Street grade, crown slope, inlet type, grade design, and tolerable bypass flow are all important factors in the selection of inlet structures. Information on the hydraulic performance of inlets is essential for effective storm drainage designs.

7.21 HYDRAULIC DESIGN

The basic hydraulic and hydrologic principles discussed in this chapter are applicable to the design of urban drainage systems. The following examples illustrate their use.

Example 7.12

Given the sketch of the urban drainage area shown in Figure 7.20, use the standard rational method to determine the pipe sizes required for the pipes from M1–M2, M2–M3, and M3 to the outfall. Assume a Manning's n of 0.013, a runoff coefficient c for all areas = 0.60, the pipe lengths and slopes shown on Table 7.17, and the IDF curve for a 10-year storm given in Figure 7.12. Solve using a spreadsheet.

Solution: Enter the given data on the spreadsheet (Table 7.17). Then enter formulas for calculating Q, pipe diameter, flow velocity, pipe flow time, and total flow time to

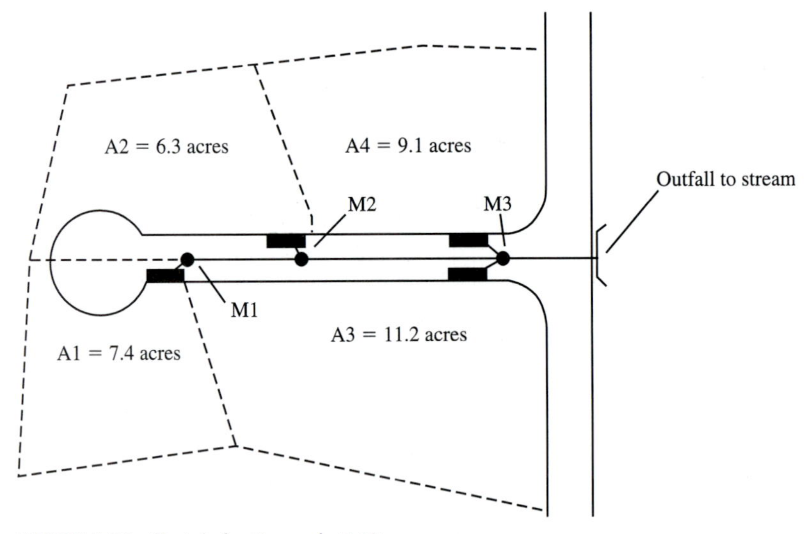

FIGURE 7.20 Sketch for Example 7.12.

TABLE 7.17 Spreadsheet Solution for Example 7.12

Pipe Section	Drainage Area	Area acres	Flow Time - min.			Rainfall Intensity *i*-in./hr	*c*	*Q* (cfs)	Pipe Length ft	Slope ft/ft	Calculated Diameter ft	Diameter Selection (ft)	Diameter Selection (in.)	Velocity	Flow Time (min)
			Inlet	Pipe	Total										
M_1–M_2	A_1	7.4	5	—	5	7	0.6	31.08	450	0.012	2.18	2.33	28	7.29	1.03
M_2–M_3	$A_1 + A_2$	13.7	5	1.03	6.03	6.7	0.6	55.07	700	0.025	2.35	2.5	30	11.22	1.04
M_3–Out	A_1 to A_4	34		1.04	7.07	6.4	0.6	130.56	200	0.025	3.25	3.67	44	12.34	0.27

manhole moving downstream to the outfall. To get pipe diameter D, solve Manning's equation for D. Then the velocity V can be determined using the relationship $V = Q/A$, and the flow time can be determined using the relationship $T = L/V$.

The solution is shown in Table 7.17. Note that calculated diameters are converted to diameters used based on standard pipe sizes.

Example 7.13

Design a storm drainage system for the Mextex drainage area shown in Figure 7.21. Note that the eight subareas are tributary to individual storm water inlets. Assume that a 10-yr frequency rainfall satisfies local design requirements. Assume clay soil to be predominant, and that average lawn slopes prevail.

Solution:

1. Prepare a drainage area map showing drainage limits, streets, impervious areas, and direction of surface flow.
2. Divide the drainage area into subareas tributary to the proposed stormwater inlets (Fig. 7.21).
3. Compute the acreage and imperviousness of each area.
4. Calculate the required capacity of each inlet, using the rational method. Assume a 5-min inlet time to be appropriate and compute inlet flows for a rainfall intensity of 7.0 in./hr. This is obtained by using the 10-yr frequency curve in Figure 7.12 with a 5-min concentration time. Appropriate c values are obtained from Figure 7.11 by entering the graph with the calculated percentage imperviousness (the percent of the inlet area covered by streets,

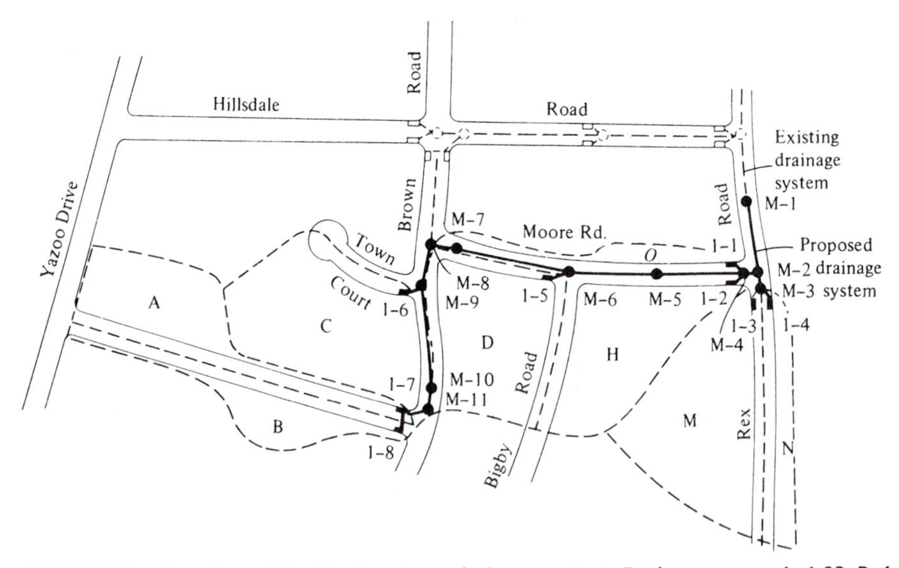

FIGURE 7.21 Plan view of the Mextex storm drainage system. Drainage areas: A, 1.93; B, 1.34; C, 3.19; D, 2.66; H, 2.25; M, 2.75; N, 1.27; O, 0.40.

TABLE 7.18 Required Storm Water Inlet Capacities for Example 7.13

(1) Inlet	(2) Area Designation	(3) Area (acres)	(4) Percent Impervious	(5) C	(6) Rainfall Intensity	Q: (3) × (5) × (6) (cfs)
I-1	O	0.40	49	0.57	7.0	1.59
I-2	H	2.25	20	0.35	7.0	5.52
I-3	M	2.75	26	0.40	7.0	7.70
I-4	N	1.27	26	0.40	7.0	3.55
I-5	D	2.66	26	0.40	7.0	7.43
I-6	C	3.19	24	0.38	7.0	8.48
I-7	A	1.93	23	0.37	7.0	4.99
I-8	B	1.34	29	0.42	7.0	3.93

sidewalks, drives, roofs, etc.), projecting up to the average lawn-slope curve, and reading c on the ordinate. Computations for the inlet flows are tabulated in Table 7.18.

5. Select the type inlets required to adequately drain the flows given in Table 7.18 adequately. The choice will be based on a knowledge of the street slopes and their relation to inlet capacities. Inlet capacity curves such as those given in Figure 7.19 would be used. For the purposes of this example, no selections will be made, but the reader should recognize that this is the next logical step, and an important one.
6. Beginning at the upstream end of the system, and moving downstream, compute the discharge to be carried by each successive length of pipe. Calculations are summarized in Table 7.19. Note that at each downstream location where a flow is introduced, a new time of concentration must be determined as well as new values for c and drainage area size. As the upstream inlet areas are combined to produce a larger tributary area at a design point, a revised c value representing this composite area must be determined. This is generally done by taking a weighted average of the individual c values of the components of the combined area. For example, for computing the flow to be carried by the pipe from M-9 to M-8, the tributary area is A + B + C = 6.46 acres, and the composite value of c is determined as

$$c = \frac{\Sigma c_i a_i}{\Sigma a_i} = \frac{0.37 \times 1.93 + 0.42 \times 1.34 + 0.38 \times 3.19}{6.46} = 0.38$$

At the design location, the value of t_c will be equal to the inlet time at I-8 plus the pipe flow time from I-8 to M-9 (see Table 7.19), which must be known to permit determining the rainfall intensity to be used in computing the runoff from composite area A + B + C.
7. Using the computed discharge values, select tentative pipe sizes for the approximate slopes given in column 8 of Table 7.19. Once the pipe sizes are known, flow velocities between input locations can be determined. Normally, velocities are approximated by computing the full-flow velocities for maximum discharge at the specified grade. These velocities are used to compute channel flow time for estimating the time of concentration. If, upon completing the hydraulic design, enough change has been made in any concentration

TABLE 7.19 Computation of Design Pipe Flows for the Storm Drainage System of Example 7.13

			Flow Time (min)						Pipe			
Pipe Section	Tributary Area	Area (acres)	Inlet	Pipe	Total	Rainfall Intensity i	c	Q (cfs)	Slope (%)	Size (in.)	Full-Flow Velocity (fps)	Length (ft)
I-8–I-7	B	1.34	5	0.10	5	7.0	0.42	3.93	1.0	15	5.2	30
I-7–M-11	A + B	3.27	5	0.13	5.10	7.0	0.39	8.93	1.0	18	5.9	46
M-11–M-10	A + B	3.27	—	0.24	5.23	—	0.39	8.93	1.0	18	5.9	85
M-10–M-9	A + B	3.27	—	0.37	5.47	—	0.39	8.93	2.0	18	8.1	178
I-6–M-9	C	3.19	5	—	5	7.0	0.38	8.48	1.0	18	5.9	40
M-9–M-8	A + B + C	6.46	—	0.21	5.80	6.9	0.38	16.90	1.8	21	8.5	110
M-8–M-7	A + B + C	6.46	—	0.11	6.01	—	0.38	16.90	1.8	21	8.5	57
M-7–M-6	A + B + C	6.46	—	0.47	6.12	—	0.38	16.90	1.6	21	8.1	230
I-5–M-6	D	2.66	5	—	5	7.0	0.40	7.43	2.0	15	7.4	19
M-6–M-5	A + B + C + D	9.12	—	0.38	6.59	6.8	0.39	24.20	2.0	24	10.0	230
M-5–M-4	A + B + C + D	9.12	—	0.42	6.97	—	0.39	24.20	1.9	24	9.8	247
I-1–M-4	O	0.40	5	—	5	7.0	0.57	1.59	3.0	15	9.0	19
I-2–M-4	H	2.25	5	—	5	7.0	0.35	5.52	3.0	15	9.0	17
M-4–M-2	A + B + C + D + O + H	11.77	—	0.05	7.39	6.6	0.39	30.3	1.5	27	9.4	29
I-3–M-3	M	2.75	5	—	5	7.0	0.40	7.70	2.0	15	7.4	15
I-4–M-3	N	1.27	5	—	5	7.0	0.40	3.55	2.0	15	7.4	20
M-3–M-2	M + N	4.02	—	—	5	7.0	0.40	11.30	1.8	18	7.8	37
M-2–M-1	A + B + C + D + O + H + M + N	15.79	—	—	7.44	6.6	0.39	40.6	1.4	30	9.8	176

time to alter the design discharge, new values of flow should be computed. Generally, this will not be the case.

8. Using the pipe sizes selected in step 7, draw a profile of the proposed drainage system. Begin the profile at the point farthest downstream, which can be an outfall into a natural channel, an artificial channel, or an existing drain, as in the case of the example. In constructing the profile, be certain that the pipes have at least the minimum required cover. Normally, 1.5–2 ft is sufficient. Pipe slopes should conform to the surface slope wherever possible. Indicate the change in invert elevation at all manholes. In this example, where there is no change in pipe size through the manhole, a drop of 0.2 ft will be used. Where the size decreases upstream through a manhole, the upstream invert will be set above the downstream invert a distance equal to the difference in the two diameters. In this way, the pipe crowns are kept at the same elevation. A portion of the profile of the drainage system in the example is given in Figure 7.22.

9. Compute the position of the hydraulic gradient along the profile of the pipe. If the gradient lies less than 1.5 ft below the ground surface, it must be lowered to preclude the possibility of surcharge during the design flow. Note that the value of 1.5 ft is arbitrarily chosen here. In practice, local standards indicate the limiting value. Hydraulic gradients may be lowered by increasing pipe sizes, decreasing head losses at structures, designing special transitions, lowering the system below ground, or a combination of these means.

Computations for a portion of the hydraulic gradient of the example are carried out in the following manner. Pipe head losses are determined by applying Manning's

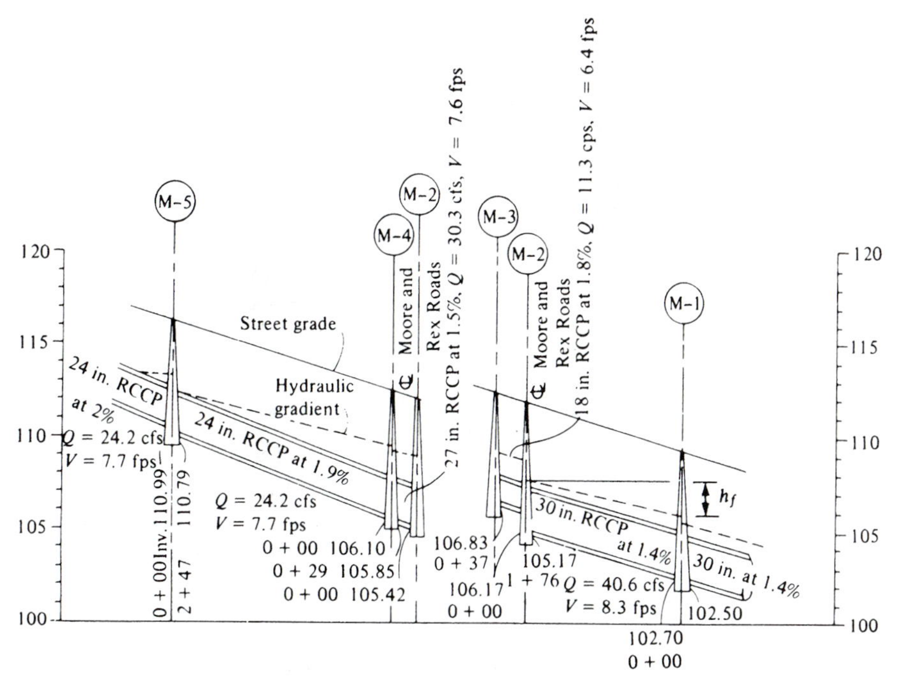

FIGURE 7.22 Profile of part of the Mextex storm drain, showing the hydraulic gradient.

equation, assuming $n = 0.013$ in this example. Head losses in the structures are determined by using the relationships defined in Figures 7.23 and 7.24. These curves were developed for surcharged pipes entering rectangular structures, but may be applied to wye branches, manholes, and junction chambers [48]. The A curve is used to find entrance and exit losses, the B curve is used to evaluate the head loss due to an increased velocity in the downstream direction. This loss is designated as the

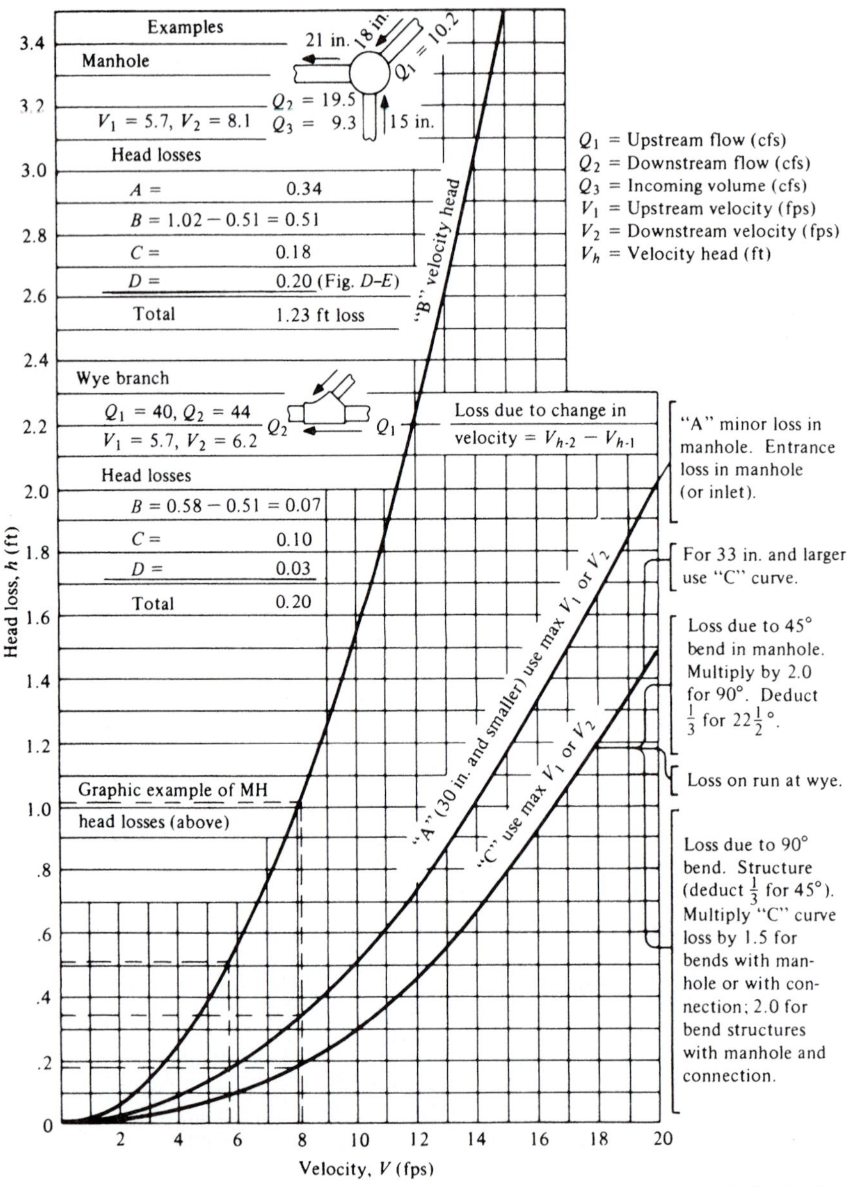

FIGURE 7.23 Types A, B, and C head losses in structures. *Source:* Courtesy of the Baltimore County, Maryland, Department of Public Works.

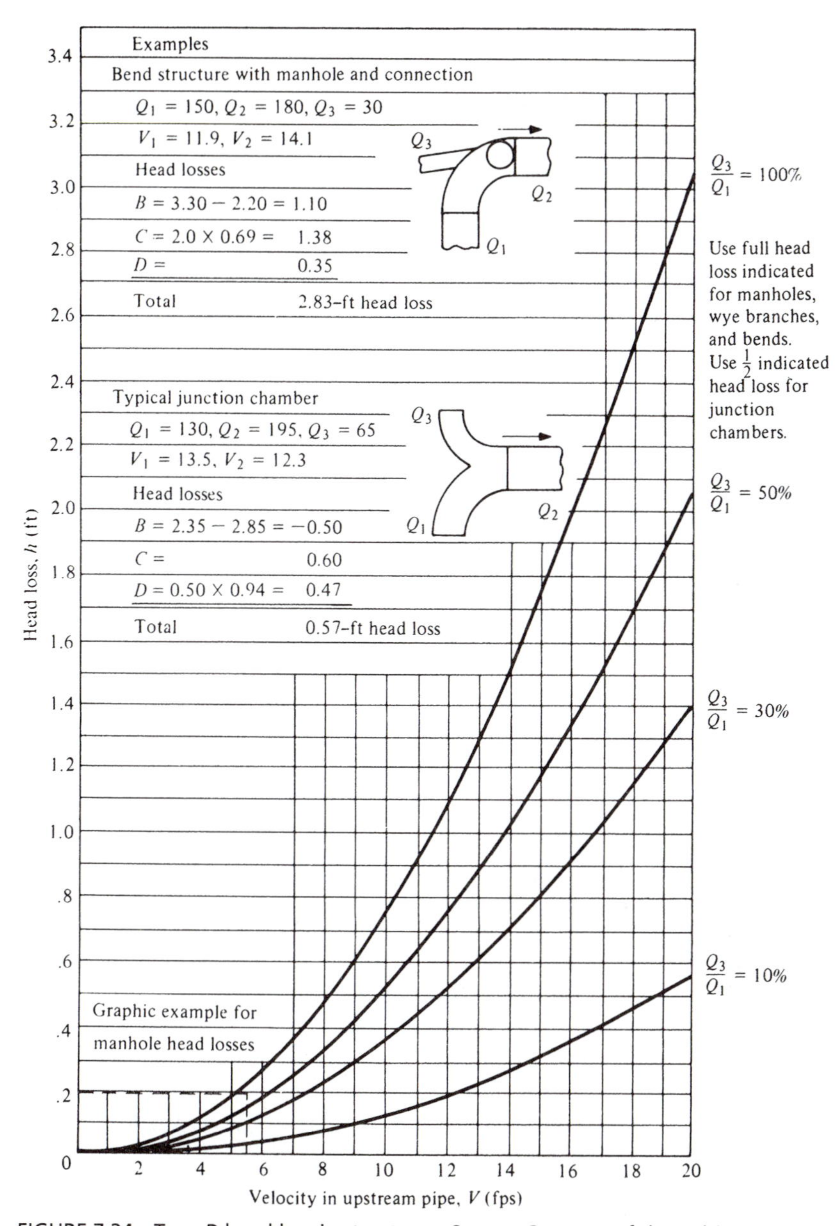

FIGURE 7.24 Type *D* head loss in structures. *Source:* Courtesy of the Baltimore County, Maryland, Department of Public Works.

difference between the head losses found for the downstream and upstream pipe $(V_{h-2} - V_{h-1})$. In cases where the greatest velocity occurs upstream, the difference will be negative and may be applied to offset other losses in the structure. The *C* loss results from a change in direction in a manhole, wye branch, or bend structure. The *D* loss is related to the impact of secondary flows entering the structure. Examples of the use of these curves are shown in Figures 7.23 and 7.24.

Computations for the hydraulic gradient shown in Figure 7.22 are as follows:

1. Begin at the elevation of the hydraulic gradient at the upstream end of the existing 30-in reinforced concrete culvert pipe (RCCP). This elevation is 105.50. The existing hydraulic gradient is shown in Figure 7.22.
2. Compute the head losses in manhole M-1 using Figures 7.23 and 7.24.

$$A \text{ loss} = 0.36 \text{ ft } (V = 8.3 \text{ fps} = Q/A)$$

$$B \text{ loss} = 0; \text{ no change in velocity, } V_1 = V_2$$

$$C \text{ loss} = 0; \text{ no change in direction}$$

$$D \text{ loss} = 0; \text{ no secondary flow}$$

$$\text{total} = 0.36 \text{ ft}$$

The hydraulic gradient rises in the manhole to an elevation of $105.50 + 0.36 = 105.86$ ft, as plotted in M-1 in Figure 7.22.

3. Compute the head loss due to friction in the 30-in. drain from M-1 to M-2. Assume that $n = 0.013$. Using Manning's equation, the head loss per linear foot of drain is

$$S = \frac{(nV)^2}{2.21R^{4/3}}$$

and from M-1 to M-2,

$$S = \frac{(0.013 \times 8.3)^2}{2.21 \times 0.534} = 0.00986$$

The total frictional head loss is therefore

$$hf = S \times L = 0.00986 \times 176 = 1.73 \text{ ft}$$

Elevation of the hydraulic gradient at the downstream end of M-2 is thus $105.86 + 1.73 = 107.59$ ft. This elevation is plotted in Figure 7.22, where the hydraulic gradient in this reach is drawn in.

4. Compute the head losses in M-2.

$$A = 0.36 \; (V = 8.3 \text{ fps})$$

$$B = 1.07 - 0.90 = 0.17 \; (V_2 = 8.3, V_1 = 7.6 = Q/A \text{ for 27-in. drain})$$

$$C = 0.20 \times 2.0 \text{ (multiply by 2 for 90° bend in manhole—see Fig.7.23)} = 0.40$$

$$D = 0.22 \text{ for } Q_3/Q_1 = 11.3/30.3 = 37\%$$

The total head loss in M-2 equals 1.15 ft and the elevation of the hydraulic gradient in M-2 is therefore $107.59 + 1.15 = 108.74$ ft.

5. Compute the friction head loss in the section of pipe from M-2 to M-3.

$$S = \frac{(0.013 \times 6.4)^2}{2.21 \times 0.272} = 0.0113$$

$$\text{hf} = 0.0113 \times 37 = 0.42 \text{ ft}$$

The elevation of the hydraulic gradient at the downstream end of M-3 is therefore 108.74 + 0.42 = 109.16 ft. Plot this point on the profile and draw the gradient from M-2 to M-3.

The hydraulic gradient in the above example was computed under the assumption of uniform flow. In closed-conduit systems, if the pipes are flowing full or the system is surcharged (the usual design flow conditions), the method will produce good results. When open conduits are used, or in partial-flow systems, the hydraulic gradient can be determined by computing surface profiles in the manner described in Section 7.2.

Computations for the remainder of the hydraulic gradient are identical to those just given. It should be noted that none of the gradients shown in Figure 7.22 come within 1.5 ft of the surface, so no revisions are needed. If they were too close to the surface, it would be necessary to modify all or a portion of the system to lower the gradient. This could be accomplished by reducing head losses, by increasing the depth of the system, or both. The choice would depend primarily on cost.

7.22 STORM WATER STORAGE

Historically, storm water was considered a resource to be disposed of quickly. The notion was to remove it from the contributing area as rapidly as possible. More recently, however, storm water flows have been looked upon as a resource to be captured and utilized for groundwater recharge, recreation, and other purposes or detained as a water quality control or peak-flow reduction measure. The principal mechanism employed is storage. This approach is also consistent with the regulatory policies in many states that require storage for water quality control in urban areas [49].

The idea behind detention storage is to reduce the rate of delivery of storm water at a discharge point. Retention is employed where no downstream releases are to be made. In this case, the volumes of water generated are ultimately infiltrated or evaporated. Storage requirements for detention and retention basins are estimated by analyzing anticipated storm events. Flows from detention basins are calculated by using standard routing techniques [13,49]. Most detention basins have a low-flow outlet and an emergency spillway or other type of control structure for dealing with high flows. Retention and detention basins may comprise parking lot storage, earth ponds, concrete tanks, depressed park lands, roofs, and capacity within the drains themselves.

Figure 7.25 illustrates a typical detention pond configuration. Note that detention basin design and analysis is founded on storage–discharge relationships. If, in Figure 7.25. Q_i represents the inflow to the pond, Q represents the outflow, and S represents storage in the pond, then a continuity equation of the form

$$Q_i - Q = dS/dt \tag{7.33}$$

can be written, where dS/dt is the rate of change of storage with respect to time. If one considers that storage increments can be calculated as the product of mean surface area of the pond at a specified depth and the depth increment, it can be stated that

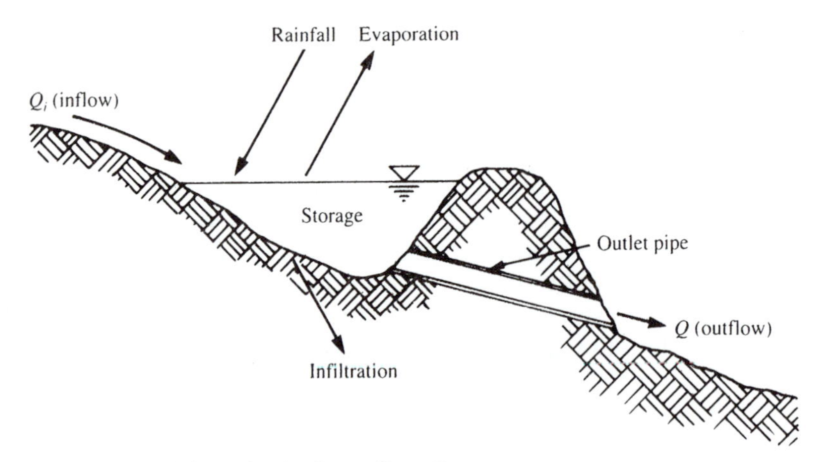

FIGURE 7.25 A detention basin configuration.

$$\frac{dS}{dt} = A(h)\frac{dh}{dt} \tag{7.34}$$

where $A(h)$ is a relationship equating pond surface area A to a reference depth (or elevation) and dh/dt is the rate of change of depth with respect to time. Depth–area relationships for reservoirs can be obtained by developing contours for the reservoir and then relating the surface areas contained within these contours to the elevations associated with them. Incremental volumes can be determined by using average end-area or other formulas. Often, a relationship of the form

$$A = Kh^m \tag{7.35}$$

can be derived relating surface area to depth. In Eq. (7.35), A is the surface area, K and m are constants for the site, and h is the depth of water in the reservoir above a reference point such as the elevation of the centerline of the outlet [2,49]. Where the outlet is a pipe, as shown in Figure 7.25, the outflow can be calculated using the orifice equation

$$Q = C_d A_o (2gh)^{0.5} \tag{7.36}$$

where Q is the outflow, C_d is a discharge coefficient, A_o is the orifice area, h is the elevation above the orifice, and g is the acceleration of gravity. Discharge coefficients for submerged pipes behaving as orifices range from about 0.62 to 1.0, depending on whether they are sharp edged or well rounded [50].

If we insert Eqs. (7.36) and (7.34) in Eq. (7.33), the result is

$$Q_i = C_d A_o (2gh)^{0.5} = A(h)\frac{dh}{dt} \tag{7.37}$$

Solving for dh/dt, we get

$$\frac{dh}{dt} = \frac{Q_i - C_d A_o (2gh)^{0.5}}{A(h)} \tag{7.38}$$

which is the governing differential equation for depth as a function of time. Specifying the right side of Eq. (7.38) as $f(h, t)$, then

$$dh/dt = f(h, t) \tag{7.39}$$

This equation can be solved numerically in several ways. The Euler method is described here and is applied to the example that follows. The procedure is simple but coarse [49,51].

If a time step of Δt is assumed, then the derivative in Eq. (7.39) can be approximated as

$$\frac{dh}{dt} = \frac{h(t + \Delta t) - h(t)}{\Delta t} \tag{7.40}$$

If the right side of Eq. (7.3) is substituted into this relationship for dh/dt, the equation can be solved for $h(t + \Delta t)$ as follows:

$$h(t + \Delta t) = h(t) + \Delta t f(h, t) \tag{7.41}$$

Note that the function $h(t)$ is always evaluated at time step t rather than at step $t + \Delta t$. Since $f(h, t)$ is quantified by Eq. (7.38), the calculation of successive values of h can be accomplished in an organized and straightforward manner. The procedure is illustrated in Example (7.14).

Example 7.14

Consider the sketch of Figure 7.25 and the input hydrograph given in column 1 of Table 7.20. In addition, consider that the discharge pipe is 8 in. in diameter (0.203 m), that C_d is 0.9, that the pond surface area is related to elevation by the expression $A = 400h^{0.7}$ (where h is the depth above the culvert centerline and A is the surface area in m^2), and the initial depth in the pond above the outlet centerline is 0.5 m.

Solution: A spreadsheet analysis is suggested. For this example, Quattro Pro was used. For ease of calculations, we calculate the product of $C_d A_o$ to be 0.02918. Substituting this value, the numerical value of the product $2g = (9.8 \text{ m/s}) \times 2 = 19.6$, and the relationship for h in Eq. (7.38), we get

$$\frac{dh}{dt} = \frac{Q_i t - 0.02918(19.6h)^{0.5}}{400h^{0.7}}$$

With this relationship, a solution for successive values of h, at quarter-hour (900 sec) time steps, can be carried out as illustrated in Table 7.20. In the table, the values in columns 1 and 2 are given, as is the initial value of depth (0.5 m) in column 3. Column headings are described in the table footnotes and follow the terminology of Eqs. (7.33) to (7.41). The calculated outflow hydrograph is plotted in Figure 7.26, along with the inflow hydrograph. It is easy to see from the plot that the inflow rates are

TABLE 7.20 Spreadsheet Detention Pond Analysis

(1) Time (hr)	(2) Q-In. (cms)	(3) $h(t)$ (m)	(4) Q-Out (cms)	(5) A-Surf (sq. m)	(6) dh/dt (m/s)	(7) $h(t + dt)$ (m)
0	0	0.5	0.09141	246.2289	−0.00037	0.165882
0.25	0.18	0.16634	0.052724	113.9614	0.001117	1.17149
0.5	0.36	1.17206	0.139954	447.018	0.000492	1.615088
0.75	0.54	1.615474	0.164309	559.59	0.000671	2.219706
1	0.72	2.220068	0.192617	699.063	0.000754	2.899041
1.25	0.9	2.89938	0.220122	842.7012	0.000807	3.625486
1.5	1.1	3.625808	0.246158	985.4699	0.000866	4.405597
1.75	0.99	4.405905	0.271349	1129.496	0.000636	4.978537
2	0.88	4.978833	0.288453	1230.408	0.000481	5.411529
2.25	0.77	5.411818	0.300734	1304.368	0.00036	5.735607
2.5	0.66	5.735891	0.309607	1358.566	0.000258	5.968013
2.75	0.55	5.968294	0.315817	1396.867	0.000168	6.119178
3	0.44	6.119456	0.319792	1421.54	8.46E-05	6.195562
3.25	0.33	6.19584	0.321781	1433.937	5.73E-06	6.200998
3.5	0.22	6.201274	0.321923	1434.817	−7.1E-05	6.137343
3.75	0.11	6.137619	0.320266	1424.492	−0.00015	6.004772
4	0	6.00505	0.316788	1402.884	−0.00023	5.801819
4.25	0	5.802097	0.311389	1369.523	−0.00023	5.597464
4.5	0	5.597744	0.305856	1335.578	−0.00023	5.391638
4.75	0	5.39192	0.300181	1301.009	−0.00023	5.184264
5	0	5.184549	0.294352	1265.778	−0.00023	4.975257
5.25	0	4.975544	0.288357	1229.839	−0.00023	4.764523
5.5	0	4.764812	0.282185	1193.142	−0.00024	4.551957
5.75	0	4.552249	0.275819	1155.628	−0.00024	4.337442
6	0	4.337736	0.269242	1117.234	−0.00024	4.120845
6.25	0	4.121142	0.262434	1077.885	−0.00024	3.902018
6.5	0	3.902319	0.255371	1037.495	−0.00025	3.680791
6.75	0	3.681094	0.248027	995.9644	−0.00025	3.456965
7	0	3.457272	0.240369	953.1768	−0.00025	3.230313
7.25	0	3.230624	0.232356	908.9929	−0.00026	3.000567
7.5	0	3.000882	0.223942	863.2453	−0.00026	2.767405
7.75	0	2.767725	0.215066	815.7293	−0.00026	2.530441
8	0	2.530766	0.205654	766.1898	−0.00027	2.289196
8.25	0	2.289527	0.195607	714.3023	−0.00027	2.043068
8.5	0	2.043406	0.184794	659.6419	−0.00028	1.791277

(1) Time sequence
(2) Inflow data
(3) Head above outlet centerline—meters
(4) Outflow calculated using orifice equation
(5) Surface area of pond calculated using power relationship
(6) dh/dt = (inflow−outflow)/(surface area)
(7) $h(t + dt) = h(t) + dt[f(h, t)]$

significantly attenuated by the detention structure. Note that the drop in outflow at time step 1 occurs because there was no inflow at $t = 0$ and there was already a positive head on the outlet of 0.5 m at the start of inflow. Accordingly, the system was draining at that time.

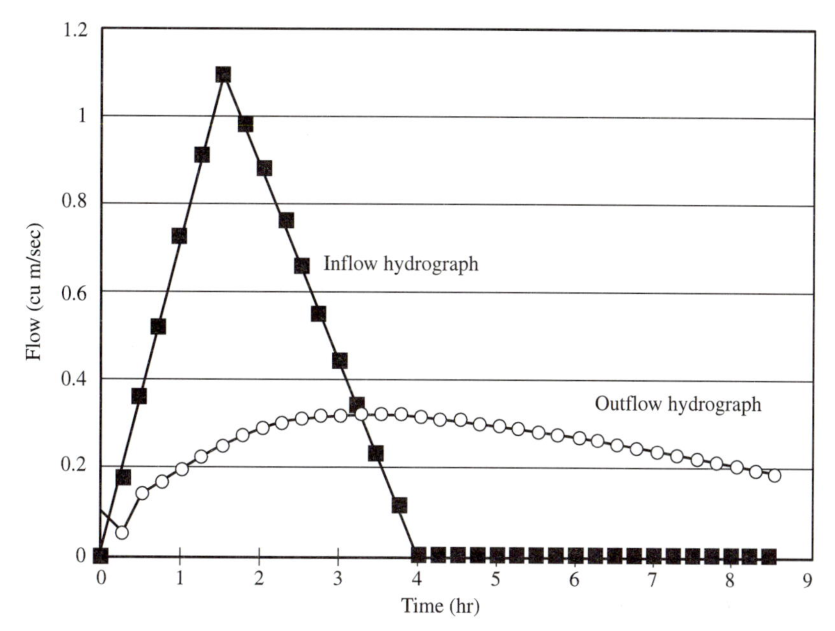

FIGURE 7.26 Detention pond hydrographs.

In the design mode, it is easy to see that for selected inflow sequences and reservoir and outlet dimensions, a determination can be made of flow attenuation, and the aforementioned parameters can be modified if the amount of attenuation is unacceptable.

PROBLEMS

7.1 Wastewater flows in a rectangular concrete channel 6.0 ft wide and 3.0 ft deep. The design flow is 25 cfs. Find the critical velocity. Also find the slope of the channel if the flow velocity is to be 2.5 fps.

7.2 A 16-in. sewer pipe flowing full is expected to carry 6.0 cfs. The n value is 0.011. The minimum flow is $\frac{1}{12}$ that of the maximum. Find the depth and velocity at minimum flow.

7.3 The sewage from an area of 400 acres is to be carried by a circular sewer at a velocity not less than 2.5 fps. Manning's $n = 0.013$. The population density is 16 persons per acre. Find the maximum hourly and minimum hourly flows. Determine the pipe size required to handle these flows and the required slope.

7.4 Water flows in a rectangular concrete channel 9.5 ft wide and 8 ft deep. The channel invert has a slope of 0.10% and the applicable n value is 0.013. The flow carried by the channel is 189 cfs. At an intersection of this channel with a canal, the depth of flow is 7.1 ft. Find the distance upstream to a point where normal depth prevails. Plot the surface profile.

7.5 A 54-in. sewer ($n = 0.013$) laid on a 0.17% grade carries a flow of 19 mgd. At a junction with a second sewer, the sewage depth is 36 in. above the invert. Plot the surface profile back to the point of uniform depth.

7.6 Determine the minimum velocity and gradient required to transport $\frac{3}{16}$-in. gravel through a 42-in.-diameter pipe, given $n = 0.013$ and $K = 0.05$.

7.7 An inverted siphon is to carry a minimum dry-weather flow of 1.5 cfs, a maximum dry-weather flow of 3.7 cfs, and a storm flow of 48.0 cfs in three pipes. Select the proper diameters to ensure velocities of 3.0 fps in all pipes. Make a detailed sketch of your design. Assume that the siphon goes under a highway with a 3.2-ft drop and is 73 ft long.

7.8 Given the following 25-yr record of 24-hr maximum annual stream flows (cfs), plot on log-probability paper these values versus the percent of years during which runoff was equal to or less than the indicated value. Find the peak flow expected on the average of (a) once every 5 yr, (b) once every 15 yr.

220	196	89	53	47
200	129	76	50	38
218	142	62	52	36
199	127	67	49	32
180	118	54	47	28

7.9 Given the following unit storm, storm pattern, and unit hydrograph, determine the composite hydrograph.

Unit storm = 1 unit of rainfall for 1 unit of time

Actual storm	(time units)	1	2	3	4
Pattern	(rainfall units)	1	3	6	4

Unit hydrograph: triangular with base length of 4 time units, time of rise of 1 time unit, and maximum ordinate of $\frac{1}{2}$ rainfall unit height.

7.10 Given a unit rainfall duration of 1 time unit, an effective precipitation of 1.5 in., the following hydrograph and storm sequence, determine (a) the unit hydrograph and (b) the hydrograph for the given storm sequence.

Storm Sequence

Time units	1	2	3	4
Precipitation (in.)	0.4	1.1	1.8	0.9

Hydrograph

Time units	1	2	3	4	4.5	5	6
Flow (cfs)	101	96	218	512	610	580	460
Time units	7	8	9	10	11	12	13
Flow (cfs)	320	200	180	100	86	60	50

7.11 Do Problem 7.10 if the storm sequence is as follows:

Time units	1	2	3	4
Precipitation—in.	0.5	1.2	1.5	0.8

7.12 Determine the discharge of a trapezoidal channel having a brick bottom and grassy sides, with the following dimensions: depth—5 ft, bottom width—11 ft, top width—16 ft. Assume $S = 0.001$.

7.13 Determine the discharge of a rectangular channel built of concrete with the following dimensions: depth—6 ft, bottom width—10 ft. Assume $S = 0.0015$.

7.14 Given the following data for a 50-year storm, estimate the peak rate of flow to a culvert entrance. Consider the rainfall intensity–duration–frequency curves of Figure 7.12 applicable. The contributing upstream drainage area is 45 acres. The overland flow distance is 180 ft and the average land slope is 2.5%. The land use of the overland flow portion of the drainage area is lawns, sandy soil. The land use for the drainage basin is 75% residential, single-family units and 25% lawns, sandy soil with an overall average slope of about 2.7%. A channel leading to the culvert is 1900 ft long with a slope of 0.012 ft/ft. The grassed channel is trapezoidal with a bottom width of 5 ft and side slopes of 2 ft vertical to 1 ft horizontal.

7.15 Given the following data for a 10-year storm, estimate the peak rate of flow to a culvert entrance. Consider the rainfall intensity–duration–frequency curves of Figure 7.12 applicable. The contributing upstream drainage area is 20 acres. The overland flow distance is 125 ft and the average land slope is 2.5%. The land use of the overland flow portion of the drainage area is lawns, sandy soil. The land use for the drainage basin is 75% residential, multiple units, detached, and 25% lawns, sandy soil with an overall average slope of about 2.7%. A channel leading to the culvert is 2000 ft long with a slope of 0.016 ft/ft. The Manning's n value for the grassed channel is 0.030. The channel is trapezoidal with a bottom width of 3 ft and side slopes of 2 ft vertical to 1 ft horizontal.

7.16 Rework solved Example 7.12 for the drainage areas as follows: A1 = 6.5 acres, A2 = 5.4 acres, A3 = 8 acres, and A4 = 9 acres.

7.17 Rework solved Example 7.12 for the following pipe slopes: M1 − M2 = 0.025 ft/ft, M2 − M3 = 0.035 ft/ft, and M3 to outfall = 0.04 ft/ft.

7.18 Rework solved Example 7.12 if the areas are as in Problem 7.16 above and the slopes are as in Problem 7.17.

7.19 Compute, by the rational method, the design flows for the pipes shown in the accompanying sketch: I-1 to M-4, M-4 to M-3, M-3 to M-2, M-2 to M-1, and M-1 to outfall. A-1 = 2.1 acres, $c_1 = 0.5$; A-2 = 3.0 acres, $c_2 = 0.4$; A-3 = 4.1 acres, $c_3 = 0.7$. Given pipe flow times are I-1 to M-3, 1 min, and M-3 to M-1, 1.5 min.

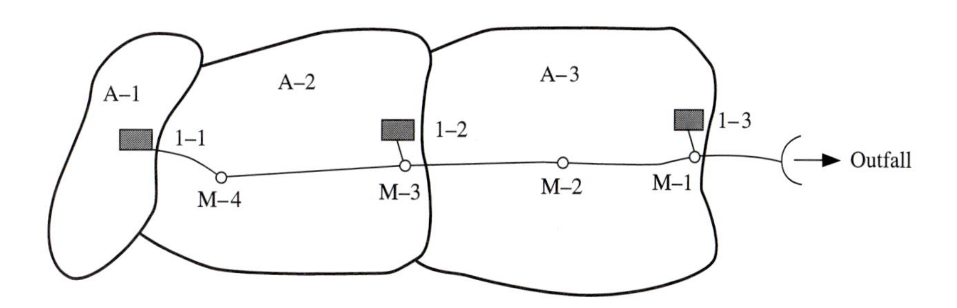

7.20 A watershed of 90 acres has a t_c of 20 min and a runoff coefficient of 0.6. Consider that a rainstorm having an average intensity of 2.0 in./hr, and a duration 40 min occurs over the watershed. Use the modified rational method to develop a storm hydrograph.

7.21 A 45-min rainstorm occurs on a 9-acre urban area having a runoff coefficient of 0.7. The time of concentration has been calculated as 20 min. Incremental rainfall depths for the storm are given in the table below. Using a spreadsheet, calculate the runoff from the storm at 5-minute intervals. Plot the resulting hydrograph.

Time Minutes	Incremental Rainfall Depth—in.
0	0
5	0.1
10	0.18
15	0.29
20	0.6
25	0.64
30	0.51
35	0.3
40	0.17
45	0.07
50	
55	0
60	0
65	0
70	0

7.22 Given the data below, design a section of sanitary sewer to carry these flows at a minimum velocity of 2.5 fps. Assume that the minimum depth of the sewer below the street must be 8.5 ft and that an n value of 0.013 is applicable. Make a plan and profile drawing. (*Note:* The invert into manhole F must be 937 ft. Assume an invert drop of 0.2 ft across each manhole.)

Manholes	Distance Between Manholes (ft)	Flow (gpm)	Street Elevations at Manholes (ft)
A to B	180	1200	A—978
B to C	300	1450	B—972
C to D	450	2200	C—964
D to E	300	2500	D—958
E to F	400	3000	E—954
			F—949

7.23 For an inlet area of 2.0 acres having an imperviousness of 0.50 and a clay soil, find the peak rate of runoff for the 5-, 10-, and 20-yr storms.

7.24 Using the manhole spacing-elevation data given in Problem 7.19 and the data below, design a storm drainage system such that manhole A is at the upper end of the drainage area and manhole F is replaced by an outfall to a stream. The outfall invert elevation is equal to 940.3 ft. Design for the 5-yr storm.

Inflow Point	Incremental Area Contributing to Inflow Point (acres)	Imperviousness of Areas (%)
A to B	3.0	41
B to C	2.7	50
C to D	5.3	55
D to E	3.6	45
Total area = 14.6 acres		

7.25 An impounding reservoir is to provide for constant withdrawal of 430 million gallons per square mile per year. The following record of monthly inflow values was selected from a critical period and was chosen as a basis for design. Find the amount of storage required in Mg/mi^2.

Month	J	F	M	A	M	J	J	A	S	O	N	D
Gauged inflow (mg/mi^2/mo)	23	48	53	87	12	9	7	1	22	34	98	28

7.26 Given the following storm pattern, unit storm, and unit hydrograph, determine the composite hydrograph.

Unit storm = 1 in effective rainfall in 1 hr

Actual Storm Pattern	Time	Volume of Effective Rainfall (in.)
	1:00	2
	2:00	2.5
	3:00	7
	4:00	3

Unit hydrograph (triangular) with base length = 5 hr, time to peak flow = 1.5 hr, and maximum ordinate = 0.4 in./hr.

7.27 Using the derived unit hydrograph from Example 7.8, calculate the storm hydrograph for the following rainstorm pattern: time unit 1, rainfall = 1 in.; time unit 2, rainfall = 3 in.; time unit 3, rainfall = 2 in. Assume the storm data are effective rainfall.

7.28 The following hydrograph was measured on a creek for a storm that produced a total of 1.9 in. of rainfall in 2 hr. Losses due to depression storage, infiltration, and interception amounted to 0.4 in.

	Time from Beginning of Storm (hr)	Discharge (ft^3/s)
	0	73
Storm begins	6	75
	12	900
	18	3400
	24	4350
	30	2900
	36	1450
	42	600
	48	450
	54	375
	60	250
	66	180
	72	135

A spillway design storm is 10 in. of effective rainfall in 8 hr. Estimate the inflow hydrograph for the design storm.

7.29 The ordinates of a 2-hr unit hydrograph are given in the following table. Derive a 4-hr unit hydrograph from the 2-hr unit hydrograph by (a) the lagging method and (b) the *S*-hydrograph method.

Time Unit	2-hr Unit Hydrograph Ordinates
0	0
1	200
2	400
3	300
4	200
5	100
6	0

7.30 Find the peak hourly flow in mgd and m^3/s for a 1000-acre urban area having the following features: domestic flows 100 gpcd, commercial flows 15 gpcd, infiltration 650 gpd/acre, and population density 20 persons per acre.

7.31 A sewer with a flow of 11 cfs enters manhole 2. The distance downstream to manhole 3 is 450 ft. The finished street surface elevation at manhole 2 is 165.0 ft; at manhole 3 it is 162.8 ft. Note that the crown of the pipe must be at least 6 ft below the street surface at each manhole. For a Manning's $n = 0.013$, find the required standard size pipe to carry the flow under (a) full flow and (b) one-half full flow. The flow velocity must be at least 2 ft/s and must not exceed 10 ft/s.

7.32 A sewer system is to be designed for the urban area of Figure 7.7. Street elevations at manhole locations are shown on the figure. It has been determined that the population density is 50 persons per acre and that the sewage contribution per capita is 100 gpd. In addition, there is an infiltration component of 658 gpad. Local regulations require that no sewers be less than 8 in in diameter. Assume an n value of 0.013.

7.33 Given an invert elevation of 105.1 for the pipe leaving M-6 in Figure 7.7, and using the pipe sizes determined in Problem 7.29, calculate the invert elevations in and out of M-5 and M-4 if the street elevation at M-5 is 108.95 ft and at M-4 is 106.82 ft.

7.34 Solve Problem 7.32 if the population density is 45 persons per acre and the infiltration component is 625 gpad.

7.35 Find the peak hourly flow in mgd and m^3/s for a 700-acre urban area having the following features: domestic flows 100 gpcd, commercial flows 18 gpcd, infiltration 650 gpd/acre, and population density 25 persons per acre.

7.36 A sewer with a flow of 8 cfs enters manhole 2. The distance downstream to manhole 3 is 380 ft. The finished street surface elevation at manhole 2 is 135.0 ft; at manhole 3 it is 133.9 ft. Note that the crown of the pipe must be at least 6 ft below the street surface at each manhole. For a Manning's $n = 0.013$, find the required standard size pipe to carry the flow under (a) full flow and (b) one-half full flow. The flow velocity must be at least 2 ft/s and must not exceed 10 ft/s.

7.37 Modify the data given in Example 7.14 so that the initial depth is 0.6 m and the outlet pipe is 10 in. in diameter. Use a spreadsheet approach to solve the problem.

7.38 Solve Problem 7.37 if the inflows given are all increased by 10%.

REFERENCES

[1] S. M. Woodward and C. J. Posey, *Hydraulics of Steady Flow in Open Channels* (New York: Wiley, 1955).

[2] G. M. Fair and J. C. Geyer, *Water Supply and Waste-Water Disposal* (New York: Wiley, 1954).

[3] K. E. Spells, *Trans. Inst. Chem. Engrs.* (London) 33 (1955).

[4] R. S. Gupta, *Hydrology and Hydraulic Systems* (Upper Saddle River, NJ: Prentice Hall, 1989).

[5] H. W. Gehm and J. L. Bregman, ed. *Handbook of Water Resources and Pollution Control* (New York: Van Nostrand Reinhold, 1976).

[6] "Design and Construction of Sanitary and Storm Sewers," *Manual of Engineering Practice No. 37* (New York: American Society of Civil Engineers, 1960).

[7] E. L. Thackston, "Notes on Sewer System Design," unpublished, Vanderbilt University, 1982.

[8] W. Viessman, Jr. and C. Welty, *Water Management: Technology and Institutions* (New York: Harper & Row, 1985).

[9] Metcalf and Eddy, Inc., *Wastewater Engineering* (New York: McGraw-Hill, 1972).

[10] J. A. Lager and W. G. Smith, *Urban Stormwater Management and Technology: An Assessment*, U.S. EPA Rep. No. EPA-670/2-74-040 (Washington, DC, 1975).

[11] V. Novotny and G. Chesters, *Handbook of Nonpoint Pollution: Sources and Management* (New York: Van Nostrand Reinhold, 1981).

[12] J. A. Lager et al., *Urban Stormwater Management and Technology, Update and Users Guide*, U.S. EPA Rep. No. EPA-600/8-77-014 (Washington, DC, 1977).

[13] W. Viessman, Jr. and G. L. Lewis, *Introduction to Hydrology*, 5th ed. (Upper Saddle River, NJ: Prentice Hall, 2003).

[14] L. K. Sherman, "Streamflow from Rainfall by the Unit Hydrograph Method," *Eng. News-Record* 108 (1932): 501–505.

[15] R. K. Linsley, Jr., M. A. Kohler, and J. L. H. Paulhus, *Applied Hydrology* (New York: McGraw-Hill, 1949), 387–411, 444–464.

[16] B. S. Barnes, "Consistency in Unitgraphs," *Proc. Am. Soc. Civil Engrs., J. Hydraulics Div.* 85 (August 1959): 39–61.

[17] C. F. Izzard, "Hydraulics of Runoff from Developed Surfaces," *Proc. Highway Res. Bd.* 26 (1946): 129–150.

[18] J. C. Schaake, *Report No. XI of the Storm Drainage Research Project*, (Baltimore: The Johns Hopkins University, Department of Sanitary Engineering, September 1963).

[19] Federal Aviation Administration, "Circular on Airport Drainage," Report A/C 050-5320-5B (Washington, DC, 1970).

[20] J. C. Y. Guo, *Urban Storm Water Design*, Water Resources Publications, LLC, (Highlands Ranch, CO, 2003).

[21] U.S. Army Waterways Experiment Station, "Models and Methods Applicable to Corps of Engineers Urban Studies," miscellaneous paper H-74-8, National Technical Information Service (August 1974).

[22] G. Fleming, *Computer Simulation Techniques in Hydrology* (New York: American Elsevier, 1975).

[23] N. H. Crawford and R. K. Linsley, Jr., "Digital Simulation in Hydrology: Stanford Watershed Model IV," Department of Civil Engineering, Stanford University, Stanford, CA, Tech. Rep. No. 39 (July 1966).

[24] J. B. Stall and M. L. Terstriep, "Storm Sewer Design—An Evaluation of the RRL Method," EPA Technology Series EPA-R2-72-068 (October 1972).

[25] M. L. Terstriep and J. B. Stall, "The Illinois Urban Drainage Area Simulator, ILLUDAS," *Illinois State Water Surv. Bull.* 58, No. 130 (1974).

[26] A. Brandstetter, "Comparative Analysis of Urban Stormwater Models," Battelle Memorial Institute (August 1974).

[27] H. G. Wenzel and M. L. Terstriep, "Sensitivity of Selected ILLUDAS Parameters," Illinois State Water Survey, Contract Rep. 178 (August 1976).

[28] J. Han and J. W. Delleur, "Development of an Extension of ILLUDAS for Continuous Simulation of Urban Runoff Quantity and Discrete Simulation of Runoff Quality" (Purdue University, July 1979).

[29] U.S. Army Corps of Engineers, "Urban Storm Water Runoff, STORM," Computer Program 723-S8-L2520, Hydrologic Engineering Center (Davis, CA, October 1974).

[30] National Technical Information Service, "TR-55, Hydrology for Small Urban Watersheds," U.S. Department of Commerce (Springfield, VA, 1986).

[31] U.S. Soil Conservation Service, "Urban Hydrology for Small Watersheds," Technical Release 55, rev., U.S. Department of Agriculture (Washington, DC, 1986).

[32] ASCE Task Committee on Urban Stormwater Software, "Microcomputer Software in Urban Hydrology," by D. F. Kibler, M. E. Jennings, G. L. Lewis, B. A. Tschantz, and S. G. Walesh, *HYDATA*, American Water Resources Association, 10, no. 5 (September 1991).

[33] C. N. Papadakis and H. C. Preul, "University of Cincinnati Urban Runoff Model," *Proc. ASCE J. Hyd. Div.* 98 (HY 10), 1789–1804 (October 1972).

[34] G. L. Lewis, "A Shopper's Guide to Urban Stormwater Software Revisited," *Proceedings, ASCE National Conference on Hydraulic Engineering* (Colorado Springs, CO 1988).

[35] Hydrocomp International, Inc., *Hydrocomp Simulation Programming Operations Manual*, 4th ed. (Palo Alto, CA, January 1976).

[36] Metcalf and Eddy, Inc., University of Florida, Gainesville, Water Resources Engineers, Inc., "Storm Water Management Model," Environmental Protection Agency, vol. 1, 1971.

[37] W. C. Huber et al., "Storm Water Management Model User's Manual, Version III," EPA-600/2-84-109a (NTIS PB84-198423), November 1981.

[38] L. A. Roesner et al., "Storm Water Management Model User's Manual, Version III: Addendum I, Extran," EPA-600/2-84-109b (NTIS PB84-198431), November 1981.

[39] W. C. Huber, J. P. Heaney, S. J. Nix, R. E. Dickinson, and D. J. Polmann, *Storm Water Management Model User's Manual, Version III*, EPA-600/2-84-109a (NTIS PB84-198423) (Cincinnati, OH: Environmental Protection Agency, November 1981).

[40] D. R. Maidment, ed., *Handbook of Hydrology* (New York: McGraw-Hill, 1993).

[41] V. K. Hagen et al., *Urban Watershed Use of Hydrologic Procedures*, National Research Council, Transportation Research Record No. 1471 (Washington, DC, 1994).

[42] R. S. Gupta, *Hydrology and Hydraulic Systems* (Prospect Heights, IL: Waveland Press, Inc., 1995).

[43] V. T. Chow et al., *Applied Hydrology*, (New York: McGraw-Hill, 1988).

[44] K. C. Patra, *Hydrology and Water Resources Engineering* (Boca Raton, FL: CRC Press, 2000).

[45] C. T. Haan et al., *Design Hydrology and Sedimentology for Small Catchments* (San Diego: Academic Press, 1994).

[46] U.S. Environmental Protection Agency, *Maximum Utilization of Water Resources in a Planned Community, Application of the Storm Water Management Model*, vol. I (Cincinnati, OH: Municipal Environmental Research Laboratory, 1979).

[47] *The Design of Storm Water Inlets*, (Baltimore: The Johns Hopkins University, Department of Sanitary Engineering and Water Resources, June 1956).

[48] *Baltimore County Department of Public Works Design Manual* (Towson, MD: Baltimore County Department of Public Works, 1998).

[49] P. B. Bedient and W. C. Huber, *Hydrology and Flood Plain Analysis* (Upper Saddle River, NJ: Prentice Hall, 2003).

[50] R. L. Daugherty, J. G. Franzini, and E. J. Finnemore, *Fluid Mechanics with Engineering Applications* (New York: McGraw-Hill, 1985).

[51] R. W. Hornbeck, *Numerical Methods* (Upper Saddle River, NJ: Prentice Hall, 1982).

CHAPTER 8

Water Quality

OBJECTIVES

The purpose of this chapter is to:

- Introduce the standards of quality that must be met by water for different uses
- Discuss the types, sources, and effects of microbial and chemical pollutants
- Acquaint the reader with laboratory tests commonly used to assess water quality

Drinking water is required to meet stringent microbiological and chemical standards of quality to prevent waterborne diseases and health risks from toxic chemicals. Hence, these subjects are discussed in considerable detail. Quality criteria are equally important to protect aquatic life in surface waters from the discharge of conventional and toxic pollutants in treated wastewater effluents. This chapter ends with brief descriptions of common laboratory tests to provide the reader with a better understanding of the meaning of these water-quality parameters.

MICROBIOLOGICAL QUALITY

A variety of pathogens are present in domestic wastewater, with the kinds and concentrations relating to the health of the contributing community. Although a reduction of pathogens occurs in conventional wastewater treatment, even after chlorination the effluent still contains persistent pathogens. (For unrestricted water reuse, filtration after chemical coagulation and chlorination with an extended contact time are necessary to remove all pathogens.) Thus, surface waters downstream from wastewater discharges may contain pathogens, and their removal is required in water supply treatment. Since testing water samples for pathogens is not practical, nonpathogenic fecal coliform bacteria are used as indicators for the potential presence of pathogens. The following sections discuss waterborne diseases and coliform bacteria as indicator organisms.

8.1 WATERBORNE DISEASES

Many human diseases are transmitted by the feces of an infected person getting into the mouth of another person. This travel from anus to mouth, referred to as the fecal–oral route, may be direct from person to person by contaminated fingers or indirect through food or water. In addition, some pathogens may reinfect by inhalation of dust or aerosol droplets, and a few (notably hookworm) can penetrate through the skin. Some of these communicable diseases are endemic in form; that is, they are native to a particular region or country. The four major groups of pathogens are viruses, bacteria, protozoans, and helminths (worms).

Pathogens Excreted in Human Feces

Viruses are obligate, intracellular parasites that replicate only in living hosts' cells. Being composed of only complex organic compounds, they lack the metabolic systems for self-reproduction. The size range of enteric viruses is 20–100 nm, about $^1/_{50}$ the size of bacterial cells, and requires electron microscopy for viewing. Human feces contains over 130 serotypes of enteric viruses. Of those, the groups listed in Table 8.1 are most likely to be transmitted by water. Persons infected by ingesting these viruses do not always become ill, but disease is a possibility in persons infected with any of the enteric viruses, particularly the hepatitis A virus. Several diseases involving the central nervous system, and more rarely the skin and heart, are caused by enteroviruses. Waterborne outbreaks of infectious hepatitis have occurred, but the most common route of transmission is by person-to-person contact. Infectious hepatitis may cause diarrhea and jaundice, and result in liver damage. Human beings are the reservoir for all of the enteric viruses.

Bacteria are microscopic single-celled plants that use soluble food and are capable of self-reproduction without sunlight. Their approximate range in size is 0.5–5 μm (500–5000 nm). The feces of healthy persons contains 1 to 1000 million of each of the following groups of bacteria per gram: enterobacteria, enterococci, lactobacilli, clostridia, bacteroides, bifidobacteria, and eubacteria. *Escherichia coli*, the common fecal coliform, is in the enterobacteria group. For many bacterial infections of the intestines, the major symptom is diarrhea. The most serious waterborne diseases are typhoid fever, paratyphoid fever, dysentery, and cholera. Typhoid and paratyphoid result in high fever and infection of the spleen, gastrointestinal tract, and blood. Dysentery causes diarrhea, bloody stools, and sometimes fever. Cholera symptoms are diarrhea, vomiting, and dehydration. While all of these diseases are debilitating and can cause death if untreated, their transmission can be controlled by pasteurization of milk, sanitary disposal of wastewater, and disinfection of water supplies.

Protozoans infecting humans are intestinal parasites that replicate in the host and exist in two forms. Trophozoites live attached to the intestinal wall where they actively feed and reproduce. At some time during the life of a trophozoite, it releases and floats through the intestines while making a morphological transformation into a cyst, or oocyst, for protection against the harsh environment outside of the host. This cyst form is infectious for other persons by the fecal–oral route of transmission. The cysts are 5–15 μm in size, significantly larger than intestinal bacteria. The most common protozoal diseases are diarrhea and dysentery (Table 8.1). *Entamoeba histolytica* causes

TABLE 8.1 Typical Pathogens Excreted in Human Feces

Pathogen Group and Name	Associated Diseases	Category for Transmissibility[a]
Virus:		
Adenoviruses	Respiratory, eye infections	I
Caliciviruses	Diarrhea	I
Enteroviruses		
Polioviruses	Aseptic meningitis, poliomyelitis	I
Echoviruses	Aseptic meningitis, diarrhea, respiratory infections	I
Coxsackie viruses	Aseptic meningitis, herpangina, myocarditis	I
Hepatitis A virus	Infectious hepatitis	I
Other viruses	Gastroenteritis, diarrhea	I
Bacterium:		
Salmonella typhi	Typhoid fever	II
Salmonella paratyphi	Paratyphoid fever	II
Other salmonellae	Gastroenteritis	II
Shigella spp.	Bacillary dysentery	II
Vibrio cholerae	Cholera	II
Other vibrios	Diarrhea	II
Yersinia enterocolitica	Gastroenteritis	II
Protozoan:		
Giardia lamblia	Diarrhea	I
Crytosporidium spp.	Diarrhea	I
Entamoeba histolytica	Amoebic dysentery	I
Helminth:		
Ancylostoma duodenale (Hookworm)	Hookworm	III
Ascaris lumbricoides (Roundworm)	Ascariasis	III
Hymenolepis nana (Dwarf tapeworm)	Hymenolepiasis	I
Necator americanus (Hookworm)	Hookworm	III
Strongyloides stercoralis (Threadworm)	Strongyloidiasis	III
Trichuris trichiura (Whipworm)	Trichuriasis	III

[a]I = Nonlatent, low infective dose
II = Nonlatent, medium to high infective dose, moderately persistent
III = Latent and persistent
Source: Adapted from R.G. Feachem, D.J. Bradley, H. Garelick, and D. D. Mara, *Sanitation and Disease, Health Aspects of Excreta and Wastewater Management; World Bank Studies in Water Supply and Sanitation 3* (Chichester: Wiley, 1983).

amoebic dysentery that is severely debilitating to the human host. (Amoebic dysentery, while common in tropical climates, is considered nontransmittable in temperate climates.) *Giardia lamblia* causes the less severe gastrointestinal infection of giardiasis, resulting in diarrhea, nausea, vomiting, and fatigue. *Cryptosporidium* species cause diarrhea with variable severity. While many infected persons have mild, nonspecific symptoms, others have prolonged diarrhea with accompanying weight loss. In persons with immunodeficiency syndrome, serious diarrhea can cause life-threatening illness. Human beings are the reservoir for both of these infectious protozoans.

Helminths are intestinal worms that (except for *Strongyloides*) do not multiply in the human host. Therefore, the worm burden in an infected person is directly related to the number of infective eggs ingested. The worm burden is also related to the severity of the infected person's disease symptoms. Eggs are excreted in the host's feces. Of the helminths listed in Table 8.1, most can be transmitted by ingestion of contaminated water or food after a latent period of several days. Hookworms live in the soil and, after molting, can infect humans by penetrating their skin. With a heavy worm infection, the symptoms can be anemia, digestive disorder, abdominal pain, and debility. Helminth eggs are commonly 40–60 μm in length and denser than water.

Human carriers exist for all enteric diseases. Thus, in communities where a disease is endemic, a proportion of the healthy persons excrete pathogens in feces. In some infections, the carrier condition may cease along with symptoms of the illness, but in others it may persist for months, years, or a lifetime. The carrier condition exists for most bacterial and viral infections, including the dreaded diseases of cholera and infectious hepatitis. The asymptomatic carriers of *Entamoeba histolytica* and *Giardia lamblia* are primarily responsible for continued transmission of these intestinal protozoans. In light helminthic infections, the human host may have only minor symptoms of illness while passing eggs in feces for more than a year.

Factors Affecting Transmission of Diseases

The transmission of waterborne diseases is influenced by latency, persistence, and infective dose of the pathogens. Latency is the period of time between excretion of a pathogen and its becoming infective to a new host. No excreted viruses, bacteria, and protozoans have a latent period. Among the helminths, only a few have eggs or larvae passed in feces that are immediately infectious to humans. The majority of helminths require a distinct latent period either for eggs to develop to the infectious stage or to pass through an intermediate to complete their life cycles. For example, *Ascaris lumbricoides* has a latency of 10 days and hookworms about 7 days. Persistence is measured by the length of time that a pathogen remains viable in the environment outside a human host. The transmission of persistent microorganisms can follow a long route—for example, through a wastewater treatment system—and still infect persons located remotely from the original host. In general, persistence increases from bacteria, the least persistent, to protozoans, to viruses, to helminths having persistence measured in months. Infective dose is the number of organisms that must be ingested to result in disease. Usually, the minimum infective doses for viruses and protozoans are low and less than for bacteria, while a single helminth egg or larva can infect. Median infective dose is that dose required to infect half of those persons exposed [1].

The transmission characteristics of pathogens can be categorized based on latency, persistence, and infective dose, as shown in the right column of Table 8.1. Category I comprises infections that have a low median infective dose (less than 100) and are infective immediately upon excretion. These infections are transmitted person to person where personal and domestic hygiene are poor. Therefore, control of these diseases requires improvements in personal cleanliness and environmental sanitation, including food preparation, water supply, and wastewater disposal. Category II comprises all bacterial diseases having a medium to high median infective dose (greater than 10,000) and are less likely to be transmitted by person-to-person contact than category I infections. In addition to the control measures given for category I, wastewater collection, treatment, and reuse are of greater importance, particularly if personal hygiene and living standards are high enough to reduce person-to-person transmission. Category III contains soil-transmitted helminths that are both latent and persistent. Their transmission is less related to personal cleanliness since these helminth eggs are not immediately infective to human beings. Most relevant is the cleanliness of vegetables grown in fields exposed to human excreta by reuse of wastewater for irrigation and sludge for fertilization. Effective wastewater treatment is necessary to remove helminth eggs, and sludge stabilization is necessary to inactivate the removed eggs.

Waterborne Diseases in the United States

Waterborne disease outbreaks are not a major cause of illness in the United States. The number of outbreaks reported during the 35-year period of 1946–1980 was 672, which affected more than 150,000 persons [2]. Based on a served population of 150 million, this is only an average illness rate of 4400 persons throughout the country or about 1 per 34,000 persons per year. During the 15-year period of 1981–1996, 156 outbreaks occurred in the 54,000 community water systems for an average of 1.9 per 10,000 systems per year, and 148 outbreaks occurred in 150,000 noncommunity systems for an average of 0.7 per 10,000 systems per year [3]. The cases of illness from the outbreaks of all water systems during the period 1981–1996 was approximately 500,000 persons, which is extraordinarily high because of outbreaks of waterborne crytosporidiosis and giardiasis in large systems in 1993 and 1994.

The causative agent was not determined in half of the waterborne disease outbreaks reported. The most common identified bacterial diseases were gastroenteritis (salmonellosis) and dysentery (shigellosis). Gastroenteritis is an inflammation of the lining membrane of the stomach and intestines, and dysentery is diarrhea with bloody stools and sometimes fever. The most serious viral disease identified as waterborne is infectious hepatitis, resulting in loss of appetite, fatigue, nausea, and pain. The most characteristic feature of the disease is a yellow color that appears in the white of the eyes and skin, hence the common name *jaundice*. None of the outbreaks of waterborne infectious hepatitis have occurred in municipal water systems. For 1991 and 1992, 34 disease outbreaks associated with drinking water were reported, affecting an estimated 17,000 persons [4]. Five of the 34 were in community systems and the remaining 29 in campgrounds, resorts, recreation areas, restaurants, and private systems. Of the 11 outbreaks for which an etiologic agent was determined, a protozoal parasite (*Giardia lamblia* or *Cryptosporidium*) was identified in seven. The remaining 4 were hepatitis A, *Shigella sonnei*, or chemicals. For 1993 and 1994, 30 outbreaks associated with drinking water were reported, affecting an estimated 405,000 persons [5]. This unusually large number of affected persons was the result of 403,000 cases of gastrointestinal illness in a large city resulting from *Cryptosporidium parvum* in the lake-water source. Oocysts were identified in both the raw and treated waters. Failure of adequate chemical coagulation and filtration was primarily attributed to inadequate operation, even though the treated water met all existing state and federal water quality standards in effect at that time. Of the 25 outbreaks for which an etiologic agent was determined, 10 were caused by *Giardia lamblia* or *Cryptosporidium parvum*, 8 were caused by chemical poisoning, 3 by *Campylobacter jejuni*, 2 by *Shigella* spp., and 1 each by *Vibrio cholerae* and *Salmonella*. Eight of the outbreaks occurred in community systems and the remainder in private homes, resorts, and other noncommunity establishments.

Giardiasis is a common waterborne protozoal disease characterized by diarrhea that usually lasts one week or more and may be accompanied by abdominal cramps, bloating, flatulence, fatigue, and weight loss. A unique feature of giardiasis is transmission to humans through beavers serving as amplifying hosts. In mountainous regions, beavers have been infected by upstream contamination with human excreta containing *G. lamblia*. After being infected, they return millions of cysts to the water for every one ingested, amplifying the number of *Giardia* cysts in clear mountain streams. Outbreaks of disease have occurred in downstream resorts and towns where the water supplies withdrawn from these streams were not properly treated to remove the cysts.

Cryptosporidiosis is mild to profuse and watery diarrhea for which no effective remedial treatment is known. Depending on the immune competency and health status of a sick person, *Cryptosporidium* can produce a continuum of deviations from normal diarrhea disease and require hospitalization. *Cryptosporidium* oocysts have been found in surface waters contaminated by runoff containing cattle and sheep feces; hence, waterborne disease in humans has been attributed to infected livestock. As with *Giardia* cysts, oocysts are persistent in nature and much more resistant to chlorination than are bacteria and viruses. Removal of oocysts in water treatment requires effective chemical coagulation and granular-media filtration. As an aside, these two parasitic diseases are also readily transmitted by contaminated food, resulting in traveler's diarrhea, and many reported cases are from person-to-person transmission, e.g., among diapered children and workers at day-care centers.

Although all of these waterborne gastrointestinal diseases can stress sensitive individuals, they are not like the dreaded bacterial diseases of cholera and typhoid that resulted in hundreds of deaths per year in the United States during the early years of the twentieth century. "During the 35-year period of 1946–1980 the rate (of deaths resulting from waterborne disease outbreaks) has been reduced to one death per year." [2] By including the deaths of 50 to 100 persons in 1993 and 1994 from waterborne outbreaks of cryptosporidiosis and giardiasis, the rate of deaths during the 55-year period of 1946–2000 is only two deaths per year for the average population over that period of approximately 200 million. The United States provides safe and dependable public water supplies for its residents.

8.2 COLIFORM BACTERIA AS INDICATOR ORGANISMS

Public water supplies, reclaimed waters, and wastewater effluents are not tested for pathogens to determine microbiological quality. Laboratory analyses for pathogens are difficult to perform, quantitatively unreliable, and, for some pathogenic microorganisms, impossible to perform. Therefore, the microbial quality is based on testing for an indicator organism, i.e., a microorganism whose presence is evidence that the water has been polluted with feces of humans or warm-blooded animals.

Nonpathogenic fecal coliform bacteria, as typified by *Escherichia coli (E. coli)*, that reside in the human intestinal tract are excreted in large numbers in feces, averaging about 50 million coliforms per gram. Untreated domestic wastewater generally contains more than 3 million coliforms per 100 ml. Pathogenic microorganisms causing enteric diseases in humans originate from fecal discharges of diseased persons. Consequently, water contaminated by fecal pollution is identified as being potentially dangerous by the presence of coliform bacteria. Hence, coliform bacteria are indicator organisms of fecal contamination and the possible presence of pathogens. Nevertheless, some genera of the coliform group of bacteria found in water and soil are not of fecal origin but grow and reproduce on organic matter outside the intestines of humans and animals. These coliforms indicate neither fecal contamination nor the possible presence of pathogens. In laboratory testing, the term *total coliforms* refers to coliform bacteria from feces, soil, or other origin, and *fecal coliforms* refers to coliform bacteria from human or warm-blooded animal feces. If a sample of drinking water tests positive for total coliforms in lauryl tryptose broth at 35°C, the growth is aseptically transferred to a culture tube containing EC medium and incubated at 44.5°C to confirm or deny the presence of fecal coliforms.

Tests for total coliforms, fecal coliforms, and *Escherichia coli* are within the capability of the majority of water microbiology laboratories. The procedures are well established and laboratory equipment and supplies are readily available from vendors. Occasionally, false positives result from inadvertent contamination of a water sample during collection. Also interference from *Klebsiella* organisms, which respond as coliform bacteria in testing, gives a false indication of contamination caused by presence of biological organic carbon in distribution piping. Approximately, 15% of *Klebsiella* are thermotolerant and can result in a false positive test for the presence of fecal coliforms [6]. Testing drinking water for *Escherichia coli* by the presence–absence test is recommended as the preferred and most specific indicator of microbiological drinking water quality. Sterile reagents and apparatus for the Colilert® technique are available, making the test for *Escherichia coli* easy to perform. (Section 8.12 describes several techniques to test for the coliform group of bacteria in drinking water, natural water, and wastewater.)

The reliability of coliform bacteria to indicate the presence of pathogens in water depends on the persistence of the pathogens relative to coliforms. For pathogenic bacteria, the die-off rate is greater than coliforms outside the intestinal tract of humans. Thus, exposure in the water environment reduces the number of pathogenic bacteria relative to coliform bacteria. Viruses, protozoal cysts, and helminth eggs are more persistent than coliform bacteria. For example, the threshold chlorine residual effective as a bactericide may not inactivate enteric viruses, is ineffective in killing protozoal cysts, and cannot harm helminth eggs. In contrast, filtration through natural sand aquifers for a sufficient distance, or granular media in a treatment plant after chemical coagulation, can entrap cysts and eggs because of their relatively large size while allowing viruses to be carried through, suspended in the water. In surface water treatment, coliforms are a reliable indicator of the safety of the processed water for human consumption, provided the treatment includes chemical coagulation and filtration to remove cysts, eggs, and suspended matter for effective chlorination of the clear water to inactivate viruses and kill bacteria. In a similar manner, coliforms can be used as an indication of water quality for reuse of reclaimed wastewater, provided the treatment processes physically remove the persistent protozoal cysts and helminth eggs.

Escherichia coli O157 : H7, which is pathogenic to humans, is an antibiotic-resistant mutant strain found in the feces of infected cattle. The primary health concerns have been with contaminated ground beef, raw milk, unpasteurized fruit juices, and person-to-person transmission. The only waterborne outbreak of *E. coli* O157 : H7 (with 243 case patients and 4 deaths) occurred in 1987 in a community with a population of 2090 [7]. Contamination of the water was attributed to wastewater entering distribution piping during repair of water-main breaks.

Drinking Water Standards

The Safe Drinking Water Act (SDWA), initially enacted in 1974, authorizes the U.S. Environmental Protection Agency (EPA) to establish comprehensive national regulations to ensure the drinking water quality of public water systems [8]. The following are the three categories of public water systems. A community water system serves at least 25 people at their primary residences (or at least 15 residences that are primary residences). Examples are a municipality, mobile home park, and homeowner subdivision. A nontransient–noncommunity water system regularly serves at least 25 of the same people for at least 6 months per year but not at their primary residences. Examples are schools, commercial facilities, or manufacturing plants that have their own water systems. A transient–noncommunity water system serves 25 or more people for at least 60 days per year but not the same people or not on a regular basis. Examples are highway rest areas, recreation areas, gas stations, and motels that have their own water available for employees and the public.

The Total Coliform Rule in the SDWA specifies a maximum contaminant level goal (MCLG) of zero for total coliforms, fecal coliforms, and *Escherichia coli*. The Surface Water Treatment Rule specifies a MCLG of zero for *Giardia lambia, Cryptosporidium* species, enteric viruses, and *Legionella*. (MCLG is a nonenforceable, health-based goal set at a level with an adequate margin of safety to ensure no adverse effect on human health.)

The maximum contaminant level (MCL) for coliforms allows for a limited number of positive samples because of inadvertent contamination, but not for fecal coliforms and *Escherichia coli*. (MCL is an enforceable standard set at a numerical value with an adequate margin of safety to ensure no adverse effect on human health.) Coliform bacteria are common in the natural environment—for example, on dirty water faucets, on the hands of the person collecting the water sample, and in dust and soil. Even with careful sampling procedures, an occasional water sample is likely to test positive for coliform bacteria of nonfecal origin. When a positive test occurs, multiple repeat samples are required to identify whether the contamination is actual or inadvertent, and the positive sample is tested further to determine if the coliforms are fecal coliforms.

The number of water samples tested for monitoring a public water supply is based on the population served. For a population under 1000, one sample per month is required; from 1000 to 100,000, one sample per 1000 population is required per month; above 100,000, the number of samples required is less than one per 1000 population.

For a population under 33,000, only one sample may test positive per month for total coliforms for no violation and, above 33,000, no more than 5.0% may test positive. Violation of this MCL requires public notification and an evaluation to determine the source of contamination and risk of contamination with pathogens.

The Safe Drinking Water Act rules also specify the disinfection treatment of public water supplies in addition to the coliform MCL of the water in the distribution system. The regulations on treatment technique are to ensure removal of pathogens more persistent than coliform bacteria during treatment. Strictly speaking, the coliform standard for drinking water applies only after proper treatment and disinfection of water supplies. (Sections 11.22–11.25 present the treatment techniques specified by the Environmental Protection Agency for disinfection of public water supplies.)

Wastewater Effluent Standards

Conventional treatment removes 99–99.9% of pathogenic microorganisms in the raw wastewater; however, the effluent still contains significant concentrations of excreted viruses, bacteria, protozoal cysts, and helminth eggs. The kinds of pathogens in the wastewater depend on the health of the contributing human population. If discharged to recreational waters, effluents are generally chlorinated at dosages in the range of 8–15 mg/l with a minimum contact time of 30 min at peak hourly flow. Satisfactory effluent disinfection is normally defined by an average fecal coliform count of 200 per 100 ml or less. This is a reduction from about 1,000,000 per 100 ml in the biologically treated effluent. The safety of disposing of chlorinated effluent by dilution in surface water is based on the argument that this reduction eliminates the great majority of bacterial pathogens and inactivates large numbers of viruses. In fact, viruses and bacteria may be harbored and protected in suspended organic matter (allowable suspended solids in a secondary effluent is 30 mg/l), and protozoal cysts and helminth eggs are resistant to chlorination. (For further discussion, refer to Section 11.26.)

Adequate disinfection for unrestricted reuse of wastewater effluents requires removal or inactivation of all pathogens. The processes following biological treatment are tertiary filtration to physically remove helminth eggs, protozoal cysts, and suspended solids, and long-term chlorination to inactivate viruses and bacteria. The tertiary scheme may be conventional coagulation–sedimentation–filtration or direct filtration without sedimentation or filter screens and membrane filtration. The subsequent disinfection requires rapid mixing of the chlorine followed by a long contact time (usually about 2 hr) in a tank simulating plug flow. The common effluent standard for unrestricted reuse for irrigation is 2.2 fecal coliforms per 100 ml and a turbidity of less than 2 NTU (nephelometric turbidity unit). (For further discussion, refer to Section 11.27.)

CHEMICAL QUALITY OF DRINKING WATER

The primary drinking water standards, which are approval limits for health, are specified for inorganic chemicals, organic chemicals, radionuclides, and turbidity. Secondary standards that recommend limits for aesthetics include inorganic chemicals, dissolved salts, corrosivity, color, and odor. Risk assessment is the process applied in developing standards for toxic chemicals.

8.3 MONITORING DRINKING WATER FOR PATHOGENS

Laboratory tests for pathogenic bacteria, viruses, and protozoans are difficult to perform and generally not quantitatively reproducible. Most utilities have neither qualified personnel nor laboratories equipped to monitor for pathogens. Some large utilities with surface water sources are testing for *Cryptosporidium* oocysts and *Giardia* cysts at a frequency of a few samples per month. Since pathogens are most likely to enter a distribution system because of a treatment breakthrough on an intermittent rather than continuous basis, conducting limited monitoring provides little value as a means of protecting public health. In the case of groundwater, the most likely pathogens are selected species of enteric viruses for which tests are impractical and, for some viruses, impossible to perform. The available methods for detection and identification of human pathogens do not produce credible data for making public health decisions.

"Past experience and data have shown that pathogen monitoring does not and cannot confirm the absolute presence or absence of infectious microorganisms in drinking water. With public health protection at stake, one obvious solution lies in treatment process optimization coupled with source water protection and infrastructure integrity practices" [9].

Testing for Enteric Viruses

Viruses of particular importance in drinking water and reclaimed water are those that infect the gastrointestinal tract of humans and are excreted with feces of infected persons. They range in size from 20 to 100 nanometers, which is about one-fiftieth the size of bacteria and less than one-thousandth the size of protozoan cysts. Testing for these viruses requires extraction, concentration, and identification [3]. The process of extraction is by pumping a large volume of water, several hundred liters, through a filter to increase the probability of virus detection since virus levels are likely to be low. The different techniques used to concentrate the eluate from the filter are microporous filters, chemical adsorption–precipitation, and dialysis. During extraction and concentration, only a portion of the viruses present in the original sample is likely captured. Therefore, to determine the precision of separation, the test procedure must be conducted on samples to which known suspensions of one or more test virus types have been added to a water sample to establish recovery efficiency.

Assay and identification of viruses in sample concentrates rely on viruses as obligate intracellular parasites to multiply in and destroy their host cells. The two major host cell systems are mammalian cell cultures of primate origin and whole animals such as sucking mice. No single universal host system exists for all enteric viruses. A measured portion of virus concentrate is spread on the surface of a cell tissue. Following incubation, the monolayer of host cells is microscopically examined to count the number of plaques. (A *plaque* is a clear area produced by viral destruction of the cells.) Virus concentration in the water sample is expressed as the number of plaque-forming units (PFUs) per liter. Further examination is required to identify the virus type creating a plaque. Precise identification involves recovering viruses from an individual plaque and inoculating them into different cell cultures and assay in mice. Virus assay and identification require a trained virologist working in a specially equipped virology laboratory facility.

Testing for *Giardia* Cysts

The procedure consists of filtration of the water, extraction from the filter material, extract concentration, and microscopic examination by immunofluorescence detection [3]. A large volume of water, several hundred liters, is pumped through a filter with a 1 μm nominal porosity. The *Giardia* cyst is oval 8 to 18 μm long and 5 to 15 μm wide containing 2 to 4 nuclei and distinctive axoneme. The retained particles and cysts are eluted, the filter extract concentrated by centrifugation, and the cysts separated by flotation. A portion is placed on a membrane filter and stained with an indirect fluorescent antibody and examined using epifluorescent microscopy. Because slide examination requires subjective judgment, extraneous organisms may be misidentified as *Giardia* cysts. As a result, greater confidence should be given to counts of cysts in which appropriate internal structures have been identified rather than counts including empty organisms.

Testing for *Cryptosporidium* Oocysts

The procedure is similar to that of *Giardia* cysts including filtration of a large volume of water for extraction, removal of particles and oocysts from the filter, concentration by centrifugation, separation of oocysts from debris, and staining with a fluorescent antibody for microscopic identification. Nevertheless, *Cryptosporidium* oocysts are more difficult to separate because they are less than one-half the size of *Giardia* cysts and more difficult to identify from similarly shaped organisms. The *Cryptosporidium* oocyst is spherical, 4–6 μm in size, containing 4 sporozoites.

Case histories reveal that boil-water advisories had been unnecessarily issued in two major cities, resulting in false public health concerns and anxiety among the residents [9]. In one case, the suspected contamination was based on "observed" oocysts in treated water in the distribution pipe network, although no cases of human disease were apparent. In another, the misidentification of oocysts resulted from algae in both raw and treated water that mimicked oocysts in microscopy. Laboratories conducting tests for *Cryptosporidium* oocysts must be audited and approved for quality assurance.

8.4 ASSESSMENT OF CHEMICAL QUALITY

Treated drinking water can contain trace amounts of toxic chemicals often so low in concentration that predicting an observable effect on human health is difficult. Reliable information on the toxicity to humans of most chemicals is difficult to obtain, since it must usually be based on uncontrolled accidental or occupational exposures. Therefore, data from long-term animal studies are applied to evaluate chronic exposure and carcinogenic risk to humans.

Risk Assessment

Risk assessment is the scientific evaluation of toxic chemicals, human exposure, and adverse health effects. Identification of a hazard results from exploratory studies on laboratory animals or case reports of actual human exposure. Based on the inference of toxicity, dose–response relationships are determined on laboratory animals between specific quantities of the substance and associated physical responses, such as growth of tumors, birth defects, or neurologic deficits. Then follows an evaluation of exposure and assessment. The purpose is to describe the magnitude and duration of exposure to human populations both in the past and anticipated in the future. Finally, a quantitative estimate of risk to humans is predicted from the expected exposure levels by applying a dose–response model. If human exposure data are not available, which is often the case, the quantitative estimate is a hypothetical risk based on dose–response data from studies of laboratory animals.

Risk assessment has been used extensively for estimating the risk of developing cancer [10] and is now being applied to assess risks to development, reproduction, and neurologic functioning [11]. Development effects include embryo and fetal death, growth retardation, and malfunctions. Developmental toxicity studies are very complex, and applying results to risk assessment remains problematic. Also highly complex is reproductive toxicity affecting any event from germ cell formation and sexual functioning in the parents through sexual maturation of the offspring. Neurotoxicity in humans ranges from cognitive, sensory, and motor impairments to immune system deficits. Numerous chemicals, including many pesticides, are known neurotoxins.

Chronic Noncarcinogenic Toxicity

For noncarcinogens and nonmutagens, human exposure should be less than the threshold level causing chronic disease. The acceptable daily intake (ADI) of a noncarcinogenic chemical is defined as the dose anticipated to be without lifetime risk to humans when taken daily, expressed in milligrams of chemical per kilogram of body weight per day. It is an empirically derived value arrived at by combining exposure knowledge and uncertainty concerning the relative risk of a chemical.

The ADI is based on toxicity data from chronic (long-term) feeding studies of laboratory animals to identify the highest no-observed-adverse-effect level (NOAEL) and the lowest-observed-adverse-effect level (LOAEL). The dosage of a toxin given to laboratory animals is assumed to be physiologically equivalent to humans on the basis of body weight. (Typical weights are 0.3 kg for a rat, 10 kg for a dog, and 70 kg for an adult human.) In other words, the intake per unit body weight of a rat or dog is equated to the intake per unit body weight of a human. After determining the NOAEL or LOAEL in animals, an uncertainty factor (safety factor) is applied to reduce the allowable human intake to account for uncertainties involved in extrapolating from animals to humans. The EPA determines the no-effect level, known as the reference dose (RfD), for chronic or lifetime exposure without significant risk to humans, including adults or sensitive subgroups such as infants, children, pregnant women, the elderly, and immunodeficient persons. Human and/or animal toxicology data are used to calculate the RfD, which is expressed in milligrams per kilogram of body weight per day, as follows:

$$\text{RfD} = \frac{\text{NOAEL or LOAEL}}{\text{uncertainty factor}} \tag{8.1}$$

The uncertainty factors used in calculating acceptable daily intakes for establishing drinking water standards are a factor of 10 when chronic human exposure data are available and are supported by chronic oral toxicity data in animal species; a factor of

100 when good chronic oral toxicity data are available in some animal species but not in humans; and a factor of 1000 with limited chronic animal toxicity data [12].

Using the RfD, the drinking water equivalent level (DWEL) is calculated as follows:

$$\text{DWEL (mg/l)} = \frac{\text{RfD} \times \text{body weight (kg)}}{\text{drinking water consumption (l/d)}} \tag{8.2}$$

The DWEL represents a lifetime exposure with no adverse health effects assuming only exposure from drinking water. For regulatory purposes, a body weight of 70 kg and drinking water consumption rate of 2 l/d are assumed for adults or, if based on effects on infants, a body weight of 10 kg and consumption of 1 l/d are applied to Eq. (8.2). The DWEL is expressed as the maximum contaminant level (MCL), an enforceable standard in drinking water regulations. For some contaminants, the MCL is reduced to account for human intake from other sources, i.e., food, beverages, and air pollution.

For example, consider the chemical aldicarb, which is a systemic insecticide with high mammalian toxicity. This chemical is used on cotton with only a few registered uses on food crops, including potatoes, peanuts, and sugar beets. By oral dose, it inhibits blood cholinesterases that are enzymes responsible for hydrolyzing esters of choline. The NOAEL on both rats and dogs was established as 0.1 mg/kg · d. From Eq. (8.1) with an uncertainty factor of 100, the ADI equals 0.035 mg/l. Then, assuming 20% of the intake apportioned to water, the allowable concentration becomes 0.007 mg/l or 7 μg/l [11].

The ADI is a judgment regarding acceptable levels of chronic exposure to a chemical and is neither an estimate of risk nor a guarantee of absolute safety [12]. The ADI concept is not recommended for evaluating the intake of organic chemicals that are suspected carcinogens.

Carcinogenic Toxicity

The development of cancer appears to have three distinct stages: initiation, promotion, and progression. Each stage is influenced by such factors as age, heredity, diet, metabolic activity, and exposure to carcinogenic chemicals. The most widely used tests for carcinogen evaluation are long-term animal bioassays.

The hazard of ingesting a chemical assessed as a confirmed or suspected carcinogen can be evaluated in terms of dose-related risk. The two major problems in assessing risk are (1) extrapolation from observed risks in relatively high exposure levels used in laboratory animal studies to the low levels of exposure in humans and (2) extrapolation of the estimated risk from laboratory animals to humans. The assessment based on high-dose animal bioassay to low-dose human exposure further suffers from a lack of basic knowledge concerning the disease process in animals and humans and the total lack of data regarding potential synergistic and antagonistic reactions of chemicals.

The estimate of risk from results of animal bioassays is made by first converting the laboratory animal dose to the physiologically equivalent human dose. One method of conversion is on the basis of relative skin areas of test animals and the human body. This is justified from the observation that effects of acute toxicity in humans on a dose per unit of body surface area are in the same range as those in experimental animals [10]. The linearized multistage mathematical model, based on the one-hit theory of cancer initiation, is generally used for low-dose risk estimation [10]. Although its application cannot be proved or disproved by current scientific data, it is considered the best available model that yields an estimate of risk representing a plausible upper limit. The actual risk is not likely to be higher than the risk predicted by this model.

For example, consider the insecticide lindane, which is related to tumorigenic effects observed in laboratory mice and neurologic impairment in humans. The risk estimate in humans at a concentration in drinking water of 1.0 μg/l and daily consumption of 1 l/d is between 3.3×10^{-6} and 8.1×10^{-6} [10]. (Most experts agree that current technologies for assessing cancer and neurotoxicity risk cannot generate a single precise estimate of human risk, and risks are best expressed in terms of ranges or confidence intervals.) The upper 95% confidence estimate of risk at the same chemical dose is from 5.6×10^{-6} to 13×10^{-6}. These risk estimates are expressed as the probability of cancer after a lifetime consumption of 1 liter of water per day containing 1.0 μg/l of chemical. The MCL for lindane is 0.0002 mg/l. With 2.0 l/d water consumption at an MCL of 0.2 μg/l, the numerical risks decrease to 0.4 of the values for 1.0 l/d and 1.0 μg/l They become 1.3×10^{-6} to 3.2×10^{-6} and 2.2×10^{-6} to 5.2×10^{-6}. The average value of 3.0×10^{-6} means that 70-yr lifetime exposure to lindane would be expected to produce one excess case of cancer for every 340,000 persons exposed.

For an example of carcinogenic risk assessment, refer to the discussion of chloroform in Section 11.20.

8.5 CHEMICAL CONTAMINANTS

The chemicals listed in Table 8.2 are currently regulated in drinking water by the EPA. Retaining the chemicals on this list and/or the associated maximum contaminant levels depends on the results of ongoing risk assessments. A chemical may be deleted when future studies lack convincing evidence of risk to human health, or the use of the chemical may be banned, resulting in significantly reduced presence in water. On the other hand, new chemicals may be added, resulting from additional risk assessments. In the future, drinking water standards are likely to be flexible, with continuous adjustments to keep up with the changing chemical environment. (A reader interested in a specific contaminant is advised to contact the EPA, or state regulatory agency, to confirm the most recent regulatory status and MCL.)

The MCL is an enforceable standard for protection of human health. Monitoring requirements are established for each contaminant on a prescribed schedule of routine sampling and check sampling to confirm the results if an MCL is exceeded. The frequency of sampling varies from every 3 years for chemicals in groundwater to every 3 months for selected chemicals in treated surface waters. Testing is performed in an approved laboratory by specified methods. The MCLG is a nonenforceable health goal. MCLs are set as close to MCLGs as feasible based on best control management, treatment technology, and other means while considering cost.

TABLE 8.2 Chemical Drinking Water Standards, Maximum Contaminant Levels in Milligrams per Liter

Inorganic Chemicals			
Antimony	0.006	Lead	TT[a]
Arsenic	0.01	Mercury (inorganic)	0.002
Barium	2.	Nickel	0.1
Beryllium	0.004	Nitrate (as N)	10.
Cadmium	0.005	Nitrite (as N)	1.
Chromium (total)	0.1	Nitrate + Nitrite (as N)	10.
Copper	TT[a]	Selenium	0.05
Cyanide	0.2	Thallium	0.002
Fluoride[b]	4.0		
Asbestos 7 million fibers/liter (longer than 10 μm)			

Volatile Organic Chemicals			
Benzene	0.005	Ethylbenzene	0.7
Carbon tetrachloride	0.005	Monochlorobenzene	0.1
Chlorobenzene	0.1	Styrene	0.1
Dichloromethane	0.005	Tetrachloroethylene	0.005
p-Dichlorobenzene	0.075	1,2,4-Trichlorobenzene	0.07
o-Dichlorobenzene	0.6	Toluene	1.
1,2-Dichloroethane	0.005	1,1,1-Trichloroethane	0.2
1,1-Dichloroethylene	0.007	1,1,2-Trichloroethane	0.005
cis-1,2-Dichloroethylene	0.07	Trichloroethylene	0.005
trans-1,2-Dichloroethylene	0.1	Vinyl chloride	0.002
Dichloromethane	0.005	Xylenes (total)	10.
1,2-Dichloropropane	0.005		

Synthetic Organic Chemicals			
Acrylamide	TT[a]	Ethylene dibromide	0.00005
Alachlor	0.002	Glyphosate	0.7
Aldicarb	0.003	Heptachlor	0.0004
Aldicarb sulfone	0.002	Heptachlor epoxide	0.0002
Aldicarb sulfoxide	0.004	Hexachlorobenzene	0.001
Atrazine	0.003	Hexachlorocyclopentadiene	0.05
Carbofuran	0.04	Lindane	0.0002
Chlordane	0.002	Methoxychlor	0.04
Dalapon	0.2	Oxamyl (Vydate)	0.2
Di(2-ethylhexyl)adipate	0.4	PAHs [benzo(a)pyrene]	0.0002
Dibromochloropropane	0.0002	Pentachlorophenol	0.001
Diethylhexyl phthalate	0.006	Picloram	0.5
Dinoseb	0.007	Polychlorinated biphenyls	0.0005
Diquat	0.2	Simazine	0.004
Endothall	0.1	Toxaphene	0.003
Endrin	0.002	2,4-D	0.07
Epichlorohydrin	TT[a]	2,4,5-TP (Silvex)	0.05
2,3,7,8-TCDD (Dioxin)	0.00000003 (3×10^{-8})		

Disinfection By-Products	
Total trihalomethanes	0.080
Five haloacetic acids	0.060

(continued)

TABLE 8.2 Continued

Radionuclides	
Radium-226 + Radium-228	5 pCi/l
Gross alpha particle activity	15 pCi/l
Beta particle and photon radioactivity	4 mrem/yr
Uranium	30 μg/l
Turbidity	
Turbidity[c]	0.3 NTU

[a]Treatment technique (TT) requires modification or improvement of water processing to reduce the contaminant concentration.
[b]Many states require public notification at least annually of fluoride in excess of 2.0 mg/l to warn consumers of potential dental fluorosis.
[c]Turbidity is a performance standard in filtration treatment of surface waters to ensure removal of *Giardia lamblia* cysts and *Cryptosporidium* spp. oocysts.

Generally, the MCLG for a carcinogenic chemical is zero. Treatment technique rather than an MCL is specified for selected chemicals. Acrylamide and epichlorohydrin are used during water treatment in flocculants to decrease turbidity. The treatment technique requirements limit the concentration of these chemicals in polymers and their application.

The regulation of specific MCLs depends on the kind of water system. All of the standards are applicable to community systems and nontransient-noncommunity systems that supply water to the same people for a long period of time, e.g., schools and factories. Transient-noncommunity systems that serve different people for a short time (e.g., campgrounds, parks, and highway rest stops) are required to meet only the MCLs of those contaminants with health effects caused by short-term exposure.

Inorganic Chemicals

The sources of trace metals are associated with the natural processes of chemical weathering and soil leaching and with human activities such as mining and manufacturing. Corrosion in distribution piping and customers' plumbing can also add trace metals to tap water.

Antimony, arsenic, barium, beryillium, cadmium, chromium, mercury, nickel, selenium, and thallium are toxic metals affecting the internal organs of the human body. *Antimony* is a trace metal used as a constituent of alloys and is rare in natural waters. Ingestion affects the blood, decreasing longevity. *Arsenic* is widely distributed in waters at low concentrations, with isolated instances of higher concentrations in well waters. It is also found in trace amounts in food. *Barium*, one of the alkaline earth metals, occurs naturally in low concentrations in most surface waters and in many treated waters. *Beryllium* is not likely to be found in natural waters in greater than trace amounts because beryllium oxides and hydroxides are relatively insoluble. Soluble beryllium sulfate is transported in the bloodstream to bone where it is found to induce bone cancer in animals. *Cadmium* can be introduced into surface waters in amounts significant to human health by improper disposal of industrial wastewaters. Nevertheless, the major sources are food, cigarette smoke, and air pollution; hence, the MCL is set so that less than 10% of the total intake is expected to be from water consumption. The health effects of cadmium can be acute, resulting from overexposure at a high concentration, or chronic, caused by accumulation in the liver and renal cortex. *Copper* is commonly found in drinking water. Trace amounts below 20 μg/l can derive from weathering rock, but the principal sources in house water supplies are from corrosion of copper service pipes and brass plumbing fixtures. As an essential element in human nutrition, copper intake is safe and adequate at 1.5 to 3 mg/day. As an indicator of corrosivity, copper has a no-action level of 1.3 mg/l in first-flush samples from household plumbing. (Refer to Section 11.31 for a discussion of corrosion of lead pipe and solder.)

Chromium is amphoteric and can exist in water in several valence states. The content in natural waters is extremely low because it is held in rocks in essentially insoluble trivalent forms. Acute systemic poisoning can result from high exposures to hexavalent chromium; trivalent is relatively innocuous. *Mercury* is a scarce element in nature and has been banned for most applications with environmental exposure, such as mercurial fungicides. The biological magnification of mercury in freshwater food fish is the most significant hazard to human health. The mercury becomes available in the food chain through the transformation of inorganic mercury to organic methylmercury by microorganisms present in the sediments of lakes and rivers. Toxicity via the oral route is related mainly to methylmercury compounds rather than to inorganic mercury salts or metallic mercury. Symptoms of methylmercury poisoning include mental disturbance, ataxia, and impairment of speech, hearing, vision, and movement. *Selenium* is a trace metal naturally occurring in soils derived from some sedimentary rocks. Surface streams and groundwater in seleniferous regions contain variable concentrations. Cattle grazing on seleniferous vegetation suffer from "blind staggers." Effects on human health have not been clearly established—a low-selenium diet is beneficial, whereas high doses can produce undesirable physical manifestations.

Lead exposure occurs through air, soil, dust, paint, food, and drinking water. Lead toxicity affects the red blood cells, nervous system, and kidneys, with young children, infants, and fetuses being most vulnerable. Depending on local conditions, the contribution of lead from drinking water can be a minor or major exposure for children. Lead is not a natural contaminant in surface waters or groundwaters and is rarely in source water. It is a corrosion by-product from high-lead solder joints in copper piping, old lead-pipe goosenecks connecting the service lines to the water main, and old brass fixtures. Lead pipe and brass fixtures with high-lead content are not installed today, and lead-free solder has replaced the old 50% lead–50% tin solder in water piping. Since dissolution of lead requires an extended contact time, lead is most likely to be present in tap water after being in the service connection piping and plumbing overnight. Therefore, the first-flush sample concentration is the highest expected, and if it is less than 0.015 mg/l, no corrective action is required.

Fluoride, the naturally occurring form of fluorine, is commonly found in trace amounts in most soil and rock. Groundwaters usually contain fluoride ion dissolved from geologic formations. Surface waters generally contain smaller amounts, under 0.3 mg/l, except when contaminated by industrial wastes. Excessive concentration of fluoride in drinking water causes the dental disease of fluorosis, also referred to as mottling. This disease in mildest form results in very slight, opaque whitish areas on some of

TABLE 8.3 Recommended Optimum Concentrations of Fluoride Based on the Annual Average of the Maximum Daily Air Temperatures

Temperature Range (°F)	Recommended Optimum (mg/l)
53.7 and below	1.2
53.8 to 58.3	1.1
58.4 to 63.8	1.0
63.9 to 70.6	0.9
70.7 to 79.2	0.8
79.3 to 90.5	0.7

the posterior teeth. With greater severity, the fluorosis is widespread and the color of the teeth is gray to black. Absence or low concentration of fluoride in drinking water results in a high incidence of dental caries in children's teeth. The optimum fluoride concentration in drinking water protects teeth from decay without causing noticeable fluorosis. Food is another source of fluoride in the diet; however, as a contributor of fluorides, many studies have shown that the average dietary intake with food is a constant amount throughout the United States. Hence, the dental effect of fluoride results primarily from the concentration in the public water supply. Since water consumption is influenced by climate, the recommended optimum concentrations listed in Table 8.3 are based on the annual average of the maximum daily air temperatures. Cities with water supplies deficient in natural fluoride have successfully provided supplemental fluoridation to optimum levels to reduce the rate of dental decay in children [13].

Nitrate is the common form of inorganic nitrogen found in water solution. In agricultural regions, heavy fertilizer application results in unused nitrate migrating down into the groundwater. As a result, groundwater withdrawn by private and public wells is likely to have measurable concentrations of nitrate, and in the same regions well waters in many rural communities can exceed the recommended limit of 10 mg/l of nitrate nitrogen. Surface waters can be contaminated by nitrogen from both discharge of municipal wastewater and drainage from agricultural lands. The health hazard of ingesting excessive nitrate in water is infant methemoglobinemia. In the intestine of an infant, nitrate can be reduced to nitrite that is absorbed into the blood, oxidizing the iron of hemoglobin. This interferes with oxygen transfer, resulting in cyanosis and giving the baby a blue color. During the first 3 months of age, infants are particularly susceptible. Incidents of infant methemoglobinemia are extremely rare since most mothers in regions of known high-nitrate drinking water use either bottled water or a liquid formula requiring no dilution. Methemoglobinemia is readily diagnosed and rapidly reversed by injecting methylene blue into the infant's blood. Healthy adults are able to consume large quantities of nitrate in drinking water without adverse effects. The principal sources of nitrate in the average adult diet are saliva and vegetables, amounting to about 130 mg/day [10]. Two liters per day at 10 mg/l equals only 20 mg/day. Justified by epidemiological evidence on the occurrence of methemoglobinemia in infants, the standard of 10 mg/l is the maximum contaminant level for water with no observed adverse health effects.

Organic Chemicals

In the evaluation of organic chemicals, the EPA groups them into three categories, depending on the evidence of carcinogenicity. Category I chemicals are probable human carcinogens, based on human or animal risk assessment, and are assigned MCLGs of zero. Category II chemicals are not regulated as human carcinogens, but the MCLGs are lower than MCLs based on inconclusive evidence of carcinogenicity. Category III chemicals causing chronic disease without evidence of carcinogenicity are given MCLs based on the acceptable daily intake.

The MCLs of chemicals are set as close to the MCLGs as feasible, including the feasibility of laboratory testing. Gas chromatography has high sensitivity and reliability in detecting concentrations down to a few micrograms per liter; therefore, for many carcinogenic and highly toxic chemicals the MCLs are set at the lowest level of quantitative measurement achievable in a good laboratory. For example, several of the volatile organic chemicals considered carcinogens have MCLGs of zero and MCLs of 0.005 mg/l, which is the practical quantitative level of measurement. This quantity of chemical in drinking water is extremely small. The amount of water theoretically consumed by a person in a lifetime of 70 years at 2 liters per day is 1,100 l (13,500 gal). For a concentration of 0.005 mg/l, the amount of organic chemical added to this quantity of water would be 3 to 7 drops, depending on the specific gravity of the chemical. Thus, the MCL allows the ingestion of only about 5 drops of a carcinogenic chemical in a person's lifetime.

Volatile organic chemicals (VOCs) are produced in large quantities for use in industrial, commercial, agricultural, and household activities. The adverse health effects of VOCs include cancer and chronic effects on the liver, kidneys, and nervous system. Because of their volatility, air stripping is a proposed method of removal from water (Section 11.35). Volatility also reduces their concentrations in surface waters. Groundwater contamination is more common because VOCs have little affinity for soils and are diminished only by dispersion and diffusion, which is often limited. Those most frequently detected in contaminated groundwaters are trichloroethylene, a degreasing solvent in metal industries and a common ingredient in household cleaning products; tetrachloroethylene, a dry-cleaning solvent and chemical intermediate in producing other compounds; carbon tetrachloride, used in the manufacture of fluorocarbons for refrigerants and solvents; 1,1,1-trichloroethane, a metal cleaner; 1,2-dichloroethane, an intermediate in the manufacture of vinyl chloride monomers; and vinyl chloride, used in the manufacture of plastics and polyvinyl chloride resins.

Many of the synthetic organic chemicals (SOCs) listed in Table 8.2 are *insecticides* and *herbicides*. Pesticides may be present in surface waters receiving runoff from either agricultural or urban areas where these chemicals are applied. Groundwaters can be contaminated by pesticides, manufacturing wastewaters, spillage, or infiltration or rainfall and irrigation water. Alachlor, aldicarb, atrazine, carbofuran, ethylene dibromide, and dibromochloropropane have been detected in drinking waters. Most pesticides can be absorbed into the human body through the lungs, skin, and gastrointestinal tract. From acute exposure, the symptoms in humans are dizziness, blurred vision, nausea, and abdominal pain. Chronic exposure of laboratory animals indicates possible neurologic and kidney effects and, for some pesticides, cancer.

Disinfection By-Products

Trihalomethanes (THMs) are produced by chlorination of surface waters and groundwaters containing natural organic substances from decaying vegetation, such as humic and fulvic acids. THMs are derivatives of methane in which three of the four hydrogen atoms have been replaced by three atoms of chlorine, bromine, or iodine. Chloroform ($CHCl_3$) is the THM most commonly found in drinking water and is usually present in the highest concentration. In some cases, brominated THMs ($CHBrCl_2$, $CHBr_2Cl$, and $CHBr_3$) dominate as a result of naturally occurring bromide in the water. Bromide ions are oxidized by aqueous chlorine to bromine, and, since it is more reactive, bromine substitutions can dramatically increase the total THM level. The general reaction producing THMs is

$$\text{Chlorine} + (\text{bromide ion or iodide ion}) + \text{precursors} \\ = \text{trihalomethanes and other halogenated compounds} \quad (8.3)$$

Five haloacetic acids (HAA5) are the next most significant fraction of disinfection by-products resulting from chlorination. Other by-products that have been detected by laboratory testing of chlorinated waters are not present in significant concentrations.

The MCLs for total THMs ($CHCl_3$, $CHBrCl_2$, $CHBr_2Cl$, and $CHBr_3$) at 80 μg/l and HAA5 at 60 μg/l have been established because of carcinogenicity in laboratory animals. The numerical values of these parameters are determined in a water system by calculating a 12-month running average value, rather than from a single test. Several water samples are collected for each 3-month period from the distribution system with 25% from the extremities of the pipe network and 75% based on population. (Different locations of sampling points are necessary because formation of disinfection by-products depend on the time of chlorine contact. Their concentrations are not instantaneously formed but continue for an extended time following the chlorination.) The values for each 3-month period are the arithmetical averages of the total THM and HAA5 concentrations of all samples tested. Finally, the contaminant levels are calculated by averaging these quarterly values with the measured levels from the previous three quarters.

The disinfectants/disinfection by-products rule also establishes MCLs of 0.01 mg/l for bromate and 1.0 mg/l for chlorite. Maximum residual disinfection levels are set at 4.0 mg/l for chlorine plus chloramines and 0.8 mg/l for chlorine dioxide.

Radionuclides

Radioactive elements decay by emitting alpha, beta, or gamma radiation caused by transformation of the nuclei to lower energy states. An alpha particle is the helium nucleus (2 protons + 2 neutrons); e.g., radon-222 decays to polonium-218 and emits helium-4. A beta particle is an electron emitted from the nucleus as a result of neutron decay; e.g., radium-228 decays to actinium-228 and emits β^-. In these processes, the helium nucleus emitted as an alpha particle or the electron ejected as a beta particle changes the parent atom into a different element. A gamma ray is a form of electromagnetic radiation; other forms are light, infrared and ultraviolet radiations, and X-rays. Gamma decay involves only energy loss and does not create a different element. Alpha, beta, and gamma radiations have different energies and masses, thus producing different effects on matter. Each is capable of knocking an electron from its orbit around the nucleus and away from the atom, which is a process referred to as ionization. Radiation is detected by ionization, and high-reactive ions taken into the human body can lead to deleterious health effects, such as cancer.

The ability to penetrate matter varies among nuclear radiations. Most alpha particles are stopped by a single thickness of paper, while most gamma rays pass through the human body, as do X-rays. Since alpha particles are stopped by short penetrations, more energy is deposited and does more damage per unit volume of matter receiving radiation.

Radioactivity in drinking water can be from natural or artificial radionuclides. Radium-226 is usually found in groundwater as a result of geologic conditions. Radioactivity from radium is widespread in surface waters because of fallout from testing of nuclear weapons. In some localities this radioactivity could be increased by small releases from nuclear power plants and industrial users of radioactive materials. The average amount of background radiation from cosmic rays and terrestrial sources is about 100 mrem/yr [14]. Only a small portion of this unavoidable background radiation comes from drinking water containing radionuclides.

The recommended allowable dose from radioisotopes in drinking water supplies is very low. If a water supply were constituted in such a way as to contain either average or likely amounts of radioactivity, a total-body dose of 0.244 mrem/yr would be accumulated [10]. This is less than 1% of background. Although the dose to bone would be considerably higher, because strontium and radium are bone seekers, even this dose constitutes less than 10% of the total average natural background [10]. Estimates were made of three possible types of risks that could be induced by this magnitude of radiation: developmental and teratogenic, genetic, and somatic. Although a developing fetus is sensitive to radiation, this low dosage delivered from drinking water during the sensitive periods of gestation is so small that no measurable effects of the radiation will be found [10]. The lowest dose level at which any effect has been reported is 3 mrem/day or 1100 mrem/yr, in contrast to 0.244 mrem/yr. For genetic risk to the general population, the maximum permissible dose of artificial radiation is 170 mrem/yr, excluding medical uses of radiation. This amounts to a 5-rem genetic dose in each 30-yr generation, which is insignificant in increasing the current incidence of genetic diseases [10]. The natural background of radiation can be estimated to cause 4.5–45 fatal cases of cancer per year per million people, depending on the risk model used to make the calculation. Less than 1% of this is attributed to radionuclides in drinking water [10].

The National Academy of Sciences concluded that the radiation associated with most water supplies is such a small proportion of the normal background that it is difficult, if not impossible, to measure any adverse health effects with certainty. In a few water supplies, however, radium can reach concentrations that pose a higher risk of bone cancer for the people exposed [10]. Because of the many uncertainties of the environmental effects, the EPA has adopted the MCLs given in Table 8.2.

The measurement unit of pCi/l is 10^{-12} curie per liter, with a curie being the activity of 1 g of radium. The rem (radiation equivalent man) is a unit of radiation dose equivalence that is numerically equal to the absorbed dose in rad multiplied by a quality factor, to describe the actual damage to tissue from the ionizing radiation. The mrem is 1/1000 of a rem. The rad is the unit of dose or radiation absorbed. One rad deposits 100 ergs of energy in 1 g of matter.

In testing, gross alpha activity that could include radium radiation is determined first. If the value is greater than 5 pCi/l, separate tests for Rd-226 and Rd-228 are conducted. Beta activity is primarily from artificial radionuclides, e.g., contamination from nuclear weapons testing. The screening test is gross beta particle activity, since the decay products of fission are beta and gamma emitters. If the measurement is less than 50 pCi/l, tritium and strontium-90 activities are determined and converted to mrem/yr units. If the gross beta activity is greater than 50 pCi/l, analyses are performed to identify the radionuclides present to determine the mrem/yr.

Turbidity

Insoluble particulates impede the passage of light through water by scattering and absorbing the rays. This interference of light passage is referred to as turbidity. The standard is a suspension of silica of specified particle size selected so that a 1.0-mg/l suspension measures as 1.0 NTU. The common method of measurement uses a photoelectric detector and nephelometry to measure the intensity of scattered light. In surface water filtration for treatment of drinking water, turbidity in the filtered water must be equal to or less than 0.3 NTU in at least 95% of the measurements taken each month.

Secondary Standards

Standards for aesthetics, listed in Table 8.4, are recommended for characteristics that render the water less desirable for use: They are not related to health hazards and are nonenforceable by the EPA.

Excessive color, foaming, or odor cause customers to question the safety of drinking water and result in complaints from users. Aluminum contributes to color in water, copper has a metallic taste and causes blue-green stain, and zinc has a metallic taste. Excessive dietary intake of silver causes skin discoloration and graying of the white of the eye. Excessive fluoride causes tooth discoloration. Chloride or sulfate ion concentrations greater than 250 mg/l, or dissolved solid concentrations greater than 500 mg/l, can have taste and laxative properties. Sodium sulfate and magnesium sulfate are laxatives, with the common names of Glauber salt and Epsom salt, respectively. The laxative effect may be noticed by travelers or new consumers drinking waters high in sulfates; however, most persons become acclimated in a relatively short time. Excessive dissolved salts can also affect the taste of coffee and tea brewed with the water. For persons on a sodium-restricted diet, which is usually 2000 mg of sodium per day, the recommended maximum concentration in

TABLE 8.4 Recommended Secondary Contaminant Standards for Aesthetics of Drinking Water

Aluminum	0.05 to 0.2 mg/l	Manganese	0.05 mg/l
Chloride	250 mg/l	Odor	3 threshold odor numbers
Color	15 color units	pH	6.5 to 8.5
Copper	1.0 mg/l	Silver	0.1 mg/l
Corrosivity	noncorrosive	Sulfate	250 mg/l
Fluoride	2 mg/l	Total dissolved solids	500 mg/l
Foaming agents	0.5 mg/l	Zinc	5 mg/l
Iron	0.3 mg/l		

their drinking water is 100 mg/l [10]. For a severely restricted diet of 500 mg of sodium per day, the recommended maximum concentration is 20 mg/l of sodium ion.

Iron and manganese above 0.3 mg/l and 0.05 mg/l, respectively, are objectionable because of brown stains imparted to laundry and porcelain and the bittersweet taste of iron. A noncorrosive water with an alkaline pH is desirable to reduce the probability of pipe corrosion contributing iron and other trace metals to the water by dissolution from water mains and plumbing.

QUALITY CRITERIA FOR SURFACE WATERS

The Clean Water Act authorizes the U.S. Environmental Protection Agency (EPA) to direct and define natural water pollution control programs [15]. The objective is to maintain the chemical, physical, and biological quality of surface waters, seawater, and groundwater by placing ecological considerations and protection of human health ahead of economic concerns. In 1972, many of the nation's waters were polluted, thus, the initial goals were to reduce discharge of pollutants and achieve an interim water quality to protect fish, shellfish, and wildlife and achieve a water quality for fishing and swimming wherever attainable. Congressional policy was to recognize and preserve the states' primary responsibility to meet these goals. In the original act, the two principle policies were to research, study, and establish water quality standards for surface waters and to prohibit the discharge of toxic pollutants in toxic amounts.

An important amendment to the act was the National Pollutant Discharge Elimination System (NPDES) permit program. Technology-based effluent limits and water quality–based limits backed by surface water quality standards were defined for treatment plants discharging pollutants through a pipe or conveyance. For compliance, the owner of the treatment plant is required to monitor and record discharge data and report any violations. For enforcement, a willful or negligent violator, or one making a false statement or representation regarding a discharge, can be fined and is also subject to possible imprisonment.

The NPDES program for wastewater treatment plants has significantly changed and expanded since initiation. Specific areas are effluent standards, water quality–based permitting, controls of toxic substances, industrial wastewater pretreatment program, new performance standards, inspection and monitoring provisions, and seawater discharge criteria. In addition, several other programs have been incorporated into the NPDES permit system, including the control of combined sewer overflows; use and disposal of waste sludge by regulating management practices and acceptable levels of toxic substances in sludge; and watershed protection strategy to integrate the NPDES to support states' basin management.

8.6 WATER QUALITY STANDARDS

States are required to develop water quality standards, on a site-specific basis, for all of their surface waters. These should

- Include provisions for restoring and maintaining the chemical, physical, and biological integrity

- Provide, where attainable, water quality for protection and propagation of fish, shellfish, and wildlife and recreation in and on the water ("fishable/swimmable")
- Consider the use and value of waters for public water supplies, propagation of fish and wildlife, recreation, agriculture and industrial purposes, and navigation

The water quality standards must meet the requirements of the Clean Water Act [15] and Water Standards Regulations [16].

Water quality standards are composed of use classifications, quality criteria, and an antidegradation policy. The classification system is based on expected beneficial water uses including drinking water supplies, recreation, and propagation of fish and wildlife. More specific uses may be designated. Designated uses should support a "fishable/swimmable" classification unless an attainability analysis shows that only subcategories of this designation can be applied that require less stringent criteria.

The second part of standards is the water quality criteria necessary to support the designated uses, which may be both numeric values and narrative statements. Numeric values are ambient levels of individual pollutants to protect water uses by designating the magnitude expressed as an allowable concentration, duration of time over which the concentration in the surface water is averaged for comparison with the numeric criterion, and frequency a criterion can be exceeded. In the development of criteria by the EPA, the emphasis has been on 126 priority toxic pollutants [15]. The criteria that have been developed are contained in individual documents and summarized in the publication "Quality Criteria for Water" [17]. Narrative statements can supplement numeric values or can be the basis for limiting pollutants where no numeric criteria exist. Examples are "free from toxic substances in toxic amounts" and "free of objectionable color, odor, taste, or turbidity."

An antidegradation policy provides three tiers of protection from degradation of water quality:

- Protects existing uses
- Protects the level of water quality necessary to support propagation of fish, shellfish, and wildlife and recreation in waters that are currently of higher quality than required to support these uses. Before water quality can be lowered, an antidegradation review must be undertaken
- Protects the quality of outstanding national resources, such as waters of state and national parks, wildlife refuges, and waters of recreational or ecological significance

8.7 POLLUTION EFFECTS ON AQUATIC LIFE

A normal, healthy stream or lake has a balance of plant and animal life represented by great species diversity. Pollution disrupts this balance, resulting in a reduction in the variety of individuals and dominance of the surviving organisms. Complete absence of species normally associated with a particular habitat reveals extreme degradation. Of course, biological diversity and population counts are meaningful only if existing communities in a polluted environment are compared to those normally present in that particular habitat. Fish are good indicators of water quality, and no perennial river can be considered in satisfactory condition unless a variety of fish can survive in it.

Being an end product of the aquatic food chain, fish reflect both satisfactory water quality and a suitable habitat for food supply, shelter, and breeding sites. Even though depletion of dissolved oxygen is commonly blamed, poisons appear to cause the most damage to plant and animal life in surface waters. The effects of toxic substances are frequently magnified by environmental conditions; for example, temperature has a direct influence on morbidity. At a given concentration of a toxic substance, a rise of 10°C generally halves the survival time of fish; poisons therefore become more lethal in rivers during the summer. Many toxic substances become more lethal with decreasing dissolved oxygen content. Also, the rate of oxygen consumption of fish is altered by the presence of poisons and their resistance to low oxygen levels can be impaired. The pH of a water within the allowable range of 6.5–9 can influence some poisons. The dissolved salt content can also influence toxicity, particularly the presence of calcium, which reduces the adverse effect of some heavy metals. For example, cadmium, copper, lead, nickel, and zinc decrease in toxicity with increased hardness in the water.

Poisonous effects on fish life also relate to the character of the watercourse, species of fish, and season of the year. During the winter fish are much more resistant because of the cold water. The rapid rise of temperature in spring and hot periods in summer create critical times when fish are susceptible to unfavorable conditions and likely to die. During spawning even slight pollution can cause damage to salmon and trout.

Acute toxic effects (24–16-hr exposure) have been studied for most toxic pollutants. In contrast, few data are available on chronic toxicity. Applying a safety factor to the median lethal concentration can result in a criterion that is either too conservative or unsafe for long-term exposure. Chronic effects often occur in the species population rather than in the individual. If eggs fail to develop or the sperm does not remain viable, the species would be eliminated from an ecosystem because of reproductive failure. Physiologic stress can make a species less competitive, resulting in a gradual population decline or absence from an area. The same phenomenon could occur if a crustacean that serves as a vital food during the larval period of a fish's life is eliminated. Finally, biological accumulation of certain toxic substances can result in acute effects in fish that are the ultimate consumers in the aquatic food chain.

8.8 CONVENTIONAL WATER POLLUTANTS

The common water pollutants are biochemical oxygen demand (BOD), suspended solids, fecal coliforms, pH, ammonia nitrogen, phosphorus, oil and grease, and chlorine residual. For these pollutants the EPA has developed water quality criteria consisting of numerical limits; their rationale is based on bioassays of aquatic organisms [17]. Bioassays are the best method for determining safe concentrations of conventional pollutants to aquatic organisms. Test species, usually fish, are exposed to various concentrations of a pollutant in water for a specified time span of 96 hr or less in laboratory tanks. The median lethal concentration (LC_{50}) is the level that kills 50% of the test organisms. The maximum allowable toxic concentration in surface waters is usually between 0.1 and 0.01 of the LC_{50} value. The uncertainty factor of 10–100 is to account for long-term exposure and other constituents already present in the river water creating additional physiologic stresses. These criteria may be modified to take into account the variability of local waters in establishing state standards.

Water quality standards associate particular numerical limits with the designated beneficial uses for specific surface waters, thus recognizing that use and criteria are interdependent. Local conditions commonly considered are natural background levels of pollutants and other constituents such as hardness, the presence or absence of sensitive aquatic species, characteristics of the biological community, temperature and weather, flow characteristics, and synergistic or antagonistic effects of combinations of pollutants. In general, EPA criteria are considered conservative estimates of pollutant concentrations that can be safely tolerated by a general ecosystem, whereas state standards address site-specific pollution problems.

The *dissolved-oxygen* standard establishes lower limits to protect propagation of fish and other aquatic life, enhance recreation and reduce the possibility of odors resulting from decomposition of organic matter, and maintain a suitable quality for water treatment. The primary pollutant associated with depletion of dissolved oxygen is carbonaceous BOD. In addition, sedimentation of suspended solids can cause a buildup of decomposing organic matter in sediments, and dissolved ammonia can contribute to oxygen depletion by nitrification. Fish vary in their oxygen requirements according to species, age, activity, temperature, and nutritional state. In general, the minimum dissolved-oxygen level needed to support a diverse population of fish is 5 mg/l. Coldwater fish require stringent limitations, 6 mg/l with a minimum of 7 mg/l at spawning times, and warm-water species, being more tolerant, need 4–5 mg/l. A typical minimum standard for water supply, recreation, and shellfish harvesting is 4 mg/l. The EPA criteria are a minimum 5.0 mg/l for freshwater aquatic life and maintenance of aerobic conditions throughout a body of water for aesthetic considerations [17].

Suspended solids interfere with the transmission of light and can settle out of suspension, covering a streambed or lake bottom. Turbid water interferes with recreational use and aesthetic enjoyment. Excess suspended solids adversely affect fish by reducing their growth rate and resistance to disease, preventing the successful development of fish eggs and larvae, and reducing the amount of food available. Settleable solids covering the bottom damage invertebrate populations and fill gravel spawning beds. The EPA criterion states that solids should not reduce the depth of the compensation point (penetration of sunlight) for photosynthetic activity by more than 10% from the seasonally established norm for aquatic life [17].

Oil and grease contaminants include a wide variety of organic compounds having different physical, chemical, and toxicological properties. Common sources are petroleum derivatives and fats from vegetable oil and meat processing. Domestic water supplies need to be virtually free from oil and grease, particularly from the tastes and odors that emanate from petroleum products. Surface waters should be free of floating fats and oils. Based on EPA criteria, individual petrochemicals should not exceed 0.01 of the median lethal concentration that kills 50% (LC_{50}) of a specific aquatic species during an exposure period of 96 hr [17].

Fecal coliform bacteria indicate the possible presence of pathogenic organisms. The correlation between coliforms and human pathogens in natural waters is not absolute, however, since these bacteria can originate from the feces of both humans and other warm-blooded animals. Coliforms from the intestinal tract of a human cannot be distinguished from those of animals. Therefore, the significance of testing in pollution surveys depends on a knowledge of the river basin and probable source of the observed fecal coliforms. The EPA criterion for fecal coliform bacteria in bathing waters is a logarithmic mean of 200 per 100 ml, based on a minimum of five samples taken over a 30-day period, with not more than 10% of the total samples exceeding 400 per 100 ml. Since shellfish may be eaten without being cooked, the strictest coliform criterion applies to shellfish cultivation and harvesting. The EPA criterion states that the mean fecal coliform concentration should not exceed 14 per 100 ml, with not more than 10% of the samples exceeding 43 per 100 ml.

Residual chlorine resulting from disinfection of wastewater effluents is very toxic to fish. When chlorine is added to wastewater, chloramines are formed by reacting with ammonia. These can be eliminated in wastewater effluents by the addition of a reducing agent such as sulfur dioxide [17]. The EPA criteria for total chlorine residual are 2.0 μg/l for salmonid fish and 10.0 μg/l for other freshwater and marine organisms.

Un-ionized ammonia is toxic to fish and other aquatic animals. When ammonia dissolves in water, a portion reacts with the water to form ammonium ions (NH_4^+) with the balance remaining as un-ionized ammonia (NH_3). The concentration of un-ionized ammonia increases with increasing pH, increases with increasing temperature, and decreases with decreasing ionic strength. The EPA criterion for freshwater aquatic life is 0.02 mg/l of un-ionized ammonia, which is based on salmonid fish [17]. This value was calculated by applying a safety factor of 10 to the lowest reported lethal concentration (LC_{50}) of 0.2 mg/l of un-ionized ammonia nitrogen for rainbow trout fry. Since un-ionized ammonia cannot be measured, its concentration in water is based on the measured concentration of total ammonia ($NH_3 + NH_4^+$). Values of total ammonia nitrogen that result in concentrations of 0.02 mg/l of un-ionized ammonia for 20 °C are 5.1 mg/l at pH 7.0, 1.6 at pH 7.5, 0.52 at pH 8.0, and 0.18 at pH 8.5 [17]. For more tolerant fish species these maximum allowable concentrations are very conservative. Interpretation of bioassay data on warm-water species suggests that the acute toxic concentration of ammonia at the gill surface is 0.4 mg/l [18]. Using a safety factor of 5 yields a no-effect un-ionized ammonia concentration of 0.08 mg/l. Values of total ammonia nitrogen that result in concentrations of 0.08 mg/l of un-ionized ammonia for 20 °C and an alkalinity of 200 mg/l are 23 mg/l at pH 7.0, 9.0 at pH 7.5, 5.0 at pH 8.0, and 2.3 at pH 8.5 [18].

Controversy over ammonia criteria results from the high concentration of ammonia nitrogen in treated effluents from municipal wastewater plants. After biological treatment without nitrification, an average domestic wastewater contains 24 mg/l of ammonia nitrogen (Table 12.1). The dilution ratio of ammonia-free water in a receiving watercourse at pH 8.0 to wastewater effluent containing 24 mg/l to protect warm-water fish is 3.8; for cold-water salmonid fish the required dilution ratio is 23.5. Ammonia can be converted to nitrate to reduce the effluent concentration by additional aeration in biological treatment. This nitrification process, usually a second stage of aeration, adds significant cost to wastewater treatment.

The *pH of surface waters* is specified for protection of fish life and to control undesirable chemical reactions, such as the dissolution of metal ions in acidic waters. Many substances increase in toxicity with changes in pH. For example, the ammonium ion is shifted to the much more poisonous form of un-ionized ammonia as the pH of water rises above neutrality. The EPA criteria for pH are 6.5–9.0 for freshwater aquatic life, 6.5–8.5 for marine aquatic life, and 5–9 for domestic water supplies [17].

Phosphate phosphorus is a key nutrient stimulating excessive plant growth—both weeds and algae—in lakes, estuaries, and slow-moving rivers. Cultural eutrophication is the accelerated fertilization of surface waters arising from phosphate pollution associated with discharge of wastewaters and agricultural drainage. Since phosphate removal is feasible by chemical precipitation in wastewater treatment, effluent permits for municipal and industrial discharges to lakes, or streams that flow into lakes, commonly limit the concentration to 1.0 mg/l or less of phosphate phosphorus; that is equivalent to about 85% or more removal from domestic wastewater. For lakes in the northern United States to be free of algal nuisances, the generally accepted upper concentration limit in impounded water when completely mixed in the spring of the year is 0.01 mg/l of orthophosphate.

Discharges of conventional pollutants from wastewater treatment plants are controlled through the National Pollutant Discharge Elimination System (NPDES) permit program (Section 9.1). Technology-based effluent limits of BOD, suspended solids, oil and grease, fecal coliforms, and pH define secondary treatment. As a general guideline, an adequate dilution ratio for secondary effluent discharged to flowing water is 20 to 1 or lower. Water quality–based effluent limits are set to ensure water quality standards are not exceeded in the receiving water. These may reduce the limits for secondary treatment where flow in the receiving watercourse does not provide adequate dilution. Water quality standards may also require limits on phosphorus to reduce eutrophication or limit un-ionized ammonia and residual chlorine to reduce toxicity.

8.9 TOXIC WATER POLLUTANTS

Numerous organic chemicals and several inorganic ions, mostly heavy metals, are classified as toxic water pollutants. To qualify as a priority toxin, a substance must be an environmental hazard and known to be present in polluted waters. Toxicity to fish and wildlife may be related to either acute or chronic effects on the organisms themselves or to humans by bioaccumulation in food fish. Persistence in the environment (including mobility and degradability) and treatability are also important factors. Currently, the EPA list of priority toxic pollutants consists of over 120 substances [15]. While some are well-documented toxic substances, others have limited data supporting their hazard in the environment.

The majority of the toxic pollutants can be categorized into 10 groups. *Halogenated aliphatics* are used in fire extinguishers, refrigerants, propellants, pesticides, and solvents. Health effects include damage to the central nervous system and liver. *Phenols* are industrial compounds used primarily in production of synthetic polymers, pigments, and pesticides and occur naturally in fossil fuels. They impart objectionable taste and odor at very low concentrations, taint fish flesh, and vary in toxicity depending on chlorination of the phenolic molecule. *Monocyclic aromatics* (excluding phenols and phthalates) are used in the manufacture of chemicals, explosives, dyes, fungicides, and herbicides. These compounds are central nervous system depressants and can cause damage to the liver and kidneys. *Ethers* are solvents for polymer plastics. They are suspected carcinogens and aquatic toxins. *Nitrosamines*, used in production of organic chemicals and rubber, are suggested carcinogens. *Phthalate esters* are used in production of polyvinylchloride and thermoplastics. They are an aquatic toxin and can be biomagnified. *Polycyclic aromatic hydrocarbons* are in pesticides, herbicides, and petroleum products. *Pesticides* of concern are those that biomagnify in the food chain and are persistent in nature; chlorinated hydrocarbons are common in this group of pesticides. Aldrin, dieldrin, and chlordane applications are already restricted by the EPA. *Polychlorinated biphenyls* (PCBs) were banned from production in 1979. They are readily assimilated by the aquatic environment and still persist in sediments and fish; PCBs were used in electric capacitors and transformers, paints, plastics, insecticides, and other industrial products. *Heavy metals* vary in toxicity, and some are subject to biomagnification.

The general goal of wastewater treatment in the NPDES permit program is to reduce the discharge of toxic pollutants to an insignificant level by enforcing an industrial wastewater pretreatment program. Toxicity reduction evaluation of a treatment plant is the first step. For municipal plants, the objectives are to evaluate operation and performance to identify and correct treatment deficiencies causing effluent toxicity; identify the toxic substances in the effluent; trace the toxic pollutants to their sources in the wastewater collection system; and implement appropriate remedial measures to reduce effluent toxicity.

Chemical Evaluation of Effluent

Gross toxicity of a wastewater effluent is evidenced in the receiving watercourse by a reduction in species diversity. Presence of a poison disrupts the normal balance of plant and animal life, which is represented by a great variety of individuals. Complete absence of species normally associated with a particular aquatic habitat is evidence of severe degradation. Obviously, a fish kill is an example associated with extreme acute toxicity from the discharge of a lethal quantity of a poison. Periodic upset of the biological process in wastewater treatment is an indication of industrial toxic chemicals being discharged to the sewer collection system and passing through the treatment plant. If the evidence of toxicity is less dramatic, the presence of toxic chemicals in a wastewater effluent may be unnoticed.

The chemical-specific approach, if toxicity is indicated by the whole effluent toxicity (WET) test, is to test effluent samples for selected toxic chemicals. This can be very costly unless the toxic chemicals likely to be present can be reduced to a reasonable number. Initial tests should be for those substances related to wastes from industries served by the sewer system. After the toxic pollutants have been quantitatively identified, the NPDES permit is modified to ensure protection of aquatic life in accordance with appropriate criteria [17]. Remedial measures must be taken by industrial pretreatment and improved municipal treatment to meet the revised effluent standards. Mandatory monitoring for toxic chemicals in the effluent is established by the NPDES discharge permit. Potential deficiencies of this analytical approach through chemical analyses are the possible presence of undetected toxic chemicals and combinations of synergistic toxic pollutants.

Effluent Biological Toxicity Testing

Biological testing by the whole effluent toxicity (WET) method defined by the EPA is required under their NPDES permit for selected municipal treatment plants [19,20]. This bioassay to determine toxicity is performed by exposing selected aquatic organisms to wastewater effluent in a controlled laboratory environment. In the static short-term WET test, effluent and uncontaminated water are placed in separate laboratory containers, test organisms are added to both containers, and then monitored for toxic effects in a controlled laboratory environment.

The containers are borosilicate glass or disposable polystyrene ranging in volume from 250 ml to 1000 ml, depending on the size of the test organism [21]. During the test period the dissolved oxygen concentration should be near saturation, temperature held at 25°C for warm-water species (12°C for cold-water species). The pH is checked periodically. The effluent is a 24-hr composite filtered through a sieve to remove suspended solids. During effluent preparation, aeration should be limited to prevent loss of volatile organic compounds. The common warm-water test organisms are fathead minnow (*Pimephales promelas*), daphnid *(Ceriodaphnia dubia)*, and a green alga *(Selenastrum capricornutum)*. The fathead minnow, a popular bait fish, feeds primarily on algae, and grows to an average length of 50 mm with a life span of less than 3 years. *Daphnia* are invertebrates that feed on algae, grow to a maximum of 4–6 mm, and have a life span of 40–60 days. Some states have developed culturing and testing methods for indigenous species.

The test result is expressed in terms of mortality after a 24-hr exposure. A 25% reduction in survival is defined as the threshold of biological significance indicating probable impairment of the receiving water. The WET test measures the aggregate toxic effect of a mixture of unknown toxic pollutants in the effluent. When significant toxicity is indicated by biological testing, a chemical evaluation of the effluent is required. This can be very costly unless toxic substances likely to be present can be reduced to a reasonable number. A release inventory of toxic pollutants from industries contributing wastewater to the sewer system is the best way to identify possible toxic substances and their sources. A proper pretreatment program for an industry requires sampling, flow measurement, laboratory testing, and reporting. After toxic pollutants have been quantitatively identified, remedial measures to reduce or eliminate effluent toxicity can be taken by additional industrial pretreatment and improved municipal wastewater treatment.

Impairment in the receiving water and effluent toxicity are related to dilution of the effluent within the receiving water. Therefore, a more definitive toxicity test is to set up a series of laboratory containers at decreasing effluent dilution, such as 100%, 50%, 25%, 12.5%, and 6.25% (a geometric series). The objective is to estimate the safe, or no-observed-effect level, that will permit normal propagation of aquatic life in the receiving water. This range of effluent dilution is to encompass the recorded one-in-10-year, 7-consecutive-day low flow in the receiving water relative to the effluent flow. The average low flow over a 7-consecutive-day period that is likely to occur once every 10 years is usually the flow in a river or stream established for the maximum allowable concentration of pollutants set by surface water standards. The dilution water used in laboratory testing is from the receiving stream above the outfall pipe or from the edge of the mixing zone. It is filtered through a plankton net to remove indigenous organisms that may attack or be confused with the test organisms. As an alternate, the dilution water may be a moderately hard mineral water.

These tests may be conducted for either acute 24-hr toxicity or chronic toxicity monitored up to 7 days. In chronic testing, food is supplied to the organism cultures, for example, flake fish food or brine shrimp for fathead minnows. Also, chronic tests may be static-renewal tests where test organisms are transferred every 24 hr to a fresh solution of the same concentration of wastewater. Based on the results, the no-observed-effect concentration is determined for effluent in the diluted test samples. This can be compared against the ratio of wastewater diluted with the 7-consecutive-day, one-in-10-year low flow.

SELECTED POLLUTION PARAMETERS

Total solids, suspended solids, BOD, chemical oxygen demand (COD), and coliform bacteria are common parameters used in water and wastewater engineering. Knowledge of testing procedures is essential to understand the meaning of these terms. For a detailed description of testing procedures for these and other substances, refer to *Standard Methods for the Examination of Water and Wastewater* [22].

8.10 TOTAL AND SUSPENDED SOLIDS

The term *total solids* refers to the residue left in a drying dish after evaporation of a sample of water or wastewater and subsequent drying in an oven (Fig. 8.1). After a measured volume is placed in a porcelain dish, the water is evaporated from the dish on a steam bath. The dish is then transferred to an oven and dried to a constant weight at 103°–105°C. The total residue is equal to the difference between the cooled weight of the dish and the original weight of the empty dish. The concentration of total solids is the weight of dry solids divided by the volume of the sample, usually expressed in milligrams per liter.

Total volatile solids are determined by igniting the dry solids at 550° ± 50°C in an electric muffle furnace. The residue remaining after burning is referred to as fixed solids, and the loss of weight on ignition is reported as volatile solids. The concentration of

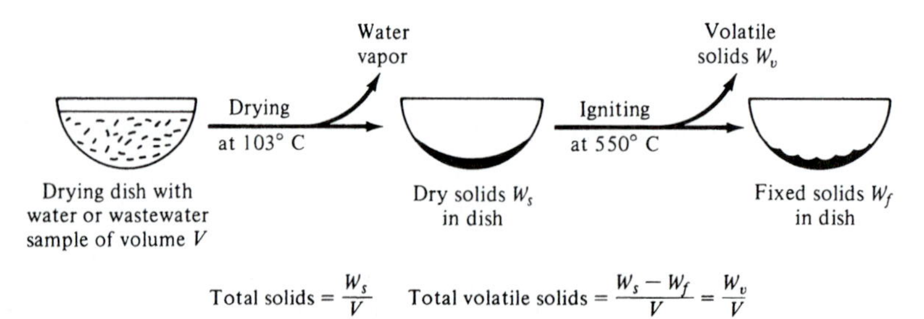

FIGURE 8.1 Diagram of laboratory procedure to determine total solids and total volatile solids concentrations of a water or wastewater sample.

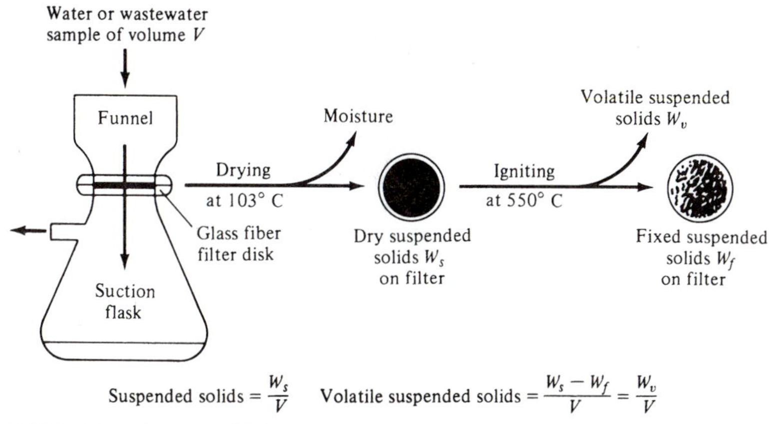

FIGURE 8.2 Diagram of laboratory procedure to determine the total suspended-solids and volatile suspended-solids concentrations of a water or wastewater sample.

total volatile solids is the weight of dry solids minus the weight of fixed solids divided by the volume of the original liquid sample. Volatile solids content also can be expressed as a percentage of the dry solids in the sample.

The term *total suspended solids* refers to the nonfilterable residue that is retained on a glass-fiber disk after filtration of a sample of water or wastewater (Fig. 8.2). A measured portion of a sample is drawn through a glass-fiber filter, retained in a funnel, by applying a vacuum to the suction flask under the filter. The filter with damp suspended solids adhering to the surface is transferred from the filtration apparatus to an aluminum or stainless steel planchet as a support. After drying at 103°–105°C in an oven, the filter with the dry suspended solids is weighed. The weight of suspended solids is equal to the difference between this weight and the original weight of the clean filter. The concentration of total suspended solids is the weight of the dry solids divided by the volume of the sample and is usually expressed in milligrams per liter.

Volatile suspended solids are determined by igniting the dry solids at 550° ± 50°C after placing the filter disk in a porcelain dish. The concentration of volatile suspended solids is the weight of dry solids minus the weight of fixed solids divided by the volume of the original liquid sample.

Dissolved solids are the solids that pass through the glass-fiber filter and are calculated from total and suspended solids analyses. Total dissolved solids equals total solids minus total suspended solids. Volatile dissolved solids equals total volatile solids minus volatile suspended solids.

8.11 BIOCHEMICAL AND CHEMICAL OXYGEN DEMANDS

Biochemical Oxygen Demand

Biochemical oxygen demand (BOD) is the quantity of oxygen used by microorganisms in the aerobic stabilization of wastewaters and polluted waters. The standard 5-day BOD value is commonly used to define the strength of municipal wastewaters, to evaluate the efficiency of treatment by measuring oxygen demand remaining in the effluent, and to determine the amount of organic pollution in surface waters.

Laboratory analyses of wastewaters and polluted waters are conducted using 300-ml BOD bottles incubated at a temperature of 20°C. The preparation of BOD tests is diagrammed in Figure 8.3. To measure the BOD of a wastewater sample, a measured portion is placed in the BOD bottle, seed microorganisms are added if needed, and the bottle is filled with aerated dilution water containing phosphate buffer and inorganic nutrients. The amount of wastewater added to a 300-ml bottle depends on the estimated strength. For example, typical amounts are 5.0 ml for untreated wastewater in the BOD range of 120–420 mg/l and 50 ml for effluents with 12–42 mg/l. Untreated municipal wastewaters and unchlorinated effluents have

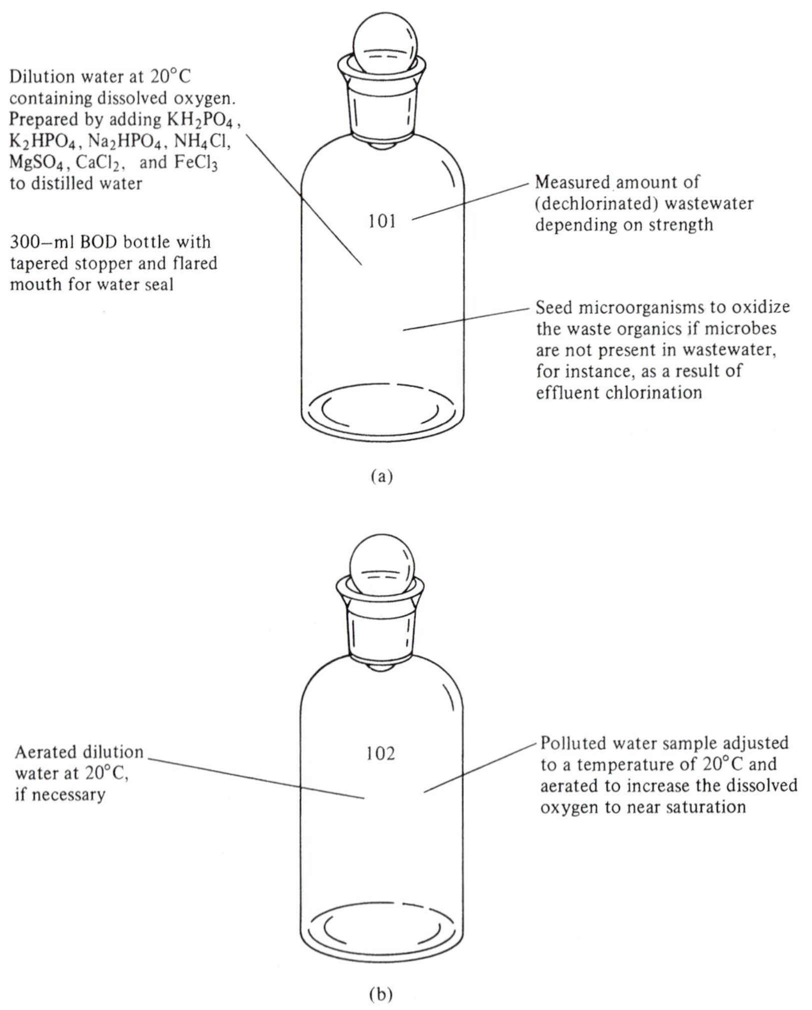

FIGURE 8.3 Preparation of biochemical oxygen demand (BOD) tests on (a) wastewater sample, (b) polluted surface water.

adequate microbial populations without seeding. Industrial wastewaters and dechlorinated effluents require a prepared seed, usually aged domestic wastewater, to perform the biological reactions. While the wastewater provides the organic matter, the dilution water provides the dissolved oxygen for aerobic decomposition. For a polluted surface water, the sample is adjusted to 20 °C and aerated to increase the dissolved oxygen to near saturation. The sample, or a portion of the sample mixed with dilution water, is then placed in the BOD bottle. A 50% mixture of sample and dilution waters is suitable for a polluted-water BOD in the range of 4–14 mg/l.

The biochemical oxygen demand exerted by diluted wastewater progresses approximately by first-order kinetics as shown in Figure 8.4. The initial depletion of dissolved oxygen is the result of carbonaceous oxygen demand resulting from organic matter degradation:

$$\text{Dissolved oxygen} + \text{organic matter} \xrightarrow[\text{and protozoans}]{\text{bacteria}} \text{carbon dioxide} + \text{biological growths} \tag{8.4}$$

If present in sufficient numbers, nitrifying bacteria exert a secondary oxygen demand by oxidation of ammonia:

$$\text{Dissolved oxygen} + \text{ammonia nitrogen} \xrightarrow[\text{bacteria}]{\text{nitrifying}} \text{nitrate nitrogen} + \text{bacterial growth} \tag{8.5}$$

Nitrification often lags several days behind the start of carbonaceous oxygen demand.

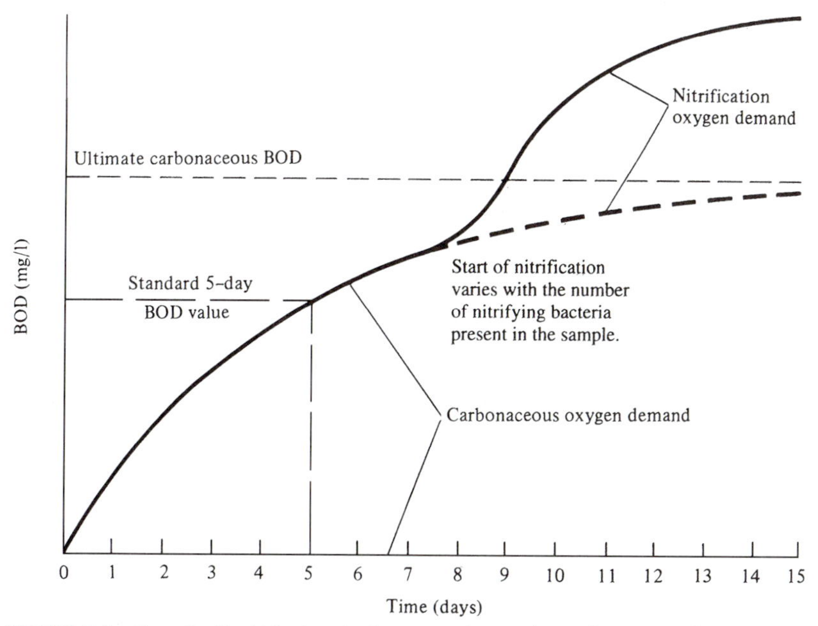

FIGURE 8.4 Hypothetical biochemical oxygen demand reaction curve showing the cabornaceous and nitrification reactions.

The carbonaceous oxygen demand curve can be expressed mathematically as

$$BOD_t = L(1 - 10^{-kt}) \quad (8.6)$$

where

BOD_t = biochemical oxygen demand at time t, mg/l

L = ultimate BOD, mg/l

k = deoxygenation rate constant, day^{-1}

t = time, days

The equation for calculating BOD from a seeded laboratory test is

$$BOD = \frac{(D_1 - D_2) - (B_1 - B_2)f}{P} \quad (8.7)$$

where

BOD = biochemical oxygen demand, mg/l

D_1 = dissolved oxygen (DO) of diluted seeded wastewater about 15 min after preparation, mg/l

D_2 = DO of wastewater after incubation, mg/l

B_1 = DO of diluted seed sample about 15 min after preparation, mg/l

B_2 = DO of seed sample after incubation, mg/l

f = ratio of seed volume in seeded wastewater test to seed volume in BOD test on seed

P = decimal fraction of wastewater sample used

$$= \frac{\text{volume of wastewater}}{\text{volume of dilution water plus wastewater}}$$

If the sample is unseeded, the relationship is

$$BOD = \frac{D_1 - D_2}{P} \quad (8.8)$$

The standard value is the BOD exerted during the first 5 days of incubation. More detailed discussions of BOD testing are presented by Hammer and Hammer [23].

Example 8.1

BOD tests were conducted on composited samples of a raw wastewater and a treated wastewater after chlorination.

1. The BOD tests for the raw wastewater were set up by pipetting 5.0 ml into each 300-ml bottle. For one pair of bottles, the test results were: The initial

dissolved oxygen (DO) was 8.4 mg/l, and after 5 days of incubation at 20°C the final DO was 3.7 mg/l. Calculate the BOD and estimate the 20-day BOD value assuming a k of 0.10 day^{-1}.

2. The treated wastewater sample was dechlorinated prior to conducting a seeded test. The BOD bottles were set up with 50.0 ml of treated wastewater and 0.5 ml of raw wastewater for seed added to each bottle. For one pair of bottles, the test results were: the initial DO was 7.6 mg/l, and the final DO was 2.9 mg/l. Calculate the BOD.

Solution:

1. Using Eq. (8.8)

$$BOD_5 = \frac{8.4 - 3.7}{5.0/300} = 282 \text{ mg/l}$$

Using Eq. (8.6)

$$L = \frac{282}{1 - 10^{-0.1 \times 5.0}} = 412 \text{ mg/l}$$

$$BOD_{20} = 412(1 - 10^{-0.1 \times 20}) = 408 \text{ mg/l}$$

2. Using Eq. (8.7)

$$BOD = \frac{(7.6 - 2.9) - (8.4 - 3.7)(0.5/5.0)}{50/300} = 25 \text{ mg/l}$$

Chemical Oxygen Demand

The *chemical oxygen demand* (COD) of wastewater or polluted water is a measure of the oxygen equivalent of the organic matter susceptible to oxidation by a strong chemical oxidant. The organic matter destroyed by the mixture of chromic and sulfuric acids is converted to CO_2 and water. The test procedure is to add measured quantities of standard potassium dichromate, sulfuric acid reagent containing silver sulfate, and a measured volume of sample into a flask. After attaching a condenser on top, this mixture is refluxed (vaporized and condensed) for 2 hr. The oxidation of organic matter converts dichromate to trivalent chromium,

$$\text{Organic matter} + Cr_2O_7^{2-} + H^+ \xrightarrow[Ag^+]{\text{heat}} CO_2 + H_2O + 2Cr^{3+} \tag{8.9}$$

After cooling, washing down the condenser, and diluting the mixture with distilled water, the excess dichromate remaining in the mixture is measured by titration with standardized ferrous ammonium sulfate. A blank sample of distilled water is carried through the same COD testing procedure as the wastewater sample. The purpose of testing a blank is to compensate for any error that can result because of the presence of extraneous organic matter in the reagents. COD is calculated from the following equation:

$$COD = \frac{(a - b)[\text{normality of Fe }(NH_4)_2(SO_4)_2]8000}{V} \tag{8.10}$$

where

$$\begin{aligned} COD &= \text{chemical oxygen demand, mg/l} \\ a &= \text{amount of ferrous ammonium sulfate titrant added to blank, ml} \\ b &= \text{amount of titrant added to sample, ml} \\ V &= \text{volume of sample, ml} \\ 8000 &= \text{multiplier to express COD in mg/l of oxygen} \end{aligned}$$

Applications of BOD and COD Testing

BOD is the most common parameter for defining the strengths of untreated and treated municipal and biodegradable industrial wastewaters. The oxygen requirement and tank sizing for aerobic treatment processes are based on BOD loadings. It is also used in quantifying the quality of an effluent from a treatment plant; the maximum allowable BOD is specified in the wastewater discharge permit. Since oxygen depletion in surface waters results from aquatic microorganisms exerting oxygen demand, the BOD test is a key measurement in the evaluation of water pollution by biodegradable wastes.

Commonly, COD is used to define the strength of industrial wastewaters that are either not readily biodegradable or contain compounds that inhibit biological activity. Frequently, laboratory wastewater treatability studies are based on COD testing rather than on BOD analyses. The COD test has the advantages of rapid analysis and reproducible results. The BOD test requires incubation for 5 days, and the results of multiple analyses on an industrial wastewater sample often show considerable scatter. Also, COD testing is becoming more popular in all applications of oxygen demand analyses as a result of simplified laboratory techniques.

The relationship between BOD and COD concentrations must be defined for each individual wastewater. Ideally, for a wastewater composed of biodegradable organic substances, the COD concentration approximates the ultimate carbonaceous BOD value. Yet this simple relationship is rarely substantiated in testing of municipal wastewaters. Many organic compounds can be oxidized chemically that are only partly biodegradable.

8.12 COLIFORM BACTERIA

The coliform group of bacteria is defined as aerobic and facultative anaerobic, non-spore forming, Gram's-stain negative rods that ferment lactose with gas production within 48 hr of incubation at 35 °C.

Fermentation Tube Technique

The basic coliform analysis, diagrammed in Fig. 8.5, is the test for total coliforms based on gas production during the fermentation of lauryl tryptose broth, which contains beef extract, peptone (protein derivatives), and lactose (milk sugar). Ten-milliliter portions of a water or wastewater sample are transferred using sterile pipettes into prepared fermentation tubes containing inverted glass vials. The inoculated tubes are

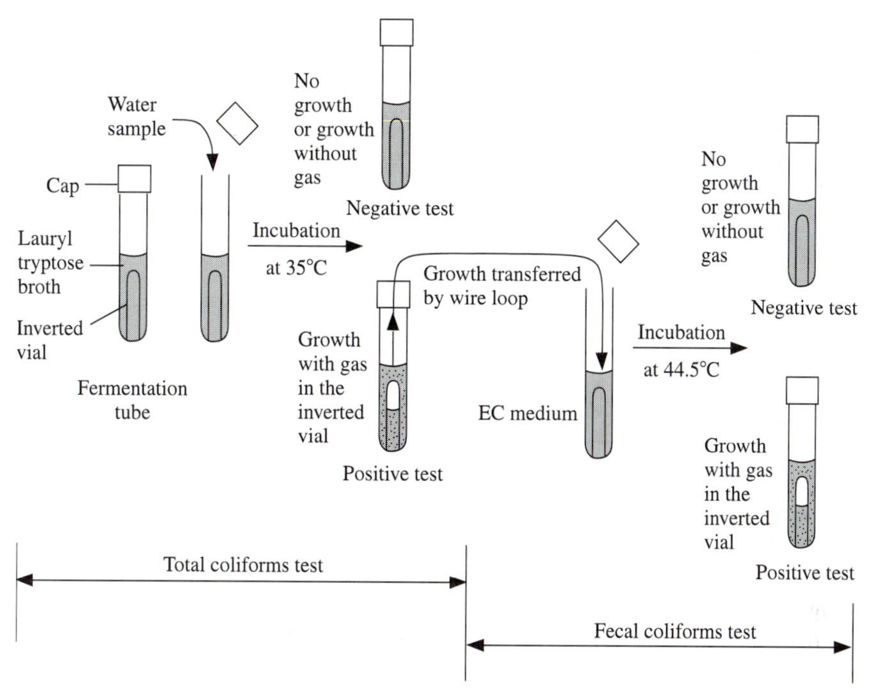

FIGURE 8.5 Diagram of the basic test for total coliforms and second-phase confirmatory test for thermotolerant fecal coliforms.

placed in a warm-air incubator at 35° ± 0.5°C for 48 hr. Growth with the production of gas, identified by the presence of a bubble in the inverted vial, is a positive test, indicating that coliform bacteria may be present. A negative reaction, either no growth or growth without gas, excludes the coliform group.

The common test for fecal coliforms is a second-phase confirmatory test following growth of coliforms in the presumptive total coliform test. A minute portion of the broth from a positive test, or a positive colony from a membrane filter, is transferred aseptically on a sterile wire loop to a fermentation tube of an EC medium containing tryptose, lactose, bile salts, and chemical buffers. The glass tube has a removable cap and an inverted vial at the bottom of the tube in the broth, as illustrated in Figure 8.5. A positive test is growth with gas after 22–26 hr at 44.5 °C. Gas production is evidence that coliforms have converted the lactose sugar to lactic acid, thus lowering the pH releasing gas. If no gas appears in the inverted vial, the test is negative and no fecal coliform bacteria were present in the positive total coliform test.

Membrane Filter Technique

The membrane filter test for coliform testing is commonly used when a large number of water or wastewater samples are routinely analyzed. This method, diagrammed in Figure 8.6, consists of drawing a measured volume of water through a filter membrane with small enough openings to take out bacteria, and then placing the filter on a growth medium in a culture dish. This technique assumes that each bacterium retained

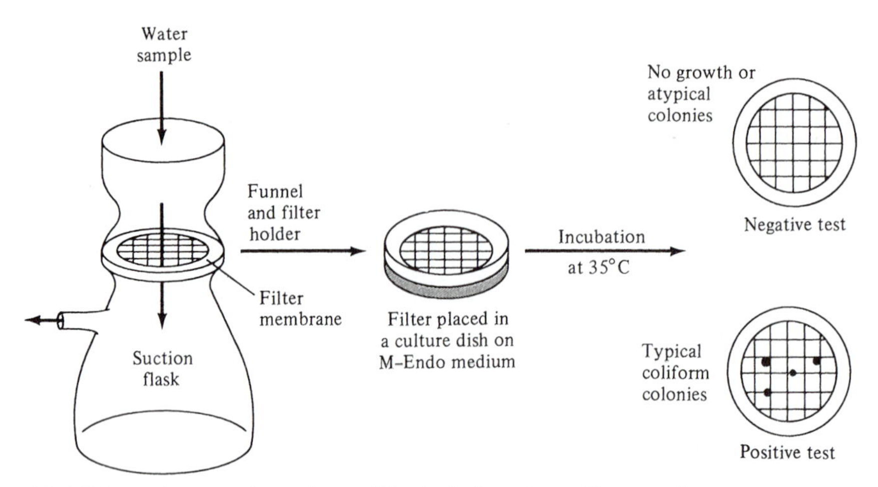

FIGURE 8.6 Diagram of membrane filter technique for coliform testing.

by the filter grows and forms a small visible colony. The number of coliforms present in a filtered sample is determined by counting the number of typical colonies and expressing this value in terms of number per 100 ml of water.

The apparatus to perform membrane filter coliform testing includes a filtration unit, sterile filter membranes, sterile absorbent pads, culture dishes, nutrient media, and forceps. The forceps for handling the filters are sterilized before each use. Rapid decontamination of filter units between successive filtrations can be accomplished using an ultraviolet sterilizer with exposure to radiation for 2 min. The filter membranes with pore openings of 0.45 μm are in 2-in-diameter disks with a grid printed on the surface for ease in counting colonies. In drinking water testing, a 100-ml sample is filtered through the membrane. A culture dish with a fitted cover is prepared by placing an absorbent pad in the bottom half of the dish and saturating it with 1.0–2.0 ml of M-Endo medium. Using the forceps, the membrane is removed from the filtration unit and placed on the pad in the culture dish. After incubation for 22–24 hr at 35° ± 0.5°C, the cover is removed from the dish and the membrane examined for growth. Typical coliform colonies are a pink to dark-red color with a green metallic surface sheen. Absence of bacterial growth on the membrane is a negative test. Occasionally, atypical colonies occur as light-colored growth without dark centers or a green sheen.

Presence–Absence Technique

The presence–absence (P–A) test can determine if coliform bacteria or *Escherichia coli* are present in a water sample without indicating the number of coliforms in a positive result. This test is intended for use in routine monitoring of drinking water immediately after treatment and in the distribution system pipe network. This test is based on the concept that the MCL for coliform bacteria is zero.

The drinking water to be tested is collected in a sterile bottle containing 15–30 mg of sodium thiosulfate, which is sufficient to neutralize and stop the disinfecting action of 10 mg/l of chlorine. The test procedure starts by vigorously shaking the sample bottle to suspend the bacteria and the particulate matter. The lid of the P–A culture bottle is aseptically removed to add 100 ml of the water sample and to add the contents of one packet of Colilert® reagent. The reagent contains bacterial nutrients and two special compounds (ONPG, O-Nitrophenyl-β-d-galactopyranoside, and MUG, 4-Methylumbelliferyl-β-d-glucuronide) that indicate the growth and presence of coliform bacteria and *Escherichia coli*. (Colilert is based on IDEXX's patented Defined Substrate Technology®.)

The culture bottle is incubated for 24 hr at 35 °C. Growth of coliform bacteria in the Colilert® uses the enzyme β-galactosidase to metabolize ONPG and change the color of the sample from clear to yellow. If a yellow color does not appear, the test is negative and no coliforms were present in the water sample. *Escherichia coli* uses the enzyme β-galactosidase to metabolize MUG and create fluorescence under 365-nanometer ultraviolet light. (Fluorescence is the emission of radiation as visible light resulting from the absorption of radiation from another source.) If the yellow-colored broth glows fluorescence with a bluish color, the test is positive for *E. coli*. Absence of a blue color is a negative test.

PROBLEMS

8.1 Define the meaning of the term *pathogen* and give the names of pathogen groups. What determines the kinds and concentrations of pathogens in wastewater? Define the meaning of fecal–oral route in the transmission of diseases.

8.2 Compare the latency, persistence, and infective dose of *Ascaris* and *Salmonella*.

8.3 Discuss the significance of the carrier condition in transmission of enteric diseases. What major waterborne diseases in the United States are commonly spread by carriers? How are these diseases amplified by animals?

8.4 Historically in the United States, the prevalent infectious diseases have been typhoid, cholera, and dysentery. How have these diseases been virtually eliminated? Currently, the prevalent infectious diseases are *giardiasis* and *cryptosporidiois*, causing diarrhea that can be life-threatening for persons with immunodeficiency syndrome. What actions are being taken to reduce the probability of waterborne transmission of these diseases? (Refer to Sections 8.1 and 11.24.)

8.5 What are the symptoms produced by the two pathogenic protozoa that are found worldwide? What is the common mode of *transmission* and in what environments is transmission most likely to occur?

8.6 Why can coliform bacteria be used as indicators of drinking water quality? Discuss the limitations of coliforms as an indicator. Why is a positive test for fecal coliforms in a public water supply considered more serious than a positive test for total coliforms?

8.7 What are the significant differences between *Escherichia coli* and *Escherichia coli* O157:H7?

8.8 What is the definition of a community water system? What is the difference between a nontransient-noncommunity water system and a transient-noncommunity system? The regulation of specific MCLs (maximum contaminant levels) depends on the kind of water system. (Refer to Section 8.2.) What kinds of contaminants are transient-noncommunity water systems required to comply with?

8.9 Outline the three phases in testing for enteric viruses in relatively unpolluted water. Since extraction and concentration capture only a portion of the viruses in a water sample, how is the precision of separation determined? When is testing for enteric viruses recommended?

8.10 In one statement, what is the general process in testing for *Giardia* cysts and *Cryptosporidium* oocysts? If the water sample is only 10 liters for testing natural stream water for *Cryptosporidium* oocysts, why is the accuracy for detection and enumeration of oocysts likely to be low? Why must laboratories conducting tests for *Cryptosporidium* oocysts be audited and approved for quality assurance?

8.11 Based on laboratory animal studies, a synthetic organic chemical has an estimated cancer risk of 1×10^{-6} after a lifetime consumption of 1 l/d of water containing 1.0 μg/l of chemical. If the MCL is 0.005 mg/l and consumption is 2 l/d, what is the calculated risk and the number of excess cases of cancer expected among persons exposed?

8.12 Discuss the health risk of finding lead in drinking water.

8.13 What is the health risk of excess nitrate ion in drinking water?

8.14 What are the most frequently detected VOCs in contaminated groundwater? What pesticide SOCs have been detected in drinking water?

8.15 List the regulated disinfection by-products. What is their source in drinking water?

8.16 Why are iron and manganese included in secondary standards for aesthetics in drinking water?

8.17 What are the objectives of the Clean Water Act?

8.18 What does the acronym NPDES refer to? Since inception, list several aspects of this permit program. How is the NPDES program monitored?

8.19 What are the technology-based standards for secondary (biological) treatment for all municipal wastewater treatment plants? (Refer to Sections 8.8 and 9.1.) When are water quality–based standards necessary for a wastewater discharge?

8.20 What are the major steps in performing the whole effluent toxicity (WET) test?

8.21 What is the procedure following a negative effluent biological toxicity test that indicates probable impairment of the receiving water?

8.22 The following data are from total solids and total volatile solids tests on a wastewater sample. Calculate the total and volatile solids concentrations in milligrams per liter.

Weight of empty dish = 68.942 g
Weight of dish plus dry solids = 69.049 g
Weight of dish plus ignited solids = 69.003 g
Volume of wastewater sample = 100 ml

8.23 Listed below are total solids and suspended-solids data on an industrial wastewater sample. Calculate the total and volatile solids, suspended solids, and dissolved solids.

Total solids data
Weight of empty dish = 85.337 g
Weight of dish plus dry solids = 85.490 g
Weight of dish plus ignited solids = 85.375 g
Volume of wastewater sample = 85 ml
Suspended-solids data
Weight of glass-fiber filter disk = 0.1400 g
Weight of disk plus dry solids = 0.1530 g
Weight of disk plus ignited solids = 0.1426 g
Volume of wastewater filtered = 200 ml

8.24 An unseeded BOD test is conducted on a polluted surface water sample by adding 100 ml to a 300-ml BOD bottle and filling with dilution water. The initial dissolved oxygen measured 8.2 mg/l, and the final concentration after 5 days of incubation at 20°C measured 2.9 mg/l. Calculate the BOD.

8.25 A BOD test was conducted on the unchlorinated effluent of a municipal treatment plant. The wastewater portion added to a 300-ml BOD bottle was 30 ml, and the dissolved oxygen values listed below were measured using a dissolved-oxygen probe. Plot a BOD-versus-time curve and determine the 5-day BOD value.

Time (days)	DO (mg/l)	Time (day)	DO (mg/l)
0	8.7	6.0	4.9
2.0	6.7	10.0	3.9
4.0	5.7	14.0	0.7

8.26 A BOD test was conducted on a raw domestic wastewater sample. The wastewater portion added to each 300-ml test bottle was 8.0 ml. The dissolved-oxygen values and incubation periods are listed below. Plot a BOD-versus-time curve and determine the 5-day BOD value.

Bottle Number	Initial DO (mg/l)	Incubation Period (days)	Final DO (mg/l)	DO Drop (mg/l)	Calculated BOD (mg/l)
1	8.4	0	8.4		
2	8.4	0	8.4		
3	8.4	1.0	6.2		
4	8.4	1.0	5.9		
5	8.4	2.0	5.2		
6	8.4	2.0	5.2		
7	8.4	3.0	4.4		
8	8.4	3.0	4.6		
9	8.4	5.0	3.8		
10	8.4	5.0	3.5		

8.27 A seeded BOD analysis was conducted on a food-processing wastewater sample. Ten-milliliter portions were used in preparing the 300-ml bottles to determine the dissolved-oxygen demand of the aged, settled wastewater seed. The seeded sample BOD bottles contained 2.7 ml of food-processing wastewater and 1.0 ml of seed wastewater. The results of this series of test bottles are listed below. Calculate the wastewater BOD values and plot a BOD–time curve. What is the 5-day BOD?

	Seed Tests		Sample Tests	
Time (days)	B_1 (mg/l)	B_2 (mg/l)	D_1 (mg/l)	D_2 (mg/l)
0	7.8	—	8.1	—
1.0	7.8	6.9	8.1	5.6
2.0	7.8	6.6	8.1	4.3
3.0	7.8	6.3	8.1	3.6
4.0	7.8	5.8	8.1	3.0
5.0	7.8	5.7	8.1	2.5
6.0	7.8	5.3	8.1	2.0
7.0	7.8	5.4	8.1	1.8

8.28 In laboratory testing for coliform bacteria, what techniques can be performed to enumerate total coliforms in treated wastewater effluent or drinking water? What test can be used to determine the presence of *Escherichia coli* in drinking water?

REFERENCES

[1] R. G. Feachem, D. J. Bradley, H. Garelick, and D. D. Mara, *Sanitation and Disease, Health Aspects of Excreta and Wastewater Management: World Bank Studies in Water Supply and Sanitation 3* (Chichester: Wiley, 1983).

[2] E. C. Lippy and S. C. Waltrip, "Waterborne Disease Outbreaks, 1946–1980: A Thirty-Five-Year Perspective," *J. Am. Water Works Assoc.* 76, no. 2 (1984): 60–67.

[3] *Waterborne Pathogens*, AWWA M48 (Denver, CO: Am. Water Works Assoc., 1999).

[4] A. C. Moore et al., "Waterborne Disease in the United States, 1991 and 1992," *J. Am. Water Works Assoc.* 86, no. 2 (1994): 87–99.

[5] M. H. Kramer et al., "Waterborne Disease: 1993 and 1994," *J. Am. Water Works Assoc.* 88, no. 3 (1996): 66–80.

[6] S. C. Edberg, M. J. Allen, and D. B. Smith, *Comparison of the Colilert Method and Standard Fecal Coliform Methods*, AWWA Research Foundation (Denver, CO: *J. Am. Water Works Assoc.,* 1994.)

[7] D. L. Swerdlow et al., "A Waterborne Outbreak in Missouri of *Escherichia coli* O157: H7 Associated with Bloody Diarrhea and Death," *Annals of Internal Medicine.* 117, no. 10 (1992): 812–819.

[8] F. W. Pontius and S. W. Clark, "Drinking Water Quality Standards, Regulations, and Goals," in *Water Quality & Treatment*, 5th ed. (New York: McGraw-Hill/Am. Water Works Assoc., 1999).

[9] M. J. Martin, J. L. Clancy, and E. W. Rice, "The Plain, Hard Truth About Pathogen Monitoring," *J. Am. Water Works Assoc.* 92, no. 9 (2000): 64–76.

[10] *Drinking Water and Health*, Vol. 1, National Academy of Sciences (Washington, DC: National Academy Press, 1977).

[11] *Drinking Water and Health*, Vol. 6, National Academy of Sciences (Washington, DC: National Academy Press, 1986).

[12] *Drinking Water and Health*, Vol. 3, National Academy of Sciences (Washington, DC: National Academy Press, 1980).

[13] D. F. Striffler, W. O. Young, and B. A. Bart, "The Prevention and Control of Dental Caries: Fluoridation," in *Dentistry, Dental Practice, and the Community* (Philadelphia: Saunders, 1983), 155–199.

[14] "Radioactivity in Drinking Water," Environmental Protection Agency, Office of Drinking Water, EPA-570/9-81-002 (January 1981).

[15] *The Clean Water Act, 25th Anniversary Edition* (Alexandria, VA: Water Environment Federation, 1997).

[16] *Water Quality Standards Handbook*, 2nd ed., Environmental Protection Agency, Office of Water, EPA-823-B-94-005a (1994).

[17] "Quality Criteria for Water," Environmental Protection Agency Office of Water Regulations and Standards, EPA-440/5-86-001(1986).

[18] D. S. Szumski, D. A. Barton, H. D. Putman, and R. C. Polta, "Evaluation of EPA Unionized Ammonia Toxicity Criteria," *J. Water Poll. Control Fed.* 54, no. 3 (1982): 281–291.

[19] *Short-Term Methods for Estimating the Chronic Toxicity of Effluents and Receiving Water to Freshwater Organisms*, 3rd ed., Environmental Protection Agency, Research and Development, EPA-600-4-91-002 (July 1994).

[20] *Evaluating Whole Effluent Toxicity Testing as an Indicator of Instream Biological Conditions* (Alexandria, VA: Water Environment Federation, Water Environment Research Foundation, 1999).

[21] "Methods of Measuring the Acute Toxicity of Effluents to Freshwater and Marine Organisms," 3rd ed., Environmental Protection Agency, Environmental Monitoring and Support Laboratory, EPA-600/4-85-013 (March 1985).

[22] *Standard Methods for the Examination of Water and Wastewater*. (Washington, DC: Am. Public Health Assoc., 1998).

[23] M. J. Hammer and M. J. Hammer, Jr., *Water and Wastewater Technology*, 5th ed. (Upper Saddle River, NJ: Prentice Hall, 2004).

CHAPTER 9

Systems for Treating Wastewater and Water

OBJECTIVES

The purpose of this chapter is to:

- Describe typical systems for wastewater and water treatment
- Summarize the roles of the different processes that comprise these treatment systems

This chapter provides an overview of wastewater and water treatment systems as an introduction to subsequent chapters that deal separately with physical, chemical, biological, and sludge treatment methods. These introductions to wastewater and water treatment processes also summarize subsequent chapters by drawing the unit processes into integrated treatment systems. *Note: Students should refer to Sections 9.1 and 9.3 while studying Chapters 10–13.*

WASTEWATER TREATMENT SYSTEMS

The purpose of municipal wastewater treatment is to prevent pollution of the receiving waters or to reclaim the water for reuse. Characteristics of a municipal wastewater depend to a considerable extent on the type of sewer collection system and industrial wastewaters entering the sewers. The degree of treatment required is determined by the beneficial uses of the receiving waters or, in the case of reclamation, the reuse application. Pollution of flowing waters and eutrophication of impounded waters are particularly troublesome in water use for water supplies and recreation.

9.1 PURPOSE OF WASTEWATER TREATMENT

Wastewaters from households and industries are collected in a sewer system and transported to the treatment plant. The treated effluent is commonly disposed of by dilution in rivers, lakes, and estuaries. Disposal to the ocean is through a submarine outfall sewer extending into deep water. Three common reuse applications are agricultural irrigation, urban landscape irrigation, and groundwater recharge.

Water quality criteria have been established by the Clean Water Act for receiving waters to define the degree of treatment required for disposal of treated wastewater by dilution to protect prescribed beneficial uses. Effluent standards prescribe the required quality of the discharge from each treatment plant. The minimum extent of processing is secondary treatment. Some cities and industries are required to install tertiary or advanced wastewater treatment processes for removal of pollutants that are resistant to conventional treatment, e.g., removal of phosphorus to retard eutrophication of receiving lakes. Stream classification documents, published by each state as required by federal law, categorize surface waters according to their most beneficial present or future use (i.e., for drinking water supplies, body-contact recreation, etc.). These publications also incorporate water quality standards that establish maximum allowable pollutant concentrations for a given watercourse under defined flow conditions.

Effluent standards under the National Pollutant Discharge Elimination System (NPDES) are used for regulatory purposes to achieve compliance with these water quality standards. Technology-based effluent limits define the minimum level of effluent quality that must be attained by secondary treatment of municipal wastewater without regard to the quality of the receiving waters.

Acceptable secondary effluent is defined in terms of biochemical oxygen demand, suspended solids, oil and grease, fecal coliform bacteria, and pH. The arithmetic mean of BOD and suspended-solids concentrations for effluent samples collected in a period of 30 consecutive days must not exceed 30 mg/l; and, during any 7-consecutive-day period, the average must not exceed 45 mg/l. Furthermore, removal efficiencies shall not be less than 85% (i.e., if an influent concentration is less than 200 mg/l, the effluent cannot exceed 15% of this value). For concentrations of oil and grease, the arithmetic mean must exceed neither 10 mg/l for any period of 30 days nor 20 mg/l for any period of 7 days. Where specified in a discharge permit, the geometric mean of fecal coliform counts for effluent samples collected in a period of 30 consecutive days must not exceed 200 per 100 ml; and the geometric mean of fecal coliform bacteria for any 7-consecutive-day period must not exceed 400 per 100 ml. The effluent values for pH shall remain within the limits 6.0–9.0. These effluent limits for biological treatment require a well-designed plant that is properly operated. Moderately loaded activated-sludge processes, biological towers, and two-stage trickling-filter plants are capable of achieving this degree of treatment. However, some designs that were popular in the past are not efficient enough, such as single-stage tricking-filter plants and high-rate activated-sludge systems. Stabilization ponds are often permitted to discharge suspended-solids concentrations of 50–70 mg/l, since these suspended solids are algae with little or no raw organic matter. Disinfection for control of bacterial populations may not be required if no threat to public health exists.

Refractory contaminants, both inorganic and organic materials, are pollutants resistant to, or totally unaffected by, conventional treatment processes. The mineral quality of wastewater depends largely on the character of the municipal water supply, but during water use numerous substances are added such as common salt (sodium chloride) and other dissolved solids. Phosphates, which occur in low concentrations in most natural waters, are increased during municipal use of water. Phosphorus and nitrogen removals in secondary treatment are only about 30%, depending on the types of processes and concentrations in the raw wastewater. Eutrophication of lakes and reservoirs, induced by disposal of treated wastewater, can be retarded by employing advanced treatment processes to remove phosphorus. Organic nitrogenous compounds decompose to ammonia and to a variable extent oxidize to nitrate in biological treatment. While nitrification of ammonia nitrogen is feasible by additional aeration of the wastewater, denitrification (conversion to nitrogen gas) requires special processes that are costly. Aside from disposal of treated wastewater, no ready solution exists for removing nutrients from runoff, either from natural land or cultivated fields.

A primary goal of the Clean Water Act is to prohibit the discharge of toxic pollutants in toxic quantities. Wastewaters from industries discharging to a municipal sewer system can contain toxic metal ions and hazardous organic chemicals. Toxic substances interfere with the operation of biological treatment processes, and many are refractory, passing through in the plant effluent, to contaminate the receiving water. Under their NPDES permit, municipalities are required to establish regulations for pretreatment of wastewaters from industries at the industrial sites to remove toxic substances. To ensure removal, industrial wastewaters entering the sewer system are monitored by the municipality, and the effluent from the municipal treatment plant is tested for biological toxicity (Section 8.9).

Water quality-based effluent limits, in contrast to technology-based effluent limits, are site-specific standards related to the water quality designation of the receiving water. These limits may apply to conventional pollutants, plant nutrients, microorganisms, toxic substances, or any other contaminant adversely affecting water quality. Under the Clean Water Act, states are required to survey and identify surface waters that do not meet specified water quality standards after application of technology-based and water quality-based limits. States are then required to establish the total maximum daily load (TMDL) of each pollutant to restore such waters to meet the quality standards and beneficial uses. Implementation of the TMDL Rule poses significant demands since the majority of impaired waters exceed their water quality criteria because of nonpoint sources, which are diffuse, transient, and highly variable runoff events from rural and urban lands. The common major causes for impairment are excess sediments, plant nutrients, and human pathogens and the lesser causes are low dissolved oxygen, toxic metals, pH, suspended solids, and pesticides.

9.2 SELECTION OF TREATMENT PROCESSES

Conventional wastewater treatment consists of preliminary processes (pumping, screening, and grit removal), primary settling to remove heavy solids and floatable materials, and secondary biological aeration to metabolize and flocculate colloidal and dissolved

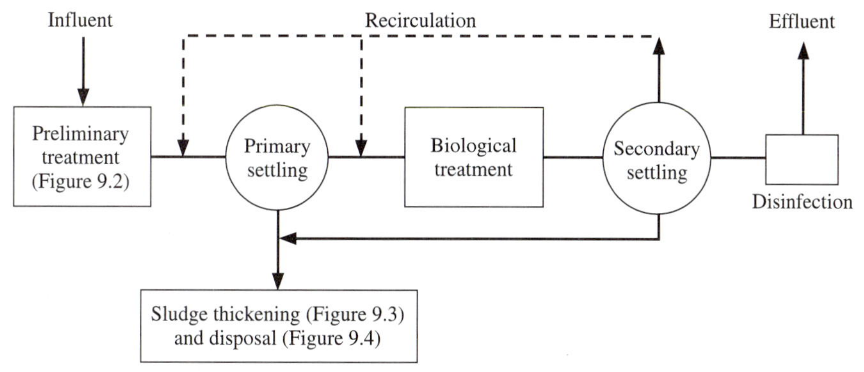

FIGURE 9.1 Schematic diagram of conventional wastewater treatment.

organics (Fig. 9.1). Waste sludge drawn from these unit operations is thickened and processed for ultimate disposal.

Preliminary Treatment Units

The following preliminary processes are used in municipal wastewater treatment: coarse screening (bar racks), medium screening, shredding of solids, flow measuring, pumping, grit removal, and preaeration. Although not common in pretreatment, flotation, flocculation, and chemical treatment are sometimes dictated by the industrial pollutants in the municipal wastewater. Flotation is used to remove fine suspensions, grease, and fats and is performed either in a separate unit or in a preaeration tank also used for grit removal. If adequate pretreatment is provided by petroleum industries and meat-processing plants, flotation units are not required at a municipal facility. Flocculation with or without chemical additions may be practiced on high-strength municipal wastewaters to provide increased primary removal and prevent excessive loads on the secondary treatment processes. Chlorination of raw wastewater is sometimes used for odor control and to improve settling characteristics of the solids.

The arrangement of preliminary treatment units varies depending on raw wastewater characteristics, subsequent treatment processes, and the preliminary steps employed. A few general rules always apply in arrangement of units. Screens are used to protect pumps and prevent solids from fouling grit-removal units and flumes. In small plants a Parshall flume is normally placed ahead of constant-speed lift pumps, but may be located after them in large plants or where variable-speed pumps are used. Grit removal should be placed ahead of the pumps when heavy loads are anticipated, although the grit chamber follows the lift pumps in most separate sanitary wastewater plants. Three possible arrangements for preliminary units are shown in Figure 9.2.

Primary Treatment Units

Primary treatment is sedimentation. In common usage, the term usually includes the preliminary treatment processes. Sedimentation of raw wastewater is usually practiced

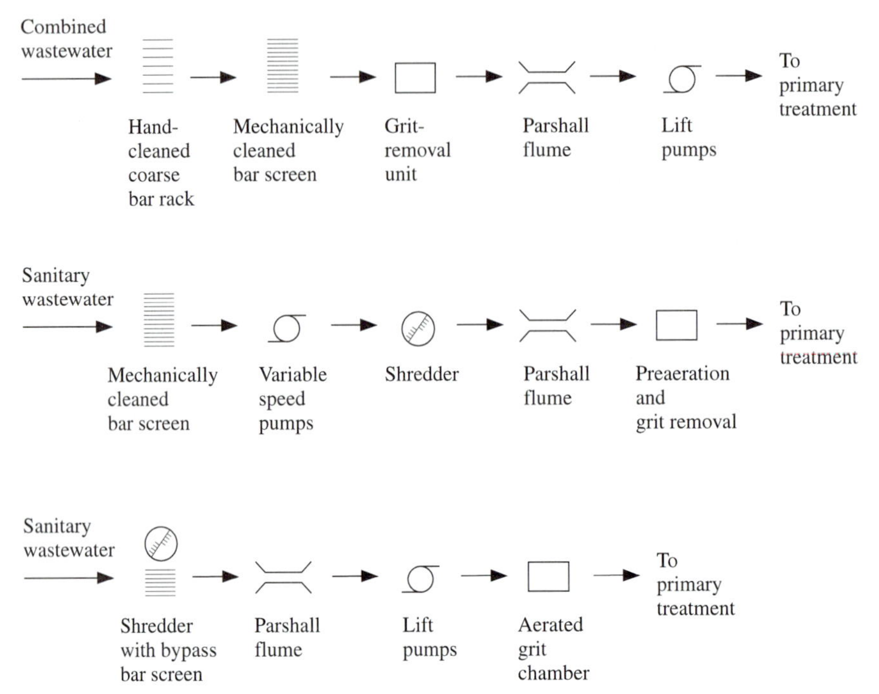

FIGURE 9.2 Possible arrangements of preliminary treatment units in municipal wastewater processes.

in all large municipal plants and must precede trickling (biological) filtration. Completely mixed activated-sludge processes can be used to treat unsettled raw wastewater; however, this method is generally used in smaller treatment plants because of the costs involved in sludge disposal and operation.

Secondary Treatment Units

Primary sedimentation removes 30%–50% of the suspended solids in raw municipal wastewater. Remaining organic matter is extracted in biological secondary treatment to the allowable effluent residual using activated-sludge processes or trickling filters. In the activated-sludge method, wastewater is fed continuously into an aerated tank, where microorganisms synthesize the organics. The resulting microbial floc (activated sludge) is settled from the aerated mixed liquor under quiescent conditions in a final clarifier and returned to the aeration tank (Fig. 9.1). The plant effluent is clear supernatant from secondary settling. Advantages of suspended-growth aeration are high-BOD removals, ability to treat high-strength wastewater, and adaptability for future use in plant conversion to advanced treatment. On the other hand, a high degree of operational control is needed, shock loads may upset the stability of the biological process, and sustained hydraulic or organic overloading results in process failure.

Trickling filters have stone or plastic media to support microbial films. These slime growths extract organics from the wastewater as it trickles over the surfaces.

Oxygen is supplied from air moving through voids in the media. Excessive biological growth washes out and is collected in the secondary clarifier. In northern climates, two-stage shallow trickling filters are needed to achieve efficient treatment. Biological towers (deep trickling filters) may be either single- or two-stage systems. Advantages of filtration are ease of operation and capacity to accept shock loads and overloading without causing complete failure.

Sludge Disposal

Primary sedimentation and secondary biological flocculation processes concentrate the waste organics into a volume of sludge significantly less than the quantity of wastewater treated. But disposal of the accumulated waste sludge is a major economic factor in wastewater treatment. The construction cost of a sludge processing facility is approximately one-third that of a treatment plant.

Flow schemes for withdrawal, holding, and thickening raw waste sludge from sedimentation tanks are illustrated in Figure 9.3. The settled solids from clarification of trickling-filter effluent are frequently returned to the plant head for removal with the

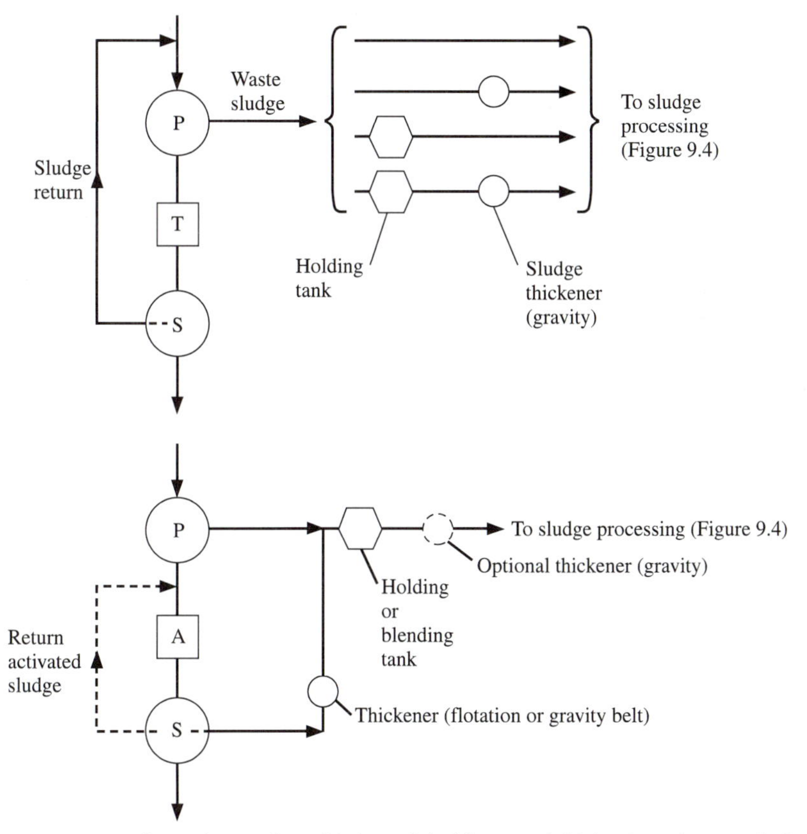

FIGURE 9.3 Flow schemes for withdrawal, holding, and thickening of waste sludge. P, primary settling; T, trickling filtration; A, activated-sludge aeration; S, secondary settling.

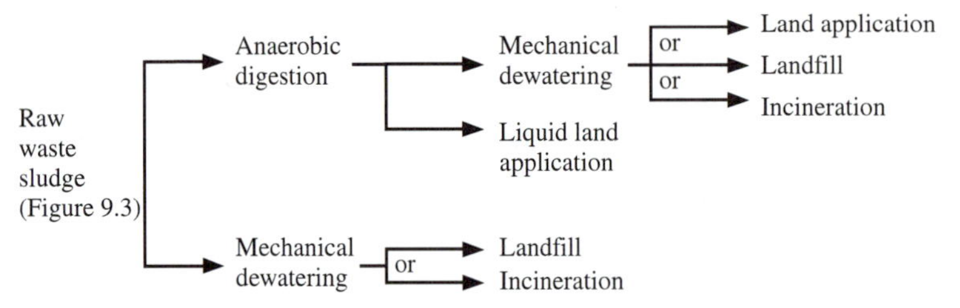

FIGURE 9.4 Common methods for processing and disposal of raw waste sludge.

primary sludge (top diagram, Fig. 9.3). Raw settlings may be stored in the primary tank bottom until processed or pumped into a holding tank for storage. The withdrawn sludge may be concentrated in a gravity thickener prior to processing. In the bottom diagram of Figure 9.3, waste-activated sludge is mixed with primary residue after withdrawal. A holding tank is commonly used in this system arrangement, along with a thickener. The waste activated is preferably thickened separately or the combined sludges may be thickened prior to processing.

Alternatives for processing and disposal of raw sludge are shown in Figure 9.4. Common methods are anaerobic digestion and mechanical dewatering by belt filter pressing or centrifugation. Conventional methods of disposal are application as a fertilizer/soil conditioner on agricultural land and landfill in a dedicated disposal site or codisposal with municipal solid waste. In small plants, anaerobically digested sludge may be applied on agricultural land by surface spreading or subsurface injection through chisel plows. Dewatered raw sludge must be covered by soil the same day if applied on the surface or buried. Therefore, the common methods are codisposal in a municipal solid waste landfill or disposal in a dedicated disposal site. Incineration, although significantly higher in cost, is sometimes the only acceptable method of sludge disposal in large urban areas. Waste-activated sludge from a plant without primary sedimentation is usually stabilized by aerobic digestion and spread on the land surface as a liquid or mechanically dewatered for application on agricultural land.

All possible sludge disposal processes for a municipal treatment plant must be given careful consideration. The method selected should be the most economical process, if it is best, with due regard to environmental conditions. Attention must be given to such factors as trucking sludge through residential areas, future use of landfill areas, ground-water pollution, air pollution, other potential public health hazards, and aesthetics.

Advanced Wastewater Treatment

Phosphorus removal is practiced in processing wastewaters that are discharged to receiving waters subject to eutrophication, including lakes, reservoirs, and slow-moving rivers acting as impoundments. Since nitrogen removal is much more difficult and costly, it is rarely performed, although nitrification to reduce the ammonia content is sometimes done to control the oxygen demand and toxicity of the plant effluent. Water reclamation is achieved in varying degrees by many plants depending on the local environmental concerns; however, only a few large-scale plants are reclaiming water to near original quality. Advanced wastewater treatment systems are discussed in Chapter 14.

Size of Municipality

Operational management and control and the necessity for sludge handling dictate the selection of wastewater treatment processes for small communities. Methods that do not require sludge disposal (stabilization ponds) or only occasional sludge withdrawals (extended aeration) are preferable for small villages and subdivisions. Towns large enough to employ a part-time operator frequently use systems that require more operational control and maintenance (e.g., oxidation ditch plants). Cities with trickling-filter and activated-sludge treatment plants employ several workers.

Listed in Table 9.1 are the common types of wastewater treatment plants built in municipalities of various sizes. Many existing plants do not conform to the listing in Table 9.1. Some are no longer popular and others were selected on the basis of unique local conditions.

Layout of Treatment Plants

The flow diagram for a two-stage trickling-filter plant is given in Figure 9.5, and an aerial view is shown in Figure 9.6. The sequence of wastewater treatment units is the influent pumping station with a mechanically cleaned bar screen, primary clarification tanks, first-stage trickling filter, intermediate clarifier, second-stage filter, and final clarifiers. The stone-media biological filters are covered with aluminum domes to prevent ice formation and to retain an adequate wastewater temperature during the winter. For optimum removal of organic matter, trickling-filter recirculation and humus return are controlled to maintain a constant hydraulic loading rate. Humus washed off the filter media collects in the intermediate and final clarifiers. Underflow from these

TABLE 9.1 Common Types of Wastewater Treatment Plants for Municipalities of Various Sizes

Subdivisions, schools, etc.:
Extended aeration (factory built)
Stabilization ponds
Towns (populations less than 2000):
Extended aeration
Contact stabilization (field erected, factory built)
Oxidation ditch
Stabilization ponds
Small cities (2000–8000 population):
Oxidation ditch
Completely mixed activated sludge without primary
Primary plus trickling filters
Primary plus rotating biological contractors
Cities (population greater than 10,000):
Primary plus trickling filters
Primary plus activated sludge

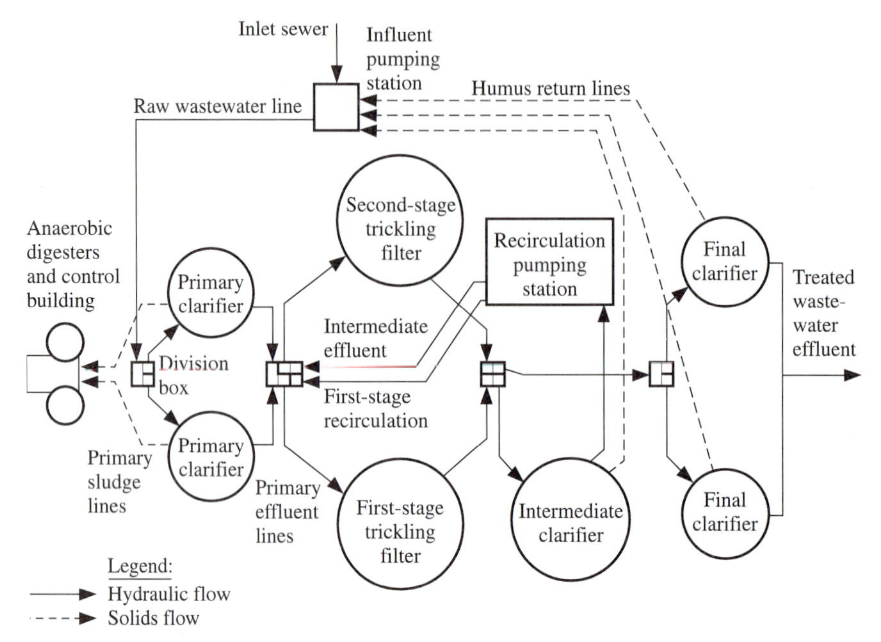

FIGURE 9.5 Plant layout for a two-stage trickling-filter plant, Sidney, NE. Average design flow of 1.0 mgd with 210 mg/l of BOD and 230 mg/l of suspended solids and maximum wet-weather flow of 2.4 mgd. *Source:* Courtesy of HDR Engineering, Inc.

FIGURE 9.6 Aerial view of two-stage filter plant shown in Figure 9.5. *Source*: Courtesy of HDR Engineering, Inc.

clarifiers to the influent pumping station is controlled to return more flow at night when raw wastewater inflow is reduced. Primary sludge containing raw organics and filter humus is withdrawn from the bottom of the clarifiers and pumped into the first-stage anaerobic digester for stabilization. The second-stage digester is for storage and thickening of the digested sludge and collection of methane gas under a floating-dome cover for use as a fuel. The digested sludge is dewatered to a cake on drying beds. The liquid sludge can also be composted with wood chips.

The sequence of wastewater treatment units in the activated-sludge plant diagrammed in Figure 9.7 is as follows: headworks consisting of mechanically cleaned bar screens, screenings compactor and vortex grit remover; influent pumping station; primary clarifiers; aeration tanks followed by final clarifiers; return activated-sludge pumping station; chlorination tank; treated water storage ponds; and dechlorination prior to discharge to the river or for irrigation. Cascade aeration increases the dissolved-oxygen level in the effluent to the river, and effluent pumping conveys the effluent to irrigation. After dewatering in a compactor, screenings and grit are hauled to a landfill. Fine-bubble diffusers in the aeration tanks provide high-efficiency oxygen transfer for suspended-growth biological treatment. Primary sludge is pumped through a grinder and blended with waste-activated sludge after it is thickened by dissolved-air flotation. The mixed sludges are stabilized by anaerobic digestion and then pumped into a storage tank. After concentrating from approximately 2% to 7% on gravity belt thickeners, the biosolids are applied to agricultural land as a fertilizer/soil conditioner.

WATER TREATMENT SYSTEMS

The purpose of a municipal water supply system is to provide potable water that is chemically and microbiologically safe for human consumption and has adequate quality for industrial users. For domestic uses, water should be free from unpleasant tastes and odors, and improved for human health (i.e., by fluoridation). Since the quality of public supplies is based primarily on drinking water standards, the special temperature and other needs of some industries are not always met by public supplies. Boiler feedwater, water used in food processing, and process water in the manufacture of textiles and paper have special quality tolerances that may require additional treatment of the municipal water at the industrial site.

9.3 WATER SOURCES

Typical water sources for municipal supplies are deep wells, shallow wells, rivers, natural lakes, and impounding reservoirs. No two sources of supply are alike, and the same origin may produce water of varying quality at different times. Water treatment processes selected must consider the raw water quality and differences in quality for each particular water source.

Municipal water quality factors of safety, temperature, appearance, taste and odor, and chemical balance are most easily and frequently satisfied by a deep-well source. Furthermore, the treatment processes employed are simplest because of the relatively uniform quality of such origin. Excessive concentrations of iron, manganese, and hardness habitually exist in well waters. Some deep-well supplies may contain hydrogen sulfide; others may have excessive concentrations of chlorides, sulfates, and carbonate. Hydrogen sulfide can be removed by aeration or other oxidation processes, but the prevalent sodium and potassium salts of anions bicarbonate, chloride, and sulfate are

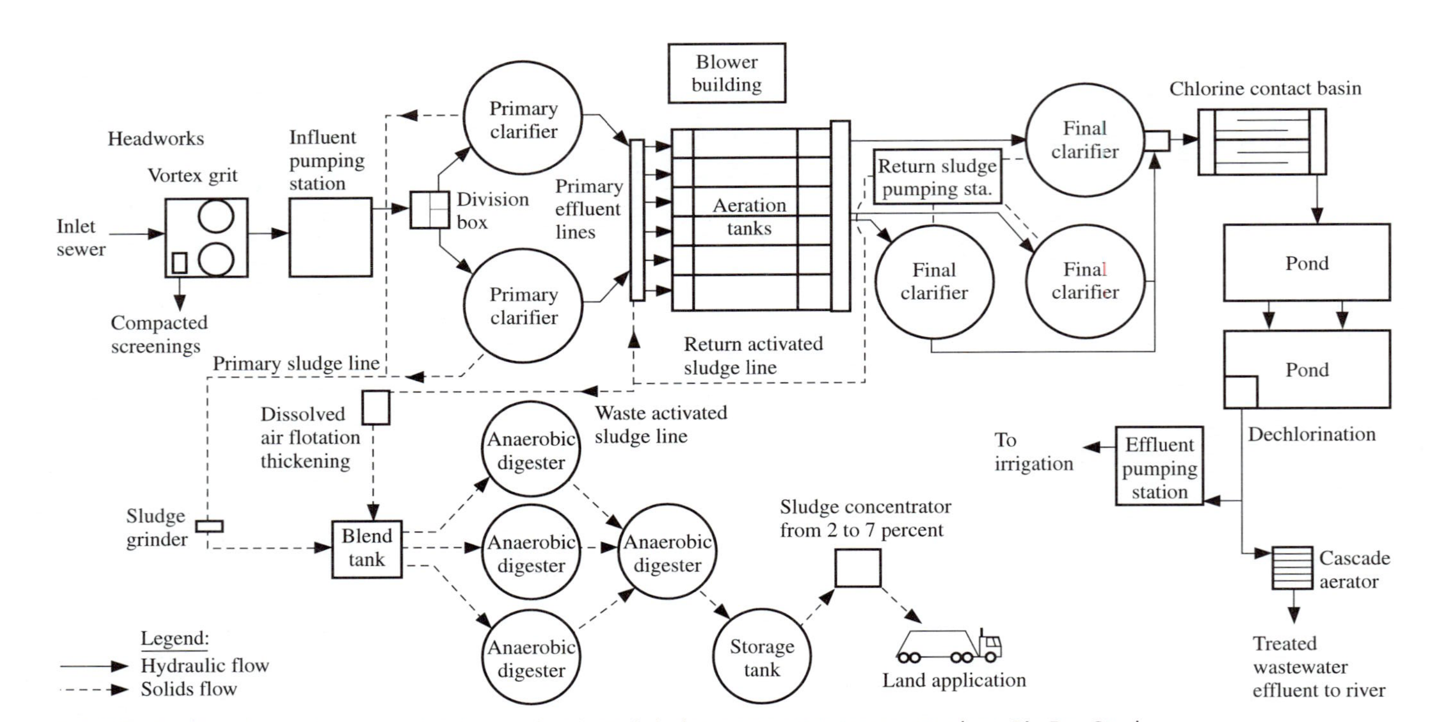

FIGURE 9.7 Plant layout for a completely mixed activated-sludge wastewater treatment plant, Big Dry Creek Water Reclamation Facility, Westminster, CO. *Source*: Courtesy of HDR Engineering, Inc.

refractory inorganics. Calcium and magnesium can be removed by precipitation softening. Excessive concentration of fluoride, responsible for mottled enamel of the teeth, has in rare cases caused a municipality to abandon a well supply source. The mineral constituents of a potential groundwater source should be carefully examined in selecting a deep-well water supply.

Shallow wells recharged by a nearby surface watercourse may have quality characteristics similar to deep wells or may relate more closely to the watercourse quality. A sand aquifer adjacent to a river may act as an effective filter for removal of organic matter and as a heat exchanger for leveling out temperature changes of the recharge water seeping into it. A case in point is the municipal water supply for Lincoln, Nebraska, located in the Platte River Valley near Ashland. The wells are located in a sand aquifer adjacent to the river channel. Chemical composition of the well water and river water is nearly identical. River water temperature varies from 80 °F in the summer to 33 °F in the winter, whereas the well water temperature remains within 3° of 56 °F year round. None of the bacterial or organic pollution existing in the river has been identified in the well water. Another shallower well field located adjacent to the Platte River in a coarser sand and gravel aquifer withdraws water of significantly poorer quality after peak pumping periods. The well water temperature increases to nearly 70 °F in late summer and bacterial counts in the raw water rise. Clogging of the well screens, attributed in part to bacterial slime growths, has occurred in some wells. Groundwater under the direct influence of surface water, which is defined as containing algae, insects, or other macroorganisms, or experiencing significant and relatively rapid shifts in water characteristics, requires disinfection under the Surface Water Treatment Rule. Therefore, disinfection for removal of pathogens requires chemical coagulation and granular-media filtration, followed by chlorination.

To predict water quality from shallow wells, careful studies of the aquifer and nature of recharge water are necessary. Quantity of the well water depends upon such things as aquifer permeability, well spacing and depth, seasonal changes in river flow, and pumping rates. The best way to evaluate these variables is by full-scale field pumping tests or by checking similar existing well fields.

Pollution and eutrophication are major concerns in selecting and treating surface water supplies. Where these are highly contaminated and difficult to treat, municipalities seek either groundwater supplies or alternate less polluted surface sources within a feasible pumping distance. However, in many regions of the United States, adequate groundwater resources are not available and high-quality surface supplies are not within economic reach. Although the majority of municipal water supply systems are from underground origins, only about one-fourth of the nation's population is served by these sources. In general, larger cities are dependent on surface supplies. Treatment of polluted waters is perhaps the greatest challenge in water supply engineering, and the importance of pollution control is realized most vividly when one ponders the problems of providing safe, palatable water from contaminated sources.

Water quality in rivers depends on the character of the watershed; pollution caused by municipalities, industries, and agricultural practices; river development such as dams; the season of the year; and climatic conditions. During spring runoff and other periods of high flows, river water may be muddy and high in taste- and odor-producing compounds. During drought flows, pollutants are often present in higher concentrations and odorous conditions can occur. During late summer, algal blooms frequently create problems. River temperature variations depend on latitude and the location of the stream headwaters. In northern states, river water is warm in the summer and cold in the winter. Water quality conditions vary from one stream to another and, in addition, each has its own peculiar characteristics. River water quality is usually deteriorating if the watershed is under development.

Water supplies from rivers normally require the most extensive treatment facilities and the greatest operational flexibility of any source. A river water treatment plant must be capable of handling day-to-day variations and the anticipated quality changes likely to occur within its useful life.

The quality of water in a lake or reservoir depends on the physical, chemical, and biological characteristics (limnology) of the body. Size, depth, climate, watershed, degree of eutrophication, and other factors influence the nature of an impoundment. The relatively quiescent retention of river water produces marked changes in quality brought about by self-purification. Common benefits derived from impounding river water include reduction of turbidity, coliform counts and usually color, and elimination of the day-to-day variations in quality.

Lakes and reservoirs are subject to seasonal changes that are particularly noticeable in eutrophic waters. In summer and winter during stratification, the hypolimnion (unmixed bottom water layer) in a eutrophic lake may contain dissolved iron and manganese and taste and odor compounds. The decrease of oxygen by bacterial activity on the bottom results in dissolution of iron and manganese and production of hydrogen sulfide and other metabolic intermediates. Algal blooms frequently occur in the epilimnion (warmer mixed surface water layer) of fertile lakes in early spring and late summer. Heavy growths of algae, particularly certain species of blue-greens, produce difficult-to-remove tastes and odors. Normally, the best quality of water is found near middepth, below the epilimnion and above the bottom. The intake for a fertile lake or reservoir should preferably be built so that water can be drawn from selected depths. Lakes that stratify experience spring and fall overturns, which occur when the temperature profile is nearly uniform and wind action stirs the lake from top to bottom. If the bottom water has become anaerobic during stratification, water drawn from all depths during the overturn will contain taste- and odor-producing compounds. The limnology of a lake or reservoir and the present and future level of eutrophication should be thoroughly understood before beginning the design of intake structures or treatment systems.

The concept of multiple barriers is important for water treatment systems. The use of several barriers is effective in preventing pathogens and contaminants from reaching water customers. The first barrier is source protection. For a surface water source, land and water use on the watershed should be controlled and wastewater contamination from human and animal sources minimized. For a groundwater source, a wellhead protection program should be adopted to prevent surface contamination from entering the aquifer and proper design and construction of wells should be implemented to prevent surface water from entering the bore holes. The most important barrier for a surface supply is water treatment, commonly consisting of chemical coagulation, sedimentation, and filtration. Treatment facilities for well water vary depending on the quality of the groundwater, which may be softening, iron and manganese removal, or no treatment. The third barrier is disinfection of the treated water by chlorination in a contact tank and an adequate protective chlorine residual maintained in the distribution system. As the final barrier, the distribution system mains and storage reservoirs are properly maintained and operated at a high enough pressure to prevent infiltration of groundwater from entering the system. Enforcement of a cross-connection-control program is essential to prevent backflow of contaminated water through service connections from industries, hospitals, and other institutions handling hazardous substances.

9.4 SELECTION OF WATER TREATMENT PROCESSES

Current pretreatment processes in municipal water treatment are screening, presedimentation or desilting, chemical addition, and aeration. Screening is practiced in pretreating surface waters. Presedimentation is regularly used to remove suspended matter from river water. Chemical treatment, in advance of in-plant coagulation, is most frequently applied to improve presedimentation, to pretreat hard-to-remove substances such as taste and odor compounds and color, and to reduce high bacterial concentrations. Conventional chemicals used with presedimentation are polymers and alum. Aeration is customarily the first step in treatment for the removal of iron and manganese from well waters and is a standard way to separate dissolved gases such as hydrogen sulfide and carbon dioxide.

Treatment processes used in water plants depend on the raw water source and quality of finished water desired. The specific chemicals selected for treatment are based on their effectiveness to perform the desired reaction and cost. For example, activated carbon, chlorine, chlorine dioxide, and potassium permanganate are all used for taste and odor control. Excess chlorination, although least expensive, can create undesired trihalomethanes; activated carbon is the most effective chemical. In surface water treatment plants, equipment for feeding two or three taste- and odor-removal chemicals is usually provided, so the operator can select the most effective and economic chemical applications. There is no fixed rule for color removal applicable to all waters. Alum coagulation with adequate pretreatment and applying oxidizing chemicals or activated carbon may provide satisfactory removal. On the other hand, a more expensive coagulant might prove to be more effective and reduce overall chemical costs (i.e., copperas or ferric salts can be substituted for alum in the coagulation process).

Perhaps the most important consideration in designing water treatment processes is to provide flexibility. The operator should have the means to change the point of application of certain chemicals. For example, chlorine feedlines are normally provided for pre-, intermediate-, and postchlorination. Multiple chemical feeders and storage tanks should be supplied so that various chemicals can be employed in the treatment process. Degradation of the raw water quality, or changes in costs of chemicals, may dictate a change in the kind of coagulant or auxiliary chemicals used in coagulation. In the case of surface water treatment plants it is desirable to provide space for the construction of additional treatment facilities. The EPA can add new contaminants to the list of regulated chemicals, lower the maximum contaminant level (MCL) of currently regulated contaminants, add new or enhanced treatment techniques, or promulgate other regulations in the Safe

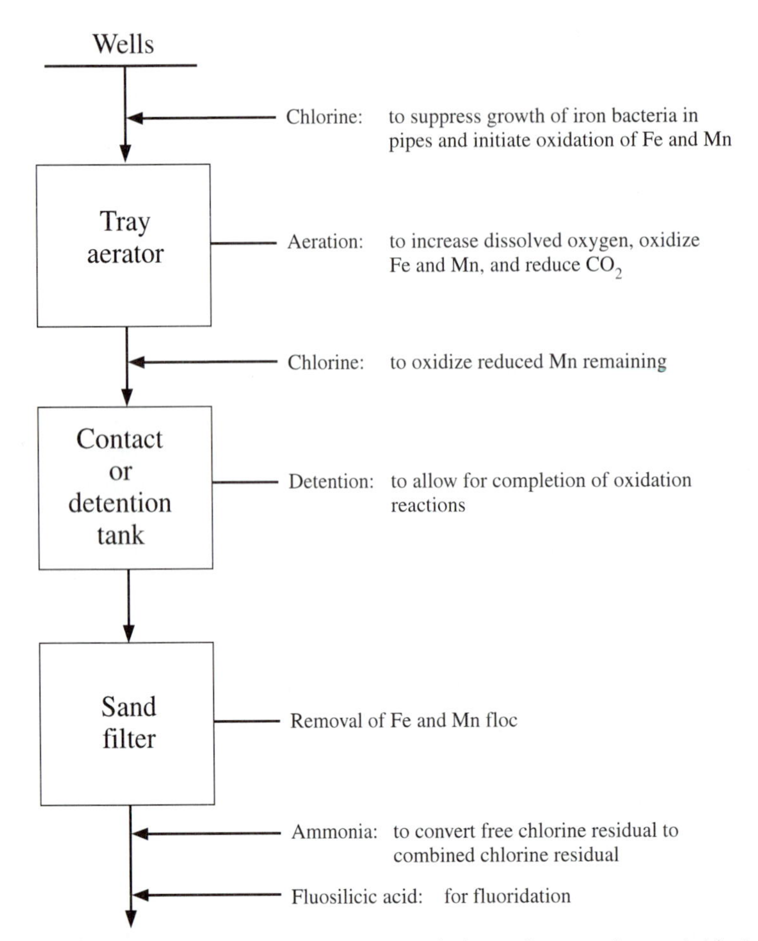

FIGURE 9.8 Iron and manganese removal plant using aeration and chlorine for oxidation.

Drinking Water Act. The flow in rivers may change due to construction of dams, channel improvements, or upstream water use. The quality of water changes due to human alteration and occupation of the watershed. Concentrations of pollutants from disposal of municipal and industrial wastewaters and agricultural land runoff may increase. Lakes can become more eutrophic.

Iron and manganese removal is the primary purpose for treatment of groundwater by the scheme shown in Figure 9.8. Prechlorination at the well site is to control growth of iron bacteria in pipes to the plant and in the tray aerator. After aeration to increase dissolved oxygen, chlorine is added as a catalyst for oxidation of dissolved iron and manganese. The metal oxides, which are not heavy enough to settle out of suspension, are removed by sand filtration. In order to maintain a chlorine residual in the pipeline to the municipality, ammonia is added to convert a portion of the free residual to combined chlorine residual. Fluosilicic acid is added for fluoridation. The finished water quality is excellent all year round because of the stable character and nearly constant temperature of the well supply.

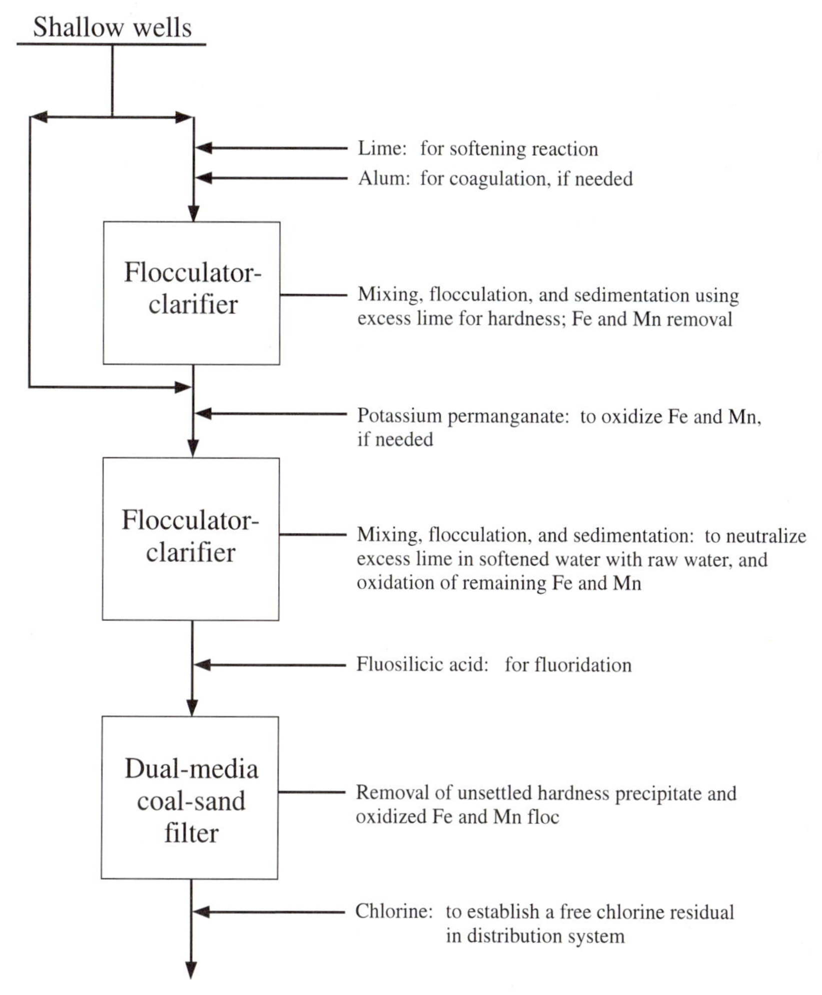

FIGURE 9.9 Plant using split treatment for partial softening and iron and manganese removal.

The treatment scheme of groundwater shown in Figure 9.9 is partial softening and iron and manganese removal. The flocculator-clarifiers, which provide chemical mixing, flocculation, and sedimentation in a compartmented tank, are operated in two stages. The first stage is for excess-lime softening of a portion of the raw water, and the second stage is for blending of the softened water containing excess lime with the bypassed flow. By this split treatment, the addition of carbon dioxide is unnecessary to neutralize the excess lime and stabilize the water. The softened water after sedimentation flows by gravity through dual-media coal–sand filters to remove suspended solids. Potassium permanganate can be added to oxidize reduced iron and manganese passing through the softening process; however, this is rarely needed. The treated water is chlorinated, stored in a large clear well, and rechlorinated as necessary to establish a free residual in the water pumped into the distribution system.

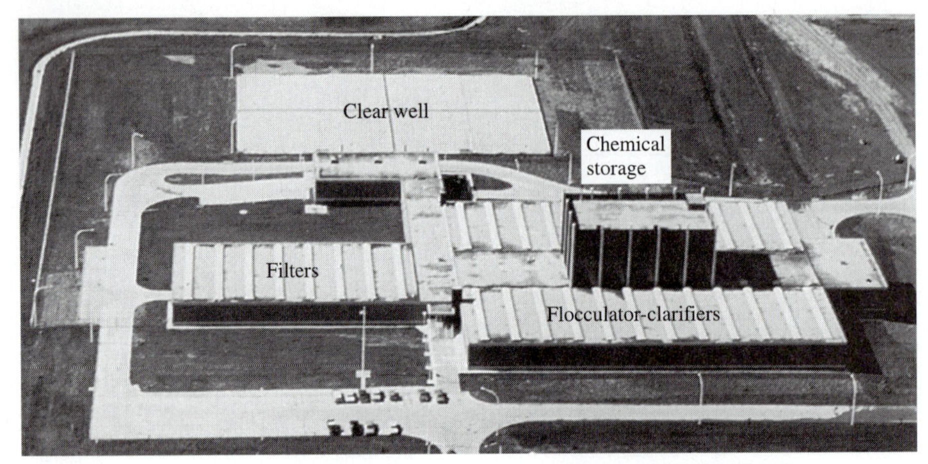

FIGURE 9.10 Aerial view of a water treatment plant. The flow diagram for this water utility is given in Figure 9.9. *Source:* Courtesy of Metropolitan Utilities District, Omaha, NE.

An aerial view of this treatment plant is shown in Figure 9.10, with the major treatment units labeled.

The treatment plant with the flow diagram in Figure 9.11 was originally constructed for processing lake water by alum coagulation, flocculation, sedimentation, and sand filtration. Since then the lake water source has become increasingly eutrophic, and additional facilities for taste and odor control have been provided. Activated carbon, chlorine dioxide, and various auxiliary chemicals for improved chemical treatment are now available to the operator during critical periods of poorer raw water quality. During most of the year the finished water is very palatable, but tastes and odors cannot be completely removed during spring and fall lake overturns.

The river water–processing scheme shown in Figure 9.12 is a very complex, flexible treatment system for a turbid and polluted stream with highly variable quality. The plant has a complex of uncovered and covered mixing and settling basins with a variety of chemicals that can be added at several points during water processing. The general treatment scheme consists of plain sedimentation for desilting; mixing followed by sedimentation, using coagulants as necessary; split treatment for partial softening and coagulation in flocculator-clarifiers; blending of the split flows followed by sedimentation; dual-media filtration; and chemical additions for chlorine residual, pH adjustment, and control of scaling. Operation of this plant is varied from day to day and from season to season depending on the raw water quality. In general, the most troublesome time of the year is during spring runoff. Little is known about the degree to which refractory inorganic or organic substances can be transported through a complex treatment system. A river water treatment plant should have process depth to prevent any possibility of short-circuiting in pollution emergencies.

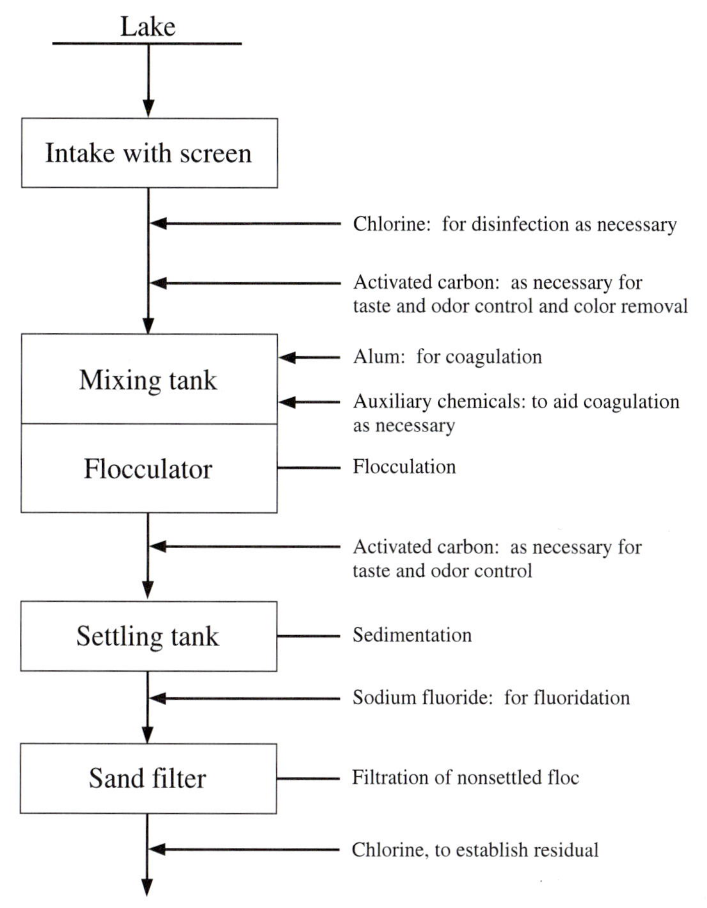

FIGURE 9.11 Chemical coagulation and partial-softening treatment plant with provisions for handling high turbidity, tastes and odors, and color.

9.5 WATER-PROCESSING SLUDGES

Chemical residues from water treatment historically were discharged to surface watercourses without treatment. Water quality regulations now require processing of these wastes to minimize environmental degradation. The method of eliminating residues is unique to each water utility because of the individuality of each treatment scheme and differing characteristics of the waste slurries generated. The two primary wastes are sludge from the settling basins following chemical coagulation, or precipitation softening, and wash water from backwashing the filters. These residues, containing matter removed from the raw water and chemicals added during processing, are relatively nonputrescible and high in mineral content. Figure 9.13 is a generalized scheme of alternatives for disposing of water-processing wastes. Filter wash water is a very dilute waste discharged for only a few minutes once or twice a day for each filter bed. Therefore, a holding tank is necessary for

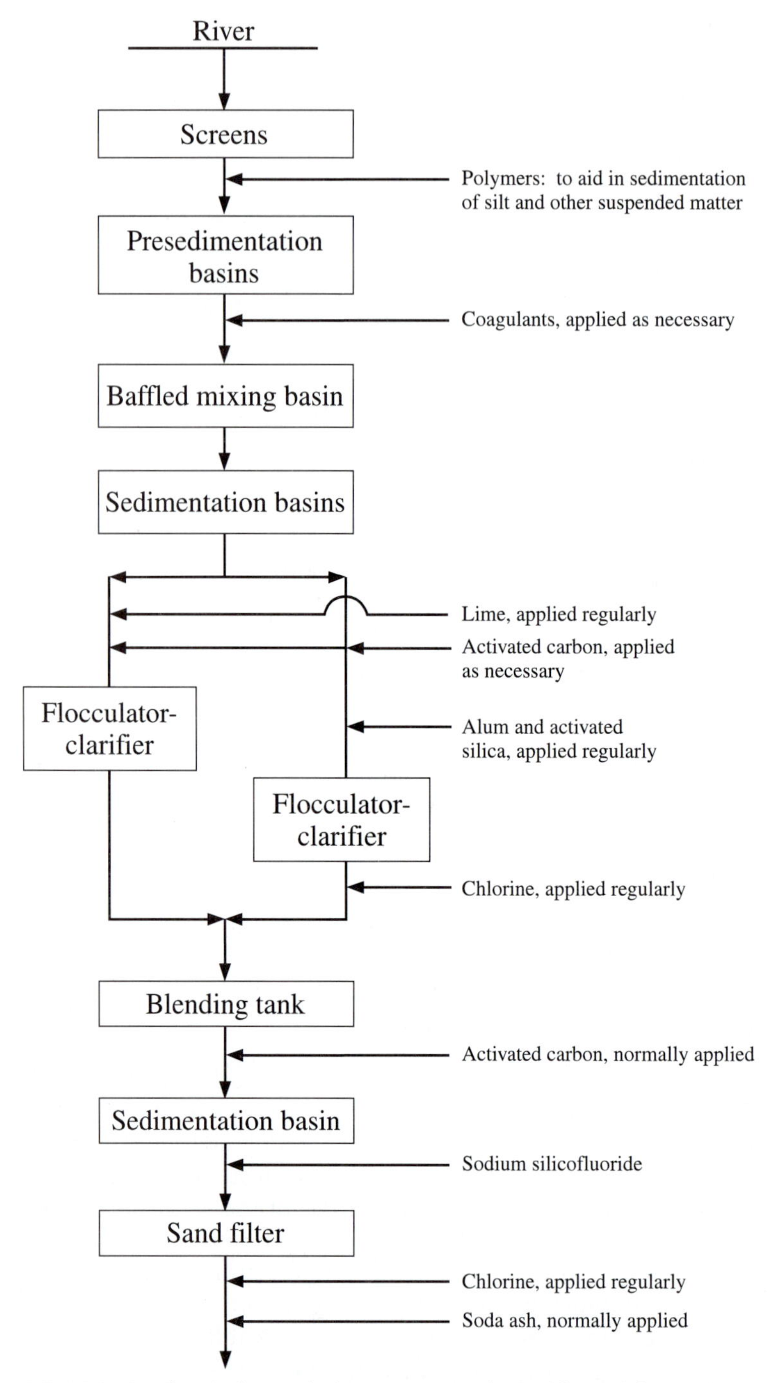

FIGURE 9.12 Chemical coagulation treatment plant with special provisions for taste and odor control.

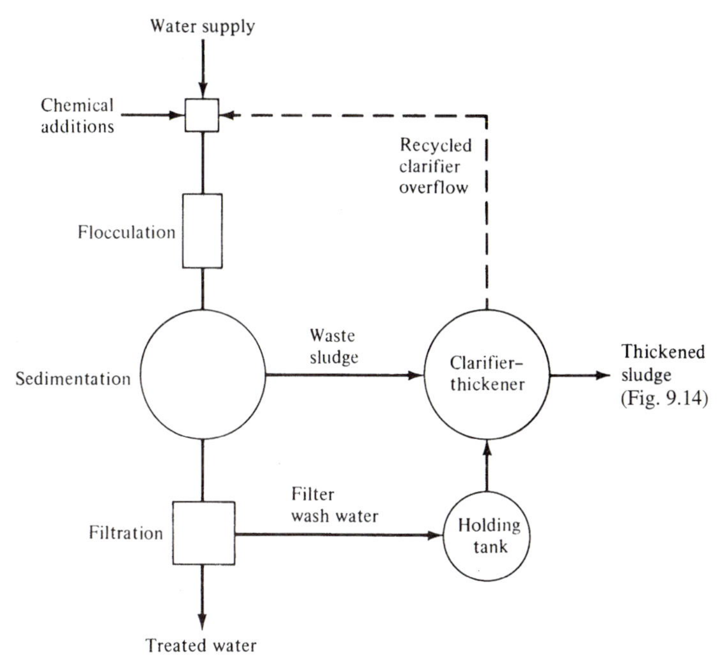

FIGURE 9.13 General flow scheme for withdrawal and gravity thickening of water-processing wastes.

flow equalization (i.e., to accept the intermittent peak discharges and release a relatively uniform effluent flow). Backwash water is settled in a clarifier-thickener with the overflow recycled to the plant influent; the underflow is withdrawn for further processing. Waste sludge from settling basins may also be concentrated further in a clarifier-thickener.

Disposal alternatives for gravity-thickened precipitates are diagrammed in Figure 9.14. In some municipalities, slurries can be drained into a sanitary sewer for processing with the municipal wastewater. Air drying is accomplished either in lagoons or on sand-drying beds. Ponding is a popular method for dewatering, thickening, and temporary storage of waste solids. Where suitable land area is available, this technique

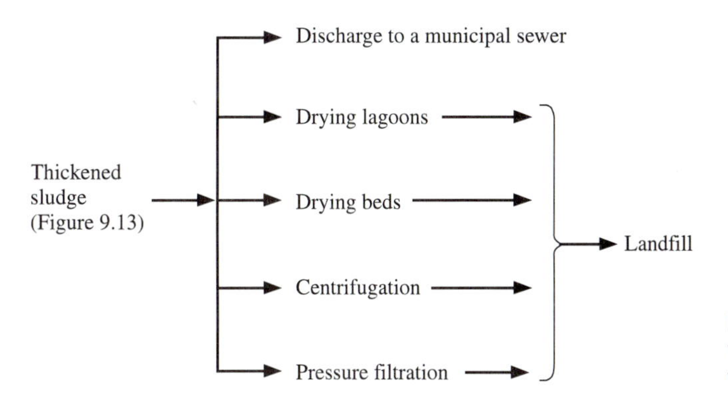

FIGURE 9.14 Common methods for dewatering and disposal of water-processing sludges.

is both inexpensive and efficient when compared to other methods. Drying beds are more costly and applicable only to small water plants. Centrifugation and pressure filtration are the two most successful mechanical dewatering processes. Residuals from air drying and concentrated cake from dewatering can be disposed of by codisposal in a municipal solid-waste landfill, burial in a dedicated landfill, or spread on agricultural land if beneficial for the soil.

CHAPTER 10

Physical Treatment Processes

OBJECTIVES

The purpose of this chapter is to:

- Describe the purposes and operation of physical processes
- Discuss the conditions under which different types of physical processes are appropriate

Physical treatment methods are used in both water treatment and wastewater processing. Except for preliminary steps, most physical processes are associated directly with chemical and biological operations. In water treatment, granular-media filtration (a physical method) must be preceded by chemical coagulation. In wastewater processing, the physical procedures of mixing and sedimentation in activated sludge are related directly to the biology of the system.

FLOW-MEASURING DEVICES

Measurement of flow is essential for operation, process control, and record keeping of water and wastewater treatment plants.

10.1 MEASUREMENT OF WATER FLOW

The flow of water through pipes under pressure can be measured by mechanical or differential head meters, such as a venturi meter, flow nozzle, or orifice meter. The drop in piezometric head between the undisturbed flow and the constriction in a differential head meter is a function of the flowrate. A venturi meter (Fig. 10.1), although more expensive than a nozzle or orifice meter, is preferred because of its lower head loss.

Piping changes upstream from a venturi meter produce nonuniform flow, causing inaccuracies in metering. In general, a straight length of pipe equal to 10–20 pipe diameters is recommended to minimize error from flow disturbances created by pipe fittings.

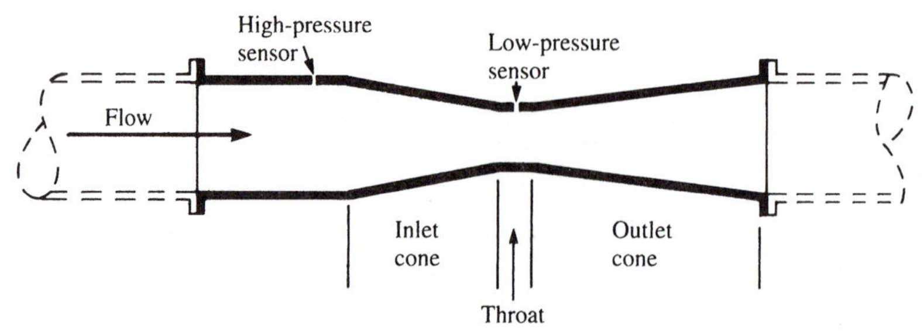

FIGURE 10.1 Venturi meter for measuring flow in a pipeline by differential pressure between inlet and throat.

Flow in a venturi meter can be calculated using the equation

$$Q = CA_2\left[2g\left(\frac{P_1 - P_2}{w}\right)\right]^{0.5} \tag{10.1}$$

where

Q = flow, cfs

C = discharge coefficient, commonly in the range of 0.90–0.98

A_2 = cross-sectional area of throat, ft^2

g = acceleration of gravity, ft/s^2

$P_1 - P_2$ = differential pressure, psf

w = specific weight of water, pcf

10.2 MEASUREMENT OF WASTEWATER FLOW

The flow of wastewater through an open channel can be measured by a weir or a venturi type flume, such as the Parshall flume [1] (Fig. 10.2). Parshall flumes have the advantages of lower head loss than a weir and smooth hydraulic flow preventing deposition of solids.

Under free-flowing (unsubmerged) conditions, the Parshall flume is a critical-depth meter that establishes a mathematical relationship between the stage h and discharge Q. For a flume with a throat width of at least 1 ft but less than 8 ft, flow can be calculated as

$$Q = 4Wh^{1.522W^{0.026}} \tag{10.2}$$

where

Q = flow, cfs

W = throat width, ft

h = upper head, ft

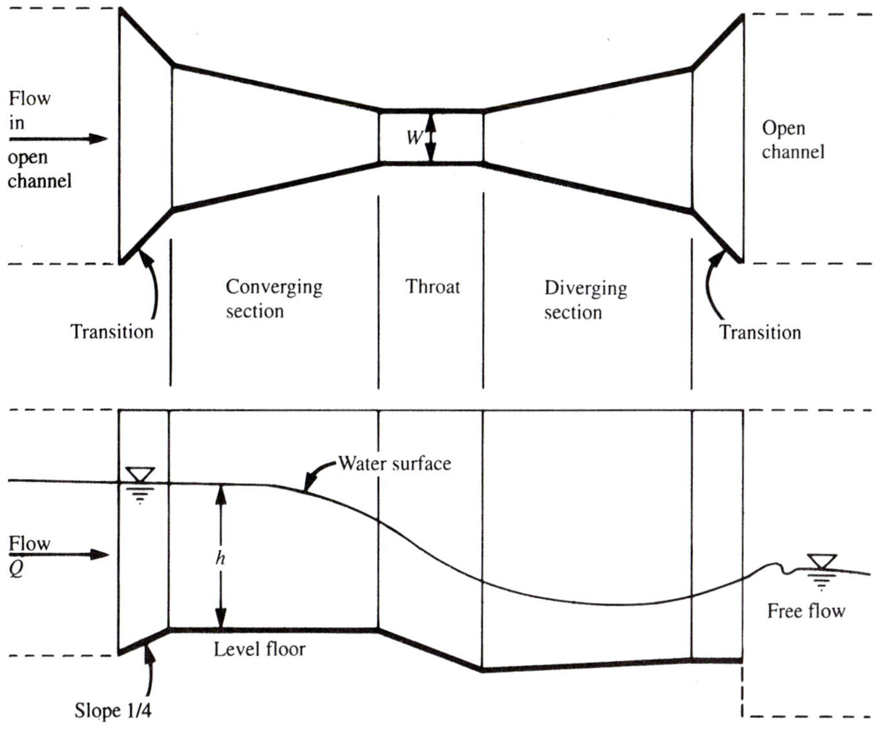

FIGURE 10.2 Parshall flume for measuring flow in an open channel by measuring the free-flowing upper head *h*.

Example 10.1

What is the calculated water flow through a venturi meter with $5^1/_{16}$-in. throat diameter when the measured pressure differential is 160 in. H_2O? Assume $C = 0.95$ and a water temperature of 50° F.

Solution: Substituting in Eq. (10.1), we obtain

$$Q = 0.95 \times \frac{\pi(0.422)^2}{4}\left(64.4\frac{160 \times 62.4}{12 \times 62.4}\right)^{0.5}$$

$$= 3.9 \text{ cfs} = 1750 \text{ gpm}$$

Example 10.2

What is the calculated wastewater flow through a Parshall flume with a throat width of 2.0 ft at the maximum free-flow head of 2.5 ft?

Solution: By Eq. (10.2),

$$Q = 4 \times 2.0 \times 2.5^{1.522 \times 2.0^{0.026}}$$

$$= 33.4 \text{ cfs} = 21.6 \text{ mgd}$$

SCREENING DEVICES

River water and wastewater in sewers frequently contain suspended and floating debris varying in size from logs to small rags. These solids can clog and damage pumps or impede the hydraulic flow in open channels and pipes. Screening is the first step in treating water containing large solids.

10.3 WATER-INTAKE SCREENS

River-water intakes are commonly located in a protected area along the shore to minimize collection of floating debris. Lake water is withdrawn below the surface to preclude interference from floating materials.

Coarse screens of vertical steel bars having openings of 1–3 in. are employed to exclude large materials. The clear openings should have sufficient total area so that the velocity through them is less than 3 fps to prevent pulling through stringy solids. These screens are available with mechanical rakes to clear accumulated material from the bars. A coarse screen can be installed ahead of a finer one used to remove leaves, twigs, small fish, and so on. The traveling screen shown in Figure 10.3 serves this purpose. Trays with sections constructed of wire mesh or slotted metal plates generally have $^3/_8$-in. openings. As water passes through the upstream side of the screen, the solids are retained and elevated by the trays. As the trays rise into the head enclosure, the solids are removed by means of water sprays. The operation of a traveling screen is intermittent, being controlled by the water head differential across it resulting from clogging.

10.4 SCREENS IN WASTEWATER TREATMENT

Coarse screens (bar racks) constructed of steel bars with clear openings not exceeding 2.5 in. are normally used to protect wastewater lift pumps. (Centrifugal pumps used for wastewater works, in sizes larger than 100 gpm, are capable of passing spheres at least 3 in. in diameter.) To facilitate cleaning, the bars are usually set in a channel inclined $22°$–$45°$ to the horizontal.

Mechanically cleaned medium screens (Fig. 10.4) with clear bar openings of $^5/_8$–$1^3/_4$ in. are commonly used instead of a manually cleaned coarse screen. The maximum velocity through the openings should not exceed 2.5 fps.

Collected screenings are generally hauled away for disposal by land burial or incineration, although they can be shredded in a grinder and returned to the wastewater flow for removal in a treatment plant.

Fine screens with openings as small as $^1/_{32}$ in. have been used in wastewater treatment, but because of their high installation and operation cost, they are rarely employed in handling municipal wastewater. Fine screens are used to take out suspended and settleable solids in pretreating certain industrial waste streams. Typical applications are the removal of coarse solids from cannery and packing-house wastewaters. Self-cleaning filter screens with 1-mm openings are used prior to spray irrigation and membrane filtration.

FIGURE 10.3 Traveling water-intake screen.
Source: Courtesy of U.S. Filter/Envirex.

10.5 SHREDDING DEVICES

A comminutor or grinder cuts solids that pass through a bar screen to about $^{1}/_{4}$–$^{3}/_{8}$ in. in size. Comminutors are installed directly in the wastewater flow channel and provided with a bypass so that the channel containing the unit can be isolated and drained for machine maintenance. In small waste treatment plants, a comminutor may be installed without being preceded by a medium screen. The bypass channel around a comminutor contains a hand-cleaned medium screen.

HYDRAULIC CHARACTERISTICS OF REACTORS

Chemical and biological reactions are performed in tanks often referred to as reactors in process engineering. The common mixed reactor in water treatment is for chemical flocculation, whereas in wastewater treatment it is for biological activated sludge. Sedimentation basins for both water and wastewater treatment are used for the physical process of solid–liquid separation. The following sections discuss the

FIGURE 10.4 Mechanically cleaned bar screen for wastewater treatment.
Source: Courtesy of U.S. Filter/Envirex.

hydraulic characteristics of complete mixing, plug flow, and plug flow with longitudinal dispersion.

10.6 RESIDENCE TIME DISTRIBUTION

The hydraulic character of a process reactor is defined by the residence time distribution of individual particles of the liquid flowing through the tank. Since routes through the reactor differ in travel times, the effluent age distribution of a nonideal reactor extends from less than to greater than the theoretical detention time, which is defined as

$$t_R = \frac{V}{Q} \tag{10.3}$$

where

t_R = theoretical mean residence time

V = volume of the reactor

Q = rate of flow through the reactor

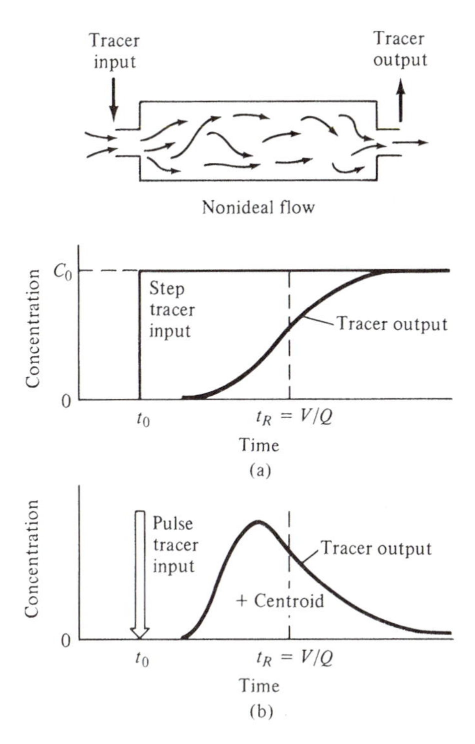

FIGURE 10.5 Output tracer distribution curves for flow through a nonideal reactor in response to (a) a continuous tracer input; (b) a pulse tracer input.

Evaluation of an actual process tank is performed by introducing a tracer to the influent during steady-state flow and measuring the concentration in the output over an extended period of time. The general shapes of residence time distributions for a dispersed plug-flow reactor are illustrated in Figure 10.5. If the tracer input is suddenly initiated and then held at a constant application rate, the tracer output curve of concentration in the effluent slowly rises and approaches the influent tracer concentration C_0 at a time interval greater than the mean residence time t_R. Applying a pulse input of dye solution results in an effluent tracer concentration curve as shown in Figure 10.5(b). The centroid of the curve is located to the left of time t_R as a result of flow dispersion with backmixing and short-circuiting of liquid in the tank because of stagnant pockets.

10.7 IDEAL REACTORS

Common ideal reactors are the plug flow reactor, completely mixed reactors, and completely mixed vessels in series. The residence time distributions of these units can be expressed in mathematical equations. Although real process tanks never precisely follow these flow patterns, many designs approximate ideal conditions with negligible error. Laboratory analyses of biological and chemical processes are often conducted using small tanks that are considered ideal reactors.

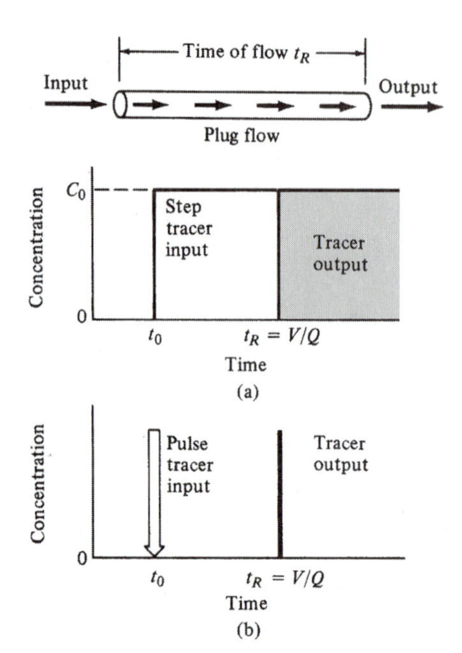

FIGURE 10.6 Output residence time distributions for an ideal plug-flow reactor in response to (a) a continuous input; (b) a pulse input.

Plug-Flow Reactor

The ideal plug-flow reactor is a tube with the particles of liquid entering and discharging in the same sequence. The residence time of each particle is equal to the mean residence time t_R, as illustrated in Figure 10.6. If a tracer with concentration C_0 is applied to the input, at time t_R the output contains the tracer at the same concentration. A pulse input is observed as a pulse in the output after time t_R. Actual treatment units that approach plug flow are long rectangular tanks used as chlorine contact chambers and in the conventional activated-sludge process.

Completely Mixed Reactor

The contents of a completely mixed reactor are uniformly and continuously redistributed. Particles entering are dispersed immediately throughout the tank and leave in proportion to their concentration in the mixing liquid. Figure 10.7(a) is the residence time distribution for a step tracer input. The rate of accumulation of tracer in the reactor, after start of the continuous feed, is equal to the input minus output, which is expressed in the following mass balance equation:

$$V\frac{dC}{dt} = QC_0 - QC \tag{10.4}$$

Rearranging and integrating between the limits of 0 and C and 0 and t gives

$$\int_0^C \frac{dC}{C_0 - C} = \frac{Q}{V}\int_0^t dt \tag{10.5}$$

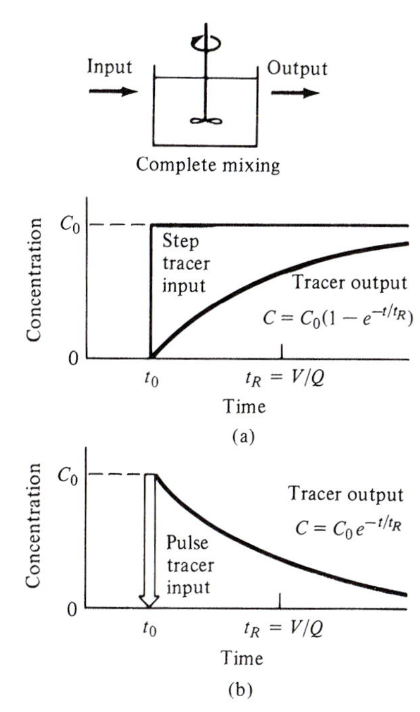

FIGURE 10.7 Output residence time distributions for an ideal completely mixed reactor in response to (a) a continuous input; (b) a pulse input.

which, when solved, yields

$$C = C_0(1 - e^{-t/t_R}) \tag{10.6}$$

Figure 10.7(b) is the residence time distribution for a pulse input of tracer in an amount such that the initial concentration after immediate mixing throughout the tank is C_0. The formula for calculating the concentration of tracer in the effluent with passage of time under steady-state flow is

$$C = C_0 e^{-t/t_R} \tag{10.7}$$

Treatment processes that closely simulate this ideal reactor are chemical mixing tanks and completely mixed activated-sludge basins. Studies of wastewater treatability and the evaluation of biological kinetics are commonly performed in laboratory containers that are completely mixed reactors.

Completely Mixed Reactors in Series

The residence time distributions for flow through a series of equal-sized ideal completely mixed reactors are plotted in Figure 10.8. These output curves for tanks in series are the response to a pulse tracer input to the first tank at time t_0. Particles entering each reactor are dispersed immediately and leave in proportion to their concentration in the mixing liquid. For the last tank in a series of n, the residence time distribution can be calculated using the generalized equation

$$C_n = C_0 \frac{(t/t_R)^{n-1}}{(n - 1)!} e^{-t/t_R} \tag{10.8}$$

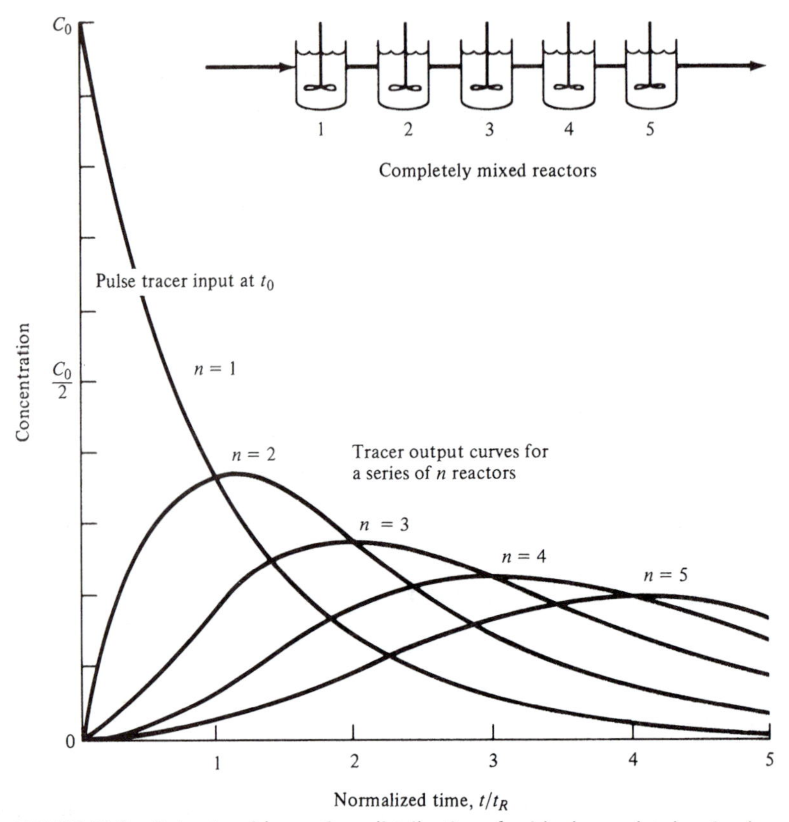

FIGURE 10.8 Output residence time distributions for ideal completely mixed reactors in series in response to a pulse tracer input.

where

C_n = effluent tracer concentration from reactor n

C_0 = influent tracer concentration in the first reactor

n = number of reactors in series

t = time elapsed since tracer input to the first reactor

t_R = theoretical mean residence time of n reactors

Substituting $n = 1$ in this equation yields Eq. (10.7).

The model of ideal reactors in series is representative in water treatment of long rectangular flocculation tanks separated into a series of compartments by baffles. In wastewater processing, compartmented aeration tanks for high-purity-oxygen activated sludge are commonly three or more reactors in series. Several completely mixed reactors in series are frequently used in laboratory studies to simulate dispersed plug flow through a rectangular aeration basin. Observe the similarity in the shapes of the curves in Figure 10.5(b) and the distribution diagrams for an n of 4 or 5 in Figure 10.8.

10.8 DISPERSED PLUG FLOW

Longitudinal flow passing through a tank has variations in flow velocities that result in axial intermixing and backmixing. The turbulence of this nonideal flow is in the range between ideal plug flow and ideal mixed flow. Dispersed plug flow, however, does not account for stagnant pockets and short-circuiting of the liquid in the tank.

The dispersion number for a particular reactor is defined as

$$d = \frac{D}{uL} \tag{10.9}$$

where

d = reactor dispersion number, dimensionless

D = longitudinal dispersion coefficient, ft^2/min

u = mean velocity of flow, fpm

L = length of the reactor, ft

Figure 10.9 illustrates tracer response curves to a pulse input for dispersed plug flow. The dispersion number for negligible mixing approaches zero, and hence plug flow. With greater intermixing, the dispersion number increases with a completely mixed reactor, represented by a value of infinity.

The spread of a tracer response curve, since it represents a continuous distribution, may be measured by calculating the variance σ^2 of the distribution about its mean. For large extents of dispersion, which are common in tracer studies of actual processing tanks, the residence time curves are often skewed. This is evidenced by an elongated curve where the concentration of tracer gradually approaches zero. For this case, the relationship between variance and the dispersion number is given by [2]

$$\sigma_\theta^2 = \frac{\sigma^2}{\bar{t}^2} = 2\frac{D}{uL} - 2\left(\frac{D}{uL}\right)^2 (1 - e^{-uL/D}) \tag{10.10}$$

where

σ_θ^2 = normalized variance, dimensionless

σ^2 = variance, min^2

$\bar{t}$ = mean residence time (time to the centroid of the distribution), min

The variance and dispersion number for an actual reactor are calculated from the effluent concentration-time data from a pulse tracer input. After dye is injected at the inlet of the tank, output samples are collected at time intervals from $^1/_2$ min to several minutes, with the greatest frequency during discharge of the dye peak. The shape of the residence time distribution is determined by graphing the concentration versus time. If

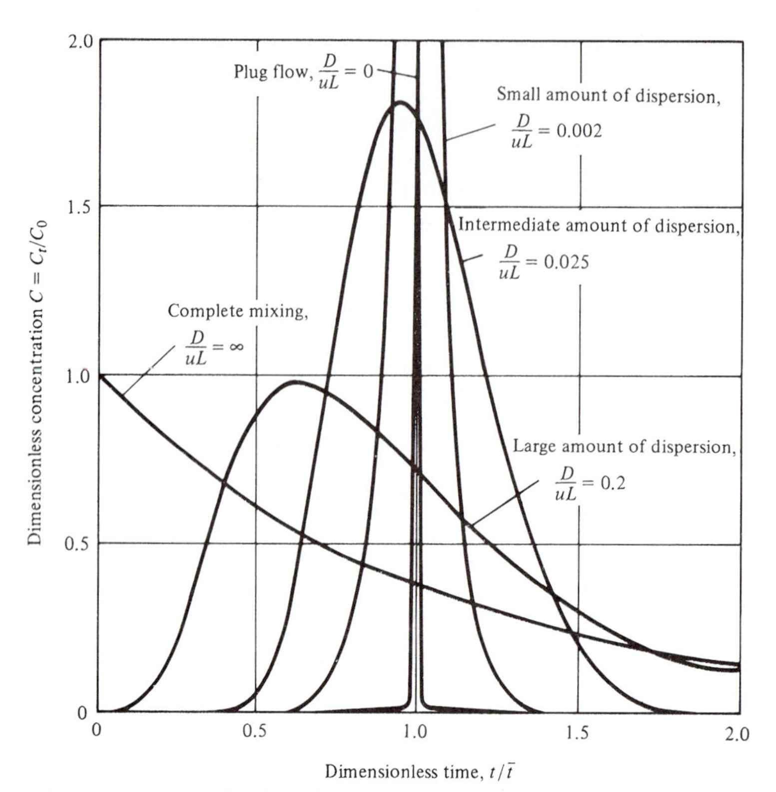

FIGURE 10.9 Normalized residence time distributions in response to a pulse tracer input and corresponding dispersion numbers for various degrees of longitudinal dispersion resulting from intermixing and backmixing of the flow through the tank.

desired, a normalized curve can be plotted by dividing the concentration values by the average of the measurements and corresponding times by the mean time. Variance of a curve is calculated from a finite number of measurements i at equal time intervals using the equation

$$\sigma^2 = \frac{\Sigma\, t_i^2 C_i}{\Sigma\, C_i} - \bar{t}^2 \tag{10.11}$$

where

$$\bar{t} = \frac{\Sigma\, t_i C_i}{\Sigma\, C_i}$$

From the second term of this equation, the time to the centroid of the distribution $\bar{t}$ is equal to the sum of time-concentration values divided by the sum of the concentrations. After calculation of the variance σ^2, the dispersion number D/uL can be computed using Eq. (10.10). Another parameter used to define a residence time distribution is the ratio of time for passage of 90% of the tracer to the time for passage of 10%.

Example 10.3

A completely mixed tank for blending chemicals into water is tested for effectiveness of hydraulic dispersion [3]. A pulse of tracer, equivalent to a concentration of 10 μg/l C_0 in the tank volume, was injected in the influent, followed by analyses of effluent samples for tracer concentrations for a period of three detention times. The experimental data are listed in the first two columns of Table 10.1. The theoretical detention time t_R of the tank is 60 s. Plot a residence time distribution for the test measurements and compare it to the theoretical curve.

TABLE 10.1 Experimental Data and Calculated Values for the Output Residence Time Distribution for the Completely Mixed Reactor in Example 10.3

Experimental Values				Theoretical Values, from Eq. (10.7)	
Measured		Calculated			
t (s)	C (μg/l)	t/t_R	C/C_0	t/t_R	C/C_0
10	4.5	0.17	0.45	0.17	0.85
20	7.6	0.33	0.76	0.33	0.72
40	6.5	0.67	0.65	0.67	0.51
60	5.6	1.00	0.56	1.00	0.37
80	3.0	1.33	0.30	1.33	0.26
100	2.1	1.67	0.21	1.67	0.19
120	1.8	2.00	0.18	2.00	0.14
140	1.2	2.33	0.12	2.33	0.10
160	0.7	2.67	0.07	2.67	0.07
180	0.4	3.00	0.04	3.00	0.05

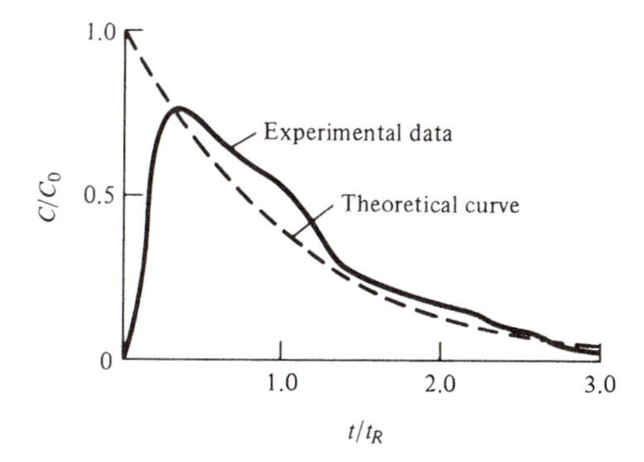

FIGURE 10.10 Plot of residence time distribution curves from data in Table 10.1 for Example 10.3.

Solution: The normalized time t/t_R and tracer concentrations C/C_0 are calculated by dividing the experimental t and C values by 60 s and 10 μg/l, respectively. The theoretical values are computed using Eq. (10.7). Plots of both the measured and theoretical curves are shown in Figure 10.10.

Example 10.4

Dispersed plug flow through a long narrow chlorination tank was analyzed by injecting a pulse input of a dye tracer in the influent. The tracer response curve is drawn in Figure 10.11, and the measured t and C values are listed in the first two columns of Table 10.2. The theoretical detention time at the test flow rate was 50 min. Calculate the dispersion number and the time ratios of t_{50}/t_R and t_{90}/t_{10}.

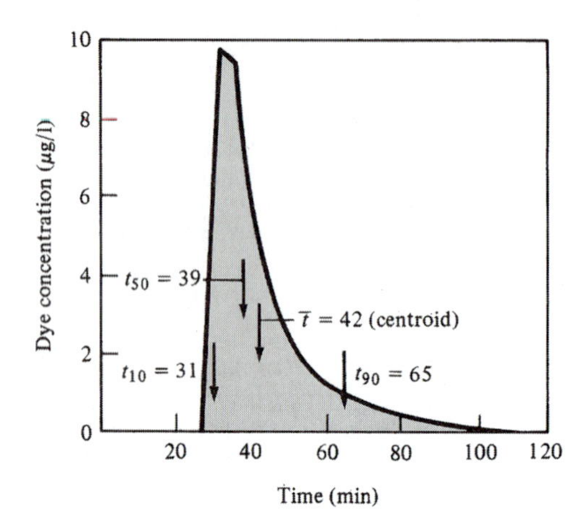

FIGURE 10.11 Tracer response curve from injecting a pulse input of dye into the influent of a long narrow chlorination tank.

TABLE 10.2 Experimental Data and Calculated Values for Hydraulic Analysis of the Long Narrow Chlorination Tank in Example 10.4

Measured Values		Calculated Values	
t (min)	C (μg/l)	tC	t^2C
25	0	0	0
30	4.0	120	3,600
35	9.7	340	11,900
40	5.8	232	9,300
45	3.6	162	7,300
50	2.3	115	5,800
55	1.9	105	5,700
60	1.4	84	5,000
65	1.1	72	4,600
70	0.8	56	3,900
75	0.6	45	3,400
80	0.4	32	2,600
85	0.3	26	2,200
90	0.2	18	1,700
95	0.2	19	1,800
100	0.1	10	1,000
Total	32.4	1440	69,800

Solution: Calculated values for $\Sigma\, tC$ and $\Sigma\, t^2C$ are given in Table 10.2. Using Eq. (10.11),

$$\bar{t} = \frac{\Sigma\, t_i C_i}{\Sigma\, C_i} = \frac{1440}{32.4} = 44.4 \text{ min}$$

$$\sigma^2 = \frac{69{,}800}{32.4} - \left(\frac{1440}{32.4}\right)^2 = 179$$

The normalized variance from Eq. (10.10) is

$$\sigma_\theta^2 = \frac{179}{(44.4)^2} = 0.091$$

and the reactor dispersion number D/uL, from a trial and solution of Eq. (10.10), equals 0.048.

From the values given in Figure 10.11 and $t_R = 50$ min,

$$t_{50}/t_R = 39/50 = 0.78$$

and

$$t_{90}/t_{10} = 65/31 = 2.1$$

MIXING AND FLOCCULATION

Chemical reactors in water treatment and biological aeration basins are mixed to produce the desired reactions and to keep the solids produced by the process in suspension. Rapid-mixing tanks are used to blend chemicals into the water. The gentle mixing in flocculation basins is to promote agglomeration of particles into large floc that can be removed from suspension by subsequent sedimentation.

10.9 RAPID MIXING

A rapid mixer provides dispersion of chemicals into water so that blending occurs in 10–30 s. Most common is mechanical mixing using a vertical-shaft impeller in a tank with stator baffles, as illustrated in Figure 10.12. The stators reduce vortexing about the

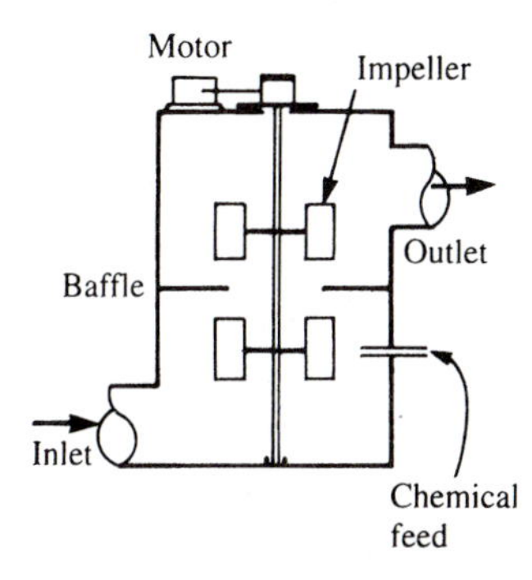

FIGURE 10.12 Impeller-type mechanical rapid mixer for dispersion of chemicals into water.

FIGURE 10.13 Static mixing elements that are inserted into a pipe for in-line blending of chemicals. *Source:* Courtesy of Koch-Glitsch, Inc.

impeller shaft. They are located to disrupt rotational flow by deflecting the water inward toward the impeller shaft. Vortexing hinders mixing and reduces the effective impeller power usage. The turbulent flow pattern in mixing is a function of the tank size and shape; number, shape, and size of impellers; kind and location of stator baffles; and power input. Recommended guidelines for designing a mechanical rapid mixer are the following: a square vessel is superior in performance to a cylindrical vessel; stator baffles are advantageous; a flat-bladed impeller performs better than a fan or propeller impeller; and chemicals introduced at the agitator blade level enhance coagulation [4].

Other methods of rapid mixing include hydraulic action, e.g., injection of chemical into the inlet of a centrifugal pump, jet injection into the water flow in a pipe, and mechanical or static in-line blending. The static mixing elements, shown in Figure 10.13, are inserted into a pipe immediately after the point of chemical injection for blending. The kind of mixer configuration selected depends on whether the blending is for dispersion, reaction, or gas–water contact.

10.10 FLOCCULATION

Flocculation is agitation of chemically treated water to induce coagulation. In this manner, very small suspended particles collide and agglomerate into larger heavier floc that settles out by gravity. Flocculation is a principal mechanism in removing turbidity from water. Floc growth depends primarily on two factors: intermolecular chemical forces and the physical action induced by agitation.

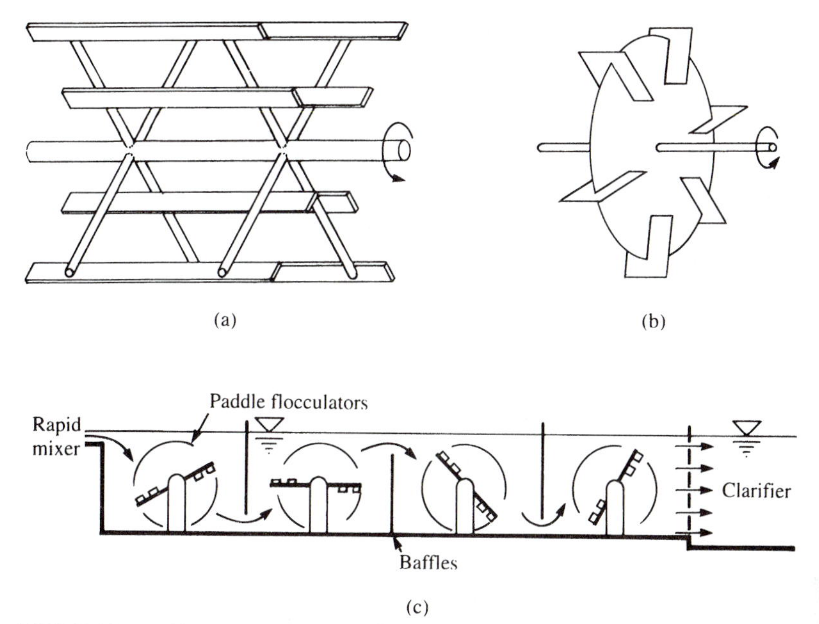

FIGURE 10.14 The common mixing devices for flocculation in water treatment are (a) the paddle flocculator, (b) the flat-blade turbine, (c) a typical flocculation tank arrangement—horizontal shaft paddles in a series of compartments separated by baffles to direct water flow through the paddle flocculators.

The common mechanical mixing devices in water treatment are paddle (reel) flocculators [Fig. 10.14(a)], flat-blade turbines [Fig. 10.14(b)], and vertical-turbine mixers. The paddle flocculator consists of a shaft with protruding steel arms that support wooden or steel blades. The paddle shafts can be located transverse to or parallel to the flow, although transverse installation is more common [Fig. 10.14(c)]. The paddle units slowly rotate, usually at 1–10 rpm, causing collision among the floc particles that are held in suspension by the gentle agitation. The result is growth of the suspended particles from colloids to settleable floc.

The energy dissipated by the flocculator as shearing forces in the water has been observed to be the controlling factor in floc growth. This energy dissipation is related to the velocity gradient, which represents the changes in velocity from point to point in the water being mixed. The relationship between velocity gradient and energy dissipated is defined as [5]

$$G = \left(\frac{P}{\mu V}\right)^{1/2} \tag{10.12}$$

where

G = velocity gradient, fps/ft or s^{-1} (s^{-1})

P = power input, ft · lb/s (N · m/s)

μ = absolute viscosity, lb · s/ft^2 (N · s/m^2 or kg/m · s)

V = volume, ft^3 (m^3)

Optimum values exist for flocculation time and velocity gradient. Below a minimum time, no flocculation occurs, and increasing the time beyond maximum floc formation does not significantly improve flocculation. Based on experience, flocculation tanks with a minimum of three compartments in series reduce the time required for flocculation. In general, the minimum and maximum values of G to promote satisfactory growth of floc are 10 and 75 fps/ft (s^{-1}), respectively, with an optimum range in water treatment with paddle flocculators of 30–60 fps/ft [6].

The power dissipated in water by a paddle flocculator can be calculated as

$$P = \frac{C_d A \rho v^3}{2} \tag{10.13}$$

where

P = power dissipated, ft · lb/s (N · m/s)

C_d = coefficient of drag

= 1.8 for flat plates

A = area of paddles, ft^2 (m^2)

ρ = density of water, lb · s^2/ft^4 (kg/m^3)

v = velocity of paddles relative to the water, ft/s (m/s)

= normally 0.50–0.75 of the velocity of paddle

For a paddle flocculator,

$$v_p = 2\pi r N \tag{10.14}$$

where

v_p = velocity of paddle blade, ft/s (m/s)

r = distance from shaft to center of paddle, ft (m)

N = rotational speed, rev/s (rev/s)

Then for a flocculation tank with n symmetrical paddle arms with blades at radii r_1 and r_2 rotating at N rev/s, the power dissipated is

$$P = \frac{n}{2} C_d A \rho (1 - k)^3 (2\pi N)^3 (r_1^3 + r_2^3) \tag{10.15}$$

where

k = ratio of water velocity to paddle velocity

= 0.25–0.50 for well-baffled tanks with paddle flocculators, with 0.30 being a common value

A = area of each paddle arm across tank

The *Recommended Standards for Water Works, Great Lakes Upper Mississippi River Board of State Sanitary Engineers* [7] recommends that rapid mixing for dispersion of chemicals in water be performed with mechanical mixing devices. The mixing detention period should not be more than 30 s, and the rapid-mix and flocculation basins shall be as close together as possible.

With regard to flocculation, the following recommendations are presented in the *Standards*:

1. Inlet and outlet design shall prevent short-circuiting and destruction of floc.
2. The flow-through velocity shall be not less than 0.5 ft/min nor greater than 1.5 ft/min with a detention time for floc formation of at least 30 min.
3. Agitators shall be driven by variable-speed drives with the peripheral speed of paddles ranging from 0.5 ft/s to 3.0 ft/s.
4. Flocculation and sedimentation basins shall be as close together as possible. The velocity of flocculated water through pipes or conduits to settling basins shall be not less than 0.5 ft/s nor greater than 1.5 ft/s. Allowances must be made to minimize turbulence at bends and changes in direction.

Example 10.5

A 25-mgd water treatment plant has a flocculation tank 64 ft long and 100 ft wide with a water depth of 16 ft. The four horizontal shafts of paddle flocculators are in compartments separated by baffles. All the paddle units have four arms with three blades, each with radii of 3.0, 4.5, and 6.0 ft measuring from the shaft to the center of the 0.80-ft-wide boards. The total length of the paddle boards across the tank is 50 ft. Assume a ratio of water velocity to paddle velocity of 0.3, rotational speed of 0.050 rev/s, C_d equal to 1.8, and water temperature of 50°F. Calculate the velocity gradient.

Solution: From a table in the appendix for water at 50°F, $\rho = 1.936\ \mathrm{lb \cdot s/ft^4}$, and $\mu = 2.735 \times 10^{-5}\ \mathrm{lb \cdot s/ft^2}$. The number of paddle arms = 16.

$$P = \frac{16}{2} 1.8(0.80 \times 50)1.936(1 - 0.3)^3 (2 \times \pi \times 0.050)^3[(3.0)^3 + (4.5)^3 + (6.0)^3]$$

$$= 3960\ \mathrm{ft \cdot lb/s}$$

$$G = \left(\frac{3960}{2.735 \times 10^{-5} \times 16 \times 64 \times 100} \right)^{1/2} = 38\ \mathrm{fps/ft}$$

SEDIMENTATION

Sedimentation (clarification) is the removal of solid particles from suspension by gravity. In water treatment, the common application of sedimentation is after chemical treatment to remove flocculated impurities and precipitates. In wastewater processing, sedimentation is used to reduce suspended solids in the influent wastewater and to remove settleable solids after biological treatment.

Design of clarifiers is based on empirical data from the performance of full-scale sedimentation tanks. Mathematical relationships to predict settling of suspensions in actual treatment processes have not been successful in design, and the application of data from laboratory settling column tests has met with only limited success. For these reasons, design criteria are derived from operation of clarifiers settling similar kinds of waters and wastewaters.

10.11 FUNDAMENTALS OF SEDIMENTATION

Settling of nonflocculent particles in dilute suspension, as defined by classical settling theory of discrete particles, has no direct application in water and wastewater sedimentation. It is, however, the basis for the hypothetical relationship between settling velocity and overflow rate. Figure 10.15 illustrates an ideal rectangular clarifier with an inlet zone for transition of influent flow to uniform horizontal flow, a sedimentation zone where the particles settle out of suspension by gravity, an outlet zone for transition of uniform flow in the sedimentation zone to rising flow for discharge, and a sludge zone where the settled particles collect. All influent discrete particles with a settling velocity greater than V_0 settle to the sludge zone and are removed. Particles with settling velocities of V_i less than V_0 are removed only if they enter the basin within a vertical distance $H_i = V_i t_0$, where t_0 is the flowing through or detention time. The velocity V_0 is defined as the overflow rate (gpd/ft^2), or as the surface settling rate (ft/s), and is calculated as

$$V_0 = \frac{Q}{A} \tag{10.16}$$

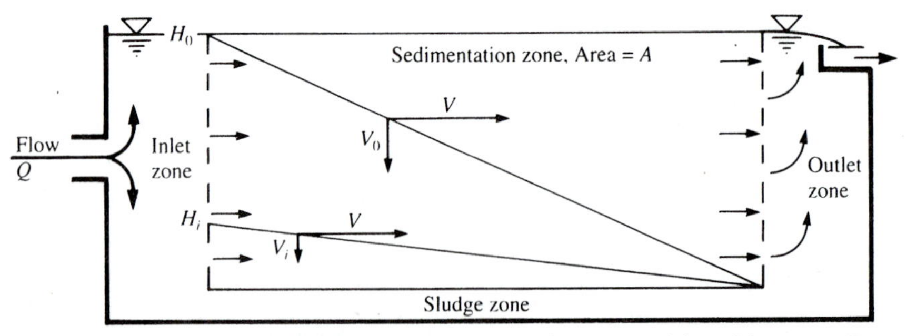

FIGURE 10.15 An ideal rectangular clarifier settling discrete particles with an overflow rate of $V_0 = Q/A$.

where

$$V_0 = \text{overflow rate, gpd/ft}^2\ (\text{m}^3/\text{m}^2 \cdot \text{d})$$
$$Q = \text{average daily flow, gpd}\ (\text{m}^3/\text{d})$$
$$A = \text{surface area of the clarifier, ft}^2\ (\text{m}^2)$$

Overflow rate is an important design criterion in the sizing of sedimentation tanks. A second criterion specified in design is liquid depth in the tank. In addition to overflow rate and depth, performance of a sedimentation tank is affected by many other factors, such as nonideal inlet and outlet conditions, presence of sludge removal equipment, and currents causing nonuniform flow and disturbance of settled sludge on the bottom.

Settling of flocculent particles results in coalescence during sedimentation, with the particles growing into larger floc as they descend. Flocculation is caused by differences in settling velocities of particles, resulting in heavy particles overtaking and coalescing with slower ones and velocity gradients within the water producing collisions among particles. The beneficial results are smaller particles growing into faster-settling floc and flocculation sweeping smaller and slower particles from suspension. The opportunity for contact among settling solids increases with depth. As a result, removal of suspended solids depends on water depth as well as overflow rate and detention time. Flocculent settling is common in clarifying both chemical and biological suspensions.

Zone settling occurs when flocculent particles in high concentration settle as a mass with a well-defined, clear interface between the surface of the settling particles and the clarified water above. Because of crowding of the particles, upward velocity of the displaced water acts to reduce settling velocity. This hindered settling reduces the differences in settling velocities of different-sized particles and increases the sweeping action of flocculation. The suspension settles as a blanket without interparticle movement; individual particles of all sizes move downward at the same velocity. Deceleration of settling occurs as hindered settling changes to *compression*. This results when subsiding particles accumulate at the bottom of the tank. Compaction slowly continues from the weight of supported particles above. Hindered settling is usually negligible in water treatment sedimentation and in clarifying of raw wastewaters, because the particles in suspension are dispersed in low concentration. Zone settling does take place in clarifying biological activated sludge in wastewater treatment. Gravity thickening in both water and wastewater sludges exhibits zone settling with hindered settling in the upper layer and compression in the bottom layer.

10.12 TYPES OF CLARIFIERS

Figure 10.16 illustrates several types of clarifiers, showing the direction of flow through the tanks. In a long rectangular clarifier, the influent is spread across the end of the tank by a baffle structure that dissipates energy to reduce the velocity of the entering water

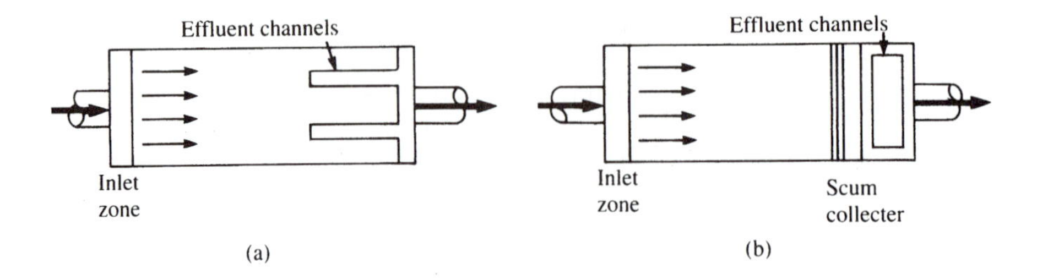

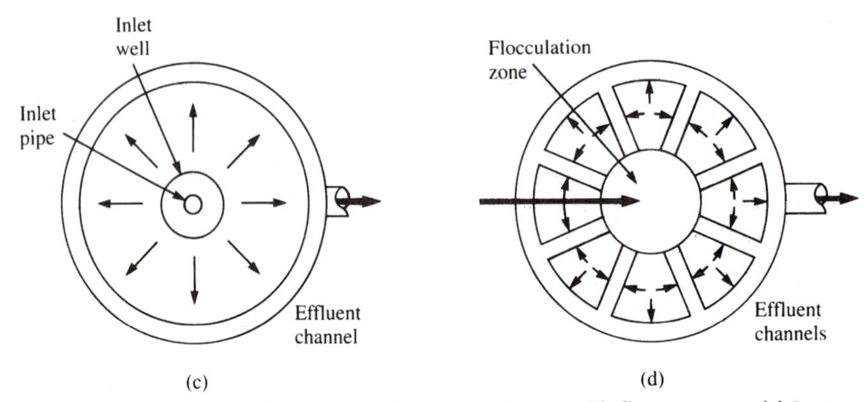

FIGURE 10.16 Diagrams of various clarifier shapes shown with flow patterns. (a) Rectangular clarifier with horizontal flow to effluent "finger" channels extending into the tank from the outlet end. (b) Rectangular clarifier with a scum collector and effluent channels located at the outlet end. (c) Circular clarifier with central feed well and radial flow to a peripheral effluent channel. (d) Circular flocculator–clarifier with water flowing up to radial effluent channels in the settling zone surrounding the submerged hood of the flocculation zone.

and uniformly distributes the flow. The effluent overflows V-notch weirs mounted along the edges of discharge channels, which convey the water out of the tank. To reduce vertical velocities as the water rises up to the discharge channels, the total effluent weir length is several times greater than the width of the tank. Rectangular clarifiers in water treatment often have long "finger" channels extending into the tank [Fig. 10.16(a)]. In contrast, sedimentation tanks in wastewater treatment require scum collectors; therefore, the effluent channels are placed across the width of the tank close to the discharge end [Fig. 10.16(b)]. Sludge is removed by flat scrapers attached to continuous chains that are supported and driven by sprocket wheels mounted on the inside walls of the tank. Rather than submerged chain-and-sprocket mechanisms, in warm climates without snow or ice the scrapers can be hung in the water from an overhead bridge that spans the width of the tank. The wheels of the bridge roll on steel rails mounted on the side walls of the clarifier.

In a circular clarifier, the influent enters through a vertical riser pipe with outlet ports discharging behind a circular inlet well. This baffle dissipates energy and directs the flow downward and radially out from the center [Fig. 10.16(c)]. The effluent channel may be attached to the outside wall with only a single peripheral weir, or the channel may be located inboard, supported from the outside wall with brackets so that water can overflow weirs located on both sides of the channel. Sludge is scraped toward a central hopper by blades attached to arms that rotate around the center of the tank. The drive is a sealed turntable, with gear and pinion running in oil, supported by a bridge or pier. Circular clarifiers are used in both water and wastewater treatment. In wastewater sedimentation, a circular shallow scum baffle is placed in front of the weir channel to prevent the overflow of floating solids, and a rotating skimmer pushes the scum into a hopper that discharges to a scum tank outside of the clarifier.

Figure 10.16(d) diagrams the plan view of a flocculator–clarifier that is illustrated in Figure 10.18. The flocculation zone is under a central cone-shaped hood that extends to near the bottom of the tank. As the water flows out from under the hood, it rises vertically to radial effluent channels. This kind of settling tank, common in water softening, is also referred to as an upflow clarifier.

Several parameters are used in sizing clarifiers. The most common is overflow rate, which is the average flow divided by the total surface area of the tank [Eq. (10.16)]. Of equal significance is the side-water depth, which is the depth of water in the tank excluding the sludge zone. Detention time, which is the volume divided by the average flow [Eq. (10.3)], although less important as a design parameter, is often specified to ensure greater water depth at the overflow rate. Solids loading rates, expressed as the maximum allowable weight of solids per unit area of tank bottom per day, apply only to gravity sludge thickening and final clarifiers of wastewater aeration systems carrying a high concentration of activated sludge.

Conservative overflow rate and sufficient depth are important to compensate for hydraulic instability caused by temperature gradients, inlet energy dissipation, outlet currents, equipment movement, and wind effects. Warm influent to a sedimentation tank containing cooler water can lead to short-circuiting, with the warm water rising to the surface and reaching the effluent channels in a fraction of the theoretical detention time. Similarly, cold water flowing into a tank containing warm water can sink, flow along the bottom, and short-circuit up to the effluent channels. When the inflow and water in the tank are the same temperature, a density current is often caused by the higher turbidity of the influent. Particles entering, being heavier than the clarified water, descend and move under the water surface toward the discharge end.

Inlet energy dissipation is important to slow down and distribute the water at the tank entrance. This is particularly critical if the water enters the inlet zone from a pipe conveying the flocculated water at the higher velocity necessary to keep the solids in suspension in the pipeline. Outlet currents are reduced by proper design of the effluent weirs and channels. In preference to flow plates, V-notch weirs provide better lateral distribution of overflow when leveling is not perfect. Channels require sufficient length and adequate spacing to reduce the approach velocity of overflow. Movement of sludge collection mechanisms through the sludge is slow to prevent resuspension of solids. Finally, wind can have a significant effect on large open sedimentation tanks by creating a surface current that moves the upper layer of water downwind.

10.13 SEDIMENTATION IN WATER TREATMENT

Surface water containing high turbidity (e.g., from a muddy river) may require sedimentation prior to chemical treatment. Presedimentation basins can have hopper bottoms or be equipped with continuous mechanical sludge-removal apparatus. The minimum recommended detention period for presedimentation is 3 hr, although in many cases this is not adequate to settle out fine suspensions that occur at certain times of the year. Chemical feeding equipment for coagulation is frequently provided ahead of presedimentation basins for periods when the raw water is too turbid to clarify adequately by plain sedimentation. Sludge withdrawn from the bottom of presedimentation basins is generally discharged back to the river.

Sedimentation following flocculation depends on the settling characteristics of the floc formed in the coagulation process. A general range for the settling velocities of floc from chemical coagulation is 2–6 ft/hr. Detention periods used in floc sedimentation range from 2 to 4 hr, and overflow rates (surface settling rates) vary from 500 to 1000 gpd/ft^2 (20–41 m^3/m$^2 \cdot$ d).

The *Recommended Standards for Water Works, Great Lakes Upper Mississippi River Board of State Sanitary Engineers* [7] recommends the following standards for sedimentation design following flocculation: a minimum detention time of 4 hr, which may be reduced by approval when equivalent effective settling is demonstrated; a maximum horizontal velocity through the settling tanks of 0.5 fpm (0.15 m/min) with the tanks designed to minimize short-circuiting; and a maximum rate of flow over the outlet weir of 20,000 gpd/ft of weir length (250 m^3/m $\cdot$ d).

Plan-view diagrams of clarifiers used in water treatment are sketched in Figure 10.16. Figure 10.17 shows a square clarifier having a sludge scraper equipped with special corner blades. The method of operation for the unit shown in Figure 10.17 is referred to as cross-flow. Feed is introduced through submerged ports along one side of the tank and flows toward the effluent weir positioned along the opposite side of the clarifier. An alternative method of feed is to pipe influent to the center column of the clarifier, distribute it radially, and collect the effluent over a peripheral weir extending around four sides of the sedimentation tank.

Where quality of the water supply does not vary greatly, such as that from a well supply, flocculator–clarifiers (solids contact units) are an efficient method of chemical treatment. One such unit, shown in Figure 10.18, combines the processes of mixing, flocculation, and sedimentation in a single-compartmented tank. Raw water and added chemicals are mixed with the slurry of previously precipitated solids to promote growth of larger crystals and agglomerated clusters that settle more readily.

The *Recommended Standards for Water Works* [7] states that flocculator–clarifiers

> are generally acceptable for combined softening and clarification where water characteristics, especially temperature, do not fluctuate rapidly, flow rates are uniform and operation is continuous. Before such units are considered as clarifiers without softening, specific approval of the reviewing authority shall be obtained. Clarifiers should be designed for the maximum uniform rate and should be adjustable to changes in flow which are less than the design rate and for changes in water characteristics.

Design criteria recommended are as follows: (1) a flocculation and mixing time of not less than 30 min based on total volume of mixing and flocculation zones; (2)

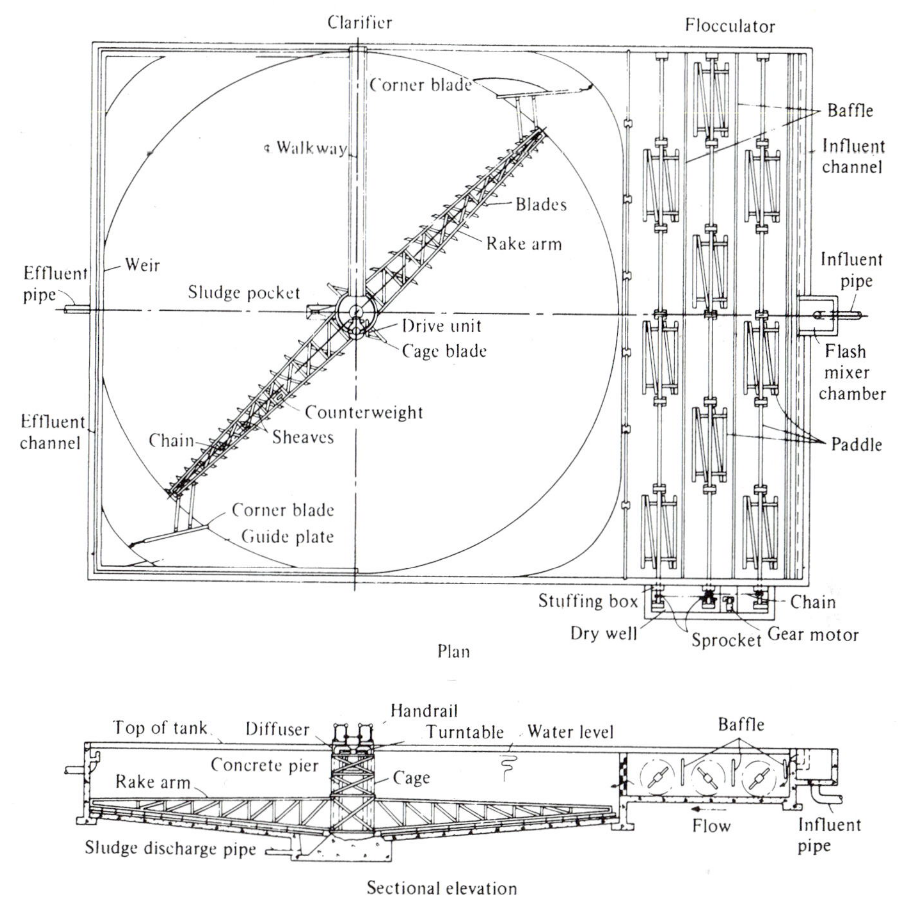

FIGURE 10.17 Flocculator and square sedimentation tank for water clarification, illustrating cross-flow operation. *Source:* Courtesy of GL+V/Dorr-Oliver Inc.

minimum detention times of 2 to 4 hr for suspended-solids contact clarifiers and softeners treating surface water and 1 to 2 hr for the suspended-solids contact softeners treating only groundwater, with the calculated detention time based on the entire volume of the flocculator–clarifier; weir loadings not exceeding 10 gpm/ft (2.1 l/m · s) for units used as clarifiers and 20 gpm/ft (4.1 l/m · s) for units used as softeners; and upflow rates not exceeding 1.0 gpm/ft^2 (0.68 l/m^2 · s) for units used as clarifiers and 1.75 gpm/ft^2 (1.19 l/m^2 · s) for units used as softeners, with the upflow velocity calculated by using the cross-sectional area under the outlet channels at the sludge separation line.

The volume of sludge removed from clarifiers in water treatment plus the quantity of backwash water from cleaning granular-media filters is generally from 4% to 6% of the processed water. The exact amount of wastewater produced depends on chemical treatment processes employed, type of sludge collection apparatus, and kind of filter backwash equipment.

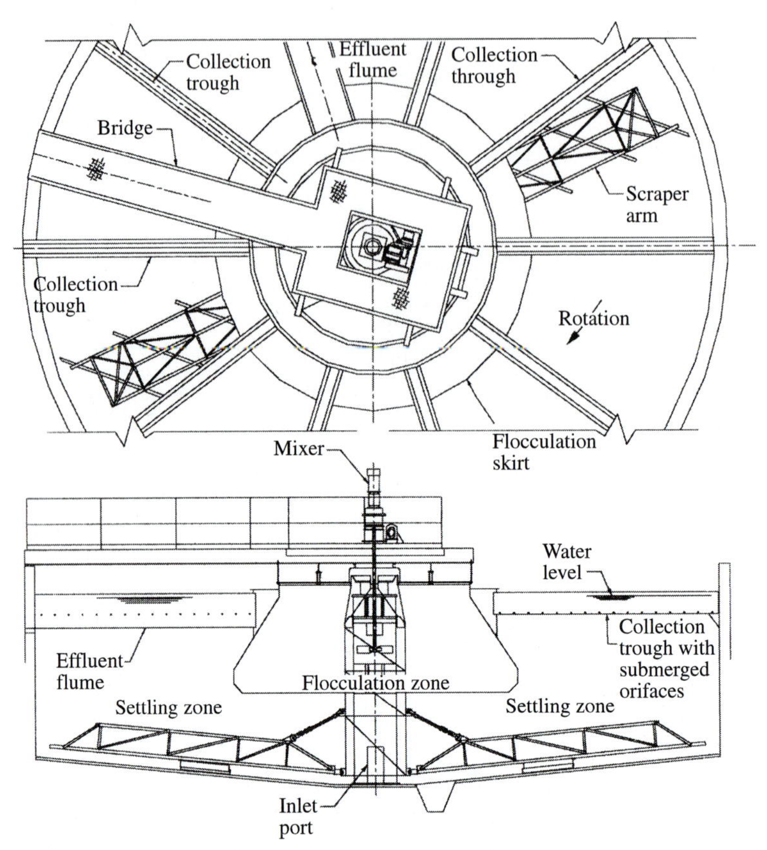

FIGURE 10.18 Flocculator–clarifier provides mixing, flocculation, and sedimentation in a compartmented concentric circular tank. *Source:* Courtesy of Walker Process Equipment Division of McNish Corp.

Example 10.6

Two rectangular clarifiers each 90 ft long, 16 ft wide, and 12 ft deep are used to settle 0.50 mil gal of water in an 8-hr operational period. Calculate the detention time, horizontal velocity, overflow rate, and weir loading, assuming multiple effluent weirs with a total length equal to three tank widths.

Solution:

$$\text{flow} = 0.50 \text{ mil gal/8 hr} = 8356 \text{ ft}^3\text{/hr} = 139 \text{ cfm}$$

$$\text{detention time} = \frac{2 \times 90 \times 16 \times 12}{8356} = 4.1 \text{ hr}$$

$$\text{horizontal velocity} = \frac{139}{2 \times 16 \times 12} = 0.36 \text{ fpm}$$

$$\text{overflow rate} = \frac{3 \times 500{,}000}{2 \times 90 \times 16} = 520 \text{ gpd/ft}^2$$

$$\text{weir loading} = \frac{3 \times 500{,}000}{3 \times 2 \times 16} = 15{,}600 \text{ gpd/ft}$$

10.14 SEDIMENTATION IN WASTEWATER TREATMENT

Clarifiers used to settle raw wastewater are referred to as *primary tanks*. Sedimentation tanks used between trickling filters in a two-stage secondary treatment system are called *intermediate clarifiers*. *Final clarifiers* are settling tanks following secondary aerobic treatment units. Placement of sedimentation tanks in single-stage trickling-filter plants, two-stage filter plants, and activated-sludge plants are illustrated in Figures 12.22, 12.23, and 12.32, respectively.

Sedimentation tanks in wastewater treatment have average overflow rates in the range of 300–1000 gpd/ft^2 and detention times between 1 and 3 hr. Side-water depths depend on settling characteristics of the waste and need for sludge storage volume in the tank bottom.

Separate design criteria are recommended for different sedimentation applications [8]. Primary clarifiers normally have an average overflow rate in the range of 600–800 gpd/ft^2 (24–32 m^3/m$^2 \cdot$ d) with a maximum recommended average rate of 1000 gpd/ft^2 (41 m^3/m$^2 \cdot$ d) and a maximum rate not exceeding 1500 gpd/ft^2 (61 m^3/m$^2 \cdot$ d) for peak hourly flow. The side-water depth should be as shallow as practicable, but not less than 7.0 ft (2.1 m). Maximum recommended weir loadings are 10,000 gpd/ft (124 m^3/m $\cdot$ d) for plants designed for an average flow of 1 mgd or less and 15,000 gpd/ft (186 m^3/m $\cdot$ d) for plants designed for average flows greater than 1 mgd.

Final clarifiers in trickling filter and rotating biological contactor plants normally have an average overflow rate of 800 gpd/ft^2 (33 m^3/m$^2 \cdot$ d) with a recommended maximum of 1200 gpd/ft^2 (49 m^3/m$^2 \cdot$ d) for peak hourly flow. The side-water depth is commonly 7.0–8.0 ft (2.1–2.4 m), with 7.0 ft being the recommended minimum. Maximum recommended weir loadings are 10,000 gpd/ft (124 m^3/m $\cdot$ d) for plants designed for average flows of 1 mgd or less and 15,000 gpd/ft (186 m^3/m $\cdot$ d) for plants designed for average flows greater than 1 mgd. Intermediate clarifiers have the same recommended side-water depths and weir loadings, but the average overflow rate is normally 1000 gpd/ft^2 (41 m^3/m$^2 \cdot$ d), with a recommended maximum of 1200 gpd/ft^2 (49 m^3/m$^2 \cdot$ d) for peak hourly flow.

Final settling tanks for activated-sludge processes are sized to account for zone settling of a flocculent suspension. Design detention times, overflow rates, and weir loadings are selected to minimize the problems with solids loadings, density currents, inlet hydraulic turbulence, and occasional poor sludge settleability. Compared to other wastewater sedimentation tanks, activated-sludge final clarifiers are deeper to accommodate the greater depth of hindered solids settling, have a lower overflow rate to reduce carryover of buoyant biological floc, and have inboard weir channels to reduce the approach velocity of the effluent.

Typical overflow rates for determining clarifier surface area based on average daily design flow are 600 gpd/ft^2 (24 m^3/m$^2 \cdot$ d) for plants smaller than 1 mgd and 800 gpd/ft^2 (33 m^3/m$^2 \cdot$ d) for larger plants. Overflow rates during the peak diurnal flows should not exceed 1200 gpd/ft^2 (49 m^3/m$^2 \cdot$ d) and 1600 gpd/ft^2 (65 m^3/m$^2 \cdot$ d) for plants smaller and larger than 1 mgd, respectively. The recommended minimum side-water depth is 10 ft (3.1 m), with greater depths for larger diameter tanks (e.g., 11 ft at 50-ft diameter and 12 ft at 100-ft diameter). Minimum detention times are 2.0–3.0 hr, depending on the values selected for overflow rate and side-water depth. The maximum recommended weir loadings are 10,000–20,000 gpd/ft (125–250 m^3/m $\cdot$ d). The allowable solids loading relates primarily to the depth of the clarifier, type of mechanical sludge-removal equipment, and characteristics of the activated sludge. Appropriate solids loadings for a properly designed clarifier are 50–60 lb/ft^2/day (250–290 kg/m$^2 \cdot$ d) for a good settling sludge and 40–50 lb/ft^2/day (200–250 kg/m$^2 \cdot$ d) for a poor settling sludge. Clarifiers with shallow water depths or high overflow rates, in comparison to optimum design, may be limited to solids loadings of 20–30 lb/ft^2/day (100–150 kg/m$^2 \cdot$ d) for adequate solids separation.

The BOD removal from raw domestic wastewater in primary settling is generally 30%–40%, with the commonly assumed design BOD removal being 35% and suspended-solids removal of 50%. Of course, the effectiveness of plain sedimentation depends on the characteristics of the wastewater as well as clarifier design. If a municipal wastewater contains an unusual amount of soluble organic matter from industrial wastes, the BOD removal could be less than 30%; conversely, if industrial wastes contribute settleable solids, the BOD removal could be greater than 40%.

Design of a clarifier, the hydraulic loading, and wastewater characteristics influence the density of the accumulated sludge withdrawn from a clarifier [9]. With proper design and operation, the sludge gravity thickens in the bottom of the tank. Hydraulic disturbances usually result in a more dilute sludge, in addition to reduced BOD and suspended-solids removal from the wastewater. Retaining the sludge solids too long can increase biological activity and expansion of the settled sludge blanket in both primary and final tanks. The higher the wastewater temperature, the greater the biological gas production and thinning of the settled sludge. In primary sedimentation, return of waste-activated sludge to the head of a plant can cause detrimental microbiological activity in the settled raw sludge, leading to foul odors, floating sludge, and thinning of the withdrawn sludge. Shortening the retention time of the sludge in the tank by more frequent withdrawal can help, but the best arrangement is separate disposal of the waste-activated sludge. In final sedimentation of activated sludge, retention of the settled sludge for too long can lead to floating solids and thinning of the sludge. The best solution is to use a scraper mechanism with uptake pipes spaced along the arm to remove the solids from the entire floor of the tank rather than plowing them to a central hopper prior to withdrawal.

Figure 10.19 pictures a typical primary settling tank. Wastewater enters at the center behind a stilling baffle and travels down and outward toward effluent weirs located on the periphery of the tank. The inlet line usually terminates near the surface, but the wastewater must travel down behind the stilling baffle before entering the actual settling zone. A stilling well reduces the velocity and imparts a downward motion to the solids, which drop to the tank floor.

(a)

(b)

FIGURE 10.19 Typical primary settling tank for wastewater treatment. (a) Circular settling tank with an inboard weir trough. (b) Stilling well and sludge-collecting mechanism in a circular clarifier. *Source:* Courtesy of Walker Process Equipment, Division of McNish Corp.

The clarifier in Figure 10.20 is designed as a final clarifier for an activated-sludge secondary. The liquid flow pattern is the same as that of a primary clarifier, but the sludge collection system is unique. Suction pipes are attached to and spaced along a V-plow-scraper mechanism rotated by a turntable above the liquid surface. The discharge elevation of the suction riser pipes in the sight well is lower than the water surface in the tank so that sludge is forced up and out of the uptake pipes by water pressure. The flow from each suction pipe is controlled by elevation adjustment of a slip tube over the riser pipe. The sludge collected in the sight well flows out through a vertical pipe centrally located in the influent wastewater pipe. A seal between the sight

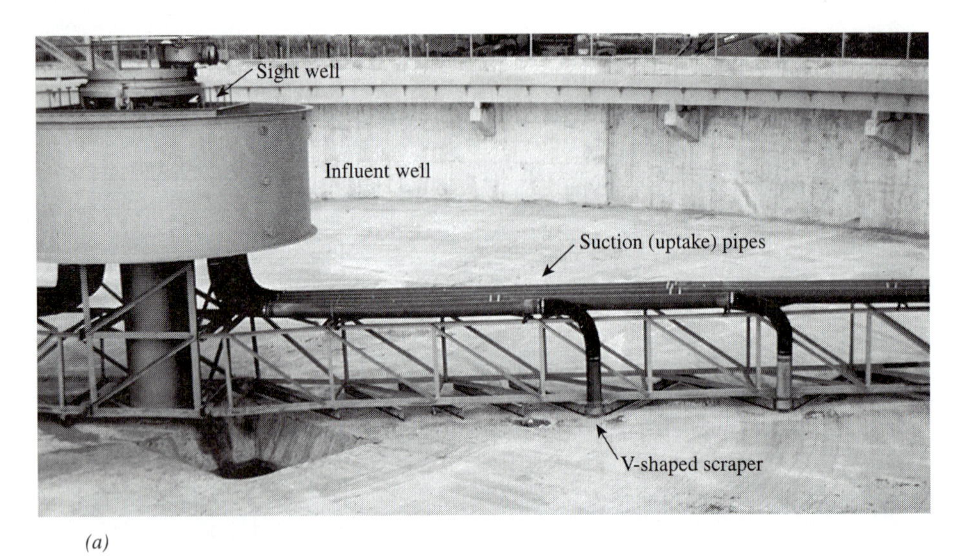

FIGURE 10.20 Final clarifier designed for use with activated-sludge processes. Return sludge is withdrawn through suction (uptake) pipes located along the collector arm for rapid return to the aeration tank. Sludge flowing from each pipe can be observed in the sight well. *Source:* Courtesy of Walker Process Equipment, Division of McNish Corporation.

well and the wastewater riser allows the well to rotate around with the collector arm without leaking sludge into the tank.

Rapid uniform withdrawal of sludge across the entire bottom of an activated-sludge final clarifier has two distinct advantages. The retention time of solids that settle near the tank's periphery is not greater than those that land near the center; thus, aging of the biological floc and subsequent floating solids due to gas production are eliminated. With a scraper collector, the residence time of settled solids depends directly on the radial distance from the sludge hopper. The second advantage is that the direction of activated-sludge return flow is essentially perpendicular to the tank bottom, rather than horizontal toward a centrally located sludge hopper. Downward flow through a sludge blanket enhances gravity settling of the floc and increases sludge density. This is an important factor when one considers that the return flow may be as great as one-half of the influent flow.

Zone settling occurs in gravity separation of flocculent biological suspensions from activated-sludge aeration in wastewater treatment and flocculent chemical suspensions from coagulated water. Figure 10.21 is a schematic diagram of the circular final clarifier in Figure 10.20. The influent, a flocculent suspension, from the biological aeration tank enters in the center and clear effluent overflows the outlet weir around the periphery of the tank. In this continuous-flow system, zone settling is represented by the clear supernatant and the hindered settling layer beneath. Compression of solids occurs near the bottom of the tank. Thickened sludge is continuously withdrawn from the tank bottom by uptake pipes and returned to the aeration tank.

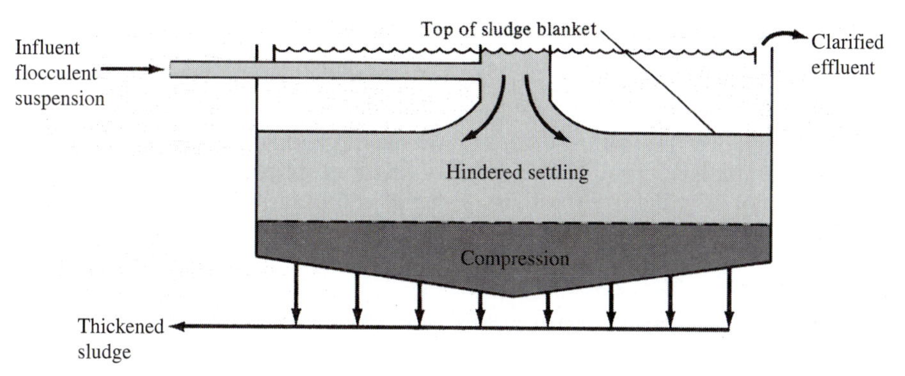

FIGURE 10.21 A schematic diagram of zone settling in the activated-sludge final clarifier illustrated in Figure 10.20.

Example 10.7

Determine the dimensions for two rectangular primary clarifiers to settle 3400 m^3/d of domestic wastewater based on the following criteria: an overflow rate of 32 $m^3/m^2 \cdot d$, a side-water depth of 2.4 m, weir loading of 125–250 $m^3/m \cdot d$, and a length/width ratio in the range of 3/1 to 5/1. Calculate the detention time and estimate the BOD removal efficiency after sizing the tanks.

Solution:

$$\text{surface area of each tank} = \frac{3400 \text{ m}^3/\text{d}}{2 \times 32 \text{ m}^3/\text{m}^2 \cdot \text{d}} = 53.1 \text{ m}^2$$

The appropriate width for a standard-sized sludge-removal mechanism is 4.0 m. Therefore, the tank length is 53.1/4.0 = 13.3 m, and the length/width ratio is 13.3/4.0 = 3.3/1, which is within the desired range of 3/1–5/1.

$$\text{detention time} = \frac{2 \times 53.1 \text{ m}^2 \times 2.4 \text{ m} \times 24 \text{ hr/d}}{3400 \text{ m}^3/\text{d}} = 1.8 \text{ hr}$$

$$\text{maximum weir length} = \frac{3400 \text{ m}^3/\text{d}}{125 \text{ m}^3/\text{m}^2 \cdot \text{d}} = 27.2 \text{ m}$$

$$\text{minimum weir length} = \frac{3400 \text{ m}^3/\text{d}}{250 \text{ m}^3/\text{m}^2 \cdot \text{d}} = 13.6 \text{ m}$$

Consider one inboard effluent channel and an end weir across the width of each tank. For this arrangement,

$$\text{weir loading} = \frac{3400\ \text{m}^3/\text{d}}{2 \times 3 \times 4\ \text{m}} = 140\ \text{m}^3/\text{m} \cdot \text{d (OK)}$$

The BOD removal efficiency for an overflow rate of $32\ \text{m}^3/\text{m}^2 \cdot \text{d} = 800\ \text{gpd/ft}^2/\text{day}$ is assumed to be 35%.

Example 10.8

Two final clarifiers of the type shown in Figure 10.20, with 100-ft diameter and 12-ft side-water depth, are provided for an activated-sludge plant designed to treat 12.5 mgd. Calculate the overflow rate and detention time based on design flow. If the aeration tank is operated at a mixed-liquor suspended-solids concentration of 4000 mg/l and a recirculation ratio of 0.5, calculate the solids loading on the clarifier.

Solution:

$$\text{surface area of clarifiers} = 2 \times \pi(50)^2 = 15{,}700\ \text{ft}^2$$

$$\text{volume of clarifiers} = 15{,}700 \times 12 \times 7.48 = 1{,}410{,}000\ \text{gal}$$

$$\text{overflow rate} = \frac{12{,}500{,}000}{15{,}700} = 800\ \text{gpd/ft}^2\ (32\ \text{m}^3/\text{m}^2 \cdot \text{d})$$

$$\text{detention time} = \frac{1{,}410{,}000 \times 24}{12{,}500{,}000} = 2.7\ \text{hr}$$

Flow from the aeration tank to the clarifier with a recirculation ratio of 0.5 equals $1.5 \times 12.5 = 18.8$ mgd.

$$\text{Solids loading} = \frac{18.8 \times 4000 \times 8.34}{15{,}700} = 40\ \text{lb/ft}^2/\text{day}\ (195\ \text{kg/m}^2 \cdot \text{d})$$

10.15 GRIT CHAMBERS IN WASTEWATER TREATMENT

Grit includes gravel, sand, and heavy particulate matter such as corn kernels, bone chips, and coffee grounds. For design purposes, grit is defined as fine sand, 0.2-mm-diameter particles with a specific gravity of 2.65 and a settling velocity of 0.075 fps.

Grit removal in municipal wastewater treatment protects mechanical equipment and pumps from abnormal abrasive wear, prevents pipe clogging by its deposition, and reduces accumulation in settling tanks and digesters. Grit chambers are commonly placed between lift pumps and primary settling tanks. Wear on the centrifugal pumps is tolerated to obtain the convenience of ground-level construction of grit chambers.

Several types of grit-removal units are used in wastewater treatment. The kind selected depends on the amount of grit in the wastewater, size of the plant, convenience of operation and maintenance, and costs of installation and operation. Standard types are channel-shaped settling tanks, aerated tanks of various shapes with hopper bottoms, clarifier-type tanks with mechanical scraper arms, and cyclone grit separators with screw-type grit washers.

Historically popular but rarely in use today, the channel-type grit chamber is a long, narrow, shallow settling channel with a proportional weir placed in the discharge end. The weir opening is shaped to keep the horizontal velocity relatively constant with varying depths of flow. To prevent scouring of the precipitated grit, the horizontal velocity is controlled at approximately 1 fps.

Square and rectangular hopper-bottom tanks with the inlet and effluent weir on opposite sides of the tank are generally used in small plants. These units are often mixed by diffused aeration to keep the organic solids in suspension while the grit settles out. Design of an aerated hopper-bottom grit tank is based on a detention time of approximately 1 min at peak hourly flow. Settled substances are removed from the hopper bottom by pumping, screw conveyor, bucket elevator, or, if possible, gravity flow.

Clarifier-type tanks are generally square, with influent and effluent weirs on opposite sides. A centrally driven collector arm scrapes the settled grit into an end hopper. The grit is then elevated into a receptacle with a chain and bucket, or screw conveyor. The clarifier may be a shallow tank or a deeper aerated chamber.

Grit taken from these kinds of grit chambers is sometimes relatively high in organic content. A counterflow grit washer that functions like a screw conveyor can be used to wash the grit and return waste organics to the plant influent. Alternatively, a special centrifugal pump can lift the slurry from a grit sump to a centrifugal cyclone. The cyclone separates the grit from the organic material and discharges it to a classifier for washing and draining. Wash water and washed-out organics from the cyclone and classifier return to the wastewater.

Preaeration of raw wastewater prior to primary sedimentation is practiced to restore freshness, scrub out entrained gases, and improve subsequent settling. The process is similar to flocculation if the detention time is not less than 45 min without chemical addition, or about 30 min with chemicals. Normally, the purpose of preaeration is freshening wastewater, not flocculation, and little or no BOD reduction occurs. The detention time in preaeration basins is generally less than 20 min. Air supplied for proper agitation ranges from 0.05 to 0.20 ft^3/gal of applied wastewater.

Processes of grit removal and preaeration can be performed in the same basin—a hopper-bottom or clarifier-type tank provided with grit-removal and washing equipment. Aeration mixing improves grit separation while freshening the wastewater.

Example 10.9

Design wastewater flows for a treatment plant are an average daily flow of 280 gpm and a peak hourly rate of 450 gpm. Estimate tank volumes and dimensions for the following grit-removal units: (1) a hopper-bottom tank with grit washer and (2) a clarifier-type unit for grit removal and preaeration.

Solution:

1. Design criteria for a hopper-bottom grit-removal tank:

$$\text{detention time} = 1 \text{ min at peak flow}$$

$$\text{volume of tank required} = 450 \times 1.0 = 450 \text{ gal} = 60 \text{ ft}^3$$

2. Design criteria for a preaeration basin with grit removal:

$$\text{detention time} = 20 \text{ min}$$

$$\text{volume of basin required} = 280 \times 20 = 5600 \text{ gal} = 750 \text{ ft}^3$$

Use a 12 × 12 ft basin 6 ft deep + freeboard. Provide a revolving clarifier mechanism, a separate grit washer, and diffused aeration along one side of the tank.

FILTRATION

Filtration is used to separate nonsettleable solids from water and wastewater by passing it through a porous medium. The most common system is filtration through a layered bed of granular media, usually a coarse anthracite coal underlain by a finer sand.

10.16 GRAVITY GRANULAR-MEDIA FILTRATION

Gravity filtration through beds of granular media is the most common method of removing colloidal impurities in water processing and tertiary treatment of wastewater.

The mechanisms involved in removing suspended solids in a granular-media filter are complex, consisting of interception, straining, flocculation, and sedimentation, as shown schematically in Figure 10.22. Initially, surface straining and interstitial removal results in accumulation of deposits in the upper portion of the filter media. Because of the reduction in pore area, the velocity of water through the remaining voids increases, shearing off pieces of captured floc and carrying impurities deeper into the filter bed. The effective zone of removal passes deeper and deeper into the filter. Turbulence and the resulting increased particle contact within the pores promotes flocculation, resulting in trapping of the larger floc particles. Eventually, clean bed depth is no longer available and breakthrough occurs, carrying solids out in the underflow and causing termination of the filter run.

Microscopic particulate matter in raw water that has not been chemically treated will pass through the relatively larger pores of a filter bed. On the other hand, suspended solids fed to a filter with excess coagulant carryover from chemical treatment produces clogging of the bed pores at the surface. Optimum filtration occurs when impurities in the water and coagulant concentration cause "in-depth" filtration. The impurities neither pass through the bed nor are all strained out on the surface, but a significant amount of flocculated solids are removed throughout the entire depth of the filter.

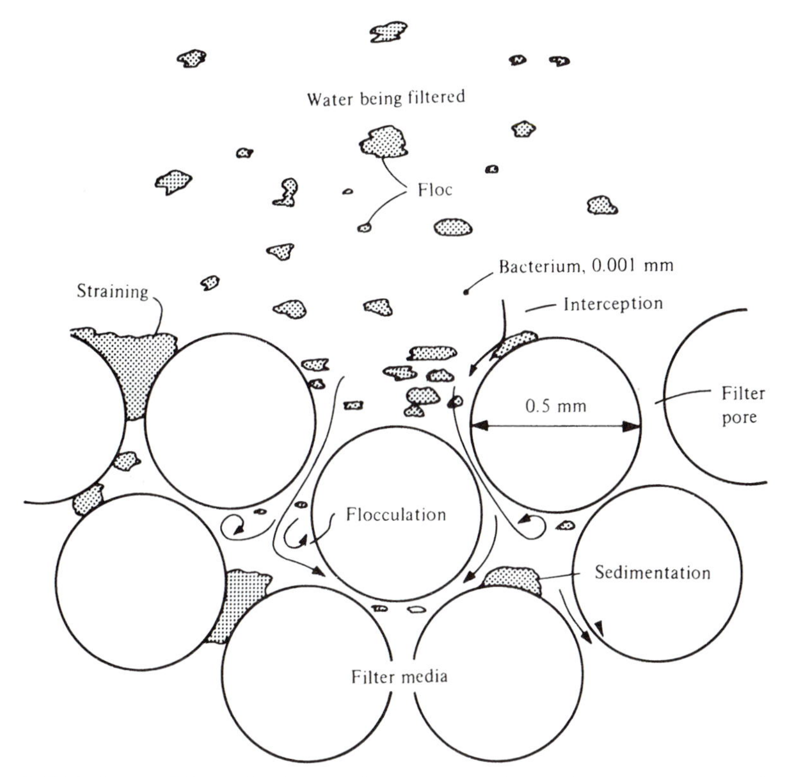

FIGURE 10.22 Schematic diagram illustrating straining, flocculation, and sedimentation actions in a granular-media filter.

Granular-media filtration of surface water for removal of *Giardia* cysts using field-scale pilot filters and testing at full-scale treatment plants has demonstrated the necessity of proper chemical pretreatment for effective removal of microscopic particles [10,11]. The dual-media pilot filters were operated at a rate of 5 gpm/ft^2 and filtered low-temperature (near 0° C–8° C), low-turbidity (0.5–1.5 NTU) waters after coagulation with either a cationic coagulant or alum and a polymer aid. The contaminant removals with proper chemical pretreatment were turbidity 84%–96%, coliform bacteria 97%–99.95%, and *Giardia* cysts 100%, meaning no cysts were detected in the filtered water. The removals without chemical pretreatment were turbidity 35%–57%, coliform bacteria 60%, and *Giardia* cysts 80%–91%. Even a lapse in chemical feed for a short time allowed a higher proportion of particles, including cysts, to pass through the filter. To ensure that large enough numbers of coliform bacteria and *Giardia* cysts were present, the raw water was seeded with these microorganisms for several of the test runs. The best indicators of proper chemical pretreatment for removal of *Giardia* cysts and other microscopic particles were the percentage reduction in turbidity and the concentration of turbidity in the filtered water. The survey of full-scale treatment plants processing waters from mountain streams containing *Giardia* cysts verified the importance of proper chemical pretreatment. *Giardia* cysts were found in the finished waters of those plants with either no chemical pretreatment or improper coagulation.

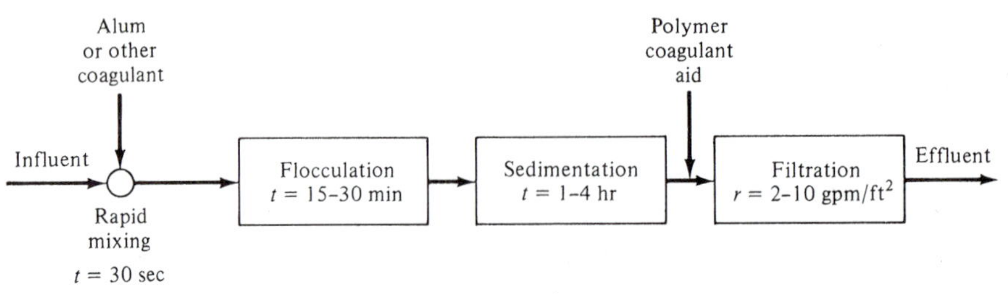

FIGURE 10.23 Flow diagram for a traditional surface water treatment system.

In filtering these low-turbidity waters (less than 1 NTU), effective chemical pretreatment is either a finished water turbidity of less than 0.1 NTU or the absence of microscopic particles [11].

Traditional Filtration

A typical scheme for processing surface supplies to drinking water quality, shown in Figure 10.23, consists of flocculation with a chemical coagulant and sedimentation prior to filtration. Under the force of gravity, often by a combination of positive head and suction from underneath, water passes downward through the media that collect the floc and particles. When the media become filled or solids break through, a filter bed is cleaned by backwashing where upward flow fluidizes the media and conveys away the impurities that have accumulated in the bed. Destruction of bacteria and viruses depends on satisfactory turbidity control to enhance the efficiency of chlorination.

Filtration rates following flocculation and sedimentation are in the range of 2–10 gpm/ft^2 (1.4–6.8 $l/m^2 \cdot s$), with 5 gpm/ft^2 (3.4 $l/m^2 \cdot s$) normally the maximum design rate.

Direct Filtration

The process of direct filtration, diagrammed in Figure 10.24, does not include sedimentation prior to filtration. The impurities removed from the water are collected and stored in the filter. Although rapid mixing of chemicals is necessary, the flocculation stage is either eliminated or reduced to a mixing time of less than 30 min. Contact flocculation of the chemically coagulated particles in the water takes place in the granular media [12]. Successful advances in direct filtration are attributed to the development of coarse-to-fine

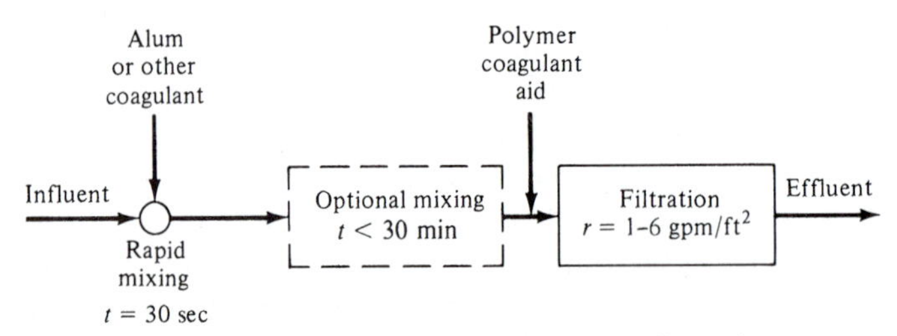

FIGURE 10.24 Flow diagram for direct filtration of a surface water supply or tertiary treatment of wastewater.

multimedia filters with greater capacity for "in-depth" filtration, improved backwashing systems using mechanical or air agitation to aid cleaning of the media, and the availability of better polymer coagulants.

Surface waters with low turbidity and color are most suitable for processing by direct filtration. Based on experiences cited in the literature, waters with less than 40 units of color, turbidity consistently below 5 units, iron and manganese concentrations of less than 0.3 and 0.05 mg/l, respectively, and algal counts below 2000/ml can be successfully processed [13]. Operational problems in direct filtration are expected when color exceeds 40 units or turbidity is greater than 15 units on a continuous basis. Potential problems can often be alleviated during a short period of time by application of additional polymer. Tertiary filtration of wastewaters containing 20–30 mg/l of suspended solids following biological treatment can be reduced to less than 5 mg/l by direct filtration. For inactivation of viruses and a high degree of bacterial disinfection, filtration of chemically conditioned wastewater precedes disinfection by chlorine.

The feasibility of filtration without prior flocculation and sedimentation relies on a comprehensive review of water quality data. The incidence of high turbidities caused by runoff from storms and blooms of algae must be evaluated. Often, pilot testing is valuable in determining efficiency of direct filtration compared to conventional treatment, design of filter media, and selection of chemical conditioning.

Filtration rates in direct filtration are usually $1-6\ \text{gpm/ft}^2$ $(0.7-4.1\ \text{l/m}^2 \cdot \text{s})$, somewhat lower than the rates following traditional pretreatment.

10.17 DESCRIPTION OF A TYPICAL GRAVITY FILTER SYSTEM

A cutaway view of a gravity filter is shown in Figure 10.25. During filtration, the water enters above the filter media through an inlet flume. After passing downward through the granular media (24–30 in. in thickness) and the supporting gravel bed, it is collected in the underdrain system and discharged through the underdrain pipe. To backwash the dirty filter, the water level is lowered to near the surface of the granular bed, and the media are scoured by either upward flow of air alone or by upward flow of air and water concurrently. Wash water entering the underdrain is distributed under the media and flows upward hydraulically, expanding the media and conveying out impurities. The turbid wash water is collected in the wash-water troughs that discharge to the outlet flume.

Filter Media

Broadly speaking, filter media should possess the following qualities: (1) coarse enough to retain large quantities of floc, (2) sufficiently fine to prevent passage of suspended solids, (3) deep enough to allow relatively long filter runs, and (4) graded to permit backwash cleaning. These attributes are not all compatible. For example, a very fine sand retains floc, which also tends to shorten the filter run, while for a coarse sand the opposite would be true. Recent trends are toward coarse sands and dual-media beds of anthracite overlying sand so that high rates of filtration can be obtained.

A filter medium is defined by effective size and uniformity coefficient. The effective size is the 10-percentile diameter; that is, 10% by weight of the filter material is less

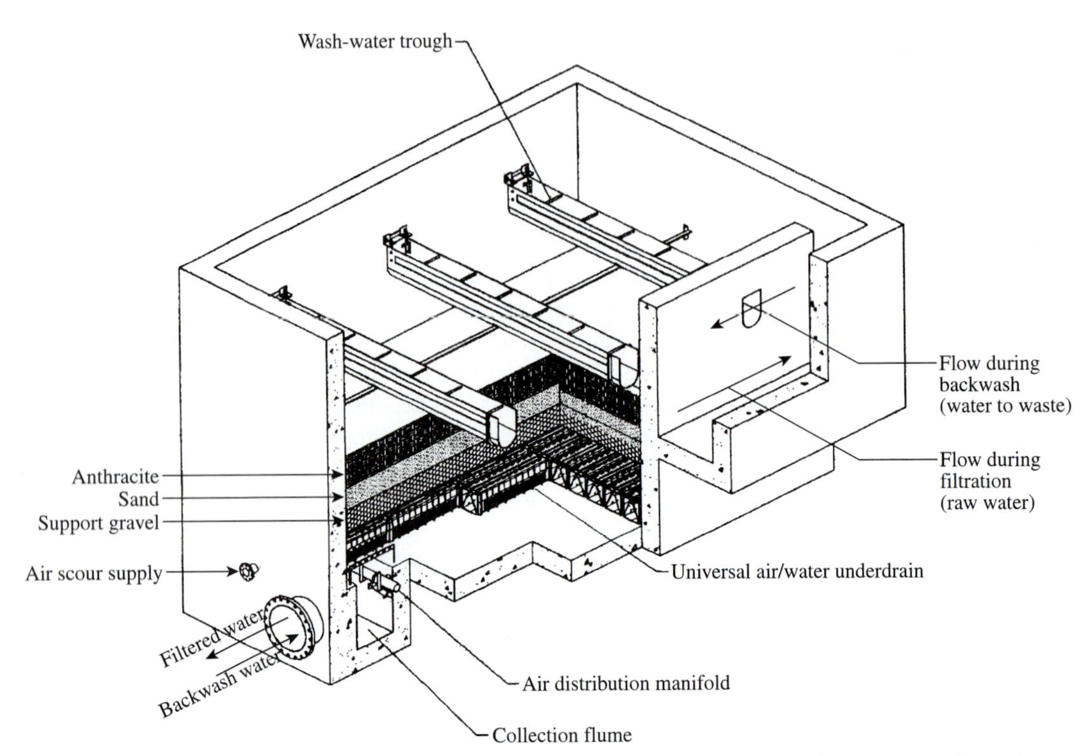

FIGURE 10.25 Cutaway view of a gravity-filter box with water and air piping, wash-water troughs, and dual-media filter supported on a plastic block underdrain. *Source:* Courtesy of the F.B. Leopold Company, Inc.

than this diameter. The uniformity coefficient is the ratio of the 60-percentile size to the 10-percentile size. In water treatment [7], the conventional sand medium has an effective size of 0.45–0.55 mm, a uniformity coefficient less than 1.7, and a bed depth of 24–30 in. For dual-media filters, as shown in Figure 10.26, the top anthracite layer has an effective size of 0.9–1.1 mm, a uniformity coefficient of less than 1.7, thickness of a few inches to two-thirds of the total filter thickness of 24–30 in., and is underlain by a sand filter layer as described above. The supporting coarse sand layer between the filter sand and the underlying gravel has an effective size of 0.8–2.0 mm and a uniformity coefficient less than 1.7. The conventional supporting gravel layers, beneath the filter to the underdrain, are [7] 2 to 3 in. of $^{1}/_{16}$–$^{3}/_{16}$ in., 2 to 3 in. of $^{3}/_{16}$–$^{1}/_{2}$ in., 3 to 5 in. of $^{1}/_{2}$–$^{3}/_{4}$ in., 3 to 5 in. of $^{3}/_{4}$–$1^{1}/_{2}$ in., and 5 to 8 in. of $1^{1}/_{2}$–$2^{1}/_{2}$ in. The coarsest layer of gravel required is determined by the kind of underdrain and size of openings for passage of filtered and backwash water.

A sand filter bed with a relatively uniform grain size can provide effective filtration throughout its depth. If the grain size gradation is too great, effective filtering is confined to the upper few inches of sand. This results because the finest sand grains accumulate on the top of the bed during stratification after backwashing. The problem of surface plugging of sand filters led to development of dual-media filters. A dual-media filter consists

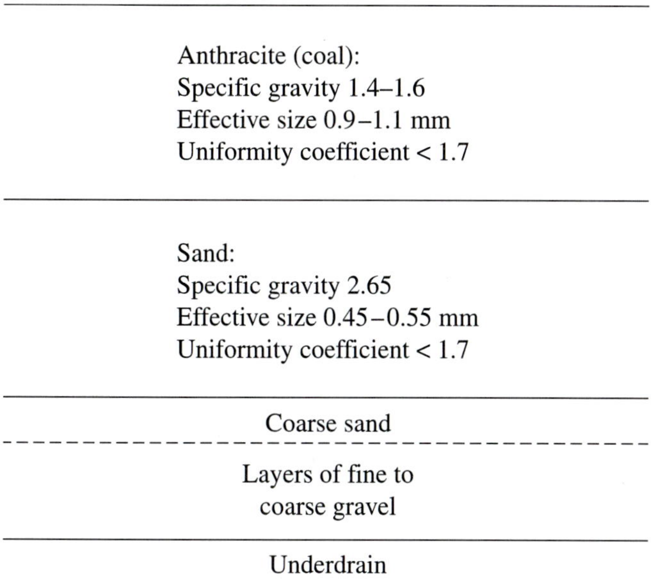

FIGURE 10.26 Cross section of the media in a coal–sand dual-media filter and supporting gravel layer showing typical grain sizes, specific gravities, effective sizes, and uniformity coefficients.

of a sand [specific gravity (sp gr), 2.65] layer topped with a bed of anthracite coal medium (1.4–1.6 sp gr). The coarser anthracite top layer has pores about 20% larger than the sand medium. These openings are capable of adsorbing and trapping particles so that floc carried over in clarified water does not accumulate prematurely on the filter surface and plug the sand filter.

Unconventional filters described by Cleasby [14] are dual media with coarse anthracite having an effective size of about 1.5 mm and a low uniformity coefficient to provide a greater volume of voids to collect impurities and extend filter runs of highly turbid surface waters and wastewaters. To avoid problems in backwashing, the recommended effective size of the underlying sand medium is 0.75 to 0.90 mm for anthracite densities of 1.45 to 1.65. The use of a single-medium coarse-sand filter is practiced in Europe. For example, sand with a size range of 0.9–1.5 mm and about 1.0 m depth has been used in surface water treatment plants. These filters are scoured with concurrent air and water at nonfluidizing velocity followed by a brief wash at high velocity with water alone. Triple-media filters comprising anthracite, sand, and garnet layers have been used for many years in the United States. Both dual- and triple-media filters are substantially better than the conventional sand filter in providing longer filter runs with a corresponding reduction in required backwash water. In comparing these two filters, however, the benefit of adding a third layer has not been well demonstrated [14].

Underdrain Systems

The purposes for the underdrain of a filter are to support the filter media, collect the filtered water, and distribute the water for backwashing and air for scouring. The number and depths of the gravel layers, both to prevent loss of the media during filtration and to contribute to uniform distribution of backwash water, depend on the

TABLE 10.3 Common Filter Underdrain Systems

Kind of Underdrain	Features
Pipe laterals with orifices	Deep gravel layer Medium head loss No air scouring
Pipe laterals with nozzles	Shallow gravel layer High head loss Air scouring
Vitrified tile block	Shallow gravel layer Medium head loss No air scouring
Plastic dual-lateral block	Deep gravel layer Low head loss Concurrent air-and-water scouring or air scouring
Plastic nozzles	Shallow or no gravel layer High head loss Air scouring

size of openings in the underdrain. Large openings require a deep gravel layer, and hence a deeper filter box, but have reduced head loss during backwashing. Smaller openings, such as the slots in nozzles, require only the finer gravel layers but have higher head loss. The underdrain systems of pipe laterals with orifices and vitrified tile blocks allow only water backwash; some underdrains permit separate air scour and water backwash; and others are designed for concurrent air-and-water scour and water backwash. Table 10.3 lists some of the common filter underdrain systems. Vitrified tile, plastic block, and nozzles are manufactured, proprietary systems. Pipe lateral underdrains can be fabricated on site. The quality of water or wastewater being filtered, type of filter media, backwash, and scouring system desired, and cost are factors that influence underdrain design [15].

The earliest underdrain was composed of pipe laterals with orifices composed of perforated pipes surrounded and overlain by graded gravel layers. If the pipe laterals were fitted with nozzles, the gravel layer could be shallower and air scouring used. Vitrified tile blocks have upper and lower laterals with orifices. Wash water enters the lower lateral, flows up through orifices into the upper lateral, and upward again through a second set of orifices for uniform distribution under the gravel layer. The disadvantage is the inability to use air scouring or air-and-water backwash. The plastic block, shown in Figures 10.27 and 10.25, has a dual-parallel lateral design for distribution of either air scouring and water backwash or concurrent air-and-water scouring and water backwash. The underdrain illustrated in Figure 10.28 consists of plastic nozzles without a gravel layer. Each nozzle is constructed of wedge-shaped segments assembled with stainless steel bolts to form media-retaining slots that are tapered larger in the direction of filtering water. For air scouring before water backwash, the nozzles have 6-in.-long plastic tubes that extend into the collecting well. Air is applied through the air holes that are sized to maintain a 3.5-in. cushion to ensure equal nozzle pressure and uniform air distribution.

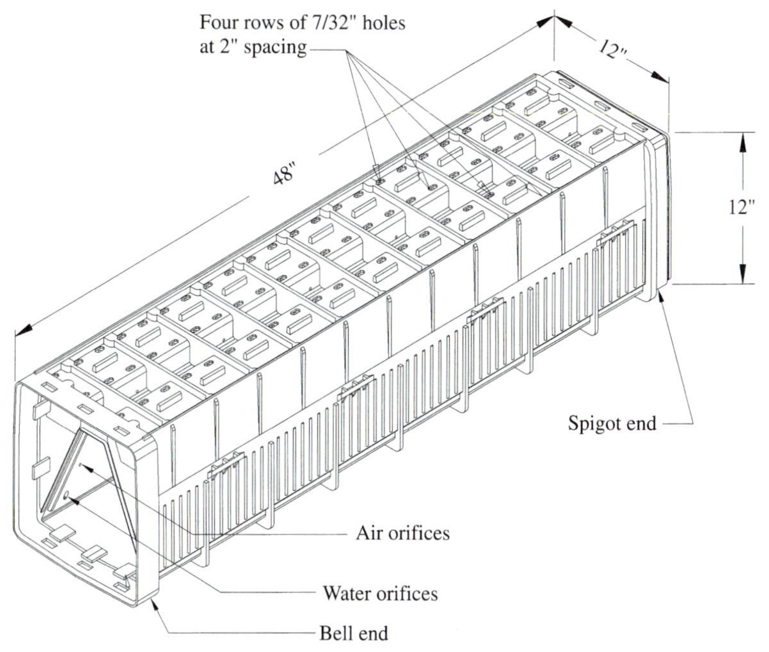

FIGURE 10.27 High-density polyethylene underdrain block with dual-parallel lateral design for either air scour followed by a water backwashing or concurrent air-and-water scouring and water backwashing. *Source:* Courtesy of the F. B. Leopold Company, Inc.

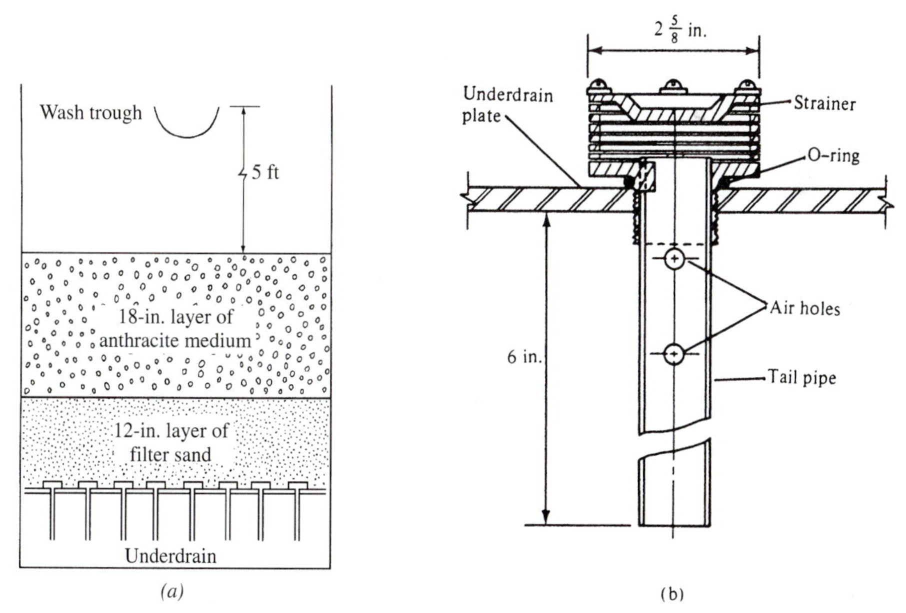

FIGURE 10.28 Underdrain system for air scouring and water backwashing of a granular-media filter. (a) Cross section of the filter. (b) Detail of the air-water nozzle. *Source:* Courtesy of General Filter Co., Ames, IA.

10.18 FLOW CONTROL THROUGH GRAVITY FILTERS

Depending on the design, the operation of a gravity filter can be controlled by the following methods: rate of discharge, constant level, influent flow splitting, and declining rate.

Rate-of-Flow Control

Traditional systems for control of gravity filtration regulate the rate of discharge from the filter underdrain. The influent enters each filter below the operating water level and is essentially unrestricted. A step-by-step description of the flow control operation follows the valve numbering illustrated in Figure 10.29. Initially, valves 1 and 4 are open and 2, 3, and 5 are closed, permitting filtration to proceed. Overflow from the settling basin is applied to the filter, and water passes through the bed into the clear well. When head loss becomes excessive, valves 1 and 4 are closed (3 remains closed), and 2 and 5 are opened to permit backwashing. Clean wash water flows into the filter underdrain system, where it is distributed upward through the filter media. Dirty wash water is collected by troughs and flows into the flume connected to a drain. At the beginning of a filter run, some filtered water can be wasted to flush out the wash water remaining in the bed. This initial wasting is accomplished by opening valve 3 when valve 1 is opened to start filtration (2, 4, and 5 are shut). Then valve 4 is opened while valve 3 is being closed. The sequence is then repeated.

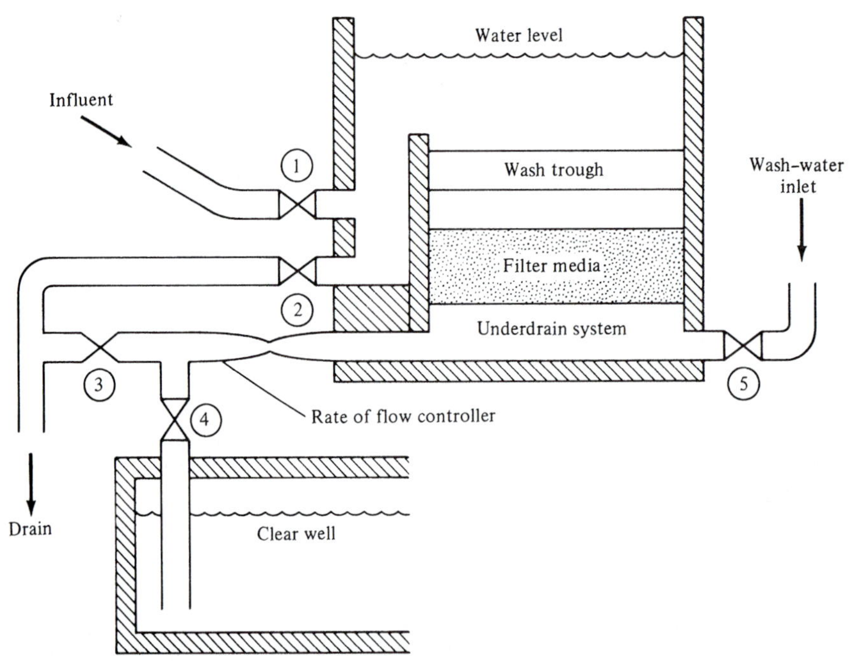

FIGURE 10.29 Diagram illustrating the operation of a traditional gravity filter system by control of the rate of flow.

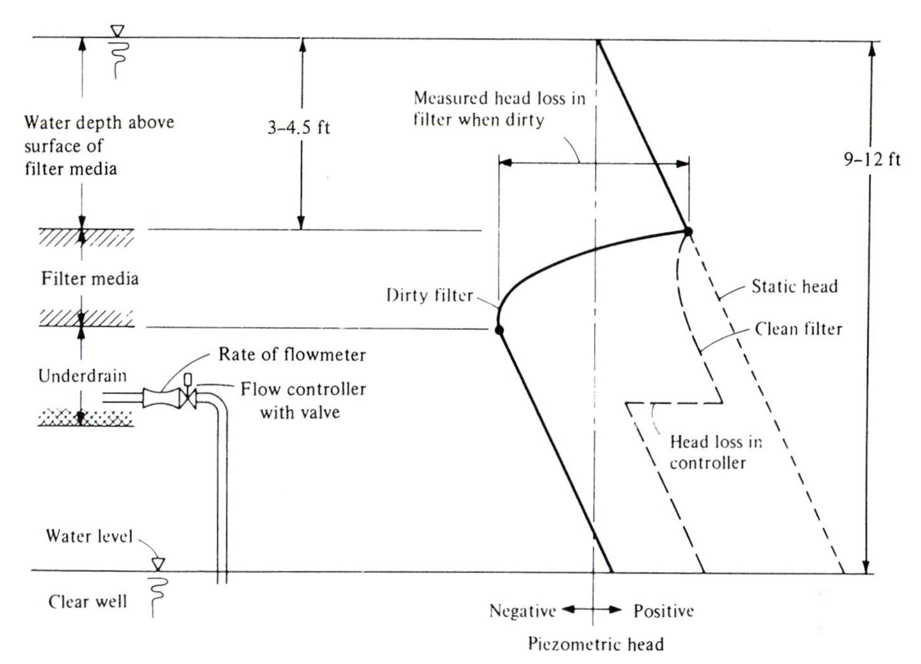

FIGURE 10.30 Piezometric head diagrams through a gravity filter system operated by control of the rate of flow.

The piezometric diagrams in Figure 10.30 illustrate the approximate hydraulic profiles through a gravity filter system operated by rate-of-flow control. During filtration the depth of water above the surface of the filter media is 3–4.5 ft (0.9–1.4 m). The filter effluent pipe is trapped in the clear well to provide a connection to water above the media and to prevent backflow of air to the bottom of the filters. The total head available for filtration is equal to the difference between the elevation of the water surface above the media and the water level in the clear well, commonly 9–12 ft (2.7–3.7 m). A rate-of-flow controller (a valve controlled by a flowmeter) in the filter effluent pipe regulates the rate of flow through a clean bed. As the pores of the filter media fill, the head loss in the bed increases and the valve automatically opens wider to maintain constant flow. The flow controller valve is wide open when the measured head loss through the filter media is 8–10 ft (2.4–3.0 m) and the bed is cleaned by backwashing. A control console for each filter unit is provided with a head-loss gauge, flowmeter, and rate controller.

The advantage of rate-of-flow control is that the design procedures are well established. The main disadvantages are the high construction and maintenance costs of this relatively complex control system. Also, manual or automatic control of plant influent is necessary to equal the outflow and to maintain constant water levels above the filter beds.

Constant-Level Control

Constant-level control, which is similar to constant-rate control, employs an effluent control valve that automatically adjusts the filter discharge to maintain a preset water level above the bed in the filter box. If the filtration rate is too high, a dropping water level throttles the valve. Conversely, as the rate decreases, resulting from accumulation of impurities in the bed, the tendency of the water level to rise opens the control valve. For uniform operation, the plant inflow must remain unchanged and the effluent water level in the clear well held constant because of the hydraulic connection to the water above the bed. In addition to cost and mechanical complexity, a potential disadvantage for this type of control system is the possibility of momentary surges in filtration rates caused by an increased plant inflow or sudden rise in water level above the operating filters when one unit is taken out of service for backwashing. Sudden filtration rate increases through partially dirty filters can result in accumulated impurities being flushed from the bed in the effluent.

Influent Flow Splitting

The typical arrangement of filtration control through flow splitting for constant-rate filtration is illustrated in Figure 10.31. While employed selectively in drinking water filtration, this system is commonly constructed for effluent filtration in advanced wastewater processing.

Plant inflow is split equally to all operating filters through an influent weir box (1) at each unit, passes through the media to the underdrain, and then into the clear well through valve 2. At minimum operating water level, the applied water is conveyed over the bed through the wash troughs and overflows into the water overlying the bed without disturbing the media. Head loss through the clean filter is h_c (Fig. 10.31). After

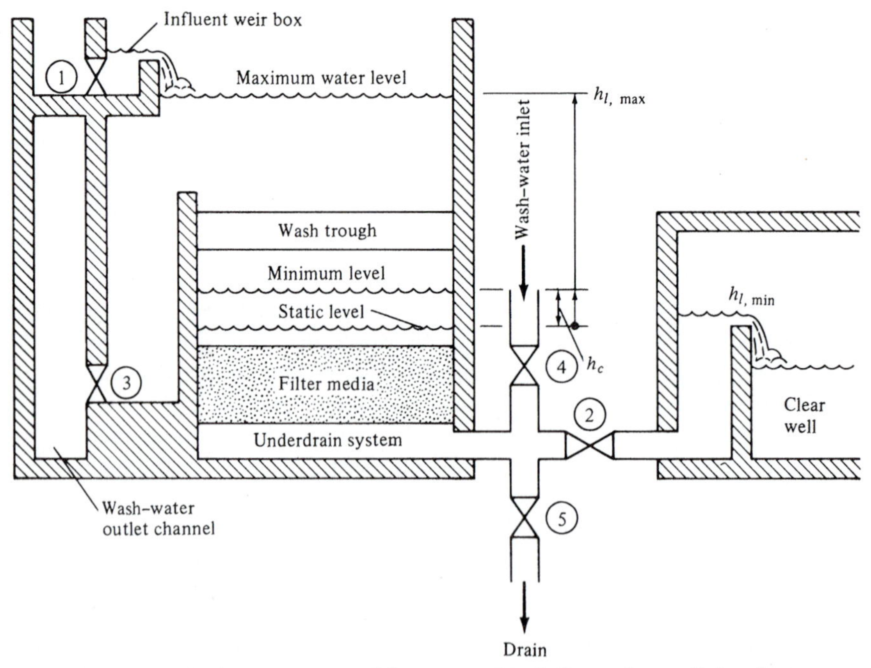

FIGURE 10.31 Typical arrangement of flow control by influent flow splitting for constant-rate filtration.

the troughs are submerged, the influent weir discharges directly into the water above the bed. To prevent accidental dewatering of the filter media, the underflow passes over an effluent weir located above the surface of the bed to enter the clear well. This arrangement eliminates the possibility of creating a negative head in the filter since the hydraulic pressure for filtration is generated only by the depth of water above static level. Constant-rate filtration is achieved without rate-of-flow controllers if the total plant inflow remains constant. Furthermore, effluent flowmeters are not necessary on the filters because the flow per filter is approximately the total plant inflow divided by the number of units in operation. As impurities fill the media, the water level above the bed rises to maintain the constant filtration rate. Thus, the head loss is apparent by the depth of water above the filter bed, which can be read by either a simple staff gauge or an automatic water-level recorder. When the water level reaches the maximum operating level, the filter is backwashed.

The sequence for cleaning the bed is started by closing influent valve 1 and opening valve 3 to the wash-water outlet channel, thus draining out the water overlying the bed and lowering the level to the wash troughs. After the close of valve 2 to the clear well, wash-water valve 4 can be opened, allowing wash water to flush impurities out of the fluidized bed. The clean media are allowed to restratify in quiescent water before closing and opening appropriate valves to restore filtration. The initial filtered water, if contaminated, can be wasted to a drain by opening valve 5. When one unit is taken out of service for backwashing, the water level gradually rises above the remaining beds until sufficient head is achieved to filter the higher flow received. Filtration rates increase slowly and smoothly with minimal effect on the filtered water quality.

The flow-splitting filtration system reduces the need for the piping and hydraulic controls required by traditional rate-of-flow control systems. The only disadvantage is the additional 5–6 ft (1.5–1.8 m) required in the depth of filter boxes to provide available head above the bed for filtration. On the other hand, eliminating the suction head under the bed removes the possibility of a negative pressure in the media with the consequent release of dissolved gases.

Declining-Rate Filtration

As illustrated in Figure 10.32, the cross-sectional view of a declining-rate filter appears similar to that of an influent flow-splitting unit. The principal differences are the location and type of influent arrangement and the provision of less available head loss.

Inflow enters a filter through valve 1 below the low-water level from a common channel (or large influent pipe) supplying a series of filters. It passes through the filter media, underdrain, and valve 2 into the clear well. Because of the common supply, the depth of water over all filters in the group being served is the same at any time. The filtration rate of any one unit depends on the head loss through that bed; consequently, the cleanest filter accepts the greatest flow and the dirtiest the least. As the rates of filtration through some beds decrease, inflow is automatically redistributed to the cleaner beds to compensate for the capacity lost by the dirtier filters. During redistribution of flow, filtration rates shift gradually among the interconnected units as the water level rises slowly, providing the needed additional hydraulic head to maintain a constant

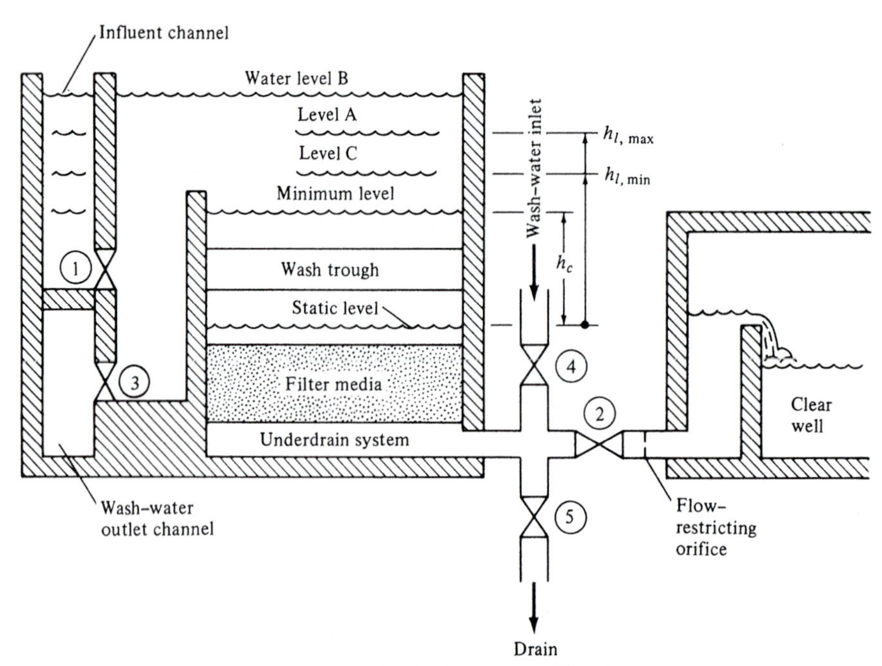

FIGURE 10.32 Typical arrangement of a declining-rate filtration system.

overall quantity of water being processed. Eventually, the operating water level reaches the maximum elevation and the dirtiest filter is backwashed to increase its rate of filtration. After this unit returns to service, the flow to all filters redistributes at a lower operating water level.

The general behavior of declining-rate filters is depicted in Figure 10.33 [16]. The upper curve gives the rate of filtration for filter 1 starting at 6 gpm/ft^2 and slowly declining to 3 gpm/ft^2, which represents the time from starting as a clean bed to backwash. In the lower diagram, the solid line shows the water level gradually rising above all filters as their head losses increase, resulting from accumulation of impurities in the granular media. At point *A* the dirtiest bed (2) is backwashed and the three remaining filters receive additional flow to compensate for this unit being removed from service for cleaning. Overall filtration capacity is retained by the automatic increase in rates through the operating filters forced by the rise in water level from point *A* to point *B*. After the clean filter 2 is returned to operation, the common water level drops to level *C*.

The vertical distance from the horizontal axis of Figure 10.33 to the water level line is the elevation of water above the effluent weir, which is the total head loss through any filter. The dashed lines apply to filter 1 starting as a clean bed to the time of backwash. Measuring upward from the horizontal axis, the vertical distances between the lines represent head losses in the underdrains and piping, clean media, and head loss caused by the accumulation of impurities in the media. The latter increases as the bed becomes dirtier, and the losses in the clean media and underdrain system decrease relative to the loss caused by solids removal as the rate of

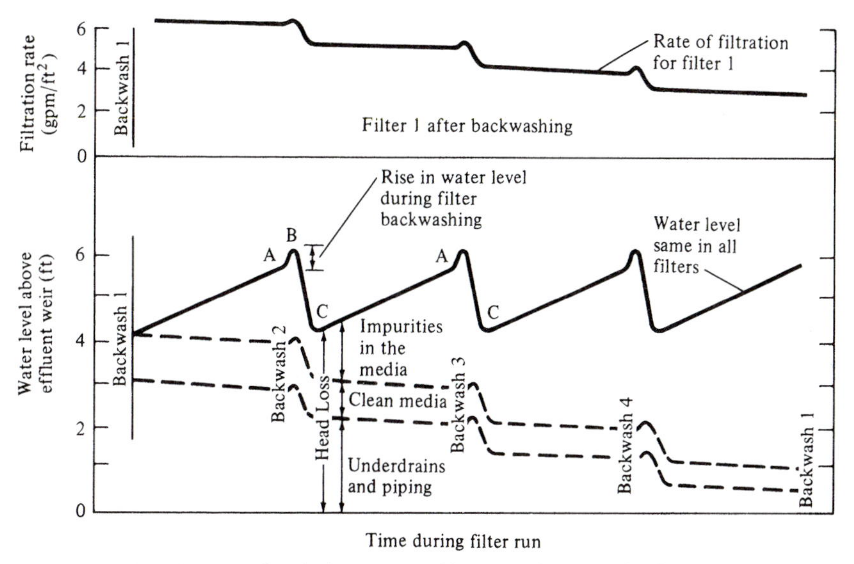

FIGURE 10.33 Diagrams for declining-rate filtration showing the filtration rate, water level, and head losses for one filter run in a plant having four filters. *Source:* From J. L. Cleasby, "Declining-Rate Filtration," *J. Am. Water Works Assoc.* 73, no. 9 (1981): 485.

filtration through filter 1 decreases. Backwashing of filters 2–4 causes the temporary rises in all of the curves.

Design considerations unique to a declining-rate filtration system are predicting water-level variations, preventing excessive rates through clean filters, and the influence of inflow variation on filtration hydraulics. Cleasby and DiBernardo [17] discuss the hydraulic considerations in declining-rate filtration in detail, including sample calculations to predict water-level variations and head losses necessary for design. The minimum water level with a head loss of h_c in Figure 10.32 results only if all beds are clean. The other water levels, labeled A–C in Figure 10.32, correspond to the labeled points in Figure 10.33. The maximum level B results when one filter is out of service for backwashing. The range between levels C and A is the normal operating zone of head loss during the time period between backwashing of filters. The recommendation for preventing an excessive filtration rate through a clean filter is to insert a flow-restricting device in the pipeline leading from the filter underdrain to the clear well. This could be an orifice, a fixed-position valve, or a two-position valve.

Advantages attributed to declining-rate filtration compared to traditional systems are that effluent quality is improved, head loss is lower for an equal length of filter run, smooth transitions in filtration rate changes are attained, negative pressure cannot occur in the bed, flowmeters and effluent control equipment are not needed, and construction and maintenance costs are lower. The disadvantages are that a greater depth is required in filter boxes, a high-water-level alarm is desirable, and, since the rate through filters cannot be manipulated, the ability of the plant to process increased flows must be considered by the operator [17].

10.19 HEAD LOSSES THROUGH FILTER MEDIA

Head loss through a clean granular-media filter is generally less than 3 ft (0.9 m). With accumulation of impurities, head loss gradually increases until the filter is backwashed, usually at 8–10 ft (2.4–3.0 m). In this section the Kozeny equation is presented for calculating head loss through a clean filter. Several other formulas not discussed here have been developed, such as those of Carman and Kozeny [18] and Rose [19]. Formulation of a relationship for head loss through dirty filters is very difficult; although several techniques have been proposed, none is considered able to provide substantially accurate predictions.

Poiseuille's equation for laminar flow through a circular capillary tube is

$$\frac{h}{l} = \frac{32\nu v}{gD^2} \tag{10.17}$$

where

h/l = head loss per unit length, ft H_2O/ft (m/m)
ν = kinematic viscosity (μ/ρ), ft^2/s (m^2/s)
v = mean velocity of flow, ft/s (m/s)
g = acceleration of gravity, ft/s^2 (m/s^2)
D = capillary tube diameter, ft (m)

Granular media can be idealized as a bundle of capillary tubes representing the pore passages of the bed. The hydraulic radius for such an ideal bed is equal to the volume of the pores divided by the wetted internal surface area of the tubular passages. For a unit volume of the bed, the pore volume is the porosity and the wetted surface area is the volume of spherical grains times the volume of a sphere ($\pi d^3/6$) divided by the area of a sphere (πd^2). Therefore,

$$\text{hydraulic radius} = \frac{\varepsilon}{(1-\varepsilon)(\pi d^2)/(\pi d^3/6)} = \frac{d\varepsilon}{6(1-\varepsilon)} \tag{10.18}$$

where

ε = porosity of the stationary filter bed, dimensionless
d = diameter of spherical grains, ft (m)

Replacing the 6 with S (the shape factor) and substituting a constant times the square of the hydraulic radius for D^2 in Poiseuille's equation yields the Kozeny equation [20] for the rate of head loss for water flow through a clean granular medium:

$$\frac{h}{l} = \frac{J\nu(1-\varepsilon)^2 VS^2}{g\varepsilon^3 d^2} \tag{10.19}$$

where

h/l = head loss per unit depth of filter bed, ft H_2O/ft (m/m)

J = constant, approximately 6 for filtration in the laminar flow region, dimensionless

ν = kinematic viscosity (μ/ρ), ft^2/s (m^2/s)

g = acceleration of gravity, ft/s^2 (m/s^2)

ε = porosity of the stationary filter bed, dimensionless

V = superficial (approach) velocity of water above the bed, ft/s (m/s)

S = shape factor ranging from 6.0 for spherical grains to 7.5 for angular grains, dimensionless

d = mean grain diameter, ft (m)

Equation (10.19) can be easily applied to determine head loss in a homogeneous granular-media bed. Filter beds in water and wastewater, however, are usually graded beds stratified as a result of backwashing with the coarsest grains on the bottom and finest on top. The grain-size distribution in a bed is defined by a sieve analysis. Table 10.4 lists the U.S. sieve series by the sieve designation number (the approximate number of meshes per inch) and the size of openings. For practical purposes, the thicknesses of substantially uniform layers in a stratified bed can be assumed to be proportional to the weights of the portions separated by the sieves. Hence, the total head loss is the sum of the head losses calculated by the Kozeny equation for successive layers based on the weight gradation from a sieve analysis of the filter material. Fair and Hatch [21] developed this concept and proposed the following equation for calculating head loss through each layer of a clean stratified filter bed:

$$\frac{h}{l} = \frac{36k\nu(1 - \varepsilon)^2 V}{g\varepsilon^3\psi^2}\sum_{i=1}^{n}\frac{P_i}{d_i^2} \tag{10.20}$$

TABLE 10.4 U.S. Sieve Series

Sieve Designation Number	Size of Opening (mm)	Sieve Designation Number	Size of Opening (mm)
200	0.074	20	0.84
140	0.105	(18)	(1.00)
100	0.149	16	1.19
70	0.210	12	1.68
50	0.297	8	2.38
40	0.42	6	3.36
30	0.59	4	4.76

where

k = constant equal to 5.0, dimensionless

ψ = sphericity of grains, ratio of surface area of equal volume spheres to the actual surface area of grains, dimensionless

P_i = fraction of total weight of filter grains in any layer i, dimensionless

d_i = geometric mean diameter of grains in layer i, ft (m)

Based on tests, Fair and Hatch [21] used a value for k of 5.0 as opposed to Kozeny's J of 6. The geometric mean grain diameter d_i (the square root of thc product of the upper and lower grain sizes) can be computed from either the upper and lower sieve size of any layer i or taken from the gradation curve of a grain size analysis. P_i is the percentage by weight of filter grains trapped between adjacent sieves. The $6/\psi$ is the same as the shape factor S. Values of ψ for filter sand are generally in the range of 0.8–0.7, rounded to angular, and for anthracite media from 0.7 to 0.4, angular to jagged.

Example 10.10 illustrates the application of Eq. (10.20).

Example 10.10

Calculate the head loss through a clean sand filter with a gradation defined by the sieve analysis given in Table 10.5 at a filtration rate of 2.7 $l/m^2 \cdot s$ (0.0027 m/s). The bed has a depth of 0.70 m with a porosity of 0.45 and the grains of sand have a sphericity of 0.75. Assume a water temperature of 10° C.

Solution: From Table 10.5, the computed value

$$\sum \frac{P_i}{d_i^2} = 2.34\ \text{mm}^{-2} = 2.34 \times 10^6\ \text{m}^{-2}$$

Appendix Table A.9 gives ν at 10° C as 1.306×10^{-6} m²/s.

TABLE 10.5 Sieve Analysis and Computation of Head Loss for Example 10.10

Sieve Designation Number	Size of Opening S (mm)	Geometric Mean Diameter $(S_1 \times S_2)^{0.5}$ (mm)	Fraction of Sand Retained	$\frac{P_i}{d_i^2}$ (mm)$^{-2}$
12	1.68			
		1.41	0.02	0.01
16	1.19			
		1.00	0.25	0.25
20	0.84			
		0.70	0.47	0.96
30	0.59			
		0.50	0.24	0.96
40	0.42			
		0.35	0.02	0.16
50	0.297			
Total			1.00	2.34

Substituting into Eq. (10.27),

$$\frac{h}{l} = \frac{36 \times 5.0 \times 1.306 \times 10^{6}\ \text{m}^2/\text{s} \times (1 - 0.45)^2\ 0.0027\ \text{m/s} \times 2.34 \times 10^{-6}\ \text{m}^{-2}}{9.807\ \text{m/s}^2\ (0.45)^3\ (0.75)^2}$$

$$= 0.89\ \text{m/m}$$

head loss through filter = 0.70 m × 0.89 m/m = 0.62 m.

10.20 BACKWASHING AND MEDIA FLUIDIZATION

Filtration can be stopped because of a low rate of filtration, passage of excess turbidity through the bed, or "air binding." As head loss increases across the bed, the lower portion of the filter is under a partial vacuum (Fig. 10.30). This negative head permits the release of dissolved gases, which tend to fill the pores of the filter, causing air binding and reducing the rate of filtration.

Under average operating conditions, granular-media filters are backwashed about once in 24 hr at a rate of about 15 gpm/ft^2 (10 l/m$^2 \cdot$ s) for a period of 5–10 min. Initial filtered water may be wasted for 3–5 min. A bed is out of operation for 10–15 min to complete the cleaning process. The amount of water used in backwashing varies from 2% to 4% of the filtered water. During backwashing the bed of filter media is expanded hydraulically about 50%, and the released impurities are conveyed in the wash water to the wash troughs.

Problems in backwashing of dual-media filters can result if the cleaning action is limited to wash-water fluidization. Nonuniform expansion and poor scouring can result in mud balls dropping through the coarser coal media and lodging on top of the sand layer. The common methods of cleaning dual-media filters use either air scouring, or air-and-water scouring prior to water backwash. For separate air scouring, the water level is lowered below the wash-water troughs to near the surface of the media and only air is introduced to mix and scour the media. Wash water is then used to fluidize the bed and purge contaminants. For air-and-water scouring, the wash cycle also begins by draining off the water above the filter. After backwash flow has started, at about one-quarter of the fluidizing rate, air is supplied and the simultaneous flow of air and water scours the bed as the wash-water level rises in the filter box. When the water surface approaches the bottom of the wash troughs, air injection is terminated and the backwash rate increases to the desired fluidization velocity to carry the impurities out of the expanded bed.

The granular media are thoroughly mixed in the agitated, turbulent flow of an expanded bed during backwashing. When the upward flow of wash water is stopped, the suspended grains settle down to form a stratified bed with the finest grains of each medium on top. In a mixed-media bed the medium of lowest density settles on top, that is, the anthracite layer above the sand bed.

Fluidization

Fluidization is defined as upward flow through a granular bed at sufficient velocity to suspend the grains in the water. During the process of fluidization, the upward flow overcomes the gravitational force on the grains, and the energy loss is due to fluid motion.

The viscous energy loss is proportional to the velocity of flow, and the kinetic energy loss is proportional to the square of the velocity.

The pressure loss through a fixed bed is a linear function of flow rate at low superficial velocities when flow is laminar. (The superficial velocity is the quantity of flow divided by the cross-sectional area of the filter.) As the flow rate increases further, the resistance of the grains to wash-water flow increases until this resistance equals the gravitational force and the grains are suspended in the water. Any further increase in upward velocity results in additional expansion of the bed while maintaining a constant pressure drop equal to the buoyant weight of the media. The characteristics of an ideal fluidized bed and departures from that behavior because of real conditions are shown in Figure 10.34 [22].

The frictional drag of grains suspended in upward flowing water is counterbalanced exactly by the pull of gravity. Therefore, the pressure drop after fluidization is equal to the buoyant weight of the grains, which can be calculated as

$$\Delta p = h\rho g = l(\rho_s - \rho)g(1 - \varepsilon) \tag{10.21}$$

where

Δp = pressure drop after fluidization, lb (N)

h = head loss (as a water column height), ft (m)

l = height of expanded bed, ft (m)

ρ = mass density of wash water, $\text{lb} \cdot \text{s}^2/\text{ft}^4$ (kg/m^3)

ρ_s = mass density of solid grains, $\text{lb} \cdot \text{s}^2/\text{ft}^4$ (kg/m^3)

ε = porosity of expanded bed, dimensionless

g = acceleration of gravity, ft/s^2 (m/s^2)

Since the quantity of filter medium remains the same whether the bed is stationary or fluidized, the volume of grains initially can be equated to the volume of grains after expansion.

$$l(1 - \varepsilon) = l_0(1 - \varepsilon_0) \tag{10.22}$$

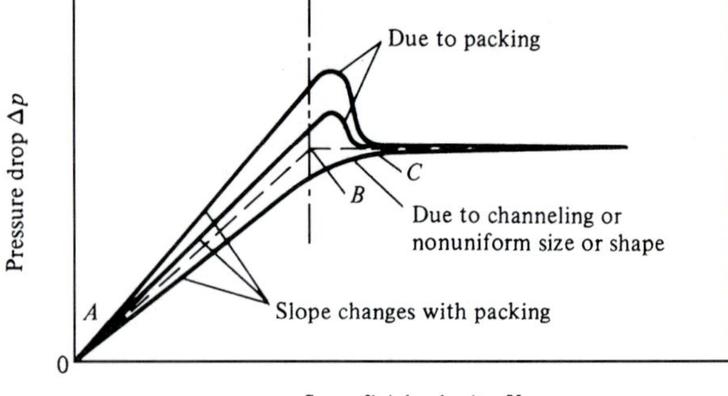

FIGURE 10.34 Characteristics of a fluidized bed of granular media. Solid curves: real bed; dashed lines: ideal bed. *Source:* From J. L. Cleasby and K. Fan, "Predicting Fluidization and Expansion of Filter Media," *J. Env. Eng. Div., Proc. Am. Soc. Civil Engrs.* 107 (EE3) (1981): 459.

where

$$l_0 = \text{height of stationary bed}$$
$$\varepsilon_0 = \text{porosity of stationary bed}$$

Minimum Fluidizing Velocity

The superficial water velocity required for the onset of fluidization is defined by point B in Figure 10.34 for an ideal bed composed of unisized spherical particles. For a graded bed, the minimum fluidizing velocity is not the same for all the grains. Therefore, the change from a stationary to an expanded bed occurs gradually, with complete fluidization at a higher superficial velocity indicated by point C. Further increase in flow causes the concentration of particles to decrease and the porosity to approach 1. Ideally, the superficial velocity equals the unhindered terminal settling velocity of a single particle. It can be measured by experimentation in a laboratory using a filter column. At gradually increasing superficial velocities, head losses through the expanding bed are measured and plotted in a form similar to Figure 10.34. The minimum fluidizing velocity is at the point where the curve becomes horizontal and subsequent head loss measurements remain constant.

The minimum fluidizing velocity can also be calculated from relationships substantiated by experimental observations. One rational approach equates the head loss determined by the Ergun equation [22,23] to the head loss using Eq. (10.21) at the point of incipient fluidization. For nonspherical particles, Wen and Yu [24] modified this relationship to account for sphericity and stationary bed porosity. The following equations resulted:

$$\text{Re}_{\text{mf}} = [(33.7)^2 + 0.0408\ \text{Ga}]^{0.5} - 33.7 \tag{10.23}$$

$$\text{Re}_{\text{mf}} = \frac{d_{\text{eq}} V_{\text{mf}} \rho}{\mu} \tag{10.24}$$

$$\text{Ga} = \frac{d_{\text{eq}}^3 \rho (\rho_s - \rho) g}{\mu^2} \tag{10.25}$$

where

Re_{mf} = Reynolds number, dimensionless

Ga = Galileo number, dimensionless

d_{eq} = grain diameter of a sphere of equal volume, ft (m)

V_{mf} = minimum fluidizing velocity (superficial velocity at the point of minimum fluidization), ft/s (m/s)

μ = absolute (dynamic) viscosity of water, $\text{lb} \cdot \text{s/ft}^2$ ($\text{kg/m} \cdot \text{s}$)

ρ = mass density of water, $\text{lb} \cdot \text{s}^2/\text{ft}^4$ (kg/m^3)

ρ_s = mass density of grains, $\text{lb} \cdot \text{s}^2/\text{ft}^4$ (kg/m^3)

In a bed with a gradation in size of the grains, a coarser particle diameter is used to calculate V_{mf} to ensure fluidizing of the entire bed. Cleasby [22] recommends substituting d_{90} sieve size (90% of the grains by weight are smaller) in Eqs. (10.24) and (10.25); the value for d_{eq} is often not conveniently available. Furthermore, to allow free movement of these coarse grains during backwashing, a 30% margin of safety is recommended so that the minimum backwash rate becomes $1.3V_{mf}$ [22].

This procedure is adequate for selecting the minimum backwash rate for a single-medium filter. For a dual-media bed, the minimum fluidizing velocity for the mixed interfacial region between the two adjacent media may be somewhat lower than the fluidizing velocity of the coarser grains of the upper layer at the interface.

Expansion of a Fluidized Bed

Calculation of bed expansion relies on selecting the best estimates for sphericity and stationary-bed porosity. For irregular shapes, indirect laboratory methods are required to determine these parameters. A technique based on the hydrodynamic behavior of the grains, proposed by Briggs et al. [25], is the dynamic shape factor defined from one of the following equations:

$$\text{DSF} = \left(\frac{V_s}{V_n}\right)^2 \tag{10.26}$$

or

$$\text{DSF} = a\left(\frac{V_s}{V_n}\right) + b\left(\frac{V_s}{V_n}\right)^2 \tag{10.27}$$

where

DSF = dynamic shape factor

V_s = unhindered terminal settling velocity of the grains as measured for a representative sample of grains, ft/s (m/s)

V_n = unhindered terminal settling velocity of equivalent volume spherical grains calculated as follows with a C_D value from Figure 10.35, ft/s (m/s)

$$= \left[\frac{4g(\rho_s - \rho)d}{3C_D\rho}\right]^{0.5}$$

a, b = constants depending on Reynolds number of the spherically equivalent grains: a and b are selected such that $a + b = 1$ and b/a is the Reynolds number

An empirical expansion correlation for uniform beds of spherical particles, including a broad range of Reynolds numbers ranging from laminar to turbulent flow, was developed by Richardson and Zaki [26]. From laboratory fluidization experiments, they found that the logarithm of superficial velocity versus the logarithm of

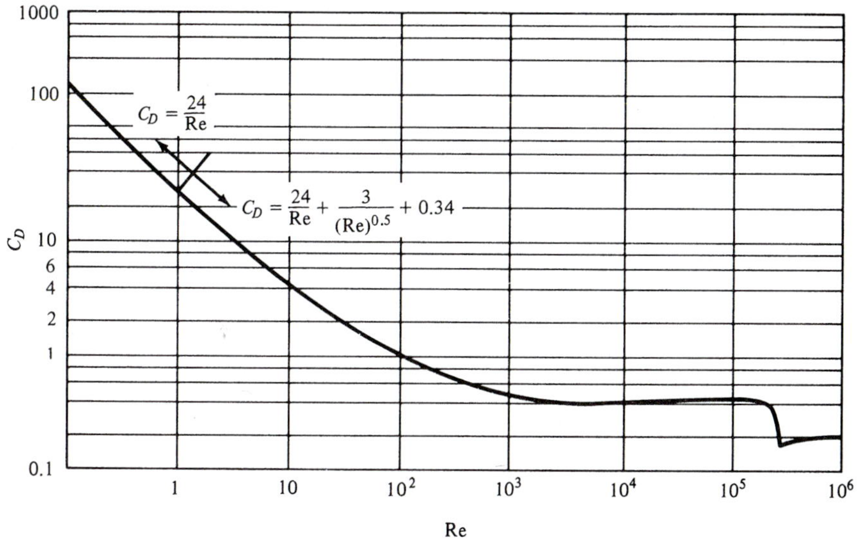

FIGURE 10.35 Drag coefficient C_D for a sphere versus Reynolds number Re.

porosity graphed as a straight line, as shown in Figure 10.36. The mathematical expression for this relationship is

$$\frac{V}{V_i} = \varepsilon^n \tag{10.28}$$

where

V = superficial velocity of water above the bed

V_i = intercept velocity at a porosity of 1 (a log porosity of zero)

ε = porosity of the expanded bed at V

n = slope of log V versus log ε plot (Fig. 10.36)

The slope n is a characteristic value for grains of a particular size, shape, and density.

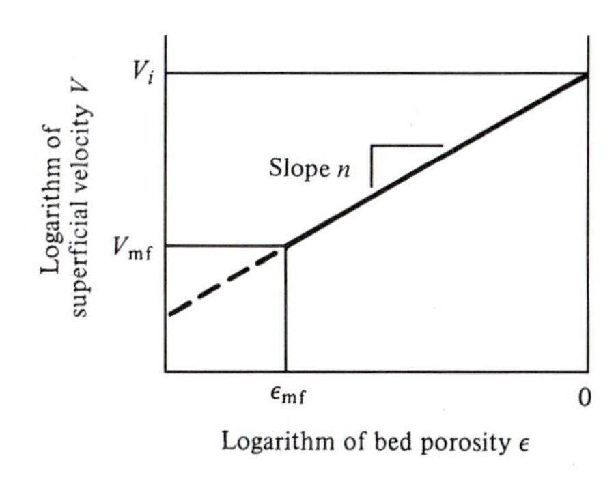

FIGURE 10.36 Relationship between superficial velocity and porosity for fluidization of an ideal bed of spherical particles developed by Richardson and Zaki [26].

Cleasby [22] developed empirical relationships for media with irregularly shaped grains by incorporating DSF in the determination of n for Eq. (10.28). This was accomplished using experimental data from fluidization of sand and anthracite beds of known physical properties. The following expression was solved for the exponent α to satisfy the expression for each size medium:

$$n_{(\text{actual})} = n_{(\text{spherical})}\text{DSF}^{\alpha} \tag{10.29}$$

A two-variable correlation of α against DSF and Re_0 yielded the desired form for α:

$$\alpha = -2.2715(\text{DSF})^{0.420}\,\text{Re}_0^{-0.441} \tag{10.30}$$

Thus, the resulting equations for the expansion coefficients are

$$n = \left(4.45 + 18\frac{d}{D}\right)\text{Re}_0^{-0.1}(\text{DSF})^{\alpha} \tag{10.31}$$

for Re_0 data from 15 to 200, and

$$n = 4.45\,\text{Re}_0^{-0.1}(\text{DSF})^{\alpha} \tag{10.32}$$

for Re_0 data from 200 to 503, where

Re_0 = Reynolds number based on unhindered terminal settling velocity V_s of grains

Values of n calculated using these formulas agreed closely with actual measured values of n at temperatures of both 25° C and 40° C. In these equations, the expansion slope n is related to Re_0, since V_s can be determined in a simple settling column and is measured in determining DSF. During fluidization experiments, the intercept velocities V_i when $\log \varepsilon = 0$ can be determined by extrapolation. Values of V_i are different from the V_s values in experiments with angular particles. The relation of V_i/V_s to DSF determined from experimental data is

$$\frac{V_i}{V_s} = 0.90(\text{DSF})^{-0.261} \tag{10.33}$$

Predicting expansion of a graded filter bed is possible if a sieve analysis is available. The technique involves dividing the material represented by the sieve analysis curve into several segments and applying the equation to determine the expanded depth of each segment at a particular flow rate. The total expanded depth is then the sum of the depths of equal segments, which are usually taken as five equal portions of 20% by weight. The middle size in each segment is used for the diameter term in the equation.

Example 10.11

Calculate the recommended minimum backwash rate for a sand filter with a d_{90} sieve opening equal to 1.00 mm (0.001 m). This grain diameter designation means that 90% of the filter sand by weight passed through a U.S. sieve number 18, which has an opening size of 1.00 mm. The water temperature is 25° C and the density of the sand is 2.65 g/cm^3 (2650 kg/m^3).

Solution: Calculation of the Reynolds number using d_{90} in Eq. (10.24):

$$\text{Re}_{mf} = \frac{0.001 \text{ m} \times V_{mf} \times 997 \text{ kg/m}^3}{0.890 \times 10^{-3} \text{ kg/m} \cdot \text{s}} = 1120 V_{mf} \text{ m/s}$$

Values for ρ and μ are for water at 25° C taken from Table A.9 in the Appendix.

Computation of the Galileo number using Eq. (10.25):

$$\text{Ga} = \frac{(0.001 \text{ m})^3 \times 997 \text{ kg/m}^3 \times [(2650 - 997) \text{ kg/m}^3] \times 9.81 \text{ m/s}^2}{(0.890 \times 10^{-3} \text{ kg/m} \cdot \text{s})^2} = 20{,}400$$

Substituting Re_{mf} and Ga into Eq. (10.23) and solving for V_{mf} yields

$$1120 V_{mf} = \{[(33.7)^2 + 0.0408 \times 20{,}400]^{0.5} - 33.7\} \text{ m/s}$$

$$V_{mf} = 0.0095 \text{ m/s} = 9.5 \text{ mm/s} = 9.5 \text{ l/m}^2 \cdot \text{s} = 14 \text{ gpm/ft}^2$$

With addition of a 30% margin of safety, the recommended backwash rate is $1.3 \times 9.5 = 12 \text{ l/m}^2 \cdot \text{s} = 18 \text{ gpm/ft}^2$.

10.21 PRESSURE FILTERS

Pressure filters have the granular media and underdrains contained in a steel tank, as illustrated in Figure 10.37. Water is pumped through the filter under pressure, and the media are washed by reversing flow through the bed, flushing out the impurities. Filtration rates are comparable to gravity filters; however, the maximum head loss can be significantly greater since it is a function of the input pump pressure rather than static water levels. Pressure filters are commonly installed in small municipal softening and iron-removal plants, in industrial water treatment processes, and for tertiary filtration of effluent from small wastewater plants.

10.22 MEMBRANE FILTRATION

Membrane filtration has been used for many years to provide high-purity water for such industries as beverages, electronics, and pharmaceuticals. In dry climates, reverse osmosis is used to reduce salinity of brackish groundwater and to desalt seawater. Currently, membrane processes are being considered in treatment of selected water sources to comply with more stringent regulatory standards for drinking water.

The application of different membrane systems is based on pore size for removal of contaminants. The fine-particle range of 0.1–10 μm (0.1–10 microns) encompasses turbidity-producing particles, *Giardia* cysts (8–18 μm by 5–16 μm), *Cryptosporidium* oocysts (4–6 μm), and large bacteria. The membrane process in the pore size range of approximately 0.7–7 μm is referred to as *microfiltration*. The molecular range of 0.001–0.1 μm includes microorganisms, colloids, and high-molecular-weight compounds.

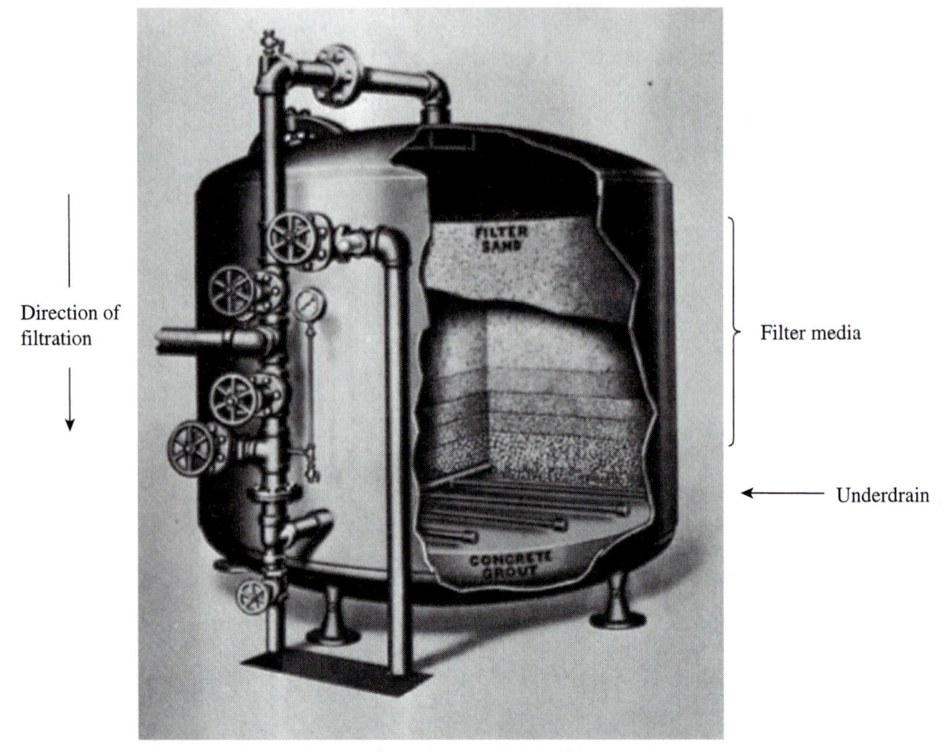

FIGURE 10.37 Pressure filter. *Source:* Courtesy of Infilco Degremont, Inc.

The membrane process in the pore size range of approximately 0.008–0.8 μm is *ultrafiltration*. The smallest range of 0.0001–0.001 includes removal of aqueous salts, dissolved organic compounds, and metal ions. The membrane processes in this range include *nanofiltration*—approximately 0.005–0.008 μm—and *reverse osmosis*—0.0001–0.007 μm.

Selection of a membrane process is determined by the desired treatment, such as turbidity reduction, disinfection, removal of organic compounds, softening, desalination, or specific ion removal (e.g., arsenic or nitrate). The cost of membrane treatment increases with the decreasing size of the contaminants to be removed. The major factors in operating costs are pretreatment and posttreatment for membrane separation and operating pressure to force water through the membrane. Microfiltration and ultrafiltration can be operated at 60 psi or less, while reverse osmosis is operated in the range of 150–800 psi, depending on recovery of product water.

Microfiltration and Ultrafiltration

These membranes for water and wastewater applications are usually hollow fibers bundled into a pressure vessel with one end potted with an epoxy resin. In the ultrafiltration module shown in Figure 10.38, the feedwater flows inside the fibers and the filtrate is collected from inside the pressure vessel by a central core tube. Concentrate is collected

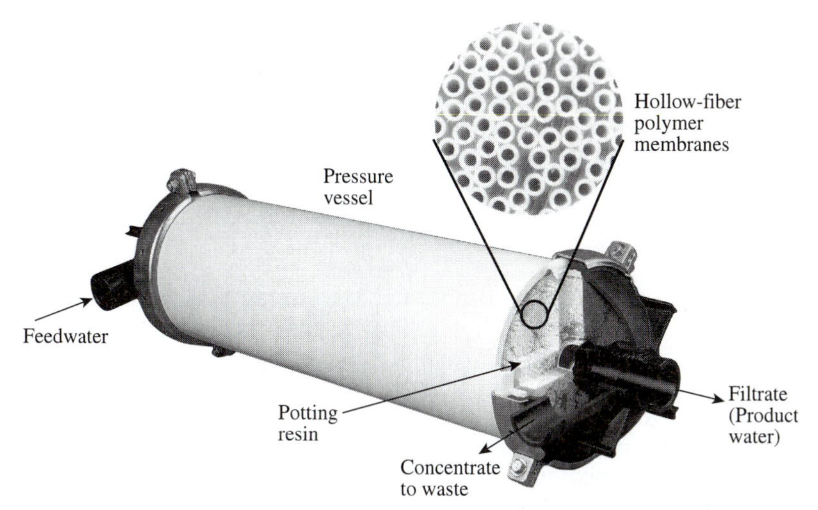

FIGURE 10.38 Ultrafiltration hollow-fiber module for removal of pathogens, including *Giardia* cysts, *Cryptosporidium* oocysts and viruses, and turbidity-producing particles. *Source:* Courtesy of Hydranautics, a Nitto Denko Corp.

from the ends of the hollow fibers and discharged to waste. The hollow-fiber membranes are available in two inside diameters, 0.8 mm and 1.2 mm for higher-turbidity feedwater. The maximum feedwater pressure is 73 psi, and maximum transmembrane pressure is 30 psi. Filtrate flow is 11–30 gpm (2.7–6.9 m^3/h). In the backwash mode, filtrate flows in reverse from the central core tube inside the pressure vessel through fibers to flush out the contaminants to waste. The water backwash cycle is every 15–60 min for 30–60 sec at 35 psi. Chemical enhanced backwash frequency is a minimum of 1 or 2 times per day for a duration of 1–10 min.

Microfiltration and ultrafiltration systems can also be constructed with hollow-fiber membranes that filter water from outside-in, transverse flow from the pressure vessel through the walls to the inside of the membranes [27]. In transverse-flow, contaminants in the feedwater build up on the outside of the hollow-fiber membrane surface and clean filtrate flows out of the inside. The advantage is the lower pressure drop for a given flow and number of fibers because of the greater surface area on the outside of the fibers than on the inside. The transmembrane pressure increases as the contaminant load increases, and a set point air-water backwash is automatically initiated. Pulses of compressed air from inside of a hollow-fiber membrane dislodge contaminants from the outside of the membrane surface. The contaminants are then carried out of the pressure vessel with either feedwater or filtrate as backwash water. Using air in the backwash sequence keeps fibers from coming together to impede cleaning. Other advantages demonstrated by the transverse-flow system are up to 95% recovery of feedwater and a significant reduction in the quantity of backwash-water production by using air-water backwash.

Pretreatment prior to membrane separation is required to remove large solids to prevent plugging of the hollow fibers. In water treatment, self-cleaning screens are often all that is necessary to protect the membranes. In wastewaters, plugging and fouling can result from organic compounds, chemical precipitates, bacterial slimes, and oils and greases. Pretreatment may require several unit processes, chemical cleaning of membranes is extensive, and membrane life can be reduced to a few years.

Reverse osmosis for reducing salinity of brackish groundwater and desalting seawater is considered a physical process within a chemical treatment system. Therefore, it is presented in Chapter 11, Section 11.34.

Example 10-12 [28]

Until 1997, the public water supply for Marquette, Michigan, was water taken from Lake Superior and then chlorinated. The lack of processing, other than chlorination, did not meet the Surface Water Treatment Rule (Section 11.24), which states that filtration is required unless the water supplier can demonstrate its drainage basin is fully controlled and the raw water quality meets stringent criteria. Under this rule, filtration was required even though the raw water turbidity averaged between 0.2 and 0.3 NTU (with a historical maximum of 5–6 NTU) from an intake on a rock bottom at a depth of 60 ft and 3000 ft from the shoreline. The recommendation for compliance was to construct a direct filtration plant (Section 10.16). The serious problem for the city was limited building space at the existing pumping station on the shore of the lake. Also, the site was surrounded by private residences and a museum, and the ground surface was bedrock.

Solution [28]: Direct filtration was not a viable option. Microfiltration was recommended because it could, along with chlorination, meet the required $C \cdot t$ (disinfectant concentration time of contact), only minimal amounts of chemicals are required, and better water quality can be produced compared to direct filtration.

Two manufacturers provided microfiltration equipment for a 6-month testing period to demonstrate the effectiveness of microfiltration during critical coldwater months. System A was microfiltration using an outside-in flow path (transverse-flow) and air-water backwash. System B was microfiltration using an inside-out flow path and water backwash. The pilot study monitored filtrate water quality by particle counts in the critical range of *Giardia* cysts and *Cryptosporidium* oocysts, turbidity, transmembrane pressure, and general operating parameters, such as backwash production and cleaning frequency. Results of the particle removal in the range of 2–10 μm in *log base 10* was 3.3 for system A and 3.8 for system B, and the turbidity was consistently in the range of 0.03–0.05 NTU for both systems. Both pilot units provided excellent filtrate water quality. Nevertheless, the air-water backwash of system A was more effective in cleaning the hollow fibers, and more effective in controlling transmembrane pressure. For systems A and B, the backwash production in percentage of filtered water were 8% and 40%, respectively, and the average cleaning frequencies expressed in days were 22 d and 2.4 d. Because of these considerations, system A was selected.

The full-scale microfiltration plant is designed based on the following: maximum membrane flux rate of 125 $l/m^2 \cdot h$, based on demonstrated pilot rates; average daily flow of 11,000 m^3/d (2.8 mgd); 8 frames with 90 modules per frame for a total of inside membrane area of 10,800 m^2 (116,000 ft^2) with one frame out of service; plant rated flow of 26,000 m^3/d (7.0 mgd) with one frame out of service, which is a flux rate of 100 $l/m^2 \cdot h$; and maximum backwash design of 3000 m^3/d (0.80 mgd).

Self-cleaning strainers (screens) are installed as pretreatment for microfiltation. Air-water backwash is discharged without treatment to Marquette Harbor under a NPDES (National Pollutant Discharge Elimination System) permit. Backwash from chemical cleaning of the membranes (sodium hydroxide and surfactant) is neutralized with acid before discharge to the sanitary sewer. The filtered water is chlorinated as it flows into the ground-level storage reservoir and rechlorinated as it is pumped into the distribution system. All operations are highly automated using computer-based control software.

PROBLEMS

10.1 A venturi water meter with a throat of 6.0 in. registers a pressure head difference of 128 in. of water between the inlet and throat. Calculate the quantity of flow through the meter using a discharge coefficient of 0.93.

10.2 Calculate the upper head in a Parshall flume with a throat width of 1.5 ft for a flow of 5.56 cfs.

10.3 The recommended minimum and maximum measurements for a Parshall flume have been experimentally determined. For a throat width of 1.5 ft, the minimum head is 0.10 ft and the maximum is 2.5 ft. Calculate the flow rates for these heads.

10.4 A wastewater bar screen is constructed with 0.25-in.-wide bars spaced 2 in apart center to center. If the approach velocity in the channel is 2.0 fps, what is the velocity through the screen openings?

10.5 Calculate points and sketch ideal output tracer curves (concentration versus normalized time) for a pulse tracer input for (a) a single completely mixed reactor and (b) a series of four equal-sized completely mixed reactors.

10.6 Dispersed plug flow through a baffled tank was analyzed by injecting a pulse of dye tracer into the influent and measuring the tracer concentrations in the effluent at time intervals of every 2 min after injection. The results are listed below. Sketch the tracer distribution curve by plotting C (mg/l) versus t (min) and locate the centroid. Calculate the dispersion number D/uL. By comparing this calculated value to those given in Figure 10.10, how is this amount of dispersion described?

t	C	t	C	t	C	t	C
0	0.00	6	0.10	12	0.40	18	0.05
2	0.00	8	0.35	14	0.22	20	0.00
4	0.00	10	0.65	16	0.11		

10.7 Dispersed plug flow through a compartmented aeration tank was analyzed by injecting a pulse of lithium chloride tracer in the influent. From the time and output concentration data listed, plot C (kg/m^3) versus t (min), the tracer response curve. Calculate the location of the centroid of the distribution $\bar{t}$, variance of the curve σ^2, normalized variance σ_θ^2, and the reactor dispersion number D/uL.

t	C	t	C	t	C	t	C
0	0	105	89.0	210	33.5	315	6.0
15	0	120	95.0	225	25.8	330	4.6
30	0	135	88.0	240	20.0	345	3.5
45	3.5	150	78.2	255	15.4	360	2.6
60	16.5	165	65.0	270	12.1	375	1.7
75	46.5	180	55.2	285	9.5	390	0.7
90	72.0	195	43.0	300	7.5	405	0

10.8 The residence time distribution of a cross-baffled serpentine chlorination tank was determined by injecting a pulse of 200 g of dissolved rhodamine dye into the influent. The tank is 9.4 m long and 6.7 m wide, with a longitudinal wall in the center. Each side has three baffle walls extending 2.5 m in from the outside wall and two extending in from the center wall to direct the serpentine flow pattern. The wastewater effluent enters through a channel near the top of the tank and discharges over an outlet weir. At the operating water depth of 2.55 m, the liquid volume is 137 m^3 (excluding the volume occupied by the concrete walls). The test was conducted at the peak hourly flow rate of 10,100 m^3/d. Effluent sampling times in minutes after injecting the dye and corresponding dye concentrations are listed below. The dye concentrations were determined by measuring absorbance with a spectrophotometer at a wavelength of 550 μm. From these data, plot the tracer response curve C (mg/l) versus t (min). Calculate and locate the mean residence time $\bar{t}$ and the theoretical mean residence time t_R. Calculate the reactor dispersion number D/uL. In order to normalize the concentration test data, the value of C_0 must be determined. A portion of the dye adsorbed onto the walls of the tank and became trapped in stagnant water in the corners and near the bottom at the inlet and outlet. Therefore, C_0 must be based on the calculated quantity of dye recovered in the effluent rather than on the amount injected. To estimate the milligrams of dye discharged, determine the area under the C-versus-t tracer response curve in milligrams · minutes/liter and multiply this value by the flow rate in liters/minute. The C_0 value is this amount of dye divided by the volume of water in the tank, which is equal to 137 m^3. Calculate the normalized value of the centroid $\bar{t}/t_R$ and the peak normalized concentration C_{peak}/C_0.

t	C	t	C	t	C	t	C
0	0.00	10	0.97	19	0.92	40	0.08
2	0.00	11	1.01	20	0.83	44	0.06
3	0.00	12	1.08	22	0.72	48	0.04
4	0.00	13	1.12	24	0.60	52	0.06
5	0.00	14	1.15	26	0.49	56	0.00
6	0.04	15	1.17	28	0.42	60	0.00
7	0.45	16	1.06	30	0.33	64	0.00
8	0.67	17	1.06	32	0.22	68	0.00
9	0.83	18	0.92	36	0.11		

10.9 A flocculation basin equipped with revolving paddles is 60 ft long (the direction of flow), 45 ft wide, and 14 ft deep and treats 10 mgd. The power input to provide paddle-blade velocities of 1.0 and 1.4 fps for the inner and outer blades, respectively, is 930 ft · lb/s. Calculate the detention time, horizontal flow-through velocity, and G (the mean velocity gradient) for a water temperature of 50° F.

10.10 A surface water treatment plant designed to process 70,000 m^3/d at a water temperature of 10° C has a four-compartment flocculation tank with a total length of 25 m, width of 12 m, and water depth of 5.0 m. The paddle flocculator in each of the four chambers between baffles has 4 blades, each 25 cm wide and 11.5 m long with the centerline of the paddles at a radius of 1.8 m. Assume the velocity of the water is 30% of the paddle velocity and the drag coefficient is 1.8. At a rotational speed of 2.0 rpm, calculate the velocity gradient.

10.11 A surface water treatment plant is being designed to process 50 mgd. The preliminary size of the flocculation tank is 96.0 ft long, with a series of 6 baffled compartments to create an over–under flow pattern. The tank width is 96.0 ft, and the water depth is 14.5 ft. Each of the six compartments has a horizontal shaft supporting six paddle flocculators with four arms on which to attach blades; the total number of units in the tank is 36. Up to five blades, each 15 ft by 0.5 ft, can be attached on each arm at radii of 6.5 ft, 5.5 ft, 4.5 ft, 3.5 ft, and 2.5 ft from the centerline of the shaft to the center of the blades. The maximum rotation of the flocculators, driven by variable-speed drives, can provide a velocity of 2.5 fps at a radius of 6.5 ft (center of the outer blade). What is the minimum number of blades needed on the flocculators for a velocity gradient of 60 fps/ft? Assume a ratio of water velocity to paddle velocity of 0.3, C_d equal to 1.8, and a water temperature of 50° F.

10.12 The settling velocity of alum floc is approximately 0.0014 fps in water at 10° C. Calculate the equivalent overflow rate in gpd/ft^2. What is the minimum detention time in hours to settle out alum floc in an ideal basin with a depth of 10 ft?

10.13 A treatment plant has straight-line processing with rapid mix chambers, flocculation tanks, and settling tanks sized according to the minimum values specified by *GLUMRB*. If the volume of the flocculation tanks is 100,000 gal, compute the design capacity of the plant. What are the total volumes of the tanks for applying chemicals to the influent and settling the flocculated water?

10.14 Each half of an in-line treatment plant, with a longitudinal cross-section as illustrated in Figure 10.39, has the following sized units: rapid mixing chamber with a volume of 855 ft^3; flocculation tank 140 ft wide, 58 ft long, and 14.5 ft of water depth; and sedimentation tank 140 ft wide, 280 ft long, and 17.0 ft of water depth. The length of the effluent weir along four weir channels and along the end of the tank is 1260 ft. Calculate the major parameters used in sizing these units based on a design flow of 40.0 mgd for each half of the plant. Compare the calculated values to the *GLUMRB Standards*.

10.15 Two rectangular clarifiers, each 30 ft long, 15 ft wide, and 10 ft deep, settle 0.40 mgd following alum coagulation. The effluent channels have a total weir length of 60 ft. Calculate the detention time, horizontal velocity of flow, and rate of flow over the outlet weir. Do these values meet the *Recommended Standards for Water Works* given in Section 10.13?

10.16 Lime precipitation of raw wastewater may be practiced for removal of phosphorus. Circular primary clarifiers with aerated flocculation wells as illustrated in Figure 10.40 were installed at a wastewater treatment plant for phosphorus precipitation and enhanced suspended-solids removal. The maximum hourly flow rate for each tank was 1.1 mgd. The lime dosage of 300 mg/l of CaO in a slurry was applied to the influent prior to entering the flocculation well. For this peak hydraulic loading, calculate the detention time based on the entire tank volume, flocculation time based on the volume of the flocculation well, overflow rate, and weir loading. Compare these values to the *Recommended Standards* for flocculator–clarifiers given in Section 10.13.

The performance of the clarifiers was very poor. The overflow was continuously milky, carrying out approximately one-third of the applied lime. Subsequent recarbonation was not

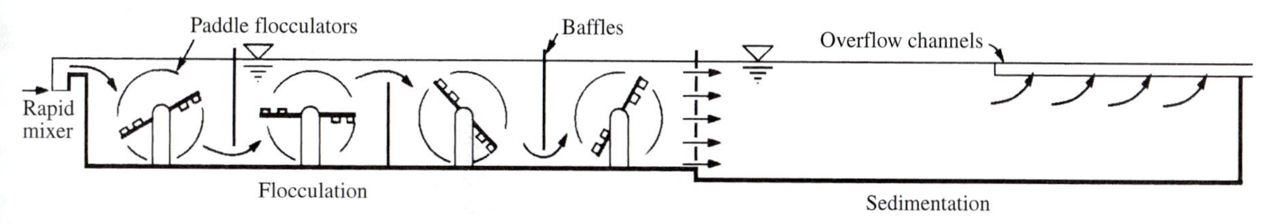

FIGURE 10.39 Illustration for Problem 10.14. In-line rapid mixing, flocculation, and sedimentation for treatment of a surface water.

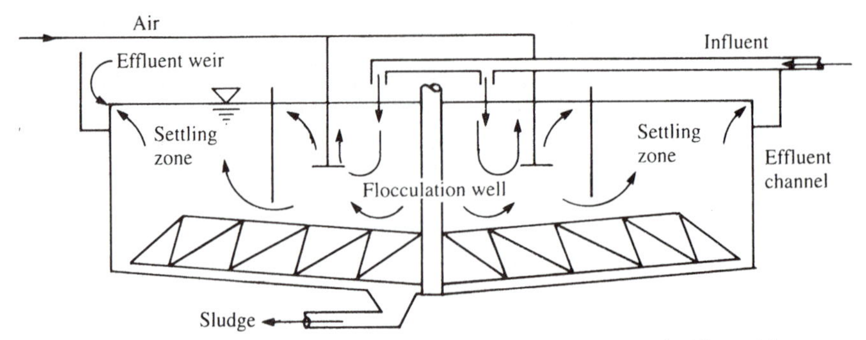

FIGURE 10.40 Illustration for Problem 10.16. A circular primary clarifier with an aerated flocculation well that receives the wastewater after the addition of lime in a rapid-mix tank. The key dimensions are diameter of the tank 40 ft, diameter of the flocculation well 20 ft, side-water depth 10 ft, and submergence of flocculation well 6.0 ft.

adequate to neutralize the pH below the range of 8.5–9.5; consequently, calcium carbonate precipitated in the following equalization tank and RBC tanks. The sludge production was significantly less than the anticipated quantity. Why was the operation of the clarifiers unsatisfactory? (Read the discussion of flocculator–clarifiers and examine Figure 10.18.)

10.17 A wastewater treatment plant has two primary clarifiers, each 20 m in diameter with a 2-m side-water depth. The effluent weirs are inboard channels set on a diameter of 18 m. For a flow of 12,900 m^3/d, calculate the overflow rate, detention time, and weir loading.

10.18 Calculate the diameter and depth of a circular sedimentation basin for a design flow of 3800 m^3/d based on an overflow rate of 0.00024 m/s and a detention time of 3 hr.

10.19 Two final clarifiers following activated-sludge processing have diameters of 60 ft and side-water depths of 11 ft. The effluent weirs are inboard channels set on a mean diameter of 55 ft. For a total flow of 3.4 mgd, calculate the overflow rate, detention time, and weir loading.

10.20 A rectangular sedimentation basin is to be designed for a flow of 1.0 mgd using a 2:1 length/width ratio, an overflow rate of 0.00077 fps, and a detention time of 3.0 hr. What are the dimensions of the basin?

10.21 Calculate the overflow rate and detention time in a primary clarifier with a diameter of 24 m and water depth of 2.3 m for a flow of 10,000 m^3/d.

10.22 A wastewater treatment plant has two rectangular primary settling tanks, each 40 ft long, 12 ft wide, and 7 ft deep. The effluent weir length in each tank is 45 ft. The average daily wastewater flow is 387,000 gal. Calculate the overflow rate and effluent weir loading. What is the estimated BOD removal?

10.23 How does the final clarifer for use with biological aeration illustrated in Figure 10.20 differ from the primary clarifier in Figure 10.19? Why are these differences important in gravity separation of suspended solids? (Refer to Figure 10.21, zone settling.) The proposed renovation of a wastewater treatment plant is considering construction of activated-sludge aeration tanks to replace existing trickling filters and keeping the existing final clarifiers in service. If you were reviewing this proposal, what would you recommend?

10.24 List the recommended design criteria for final clarifiers for an activated-sludge process with a design capacity of 20 mgd. Assuming four identical circular clarifiers, calculate the surface area, diameter, side-water depth, detention time, and weir loading based on an inboard weir channel. If the flow into each tank is 6.5 mgd (5.0 mgd plus 30% recirculation flow) and the suspended-solids concentration equals 2500 mg/l, calculate the solids loading. Is this solids loading satisfactory?

10.25 An aerated clarifier-type unit for grit removal and preaeration is 12 ft square with an 8-ft liquid depth. The wastewater flow is 0.80 mgd with an estimated grit volume of 3 ft^3/mil gal. A separate hopper-bottomed grit storage tank has a usable volume of 3 yd^3. Compute the detention time in the aerated unit and the estimated length of time required to fill the storage tank with grit.

10.26 Why must granular-media filtration in water treatment be preceded by chemical coagulation?

10.27 What are the limitations in applying direct filtration in water processing?

10.28 What is the major advantage of a dual-media coal-sand filter compared to a conventional sand filter?

10.29 Of the underdrains discussed, which kinds can use air scouring, air-and-water scouring, and no air scouring for filter backwashing?

10.30 In a gravity filter, how can the head loss gauge record a 9-ft loss when the water depth above the filter media surface is only 3.5 ft?

10.31 Refer to Figure 10.30. Assume the water depth above the surface of the filter media is 4 ft, the filter media is 2 ft deep, and the measured head loss as diagrammed in the dirty filter is 8 ft. This traditional flow-control system, by regulating the discharge from the underdrain, was installed at a treatment plant where the influent was cold water transported in a pipeline from a mountain reservoir at a higher elevation. Being delivered under pressure, the influent contained dissolved gases released during processing by depressurization and warming of the water. The dissolved gases came out of solution under the negative piezometric head in the filter media and formed tiny bubbles. This "air binding" (Section 10.20) increased the head loss and, at the start of backwashing, disturbed the underdrain media by eruption of the accumulated gas bubbles. A plant expansion is being proposed. What change in design of the filtration system would you recommend to reduce the problem of air binding?

10.32 What are the principal similarities and differences between declining-rate filtration and influent flow-splitting filtration?

10.33 Calculate the initial head loss through a filter with 18 in. of uniform sand having a porosity of 0.42 and a grain diameter of 1.6×10^{-3} ft. Assume spherical particles and a water temperature of 50° F. The filtration rate is 2.5 gpm/ft^2. Use the Kozeny equation.

10.34 Calculate the initial head loss through a dual-media filter consisting of a 12-in layer of uniform anthracite with a grain diameter of 2.0 mm and a 24-in. layer of uniform sand with a grain diameter of 1.0 mm at a filtration rate of 4.0 gpm/ft^2. The porosity of both media is 0.40, the shape factor for the anthracite is 7.5, and the shape factor for the sand is 6.0. Assume a water temperature of 60° F.

10.35 Calculate the initial head loss through a dual-media filter consisting of a 0.30-m layer of uniform anthracite with a grain diameter of 1.0 mm and a 0.30-m layer of uniform sand with a grain diameter of 0.50 mm at a filtration rate of 2.7 l/$m^2 \cdot$s. The porosity of both media is 0.42, the shape factor for the anthracite is 7.5, and the shape factor for the sand is 6.0. Assume a water temperature of 10° C.

10.36 Calculate the head loss through a clean sand filter with a gradation as given by the sieve analysis below. The filtration rate is 2.7 l/$m^2 \cdot$s, and the water temperature is 10° C. The filter depth is 0.70 m with a porosity of 0.45, and the sand grains have a sphericity of 0.75.

Sieve designation number	12	16	20	30	40	50
Fraction of sand retained	0	0.05	0.22	0.51	0.20	0.02

10.37 Calculate the recommended minimum backwash rate for a sand filter with a d_{90} sieve opening equal to 0.84 mm. (Ninety percent of the filter sand by weight passed through a sieve number 20.) The water temperature is 10° C and the density of the sand is 2650 kg/m^3.

10.38 Calculate the recommended minimum backwash rate for a single-medium, coarse-sand filter for tertiary filtration of reclaimed water at a temperature of 20° C. The grain diameter of a sphere of equal volume is 1.2 mm, density of the sand is 2650 kg/m^3, filter depth is 1.0 m, and porosity is 0.40. Also calculate the pressure drop through the filter during fluidization.

10.39 A dual-media filter consists of a 0.50-m layer of anthracite and 0.30-m layer of sand. The anthracite medium has a specific gravity of 1.67, porosity of 0.50, and a d_{90} grain size of 1.50 mm. The sand medium has a specific gravity of 2.65, a porosity of 0.40, and a d_{90} grain size of 0.90 mm. The water temperature is 20° C. (a) Calculate the recommended minimum backwash rate for the filter. (b) Calculate the pressure drop through the filter during fluidization.

REFERENCES

[1] L. D. Benefield and J.F. Judkins, Jr., *Treatment Plant Hydraulics for Environmental Engineers* (Upper Saddle River NJ: Prentice-Hall, 1984).

[2] O. Levenspiel, *Chemical Reaction Engineering* (New York: Wiley, 1972), 276.

[3] D. M. Marske and J. D. Boyle, "Chlorine Contact Chamber Design—A Field Evaluation," *Water and Sewage Works* 120, no. 1 (January 1973): 70–77.

[4] A. Amirtharajah, "Design of Rapid Mix Units," in *Water Treatment Plant Design*, R. Sanks, ed. (Ann Arbor, MI: Ann Arbor Science, 1978), 131–147.

[5] T. R. Camp and P. C. Stein, "Velocity Gradients and Internal Work in Fluid Motion," *J. Boston Soc. Civil Engrs.* 30 (1943): 219.

[6] G. M. Fair, J. C. Geyer, and D. A. Okun, *Water and Wastewater Engineering*, Vol. 2 (New York: Wiley, 1968).

[7] *Recommended Standards for Water Works, Great Lakes Upper Mississippi River Board of State Sanitary Engineers* (Albany, NY: Health Research Inc., 1992).

[8] *Recommended Standards for Wastewater Facilities, Great Lakes Upper Mississippi River Board of State Public Health & Environmental Managers* (Albany, NY: Health Research Inc., 1990).

[9] "Design Manual, Dewatering Municipal Wastewater Sludges," U.S. Environmental Protection Agency, Office of Research and Development, Center for Environmental Research Information, EPA/625/1-87/014 (September 1987).

[10] R. R. Mosher and D. W. Hendricks, "Rapid Rate Filtration of Low Turbidity Water Using Field-Scale Pilot Filters," *J. Am. Water Works Assoc.* 78, no. 12 (1986): 42–51.

[11] D. W. Hendricks et al., *Filtration of Giardia Cysts and Other Particles Under Treatment Plant Conditions* (Denver: Am. Water Works Assoc. Research Foundation, 1988).

[12] R. L. Culp, "Direct Filtration," *J. Am. Water Works Assoc.* 69, no. 7 (1977): 375–378.

[13] Committee Report, "The Status of Direct Filtration," *J. Am. Water Works Assoc.* 72, no. 7 (1980): 405–411.

[14] J. L. Cleasby, "Unconventional Filtration Rates, Media, and Backwashing Techniques," in *Innovations in the Water and Wastewater Fields*, E. A. Glysson et al., eds. (Boston: Butterworths, 1985), 1–23.

[15] R. D. G. Monk, "Design Options for Water Filtration," *J. Am. Water Works Assoc.* 79, no. 9 (1987): 93–106.

[16] J. L. Cleasby, "Declining-Rate Filtration," *J. Am. Water Works Assoc.* 73, no. 9 (1981): 484–489.

[17] J. L. Cleasby and L. DiBernardo, "Hydraulic Considerations in Declining-Rate Filtration," *J. Env. Eng. Div., Proc. Am. Soc. Civil Engrs.* 106(EE6) (1980): 1043–1055.

[18] L. G. Rich, *Unit Operations of Sanitary Engineering* (New York: Wiley, 1961), 137–146.

[19] H. E. Rose, "On the Resistance Coefficient-Reynolds Number Relationship for Fluid Flow through a Bed of Granular Material," *Proc. Inst. Mech. Engrs.* 153 (1945): 154–168; 160 (1949): 492–511.

[20] T. R. Camp, "Theory of Water Filtration," *J. San. Eng. Div., Proc. Am. Soc. Civil Engrs.* 90(SA4) (1964): 1–30.

[21] G. M. Fair and L. P. Hatch, "Fundamental Factors Governing the Streamline Flow of Water through Sand," *J. Am. Water Works Assoc.* 25, no. 11 (1933): 1551–1565.

[22] J. L. Cleasby and K. Fan, "Predicting Fluidization and Expansion of Filter Media," *J. Env. Eng. Div., Proc. Am. Soc. Civil Engrs.* 107(EE3) (1981): 455–471.

[23] S. Ergun, "Fluid Flow through Packed Columns," *Chem. Eng. Progress* 48, no. 2 (1952): 89–94.

[24] C. Y. Wen and Y. H. Yu, "Mechanics of Fluidization," *Chem. Eng. Progress Symposium Series* 62 (1966): 100–111.

[25] L. I. Briggs, D. S. McCulloch, and F. Moser, "The Hydraulic Shape of Sand Particles," *J. Sedimentary Petrol.* 32 (1962): 645–657.

[26] J. F. Richardson and W. N. Zaki, "Sedimentation and Fluidization, Part I," *Trans. Institution of Chem. Engrs.* 32 (1954): 35–53.

[27] *Handbook of Public Water Systems*, 2nd ed., "Membrane Treatment," (New York: John Wiley & Sons, Inc., for HDR Engineering Inc., 2001).

[28] W. A. Kelley and R. A. Olson, "Selecting MF to Satisfy Regulations," *J. Am. Water Works Assoc.* 91, no. 6 (1999): 52–63.

CHAPTER 11

Chemical Treatment Processes

OBJECTIVES

The purpose of this chapter is to:

- Review the basic concepts of the chemistry involved in chemical treatment processes
- Discuss types of chemical processes and their roles within treatment systems

Chemical processes are the most important unit operations in treatment of drinking water supplies. Surface waters require chemical coagulation to remove turbidity, color, and taste- and odor-producing compounds. In well water treatment, hardness can be reduced by lime–soda ash softening and iron and manganese removed by chemical oxidation. Distillation and reverse osmosis are employed to convert saltwater and brackish groundwater to fresh water. To meet chemical drinking water standards, organic compounds can be removed by activated-carbon adsorption. Other chemical processes as health benefits are fluoridation and corrosion control to reduce the dissolution of lead into drinking water.

Selected chemical processes also are applied in wastewater treatment, water reclamation, and conditioning of waste sludges. Chlorine disinfection of both drinking water and wastewater effluents is a major common chemical treatment. In water reclamation, wastewater can be clarified by chemical coagulation or lime precipitation. Polymers are commonly applied to wastewater sludges to allow release of water during thickening and dewatering. As a review of principles related to chemical treatment processes, the first section of this chapter discusses chemical considerations.

CHEMICAL CONSIDERATIONS

The purpose of these initial sections is to refresh students on selected fundamental concepts, including definitions, compounds, units of expression, the bicarbonate–carbonate system, chemical equilibria, and process kinetics. Basic chemistry is also presented as introductory material for each specific unit process discussed in the chapter. These brief comments are not intended to replace formal chemistry courses prerequisite to sanitary engineering.

11.1 INORGANIC CHEMICALS AND COMPOUNDS

Definitions

Chemical elements and their atomic weights are given in Table A.7 in the Appendix. *Atomic weight* is the weight of an element relative to that of carbon-12, which has an atomic weight of 12. *Valence* is the combining power of an element relative to that of the hydrogen atom, with an assigned value of 1. Thus, an element with a valence of 2+ can replace two hydrogen atoms in a compound or, in the case of a 2− valence, can react with two hydrogen atoms. Equivalent or combining weight of an element is equal to its atomic weight divided by the valence. For example, the equivalent weight of calcium (Ca^{2+}) equals 40.0 g divided by 2, or 20.0 g.

The *molecular weight* of a compound equals the sum of the weights of the combining elements and is conventionally expressed in grams. *Equivalent weight* is the molecular weight divided by the number of positive or negative electrical charges resulting from dissolution of the compound. Consider sulfuric acid, with a molecular weight of 98.1 g. *Ionization* releases two H^+ ions and one SO_4^{2-} radical; therefore, the equivalent weight of sulfuric acid is 98.1 divided by 2, or 49.0 g.

Chemicals Applied in Treatment

The common inorganic compounds used in water and wastewater processing are listed in Table 11.1. Given are the name, formula, common usage, molecular weight, and equivalent weight when appropriate. Common names and purity of commercial-grade chemicals are presented in the sections dealing with specific chemical treatment processes.

Units of Expression

The concentration of ions or chemicals in solution is normally expressed as weight of the element or compound in milligrams per liter of water, abbreviated as mg/l. Occasionally, the term *parts per million* (ppm) is used rather than mg/l. These are identical in meaning, since 1 mg/1,000,000 ml is essentially the same as 1 part by weight to 1 million parts for low concentrations. Chemical dosages may be expressed in units of pounds per million gallons or rarely as grains per gallon. To convert milligrams per liter to pounds per million gallons, multiply by 8.34, which is the weight of 1 gal of water. In other words,

$$1.0 \text{ mg/l} = \frac{1 \text{ gal by weight}}{1{,}000{,}000 \text{ gal}} = 8.34 \text{ lb/mil gal}$$

One pound contains 7000 grains, and 1.0 grain per gallon (gpg) equals 17.1 mg/l.

TABLE 11.1 Common Chemicals in Water and Wastewater Processing

Name	Formula	Common Application	Molecular Weight	Equivalent Weight
Activated carbon	C	Taste and odor control	12.0	n.a.[a]
Aluminum sulfate	$Al_2(SO_4)_3 \cdot 14.3\ H_2O$	Coagulation	600	100
Ammonia	NH_3	Chloramine disinfection	17.0	n.a.
Ammonium sulfate	$(NH_4)_2SO_4$	Coagulation	132	66.1
Calcium hydroxide	$Ca(OH)_2$	Softening	74.1	37.0
Calcium hypochlorite	$Ca(ClO)_2 \cdot 2\ H_2O$	Disinfection	179	n.a.
Calcium oxide	CaO	Softening	56.1	28.0
Carbon dioxide	CO_2	Recarbonation	44.0	22.0
Chlorine	Cl_2	Disinfection	71.0	n.a.
Chlorine dioxide	ClO_2	Taste and odor control	67.0	n.a.
Copper sulfate	$CuSO_4$	Algae control	160	79.8
Ferric chloride	$FeCl_3$	Coagulation	162	54.1
Ferric sulfate	$Fe_2(SO_4)_3$	Coagulation	400	66.7
Ferrous sulfate	$FeSO_4 \cdot 7\ H_2O$	Coagulation	278	139
Fluosilicic acid	H_2SiF_6	Fluoridation	144	n.a.
Magnesium hydroxide	$Mg(OH)_2$	Defluoridation	58.3	29.2
Oxygen	O_2	Aeration	32.0	16.0
Potassium permanganate	$KMnO_4$	Oxidation	158	n.a.
Sodium aluminate	$NaAlO_2$	Coagulation	82.0	n.a.
Sodium bicarbonate	$NaHCO_3$	pH adjustment	84.0	84.0
Sodium carbonate	Na_2CO_3	Softening	106	53.0
Sodium chloride	$NaCl$	Ion exchanger regeneration	58.4	58.4
Sodium fluoride	NaF	Fluoridation	42.0	n.a.
Sodium fluosilicate	Na_2SiF_6	Fluoridation	188	n.a.
Sodium hexametaphosphate	$(NaPO_3)_n$	Corrosion control	n.a.	n.a.
Sodium hydroxide	$NaOH$	pH adjustment	40.0	40.0
Sodium hypochlorite	$NaClO$	Disinfection	74.4	n.a
Sodium silicate	Na_4SiO_4	Coagulation aid	184	n.a.
Sodium thiosulfate	$Na_2S_2O_3$	Dechlorination	158	n.a.
Sulfur dioxide	SO_2	Dechlorination	64.1	n.a.
Sulfuric acid	H_2SO_4	pH adjustment	98.1	49.0

[a] Note applicable.

Elemental concentrations expressed in units of mg/l can usually be interpreted to mean that the solution contains the stated number of milligrams of that particular element. For example, a water containing 1.0 mg/l of fluoride means that there is 1.0 mg of F ion by weight per liter. However, in some cases, the concentration given in milligrams of weight does not relate to the specific element whose concentration is being expressed. For example, hardness, which is a measure of the calcium ion and magnesium ion content of a water, is given in weight units of calcium carbonate. This facilitates treating hardness as a single value rather than two concentrations expressed in different weight units, one for Ca^{2+} and the other for Mg^{2+}. The alkalinity of a water may consist of one or more of the following ionic forms: OH^-, CO_3^{2-}, and HCO_3^-. For commonality, the concentrations of these various radicals are given in mg/l as $CaCO_3$. According to *Standard Methods* [1], all nitrogen compounds—ammonia, nitrate, and organic nitrogen—are expressed in units of mg/l as nitrogen and phosphates are given in mg/l as phosphorus.

The term *milliequivalents per liter* (meq/l) expresses the concentration of a dissolved substance in terms of its combining weight. Milliequivalents are calculated from milligrams per liter for elemental ions by Eq. (11.1) and for radicals or compounds by Eq. (11.2):

$$\text{meq/l} = \text{mg/l} \times \frac{\text{valence}}{\text{atomic weight}} = \frac{\text{mg/l}}{\text{equivalent weight}} \tag{11.1}$$

$$\text{meq/l} = \text{mg/l} \times \frac{\text{electrical charge}}{\text{molecular weight}} = \frac{\text{mg/l}}{\text{equivalent weight}} \tag{11.2}$$

Milliequivalents-per-Liter Bar Graph

Results of a water analysis are normally expressed in milligrams per liter and reported in tabular form. For better visualization of the chemical composition, these data can be expressed in milliequivalents per liter to permit graphical presentation, as illustrated in Figure 11.1. The top row of the bar graph consists of major cations arranged in the order of calcium, magnesium, sodium, and potassium. Anions in the bottom row are aligned in the sequence of carbonate (if present), bicarbonate, sulfate, and chloride. The sum of the positive milliequivalents per liter must equal the sum of the negative values for a water in equilibrium. Hypothetical combinations of positive and negative ions can be written from a bar graph. These combinations are useful in evaluating a water for lime–soda ash softening.

Table 11.2 lists basic data for selected elements, radicals, and compounds. Included are equivalent weights for calculating milliequivalents per liter for use in bar graph presentations and chemical equations.

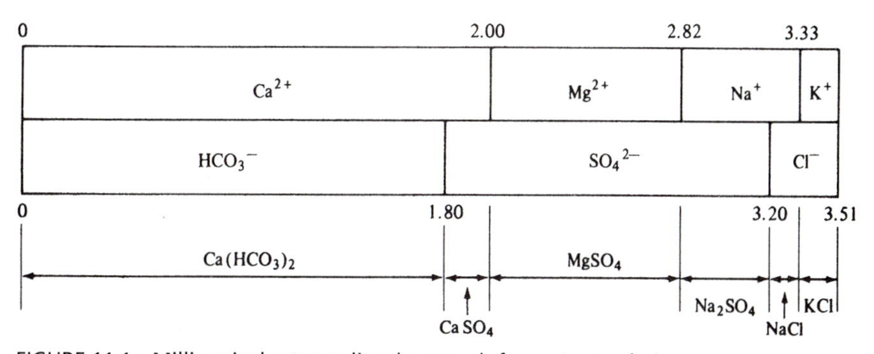

FIGURE 11.1 Milliequivalents-per-liter bar graph for water analysis.

TABLE 11.2 Data on Selected Elements, Radicals, and Compounds

Name	Symbol or Formula	Atomic or Molecular Weight	Equivalent Weight
Aluminum	Al^{3+}	27.0	9.0
Calcium	Ca^{2+}	40.1	20.0
Carbon	C	12.0	
Hydrogen	H^{+}	1.0	1.0
Magnesium	Mg^{2+}	24.3	12.2
Manganese	Mn^{2+}	54.9	27.5
Nitrogen	N	14.0	
Oxygen	O	16.0	
Phosphorus	P	31.0	
Sodium	Na^{+}	23.0	23.0
Ammonium	NH_4^{+}	18.0	18.0
Bicarbonate	HCO_3^{-}	61.0	61.0
Carbonate	CO_3^{2-}	60.0	30.0
Hydroxyl	OH^{-}	17.0	17.0
Hypochlorite	OCl^{-}	51.5	51.5
Nitrate	NO_3^{-}	62.0	62.0
Orthophosphate	PO_4^{3-}	95.0	31.7
Sulfate	SO_4^{2-}	96.0	48.0
Aluminum hydroxide	$Al(OH)_3$	78.0	26.0
Calcium bicarbonate	$Ca(HCO_3)_2$	162	81.0
Calcium carbonate	$CaCO_3$	100	50.0
Calcium sulfate	$CaSO_4$	136	68.0
Carbon dioxide	CO_2	44.0	22.0
Ferric hydroxide	$Fe(OH)_3$	107	35.6
Hydrochloric acid	HCl	36.5	36.5
Magnesium carbonate	$MgCO_3$	84.3	42.1
Magnesium hydroxide	$Mg(OH)_2$	58.3	29.1
Magnesium sulfate	$MgSO_4$	120	60.1
Sodium sulfate	Na_2SO_4	142	71.0

Example 11.1

The results of a water analysis are calcium 40.0 mg/l, magnesium 10.0 mg/l, sodium 11.7 mg/l, potassium 7.0 mg/l, bicarbonate 110 mg/l, sulfate 67.2 mg/l, and chloride 11.0 mg/l. Draw a milliequivalents-per-liter bar graph and list the hypothetical combinations. Express the hardness and alkalinity in units of mg/l as $CaCO_3$.

Solution: Using Eqs. (11.1) and (11.2), we have the data in Table 11.3. From the bar graph of these data in Figure 11.1, the hypothetical chemical combinations are

1.80 meq/l of $Ca(HCO_3)_2$ 0.38 meq/l of Na_2SO_4

0.20 meq/l of $CaSO_4$ 0.13 meq/l of NaCl

TABLE 11.3 Water Analysis Data for Example 11.1

Component	mg/l	Equivalent Weight	meq/l
Ca^{2+}	40.0	20.0	2.00
Mg^{2+}	10.0	12.2	0.82
Na^{+}	11.7	23.0	0.51
K^{+}	7.0	39.1	0.18
			Total cations = 3.51
HCO^{3-}	110	61.0	1.80
SO_4^{2-}	67.2	48.0	1.40
Cl^{-}	11.0	35.5	0.31
			Total anions = 3.51

0.82 meq/l of $MgSO_4$ 0.18 meq/l of KCl

Hardness is the sum of the Ca^{2+} and Mg^{2+} concentrations expressed in mg/l as $CaCO_3$ and alkalinity equals the bicarbonate content. Thus,

$$\text{hardness} = 2.82 \text{ meq/l} \times 50 \frac{\text{mg/l of } CaCO_3}{\text{meq/l}} = 141 \text{ mg/l}$$

$$\text{alkalinity} = 1.80 \times 50 = 90 \text{ mg/l}$$

11.2 HYDROGEN ION CONCENTRATION

The hydrogen ion activity (i.e., intensity of the acid or alkaline condition of a solution) is expressed by the term pH, which is defined as

$$pH = \log \frac{1}{[H^+]} \tag{11.3}$$

Water dissociates to only a slight degree, yielding hydrogen ions equal to 10^{-7} mole/l; thus pure water has a pH of 7. It is also neutral since 10^{-7} mole/l of hydroxyl ion is produced simultaneously:

$$H_2O \rightleftharpoons H^+ \, OH^- \tag{11.4}$$

When an acid is added to water, the hydrogen ion concentration increases, resulting in a lower pH value. Addition of an alkali reduces the number of free hydrogen ions, causing an increase in pH, because OH^- ions unite with H^+ ions. The pH scale is acidic from 0 to 7 and basic from 7 to 14.

The chemical equilibrium of water can be shifted by changing the hydrogen ion activity in solution. Thus pH adjustment is used to optimize coagulation, softening, and disinfection reactions, and for corrosion control. In wastewater treatment, pH must be maintained in a range favorable for biological activity.

11.3 ALKALINITY AND PH RELATIONSHIPS

Alkalinity is a measure of water's capacity to absorb hydrogen ions without significant pH change (i.e., to neutralize acids). It is determined in the laboratory by titrating a water sample with a standardized sulfuric acid solution. The three chemical forms that contribute to alkalinity are bicarbonates, carbonates, and hydroxides that originate from the salts of weak acids and strong bases. Bicarbonates represent the major form since they originate naturally from reactions of carbon dioxide in water.

Below pH 4.5, dissolved carbon dioxide is in equilibrium with carbonic acid in solution; thus no alkalinity exists. Between pH 4.5 and 8.3, the balance shown in Eq. (11.5) shifts to the right, reducing the CO_2 and creating HCO_3^- ions. Above 8.3, the bicarbonates are converted to carbonate ions. Hydroxide appears at a pH greater than 9.5 and reacts with carbon dioxide to yield both bicarbonates and carbonates [Eq. (11.6)]. The maximum CO_3^{2-} concentration for dilute solutions is in the pH range 10–11. Figure 11.2 shows the relationship between carbon dioxide and the three forms of alkalinity with respect to pH calculated for water having a total alkalinity of 100 mg. at 25°C. Temperature, total alkalinity, and the presence of other ionic species influence alkalinity–pH relationships; nevertheless, Figure 11.2 is a realistic representation of most natural waters.

$$CO_2 + H_2O \rightleftharpoons H_2CO_3 \rightleftharpoons H^+ + HCO_3^- \tag{11.5}$$

$$CO_2 + OH^- \rightleftharpoons HCO_3^- \rightleftharpoons H^+ + CO_3^{2-} \tag{11.6}$$

Substances that offer resistance to change in pH as acids or bases are added to a solution are referred to as *buffers*. Since the pH falls between 6 and 9 for most natural waters and wastewaters, the primary buffer is the bicarbonate–carbonate system. When

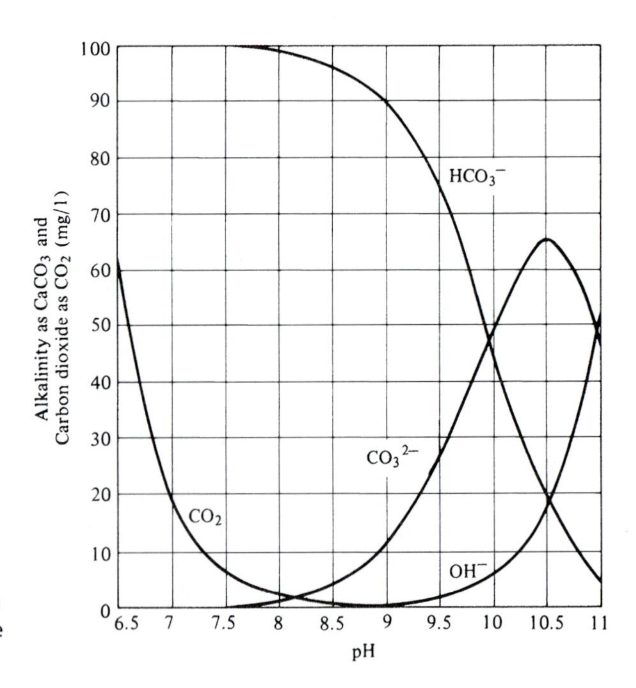

FIGURE 11.2 Carbon dioxide and various forms of alkalinity relative to pH in water at 25°C.

acid is added, a portion of the H^+ ions is combined with HCO_3^- to form un-ionized H_2CO_3; only the H^+ remaining free affect pH. If a base is added, the OH^- ions react with free H^+, increasing the pH. However, some of the latter are replaced by a shift of HCO_3^- to CO_3^{2-}, attenuating the change in hydrogen ion concentration. Both chemical reactions and biological processes depend on this natural buffering action to control pH changes. Sodium carbonate or calcium hydroxide can be added if naturally existing alkalinity is insufficient, such as in the coagulation of water where the chemicals added react with and destroy alkalinity.

Precipitation softening is easily understood by referring to the pH–alkalinity relationship illustrated in Figure 11.2. Calcium and magnesium ions are soluble when associated with bicarbonate anions. But, if the pH of a hard water is increased, insoluble precipitates of $CaCO_3$ and $Mg(OH)_2$ are formed. This is accomplished in water treatment by adding lime to raise the pH level. At a value of about 10, hydroxyl ions convert bicarbonates to carbonates to allow the formation of calcium carbonate precipitate as follows:

$$\underset{\text{bicarbonate hardness}}{Ca(HCO_3)_2} + \underset{\text{lime slurry}}{Ca(OH)_2} \longrightarrow \underset{\text{solid precipitate}}{2CaCO_3} + 2H_2O \tag{11.7}$$

11.4 CHEMICAL EQUILIBRIA

Many chemical reactions are reversible to some degree, and the concentrations of reactants and products determine the final state of equilibrium. For the general reaction expressed by Eq. (11.8), increase in either A or B shifts the equilibrium to the right, whereas a larger concentration of either C or D drives it to the left. A reaction in true equilibrium can be expressed by the *mass-action formula*, Eq. (11.9):

$$a\text{A} + b\text{B} \rightleftharpoons c\text{C} + d\text{D} \tag{11.8}$$

$$\frac{[\text{C}]^c[\text{D}]^d}{[\text{A}]^a[\text{B}]^b} = K \tag{11.9}$$

where

A, B = reactants

C, D = products

[ ] = molar concentrations

K = equilibrium constant

Strong acids and bases in dilute solutions approach 100% ionization, while weak acids and bases are poorly ionized. The degree of ionization of the latter is expressed by the mass-action equation. For example, Eqs. (11.10–11.13) are, for carbonic acid,

$$H_2CO_3 \rightleftharpoons H^+ + HCO_3^- \tag{11.10}$$

$$\frac{[H^+][HCO_3^-]}{[H_2CO_3]} = K_1 = 4.45 \times 10^{-7} \text{ at } 25°\text{C} \tag{11.11}$$

$$HCO_3^- \rightleftharpoons H^+ + CO_3^{2-} \tag{11.12}$$

$$\frac{[H^+][CO_3^{2-}]}{[HCO_3^-]} = K_2 = 4.69 \times 10^{-11} \text{ at } 25°C \tag{11.13}$$

The foregoing characterizes homogeneous chemical equilibria where all reactants and products occur in the same physical state. Heterogeneous equilibrium exists between a substance in two or more physical states. For example, at greater than pH 10, solid calcium carbonate in water reaches a stability with the calcium and carbonate ions in solution,

$$CaCO_3 \rightleftharpoons Ca^{2+} + CO_3^{2-} \tag{11.14}$$

Equilibrium between crystals of a compound in the solid state and its ions in solution can be treated mathematically as if the equilibrium is homogeneous in nature. For Eq. (11.14), the expression is

$$\frac{[Ca^{2+}][CO_3^{2-}]}{[CaCO_3]} = K \tag{11.15}$$

Concentration of a solid substance can be treated as a constant K_s in mass-action equilibrium; therefore, $[CaCO_3]$ can be assumed equal to KK_s, and then

$$[Ca^{2+}][CO_3^{2-}] = KK_s = K_{sp} = 5 \times 10^{-9} \text{ at } 25°C \tag{11.16}$$

The constant K_{sp} is called the *solubility-product constant.*

If the product of the ionic molar concentrations is less than the solubility-product constant, the solution is unsaturated. Conversely, a supersaturated solution contains an $[A^+][B^-]$ value greater than K_{sp}. In this case, crystals form and precipitation progresses until the ionic concentrations are reduced equal to those of a saturated solution.

Other ions in solution affect the solubility of a substance by either the common-ion effect or the secondary-salt effect. The common-ion effect is a repression of solubility of a substance in the presence of an excess of one of the solubility-product ions. For example, in pure water the solubility of $CaCO_3$ is about 13 mg/l, whereas in a solution containing 100 mg/l of carbonate alkalinity, the solubility is only 0.5 mg/l. The secondary-salt effect is the increase in solubility of slightly soluble salts when other salts in solution do not have an ion in common with the slightly soluble substance. For example, $CaCO_3$ is several times more soluble in seawater than in fresh water.

11.5 WAYS OF SHIFTING CHEMICAL EQUILIBRIA

Chemical reactions in water and wastewater treatment rely on shifting of homogeneous or heterogeneous equilibria to achieve the desired results. The most common methods for completing reactions are by formation of insoluble substances, weakly ionized compounds, gaseous end products, and oxidation and reduction.

The best example of shifting equilibrium to form precipitates is lime–soda ash softening. Calcium is removed from solution by adding lime, as shown in Eq. (11.7). If insufficient alkalinity is available to complete this reaction, sodium carbonate (soda ash) is also applied. However, magnesium hardness must be removed by forming $Mg(OH)_2$ since $MgCO_3$ is relatively soluble. This is effected by addition of excess lime to increase the value of $[Mg^{2+}][OH^-]^2$ above the solubility product of magnesium hydroxide, which equals 9×10^{-12}.

$$MgCO_3 + Ca(OH)_2 \xrightarrow{\text{excess } OH^-} CaCO_3\downarrow + Mg(OH)_2\downarrow \quad (11.17)$$

A common example of destroying equilibrium by forming a poorly ionized compound is neutralization of acid or caustic wastes. Here, the combining of hydrogen ions and hydroxyl ions forms poorly ionized water and a soluble salt. For example,

$$2H^+ + SO_4^{2-} + 2Na^+ + 2OH^- \rightarrow 2H_2O + 2Na^+ + SO_4^{2-} \quad (11.18)$$

Reactions involving production of a gaseous product go to practical completion if the gas escapes from solution. One illustration is breakpoint chlorination, which oxidizes ammonia to nitrogen and nitrous oxide gases.

Oxidation and reduction is a very positive method of sending reactions to completion, since one or more of the ions involved in the equilibrium is destroyed. A practical example in water treatment is removal of soluble iron from solution by oxidation using potassium permanganate. In this reaction [Eq. (11.19)], the iron gains one positive charge while the manganese in the permanganate ion is reduced from a valence of 7+ to a valence of 4+, forming manganese dioxide.

$$Fe(HCO_3)_2 + KMnO_4 \rightarrow Fe(OH)_3\downarrow + MnO_2\downarrow \quad (11.19)$$

11.6 CHEMICAL PROCESS KINETICS

Chemical reactions are classified on the basis of stoichiometry, which defines the number of moles of a substance entering into a reaction and the number of moles of products, and process kinetics that describe the rate of reaction. The types discussed in this section are irreversible reactions occurring in one phase where the reactants are distributed uniformly throughout the liquid. Heterogeneous processes involve the presence of a solid-phase catalyst, such as adsorption onto a granular medium. In irreversible reactions, the stoichiometric combination of reactants leads to nearly complete conversion to products. Most reactions in water and wastewater are sufficiently irreversible to allow this assumption for purposes of kinetic interpretation of experimental data. The type of reactor containing a reaction influences the degree of completion; therefore, process kinetics must also incorporate the hydraulic characteristics of reactors discussed in Section 10.3.

Reaction Rates

Zero-Order Reactions These reactions proceed at a rate independent of the concentration of any reactant or product. Consider conversion of a single reactant to a single product, represented as

$$A \text{ (reactant)} \rightarrow P \text{ (product)}$$

If C represents the concentration of A at any time t, the disappearance of A is expressed as

$$\frac{dC}{dt} = -k \tag{11.20}$$

where

$$\frac{dC}{dt} = \text{rate of change in concentration of A with time}$$
$$k = \text{reaction-rate constant}$$

The negative sign in front of k means that the concentration of A decreases with time. Integrating Eq. (11.20) and rearranging yields

$$C = C_0 - kt \tag{11.21}$$

where

$$C = \text{concentration of A at any time } t$$
$$C_0 = \text{initial concentration of A}$$

and C_0 becomes the constant of integration if we let $C = C_0$ when $t = 0$.

A plot of a zero-order reaction is shown in Figure 11.3(a) by the straight line with a constant slope equal to $-k$.

First-Order Reactions These reactions proceed at a rate directly proportional to the concentration of one reactant. Again consider conversion of a single reactant to a single product:

$$\text{A (reactant)} \longrightarrow \text{P (product)}$$

If C represents the concentration of A at any time t, the disappearance of A is expressed

$$\frac{dC}{dt} = -kC \tag{11.22}$$

Integrating Eq. (11.22) and letting $C = C_0$ at $t = 0$ gives

$$\ln\frac{C_0}{C} = kt \quad \text{or} \quad \log\frac{C_0}{C} = \frac{kt}{2.30} \tag{11.23}$$

Plotting a first-order reaction on arithmetic paper results in a curve, as shown by the dashed line in Figure 11.3(a), with the slope of a tangent at any point equal to $-kC$. With passage of time from the start of the reaction, the rate decreases with the decreasing concentration of A remaining; this is reflected by the reduction in tangent slopes along the curve. First-order reactions can be linearized by graphing on semilogarithmic paper, as illustrated in Figure 11.3(b).

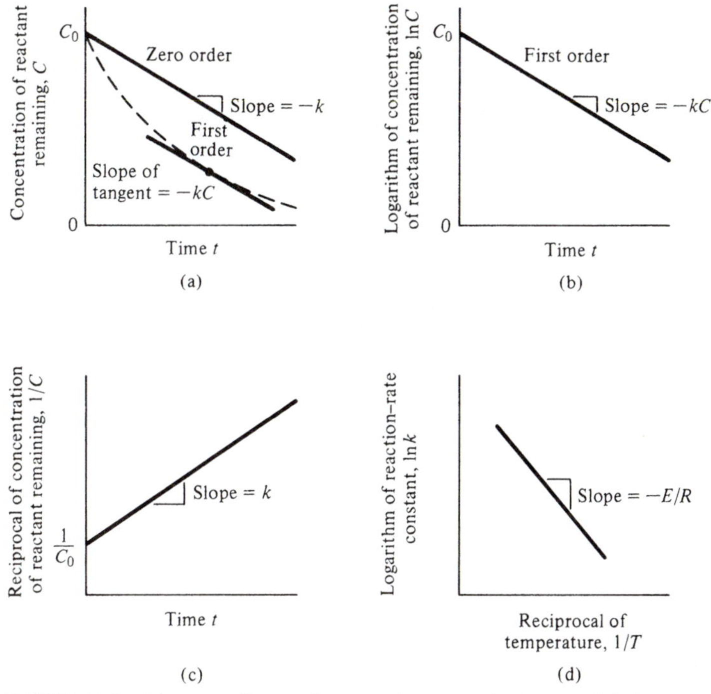

FIGURE 11.3 Diagrams illustrating reaction rates for irreversible homogeneous reactions occurring in one phase. (a) Arithmetic graphs of a zero-order reaction as a solid line and first-order reaction as a dashed line. (b) Semilogarithmic plot of a first-order reaction. (c) Plot of a second-order process. (d) Diagram showing the effect of temperature on reaction-rate plotting ln k versus $1/T$ based on the Arrhenius equations [Eqs. (11.26) and (11.27)].

Second-Order Reactions These reactions proceed at a rate proportional to the second power of a single reactant being converted to a single product,

$$2\text{A (reactant)} \rightarrow \text{P (product)}$$

The rate of disappearance of A is described by the rate equation

$$\frac{dC}{dt} = -kC^2 \tag{11.24}$$

The integrated form for a second-order reaction is

$$\frac{1}{C} - \frac{1}{C_0} = kt \tag{11.25}$$

A plot of Eq. (11.25) is shown in Figure 11.3(c). On arithmetic paper, the reciprocal of the concentration of reactant remaining versus time plot is a straight line with a slope equal to k.

Effect of Temperature on Reaction Rate

The rates of simple chemical reactions accelerate with increasing temperature of the liquid in a reactor, provided the higher temperature does not alter a reactant or catalyst. In 1889 Arrhenius examined the available data on the effect of temperature on the rates of chemical reactions and proposed the following equation:

$$\frac{d(\ln k)}{dT} = \frac{E}{RT^2} \tag{11.26}$$

where

k = reaction rate constant

T = absolute temperature, K

E = a constant characteristic of the reaction called activation energy, cal/mole

R = ideal gas constant (1.987 cal/K/mole)

Integration of this equation between T_1 and T_2, corresponding to reaction-rate constants k_1 and k_2, respectively, gives the relationship

$$\ln\frac{k_2}{k_1} = \frac{E(T_2 - T_1)}{RT_1T_2} \tag{11.27}$$

If experimental data are available, then E/R can be determined from the slope of a plot of ln k versus $1/T$ [Fig. 11.3(d)]. In turn, the activation energy E for the reaction can be calculated by dividing the value of the slope by the ideal gas constant R.

The integrated form of Eq. (11.27) is

$$\frac{k_2}{k_1} = (e^{E/RT_1T_2})^{T_2-T_1} \tag{11.28}$$

For water and wastewater reactions near ambient temperature, the quantity E/RT_1T_2 located on the right side of Eq. (11.28) can be assumed constant for practical temperature ranges. Replacing the term in brackets with a temperature coefficient Θ yields

$$\frac{k_2}{k_1} = \Theta^{T_2-T_1} \tag{11.29}$$

where

T = temperature, °C

which is commonly used to adjust the value of a rate constant for a temperature change. Equation (11.29) is applied to both chemical and biological processes, although in some cases linearity may be limited to a narrow temperature range. A common value for the temperature coefficient is 1.072, which doubles the reaction-rate constant with a 10°C temperature rise.

Mass Balance Analysis

Reaction-rate equations are based on batch reactor analyses in which the reactants are added to a stirred container and the change in concentrations measured with respect to time. In continuous-flow reactors, the extent of a reaction depends on how the hydraulic characteristics of the tank affect reaction time. For steady-state conditions of uniform flow and reactant concentrations, mass balance analyses are used to calculate the changes that occur between the influent and effluent of a reactor. Mathematical expressions can be derived for process reactions to predict the degree of reaction completion in a particular system or to compute the mean residence time (mean hydraulic detention time) needed for a specific degree of reaction completion.

Mass balance analysis of ideal plug flow is the same as reaction analysis of a stirred batch reactor, since the liquid flows through an ideal tubular reactor without longitudinal mixing. The horizontal time scales in Figure 11.3 represent time of passage, or travel distance along the reactor at a known velocity of flow. The plug-flow equations relating mean hydraulic detention time to reaction-rate constants for zero-, first-, and second-order reactions are given in Table 11.4. These were derived by rearranging and solving Eqs. (11.21), (11.23), and (11.25) for t.

Analysis of an ideal completely mixed reactor assumes steady-state conditions, so that the reactants flow continuously into the tank and products are continuously discharged at time t later. The concentrations of reactants and products are uniform throughout the mixing liquid, and the effluent has identical concentrations of reactants and products. Considering both the rate of reaction within the reactor and hydraulic characteristics of the system, the mass balance is

$$V\left(\frac{dC}{dt}\right) = QC_0 - QC_t - V(\text{rate of reaction}) \tag{11.30}$$

TABLE 11.4 Mean Hydraulic Detention Times (Mean Residence Times) for Reactions of Different Order in Plug-Flow and Completely Mixed Reactors

Reaction Order	Equations for Mean Hydraulic Detention Times: Ideal Plug Flow	Ideal Completely Mixed Flow
0	$\frac{1}{k}(C_0 - C_t)$	$\frac{1}{k}(C_0 - C_t)$
1	$\frac{1}{k}\left[\ln\left(\frac{C_0}{C_t}\right)\right]$	$\frac{1}{k}\left(\frac{C_0}{C_t} - 1\right)$
2	$\frac{1}{kC_0}\left(\frac{C_0}{C_t} - 1\right)$	$\frac{1}{kC_t}\left(\frac{C_0}{C_t} - 1\right)$

where the change of mass of reactant A within the reactor equals the mass input minus the mass output and minus the decrease in mass caused by reaction in the reactor.

For a first-order reaction rate [Eq. (11.22)], the mass balance becomes

$$V\left(\frac{dC}{dt}\right) = QC_0 - QC_t - V(kC_t) \tag{11.31}$$

Under steady-state conditions, the rate of change in the mass of reactant within the tank is zero. Hence,

$$0 = QC_0 - QC_t - V(kC_t) \tag{11.32}$$

Rearranging Eq. (11.32) and solving for V/Q, which equals t, gives

$$\frac{V}{Q} = t = \frac{1}{k}\left(\frac{C_0}{C_t} - 1\right) \tag{11.33}$$

where

V = volume of the reactor

Q = rate of flow through the reactor

t = mean hydraulic detention time (mean residence time)

k = reaction rate constant

C_0 = influent concentration of reactant A

C_t = effluent concentration of reactant A

Thus, the reaction-rate constant and mean hydraulic detention time are expressed mathematically with the influent and effluent concentrations of reactant A for a first-order reaction.

The equations for zero- and second-order reactions can be derived in a similar manner. Table 11.4 lists the equations for mean hydraulic detention times of ideal completely mixed-flow reactors based on zero-, first-, and second-order reactions. Mathematical expressions for more complex chemical process kinetics are given by others [2,3].

11.7 COLLOIDAL DISPERSIONS

Colloidal dispersions in water consist of discrete particles held in suspension by their extremely small size (1–200 nm), state of hydration (chemical combination with water), and surface electric charge. The size of particles is the most significant property responsible for the stability of a *sol* (a colloidal dispersion in a liquid). With larger particles, the ratio of surface area to mass is low, and mass effects, such as sedimentation by gravity forces, predominate. For colloids, the ratio of surface area to mass is high, and surface phenomena, such as electrostatic repulsion and hydration, become important.

There are two types of colloids—hydrophilic and hydrophobic. *Hydrophilic* colloids are readily dispersed in water, and their stability (lack of tendency to agglomerate) depends on a marked affinity for water rather than on the slight charge (usually negative) that they possess. Examples of hydrophilic colloidal materials are soaps, soluble starch, soluble proteins, and synthetic detergents.

Hydrophobic colloids possess no affinity for water and owe their stability to the electric charge they possess. Metal oxide colloids, most of which are positively charged, are examples of hydrophobic sols. A charge on the colloid is gained by adsorbing positive ions from the water solution. Electrostatic repulsion between the charged colloidal particles produces a stable sol.

The concept of *zeta potential* is derived from the diffuse double-layer theory applied to hydrophobic colloids (Fig. 11.4). A fixed covering of positive ions is attracted to the negatively charged particle by electrostatic attraction. This stationary zone of positive ions is referred to as the *Stern layer*, which is surrounded by a movable,

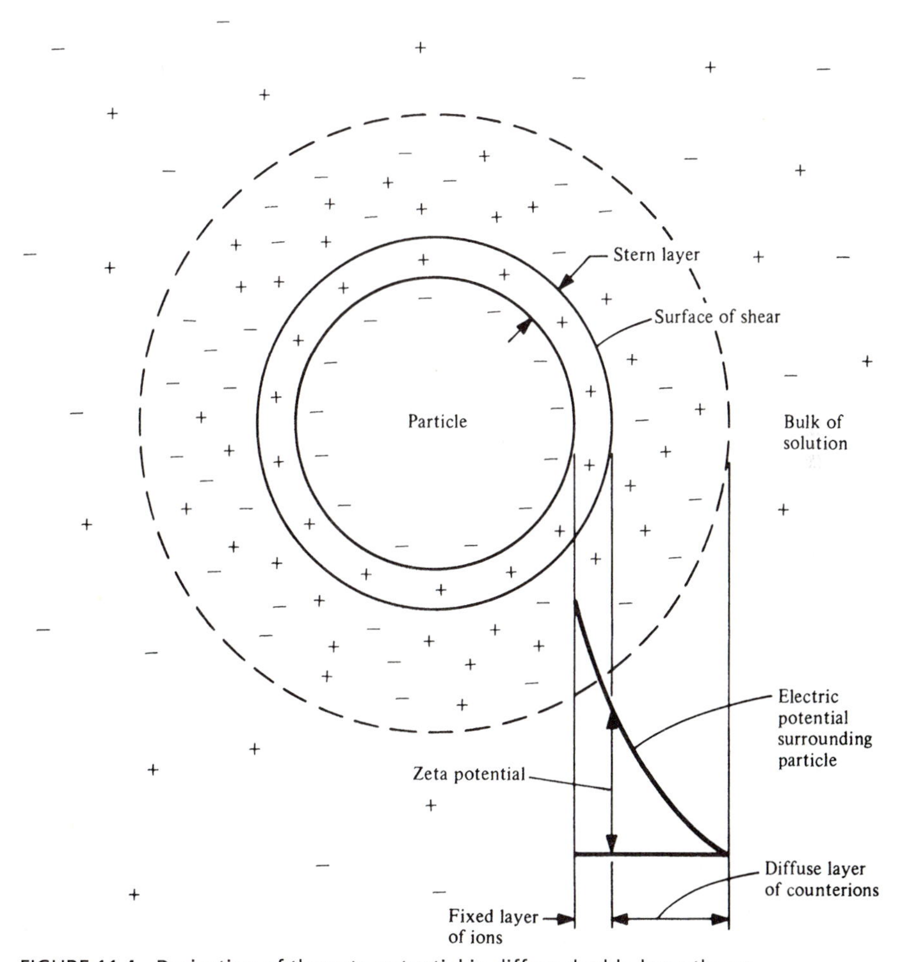

FIGURE 11.4 Derivation of the zeta potential in diffuse double-layer theory.

diffuse layer of counterions. The concentration of these positive ions in the diffuse zone decreases as it extends into the surrounding bulk of electroneutral solution. Zeta potential is the magnitude of the charge at the surface of shear. The boundary surface between the fixed ion layer and the solution serves as a shear plane when the particle undergoes movement relative to the solution. The zeta potential magnitude can be estimated from electrophoretic measurement of particle mobility in an electric field.

A colloidal suspension is defined as stable when the dispersion shows little or no tendency to aggregate [Fig. 11.5(a)]. The repulsive force of the charged double layer disperses particles and prevents aggregation; thus, particles with a high zeta potential produce a stable sol. Factors tending to destabilize a sol are van der Waals forces of attraction and Brownian movement. *Van der Waals forces* are the molecular cohesive forces of attraction that increase in intensity as particles approach each other. These forces are negligible when the particles are slightly separated but become dominant when particles contact. *Brownian movement* is the random motion of colloids caused by their bombardment by molecules of the dispersion medium. This movement has a destabilizing effect on a sol because aggregation may result.

Destabilization of hydrophobic colloids can be accomplished by adding electrolytes to the solution [Fig. 11.5(b)]. Counterions of the electrolyte suppress the

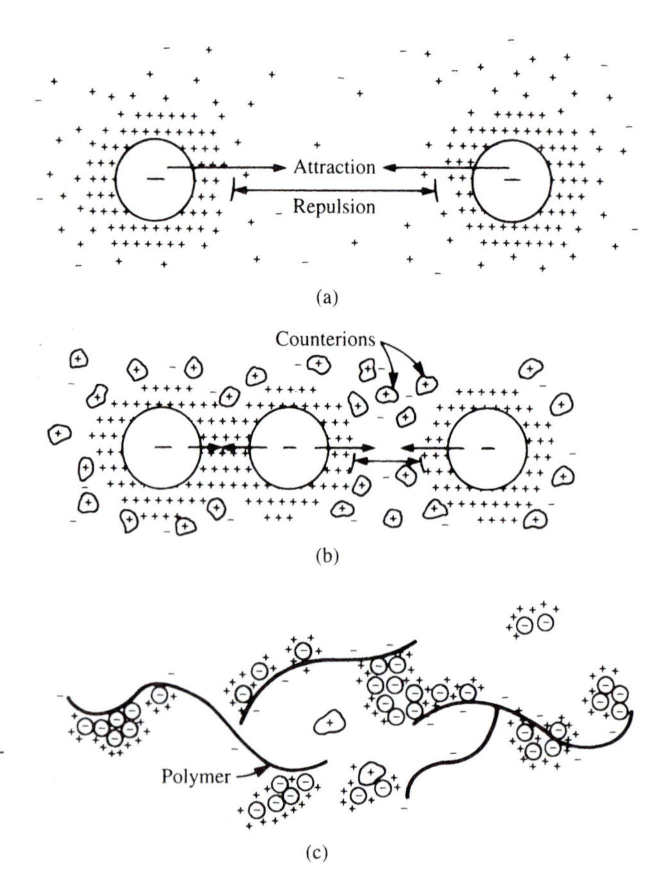

FIGURE 11.5 Schematic representations of coagulation and bridging of colloids. (a) A stable suspension of particles where forces of repulsion exceed the forces of attraction. (b) Destabilization and coagulation caused by counterions of a coagulant suppressing the double-layer charges. (c) Agglomeration of destabilized particles by attaching of coagulant ions and bridging of polymers.

double-layer charge of the colloids sufficiently to permit particles to contact. As the particles meet, van der Waals forces of attraction become dominant and aggregation results. Electrolytes found to be most effective are multivalent ions of opposite charge to that of the colloidal particles. Sols can also be destabilized by cationic polymers that act much the same as neutral salts in suppressing the diffuse double layer. The primary mechanism of polymers appears to be bridging. The long polymer molecule attaches to absorbent surfaces of colloidal particles by chemical or physical interactions, resulting in aggregation [Fig. 11.5(c)]. The destabilizing action of hydrolyzed metal ions (i.e., aluminum and iron salts) appears to fall into an intermediate category between simple ions and cationic polymers. Highly charged, soluble hydrolysis products of these metal salts reduce the repulsive forces between colloids by compressing the double-layer charge, bringing on coagulation. Hydrolyzed metal ions are also adsorbed on the colloids, creating bridges between the particles.

In contrast to the electrostatic nature of hydrophobic colloids, the stability of hydrophilic colloids is related to their state of hydration (i.e., their marked affinity for water). Chemical coagulation does not materially affect the degree of hydration of colloids. Therefore, hydrophilic colloids are extremely difficult to coagulate, and heavy doses of coagulant salts, often 10–20 times the amount used in conventional water treatment, are needed for destabilization.

WATER COAGULATION

Coagulation is the process of adding chemicals to surface waters to collect particulate matter and colloids into clusters that can be removed from solution by subsequent sedimentation and filtration through granular media. In addition to improving aesthetics, the purpose of clarification is to remove substances that interfere with the disinfection of drinking water by chlorine.

11.8 COAGULATION PROCESS

Two basic mechanisms have been defined in destabilizing a colloidal suspension to produce floc for subsequent solids-separation processes: *Coagulation* reduces the net electrical repulsive forces at particle surfaces by adding coagulant chemicals, whereas *flocculation* is agglomeration of the destabilized particles by chemical joining and bridging.

The primary purpose of chemical treatment is to agglomerate particulate matter and colloids into floc that can be separated from the water by sedimentation and filtration. In water treatment, coagulation and flocculation are used to destabilize turbidity, color, odor-producing compounds, pathogens, and other contaminants in surface waters. In wastewater reclamation, coagulation precedes tertiary filtration necessary to clarify a biologically treated effluent for effective chemical disinfection.

Traditionally, environmental engineers have not restricted the use of the terms *coagulation* and *flocculation* to describing chemical mechanisms only. Common use of these terms refers to both chemical and physical processes in treatment, possibly because the complex reactions that take place in chemical coagulation–flocculation are only partially understood. Even more important is that engineers tend to associate coagulation with the operational units (rapid mixing and flocculation) used in chemical treatment. Hence, the term *coagulation* in common usage refers to the series of chemical and mechanical operations by which coagulants are applied and made effective. These operations are customarily considered to comprise two distinct phases: (1) rapid mixing to disperse coagulant chemicals by violent agitation into the water being treated and (2) flocculation to agglomerate small particles into well-defined floc by gentle agitation for a much longer time [4]. (Refer to Section 10.4 for a description of rapid mixing and flocculation processes.)

The common unit operations and chemical additions in the treatment of surface waters for a potable supply are diagrammed in Figure 11.6. Coagulant chemicals are added by rapid mixing, and a coagulant aid, usually a polymer, is blended into the destabilized water prior to or during flocculation. After filtration, the turbidity must be equal to or less than 0.3 NTU to ensure satisfactory disinfection without creating undesirable by-products, such as trihalomethanes. Activated carbon is applied to adsorb taste- and odor-producing compounds and both natural and synthetic organic compounds. Fluoride is to achieve an optimum concentration as a public health measure. If the formation of trihalomethanes is not a problem, prechlorination may be used for disinfection of the raw water and/or for control of tastes and odors. Chemical treatment varies with the season of the year, and, for river waters, operational flexibility is required to handle the day-to-day variations.

The removal of contaminants by coagulation depends on their nature and concentration; use of both coagulants and coagulant aids; and other characteristics of the water, including pH, temperature, and ionic strength. Because of the complex nature of coagulation reactions, chemical treatment is based on empirical data derived from laboratory and field studies.

The *jar test* is widely used to simulate a full-scale coagulation–flocculation process to determine optimum chemical dosages. The apparatus consists of six agitator paddles coupled to operate at the same rotational speed, which can be varied from 10 to 100 rpm.

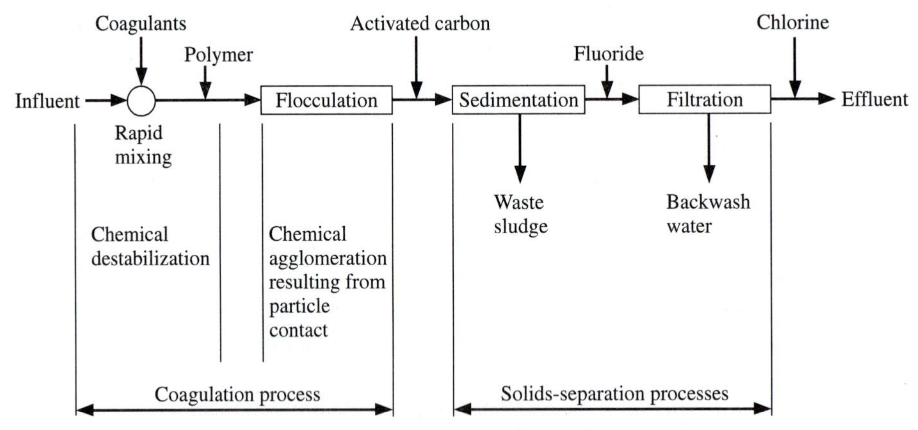

FIGURE 11.6 Schematic diagram of the coagulation process as a part of the scheme for treating surface water for a potable supply.

The laboratory containers are 1- to 2-liter beakers or square jars. The general procedure for conducting a jar test is as follows:

1. Fill six 1- or 2-liter beakers with a measured amount of the water to be treated.
2. Add the coagulant and/or other chemicals to each sample.
3. Flash-mix the samples by agitating at maximum speed (100 rpm) for 1 min.
4. Flocculate the samples at a stirring rate of 20–70 rpm for 10–30 min. Record the time of floc appearance for each beaker.
5. Stop the agitation and record the nature of the floc, clarity of supernatant, and settling characteristics of the floc.

Experiments can be conducted to evaluate the effectiveness of different coagulants and optimum dosage for destabilization, optimum pH, effectiveness of various coagulant aids and best dosage for floc formation, and the most effective sequence of chemical applications [5]. Ordinarily, a full-scale plant provides better results than the laboratory jar test for the same chemical dosages.

11.9 COAGULANTS

The most widely used coagulants for water and wastewater treatment are aluminum and iron salts. The common metal salt is aluminum sulfate, which is a good coagulant for water containing organic matter. Iron coagulants are effective over a wider pH range and are generally more effective in removing color from water, but they usually cost more. For some waters, cationic polymers are effective as a primary coagulant, but polymers are more commonly applied as coagulant aids. The final choice of coagulants and chemical aids is based on being able to achieve the contaminant removals and low turbidity desired in the filtered water at least cost.

The following paragraphs discuss various coagulants and present the theoretical chemical reactions. These traditional coagulation reactions yield only approximate results. The molecular and equivalent weights of chemical compounds listed in Tables 11.1 and 11.2 are useful in solving numerical problems.

Aluminum Sulfate (Filter Alum)

Aluminum sulfate is the standard coagulant used in water treatment. The commercial product strength ranges from 15% to 22% as Al_2O_3 with a hydration of about 14 moles of water. A formula used for filter alum is $Al_2(SO_4)_3 \cdot 14.3\, H_2O$ with a molecular weight of 600. The material is commonly shipped and fed in a dry granular form, although it is available as a powder or liquid alum syrup.

Aluminum sulfate reacts with natural alkalinity in water to form aluminum hydroxide floc.

$$Al_2(SO_4)_3 \cdot 14.3\, H_2O + 3Ca(HCO_3)_2 \rightarrow 2\, Al(OH)_3\downarrow + 3\, CaSO_4 + 14.3\, H_2O + 6\, CO_2 \qquad (11.34)$$

Each mg/l of alum decreases water alkalinity by 0.50 mg/l (as $CaCO_3$) and produces 0.44 mg/l of carbon dioxide. Production of carbon dioxide is undesirable since this increases the corrosiveness of water.

If water does not contain sufficient alkalinity to react with the alum, lime or soda ash is fed to provide the necessary alkalinity.

$$Al_2(SO_4)_3 \cdot 14.3H_2O + 3Ca(OH)_2 \rightarrow 2Al(OH)_3\downarrow + 3CaSO_4 + 14.3H_2O \quad (11.35)$$

$$Al_2(SO_4)_3 \cdot 14.3H_2O + 3Na_2CO_3 + 3H_2O \rightarrow 2Al(OH)_3\downarrow + 3Na_2SO_4 + 3CO_2 + 14.3H_2O \quad (11.36)$$

An advantage of using sodium carbonate (soda ash) is that unlike lime it does not increase water hardness, only corrosiveness. Lime, which is more popular, is less expensive than soda ash.

The dosage of alum used in water treatment is in the range of 5–50 mg/l. The effective pH range for alum coagulation is 5.5–8.0. Alum is preferred in treating relatively high-quality surface waters because it is the only chemical needed for coagulation.

Ferrous Sulfate (Copperas)

Commercial ferrous sulfate has a strength of 55% $FeSO_4$ and is supplied as green crystal or granule for dry feeding. Ferrous sulfate reacts with natural alkalinity [Eq. (11.37)], but the response is much slower than that between alum and natural alkalinity. Lime is generally added to raise the pH to the point where ferrous ions are precipitated as ferric hydroxide by the caustic alkalinity [Eq. (11.38)].

$$2FeSO_4 \cdot 7H_2O + 2Ca(HCO_3)_2 + 0.5O_2 \rightarrow 2Fe(OH)_3\downarrow + 2CaSO_4 + 4CO_2 + 13H_2O \quad (11.37)$$

$$2FeSO_4 \cdot 7H_2O + 2Ca(OH)_2 + 0.5O_2 \rightarrow 2Fe(OH)_3\downarrow + 2CaSO_4 + 13H_2O \quad (11.38)$$

Treatment using ferrous sulfate and lime adds some hardness but no corrosiveness to the water. This process is usually cheaper than alum coagulation, but the dosing operation with two chemicals is more difficult. If excess lime is used, the water may require treatment for stabilization.

Chlorinated copperas treatment is a second method of using ferrous sulfate. In this process chlorine is used to oxidize the ferrous sulfate to ferric sulfate.

$$3FeSO_4 \cdot 7H_2O + 1.5Cl_2 \rightarrow Fe_2(SO_4)_3 + FeCl_3 + 21H_2O \quad (11.39)$$

Oxidation is generally performed by adding ferrous sulfate to the discharge from a solution feed chlorinator. Theoretically, 1.0 lb of chlorine oxidizes 7.8 lb of copperas. In practice, a chlorine feed slightly in excess of the theoretical amount produces good results.

Ferric sulfate and the ferric chloride react with natural alkalinity or lime, as illustrated by the following reactions with ferric chloride:

$$2FeCl_3 + 3Ca(HCO_3)_2 \rightarrow 2Fe(OH)_3\downarrow + 3CaCl_2 + 6CO_2 \tag{11.40}$$

$$2FeCl_3 + 3Ca(OH)_2 \rightarrow 2Fe(OH)_3\downarrow + 3CaCl_2 \tag{11.41}$$

Color in water is generally not affected by copperas and lime treatment, whereas chlorinated copperas is effective in the removal of color.

Ferric Salts

Ferric sulfate and ferric chloride are available as coagulants under a variety of trade names. The reactions of these salts with natural alkalinity and with lime are noted in Eqs. (11.40) and (11.41). Advantages of the ferric coagulants are that (1) coagulation is possible over a wider pH range, generally pH 4–9 for most waters; (2) the precipitate produced is a heavy quick-settling floc; and (3) they are more effective in the removal of color, taste, and odor compounds.

Ferric sulfate is available in crystalline form and may be fed using dry or liquid feeders. Ferric sulfate, although not as aggressive as ferric chloride or chlorinated copperas, must be handled by corrosive-resistant equipment.

Ferric chloride is supplied in either crystalline or liquid form. Although ferric chloride can be used in water treatment, its most frequent application is in wastewater treatment (e.g., as a waste-sludge-conditioning chemical in combination with lime prior to mechanical dewatering).

Sodium Aluminate

Sodium aluminate is essentially alumina dissolved in sodium hydroxide. The principal use of sodium aluminate is as an additional coagulant with aluminum sulfate, generally in the treatment of boiler water. The limited employment of sodium aluminate is dictated by its high cost. Sodium aluminate is alkaline in its reactions, instead of acidic like other coagulants. Reactions of sodium aluminate with aluminum sulfate and carbon dioxide are

$$\begin{aligned}6NaAlO_2 &+ Al_2(SO_4)_3 \cdot 14.3H_2O \\ &\rightarrow 8Al(OH)_3\downarrow + 3Na_2SO_4 + 2.3H_2O\end{aligned} \tag{11.42}$$

$$2NaAlO_2 + CO_2 + 3H_2O \rightarrow 2Al(OH)_3\downarrow + Na_2CO_3 \tag{11.43}$$

Polymers

Cationic polymers can be effective for coagulation, without hydrolyzing metals, by producing destabilization through charge neutralization and interparticle bridging. In contrast to metal coagulants, rapid mixing for low-molecular-weight cationic polymers is performed at reduced turbulence relative to metal coagulants to prevent breakup of the more fragile polymer floc. Feed control of polymer coagulants must be precise since effective dosage has a narrow band; overdosing or underdosing results in restabilization of the colloidal particles. Because polymers do not affect pH, they are advantageous for coagulating low-alkalinity waters. Another potential advantage of polymers relative to metal salts is reduced sludge production. Dosages of cationic polymers for coagulation are commonly 0.5–1.5 mg/l, which is much less than the dosages of metal coagulants.

Acids and Alkalies

Acids and alkalies are added to water to adjust the pH for optimum coagulation. Typical acids used to lower the pH are sulfuric and phosphoric.

Alkalies used to raise the pH are lime, sodium hydroxide, and soda ash. Hydrated lime, with about 70% available CaO, is suitable for dry feeding but costs more than quicklime, which is 90% CaO. The latter must be slaked (combined with water) and fed as a lime slurry. Soda ash is 98% sodium carbonate and can be fed dry but is more expensive than lime. Sodium hydroxide is purchased and fed as a concentrated solution.

Example 11.2

A surface water is coagulated with a dosage of 30 mg/l of ferrous sulfate and an equivalent dosage of lime. (a) How many pounds of ferrous sulfate are needed per mil gal of water treated? (b) How many pounds of hydrated lime are required, assuming a purity of 70% CaO? (c) How many pounds of $Fe(OH)_3$ sludge are produced per mil gal of water treated?

Solution: Converting the dosage of ferrous sulfate from mg/l to lb/mil gal, one obtains

$$30 \times 8.34 = 250 \text{ lb/mil gal}$$

By Eq.(11.38) and molecular weights from Tables 11.1 and 11.2,

$$\underset{FeSO_4 \cdot 7H_2O}{2 \times 278 \text{ lb}} + \underset{Ca(OH)_2}{2 \times 74 \text{ lb}} \rightarrow \underset{Fe(OH)_3}{2 \times 107 \text{ lb}}$$

Therefore,

$$\frac{556}{250 \text{ lb}} = \frac{148}{Y \text{ lb } Ca(OH)_2} = \frac{214}{Z \text{ lb } Fe(OH)_3}$$

Solving for the lime dosage yields

$$Y \text{ lb } Ca(OH)_2 = \frac{250 \times 148}{556} = 66.5 \text{ lb/mil gal } (8.0 \text{ mg/l})$$

$$\text{lb } 70\% \text{ CaO} = 66.5 \times \frac{56}{74} \times \frac{1}{0.70} = 72 \text{ lb/mil gal } (8.6 \text{ mg/l})$$

The $Fe(OH)_3$ sludge production is

$$Z \text{ lb } Fe(OH)_3 = \frac{250 \times 214}{556} = 96 \text{ lb/mil gal } (12 \text{ mg/l})$$

Alternative solution for lime dosage: One equivalent weight of ferrous sulfate (139) reacts with 1 equivalent weight of 70% CaO (28 ÷ 0.70 = 40). Therefore,

$$\text{lime dosage} = 30 \times \frac{40}{139} \times 8.34 = 72 \text{ lb/mil gal}$$

Example 11.3

Dosage of alum with the alum–lime coagulation of a water is 50 mg/l. It is desired to react only 10 mg/l (as $CaCO_3$) of the natural alkalinity with the alum. Based on the theoretical Eqs. (11.34) and (11.35), what dosage of lime is required, in addition to 10 mg/l of natural alkalinity, to react with the alum dosage?

Solution: Using equivalent weights, the alum that reacts with 10 mg/l of natural alkalinity is $10 \times (100/50) = 20$ mg/l. The amount of alum remaining to react with lime is $50 - 20 = 30$ mg/l. The lime dosage required to react with 30 mg/l of alum is $30 \times (28/100) = 8.4$ mg/l as CaO.

11.10 POLYMERS

Synthetic polymers are water-soluble high-molecular-weight organic compounds that have multiple electrical charges along a molecular chain of carbon atoms. If the ionizable groups have a positive charge, the compound is referred to as a cationic polymer. If ionizable groups have a negative charge, it is an anionic polymer. If no charges are exhibited, it is a nonionic polymer.

In drinking water treatment, polymers are extensively used as coagulant aids with aluminum and iron coagulants in treatment of turbid waters to build larger floc prior to sedimentation and filtration.

In wastewater treatment, they can be applied in activated-sludge processing after aeration to remedy overflow of biological floc from the final clarifier, usually as a result of plant overloading or poor settleability of filamentous growths and poorly agglomerated floc. Polymers are the flocculating chemicals in the agglomeration of organic solids to release bound water for sludge thickening on a gravity belt and for dewatering by a belt press or centrifuge.

In water and wastewater coagulation, anionic and nonionic polymers are effective coagulant aids. After destabilizing the colloidal suspension by hydrolyzing metals such as alum, polymers promote larger and tougher floc by a bridging mechanism (Fig. 11.5). Generally, nonionic polymers are more effective in water containing higher concentrations of divalent cations, i.e., Ca^{2+} and Mg^{2+}. Common dosages are 0.1–0.5 mg/l, to aid alum coagulation. The combination of polymer and alum can reduce the alum dosage that would be required without the polymer. Since a wide selection of proprietary polymers is available, the brand name of the selected polymer should be determined and verified by laboratory and full-scale testing to ensure it is the best aid at least cost. The effectiveness of a polymer also depends on the point of application. The mixing must be adequate for complete blending but not so turbulent as to hinder particle bridging.

Several kinds of polymers are available. *Dry polymers* are high-molecular-weight, powder or granular, available in bags or bulk, and 100% active. They are diluted with water to the desired concentration in the range of 0.10%–0.25% solution by weight to allow accurate metering. In order to uncoil and activate the polymer molecules, the solution is aged in a mixing tank. *Emulsion polymers* are high-molecular-weight, available as a liquid in an oil medium, and 35%–50% active by weight. The minimum dilution ratio is 100 to 1 and may be as high as 1000 to 1. Because of the hydrophobic nature of oil, emulsion polymers tend to agglomerate when diluted with water. *Solution polymers* are low-molecular-weight, supplied in water solution, 10%–100% active, and are the easiest to blend into a diluted solution. *Mannich polymers* are high-molecular-weight, water soluble, available in bulk, and 6%–10% active. The typical water dilution for application is 3 to 1. Solution and mannich polymers are easily miscible in water and all three liquid polymers require less aging time than dry polymers. Process application, size of treatment plant, selection of polymer feed system, price based on active polymer weight, and availability dictate polymer selection.

WATER SOFTENING

Hardness in water is caused by the ions of calcium and magnesium. Although ions of iron, manganese, strontium, and aluminum also produce hardness, they are not present in significant quantities in natural waters.

A single criterion for maximum hardness in public water supplies is not possible. Water hardness is largely the result of geological formations of the water source. Public acceptance of hardness varies from community to community, consumer sensitivity being related to the degree of hardness to which the consumer is accustomed.

Hardness of more than 300–500 mg/l as $CaCO_3$ is considered excessive for a public water supply and results in high soap consumption as well as objectionable scale in plumbing fixtures and pipes. Many consumers object to water harder than 150 mg/l, a moderate figure being 60–120 mg/l. The magnesium hardness should be less than 40 mg/l to prevent magnesium hydroxide scale in household water heaters. For chemical equilibrium of the bicarbonate–carbonate buffer, the total alkalinity should be 60–70 mg/l as $CaCO_3$.

11.11 CHEMISTRY OF LIME–SODA ASH PROCESS

The lime–soda water-softening process uses lime, $Ca(OH)_2$, and soda ash, Na_2CO_3, to precipitate hardness from solution. Carbon dioxide and carbonate hardness (calcium and magnesium bicarbonate) are complexed by lime. Noncarbonate hardness (calcium and magnesium sulfates or chlorides) requires the addition of soda ash for precipitation.

The following are chemical reactions in lime–soda ash treatment:
Reaction with CO_2 and precipitation of Ca^{2+} from lime.

$$CO_2 + Ca(OH)_2 = CaCO_3\downarrow + H_2O \tag{11.44}$$

Precipitation of bicarbonate Ca^{2+} and Ca^{2+} from lime.

$$Ca(HCO_3)_2 + Ca(OH)_2 = 2CaCO_3\downarrow + 2H_2O \tag{11.45}$$

Conversion of HCO_3^- to CO_3^{2-} and precipitation of Ca^{2+} from lime.

$$Mg(HCO_3)_2 + Ca(OH)_2 = CaCO_3\downarrow + MgCO_3 + 2\,H_2O \quad (11.46)$$

Precipitation of carbonate Mg^{2+} and Ca^{2+} from lime.

$$MgCO_3 + Ca(OH)_2 = Mg(OH)_2\downarrow + CaCO_3\downarrow \quad (11.47)$$

Precipitation of noncarbonate Mg^{2+}, leaving Ca^{2+} from lime in solution.

$$MgSO_4 + Ca(OH)_2 = Mg(OH)_2\downarrow + CaSO_4 \quad (11.48a)$$

$$MgCl_2 + Ca(OH)_2 = Mg(OH)_2\downarrow + CaCl_2 \quad (11.48b)$$

Precipitation of noncarbonate Ca^{2+} by addition of soda ash.

$$CaSO_4 + Na_2CO_3 = CaCO_3\downarrow + Na_2SO_4 \quad (11.49a)$$

$$CaCl_2 + Na_2CO_3 = CaCO_3\downarrow + 2\,NaCl \quad (11.49b)$$

These equations give all the reactions taking place in softening water containing both carbonate and noncarbonate hardness, by additions of both lime and soda ash. The carbon dioxide in Eq. (11.44) is not hardness as such, but it consumes lime and must therefore be considered in calculating the amount required. Equations (11.45–11.47) demonstrate removal of carbonate hardness by lime. Note that only 1 mole of lime is needed for each mole of calcium alkalinity, whereas 2 moles are required for each mole of magnesium alkalinity [Eqs. (11.46) and (11.47)]. Equation (11.48) shows the removal of magnesium noncarbonate hardness by lime. No softening results from this reaction because 1 mole of calcium noncarbonate hardness is formed for each mole of magnesium salt present. Equation (11.49) is for removal of noncarbonate calcium originally present in the water and also that formed as stated in Eq. (11.48).

Precipitation softening cannot produce a water completely free of hardness because of the solubility of calcium carbonate and magnesium hydroxide. Furthermore, completion of the chemical reactions is limited by physical considerations, such as adequate mixing and limited detention time in settling basins. Therefore, the minimum practical limits of precipitation softening are 30 mg/l of $CaCO_3$ and 10 mg/l of $Mg(OH)_2$ expressed as $CaCO_3$. Hardness levels of 80–100 mg/l are generally considered acceptable for a public water supply, but the magnesium content should not exceed 40 mg/l as $CaCO_3$ in a softened municipal water.

There are several advantages of lime softening in water treatment. The most obvious is that the total dissolved solids may be significantly reduced; hardness is taken out of solution; and the lime added is also removed. When soda ash is applied, sodium ions remain in the finished water; however, noncarbonate hardness requiring the addition of soda ash is generally a small portion of the total hardness. Lime also precipitates the soluble iron and manganese often found in groundwaters. In processing surface waters, excess lime treatment provides disinfection and aids in coagulation for removal of turbidity.

11.12 PROCESS VARIATIONS IN LIME–SODA ASH SOFTENING

Three different basic schemes are used to provide a finished water with the desired hardness: excess-lime treatment, selective calcium removal, and split treatment.

Excess-Lime Treatment

Carbonate hardness associated with the calcium ion can be effectively removed to the practical limit of $CaCO_3$ solubility by stoichiometric additions of lime [Eq. (11.45)]. Precipitation of the magnesium ions [Eqs. (11.47) and (11.48)] calls for a surplus of approximately 35 mg/l of CaO (1.25 meq/l) above stoichiometric requirements. The practice of excess-lime treatment reduces the total hardness to about 40 mg/l (i.e., 30 mg/l of $CaCO_3$ and 10 mg/l of magnesium hardness).

After excess-lime treatment, the water is scale forming and must be neutralized to remove caustic alkalinity. Recarbonation and soda ash are regularly used to stabilize the water. Carbon dioxide neutralizes excess lime as follows:

$$Ca(OH)_2 + CO_2 = CaCO_3\downarrow + H_2O \tag{11.50}$$

This reaction precipitates calcium hardness and reduces the pH from near 11 to about 10.2. Further recarbonation of the clarified water converts a portion of the remaining carbonate ions to bicarbonate by the reaction

$$CaCO_3 + CO_2 + H_2O = Ca(HCO_3)_2 \tag{11.51}$$

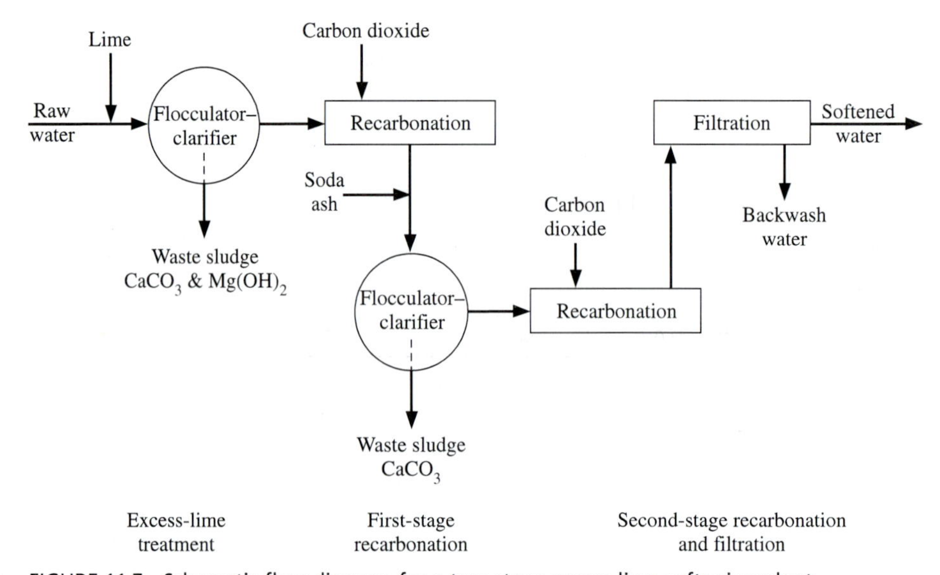

FIGURE 11.7 Schematic flow diagram for a two-stage excess-lime softening plant.

The final pH is in the range 9.5–8.5, depending on the desired carbonate-to-bicarbonate ratio (Fig. 11.2).

The equipment required for generating carbon dioxide gas consists of a furnace to burn the fuel (coke, coal, gas, or oil), a scrubber to remove soot and other impurities from the gas, and a compressor for forcing the gas into the water. Commercial liquid carbon dioxide is also used if the consumption is sufficiently low to be cost effective. Receiving and storing bulk quantities of commercial gas and the ease of feeding reduce equipment and operating costs compared to on-site gas generation for smaller facilities.

A two-stage system is preferred for excess-lime treatment as diagrammed in Figure 11.7. Lime is applied in first-stage mixing and sedimentation to precipitate both calcium and magnesium. Then carbon dioxide is applied to neutralize the excess lime [Eq. (11.50)], and soda ash is added to reduce noncarbonate hardness. Solids formed in these reactions are removed by secondary settling and subsequent filtration. Recarbonation immediately ahead of the filters may be used to prevent scaling of the media [Eq. (11.51)].

Example 11.4

Water defined by the following analysis is to be softened by excess-lime treatment in a two-stage system (Fig. 11.7):

$$CO_2 = 8.8 \text{ mg/l as } CO_2 \qquad \text{Alk}(HCO_3^-) = 115 \text{ mg/l as } CaCO_3$$

$$Ca^{2+} = 70 \text{ mg/l} \qquad SO_4^{2-} = 96 \text{ mg/l}$$

$$Mg^{2+} = 9.7 \text{ mg/l} \qquad Cl^- = 10.6 \text{ mg/l}$$

$$Na^+ = 6.9 \text{ mg/l}$$

The practical limits of removal can be assumed to be 30 mg/l of $CaCO_3$ and 10 mg/l of $Mg(OH)_2$, expressed as $CaCO_3$. Sketch a meq/l bar graph and list the hypothetical combinations of chemical compounds in the raw water. Calculate the quantity of softening chemicals required in pounds per million gallons of water treated and the theoretical quantity of carbon dioxide needed to provide a finished water with one-half of the alkalinity converted to bicarbonate ion. Draw a bar graph for the softened water after recarbonation and filtration.

Solution:

Component	mg/l	Equivalent Weight	meq/l
CO_2	8.8	22.0	0.40
Ca^{2+}	70	20.0	3.50
Mg^{2+}	9.7	12.2	0.80
Na^+	6.9	23.0	0.30
Alk	115	50.0	2.30
SO_4^{2-}	96	48.0	2.00
Cl^-	10.6	35.5	0.30

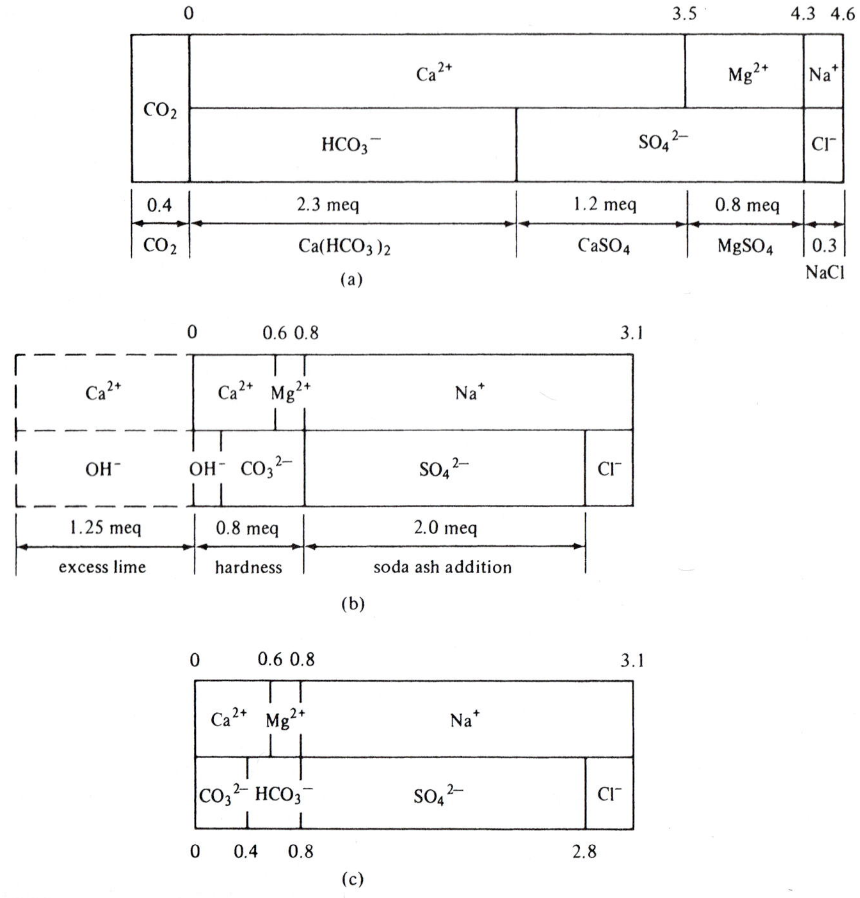

FIGURE 11.8 Milliequivalent bar graphs for Example 11.4. (a) Bar graph and hypothetical chemical combinations in the raw water. (b) Bar graph of the water after lime and soda ash additions and settling but before recarbonation. (c) Bar graph of the water after two-stage recarbonation and final filtration.

The meq/l bar graph of the raw water is shown in Figure 11.8(a), and the hypothetical combinations are listed.

Component	**meq/l**	**Lime**	**Soda Ash**
CO_2	0.4	0.4	0
$Ca(HCO_3)_2$	2.3	2.3	0
$CaSO_4$	1.2	0	1.2
$MgSO_4$	0.8	0.8	0.8
		3.5	2.0

$$\text{lime required} = \text{stoichiometric quantity} + \text{excess lime}$$
$$= 3.5 \times 28 + 35 = 133 \text{ mg/l of CaO} = 1100 \text{ lb/mil gal}$$

$$\text{soda ash required} = 2.0 \times 53 = 106 \text{ mg/l of } Na_2CO_3 = 900 \text{ lb/mil gal}$$

A hypothetical bar graph for the water after addition of softening chemicals and first-stage sedimentation is shown in Figure 11.8(b). The dashed box is the excess-lime addition, 35 mg/l of CaO = 1.25 meq/l. The 0.6 meq/l of Ca^{2+} (30 mg/l as $CaCO_3$) and 0.20 meq/l of Mg^{2+} (10 mg/l as $CaCO_3$) are the practical limits of hardness reduction. The 2.0 meq/l of Na_2SO_4 results from the addition of soda ash. Alkalinity consists of 0.20 meq/l of OH^- associated with $Mg(OH)_2$ and 0.60 meq/l of CO_3^{2-} related to $CaCO_3$.

Recarbonation converts the excess hydroxyl ion to carbonate ion; using the relationship in Eq. (11.50) and 22.0 as the equivalent weight of carbon dioxide is $(1.25 + 0.2)22.0 = 31.9$ mg/l of CO_2. After second-stage processing, final recarbonation converts one-half of the remaining alkalinity to bicarbonate ion by Eq. (11.51), giving $0.5 \times 0.8 \times 22.0 = 8.8$ mg/l of CO_2. Therefore the total carbon dioxide reacted is $(31.9 + 8.8)8.34 = 340$ lb/mil gal of CO_2.

The bar graph of the finished water is shown in Figure 11.8(c).

Selective Calcium Removal

Waters with a magnesium hardness of less than 40 mg/l as $CaCO_3$ can be softened by removing only a portion of the calcium hardness. The processing scheme can be a single-stage system of mixing, sedimentation, recarbonation, and filtration, as diagrammed in Figure 11.9. Enough lime is added to the raw water to precipitate calcium hardness without providing any excess for magnesium removal. Soda ash may be required, depending on the amount of noncarbonate hardness. Recarbonation is usually practiced to reduce scaling of the filter media and produce a stable effluent.

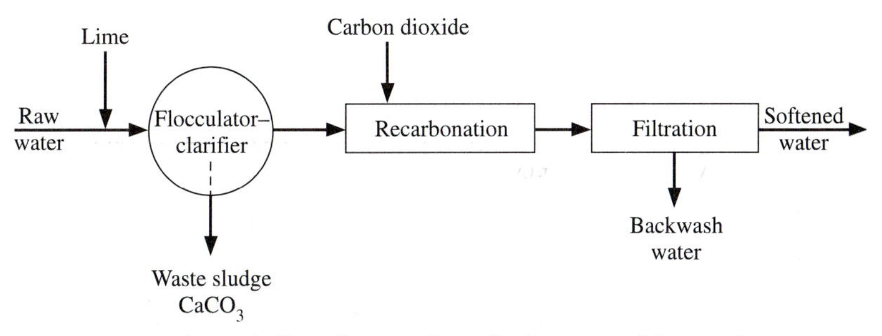

FIGURE 11.9 Schematic flow diagram for a single-stage calcium-carbonate softening plant.

Example 11.5

Determine the chemical dosages needed for selective calcium softening of the water described in Example 11.4. Draw a bar graph of the processed water.

Solution: The hypothetical combinations of concern in calcium precipitation are CO_2, $Ca(HCO_3)_2$, and $CaSO_4$ [Fig. 11.8(a)].

Component	meq/l	Lime	Soda Ash
CO_2	0.4	0.4	0
$Ca(HCO_3)_2$	2.3	2.3	0
$CaSO_4$	1.2	0	1.2
		2.7	1.2

$$\text{lime required} = 2.7 \times 28 = 76 \text{ mg/l of CaO} = 630 \text{ lb/mil gal}$$

$$\text{soda ash required} = 1.2 \times 53 = 64 \text{ mg/l of } Na_2CO_3 = 530 \text{ lb/mil gal}$$

Final hardness is all the Mg^{2+} in the raw water plus the practical limit of $CaCO_3$ removal (30 mg/l or 0.60 meq/l), $(0.8 + 0.6)50 = 70$ mg/l as $CaCO_3$. The bar graph of the softened water is shown in Figure 11.10. Recarbonation would be desirable to stabilize the water by converting a portion of the carbonate alkalinity to bicarbonate.

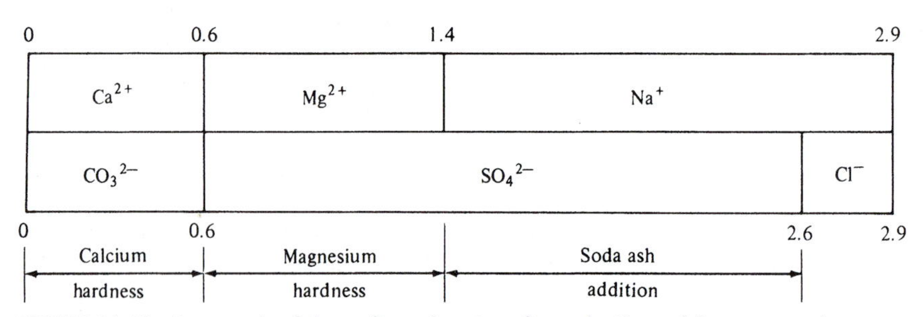

FIGURE 11.10 Bar graph of the softened water after selective calcium removal.

Split Treatment

Split treatment consists of treating a portion of the raw water by excess lime and then neutralizing the excess lime in the treated flow with the remaining portion of raw water. When split treatment is used, any desired hardness level above 40 mg/l is attainable. Since hardness levels of 80–100 mg/l are generally considered acceptable, split treatment can result in considerable chemical savings. Recarbonation is not customarily required.

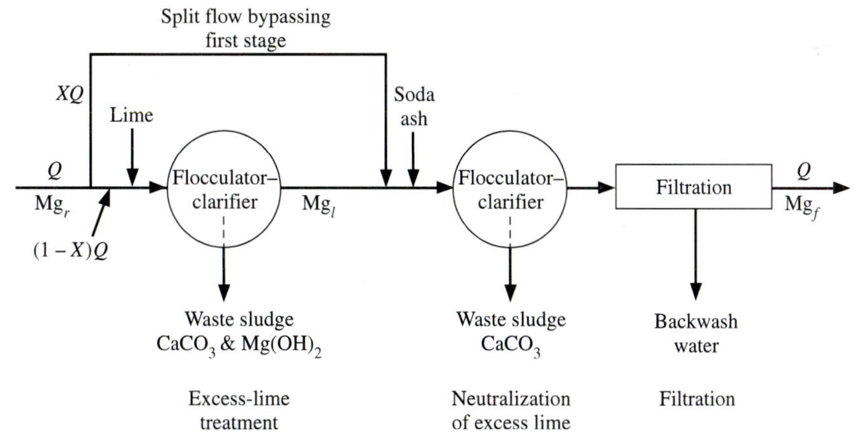

FIGURE 11.11 Schematic flow diagram for a spilt-treatment softening plant.

Split treatment is particularly advantageous on well waters. In softening surface waters, where taste, odor, and color may be problems, two stages of processing for the total flow are usually preferred over split treatment.

The flow pattern of a typical two-stage split-treatment plant is shown in Figure 11.11. If X is the ratio of the bypassed flow to the total quantity Q, then bypassed flow is XQ, and that through the first stage is $Q - XQ$ or $(1 - X)Q$. The magnesium content leaving the first stage (designated Mg_1) will be less than 10 mg/l (as $CaCO_3$). Magnesium in the bypass will be the same as that in the raw water (designated as Mg_r). Permissible magnesium in finished water (designated Mg_f) is about 50 mg/l as $CaCO_3$. The bypass flow fraction can be calculated for any desired level of magnesium by using the formula

$$X = \frac{Mg_f - Mg_1}{Mg_r - Mg_1} \tag{11.52}$$

The bypass flow is usually 30%–50%, depending on the limiting concentration of about 50 mg/l magnesium hardness in the finished water, with a total hardness of 80–100 mg/l.

The first stage in split treatment is excess-lime treatment to precipitate both calcium carbonate and magnesium hydroxide (Fig. 11.11). However, soda ash is not applied until the combined lime-treated flow $(1 - X)Q$ and bypassed flow XQ enter the second stage. In second-stage flocculation and sedimentation, the excess lime in the treated flow reacts with the carbon dioxide and calcium bicarbonate to precipitate as calcium carbonate with no precipitation of magnesium hydroxide. Blending of these two flows neutralizes the pH and eliminates the need for recarbonation. Suspended solids in the overflow of the flocculator–clarifier are removed by filtration.

Split-treatment calculations are illustrated in Examples 11.6 and 11.7. The solution in Example 11.6 is theoretically precise and calculations are in the same sequence

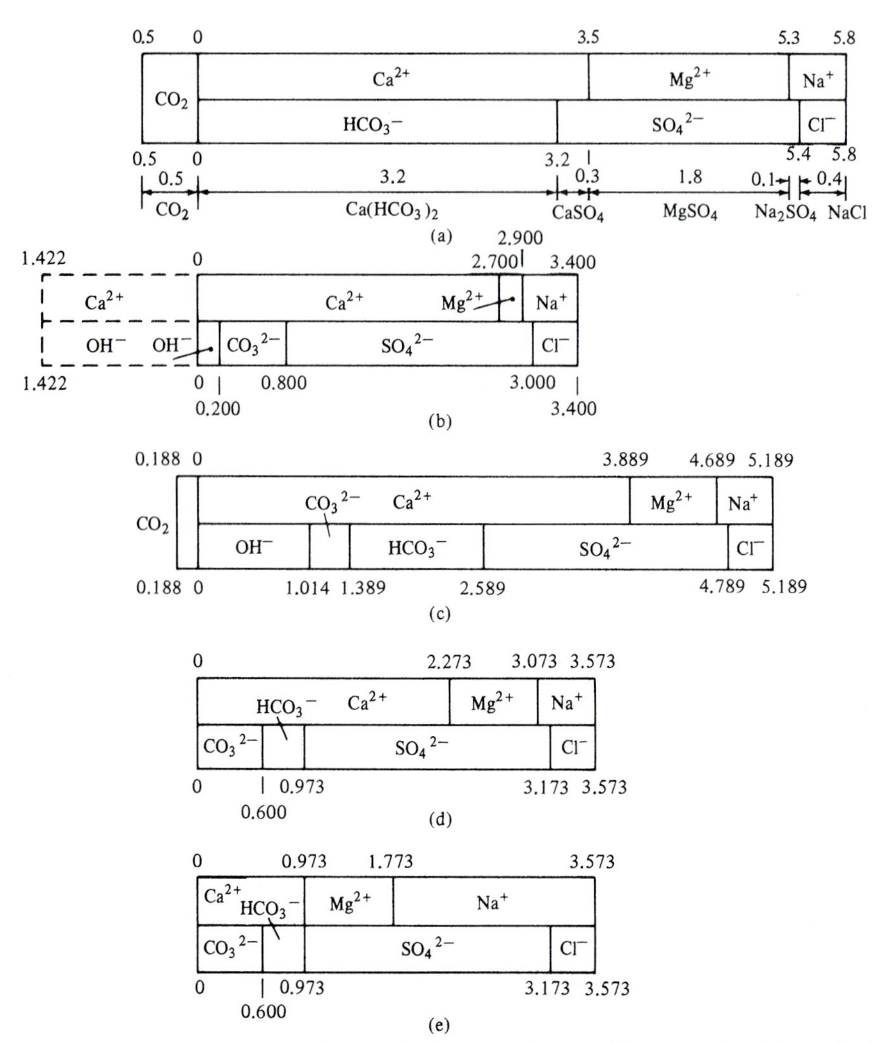

FIGURE 11.12 Milliequivalent bar graphs for Example 11.6. (a) Bar graph and hypothetical chemical combinations in the raw water. (b) Bar graph of the water after first-stage treatment with lime. (c) Unreacted combination of the first-stage effluent times 0.625 plus the bypassed flow times 0.375. (d) Hypothetical graph after reaction of the excess lime in the combined flows. (e) Bar graph of finished water after second-stage treatment with soda ash, sedimentation, and final filtration.

as the treatment processes. Refer to the bar graphs in Figure 11.12 in the order from top to bottom: (a) the raw water and bypassed water, (b) settled overflow from first-stage lime treatment, (c) combined bar graphs of treated and bypass flows entering the second stage, (d) after reaction of excess lime in the combined flows and removal of calcium carbonate precipitate, and (e) finished water after second-stage treatment with soda ash and final granular-media filtration.

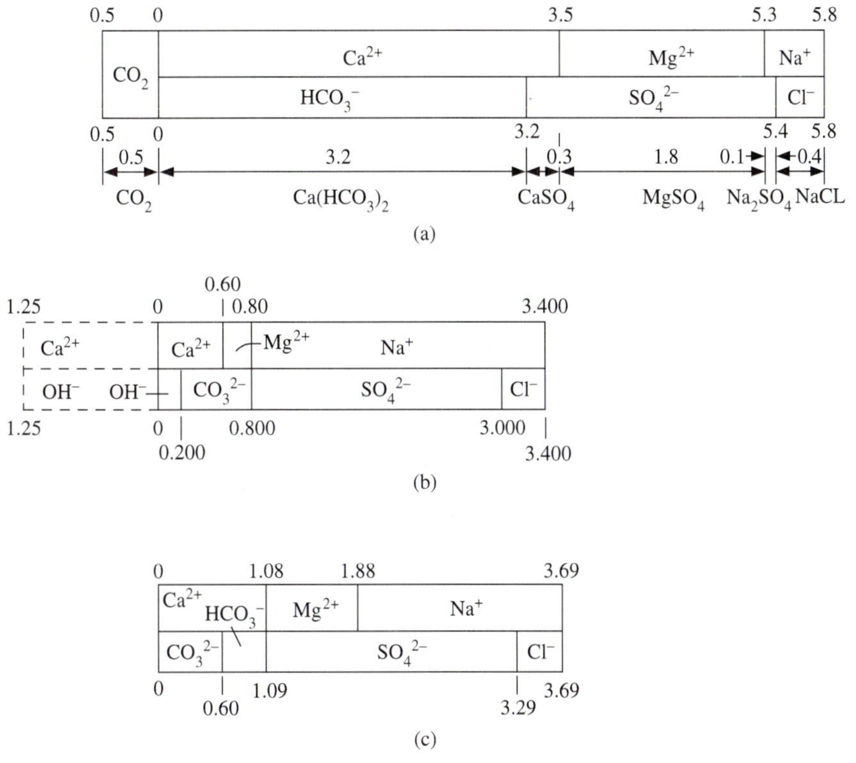

FIGURE 11.13 Milliequivalent bar graphs for Example 11.7. (a) Bar graph and hypothetical chemical combinations in the raw water. (b) Bar graph of the water after excess-lime–soda ash softening. (c) Bar graph of finished water after second-stage blending, sedimentation, and final filtration.

The solution in Example 11.7 is a somewhat simplified method of calculations for split treatment of the same raw water as in Example 11.6. Although the results are approximately the same, the process is not theoretically correct. The calculated chemical dosages for the first stage are identical to those for separate excess-lime treatment. The bar graphs in Figure 11.13 summarize the results: (a) the raw water and bypassed water as in Figure 11.12, (b) the first stage after excess-lime treatment, with additions of both lime and soda ash, and (c) the finished water after combining the treated and bypassed flows. This bar graph differs only slightly from the bar graph of finished water from Example 11.6.

Coagulation and Softening

Addition of a coagulant or a coagulant aid may result in more efficient removal of the hardness precipitates formed in lime–soda softening. Alum is the prevalent coagulant.

Lime softening is often used to treat hard surface waters. Improved coagulation and turbidity removal usually result from lime–soda softening, as compared to simple coagulation, because of the greater quantity of precipitate formed in chemical treatment processes.

Example 11.6

Consider the split-treatment softening of water described by the following analysis. Criteria for the finished water are a maximum permissible magnesium hardness of 40 mg/l as $CaCO_3$ and calcium hardness in the range 40–60 mg/l.

$$CO_2 = 0.5 \text{ meq/l} \qquad HCO_3^- = 3.2 \text{ meq/l}$$

$$Ca^{2+} = 3.5 \text{ meq/l} \qquad SO_4^{2-} = 2.2 \text{ meq/l}$$

$$Mg^{2+} = 1.8 \text{ meq/l} \qquad Cl^- = 0.4 \text{ meq/l}$$

$$Na^+ = 0.5 \text{ meq/l}$$

Solution: Solving for the fraction of bypassed flow using Eq. (11.52) gives

$$X = \frac{40 - 10}{(1.8)(50) - 10} = \frac{30}{80} = 0.375$$

The first-stage flow is therefore

$$1 - X = 1 - 0.375 = 0.625$$

The lime and soda ash required for treatment are as follows:

		Lime		
Component	**meq/l**	**First-Stage Flow**	**Bypassed Flow**	**Soda Ash**
CO_2	0.5	$0.625 \times 0.5 = 0.313$	$0.375 \times 0.5 = 0.188$	0
$Ca(HCO_3)_2$	3.2	$0.625 \times 3.2 = 2.000$	$0.375 \times 3.2 = 1.200$	0
$CaSO_4$	0.3	0	0	0.3
$MgSO_4$	1.8	$0.625 \times 1.0 = 0.625$	0	1.0
		2.938	1.388	1.300

The lime dose for $MgSO_4$ is for only 1.0 meq/l of the 1.8 meq/l, since 0.8 meq/l of Mg (40 mg/l of magnesium hardness) is the allowable concentration in the finished water.

The lime addition to the first stage is

$$2.938 + 1.388 = 4.326 \text{ meq/l} = 121 \text{ mg/l of CaO}$$

The soda ash addition to the second stage is

$$1.300 \text{ meq/l} = 69 \text{ mg/l of } Na_2CO_3$$

The hypothetical bar graphs of the water at various stages of treatment are shown in Figure 11.12. The first bar graph showing the hypothetical combinations is for the raw water. All of the lime is applied to this water in the first stage. Since only $0.625Q$ of the influent is treated, the lime addition relative to the full bar graph is the actual amount of 4.326 meq/l divided by 0.625 to equal 6.922 meq/l. Reacting with this applied lime are 0.5 meq/l of CO_2 [Eq. (11.44)], 3.2 of $Ca(HCO_3)_2$ [Eq. (11.45)], and 1.8

of $MgSO_4$ [Eq. (11.48a)]. The excess lime remaining after these reactions equals $6.922 - (0.50 + 3.20 + 1.80) = 1.422$ meq/l, which is greater than the required minimum of 1.25 meq/l for $Mg(OH)_2$ precipitation. The 0.3 meq/l of $CaSO_4$ do not react, and the 1.8 meq/l of $MgSO_4$ are converted to $CaSO_4$ [Eq. (11.48a)]. The practical limits of removal are assumed to be 0.60 meq/l of $CaCO_3$ and 0.20 meq/l of $Mg(OH)_2$. Figure 11.12(b) is a bar graph for the settled effluent from the first stage.

The next step is the reaction between the first-stage effluent and the untreated bypassed flow. Figure 11.12(c) is a bar graph of the combined flow without considering any reactions. It was drawn by adding Figure 11.12(a) times 0.375 to Figure 11.12(b) times 0.625. The 1.014 meq/l of $Ca(OH)_2$ reacts with the CO_2 and $Ca(HCO_3)_2$ to form $CaCO_3$, which precipitates, leaving 0.60 meq/l of $CaCO_3$ in solution. Figure 11.12(d) is a hypothetical bar graph after reaction of the excess lime in the first-stage effluent with the CO_2 and $Ca(HCO_3)_2$ in the bypassed flow.

In the second stage, 1.30 meq/l of soda ash (Na_2CO_3) are added and react with 1.30 meq/l of $CaSO_4$ to precipitate 1.30 meq/l of $CaCO_3$ in accordance with Eq. (11.49a). The net effect in the bar graph is the replacement of 1.30 meq/l of Ca with Na. The finished water after sedimentation and filtration is graphed in Figure 11.12(e).

The finished water has the following calculated alkalinity and hardness values:

$$\text{alkalinity} = 0.973 \times 50 = 49 \text{ mg/l}$$

$$\text{calcium hardness} = 0.973 \times 50 = 49 \text{ mg/l}$$

$$\text{magnesium hardness} = 0.800 \times 50 = 40 \text{ mg/l}$$

$$\text{total hardness} = 49 + 40 = 89 \text{ mg/l}$$

Example 11.7

This is an alternate solution to the split-treatment softening of the water described in Example 11.6. Although less precise, these calculations result in similar answers.

Solution: From Example 11.6, $X = 0.375$ and $(1 - X) = 0.625$.

From the meq/l bar graph of the raw water in Figure 11.13, the hypothetical combinations are $Ca(HCO_3)_2$, $CaSO_4$, and $MgSO_4$. Calcium hardness $= 3.5 \times 50 = 175$ mg/l as $CaCO_3$, and magnesium hardness $= 1.8 \times 50 = 90$ mg/l.

The lime and soda ash for excess-lime softening are as follows:

Component	meq/l	Applicable Equation	Lime meq/l	Soda Ash meq/l
CO_2	0.5	(11.44)	0.50	
$Ca(HCO_3)_2$	3.2	(11.45)	3.20	
$CaSO_4$	0.3	(11.49a)		0.30
$MgSO_4$	1.8	(11.48a) & (11.49a)	1.80	1.80
			5.50	2.10

Flow passing through the first stage is processed as excess-lime treatment. However, the chemical additions are reduced by multiplying by 0.625, since this is the fraction of raw water being treated.

$$\text{lime required} = \text{stoichiometric quantity} + \text{excess lime}$$
$$= 0.625[(5.50 \times 28) + 35] = 118 \text{ mg/l of CaO}$$
$$\text{soda ash required} = 0.625(2.10 \times 53) = 69 \text{ mg/l of } Na_2CO_3$$

Figure 11.13(a) is the bar graph of the bypassed flow, and Figure 11.13(b) is the bar graph of the water after first-stage excess-lime treatment. These two are blended in the second stage, where excess lime reacts with the untreated water. The amount of excess hydroxide ion in the mixed flows is equal to

$$0.625(1.25 + 0.20) = 0.906 \text{ meq/l of } OH^-$$

The other components of interest in the blended water are

$$CO_2 = 0.375 \times 0.50 = 0.188 \text{ meq/l}$$
$$Ca(HCO_3)_2 = 0.375 \times 3.20 = 1.20 \text{ meq/l}$$

First, the carbon dioxide is eliminated by the excess lime,

$$0.906 - 0.188 = 0.718 \text{ meq/l of } OH^- \text{ remaining}$$

Then, the balance of the hydroxide ion reacts with calcium bicarbonate, reducing it to

$$1.20 - 0.718 = 0.48 \text{ meq/l of } Ca(HCO_3)_2$$

Final calcium hardness equals this remainder plus the limit of calcium carbonate removal,

$$0.48 + 0.60 = 1.08 \text{ meq/l} = 54 \text{ mg/l as } CaCO_3$$

and magnesium hardness in the finished water is

$$0.375 \times 1.80 + 0.625 \times 0.20 = 0.80 \text{ meq/l}$$
$$= 40 \text{ mg/l as } CaCO_3$$

for a total hardness of 94 mg/l.

The sulfate ion and chloride ion concentrations are unchanged from the raw water, 2.20 meq/l and 0.40 meq/l, respectively. The carbonate ion concentration is at the practical limit of $CaCO_3$ removal equal to 0.60 meq/l.

Figure 11.13(c) is the bar graph of the finished water after second-state sedimentation and filtration.

11.13 CATION EXCHANGE SOFTENING

The hardness-producing elements of calcium and magnesium are removed and replaced with sodium by a cation resin. Ion exchange reactions for softening may be written

$$Na_2R + \left.\begin{matrix} Ca \\ Mg \end{matrix}\right\} \left\{\begin{matrix} (HCO_3)_2 \\ SO_4 \\ Cl_2 \end{matrix}\right. \longrightarrow \left.\begin{matrix} Ca \\ Mg \end{matrix}\right\} R + \left\{\begin{matrix} 2\,NaHCO_3 \\ Na_2SO_4 \\ 2\,NaCl \end{matrix}\right. \qquad (11.53)$$

where R represents the exchange resin. They show that if a water containing calcium and magnesium is passed through an ion exchanger, these metals are taken up by the resin, which simultaneously gives up sodium in exchange.

After the ability of the bed to produce soft water has been exhausted, the unit is removed from service and backwashed with a solution of sodium chloride. This removes the calcium and magnesium in the form of their soluble chlorides and at the same time restores the resin to its original sodium condition. The bed is rinsed free of undesirable salts and returned to service. The governing reaction may be written

$$\left.\begin{matrix} Ca \\ Mg \end{matrix}\right\} R + 2\,NaCl \longrightarrow Na_2R + \left.\begin{matrix} Ca \\ Mg \end{matrix}\right\} Cl_2 \tag{11.54}$$

A majority of ion exchange softeners are the pressure type, with either manual or automatic controls. They normally operate at rates of 6–8 gpm/ft^2 of surface filter area. A water meter is usually employed on the inlet or outlet side. For manual-type operations, this meter can be set to turn on a light or sound an alarm at the end of the softening run.

About 8.5 lb of salt is required to regenerate 1 ft^3 of resin and remove approximately 4 lb of hardness in a commercial unit. The reduction is directly related to the amount of cations present in the raw water and the amount of salt used to regenerate the resin bed.

IRON AND MANGANESE REMOVAL

Iron and manganese in concentrations greater than 0.3 mg/l of iron and 0.05 mg/l of manganese stain plumbing fixtures and laundered clothes. Although discoloration from precipitates is the most serious problem associated with water supplies having excessive iron and manganese, foul tastes and odors can be produced by growth of iron bacteria in water distribution mains. These filamentous bacteria, using reduced iron as an energy source, precipitate it, causing pipe encrustations. Decay of the accumulated bacterial slimes creates offensive tastes and odors.

Dissolved iron and manganese are often found in groundwater from wells located in shale, sandstone, and alluvial deposits. Impounded surface water supplies may also have troubles with iron and manganese. An anaerobic hypolimnion (stagnant bottom-water layer) in a reservoir dissolves precipitated iron and manganese from the bottom muds, and during periods of overturn these minerals are dispersed throughout the entire depth.

11.14 CHEMISTRY OF IRON AND MANGANESE

Iron (II) (Fe^{2+}) and manganese (II) (Mn^{2+}) are chemically reduced, soluble forms that exist in a reducing environment (absence of dissolved oxygen and low pH). These conditions exist in groundwater and anaerobic reservoir water. When it is pumped from underground or an anaerobic hypolimnion, carbon dioxide and hydrogen sulfide are released, raising the pH. In addition, the water is exposed to air, creating an oxidizing environment. The reduced iron and manganese start transforming to their stable, oxidized, insoluble forms of iron (III) (Fe^{3+}) and manganese (IV) (Mn^{4+}).

The rate of oxidation of iron and manganese depends on the type and concentration of the oxidizing agent, pH, alkalinity, organic content, and presence of catalysts [6].

Oxygen, chlorine, and potassium permanganate are the most frequent oxidizing agents. The natural reaction by oxygen is enhanced in water treatment by using spray nozzles or waterfall-type aerators. Chlorine and potassium permanganate ($KMnO_4$) are the chemicals commonly used in iron- and manganese-removal plants. Oxidation reactions using potassium permanganate are

$$3Fe^{2+} + MnO_4^- \rightarrow 3Fe^{3+} + MnO_2 \tag{11.55}$$

$$3Mn^{2+} + 2MnO_4^- \rightarrow 5MnO_2 \tag{11.56}$$

Rates of oxidation of the ions depend on the pH and bicarbonate ion concentration. The pH for oxidation of iron should be 7.5 or higher; manganese oxidizes readily at pH 9.5 or higher. Organic substances (i.e., humic or tannic acids) can create complexes with iron (II) and manganese (II) ions, holding them in the soluble state at higher pH levels. If a large concentration of organic matter is present, iron can be held in solution at pH levels of up to 9.5.

Manganese oxides are catalytic in the oxidation of manganese. Tray aerators frequently contain coke or stone contact beds through which the water percolates. These media develop and support a catalytic coating of manganese oxides. After chemical-oxidation pretreatment to remove manganese, the grains of sand in a filter bed become coated with manganese oxide. When a new sand filter is initially put into operation, reduced soluble manganese (II) can pass through until the filter is "conditioned" by coating the sand grains with catalytic manganese (IV) oxides from removal of manganese in the water.

11.15 PREVENTIVE TREATMENT

When an industry or a city is confronted with iron and manganese problems, solutions are difficult and probably costly. Treatment of the water supply is the only permanent answer. Control and preventive measures can be employed with expectation of reasonable success.

Phosphate chemicals may be effective in sequestering iron and manganese in well water supplies [6]. The sequestering chemical is added directly into the groundwater pumped from the well prior to any unintentional aeration. When applied at the proper dosage, before oxidation of the iron and manganese occurs, polyphosphates tend to hold the metals in solution and suspension, preventing destabilization and, thus, stopping agglomeration of the individual tiny particles of iron and manganese oxides. The concept is that the sequestered metals will pass through the distribution system without creating discolored water. Nevertheless, oxidized particles often settle out and collect in water mains at times when the velocities of flow in the pipes are low and in storage reservoirs during quiescent periods. Then, when the velocities of flow increase and when water in storage is agitated, these particles are resuspended in the water at much higher concentrations than in the raw water. The scouring and mixing flows can be caused by the dramatic increase in water consumption (e.g., because of lawn watering in the spring of the year), operation of well or booster pumps in the distribution system that have not been used for several weeks or months, and sudden withdrawals of water from hydrants.

Growth of iron bacteria also gravitates the adverse effects of iron and manganese. Some of the iron bacteria are autotrophs that oxidize soluble iron (II) to insoluble iron (III) for energy and use carbon dioxide as a carbon source. Filamentous *Crenothix* grow in sheaths impregnated with ferric hydroxide that form reddish-colored slimes. *Crenothix* infestations in distribution systems are often widespread, since these bacteria produce and release vast numbers of spores that can be carried throughout the pipe network. *Leptothrix*, a member of the genus *Sphaerotilus*, grow on iron (II) and organic matter. Their growth may be accompanied by a musty odor, which changes to a foul odor upon their death. Also associated with decay of the slime growth, sulfate-reducing bacteria can release hydrogen sulfide, giving the water a rotten-egg odor. Manganese oxidation is also attributed to several species of iron bacteria.

No easy or inexpensive way exists for controlling the chemical and bacterial oxidation of iron and manganese. Periodic flushing of small distribution pipes can be effective in removing accumulations of oxide particles; however, elimination of iron bacteria is generally impossible. Heavy chlorination of isolated sections of water mains followed by flushing may be effective for a limited time. If chlorine is continuously added for disinfection, the rate of oxidation of iron and manganese is increased and the problem of colored water is likely to become more severe. The only permanent solution is treatment to remove iron and manganese from the raw water before distribution in the pipe network.

11.16 IRON AND MANGANESE REMOVAL PROCESSES

Aeration–Sedimentation–Filtration

The simplest form of oxidation treatment uses plain aeration. The units most commonly employed are the tray type, where a vertical riser pipe distributes the water on top of a series of trays, from which it then drips and spatters down through a stack of three or four of them. Soluble iron is readily oxidized by the following reaction:

$$2Fe(HCO_3)_2 + 0.5O_2 + H_2O = 2Fe(OH)_3 + 4CO_2 \tag{11.57}$$

Manganese cannot be oxidized as easily as iron, and aeration alone is generally not effective. If, however, the pH is increased to 8.5 or higher (by the addition of lime, soda ash, or caustic soda), and if aeration is accompanied by contact with coke beds coated with oxides in the aerator, catalytic oxidation of the manganese occurs.

In plants using the aeration–sedimentation–filtration process, most of the oxidized iron and manganese is removed by a granular-media filter. Flocculant metal oxides are not heavy enough to settle out in the settling basin. The main function of the basin is to allow sufficient reaction time for oxidation to proceed to near completion.

Aeration–Chemical Oxidation–Sedimentation–Filtration

This sequence of processes is the usual method for removing iron and manganese from well water without softening treatment. Contact tray aeration is designed to displace dissolved gases (i.e., carbon dioxide) and initiate oxidation of the reduced iron and manganese.

Either chlorine or potassium permanganate can chemically oxidize the manganese. When chlorine is utilized, a free available chlorine residual is maintained throughout the treatment process. The rate of manganese (II) oxidation by chlorine depends on the pH, the chlorine dosage, mixing conditions, and other factors. Copper sulfate is a catalyst in the oxidation of manganese.

Theoretically, 1 mg/l of potassium permanganate will oxidize 1.06 mg/l of iron or 0.52 mg/l of manganese [Eqs. (11.55) and (11.56)]. In actual practice, however, the permanganate necessary for oxidation of the soluble manganese is less than the theoretical requirement. One main advantage of potassium permanganate oxidation is the high rate of the reaction, many times faster than for chlorine. Also, the rate of reaction is relatively independent of the hydrogen ion concentration within a pH range of 5–9.

Filtration following chemical oxidation and sedimentation is very important. Practice has shown that filters pass manganese unless grains of the media are coated with metal oxides. This covering develops naturally during filtration of manganese-bearing water. The coating serves as a catalyst for the oxidation and removal of manganese.

Water Softening

Lime–soda ash softening takes out iron and manganese. If split treatment is employed, potassium permanganate can oxidize the iron and manganese in water, bypassing the first-stage excess-lime treatment. Lime–soda ash softening should be given careful consideration as a possible process for treating a hard water requiring iron and manganese elimination.

Lime treatment has been used to remove organically bound iron and manganese from surface water. The process scheme aeration–coagulation–lime treatment–sedimentation–filtration can treat surface waters containing color, turbidity, and organically bound iron and manganese.

Manganese Zeolite Process

Manganese zeolite is made by coating natural greensand (glauconite) zeolite with oxides. Manganese dioxide removes soluble iron and manganese until it becomes degenerated [Eq. (11.58)]. The filter is regenerated by using potassium permanganate [Eq. (11.59)].

$$Z\text{—}MnO_2 + \begin{cases} Fe^{2+} \\ Mn^{2+} \end{cases} \rightarrow Z\text{—}Mn_2O_3 + \begin{cases} Fe^{3+} \\ Mn^{3+} \\ Mn^{4+} \end{cases} \tag{11.58}$$

$$Z\text{—}Mn_2O_3 + KMnO_4 \rightarrow Z\text{—}MnO_2 \tag{11.59}$$

Manganese zeolite filters are pressure filters. Greensand is not practical for gravity filtration because of its small effective size (0.3 mm) resulting in high head loss during filtration.

Disadvantages of the regenerative-batch process are the possibility of soluble-manganese leakage when the bed is nearly degenerated and the waste of excess potassium permanganate needed to regenerate the greensand. These two drawbacks have been substantially overcome by continuously supplying a feed of potassium permanganate solution ahead of a dual-media filter of anthracite and manganese zeolite. Figure 11.14 is

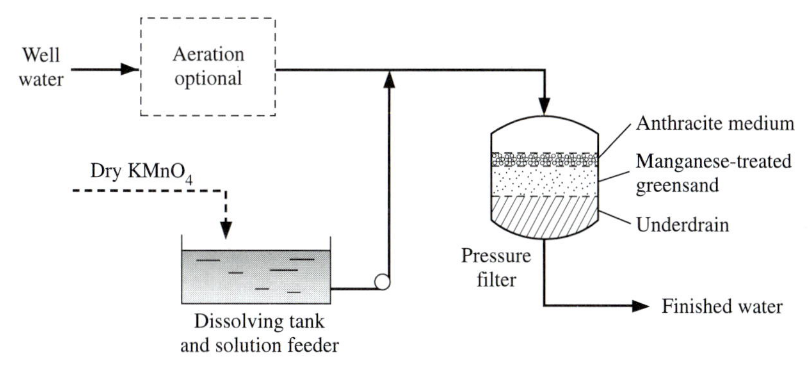

FIGURE 11.14 Schematic flow diagram for removal of iron and manganese from groundwater by the manganese-zeolite process using a dual-media pressure filter with manganese-treated greensand.

a schematic flow diagram of a continuous-flow system ahead of a pressure filter. The anthracite filter media remove most insolubles, thereby reducing the problem of plugging the greensand. A continuous feed of permanganate reduces the frequency of greensand regeneration. When the permanganate feed is less than the reduced iron and manganese in the water, excess iron and manganese are oxidized by the greensand. If a surplus is applied, it regenerates the greensand.

Example 11.8

A well water supply contains 3.2 mg/l of iron and 0.8 mg/l of manganese at pH 7.8. Estimate the dosage of potassium permanganate required for iron and manganese oxidation.

Solution: The theoretical potassium permanganate dosages are 1.0 mg/1per 1.06 mg/l of iron and 1.0 mg/1 per 0.52 mg/1 of manganese. Therefore,

$$KMnO_4 \text{ required} = \frac{3.2 \times 1.0}{1.06} + \frac{0.8 \times 1.0}{0.52} = 4.6 \text{ mg/l}$$

CHEMICAL DISINFECTION AND BY-PRODUCT FORMATION

Chlorine is the common chemical used for disinfection of water and wastewater. One alternative to chlorine is chlorine dioxide, although it is usually applied for taste and odor control. Another alternative is ozone, but because of high cost it is rarely applied solely for disinfection. Other benefits attributed to ozonation are trihalomethane reduction, taste and odor control, and improved coagulation.

11.17 CHEMISTRY OF CHLORINATION

Chlorine gas is soluble in water (7160 mg/l at 20°C and 1 atm) and hydrolizes rapidly to form hypochlorous acid:

$$Cl_2 + H_2O \rightleftharpoons HOCl + H^+ + Cl^- \tag{11.60}$$

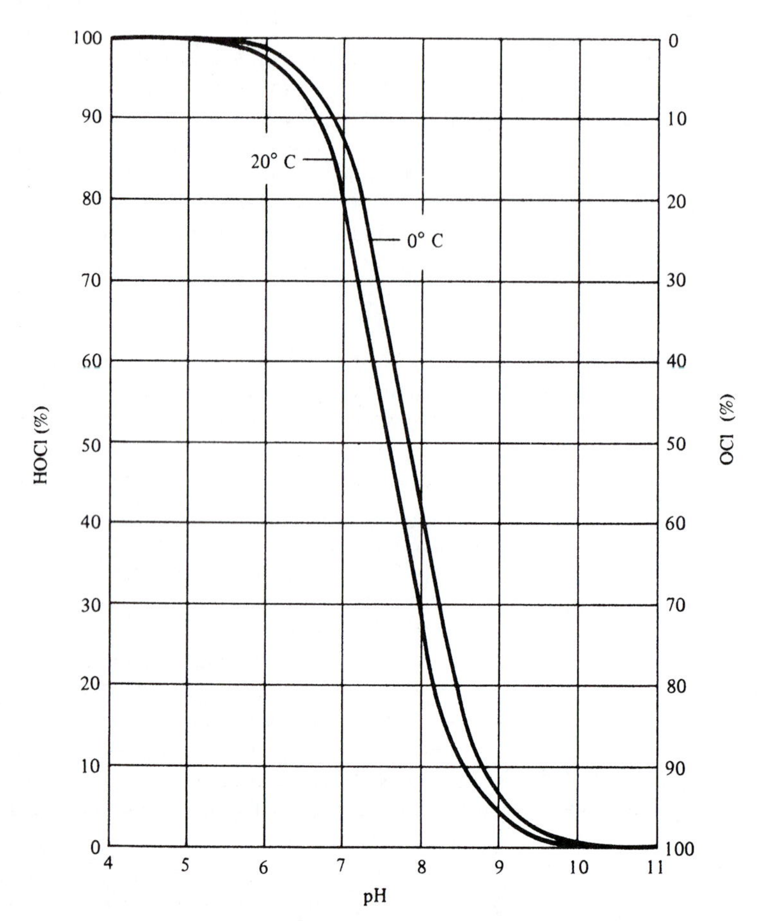

FIGURE 11.15 Effect of pH on the portions of hypochlorous acid (HOCl) and hypochlorite ion (OCl^-) present in water.

Hydrolysis goes virtually to completion at pH values and concentrations normally experienced in water treatment and waste treatment operations.

Figure 11.15 shows the relationship between HOCl and OCl^- at various pH levels. Hypochlorous acid ionizes according to the following equation:

$$HOCl \rightleftharpoons H^+ + OCl^- \tag{11.61}$$

$$\frac{[H^+][OCl^-]}{[HOCl]} = K_i \tag{11.62}$$

The dissociation rate from hypochlorous acid to hypochlorite ion is sufficiently rapid so that equilibrium is maintained even though the former is being continuously consumed. If a reducing agent is put into water containing free available chlorine, the unconsumed residual redistributes itself between HOCl and OCl^-.

Chlorine reacts with ammonia in water to form chloramines as follows:

$$HOCl + NH_3 \rightarrow H_2O + NH_2Cl \quad \text{(monochloramine)} \tag{11.63}$$

$$HOCl + NH_2Cl \rightarrow H_2O + NHCl_2 \quad \text{(dichloramine)} \tag{11.64}$$

$$HOCl + NHCl_2 \rightarrow H_2O + NCl_3 \quad \text{(trichloramine)} \tag{11.65}$$

The chloramines formed depend on the pH of the water, the amount of ammonia available, and the temperature. In the pH range 4.5–8.5, monochloramine and dichloramine are formed. At room temperature, monochloramine exists alone above pH 8.5 and dichloramine occurs alone at pH 4.5. Below pH 4.4, trichloramine is produced.

Free available residual chlorine is that residual chlorine existing in water as hypochlorous acid or hypochlorite ion. *Combined available residual chlorine* is that residual existing in chemical combination with ammonia (chloramines) or organic nitrogen compounds. *Chlorine demand* is the difference between the amount added to a water and the quantity of free and combined available chlorine remaining at the end of a specified contact period.

When chlorine is added to water containing reducing agents and ammonia, residuals develop that yield a curve similar to that in Figure 11.16. Chlorine reacts first with reducing agents present and develops no measurable residual, as shown by the portion of the curve extending from A to B. The chlorine dosage at B is the amount required to meet the demand exerted by the reducing agents (those common to water and wastewater include nitrites, ferrous ions, and hydrogen sulfide).

The addition of chlorine in excess of that required up to point B results in the formation of chloramines. Monochloramines and dichloramines are usually considered together because there is little control over which will be formed. The quantities of each are determined primarily by pH. Chloramines thus established show an available chlorine residual and are effective as disinfectants. When all the ammonia has been reacted with, a free available chlorine residual begins to develop (point C on the curve). As the free available chlorine residual increases, the previously produced chloramines are oxidized. This results in the creation of oxidized nitrogen compounds, such as nitrous oxide, nitrogen,

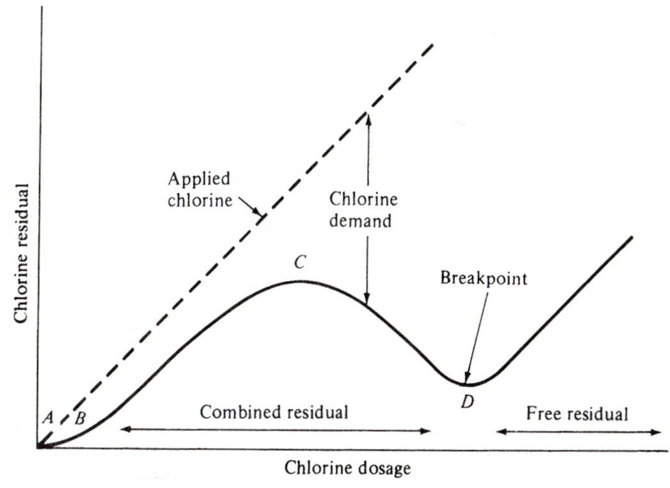

FIGURE 11.16 Chlorine residual curve for breakpoint chlorination.

and nitrogen trichloride, which in turn reduce the chlorine residual, as seen on the curve between C and D.

Once most of the chloramines are oxidized, additional chlorine applied to the water creates an equal residual, as indicated by the rising curve at point D. Point D is generally referred to as the *breakpoint;* beyond it, all added residual is free available chlorine. Some resistant chloramines can still be present beyond D, but their relative importance is small.

Hypochlorites (salts of hypochlorous acid) may be used for chlorination at small installations such as swimming pools and in emergencies. Since hypochlorites are more expensive, liquid chlorine is applied in most water treatment plants in the United States. Calcium hypochlorite, $Ca(OCl)_2$, is available commercially in granular and powdered forms that contain about 70% available chlorine. Sodium hypochlorite (NaOCl) is handled in liquid form at concentrations between 5% and 15% available chlorine. These salts in water solution yield the hypochlorite ion directly.

Feeding of chlorine involves controlled dissolution of the gas into a carrier water supply for delivery to the point of application and blending with the water or wastewater being chlorinated. Direct feed of chlorine gas into a pipe or channel is not practiced for safety reasons, e.g., the danger is pipes leaking gas outside the controlled environment of the chlorine room. Figure 11.17 is a schematic diagram of a vacuum chlorinator and ejector suitable for a small system. Chlorine is shipped as a liquid in pressurized steel cylinders ranging in size from 100 1b to 1 ton. As gas is released, the liquid vaporizes, yielding about 450 volumes of gas per volume of liquid. To start the flow of gas from the cylinder, water under pressure is pumped at high velocity through the ejector throat to draw a vacuum on the regulator. The safety features of the regulator are that the gas cannot be

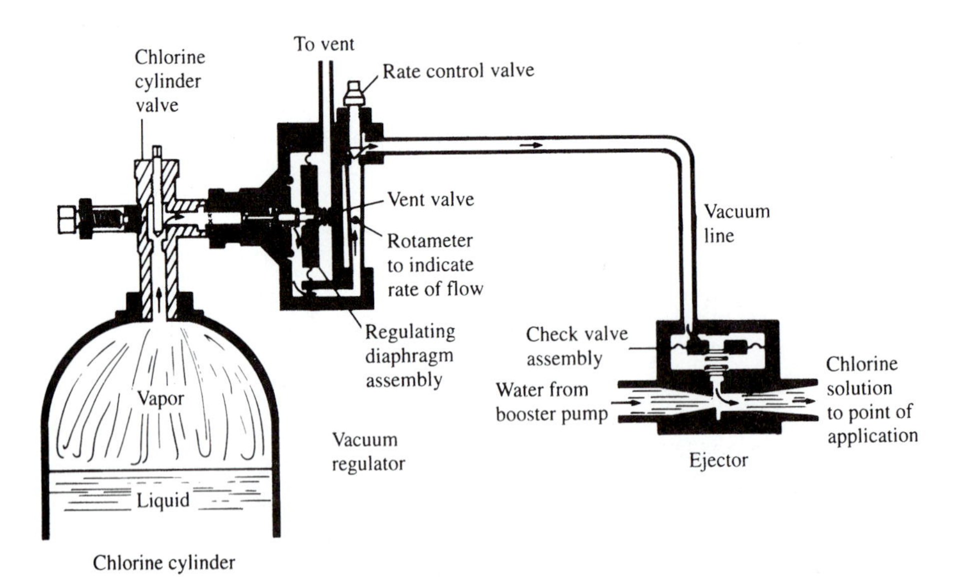

FIGURE 11.17 Gas chlorinator with a vacuum regulator mounted on a cylinder and an ejector to dissolve the chlorine in water for conveying a strong solution to the point of application. *Source:* Courtesy of Severn Trent Services, Capitol Controls.

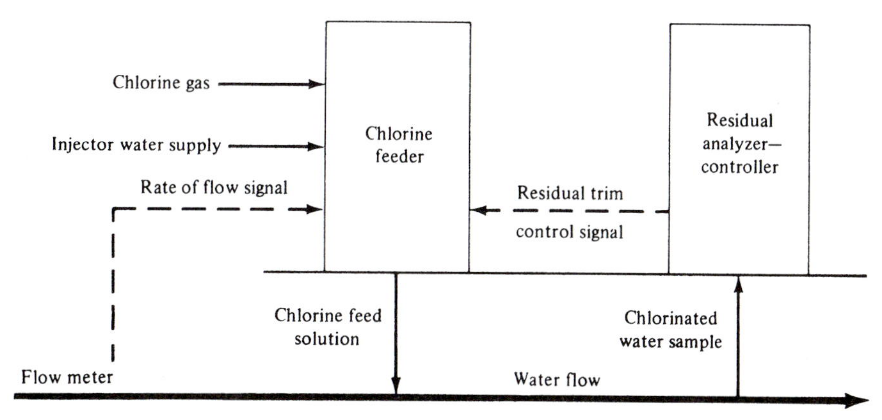

FIGURE 11.18 Automatic proportional-control system for feeding to a constant preestablished chlorine residual.

released except under vacuum, and the regulator case is vented to release the gas outside the room if the diaphragm leaks. The rate of flow is controlled by the rate-control valve on the regulator, and the flow can be observed by the rotameter. The ejector dissolves the gas in water, and this concentrated solution is piped to the water being treated. Moist chlorine gas is extremely corrosive, so piping and dosing equipment are nonmetallic or a special alloy. The yellow-green chlorine gas is poisonous, causing respiratory and eye irritation at very low concentrations and physiologic damage at high doses. Chlorine feeding rooms and storage areas must be kept cool and ventilated.

A manual-control chlorinator, as illustrated in Figure 11.17, is appropriate for a small treatment plant or well water supply discharging water at a nearly constant rate of flow. For large facilities, controlling the chlorine rate of feed based on water flow rate or chlorine residual is important for reliable performance and economical operation. The automatic proportional-control system illustrated in Figure 11.18 adjusts the feed rate to maintain a constant preset dosage for all rates of flow. The chlorine feeder is responsive to signals from both the flowmeter transmitter and the chlorine residual analyzer. In this manner rate-of-flow measurement is the primary feed regulator, and residual monitoring trims the dosage. At some installations, it may be satisfactory to proportion feed to flow and thus apply a constant preset dosage without residual monitoring. However, this type of regulator is only satisfactory where chlorine demand and flow are reasonably constant and an operator is available to make adjustments as necessary.

11.18 CHLORINE DIOXIDE

Chlorine dioxide has had limited application in the United States as a water disinfectant; meanwhile, it has been used for taste and odor control. It is manufactured at a water treatment plant by mixing solutions of sodium chlorite and chlorine in controlled proportions, as shown by Eq. (11.66) [7]:

$$2NaClO_2 + Cl_2 \longrightarrow 2ClO_2 + 2NaCl \qquad (11.66)$$

By maintenance of the pH in the reactor at about 3.5, the production of ClO_2 is optimized with minimum residuals of unreacted chlorine or chlorite. A mineral acid such as HCl or H_2SO_4 can be added to reduce the pH in the reactor if necessary. Under good control, chlorine dioxide yields of 85%–90% are expected with solution concentrations between 500 and 2000 mg/l.

Disinfection using chlorine dioxide has several advantages. It is a strong bactericide and viricide over a wide pH range and forms a residual capable of persisting in the distribution system. Equally important in treatment of surface waters, chlorine dioxide reacts with neither nitrogenous compounds to form chloramines nor humic acids to form trihalomethanes. The greatest potential disadvantage is creation of chlorate and chlorite residuals, which are toxins. For this reason, the recommended limit for these residuals in drinking water is 1.0 mg/l. Second, the high cost of sodium chlorite makes ClO_2 disinfection more expensive than chlorination.

11.19 OZONE

Ozone is a strong oxidizing gas that reacts with most organic and many inorganic molecules. It is more reactive than chlorine. The reactions are rapid in inactivating microorganisms, oxidizing iron, manganese, sulfide, and nitrite, and slower in oxidizing organic compounds like humic and fulvic substances, pesticides, and volatile organic compounds. Unlike chlorine, it does not react with water to produce disinfecting species but decomposes in water to produce oxygen and hydroxyl free radicals. Since it does not produce a disinfecting residual, chlorine is added to treated potable water before distribution for a protective residual. The half-life of ozone in water is approximately 10–30 min and shorter above pH 8; therefore, it must be generated on site.

An ozonation system consists of (1) air preparation or oxygen feed, (2) electric power supply, (3) ozone generation, (4) ozone contacting, and (5) ozone contactor exhaust gas destruction. Ambient air is dried to prevent fouling of the ozone production tubes and to reduce corrosion. A common system uses desiccant dryers in conjunction with compression and refrigerant dryers. The voltage or frequency of the electrical supply is varied to control the rate of ozone production, which requires a specialized power source normally supplied by the generator manufacturer. Generators for water treatment usually employ a corona discharge cell. The cell consists of two electrodes separated by a discharge gap and a dielectric material across which high-voltage potentials are maintained. As the air or oxygen flows between the electrodes, ozone is produced. If ambient air is used, the concentration of ozone is 1%–3.5% by weight. This is an adequate concentration to dissolve enough ozone to attain the concentration–contact time necessary in water processing. When pure oxygen is used, the concentration of ozone is approximately doubled. The common ozone contactor has two or three compartments in series with porous diffusers under a 16-ft water column, as sketched in Figure 11.19. The water flows downward in each compartment counter to the rising fine bubbles of ozonated air. The covered tank increases the partial pressure of ozone and collects the head gases for disposal. Ozone in the exhaust gas must be destroyed or removed by

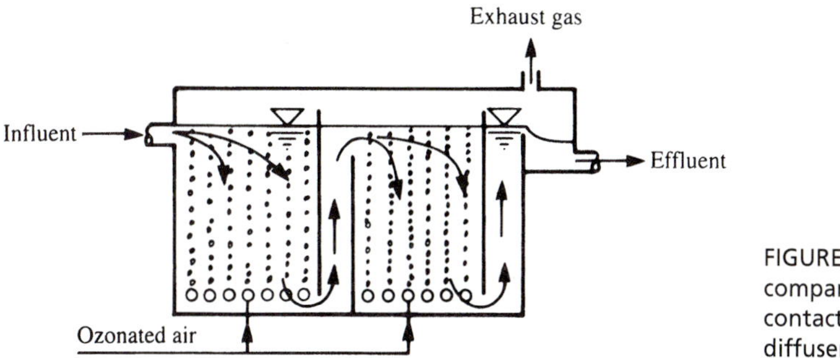

FIGURE 11.19 Two-compartment ozone contactor with porous diffusers.

recycling prior to venting, since the concentration of ozone exceeds air quality standards. When the ozone is generated from air, reducing ozone by the process of thermal–catalytic destruction is less expensive than recirculating the exhaust gas through the air preparation system.

Ozone is rarely applied solely for disinfection because of the high cost relative to chlorine. In most cases, its application is for a combination of operations incorporating taste, odor, or color control, which is a common application; oxidation of humic organic substances that are precursors in the formation of trihalomethanes; destabilization of colloids; and inactivation of microorganisms. The only by-products that have been identified with ozonation of water are detectable levels of aldehydes. The increased interest in application of ozone in potable water treatment relates to this absence of health-related by-products and the potential for destruction of many trace organic compounds.

11.20 DISINFECTION BY-PRODUCTS

Chlorination inactivates pathogenic microorganisms and oxidizes many organic molecules to carbon dioxide. In the treatment of surface waters and groundwaters containing humic substances, however, it also produces chlorinated by-products and incompletely oxidized compounds that are risks to human health. Studies involving chlorination of humic and fulvic acid precursors, isolated from waters, have improved the understanding of by-product formation during disinfection of water supplies. Many of the specific chemical structures have been characterized and vary with chlorine-to-carbon (Cl/C) ratio, pH, time of reaction, and other factors [8]. The principal by-products at high Cl/C molar ratios of 3:1 or 4:1 are volatile hydrophobic compounds, mainly chloroform. In addition, a large variety of nonvolatile hydrophilic compounds are produced, including chlorinated and unchlorinated aromatic and aliphatic compounds. These appear to increase at lower pH and Cl/C molar ratios less than 1:1. At lower Cl/C ratios, which more closely approximate typical drinking water disinfection conditions, the humic acid precursors appear to support the formation of unchlorinated by-products, such as monobasic and dibasic aliphatic acids, to a greater extent than do fulvic acid precursors. Increasing the Cl/C ratio appears to drive both kinds of precursors toward chloroform production and a larger fraction of identifiable products. Nevertheless, the by-products represent only a small fraction of the initial organic matter [8].

An extensive study on the occurrence of disinfection by-products in 35 water treatment plants processing surface waters revealed that trihalomethanes (mainly chloroform, bromodichloromethane, and dibromochloromethane) accounted for about 50% of the total by-products on a weight basis [9]. Five haloacetic acids were the next most significant fraction, accounting for about 25%, and aldehydes accounted for about 7%. Of the remaining by-products, none was present in a significant concentration. The median total trihalomethane concentration was 39 μg/l, and the median haloacetic acids concentration was 19 μg/l [9].

The evidence that trihalomethanes and haloacetic acids are capable of inducing cancer gives rise to a complex set of issues in risk assessment [8,10]. Evidence that trihalomethanes are intrinsically carcinogenic to humans is lacking. Some epidemiologic studies have associated increased cancer risk with chlorination of drinking waters, but most studies have been seriously hampered by universal exposures, such as diet and smoking, and geographic migration in large populations. On the other hand, trihalomethanes are not the only mutagenic and potentially carcinogenic by-products. Therefore, calculation of the levels of risk to humans, who consume drinking water containing these compounds, must rely on data that demonstrate the carcinogenicity of individual trihalomethanes in experimental animals [8].

The carcinogenicity of chloroform in laboratory animals was studied by several researchers. The results of one study are shown in Table 11.5 [8]. Male rats were administered chloroform for 104 weeks in drinking water containing 0, 200, 400, 900, and 1800 mg/l. Based on water intakes and body weights, the average doses were 0, 19, 38, 81, and 160 mg/kg body weight per day. The rats consuming the high-dose rate had a 14% incidence of adenomas and adenocarcinomas of the renal tubules, which is the 7/50 tumor rate shown in Table 11.5. The control group had a 1% (4/301) incidence of tumors. Based on these laboratory data, the carcinogenic risk for chloroform estimated for humans from the linearized multistage mathematical model is 5.16×10^{-8} for a lifetime risk and 8.9×10^{-8} cancer risk for the upper 95% confidence level. This risk estimate expresses the probability of cancer in a person weighing 70 kg who has consumed 1 l of water per day containing 1 μg/l of chloroform for a lifetime of 70 years. At 2 l/d with a maximum contaminant level of 100 μg/l, the risk at the upper 95% confidence level increases to 1.78×10^{-5}, and, therefore, lifetime exposure would be expected to produce one excess case of cancer for every 56,000 persons exposed for their lifetimes. If the maximum contaminant level were lowered to 25 μg/l the risk decreases to

TABLE 11.5 Tumor Incidence in Male Rats Fed Chloroform in Drinking Water

Animal	Sex	Tumor Site	Dose (mg/kg·bw/d)	Tumor Rates
Obsorne–Mendel rats	Male	Kidney	0	4/301
			19	4/313
			38	4/148
			81	3/48
			160	7/50

Source: Drinking Water and Health, Disinfectants and Disinfection By-products, National Research Council (Washington, DC: National Academy Press, 1987), p. 131

4.45×10^{-5} and lifetime exposure would be expected to produce one excess case of cancer for every 225,000 persons exposed. A trihalomethane standard of 25 μg/l or less would severely limit the use of free chlorine for drinking water disinfection.

11.21 CONTROL OF DISINFECTION BY-PRODUCTS

Chlorination is the most common process for disinfecting and establishing a protective residual in water treatment because of its relatively low cost, ease of application, proven reliability, and residual detectability. At the same time, chlorine reacts with humic substances commonly found in raw surface waters to form trihalomethanes and haloacetic acids as by-products. Decaying vegetation produces the humic substances—humic and fulvic acids—referred to as precursors.

Alternatives for reducing the production of trihalomethanes are to (1) change the point of chlorine application, (2) improve the removal of precursors prior to chlorination, (3) use an alternative disinfectant, and (4) reduce the free chlorine residual. Moving the point of chlorine application to later stages in water treatment is the easiest method for reducing by-product formation. Prior to the mid-1970s, the common practice for taste and odor control and disinfection was breakpoint chlorination of raw surface waters as the first step in processing. Now, chlorine is not added until after coagulation and sedimentation or, if previous control of microbial populations is not necessary, after filtration. The benefits of delayed chlorination are reduction in the required dosage and prior removal of precursors.

Minimizing precursor concentrations prior to chlorination complements the alternative of delaying and reducing chlorine additions. Optimizing chemical coagulation also has the benefits of reducing turbidity and removing other organic compounds. Powdered activated carbon applied in the early stages of treatment may adsorb humic substances. Where by-product formation cannot be controlled by improved clarification and filtration, predisinfection using chlorine dioxide or ozone may be considered. Both of these chemicals provide taste and odor control and disinfection, and ozone can improve flocculation and filtration.

The alternative disinfectants are ozone, chlorine dioxide, and chloramines. If ozone is used as the primary disinfectant, secondary chlorination is needed to provide a disinfectant residual in the water entering the distribution system. The major disadvantage of ozonation is high cost. Chlorine dioxide does not produce trihalomethanes, but a major portion can revert to the chlorite ion, which is limited to a maximum contaminant level of 1.0 mg/l.

A chlorine residual in the distribution pipe network of 0.05 mg/l to greater than 0.1 mg/l is common, particularly for treated surface water supplies. The major reasons for carrying a chlorine residual are to minimize regrowth of bacteria present in biofilms inside the pipes and to reduce the risk of pathogen contamination from external sources (i.e., a protective residual). Nevertheless, if retention time in the distribution system results in an undesirable increase in by-product formation, the free chlorine residual can be converted to a combined residual by addition of ammonia. Although chloramine is a significantly weaker disinfectant than free chlorine, establishing a combined available chlorine residual can control regrowth of bacteria and reduce the rate of by-product formation. However, it is much less effective as a protective residual.

11.22 DISINFECTION/DISINFECTION BY-PRODUCTS RULE

The conflicting requirements to provide effective disinfection and to reduce adverse health effects of disinfection by-products lead to the Disinfection/Disinfection By-Products (D/DBP) rule. Under this EPA rule, public water systems are required to limit trihalomethanes, five haloacetic acids, and maintain a residual chlorine in the distribution system [11]. (Section 8.5 includes MCLs of these chemical contaminants established by the Stage 1 D/DBP rule.) The rule also dictates detailed regulations on sampling and monitoring for compliance of by-product concentrations in the distribution system. Future revision and expansion of this rule is expected.

Water treatment plants providing chemical coagulation and filtration must also achieve prescribed reductions of total organic carbon (TOC), depending on source water quality. This contaminant limit is a treatment technique requiring modification or improvement of water processing to reduce TOC concentration in the treated water, which is likely to reduce the precursors in disinfection by-product formation. Although this rule has no exceptions, alternative compliance criteria exist. These criteria are [12]:

- The system's source water TOC is less than 2.0 mg/L.
- The system's treated water TOC is less than 2.0 mg/L.
- The system's source water TOC is less than 4.0 mg/l, the source water alkalinity is greater than 60 mg/l (as $CaCO_3$), and the system's DBP levels for TTHM (total trihalomethanes) and HAA5 (5 haloacetic acids) are less than 40 μg/L and 30 μg/L, respectively.
- The system is using only chlorine as its disinfectant and the DBP (disinfection by-products) levels for TTHM and HAA5 are less than 40 μg/L and 30 μg/L, respectively.
- The system's source water specific ultraviolet absorbance (SUVA) prior to any treatment is less than 2.0 L/mg-m.
- The system's treated SUVA is less than 2.0 L/mg-m.

Total organic carbon is the covalently bonded carbon in a wide variety of organic compounds in runoff containing decaying vegetation, contributed by domestic and industrial water usage, disinfection by-products from chlorination, and micobiological products from wastewater treatment. The use of TOC as a surrogate (composite parameter) for quality of drinking water is considered to establish a measure of safety even when individual contaminants cannot be identified [13]. Removing TOC from a drinking water supply reduces the concentration of potentially hazardous, although unidentified, compounds.

DISINFECTION OF POTABLE WATER

The Environmental Protection Agency has established maximum contaminant level goals of zero for *Giardia lamblia, Cryptosporidium* species, enteric viruses, and *Legionella* for public water supplies. Since routine tests cannot be used to determine the presence of these microorganisms, treatment techniques have been established to ensure their removal and inactivation during water processing. Furthermore, because *G. lamblia* cysts, *Cryptosporidium* oocysts, and viruses represent the most persistent pathogens, treatment for their removal ensures the absence of other pathogens. Although some water supplies may not contain significant numbers of these pathogens, demonstrating their absence by water quality monitoring is not feasible and cannot be used in lieu of applying the specified treatment techniques. The three categories of water supplies are (1) surface water open to the atmosphere and subject to surface runoff, (2) groundwater under the direct influence of surface water (i.e., containing algae, insects, or other macroorganisms, or experiencing significant and relatively rapid shifts in water characteristics), and (3) groundwater.

11.23 CONCEPT OF THE $C \cdot t$ PRODUCT

Chemical inactivation of a specific species of microorganism is a function of disinfectant concentration and contact time. Other important factors are the kind of disinfectant, temperature, pH, viability of the microorganisms, and presence of suspended organic matter.

The $C \cdot t$ concept is expressed as

$$k = C^n \times t \quad (\text{or } C^n \cdot t) \tag{11.67}$$

where

k = a constant for a specific microorganism exposed under specific conditions

n = a constant

C = disinfectant concentration, mg/l

t = contact time required to inactivate a specified percentage of the microorganisms, min

To apply this equation, results of several individual experiments with different disinfectant concentrations under identical conditions are plotted on double-logarithmic paper. For a selected degree of inactivation (e.g., 2.0 log or 99%), the t-versus-C plot produces a straight line with a slope of n. The $C \cdot t$ value remains constant when $n = 1$, regardless of disinfectant concentration; also, the concentration and contact time are of equal importance. The n values determined from experimental analyses on a variety of microorganisms were in a broad range (0.5–2.0), with the majority between 0.8 and 1.2, averaging close to 1.0 [14]. Thus, for engineering practice, the $C \cdot t$ values are based on the assumption that $n = 1.0$.

The kind of disinfectant (chlorine, chlorine dioxide, chloramine, or ozone) has individual characteristics that result in different $C \cdot t$ values for the same conditions. All are influenced by water temperature such that a two- or threefold increase in inactivation rates results from a 10°C increase. Of the chlorine disinfectants, free chlorine is most influenced by pH because of the dissociation of HOCl to OCl^- (Fig. 11.15). Hypochlorous acid, the stronger disinfectant species, is present in a proportion of about 98% below pH 6 and the hypochlorite ion, the weaker disinfectant species, is present in a proportion of 99% above pH 10. At intermediate pH values, the two chemical species are in equilibrium, and as one is consumed more rapidly in reactions, the proportions remain the same, with the concentration of each being decreased accordingly. In general, the order of resistance to chemical disinfection from least to greatest is bacteria, viruses, protozoal cysts, helminth eggs. Each category has a variety of species encompassing a wide diversity of sizes, life cycles, and other biological characteristics, including resistance to chemical disinfectants. Even within the same species, resistance can vary, e.g., between those cultured in the laboratory and those found naturally in the environment. Protection by clumping, or being protected by organic matter, affects the kinetics of disinfection, thereby extending the required contact time. Consequently, turbidity reduction in water treatment and suspended-solids removal in wastewater treatment are important before applying disinfecting chemicals.

Extensive evaluations of the inactivation of microorganisms by chemicals have been documented in several studies. The $C \cdot t$ values for 99% inactivation in Table 11.6 and summarized in the following statements are from Hoff [15]. The $C \cdot t$ value for 99% inactivation of *Escherichia coli* by free chlorine at 5°C and pH 6 averaged 0.045 (mg/l)·min. The $C \cdot t$ values for 99% inactivation of poliovirus 1 by free chlorine at 5°C and pH 6–7 averaged 1.1 and 2.0 (mg/l)·min. in different studies; at pH 10 the average was 10.5 (mg/l)·min. The $C \cdot t$ values for 99% inactivation of *G. lamblia* (infective in humans) by free chlorine at 5°C and pH 6 were 65–150 (mg/l)·min. For *G. muris* (infective in mice), the 99% values in different studies ranged from 68 (mg/l)·min. at 3°C and pH 6.5 to greater than 150 (mg/l)·min. at 5°C and pH 6.

Animal infectivity data for cysts of *G. lamblia* were collected in studies by Hibler [16]. Isolates of *G. lamblia* acquired from several human sources were maintained by passage in Mongolian gerbils. Clean cysts from these animals at a concentration of 1000 cysts/ml were exposed to selected free-chlorine concentrations for specified $C \cdot t$ values from 0.5° to 5°C and at pH values of 6, 7, and 8. At specified time intervals for each temperature and pH condition, the chlorine activity was chemically arrested and the cyst suspension was concentrated and washed. Five gerbils were fed 50,000 cysts that had been exposed to the chlorine solution. An equal

TABLE 11.6 Summary of $C \cdot t$ Value Ranges for 99% Inactivation of Various Microorganisms by Disinfectants at 5°C

	Disinfectant			
Microorganism	Free Chlorine pH 6–7	Preformed Chloramine pH 8–9	Chlorine Dioxide pH 6–7	Ozone pH 6–7
E.coli	0.034–0.05	95–180	0.4–0.75	0.02
Polio 1	1.1–2.5	770–3700	0.2–6.7	0.1–0.2
Rotavirus	0.01–0.05	3800–6500	0.2–2.1	0.006–0.06
G.lamblia cysts	47–>150	—	—	0.5–0.6
G.muris cysts	30–630	—	7.2–19	1.8–2.0

Source: J. C. Hoff, Project Summary, "Inactivation of Microbial Agents by Chemical Disinfectants," EPA/600/S2-86/067 (September 1986), p. 5.

number of positive control animals were each orally inoculated with 50 unchlorinated cysts maintained at the same temperature and pH. After 6–7 days, the gerbils were examined to determine the number per group infected. A separate infectivity study of gerbils demonstrated that approximately five cysts constituted an infective dose, which is 0.01% of the 50,000 dose of treated cysts fed to the test animals. If all five gerbils became infected, the $C \cdot t$ during chlorination of the cyst culture produced less than 99.99% (4.0 log) inactivation, and if no gerbils were infected, the inactivation was greater than 99.99%. If some of the group were infected and others were not, the $C \cdot t$ value was assumed to have produced 99.99% inactivation for the specified temperature and pH conditions.

A regression analysis of the animal infectivity data yielded the equation

$$C \cdot t = 0.985 C^{0.176} \mathrm{pH}^{2.75} T^{-0.147} \quad (11.68)$$

where

$C \cdot t$ = chlorine concentration times contact time for 99.99% inactivation of *G. lamblia* cysts, (mg/l) · min

C = free chlorine concentration, mg/l

T = temperature, °C

Equation (11.68) was derived from experimental results for C in the range of 0.44–4.23 mg/l, pH of 6–8, and T of 0.5°–5°C. Extrapolating these values to calculate estimated $C \cdot t$ values can be performed by assuming first-order kinetics. For example, $C \cdot t$ values for 99.9%, 99%, and 90% would be three-quarters, one-half, and one-quarter of the calculated value for 99.99%, respectively. For temperature correction, the reaction rate can be assumed to double (a twofold increase) for a 10°C temperature increase. For example, the $C \cdot t$ values at 15°C and 25°C would be one-half and one-quarter of the calculated value at 5°C, respectively. For intermediate temperatures, use Eq. (11.29) to calculate the change in reaction rate. Extrapolated values are subject to error and should be considered only as estimated values.

Example 11.9

Using Eq. (11.68), calculate the $C \cdot t$ value for *G. lamblia* inactivation for a chlorine concentration of 2.0 mg/l, pH 7.0, and temperature of 5.0°C. Extrapolate this value to 99% and 99.9% inactivation and compare the results with the values given in Tables 11.6 and 11.7, respectively.

Solution: For 99.99%

$$C \cdot t = 0.985(2.0)^{0.176}(7.0)^{2.75}(5.0)^{-0.147}$$
$$= 185 \text{ (mg/l)} \cdot \text{min}$$

For 99%

$$C \cdot t = (1/2)185 = 93 \text{ (mg/l)} \cdot \text{min}$$

TABLE 11.7 $C \cdot t$ Values for 99.9% (3.0 Log) Inactivation of *Giardia Lamblia* Cysts by Free Chlorine at Various Temperatures and pH Values.

Free Residual Chlorine (mg/l)	pH	Water Temperature 0.5°C [(mg/l) · min]	5°C [(mg/l) · min]	10°C [(mg/l) · min]	20°C [(mg/l) · min]
≤0.4	6.5	163	117	88	44
	7.0	195	139	104	52
	7.5	237	166	125	62
	8.0	277	198	149	74
1.0	6.5	176	125	94	47
	7.0	210	149	112	56
	7.5	253	179	134	67
	8.0	304	216	162	81
2.0	6.5	197	138	104	52
	7.0	236	165	124	62
	7.5	286	200	150	75
	8.0	346	243	182	91
3.0	6.5	217	151	113	57
	7.0	261	182	137	68
	7.5	316	221	166	83
	8.0	382	268	201	101

Source: Adapted from *Guidance Manual for Compliance with the Filtration and Disinfection Requirements for Public Water Systems Using Surface Water Sources* (Environmental Protection Agency, 1991).

The value of 93 (mg/l) · min is within the range of $C \cdot t$ values of 47–>150 (mg/l) · min given in Table 11.6 for 99% inactivation of *G. lamblia* by free chlorine at 5°C in the pH range of 6–7.

For 99.9%,

$$C \cdot t = (3/4)185 = 139 \text{ (mg/l)} \cdot \text{min}$$

This value is slightly (16%) less than the $C \cdot t$ value of 165 (mg/l) · min given in Table 11.7 for 2.0 mg/l, pH 7.0, and 5°C.

11.24 SURFACE WATER DISINFECTION

The EPA Surface Water Treatment Rule [17] requires at least 99.9% (3.0 log) removal of *Giardia lamblia* cysts, 99% (2.0 log) removal of *Cryptosporidium* species oocysts, and 99.99% (4.0 log) removal or inactivation of enteric viruses. These performance standards are generally achieved in well-operated conventional plants processing water by coagulation, sedimentation, filtration, and disinfection. Since *Giardia* cysts and *Cryptosporidium* oocysts are very resistant to chlorine, the intent of this rule is to promote coagulation–filtration treatment for physical removal of a substantial portion of the *Giardia* cysts and *Cryptosporidium* oocysts.

Treatment technique of surface water and groundwater under the direct influence of surface water is required to include filtration unless the source water meets stringent water quality criteria. The fecal coliform concentration in the raw water must be equal to or less than 20 per 100 ml, or total coliform concentration equal to or less than 100 per 100 ml, in at least 90% of the samples tested. The turbidity level cannot exceed 5 NTU except for an unexpected event, but the number of events cannot exceed two in the preceding 12 months. A comprehensive watershed control program is mandated to minimize the potential for contamination by *Giardia, Cryptosporidium*, and viruses, and a defined water quality monitoring schedule must be instituted.

The disinfection requirement for unfiltered public water supplies is at least 99.9% (3.0 log) removal of *Giardia* cysts, 99% (2.0 log), removal of *Cryptosporidium* oocysts, and 99.99% (4.0 log) inactivation of enteric viruses. The calculated $C \cdot t$ value must be equal to or greater than those listed in Table 11.7. Only the $C \cdot t$ for inactivation of *Giardia* cysts is necessary since the $C \cdot t$ for a 4.0 log inactivation of viruses is a much lower value, equivalent to approximately a 0.5 log inactivation of cysts. Disinfection by chemicals other than chlorine must be demonstrated to be equally effective. The free-plus-combined chlorine residual in the water entering the distribution system cannot be less than 0.2 mg/l for more than 4 hr, and the chlorine residual in the distribution system must be detectable in at least 95% of the samples tested each month. Where the residual is undetectable, a measurement of heterotrophic bacteria by a plate count of equal to or less than 500 per ml is deemed to be equivalent to a detectable chlorine residual.

The procedure for calculating the $C \cdot t$ for a treatment system is described in the Guidance Manual [17]. The $C \cdot t$ is the summation of the calculated $C \cdot t$ values before the water arrives at the first customer with the C being the free chlorine residual measured at the end of each chlorination segment, in milligrams per liter, and the t being the calculated contact time of the segment, in minutes. For example, if chlorine were added at the pumping station as the water entered a pipeline and again as the water was discharged into a contact tank, the $C \cdot t$ would be the chlorine residual measured in the discharge of the pipeline multiplied by the contact time in transit plus the residual at the discharge of the contact tank multiplied by the t_{10} time. The contact time in the pipeline is the theoretical detention time, assuming plug flow. In chlorine contact tanks or storage reservoirs, the

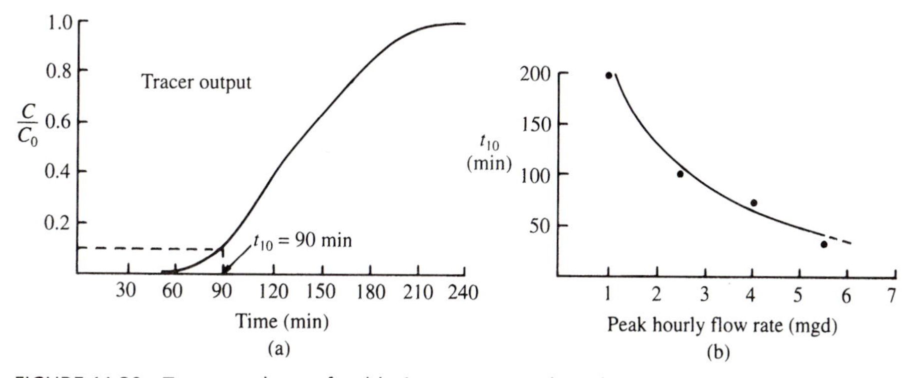

FIGURE 11.20 Tracer analyses of a chlorine contact tank to determine t_{10} times at peak hourly flow for calculating $C \cdot t$ values. (a) Normalized tracer output for a test applying a continuous tracer input. (b) A plot of the t_{10} times for four tracer tests at different flow rates to draw a curve to determine t for calculating $C \cdot t$ values.

contact time can be determined by tracer studies. (Refer to Sections 10.6–10.8.) The tracer tests should be conducted for at least four different flow rates. The residual tracer concentrations C measured at time intervals are normalized by dividing by the applied tracer concentration C_0 and plotted versus time, as in Figure 11.20(a). The contact time t_{10} is the time for 10% of the tracer to pass through the contact tank or reservoir. Figure 11.20(b) is a plot of t_{10} times versus flow rate that can be used to determine t for daily $C \cdot t$ calculations. The contact time t used in computing the $C \cdot t$ is the peak hourly flow for that day.

Surface water plants that provide filtration are required to achieve at least 99.9% removal of *Giardia* cysts, 99% *Cryptosporidium* oocysts, and 99.99% removal or inactivation of enteric viruses by the total treatment system, including coagulation, sedimentation (if provided), granular-media filtration, and chemical disinfection. For filtration to be considered effective in pathogen removal, the turbidity in the combined filtered water must be equal to or less than 0.3 NTU in at least 95% of the measurements taken each month. The combined filtered water must not exceed 1 NTU. For filtration systems serving more than 10,000 people, turbidity must be monitored in the filtered water from each individual filter in order to identify poor performance so corrective action can be taken. In systems serving less than 10,000 people, turbidity in the combined filtered water must be measured at least every 4 hr.

The recommended removals of pathogens achievable in conventional treatment (coagulation, flocculation, sedimentation, and filtration) are a 2.5 log reduction of *Giardia* cysts and a 2.0 log reduction of viruses [17]. Therefore, only 0.5 log of *Giardia* and 2.0 log of viruses need to be inactivated by chemical disinfection. For direct filtration (coagulation and filtration excluding sedimentation), the recommended removals are a 2.0 log reduction of cysts and a 1.0 log reduction of viruses, which leaves 1.0 log of cysts and 3.0 log of viruses for inactivation. Table 11.8 lists the $C \cdot t$ values for 90% (1.0 log) inactivation of *Giardia* cysts by various chemical disinfectants; the $C \cdot t$ values for 0.5 log inactivation are equal to one-half the 1.0 log inactivation values. Recommended $C \cdot t$ values for achieving different levels of virus inactivation are given in Table 11.9. For free chlorine, $C \cdot t$ values recommended for *Giardia* inactivation also provide adequate virus inactivation. For other disinfectants, this may not be true.

TABLE 11.8 $C \cdot t$ Values for 90% (1.0 Log) Inactivation of *Giardia Lamblia* Cysts[a]

		Water Temperature			
	pH	0.5°C [(mg/l) · min]	5°C [(mg/l) · min]	10°C [(mg/l) · min]	15°C [(mg/l) · min]
Free chlorine[b]	6	49	35	26	19
	7	70	50	37	28
	8	101	72	54	36
	9	146	146	78	59
Preformed chloramine		1300	730	620	500
Chlorine dioxide		21	8.4	7.4	6.3
Ozone		0.97	0.63	0.48	0.32

[a] $C \cdot t$ values for 0.5-log inactivation are one-half those shown in table.
[b] Free chlorine values are based on a residual of 1.0 mg/l.
Source: Adapted from *Guidance Manual for Compliance with the Filtration and Disinfection Requirements for Public Water Systems Using Surface Water Sources* (Environmental Protection Agency, 1991).

TABLE 11.9 $C \cdot t$ Values for Inactivation of Viruses at Various Temperatures and pH 6–9[a]

	Log Inactivation	Water Temperature 0.5°C [(mg/l) · min]	5°C [(mg/l) · min]	10°C [(mg/l) · min]	15°C [(mg/l) · min]	20°C [(mg/l) · min]
Free chlorine	2.0	6	4	3	2	1
	3.0	9	6	4	3	2
	4.0	12	8	6	4	3
Preformed chloramine	2.0	1200	860	640	430	320
	3.0	2100	1400	1100	710	530
Chlorine dioxide	2.0	8.4	5.6	4.2	2.8	2.1
	3.0	25.6	17.1	12.8	8.6	6.4
Ozone	2.0	0.9	0.6	0.5	0.3	0.2
	3.0	1.4	0.9	0.8	0.5	0.4

[a] $C \cdot t$ values for free chlorine, ozone, and chlorine dioxide include safety (uncertainty) factors. The chloramine values are based on laboratory data using preformed chloramine to inactivate hepatitis A and do not include a safety factor.

Source: Adapted from *Guidance Manual for Compliance with the Filtration and Disinfection Requirements for Public Water Systems Using Surface Water Sources* (Environmental Protection Agency, 1991).

The residual chlorine (free chlorine plus chloramine) in the water entering the distribution system cannot be less than 0.2 mg/l for more than 4 hr, and a residual in the distribution system must be detectable in at least 95% of the samples tested each month. Where the residual is undetectable, a measurement of heterotrophic plate count equal to or less than 500 per ml is deemed equivalent to a detectable residual.

Example 11.10

A conventional surface water plant with coagulation, flocculation, sedimentation, and filtration produces a filtered water with a turbidity less than 0.3 NTU, pH 7, and temperature 5°C on a day when the peak hourly flow is 3.0 mgd. After filtration, the water is chlorinated in a baffled reservoir with hydraulic characteristics as shown in Figure 11.20. What is the required disinfection of the filtered water if free chlorine is used?

Solution: For conventional treatment, a 2.5-log removal of *Giardia* cysts is achieved by filtration, leaving 0.5-log inactivation required by disinfection.

The required $C \cdot t$ value for 1.0-log inactivation from Table 11.8 is 50 (mg/l) · min at pH 7 and 5°C. (Note that this value is based on a free residual of 1.0 mg/l. If a $C \cdot t$ value for inactivation were given for a lower chlorine residual, it would be lower. In other words, $C \cdot t$ values at a residual of 1.0 mg/l are conservative for postchlorination, since levels above about 0.5 mg/l in drinking water are intolerable for most customers.)

$$C \cdot t \text{ for 0.5-log inactivation} = (1/2)50 = 25 \text{ (mg/l)} \cdot \text{min}$$

From Figure 11.20(b), for a peak hourly flow rate of 3.0 mgd, the t_{10} is 90 min. Therefore, the required free chlorine residual in the effluent from the baffled reservoir is

$$C = \frac{25 \text{ (mg/l)} \cdot \text{min}}{90 \text{ min}} = 0.28 \text{ mg/l}$$

This chlorine residual also satisfies the requirement of a minimum residual of 0.2 mg/l for water entering the distribution system.

11.25 GROUNDWATER DISINFECTION

The Safe Drinking Water Act states:

> ... the Administrator (of EPA) shall also promulgate national primary drinking water regulations requiring disinfection as a treatment technique for all public water systems, including surface water systems and, as necessary, (for) groundwater systems [18].

The first two of three specific categories of drinking water are surface water and groundwater under the direct influence of surface water that are both covered by the Surface Water Treatment Rule. The third category is groundwater not under the direct influence of surface water, for which the need for disinfection is currently determined based on state and local requirements.

The primary pathogens of concern in the contamination of groundwater are fecal viruses, since large pathogens such as *Giardia* cysts and *Cryptosporidium* oocysts are removed by natural filtration through the vadose zone and aquifer. Testing for the presence of pathogenic viruses at low concentrations and identifying their species involve complex and difficult procedures (Section 8.3). Furthermore, no indicative biological organism or substance has been identified to reliably confirm or deny the presence of enteric viruses. Thus, direct monitoring of well water to indicate the presence of enteric viruses (without concurrent coliform bacteria) is not feasible.

Scientific soundness is the foundation for disinfection standards for surface water and groundwater under the direct influence of surface water. Outbreaks of waterborne disease, typified by giardiasis, occurred in water systems with inadequately treated surface waters. *Giardia* cysts and *Cryptosporidium* oocysts were found to be present in many surface waters used as sources for public water supplies. While plain chlorination was inadequate, full-scale and pilot-plant studies clearly demonstrated that treatment by chemical coagulation and filtration followed by chlorination was adequate for cyst and virus removal. Furthermore, the adequacy of treatment could be monitored by measuring turbidity, which is an easy, reliable test.

The Centers for Disease Control and Prevention (CDC), during the 6-year period 1986–1992, reported 110 outbreaks of illness from consuming drinking water from a groundwater source. Only one of these 110 outbreaks was attributed to untreated groundwater in a community supply; the remainder were caused from contamination entering the distribution system [19]. Although outbreaks in transient and noncommunity systems may be underreported, outbreaks in community systems under closer surveillance are more likely to be reported to the CDC.

Some writers, in the absence of significant incidence of waterborne outbreaks, have proposed the theory of endemic disease resulting from public water supplies [20]. (In this context, endemic means constantly present at a significant level affecting individuals within a community rather than a large number of persons; not an epidemic.) Nevertheless, no epidemiological evidence has been established to implicate untreated groundwaters in public water supply systems in endemic infectious disease. For other medical scientists, extending the waterborne-diseases model beyond cholera and typhoid to endemic diarrheal disease without any proper scientific basis gives it exaggerated importance [21]. In the United States, almost all of the transmission of enteric viruses is through person-to-person transmission, fomites, and other close contact. The most effective routes of transmission are probably droplets or ocular inoculation, not necessarily direct ingestion.

Chlorination of groundwater supplies is the addition of chlorine to establish a "protective" residual of 0.2–0.6 mg/l of free chlorine as the water enters the distribution system. Public acceptance of chlorine residual varies, although most consumers object to a concentration greater than 0.5 mg/l. This chlorination is not disinfection based on the $C \cdot t$ concept as applied by the Surface Water Treatment Rule. Well water not requiring treatment is commonly pumped directly into the pipe network without chlorination. In contrast, if groundwater is processed in a treatment plant, the water is always chlorinated prior to distribution.

The practice of residual chlorination greatly reduces the probability of positive coliform tests, since coliform bacteria are rapidly inactivated by a free chlorine residual. Based on the $C \cdot t$ value for E. coli in Table 11.6, inactivation time in a free chlorine residual of 0.5 mg/l in water at 5°C is less than 0.1 min. For pathogens, the inactivation times are greater: Polio virus is 2–5 min and *Giardia* cysts 100–300 min.

A rational assessment of the degree of protection of health provided by maintaining a chlorine residual in a distribution system supplied by groundwater is very difficult. (A chlorine residual in system-supplied surface waters is important to suppress the growth of numerous species of heterotrophic bacteria that survive treatment and regrow in the pipe network.) If the water in the system is contaminated with only 0.1% of wastewater containing low concentrations of bacterial and viral pathogens, a chlorine residual of 0.2–0.4 mg/l provides some protection. This chlorine residual, however, is no guarantee of protection if the percentage of wastewater is greater or concentrations of pathogens are high. The chlorine demand of the contaminating wastewater rapidly neutralizes the chlorine residual, and the pathogens survive.

A potential disadvantage of residual chlorination of uncontaminated groundwater entering a distribution system is interference with coliform testing as an indicator of fecal pollution entering the system. If the supply is unchlorinated, the presence of coliforms in the water is a warning of the possibility of backflow contamination. Conversely, if the supply has a chlorine residual, the coliforms can be inactivated and not indicate the presence of more persistent viruses and protozoal cysts. The water utility personnel may never suspect that backflow at a low level of contamination is occurring in the absence of positive coliform tests.

The best protection against contamination of water in a distribution system is to prevent backflow by installation of backflow preventers in high-risk service connections (e.g., hospitals and mortuaries) and enforcement of a plumbing code for all residential, commercial, and industrial buildings. In case of an unprotected system outlet, back-siphonage is prevented by maintaining adequate pressure in the supply mains to prevent reversal of flow. The risk of backflow is remote in a properly designed and operated water distribution system. A chlorine residual is not a substitute for a well-maintained system with backflow protection.

Groundwater supplies are rarely treated based on the $C \cdot t$ concept solely for disinfection, although some systems may unintentionally achieve this level of disinfection. Examples are disinfection by lime–soda ash softening, iron and manganese removal using potassium permanganate or chlorine, and contact time in a long transmission main to the community from a well field where chlorine is applied to the water. Based on the $C \cdot t$ concept, the only direct application of disinfection would be to treat a groundwater known to be contaminated with pathogenic bacteria and viruses as indicated by the presence of fecal coliforms or because the well has been identified as the source of an outbreak of waterborne disease. The appropriate level of disinfection, between 99% and 99.99% virus inactivation, depends on the specific conditions as determined by the reviewing authority. The $C \cdot t$ values for inactivation of 99% (2.0 log) and 99.99% (4.0 log), which include safety factors (uncertainty factors), are listed in Table 11.9 for various temperatures for a pH range of 6–9.

Methods of disinfection less common than chlorination in water disinfection are not feasible. Ozonation costs much more for installation, operation, and maintenance. Ultraviolet light disinfection is not a proven technology in drinking water treatment. Disinfection reliability is a serious problem, since radiation has not been tested under field conditions for inactivation of viruses. Furthermore, neither of these processes can provide residual disinfection in water entering the distribution system. Therefore, if a disinfectant residual is required in the distributed water, chlorination would be required.

Construction of a drinking water well is a deliberate step-by-step process to ensure the best possible water quality and satisfactory quantity. Sites of potentially suitable hydrogeology are selected where the well location is isolated as far as possible from potential sources of contamination. At the preferred sites, a small-diameter monitoring well is drilled to collect soil samples to record the soil profile and to collect groundwater samples for quality testing. If suitable, the well is constructed in accordance with engineering standards specified for casing and screen, seal around the upper casing to prevent contamination from surface water, column pipe and turbine pump, and well house. Drinking water wells must be designed by a licensed professional engineer, and design, siting, field work, and testing supervised and approved by the appropriate state agency. Existing wells are protected against contamination by a state wellhead protection program, and periodic sanitary surveys are conducted by state personnel to ensure that no potential sources of contamination are located in the proximity of a well.

Example 11.11

The transmission main from the well field to the community is 2500 ft, the velocity of flow at peak pumping capacity is 3.5 fps, water temperature is 10°C, and pH is 7.5. If a free chlorine residual is applied at the well field and is 0.5 mg/l entering the distribution system, what is the log inactivation of viruses based on the data given in Table 11.9?

Solution:

$$\text{travel time} = \frac{2500}{3.5 \times 60} = 11.9 \text{ min}$$

$$C \cdot t = 11.9 \times 0.50 = 6.0 \text{ (mg/l)} \cdot \text{min}$$

From Table 11.9, log inactivation = 4.0 = 99.99%

Example 11.12

Sketch a plan for a continuous-flow chlorine contact tank housed in a building to provide a 4 log (99.99%) virus reduction at 10°C for a well with a capacity of 400 gpm. Assume that the contact time is the time for 10% of the flow to pass through the contact tank as required in the Surface Water Treatment Rule (Fig. 11.20).

Solution: From Table 11.9, the $C \cdot t$ for a 4 log reduction at 10°C equals 6 (mg/l) · min. Based on a tolerable chlorine residual of 0.5 mg/l in the drinking water,

$$\text{contact time} = 6/0.5 = 12 \text{ min}$$

A continuous-flow chlorine contact tank requires a much greater detention time because of short-circuiting and backmixing and the requirement that no more than 10% of the water can receive less than the required contact time of 12 min. Using a long and narrow chlorine contact channel with a length to width ratio of at least 40 to

1, the t_{10} time required for design is assumed to be 3 times greater than the theoretical detention time. Therefore,

$$\text{design detention time} = 3 \times 12 = 36 \text{ min}$$

and

$$\text{design water volume} = 400 \times 36 = 14{,}400 \text{ gal}$$
$$= 1925 \text{ ft}^3$$

A plan-view sketch of the disinfection building is shown in Figure 11.21. The channel is constructed as a serpentine open channel. Using a channel width of 3.5 ft and length of 157 ft, the liquid volume is 1924 ft^3 and the length-to-width ratio equals 45 to 1.

The treatment building is sized at 32 ft by 50 ft for a channel folded 4 times, chlorination equipment, inlet chamber with a vertical riser pipe from the well, outlet chamber for a vertical-turbine pump to discharge disinfected water to the distribution system, pump controls, laboratory bench, and storage area for supplies and tools. Water must be retained in the channel at all times to supply upon demand. The building must be secure and have heating, lighting, ventilation, and utilities.

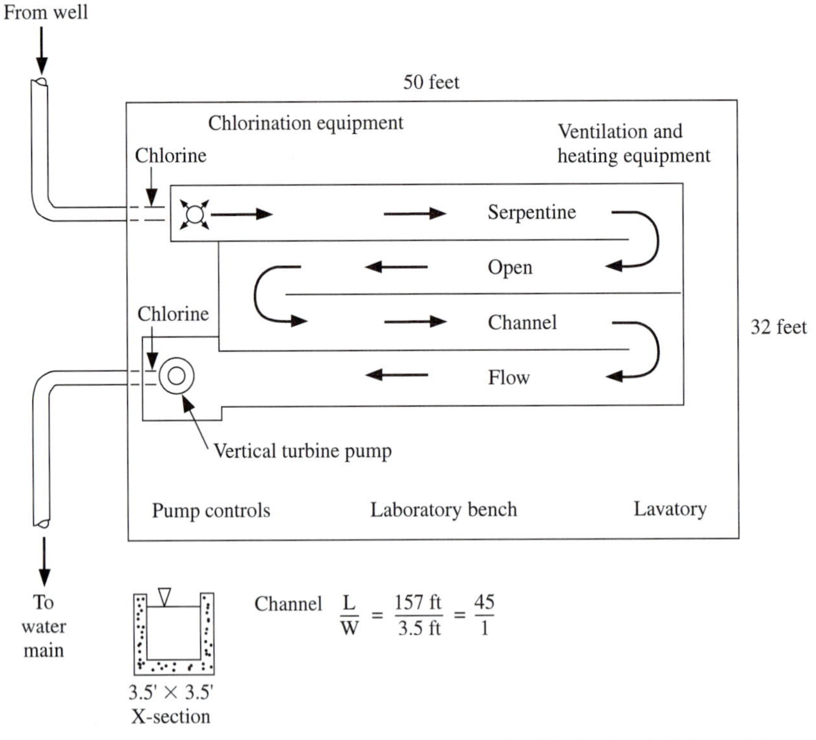

FIGURE 11.21 Plan-view sketch for Example 11.12 of a disinfection building with a continuous flow chlorine contact tank to disinfect groundwater pumped from a drinking water well with a capacity of 400 gpm.

DISINFECTION OF WASTEWATER

The degree of wastewater disinfection depends on the uses of the receiving watercourse or direct reuse of the wastewater. In general, conventionally treated wastewaters discharged to recreational waters with adequate dilution are given only plain chlorination. Reclaiming water for reuse involving human contact or protection of sensitive aquatic organisms for a food source (e.g., shellfish) requires filtration prior to chemical disinfection.

11.26 CONVENTIONAL EFFLUENT DISINFECTION

The kinds and numbers of pathogens in wastewater depend on the health of the contributing population. Although the concentrations of fecal microorganisms are reduced progressively by each stage in wastewater treatment, the effluent still contains a remnant of those contributed to the wastewater in human excreta. The removal of microorganisms by conventional treatment processes are summarized in Table 11.10 [22]. Helminth eggs are most likely to be removed by primary sedimentation because of their large size. In contrast, bacteria and protozoa removals are generally greater in trickling filtration. Apparently they adhere to the fixed-film biological growths coating the filter media. Removals in the activated-sludge processing rank low because of the variability in operational control. If the suspended biological floc under aeration settle out of suspension, removals of all the kinds of microorganisms can be as high as 99%. When the activated-sludge floc are carried out in the effluent, removals of microorganisms are greatly reduced. Even though sedimentation and biological treatment are able to remove 99% to 99.9% of microorganisms, the effluent still can contain significant concentrations of fecal viruses, bacteria, protozoal cysts, and helminth eggs.

Chlorination Wastewater effluents can be chlorinated to inactivate pathogens in an attempt to protect public health where discharges enter surface waters used for body-contact recreation or as municipal water supplies. Disinfection is accomplished by rapid mixing of the chlorine solution with the wastewater followed by contact with the chloramines formed when the chlorine reacts with ammonia present in the wastewater [Eqs.(11.63) and (11.64)]. Acceptable disinfection is defined by the reduction of fecal coliforms to either less than 1000 per 100 ml or less than 200 per 100 ml, depending on

TABLE 11.10 Removal of Microorganisms in Conventional Wastewater Treatment

Kind of Microorganism	Primary Sedimentation (%)	Trickling Filtration (%)	Activated Sludge (%)	Postchlorination (%)
Bacteria	0–90	0–99	0–99	99–99.99
Viruses	0–90	0–90	0–90	—
Protozoa	0–90	0–99	0–90	—
Helminths	0–99	0–90	0–90	0

Source: Adapted from R.G. Feachem et al., *Sanitation and Disease, Health Aspects of Excreta and Wastewater Management* (Chichester: Wiley, 1983).

the state standard. Some standards are more stringent and require pretreatment by granular-media filtration. The requirement for effluent disinfection may also be seasonal, depending on recreational water use. Since biologically treated wastewater contains approximately 1,000,000 coliforms per 100 ml, oxidative reduction of fecal coliforms from this large number to 200–1000 per 100 ml destroys most bacteria (Table 11.10). Nevertheless, viruses and protozoal cysts are more resistant to chlorination than are bacteria, and helminth eggs are unharmed. Furthermore, the presence of suspended solids inhibits disinfection by harboring viruses within floc material, shielding them from the action of the chlorine.

An efficient chlorination system provides initial contact of the chlorine solution with the wastewater during mixing, followed by contact time with chloramines in a plug-flow tank for a minimum of 30 min at the peak hourly flow. Rapid blending can be accomplished by applying the chlorine into a pressure pipe conveying turbulent flow or into a channel immediately upstream from a mechanical mixer. Figure 11.22 illustrates the importance of longitudinal baffling in a chlorine contact tank to provide a long narrow flow stream to approach plug flow. Cross baffling with a wider flow stream creates short-circuiting of flow through the center with stagnant corners and excessive dispersion resulting from intermixing and backmixing. The tracer response curves are normalized residence time distributions in response to a pulse input of dye. The y-axis is dimensionless concentration C_t/C_0 (effluent concentration/influent concentration) and the x-axis is dimensionless time t/t_R (time elapsed since tracer input/theoretical residence time). By

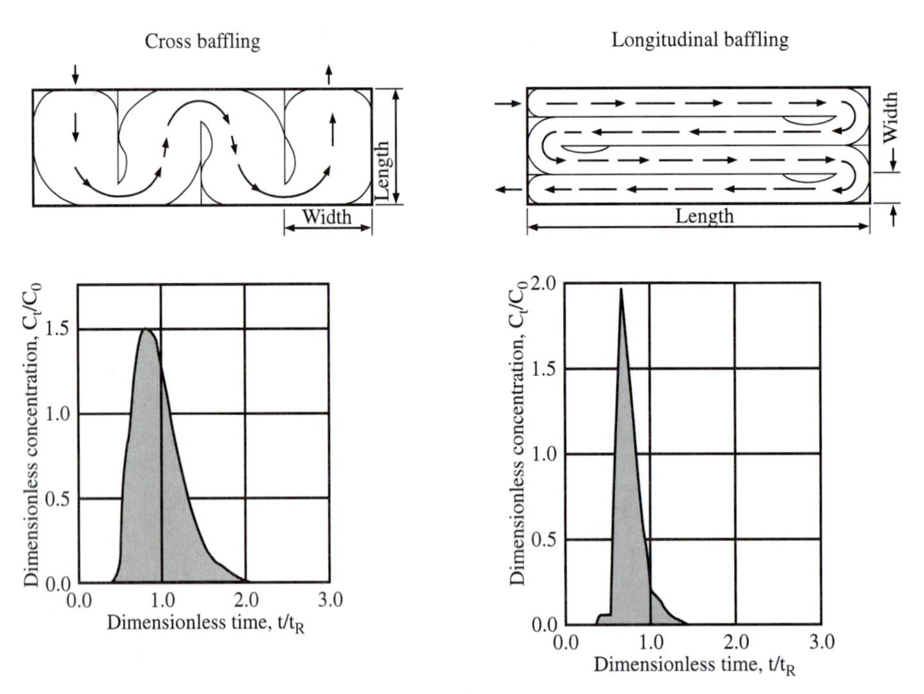

FIGURE 11.22 Plan views of chlorine contact tanks with cross baffling and longitudinal baffling for serpentine flow and corresponding output residence time distributions in response to pulse tracer inputs in field analyses of full-scale tanks.

comparing these tracer response curves to those in Figure 10.9, the curve for cross baffling exhibits an intermediate or a large amount of dispersion, while the curve for longitudinal baffling exhibits a small amount of dispersion, much closer to the desired plug flow.

Control of chlorine dosage is extremely important for proper operation. A well-designed unit provides adequate disinfection of a secondary effluent with a dosage of 8–15 mg/l. Automatic residual monitoring and feedback control is necessary to prevent either inadequate disinfection or excessive chlorination resulting in discharge of an effluent toxic to aquatic life. To protect receiving streams, some regulatory agencies have specified maximum chlorine residuals in undiluted effluents of 0.1–0.5 mg/l. Dechlorination may be required to detoxify a discharge after chlorination. This is usually done by adding sulfur dioxide in aqueous solution at the discharge end of the chlorination tank. The oxidation–reduction reaction [Eq.(11.69)] between the chloramine residual and the sulfur dioxide is very rapid. Any excess sulfur dioxide applied reduces dissolved oxygen [Eq. (11.70)].

$$NH_2Cl + SO_2 + 2H_2O = NH_4^+ + SO_4^{2-} + Cl^- + 2H^+ \tag{11.69}$$

$$2SO_2 + O_2 + 2H_2O = 2SO_4^{2-} + 4H^+ \tag{11.70}$$

Ultraviolet Radiation

The energy waves of ultraviolet (UV) radiation are in the range of electromagnetic waves 100 to 400 nm long, between the X ray and visible light spectrums. For disinfection, the optimum UV range is 245 to 285 nm; therefore, optimum UV lamps emit maximum energy output at a wavelength of 253.7 nm. UV rays inactivate microorganisms by penetrating the cell walls, altering molecular compounds essential to cell function. Based on available data, bacteria and viruses are inactivated at significantly lower UV doses than larger protozoa, such as *Giardia* and *Cryptosporidium*. UV radiation quickly dissipates in water by being reflected or absorbed by material in the water without producing disinfection by-products [12].

The UV dose for inactivation of microorganisms is directly related to radiation intensity and time of exposure, so that high-intensity UV energy over a short period of time is as effective as lower-intensity UV energy at a proportionally longer period of time. In the application of UV disinfection, effective dose is determined by site-specific data related to water quality and permitted microorganism concentrations. UV transmittance is adversely affected by turbidity, color, and suspended solids and defines the quantity of UV equipment required to meet the disinfection limits. The concentration and size of suspended solids in the wastewater is particularly important because bacteria and viruses contained within the clusters are shielded from the UV radiation. In contrast, chemical water quality parameters like pH, temperature, alkalinity, and total organic carbon do not influence the effectiveness of UV disinfection. Nevertheless, chemical and biological films reduce transmittance and cause problems with keeping the lamp sleeves clean. As a result of all of these complex factors, specific design parameters vary for individual waters and should be determined empirically for each application [12].

UV lamps are powered by electricity and operated in much the same way as fluorescent lamps. The difference is that the fluorescent lamp bulb is coated with phosphorus to convert the UV radiation to visible light. The UV lamps are not coated so UV radiation is generated. Lamps used in UV disinfection are commonly quartz tubes filled with an inert gas, such as argon, and small quantities of mercury. Both low-pressure and medium-pressure lamps are available. Low-pressure lamps emit their maximum energy output at a wavelength of 253.7 nm, and medium-pressure lamps emit energy output with wavelengths ranging for 180 to 1370 nm. Therefore, more low-pressure lamps are required for an equivalent dosage. Low-pressure lamps are enclosed commonly in a quartz sleeve to separate the water from the lamp surface, which is required to maintain the lamp temperature operating near its optimum of 40°C. Ballasts are transformers (electronic or electromagnetic) to control the power to the lamps and should operate at 60°C to prevent premature failure.

Figure 11.23(a) illustrates an installation of UV radiation for disinfection of an effluent from conventional biological or tertiary wastewater treatment and (b) shows a frame with 8 lamps. The geometry of the open channel and the installation of multiple frames set at optimum spacing in the channel are designed for dispersed plug flow to provide radial mixing so the flow is uniformly distributed through the varying regions of UV energy intensity. The microprocessor in the system control center adjusts UV intensity in response to changes in wastewater quality and rate of flow to conserve energy and extend lamp life. The ballasts enclosed in the tops of the frames operate at multiple power settings between 60% and 100%. The system control can also turn frames of lamps on and off in relation to the water flow. The dose pacing program is based on flow signal, UV transmittance value, and lamp age. The automatic sleeve cleaning system is a combination of mechanical cleaning by pulling the wiper collars over the lamp sleeves and chemical cleaning by applying a food-grade chemical gel. This cleaning system reduces the amount of operator maintenance.

11.27 TERTIARY EFFLUENT DISINFECTION

The inability of gravity sedimentation to remove small suspended solids following biological aeration is a major limitation of conventional treatment. Tertiary granular-media filtration preceded by chemical coagulation can be used for effluent clarification to remove tiny particles, like helminth eggs, and colloidal solids that interfere with the disinfecting action of chlorine.

Tertiary treatment and disinfection processes are similar to those applied in surface water treatment for potable supplies: (1) rapid mixing of chemicals, flocculation, sedimentation, and granular-media filtration or (2) direct filtration after chemical mixing without sedimentation (Section 14.5). The flocculation–sedimentation–filtration system is recommended for tertiary treatment of a conventional biologically treated effluent with BOD and suspended solids of 30 mg/l. For direct filtration, the biologically treated wastewater must be of consistently higher quality, which can be achieved only by extended aeration in an activated-sludge process. The state of California regulatory guideline for direct filtration is a settled wastewater turbidity of less than 5 NTU (Section 14.15).

The effluent chlorination system following filtration must be properly designed and operated to ensure effective disinfection. Rapid mixing is used to blend the chlorine solution with the filtered wastewater, and chlorine contact is by plug flow through the

FIGURE 11.23 Disinfection of wastewater effluent by ultraviolet (UV) radiation. (a) Multiple frames of lamps are suspended in the channel with a fixed depth of flow to ensure continuous submergence. Each frame can be electrically unplugged and removed from the channel for maintenance and a spare frame inserted. (b) Each frame of lamps has a sealed enclosure on top containing electrical wiring and electronic ballast. An automatic mechanical and chemical self-cleaning system, shown as wiper collars in the center of the rack of lamps, is transported across on the sleeves surrounding the lamps to remove chemical and biological films that interfere with UV transmission. *Source:* Courtesy of Trojan Technologies, Inc., Ontario, Canada.

tank with a minimum of short-circuiting and backmixing in a long narrow channel. Common design specifications are a theoretical detention time of at least 2.0 hr and an actual modal contact time of at least 1.5 hr. A chlorine residual in the range of 5–10 mg/l is maintained by automatic residual monitoring and feedback control. This provides a chloramine $C \cdot t$ in the range of 360–720 mg/l · min based on a t_{10} equal to 60% of the theoretical detention time. If dechlorination of the effluent is necessary, a solution of sulfur dioxide can be added at the discharge end of the chlorination tank.

TASTE AND ODOR

Surface waters contain tastes and odors associated with decaying organic matter, biological growths, and chemicals originating from industrial waste discharges. Land drainage from snowmelt and spring rains contains pollutants that are difficult to remove, especially at the low water temperatures that occur in northern climates. During the summer, lake and reservoir supplies may be plagued by blooms of algae that impart compounds producing fishy, grassy, or other foul odors. Actinomycetes are mold-like bacteria that create an earthy odor. Well waters may contain dissolved gases, such as hydrogen sulfide, and inorganic salts or metal ions that flavor the water. Tastes in groundwater can be identified, but defining specific odor-bearing substances in surface waters is usually an impossible task. Therefore, the best practical treatment for each water supply is determined by experiment and experience.

11.28 CONTROL OF TASTE AND ODOR

Water treatment plant designs should allow maximum flexibility of operations for control of tastes and odors. Problems may vary considerably during different seasons of the year, and future adverse changes in water quality of a supply are difficult to predict. Breakpoint chlorination and treatment with activated carbon are common control techniques for surface water supplies, while aeration is applied most frequently in groundwater processing. Preventive measures, such as selection of a source with the best water quality, should be given primary consideration. Regulatory controls should be used to reduce contaminants from waste discharges that enter surface and underground waters. Algal blooms may be suppressed in reservoirs by regular application of copper sulfate or approved herbicide.

Oxidative Methods

Chlorine dioxide, potassium permanganate, and ozone are strong oxidants capable of destroying many odorous compounds. These chemicals rather than heavy chlorination are favored since they do not form significant by-products (Section 11.21). Occasionally, a combination of chemicals can be used to achieve maximum control. The use of ozone is limited in the United States.

Powdered Activated Carbon

Adsorption on activated carbon is the most effective means of taste and odor removal. It is primarily a surface phenomenon; that is, one substance is attracted to the surface of another. The larger the surface area of an adsorber, the greater its power. Carbon for water treatment is rated in terms of square meters of surface area per gram. One pound of activated carbon has an estimated surface area of 100 acres. Besides controlling tastes and odors, powdered carbon aids in sludge stabilization, improved floc formation, and reduction in non-odor-causing organics.

Carbon is fed to water either as a dry powder or as a wet slurry. The latter has the advantage of being cleaner to handle and assures complete effectiveness by thoroughly wetting the carbon. Although granular carbon adsorption beds are used extensively to purify product water in the food and beverage industries, their application in municipal water processing is limited by economic considerations.

Activated carbon can be introduced at any stage of water processing ahead of filtration. Although adsorption is nearly instantaneous, a contact time of 15 min or more is desirable before sedimentation or filtration. The best point of application is generally determined by trial and error based on previous experience. Carbon is often fed during flocculation or just prior to filtration. Since carbon adsorbs chlorine, these two chemicals should not be applied simultaneously or in close sequence.

Aeration

Air stripping is effective for removing dissolved gases and highly volatile odorous compounds. Aeration as a first step in processing well water may achieve any of the following: removal of hydrogen sulfide, reduction of dissolved carbon dioxide, and addition of dissolved oxygen for oxidation of iron and manganese. Aeration is rarely effective in processing surface waters, since the odor-producing substances are generally nonvolatile.

FLUORIDATION

The incidence of dental caries relative to the concentration of fluoride in drinking water is well established [23]. Low levels of fluoride result in increasing incidence of caries, while excessive fluoride results in mottled tooth enamel. At the optimum concentration for the local climate, consumers realize maximum reduction in tooth decay with no esthetically significant dental fluorosis. Recommended limits given in Table 11.11 are based on air temperature, since this influences the amount of water ingested by people. Medical studies have also indicated that fluoride benefits older persons by reducing the prevalence of osteoporosis and hardening of the arteries. Most public water supplies add fluoride ion to achieve an optimum level as a public health measure.

TABLE 11.11 Recommended Fluoride Limits for Public Drinking Water Supplies

Annual Average of Maximum Daily Air Temperatures Based on Temperature Data Obtained for a Minimum of 5 Yr (°F)	Fluoride Ion Concentrations (mg/l) Recommended Limits		
	Lower	Optimum	Upper
50.0–53.7	0.9	1.2	1.7
53.8–58.3	0.8	1.1	1.5
58.4–63.8	0.8	1.0	1.3
63.9–70.6	0.7	0.9	1.2
70.7–79.2	0.7	0.8	1.0
79.3–90.5	0.6	0.7	0.8

Source: National Primary Drinking Water Regulations, U.S. Environmental Protection Agency.

11.29 FLUORIDATION

The fluoride compounds most commonly applied in water treatment are sodium fluoride, sodium silicofluoride, and fluosilicic acid (Table 11.12). Sodium silicofluoride is commercially available in various gradations for application by dry feeders. Sodium fluoride is also popular, particularly the crystalline type where manual handling is involved. Fluosilicic acid is a strong corrosive acid that must be handled with care. It is preferred in waterworks where it can be applied by liquid feeders without prior dilution.

Design of a fluoridation system varies with size and type of water facility, chemical selection, and availability of operating personnel. Small utilities often choose liquid feeders to apply solutions of NaF or Na_2SiF_6 that are prepared in batches. The solution tank may be placed on a platform scale for the convenience of weighing during preparation and feed. Saturator tanks containing a bed of sodium fluoride crystals yield a solution of about 4.0% (18,000 mg/l of F). Larger water plants use either gravimetric dry feeders to apply chemicals or solution feeders to inject full-strength H_2SiF_6 directly from the shipping drum. Automatic control systems use flow meters and recorders to adjust feed rate.

Application of fluoride is best in a channel or water main coming from the filters, or directly to the clear well. If applied prior to filtration, losses could occur due to reactions with other chemicals, such as coagulation with heavy alum doses or lime softening. If no treatment plant exists, fluoride can be injected into mains carrying water to the distribution system. This may be a single point or several separate fluoride feeding installations where wells supply water at different points in the piping network.

Example 11.13

The fluoride ion concentration in a water supply is increased from 0.30 mg/l to 1.00 mg/l by applying 98% pure sodium silicofluoride. How many pounds of chemical are required per million gallons of water?

Solution: From Table 11.12, Na_2SiF_6 is 61% F.

$$\text{dosage} = \frac{(1.00 - 0.30) \times 8.34}{0.98 \times 0.61} = 9.77 \text{ lb/mil gal}$$

TABLE 11.12 Common Chemicals Used in the Fluoridation of Drinking Water

	Sodium Fluoride	Sodium Silicofluoride	Fluosilicic Acid
Formula	NaF	Na_2SiF_6	H_2SiF_6
Fluoride ion (%)	45	61	79
Molecular weight	42	188	144
Commercial purity (%)	90–98	98–99	22–30
Commercial form	Powder or crystal	Powder or fine crystal	Liquid

CORROSION AND CORROSION CONTROL

Internal corrosion of piping and valves is a serious problem in many water distribution systems. A national survey of the chemical composition of water in 130 systems throughout the United States revealed that 17% had highly corrosive water and approximately 50% more had moderately aggressive water [24]. In addition to economic loss, corrosive waters have the potential of degrading water quality by the dissolution of metals from the distribution system and household plumbing.

11.30 ELECTROCHEMICAL MECHANISM OF IRON CORROSION

The primary mechanism in metal dissolution is a chemical process accompanied by the passage of an electric current. In aerated water, the following complementary electrode process is capable of absorbing electrons and therefore acting as a cathode.

$$O_2 + 2\,H_2O + 4e^- \longrightarrow 4\,OH^- \tag{11.71}$$

As shown in Figure 11.24, the principal reaction at the lower potential, the anode, is dissolution of the iron-releasing iron ions into solution. These are oxidized in the presence of hydroxide ions and dissolved oxygen to form ferric oxides that are only slightly soluble. Thus, rather than the ferrous ions being carried away in the water, leaving a clean surface, as in the case of corrosion in deaerated water, the iron oxides collect around the anode. Nodules created by this precipitation reduce the diffusion of oxygen and strengthen the anodic character of the surface covered by the oxide nodules. The result is pitting and the ultimate perforation of the pipe wall by corrosive action.

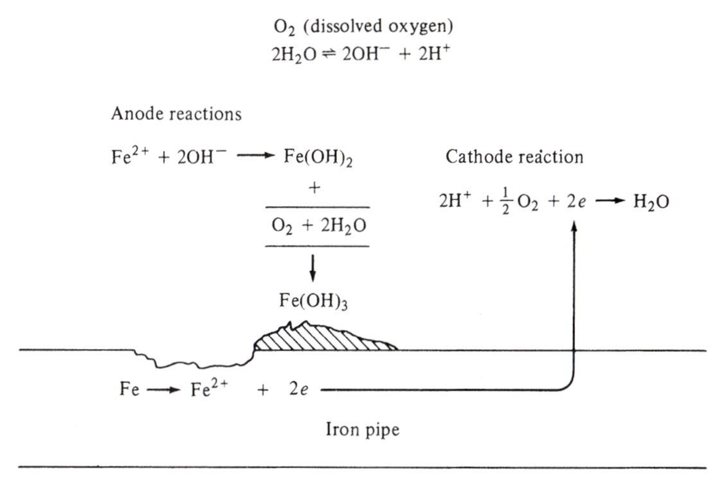

FIGURE 11.24 Electrochemical mechanism for corrosion of iron exposed to aerated water. Metal dissolution is the result of iron oxidation at the anode coupled with reduction of oxygen at the cathode.

Corrosion is generally considered to be limited by formation of protective films that coat interior pipe surfaces [25]. In potable water systems the carbonate–carbon dioxide equilibrium can be shifted and controlled by chemical treatment to deposit and maintain a calcium carbonate film. The mechanically applied cement lining in new ductile iron pipe is preserved by this chemical stabilization. After establishing a water slightly oversaturated with calcium carbonate, polyphosphates are applied at a dosage of about 2 mg/l to inhibit crystal nucleation. Thereby, the protective carbonate scale does not dissolve and the excess calcium ions cannot precipitate.

The primary chemical characteristics of water quality that influence corrosion of iron pipe are pH, alkalinity, and bicarbonate–carbonate buffer capacity [26]. In the pH range of 7 to 9, dissolution of iron and precipitation of nodules of iron hydroxides generally increase with increasing pH. Increased alkalinity generally lowers dissolution of iron, and higher buffer capacity attenuates pH changes from corrosion reactions at anodic and cathodic areas. As illustrated in the chemical reactions in Figure 11.24, dissolved oxygen is an important electron acceptor in the dissolution of iron and in the precipitation of iron hydroxide. Scale is actually composed of many heterogeneous compounds, including calcite ($CaCO_3$, a protective scale), siderite ($FeCO_3$), and green rust (iron compounds containing both ferrous and ferric iron plus other anions). Other parameters potentially contributing to corrosion include age of pipe, biological activity, chlorine residual, water temperature, and chloride concentration.

Iron corrosion is a very complex process highly variable among different water distribution pipe networks. None of the mathematical corrosion indexes can either universally indicate or resolve corrosion problems. The Langelier index (also referred to as the Saturation Index, SI) has been improperly applied as the cure-all method for solving corrosion problems since it was first proposed in 1936 [26]. Although successful sometimes, it is not a universal method for controlling corrosion. Another method, the Larson index, has provided mixed results. Phosphate inhibitors are added for corrosion control, for sequestering soluble iron to reduce "red water," and for protective $CaCO_3$ scale formation. Along with a comprehensive list of references on iron pipe corrosion, L. S. McNeill and M. Edwards [26] provide a summary of key factors that utilities must evaluate in order to mitigate iron corrosion problems.

11.31 CORROSION OF LEAD PIPE AND SOLDER

Lead in excess of 5 μg/l in raw water sources is rare. In tap water, it is a corrosion by-product from customer service lines, household piping, and plumbing fixtures. The sources are solder containing 50% lead and 50% tin joining copper pipes (which is no longer being used), old lead goosenecks that were used to connect the service line to the water main, and brass fixtures containing 3%–8% lead. The highest lead concentration is in water that has been in contact with the service line and plumbing for several hours, while the content in flowing water is undetectable. Therefore, first-flush sampling is recommended in monitoring for compliance with the allowable limit of 15 μg/l at the customer's tap. Aging of copper plumbing reduces the dissolution of lead from solder to a level below 15 μg/l, usually after 5 years. Chemical characteristics that increase dissolution of lead are low alkalinity, acidic pH, and high temperature.

Recommended methods of controlling excessive lead concentrations are corrosion-control treatment and replacement of long lead service lines.

Lead corrosion control is complicated because of the many interdependent reactions that occur in the formation of protective films on lead surfaces. The effect of pH on lead solubility is very strong [27]. Based on this, one recommendation to reduce lead corrosion has been to increase the pH to 8.0 or above. Nevertheless, dissolution in actual service lines is often lower than the theoretically predicted concentration because of protective coatings and other factors. The formation of an effective film of lead carbonates (e.g., $PbCO_3$) depends on pH and alkalinity, which has resulted in a second recommendation of chemical addition for supplementing alkalinity. But because the relationship is complex, increasing the carbonate content (except in very low alkalinity waters) can be complicated by the precipitation of calcium carbonate. Although $CaCO_3$ deposition is accepted practice to reduce iron corrosion, the adequacy of protection from lead leaching has been questioned by the argument that sloughing off of the deposit or incomplete coating of the pipe could result in periodic dissolution of lead in high concentrations. Despite this, in well waters high in calcium and alkalinity, service lines with lead goosenecks and copper lines joined with lead solder have shown negligible lead content in first-flush samples.

The use of phosphate compounds (orthophosphates and polyphosphates) as a remedial measure to reduce lead levels is controversial. Recent experimental studies concluded that polyphosphates cannot be recommended for lead corrosion control [28]. Orthophosphate reduced soluble lead in most cases by about 70% except in new pipes, where lead release was observed to be comparable to the same system without orthophosphate addition. Dosing with hexametaphosphate generally increased soluble lead release over a range of water qualities. Polyphosphate cannot be recommended for lead corrosion control without extensive testing that provides evidence contradicting this research [28].

Verifying the effectiveness of a lead corrosion control program is time-consuming and complicated by changing conditions. The most likely measures are adjustment of pH or supplementing alkalinity, or some combination of these. The only way to measure the success of corrective treatment is to monitor changes in lead concentrations in selected first-flush tap waters. Because formation of protective films is a slow process, changing field conditions can occur that are independent of the treatment. For example, the most significance source of lead at the tap is lead-based solder in copper service lines [29]. Aging of the solder rapidly decreases lead dissolution. In fact, the most important measure to control lead in drinking water has been to require the use of lead-free solder and plumbing fixtures in public water supplies.

11.32 CORROSION OF SEWER PIPES

Corrosion in sewers is the destruction of pipe materials by chemical action. Sewer corrosion can result from biological production of sulfuric acid and is caused by strong industrial wastes unless they are neutralized prior to disposal. Most municipal sewer ordinances prohibit industrial wastes having a pH less than 5.5 or higher than 9.0 or having other corrosive effects.

Crown corrosion in sanitary sewers is most prevalent in warm climates where they are laid on flat grades or where the sulfur content of the wastewater is high. Biological activity in wastewater in a sewer creates anaerobic conditions, producing hydrogen sulfide. Condensation moisture on the crown and walls of the sewer pipe absorbs hydrogen sulfide and oxygen from the atmosphere in the sewer. The sulfur-oxidizing bacteria *Thiobacillus* form sulfuric acid in the moisture of condensation:

$$H_2S + O_2 \xrightarrow{\textit{Thiobacillus}} H_2SO_4 \qquad (11.72)$$

In concrete sewers, sulfuric acid reacts with lime to form calcium sulfate, which lacks structural strength. If the concrete is sufficiently weakened, the pipe can collapse under, heavy overburden loads.

The best protection for sanitary sewers is a corrosion-resistant pipe material such as vitrified clay or plastic. In large sewers, where size and economics dictate concrete pipe, sacrificial concrete can be placed in the crown of the pipe. Crown corrosion can be retarded by ventilation or by chlorinating the wastewater to control hydrogen sulfide generation. Recent advances for the protection of concrete pipe include the development of synthetic coatings and linings.

REDUCTION OF DISSOLVED SALTS

The common processes for desalination of seawater are distillation and reverse osmosis; for desalting brackish groundwater, the processes are reverse osmosis and electrodialysis. The selection of process is based on size of plant, sources of energy, and capital and operating costs.

11.33 DISTILLATION OF SEAWATER

Typical seawater has a salinity (total dissolved solids) of 35,000 mg/l, of which 30,000 mg/l is NaCl. The generally accepted quality standards for drinking water are 500 mg/l of total dissolved solids and 200 mg/l of chloride. Distillation is cost competitive for desalination of feedwater with a high salt content since the process operates virtually independent of influent solids concentration. Moreover, a product purity of less than 100 mg/l is easily attained [30].

Distillation involves heating feedwater to the boiling point and then into steam to form water vapor, which is then condensed to yield a salt-free water. The principal commercial processes are multistage flash distillation and thin-film multiple-effect evaporation. The prefix *multi-* in the names means that a series of evaporation–condensation units is employed to obtain multiple reuse of the energy content of the heated steam. There may be as many as 15–25 stages.

Multistage flash distillation is illustrated schematically in Figure 11.25. Seawater entering the plant is initially heated in a heat recovery unit in which the hot desalted product water and waste brine discharge are cooled. The warmed seawater is then blended with recycled brine and passed through a series of condenser tubes in the evaporator chambers for further heating. In the process of condensing the steam to

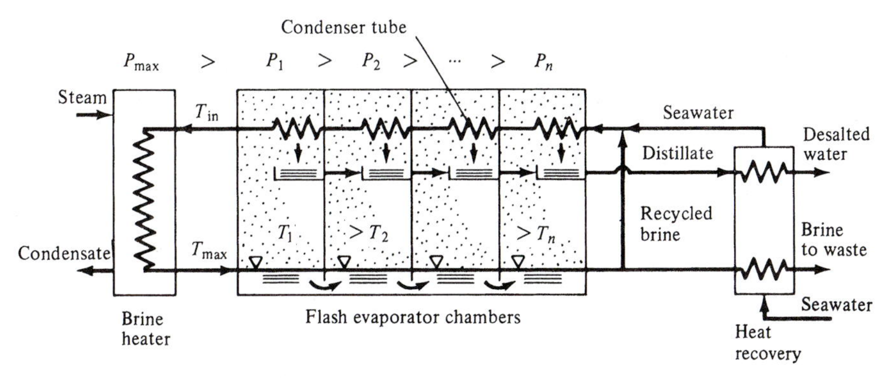

FIGURE 11.25 Multistage flash distillation.

distillate (the desalted water), the temperature of the brine is increased in stages to temperature T_{in} as it enters the brine heater. In this unit, the temperature of the brine feed is raised by external thermal energy to just below the saturation temperature T_{max} under a pressure P_{max}. The hot brine is then discharged through the series of stages each at a reduced pressure compared with the previous stage. The pressure $P_{max} > P_1, P_1 > P_2$, and so forth. Because of the reduction in pressure, a portion of the heated feed flashes to vapor in each stage to obtain equilibrium with the vapor condition prevailing in each individual stage. This results in a temperature drop in each stage (e.g., T_{max} drops to T_1, and T_1 drops to T_2), arriving at the minimum temperature T_n in the last stage. The total temperature drop is usually from a T_{max} of 250°F to a T_n of 100°F. The fraction of recirculating brine that can be flashed with each cycle is restricted to 0.10–0.15. Therefore, to produce the desired rate of distillate requires a minimum brine recirculation rate in the range of 10, which is 6.6 times the production rate of desalted water. The brine wasted from the recycling feed line may contain approximately 70,000 mg/l for seawater input of 35,000 mg/l.

The controlling parameters for output of desalted water and energy consumption are the temperature drop allowed in each stage, the difference between the brine inlet temperature to the first stage and the discharge temperature at the last (overall flash range), and the stage heat transfer coefficients. The principal unavoidable heat losses result from imperfect heat transfer by the condenser tubes and heat exchangers. Other losses include poor venting resulting in vapor blanketing of the condenser tube surfaces and tube fouling as a result of scale formation. Energy consumption in distillation is always well in excess of the ideal theoretical minimum. For a particular installation, efficiency is related to design factors such as the number of stages.

Thin-film multiple-effect evaporation is the second process widely used for distillation of seawater. The steam generated in one effect condenses on the outside of long vertical tubes in the next effect, evaporating more water from a film of brine that runs down the inside of the tube. Tracing this on Figure 11.26, prime steam enters the shell of the first effect where it condenses on the outside of the tubes. The latent heat of condensation furnishes the energy required to evaporate a portion of the brine feed within the tubes. The partially concentrated brine proceeds to the second effect, which

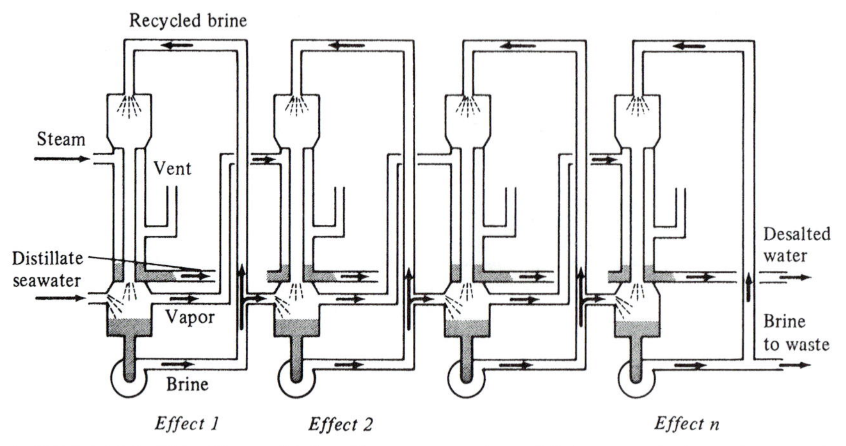

FIGURE 11.26 Thin-film multiple-effect evaporation with long vertical condenser tubes.

operates at a slightly lower pressure. The vapor leaving the first effect condenses on the tubing of the second effect, causing further evaporation of water from the brine. This process continues from effect to effect until the lowest pressure vapor is condensed in a final condenser by giving up its latent heat to circulating cooling water. The combined condensate from all effects constitutes the product water from the plant. Using this same principle, several variations may be designed into the multiple-effect process. One modification is the use of horizontal rather than vertical tubes with the steam inside the tubes and the evaporating brine flowing in a film on the outside tube surfaces.

11.34 REVERSE OSMOSIS

Reverse osmosis is the most common process for reducing the salinity of brackish groundwater. It is the forced passage of water through a membrane against the natural osmotic pressure to accomplish separation of water from a solution of dissolved salts. The process of osmosis is illustrated in Figure 11.27, where a thin membrane separates waters with different salt concentrations. In direct osmosis, water naturally flows from the side of lower salt concentration through the membrane to the solution of higher concentration, attempting to equalize the salt content; the membrane allows water flow while blocking the passage of salt ions. If pressure is applied to the side of higher salt content, this flow of water can be prevented at a pressure termed the *osmotic pressure* of the salt solution. In reverse osmosis, the water is forced by high pressure from a salt solution through the membrane into fresh water, separating desalted water from the saline solution. The rate of flow through a reverse-osmosis membrane is directly proportional to the effective pressure (i.e., to the difference between the applied and osmotic pressures). In practice, operating pressures vary between 350 and 1500 psi, with a typical range of 600–800 psi. The quantity of product water is 70%–90% for a feed of brackish groundwater and about 30% for a feed of seawater.

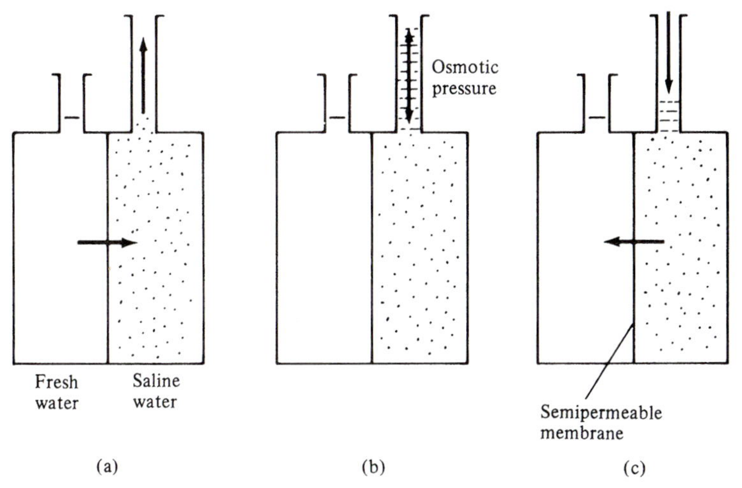

FIGURE 11.27 Illustrations describing the process of reverse osmosis to remove dissolved salts from water. (a) Direct osmosis. (b) Osmotic equilibrium. (c) Reverse osmosis.

The two common membrane materials are cellulose acetate and aromatic polyamide. Cellulose acetate membranes have a high flow rate per unit area and are commonly used to form tubes of spiral-wound flat sheets. In contrast, polyamide membranes have a lower specific flow rate and are manufactured in the form of hollow fibers to achieve the maximum surface area per unit volume, which is about 15 times that of spiral-wound membranes. By assembling the membranes in modular units, a large membrane surface area can be compacted into a cylindrical pressure vessel fitted with an inlet for the saline feedwater and outlets for the product water (permeate) and reject brine.

A *spiral-wound module* (Fig. 11.28) is made up of large membrane sheets covering both sides of a flat sheet of porous backing material that collects the permeate (product

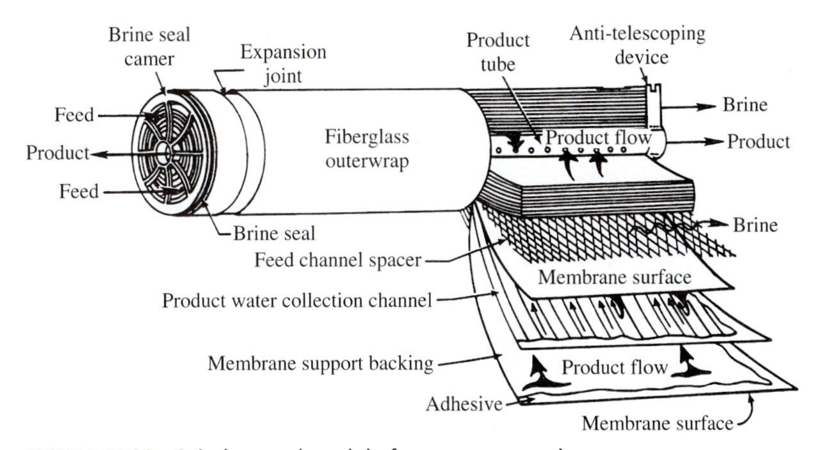

FIGURE 11.28 Spiral-wound module for reverse osmosis.

water). The membranes are sealed on the two long edges and one end to form an envelope enclosing the permeate collector. The other end of the membrane envelope is sealed to a perforated tube, which receives and carries away the permeate from the collectors. Several of these membrane envelopes, with mesh spacers in between for brine flow, are rolled up to form a spiral-type module. Operation of a spiral-wound module is diagrammed in Figure 11.28. Saline water enters the end of the module through the voids between the membrane envelopes provided by the spacers. Under high pressure, water is forced from the brine in the spacer voids through the membranes and conveyed by the enclosed porous permeate collectors to the perforated tube in the center of the module. Brine reject discharges from the spacer voids at the outlet end of the module.

A *hollow-fiber module* is a pressure vessel containing a very large number of microfiber membranes densely packed in a U-bundle with their openings secured in an end block of the module. The hollow fibers have an outside diameter of 85–100 μm and an inside diameter of 42 μm. Because of their small diameter and thick wall, these tubes can withstand the high reverse-osmosis pressure required to force water from the surrounding brine into the hollow cores of the fibers. The process flow is shown in Figure 11.29. Saline water enters the module through a central perforated feed tube and flows radially through the fiber bundle toward the outer shell of the cylinder. Under high pressure, water enters the hollow fibers and exits from their open ends at the discharge end of the module. Reject brine arriving at the outer shell of the cylinder is collected by a flow screen and conveyed from the module.

The potential for membrane fouling and scaling must be considered in both the design and operation of a reverse-osmosis system. The feedwater must be clear and low in chemical ions that can cause scale (e.g., calcium). In groundwater, silt and iron oxides

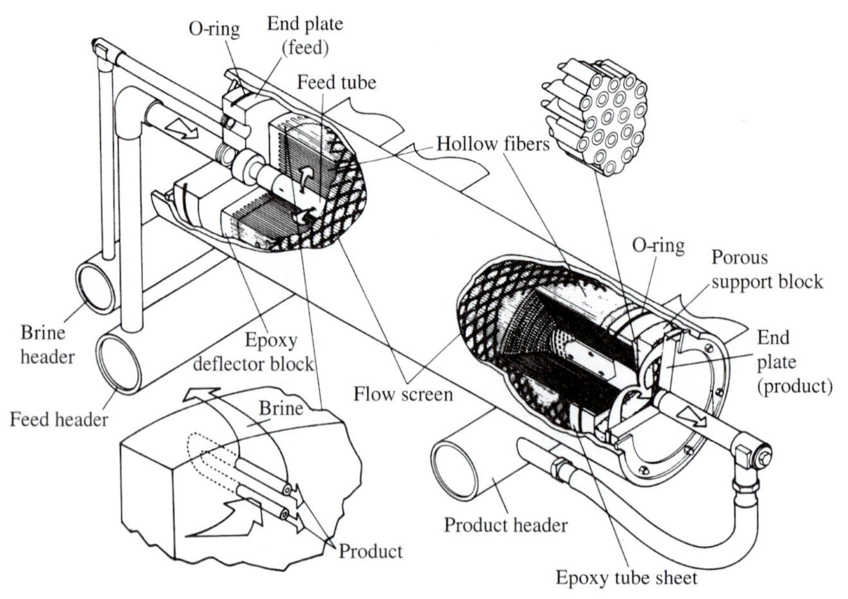

FIGURE 11.29 Hollow-fiber module for reverse osmosis. *Source:* Courtesy of Permasep Products, E.I. du Pont de Nemours & Co.

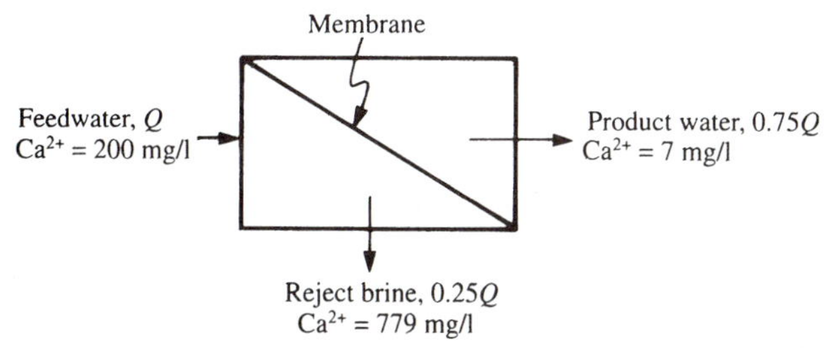

FIGURE 11.30 Schematic diagram of a reverse-osmosis module illustrating the concentration of calcium ion in the reject brine relative to the feedwater and product water for 75% recovery.

are the two common contaminants resulting from poor well construction. If the gravel pack and screen are not properly designed and placed, fine sand and silt from the aquifer can be carried out in the well water. Since many groundwaters are aggressive, the well casing, pump column, and transmission piping must be constructed of noncorrosive materials to prevent formation of rust particles.

The three most common chemical scalants are calcium sulfate, calcium carbonate, and silicon dioxide. As shown schematically in Figure 11.30, the reject brine is in contact with the membrane. Therefore, the scaling potential is determined by the concentration of ions in the brine, which in turn is controlled by the recovery of product water. For example, if the concentration of calcium ion in feedwater is 200 mg/l and the recovery of product water is 75% with 7 mg/l, the reject brine contains 779 mg/l of calcium ion (4 times the concentration in the feedwater). The calcium concentration can be lowered by reducing the permeate recovery and wasting more of the feedwater as brine. If a high recovery is necessary, such as 90%, the feedwater can be pretreated by softening to reduce the calcium content. Regardless of the methods used to reduce the calcium concentration, sodium hexametaphosphate or a polymeric antiscalant is commonly applied to inhibit the formation of calcium sulfate [31].

Calcium carbonate precipitation is controlled by acidifying the feedwater to convert bicarbonate ion to carbon dioxide, thus making the water corrosive. Since the reverse-osmosis process is a closed system, the carbon dioxide cannot escape and appears in the permeate and reject brine. Silica must be removed by pretreatment of the feedwater by chemical coagulation and filtration. High feedwater temperature deteriorates membrane material; hence, hot groundwaters require cooling prior to processing.

The product water from reverse osmosis is highly corrosive, with low pH and high concentration of carbon dioxide. The three methods of stabilization are degasification (decarbonation), addition of lime or soda ash for neutralization and increase in alkalinity, and blending with raw water. Degasification can be performed in a packed column with the water sprayed in at the top and percolating through the media against a countercurrent of air. Carbon dioxide is reduced to less than 10 mg/l and the pH raised to near neutrality. Addition of lime slurry [$Ca(OH)_2$] neutralizes the carbon dioxide, increases the calcium concentration, and raises the pH. Soda ash (Na_2CO_3) addition increases alkalinity and raises the pH. It also adds sodium ion, which is undesirable, both for corrosion control and for consumption as drinking water by persons with hypertension (high blood pressure). Nevertheless, soda ash is often applied in small systems since it is much easier to feed; a solution of soda ash is a clear liquid, whereas lime is a milky slurry. Blending permeate (after degasification, if performed) with raw water is often an economical method of partial or, in some cases, complete stabilization. The blending ratio depends on the chemical characteristics of the two waters. Since the fluoride ion content is reduced in desalting, supplemental fluoridation of the stabilized water to the optimum level is recommended.

This limited discussion and the calculations in Example 11.14 are only an overview of reverse osmosis. The presentation by Ko and Guy [32] on brackish and seawater desalting expands the topics of membranes, modules, pretreatment, and design and operation. In the same book, Ridgway [33] reports the results of research studies on biological fouling of membrane surfaces.

Example 11.14

A reverse-osmosis plant treats warm brackish groundwater with total dissolved solids of 2600–2700 mg/l and pH 6.8–7.2. Since the well water is free of silt, iron, and manganese, and low in silica, no granular-media filtration is required. The first step in pretreatment is to cool the water in a heat exchanger to 35°C, when necessary, for longer membrane life. Next, the water is acidified to pH 5.8 with 150 mg/l of sulfuric acid to prevent $CaCO_3$ scale formation, and 10 mg/l of hexametaphosphate are added to prevent $CaSO_4$ formation. Cartridge filters remove particles down to 5 microns in size; the average life of the replaceable filter elements is 12 weeks. To force the pretreated water through the membranes, a high-pressure pump at each reverse-osmosis unit increases the pressure to 370 psi. Each unit has 13 modules, with 9 in the first stage and 4 in the second stage. The feedwater applied to the first stage produces 50% permeate and 50% brine. The first-stage brine is applied to the second stage to again produce 50% permeate and 50% brine, which is rejected. Therefore, the total recovery of product water is 75% of the feedwater, and the reject brine is 25%. The product water is stripped to remove carbon dioxide in packed countercurrent columns, raising the pH from 5.8 to about 7.0. It is then stabilized by adding approximately 10 mg/l of soda ash to raise the pH to 8.2–8.5. The fluoride ion concentration is increased from 0.2 mg/l to the optimum of 0.8 mg/l by adding fluosilicic acid. The finished water has a total dissolved-solids concentration of 250–350 mg/l, alkalinity of 80–100 mg/l, and calcium ion concentration of approximately 7 mg/l.

Trace the chemical changes that occur in the water during treatment. The milliequivalents-per-liter bar graph of the untreated groundwater is shown in Figure 11.31(a).

Solution: Adding 150 mg/l of sulfuric acid for acidification to a pH of 5.8 converts bicarbonate ion to carbon dioxide and increases the sulfate ion content as shown in Figure 11.31(b).

If the water recovery is 75%, the concentrations of ions in the reject brine are about 4 times the concentrations in the acidified feedwater. Therefore, the contents are calcium ion 860 mg/l (0.0215 moles/l), sulfate ion 2270 mg/l (0.0236 moles/l), alkalinity 228 mg/l (4.56 meq/l), total dissolved solids 10,200 mg/l, and ionic strength 0.226. The scaling potential of $CaCO_3$ can be estimated by calculating the Langelier saturation index by using the following equation:

$$SI = pH - pH_s = pH - [(pK_2' - pK_s') + pCa^{2+} + pAlk] \qquad (11.73)$$

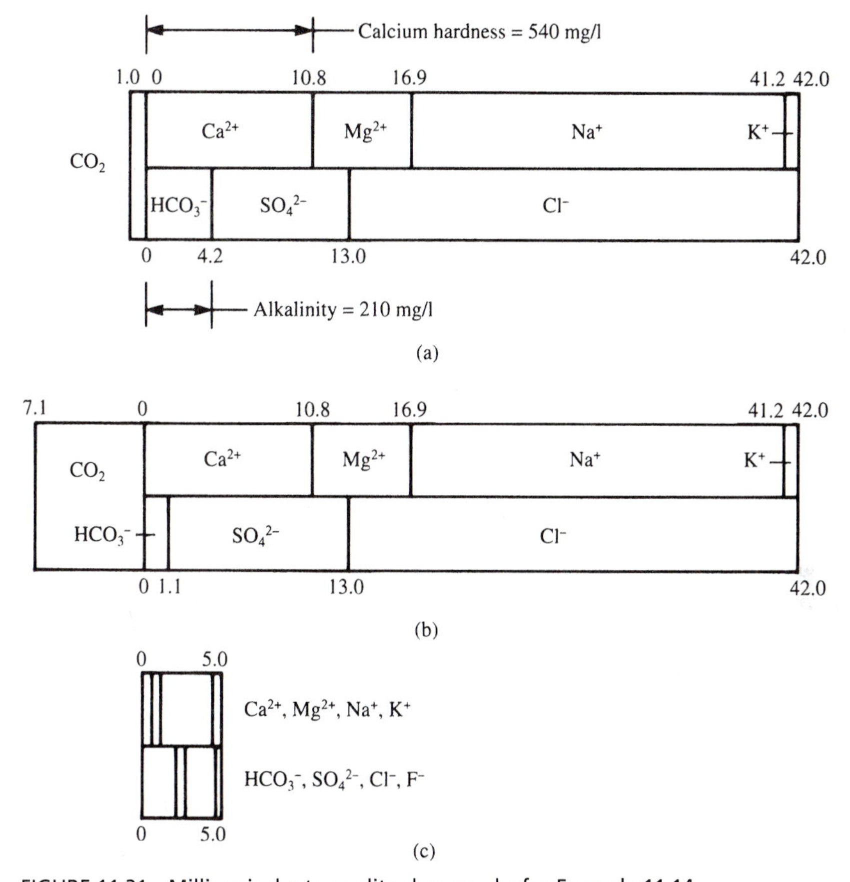

FIGURE 11.31 Milliequivalents-per-liter bar graphs for Example 11.14. (a) Untreated groundwater with a total dissolved-solids concentration of 2600 mg/l. (b) Feedwater after acidification to pH 5.8. (c) Product water after post treatment with a total dissolved-solids concentration of 350 mg/l.

where

$$
\begin{aligned}
\text{pH} &= \text{measured pH of the water} \\
\text{pH}_s &= \text{pH at CaCO}_3 \text{ saturation (equilibrium)} \\
\text{p}K_2' - \text{p}K_s' &= \text{constants based on ionic strength and temperature} \\
\text{pCa}^{2+} &= \text{negative logarithm of the calcium ion concentration, moles/liter} \\
\text{pAlk} &= \text{negative logarithm of the total alkalinity, equivalents/liter} \\
\text{SI} &= 5.8 - [2.1 + \text{p}(1/0.0215) + \text{p}(1000/4.56)] \\
&= -0.3 \quad \text{(non-scale-forming water)}
\end{aligned}
$$

The $pK_2' - pK_s'$ value for ionic strength is less than 0.020 and total dissolved solids are less than 800 mg/l. The value of 2.10 used in the above calculation was taken from a Langelier diagram in [34] for a total dissolved-solids content of 10,200 mg/l.

The potential for $CaSO_4$ scale is estimated by calculating the product of the ionic molar concentrations of the calcium and sulfate ions and comparing the result to the solubility-product constant. For the acidified brine,

$$[Ca^{2+}][SO_4^{2-}] = 0.0215 \times 0.0236 = 0.51 \times 10^{-3}$$

The estimated K_{sp} for a brine after addition of sodium hexametaphosphate is 1.0×10^{-3}; hence, this brine is unsaturated and non-scale-forming.

Figure 11.31(c) is the approximate bar graph of the finished water with a total dissolved-solids concentration of 350 mg/l after carbon dioxide stripping, soda ash addition, and fluoridation.

VOLATILE ORGANIC CHEMICAL REMOVAL

The two processes for removal of volatile organic chemicals (VOCs) are stripping by aeration and granular activated carbon (GAC) adsorption. Because of their volatility, these chemicals are rarely found in surface waters. However, VOCs are stable in groundwaters contaminated by leaching of chemicals from industrial discharges, improper chemical use, and spillage. The maximum contaminant levels for VOCs are in the range of 2–10,000 μg/l, with a most common value of 5 μg/l (Table 8.2). Because of these extremely low allowable concentrations, air stripping in a countercurrent packed tower is the only aeration method satisfactory for drinking water treatment. In cold climates the process may not be feasible because of poor removal at low temperatures and the possibility of ice formation on the tower packing. The most costly process of GAC adsorption may replace air stripping or be applied as a second stage following partial removal by aeration.

11.35 DESIGN OF AIR-STRIPPING TOWERS

The water is sprayed on the top of the packing and passes down through the column while air is blown countercurrent up through the tower. The packing can be random or stacked lightweight plastic media. As the water spreads over the surfaces of the packing, a large area of water is exposed for mass transfer to the flow of air. The VOCs can move freely toward equilibrium between liquid and gas phases. For air stripping very dilute solutions, this equilibrium can be expressed by Henry's law as

$$C_{GM}^{E} = HC_{LM}^{E} \tag{11.74}$$

where

C_{GM}^{E} = gas-phase molar concentration in equilibrium with the liquid-phase concentration, kmol/m^3

C_{LM}^{E} = liquid-phase molar concentration in equilibrium with the gas-phase concentration, $kmol/m^3$

H = Henry's law constant, mass concentration/mass concentration (dimensionless)

For efficient air stripping, the equilibrium between the liquid and gas phases is continuously destabilized by replenishing the air exhausted from the top with contaminant-free air entering at the bottom. By the time the water discharges from the column bottom, the contaminant in the liquid phase is reduced to a very low concentration. The higher the value of Henry's law constant, the more readily a VOC is air stripped from water.

The rate of mass transfer of a VOC from water to air is proportional to the difference between the equilibrium concentration in solution and the existing concentration in solution.

$$J = K_L a(C_{LM}^{E} - C_{LM}) \tag{11.75}$$

where

J = rate of mass transfer per unit volume of packing, $kmol/m^3 \cdot s$

$K_L a$ = overall mass transfer coefficient, s^{-1}

C_{LM}^{E} = molar concentration in liquid phase in equilibrium with the gas-phase concentration, $kmol/m^3$

C_{LM} = average molar concentration in liquid phase, $kmol/m^3$

The value of $K_L a$ depends on the geometry of the tower and packing, operation of the air-stripping system (e.g., the air-to-water ratio), and temperature.

The design of an air-stripping packed column for steady-state mass transfer is based on the following relationship:

$$Z = (\text{HTU})(\text{NTU}) \tag{11.76}$$

where

Z = depth of packing, m

HTU = height of a transfer unit, m

NTU = number of transfer units (dimensionless)

The height of a transfer unit (HTU) characterizes the mass transfer efficiency from the liquid to the gas phase.

$$\text{HTU} = L/\rho_L \cdot K_L a \tag{11.77}$$

where

$$L = \text{water mass loading rate, kg/m}^2 \cdot \text{s}$$
$$\rho_L = \text{water density, kg/m}^3$$
$$L/\rho_L = \text{volumetric loading rate, m}^3\text{/m}^2 \cdot \text{s}$$
$$K_L a = \text{overall mass transfer coefficient, s}^{-1}$$

The number of mass transfer units (NTU) corresponds to the difficulty in removing the VOC from the liquid phase.

$$\text{NTU} = \left(\frac{S}{S - 1}\right) \ln \frac{(C_{\text{in}}/C_{\text{out}})(S - 1) + 1}{S} \tag{11.78}$$

where

$$S = \text{stripping factor, mol/mol (dimensionless)}$$
$$C_{\text{in}} = \text{VOC concentration in influent, kg/m}^3$$
$$C_{\text{out}} = \text{VOC concentration in effluent, kg/m}^3$$

The stripping factor is defined as

$$S = H(Q_A/Q_w) \tag{11.79}$$

where

H = Henry's law constant, mass concentration/mass concentration (dimensionless)

Q_A = volumetric airflow rate, m^3/s

Q_W = volumetric water loading rate, m^3/s

Several mathematical models to calculate mass transfer coefficients have been proposed based on the two-film theory, which assumes that overall resistance to mass transfer is the sum of liquid- and gas-phase resistances. Lamarche and Droste [35] evaluated these models in packed-column air stripping of 6 VOCs. They also presented a laboratory technique for determining the value of Henry's law constant for VOCs at different temperatures.

The procedure for designing an air-stripping tower starts with the selection of a packing. A stripping factor is selected between 2 and 5, if high removal efficiency is required, and, based on this value, the air-to-water ratio is calculated. An allowable air pressure drop is selected. Data on the air pressure drop through a particular packing are generally available from the manufacturer. Operation at a high-pressure drop allows a smaller volume of packing, reducing the construction cost but increasing the operational costs. Various combinations of pressure drops and air-to-water ratios can be calculated to find the most cost-effective choice. Henry's law constant for the anticipated operating temperature is taken from the literature or determined by laboratory analysis. Selection of a mass transfer coefficient should preferably be from pilot-plant studies or based on experience in full-scale performance. Using these data, along with influent and effluent VOC concentrations, the required depth of packing can be calculated from Eqs. (11.76–11.79). The surface area of the packing is calculated from the quantity of water to be treated and the water loading.

Example 11.15

Determine the depth of packing and surface area for a countercurrent air-stripping tower to reduce the trichloroethylene from 200 μg/l to 2 μg/l (99% removal). The design water flow rate is 76 l/s, and the lowest operating temperature anticipated is 10°C, based on groundwater temperature and cooling in the tower. Henry's law constant for trichloroethylene is 0.30 at 10°C. The manufacturer of the proprietary random packing recommends a mass transfer coefficient of 0.017 s^{-1} based on pilot studies and an allowable pressure drop of 50 $N/m^2 \cdot m$.

Solution: After discussions with the client and the manufacturer of the packing, the designer selected a loading of 10 $l/m^2 \cdot s$ ($kg/m^2 \cdot s$) and a stripping factor of 3.6. Using Eq. (11.79),

$$\text{air-to-water ratio } (Q_A/Q_W) = 3.6/0.30 = 12 \text{ m}^3/\text{m}^3$$

$$\text{volumetric airflow rate} = 0.076 \text{ m}^3/\text{s} \times 12 \text{ m}^3/\text{m}^3 = 0.91 \text{ m}^3/\text{s}$$

Using Eq. (11.77),

$$\text{HTU} = \frac{10 \text{ kg/m}^2 \cdot \text{s}}{(1000 \text{ kg/m}^3)(0.017/\text{s})} = 0.59 \text{ m}$$

Using Eq. (11.78),

$$\text{NTU} = \left(\frac{3.6}{3.6 - 1}\right) \ln \frac{(200/2)(3.6 - 1) + 1}{3.6} = 5.9$$

Using Eq. (11.76),

$$Z \text{ (depth of packing)} = 0.59 \times 5.9 = 3.5 \text{ m}$$

$$\text{surface area of packing} = \frac{76 \text{ l/s}}{10 \text{ l/m}^2 \cdot \text{s}} = 7.6 \text{ m}^2$$

SYNTHETIC ORGANIC CHEMICAL REMOVAL

Synthetic organic chemicals (SOCs) include pesticides (herbicides and insecticides), volatile organic chemicals, and trihalomethanes. Trace concentrations of pesticides are found in runoff from agricultural lands, occasionally in groundwaters under agricultural lands, and in groundwaters contaminated by seepage from improper disposal of industrial wastes and spillage of chemicals. Conventional water treatment provides limited removal of organic chemicals. If adsorbed on particles or associated with large hydrophobic molecules, they can be taken out by coagulation–sedimentation–filtration. However, dissolved organic chemicals rarely adsorb to metal hydroxides and polymers, resulting in negligible removal. Adjustment of pH, changing coagulants or coagulant aids, and application of powdered activated carbon are options to be considered for improved treatment. In surface water treatment, these process variables are generally successful for greater removal of natural organic precursors to reduce subsequent formation of disinfection by-products.

11.36 ACTIVATED CARBON ADSORPTION

Activated carbon can be made from a variety of carbonaceous raw materials. Processing is dehydration and carbonization by slow heating in the absence of air followed by chemical activation to produce a highly porous structure. Powdered activated carbon (PAC) for water treatment, which has good characteristics for adsorption of taste and odor compounds, is commonly made from lignin or lignite. Granular activated carbon (GAC) made from coal has the best physical properties of density, particle size, abrasion resistance, and ash content. These characteristics are essential, since GAC is subject to filter backwashing, conveyance as a slurry, and heat reactivation.

The activation process in manufacture creates a highly porous surface on the carbon particles with macropores and micropores down to molecular dimensions. Organic contaminants are adsorbed by attraction to and accumulation in pores of appropriate size; thus the pore structure is extremely important in determining adsorptive properties for particular compounds. In general, GAC most readily adsorbs branch-chained high-molecular-weight organic chemicals with low solubility. These include pesticides, volatile organic chemicals, and trihalomethanes. Macropores are large enough for colonies of bacteria to grow and proliferate if biodegradable organic compounds are in the water. The benefits of microbial growth or potential risks to water quality are not well understood. GAC is reactivated thermally at a furnace temperature and retention time based on the volatility of the adsorbed chemicals. Two percent to 5% of the carbon is lost during each reactivation and must be replaced with fresh carbon.

Powdered activated carbon is a fine powder applied in a water slurry, which can be added at any location in the treatment process ahead of filtration. At the point of application, the mixing must be adequate to ensure dispersion and the contact time long enough for adsorption. The dosage for normal taste and odor control is usually up to 5 mg/l with a contact time of 10–15 min. Although PAC is an effective adsorber of organic compounds that cause taste and odor, this success is not repeated by the adsorption of SOCs. Poor adsorption is attributed to the pore structure of the PAC, short contact time between the carbon particles and the dissolved organic chemicals, and interference by adsorption of other organic compounds.

Efficiency in removal of SOCs requires a granular activated carbon filter to ensure close contact between the water and carbon for a sufficient time for adsorption to occur. Design of a GAC system requires pilot-plant tests for selection of the best carbon, determination of the required contact time, the effects of influent water quality variations, and to establish the carbon loss during reactivation.

Pilot column tests make it possible to [36]:

- Determine treatability
- Select the best carbon for the specific purpose based on performance
- Establish the required carbon dosage that, together with laboratory tests of reactivation, will determine the capacity of the carbon reactivation furnace or the necessary carbon replacement costs
- Determine the effects of influent water quality variations on plant operation

11.37 GRANULAR ACTIVATED CARBON SYSTEMS

The majority of new systems in water treatment to remove SOCs will use separate deep-bed contactors. GAC facilities require the following system components [36]:

- Carbon contactors for the water to be treated for the length of time required to obtain the necessary removal of organics
- Reactivation or replacement of spent carbon
- Transport of makeup or reactivated carbon into contactors
- Transport of spent carbon from the contactors to reactivation or hauling facilities
- Facilities to backwash the GAC beds

A fixed-bed contactor has a GAC bed that remains stationary (fixed) during operation. Although the bed can be designed for downflow or upflow, downflow operation with provision for backwashing is more common. The GAC is not reactivated until chemical breakthrough; then the entire bed is removed and replaced. The design of a fixed-bed contactor is similar to a gravity granular-media filter or a pressure filter. In a surface water treatment plant, the conventional filter is retained for removal of turbidity, and the fixed-bed contactor added as a second stage. A postfilter contactor designed specifically to adsorb SOCs (without the necessity of turbidity removal) provides better use of the adsorptive capacity of the GAC and allows longer contact times. Adsorptive capacity can be significantly reduced by organic contaminants in an unfiltered water. Also, GAC suitable for filtration may not be optimum for adsorbing the contaminating organic chemicals. Contact time is expressed as empty bed contact time, calculated by dividing the volume of the bed by the flow rate. In conventional filtration, the empty bed contact time is usually 3–9 min, while in a GAC contactor it is 15–30 min or greater. A contactor following filtration also reduces the quantity of backwash water required and can be designed for easier removal of spent carbon and replacement. A gravity contactor is appropriate for removal of chemicals from a groundwater supply in a large plant. Pretreatment may be necessary to remove contaminants that can interfere with filtration through the GAC bed, for example, iron oxide deposits and growth of iron bacteria. For individual wells, the contactor may be a pressure vessel with discharge pressure from the well pump forcing the water through the bed.

PROBLEMS

11.1 (a) Using atomic weights from the table of elements given in Table A.7, calculate the molecular and equivalent weights of alum (aluminum sulfate), ferric sulfate, and soda ash (sodium carbonate). The formulas of these compounds are given in Table 11.1. (b) Using atomic weights, compute the equivalent weights of the ammonium ion, bicarbonate ion, calcium carbonate, and carbon dioxide. Values are given in Table 11.2.

11.2 (a) Water contains 38 mg/l of calcium ion and 10 mg/l of magnesium ion. Express the hardness as mg/l of $CaCO_3$. (b) Alkalinity in water consists of 120 mg/l of bicarbonate ion and 15 mg/l of carbonate ion. Express the alkalinity in units of mg/l of $CaCO_3$.

11.3 Draw a milliequivalents-per-liter bar graph and list the hypothetical combinations for the following analysis of a soft water:

Ca^{2+} = 36 mg/l HCO_3^- = 208 mg/l

Mg^{2+} = 14 mg/l SO_4^{2-} = 14 mg/l

Na^+ = 43 mg/l Cl^- = 44 mg/l

K^+ = 7 mg/l

11.4 Draw a milliequivalents-per-liter bar graph and list the hypothetical combinations for the following analysis of a groundwater:

Ca^{2+} = 94 mg/l HCO_3^- = 317 mg/l

Mg^{2+} = 24 mg/l SO_4^{2-} = 67 mg/l

Na^+ = 14 mg/l Cl^- = 24 mg/l

11.5 Draw a milliequivalents-per-liter bar graph for the following water analysis:

calcium hardness = 185 mg/l   alkalinity = 200 mg/l

magnesium hardness = 50 mg/l   sulfate ion = 58 mg/l

sodium ion = 23 mg/l   chloride ion = 36 mg/l

potassium ion = 20 mg/l   pH = 7.7

11.6 Calculate the pH of a solution of pure water containing 1.0 mg/l of sulfuric acid.

11.7 What is the dominant form of alkalinity in a natural water at pH 7? What are the forms present at pH 10.5?

11.8 What parameter dictates the rate of decrease in concentration of remaining reactant with time in (a) zero-order kinetics and (b) first-order kinetics?

11.9 The kinetics of a chemical reaction were analyzed by laboratory experiment. Lime was added to a water sample to precipitate reactant A as product P. While the water was continuously stirred, portions were withdrawn at 10-min intervals to measure the amount of A remaining. The data collected were as follows: t = 0, C_0 of A = 100 mg/l; t = 10 min, C of A remaining = 55 mg/l; t = 20 min, C = 22 mg/l; t = 30, C = 8; and t = 40, C = 5. Plot graphs as shown in Figure 11.3(a) and (b). What are the kinetics of the reaction? What is the value of the reaction-rate constant?

11.10 The number of coliform bacteria is reduced from an initial concentration of 2,000,000 per 100 ml to 400 per 100 ml in a long, narrow chlorination tank under a steady wastewater flow with a hydraulic detention time of 30 min. Assuming first-order kinetics and ideal plug flow, calculate the reaction-rate constant.

11.11 Alternative reactor systems are being considered to reduce the reactant in a steady liquid flow from an initial concentration of 100 mg/l to a final concentration of 20 mg/l. Assuming a first-order reaction-rate constant of 0.80 day^{-1}, calculate the hydraulic detention time required for each of the following reactor systems: (a) one plug flow reactor, (b) one completely mixed reactor, and (c) two equal-volume completely mixed reactors in series.

11.12 Define the meanings of the terms *coagulation* and *flocculation* in reference to destabilization of colloidal suspensions. When these terms are used by an environmental engineer in reference to water treatment processes, what are their meanings?

11.13 The results from a jar test for coagulation of a turbid alkaline raw water are given in the table. Each jar contained 1000 ml of water. The aluminum sulfate solution used for chemical addition had such strength that each milliliter of the solution added to a jar of water produced a concentration of 8.0 mg/l of aluminum sulfate. Based on the jar test results, what is the most economical dosage of aluminum sulfate in mg/l?

Jar	Aluminum Sulfate Solution (ml)	Floc Formation
1	1	None
2	2	Smoky
3	3	Fair
4	4	Good
5	5	Good
6	6	Very heavy

If another jar had been filled with freshly distilled water and dosed with 5 ml of aluminum sulfate solution, what would have been the degree of floc formation?

11.14 In the coagulation reaction, commercial aluminum sulfate (alum) reacts with natural alkalinity or can be reacted with lime or soda ash if the water is deficient in alkalinity. Based on Eqs. (11.34–11.36), calculate the milligrams-per-liter amounts of alkalinity, lime as CaO, and soda ash as Na_2CO_3 that react with 1.0 mg/l of alum.

11.15 The removal of *Giardia* cysts from a soft, cold, low-turbidity water requires 15 mg/l of alum plus 0.10 mg/l of anionic polymer. (a) How many milligrams per liter of natural alkalinity are consumed in the coagulation reaction? How much carbon dioxide is released by this reaction? (b) What is the stoichiometric dosage of soda ash to react with the 15 mg/l of alum? This reduces loss of alkalinity but still produces carbon dioxide. How much carbon dioxide is released by this reaction? (c) Would a stoichiometric dosage of lime slurry be better than soda ash? Suggest a reason why lime slurry would not be used. How can carbon dioxide be removed from water?

11.16 A ferrous sulfate dosage of 40 mg/l and an equivalent dosage of lime are used to coagulate a water. (a) How many pounds of ferrous sulfate per million gallons are used? (b) How many pounds of hydrated lime per million gallons are used, assuming a purity of 70% CaO? (c) How many pounds of ferric hydroxide are theoretically produced per million gallons of water treated?

11.17 Treatment of a water supply requires 60 mg/l of ferric chloride as a coagulant. The natural alkalinity of the water is 40 mg/l. Based on theoretical chemical reactions, what dosage of lime as CaO is required to react with the ferric chloride after the natural alkalinity is exhausted?

11.18 The data listed below are from a pilot-plant study to determine turbidity and *Giardia* cyst removal from a cold, low-turbidity water (less than 1°C and 0.5 NTU in winter) by direct filtration using a cationic polymer as the coagulant [Reference 11, Chapter 10]. The filter was a dual-media coal-sand bed, 2 ft by 2 ft square, operated for most test runs at a loading of approximately 12 $m^3/m^2 \cdot h$ (4.9 gpm/ft^2) for durations in the range of 3–22 h. During selected filter runs, *Giardia* cysts and coliform bacteria were injected into the raw water for 40–60 min and tested for presence in the filtered water. Calculate the percentages for *Giardia* cyst and coliform removals. Plot turbidity versus polymer dosage for both the raw water and filtered water on the same diagram. What is the least dosage of polymer for maximum turbidity removal? What appears to be an acceptable turbidity in the effluent to ensure 98%–99% (virtually 100%) *Giardia* cyst removal?

Filter Loading ($m^3/m^2 \cdot h$)	Water Temp. (°C)	Polymer Dosage (mg/l)	Turbidity		*Giardia lamblia*		Coliforms	
			Inf. (NTU)	Eff. (NTU)	Inf. (Cysts/l)	Eff. (Cysts/l)	Inf. (Org/100 ml)	Eff. (Org/100 ml)
12.3	2.7	0	0.46	0.30				
8.0	0.3	0	0.45	0.22	340	69		
12.1	1.9	5	0.48	0.30				
12.2	0.3	10	0.60	0.05				
12.3	0.3	10	0.47	0.03				
11.4	0.3	12	0.45	0.03			360	9
12.3	0.3	12	0.48	0.04				
8.7	1.9	13	0.61	0.06	270	0.4	3900	580
12.3	0.3	13	0.47	0.03				
12.2	0.3	14	0.49	0.03				
11.9	8.3	18	0.65	0.07	3.2	0	8700	190
12.5	7.6	24	0.43	0.02	410	0	5800	3
12.3	0.2	24	0.48	0.02				

11.19 Presedimentation reduces the turbidity of a raw river water from 1500 mg/l suspended solids to 200 mg/l. How many pounds of dry solids are removed per million gallons? If the settled sludge has a concentration of 8% solids and a specific gravity of 1.03, calculate the sludge volume produced per million gallons of river water processed.

11.20 Sketch a preliminary process flow diagram for a water treatment plant to clarify and disinfect a turbid surface water at a design flow of 50 mgd. Use two identical, parallel, and separate processing lines with rapid mixing, flocculation, sedimentation, filtration, and clear-well storage. The flocculation and sedimentation processes for each line are in the same large rectangular concrete tank with paddle flocculators in baffled compartments ahead of the sedimentation section with effluent "finger" channels extending into the tank from the outlet end [Figs. 10.14 and 10.16(a)]. For each line, use four gravity dual-media coal-sand filters with deep filter boxes to prevent "air binding" and flow control by influent flow splitting for constant-rate filtration (Fig. 10.31). The preferred filter bottom is the plastic dual-lateral block underdrain illustrated in Figure 10.27.

On the flow diagram, show the chemicals to be added with alternate points of application. The raw water has a turbidity ranging from 10 to 40 NTU, and in the spring the water contains natural organic matter that creates bad taste and odor and forms trihalomethanes with prechlorination. The fluoride concentration is less than optimum. The treatment plant must meet the EPA rule for surface water disinfection as discussed in Section 11.24. Sketch a longitudinal cross-sectional view of the flocculation–sedimentation tank. List the design criteria for sizing the flocculation section and specifying paddle flocculators, sizing the sedimentation section and effluent channels, and sizing the filters. Sketch a plan view of the clear well to ensure compliance with the EPA disinfection rule.

11.21 Flocculator–clarifiers similar to the one illustrated in Figure 10.18 are proposed for precipitation in lime–soda ash softening of a groundwater. The inside diameter of the tank is 40 ft and side-water depth is 10 ft. The cone-shaped skirt is 12 ft in diameter at the water surface and 24 ft in diameter at the bottom, which is at a depth of 8.0 ft below the water surface. The design flow for each flocculator–clarifier is 750,000 gpd. Does this proposed design meet the *GLUMRB Standards* for detention time based on total volume, detention time for mixing and flocculation volume, and upflow rate based on the open-water surface area? {The cone shaped shirt can be considered to be the geometric form of a conical tank where: $\text{Area}_1 = \pi \times r_1^2$, $\text{Area}_2 = \pi \times r_2^2$, $V = \frac{1}{3}\text{depth}\,[A_1 + A_2 + (A_1 \times A_2)^{0.5}]$.}

11.22 The water defined by the analysis given below is to be softened by excess-lime treatment. (a) Sketch an meq/l bar graph. (b) Calculate the softening chemicals required. (c) Draw a bar graph for the softened water after recarbonation and filtration, assuming that 80% of the alkalinity is in the bicarbonate form.

$$CO_2 = 8.8 \text{ mg/l} \qquad \text{Alk}(HCO_3^-) = 135 \text{ mg/l}$$
$$Ca^{2+} = 40.0 \text{ mg/l} \qquad SO_4^{2-} = 29.0 \text{ mg/l}$$
$$Mg^{2+} = 14.7 \text{ mg/l} \qquad Cl^- = 17.8 \text{ mg/l}$$
$$Na^+ = 13.7 \text{ mg/l}$$

11.23 Settled water after excess-lime treatment, before recarbonation and filtration, contains 35 mg/l of CaO excess lime in the form of hydroxyl ion, 30 mg/l of $CaCO_3$ as carbonate ion, and 10 mg/l as $CaCO_3$ of $Mg(OH)_2$ in the form of hydroxyl ion. First-stage recarbonation precipitates the excess lime as $CaCO_3$ for removal by sedimentation, and second-stage recarbonation converts a portion of the remaining alkalinity to bicarbonate ion. Calculate the carbon dioxide needed to neutralize the excess lime and convert one-half of the alkalinity in the finished water to the bicarbonate form. Assume an excess of 20% of the calculated CO_2 is required to account for unabsorbed gas escaping from the recarbonation chamber.

11.24 For the water analysis given in Problem 11.5, calculate the additions of CaO, Na_2CO_3, and CO_2 needed for excess-lime softening. Sketch a bar graph for the finished water after two-stage precipitation softening by excess-lime treatment with intermediate and final recarbonation (Fig. 11.7). Assume that three-quarters of the alkalinity in the finished water is in the bicarbonate form.

11.25 For the water analysis given in Problem 11.5, calculate the lime dosage required for selective calcium removal. The process flow scheme is a single-stage system of mixing, sedimentation, and filtration. Draw a bar graph for the finished water. Is this softening process recommended for this water?

11.26 A groundwater supply has the following analysis:

calcium	= 94 mg/l	bicarbonate	= 317 mg/l
magnesium	= 24 mg/l	sulfate	= 67 mg/l
sodium	= 14 mg/l	chloride	= 24 mg/l

1. Calculate the quantities of lime and soda ash for excess-lime softening, and the carbon dioxide reacted for neutralization by two-stage recarbonation. Assume the practical limits of hardness removal for calcium as 30 mg/l as $CaCO_3$ and magnesium as 10 mg/l as $CaCO_3$, and three-quarters of the final alkalinity is converted to bicarbonate. Calculate the finished hardness and sketch the bar graphs for the raw and finished waters.
2. Calculate the chemical dosages for split-treatment softening, assuming a permissible magnesium hardness in the finished water of 40 mg/l. Follow the methods of solution in Example 11.6. Draw the bar graphs, after first-stage lime treatment, of the unreacted combined first-stage effluent and bypassed flow, and the finished water after reaction of soda ash. Calculate the finished water hardness.
3. Calculate the chemical dosages for split-treatment softening using the method of solution in Example 11.7. Draw the bar graph after first-stage excess-lime–soda ash treatment and the finished water after reaction of the excess lime with the bypassed flow. Calculate the finished water hardness.

11.27 Compute the lime dosage needed for selective calcium-removal softening of the water described by the following analysis. What is the finished water hardness?

$$\begin{aligned} Ca^{2+} &= 63 \text{ mg/l} & CO_3^{2-} &= 16 \text{ mg/l} \\ Mg^{2+} &= 15 \text{ mg/l} & HCO_3^{-} &= 189 \text{ mg/l} \\ Na^{+} &= 20 \text{ mg/l} & SO_4^{2-} &= 80 \text{ mg/l} \\ K^{+} &= 10 \text{ mg/l} & Cl^{-} &= 10 \text{ mg/l} \end{aligned}$$

11.28 Consider the split-treatment softening of water described by the analysis below. Use the criteria for the finished water as given in Example 11.6. Draw a bar graph for the finished water after second-stage treatment with soda ash, sedimentation, and final filtration.

$$\begin{aligned} CO_2 &= 15 \text{ mg/l as } CO_2 & HCO_3^{-} &= 200 \text{ mg/l as } CaCO_3 \\ Ca^{2+} &= 60 \text{ mg/l} & SO_4^{2-} &= 96 \text{ mg/l} \\ Mg^{2+} &= 24 \text{ mg/l} & Cl^{-} &= 35 \text{ mg/l} \\ Na^{+} &= 46 \text{ mg/l} & & \end{aligned}$$

11.29 Reconsider split-treatment softening of the water with the analysis as given in Problem 11.28. Use the alternate method of solution as given in Example 11.7. How does this finished water bar graph compare to the finished water bar graph determined in Problem 11.28? (Comment: The equilibrium concentrations for $CO_3^{=}$ and HCO_3^{-} ions determined in these calculations are only approximate. In actual chemical reactions, the relationship of these ions depends on final pH.)

11.30 Sketch a preliminary process flow diagram for a split-treatment lime–soda ash water treatment plant to soften a design flow of 60 mgd. Use six equal-sized flocculator–clarifiers (Fig. 10.18). The first-stage flow for excess-lime treatment is expected to be no greater than 50% of the raw water. Use eight gravity dual-media coal–sand filters with traditional flow control using rate-of-flow controllers (Fig. 10.29). The preferred backwashing system is air scouring prior to water backwash. The clear-well capacity is 6.0 mil gal. The raw water has a hardness of approximately 230 mg/l, iron in the range of 0.2–0.3 mg/l, and a less than optimum concentration of fluoride. On the flow diagram, show the chemicals being added and their points of application. Sketch a cross-sectional view of the filter box showing the wash-water troughs, filter media, and underdrain system. List the design criteria for sizing the flocculator–clarifiers and filters.

11.31 The ionic characteristics of a fossil groundwater in an arid region are listed below. Draw a milliequivalents-per-liter bar graph and calculate total hardness and alkalinity. One recommendation for improving the quality of the water for domestic use is lime–soda ash softening to reduce hardness and total dissolved solids (TDS). Calculate the lime and soda ash additions for excess-lime treatment and draw the final bar graph after recarbonation. Calculate the theoretical total dissolved solids content by summing the weights of the ions (or hypothetical combinations) in the softened water. Was the recommendation of lime–soda ash softening appropriate?

$$Ca^{2+} = 108 \text{ mg/l} \qquad HCO_3^- = 146 \text{ mg/l} \qquad TDS = 900 \text{ mg/l}$$
$$Mg^{2+} = 44 \text{ mg/l} \qquad SO_4^{2-} = 110 \text{ mg/l}$$
$$Na^+ = 138 \text{ mg/l} \qquad Cl^- = 366 \text{ mg/l}$$

11.32 Sketch a meq/1 bar graph of the water described in Problem 11.22 after it is softened to zero hardness by cation exchange softening.

11.33 Consider the ion exchange softening of water described in Example 11.4. If 0.3 lb of NaCl is required to regenerate the resin bed per 1000 grains of hardness removed, calculate the salt required per million gallons of water softened. Sketch a meq/l graph for the ion-exchange-softened water. How does finished water from ion exchange softening differ from finished water produced in lime–soda softening?

11.34 A small community has used an unchlorinated well water supply containing approximately 0.3 mg/l of iron and manganese for several years without any apparent iron and manganese problems. A health official suggested that the town install chlorination equipment to disinfect the water and provide a chlorine residual in the distribution system. After initiating chlorination, consumers complained about water staining washed clothes and bathroom fixtures. Explain what is occurring due to chlorination.

11.35 The iron and manganese removal process for the well supply of a small community is mechanical aeration, the addition of potassium permanganate followed by detention in a contact tank, pressure filtration, and postchlorination. The construction specifications called for manganese-treated greensand; however, the actual filter medium provided was plain sand. Customers often complained that the treated water caused staining of bathroom fixtures and laundry. The common response of the plant operator was to increase the chemical dosage, which did not seem to improve the situation. The operator even tried prechlorination of the water in combination with potassium permanganate addition, but that appeared to increase the staining characteristics of the treated water. Discuss the most probable cause of the poor-quality finished water and your recommendations for improvement.

11.36 Untreated well water contains 1.2 mg/l of iron and 0.8 mg/l of manganese at a pH of 7.5. Calculate the theoretical dosage of potassium permanganate required for iron and manganese oxidation.

11.37 Iron and manganese are removed by aeration, chlorine oxidation, sedimentation, and sand filtration from a well water supply. The plant processes 3000 m^3/d with application of 2.5 mg/l chlorine. Calculate the usage in kilograms per month for each of the following chemicals: chlorine gas, 70% granular calcium hypochlorite, and 12% sodium hypochlorite solution. At the same pH of the chlorinated water, do all of these chemicals form the same kind of chlorine residual?

11.38 The results of a chlorine demand test on a raw water at 20°C are given in the following table.

Sample	Chlorine Dosage (mg/l)	Residual Chlorine After 10 Min of Contact (mg/l)
1	0.20	0.19
2	0.40	0.37
3	0.60	0.51
4	0.80	0.50
5	1.00	0.20
6	1.20	0.40
7	1.40	0.60
8	1.60	0.80

1. Sketch the chlorine demand curve.
2. What is the breakpoint chlorine dosage?
3. What is the chlorine demand at a chlorine dosage of 1.20 mg/1?

11.39 The practice of combined residual chlorination involves feeding both chlorine and anhydrous ammonia. Calculate the stoichiometric ratio of chlorine feed to ammonia feed for combined chlorination.

11.40 List the possible applications of ozone in water treatment. If ozone is applied for disinfection, may the use of chlorine be eliminated?

11.41 What is the suspected health risk of trihalomethanes (THMs) in drinking water, and how was this risk demonstrated? What is the origin of THMs in treated water? If the finished water from a river water treatment plant contains an excessive concentration of these chemicals during spring runoff, what remedial actions can be taken to reduce their formation?

11.42 Define the meaning of the $C \cdot t$ product. What factors affect the $C \cdot t$ value used in design and operation of a drinking water disinfection system? With reference to Table 11.6, what kind of microorganism is most readily inactivated by free chlorine? What kind is the most difficult to inactivate? List the tabulated disinfectants in the order of most effective to least effective in disinfecting action.

11.43 What is the disease in humans caused by *Giardia lamblia* and *Cryptosporidium* species? In what manner do these protozoa infect humans and how are they transmitted to other humans by water? What are other modes of transmission? Describe the waterborne sources of these organisms. (Refer to Section 8.1.)

11.44 Using Eq. (11.68), calculate the $C \cdot t$ value for *G. lamblia* inactivation at a free chlorine concentration of 0.80 mg/l, pH equal to 8.0, and temperature of 10°C. Correct this value for 5°C, using Eq. (11.29), and extrapolate to 99% inactivation. How does this result compare to the value in Table 11.6? Extrapolate the calculated value at 10°C to 99.9% inactivation and compare the result with the value in Table 11.7.

11.45 From Table 11.7, what is the $C \cdot t$ value for 99.9% (3.0-log) inactivation of *Giardia* cysts at a free chlorine residual of 1.0 mg/l, temperature of 10°C, and pH 7.5? How does a water temperature increase to 20°C affect the $C \cdot t$? A decrease to 5°C?

11.46 A surface water treatment plant at a winter resort city has been designed to process cold, low-turbidity water by direct filtration based on the pilot-plant study described in Problem 11.18. The filtered water has a turbidity of less than 0.1 NTU, temperature of 0.5°C, and pH of 7.4

during the period of highest water consumption in the winter. Tracer analyses of the clear-well reservoir are illustrated in Figure 11.20. At the critical hourly flow of 3.0 mgd, t_{10} = 90 min. To comply with the EPA surface water disinfection regulation, what free chlorine residual must be maintained in filtered water at the outlet of the clear well?

11.47 Figure 11.32 illustrates surface water schemes of conventional water treatment and direct filtration treatment. Review data provided in Section 11.24. Fill in the names of unit processes in the boxes and the numerical values of log removals required by the EPA in the blank spaces.

11.48 A surface water treatment plant with coagulation, sedimentation, filtration, and effluent chlorination is being evaluated for compliance with the EPA surface water disinfection regulation. The critical time of the year is during peak demand in the summer when the finished water temperature is 15°C or greater, pH is 7.5 or less, and the turbidity is 0.3–0.4 NTU. Prechlorination cannot be practiced because of trihalomethane formation. Data from the tracer analysis at peak hourly flow through the clear well, which has been modified by installation of baffles to reduce short-circuiting, are given in Problem 10.6. After the clear well and prior to the first service connection, the finished water is transported through a pipeline for 4000 ft at a velocity of 5 fps. What chlorine residual must be present in the water at the outlet of the pipeline?

11.49 A groundwater supply has been designated as vulnerable to fecal contamination, and the state has specified a disinfection level of 99.99% (4.0-log) virus inactivation. The peak hourly pumping rate is 2000 gpm from the well field through a 4200-ft pipe with a 16-in.

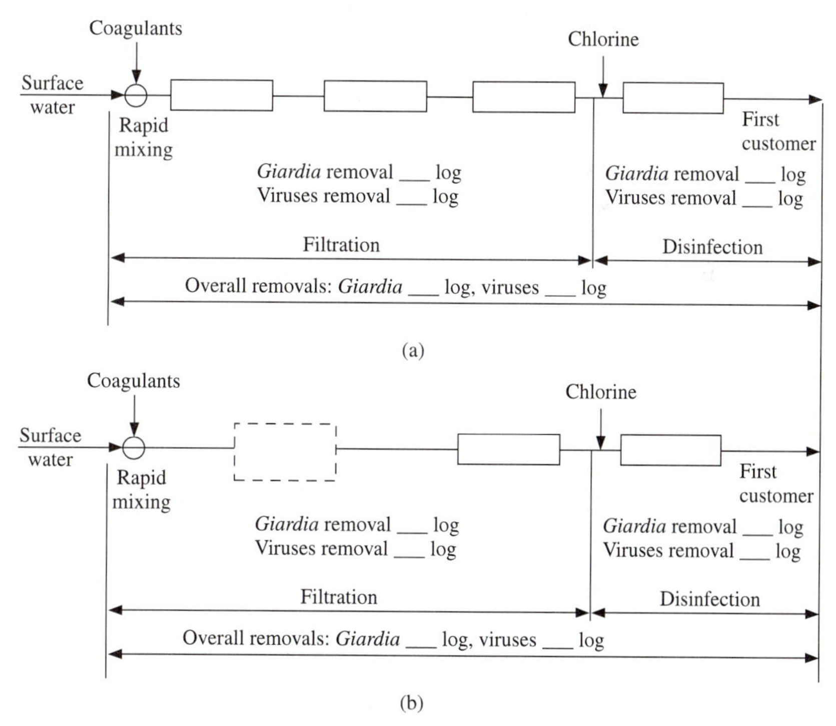

FIGURE 11.32 Surface-water treatment schemes listing the expected removals of *Giardia* and viruses by (a) conventional water treatment; (b) direct filtration treatment.

diameter to a reservoir in the town. The water temperature is 10°C. What chlorine residual is required in the water at the outlet of the pipeline?

11.50 Groundwater from a deep well is pumped at 400 gpm directly into the pipe network of a city. The well pump and pipe fittings are located in a well house. During the seasons of maximum usage, the pump operates for an average of 18 hr per day, with chlorine feed controlled by pump operation. The following chlorination systems are being considered to apply 0.50 mg/l of chlorine to the pump discharge: (a) liquid chlorine from a 100-lb pressurized cylinder through a solution feed chlorinator; (b) sodium hypochlorite solution with 10% available chlorine (by weight) fed from a 200-gal storage tank by a diaphragm pump; (c) powdered calcium hypochlorite with 70% available chlorine in a 50-gal tank and fed by a diaphragm pump. Calculate the quantity of chlorine applied per 18-hr day. Calculate the number of days each system can feed chlorine before requiring renewal. Which system would you recommend for a cold climate? For a warm climate?

11.51 A groundwater treatment plant adds chlorine to oxidize iron and manganese for removal by filtration. The groundwater temperature during peak summer usage is 10°C and the pH is 7.8. Based on a tracer study, the t_{10} time from chlorine addition in the contact tank through filtration is 1.8 min. The t_{10} time in the clear well during peak pumpage is 8.0 min. The chlorine residual in the filter effluent is 0.8 mg/l and in the clear-well effluent is 0.4 mg/l. Is the disinfection adequate for 3.0-log inactivation of viruses?

11.52 What determines the occurrence and concentrations of different kinds of pathogens present in a municipal wastewater? In general terms, to what degree are different kinds of pathogens removed in conventional wastewater treatment without effluent chlorination? With effluent chlorination?

11.53 A conventional activated-sludge treatment plant with effluent chlorination consistently produces a treated wastewater with a BOD less than 20 mg/l, suspended solids less than 30 mg/l, and fecal coliform count less than 200 organisms per 100 ml. The effluent has been described by the plant superintendent as clear "sparkling" water. You have been asked to evaluate the feasibility of using the effluent to irrigate the city's public park with playgrounds and to apply to the state department of environmental control for permission to reuse the wastewater for this application. To introduce the idea and to allay the fears of an environmental citizens group, you have been asked to present a preliminary assessment of the situation to the city council. Outline the subjects that you would discuss in your presentation. (Refer to Sections 8.1, 11.26, 11.27, 14.5, 14.15, and 14.19.)

11.54 What dosage of commercial fluosilicic acid is needed to increase the fluoride ion concentration from 0.3 to 1.0 mg/l? Use fluosilicic acid data from Table 11.12 and express the answers as mg/l and lb/mil gal.

11.55 A 4.0% sodium fluoride solution is applied to increase the fluoride concentration from 0.4 to 1.0 mg/l in a municipal water supply. (a) What is the application rate of NaF solution in gal/mil gal? (b) How many pounds of commercial-grade sodium fluoride are needed per million gallons?

11.56 What is the health risk of dietary intake of lead? (Refer to Section 8.5.) Why are first-flush samples collected from consumers' faucets used to assess lead contamination of drinking water? What are recommended methods of controlling excessive lead concentrations?

11.57 List three possible methods for controlling crown corrosion in a large concrete sanitary sewer.

11.58 A distilled seawater contains 25 mg/l of sodium chlorine. (Negligible amounts of magnesium, calcium, potassium, and sulfate are also present.) To stabilize this corrosive water, sufficient lime is applied to add 15 mg/l of calcium ion, and the hydroxide ion is converted to bicarbonate ion by applying carbon dioxide. (The equilibrium pH for stabilization is generally 8.5–8.7.)

The desalinated stabilized water is transported to a blending plant where brackish groundwater is added prior to delivery to the water distribution system. The criterion for blending is a product water with not more than 250 mg/l chloride, 250 mg/l sulfate, or 100 mg/l of sodium. Based on the analysis of the groundwater listed below, what percentage of the blended water can be groundwater? Sketch milliequivalents-per-liter bar graphs for the distilled seawater, stabilized distilled water, and blended water.

$$Ca^{2+} = 220\ \text{mg/l} \qquad Na^{+} = 580\ \text{mg/l} \qquad SO_4^{\ 2-} = 420\ \text{mg/l}$$
$$Mg^{2+} = 74\ \text{mg/l} \qquad HCO_3^{\ -} = 210\ \text{mg/l} \qquad Cl^{-} = 1030\ \text{mg/l}$$

11.59 Review Example 11.14. Evaluate the feasibility of operating this reverse-osmosis plant at a water recovery of 90%. (The $pK_2' - pK_s'$ value for a total dissolved-solids concentration above 10,000 mg/l can be assumed to remain essentially constant at 2.1.) If 90% recovery were to be performed, what process changes would be required?

11.60 The quality characteristics of a deep-well water in an arid region with a hot climate are

calcium hardness	= 270 mg/l	pH	= 8.1
magnesium hardness	= 180 mg/l	alkalinity	= 120 mg/l
sodium ion	= 138 mg/l	sulfate ion	= 110 mg/l
iron ion	= 0.40 mg/l	chloride ion	= 344 mg/l
manganese ion	= 0.15 mg/l	fluoride ion	= 1.4 mg/l
total dissolved solids	= 830 mg/l	nitrate nitrogen	= 15 mg/l
temperature	= 27°C		

1. Is the water quality satisfactory for a municipal supply without treatment? Comment on any quality problems with respect to health and aesthetic standards.
2. Propose a treatment scheme that will provide a water quality adequate to meet both health and aesthetic standards. Sketch a flow diagram showing the unit processes, chemical additions, and sources of wastes for disposal.
3. Sketch an approximate bar graph of the treated water.

11.61 In Problem 11.31, lime–soda ash softening of the groundwater did not produce a treated water of drinking quality. Sketch a preliminary process flow diagram to treat this groundwater for potable use. On the flow diagram, show the chemicals being added and their points of application. Consider also the disposal of any process wastewaters.

11.62 What categories of chemicals are included in synthetic organic chemicals? In treatment of surface waters, how effective for removal of SOCs are conventional coagulation processes and the addition of powdered activated carbon for taste and odor control? What are the limitations in using aeration as a method for removal of VOCs from contaminated well water?

11.63 Determine depths of packing and surface areas between 4 m and 8 m for a countercurrent air-stripping tower to treat well water to reduce tetrachloroethylene from 100 μg/l to 2 μg/l, trichlorethylene from 25 to 2 μg/l, and *cis*-1,2-dichloroethylene from 70 to 2 μg/l. The design flow rate is 44 l/s, and the lowest temperature anticipated is 7°C, based on groundwater temperature and cooling in the tower. The following data are based on pilot-scale packed-column tests: (1) Henry's law constants for tetrachloroethylene, trichloroethylene, and *cis*-1,2,-dichloroethylene are 0.30, 0.21, and 0.094 at 7°C, respectively. (2) The mass transfer coefficient for the most efficient and cost-effective commercial packing is 0.015 s^{-1}. (3) The optimum air-to-water ratio is 20:1.

11.64 Outlined below is the sequence of unit operations and chemical additions used in the treatment of a well water supply. Briefly state the function or purpose of each unit process and the reason for each chemical addition:

1. Mixing and flocculation with the addition of lime
2. Sedimentation
3. Recarbonation
4. Granular-media filtration
5. Postchlorination

11.65 Outlined below is the sequence of unit operations and chemical additions used in the treatment of a well water supply. Briefly state the function or purpose of each unit process and the reason for each chemical addition:

1. Prechlorination at the wells
2. Aeration over a tray aerator
3. Rechlorination
4. Detention in a settling basin
5. Granular-media filtration
6. Addition of anhydrous ammonia

11.66 Outlined below is the sequence of unit operations and chemical additions used in the treatment of a well water supply. Briefly state the function or purpose of each unit process and the reason for each chemical addition:

1. Prechlorination at the wells
2. Mixing–flocculation–sedimentation in flocculator–clarifiers using split treatment with lime and alum added to one leg and potassium permanganate to the other leg
3. Granular-media filtration
4. Postchlorination

11.67 Consider the following sequence of unit operations and chemical additions used in the treatment of a river water supply. Briefly state the function or purpose of each unit process and the reason for each chemical addition:

1. Presedimentation with polymer addition
2. Activated carbon available when needed
3. Mixing and flocculation with the addition of alum and polymer
4. Sedimentation
5. Addition of activated carbon
6. Granular-media filtration
7. Postchlorination

11.68 Outlined below is the sequence of unit operations and chemical additions used in the treatment of a reservoir water supply. Briefly state the function or purpose of each unit process and the reason for each chemical addition:

1. Intermittent applications of copper sulfate to the reservoir during summer and fall
2. Chlorine dioxide available when needed
3. Mixing and flocculation with the addition of alum and polymer

4. Sedimentation
5. Addition of activated carbon
6. Granular-media filtration
7. Postchlorination

11.69 Outlined below is the sequence of unit operations and chemical additions used in the treatment of a brackish groundwater. Briefly state the function or purpose of each unit process and the reason for each chemical addition:

1. Acidification with sulfuric acid
2. Addition of sodium hexametaphosphate
3. Cartridge filtration
4. High-pressure pumps
5. Reverse-osmosis modules
6. Degasification using stripping towers
7. Addition of sodium hydroxide
8. Addition of chlorine

REFERENCES

[1] *Standard Methods for the Examination of Water and Wastewater* (Washington, DC: Am. Public Health Assoc., 1998).

[2] V. L. Snoeyink and D. Jenkins, *Water Chemistry* (New York: Wiley, 1980).

[3] W. J. Weber, Jr., *Physicochemical Processes for Water Quality Control* (New York: Wiley-Interscience, 1972).

[4] Committee Report, "Coagulation as an Integrated Water Treatment Process," *J. Am. Water Works Assoc.* 81, no. 11 (1989): 72–78.

[5] H. E. Hudson, Jr., "Jar Testing and Utilization of Jar Test Data," *Water Clarification Processes Practical Design and Evaluation* (New York: Van Nostrand Reinhold, 1981), Chap. 3.

[6] Committee Report, "Research Needs for the Treatment of Iron and Manganese," *J. Am. Water Works Assoc.* 79, no. 9 (1987): 119–122.

[7] Committee Report, "Disinfection," *J. Am. Water Works Assoc.* 74, no. 7 (1982): 376–379.

[8] National Research Council, *Drinking Water and Health, Disinfectants and Disinfection By-Products*, Vol. 7 (Washington, DC: National Academy Press, 1987).

[9] S. W. Krasner, M. J. McGuire, J. G. Jacangelo, N. L. Patania, K. M. Reagan, and E. M. Aieta, "The Occurrence of Disinfection By-Products in U.S. Drinking Water," *J. Am. Water Works Assoc.* 81, no. 8 (1989): 41–53.

[10] P. D. Cohn, M. Cox, and P. S. Berger, "Health and Aesthetic Aspects of Water Quality," in *Water Quality & Treatment, A Handbook of Community Water Supplies*, 5th ed. (New York: McGraw-Hill, Inc. for *Am. Water Works Assoc.* 1999).

[11] F. W. Pontius and W. R. Diamond, "Complying With the Stage 1 D/DBP," *Am. Water Works Assoc.* 91, no. 4 (1999): 16–32.

[12] *Handbook of Public Water Systems*, 2nd ed., "Criteria and Standards for Improved Potable Water Quality," (New York: John Wiley & Sons, for HDR Engineering, Inc., 2001).

[13] *Issues in Potable Reuse: The Viability of Augmenting Drinking Water Supplies with Reclaimed Water*, (Washington, DC: National Research Council, National Academy Press, 1998).

[14] C. N. Hass and S. B. Karra, "Kinetics of Microbial Inactivation by Chlorine, Review of Results in Demand Free Systems," *Water Res.* 18 (1984): 1443–1449.

[15] J. C. Hoff, Project Summary, "Inactivation of Microbial Agents by Chemical Disinfectants," Environmental Protection Agency, Water Engineering Research Laboratory, EPA/600/S2-86/067 (September 1986).

[16] C. P. Hibler, C. M. Hancock, L. M. Perger, J. G. Wegrzn, and K. D. Swabbly, *Inactivation of Giardia Cysts with Chlorine at 0.5° C to 5.0° C* (Denver: Research Foundation, Am. Water Works Assoc., 1987).

[17] *Guidance Manual for Compliance with the Filtration and Disinfection Requirements for Public Water Systems Using Surface Water Sources*, (Washington, DC; Office of Drinking Water, U.S. Environmental Protection Agency, 1991).

[18] Amendment to paragraph (8) of section 1412(b) (42 U.S.C. 300g-l(b)(8)) of the Safe Drinking Water Act SEC.107. GROUND WATER DISINFECTION (August 1, 1996).

[19] Centers for Disease Control and Prevention, *Surveillance Summaries for Waterborne Disease Outbreaks*, U.S. Department of Health and Human Services publications.

[20] J. L. Melnick, "Etiologic Virus Diseases," *Monographs in Virology*, 15 (Karger, Basel, 1984): 1–16.

[21] R. A. J. Arthur, "Hanging on the Lifeline," Water & Environment International, 5, no. 41 *International Trade Publications* (Surrey, England, 1996): 12–13 and 35–36.

[22] R. G. Feachem, D. J. Bradley, H. Garelick, and D. D. Mara, *Sanitation and Disease, Health Aspects of Excreta and Wastewater Management; World Bank Studies in Water Supply and Sanitation 3* (Chichester: Wiley, 1983).

[23] D. F. Striffler, W. O. Young, and B. A. Burt, "The Prevention and Control of Dental Caries: Fluoridation," in *Dentistry, Dental Practice, and the Community* (Philadelphia: Saunders, 1983), 155–199.

[24] J. R. Millette, A. F. Hammonds, M. F. Pansing, E. C. Hanson, and P. J. Clark, "Aggressive Water: Assessing the Extent of the Problem," *J. Am. Water Works Assoc.* 72, no. 5 (1980): 262–266.

[25] D. T. Merrill and R. L. Sanks, "Corrosion Control by Deposition of $CaCO_3$ Films," *J. Am. Water Works Assoc.* 69, no. 11 (1977): 592–599; 69, no. 12 (1977): 634–640; 70, no. 1 (1978): 12–18.

[26] L. S. McNeill and M. Edwards, "Iron Pipe Corrosion in Distribution Systems," *Am. Water Works Assoc.* 93, no. 7 (2001): 88–100.

[27] M. R. Schock, "Understanding Corrosion Control Strategies for Lead," *J. Am. Water Works Assoc.* 81, no. 7 (1989): 88–100.

[28] M. Edwards and L. S. McNeill, "Effects of Phosphate Inhibitors on Lead Release from Pipes," *J. Am. Water Works Assoc.* 94, no. 1 (2002): 79–90.

[29] R. G. Lee, W. C. Becker, and D. W. Collins, "Lead at the Tap: Sources and Control," *J. Am. Water Works Assoc.* 81, no. 7 (1989): 52–62.

[30] A. Porteous, *Saline Water Distillation Processes* (London: Longman, 1975).

[31] R. L. Reitz, "Pretreatment for Potable Water Production by Membrane Processes," in *AWWA Seminar Proceedings, Membrane Processes: Principles and Practices* (Denver: Am. Water Works Assoc., 1988), 1–11.

[32] A. Ko and D. B. Guy, "Brackish and Seawater Desalting," in *Reverse Osmosis Technology, Application for High-Purity-Water Production* (New York: Marcel Dekker, 1988), 185–277.

[33] H. F. Ridgway, Jr., "Microbial Adhesion and Biofouling of Reverse Osmosis Membranes," in *Reverse Osmosis Technology, Application for High-Purity-Water Production* (New York: Marcel Dekker, 1988), 429–481.

[34] Degremont, *Water Treatment Handbook*, 5th ed. (New York: Halsted Press, 1979), 878.

[35] P. L. Lamarche and R. L. Droste, "Air-Stripping Mass Transfer Correlation for Volatile Organics," *J. Am. Water Works Assoc.* 81, no. 1 (1989): 78–89.

[36] *Handbook of Public Water Systems*, 2nd ed., "Activated Carbon Treatment," (New York: John Wiley & Sons, Inc., for HDR Engineering, Inc., 2001).

CHAPTER 12

Biological Treatment Processes

OBJECTIVES

The purpose of this chapter is to:

- Discuss the types and characteristics of microorganisms used in biological treatment processes
- Describe the purposes for and operation of various types of biological treatment processes

Biological processes are the most important unit operations in wastewater treatment. Therefore, biological considerations are discussed first, including kinds of microorganisms, their metabolism, and growth kinetics. Because the success of biological processes depends on the environment provided by treatment units, design engineers need a basic understanding of factors affecting the growth of mixed cultures. Subsequent sections present descriptions, operations, and design criteria for trickling filters, activated-sludge processes, and stabilization ponds. The final sections briefly present odor control, household septic tank systems, and ocean outfalls.

Wastewater treatment also includes physical processes, such as sedimentation and chemical disinfection, both of which were covered in previous chapters. The design engineer must select the best method of sludge disposal or reuse compatible with a selected wastewater treatment system. Processing of sludges is presented in Chapter 13.

BIOLOGICAL CONSIDERATIONS

Biological treatment systems are "living" systems that rely on mixed biological cultures to break down waste organics and remove organic matter from solution. Domestic wastewater supplies the biological food, growth nutrients, and inoculum. A treatment unit provides a controlled environment for the desired biological process. Historically, civil engineers designed treatment systems on the basis of empirical rules. This practice has led to failures in sanitary design—not unsuccessful in the sense of collapse of a structure, but deficient in that the biological process did not function properly. Understanding the biological processes involved in wastewater treatment is essential to a design engineer.

12.1 BACTERIA AND FUNGI

Bacteria (singular, bacterium) are the simplest forms of plant life that use soluble food and are capable of self-reproduction. Bacteria are fundamental microorganisms in the stabilization of organic wastes and therefore are of basic importance in biological treatment. Individual bacterial cells range in size from approximately 0.5 to 5 μm in rod, sphere, and spiral shapes and occur in a variety of forms: individual, pairs, packets, and chains.

Bacteria reproduce by binary fission (the mature cell divides into two new cells). In most species, the process of reproduction—growth, maturation, and fission—occurs in 20–30 min under ideal environmental conditions. Certain bacterial species form spores as a means of survival under adverse environmental conditions. Their tough coating is resistant to heat, lack of moisture, and loss of food supply. Fortunately, only one spore-forming bacterium, *Bacillus anthracis*,* is pathogenic to humans. As the result of stringent public health measures, incidents of anthrax in humans are rare.

Based on nutritive requirements, bacteria are classified as heterotrophic or autotrophic bacteria, although several species may function both heterotrophically and autotrophically.

Heterotrophic bacteria use organic compounds as an energy and carbon source for synthesis. Another term used instead of heterotroph is *saprophyte*, which refers to an organism that lives on dead or decaying organic matter. The heterotrophic bacteria are grouped into three classifications, depending on their action toward free oxygen. *Aerobes* require free dissolved oxygen to live and multiply. *Anaerobes* oxidize organic matter in the complete absence of dissolved oxygen. *Facultative bacteria* are a class of bacteria that use free dissolved oxygen when available but can also respire and multiply in its absence. *Escherichia coli*, a fecal coliform, is a facultative bacterium.

Autotrophic bacteria use carbon dioxide as a carbon source and oxidize inorganic compounds for energy. Autotrophs of greatest significance in sanitary engineering are the nitrifying, sulfur, and iron bacteria. Nitrifying bacteria perform the following reactions:

$$NH_3 \text{ (ammonia)} + \text{oxygen} \xrightarrow{\textit{Nitrosomonas}} NO_2^- \text{ (nitrite)} + \text{energy} \qquad (12.1)$$

$$NO_2^- \text{ (nitrite)} + \text{oxygen} \xrightarrow{\textit{Nitrobacter}} NO_3^- \text{ (nitrate)} + \text{energy} \qquad (12.2)$$

Autotrophic sulfur bacteria, *Thiobacillus*, convert hydrogen sulfide to sulfuric acid [Eq. (12.3)]. This bacterial production of sulfuric acid occurs in the moisture of condensation on side walls and crowns of sewers conveying septic wastewater. Since

*Bacteria are named using a binominal system; that is, each species is given a name consisting of two words. The first word is the genus and the second, the name of the species.

thiobacilli can tolerate pH levels less than 1.0, sanitary sewers constructed on flat grades in warm climates should be built using corrosion-resistant materials.

$$H_2S(\text{hydrogen sulfide}) + \text{oxygen} \rightarrow H_2SO_4 + \text{energy} \quad (12.3)$$

True iron bacteria are autotrophs that oxidize inorganic ferrous iron as a source of energy. These filamentous bacteria occur in iron-bearing waters and deposit the oxidized iron, $Fe(OH)_3$, in their sheath [Eq. (12.4)]. All species of the iron bacteria *Leptothrix* and *Crenothrix* may not be strictly autotrophic; however, they are truly iron-accumulating bacteria and thrive in water pipes conveying water containing dissolved iron and form yellow- or reddish-colored slimes. When mature bacteria die, they may decompose, imparting foul tastes and odors to water.

$$Fe^{2+}\ (\text{ferrous}) + \text{oxygen} \rightarrow Fe^{3+}\ (\text{ferric}) + \text{energy} \quad (12.4)$$

Fungi (singular, fungus) is a common term used to refer to microscopic nonphotosynthetic plants, including yeasts and molds. The most important group of yeasts for industrial fermentations is the genus *Saccharomyces. Saccharomyces cerevisiae* is the common yeast used by bakers, distillers, and brewers. *Saccharomyces cerevisiae* is single celled, commonly 5–10 μm in size, and reproduces by budding, in which large, mature cells divide, each producing one or more daughter cells. Under anaerobic conditions, this yeast produces alcohol as an end product. *Saccharomyces cerevisiae* is facultative and performs the following reactions:

$$\textit{Anaerobic:} \quad \text{Sugar} \rightarrow \text{alcohol} + CO_2 + \text{energy} \quad (12.5)$$

$$\textit{Aerobic:} \quad \text{Sugar} + \text{oxygen} \rightarrow CO_2 + \text{energy} \quad (12.6)$$

Energy yield in the aerobic reaction is much greater than in the anaerobic fermentation.

Molds are saprophytic or parasitic filamentous fungi that resemble higher plants in structure, composed of branched, filamentous, threadlike growths called hyphae. Molds are nonphotosynthetic, multicellular, heterotrophic, aerobic; reproduce by spore formation; and grow best in low-pH solutions (pH 2–5) high in sugar content. Molds are undesirable growths in activated sludge and can be created by low-pH conditions. The operation of an activated-sludge wastewater treatment system relies on gravity separation of microorganisms from the wastewater effluent. A large growth of molds creates a filamentous activated sludge that does not settle easily.

12.2 ALGAE

Algae (singular, alga) are microscopic photosynthetic plants. The process of photosynthesis is illustrated by the equation

$$CO_2 + 2H_2O \underset{\text{dark reaction}}{\overset{\text{sunlight}}{\rightleftharpoons}} \text{new cell tissue} + O_2 + H_2O \quad (12.7)$$

The overall effect of this reaction is to produce new plant life, thereby increasing the number of algae. By-product oxygen results from the biochemical conversion of water.

Algae are autotrophic, using carbon dioxide (or bicarbonates in solution) as a carbon source. The nutrients of phosphorus (as phosphate) and nitrogen (as ammonia, nitrite, or nitrate) are necessary for growth. Certain species of blue-green algae are able to fix atmospheric nitrogen. In addition, certain trace nutrients are required, such as magnesium, sulfur, boron, cobalt, molybdenum, calcium, potassium, iron, manganese, zinc, and copper. In natural waters, the nutrients most frequently limiting algal growth are inorganic phosphorus and nitrogen.

Energy for photosynthesis is derived from sunlight. Photosynthetic pigments biochemically convert energy in the sun's rays to useful energy for plant synthesis. The most common pigment is chlorophyll, which is green in color. Other pigments or combinations of pigments result in algae of a variety of colors, such as blue-green, yellowish green, brown, and red. In the prolonged absence of sunlight, the algae perform a dark reaction—for practical purposes the reverse of synthesis. In the dark reaction, the algae degrade stored food or their own protoplasm for energy to perform essential biochemical reactions for survival. The rate of this endogenous reaction is significantly slower than the photosynthetic reaction.

Algae grow in abundance in stabilization ponds rich in inorganic nutrients and carbon dioxide released from bacterial decomposition of waste organics. Green algae *Chlorella* are commonly found in oxidation ponds. Certain genera of algae are identified with clean water, such as *Navicula*. Descriptions and pictorial representations of algae occurring in water and wastewater are given in *Standard Methods for the Examination of Water and Wastewater* [1].

12.3 PROTOZOANS AND HIGHER ANIMALS

Protozoans are single-celled animals that reproduce by binary fission. The protozoans of significance in biological treatment systems are strict aerobes found in activated sludge, trickling filters, and oxidation ponds. These microscopic animals have complex digestive systems and use solid organic matter as an energy and carbon source. Protozoans are a vital link in the aquatic chain because they ingest bacteria and algae.

Protozoans with cilia may be categorized as free swimming and stalked. Free-swimming forms move rapidly in the water, ingesting organic matter at a very high rate. The stalked forms attach by a stalk to particles of matter and use cilia to propel their head about and bring in food. Another group of protozoans move by flagella. Long hairlike strands (flagella) move with a whiplike action, providing motility. *Amoeba* move and ingest food through the action of a mobile protoplasm.

Rotifers are the simplest multicelled animals. They are strict aerobes and metabolize solid food. A typical rotifer uses the cilia around its head for catching food. The name *rotifer* is derived from the apparent rotating motion of the cilia on its head. Rotifers are indicators of low pollutional waters and are regularly found in streams and lakes.

Crustaceans are multicellular animals with branched swimming feet or a shell-like covering, with a variety of appendages (antennae). The two most common crustaceans of interest are *Daphnia* and *Cyclops*. Crustaceans are strict aerobes and ingest microscopic plants. The zooplankton population in a lake includes a wide selection of crustaceans that serve as food for fishes.

12.4 METABOLISM, ENERGY, AND SYNTHESIS

Metabolism (catabolism) is the biochemical process (a series of biochemical oxidation–reduction reactions) performed by living organisms to yield energy for synthesis, motility, and respiration to remain viable. In standard usage, metabolism implies both catabolism and anabolism—that is, both degradation and assimilative reactions.

The metabolism of autotrophic bacteria is illustrated in Eqs. (12.1) through (12.4). In these reactions, the reduced inorganic compounds are oxidized, yielding energy for synthesis of carbon from carbon dioxide, producing organic cell tissue. [In the case of algae, Eq. (12.7), the carbon source is carbon dioxide, but the energy is from sunlight.]

In heterotrophic metabolism, organic matter is the substrate (food) used as an energy source. However, the majority of organic matter in wastewater is in the form of large molecules that cannot penetrate the bacterial cell membrane. The bacteria, in order to metabolize high-molecular-weight substances, must be capable of hydrolyzing the large molecules into diffusible fractions for assimilation into their cells. Therefore, the first biochemical reactions are hydrolysis* of complex carbohydrates into soluble sugar units, protein into amino acids, and insoluble fats into fatty acids. Under aerobic conditions, the reduced soluble organic compounds are oxidized to end products of carbon dioxide and water [Eq. (12.8)]. Under anaerobic conditions, soluble organics are decomposed to intermediate end products, such as organic acids and alcohols, along with the production of carbon dioxide and water [Eq. (12.9)]. Many intermediates, such as butyric acid, mercaptans (organic compounds with —SH radicals), and hydrogen sulfide have foul odors.

Under anaerobic conditions, if excess organic acids are produced, the pH of the solution will drop sufficiently to "pickle" the fermentation process. This is the principle used for preservation of silage. Bacteria produce an overabundance of organic acids in the anaerobic decomposition of the green fodder stored in the silo, inhibit further bacterial decomposition, and preserve the food value of the fodder. However, if proper environmental conditions exist to prevent excess acidity from the production of organic acid intermediates, populations of acid-splitting, methane-forming bacteria will develop and use the organic acids as substrate [Eq. (12.10)]. The combined biological processes of anaerobic decomposition of raw organic matter to soluble organic intermediates and the gasification of the intermediates to carbon dioxide and methane is referred to as digestion.

Aerobic: $$\text{Organics} + \text{oxygen} \rightarrow CO_2 + H_2O + \text{energy} \tag{12.8}$$

Anaerobic: $$\text{Organics} \rightarrow \text{intermediates} + CO_2 + H_2O + \text{energy} \tag{12.9}$$

$$\text{Organic acid intermediates} \rightarrow CH_4 + CO_2 + \text{energy} \tag{12.10}$$

The growth and survival of nonphotosynthetic microorganisms are dependent on their ability to obtain energy from the metabolism of substrate. Biochemical metabolic processes of heterotrophs are energy-yielding, oxidation–reduction reactions in which reduced organic compounds serve as hydrogen donors and oxidized organic or inorganic compounds act as hydrogen acceptors. *Oxidation* is the addition of oxygen,

*Hydrolysis is the addition of water to split a bond between chemical units.

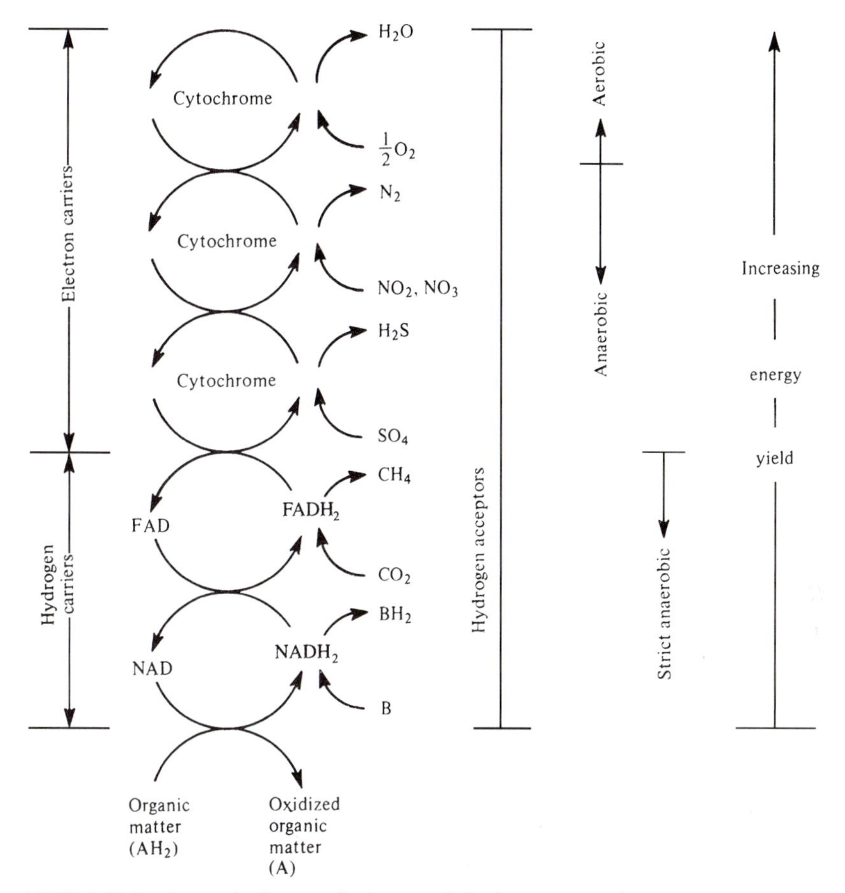

FIGURE 12.1 General scheme of substrate dehydrogenation for energy yield. (FAD: flavin adenine dinucleotide; NAD: nicotinamide adenine dinucleotide.)

removal of hydrogen, or removal of electrons. *Reduction* is the removal of oxygen, addition of hydrogen, or addition of electrons.

The simplified diagram of substrate dehydrogenation shown in Figure 12.1 is intended to illustrate the general relationship between energy yields of aerobic and anaerobic metabolism. Enzymatic processes of hydrogen transfer and methods of biologically conserving energy released are beyond the scope of this discussion. For students to understand fully the mechanisms illustrated in Figure 12.1, a knowledge of the biochemistry of microorganisms is necessary [2].

Energy stored in organic matter (AH_2) is released in the process of biological oxidation by dehydrogenation of substrate followed by transfer of hydrogen, or electrons, to an ultimate acceptor. The higher the ultimate hydrogen acceptor is on the energy (electromotive) scale, the greater will be the energy yield from oxidation of 1 mole of a given substrate. Aerobic metabolism using oxygen as the ultimate hydrogen acceptor yields the greatest amount of energy. Aerobic respiration can be traced in Figure 12.1 from reduced organic matter (AH_2) at the bottom, through the hydrogen and electron carriers, to oxygen. Facultative respiration, using oxygen bound in nitrates and sulfates, yields less energy than aerobic metabolism. The least energy yield results from strict anaerobic respiration, where the oxidation of AH_2 is coupled with reduction of B (an oxidized organic compound) to BH_2 (a reduced organic compound). The preferential use of hydrogen acceptors based on energy yield in a mixed bacterial culture is illustrated by the following equations:

Aerobic $$AH_2 + O_2 \rightarrow CO_2 + H_2O + \text{energy} \quad (12.11)$$

$$AH_2 + NO_3^- \rightarrow N_2 + H_2O + \quad (12.12)$$

(Faculative) $$AH_2 + SO_4^{2-} \rightarrow H_2S + H_2O + \quad (12.13)$$

$$AH_2 + CO_2 \rightarrow CH_4 + H_2O + \quad (12.14)$$

Anaerobic $$AH_2 + B \rightarrow BH_2 + A + \text{energy} \quad (12.15)$$

decreasing energy yield

Hydrogen acceptors are used in the sequence of dissolved oxygen first, followed by nitrates, sulfates, and oxidized organic compounds, in this general order. Thus hydrogen sulfide odor formation follows nitrate reduction and precedes methane formation.

The biochemical reactions in Figure 12.1 are performed by oxidation–reduction enzymes. Enzymes are organic catalysts that perform biochemical reactions at temperatures and chemical conditions compatible with biological life. The coenzyme component of the enzyme determines what chemical reaction will occur. Coenzymes nicotinamide adenine dinucleotide (NAD) and flavin adenine dinucleotide (FAD) are responsible for hydrogen transfer. Cytochromes are respiratory pigments that can undergo oxidation and reduction and serve as hydrogen carriers.

Synthesis (anabolism) is the biochemical process of substrate utilization to form new protoplasm for growth and reproduction. Microorganisms process organic matter to create new cells. The cellular protoplasm formed is a combination of hundreds of complex organic compounds, including proteins, carbohydrates, nucleic acids, and lipids. Major elements in biological cells are carbon, hydrogen, oxygen, nitrogen, and phosphorus. On a dry-weight basis, protoplasm is 10%–12% nitrogen and approximately 2.5% phosphorus; the remainder is carbon, hydrogen, oxygen, and trace elements.

Relationships between metabolism, energy, and synthesis are important in understanding biological treatment systems. The primary product of metabolism is energy, and the chief use of this energy is for synthesis. Energy release and synthesis are coupled biochemical processes that cannot be separated. The maximum rate of synthesis occurs simultaneously with the maximum rate of energy yield (maximum rate of metabolism). Therefore, in heterotrophic metabolism of wastewater organics, maximum rate of removal of organic matter for a given population of microorganisms occurs during maximum biological growth. Conversely, the lowest rate of removal of organic matter occurs when growth ceases.

The major limitation of anaerobic growth is energy, owing to the fact that in anaerobic decomposition a low energy yield per unit of substrate results from an incomplete reaction (Fig. 12.2). In other words, the limiting factor in anaerobic metabolism is a lack of hydrogen acceptors. When the supply of biologically available energy is exhausted, the processes of metabolism and synthesis cease.

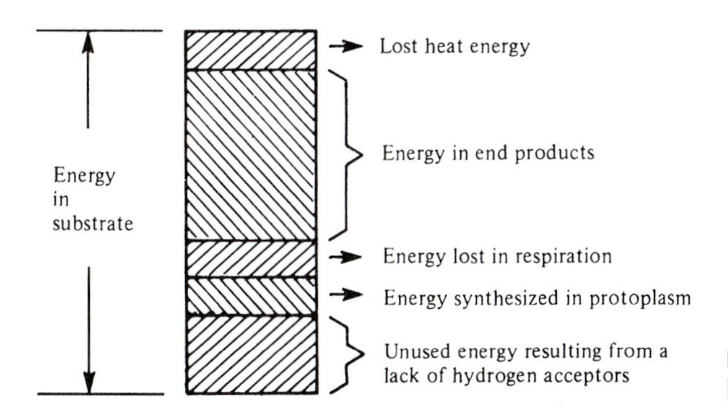

FIGURE 12.2 Energy conversions in anaerobic metabolism.

Aerobic metabolism is the antithesis of anaerobiosis, biologically available carbon being the limiting factor (Fig. 12.3). Abundance of oxygen creates no shortage of hydrogen acceptors. But the supply of substrate carbon is rapidly exhausted through respiration of carbon dioxide and synthesis into new cells.

The energy conversion diagrams shown schematically in Figures 12.2 and 12.3 illustrate the major features of anaerobic and aerobic metabolism. An anaerobic process has the following characteristics: incomplete metabolism, small quantity of biological growth, and production of high-energy products such as acetic acid and methane. An aerobic process results in complete metabolism and synthesis of the substrate, ending in a large quantity of biological growth.

12.5 ENZYME KINETICS

The reaction between an enzyme E and substrate S is postulated to be

$$\mathrm{E} + \mathrm{S} \underset{k_2}{\overset{k_1}{\rightleftharpoons}} \mathrm{ES} \overset{k_3}{\rightarrow} \mathrm{E} + \text{products} \tag{12.16}$$

In the first step, E combines with S to form ES by a reversible reaction where k_1 is the rate constant for formation of ES, and k_2 is the rate constant for dissociation of ES to E and S. After combining to form ES, S is converted to products in the second step and E is released for combination with more S. The rate of conversion of ES to products is

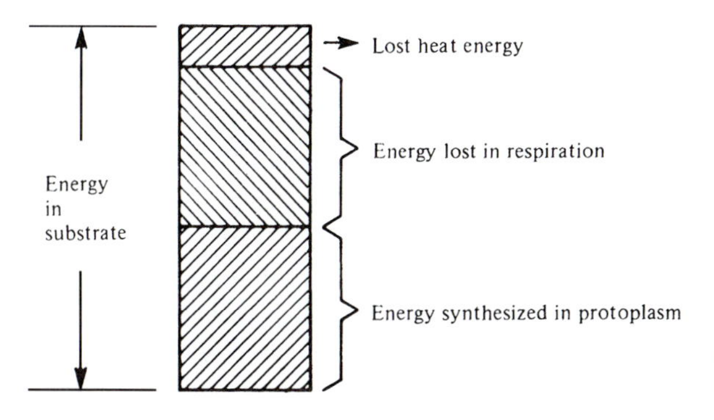

FIGURE 12.3 Energy conversions in aerobic metabolism.

represented by k_3. An enzyme-catalyzed reaction can be experimentally performed by placing a small quantity of an enzyme in a substrate solution and measuring the decomposition of substrate with time. For example, sucrose is split into its two separate sugar rings of glucose and fructose by the enzyme *invertase* that performs this hydrolysis. The reaction in the form of Eq. (12.16) is written

$$\text{Sucrose} + \textit{invertase} \rightleftharpoons \text{sucrose-}\textit{invertase}$$
$$\rightarrow \text{glucose} + \text{fructose} + \textit{invertase}$$

The first satisfactory mathematical analysis of the effect of substrate concentration on the rate of enzyme-catalyzed reactions was made in 1913 by Michaelis and Menten [3]. The derivation of their equation is based on Eq. (12.17), that is, the rate of decomposition of substrate is proportional to the concentration of the intermediate enzyme–substrate complex. For the reversible reaction $E + S \rightleftharpoons ES$, the dissociation constant of ES, defined as K_m, can be written as

$$K_m = \frac{(E - ES)S}{ES} \tag{12.17}$$

Rearranging the equation,

$$(ES) = \frac{(E)(S)}{K_m + S} \tag{12.18}$$

with k_3 the rate constant for decomposition of ES, the measured rate of decomposition of substrate r equals k_3 (ES), and

$$r = \frac{k_3(E)(S)}{K_m + S} \tag{12.19}$$

The maximum rate of decomposition r_m occurs when ES is at its maximum concentration—that is, when all of the enzyme is combined with substrate and $ES = E$. Therefore,

$$r_m = k_3(ES) = k_3(E) \tag{12.20}$$

Substituting r_m for $k_3(E)$ in Eq. (12.19) produces the Michaelis–Menten equation:

$$r = r_m\left(\frac{S}{K_m + S}\right) \tag{12.21}$$

or

$$K_m = S\left(\frac{r_m}{r} - 1\right) \tag{12.22}$$

Since K_m and r_m are constants, Eq. (12.21) is a rectangular hyperbola, as shown in Figure 12.4. Data from enzyme-catalyzed experiments by Michaelis and Menten are graphed to form this diphasic curve. From Eq. (12.22), when $r_m/r = 2$, the measured rate r is one-half the value of the limiting rate r_m and $K_m = S$. Therefore, the substrate

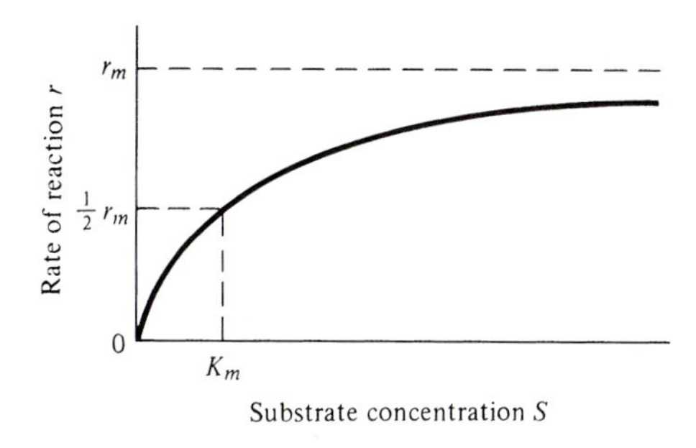

FIGURE 12.4 Theoretical curve of reaction rate versus substrate concentration for an enzyme-catalyzed reaction [Eq. (12.21)].

concentration for attaining the half-maximum reaction rate is a characteristic constant K_m of an enzyme-catalyzed reaction, which is termed the *saturation constant.*

The substrate saturation phenomenon is a unique characteristic of enzymatic reactions and is the mathematical basis of kinetics in microbiology. For further explanation of this process, consider a batch experiment starting with a large amount of substrate relative to the concentration of enzyme. From Eq. (12.21), for a large S, K_m can be neglected and $r = r_m$. This can also be seen in Figure 12.4. The magnitude of r_m depends on the maximum rate at which the enzyme present can decompose the substrate. The rate is zero order; that is, the reaction proceeds at a rate independent of the concentration of substrate. As the quantity of substrate decreases, the concentration of enzyme remaining constant, the rate of reaction becomes increasingly dependent on the remaining substrate concentration. The rate at $\frac{1}{2} r_m$ corresponds to a substrate concentration equal to the saturation constant K_m. Below this substrate concentration, some of the enzyme molecules are not combined with substrate, resulting in a first-order reaction, which proceeds at a rate directly proportional to the substrate concentration. When S is much smaller than K_m, S can be neglected in Eq. (12.21) and $r = r_m S/K_m$.

12.6 GROWTH KINETICS OF PURE BACTERIAL CULTURES

A pure culture can be grown in a laboratory reactor by inoculating a sterile liquid medium with a small number of viable bacteria of a single species. The characteristic growth pattern for bacteria in such a batch culture is sketched in Figure 12.5. After a short lag period for adaptation to the new environment, the bacteria reproduce by binary fission, exponentially increasing the number of viable cells and biomass in the culture medium. The existence of excess substrate promotes this maximum rate of growth. The rate of metabolism in the *exponential growth phase* is limited only by the microorganisms' ability to process the substrate. With X representing the concentration of biomass and μ a proportionality constant, the biomass growth rate can be expressed as

$$\left(\frac{dX}{dt}\right)_g = \mu X \tag{12.23}$$

Dividing both sides of Eq. (12.23) by X yields

$$\mu = \frac{(dX/dt)_g}{X} \tag{12.24}$$

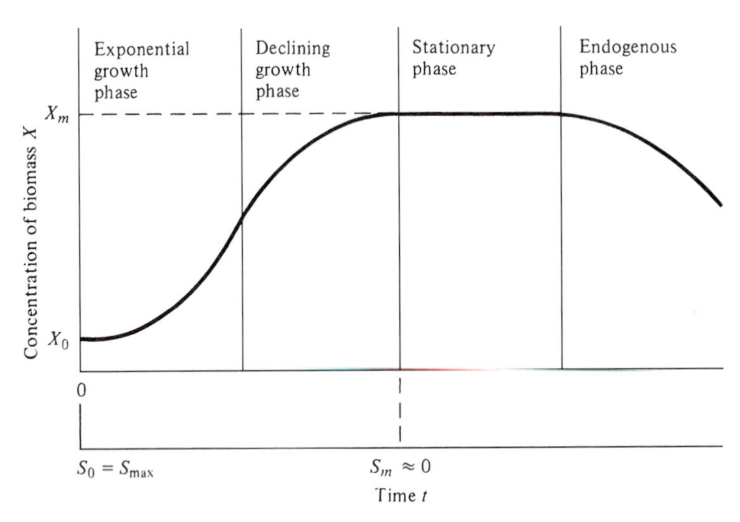

FIGURE 12.5 Characteristic growth phases of a pure culture of bacteria.

where

μ = specific growth rate (rate of growth per unit of biomass), time^{-1}

$(dX/dt)_g$ = biomass growth rate, mass/unit volume · time

X = concentration of biomass, mass/unit volume

The *declining growth phase* is caused by an increasing shortage of substrate. The rate of reproduction decreases until the number of viable bacteria is stationary, which occurs when the rate of reproduction is equal to the rate of death. The total biomass exceeds the mass of viable cells since many of the microorganisms stopped reproducing owing to substrate-limiting conditions. By laboratory experimentation, Monod [4] studied the growth of bacteria in batch cultures. He found that growth was a function of both the concentration of microorganisms and the concentration of the growth-limiting substrate. The mathematical relationship proposed by Monod between the residual concentration of the growth-limiting substrate and the specific growth rate of biomass is the hyperbolic equation

$$\mu = \mu_m\left(\frac{S}{K_s + S}\right) \tag{12.25}$$

where

μ = specific growth rate, time^{-1}

μ_m = maximum specific growth rate (at a concentration of the growth-limiting substrate at or above saturation), time^{-1}

S = concentration of growth-limiting substrate in solution, mass/unit volume

K_s = saturation constant (equal to the limiting substrate concentration at one-half the maximum growth rate), mass/unit volume

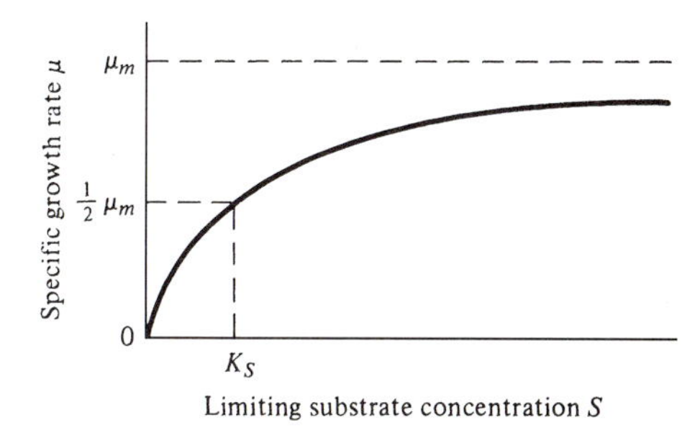

FIGURE 12.6 Specific growth rate versus substrate concentration for the exponential and declining growth phases of a bacterial culture.

As shown in Figure 12.6, Eq. (12.25) is similar in form to the Michaelis–Menten equation for enzyme-catalyzed reactions [Eq. (12.21) and Fig. 12.4]. The constant K_s, similar in function to K_m in Eq. (12.22), is equal to the concentration of substrate S when the specific growth rate μ equals one-half of the maximum growth rate μ_m.

The specific growth-rate relationship in Eq. (12.25) can be substituted for the proportionality constant in Eq. (12.23). The result is the following expression for biomass growth rate in a substrate-limiting solution:

$$\left(\frac{dX}{dt}\right)_g = \frac{\mu_m XS}{K_s + S} \tag{12.26}$$

where

$(dX/dt)_g$ = biomass growth rate, mass/unit volume · time

μ_m = maximum specific growth rate, time^{-1}

X = concentration of biomass, mass/unit volume

S = concentration of growth-limiting substrate, mass/unit volume

K_s = saturation constant, mass/unit volume

The *growth yield Y* is defined as the incremental increase in biomass resulting from metabolism of an incremental amount of substrate [5]. The growth yield in a batch culture (Fig. 12.5) is the biomass increase during the exponential and declining growth phases ($X_m - X_0$) relative to the substrate used ($S_0 - S_m$). Since growth is limited by depletion of the substrate, S_m is assumed to be zero. Therefore,

$$X_m - X_0 = YS_0 \tag{12.27}$$

If a series of batch cultures are grown starting with different initial substrate concentrations, a plot of X_m values versus their respective S_0 values is a straight line with slope Y (Fig. 12.7). For bacterial cultures, when conditions are maintained constant, the growth yield is a constant reproducible value [4].

Growth yield can also be expressed in derivative form as

$$\left(\frac{dX}{dt}\right)_g = Y\left(\frac{dS}{dt}\right)_u \tag{12.28}$$

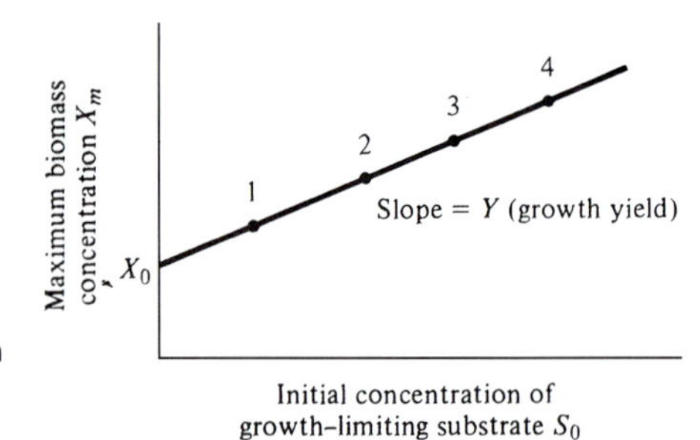

FIGURE 12.7 Growth yield for a series of four batch cultures (Fig. 12.5) is determined by plotting X_m versus S_0.

Substituting this into Eq. (12.26) results in an equation that defines the rate of substrate utilization in a solution in which the biomass growth rate is limited by the low concentration of substrate:

$$\left(\frac{dS}{dt}\right)_u = \frac{\mu_m XS}{Y(K_s + S)} \tag{12.29}$$

where

$(dS/dt)_u$ = substrate utilization rate, time^{-1}
μ_m = maximum specific growth rate, time^{-1}
X = concentration of biomass, mass/unit volume
S = concentration of growth-limiting substrate, mass/unit volume
Y = growth yield, mass/mass
K_s = saturation constant, mass/unit volume

In the *endogenous growth phase* (Fig. 12.5), viable bacteria are competing for the small amount of substrate still in solution. The rate of metabolism is decreasing at an increasing rate, resulting in a rapid decrease in the number of viable cells. Starvation occurs such that the rate of death exceeds the rate of reproduction. The total biomass decreases as cells utilize their own protoplasm as an energy source. Cells become old, die, and lyse, releasing nutrients back into solution. The action of cell lysis decreases both the number and mass of microorganisms.

The rate of biomass decrease during endogenous respiration is proportional to the biomass present. Thus,

$$\left(\frac{dX}{dt}\right)_d = -k_d X \tag{12.30}$$

where

$(dX/dt)_d$ = biomass decay rate, mass/unit volume · time
k_d = microbial decay coefficient, time^{-1}
X = concentration of biomass, mass/unit volume

To determine *net biomass growth rate* during the endogenous growth phase, Eq. (12.30) is combined with Eq. (12.26) and Eq. (12.30) with Eq. (12.28), resulting in the following equations:

$$\left(\frac{dX}{dt}\right)_g^{\text{net}} = \frac{\mu_m XS}{K_s + S} - k_d X \tag{12.31}$$

$$\left(\frac{dX}{dt}\right)_g^{\text{net}} = Y\left(\frac{dS}{dt}\right)_u - k_d X \tag{12.32}$$

The Monod relationship [Eq. (12.25)] modified to yield *net specific growth rate* is

$$\mu_{\text{net}} = \mu_m \frac{S}{K_S + S} - k_d \tag{12.33}$$

and the observed growth yield Y_{obs} accounting for the effect of endogenous respiration on the net biomass growth rate from the relationship in Eq. (12.28) is

$$Y_{\text{obs}} = \frac{(dX/dt)_g^{\text{net}}}{(dS/dt)_u} \tag{12.34}$$

12.7 BIOLOGICAL GROWTH IN WASTEWATER TREATMENT

Both fixed-film growth and suspended-solids growth systems are used in biological treatment of wastewaters. In a fixed-growth process, organic matter is removed from wastewater as it flows over a biological film (slime layer) attached to a filter medium. Trickling filters use a variety of media, including stones, crushed rock, small plastic cylinders, and plastic vertical-sheet packing. A rotating biological contactor consists of a series of large-diameter plastic disks that slowly rotate in tanks conveying wastewater flow. In a suspended-growth process, active biological solids are mixed with the wastewater and held in suspension by aeration as the organic matter is taken out of solution by the microbial floc. The common name for this suspended-growth process is activated sludge.

Wastewater treatment relies on a mixed biological culture consisting of a variety of bacteria and protozoans. Microorganisms in the raw wastewater provide continuous inocula for the treatment process. The heterogeneous substrate content of a wastewater is expressed either as BOD or COD. The purpose of the treatment unit is to hold the biological culture in a controlled environment to promote growth of the microorganisms for extraction of colloidal and dissolved organics from solution.

The batch-culture growth pattern shown in Figure 12.5 is not directly applicable to biological treatment processes that are continuous-flow systems. For example, an activated-sludge system is fed continuously, and excess microorganisms are withdrawn, either continuously or intermittently, to maintain the desired mass of microorganisms for metabolizing incoming organic wastes. A schematic diagram (Figure 12.8) illustrates the flow pattern for organic matter and microorganisms in an activated-sludge system. Influent wastewater is aerated with a mixed culture of microorganisms for a sufficient period of time to permit synthesis of the waste organics into biological cells.

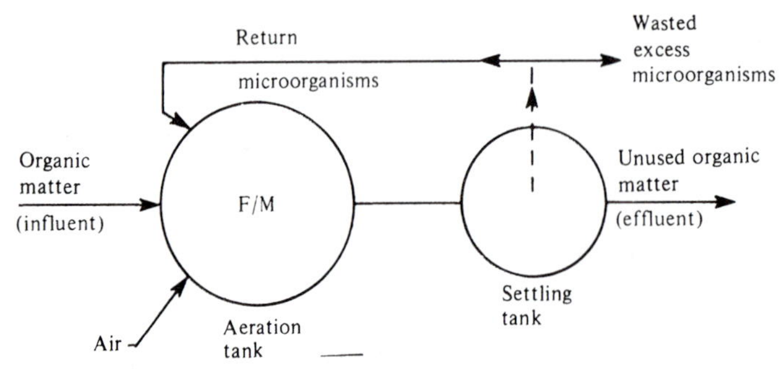

FIGURE 12.8 Schematic diagram of a continuous-flow activated-sludge process. (F/M: food/microorganism ratio.)

The microorganisms are then settled out of solution, removed from the bottom of the settling tank, and returned to the aeration tank to metabolize additional waste organics. Unused organic matter and nonsettleable microorganisms pass out in the system effluent. Metabolism of the organic matter results in an increased mass of microorganisms in the system. Excess microorganisms are removed (wasted) from the system to maintain proper balance between food supply and mass of microorganisms in the aeration tank. This balance is referred to as the *food-to-microorganism ratio* (F/M).

The F/M ratio maintained in the aeration tank defines the operation of an activated-sludge system. At a high F/M ratio, microorganisms are in the exponential growth phase, characterized by excess food and maximum rate of metabolism (Fig. 12.9). Although the exponential growth phase is desirable for maximum rate of organic matter removal, distinct disadvantages make it undesirable for operation of an activated-sludge system. The microorganisms are in dispersed growth such that they do not settle out of solution by gravity. Consequently, the settling tank is not effective in separating microorganisms from the effluent for return to the aeration tank. Second, there is excess unused organic matter in solution that cannot be

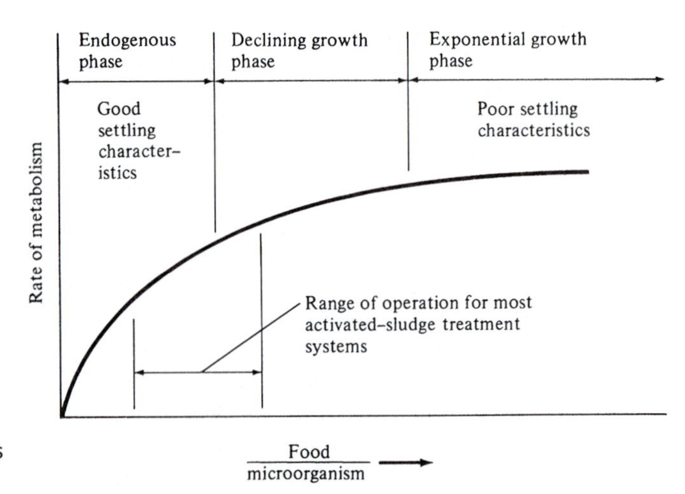

FIGURE 12.9 Rate of metabolism versus increasing food/microorganism ratio.

removed by sedimentation and passes out in the effluent. Operation at a high F/M ratio results in poor BOD removal efficiency.

At a low F/M ratio, overall metabolic activity in the aeration tank may be considered endogenous. Although initially there is rapid growth when the influent food and return microorganisms are mixed, competition for the small amount of food made available to the large mass of microorganisms results in near-starvation conditions for the majority of microorganisms within a short period of time. Under these conditions, continued aeration results in autooxidation of the microbial mass through cell lysis and resynthesis and also through the predator–prey activity where bacteria are consumed by the protozoans. Although the rate of metabolism is relatively low in the endogenous phase, metabolism of the organics is nearly complete and the microorganisms flocculate rapidly and settle out of solution by gravity. The good settling characteristics exhibited by activated sludge in the endogenous phase make operation in this growth period desirable where a high BOD removal efficiency is desired. Figure 12.9 summarizes the previous discussion and shows the range of operation for most activated-sludge treatment systems to be between the declining growth phase and the endogenous phase.

12.8 FACTORS AFFECTING GROWTH

Several factors affect the growth of microorganisms. The most important are temperature, pH, availability of nutrients, oxygen supply, presence of toxins, types of substrate, and, in the case of photosynthetic plants, sunlight. Growth with respect to both aerobic and anaerobic conditions and to the need for essential nutrients was discussed in Section 12.4.

Bacteria are classified as psychrophilic, mesophilic, or thermophilic, depending on their optimum temperature range for growth. Of least significance to sanitary engineers are the *psychrophilic* (cold-loving) bacteria, which grow best at temperatures slightly above freezing (4°–10°C).

Thermophilic (heat-loving) bacteria like an optimum temperature range of 50°–55°C. They are significant in sludge composting and attempts have been made to use a thermophilic temperature range for the anaerobic digestion of waste sludge. Thermophilic digestion has usually not been successful in practice because thermophilic bacteria are sensitive to small temperature changes, and it is difficult to maintain the required high operating temperature in a digestion tank.

Mesophilic (moderation-loving) bacteria grow best in the temperature range 20°–40°C. The vast majority of biological treatment systems operate in the mesophilic temperature range. Anaerobic digestion tanks are normally heated to near the optimum level of 35°C (95°F). Aeration tanks and trickling filters operate at the temperature of the wastewater as modified by that of the air. Generally, this is in the range of 10°–20°C (50°–68°F) within outer maximums at a low of 5°C (41°F) and a high of 25°C (77°F). Higher wastewater temperature increases biological activity in the treatment process and can cause operating problems. At high temperatures, odor problems may be more pronounced at a wastewater plant.

The rate of biological activity in the mesophilic range between 5° and 35°C doubles for every 10°–15°C temperature rise (Fig. 12.10). The common mathematical

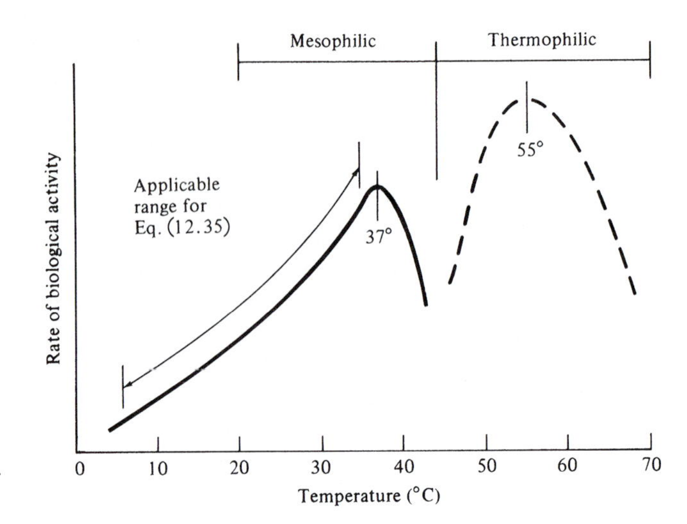

FIGURE 12.10 General effect of temperature on biological activity.

expression for relating the change in the reaction-rate constant with temperature, developed in Section 11.6, is

$$K = K_{20}\Theta^{T-20} \tag{12.35}$$

where

K = reaction-rate constant at temperature T

K_{20} = reaction-rate constant at 20°C

Θ = temperature coefficient

T = temperature of biological reaction, °C

The value of Θ is 1.072 if the rate of biological activity doubles with a 10°C temperature rise; the value of Θ is 1.047 if the rate doubles in 15°C. Above 40°C, mesophilic bacterial metabolism drops off sharply and thermophilic growth starts. Thermophilic bacteria have a range of approximately 45°–75°C, with an optimum near 55°C.

Cold wastewater can reduce BOD removal efficiency of biological processes. The efficiency of trickling filters is definitely decreased during cold weather and increased during warm periods. Trickling filters operating at 5°–10°C have poor BOD removals. However, low-loaded extended aeration systems operating at the same temperatures show good efficiencies. Extended aeration at a reduced BOD loading and resultant long aeration time compensates for the low metabolism rate of microorganisms. The biological treatment system most affected by cold winter temperature is the stabilization pond. Heat in the wastewater is not adequate to prevent formation of an ice cover in northern climates during winter.

The hydrogen ion concentration of the culture medium has a direct influence on microbial growth. Most biological treatment systems operate best in a neutral environment. The general range for operation of activated-sludge systems is between pH 6.5 and 8.5. At pH 9.0 and above, microbial activity is inhibited. Below 6.5, fungi are favored over bacteria in the competition for food. The methane-forming bacteria in anaerobic digestion have a much smaller pH tolerance range. General limits for anaerobic digestion are pH 6.7–7.4, with optimum operation at pH 7.0–7.1.

Biological treatment systems are adversely affected by toxic substances. Industrial wastes from metal-finishing industries usually contain toxic metal ions, such as nickel and chromium. Phenol is an extremely toxic compound found in chemical industry wastes. These and other inhibiting compounds are commonly removed by pretreatment at the industrial site prior to disposal of the industrial wastes to a municipal sewer.

Environmental conditions that adversely affect the desired microbial growth in an activated-sludge aeration tank can cause production of sludge with poor settling characteristics. This condition, resulting in excessive carryover of activated-sludge floc in the clarifier effluent (referred to as sludge bulking), is associated with filamentous growths and pinpoint floc with poor settleability.

12.9 POPULATION DYNAMICS

Previous sections described the important characteristics of each group of microorganisms (bacteria, fungi, algae, and protozoans) independently. In biological waste treatment systems, however, the naturally occurring cultures are mixtures of bacteria growing in mutual association and with other microscopic plants and animals. A general knowledge of the relationships, both cooperative and competitive, between various microbial populations in mixed cultures is essential to understanding biological treatment processes.

When organic matter is made available to a mixed population of microorganisms, competition arises for this food between the various species. Primary feeders that are most competitive become the dominant microorganisms. Under normal operating conditions, bacteria are the dominant primary feeders in activated sludge (Fig. 12.11). Saprobic protozoans, those that feed on dead organic matter (e.g., *Euglena*), are not effective competitors against bacteria.

Species of dominant primary bacteria depend chiefly on the nature of the organic waste and environmental conditions in aeration tanks. Conditions adverse to bacteria, such as acid pH, low dissolved oxygen, and nutrient shortage, can produce a predominance of filamentous fungi, resulting in sludge bulking. These abnormal circumstances are rare in municipal activated-sludge systems treating wastewater composed chiefly of domestic wastewater. However, when bulking in an activated-sludge system does occur, sanitary engineers must be prepared to find the cause and recommend corrective action.

Primary bacteria in an activated-sludge system are maintained in the declining or endogenous growth phases. Under these conditions, the primary bacteria die and lyse, releasing their cell contents to solution. In this process, raw organic matter is synthesized and resynthesized by various groups of bacteria.

Holozoic protozoans, which feed on living organic matter, are common in activated sludge. They grow in association with the bacteria in a prey–predator relationship; that is, the bacteria (plants) synthesize the organic matter, and the protozoans (animals) consume the bacteria (Fig. 12.11). For a single reproduction a protozoan consumes thousands of bacteria, with two major beneficial effects of the prey–predator action. First, removal of the bacteria stimulates further bacterial

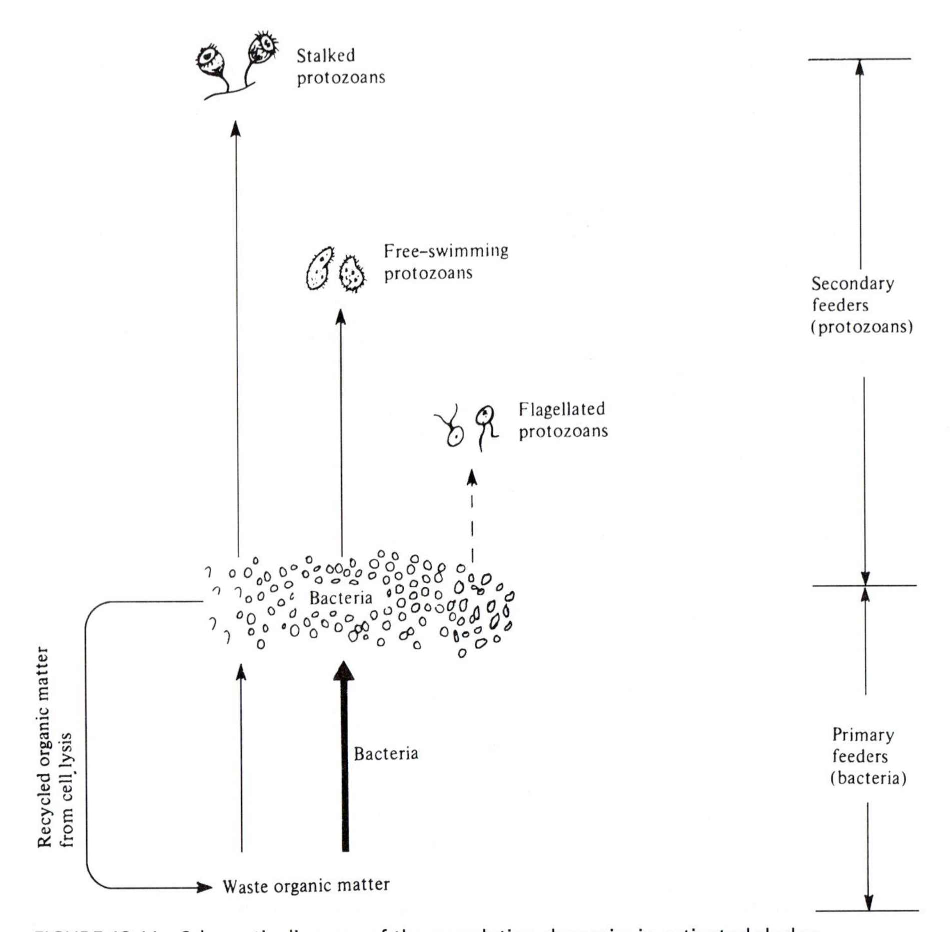

FIGURE 12.11 Schematic diagram of the population dynamics in activated sludge.

growth, resulting in accelerated extraction of organic matter from solution. Second, the flocculation characteristics of activated sludge are improved by reducing the number of free bacteria in solution, and a biological floc with improved settling characteristics results.

Competition for food also occurs between the secondary feeders. In a solution with high bacterial populations, free-swimming protozoans are dominant, but when food becomes scarce, stalked protozoans increase in numbers. Stalked protozoans do not require as much energy as free-swimming protozoans; therefore, they compete more effectively in a system with low bacterial concentrations. The photomicrographs shown in Figure 12.12 illustrate the appearance of a healthy activated sludge.

The process of anaerobic digestion is carried out by a wide variety of bacteria, which can be categorized into two main groups, acid-forming bacteria and methane-forming bacteria. (Protozoans do not function in the digester's strict anaerobic environment.) The acid formers are facultative or anaerobic bacteria that metabolize organic matter, forming organic acids as an end product, along with carbon dioxide

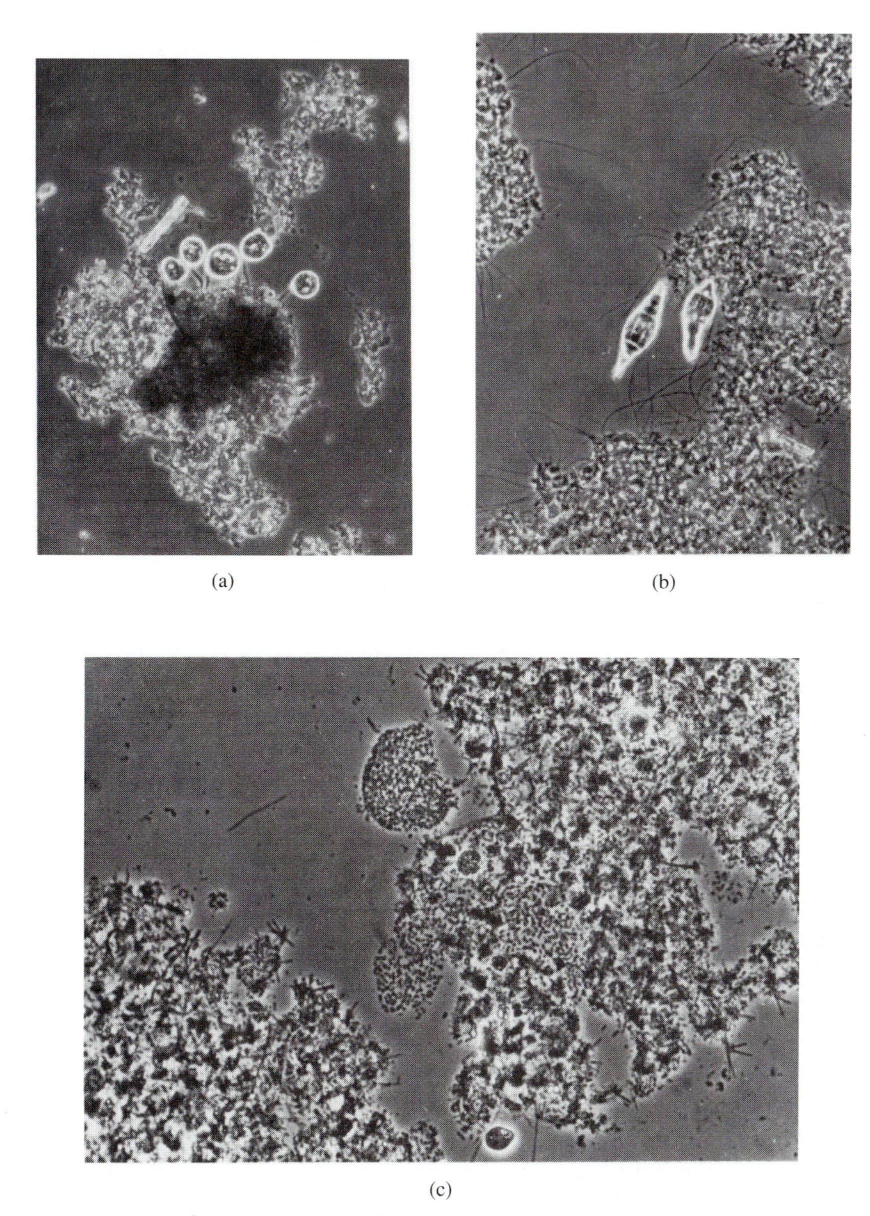

FIGURE 12.12 Photomicrographs of activated sludge. (a) Activated-sludge floc with stalked protozoans (100×). (b) Rotifers in activated sludge (100×). (c) Activated-sludge floc showing clusters of bacterial cells. Lower center: flagellated protozoan (400×).

and methane (associated with oxidation of fats to organic acids). Acid-splitting methane formers use organic acids as substrate and produce gaseous end products of carbon dioxide and methane. These methane bacteria are strict anaerobes inactivated by the presence of dissolved oxygen and inhibited by the presence of oxidized compounds. The growth medium must contain a reducing agent such as hydrogen

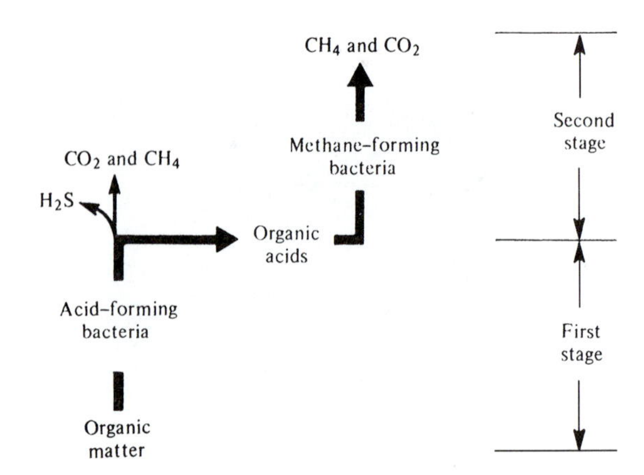

FIGURE 12.13 Simplified diagram of population dynamics in anaerobic digestion.

sulfide. Acid-splitting methane bacteria are sensitive to pH changes and other environmental conditions.

A simplified diagram (Fig. 12.13) portrays the relationship between the two bacterial stages in anaerobic digestion of organic matter. Both major groups of bacteria must cooperate to perform the overall gasification of organic matter. The first stage creates organic acids for the second stage, where these organic acids are converted to gas, preventing excess acid accumulation. In addition to producing food for the methane bacteria, acid formers also reduce the environment to one of strict anaerobiosis by using the oxidized compounds and excreting reducing agents.

Problems in operating anaerobic treatment systems result when an imbalance occurs in the population dynamics. For example, if a sudden excess of organic matter is fed to a digester, acid formers very rapidly process this food, developing excess organic acids. The methane formers, whose population had been limited by a previous lower organic acid supply, are unable to metabolize the organic acids fast enough to prevent a drop in pH. When the pH drops, the methane bacteria are affected first, further reducing their capacity to break down the acids. Under severe or prolonged overloading, the contents of the digester "pickles" in excess acids, and all bacterial activity is inhibited. In addition to organic overloading, the digestion process can be upset by a sudden increase in temperature, a significant shift in the type of substrate, or additions of toxic or inhibiting substances from industrial wastes.

A unique relationship exists between bacteria and algae in wastewater stabilization ponds (Fig. 12.14). The bacteria metabolize organic matter for reproduction, releasing soluble nitrogen and phosphorus nutrients and carbon dioxide. Algae use these inorganic compounds, along with the energy from sunlight, for synthesis, releasing oxygen. The resulting dissolved oxygen in the pond water is taken up by the bacteria, thus closing the cycle. This type of association between organisms is referred to as *symbiosis*, a relationship where two or more species live together for mutual benefit such that the association stimulates more vigorous growth of each species than if the growths were separate. The growth of algae replaces in part the organic matter decomposed by the bacteria. For this reason, ponds do not always appear to provide satisfactory removal of organic matter. A variety of predators (protozoans, rotifers, and

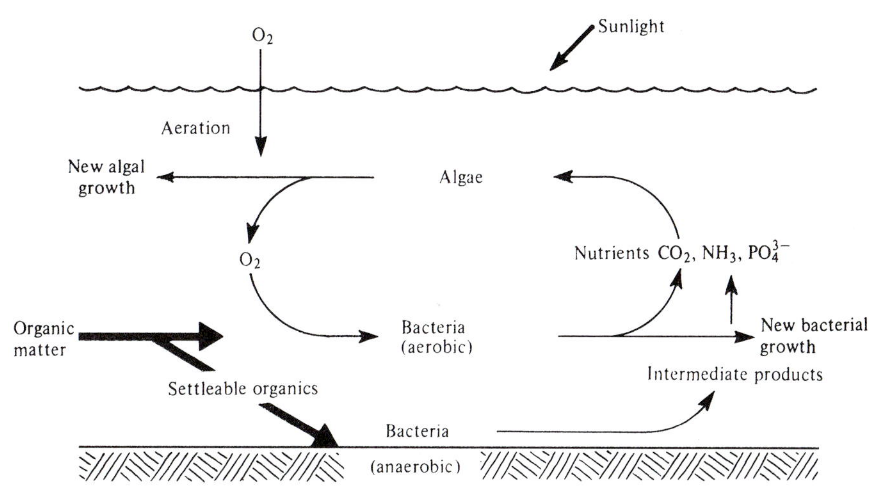

FIGURE 12.14 Schematic diagram of microbiological activity in a wastewater stabilization pond showing the symbiotic relationship between bacteria and algae and decomposition of organic matter by both aerobic and anaerobic bacteria.

higher animals) that feed on the plant growth (algae and bacteria) are also present in pond water.

At the liquid depth commonly used in stabilization-pond design, bottom waters may become anaerobic while the surface remains aerobic. In terms of general oxygen conditions, these lagoons are commonly referred to as *facultative stabilization ponds* (Fig. 12.14). During periods when the dissolved oxygen is at less than saturation level, the surface water is aerated through wind action. During the winter, both bacterial metabolism and algal synthesis are slowed by cold temperatures. The lagoon generally remains aerobic, even under a transparent ice cover. If the sunlight is blocked by a snow cover, the algae cannot produce oxygen, and the lagoon becomes anaerobic. The result is odorous conditions during the spring thaw until the algae becomes reestablished. This may take from a few days to weeks, depending on climatic conditions and the amount of organic matter accumulated in the lagoon during the winter.

CHARACTERISTICS OF WASTEWATER

Wastewater is defined as liquid wastes collected in a sewer system and conveyed to a treatment plant for processing. In most communities, storm runoff water is collected in a separate storm sewer system and conveyed to the nearest watercourse for disposal without treatment. A few large cities have a combined wastewater collection system where both stormwater and sanitary wastes are collected in the same pipe system. The dry-weather flow in the combined sewers is collected for treatment, but during storms the wastewater flow in excess of plant capacity may be stored for treatment later or, in some situations, is bypassed directly to the receiving watercourse.

Sanitary or domestic wastewater refers to liquid material collected from residences, business buildings, and institutions. The term *industrial wastes* refers to that from manufacturing plants. *Municipal wastewater* is a general term applied to liquid treated in a municipal treatment plant. Municipal wastewaters from towns frequently contain industrial effluents from dairies, laundries, bakeries, and factories, and those from large cities may have wastes from major industries, such as chemical manufacturing, breweries, meat processing, and metal processing.

12.10 FLOW AND STRENGTH VARIATIONS

The quantity of municipal wastewater flow varies from 50 to over 250 gal per capita per day (gpcd), depending on sewer uses of the community. A common value for sanitary flow is 120 gpcd (450 l/d). Per capita contribution of organic matter in domestic wastewater is approximately 0.24 lb (109 g) of suspended solids per day and 0.20 lb (91 g) of BOD per day in communities where a substantial portion of the household kitchen wastes are discharged to the sewer system through garbage grinders. These values are the population equivalents used to convert total pounds of solids or BOD of industrial wastewaters to equivalent population.

The actual quantities of flow and organic matter can differ significantly from these common values. The frequency distributions for per capita contributions of flow, suspended solids, and BOD based on analyses of data from 100 cities in Illinois, Indiana, Ohio, and Minnesota are graphed in Figure 12.15 [6]. The data show wide variations. Larger cities have greater waste contributions per capita. For cities with populations of greater than 100,000, mean flow was 194 gpcd, mean suspended solids were 0.40 lb/capita/day, and BOD was 0.30 lb/capita/day. For cities with populations of less than 10,000, the values were 140 gpcd, 0.15 lb of suspended solids, and 0.14 lb of BOD. Geographic location and climate influence waste contributions, particularly flow as related to the amount of infiltration and inflow. From a study of 700 Texas communities having populations of less than 10,000, the average contributions per capita per day were 89 gal, 0.21 lb of suspended solids, and 0.16 lb of BOD [7].

The flows and loads entering a wastewater treatment plant during the year vary with climatic conditions, industrial production, domestic water use, and other factors. In a typical industrial community in a moderate climate, the summer average daily flow frequently exceeds winter flows by 20%–30%. Munksgaard and Young [8] analyzed operational data from 11 municipal plants in the range of 0.25–40 mgd to determine peak flows and loads. All the cities had separate sanitary sewers receiving less than 20% of the wastewater from industrial sources. The wastewater flows included normal infiltration expected in humid climatic areas with an annual precipitation in the range of 20–40 in. The peaking ratio equations derived for the average annual peak month and average annual peak day are

$$Q_m = Q\left(\frac{1.26}{Q^{0.0101}}\right) \tag{12.36}$$

$$B_m = B\left(\frac{1.91}{B^{0.0430}}\right) \tag{12.37}$$

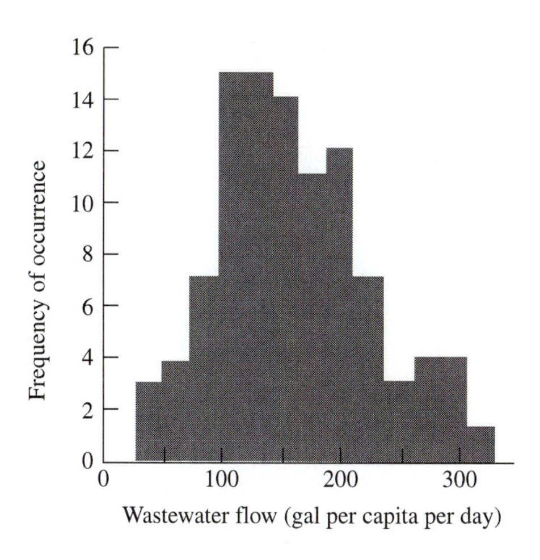

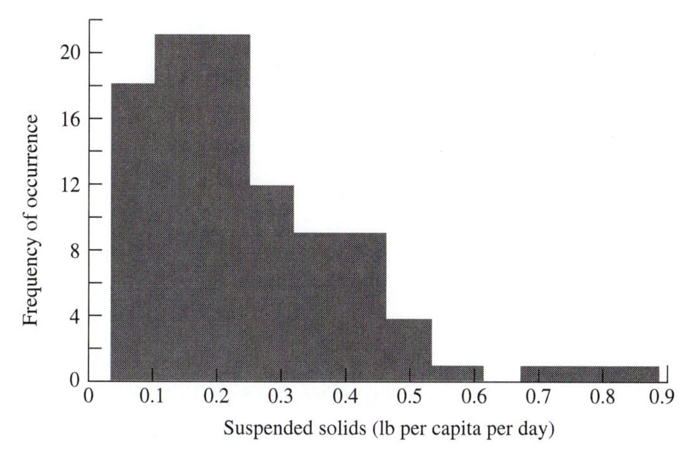

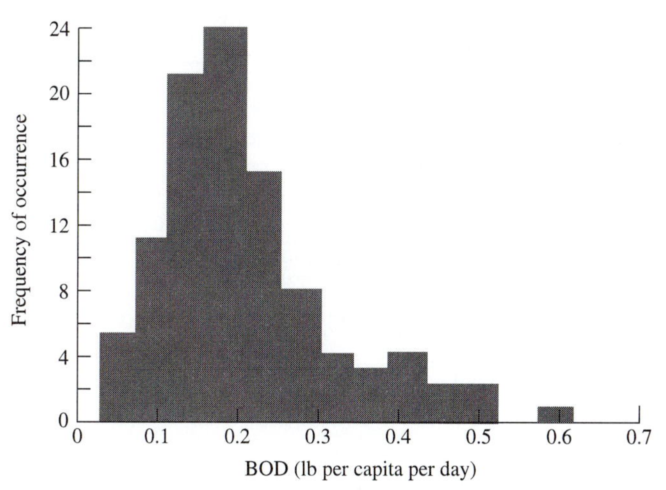

FIGURE 12.15 Histograms of average wastewater flow and quantities of suspended solids and BOD contributed per capita based on data from 100 cities in illinois, Indiana, Ohio, and Minnesota. *Source:* From D. H. Stoltenberg, "Midwestern Wastewater Characteristics," *Public Works* 111, no. 1 (1980): 52–53.

$$S_m = S\left(\frac{2.18}{S^{0.0517}}\right) \tag{12.38}$$

$$Q_d = Q\left(\frac{1.96}{Q^{0.0360}}\right) \tag{12.39}$$

$$B_d = B\left(\frac{4.08}{B^{0.0732}}\right) \tag{12.40}$$

$$S_d = S\left(\frac{5.98}{S^{0.0716}}\right) \tag{12.41}$$

where

Q = average annual wastewater flow, mgd

B = average annual BOD load, lb/day

S = average annual suspended-solids load, lb/day

Q_m, B_m, S_m = average flow, BOD, and suspended-solids values during the peak month

Q_d, B_d, S_d = average flow, BOD, and suspended-solids values for the peak day

Diurnal flow variations depend primarily on the size of a municipality and industrial flows. Hourly flow rates range from a minimum of 20% to a maximum of 250% or more of the average daily rate for small communities and from 50% to 200% for larger cities. The wastewater flow diagrams in Figure 12.16 exemplify hourly flow variations for two municipalities of different sizes.

Example 12.1

The sanitary sewer system in a municipality located in a humid climate receives over three-quarters of the wastewater discharges from domestic and commercial sources. The average annual wastewater flow is 10.0 mgd, containing 16,700 lb of BOD and 20,000 lb of suspended solids. Calculate the average daily wastewater flow, BOD load, and suspended-solids load during the peak month of the year.

Solution: Applying Eqs. (12.36) through (12.38),

$$Q_m = 10\left(\frac{1.26}{10^{0.0101}}\right) = 12.3 \text{ mgd}$$

$$B_m = 16{,}700\left(\frac{1.91}{16{,}700^{0.0430}}\right) = 21{,}000 \text{ lb/day}$$

$$S_m = 20{,}000\left(\frac{2.18}{20{,}000^{0.0517}}\right) = 26{,}100 \text{ lb/day}$$

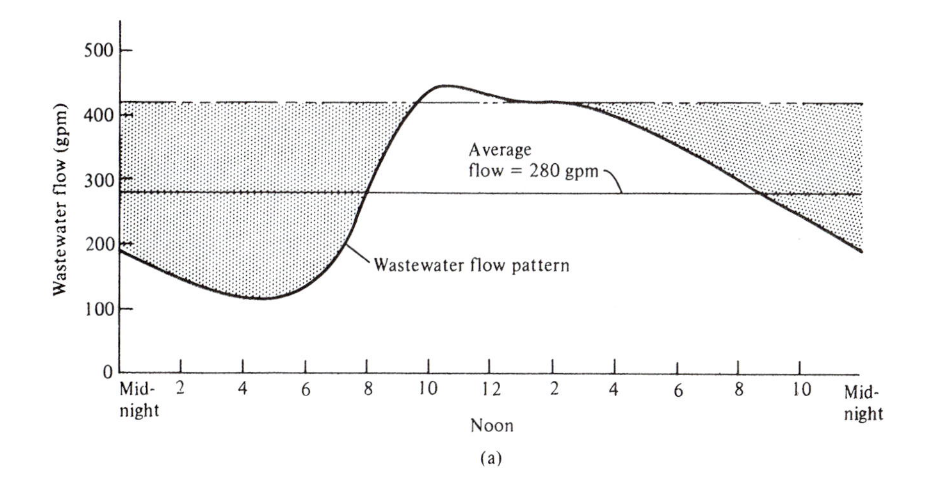

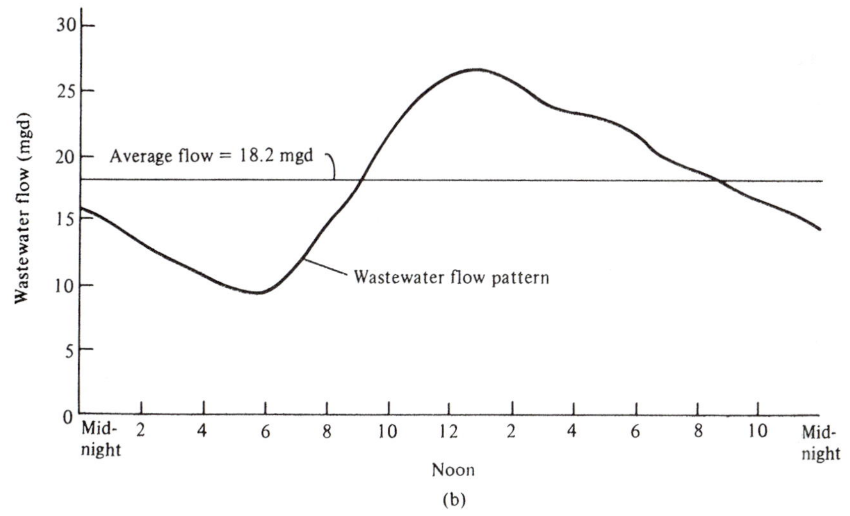

FIGURE 12.16 Diagrams of municipal wastewater flow showing hourly variations. (a) Flow diagram from a town with a population of 4500. (Shaded area is the typical recirculation flow for a high-rate trickling-filter plant at R = 0.5.) (b) Flow diagram from a city with a population of 150,000.

12.11 COMPOSITION OF WASTEWATER

The data in Table 12.1 represent the approximate composition of domestic wastewater before and after treatment. BOD and suspended solids (nonfiltrable residue) are the two most important parameters used to define the characteristics of a domestic wastewater. A suspended-solids concentration of 240 mg/l is equivalent to 0.24 lb of suspended solids in 120 gal, and 200 mg/l of BOD is equivalent to 0.20 lb of BOD in 120 gal. Reduction of suspended solids and BOD in primary sedimentation is approximately 50% and 35%, respectively. Approximately 70% of the suspended solids are volatile, defined as those lost upon ignition at 550°C.

TABLE 12.1 Approximate Composition of an Average Domestic Wastewater (mg/l)

	Before Sedimentation	After Sedimentation	Biologically Treated
Total solids	800	680	530
Total volatile solids	440	340	220
Suspended solids	240	120	30
Volatile suspended solids	180	100	20
BOD	200	130	30
Ammonia nitrogen as N	22	22	24
Total nitrogen as N	35	30	26
Soluble phosphorus as P	4	4	4
Total phosphorus as P	7	6	5

Total solids (residue on evaporation) include organic matter and dissolved salts; the concentration of the latter is dependent to a considerable extent on the hardness of the municipal water. The concentration of nitrogen in domestic waste is directly related to the concentration of organic matter (BOD). Approximately 60% of the total nitrogen is in solution as ammonia. If raw wastewater has been retained for a long time in collector sewers, a greater percentage of ammonia nitrogen results from deamination of the proteins and urea in wastewater. Seven milligrams per liter of phosphorus is approximately equivalent to a 2-lb phosphorus contribution per capita per year.

The surplus of nitrogen and phosphorus in biologically treated wastewater reveals that domestic wastewater contains nutrients in excess of biological needs. The approximate BOD/nitrogen/phosphorus (BOD/N/P) weight ratio required for biological treatment is 100/5/1. The exact BOD/N/P ratio needed for treatment depends on the process and the biological availability of the nitrogen and phosphorus compounds in the wastewater. A minimum of 100/6/1.5 is commonly related to treatment of unsettled sanitary wastewater, while 100/3/0.7 is generally adequate for wastewater where the nitrogen and phosphorus are in soluble forms. The average domestic wastewater listed in Table 12.1 has a ratio of 100/17/3 before sedimentation and 100/23/5 after sedimentation, both of which are in excess of the minimum 100/6/1.5. For biological treatment of industrial wastewaters deficient in nutrients, soluble phosphorus can be supplied by adding H_3PO_4 and soluble nitrogen by adding NH_4NO_3.

Biodegradable organic matter in wastewater is generally classified in three categories: carbohydrates, proteins, and fats. *Carbohydrates* are hydrates of carbon with the empirical formula $C_nH_{2n}O_n$ or $C_n\,(H_2O)_n$. The simplest carbohydrate unit is known as a *monosaccharide*, although few monosaccharides occur naturally. Glucose is a common monosaccharide in the structure of polysaccharides. *Disaccharides* are composed of two monosaccharide units. Sucrose (table sugar) is glucose plus fructose. Common milk sugar is lactose, consisting of glucose plus galactose. *Polysaccharides* are long chains of monosaccharides, such as cellulose, starch, and glycogen. Cellulose is the common polysaccharide in wood, cotton, paper, and plant tissues. Starches are primary nutrient polysaccharides for plant growth and are abundant in potatoes, rice, wheat, corn, and other plant forms.

Proteins in simple form are long-chain molecules composed of amino acids connected by peptide bonds and are important in both the structural (e.g., muscle tissue) and dynamic aspects (e.g., enzymes) of living matter. Twenty-one common amino acids when linked together in long peptide chains form a majority of simple proteins found in nature. A mixture of proteins as a bacterial substrate is an excellent growth medium, since proteins contain all the essential nutrients. On the other hand, pure carbohydrates are unsuitable as a growth medium since they do not contain the nitrogen and phosphorus essential for synthesis.

Lipids, together with carbohydrates and proteins, form the bulk of organic matter of living cells. The term refers to a heterogeneous collection of biochemical substances having the mutual property of being soluble to varying degrees in organic solvents (e.g., ether, ethanol, hexane, and acetone) while being only sparingly soluble in water. Lipids may be grouped according to their shared chemical and physical properties as fats, oils, and waxes. A simple fat, when broken down by hydrolytic action, yields fatty acids. In sanitary engineering, the word *fats* in current usage apparently conveys the meaning of lipids. The term *grease* applies to a wide variety of organic substances in the lipid category that are extracted from aqueous solution or suspension by trichlorotrifluoroethane.

Actually, not all biodegradable organic matter can be classed into these three simple groupings. Many natural compounds have structures that are combinations of carbohydrates, proteins, and fats, such as lipoproteins and nucleoproteins.

Approximately 20%–40% of the organic matter in wastewater appears to be nonbiodegradable. Several organic compounds, although biodegradable in the sense that specific bacteria can break them down, must be considered by sanitary engineers as partially biodegradable because of time limitations in waste treatment processes. For example, lignin, a polymeric noncarbohydrate material associated with cellulose in wood fiber, is for all practical purposes nonbiodegradable. Cellulose itself is not readily available to the general population of domestic wastewater bacteria. Saturated hydrocarbons are a problem in treatment because of their physical properties and resistance to bacterial action. Alkyl benzene sulfanate (ABS synthetic detergent) is only sparingly biodegradable in wastewater treatment.

Example 12.2

Domestic wastewater contains 0.24 lb of suspended solids and 0.20 lb of BOD per 120 gal.

1. Using these values, calculate the suspended-solids and BOD concentrations in milligrams per liter.
2. What is the BOD equivalent population for an industry that discharges 0.10 mgd of wastewater with an average BOD of 450 mg/l? What is the hydraulic equivalent population of this wastewater?

Solution:

1.

$$\text{BOD} = \frac{0.20 \text{ lb}}{120 \text{ gal}} \times \frac{1{,}000{,}000 \text{ gal}}{8.34 \text{ lb}} = 200 \text{ mg/l}$$

$$\text{suspended solids} = \frac{0.24 \text{ lb}}{120 \text{ gal}} \times \frac{1{,}000{,}000 \text{ gal}}{8.34 \text{ lb}} = 240 \text{ mg/l}$$

2. $$\begin{pmatrix}\text{BOD equivalent}\\ \text{population}\end{pmatrix} = \frac{0.10 \text{ mil gal} \times 450 \text{ mg/l} \times 8.34 \text{ lb/mil gal}}{0.20 \text{ lb/person}} \frac{}{\text{mg/l}} = 1900$$

$$\begin{pmatrix}\text{hydraulic equivalent}\\ \text{population}\end{pmatrix} = \frac{100{,}000 \text{ gal/day}}{120 \text{ gal/person}} = 830$$

Example 12.3

Wastewater from soluble coffee manufacturing is treated jointly with domestic wastewater in an activated-sludge process without primary settling. The mixture is 40% coffee wastewater and 60% domestic wastewater by volume. The characteristics of the coffee wastewater are 840 mg/l of BOD, 6.0 mg/l of total nitrogen, and 2.0 mg/l of total phosphorus. The domestic wastewater characteristics are as listed in Table 12.1.

1. If the required BOD/N/P weight ratio is assumed to be 100/6.0/1.5 for activated-sludge processing, are the nitrogen and phosphorus concentrations in the combined wastewater adequate?
2. If the nutrient content is not adequate, how many milligrams per liter of pure NH_4NO_3 and H_3PO_4 must be added to the wastewater?

Solution:

1. $$\text{BOD} = 0.40 \times 840 + 0.60 \times 200 = 456 \text{ mg/l}$$

$$\text{N} = 0.40 \times 6.0 + 0.60 \times 35 = 23.4 \text{ mg/l}$$

$$\text{P} = 0.40 \times 2.0 + 0.60 \times 7.0 = 5.0 \text{ mg/l}$$

	Available	Required	Need
BOD	456/4.56 = 100	100	—
N	23.4/4.56 = 5.1	6.0	0.9
P	5.0/4.56 = 1.1	1.5	0.4

2. $$NH_4NO_3 = 0.9 \times \frac{456}{100} \times \frac{80}{28} = 12 \text{ mg/l}$$

$$H_3PO_4 = 0.4 \times \frac{456}{100} \times \frac{98}{31} = 5.8 \text{ mg/l}$$

TRICKLING (BIOLOGICAL) FILTERS

Trickling filters are fixed-growth biological beds where wastewater is spread on the surface of media supporting microbial growths. Although the term *filter* is the accepted designation for this unit, no physical filtration occurs; contaminants are removed by biological action. The media placed in a tank under the wastewater distributor can be crushed rock, plastic-sheet packing formed into modules, or random plastic packing of various shapes. Trickling filters are preceded by primary clarifiers to remove settleable solids and are followed by final clarifiers to collect microbiological growths that slough from the media. The primary reasons for the popularity of trickling filters are their simplicity, low operating cost, and production of a waste sludge that is easy to process.

12.12 BIOLOGICAL PROCESS IN TRICKLING FILTRATION

The biological slime layers on filter media consist of bacteria, protozoans, and fungi. Frequently, noticeable populations of larger organisms such as sludge worms and rotifers are present. The top surface of a bed exposed to sunlight is coated with algae, and the lower portion of a deep filter can support nitrifying bacteria. Microorganisms near the surface of the bed where the food concentration is high are in a rapid growth phase, while the microorganisms near the bottom are in a state of starvation.

Figure 12.17 is a schematic diagram illustrating the general biological process at the surface of the medium. Although classified as aerobic treatment, the microbial film is aerobic to a depth of only 0.1–0.2 mm. The zone adjacent to the medium is anaerobic. As the wastewater flows over the microbial film, the soluble organic matter is metabolized and the colloidal organic matter is adsorbed on the film. Dissolved oxygen taken up in the liquid layer is replenished by reoxygenation from the surrounding air. Continuous passage of air through the bed is essential to prevent undesirable anaerobic

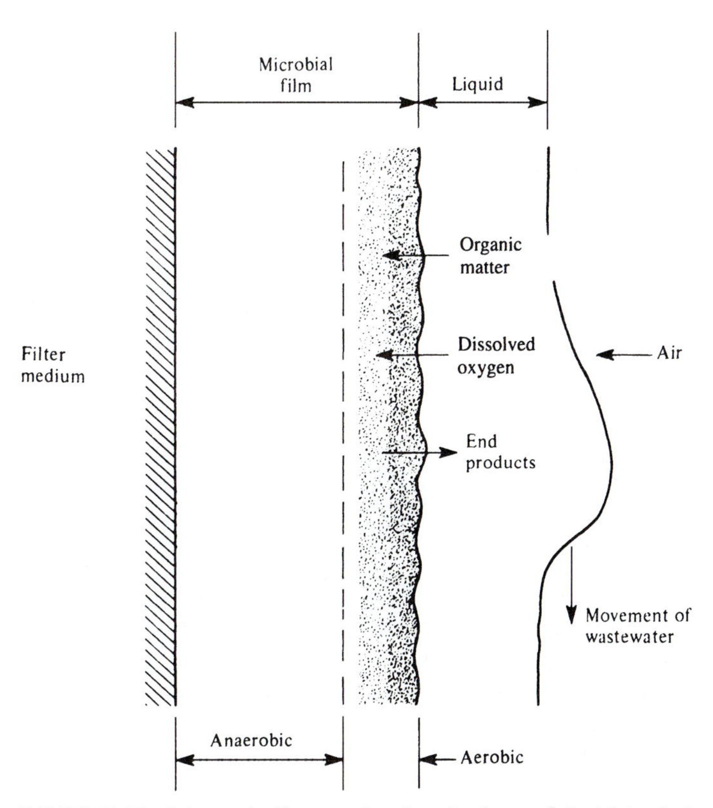

FIGURE 12.17 Schematic diagram showing the form of the biological process in a trickling filter.

conditions. The actual biological process is more complex than this simplified description and is poorly understood. Physical characteristics, such as media configuration, bed depth, and hydraulic loading, strongly influence the process. Several forms of manufactured media provide a high specific surface area with a corresponding large percentage of void volume. This permits substantial biological growth without inhibiting passage of air through the bed. A uniform medium also allows even loading distribution, and lightweight packing permits the construction of deep beds.

Problems with the biological process that can occur include poor effluent quality and emission of offensive odors, both associated with organic loading, industrial wastes, and cold-weather operation. The anaerobic zone of the film adjacent to the medium surface can produce odorous metabolic end products. If they are not oxidized while moving through the aerobic zone, they can be released and carried out in the airflow through the bed. Some industrial wastes have characteristic odors that are not easily oxidized by the microbial film. Cooling of the applied wastewater reduces the biological activity, allowing more organic matter to pass through the bed. In northern climates, the walls of deep filters can be insulated and covers can be placed over the tops to reduce cooling. Filter flies, *Psychoda*, are a nuisance around filters during warm weather. They breed on the inside of the retaining walls and on the surface of the media around the margin of the bed.

12.13 TRICKLING-FILTER OPERATION AND FILTER MEDIA REQUIREMENTS

A cutaway view of a shallow trickling filter is shown in Figure 12.18. The major components are a rotary distributor, underdrain system, and filter medium. The distributor spreads the wastewater at a uniform hydraulic load per unit area on the surface of the bed. The arms are driven by the reaction of wastewater flowing out of the distributor nozzles, which usually requires a pressure head of 24 in. measured at the center column that supports the arms. The underdrain system, often vitrified clay blocks with entrance holes to drainage channels, carries away the effluent and permits circulation of air

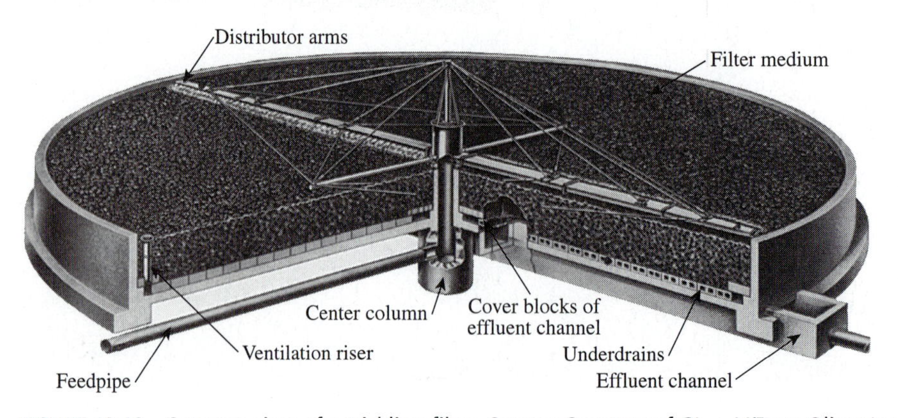

FIGURE 12.18 Cutaway view of a trickling filter. *Source:* Courtesy of GL + V/Dorr-Oliver Inc.

through the bed. The need for free passage of air controls the size of openings in the underdrain. The filter media provide a surface for biological growth and voids for passage of air and water.

Traditionally, the common filter media have been crushed rock, slag, or field stone, since these are durable, insoluble, resistant to spalling, and often locally available. Stones, however, have the disadvantage of occupying the majority of the volume in a filter bed, thus reducing the void spaces for passage of air and limiting the surface area per unit volume for biological growth. Several forms of chemical-resistant plastic media are available that have much greater surface area per unit volume and a large percentage of free space. For shallow filters, the common types are random packing and high-density cross-flow media manufactured in handleable modules that can be cut and fitted into a circular tank. The common random packing is small (2–4 in.) cylinders with perforated walls and internal ribs made of plastic, as shown in Figure 12.19. The specific surface area is 30–40 ft^2/ft^3 (100–130 m^2/m^3) with a void space of 91%–94%. The packing is placed by dumping it into a filter tank on top of the underdrains. Because of the random placement, the bed is efficient in distributing the applied wastewater to the media surfaces.

Biological towers are trickling filters with deep beds of 10–20 ft, usually in circular tanks with rotary distributors. The modules of packing with differing internal configurations are manufactured of polyvinyl chloride (PVC) in bundles 2 ft wide, 4 ft long, and 2 ft high. The module in Figure 12.20(a) is constructed of corrugated sheets bonded between flat sheets. By preventing clear vertical openings, the wastewater passing down through the packing is distributed over the surfaces of the media. The specific surface is 26–43 ft^2/ft^3 (85–140 m^2/m^3), varying with manufacturer; the void space is about 95%. Cross-flow packing in Figure 12.20(b) is made of ridged corrugated sheets with ridges on adjacent sheets at 45° or 60° angles to each other and bonded

FIGURE 12.19 Random packing for both shallow and deep trickling filters. Each element is a plastic cylinder (3.5 × 3.5 in) with perforated walls and internal ribs.

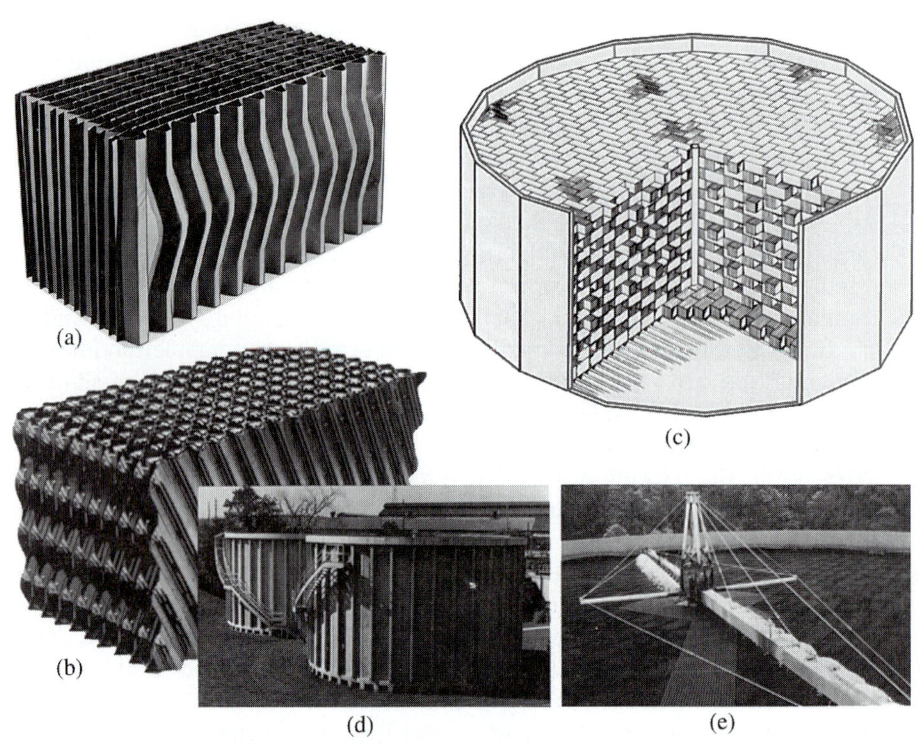

FIGURE 12.20 Biological tower media and construction. (a) Vertical-flow module with corrugated sheets bonded between flat sheets. (b) Cross-flow module with corrugated sheets assembled with adjacent sheets in a crossed pattern. (c) Cutaway view showing the construction of a circular biological tower. (d) Picture of two biological towers. (e) Distributor arm spreading wastewater on surface of tower media. *Source:* Courtesy of TLB Corporation, Newington, CT.

together where the ridges contact. As wastewater flows down through the media, each contact point permits flow splitting and combining. Cross-flow wets the surfaces completely and slows the flow rate, resulting in an increased hydraulic residence time in the bed. The specific surface of low-density, cross-flow media is 27 ft^2/ft^3 (90 m^2/m^3), and high-density is 42 ft^2/ft^3 (140 m^2/m^3).

The modules have sufficient strength to support the packing with attached wet biological growth in towers of 20 ft in height. The media bundles are stacked to interlock for structural stability and can be cut to fit the edge modules in a circular tower equipped with a rotary distributor [Fig. 12.20(c), (d), (e)].

12.14 TRICKLING-FILTER SECONDARY SYSTEMS

A trickling-filter secondary treatment system includes a final settling tank to remove biological growths that are washed off the filter media. These sloughed solids are commonly disposed of through a drain line from the bottom of the final clarifier to the head end of the plant. This return sludge flow is mixed with the raw wastewater and settled in the primary clarifier.

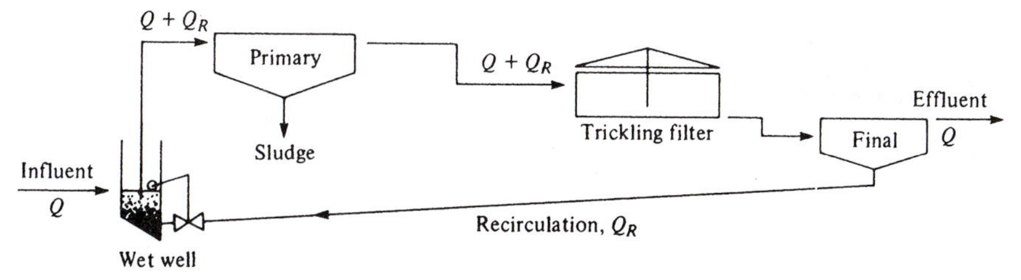

FIGURE 12.21 Profile of a typical single-stage trickling-filter plant with recirculation of underflow from the final clarifier to the head of the plant.

The raw wastewater in trickling filtration is diluted with recirculated flow so that it is reduced in strength and passes through the filter more than once. A typical filter design is illustrated in Figure 12.21. A return line from the final clarifier serves a dual function as a sludge return and a recirculation line. The combined flow ($Q + Q_R$) through the filter is always sufficient to maintain a minimum hydraulic loading on the media and sufficient flow to turn the distributor arms.

The two most common recirculation patterns used in trickling filter systems are show in Figure 12.22. The recirculation ratio is the ratio of recirculated flow to the quantity of raw wastewater. A common range for recirculation ratio values is 0.5–3.0. Recirculation may be done (1) only during periods of low wastewater flow, (2) at a rate proportional to raw-wastewater flow, (3) at a constant rate at all times, or (4) at two or more constant rates predetermined automatically or by manual control.

The gravity-flow-recirculation sludge-return system illustrated in Figures 12.21 and 12.22 is common for treatment of average-strength wastewater. The rate of recirculation flow is generally automatically regulated by the rate of raw-wastewater flow into the clear well. In this manner, the rate of return is increased during periods of low

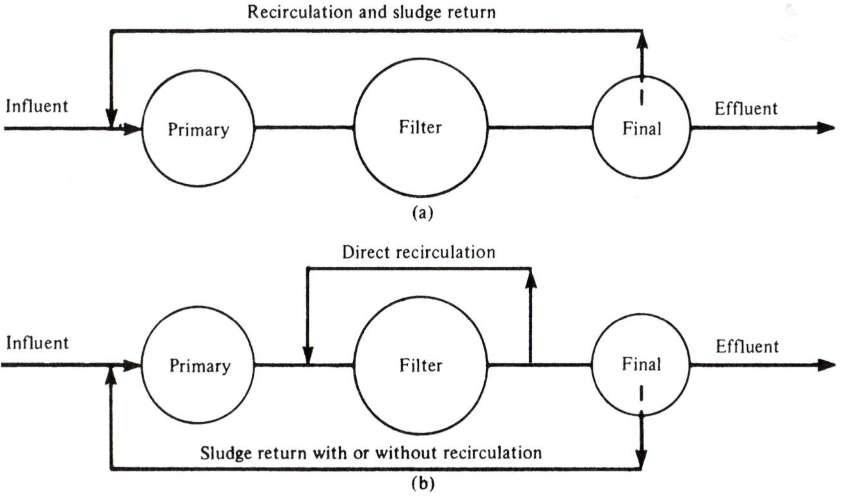

FIGURE 12.22 Typical recirculation patterns for single-stage trickling-filter plants. (a) Recirculation with sludge return. (b) Direct recirculation around the filter.

raw-wastewater flow and reduced, or even stopped, during high-flow periods. This kind of recirculation operation is shown graphically in Figure 12.16(a). The shaded area represents recirculated flow. The average raw-wastewater flow is 280 gpm. Flow through the plant is 420 gpm, equal to $Q + Q_R$, except from 10 A.M. to 3 P.M., when the raw-wastewater flow exceeds 420 gpm. The recirculation ratio for this illustrated flow recirculation pattern is 0.5.

Direct recirculation, depicted in Figure 12.22(b), is frequently used in the treatment of stronger wastewaters, where recirculation ratios of 2–3 are desirable. If high rates are used in pattern (a), the primary clarifier must be sized for the increased flow rate created by the greater volume. In other words, if recirculation flow is routed through a clarifier during the peak hourly flows of the raw wastewater, the clarifier must be increased in size to prevent disturbance of the settled solids and resultant loss of removal efficiency in the sedimentation tank. Consider the flow diagram in Figure 12.16(a), and assume that the shaded area is the flow in the sludge-return line shown in Figure 12.22(b). Then apply a direct recirculation of 420 gpm around the filter using constant-speed pumps. The resultant ratio for the trickling filter is 2.0:0.5 from indirect recirculation (from the final to the head of the primary) and 1.5 from direct recirculation.

Two-stage trickling-filter secondary systems have two filters in series, usually with an intermediate settling tank. Two typical flow diagrams for two-stage filter installations are sketched in Figure 12.23. Two-stage filters are used where a high-quality effluent is required, for treatment of strong wastewater, and to compensate for lower bacterial activity in treating cold wastewater.

Deep filters (biological towers) for treating municipal wastewater are generally single-stage units following primary sedimentation, although two stages are installed when processing strong wastewater resulting from industrial discharges. Direct recirculation is employed to maintain the desired flow through the tower. Sometimes a portion of the recirculation flow is drawn from the bottom of the final clarifier to develop

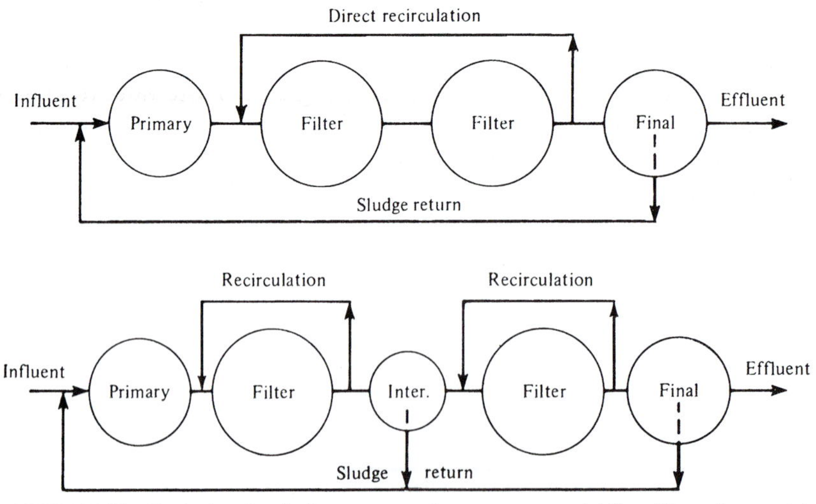

FIGURE 12.23 Typical recirculation patterns for two-stage trickling-filter plants without and with intermediate sedimentation.

a microbial floc in the wastewater circulating through the tower. Waste sludge accumulated in the final clarifier can be returned to the plant influent for settling in the primary clarifier.

12.15 EFFICIENCY EQUATIONS FOR STONE-MEDIA TRICKLING FILTERS

The BOD load on a trickling filter is calculated using the raw BOD in the primary effluent applied to the filter, without regard to the BOD in the recirculated flow. BOD loadings are expressed in terms of pounds of BOD applied per unit of volume per day. Common values are 20–40 lb/1000 ft^3/day (320–640 g/$m^3 \cdot$ d) for single-stage filters and 40–60 lb/1000 ft^3/day (640–960 g/$m^3 \cdot$ d) for two-stage filters based on the total media volume of both filters. The loadings apply to treatment of domestic wastewater (approximately 200 mg/l BOD) in the range of 15°–20°C (59°–68°F), with a minimum operating recirculation ratio of 0.5.

The hydraulic load is computed from the raw-wastewater flow plus recirculated flow. Hydraulic loadings are expressed in terms of average flow in gpm applied per square foot of surface area per day. Recirculation flow is required to maintain an open, well-aerated bed by preventing excessive accumulation of biological growth in the voids and impeding the passage of water and air. The minimum recommended hydraulic loading is 0.16 gpm/ft^2 (9.4 $m^3/m^2 \cdot$ d). The maximum recommended is 0.48 gpm/ft^2 (28 $m^3/m^2 \cdot$ d). Above this, the flushing action is excessive and contact time of the wastewater with the filter media becomes too short. The recirculation ratio for this range of hydraulic loadings is usually between 0.5 and 3.0.

Trickling filters have a bed depth of 5–7 ft (1.5–2.1 m) for most efficient BOD removal per unit volume of stone or slag media. In the early development of filters, they were constructed as deep as 8 ft and as shallow as 3 ft. Experience showed that two 3-ft-deep filters in series were no more efficient than one 6-ft-deep filter, and that the filter media below 6 ft in a bed did not result in significant increased BOD removal.

General practice in trickling-filter design has been to use empirical relationships to find the required filter volume for a desired degree of wastewater treatment. Several of these associations have been developed from operational data collected at existing treatment plants. One of the first evolved was the National Research Council (NRC) formula, based on data collected from filter plants at military installations in the United States in the early 1940s [9].

The NRC formula for a single-stage trickling filter is

$$E = \frac{100}{1 + 0.0561\,(w/VF)^{0.5}} \tag{12.42}$$

where

E = BOD removal at 20°C, %

w = BOD load applied, lb/day

V = volume of filter media, $ft^3 \times 10^{-3}$

F = recirculation factor

w/V = BOD loading, $lb/1000\ ft^3/day$

The recirculation factor is calculated from the formula

$$F = \frac{1 + R}{(1 + 0.1R)^2} \tag{12.43}$$

where R is the recirculation ratio (ratio of recirculation flow to raw wastewater flow).

The NRC formula for the second stage of a two-stage filter is

$$E_2 = \frac{100}{1 + [0.0561/(1 - E_1)](w_2/VF)^{0.5}} \tag{12.44}$$

where

E_2 = BOD removal of the second stage at 20°C, %

E_1 = fraction of BOD removed in the first stage

w_2 = BOD load applied to the second stage, lb/day

w_2/V = BOD loading, $lb/1000\ ft^3/day$

The effect of wastewater temperature on stone-filled trickling-filter efficiency may be expressed as

$$E = E_{20}\, 1.035^{T-20} \tag{12.45}$$

where

E = BOD removal efficiency at temperature T in °C

E_{20} = BOD removal efficiency at 20°C

The BOD removal efficiencies computed by the NRC formulas include final settling of the filter effluent. In the empirical development of these formulas, the field procedure used in collecting data sampled the filter influent and final clarifier effluent. Therefore, in evaluating the efficiency of a trickling-filter secondary, the overflow rate and detention time of the final clarifier should be examined for adequacy of design.

For a two-stage filter secondary without an intermediate settling tank (Fig. 12.23), the NRC formulas cannot be used to determine the efficiency of the first stage. In this case it is common to assume that the first-stage efficiency is 50% and find the efficiency of the second stage from Eq. (12.44).

Example 12.4

Calculate the BOD loading, hydraulic loading, BOD removal efficiency, and effluent BOD concentration of a single-stage trickling filter based on the following data:

$$\text{wastewater flow pattern} = \text{as shown in Figure 12.16(a)}$$

$$\text{recirculation rate} = \text{as shown in Figure 12.16(a)}$$

$$\text{settled wastewater BOD (primary effluent)} = 130 \text{ mg/l}$$

$$\text{diameter of filter} = 18.0 \text{ m}$$

$$\text{depth of media} = 2.1 \text{ m}$$

$$\text{wastewater temperature} = 18°\text{C}$$

Solution:

$$\text{raw-wastewater flow} = 280 \text{ gpm} = 1530 \text{ m}^3\text{/d}$$

$$\text{recirculation flow} = 0.50 \times 1530 = 765 \text{ m}^3\text{/d}$$

$$\text{BOD load} = 1530 \text{ m}^3\text{/d} \times 130 \text{ mg/l} \times \frac{\text{kg/m}^3}{1000 \text{ mg/l}} = 200 \text{ kg/d}$$

$$\text{surface area of filter} = \pi(18.0)^2/4 = 254 \text{ m}^2$$

$$\text{volume of media} = 254 \times 2.1 = 533 \text{ m}^3$$

$$\text{BOD loading} = \frac{200{,}000 \text{ g}}{533 \text{ m}^3} = 375 \text{ g/m}^3 \cdot \text{d} = 23.5 \text{ lb/1000 ft}^3\text{/day}$$

$$\text{hydraulic loading} = \frac{1530 \text{ m}^3\text{/d} + 765 \text{ m}^3\text{/d}}{254 \text{ m}^2} = 9.04 \text{ m}^3\text{/m}^2 \cdot \text{d} = 0.15 \text{ gpm/ft}^2 \cdot \text{d}$$

By Eqs. (12.43) and (12.42),

$$F = \frac{1 + 0.5}{(1 + 0.1 \times 0.5)^2} = 1.36$$

$$E_{20} = \frac{100}{1 + 0.0561(23.5/1.36)^{0.5}} = 81.1\%$$

Using Eq. (12.45),

$$E_{18} = 81.1 \times 1.035^{18-20} = 75.7\%$$

$$\text{Effluent BOD} = 130[(100 - 75.7)/100] = 32 \text{ mg/l}$$

Example 12.5

The design flow for a new two-stage trickling-filter plant is 1.2 mgd, with an average BOD concentration of 315 mg/l. Determine the dimensions of the sedimentation tanks

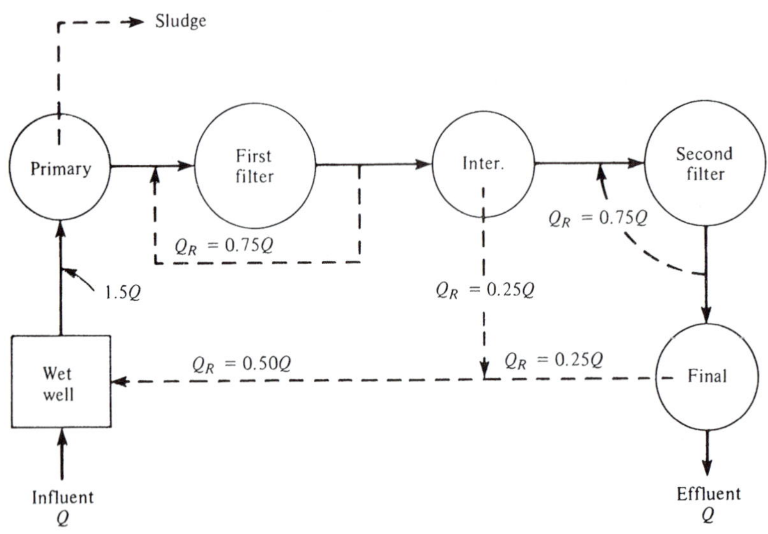

FIGURE 12.24 Flow scheme of the two-stage trickling-filter plant for Example 12.5.

and trickling filters (surface areas and depths) for the flow scheme shown in Figure 12.24. Calculate the volume of filter media based on a loading of 35 lb of BOD/1000 ft^3/day, and divide the resulting volume equally between the primary and secondary filters. Estimate the BOD concentration in the plant effluent at 20°C.

Solution:

Primary Tank

Criteria: (1) 500-gpd/ft^2 overflow rate based on raw Q or 750 gpd/ft^2 based on Q plus recirculation flow; (2) minimum depth of 7 ft. However, if accumulated sludge is to be retained in the bottom of the tank, increase the depth to accommodate the necessary sludge storage volume.

$$\text{area required} = \frac{1{,}200{,}000}{500} = 2400 \text{ ft}^2$$

or

$$\text{area required} = \frac{1.5 \times 1{,}200{,}000}{750} = 2400 \text{ ft}^2 \quad \text{(Use)}$$

Estimate the daily sludge accumulation at 4% solids, assuming a sludge solids accumulation equal to 90% of the BOD load:

$$\text{volume} = \frac{0.9 \times 450 \times 1.2 \times 8.34}{0.04 \times 62.4} = 1620 \text{ ft}^3$$

$$\text{depth of sludge} = \frac{1620}{2400} = 0.7 \text{ ft}$$

Provide a side-wall depth of 8 ft plus freeboard.

$$\text{primary BOD removal} = 35\%$$

Trickling Filters

Criteria: (1) 35 lb/1000 ft^3/day BOD loading; (2) 0.16–0.48 gpm/ft^2 hydraulic loading.

$$\text{volume required} = \frac{0.65 \times 315 \times 1.2 \times 8.34}{0.035} = 58{,}500 \text{ ft}^3$$

$$\text{volume of each filter} = 29{,}300 \text{ ft}^3$$

Try 6-ft depth:

$$\text{area} = \frac{29{,}300}{6.0} = 4880 \text{ ft}^2$$

Check the hydraulic loading:

$$\frac{(1.5 + 0.75)1{,}200{,}000}{4880 \times 1440} = 0.38 \text{ gpm/ft}^2 \quad \text{(OK)}$$

Use 6-ft-deep filters with a 4880-ft^2 area.

Intermediate Settling Tank

Criteria: (1) 1000-gpd/ft^2 overflow rate; (2) minimum depth of 7 ft.

$$\text{area required} = \frac{1.25 \times 1{,}200{,}000}{1000} = 1500 \text{ ft}^2$$

Use a side-wall depth of 7 ft plus freeboard.

Final Settling Tank

Criteria: (1) 800-gpd/ft^2 overflow rate; (2) minimum depth of 7 ft.

$$\text{area required} = \frac{1{,}200{,}000}{800} = 1500 \text{ ft}^2$$

Use a side-wall depth of 7 ft plus freeboard.

Calculation of BOD removal efficiency

$$\text{primary tank} = 35\%$$

First-stage filter:

$$\text{BOD loading} = \frac{0.65 \times 315 \times 1.2 \times 8.34}{29.3} = 70 \text{ lb/1000 ft}^3\text{/day}$$

$$R = \frac{0.50Q + 0.75Q}{Q} = 1.25$$

$$E_1 = \frac{100}{1 + 0.0561(70/1.25)^{0.5}} = 70\%$$

Second-stage filter:

$$\text{BOD loading} = 0.30 \times 70 = 21 \text{ lb/1000 ft}^3\text{/day}$$

$$R = \frac{0.25Q + 0.75Q}{Q} = 1.0$$

$$F = \frac{1 + 1.0}{(1 + 0.1 \times 1.0)^2} = 1.65$$

By Eq. (12.44),

$$E_2 = \frac{100}{1 + [0.0561/(1 - 0.70)](21/1.65)^{0.5}} = 60\%$$

The plant efficiency is

$$E = 100 - 100[(1 - 0.35)(1 - 0.70)(1 - 0.60)] = 92.2\%$$

The estimated effluent BOD is $0.078 \times 315 = 25$ mg/l.

12.16 EFFICIENCY EQUATIONS FOR PLASTIC-MEDIA TRICKLING FILTERS

The hydraulic profile for typical design of a biological tower is shown in Figure 12.25. Primary clarification is required to remove settleable and floating solids prior to filtration. Direct recirculation of the tower underflow is blended with the clarified raw wastewater to provide dilution and a greater flow through the media. In this way, the deep biological bed is more effectively used by vertical distribution of the BOD load throughout the depth of the media, and treatment is improved by passage of the wastewater through the filter more than once.

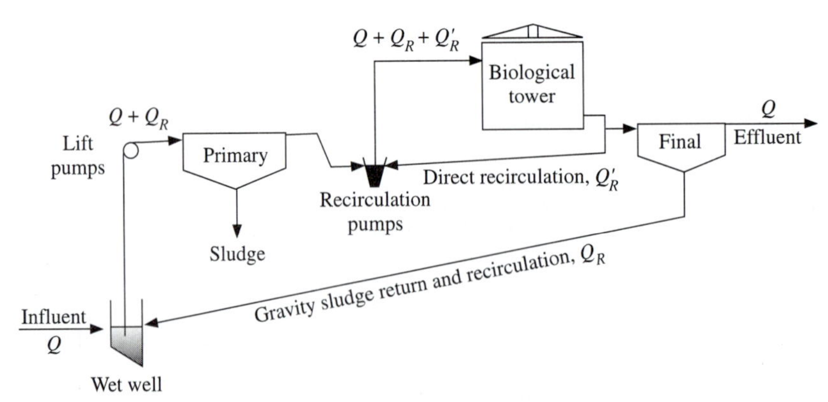

FIGURE 12.25 Profile of a typical biological tower with direct recirculation through tower and recirculation of underflow from the final clarifier to the inlet of the plant.

BOD loadings on biological towers with plastic media are usually 50 lb/ 1000 ft^3/day (800 $g/m^3 \cdot d$) or greater with surface hydraulic loadings of 1.0 gpm/ft^2 (60 $m^3/m^2 \cdot d$, 0.68 $l/m^2 \cdot s$) or greater. The design loading selected for treatment of a particular wastewater depends on BOD concentration, biodegradability, temperature, type and depth of media, and the ratio and pattern of wastewater circulation.

Plastic media, such as vertical-flow and cross-flow modules (Fig. 12.20) and random packing (Fig. 12.19), are manufactured with uniform geometric features of specific surface and shape. The objectives in design of these media are to provide a surface area that supports a continuous biological growth and distribution of the applied wastewater so that it flows uniformly over the biological growth. Based on this "definable" physical–biological system, several efficiency equations have been developed, modified, and refined in an attempt to find a suitable mathematical model. Nevertheless, no universal model exists to precisely describe substrate removal in a trickling filter. The following equations define a common model used in environmental engineering practice.

The residence time of applied wastewater in a filter bed is an important factor in BOD removal. The mean contact time between the wastewater and the biological film on the surface of the media can be related to filter depth, hydraulic loading, and the geometry of the media as

$$t = \frac{CD}{Q^n} \tag{12.46}$$

where

t = mean residence time, min

D = depth of media, ft (m)

Q = hydraulic loading, gpm/ft^2 ($m^3/m^2 \cdot h$)

C, n = constants related to specific surface and configuration of media

Exponent n and constant C are determined from pilot-plant studies using an experimental filter column packed with synthetic media to the depth of a full-scale unit. A minimum of three ports at different depths are installed to collect samples of the wastewater percolating down through the packing. Residence time is measured by adding a pulse tracer input, such as fluorescent dye, and observing the concentration of tracer in the effluent with respect to time. The mean residence time is when 50% of the tracer has passed out of the filter column. Since the thickness of the biological film affects the residence time, the experimental filter is acclimated to an applied wastewater feed for a sufficient period of time to establish dynamic equilibrium between the organic loading and microbial mass. Values of n and C are determined from a graph of the t and Q data plotted on logarithmic paper. From Eq. (12.46), t is proportional to Q^{-n}; therefore,

$$n = -\frac{\Delta \log t}{\Delta \log Q} \tag{12.47}$$

and $C = t/D$ when $Q = 1.0$.

The removal of soluble BOD in a plastic-media trickling filter, where the media supports a continuous biological growth and the wastewater is uniformly distributed, is as follows, based on first-order kinetics:

$$\frac{S_e}{S_0} = e^{-KD/Q^n} \tag{12.48}$$

where

S_e = soluble (filtered) BOD in effluent, mg/l

S_0 = soluble (filtered) BOD in influent, mg/l

e = 2.718 (the Napierian base)

K = reaction-rate constant, gal/min/ft^3

D = depth of media, ft

Q = hydraulic loading, gpm/ft^2

n = empirical flow constant

This equation can be rewritten to include the specific surface area of the media by substituting $k_{20}A_s$ for K:

$$\frac{S_e}{S_0} = e^{-k_{20}A_sD/Q^n} \tag{12.49}$$

where

S_e = soluble (filtered) BOD in effluent, mg/l

S_0 = soluble (filtered) BOD in influent, mg/l

k_{20} = reaction-rate coefficient at 20°C, (gpm/ft^2)$^{0.5}$ [(l/m$^2 \cdot$ s)$^{0.5}$]

A_s = specific surface area of media, ft^2/ft^3 (m^2/m^3)

D = depth of media, ft (m)

Q = hydraulic loading, gpm/ft^2 (l/m$^2 \cdot$ s)

n = empirical flow constant, normally selected as 0.5 for vertical-flow and cross-flow media

The reaction-rate coefficient is corrected for temperature by the relationship

$$k = k_{20}\Theta^{T-20} \tag{12.50}$$

where

k = reaction-rate coefficient at temperature T, °C

k_{20} = reaction-rate coefficient at 20°C

Θ = temperature coefficient, normally selected as 1.035

T = wastewater temperature, °C

Equation (12.49) can be rewritten and include Eq. (12.50) as follows:

$$\ln\frac{S_0}{S_e} = \frac{k_{20}A_sD\Theta^{T-20}}{Q_n} \tag{12.51}$$

This is a linear equation convenient for analysis of experimental data. The value of k_{20} is determined graphically by plotting ln S_0/S_e versus $A_sD\Theta^{T-20}/Q^n$ on arithmetic paper. The value of k is the slope of a straight line originating from the origin and drawn as a best fit through the plotted data.

The BOD concentrations of the applied wastewater before and after dilution with clear recirculation flow is

$$S_0 = \frac{S_p + RS_e}{1 + R} \tag{12.52}$$

where

S_0 = soluble BOD in influent after dilution with recirculated flow, mg/l

S_p = soluble BOD in primary effluent before dilution with recirculated flow, mg/l

S_e = soluble BOD in effluent, mg/l

R = recirculation ratio, recirculated flow/primary effluent flow, Q_R/Q

Combining Eqs. (12.49), (12.50), and (12.52):

$$\frac{S_e}{S_p} = \frac{e^{-k_{20}\Theta^{T-20}A_sD/(Q_p\,(1+R))^n}}{(1 + R) - Re^{-k_{20}\Theta^{T-20}A_sD/(Q_p\,(1+R))^n}} \tag{12.53}$$

where

Q_p = hydraulic loading of primary effluent without recirculation flow, gpm/ft^2

This mathematical model is a simplification of complex biological–physical interactions that occur in trickling filters. The values for the reaction-rate coefficient k_{20} vary with media of different configurations with the same specific surface areas. The coefficient also varies with wastewater treatability and depth of media. The approximate range for k_{20} values for vertical-flow media is 0.0008–0.0016 (gpm/ft^2)$^{0.5}$ [0.0010–0.0020 (l/m$^2\cdot$s)$^{0.5}$], and for cross-flow media it is 0.0014–0.0023 (gpm/ft^2)$^{0.5}$ [0.0017–0.0028 (l/m$^2\cdot$s)$^{0.5}$]. The k_{20} values for cross-flow media are greater than those for vertical-flow media, which is attributed to longer contact time and better wastewater flow distribution. Often manufacturers provide field data to verify reaction-rate coefficients for their media; however, where feasible, pilot studies using selected media are recommended to determine the k_{20} for design.

Example 12.6 illustrates the application of these equations to determine the values of the reaction-rate constant K_{20} and the empirical flow constant n for random packing in a pilot-plant study. Example 12.7 presents the evaluation of data from a field study to determine the k_{20} for high-density cross-flow media in a shallow trickling filter.

Example 12.6

Balakrishnan, Eckenfelder, and Brown [10] demonstrated the feasibility of determining the constants for Eqs. (12.46) and (12.48), using data from a pilot-plant study.

The trickling filter was a 20-in.-diameter, 9-ft-long fiberglass cylinder with an air sparger to provide a uniform distribution of air from the bottom and distribution plates on top for uniform hydraulic loading. The filter was filled to a depth of 8 ft with 1.5-in.-diameter cylindrical random packing with a specific surface of 40 ft^2/ft^3 and 96% void space. Eight sampling ports were located at 1-ft intervals from the top of the filter bed.

The feed was settled domestic wastewater, containing filtered (soluble) BOD concentrations in the range of 65–90 mg/l, applied at hydraulic loadings of 0.20, 0.30, and 0.43 gpm/ft^2. After acclimation at each loading rate, samples were collected at various depths in the filter for laboratory analysis. The wastewater samples were settled for 30 min and filtered through Whatman No. 42 filter paper prior to BOD testing.

Solution: Mean residence times of the wastewater in the filter were determined by measuring the time for concentrated doses of salt solution to flush through the column. The results, with and without the presence of biological slime growth, are graphed in Figure 12.26. The constant n for Eq. (12.46) is the negative slope of the line, and C is the intercept of the line at the vertical axis divided by the depth of media. With a biological growth coating the packing, the mean residence time was

$$t = \frac{11D}{Q^{0.045}}$$

The mean residence time of clean packing, without slime, was

$$t = \frac{1.9D}{Q^{0.43}}$$

A comparison of these equations shows the significant effect that slime growth has on the mean residence time.

Soluble BOD removal data are plotted in Figure 12.27 as the logarithm of BOD remaining, S_e/S_0, versus the depth D in the filter packing. From Eq. (12.48),

$$\log\left(\frac{S_e}{S_0}\right) = -\left(\frac{K}{2.3}\right)\left(\frac{D}{Q^n}\right) \tag{12.54}$$

Based on this equation, the slope of each line drawn through $\log(S_e/S_0)$ versus D data for a specific hydraulic loading is defined as

$$\text{slope} = \frac{\log(S_e/S_0)}{D} = -\left(\frac{K}{2.3}\right)Q^{-n} \tag{12.55}$$

The slopes in Figure 12.27(a) for hydraulic loadings of 0.20, 0.30, and 0.43 gpm/ft^2 are 0.059, 0.051, and 0.045, respectively.

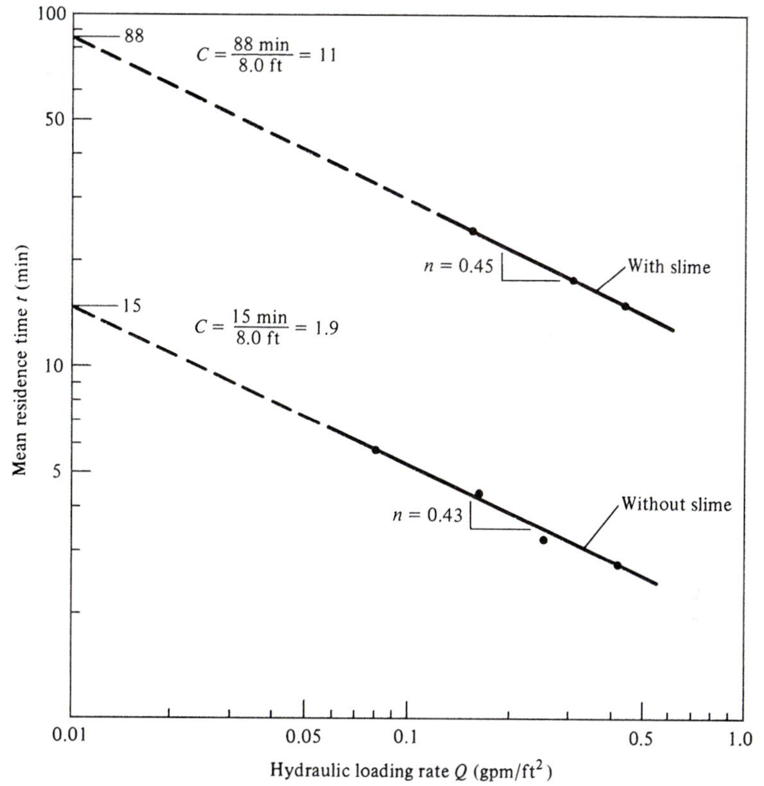

FIGURE 12.26 Relationship between hydraulic loading and mean residence time for Example 12.6. *Source:* Adapted from S. Balakrishnan, W. W. Eckenfelder, and C. Brown, "Organics Removal by a Selected Trickling Filter Media," *Water and Wastes Eng.* 6, no. 1 (1969): A.22–A.25.

Taking the logarithm of both sides of Eq. (12.55) yields

$$\log(\text{slope}) = n \log Q - \log(K/2.3) \tag{12.56}$$

which is the equation for a straight line of slope n. Therefore, the slope of a line drawn through log(slope) versus log Q data is the value of the constant n. From the logarithmic plot in Figure 12.27(b), $n = 0.39$ [based on the slopes and hydraulic loadings given in Fig. 12.27(a)].

Finally, the constant K is determined by plotting $\log(S_e/S_0)$ versus D/Q^n and determining the slope of the line drawn through the data, which from Eq. (12.54) equals $-(K/2.3)$. For the S_e/S_0 and $D/Q^{0.39}$ data plotted in Figure 12.27(c), the slope of the line is 0.0324, and $K = 2.3 \times 0.0324 = 0.0745\ \text{min}^{-1}$.

The average operating temperature of the pilot filter was 14°C. Correcting K to 20°C using Eq. (12.50), the reaction-rate constant at 20°C is

$$K_{20} = \frac{K}{(1.035)^{T-20}} = \frac{0.0745}{(1.035)^{14-20}} = 0.092\ \text{min}^{-1}$$

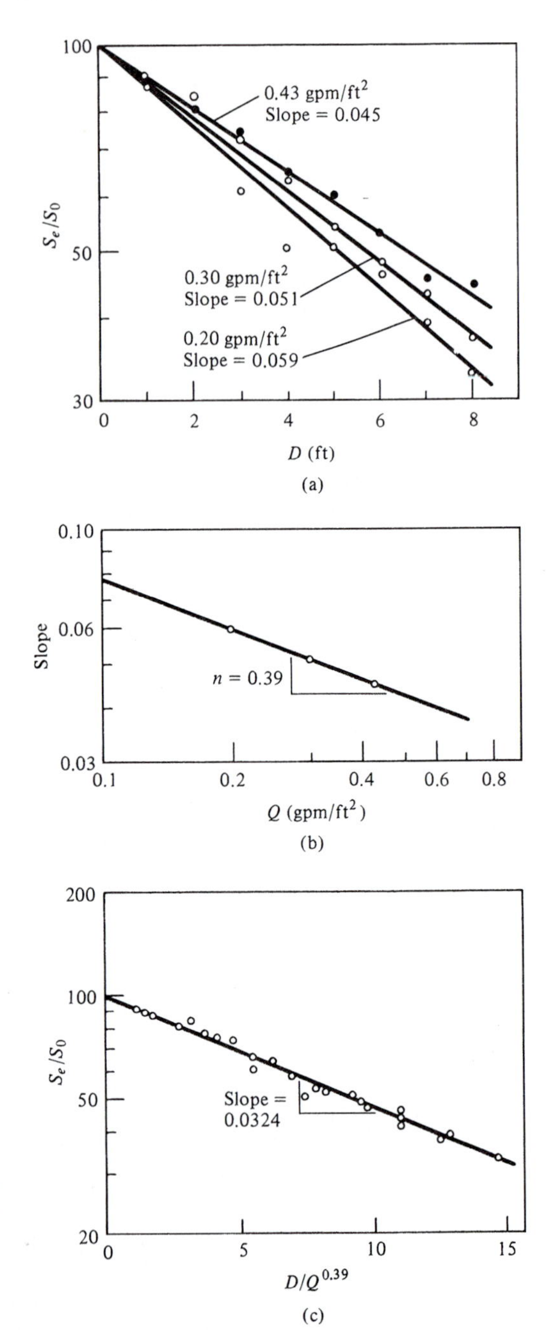

FIGURE 12.27 Plotted data for Example 12.6. (a) Relationship between filter depth and fraction of BOD remaining at various hydraulic loads. (b) Diagram for determination of *n*. (c) Diagram for determination of *K*. *Source:* Adapted from S. Balakrishnan, W. W. Eckenfelder, and C. Brown, "Organics Removal by a Selected Trickling Filter Media," *Water and Wastes Eng.* 6, no. 1 (1969): A.22–A.25.

Substituting values of $K = 0.092$ and $n = 0.39$ in Eq. (12.48), the soluble BOD removal from a settled domestic wastewater for the specific filter packing tested is

$$\frac{S_e}{S_0} = e^{-0.092D/Q^{0.39}} \tag{12.57}$$

Example 12.7

Drury, Carmona, and Delgadillo [11] describe the rehabilitation of an old shallow stone-media trickling filter by installing plastic media to solve the problem of ponding due to plugging of voids in the bed.

High-density, cross-flow media with a specific surface area of 138 m^2/m^3 (42 ft^2/ft^3) and a depth of only 1.02 m (3.33 ft) was installed in the trickling-filter tank, which was 21.3 m (70 ft) in diameter. Before undertaking costly rehabilitation of the final settling tank, a study was performed to evaluate performance of the media and to establish a suitable loading for operation. The trickling filter was operated at a constant flow rate throughout the day (7 A.M.–midnight), with the primary effluent diluted by trickling-filter effluent returned to the wet well; the recirculation ratio ranged from 0.8 to 1.1. During the 8-hr period from 8 A.M.–4 P.M., the influent and effluent were composited and refrigerated by automatic samplers at 15-min intervals. The effluent sample was allowed to settle for 1 hr in a laboratory container to simulate clarification in a final settling tank, and the supernatant was decanted for analysis. The samples were tested for total BOD and dissolved BOD after filtration through a glass-fiber filter. The test results for the winter evaluation period are given in Table 12.2.

Solution: For each set of Q, T, S_0, and S_e, calculate values for $\ln S_0/S_e$ and $A_sD\Theta^{T-20}/Q^n$. The following are sample calculations for the first set of data:

$$\ln\frac{S_0}{S_e} = \ln\frac{44}{18} = 0.89$$

$A_s = 138\ m^2/m^3$, $D = 1.02$ m, $\Theta = 1.035$, and $n = 0.5$:

$$\frac{A_sD\Theta^{T-20}}{Q^n} = \frac{138 \times 1.02(1.035)^{20.0-20}}{(0.49)^{0.5}} = 201$$

TABLE 12.2 Sampling Data and Calculated Values for Determination of Reaction-Rate Coefficient k_{20} in Example 12.7

Q ($l/m^2 \cdot s$)	T (°C)	Influent BOD (mg/l)	S_0 (mg/l)	Effluent BOD (mg/l)	S_e (mg/l)	$\text{LN}\frac{S_0}{S_e}$	$\frac{A_sD\theta^{T-20}}{Q^n}$ $[(l/m^2 \cdot s)^{-0.5}]$
0.49	20.0	75	44	32	18	0.89	201
0.54	20.5	50	29	23	13	0.80	195
0.49	20.0	73	38	27	14	1.00	201
0.49	18.5	53	22	22	9.5	0.84	191
0.59	18.5	64	34	32	17	0.69	174
0.49	19.0	45	30	24	12	0.92	195
0.31	23.0	80	33	16	9.2	1.28	280
0.35	21.5	66	37	18	12	1.13	251

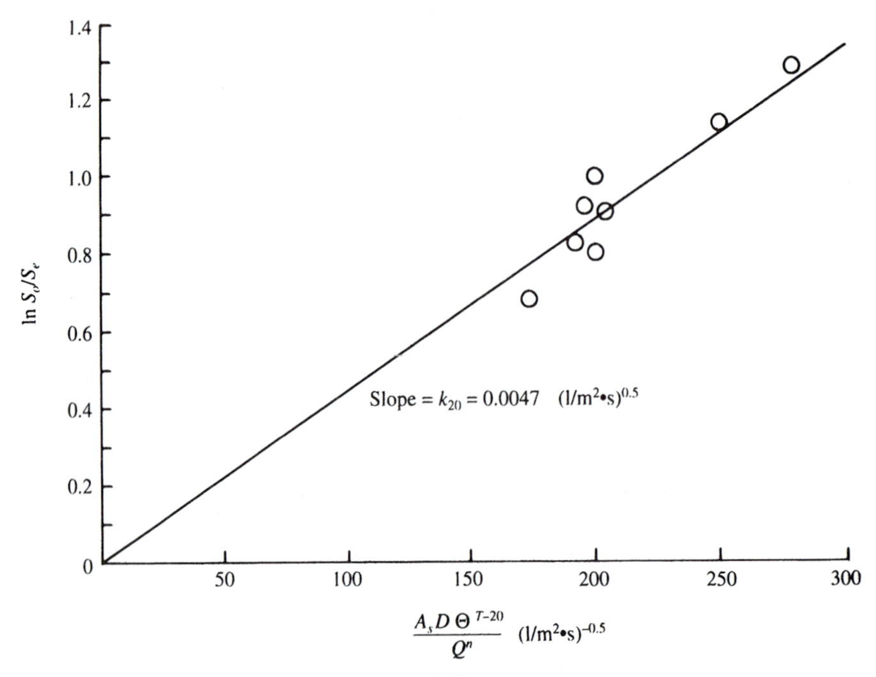

FIGURE 12.28 Plot of ln S_0/S_e versus $A_s D\Theta^{T-20}/Q^n$ to determine k_{20} for Example 12.7.

The calculated values for determining k_{20} are listed in Table 12.2 and plotted in Figure 12.28. The slope of the best-fit line through the plotted points is the reaction-rate coefficient k_{20} equal to 0.0047 $(l/m^2 \cdot s)^{0.5}$.

This field evaluation demonstrated that high-density cross-flow media can be effectively used in shallow filters at high BOD loadings. The k_{20} was significantly greater than expected, based on other studies conducted of towers with much deeper media. Although the reason for this is not known, the authors speculated that the high BOD removal efficiency of a shallow bed may be related to improved oxygen transfer.

An allowable loading on the filter used in the study with high-density cross-flow media can be calculated based on the following parameters:

$$\text{Depth of media} = 1.02 \text{ m}$$

$$\text{Specific surface area} = 138 \text{ m}^2/\text{m}^3$$

$$k_{20} = 0.0047\ (l/m^2 \cdot s)^{0.5}$$

$$\text{Wastewater temperature} = 10°\text{C}$$

$$\text{Influent BOD with recirculation flow} = 70 \text{ mg/l}$$

$$\text{Soluble influent BOD} = 35 \text{ mg/l}$$

$$\text{Allowable effluent BOD} = 25 \text{ mg/l}$$

$$\text{Soluble effluent BOD} = 12.5 \text{ mg/l}$$

Rearranging Eq. (12.51),

$$Q = \left(\frac{k_{20}A_s D\Theta^{T-20}}{\ln(S_0/S_e)}\right)^2 \tag{12.58}$$

$$= \left(\frac{0.0047 \times 138 \times 1.02(1.035)^{10-20}}{\ln(35/12.5)}\right)^2 = 0.208 \text{ l/m}^2 \cdot \text{s}$$

Allowable BOD loading is

$$\left(Q\frac{1}{\text{m}^2 \cdot \text{s}}\right)\left(86{,}400\frac{\text{s}}{\text{d}}\right)\left(\text{BOD}\frac{\text{mg}}{\text{l}}\right)\left(\frac{\text{g}}{1000 \text{ mg}}\right)\left(\frac{1}{A_s}\frac{\text{m}^2}{\text{m}^3}\right) \tag{12.59}$$

$$= \frac{0.208 \times 86{,}400 \times 70}{1000 \times 1.02} = 1200 \text{ g/m}^3 \cdot \text{d} \ (77 \text{ lb/1000 ft}^3\text{/day})$$

12.17 COMBINED TRICKLING-FILTER AND ACTIVATED-SLUDGE PROCESSES

Trickling filters alone may not be able to meet stringent effluent standards occasionally specified for treatment. Combining trickling filtration with activated-sludge processes can increase removal efficiency. The combined system consists of a deep trickling filter (biological tower) followed by an aeration tank with sludge recirculation from a final clarifier. To ensure adequate aeration, the filter media are either cross-flow or vertical-flow plastic modules with a high percentage of void space. The aeration tank and final clarifier is an activated-sludge system often referred to as a solids contact process. The final clarifier is either a conventional activated-sludge sedimentation tank or one with a flocculation zone inside a large inlet baffle.

The trickling-filter activated-sludge process can be designed to operate with various flow patterns, as illustrated in the composite flow diagram in Figure 12.29. By providing direct recirculation for adequate hydraulic loading, the trickling filter can be operated independent of the second-stage aeration process. In this arrangement, the filter is called a *roughing* filter, and the system is suitable to treat variable-strength wastewaters. The filter absorbs shock loads to stabilize loading on the subsequent activated-sludge process.

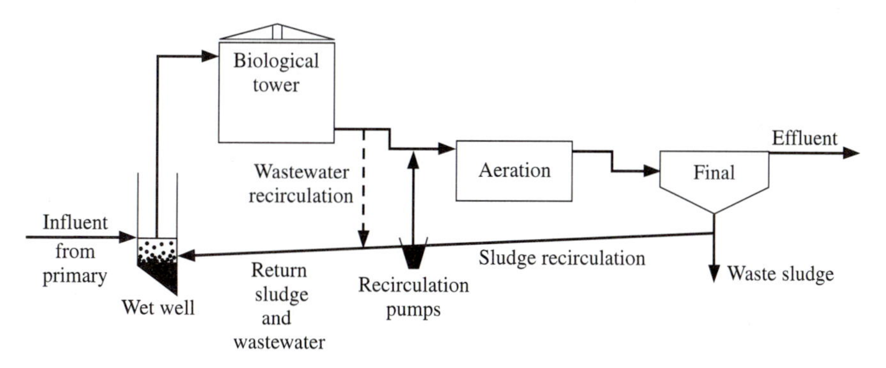

FIGURE 12.29 Profile of a combined trickling-filter activated-sludge process.

The other common flow pattern includes recirculation of a portion of the settled biological solids from the final clarifier to the trickling-filter influent to form an activated sludge that recycles through both the filter and the aeration tank. [This process is referred to by several names: activated biological filtration (ABF), activated biofiltration–activated-sludge, and trickling filtration–solids contact process.] This composite system has the characteristics of both fixed-film and suspended-growth processes. The mixed liquor in the aeration tank enhances removal of nonsettleable and dissolved BOD in the filter effluent. Because of good process stability, a high-quality effluent can be consistently produced. In a way, the biofilter in this process is acting as a static aerator; therefore, by increasing the hydraulic loading by recirculation, the transfer of oxygen is increased. However, if excess sludge solids are recirculated through the biofilter, the oxygen demand exceeds the transfer capability of the media and limits BOD removal.

Equations to model the activated biofiltration–activated-sludge process have not been successfully formulated because of the difficulty in analyzing the system. In design, removal by the biofilter must be determined first so that the aeration tank and associated aeration equipment can be properly sized. Arora and Umphres [12] evaluated the operation of 17 activated biofiltration–activated-sludge plants with redwood-media filters. The following are the ranges of various design and operational parameters of these treatment systems: percentage of the total volume (filter media plus aeration tank) that is aeration tank volume, 40%–75%; system BOD loading based on total volume, 20–50 lb/1000 ft^3/day (320–800 $g/m^3 \cdot d$); biofilter loading, 50–110 lb/1000 ft^3/day (800–1800 $g/m^3 \cdot d$); system food/microorganism ratio, 0.2–1.0 with the median at 0.5; and effluent BOD and effluent suspended solids, 5–25 mg/l. The operating personnel at most of the plants believed the system was more stable than conventional activated sludge and was better able to absorb shock loads.

12.18 DESCRIPTION OF ROTATING BIOLOGICAL CONTACTOR MEDIA AND PROCESS

A rotating biological contactor (RBC) consists of a shaft of circular plastic disks 12 ft in diameter revolving at 40% submergence in a contour-bottom tank. The nominal spacing between disks is 0.50–0.75 in. so that during submergence the wastewater can enter between the surfaces. When rotated out of the tank, air enters the spaces while the liquid trickles out over films of biological growth attached to the media. Alternating exposure to organics in the wastewater and oxygen in the air is similar to dosing a trickling filter with a rotating distributor. Excess microbial solids are hydraulically scoured from the media and carried out in the process effluent for gravity separation in a final clarifier. A rotating biological contactor system has the advantages of low power consumption and good process stability. Nevertheless, new RBC installations are rare because of cost relative to trickling filters.

The arrangement for secondary treatment of municipal wastewater by the RBC process, as shown in Figure 12.30, is similar to the flow diagram of a trickling-filter plant. The biological process is preceded by primary clarifiers and followed by secondary clarifiers. These settling tanks are sized using the same design criteria as for clarifiers in a trickling-filter plant. Settled solids that accumulate in the final clarifier

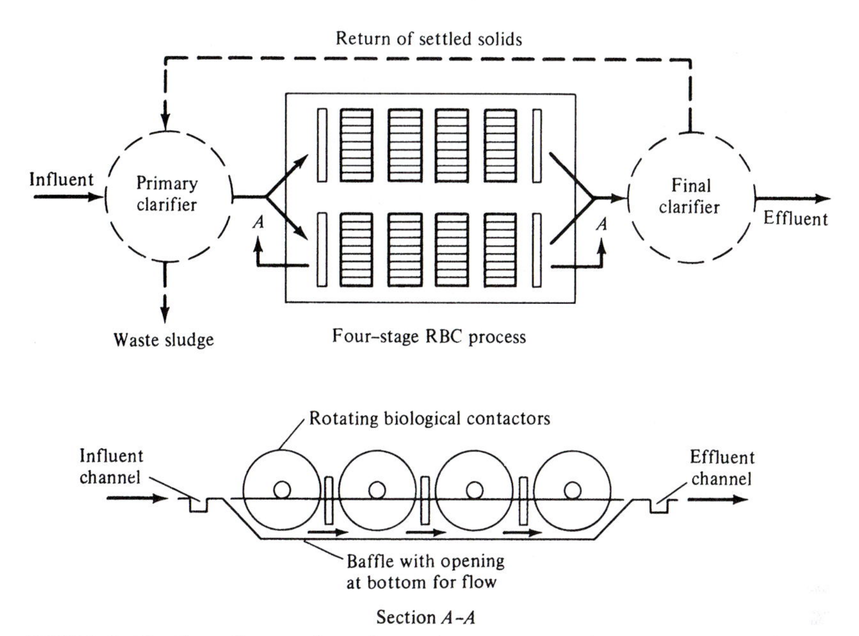

FIGURE 12.30 Flow diagram for a domestic wastewater plant using rotating biological contactors for secondary treatment.

are returned to the head of the plant for settlement with the raw-wastewater solids; thus, waste sludge is withdrawn only from the primary clarifiers. Since recirculation does not improve performance, the return flow is not designed to recycle wastewater through the RBC units.

Wastewater after sedimentation is applied to the first stage of a series of RBC chambers separated by baffles. In processing a wastewater that is essentially domestic in characteristics, a series of four stages is usually employed to ensure adequate BOD reduction; additional stages can be added to initiate nitrification. Each stage acts as a completely mixed chamber, and the slow movement of wastewater through the system simulates plug flow. Biological solids sheared from the disk surfaces are hydraulically transported under the baffles separating the chambers and conveyed from the system suspended in the effluent.

Rotating biological contactors are sensitive to cold and must be protected from normal outdoor weather conditions—that is, precipitation, wind, and intense sunshine. Standard design procedure is either to enclose individual stages under insulated plastic covers or to house a series of units in a suitable building with adequate ventilation.

Manufacturers' literature usually suggests design criteria based on operational performance of their full-scale installations. Typical recommendations follow for secondary treatment of domestic wastewater to yield an effluent of less than 30 mg/l of BOD and 30 mg/l of suspended solids:

1. Average loading based on total RBC surface area should be 1.5 lb/1000 ft^2/day of soluble BOD or 3.0 lb/1000 ft^2/day of total BOD (7.5 $g/m^2 \cdot d$ of soluble BOD or 15 $g/m^2 \cdot d$ of total BOD).

2. Maximum loading on the first stage should be 6 lb/1000 ft^2/day of soluble BOD or 12 lb/1000 ft^2/day of total BOD (30 $g/m^2 \cdot d$ of soluble BOD or 60 $g/m^2 \cdot d$ of total BOD).
3. A temperature correction for additional RBC surface area of 15% should be made for each 5°F below a design wastewater temperature of 55°F (15% for each 2.8°C below 13°C).

These recommendations are consistent with observed satisfactory performance at larger installations. For small plants, the safe upper limit is an average soluble BOD loading of approximately 1.0 lb/1000 ft^2/day (5 $g/m^2 \cdot d$). Mathematical models to predict the performance of the RBC process reliably are still in the developmental stage [13].

Example 12.8

Calculate the RBC area required for secondary treatment of a raw domestic wastewater having 230 mg/l of BOD. The design flow is 2.0 mgd at a temperature of 50°F. The effluent quality specified is 30 mg/l of BOD.

Solution:

$$\text{primary effluent BOD concentration} = 0.65 \times 230 = 150 \text{ mg/l}$$

An appropriate average BOD loading to achieve an effluent BOD of 30 mg/l is 3.0 lb/1000 ft^2/day at 55°F. Therefore,

$$\text{RBC area at 55°F} = \frac{2.0 \times 150 \times 8.34}{3.0} = \frac{2500}{3.0} = 834{,}000 \text{ ft}^2$$

Correcting for a temperature of 50°F by a 15% area increase per 5°F,

$$\text{RBC area at 50°F} = 1.15 \times 834{,}000 = 959{,}000 \text{ ft}^2$$

Use RBC shafts manufactured with a nominal surface area of 60,000 ft^2 with 12-ft-diameter disks for installation in tanks with a length of 17 ft 4 in. Install four rows of four stages for a total of $16 \times 60{,}000 = 960{,}000$ ft^2.

BOD loading on the first stage based on a temperature of 55°F should not exceed 12 lb/1000 ft^2/day.

$$\text{first-stage loading at 55°F} = \frac{2500 \times 4}{(960/1.15)} = 12.0 \text{ lb/1000 ft}^2\text{/day} \quad \text{(OK)}$$

ACTIVATED SLUDGE

Activated-sludge processes are used for both secondary treatment and complete aerobic treatment without primary sedimentation. Wastewater is fed continuously into an aerated tank, where the microorganisms metabolize and biologically flocculate the organics (Fig. 12.8). Microorganisms (activated sludge) are settled from the aerated mixed liquor under quiescent conditions in the final clarifier and returned to the aeration tank. Clear supernatant from the final settling tank is the plant effluent.

The primary feeders in activated sludge are bacteria; secondary feeders are holozoic protozoans (Fig. 12.11). Microbial growth in the mixed liquor is maintained in the declining or endogenous growth phase to ensure good settling characteristics (Fig. 12.9). Synthesis of the waste organics results in a buildup of the microbial mass in the system. Excess activated sludge is wasted from the system to maintain the proper food/microorganism ratio (F/M) and sludge age to ensure optimum operation.

Activated sludge is truly an aerobic treatment process because the biological floc are suspended in a liquid medium containing dissolved oxygen. Aerobic conditions must be maintained in the aeration tank; however, in the final clarifier, the dissolved-oxygen concentration can become extremely low. Dissolved oxygen extracted from the mixed liquor is replenished by air supplied to the aeration tank.

12.19 BOD LOADINGS AND AERATION PERIODS

General loading and operational parameters for the activated-sludge processes used in treatment of municipal wastewater in cool humid climatic regions are listed in Table 12.3.

"Allowable BOD loadings and performance of aeration processes depend on wastewater temperature, yet, few data from field studies of treatment plants are available to establish reliable design criteria based on temperature. Most of the design data listed in books [such as Table 12.3] are based on the operation of activated-sludge systems in cool humid climatic regions where wastewater temperature is in the range of 10° to 20°C (58° to 68°F). In this temperature climate, the concern is cooling of the wastewater below 10°C in winter. In a hot dry climate, the concern is the effect of warm water supplies (particularly where the drinking water is warm groundwater or from distillation of seawater) that increase wastewater temperature to the range of 20° to 30°C or higher" [14].

The BOD load on an aeration tank is calculated using the BOD in the influent wastewater without regard to that in the return sludge flow. BOD loadings are expressed in terms of pounds of BOD applied per day per 1000 ft^3 of liquid volume in the aeration tank and in terms of pounds of BOD applied/day/lb of mixed-liquor suspended solids (MLSS) in the aeration tank. The latter, the F/M ratio, is expressed

TABLE 12.3 General Loading and Operational Parameters for Activated-Sludge Processes

Process	BOD Loading: lb BOD/1000 ft^3/day[a]	BOD Loading: lb BOD/day/lb of MLSS	Sludge Age (days)	Aeration Period (hr)	Average Return Sludge Rates (%)
Step aeration	30–50	0.2–0.5	5–15	5.0–7.0	50
Conventional (tapered aeration)	30–40	0.2–0.5	5–15	6.0–7.5	30
Contact stabilization	30–50	0.2–0.5	5–15	6.0–9.0	100
Extended aeration	10–30	0.05–0.2	20+	20–30	100
High-purity oxygen	120+	0.6–1.5	5–10	1.0–3.0	30

[a] 1.0 lb/1000 ft^3/day = 16 g/m^3 · d.

by some authors in terms of lb of BOD applied/day/lb of volatile mixed-liquor suspended solids (MLVSS).

The aeration period is the detention time of the raw-wastewater flow in the aeration tank expressed in hours. It is calculated by dividing the tank volume by the daily average flow without regard to return sludge. The activated sludge returned is expressed as a percentage of the raw-wastewater influent. For example, if the return sludge rate is 20% and the raw-wastewater flow into the plant is 10 mgd, the return sludge is 2.0 mgd.

BOD loadings per unit volume of aeration tank vary from greater than 50 to 10 lb of BOD/1000 ft^3/day, while the aeration periods correspondingly vary from 5 to 24 hr. The relationship between volumetric BOD loading and aeration period is directly related to BOD concentration in the wastewater. For example, converting the average BOD concentration of 200 mg/l into units of pounds per 1000 ft^3 yields a concentration of

$$200\ \text{mg/l} \times \frac{62.4\ \text{lb}/1000\ \text{ft}^3}{1000\ \text{mg/l}} = 12.5\ \text{lb}/1000\ \text{ft}^3$$

Therefore, 200 mg/l wastewater applied to an extended aeration system with a 24-hr (1-day) aeration period results in a BOD loading of 12.5 lb/1000 ft^3/day. If a high-rate aeration period of 6.0 hr is considered, the BOD loading becomes

$$12.5\ \text{lb}/1000\ \text{ft}^3/\text{day} \times \frac{24\ \text{hr}}{6.0\ \text{hr}} = 50\ \text{lb}/1000\ \text{ft}^3/\text{day}$$

Sludge age (mean cell residence time) relates the quantity of microbial solids in an activated-sludge process to the quantity of solids lost in the effluent and excess solids withdrawn in the waste sludge. Equation (12.60) establishes the sludge age in days on the basis of MLSS in the aeration tank relative to SS discharged in the effluent and SS in the waste sludge withdrawn daily:

$$\text{Sludge age} = \frac{\text{MLSS} \times V}{\text{SS}_e \times Q_e + \text{SS}_w \times Q_w} \tag{12.60}$$

where

sludge age = mean cell residence time, days

MLSS = mixed-liquor suspended solids, mg/l

V = volume of the aeration tank, mil gal (m^3/d)

SS_e = suspended solids in effluent, mg/l

SS_w = suspended solids in waste sludge, mg/l

Q_e = quantity of effluent wastewater, mgd (m^3/d)

Q_w = quantity of waste sludge, mgd (m^3/d)

Sludge age is also calculated using the MLVSS (volatile portion of the MLSS) and the VSS (volatile suspended solids) in the effluent and waste sludge. The argument is that the volatile portion of the suspended solids is more representative of the microbial masses, and thus the sludge age expresses the residence time of the microbial cells in the system more realistically.

The suspended-solids concentration maintained in the MLSS of conventional and step-aeration processes ranges from 1500 to 3000 mg/l. The concentration held in the operation of a particular system depends on the desired F/M and sludge age for the applied BOD load. High-rate completely mixed processes generally operate with higher MLSS concentrations of 3000–4000 mg/l. Because of the variety of extended aeration processes, MLSS values encompass the entire range of 1000 to greater than 5000 mg/l.

Solids retention in an activated-sludge system is measured in days, whereas the liquid aeration period is in hours. For example, a conventional activated-sludge process with an MLSS of 2500 mg/l in the aeration tank, treating an average domestic wastewater and operating at a 6-hr aeration period, has a sludge age of approximately 7 days. The suspended solids are cycled in the system from final clarifier back to aeration tank, while the liquid flows through the aeration tank and clarifier.

Effluent quality from well-operated activated-sludge processes in the BOD loading range of 30–50 lb BOD/1000 ft^3/day can reliably meet the secondary standards of average maximum BOD of 30 mg/l and suspended solids of 30 mg/l with the temperature of mixed liquor at 10°–20°C (50°–68°F). At loadings lower in the listed range, or mixed-liquor temperature in the upper range, the effluent quality is more likely to be nearer 20 mg/l BOD and 20 mg/l suspended solids. Biological activity doubles (or halves) for every 10°–15°C temperature change. Therefore, for processes in the loading range of 30–50 lb BOD/1000 ft^3/day, reducing the mixed-liquor temperature to 5°–10°C can adversely affect effluent quality. Conversely, in the range of 15°–25°C, the quality of the effluent is likely to improve, or the loading can be increased with no detriment to effluent quality. Because extended aeration systems operate in a lower range of 10–30 lb BOD/1000 ft^3/day, a decrease or an increase in mixed-liquor temperature has less influence on effluent quality. Selection of aeration equipment is to some extent dictated by this relationship between allowable BOD loading and operating temperature. In a cool climate, submerged diffused aeration is common to reduce cooling of the mixed liquor in winter operation. In a warm climate, surface aerators that spray the mixed liquor in the air to absorb oxygen can be used since cooling is not a major consideration.

The wide range of aeration periods and BOD loadings used in activated-sludge processes tends to contrast one process with another. Also, the variety of physical features, such as the aeration tank size and shape, used in the various processes tends to accent the differences. Actually all activated-sludge processes are biologically similar, as seen in the generalized activated-sludge diagram of Figure 12.31. BOD is removed in the process by assimilative respiration of microorganisms, and the new cell growth is reduced by endogenous respiration. Excess microbial growth is withdrawn from the system by wasting activated sludge. Oxygen is added to the process to maintain aerobic biological activity.

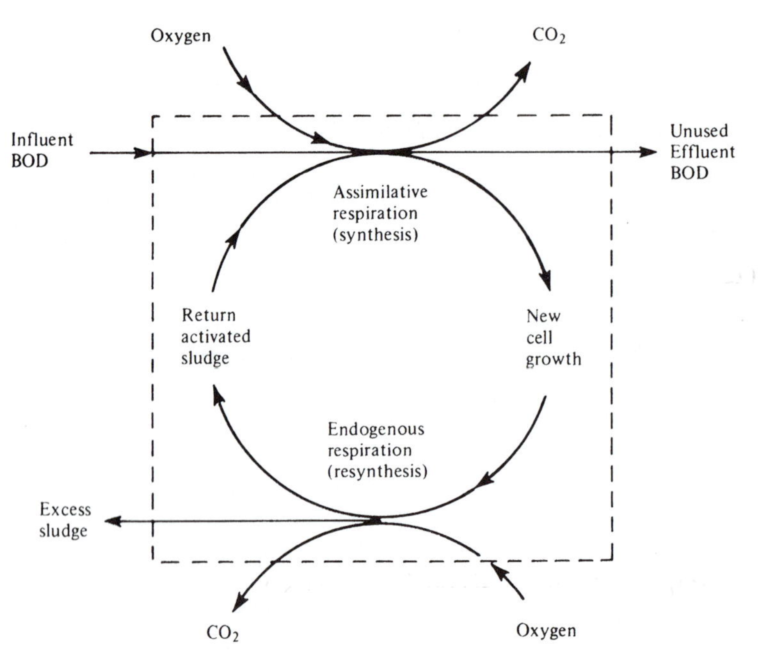

FIGURE 12.31 Generalized biological process reactions in the activated-sludge process.

Example 12.9

Data from a field study on a step-aeration activated-sludge secondary are as follows:

$$\text{aeration tank volume} = 120{,}000 \text{ ft}^3 = 0.898 \text{ mil gal}$$

$$\text{settled wastewater flow} = 3.67 \text{ mgd}$$

$$\text{return sludge flow} = 1.27 \text{ mgd}$$

$$\text{waste sludge flow} = 18{,}900 \text{ gpd} = 0.0189 \text{ mgd}$$

$$\text{MLSS in aeration tank} = 2350 \text{ mg/l}$$

$$\text{SS in waste sludge} = 11{,}000 \text{ mg/l}$$

$$\text{influent wastewater BOD} = 128 \text{ mg/l}$$

$$\text{effluent wastewater BOD} = 22 \text{ mg/l}$$

$$\text{effluent SS} = 26 \text{ mg/l}$$

Using these data, calculate the loading and operational parameters listed in Table 12.3 and the excess sludge production in pounds of excess suspended solids per pound of BOD applied.

Solution:

$$\text{BOD load} = 3.67 \text{ mgd} \times 128 \text{ mg/l} \times 8.34 = 3920 \text{ lb/day}$$

$$\text{MLSS in aeration tank} = 0.898 \text{ mil gal} \times 2350 \text{ mg/l} \times 8.34 = 17{,}600 \text{ lb}$$

$$\text{BOD loading} = 3920/120 = 32.7 \text{ lb/day/1000 ft}^3$$

$$\text{BOD loading} = 3920/17{,}600 = 0.22 \text{ lb/day/lb of MLSS}$$

Using Eq. (12.60),

$$\text{sludge age} = \frac{2350 \times 0.898}{26 \times 3.67 + 11{,}000 \times 0.0189} = 7.0 \text{ days}$$

$$\text{aeration period} = \frac{0.898 \times 24}{3.67} = 5.9 \text{ hr}$$

$$\text{return sludge rate} = \frac{1.27 \times 100}{3.67} = 35\%$$

$$\text{BOD efficiency} = \frac{(128 - 22)100}{128} = 83\%$$

$$\text{sludge production} = \frac{0.0189 \text{ mgd} \times 11{,}000 \text{ mg/l} \times 8.34}{3920}$$

$$= 0.44 \text{ lb SS wasted/lb BOD applied}$$

12.20 OPERATION OF ACTIVATED-SLUDGE PROCESSES

Operation of an activated-sludge treatment plant is regulated by (1) the quantity of air supplied to the aeration basin, (2) the rate of activated-sludge recirculation, and (3) the amount of excess sludge withdrawn from the system. Sludge wasting is used to establish the desired concentration of MLSS, food/microorganism ratio, and sludge age.

Field observations for monitoring an aeration system are the rates of wastewater influent, excess sludge wasting, and sludge recirculation; the dissolved-oxygen concentration in the mixed liquor; and the depth of the sludge blanket in the final clarifier. Laboratory tests are used to determine influent and effluent BOD, the concentration of suspended solids in the return sludge, and the concentration of MLSS in the aeration tank. From these data, BOD loadings, the aeration period, the return sludge rate, and the BOD removal efficiency can be calculated. The final clarifier operation is observed by testing for the concentration of suspended solids in the effluent and calculating the overflow rate and solids loading.

The degree of treatment achieved in an activated-sludge process depends directly on the settleability of the suspended solids in the final clarifier. If the biological floc agglomerate and settle rapidly by gravity, the overflow is a clear supernatant. Conversely, poorly flocculated particles (pin floc) and buoyant filamentous growths that do not separate by gravity contribute to BOD and suspended solids in the system effluent.

TABLE 12.4 Factors That Can Adversely Affect Settleability of Activated Sludge

Biological Factors
Species of dominant microorganisms (filamentous)
Ineffective biological flocculation
Denitrification in final clarifier (floating solids)
Excessive volumetric and food/microorganism loadings
Mixed-liquor suspended-solids concentration
Unsteady-state conditions (nonuniform feed rate and discontinuous wasting of excess activated sludge)
Chemical Factors
Lack of nutrients
Presence of toxins
Kinds of organic matter
Insufficient aeration
Low temperature
Physical Factors
Excessive agitation during aeration resulting in shearing of floc
Ineffective final clarification: inadequate rate of return sludge, excessive overflow rate or solids loading, or hydraulic turbulence

Excessive carryover of floc resulting in inefficient operation is referred to as sludge bulking. This can be caused by any one or a combination of the biological, chemical, and physical factors listed in Table 12.4. If an activated-sludge process is not functioning properly, the loadings on the aeration tank and final clarifier are calculated and compared to established design criteria. Next, operational procedures are reviewed to ensure proper aeration, sludge recirculation, and sludge wasting. Special laboratory tests can be performed to determine detrimental chemical characteristics of the wastewater, such as a lack of nutrients or the presence of toxins. Microscopic examination of the activated sludge can reveal excessive filamentous growth [15].

12.21 ACTIVATED-SLUDGE TREATMENT SYSTEMS

Conventional Activated-Sludge Process

The conventional process diagrammed in Figure 12.32(a) is an outgrowth of the earliest activated-sludge systems constructed, used for secondary treatment of domestic wastewater. The aeration basin is a long rectangular tank with air diffusers on one side of the tank bottom to provide aeration and mixing. Settled raw wastewater and return activated sludge enter the head of the tank and flow down its length in a spiral flow pattern. An air supply is tapered along the length of the tank to provide a greater amount of diffused air near the head where the rate of biological metabolism and resultant oxygen demand are the greatest. A conventional activated-sludge aeration tank is shown in Figure 12.33.

The conventional activated-sludge process uses bubble air diffusers set at a depth of 8 ft or more to provide adequate oxygen transfer and deep mixing. Several different

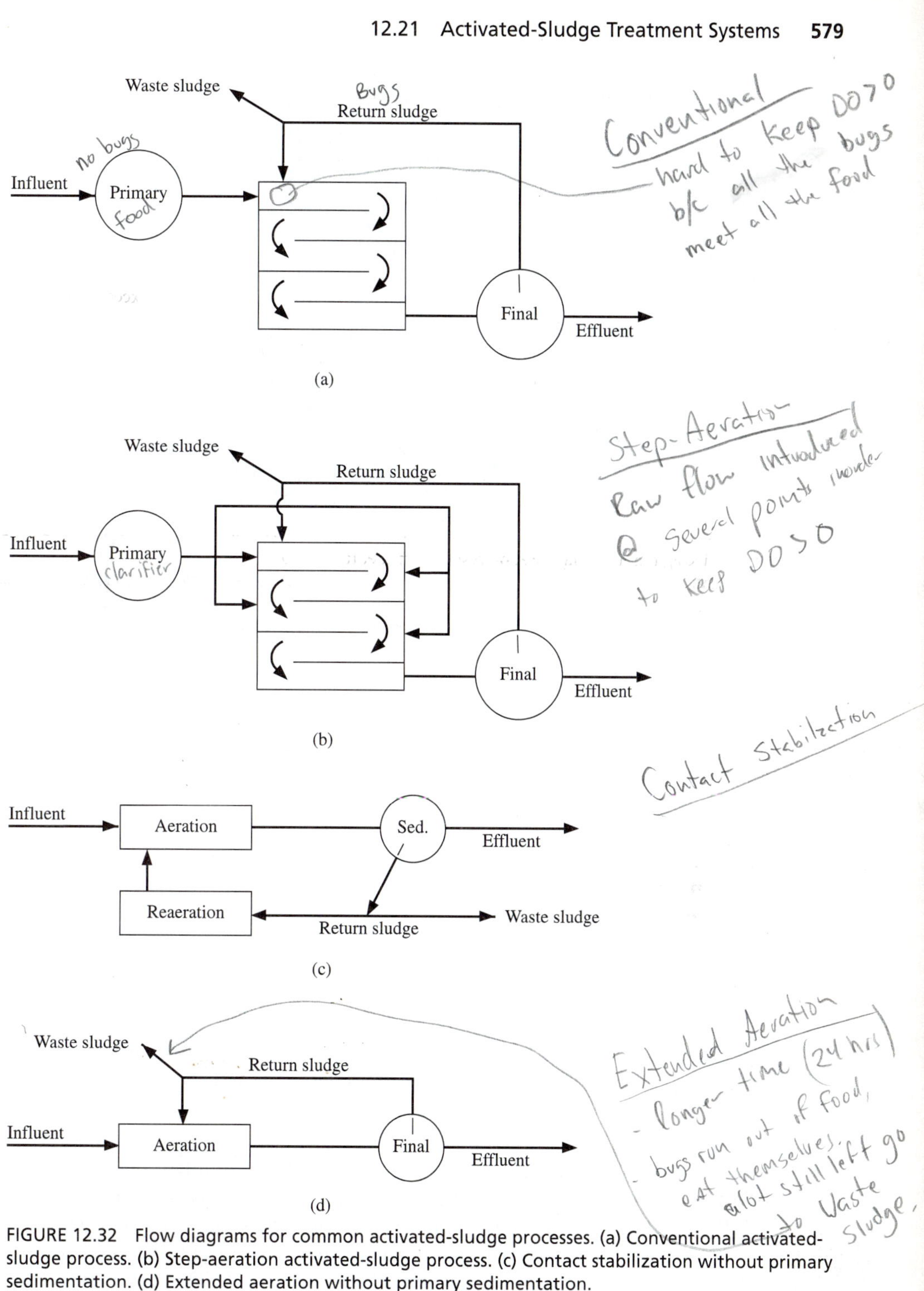

FIGURE 12.32 Flow diagrams for common activated-sludge processes. (a) Conventional activated-sludge process. (b) Step-aeration activated-sludge process. (c) Contact stabilization without primary sedimentation. (d) Extended aeration without primary sedimentation.

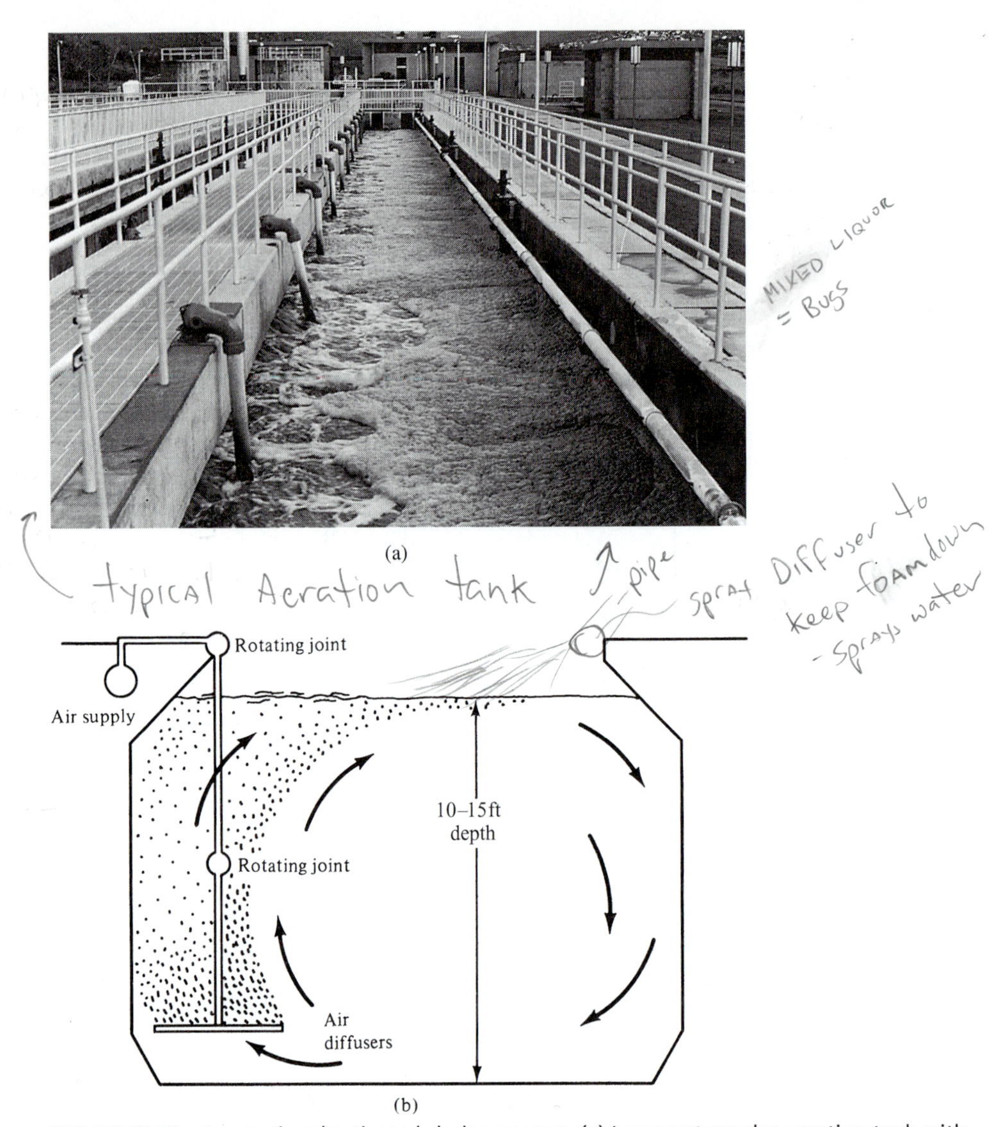

FIGURE 12.33 Conventional activated-sludge process. (a) Long rectangular aeration tank with submerged coarse-bubble diffusers along one side (Santee, CA). (b) Cross section of a typical aeration tank illustrating the spiral flow pattern created by aeration along one side.

bubble diffusers are manufactured; common kinds are stainless-steel or hollow-cylinder porous tubes 1–2 ft in length or porous disks about 6 in. in diameter. These individual diffusers are attached along a submerged air header about 10 ft in length attached to an air-supply hanger pipe. For maintenance of the diffusers, the hanger pipe can be designed with rotating joints (a swing-diffuser arm) so the header can be retracted using a portable jack. The tops of swing-diffuser hanger arms can be seen in Figure 2.33(a) along the aeration tank.

Step-Aeration Activated-Sludge Process

The *step-aeration process* [Fig. 12.32(b)] is a modification of the conventional process. Instead of introducing all raw wastewater at the tank head, raw flow is introduced at several points along the tank length. Stepping the influent load along the tank produces a more uniform oxygen demand throughout. While tapered aeration attempts to supply air to match oxygen demand along the length of the tank, step loading provides a more uniform oxygen demand for an evenly distributed air supply.

Both step-aeration and conventional processes can use fine-bubble aeration. Fine-bubble diffusers produce bubbles with a diameter of approximately 2–5 mm (0.08–0.20 in.) in clean water. The three general categories of fine-pore media are ceramics, porous plastics, and perforated membranes. As illustrated in Figure 12.34, individual diffusers are mounted on holders attached to air piping on the tank bottom. Each membrane or ceramic disc, either 9 in. or 7 in. in diameter, is sealed to a holder by a screw-on retainer ring with an O-ring seal. With the diffusers over the entire floor area, the rising streams of fine bubbles mix and aerate the mixed liquor uniformly, keeping the microbial floc in suspension. The benefit of fine-bubble aeration is a power savings of 40%–60% when compared to coarse-bubble or mechanically aerated activated-sludge processes [16]. As a result of cost savings and system performance, use of fine-bubble aeration is now common in activated-sludge processes, particularly with automated control. Automated aeration control is the manipulation of the aeration rate by computer to match the dynamic oxygen demand and maintain the desired residual dissolved-oxygen concentration in the mixed liquor.

Example 12.10

A step-aeration activated-sludge process is being sized for a settled wastewater flow of 7.40 mgd (989,000 ft^3/d) containing 7900 lb of BOD. The design maximum BOD loading is 40 lb/1000 ft^3/day, and the design minimum aeration period is 6.0 hr. (1) Calculate the dimensions for 4 identical aeration tanks. (2) Calculate the dimensions for 4 circular final clarifiers. (3) If the proposed minimum operating MLSS is 2000 mg/l, what is the calculated F/M at design loading?

Solution:

1.

$$V \text{ (based on BOD loading)} = \frac{7900 \times 1000}{40} = 198{,}000 \text{ ft}^3$$

$$V \text{ (based on aeration period)} = \frac{7{,}400{,}000 \times 6.0}{24 \times 7.48} = 247{,}000 \text{ ft}^3$$

Use 247,000 ft^3 with an aeration period of 6.0 hr, which results in a BOD loading of 32 lb/1000 ft^3/day. Install 4 aeration tanks with 13 ft liquid depth and 24 ft width for fine-bubble aeration.

$$\text{length of each tank} = \frac{247{,}000}{4 \times 13 \times 24} = 198 \text{ ft}$$

DIFFUSER = - smaller bubbles
- More surface AREA
- longer travel time to surface

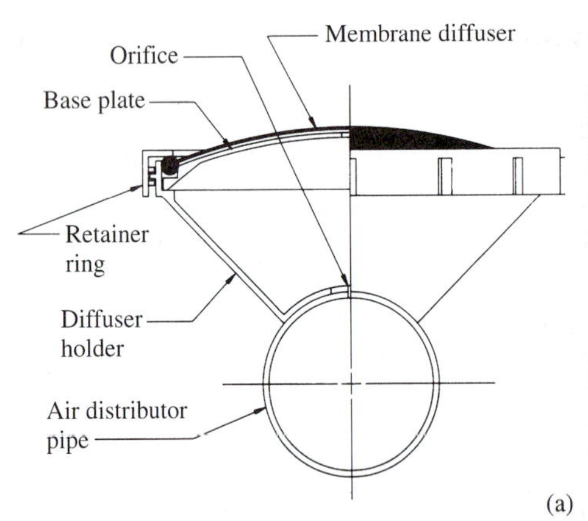

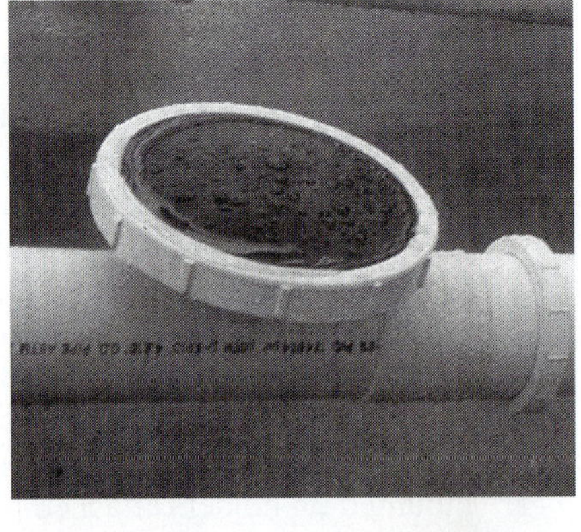

(a)

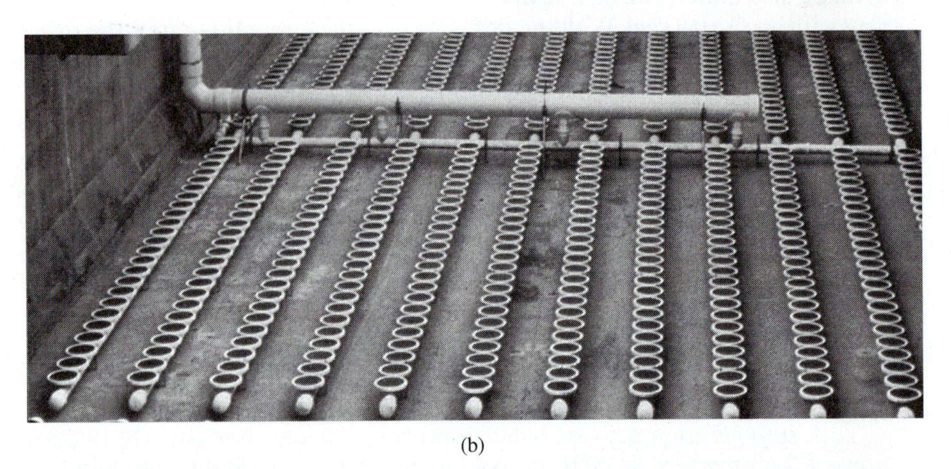

(b)

(c)

FIGURE 12.34 Fine-bubble diffuser for wastewater aeration. (a) A disc diffuser mounted on top of an air distributor pipe. (b) A grid of diffusers attached to air pipes mounted on the floor of an aeration tank. (c) Long rectangular aeration tank with uniform mixing and oxygenation by a grid of fine-bubble diffusers. *Source:* Courtesy of Sanitaire, a division of ITT Industries, Inc.

2. From Section 10.16, use an overflow rate of 800 gpd/ft^2 and side-water depth of 11 ft to size 4 circular clarifiers.

$$\text{surface area} = \frac{7{,}400{,}000}{4 \times 800} = 2310 \text{ ft}^2 \quad (\text{diameter} = 54 \text{ ft})$$

$$\text{detention time} = \frac{2300 \times 11 \times 24}{989{,}000/4} = 2.5 \text{ hr}$$

3. $$\text{F/M} = \frac{7900}{2000 \times 1.85 \times 8.34} = 0.26 \text{ lb BOD/day/lb of MLSS}$$

Contact-Stabilization Activated-Sludge Process

This process [Fig. 12.32(c)] provides for reaeration of the return activated sludge from the final clarifier, allowing this process to have a smaller aeration tank. The sequence of aeration–sedimentation–reaeration has been used as a secondary treatment process in large plants but is rare in new design. Current use is of complete aerobic treatment without primary sedimentation in factory-built, field-erected plants with capacities of 0.05–0.5 mgd, as pictured in Figure 12.35. Using common walls for economical construction, the plant consists of two concentric circular tanks about 14 ft deep with the inner shell 15–30 ft in diameter and the outer tank 30–70 ft in diameter. The doughnut-shaped space between the two tanks is divided into three chambers for aeration, reaeration, and aerobic digestion. The circular chamber in the center is the final settling tank. The plant can also be segmented and constructed for extended aeration or step-aeration processes.

The sequence of operation for contact stabilization is aeration of the raw wastewater with return activated sludge, sedimentation to overflow clarified effluent, and reaeration of the settling tank underflow with a portion wasted to the aerobic digester. Supernatant drawn from the digester is returned to the aeration chamber. Periodically, aeration to the aerobic digester is stopped and suspended solids allowed to settle for withdrawal of gravity-thickened sludge for disposal.

Example 12.11

A contact-stabilization plant, similar to the one diagrammed in Figure 12.35(a), has compartments with the following liquid volumes:

$$\text{aeration chamber} = 85 \text{ m}^3$$

$$\text{reaeration chamber} = 173 \text{ m}^3$$

$$\text{aerobic digester} = 153 \text{ m}^3$$

$$\text{sedimentation tank} = 122 \text{ m}^3 \text{ (30.7-m}^2 \text{ surface area)}$$

If the plant is designed for an equivalent population of 2000 persons, calculate the BOD loading, aeration periods, and detention times.

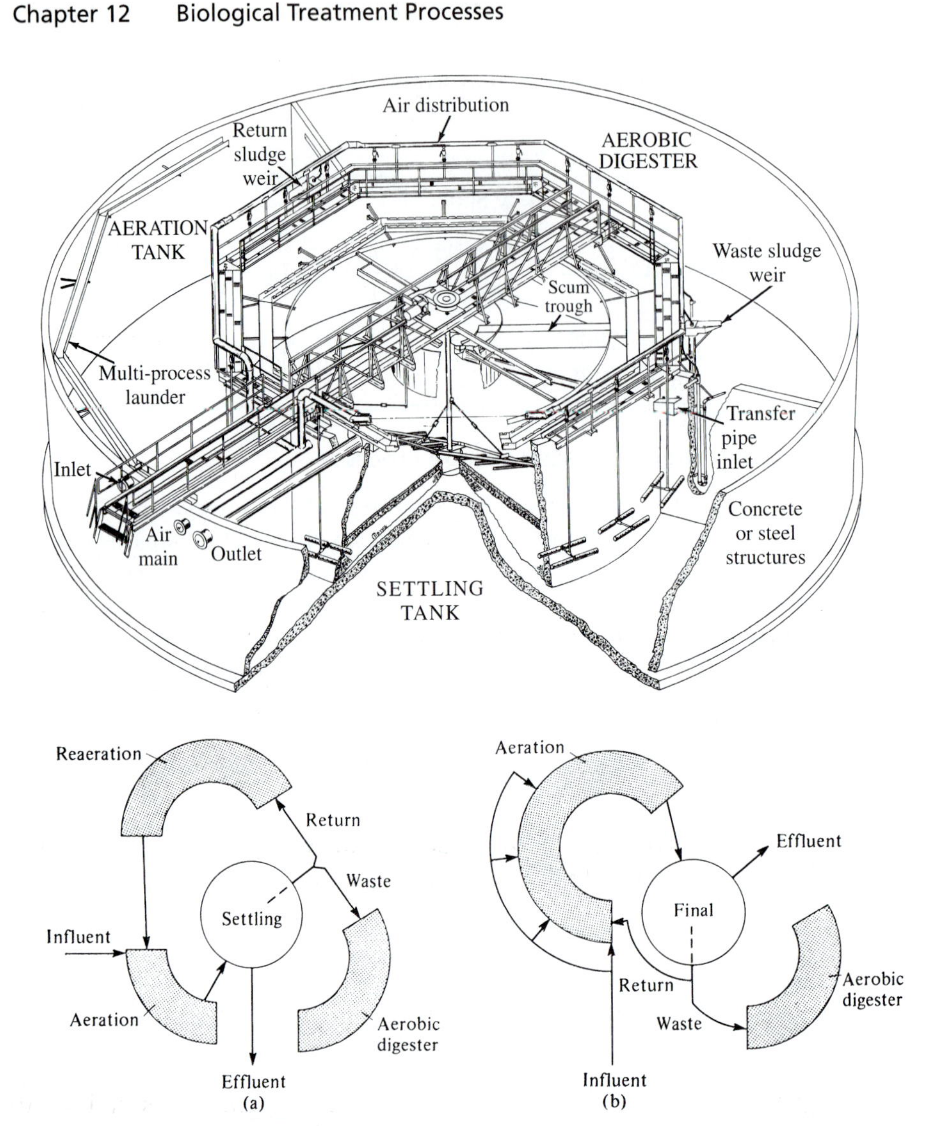

FIGURE 12.35 Field-erected circular steel wastewater treatment plant for extended-aeration, step-aeration, or contact stabilization processes. Drawing shows a cutaway view of the aeration tank and clarifier. (a) Contact stabilization (b) Extended or step aeration. *Source:* Courtesy of Sanitaire, a division of ITT Industries, Inc.

Solution:

$$\text{hydraulic load} = 2000 \times 450 \text{ l/person}$$
$$= 900{,}000 \text{ l/d} = 900 \text{ m}^3\text{/d}$$
$$\text{BOD load} = 2000 \times 91 \text{ g/person}$$
$$= 182{,}000 \text{ g/d}$$

$$\text{BOD loading on aeration tanks} = \frac{182{,}000}{85 + 173} = 705\ \text{g/m}^3 \cdot \text{d}$$

$$\text{aeration period (based on raw wastewater flow)} = \frac{85 \times 24}{900} = 2.3\ \text{h}$$

$$\text{reaeration period (based on raw wastewater flow)} = \frac{173 \times 24}{900} = 4.6\ \text{h}$$

The detention time for sedimentation (assuming 100% recirculation flow) is

$$\frac{122 \times 24}{2 \times 900} = 1.6\ \text{h}$$

The overflow rate on final clarifier (based on effluent flow) is

$$\frac{900}{30.7} = 29.3\ \text{m}^3/\text{m}^2 \cdot \text{d}$$

Extended-Aeration Activated-Sludge Process

The extended-aeration process [Fig. 12.32(d)] is used primarily to treat wastewater flows from residential communities and small municipalities. The aeration period is 24 hr or greater, with complete mixing of the aeration tank and, because of low BOD loading, the activated-sludge process operates in the endogenous growth phase. As a result, the biological process is very stable and can accept variable loading. Waste sludge is discharged to an aerobic digester for stabilization prior to disposal. Final settling tanks are conservatively sized for a long detention time and a low overflow rate, generally in the range of 200–600 gpd/ft^2 for aeration tank volumes in the range of 5000–150,000 gal.

A well-known extended-aeration process is the closed-loop reactor, or oxidation ditch, aerated and mixed by horizontal rotors, as illustrated in Figure 12.36. The modern reactor is an elongated oval with vertical walls and a center dividing wall; earlier ditches had sloping side walls with a center island. In reactor design, the wastewater depth is up to 16 ft with 2-ft freeboard, and channel width is up to 31 ft, with the horizontal rotors spanning the full width of the channel. The flow diagram in Figure 12.36(a) shows parallel operation of two reactors, which can be changed to series operation by adjusting slide gates. Also, if one reactor is to be taken out of service temporarily for inspection, operation of the plant can continue, although at higher volumetric loading.

Dimensions of the reactor must conform to design criteria established by the manufacturer of the horizontal rotors. For example, for the horizontal rotor with individual blades illustrated in Figure 12.36(b) (Lakeside's *Magna Rotor*), the maximum liquid depth of channel is 16 ft, rotor diameter is 42 in. with minimum design immersion of 5 in., and available length is 5–30 ft. Manufacturers also provide design data on rate of oxygen transfer and installation requirements. As pictured, rotor covers are available to contain spray and to reduce cooling of the mixed liquor in low-temperature operation.

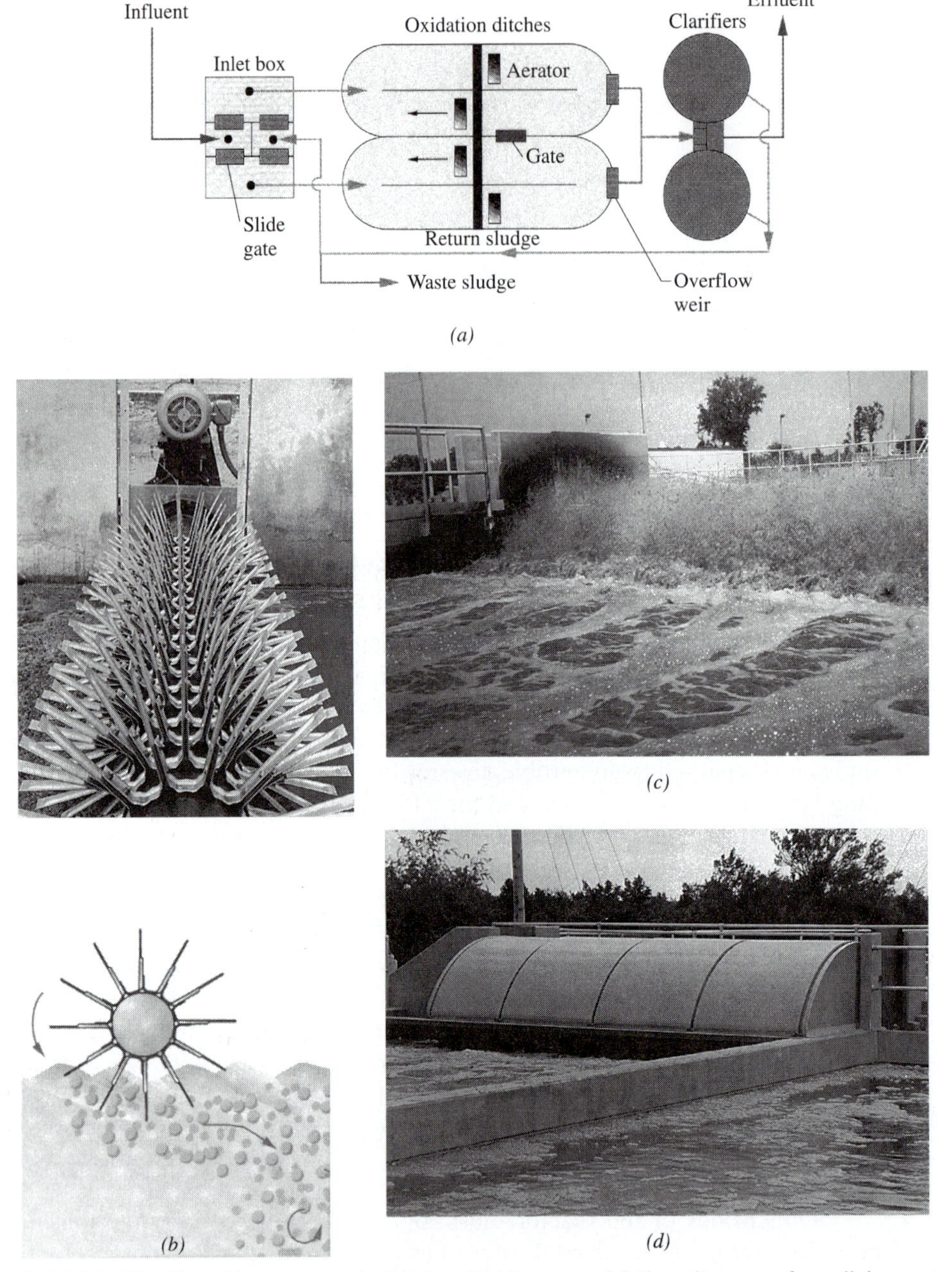

FIGURE 12.36 Closed-loop reactor (oxidation ditch) process. (a) Flow diagram of parallel operation of two reactors and clarifiers. Series operation is possible by closing two slide gates and the overflow gate of one reactor and opening the gate between ditches. (b) The horizontal rotor for aeration and moving the mixed liquor around the ditch. (c) Rotor operation showing aeration and mixing. (d) Rotor cover to contain spray and reduce icing in a cold climate. *Source:* Courtesy of Lakeside Equipment Corporation.

The *Carrousel®* system, similar in operation to an oxidation ditch, is a deep closed-loop aeration tank with vertical walls, as shown in the aerial view in Figure 12.37. However, in contrast, the aerators are vertical-shaft inverted open cones suspended from platforms constructed over the ends of the aeration channel. When treating unsettled municipal wastewater, the aeration period is generally 24 hr and the operating food/microorganism ratio less than 0.10 lb BOD/day/lb MLSS to operate at a long sludge age. Under these conditions with a warm mixed liquor, viable populations of nitrifying bacteria can be maintained in the activated sludge for nitrification.

The extended-aeration process can also be performed in a rectangular aeration tank aerated and mixed by mechanical aerators mounted on platforms supported by columns, as pictured in Figure 12.38. In order to have complete mixing, the length of the aeration tank is usually no greater than twice the width, with either two or four aerators. Mechanical aeration is often preferred to diffused aeration when the climate is sufficiently warm to prevent excessive cooling of the mixed liquor.

Example 12.12

Determine a preliminary layout for a closed-loop-reactor plant to treat an unsettled wastewater flow of 240,000 gpd (32,100 ft^3/d) with 400 lb of BOD (200 mg/l). (1) Calculate the dimensions for two oval closed-loop reactors with vertical walls, assuming a liquid depth of 5.0 ft. (2) Calculate the dimensions for two circular final clarifiers. (3) If the proposed minimum operating MLSS is 3000 mg/l, what is the operating F/M?

Solution: Use a layout of two reactors, as shown in Figure 12.36. For a typical domestic wastewater with a BOD of 200 mg/l, assume a design aeration period of 24 hr.

1. $V\,(\text{each reactor}) = 120{,}000 \text{ gal} = 16{,}000 \text{ ft}^3$

Assuming a liquid depth of 5.0 ft and channel width of 12 ft, each aeration tank would be 24 ft wide, with a straight length of 92 ft and total length, including circular ends, of 115 ft. Based on these dimensions, two 15-ft horizontal rotors could be installed in each tank.

The width and depth of the channel have to be confirmed by the manufacturer's recommendations for the installation of horizontal rotors for adequate aeration and velocity of flow in the channel. The rate of oxygen transfer must consider oxygen demand for both carbonaceous organic matter and nitrification. Because of the long sludge age, unintentional nitrification may occur, particularly in a warm mixed liquor.

2. Use an overflow rate of 400 gpd/ft^2 and side-water depth of 10 ft to size two circular clarifiers.

$$\text{surface area} = \frac{240{,}000}{2 \times 400} = 300 \text{ ft}^2 \quad (\text{diameter} = 20 \text{ ft})$$

$$\text{detention time} = \frac{300 \times 10 \times 24}{32{,}100/2} = 4.5 \text{ hr}$$

3. $$\text{F/M} = \frac{400}{3000 \times 0.24 \times 8.34} = 0.06 \text{ lb BOD/day/lb of MLSS}$$

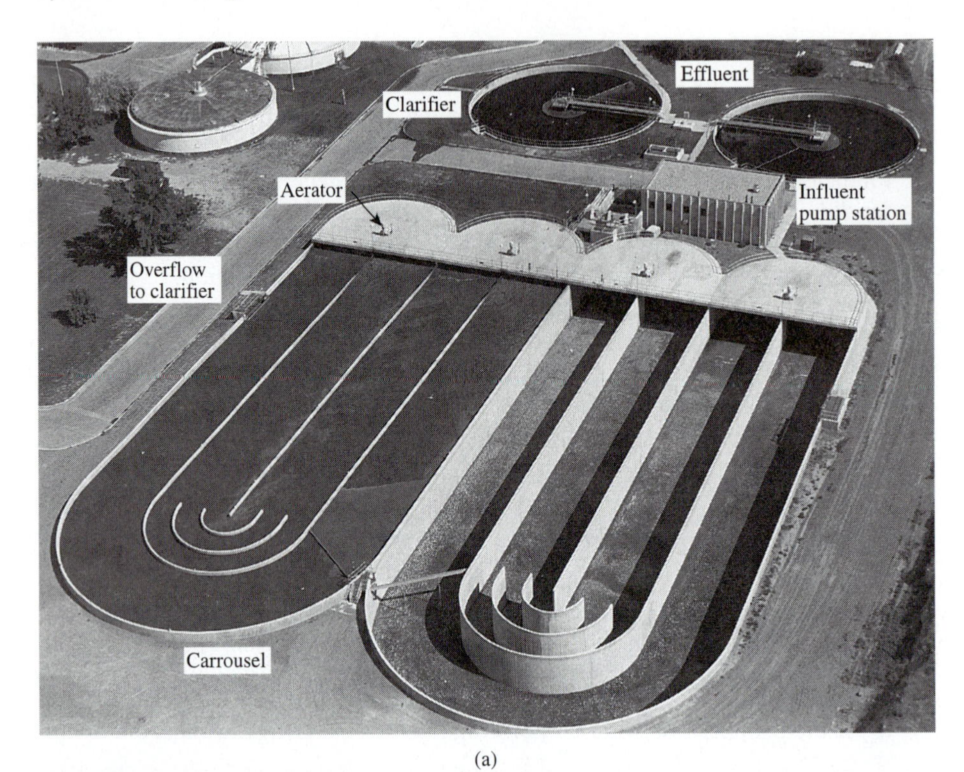

(a)

(b) (c)

FIGURE 12.37 Aerial view of a *Carrousel*® activated-sludge plant (a) Aeration and mixing is in oval serpentine aeration tanks with vertical walls followed by clarifiers for separation and return of activated sludge. (b) Low-speed vertical-shaft aerators are supported from a platform holding the drive motors and gear boxes. (c) Column-supported clarifier mechanisms have uptake pipes for rapid return of activated sludge. *Source:* Courtesy of EIMCO Process Equipment.

FIGURE 12.38 Platform-mounted low-speed mechanical aerator with an inverted open cone for high oxygen-transfer efficiency in an extended-aeration activated-sludge process. *Source:* Courtesy of EIMCO Process Equipment.

High-Purity-Oxygen Activated-Sludge Process

This process uses oxygen gas generated by cryogenic air separation or pressure-swing adsorption processes. A typical aeration tank, shown schematically in Figure 12.39, is divided into three or four stages by means of baffles to stimulate plug flow and is covered with a gas-tight enclosure. Raw wastewater, return-activated sludge, and oxygen gas under a slight pressure are introduced to the first stage and flow concurrently through succeeding sections. Oxygen can be mixed with the tank contents by injection through a hollow shaft to a rotating sparger device, or a surface aerator installed on top of the mixer turbine shaft to contact oxygen gas with the mixed liquor. Successive aeration chambers are connected to each other so that the liquid flows through submerged ports, and head gases pass freely from one stage to the next with only a slight pressure drop. Exhausted waste gas is a mixture of carbon dioxide, nitrogen, and about 10%–20% of the applied oxygen. Effluent mixed liquor is settled in either a scraper-type or rapid-sludge-return clarifier, and the activated sludge is returned to the aeration tank.

Compared to air activated-sludge processes, high-purity-oxygen activated sludge has several advantages that are attributed to its higher oxygenation capacity [17]. If all of the nitrogen in air is displaced by oxygen, the partial pressure of oxygen is 100%, resulting in a fivefold increase in the saturation value of dissolved oxygen in water.

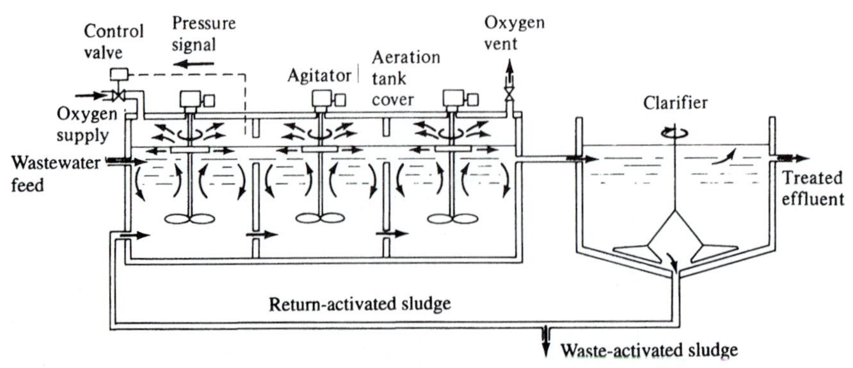

FIGURE 12.39 Schematic diagram of a high-purity-oxygen activated-sludge process with surface aerators in three stages.

High efficiency is possible at increased BOD loads and reduced aeration periods by maintaining the food/microorganism ratio with MLSS concentrations of 4000–8000 mg/l. Even though the process simulates plug flow, shock organic loads do not produce instability, since extra oxygen is supplied to the first stage automatically on demand. Emission of foul odors is virtually eliminated because of the highly aerobic environment and reduced volume of exhaust gases. Covered tanks also help to reduce the cooling of the wastewater during cold-weather operation.

Example 12.13

A municipality has an average daily wastewater flow of 280 gpm with a peak hourly rate of 450 gpm. The average BOD concentration is 200 mg/l, except during several weeks when a seasonal industry increases the mean BOD to 250 mg/l. A high-purity-oxygen system as in Figure 12.39 without primary clarification of the raw wastewater is being considered. Calculate (1) the volume of aeration tank capacity required, and (2) surface area and depth for a final settling tank. Recommended design criteria are a maximum BOD loading of 160 lb/1000 ft^3/day, a largest food/microorganism ratio of 0.5 lb of BOD/day/lb of MLVSS, an operating MLSS concentration of 5500 mg/l (MLVSS of 4200 mg/l), and a highest overflow rate of 1200 gpd/ft^2 during peak flow.

Solution:

1. $$\begin{pmatrix}\text{aeration tank volume}\\ \text{required at 250 mg/l of BOD}\end{pmatrix} = \frac{280 \times 1440 \times 250 \times 8.34}{1{,}000{,}000 \times 0.160} = 5250 \text{ ft}^3$$

$$\begin{pmatrix}\text{aeration period}\\ \text{at average flow}\end{pmatrix} = \frac{5250 \times 7.48 \times 24}{280 \times 1440} = 2.34 \text{ hr} \quad \text{(OK)}$$

$$\begin{pmatrix}\text{BOD load at}\\ \text{200 mg/l of BOD}\end{pmatrix} = \frac{280 \times 1440 \times 200 \times 8.34}{1{,}000{,}000 \times 5.25}$$

$$= 128 \text{ lb/1000 ft}^3\text{/day} \quad \text{(OK)}$$

Check the F/M ratio for both BOD loadings (128 and 160 lb/1000 ft^3/day) at an MLVSS concentration of 4200 mg/l.

$$\text{At}\ \frac{128 \text{ lb of BOD/day}}{1000 \text{ ft}^3}:\frac{F}{M} = \frac{128/1000}{\dfrac{4200 \times 62.4}{1,000,000}} = 0.49\frac{\text{lb of BOD/day}}{\text{lb of MLVSS}} \quad \text{(OK)}$$

$$\text{At}\ \frac{160 \text{ lb of BOD/day}}{1000 \text{ ft}^3}:\frac{F}{M} = \frac{160/1000}{\dfrac{4200 \times 62.4}{1,000,000}} = 0.61 \quad \text{(slightly greater than 0.5)}$$

Therefore, use a three-stage aeration basin with a total volume of 5250 ft^3.

2. $$\left(\begin{array}{c}\text{final clarifier surface area}\\ \text{required based on peak flow}\end{array}\right) = \frac{450 \times 1440}{1200} = 540 \text{ ft}^2$$

$$\text{overflow rate at average flow} = \frac{280 \times 1440}{540} = 747 \text{ gpd/ft}^2$$

Assume the additional final clarifier design parameters of 8.0 ft minimum depth and 2.5 hr as the minimum detention time. Then the clarifier depth required for an overflow rate of 747 gpd/ft^2 is

$$\text{depth} = \frac{747 \text{ gal}}{\text{day} \times \text{ft}^2} \times \frac{\text{ft}^3}{7.48 \text{ gal}} \times 2.5 \text{ hr} \times \frac{\text{day}}{24 \text{ hr}} = 10.4 \text{ ft}$$

12.22 KINETICS MODEL OF THE ACTIVATED-SLUDGE PROCESS

The principles of growth kinetics from pure culture microbiology can be applied in suspended-growth wastewater treatment even though the processes differ significantly in relation to the method of growth (batch as compared to continuous flow), microbial population (pure culture as opposed to mixed culture), and substrate content (uniform as opposed to a variety of organics). The shape of the wastewater processing curve in Figure 12.9 is similar to Monod's growth rate curve (Fig. 12.6) and the Michaelis–Menten rate curve for an enzyme-catalyzed reaction (Fig. 12.4). When plotted against substrate concentration, the reaction rates of enzyme, pure culture, and mixed culture reactions all follow the shape of a rectangular hyperbola where the substrate at one-half the maximum reaction rate is a characteristic constant of the reaction.

The following mathematical equations apply to completely mixed activated-sludge systems operating in a substrate-limiting condition, i.e., at a low F/M ratio. The fundamental relationships of growth kinetics used in the derivation of these equations are presented in Section 12.6.

The flow scheme for a completely mixed activated-sludge process is shown in Figure 12.40,

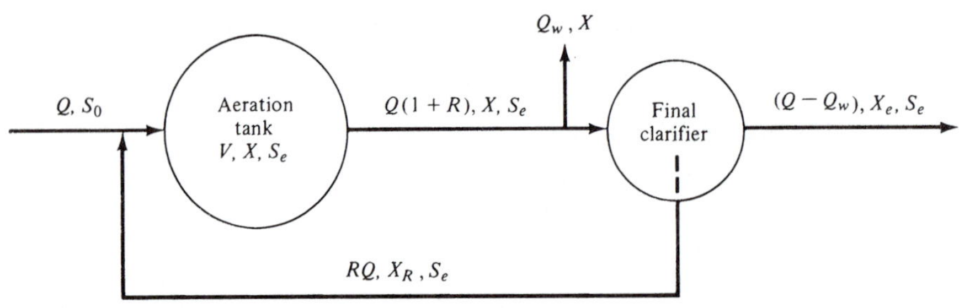

FIGURE 12.40 Flow scheme for the completely mixed activated-sludge system used in the derivation of the kinetics equations.

where

Q = rate of influent flow

Q_w = rate of excess sludge wasting from aeration tank

$Q - Q_w$ = rate of effluent flow

R = recirculation ratio (Q_R/Q)

RQ = rate of sludge recirculation

$Q(1 + R)$ = rate of flow from aeration tank

V = volume of aeration tank

X = concentration of biomass in aeration tank (MLVSS)

X_R = concentration of biomass in recirculating sludge (VSS)

X_e = concentration of biomass in effluent (VSS)

S_0 = concentration of substrate in influent flow (soluble BOD or COD)

S_e = concentration of substrate in effluent flow, recirculating sludge, and aeration tank (soluble BOD or COD)

The following conditions are assumed in the formulation of the mass balance equations:

1. Flows, biomass concentrations, and substrate concentrations are in a steady state.
2. All substrates are soluble (filtered BOD or COD).
3. The substrate concentration in the aeration tank equals the substrate concentration in the effluent after treatment.
4. Biological activity occurs only in the aeration tank.
5. No microorganisms are present in the influent wastewater.

6. The mean cell residence time is calculated based on the biomass in the aeration tank.
7. Excess activated sludge is wasted from the aeration tank rather than from the sludge recirculation line.
8. The aeration tank is complete mixing.

The aeration period (liquid detention time) is defined as

$$\theta = \frac{V}{Q} \tag{12.61}$$

where

θ = aeration period, time

The mean cell residence time (sludge age) is the biomass (MLVSS) in the aeration tank divided by the sum of the biomasses in the waste sludge and effluent.

$$\theta_c = \frac{VX}{Q_w X + (Q - Q_w)X_e} \tag{12.62}$$

where

θ_c = mean cell residence time, time

The definition of specific growth rate μ from Eq. (12.24) is the rate of growth per unit of biomass (time^{-1}). The inverse of μ is the biomass divided by the rate of growth, $X/(dX/dt)_g$. Therefore, under steady-state conditions, the mean cell residence time is

$$\theta_c = \frac{X}{(dX/dt)_g} = \frac{1}{\mu} \tag{12.63}$$

A mass balance for biomass around the entire activated-sludge system shown in Figure 12.40 is

$$\begin{pmatrix}\text{net rate of change}\\ \text{of biomass in system}\end{pmatrix} = \begin{pmatrix}\text{net rate of growth}\\ \text{of biomass in system}\end{pmatrix} - \begin{pmatrix}\text{rate of loss}\\ \text{of biomass from system}\end{pmatrix}$$

Using the notation in Figure 12.40,

$$\left(\frac{dX}{dt}\right)V = \left(\frac{dX}{dt}\right)_g V - [Q_w X + (Q - Q_w)X_e] \tag{12.64}$$

At steady state, the rate of biomass growth equals the rate of biomass loss; hence, $(dX/dt)V = 0$. Setting Eq. (12.64) equal to zero and substituting the endogenous rate of growth from Eq. (12.32) for $(dX/dt)_g$ yields

$$\frac{1}{\theta_c} = Y\frac{(dS/dt)_u}{X} - k_d \tag{12.65}$$

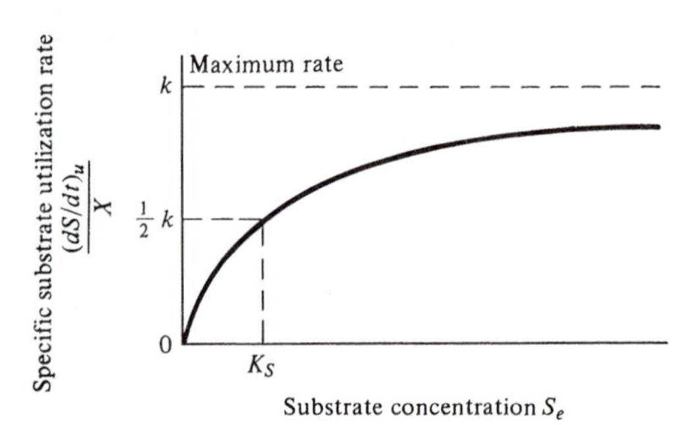

FIGURE 12.41 Specific substrate utilization rate versus substrate concentration surrounding the microorganisms based on Eq. (12.66).

Lawrence and McCarty [18] related the rate of substrate utilization both to the concentration of microorganisms in the aeration tank and to the concentration of substrate surrounding the organisms. The equation is

$$\left(\frac{dS}{dt}\right)_u = k\frac{XS_e}{K_S + S_e} \tag{12.66}$$

where

k = maximum rate of substrate utilization per unit mass of biomass, time^{-1}

X = concentration of biomass, mass/unit volume

S = concentration of substrate surrounding the microorganisms, mass/unit volume

K_S = saturation constant, equal to the substrate concentration when $(dS/dt)_u = k/2$, mass/unit volume

Equation (12.66), graphed in Figure 12.41, indicates that the functional relationship between substrate utilization rate and substrate concentration is continuous over the entire range of substrate concentrations. It is similar in form to Eq. (12.29), with k replacing μ_m/Y.

Substituting Eq. (12.66) into Eq. (12.65) incorporates the term S_e,

$$\frac{1}{\theta_c} = Y\frac{kS_e}{K_s + S_e} - k_d \tag{12.67}$$

Rearranging, the formula for S_e (effluent substrate concentration) is

$$S_e = \frac{K_s(1 + k_d\theta_c)}{\theta_c(Yk - k_d) - 1} \tag{12.68}$$

The specific substrate utilization rate is defined as the substrate utilization rate divided by the concentration of biomass in the aeration tank. Hence,

$$U = \frac{(dS/dt)_u}{X} \tag{12.69}$$

where

U = specific substrate utilization rate, time^{-1}

It can be calculated from experimental data using the following formula:

$$U = \frac{Q(S_0 - S_e)}{VX} = \frac{S_0 - S_e}{\theta X} \tag{12.70}$$

Substituting the mathematical expression for specific substrate utilization rate [Eq. (12.69)] into Eq. (12.65) gives the relationship

$$\frac{1}{\theta_c} = YU - k_d \tag{12.71}$$

where

θ_c = mean cell residence time, time

Y = growth yield, biomass increase-substrate metabolized, unitless

U = specific substrate utilization rate, time^{-1}

k_d = microbial decay coefficient, time^{-1}

A mass balance for substrate entering and leaving the aeration tank, as diagrammed in Figure 12.40, is

$$\begin{pmatrix}\text{net rate of change}\\ \text{of substrate}\\ \text{in aeration tank}\end{pmatrix} = \begin{pmatrix}\text{rate of substrate}\\ \text{entering}\\ \text{aeration tank}\end{pmatrix} - \begin{pmatrix}\text{rate of substrate}\\ \text{utilization}\\ \text{in aeration tank}\end{pmatrix} - \begin{pmatrix}\text{rate of substrate}\\ \text{leaving}\\ \text{aeration tank}\end{pmatrix}$$

Using the notation in Figure 12.40,

$$\left(\frac{dS}{dt}\right)V = QS_0 + RQS_e - \left(\frac{dS}{dt}\right)_u V - Q(1 + R)S_e \tag{12.72}$$

At steady state, the rate of substrate entering the aeration tank equals the rate of substrate removal; hence, $(dS/dt)V = 0$. Setting Eq. (12.72) equal to zero and solving for the substrate utilization rate results in

$$\left(\frac{dS}{dt}\right)_u = \frac{Q(S_0 - S_e)}{V} = \frac{S_0 - S_e}{\theta} \tag{12.73}$$

Dividing both sides by X, Eq. (12.73) becomes

$$\frac{(dS/dt)_u}{X} = \frac{Q(S_0 - S_e)}{XV} \tag{12.74}$$

Substituting this expression for specific substrate utilization into Eq. (12.65) and rearranging, the equation for V (volume of the aeration tank) is

$$V = \frac{\theta_c YQ(S_0 - S_e)}{X(1 + k_d\theta_c)} \tag{12.75}$$

Equating the terms for $(dS/dt)_u$ from Eqs. (12.73) and (12.66), and dividing by X yields

$$\frac{S_0 - S_e}{\theta X} = k\frac{S}{K_s + S} \tag{12.76}$$

By inverting and linearizing Eq. (12.76),

$$\frac{X\theta}{S_0 - S_e} = \left(\frac{K_s}{k}\right)\left(\frac{1}{S_e}\right) + \frac{1}{k} \tag{12.77}$$

Substituting Eq. (12.70) for the left side of Eq. (12.77) gives

$$\frac{1}{U} = \left(\frac{K_s}{k}\right)\left(\frac{1}{S_e}\right) + \frac{1}{k} \tag{12.78}$$

where

U = specific substrate utilization rate, time^{-1}

K_s = saturation constant, mass/unit volume

k = maximum rate of substrate utilization per unit mass of biomass, time^{-1}

S_e = concentration of substrate in effluent, mass/unit volume

This presentation was limited to the kinetics model for completely mixed activated sludge. For a more detailed mathematical analysis of this system and discussions of models for other biological processes, the reader is referred to books by Benefield and Randall [19] and Grady and Lim [20].

12.23 LABORATORY DETERMINATION OF KINETIC CONSTANTS

The numerical values for the following kinetic constants must be determined by laboratory experiments before the model described in Section 12.22 can be used for design of completely mixed activated sludge:

Y = growth yield, mg VSS/mg BOD (or mg COD)

k_d = microbial decay coefficient, d^{-1}

K_s = saturation constant, mg/l of BOD (or COD)

k = maximum rate of substrate utilization per unit mass of biomass, d^{-1}

For design, additional required data include settling characteristics of the activated sludge and oxygen uptake rates during aeration of the wastewater.

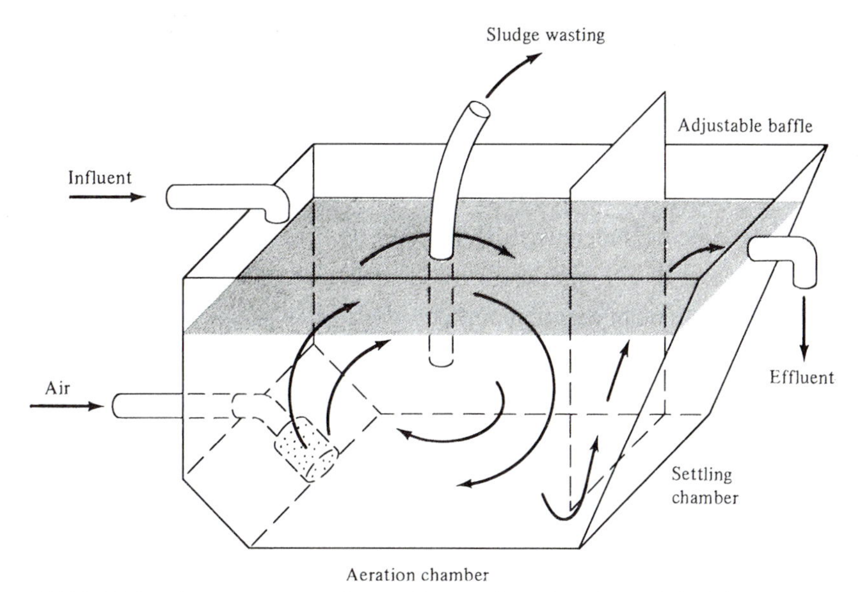

FIGURE 12.42 Bench-scale, continuous-flow, activated-sludge unit for laboratory testing of wastewater to determine kinetic constants.

A bench-scale activated-sludge unit as illustrated in Figure 12.42 is commonly used for laboratory evaluations. Operating conditions are the same as those assumed in the derivation of the mass balance equations, which are listed at the beginning of Section 12.22. Essential for satisfactory results are a continuous influent flow rate with a constant substrate concentration and complete mixing in the aeration tank. To collect sufficient data, the unit is operated at several mean cell residence times in the range of 3–20 days. Wasting the same quantity of suspended solids from the aeration tank maintains a constant θ_c, concentration of MLVSS, and F/M ratio. Ideally, the waste activated sludge is pumped out continuously; however, withdrawing batches of mixed liquor from the aeration tank at uniform time intervals may be practiced. Settleability of the activated sludge is tested by withdrawing a sample of the aerating mixed liquor and placing it in a settleometer; it may then be replaced or discarded as waste sludge. The length of time required for a test period, after steady-state conditions are established, is at least twice the mean cell residence time, with four residence times preferred. Temperature, pH, and dissolved oxygen concentration are held constant throughout the series of test runs.

Laboratory tests and operating conditions are generally recorded daily. The data required for calculating the operating parameters are as follows:

$$Q = \text{rate of influent flow, l/d}$$

$$Q_w = \text{rate of sludge wasting, l/d}$$

$$Q - Q_w = \text{rate of effluent flow, l/d}$$

$$V = \text{volume of aeration tank, l}$$

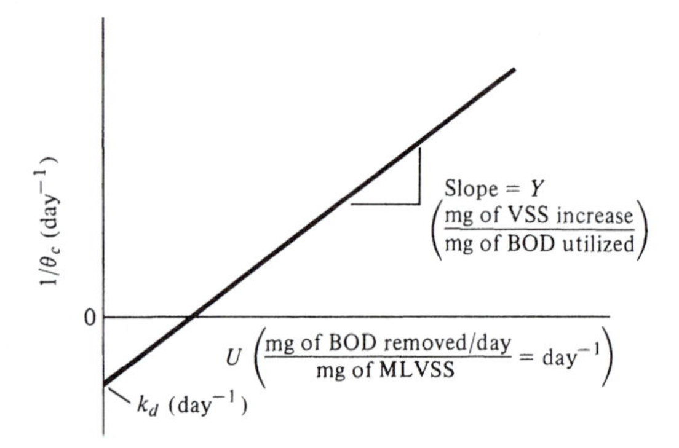

FIGURE 12.43 Plot of the inverse of the mean cell residence time versus the specific substrate utilization rate to determine Y and k_d based on Eq. (12.71).

X = concentration of MLVSS in the aeration tank, mg/l

X_e = concentration of VSS in effluent, mg/l

S_0 = soluble BOD (or COD) in influent, mg/l

S_e = soluble BOD (or COD) in effluent, mg/l

From these data, the aeration period θ is calculated using Eq. (12.61), the specific substrate utilization rate U is calculated from Eq. (12.70), and the mean cell residence time θ_c from Eq. (12.62).

Values of $1/\theta_c$ and U for each test period are plotted as shown in Figure 12.43. Based on Eq. (12.71), the slope of the line is equal to the growth yield Y and the intercept with vertical axis is the microbial decay coefficient k_d.

A plot of $1/U$ and $1/S_e$ for each test period is shown in Figure 12.44. Based on Eq. (12.78), the slope of the line is equal to the saturation constant over the maximum rate of substrate utilization per unit mass of biomass K_s/k, and the intercept with the vertical axis is equal to $1/k$.

The general range of values for kinetic constants for completely mixed activated sludge treating municipal wastewaters is listed in Table 12.5.

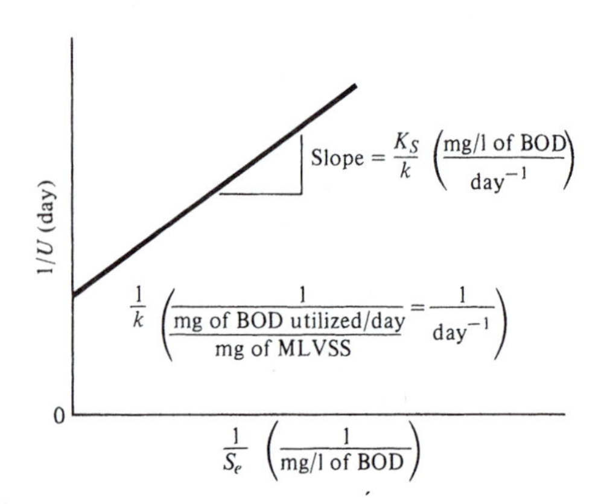

FIGURE 12.44 Plot of the inverse of the specific substrate utilization rate versus the inverse of concentration of substrate in the effluent to determine k and K_S based on Eq. (12.78).

TABLE 12.5 General Ranges of Magnitude for Kinetic Constants for Completely Mixed Activated-Sludge Processes Treating Municipal Wastewater at Approximately 20° C

Constant	Units	Range
Y	mg VSS/mg BOD	0.4–0.8
Y	mg VSS/mg COD	0.3–0.4
k_d	d^{-1}	0.04–0.08
K_S	mg/l of BOD	25–100
K_S	mg/l of COD	25–100
k	d^{-1}	4–8

Example 12.14

A municipal wastewater was tested to determine the kinetic constants using a laboratory apparatus similar to the bench-scale unit shown in Figure 12.42. The volume of the aeration chamber was 10 l. Wastewater feed was established at a constant rate of 34.3 l/d to provide a 7.0-hr aeration period for all of the test runs. A measured volume of sludge was wasted once a day from the tank. Determine the values for Y, k_d, k, and K_S from the following laboratory data.

Q (l/d)	S_0 (mg/l)	Q_w (l/d)	X (mg/l)	S_e (mg/l)	X_e (mg/l)
34.3	126	0.25	1730	5.2	9.4
34.3	126	0.35	1500	7.3	8.0
34.3	126	0.80	968	10.5	8.4
34.3	126	0.90	848	11.5	7.9

Solution: The following calculations are for the first test period. Using Eq. (12.62),

$$\theta_c = \frac{10\,\text{l} \times 1730\ \text{mg/l}}{0.25\ \text{l/d} \times 1730\ \text{mg/l} + (34.3\ \text{l/d} - 0.25\ \text{l/d})9.4\ \text{mg/l}} = 23\ \text{d}$$

The biomass growth rate per liter of aeration tank volume is

$$\left(\frac{dX}{dt}\right)_g = \frac{Q_w X + (Q - Q_w)X_e}{V}$$

$$\left(\frac{dX}{dt}\right)_g = \frac{0.25\ \text{l/d} \times 1730\ \text{mg/l} + (34.3\ \text{l/d} - 0.25\ \text{l/d})9.4\ \text{mg/l}}{10\ \text{l}} = 75\ \text{mg/l} \cdot \text{d of VSS}$$

Checking the θ_c by Eq. (12.63) gives

$$\theta_c = \frac{1730\ \text{mg/l}}{75\ \text{mg/l}} = 23\ \text{d}$$

The soluble BOD utilization rate using Eq. (12.73) is

$$\left(\frac{dS}{dt}\right)_u = \frac{(34.3\ \text{l/d})(126\ \text{mg/l} - 5.2\ \text{mg/l})}{101} = 414\ \text{mg/l} \cdot \text{d}$$

The specific substrate utilization rate as defined by Eq. (12.69) is

$$U = \frac{414\ \text{mg/l} \cdot \text{d}}{1730\ \text{mg/l}} = 0.24\ \text{d}^{-1}$$

The following values are calculated for plotting the data:

$$\frac{1}{\theta_c} = \frac{1}{23\ \text{d}} = 0.043\ \text{d}^{-1}$$

$$\frac{1}{U} = \frac{1}{0.24\ \text{d}^{-1}} = 4.2\ \text{d}$$

$$\frac{1}{S_e} = \frac{1}{5.2\ \text{mg/l}} = 0.19\ (\text{mg/l})^{-1}$$

The calculated data for all of the test runs are listed in Table 12.6.

The plot in Figure 12.45 is to determine the values of Y and k_d. From the slope of the line, the growth yield $Y = 0.35$ mg VSS/mg BOD. The intercept on the vertical axis is a microbial decay coefficient of $k_d = 0.04\ \text{d}^{-1}$.

TABLE 12.6 Data Calculated for Example 12.14 and Plotted in Figures 12.45 and 12.46 to Determine Kinetic Constants

θ_c (d)	$\left(\frac{dX}{dt}\right)_g$ (mg/l·d)	$\left(\frac{dS}{dt}\right)_u$ (mg/l·d)	U (d^{-1})	$1/\theta_c$ (d^{-1})	1/U (d)	$1/S_e$ [(mg/l)$^{-1}$]
23	75	414	0.24	0.0444	4.2	0.19
19	80	407	0.27	0.053	3.7	0.14
9.2	106	396	0.41	0.11	2.4	0.095
8.3	103	392	0.46	0.12	2.2	0.087

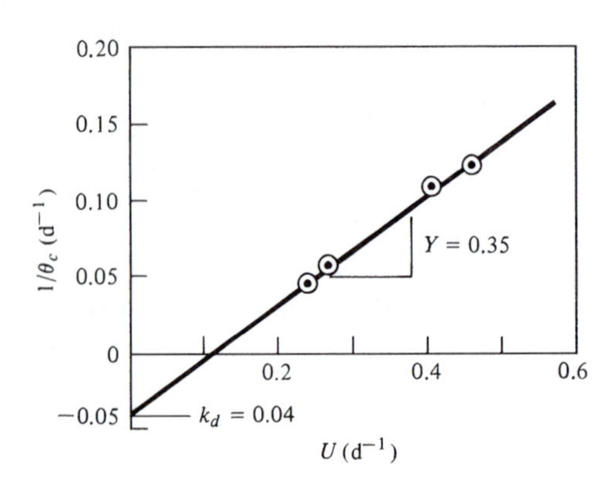

FIGURE 12.45 Plot of $1/\theta_c$ versus U to determine Y and k_d for Example 12.14.

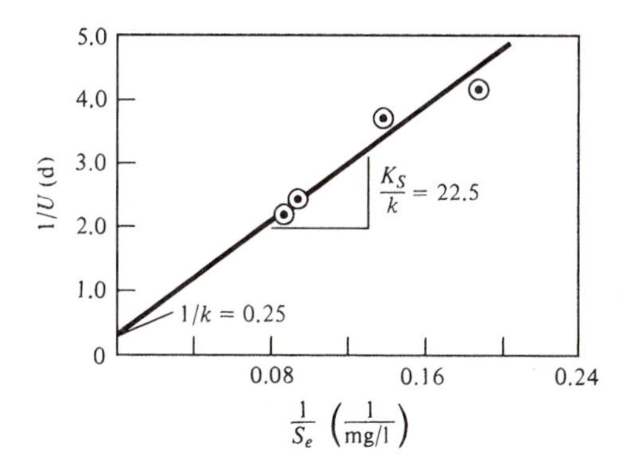

FIGURE 12.46 Plot of $1/U$ versus $1/S_e$ to determine k and K_s for Example 12.14.

The plot in Figure 12.46 is to determine the values of k and K_s. The intercept on the vertical axis is at $1/k = 0.25$ d; therefore, the maximum rate of substrate utilization per unit mass of biomass k is 4.0 d^{-1}. The slope of the line K_s/k is 22.5; hence,

$$K_s = 22.5 \text{ d} \cdot \text{mg/l} \times 4.0 \text{ d}^{-1} = 90 \text{ mg/l of BOD}$$

12.24 APPLICATION OF THE KINETICS MODEL IN PROCESS DESIGN

The equations developed in Section 12.22 can be applied in the design of a completely mixed activated-sludge process based on kinetic constants determined by laboratory testing. Since the model is based on steady-state conditions that do not exist in full-scale systems, the selection of design parameters such as the mean cell residence time must account for the diurnal and random variations in wastewater loadings. Allowances must also be made for characteristics of the actual wastewater not taken into consideration in the theoretical equations. Rather than consisting solely of a soluble substrate, municipal wastewater contains suspended solids and an abundance of biological organisms. For example, the BOD of a treatment plant effluent includes oxygen demand from both the soluble organic matter and the volatile suspended solids. During the laboratory evaluation of wastewater treatability, both total BOD and filtered BOD analyses can be performed on the effluent from the bench-scale unit to correlate total and soluble substrate. Also, an evaluation of sludge settleability is necessary to establish design criteria for the final clarifier to ensure good gravity separation of the biological suspended solids.

The first step in process design is to select the desired concentration of effluent soluble BOD based on the allowable total effluent BOD and the anticipated performance in final clarification. Soluble BOD removal efficiency of the system is calculated by the following formula:

$$E = \frac{(S_0 - S_e)100}{S_0} \tag{12.79}$$

where

E = efficiency of soluble BOD removal, %

S_0 = influent soluble BOD concentration, mg/l

S_e = effluent soluble BOD concentration, mg/l

The recommended loading criterion for completely mixed activated sludge is the mean cell residence time defined by Eq. (12.62):

$$\theta_c = \frac{VX}{Q_w X + (Q - Q_w)X_e}$$

where

θ_c = mean cell residence time d

V = volume of aeration tank, m^3

Q = influent wastewater flow, m^3/d

Q_w = rate of excess sludge wasting, m^3/d

X = concentration of MLVSS in aeration tank, mg/l

X_e = concentration of VSS in effluent, mg/l

A completely mixed activated-sludge process with an aeration period of 5–7 hr has a short θ_c of 3–5 d during the period of peak diurnal flow and is likely to exceed the effluent standards of 30 mg/l of total BOD and 30 mg/l of suspended solids. Under dispersed plug flow in long tanks (step aeration and conventional aeration), θ_c is in the range of 5–15 d and can provide a process efficiency resulting in an effluent of satisfactory uniform quality. A completely mixed extended-aeration process produces a high-quality effluent because of the long aeration period of 20–30 hr and θ_c of 20+ d. Selection of the mean cell residence time in design takes into consideration such factors as process efficiency, treatment reliability, and load variations.

Another common loading criterion is the food/microorganism ratio (F/M), which is defined for the kinetics model as

$$\text{F/M} = \frac{QS_0}{VX} = \frac{S_0}{\theta X} \tag{12.80}$$

where

F/M = food/microorganism ratio, g/d of soluble BOD applied per g of MLVSS in the aeration tank

Rearranging Eq. (12.70) allows one to express the specific substrate utilization rate as

$$U = \frac{QS_0}{VX} = \frac{S_0 - S_e}{S_0} \tag{12.81}$$

Then substituting the appropriate terms from Eqs. (12.79) and (12.80) gives a relationship between the food/microorganism ratio and the specific substrate utilization rate:

$$U = \frac{(\text{F/M})E}{100} \tag{12.82}$$

Replacing U in Eq. (12.71) with Eq. (12.82) relates θ_c and F/M such that

$$\frac{1}{\theta_c} = \frac{Y(\text{F/M})E}{100} - k_d \tag{12.83}$$

where

θ_c = mean cell residence time, d

Y = growth yield, unitless

F/M = food/microorganism ratio, g/d of soluble BOD per g of MLVSS

E = soluble BOD removal, %

k_d = microbial decay coefficient, d^{-1}

After establishing the desired effluent quality S_e and mean cell residence time θ_c, or the F/M, the required aeration tank volume can be calculated from Eq. (12.75):

$$V = \frac{\theta_c Y Q(S_0 - S_e)}{X(1 + k_d \theta_c)}$$

The choice of design flow Q and soluble influent BOD S_0 depends on the anticipated flow and strength variations, as discussed in Sections 12.10 and 12.11. Values of Y and k_d are determined from laboratory testing. Thus, the only remaining design parameter is X, the mixed-liquor volatile suspended solids to be maintained in the aeration tank.

Selection of an MLVSS concentration is based on a number of considerations, the most important of which are (1) the ability of the final clarifier to provide gravity separation of the activated-sludge suspended solids and (2) the oxygen transfer capacity of the aeration system. At a low design MLVSS the aeration period is long, resulting in an extended process time. A concentration of MLVSS that is too high produces a process characterized by poor soluble BOD removal efficiency and high suspended-solids concentration in the effluent, resulting from the limited aeration period and poor settleability of the microbial floc. In general, conventional secondary activated-sludge systems processing municipal wastewaters operate in the MLSS range of 1500–3000 mg/l with 70%–80% being volatile solids. In a typical design, therefore, the optimum MLVSS is within the range of 1200–2400 mg/l.

Waste sludge production in terms of volatile solids can be calculated based on the kinetics model. The observed growth yield Y_{obs}, as defined in Eq. (12.34), can be obtained by first substituting Eq. (12.32) for $(dX/dt)_g^{net}$ in the numerator and then replacing $(dS/dt)_u$ with UX from Eq. (12.69):

$$Y_{obs} = \frac{YU - k_d}{U} \tag{12.84}$$

Substituting the relationship from Eq. (12.71) for U in Eq. (12.84) gives

$$Y_{obs} = \frac{Y}{1 + \theta_c k_d} \quad (12.85)$$

where

Y_{obs} = observed growth yield, g of MLVSS/g of soluble BOD

Using this equation, Lawrence and McCarty [18] expressed the production of excess biomass in the waste-activated sludge as

$$P_x = \frac{YQ(S_0 - S_e)}{1 + \theta_c k_d} \quad (12.86)$$

where

P_x = volatile solids in waste sludge, g/d

Example 12.15

A completely mixed activated-sludge process is being designed for a wastewater flow of 10,000 m^3/d (2.64 mgd) using the kinetics equations. The influent BOD of 120 mg/l is essentially all soluble and the design effluent soluble BOD is 7 mg/l, which is based on a total effluent BOD of 20 mg/l. For sizing the aeration tank, the mean cell residence time is selected to be 10 d and the MLVSS 2000 mg/l. The kinetic constants from a bench-scale treatability study are as follows: $Y = 0.60$ mg VSS/mg BOD, $k_d = 0.06\ d^{-1}$, $K_s = 60$ mg/l of BOD, and $k = 5.0\ d^{-1}$.

Solution: From Eq. (12.79), the required soluble BOD efficiency is

$$E = \frac{(120 - 7)100}{120} = 94\%$$

Rearranging Eq. (12.83), one obtains the food/microorganism ratio for $\theta_c = 10$ d and $E = 94\%$:

$$\text{F/M} = \frac{(1/\theta_c + k_d)100}{YE} = \frac{(1/10 + 0.06)100}{0.60 \times 94} = 0.28 \frac{\text{g of soluble BOD}}{\text{g of MLVSS}}$$

The volume of the aeration tank, based on Eq. (12.75), is

$$V = \frac{10 \times 0.60 \times 10{,}000(120 - 7)}{2000(1 + 0.06 \times 10)} = 2100\ m^3\ (74{,}800\ ft^3)$$

$$\theta = \frac{2100 \times 24}{10{,}000} = 5.0\ hr$$

From Eq. (12.68), the soluble BOD is

$$S_e = \frac{60(1 + 0.06 \times 10)}{10(0.60 \times 5.0 - 0.06) - 1} = 3.4 \text{ mg/l}$$

From Eq. (12.86), the excess volatile solids in the waste sludge are

$$P_x = \frac{0.60 \times 10{,}000(120 - 7)}{1 + 10 \times 0.06} = 420{,}000 \text{ g/d} = 420 \text{ kg/d}$$

12.25 OXYGEN TRANSFER AND OXYGENATION REQUIREMENTS

In activated-sludge processes, oxygen is supplied to the microorganisms by dispersing air into the mixed liquor by either diffused-air or mechanical surface aeration. Diffused-air systems use a variety of fine- and coarse-bubble diffusers. The two kinds of mechanical aerators, differentiated by the plane of rotation, are horizontal rotors and impellers mounted on vertical shafts.

The commonly accepted oxygen transfer scheme is diagrammed in Figure 12.47. Oxygen is dissolved in solution and then extracted from solution by the biological cells. Direct oxygen transfer from bubble to cell is possible if the microorganisms are adsorbed on the bubble surface. Bennett and Kempe [21] demonstrated direct oxygen transfer in a laboratory fermenter using a culture of *Pseudomonas ovalis* converting glucose to gluconic acid. The extent of direct oxygen transfer in activated-sludge systems is not known; however, it is generally felt to be secondary to oxygen transfer through the intermediate dissolved-oxygen phase.

The rate of oxygen transfer from air bubbles to dissolved oxygen in an aeration tank is expressed as

$$\frac{dc}{dt} = \alpha K_L a(\beta C_s - C_t) \tag{12.87}$$

where

dc/dt = rate of oxygen transfer, mg/l/hr

α = oxygen transfer coefficient of the wastewater

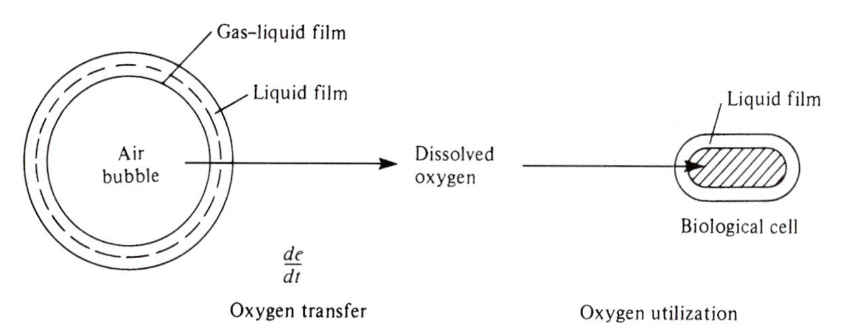

FIGURE 12.47 Schematic diagram of oxygen transfer in activated sludge.

β = oxygen saturation coefficient of the wastewater

K_La = oxygen transfer coefficient, hr^{-1}

C_s = oxygen concentration at saturation, mg/l

C_t = oxygen concentration in the liquid, mg/l

$\beta C_s - C_t$ = dissolved-oxygen deficit, mg/l

Equation (12.87) without the α and β coefficients applies to clean water. The factors α and β depend on the characteristics of the wastewater being aerated, primarily the concentration of dissolved solids; K_La depends on the temperature and the aeration system features, such as the type of diffuser, depth of aerator, type of mixer, and tank geometry. In general, the rate of oxygen transfer increases with decreasing bubble size, longer contact time, and added turbulence. Methods for determining the coefficients K_La, α, and β are discussed in Section 12.26.

The rate of dissolved-oxygen utilization by microorganisms in an activated-sludge system can be determined by placing a sample of mixed liquor in a closed container and measuring the dissolved-oxygen depletion with respect to time. The slope of the resultant curve r is the oxygen utilization rate. Figure 12.48 is a dissolved-oxygen depletion curve for a mixed liquor from a high-rate activated-sludge aeration tank. The r value depends on the microorganisms' ability to metabolize the waste organics based on such factors as the food/microorganism ratio, mixing conditions, and temperature.

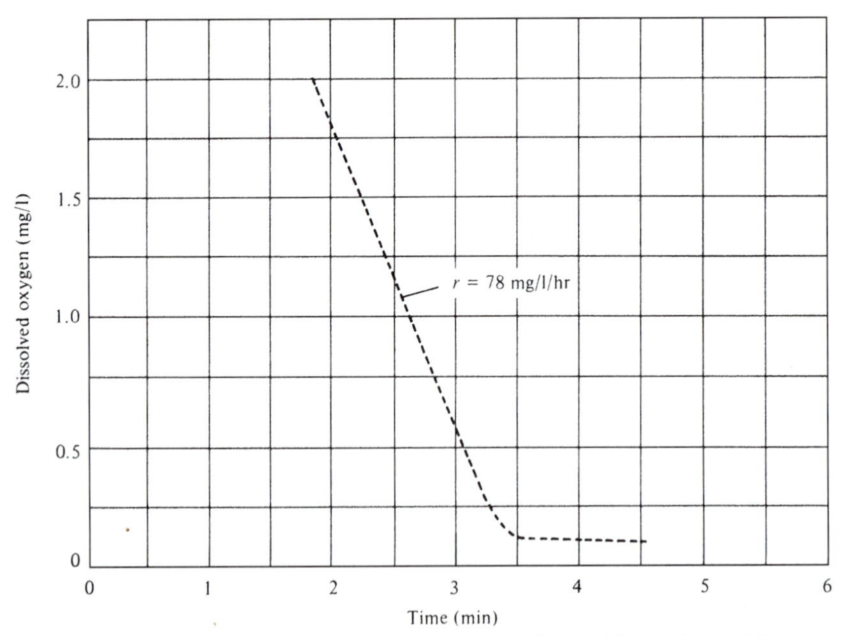

FIGURE 12.48 Oxygen utilization curve for a sample of mixed liquor from a high-rate activated-sludge aeration basin.

A general range for r in the mixed liquor of conventional and high-rate activated-sludge systems is 30–100 mg/l/hr.

Under steady-state conditions of oxygen transfer in an activated-sludge system, the rate of oxygen transfer to dissolved oxygen (dc/dt) is equal to the rate of oxygen utilization (r). Substituting r in Eq. (12.87) for dc/dt and rearranging yields the following relationship:

$$\alpha K_L a = \frac{r}{\beta C_s - C_t} \tag{12.88}$$

where

r = oxygen utilization rate by microorganisms in activated sludge, mg/l/hr

The rate of aerobic microbial metabolism is independent of the dissolved-oxygen concentration above a critical (minimum) value. Below the critical value, the rate is reduced by the limitation of oxygen required for respiration. Critical dissolved-oxygen concentrations reported in the literature for activated-sludge systems range from 0.2 to 2.0 mg/l, depending on the type of activated-sludge process and characteristics of the wastewater. The most frequently referenced critical dissolved-oxygen value for conventional and high-rate aeration basins is 0.5 mg/l.

The aerator power required to satisfy microbial oxygen demand and to provide adequate mixing in an aeration tank depends on the type of activated-sludge process, BOD loading, and oxygen transfer efficiency of the aeration equipment. In the design of any activated-sludge system, the power requirements should be based on proven performance of the aeration equipment. The capacity of the aeration equipment must furnish sufficient air to meet peak BOD loads without the dissolved-oxygen concentration dropping below the critical level for aerobic metabolism.

Aeration systems are compared on the basis of mass of gaseous oxygen transferred to dissolved oxygen per unit of energy expended, pounds of oxygen per horsepower-hour (kilogram per kilowatt-hour). Oxygen transfer efficiency is expressed as the percentage of the mass of oxygen dissolved in the water relative to the applied mass of gaseous oxygen. For the purpose of comparison, the values specified for efficiency are based on operation in clean water with zero dissolved-oxygen concentration and standard conditions of 20°C and 1 atm pressure. Table 12.7 lists oxygen transfer efficiencies and oxygen transfer rates for different kinds of aeration systems. Of course, the selection of an aeration system in process design must also consider other factors, including the flexibility and reliability of operation, effective mixing, and the maintenance of equipment. An economic analysis encompasses capital, operation, and maintenance costs.

The amount of dissolved oxygen needed for treatment of a wastewater depends on the carbonaceous and nitrogenous oxygen demands that are satisfied. For biological oxidation of carbonaceous matter, the oxygen requirement varies from approximately 0.8 to 1.6 times the BOD of the applied wastewater for a corresponding food/microorganism loading range of extended aeration to high-rate aeration. The oxygen requirement for nitrification is 4.6 times the ammonia nitrogen oxidized to nitrate.

The *Recommended Standards for Wastewater Facilities, Great Lakes—Upper Mississippi River Board of State Public Health & Environmental Managers* [22] considers

TABLE 12.7 Oxygen Transfer Data for Air Aeration Systems in Clean Water at 15-Ft Submergence

System	Oxygen Transfer Efficiency (%)	Oxygen Transfer Rate (lb/hp · hr)[a]
Fine-bubble diffusers, total floor coverage	20–32[b]	4.0–6.5
Fine-bubble diffusers, side-wall installation	11–15[b]	2.2–3.0
Jet aerators (fine bubble)	22–27[b]	4.0–5.0
Static aerators (medium-sized bubble)	12–14[b]	2.3–2.8
Mechanical surface aerators	—	2.5–3.5[c]
Coarse-bubble diffusers, wide-band pattern	6–8[c]	1.2–1.6
Coarse-bubble diffusers, narrow-band pattern	4–6[c]	0.8–1.2

[a] 1.0 lb/hp · hr = 0.61 kg/kW · h
[b] From manufacturers' bulletins and technical reports
[c] Common ranges for variations of these systems
Source: Proceedings, Workshop Toward an Oxygen Transfer-Standard, Environmental Protection Agency, EPA 600/9-78-021 (April 1979): 13.

the following as minimum normal air requirements for diffused-air systems: conventional, step aeration, and contact stabilization 1500 ft^3 of air applied per lb of BOD aeration tank load; modified or high-rate 400–1500 ft^3/lb of BOD load; and extended aeration 2000 ft^3/lb of BOD load. These demands assume that the aeration equipment is capable of transferring at least 1.0 lb of oxygen to the mixed liquor per pound of BOD aeration tank loading. In any case, aeration equipment shall be capable of maintaining a minimum of 2.0 mg/l of dissolved oxygen in the mixed liquor at all times and ensuring thorough mixing of the mixed liquor.

Example 12.16

The following data were collected during field evaluation of a completely mixed activated-sludge secondary treating of municipal wastewater. The aeration basin, with a diameter of 80 ft and a liquid depth of 17 ft, was mixed with four turbine mixers mounted above air sparge rings. Twenty-four-hour composite BOD analyses were run on the aeration basin influent, final clarifier effluent, and waste-activated sludge. The oxygen-utilization rate in the aeration basin was measured each hour throughout the 24-hr sampling period and individual values averaged for oxygen utilization rate of the mixed liquor.

$$\text{influent wastewater flow} = 6.52 \text{ mgd}$$

$$\text{waste-activated sludge} = 15{,}000 \text{ gpd}$$

$$\text{influent BOD} = 125 \text{ mg/l}$$

$$\text{effluent BOD} = 18 \text{ mg/l}$$

$$\text{waste sludge BOD} = 5300 \text{ mg/l}$$

$$\text{air supplied (20°C and 760 mm)} = 1650 \text{ cfm}$$

$$\text{minimum DO in mixed liquor} = 0.8 \text{ mg/l}$$

$$\text{average DO in mixed liquor} = 1.1 \text{ mg/l}$$

$$\text{temperature of mixed liquor} = 24°\text{C}$$

$$\text{oxygen utilization rate of mixed liquor} = 74 \text{ mg/l/hr}$$

$$\text{beta factor of mixed liquor} = 0.9$$

Use these data to calculate the following:

1. Pounds of BOD load.
2. Cubic feet of air applied per lb of BOD load.
3. Pounds of oxygen utilized per lb of BOD.
4. Oxygen transfer efficiency.
5. $\alpha K_L a$.

Solution:

1.

$$\text{lb of BOD load} = 6.52 \text{ mgd} \times 125 \text{ mg/l} \times 8.34 = 6800 \text{ lb/day}$$

$$\text{volume aeration tank} = \pi(40)^2 17 = 85{,}500 \text{ ft}^3$$

$$\text{BOD loading} = 79.5 \text{ lb of BOD/1000 ft}^3\text{/day}$$

2.

$$\text{air applied} = 1650 \frac{\text{ft}^3}{\text{min}} \times 1440 \frac{\text{min}}{\text{day}} = 2{,}380{,}000 \text{ ft}^3$$

$$\frac{\text{ft}^3 \text{ of air applied}}{\text{lb of BOD load}} = \frac{2{,}380{,}000}{6800} = 350 \frac{\text{ft}^3}{\text{lb of BOD}}$$

3.

$$\text{lb of oxygen utilized} = r \times \text{volume of aeration tank} \times \text{time}$$

$$= 74 \frac{\text{mg/l}}{\text{hr}} \times 28.3 \frac{1}{\text{ft}^3} \times 85{,}500 \text{ ft}^3 \times \frac{\text{lb}}{453{,}600 \text{ mg}} \times 24 \frac{\text{hr}}{\text{day}}$$

$$= 9420 \text{ lb/day}$$

$$\text{lb of BOD satisfied} = \text{lb of BOD removed} - \text{lb of BOD wasted}$$

$$= 6.52(125 - 18)8.34 - 0.015 \times 5300 \times 8.34$$

$$= 5190 \text{ lb/day}$$

$$\frac{\text{lb of oxygen utilized}}{\text{lb of BOD satisfied}} = \frac{9420}{5190} = 1.82$$

$$\frac{\text{lb of oxygen utilized}}{\text{lb of BOD applied}} = \frac{9420}{6800} = 1.39$$

4. $$\text{lb of oxygen applied} = 2{,}380{,}000\frac{\text{ft}^3}{\text{day}} \times 0.0174\frac{\text{lb of oxygen}}{\text{ft}^3}$$

$$= 41{,}400 \text{ lb/day}$$

$$\text{oxygen transfer efficiency} = \frac{9420}{41{,}400} \times 100 = 22.8\%$$

5. From Eq. (12.88),

$$\alpha K_L a \text{ at } 24°\text{C} = \frac{74}{0.9 \times 8.5 - 1.1} = 11 \text{ hr}^{-1}$$

12.26 DETERMINATION OF OXYGEN TRANSFER COEFFICIENTS

In order to apply Eq. (12.87) in calculating the mass transfer of oxygen in the design of an activated-sludge process, the coefficients $K_L a$, α, and β must be experimentally determined.

The rate of oxygen transfer in clean water is defined as

$$\frac{dc}{dt} = K_L a(C_s - C_t) \tag{12.89}$$

where

dc/dt = rate of oxygen transfer, mg/l/hr

$K_L a$ = oxygen transfer coefficient, hr^{-1}

C_s = oxygen concentration at saturation, mg/l

C_t = oxygen concentration in liquid, mg/l

The rate of oxygen dissolution is proportional to the dissolved-oxygen deficit $(C_s - C_t)$ and the area of the air–water interface per unit volume of water. $K_L a$ is the overall coefficient that incorporates the interfacial area a of diffusion and the liquid film coefficient K_L. The value of $K_L a$ depends on the hydrodynamics and turbulence at the interface between the air bubbles and the liquid; hence, it depends on the aeration system, geometry of the aeration tank, liquid characteristics, and temperature.

The efficiency of an aeration system to transfer oxygen is measured by conducting a non-steady-state test on a full-scale aeration basin or test tank using clean water [23, 24]. The clean aeration tank is filled with tap water at a temperature as close to 20°C as possible. Next, a cobalt chloride catalyst is dissolved in a small amount of warm water and added to the aeration tank; the concentration must be high enough to assure catalyzation of all of the sodium sulfite added. After operating the aerator for 20–30 min to achieve a steady-state mixing condition, the sodium sulfite is added to deoxygenate the water in the aeration tank as follows:

$$2\,Na_2SO_3 + O_2 \xrightarrow{\text{cobalt}} 2\,Na_2SO_4 \tag{12.90}$$

The sulfite addition is in excess of the theoretical requirement (7.88 mg/l of pure sodium sulfite per 1.0 mg/l of DO concentration) to allow a time lag for mixing before the dissolved oxygen starts to rise above zero. Simultaneously, sampling is initiated at several points in the aeration tank when the DO concentration begins to rise from zero and is continued at 1–3-min intervals, or at approximately every 1.0 mg/l increase in dissolved oxygen. At least six samples are collected at each point between the levels of 10% and 80% DO saturation. Water for sampling is continuously withdrawn by submersible pumps with sufficient capacity to limit the detention time between pump and sample outlet to 5–10 sec. Although oxygen concentrations are monitored and recorded by DO probes and meters, the standard test for dissolved oxygen is by the Winkler titration method. Three replicate tests are normally conducted to determine the aeration efficiency for each operating condition.

Dissolved-oxygen data from each sampling point are plotted to determine the $K_L a$ value based on the following relationship, derived from Eq. (12.89):

$$K_L a = \ln\left(\frac{C_s - C_2}{C_s - C_1}\right) \Big/ (t_2 - t_1) \tag{12.91}$$

Nonparallel slopes of the plots from different sampling points indicate poor mixing, and $K_L a$ values that differ significantly from the others are discarded. The saturation concentration C_s is the theoretical value at the temperature of the water during the test (Table A.10). In the case of diffused-air systems, a correction factor for submergence of the bubbles is included in the pressure correction; this is normally taken as the pressure at one-half the depth of submergence of the diffusers. The $K_L a$ at test temperature in degrees Celsius is corrected to 20°C by the relationship

$$(K_L a)_{20} = (K_L a)_T \Theta^{20-T} \tag{12.92}$$

where Θ is commonly assumed to be 1.024. (The observed range is 1.01–1.05.)

The mass of oxygen dissolved in the water contained in a test tank per unit time at standard conditions (20°C, 1 atm pressure, and zero DO) is calculated as

$$N = 10^{-6} (K_L a)_{20} (C_s)_{20} W \tag{12.93}$$

where

N = rate of oxygen dissolution, lb/hr

$(K_L a)_{20}$ = oxygen transfer coefficient at 20°C, hr^{-1}

$(C_s)_{20}$ = oxygen concentration at saturation at 20°C, mg/l

W = weight of water in the test tank, lb

$10^{-6} \sim 1$ mg/l $\sim$ 1 mg/1,000,000 mg

Oxygen transfer efficiency E in a diffused-aeration system is computed by

$$E = \frac{N \times 100}{A} \tag{12.94}$$

where

$$E = \text{oxygen transfer efficiency, \%}$$
$$A = \text{oxygen applied (standard conditions), lb/hr}$$

Determination of the applied oxygen requires accurate measurement of the air flow rate and adjustment of the observed rate to standard conditions of 20°C and 1 atm of pressure.

The oxygen transfer rate can be calculated for both diffused-air and mechanical aeration systems by the relationship

$$R_0 = \frac{N}{P} \tag{12.95}$$

where

R_0 = rate of oxygen transfer at standard conditions (20°C, 1 atm pressure, and zero DO), lb/hp · hr

P = power input, hp

The rate of oxygen transfer to a wastewater requires determining the alpha and beta coefficients in Eq. (12.87). The alpha coefficient is defined as the ratio of the oxygen transfer coefficient in wastewater to that in clean water,

$$\alpha = \frac{K_L a \text{ in wastewater}}{K_L a \text{ in clean water}} \tag{12.96}$$

The value of α is influenced by many conditions related to both the characteristics of the wastewater (temperature, soluble BOD, and concentration of suspended solids) and the aeration equipment (type of diffuser or mechanical aerator, mixing intensity, and aeration tank configuration). The magnitude can even change between the influent and effluent ends of a plug-flow aeration tank resulting from stabilization of the wastewater. Even though the most reliable method of measuring α is under field design conditions, it is often determined using a bench-scale aeration tank. Different laboratory units are designed to simulate diffused, mechanical-surface, and submerged-turbine aeration systems. The procedure involves conducting tests for $K_L a$ in the model aeration tank for both tap water and wastewater [23, 24]. Deoxygenation is performed by stripping the liquid in the tank with nitrogen gas. Conducting tests to determine alpha requires considerable expertise in oxygen transfer processes and laboratory techniques.

Alpha coefficients for municipal wastewater are generally in the range of 0.7–0.9; nevertheless, fine-bubble diffusers can have a value as low as 0.4 and mechanical aerators as high as 1.1.

The beta coefficient in Eq. (12.87) is defined as the ratio of the DO saturation concentration in the wastewater to that in clean water,

$$\beta = \frac{\text{DO saturation concentration in wastewater}}{\text{DO saturation concentration in clean water}} \tag{12.97}$$

The value of β is influenced by wastewater constituents, including dissolved salts, organics, and gases. To determine the saturation concentration in a wastewater, a settled sample is aerated by vigorous hand mixing for several minutes in a half-full jar. The temperature and dissolved oxygen are then both measured, usually with a calibrated DO probe, for several minutes to ensure stable readings. Saturation for clean water is the theoretical value for the same temperature and is corrected for the barometric pressure (Table A.10). These two values are used in Eq. (12.97) to calculate β.

The magnitude of the beta coefficient for municipal wastewater typically equals 0.9 and is seldom less than 0.8.

Example 12.17

A coarse-bubble, diffused-aeration system was tested in a large tank to determine the oxygen transfer coefficient by the non-steady-state procedure in clean water. The diffusers were submerged 8.0 ft, and the atmospheric pressure on the day of the test was 720 mm Hg. Dissolved oxygen data from a sampling point at a depth of 4.0 ft are given in columns 1 and 3 of Table 12.8. The water temperature was 17°C.

The test tank was circular, with a diameter of 20.0 ft. The air supply during the test was 310 cfm (20°C and 760 mm) and the power input was 15 hp.

Determine the value of K_La and, based on this value and the above operating data, calculate the oxygen transfer efficiency and oxygen transfer rate.

Solution: The pressure at one-half the depth of submergence of the diffusers is

$$\frac{4.0 \text{ ft} \times 305 \text{ mm/ft}}{13.6 \text{ mm } H_2O/1.0 \text{ mm Hg}} = 90 \text{ mm Hg}$$

Therefore, the barometric pressure at mid-depth is equal to 720 + 90 = 810 mm Hg.

Using Table A.10,

$$C_s' = C_s \frac{P - p}{760 - p} = 9.7 \times \frac{810 - 15}{760 - 15} = 10.4 \text{ mg/l}$$

This value for the oxygen concentration at saturation (17°C and 4.0 ft submergence) is entered in column 2 of Table 12.8 and the $C_s - C_t$ and $\ln(C_s - C_t)$ calculated.

TABLE 12.8 Sampling Data and Calculated Values for Determination of the Oxygen Transfer Coefficient in Example 12.7

t (min)	C_s (mg/l)	C_t (mg/l)	$C_s - C_t$ (mg/l)	$\ln(C_s - C_t)$
1.2	10.4	2.3	8.1	2.1
2.8	10.4	4.6	5.8	1.8
5.0	10.4	6.7	3.7	1.3
8.0	10.4	8.2	2.2	0.8
12.0	10.4	9.4	1.0	0
16.0	10.4	10.0	0.4	−0.9
22.0	10.4	10.2	0.2	−1.6

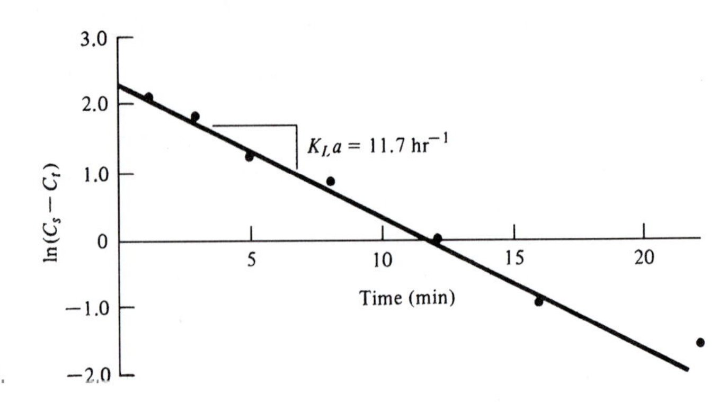

FIGURE 12.49 Plot of data from Table 12.8 to determine $K_L a$ for Example 12.17.

The $\ln(C_s - C_t)$ versus time data are plotted in Figure 12.49. The slope of a straight line of best fit through the points between 20% and 80% DO saturation is K_La, which is calculated using Eq. (12.91) as

$$K_La = \frac{(2.0 - 0)60}{11.8 - 1.5} = 11.7\ \text{hr}^{-1}$$

Correcting K_La to 20°C by Eq. (12.92),

$$(K_La)_{20} = 11.7(1.024)^{20-17} = 12.6\ \text{hr}^{-1}$$

The following parameters are calculated in order to use Eq. (12.93):

$$(C_s)_{20} = 9.2 \times \frac{810 - 18}{760 - 18} = 9.8\ \text{mg/l}$$

$$W = \pi(10)^2 \times 8.0 \times 62.4 = 157{,}000\ \text{lb}$$

Substituting into Eq. (12.93),

$$N = 10^{-6} \times 12.6 \times 9.8 \times 157{,}000 = 19.4\ \text{lb/hr}$$

Calculating oxygen transfer efficiency by Eq. (12.94),

$$E = \frac{19.4\ \text{lb/hr} \times 100}{310\ \text{ft}^3/\text{min} \times 60\ \text{min/hr} \times 0.0174\ \text{lb of oxygen/ft}^3} = \frac{19.4 \times 100}{323} = 6.0\%$$

The rate of oxygen transfer from Eq. (12.95) is

$$R_0 = \frac{19.4}{15} = 1.3\ \text{lb/hp} \cdot \text{hr}$$

Example 12.18

A step-aeration activated-sludge process is being designed using the diffused-aeration system described in Example 12.17. The design criteria for the period of critical oxygen demand in the aeration basin are as follows:

$$\text{BOD loading} = 3.3\ \text{lb BOD/1000 ft}^3\text{/hr}$$

$$\text{oxygen transfer requirement} = 1.0\ \text{lb of oxygen/1.0 lb of BOD applied}$$

temperature of mixed liquor = 14°C

minimum allowable DO = 2.0 mg/l

beta coefficient = 0.70

alpha coefficient = 0.90

oxygen transfer coefficient $(K_L a)_{20}$ = 12.0 hr^{-1}

oxygen transfer efficiency E = 6.0%

pressure at mid-depth of diffusers = 810 mm Hg

Compare the rate of oxygen transfer dc/dt to the rate of oxygen utilization r, and calculate the volume of standard air required per pound of BOD applied. (One cubic foot of air at standard temperature and pressure, 20°C and 760 mm, contains 0.0174 lb oxygen.)

Solution: Using Table A.10 for 14°C,

$$C_s = 10.4 \times \frac{810 - 12}{760 - 12} = 11.1 \text{ mg/l}$$

Correcting $(K_L a)_{20}$ for 14°C,

$$K_L a \text{ at } 14°\text{C} = (K_L a)_{20}\, 1.024^{T-20} = 12.0(1.024)^{-6} = 10.4 \text{ hr}^{-1}$$

Substituting into Eq. (12.87), the rate of oxygen transfer is

$$\frac{dc}{dt} = 0.70 \times 10.4(0.90 \times 11.1 - 2.0) = 58 \text{ mg/l/hr}$$

Assuming 1.0 lb of oxygen utilized per 1.0 lb of BOD applied, the rate of dissolved-oxygen utilization is

$$r = \frac{3.3 \text{ lb}}{1000 \text{ ft}^3 \cdot \text{hr}} \times \frac{453{,}600 \text{ mg}}{\text{lb}} \times \frac{\text{ft}^3}{28.3 \text{ l}} = 53 \text{ mg/l/hr}$$

The oxygen transfer rate of the aeration system is adequate since dc/dt exceeds r. (Refer to Fig. 12.47.)

The efficiency of oxygen transfer is proportional to the rate of oxygen transfer; therefore,

$$E_{\text{actual}} = E\frac{\alpha(K_L a)(\beta C_s - C_t)}{(K_L a)_{20}(C_s)_{20}}$$

$$= 6.0 \times \frac{0.70 \times 10.4(0.90 \times 11.1 - 2.0)}{12.6 \times 9.8} = 2.8\% \qquad (12.98)$$

The volume of standard air required at an actual oxygen transfer efficiency of 2.8% per pound of BOD load, assuming 1.0 lb of oxygen utilized per 1.0 lb of BOD applied, is

$$\frac{1.0 \text{ lb of oxygen}}{0.0174 \text{ lb of oxygen/ft}^3 \times 0.028} = 2000 \text{ ft}^3\text{/lb of BOD applied}$$

STABILIZATION PONDS

Domestic wastewater can be effectively stabilized by the natural biological processes that occur in shallow ponds. Those treating raw wastewater are referred to as facultative ponds, lagoons, or oxidation ponds. Where small ponds are installed after secondary treatment, they are referred to as tertiary, maturation, or polishing ponds. Their purpose is to further reduce suspended solids, BOD, fecal microorganisms, and ammonia in the plant effluent.

Facultative ponds have a light BOD loading of 0.1–0.3 lb/1000 ft^3/day, a normal operating water depth of 5 ft, and a long retention time of 50–150 days. Small ponds may be designed for complete retention with water loss only by evaporation. Tertiary polishing ponds generally have a retention time of only 10–15 days and are shallower, with water depths of 2–3 ft for better mixing and sunlight penetration.

A wide variety of microscopic plants and animals find the environment a suitable habitat. Waste organics are metabolized by bacteria and saprobic protozoans as primary feeders. Secondary feeders include protozoans and higher animal forms, such as rotifers and crustaceans. When the pond bottom is anaerobic, biological activity results in digestion of the settled solids. Nutrients released by bacteria are used by algae in photosynthesis. The overall process in a facultative stabilization pond (Fig. 12.14) is the sum of individual reactions of the bacteria, protozoans, and algae.

12.27 DESCRIPTION OF A FACULTATIVE POND

A stabilization pond is a flat-bottomed pond enclosed by an earth dike (Fig. 12.50). It can be round, square, or rectangular, with a length not greater than three times the width. The operating liquid depth has a range of 2–5 ft, with 3 ft of dike freeboard. A minimum of about 2 ft is required to prevent the growth of rooted aquatic plants. Operating depths greater than 5 ft can create odorous conditions because of anaerobiosis on the bottom.

Influent lines discharge near the center of the pond and the effluent usually overflows in a corner on the windward side to minimize short-circuiting. The overflow is generally a manhole or box structure with multiple-valved draw-off lines to offer flexible operation. Where the lagoon area required exceeds 6 acres, it is good practice to have multiple cells that can be operated individually, in series, or in parallel. If the soil is pervious, bottom and dikes should be sealed to prevent groundwater pollution. A commonly used sealing agent is bentonite (clay). Dikes and areas surrounding the ponds are seeded with grass, graded to prevent runoff from entering the ponds, and fenced to preclude livestock and discourage trespassing. In the case of a multiple-pond installation, the sequence of pond operation and liquid operating depth are regulated to provide control of the treatment system. Operating ponds in series generally increases BOD reduction by preventing short-circuiting. Conversely, parallel operation may be desirable to distribute the raw BOD load and avoid potential odor problems.

Where discharges of pond effluent in the winter result in pollution of the receiving stream, the operating level can be lowered before ice formation and

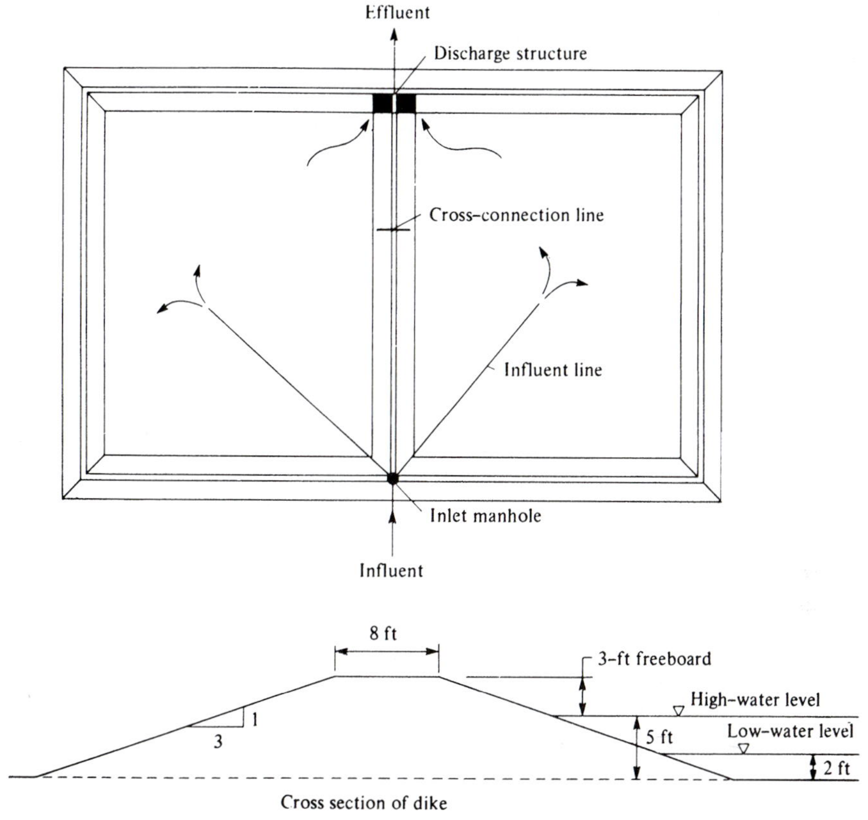

FIGURE 12.50 Two-cell stabilization pond.

gradually increased to 5 ft by retention of winter flows. The elevation can then be slowly lowered in the spring when the dilution flow in the receiving stream is high. Shallow operation can be maintained during the spring with gradually increasing depths to prevent emergent vegetation. In the fall, the level can again be lowered to hold winter flows.

Most stabilization ponds emit odors occasionally. This is the primary reason for locating them as far as practicable from present or future developed areas and on the leeward side so that prevailing winds are in the direction of uninhabited areas. Lagoons treating only domestic wastewater normally operate odor-free. Only for a short period of time in the early spring, when the ice melts and the algae are not flourishing, are offensive odors discharged. Lagoons treating certain industrial wastes, in combination with domestic wastewater, are often noted for their persistent obnoxious odors. The cause of these odors is most likely continuous or periodic overloading from industrial waste discharges or the odorous nature of the industrial waste itself. Unpleasant odors can be emitted from lagoons of small municipalities where poultry-processing wastes, slaughterhouse wastes, or creamery wastes are discharged to the municipal sewer without pretreatment.

12.28 BOD LOADINGS OF FACULTATIVE PONDS

Many efficiency equations have been proposed for modeling BOD removal in facultative ponds [25]. BOD reduction in a primary cell appears to follow a simple plug-flow hydraulic model with a first-order reaction rate; however, verification by comparison with actual performance data has been only marginally successful. More complex equations that include factors such as water temperature, light intensity, and dispersion are no more successful in predicting the efficiency of stabilization ponds. As a result, design of facultative ponds is commonly dictated by empirical rules based on the observation of pond performance in a region. Design guidelines may specify the maximum allowable BOD loading on primary cells, number of cells required for different-size pond systems, overall wastewater retention time, maximum and minimum water depths, and winter storage for a specified number of months. Example 12.20 illustrates the application of a typical set of design guidelines.

BOD loadings are usually expressed in pounds of BOD applied per acre of surface area per day (kg/ha · d or $g/m^2 \cdot d$). In northern states, loadings on primary cells are 25–35 lb BOD/acre/day (2.8–3.9 $g/m^2 \cdot d$) to minimize odor nuisance in the spring of the year. In southern states, design loadings are about 40–50 lb BOD/acre/day. Total retention time for primary plus secondary cells is 3–6 months to allow seasonal storage and controlled discharge. Series operation reduces short-circuiting and, as a result, improves BOD and fecal coliform removals.

The degree of stabilization produced in a pond is influenced by climatic conditions. During warm, sunny weather, decomposition and photosynthetic processes flourish, resulting in rapid and complete stabilization of the waste organics. The pond water becomes supersaturated with dissolved oxygen during the afternoon. Suspended solids and BOD in the pond effluent are primarily from the algae. BOD reductions in the summer usually exceed 95%. During cold weather under ice cover, biological activity is extremely slow and the lagoon-treatment process is, for practical purposes, reduced to sedimentation. Anaerobic conditions can occur from a lack of reaeration by wind action and photosynthesis. Suspended solids and BOD in the pond effluent include organics from raw wastes and intermediate organics issuing from incomplete anaerobic metabolism. Under winter ice cover, BOD reductions are generally about 50%.

Warm pond water that is rich in plant nutrients, when exposed to sunlight, supports an abundance of algae, giving the water a green color. Because wastewater contains significant concentrations of carbon dioxide, inorganic nitrogen, and soluble phosphorus, the growth of algae is usually limited by shading caused by the high turbidity from the algae in suspension. Higher forms of aquatic animal life, i.e., zooplankton, that graze on the algae are also present. As a result of this natural biological activity, pond water during summer months contains 50–80 mg/l of suspended solids. Although this exceeds the effluent standard of 30 mg/l of suspended solids, most regulatory agencies have established higher limits for stabilization pond effluent, typically 60 mg/l, to allow summer and fall discharge. Where the effluent standard prohibits discharge to a watercourse, irrigation of nearby agricultural land appears to be the best solution. In a semiarid climate where evaporation exceeds rainfall by a wide margin, ponds serving a small community can be constructed large enough to provide complete retention. However, the land area for zero-discharge ponds is often prohibitive.

12.29 ADVANTAGES AND DISADVANTAGES OF STABILIZATION PONDS

A general list of items to be considered before selecting the stabilization pond process for treatment of a municipal wastewater is offered here. In general, stabilization ponds are suitable for small towns that do not anticipate extensive industrial expansion and where land with suitable topography and soil conditions is available for siting.

Advantages

1. The initial cost is probably lower than that of a mechanical plant.
2. Operating costs are lower.
3. Regulation of effluent discharge is possible, thus providing control of pollution during critical times of the year.
4. Treatment system is not significantly influenced by a leaky sewer system that collects storm water.

Disadvantages

1. Extensive land area is required for siting.
2. The assimilative capacity for certain industrial wastes is poor.
3. Odor problems are possible.
4. The expansion of town and new developments may encroach on the lagoon site.
5. The effluent quality generally cannot meet the standard for suspended solids of 30 mg/l.

Example 12.19

Design population for a town is 1200 persons, and the anticipated industrial load is 20,000 gpd at 1000 mg/l of BOD from a milk-processing plant. Calculate the surface area required for a stabilization pond system, and estimate the number of days of winter storage available. Assume the following:

1. Wastewater flow of 100 gpcd with 0.17 lb of BOD per capita.
2. Design BOD loading of 25 lb of BOD/acre/day.
3. Water loss from evaporation and seepage of 60 in./yr.
4. Annual rainfall of 20 in./yr.

Solution:

BOD load (domestic + industrial)

$$= 1200 \times 0.17 + 0.020 \times 1000 \times 8.34 = 371 \text{ lb of BOD/day}$$

$$\text{stabilization pond area required} = \frac{371}{25} = 14.8 \text{ acres (use two ponds)}$$

volume available for winter storage between 2- and 5-ft depths

$$= (5 - 2)14.8 \times 43{,}560 = 1{,}930{,}000 \text{ ft}^3$$

water loss per day (evaporation + seepage − rainfall)

$$= \frac{(60 - 20)14.8 \times 43{,}560}{12 \times 365} = 5890\ \text{ft}^3/\text{day}$$

$$\text{wastewater influent per day} = \frac{1200 \times 100 + 20{,}000}{7.48} = 18{,}700\ \text{ft}^3/\text{day}$$

$$\text{winter storage available} = \frac{1{,}930{,}000}{18{,}700 - 5900} = 150\ \text{days}$$

Example 12.20

The design criteria for stabilization ponds specified by the state regulatory agency are that (1) the BOD loading in the primary cells shall not exceed 25 lb BOD/acre/day; (2) the minimum total water volume, based on influent flow and all cells at a water depth of 5 ft, shall not be less than 120 days; (3) the volume for winter storage between the water depths of 1.5 and 5.0 ft shall be sufficient so that no discharge is necessary for a 4-month period; and (4) the pond system shall have at least two primary cells and one or more secondary cells that cannot receive raw wastewater. Size the ponds for a community with an average daily wastewater discharge of 160,000 gpd with a BOD of 220 mg/l. Based on available data, the net water loss (evaporation plus seepage minus precipitation) during the storage months is 1.0 in./month.

Solution:

$$\text{area of primary ponds} = \frac{0.160 \times 220 \times 8.34}{25}$$

$$= 11.7\ \text{acres}$$

$$\text{construct two primary cells each} = 5.9\ \text{acres}$$

$$\text{minimum total water volume required} = 120 \times 0.160 = 19.2\ \text{mil gal}$$

$$\text{volume of primary cells} = 11.7 \times 5 \times 0.326$$

$$= 19.1\ \text{mil gal} \quad \text{(OK)}$$

$$\text{wastewater inflow during 4 months} = 0.160 \times 3.07 \times 4 \times 30 = 58.9\ \text{acre-ft}$$

$$\text{pond area required for storage} = \frac{\text{inflow} - \text{water loss}}{\text{difference in water levels}}$$

$$\text{pond area} = \frac{58.9 - \text{pond area} \times {}^4\!/_{12}}{5.0 - 1.5}$$

$$\text{pond area} = 15.4\ \text{acres}$$

$$\text{area of secondary cell} = 15.4 - 2 \times 5.9 = 3.6\ \text{acres}$$

12.30 COMPLETELY MIXED AERATED LAGOONS

Aerated ponds for pretreatment of industrial wastes or first-stage treatment of municipal wastewaters are commonly completely mixed lagoons 8–12 ft deep with floating or platform-mounted mechanical aeration units. A floating aerator consists of a motor-driven impeller mounted on a doughnut-shaped float with a submerged intake cone (Fig. 12.51). Inspection and maintenance are performed using a boat, or by disconnecting the restraining cables and pulling the unit to the edge of the lagoon. Platform-mounted aerators are placed on piles or piers extending into the pond bottom (Fig. 12.38). The impeller is held beneath the liquid surface by a short shaft connected to the motor mounted on the platform. A bridge may be constructed from the lagoon dike to the aerator for ease of inspection and maintenance.

Complete mixing and adequate aeration are essential environmental conditions for a lagoon biota. Selection and design of mixing equipment depend on manufacturers' laboratory test data and field experience. Aerators are spaced to provide uniform blending for the dispersion of dissolved oxygen and suspension of microbial solids. Their oxygen transfer capability must be able to satisfy the BOD demand of the waste while retaining a residual dissolved-oxygen concentration. Figure 12.52 illustrates the general relationships between power required for mixing and that required for aeration. Only one detention time exists for a given wastewater strength where both stirring and aeration functions are at optimum. Thus, deviations in loadings should be considered in the design selection and operational control of mechanical aeration units.

Organic stabilization depends on suspended microbial floc developed within the basin, since no provision is made for settling and returning activated sludge. BOD removal is a function of detention time, temperature, and nature of the waste, primarily biodegradability and nutrient content. The common relationships are

$$\frac{L_e}{L_0} = \frac{1}{1 + kt} \tag{12.99}$$

$$k_T = k_{20^\circ C}\Theta^{T-20} \tag{12.100}$$

where

L_e = effluent BOD, mg/l

L_0 = influent BOD, mg/l

k = BOD-removal-rate constant to base e, day^{-1}

t = detention time, days

T = temperature, °C

Θ = temperature coefficient

The value of k relates to degradability of the waste organics, temperature, and completeness of aeration mixing. At 20°C, k values have been found to range from 0.3 to over 1.0; the precise value for a particular waste must be determined experimentally.

(a)

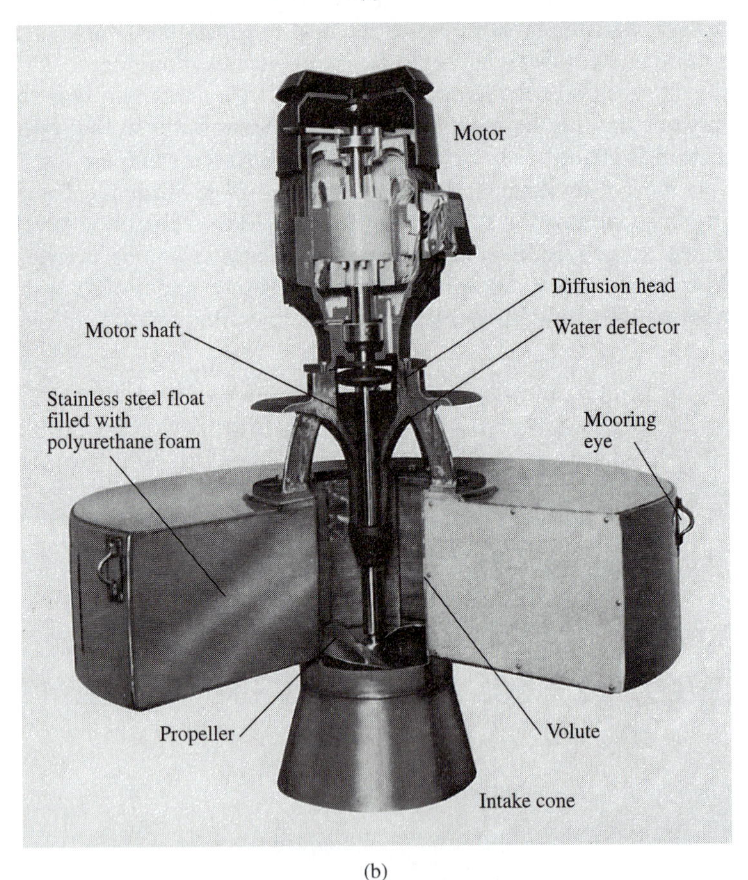

(b)

FIGURE 12.51 Floating aerator. (a) Picture of an aerator in operation. (b) Cutaway section showing aerator design with propeller directly connected to motor that is fastened to the float. Water drawn up through the intake cone is deflected by the diffusion head for aeration by dispersion. *Source:* Courtesy of Aqua-Aerobic Systems, Inc., Rockford, IL.

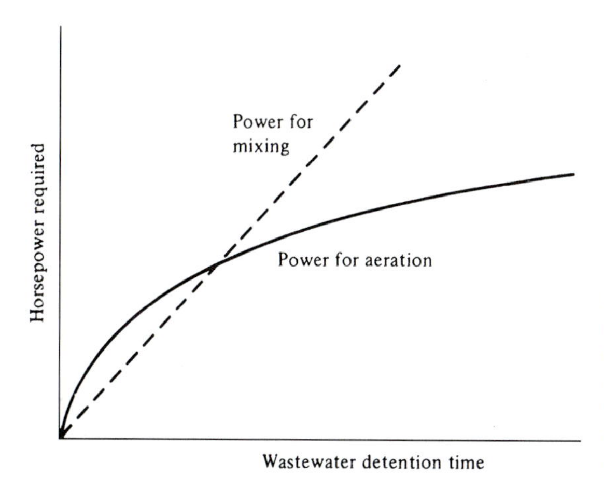

FIGURE 12.52 General relationships between the power required for mixing and that needed for aeration relative to wastewater detention time for mechanical aerators in completely mixed lagoons.

The coefficient Θ is a function of biodegradability and generally falls between 1.035 and 1.075, with 1.035 the most common value.

Biological oxygen utilization is equal to assimilative plus endogenous respiration, as in the activated-sludge process. However, with the low concentration of microbial suspended solids in the aerating wastewater, oxygen uptake can be simply related to BOD removal by the relationship

$$\text{lb of oxygen/day} = a \times \text{lb of BOD removed/day} \tag{12.101}$$

The magnitude of a is determined by laboratory or field testing the particular wastewater to be treated. Values are from 0.5 to 2.0, 1.0 being typical.

Oxygen transferred by surface-aeration units can be computed by

$$R = R_0 \frac{\beta C_s - C_t}{9.2} \times 1.02^{T-20}(\alpha) \tag{12.102}$$

where

R = actual rate of oxygen transfer, lb of oxygen/hp · hr

R_0 = rate of oxygen transfer of manufacturer's unit under standard conditions(water at 20°C, 1 atm pressure, and zero dissolved oxygen), lb of oxygen/hp · hr

β = oxygen saturation coefficient of the wastewater

C_s = oxygen concentration at saturation, mg/l

C_t = oxygen concentration existing in liquid, mg/l

T = temperature of lagoon liquid, °C

α = oxygen transfer coefficient

9.2 = saturation oxygen concentration of pure water at 20°C, mg/l

The manufacturer's oxygen transfer rate R_0 is guaranteed performance based on aerator test data stated in terms of standard conditions. The α and β factors are discussed in Section 12.26.

A facultative, aerated lagoon results if insufficient mixing permits deposition of suspended solids. BOD removal cannot be predicted with certainty in a nonhomogeneous system. Anaerobic decomposition of the settled sludge can cause emission of foul odors, particularly in treating certain industrial wastes. Facultative conditions often result from overloaded completely mixed lagoons or may derive from poorly designed systems with inadequate mixing. To ensure odor-free operation, pond contents must be thoroughly stirred, with dissolved oxygen available throughout the liquid. Aeration equipment installed should be of proven performance and purchased from a reputable manufacturer.

Significant increases in effluent BOD can occur from reducing detention time since the biological process is time dependent. Cooling-water discharges and shock loads of relatively uncontaminated water, for example, storm runoff, should be diverted around the lagoon to the secondary ponds. Sudden, large inputs of biodegradable or toxic wastes resulting from industrial spills can also upset the process. Pretreatment and control systems at industrial sites should be furnished to prevent taxing the lagoon's equalizing capacity. Biodegradability studies are essential for municipal wastewater containing measurable amounts of industrial discharges to determine such design parameters as BOD removal-rate constant, influence of temperature, nutrient requirements, oxygen utilization, and sludge production.

Example 12.21

Size an aerated lagoon to treat a domestic plus industrial waste flow of 0.30 mgd with an average BOD of 600 mg/l (1500 lb of BOD/day). The temperature extremes anticipated for the lagoon contents range from 10°C in winter to 35°C in summer. Minimum BOD reduction through the lagoon should be 75%. The surface aerators to be installed carry a manufacturer's guarantee to transfer 2.5 lb of oxygen/hp · hr under standard conditions. During laboratory treatability studies, the wastewater exhibited the following characteristics: $k_{20°C} = 0.68$ per day, $\Theta = 1.047$, $\alpha = 0.9$, and $\beta = 0.8$.

Solution: The required detention time at a critical temperature of 10°C is found using Eqs. (12.99) and (12.100):

$$k_{10°C} = 0.68 \times 1.047^{10-20} = 0.43 \text{ per day}$$

$$\frac{L_e}{L_0} = 1 - 0.75 = \frac{1}{1 + 0.43t}$$

$$t = 7.0 \text{ days}$$

$$\text{lagoon volume} = 0.30 \text{ mgd} \times 7.0 \text{ days} = 2.1 \text{ mil gal} = 280{,}000 \text{ ft}^3$$

Use a 10-ft depth with earth sides sloped appropriately for soil conditions. Oxygen utilization using $a = 0.8$:

At 10°C:

$$\text{BOD removal} = 0.75 \times 1500 = 1120 \text{ lb of BOD/day}$$

$$\text{oxygen required} = \frac{1120 \times 0.8}{24} = 37 \text{ lb of oxygen/hr}$$

At 35°C:

$$k = 0.68 \times 1.047^{35-20} = 1.35 \text{ per day}$$

$$\text{BOD removal} = 1500 - \frac{1500}{1 + 1.35 \times 7.0} = 1360 \text{ lb of BOD/day}$$

$$\text{oxygen required} = \frac{1360 \times 0.8}{24} = 45 \text{ lb of oxygen/hr}$$

Aerator power requirements using Eq. (12.102) at a minimum of 2.0 mg/l of dissolved oxygen (C_s = 11.3 mg/l at 10°C and 7.1 mg/l at 35°C):

$$R_{10°C} = \frac{2.5(0.8 \times 11.3 - 2.0)}{9.2} 1.02^{10-20} \times 0.9 = 1.4 \text{ lb of oxygen/hp} \cdot \text{hr}$$

$$\text{power required} = \frac{37 \text{ lb of oxygen/hr}}{1.4 \text{ lb of oxygen/hp} \cdot \text{hr}} = 26 \text{ hp}$$

$$R_{35°C} = \frac{2.5(0.8 \times 7.1 - 2.0)}{9.2} 1.02^{35-20} \times 0.9 = 1.2 \text{ lb of oxygen/hp} \cdot \text{hr}$$

$$\text{power required} = \frac{45}{1.2} = 38 \text{ hp}$$

Power requirements at 35°C control design: Use four 10-hp surface aerators.

ODOR CONTROL

Increased urbanization has resulted in wastewater treatment plants being situated in close proximity to housing areas and commercial developments. This has caused complaints about odors and, in serious situations, led to lawsuits against municipalities operating the disposal systems. Although the problem of foul odors emitting from treatment plants is not new, only in recent years have political and legal pressures forced processing facilities to consider abatement.

12.31 SOURCES OF ODORS IN WASTEWATER TREATMENT

Principal odors are hydrogen sulfide and organic compounds generated by anaerobic decomposition. The latter include mercaptans, indole, skatole, amines, fatty acids, and many other volatile organics. Often, industrial wastes in a municipal sewer create odors inherent in the raw materials being processed or the manufactured products (poultry processing, slaughtering and rendering, tanning, and manufacture of volatile chemicals).

With the exception of hydrogen sulfide, a specific odor-producing substance is very difficult to identify. Weather conditions, such as temperature and wind velocity, influence the intensity and prevalence of emissions.

Frequently, the initial evolution of malodors is from septic wastewater in the sewer collection system. Flat sewer grades, warm temperatures, and high-strength wastes lead to anaerobiosis. The first sources at the treatment plant are the wet well and grit chamber. Turbulent flow and preaeration of raw waste strip dissolved gases and volatile organics, discharging them into the atmosphere. Odors also may arise from the liquid held in primary clarifiers, particularly if excess activated sludge is returned to the head of the plant, resulting in an active microbial seed being mixed with the settleable raw organic matter. Sludge taken from these tanks has an obnoxious smell. Pumping it into uncovered holding tanks releases the scent previously confined under the water cover. Polymers do not neutralize the olfactory compounds prior to mechanical dewatering; therefore, the air drawn through the sludge cake picks up volatile compounds and carries them to the atmosphere. The use of ferric chloride and lime for conditioning chemicals significantly reduces odors, but for most municipal wastes, polymers provide more economical operation. The process of anaerobic digestion takes place in enclosed tanks, while digested sludge is dewatered either mechanically or on drying beds. The smell of well-digested sludge is earthy, but if the digestion process is not complete, intermediate aromatic compounds may be released during drying.

Secondary biological processes also yield odors, particularly stone-media trickling filters. Although referred to as aerobic devices, filters are actually facultative, since the microbial films are aerobic on the surface and anaerobic adjacent to the medium. Because of this potential for anaerobic decomposition, filters under heavy organic loading can reek. Odors are not as likely to be created in biological towers because of thinner biological films and improved aeration. Activated-sludge processes yield a relatively inoffensive musty odor carried by the air passing through the mixed liquor. Foul smells are rare, since microbial flocs in the aeration basin are completely surrounded by liquid containing dissolved oxygen.

12.32 METHODS OF ODOR CONTROL

Modern treatment plant design incorporates odor prevention [26]. In siting plants, providing a reasonable buffer zone is prudent to prevent encroachment of other activities in which people will be offended by the presence of a treatment plant.

The first step in analyzing an existing problem is to determine the cause of odorousness and attempt to isolate the sources. Special attention must be paid to industrial wastes entering the sewer system. Overloading often increases malodors; however, expansion of facilities is no guarantee that the situation will change dramatically. Foul emissions can be given off by properly loaded units if the design is poor, if they are not maintained properly, or when the waste includes organics with an inherent smell. Typical problems are hydrogen sulfide and other volatile odorous compounds being stripped out of solution and dispersed in air from the wet well, grit chamber, or primary clarifiers. Warm septic wastewater can release extremely offensive odors. Anaerobiosis can occur in poorly vented trickling filters. In sludge processing, foul odors are most likely to be released from raw sludge in holding tanks and in ventilation exhaust from buildings housing mechanical dewatering equipment. The sources of odors can be difficult to locate precisely. For example, convection can lift odorous air up and transport it a considerable distance before it descends.

Chemicals can sometimes be used to oxidize odorous compounds, particularly hydrogen sulfide. Chlorination of the wastewater in main sewers, or prior to primary settling, may prove beneficial. Using lime and ferric chloride as chemical sludge conditioners reduces bacterial activity and oxidizes many products of anaerobic decomposition.

The best method for preventing odorous emissions is to contain the foul air and process it through an air pollution control system. Fiberglass and aluminum covers and domes can be constructed over grit chambers, primary clarifiers, sludge-holding tanks, sludge conveyors, and trickling filters for containment. For a deep wet well, an exhaust pipe can be installed to withdraw air from above the wastewater inflow. Exhaust hoods are commonly placed over sludge dewatering equipment, e.g., belt filter presses and centrifuges, to reduce indoor air pollution and to treat the released air. In activated-sludge plants, contained or exhausted foul air can be cleansed effectively by using it as a portion of the air supply to the aeration tanks [27].

The common air pollution control system is a countercurrent packed-tower scrubber where a chemical solution, usually hypochlorite or permanganate, is used to oxidize airborne odorous compounds. The foul air flows up through the packing, passes through a mist eliminator, and exhausts to the atmosphere. The scrubbing chemical solution is sprayed down on the packing, flows over the packing, and is collected in the bottom of the tank for recirculation. The purpose of the packing is turbulent mixing of the solution and air to increase the rate of gas–liquid mass transfer. Fresh chemical is automatically added to maintain solution strength and a small amount of spent solution is wasted. While some contaminants like hydrogen sulfide are readily adsorbed, organic compounds such as amines and aldehydes are less effectively removed. The main advantages of the packed-tower scrubber are the ability to process large air flows in an economically sized system and to treat effectively a rapid increase in concentration of hydrogen sulfide. A similar process is a mist scrubber, which does not contain packing. A strong chemical solution is introduced through air-atomizing nozzles to create very fine droplets that adsorb airborne odors by gas–liquid contact.

An activated-carbon adsorber is a bed through which odorous air is passed to remove contaminants. After degradation, the carbon has to be replaced or regenerated. Since activated carbon is most effective in removing many organic molecules, an adsorber may follow a packed-tower scrubber to remove organic sulfur and volatile organic compounds.

A biofilter is a bed of organic bulk material used to adsorb and biologically oxidize airborne contaminants, including sulfur compounds, ammonia, and hydrocarbons. After collection from various treatment processes, the foul air is blown out of perforated pipes buried in the bed. Biofiltration can remove a wide variety of air contaminants to a level not achievable with packed-tower scrubbing or other absorption systems with chemical treatment. The disadvantages are the required space, extensive piping, and energy required to force the foul air through the bed.

INDIVIDUAL ON-SITE WASTEWATER DISPOSAL

12.33 SEPTIC TANK–ABSORPTION FIELD SYSTEM

Approximately 30% of the population in the United States live in unsewered areas and rely on on-site systems for wastewater treatment and disposal. Almost one-third of the housing units use septic tanks or cesspools; the majority of the remainder, usually in remote locations, use pit privies.

The installation of a septic tank and absorption field, sketched in Figure 12.53, has the advantages of low cost and underground disposal of effluent. The septic tank is an underground concrete box sized for a liquid detention time of approximately 2 days. With garbage grinders and automatic washers, the recommended minimum capacity is 750 gal (2.8 m^3) for a two-bedroom house, 900 gal (3.4 m^3) for three bedrooms, 1000 gal (3.8 m^3) for four bedrooms, and 250 gal (1.0 m^3) for each additional bedroom. Inspection and cleaning ports are accessible for maintenance by removing the earth cover of about 1 ft. Inlet and outlet pipe tees, or baffles, prevent clogging of the openings with scum that accumulates on the liquid surface. The functions of a septic tank are settling of solids, flotation of grease, and anaerobic decomposition and storage of sludge. Retention of large solids and grease is essential to prevent plugging of the absorption field. Under normal loading, the accumulated sludge (septage) must be pumped from the tank every 3–5 yr.

A typical absorption field consists of looped or lateral trenches 18–24 in. wide and at least 18 in. deep. Drain tile or perforated pipe in an envelope of gravel distributes the wastewater uniformly over the trench bottom. Organics decompose in the aerobic-facultative environment of the bed and the water seeps downward into the soil. Air enters the gravel bed through the backfill covering the trenches and by ventilation through the drain tile from the house plumbing stack. The percolation area required depends on soil permeability. For a four-bedroom dwelling, the area needed is in the range of 300–1300 ft^2 (30–120 m^2). Trench area for a particular location can be determined by subsurface soil exploration and percolation tests; however, most state environmental control agencies and county health departments have guidelines for installations based on local conditions.

The most frequent complaints in the operation of septic tank–absorption field systems involve plumbing stoppages and odorous seepage appearing at or near the ground surface. Plugging of the influent is often the result of excessive accumulation of solids due to either overloading or neglecting to pump out the sludge every few years.

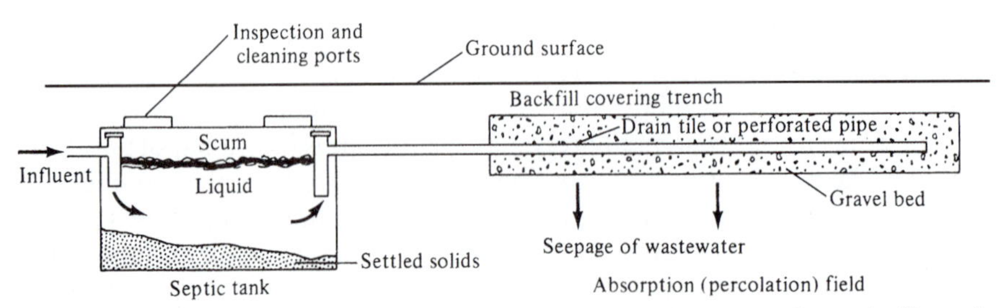

FIGURE 12.53 A typical septic tank–absorption field system for on site disposal of household wastewater.

When cleaning a tank, a small amount of the black digesting sludge should be left in the tank to ensure adequate bacterial seeding to continue solids digestion. Obstruction of percolation can result for any of the following reasons: construction in soils of low permeability that are inadequate for the seepage, high-water-table conditions that saturate the soil profile during the wet-weather season, inadequate design resulting in hydraulic and organic overloading, and improper cleaning of the septic tank.

Soils with a high infiltration rate, although advantageous for effluent disposal, often do not have the capacity to absorb contaminants. The result can be serious groundwater pollution, either limited to the area immediately surrounding an absorption field or widespread, depending on the density of housing units. Nearby wells can be polluted with disease-producing microorganisms, particularly viruses. The most common pollutants, however, are mobile substances like detergents, and ions such as chloride and nitrate.

MARINE WASTEWATER DISPOSAL

12.34 OCEAN OUTFALLS

An ocean outfall is a pipeline that extends thousands of feet from the shore to relatively deep water. At the end of the pipe, a diffuser discharges the wastewater through a series of ports spaced to provide initial dilution. After reaching the sea surface, or an intermediate equilibrium level, this mixed plume tends to move with the prevailing currents to provide secondary dispersion. Data collected for design of an outfall include physical, chemical, biological, and geological conditions. Of particular concern is a comprehensive oceanographic study to predict dilution and diffusion of the wastewater.

The general water quality objective in marine disposal is to maintain the indigenous marine life and a healthy and diverse marine community. Relevant considerations are contamination of shellfish with pathogens, toxicity of aquatic life, accumulation of sediments that impair benthic life, and aesthetics of the ocean surface.

The beneficial uses of ocean waters to be protected are water contact and noncontact recreation, commercial and sport fishing, marine habitat, shellfish harvesting (mussels, clams, and oysters), and industrial water supply.

Effluent requirements, based on samples collected from the outfall pipeline, limit both major constituents and toxins for protection of aquatic life and human health. As listed in Table 12.9, the major constituents limited by the state of California are grease and oil, suspended solids, settleable solids, turbidity, pH, and acute toxicity [28]. The limitations of specific toxins, mostly heavy metals and pesticides, are listed separately, and BOD is included for domestic wastewater. For municipal wastewaters, these limitations can be achieved by pretreatment of industrial wastewaters at industrial sites and conventional biological treatment with chlorination for effluent disinfection. Fish bioassays for acute toxicity are conducted with the threespine stickleback. The test species for chronic toxicity bioassays preferably include a fish, an invertebrate (shrimp or oyster), and an aquatic plant (red algae or giant kelp). Effluent quality limitations for chemicals for protection of human health are based on calculated initial dilution as determined from a mathematical model. The characteristics of the outfall for inputs to the model include length of diffuser, number and spacing of ports, port diameter and

TABLE 12.9 Effluent Quality Limits of Major Wastewater Constituents for Ocean Discharge to Protect Marine Aquatic Life

Parameter	Monthly (30-Day Average)	Weekly (7-Day Average)	Maximum at Any Time
Grease and oil, mg/l	25	40	75
Suspended solids, mg/l	60 with a minimum removal of 75%		
Settleable solids, ml/l	1.0	1.5	3.0
Turbidity, NTU	75	100	225
pH	within limits of 6.0–9.0 at all times		
Acute toxicity, TUa[a]	1.5	2.0	2.5

[a] $TUa = \frac{100}{96\text{-hr } LC_{50}}$

where

TUa = toxicity units acute

LC = lethal concentration, 50%

Source: California Ocean Plan, Ocean Waters of California, State Water Resources Control Board (Sacramento, CA: 2001).

angle from the horizontal, average depth of ports under mean sea level, and rate of wastewater discharge.

Water quality in the zone of initial dilution (Fig. 12.54) is determined from samples collected at boat stations. The physical characteristics of the ocean water in this zone are no visible floating particulates, grease or oil, and no aesthetically undesirable discoloration of the ocean surface. Natural light penetration shall not degrade biological benthic communities. Chemical changes are not to decrease dissolved oxygen more than 10%, lower the pH more than 0.2 units, increase dissolved sulfide concentration, increase the concentration of degrading substances in sediments, or increase nutrients causing objectionable aquatic plant growths. The biological water quality changes shall not degrade vertebrate, invertebrate, or plant species; alter the natural taste, odor, or color of fish or shellfish for human

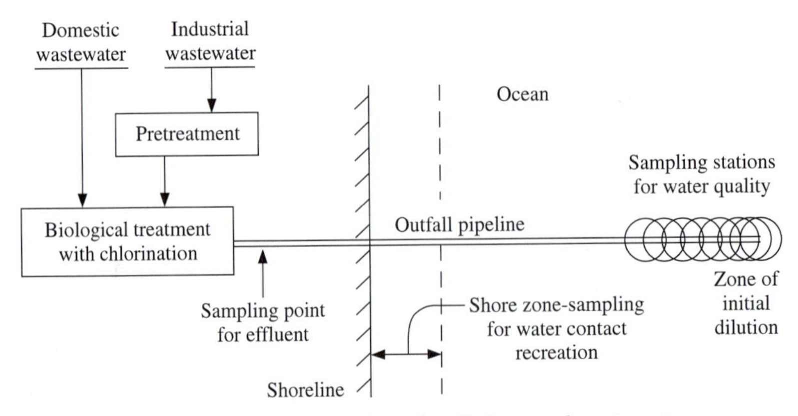

FIGURE 12.54 Schematic diagram of marine discharge of wastewater.

consumption; or increase the bioaccumulation of toxins in fish or shellfish to a level harmful to human health.

Water quality in the shore zone for water-contact recreation is monitored by collecting samples for coliform testing from the shoreline into the ocean for a distance of 1000 ft or to the 30-ft depth contour, whichever is further. The typical standard is an average of less than 1000 total coliforms per 100 ml in any 30-day period, with no single sample exceeding 10,000 per 100 ml. For fecal coliforms, the standard is not to exceed a geometric mean of 200 per 100 ml, with no more than 10% exceeding 400 per 100 ml in a 60-day period [28]. In areas where shellfish are being harvested for human consumption, the median total coliform concentration is not to exceed 70 per 100 ml, with no more than 10% exceeding 230 per 100 ml.

PROBLEMS

12.1 (a) How do autotrophic bacteria gain energy? (b) Why do some bacteria convert ammonia to nitrate [Eqs. (12.1) and (12.2)], whereas others reduce nitrate to nitrogen gas [Eq. (12.12)]? (c) Hydrogen sulfide emitted by septic wastewater and sludge is too weak an acid to cause significant corrosion, yet when hydrogen sulfide is present, concrete deteriorates and iron corrodes. Explain this phenomenon. (d) What promotes the growth of iron bacteria in water pipes distributing well water? How do these bacteria contribute foul tastes and odors to the water?

12.2 List the major nutritional and environmental conditions necessary to culture algae in a laboratory container.

12.3 A fresh wastewater containing nitrate ions, sulfate ions, and dissolved oxygen is placed in a sealed jar without air. In what sequence are these oxidized compounds reduced by the bacteria? Why is dissolved oxygen the preferred hydrogen acceptor? When do obnoxious odors appear?

12.4 Discuss the relationships among metabolism, energy, and synthesis and the effect of these on growth under aerobic and anaerobic conditions. (*a*) Include comments on growth rates, extent of metabolism, and limiting factors under the two environments. (*b*) Why is bacterial synthesis for the same quantity of substrate greater in an aerobic environment than under anaerobiosis?

12.5 What are the relationships between biomass and substrate in the exponential growth and declining growth phases of a pure bacterial culture? List the characteristics of the endogenous growth phase.

12.6 What is the mathematical relationship for growth kinetics of a pure bacterial culture under substrate-limited growth conditions proposed by Monod and graphed in Figure 12.6? How does this relate to enzymatic reactions as defined by Michaelis and Menten?

12.7 A series of fermentation tubes containing varying concentrations of glucose in a nutrient broth were inoculated with a pure bacterial culture. The concentrations of cells in the broth media were determined after 16 hr of incubation at 37°C. Rates of growth and initial glucose concentrations are listed below. Plot growth rate expressed as cell divisions per hour versus initial glucose concentration. Estimate the maximum growth rate and the saturation constant (glucose concentration at one-half the maximum growth rate). Write an equation in the form of Eq. (12.25), and draw the curve for this equation on the graph with the plotted data. Does the growth of this pure culture appear to be a hyperbolic function as defined by the Monod relationship?

Glucose (moles × 10^{-4})	Cell Divisions (per hr)	Glucose (moles × 10^{-4})	Cell Divisions (per hr)
0.1	0.23	0.8	0.94
0.1	0.28	1.6	1.06
0.2	0.32	3.2	1.15
0.4	0.71		

12.8 How is growth yield determined in pure bacterial cultures?

12.9 (a) How does temperature affect biological processes? (b) The rate of BOD reduction in aeration of a synthetic wastewater decreased 25% when the temperature of the laboratory fermentation tank was lowered from 20° to 15°C. Using Eq. (12.35), calculate the temperature coefficient. How many degrees of temperature drop are required to reduce the rate of BOD reduction by one-half?

12.10 State the two main reasons why an activated-sludge system is operated at a relatively low food/microorganism ratio.

12.11 Why are bacteria rather than protozoans the primary feeders in activated sludge?

12.12 Why is impending failure of an anaerobic digestion process forecast by an increase in the percentage of carbon dioxide in the gas produced?

12.13 (a) Describe the role of algae in biological stabilization of wastewater in a stabilization pond. (b) Describe what is wrong with the following statement: "The key to waste stabilization in a pond is the algal growth; overloading kills the algae, whereas they thrive on a reasonable supply of organic matter."

12.14 The average annual wastewater flow of a municipality is 24 mgd, with an average BOD concentration of 200 mg/l and suspended-solids concentration of 220 mg/l. Based on the equations in Section 12.10, estimate the average annual peak month and average annual peak day of wastewater flows, BOD loads, and suspended-solids loads.

12.15 The average wastewater flow and strength anticipated during the maximum month of the year are often used for design. If the annual average wastewater flow and characteristics are estimated to be 32,000 m^3/d with 250 mg/l of BOD and 280 mg/l of suspended solids, what are the values likely to be for the peak month, using the equations in Section 12.10?

12.16 The wastewater from a synthetic textile manufacturing plant is discharged to a municipal treatment plant for processing with domestic wastewater. The characteristics of the synthetic textile wastewater are 1500 mg/l BOD, 2000 mg/l SS, 30 mg/l nitrogen, and no phosphorus. The characteristics of the domestic wastewater are listed in Table 12.1. If the required BOD/N/P weight ratio is 100/5/1, what is the minimum quantity of domestic wastewater required per 1000 gal of textile wastewater to provide adequate nutrients for biological treatment?

12.17 The wastewater from the manufacture of synthetic chemicals produces a wastewater flow of 779 m^3/d containing 4300 mg/l of BOD, 1200 mg/l of suspended solids, 70 mg/l of nitrogen, and negligible phosphorus. You are asked to recommend dosages of pure NH_4NO_3 and H_3PO_4 to be applied for activated-sludge treatment. After the biological process is stabilized, how would you determine if your recommended dosages were correct?

12.18 A pharmaceutical wastewater needs a minimum BOD/N/P of 100/3.0/0.7 for biological treatment. The wastewater characteristics are BOD = 740 mg/l, suspended solids = 250 mg/l, soluble nitrogen = 24 mg/l, and soluble phosphorous = 12 mg/l. What ammonium nitrate and phosphoric acid additions would you recommend to ensure adequate nutrients for a daily flow of 28 m^3/d?

12.19 A dairy that processes about 250,000 lb of milk daily produces an average of 65,100 gpd of wastewater with a BOD of 1400 mg/l. The principal operations are bottling milk and making ice cream, with limited production of cottage cheese. Compute the flow and BOD per 1000 lb of milk received, and the BOD and hydraulic equivalent populations of the daily wastewater discharge.

12.20 A meat-processing plant discharges 10,000 m^3/d of wastewater containing 1300 mg/l of BOD, 960 mg/l of suspended solids, 2500 mg/l of COD, and 460 mg/l of grease. Calculate the BOD equivalent population and the hydraulic equivalent population.

12.21 The domestic and industrial waste from a community consists of 100 gpcd from 7500 persons; 65,000 gpd from a milk-processing plant with a BOD of 1400 mg/l; and 90,000 gpd containing 450 lb of BOD from potato-chip manufacturing. Calculate the combined wastewater flow, BOD concentration in the composite waste, and BOD equivalent population.

12.22 The combined wastewater flow from a community includes domestic waste from a sewered population of 2000 and industrial wastes from a dairy and a poultry-dressing plant. The poultry plant discharges 125 m^3/d and 136 kg of BOD/d. The dairy produces a flow of 190 m^3/d with a BOD concentration of 900 mg/l. Estimate the total combined wastewater flow from the community and the BOD concentration in the composite discharge.

12.23 A town with a sewered population of 4000 has a daily wastewater flow (including industrial wastewaters) of 1600 m^3 and an average BOD of 280 mg/l. The industrial discharges to the municipal sewers are 60 m^3 at 1800 mg/l BOD from a meat-processing plant and 100 m^3 at 400 mg/l BOD from a soup-canning plant. Determine the contribution of domestic flow in liters per person and the BOD in grams per person based on the town's wastewater, excluding the industrial wastewaters.

12.24 The municipal wastewater flow from a town is 1890 m^3/d with an average BOD of 280 mg/l. Assuming 35% BOD removal in the primary, calculate the size required for one single-stage trickling filter. Use a depth of 2.1 m, a BOD loading of 480 $g/m^3 \cdot d$, and minimum required hydraulic loading. Compute the BOD concentration in the effluent at 20°C and 15°C, using the NRC formula and temperature correction relationship in Eq. (12.25).

12.25 A single-stage trickling-filter plant (Fig. 12.21) is proposed for treating a dilute wastewater with a BOD concentration of 170 mg/l. The plant is located in a warm climate and the minimum wastewater temperature anticipated is 16°C. Using a recirculation ratio of 0.5, what is the maximum allowable BOD loading for a stone-media filter to achieve an average effluent BOD of 30 mg/l?

12.26 The sizing of primary clarifiers is generally based on the average weekday flow during the time of the year of greatest flow. If the plant is a trickling-filter system, gravity return of underflow from the final clarifier to the wet well is performed for recirculation during periods of low influent flow (Fig. 12.21). This recirculation flow, usually $0.5Q$ or less as diagrammed in Figure 12.16(a), is necessary to maintain rotation of the distributor arm at night and to provide adequate hydraulic loading of the filter media. Should the size of primary clarifiers be increased to account for this low-flow recirculation? Explain.

12.27 Estimate the effluent BOD for the two-stage trickling-filter plant designed in Example 12.5 for a wastewater flow of 1.2 mgd with a BOD concentration of 250 mg/l.

12.28 Determine the NRC BOD removal efficiency for a two-stage trickling-filter plant based on the following: primary clarification with 35% BOD reduction, first-stage filters loaded at 80 lb/1000 ft^3/day, intermediate settling, second-stage filters sized identical to the first-stage units, an operating recirculation ratio of 1.0 for all filters, final clarification of the effluent, and a wastewater temperature of 18°C.

12.29 An existing single-stage trickling-filter plant cannot meet the effluent limitation of 30 mg/l BOD during cold weather. In operation under the design loading with the wastewater temperature at 15°C, the plant influent BOD is 240 mg/l, the primary effluent BOD equals 155 mg/l, and the plant effluent averages 55 mg/l. The proposed modification is the addition of a second-stage trickling filter and final clarifier to reduce the effluent BOD from 55 to 30 mg/l with a wastewater temperature of 15°C. Under normal operation, the recirculation flow in the second-stage filtration is to be one-half of the average wastewater flow entering the plant. Calculate the design BOD loading that should be used to determine the volume of stone media needed in the proposed second-stage filter.

12.30 The following are design data for the town of Nancy, with a sewered population of 7600. Design flows are as follows: the average daily flow is 0.84 mgd, the peak hourly flow is 1.25 mgd, and the minimum hourly flow is 0.12 mgd. Design average BOD equals 1740 lb/day and average suspended solids equal 1530 lb/day. Calculate the following: equivalent population based on 0.20 lb of BOD per capita; design flows in units of gpm, ft^3/sec, m^3/min, and m^3/d; mean BOD and SS concentrations in mg/l.

12.31 Consider the feasibility of treating the wastewater from the town of Nancy (Problem 12.30) in a single-stage stone-media trickling-filter plant as diagrammed in Figure 12.21. The treated wastewater is required to meet the effluent standards of 30 mg/l of BOD and 30 mg/l of suspended solids at a wastewater temperature of 15°C. Assume a primary BOD removal efficiency of 35%. Size the trickling filters based on a BOD loading of 35 lb/1000 ft^3/day, recirculation ratio of 0.5, and a stone-media depth of 7.0 ft. Calculate the volume of filter media required, filter surface area, hydraulic loading with recirculation in gallons per minute per square foot, and effluent BOD concentration using the NRC formula.

12.32 Consider the feasibility of treating the wastewater from the town of Nancy (Problem 12.30) in a two-stage stone-media trickling-filter plant with intermediate clarification. The treated wastewater is required to meet the effluent standards of 30 mg/l of BOD and 30 mg/l of suspended solids at a wastewater temperature of 15°C. Assume a primary BOD removal efficiency of 35%. Calculate the volume of filter media required at a BOD loading of 35 lb/1000 ft^3/day and divide the resulting volume equally between the first-stage and second-stage filters. Use a filter media depth of 7.0 ft. Assuming a recirculation ratio of 0.5 for both filters, determine the effluent BOD concentration using the NRC formulas.

12.33 Consider the feasibility of treating the wastewater from the town of Nancy (Problem 12.30) in a single-stage trickling-filter plant, as diagrammed in Figure 12.21, using high-density cross-flow media with a specific surface area of 42 ft^2/ft^3. Use the same filter surface area as calculated for stone-media filtration in Problem 12.31 ($A = 4600\ ft^2$) and a recirculation ratio of 0.5. The design wastewater temperature is 15°C. As determined from laboratory testing, the soluble (filtered) BOD of the primary effluent is 100 mg/l or less, compared to the unsettled unfiltered BOD of 160 mg/l. The soluble effluent BOD concentration of trickling-filter plants treating similar wastewaters is approximately 50% of the unfiltered BOD concentration. From pilot-plant studies and full-scale experience, the manufacturer of the cross-flow media recommends a reaction-rate coefficient at 20°C of 0.0030 $(gpm/ft^2)^{0.5}$ for a media depth of 6.6 ft (2.0 m). From these data, calculate the soluble effluent BOD and double this value for the estimated effluent BOD.

12.34 A single-stage, trickling-filter plant consists of a primary clarifier, a trickling filter 70 ft in diameter with a 7-ft depth of random packing, and a final clarifier. The raw-wastewater flow is 0.80 mgd with 200 mg/l of BOD and a temperature of 15°C. The constants for the random plastic media are an n of 0.44 and K_{20} of 0.090 min^{-1}. Assuming 35% BOD removal in the primary, calculate the effluent BOD concentration (a) without recycle and (b) with an indirect recirculation to the wet well of 0.40 mgd and direct recirculation of 0.80 mgd.

12.35 A single-stage, high-rate trickling-filter plant treating wastewater with 200 mg/l of BOD cannot meet the effluent standard of less than 30 mg/l of BOD. Primary settling removes 35% of the raw-wastewater BOD and the 7-ft-deep filters have an efficiency of 75% under a loading of 40 lb/1000 ft^3/day of BOD and a hydraulic loading of 0.18 gpm/ft^2 without recirculation. This provides average effluent BODs of 30–35 mg/l during the summer and 40–45 mg/l during the winter. One proposal is to replace the stone media with the random packing described in Example 12.6, which has constants $n = 0.39$ and $K_{20} = 0.120\ min^{-1}$; cover the filters with fiberglass domes to prevent the wastewater from cooling below 16°C in winter; and modify the plant to provide both direct and indirect recirculation. Calculate the effluent BOD at 16°C for the filters packed with the random plastic media at $R = 1.0$.

12.36 Using Eq. (12.57) developed in Example 12.6, plot the filtered BOD remaining versus depth in a filter with 20 ft of random packing. Assume an applied filtered BOD of 100 mg/l and a hydraulic loading of 0.35 gpm/ft^2.

12.37 A pilot-scale study was conducted to determine the reaction-rate coefficient for cross-flow media treating a settled municipal wastewater. The tower was 1.2 m by 1.2 m square with a 6.0-m depth of media, which had a specific surface area equal to 98 m^2/m^3. From the following data, plot a graph as shown in Figure 12.27 and determine k_{20}. Assume $n = 0.5$ and $\Theta = 1.035$.

		Influent		**Effluent**	
Q ($l/m^2 \cdot s$)	T (°C)	**BOD (mg/l)**	S_0 **(mg/l)**	**BOD (mg/l)**	S_e **(mg/l)**
1.03	18	92	54	25	14
0.64	23	90	52	18	10
0.47	21	104	54	17	9
1.06	19	72	48	23	12
0.97	24	77	43	20	10
0.70	20	85	45	17	11

12.38 A biological tower has vertical-flow packing with a $k_{20} = 0.0014\ (gpm/ft^2)^{0.5}$, $n = 0.50$, and $A_s = 36\ ft^2/ft^3$. The tower is cylindrical, with a diameter of 32 ft and depth of packing of 16 ft. The primary effluent is 0.60 mgd with 80 mg/l of soluble BOD at a temperature of 14°C. Direct recirculation is 400 gpm and the final clarifier is properly sized. Calculate the soluble BOD loading, hydraulic loading, and soluble effluent BOD using Eq. (12.53).

12.39 A biological tower with high-density, cross-flow packing has a surface area of 480 ft^2 and a depth of 12.0 ft. The packing has a specific surface area of 42 ft^2/ft^3, $n = 0.5$, and $k_{20} = 0.0035\ (gpm/ft^2)^{0.5}$. The effluent from the primary clarifier is 600,000 gpd with a total BOD of 126 mg/l and soluble BOD of 62 mg/l. The wastewater temperature is 18°C and the recirculation ratio is 1.0. Calculate the effluent soluble BOD and estimate the effluent total BOD, assuming the ratio of soluble BOD to total BOD is the same as in the primary effluent.

12.40 A stone-media trickling filter with a depth of 7.0 ft cannot produce an effluent with an average BOD equal to or less than 30 mg/l, as specified in the discharge permit. From a field study, the operational data for the filter and final clarifier were as follows: hydraulic loading on the filter from primary clarifier = 0.35 gpm/ft^2; settled wastewater from the primary clarifier BOD = 126 mg/l and soluble BOD = 68 mg/l; final clarified effluent BOD = 48 mg/l and soluble BOD = 26 mg/l; wastewater temperature = 18°C;

and recirculation ratio of 1.0. Since the stone media were in poor condition and causing plugging of the voids, one of the recommendations for remedial action was to install high-density cross-flow packing with a specific surface area = 42 ft^2/ft^3, n = 0.50, and k_{20} = 0.0035 $(gpm/ft^2)^{0.5}$. Using Eq. (12.53), calculate the effluent BOD for cross-flow packing based on field-study data for hydraulic loading, influent soluble BOD, temperature, and recirculation ratio. Assume that the effluent BOD would have the same fraction of soluble BOD as in the field study, 26/48 = 0.54.

12.41 Calculate the required surface area and the minimum required recirculation ratio based on the following: cross-flow media with k_{20} = 0.0018 $(gpm/ft^2)^{0.5}$, A_s = 42 ft^2/ft^3, n = 0.5, and D = 20 ft; settled wastewater flow Q_p = 0.50 mgd; minimum wetting rate = 0.7 gpm/ft^2; influent BOD = 162 mg/l, S_p = 80 mg/l, and temperature = 15°C; and effluent BOD = 20 mg/l and S_e = 10 mg/l. At the calculated surface area, what is the BOD loading in pounds per 1000 ft^3 per day?

12.42 Calculate the RBC surface area required for secondary treatment of wastewater for the town of Nancy (Problem 12.30). Assume 35% BOD removal in the primary, a required effluent quality of 30 mg/l of BOD, and wastewater temperature of 50°F.

12.43 A community with a wastewater flow during the peak month of the year of 2.0 mgd at 15°C is considering constructing a new treatment plant. The composition of the raw wastewater is as shown in Table 12.1, and the effluent limits are 30 mg/l of BOD, 30 mg/l of suspended solids, and fecal coliform count of less than 200 per 100 ml. Sketch flow diagrams for the treatment systems listed below, showing the arrangement of unit processes (i.e., tanks, pumping stations, and division boxes) and major pipelines (i.e., for wastewater, recycle, and sludge). Assume sludge stabilization is by anaerobic digestion and dewatering by belt filter presses. Duplication of treatment units is necessary, so if any unit is out of service for maintenance the wastewater can still be processed through the plant, although the effluent quality may be reduced. No pipelines are allowed to bypass raw or settled wastewater to the plant outlet. Standby generators are provided for emergency operation during electrical power outage. (For examples of plant layouts, refer to Figs. 9.5 and 9.7.) List numerical design guidelines for sizing each treatment unit, except pumps and sludge processing.

(a) Preliminary treatment with constant-speed pumps
(b) Preliminary treatment with variable-speed pumps
(c) Two-stage trickling-filter plant with stone-media filters
(d) Trickling-filter plant with single-stage biological towers

12.44 The following are average operating data from a conventional activated-sludge secondary: wastewater flow = 7.7 mgd, volume of aeration tanks = 300,000 ft^3 = 2.24 mil gal, influent total solids = 600 mg/l, influent suspended solids = 120 mg/l, influent BOD = 173 mg/l, effluent total solids = 500 mg/l, effluent suspended solids = 22 mg/l, effluent BOD = 20 mg/l, mixed-liquor suspended solids = 2500 mg/l, recirculated sludge flow = 2.7 mgd, waste sludge quantity = 54,000 gpd, and suspended solids in waste sludge = 9800 mg/l. Based on these data, calculate the following: aeration period; BOD loading in lb/1000 ft^3/day; F/M ratio in lb BOD/day/lb MLSS; total solids, suspended solids, and BOD removal efficiencies; sludge age; and return sludge rate.

12.45 The general loading and operational parameters for activated-sludge processes in Table 12.3 are for continental climate with cold, snowy winters and short, warm, humid summers, and for a temperate climate with mild rainy winters and warm, humid summers or mild, rainy summers. In the northern United States, the concern is wastewater cooling below 10°C toward 5°C in activated-sludge processes; hence, aeration is by submerged air diffusers to prevent cooling (e.g., Fig. 12.33 and 12.34). In the southern United States, where the air temperature stays above freezing, aeration may be by mechanical surface aerators (e.g., Fig. 12.38). For regions with moderate air temperatures, mechanical aeration systems that have housing or covers over the aerators can minimize cooling of the mixed liquor (e.g., Fig. 12.36 and 12.37).

Wastewater temperatures of 20°–30°C in hot, arid climates have significant influence on the sizing of an activated-sludge aeration tank and the capacity of aeration supply. (a) Commonly recommended loading criteria for a secondary step-aeration activated-sludge process [Fig. 12.32(b)] for a minimum wastewater temperature of 10°C is a maximum of 40 lb BOD/1000 ft^3/day (640 kg BOD/$m^3 \cdot d$), a minimum aeration period of 6.0 hr, and an operating F/M of 0.26 lb BOD/day/lb of MLSS (0.26 kg BOD/kg MLSS $\cdot$ d). Using Eq. (12.35), calculate the rate of biological activity increase for a wastewater temperature increase from 15° to 25°C, assuming doubling with a 15°C temperature rise ($\theta = 1.047$). For the rise in biological activity, calculate new values of volumetric BOD loading, aeration period, and F/M ratio. (b) Higher wastewater temperature also affects oxygen transfer in the mixed liquor because the rate of transfer is directly related to the dissolved oxygen deficit [Eq. (12.87)]. Calculate the values of $\beta C_s - C_t$ at 15°C and 25°C, assuming $\beta = 0.9$ and $C_t = 2.0$ mg/l, which is a common minimum dissolved-oxygen concentration recommended in design. Determine C_s from Table A.10 for zero chloride concentration. What is the percentage reduction in the rate of oxygen transfer when the temperature of the mixed liquor increases from 15°C to 25°C? (c) "At wastewater temperatures lower than 15°C, nitrification (oxidation of ammonia to nitrate) is limited in activated-sludge aeration. In contrast nitrification during aeration of wastewater with a temperature higher than 25°C cannot be prevented" [14]. For the wastewater characteristics in Table 12.1, calculate the BOD oxygen demand, which is 1 lb of oxygen per lb of BOD (Section 12.25) and the nitrogen oxygen demand, which is 4.6 lb of oxygen per lb of nitrogen oxidized to nitrate (Section 14.11). Based on these data, how must oxygen demand should be considered in design for aeration of a warm wastewater?

12.46 Data from a field study on a step-feed activated-sludge secondary are as follows:

Aeration tank volume = 2800 m^3

Settled wastewater flow = 13,900 m^3/d

Wastewater temperature = 20°C

Return sludge flow = 4800 m^3/d

Waste sludge flow = 72 m^3/d

MLSS in aeration tank = 2350 mg/l

SS in waste sludge = 14,000 mg/l

Influent wastewater BOD = 128 mg/l

Effluent wastewater BOD = 22 mg/l

Effluent SS = 26 mg/l

Using these data, calculate the BOD loading and F/M ratio, operational parameters of sludge age, aeration period and return sludge rate, and BOD removal efficiency. Calculate the excess sludge production in kilograms of excess suspended solids per kilogram of BOD applied and compare this value to that given in Figure 13.1.

12.47 A conventional activated-sludge system treats 11,000 m^3/d of wastewater with a BOD of 180 mg/l in an aeration tank with a volume of 3400 m^3. The operating conditions are an effluent suspended solids of 20 mg/l, an MLSS concentration maintained in the aeration tank of 2500 mg/l, and an activated-sludge wasting rate of 160 m^3/d containing 8000 mg/l. From these data, calculate the aeration period, volumetric BOD loading, F/M loading, and sludge age.

12.48 Determine the activated-sludge aeration volume required to treat 2.64 mgd with a BOD of 120 mg/l based on the criteria of a maximum BOD loading of 40 lb/1000 ft^3/day and a minimum aeration period of 5.0 hr. Assuming an operating F/M of 0.20 lb BOD/day per lb of MLSS, calculate the MLSS to be maintained in the aeration tank. Estimate the operating sludge age (mean cell residence time), assuming an effluent suspended solids of 30 mg/l and the daily amount of waste-sludge solids from Figure 13.1. Determine the diameter and side-water depth of two identical final clarifiers of the type shown in Figure 10.20 for this activated-sludge system.

12.49 Determine the volume of three identical activated-sludge tanks following primary clarification to aerate 36,000 m^3/d with a BOD concentration of 180 mg/l at a BOD loading of 560 $g/m^3 \cdot d$. What is the aeration time? For an F/M of 0.35, what MLSS concentration should be maintained in the aeration tanks? Estimate the operating sludge age, assuming an effluent suspended solids of 30 mg/l and the daily amount of waste-activated sludge solids from Figure 13.1. If the waste sludge has a concentration of 10,000 mg/l, what is the calculated volume of sludge to be wasted each day? Determine the dimensions of three identical rectangular clarifiers (length, width, and side-water depth) with a 2:1 length-to-width ratio.

12.50 A treatment plant has two oxidation-ditch activated-sludge systems, as illustrated in Figure 12.36. Each ditch has a liquid volume of 35,000 ft^3 and is equipped with 2 horizontal-rotor aerators with capacity to transfer 1150 lb of oxygen per day at normal submergence. Each system has a clarifier of the type illustrated in Figure 10.20 with a diameter of 30 ft, a 9.0 ft side-water depth, and a single weir set on a diameter of 30 ft. The effluent chlorination tank has a volume of 2200 ft^3. The design flow is 0.54 mgd: 0.35 domestic and commercial, 0.02 industrial, and 0.17 infiltration and inflow. The design BOD and suspended solids are both 740 lb/day. During heavy rains, the peak hydraulic loading anticipated is 800 gpm for 2–3 hr. Calculate the (a) BOD concentration at design flow and (b) at design flow, the aeration period, volumetric BOD loading, and F/M at 2500 mg/l. How do these compare with values listed in Table 12.3? (c) How many pounds of oxygen can be transferred to the mixed liquor per pound of BOD aeration tank loading? (d) Calculate the overflow rate and weir loading at the peak hydraulic loading. How do these values compare with recommended design values? (e) What is the detention in the chlorination tank at peak hydraulic flow? At twice the design flow? How do these values compare with the recommended detention time for chlorination of wastewater effluent?

12.51 The high-purity-oxygen process illustrated in Figure 12.39 is being considered for treatment of an unsettled domestic wastewater flow of 1.40 mgd with an average BOD of 200 mg/l. Compute the aeration volume required based on a maximum allowable BOD loading of 130 lb/1000 ft^3/day and minimum aeration period of 1.8 hr. For this system, what is the F/M ratio in terms of lb of BOD/day/lb of MLVSS, assuming an MLSS of 5500 mg/l, that is 75% volatile?

12.52 Size an extended aeration system for the town of Nancy (Problem 12.30). Provide two identical activated-sludge aeration tanks with diffused aeration and final clarifiers. The treated wastewater is required to meet the effluent standards of 30 mg/l of BOD and 30 mg/l of suspended solids at a wastewater temperature of 15°C. The sludge is to be stabilized by aerobic digestion and dewatered on drying beds.

12.53 Refer to the instructions given in Problem 12.43. (a) Sketch a flow diagram showing the arrangement of unit processes and major pipelines for a treatment system of primary sedimentation followed by a step-aeration secondary and sludge stabilization by anaerobic digestion. (b) Sketch a flow diagram for a treatment system of extended aeration with aerobic sludge digestion. For both systems, list numerical design guidelines for sizing the unit processes, except preliminary treatment and sludge processing.

12.54 List the major assumptions made in the derivation of the mathematical model for biological kinetics of the activated-sludge process. What are the limitations in applying the kinetics equations given in Section 12.22?

12.55 A municipal wastewater containing both domestic and food-processing wastewaters was tested to determine the kinetic constants using a laboratory apparatus similar to the bench-scale unit shown in Figure 12.42. The volume of the aeration chamber was 10 l, and the wastewater feed was established at a constant rate of 30.0 l/d to provide an 8.0-hr aeration period for all of the test runs. A measured volume of sludge was wasted once a day from the tank. Determine Y, k_d, k, and K_s from the following laboratory data using the procedure in Example 12.14:

Q (l/d)	S_0 (mg/l)	Q_w (l/d)	X (mg/l)	S_e (mg/l)	X_e (mg/l)
30.0	150	0.31	2460	2.5	7.7
30.0	150	0.53	1690	3.3	6.5
30.0	150	0.98	1320	4.4	5.6
30.0	150	1.27	1080	5.9	5.0

12.56 Example 12.15 illustrates the application of the kinetics model in sizing the aeration volume required for a completely mixed activated-sludge process. What operating parameters must be assumed in the design procedure?

12.57 A completely mixed activated-sludge process is being designed for a wastewater flow of 3.0 mgd using the kinetics equations. The influent BOD of 180 mg/l is essentially all soluble and the design effluent soluble BOD is 10 mg/l. For sizing the aeration volume, the mean cell residence time is selected to be 8.0 days and the MLVSS 2500 mg/l. The kinetic constants from a bench-scale treatability study are as follows: $Y = 0.60$ lb VSS/lb BOD, $k_d = 0.06\ \text{day}^{-1}$, $K_s = 60$ mg/l, and $k = 5.0\ \text{day}^{-1}$.

12.58 The aeration tank for a completely mixed aeration process is being sized for a design wastewater flow of 7500 m^3/d. The influent BOD is 130 mg/l with a soluble BOD of 90 mg/l. The design effluent BOD is 20 mg/l with a soluble BOD of 7.0 mg/l. Recommended design parameters are a sludge age of 10 d and volatile MLSS of 1400 mg/l. Selection of these values takes into account the anticipated variations in wastewater flows and strengths. The kinetic constants from a bench-scale treatability study are $Y = 0.60$ mg VSS/mg soluble BOD and $k_d = 0.06$ per day. Calculate the volume of the aeration tank, aeration period, food/microorganism ratio, and excess biomass in the waste-activated sludge.

12.59 A step-aeration activated-sludge system at a loading of 40 lb of BOD/1000 ft^3/day requires an air supply of 1200 ft^3/lb of BOD applied to maintain an adequate dissolved-oxygen level. The measured average oxygen utilization of the mixed liquor is 36 mg/l/hr. Calculate the oxygen transfer efficiency. (Assume 0.0174 lb of oxygen per cubic foot of air, which is the amount at 20°C and 760 mm pressure.)

12.60 An air supply of 1000 ft^3 of air is required per pound of BOD applied to a diffused aeration basin to maintain a minimum DO of 2.0 mg/l. Assuming that the installed aeration equipment is capable of transferring 1.0 lb of oxygen to dissolved oxygen per pound of

BOD applied, calculate the oxygen transfer efficiency of the system. (One cubic foot of air at standard temperature and pressure contains 0.0174 lb of oxygen.)

12.61 A 10-hp surface aerator was tested in a tank filled with 9200 ft^3 of tap water at 22°C by the non-steady-state procedure. The dissolved-oxygen saturation was assumed to be the standard value of 8.8 mg/l from Table A.10. Based on the following time and dissolved-oxygen data, determine the value of K_La corrected to 20°C. Also calculate the oxygen transfer rate in pounds per horsepower-hour.

t (min)	C_t (mg/l)	t (min)	C_t (mg/l)
0	0	8.0	5.5
2.0	3.0	11.0	6.3
4.0	4.3	14.0	7.1
6.0	5.0	17.0	7.6

12.62 A 40-hp surface aerator was tested in a 120-ft-diameter tank with a water depth of 8.0 ft. Cobalt chloride catalyst and sodium sulfite were added to the tap water in the tank to remove the dissolved oxygen. The water temperature was 21.9°C, and the dissolved-oxygen saturation was 8.7 mg/l. During the test, the average electric power usage was 33.1 kW, which at 90% efficiency is equivalent to 40.0 hp. From the following time and dissolved-oxygen data, determine the value of K_La corrected to 20°C, and calculate the oxygen transfer rate in pounds per horsepower-hour.

t (min)	C_t (mg/l)	t (min)	C_t (mg/l)
0	0	23	6.5
2	1.0	28	7.1
4	1.9	33	7.5
6	2.7	38	7.7
8	3.1	43	7.9
13	4.9	48	8.0
18	5.9	58	8.3

12.63 Calculate the surface area required for a stabilization pond to serve a domestic population of 1000. Assume 80 gpcd at 210 mg/l of BOD. Use a design loading of 20 lb of BOD/acre/day. If the average liquid depth is 4 ft, calculate the retention time of the wastewater based on influent flow. The effluent is spread on grassland by spray irrigation at a rate of 2.0 in./week (54,300 gal/acre/wk). Compute the land area required for land disposal. In these computations, assume no evaporation or seepage losses from the ponds.

12.64 Facultative stabilization ponds with a total surface area of 6.0 ha (1 ha = 10,000 m^2) serve a community with a waste discharge of 530 m^3/d at a BOD of 280 mg/l. Calculate the BOD loading and days of winter storage available between the 0.6- and 1.5-m depths, assuming a daily water loss of 0.30 cm by evaporation and seepage.

12.65 Stabilization pond computations are required for the town of Nancy (Problem 12.30). (a) Calculate the lagoon area required for a design loading of 40 lb of BOD/acre/day. (b) Determine the percentage of the average design flow that appears as effluent from the lagoons, assuming a water loss from the ponds of 60 in/yr (seepage plus evaporation minus precipitation). (c) Using the water-balance data from part (b), calculate the BOD removal efficiency if the average BOD concentrations in the influent and effluent are 250 mg/l and 25 mg/l, respectively. (d) How many acres of cropland are needed to dispose of the effluent by spray irrigation if the application rate is 2.0 in/week year-round?

12.66 Design a layout of stabilization ponds for the town of Nancy (Problem 12.30) based on the following criteria: (1) BOD loading in the primary cells cannot exceed 25 lb/acre/day, (2) minimum total water volume with the ponds full to a 5.0-ft depth in primary cells and 8.0-ft depth in secondary cells cannot be less than 120 days times the design flow, (3) volume for winter storage above water depths of 1.5 ft should be sufficient so that no discharge is necessary for a 4-month period when the net water loss (evaporation plus seepage minus precipitation) is 1.0 in./month, and (4) the system should have at least two primary cells and at least two secondary cells that cannot receive raw wastewater.

12.67 The wastewater flow from a small town is 240 m^3/d with a BOD of 180 mg/l and SS of 210 mg/l. Size and sketch a layout for a stabilization pond arrangement consisting of two identical primary cells, which can be operated in parallel, and one secondary cell. The primary cells can have a maximum water depth of 1.5 m, and the secondary cell can have a maximum depth of 2.5 m. The minimum operating water depth in all cells is 0.5 m. Determine the dimensions of the ponds based on the following criteria: (1) The BOD loading on the primary cells cannot exceed 4.0 $g/m^2 \cdot d$. (2) The water storage capacity considering all three cells must be at least 120 d between the minimum and maximum water levels, with allowance for a water loss of 2.0 mm/d by evaporation and seepage.

12.68 A completely mixed aerated lagoon is being considered for pretreatment of a strong industrial wastewater with $k = 0.70$ at 20°C and $\Theta = 1.035$, using a detention time of 4 days. What is the BOD reduction at 20°C based on Eqs. (12.99) and (12.100)? If the wastewater temperature is 10°C, compute the detention time required to achieve the same degree of treatment.

12.69 A manufacturer's specified oxygen transfer capacity of a surface-aeration unit is 3.0 lb of oxygen/hp · hr. Using Eq. (12.102), calculate the oxygen transfer capability of this unit for an $\alpha = 0.9$, $\beta = 0.8$, a temperature of 20°C, and a dissolved-oxygen level of 2.0 mg/l in the lagoon water.

12.70 The rate of oxygen transfer for a surface aerator is specified by the manufacturer as 4.2 lb/hp · hr at standard conditions. Calculate the actual rate of transfer for a 40-hp aerator in a lagoon using a beta coefficient = 0.8, alpha coefficient = 0.9, and residual oxygen concentration = 2.0 mg/l at wastewater temperatures of 25°C and 15°C. If the efficiency of the electric drive motor is 90%, what is the electric power usage for the 40-hp aerator?

12.71 An aerated lagoon with a 10-ft depth and liquid volume of 175,000 cu ft treated an average of 0.20 mgd of wastewater with a BOD of 550 mg/l (920 lb of BOD/day). The anticipated temperature extremes of the aerating wastewater range from 10°C in winter to 30°C in summer. The two 15-hp surface aerators recommended for adequate mixing and oxygenation have the manufacturer's guarantee to provide an oxygen transfer of 2.5 lb/hp · hr under standard conditions. Based on laboratory treatability studies, the wastewater has the following characteristics: $k_{20} = 0.68$ per day, $\Theta = 1.047$, $\alpha = 0.9$, $\beta = 0.8$, and an oxygen utilization rate of 1.0 lb of oxygen to satisfy 0.8 lb of BOD removal. Is the lagoon adequately sized? For a minimum BOD reduction of 75%, is the aeration capacity adequate?

12.72 A completely mixed aerated lagoon with a volume of 1.0 mil gal is to treat a daily wastewater flow of 250,000 gal with a BOD of 400 mg/l. The liquid temperature ranges from 4°C in the winter to 30°C in the summer. There are four 5.0-hp surface aerators rated at 2.0 lb of oxygen/hp · hr. The wastewater characteristics are $\Theta = 1.035$, $\alpha = 0.9$, $\beta = 0.9$, $k = 0.80$ per day at 20°C, and $a = 1.0$ lb of oxygen utilized/lb of BOD removed. The designer states that this system will remove at least 400 lb of BOD/day and maintain a dissolved-oxygen concentration greater than 2.0 mg/l. Verify or disprove these claims by appropriate calculations.

12.73 Design a layout for an aerated lagoon followed by two secondary facultative ponds to treat a combined food-processing and domestic wastewater. The design flow is 200,000 gpd containing 1100 lb of BOD. The water temperatures in the aerated lagoon are expected to range from 2°C in winter to 25°C in summer. Determine the size and shape of the aerated lagoon and the number and horsepower of the surface aerators based on the following parameters: $\alpha = 0.9$, $\beta = 0.8$, $k = 0.90$ per day at 20°C, $a = 1.0$ lb of oxygen required per lb of BOD removed, a minimum of 80% BOD removal, and an aerator oxygen transfer rate $R_0 = 2.33$ lb of oxygen/hp · hr. The BOD loading on the two facultative ponds receiving the aerated wastewater is limited to 20 lb BOD/acre/day, and the storage volume of the ponds between the minimum water depth of 1.5 ft and maximum of 5.0 ft must be equal to or greater than 4 months of wastewater discharge at design flow.

12.74 The superintendent of a conventional activated-sludge plant receives complaints from nearby residents and businesses about odors, particularly in the summer. After a comprehensive field investigation, the major sources were found to be emissions of predominantly hydrogen sulfide from the grit chamber and primary clarifiers. Apparently, in the summer, convection lifts the foul air over the bushes and trees in the buffer zone and in the summer people are more likely to be outside or have their windows open. You are asked to recommend alternative schemes for containing and cleansing the foul air before release to the atmosphere.

12.75 Recommend the minimum wastewater treatment processes to produce an effluent to meet the standards specified by the state of California for ocean discharge. What steps should be taken if toxicity limits are exceeded?

REFERENCES

[1] *Standard Methods for the Examination of Water and Wastewater* (Washington, DC: Am. Public Health Assoc., 1995).

[2] G. J. Tortora, B. R. Funke, and C. L. Case, *Microbiology* (Menlo Park, CA: Benjamin/Cummings, 1982).

[3] L. Michaelis and M. L. Menten, "Die Kinetik der Invertinwirkung," *Biochemische Zeitschrift*, Neunundvierzigster Band (Vol. 49) (Berlin: Springer-Verlag, 1913): 333–369.

[4] J. Monod, "The Growth of Bacterial Cultures," *Ann. Rev. Microbiol.* 3 (1949): 371–393.

[5] S. J. Pirt, *Principles of Microbe and Cell Cultivation* (New York: Halsted Press Book, Wiley, 1975).

[6] D. H. Stoltenberg, "Midwestern Wastewater Characteristics," *Public Works* 3, no. 1 (1980): 52–53.

[7] N. W. Classen, "Per Capital Wastewater Contributions," *Public Works* 98, no. 5 (1967): 81–83.

[8] D. G. Munksgaard and J. C. Young, "Flow and Load Variations at Wastewater Treatment Plants," *J. Water Poll. Control Fed.* 52, no. 8 (1980): 2131–2144.

[9] "Sewage Treatment at Military Installations," Report of the Subcommittee on Sewage Treatment in Military Installations, National Research Council, *Sewage Works J.* 18, no. 5 (1946): 787–1028.

[10] S. Balakrishnan, W. W. Eckenfelder, and C. Brown, "Organics Removal by a Selected Trickling Filter Media," *Water and Wastes Eng.* 6, no. 1 (1969): A.22–A.25.

[11] D. D. Drury, J. Carmona III, and A. Delgadillo, "Evaluation of High Density Cross Flow Media for Rehabilitating an Existing Trickling Filter," *J. Water Poll. Control Fed.* 58, no. 5 (1986): 364–367.

[12] M. L. Arora and M. B. Umphres, "Evaluation of Activated Biofiltration and Activated Biofiltration/Activated Sludge Technologies," *J. Water Poll. Control Fed.* 59, no. 4 (1987): 183–190.

[13] Proceedings, First National Symposium on Rotating Biological Contactor Technology (Pittsburgh: University of Pittsburgh, 1980).

[14] M. J. Hammer, *Wastewater Treatment in Dry Climates (Desert or Desert with Some Rain)* (Alexandria, VA: Water Environment Federation, 2001).

[15] "The Causes and Control of Activated Sludge Bulking and Foaming," U.S. Environmental Protection Agency, Center for Environmental Research Information, EPA 625/8-87-012 (Cincinnati, OH: July 1987).

[16] *Design Manual, Fine Pore Aeration Systems*, EPA/625/1-89/023, U.S. Environmental Protection Agency, Technology Transfer (September 1989).

[17] *Oxygen Activated-Sludge Wastewater Treatment Systems*, U.S. Environmental Protection Agency, Technology Transfer (August 1973).

[18] A. W. Lawrence and P. L. McCarty, "Unified Basis for Biological Treatment Design and Operation," *Proc. Am. Soc. Civil Engrs., J. San. Eng. Div.* 96 (SA3) (1970): 757–778.

[19] L. D. Benefield and C. W. Randall, *Biological Process Design for Wastewater Treatment* (Upper Saddle River, NJ: Prentice Hall, 1980).

[20] C. P. Grady, Jr., and H. C. Lim, *Biological Wastewater Treatment* (New York: Dekker, 1980).

[21] G. F. Bennett and L. L. Kempe, "Oxygen Transfer in Biological Systems," *Proc. 20th Industrial Waste Conf., Purdue Univ. Ext. Service* 49, no. 4 (1965): 435–447.

[22] *Recommended Standards for Wastewater Facilities, Great Lakes—Upper Mississippi River Board of State Public Health & Environmental Managers* (Albany, NY: Health Research, Inc., Health Education Services Div., 1990).

[23] American Society of Civil Engineers, *ASCE Standard: Measurement of Oxygen Transfer in Clean Water* (New York: Am. Soc. Civil Eng., 1984).

[24] "Development of Standard Procedures for Evaluating Oxygen Transfer Devices," U.S. Environmental Protection Agency, Municipal Environmental Research Laboratory, EPA 600/2-83-002 (Cincinnati, OH: 1983).

[25] E. J. Middlebrooks, "Design Equations for BOD Removal in Facultative Ponds," *Water Science Tech.* 19, no. 12 (1987): 187–193.

[26] "Odor and Corrosion Control in Sanitary Sewerage Systems and Treatment Plants," U.S. Environmental Protection Agency, Technology Transfer, Center for Environmental Research Information, EPA/625/1-85/018 (Cincinnati, OH: 1985).

[27] R. P. G. Bowker, "Cleaning the Air on an Overlooked Odor Control Technique," *Water Environment & Treatment, Water Environment Federation*, 11, no. 2 (1999): 30–35.

[28] *California Ocean Plan, Water Quality Control Plan, Ocean Waters of California*, State Water Resources Control Board (Sacramento, CA: California Environmental Protection Agency, 2001).

CHAPTER 13

Processing of Sludges

OBJECTIVES

The purpose of this chapter is to:

- Describe the differing characteristics and quantities of sludges from wastewater processing and water treatment.
- Discuss methods of processing sludges and the conditions under which each method is appropriate.

Combinations of physical, chemical, and biological processes are employed in handling sludges. While the purpose in treating water and wastewater is to remove impurities from dilute solution and consolidate them into a smaller volume of liquid, the objective of processing sludge is to extract water from the solids and dispose of the dewatered residue. Furthermore, the relationships among the processes must be well understood, as disposal schemes involve sequencing of operations. For example, gravity thickening, anaerobic digestion, chemical conditioning, and mechanical dewatering form a physical–biological–chemical–physical sequence. Discussions on the characteristics of wastewater sludges and water treatment plant residues are isolated in different sections. Generalized process flow diagrams are also separated, although descriptions of individual unit operations are directed toward both water and wastewater sludges. Comments relating to the applicability of each process are included.

SOURCES, CHARACTERISTICS, AND QUANTITIES OF WASTE SLUDGES

Mathematical relationships for estimating the specific gravity and computing the sludge volume appear first, as they are fundamental calculations applicable to all sludges. Then the characteristics and methods for estimating sludge quantities are presented separately for wastewater and water treatment plant residues.

13.1 WEIGHT AND VOLUME RELATIONSHIPS

The majority of sludge solids from biological wastewater processing are organic, with a 60%–80% volatile fraction. The concentration of suspended solids in a liquid sludge is determined by straining a measured sample through a glass-fiber filter. Nonfilterable residue, expressed in milligrams per liter, is the solids content. Since the filterable portion of a sludge is very small, sludge solids are often determined by total residue on evaporation (i.e., the total deposit remaining in a dish after evaporation of water from the sample and subsequent drying in an oven at 103°C). Volatile solids are determined by igniting the dried residue at 550°C in a muffle furnace. Loss of weight upon ignition is reported as milligrams per liter of volatile solids, and the inerts remaining after burning as fixed solids. Waste from chemical coagulation of a surface water contains both organic matter removed from the raw water and mineral content derived from the chemical coagulants. Most solids are nonfilterable and have a volatile fraction of 20%–40%. Precipitate from treated well water is essentially mineral.

The specific gravity of solid matter in a sludge can be computed from the relationship

$$\frac{W_s}{S_s \gamma} = \frac{W_f}{S_f \gamma} + \frac{W_v}{S_v \gamma} \tag{13.1}$$

where

W_s = weight of dry solids, lb

S_s = specific gravity of solids

γ = unit weight of water, lb/ft^3 (lb/gal)

W_f = weight of fixed solids (nonvolatile), lb

S_f = specific gravity of fixed solids

W_v = weight of volatile solids, lb

S_v = specific gravity of volatile solids

The specific gravity of organic matter is 1.2–1.4, while the solids in chemically coagulated water vary from 1.5 to 2.5. The value for a solids slurry is calculated from

$$S = \frac{W_w + W_s}{(W_w/1.00) + (W_s/S_s)} \tag{13.2}$$

where

S = specific gravity of wet sludge

W_w = weight of water, lb

W_s = weight of dry solids, lb

S_s = specific gravity of dry solids

Consider a waste biological sludge of 10% solids with a volatile fraction of 70%. Their specific gravity can be estimated using Eq. (13.1) by assuming values of 2.5 for the fixed matter and 1.0 for the volatile residue.

$$\frac{1.00}{S_s} = \frac{0.30}{2.5} + \frac{0.70}{1.0} = 0.82$$

$$S_s = \frac{1}{0.82} = 1.22$$

Then the specific gravity of the wet sludge, by Eq. (13.2), is 1.02:

$$S = \frac{90 + 10}{(90/1.00) + (10/1.22)} = 1.02$$

These calculations demonstrate that for organic sludges of less than 10% solids the specific gravity may be assumed to be 1.00 without introducing significant error. Example 13.1 illustrates that even for mineral residue a high concentration of precipitate is required to increase the specific gravity of a slurry above 1.0.

The volume of waste sludge for a given amount of dry matter and concentration of solids is given by

$$V = \frac{W_s}{(s/100)\gamma S} = \frac{W_s}{[(100 - p)/100]\gamma S} \tag{13.3}$$

where

V = volume of sludge, ft^3 (gal) [m^3]

W_s = weight of dry solids, lb [kg]

s = solids content, %

γ = unit weight of water, 62.4 lb/ft^3 (8.34 lb/gal)[1000 kg/m^3]

S = specific gravity of wet sludge

p = water content, %

In this formula the volume of a sludge is indirectly proportional to the solids content. Thus, if a waste is thickened from 2% to 4% solids, the volume is reduced by one-half, and if consolidation is continued to a concentration of 8%, the quantity of wet sludge is only one-fourth of the original amount. During this concentration process, water content is reduced from 98% to 92%. In applying Eq. (13.3), specific gravity of the sludge S is normally taken as 1.0 and therefore not included in computations, as demonstrated in Example 13.2.

Example 13.1

Coagulation of a surface water using alum produces 10,000 lb (4540 kg) of dry solids/day, of which 20% are volatile. Both the settled sludge following coagulation and filter backwash water are concentrated in clarifier–thickeners to a solids density of 2.5%. Centrifugation can be used to increase the concentration to 20%, a consistency similar to soft wet clay, or the clarifier–thickener underflow can be dewatered to a 40% cake by pressure filtration. (1) Estimate the specific gravities of the thickened sludge, concentrate from centrifugation, and filter cake. (2) Calculate the daily sludge volumes from each process.

Solution:

1. Applying Eq. (13.1), one has

$$\frac{1.00}{S_s} = \frac{0.80}{2.50} + \frac{0.20}{1.00} = 0.52 \quad S_s = \frac{1}{0.52} = 1.9$$

From Eq. (13.2),

$$S(\text{thickened sludge}) = \frac{97.5 + 2.5}{(97.5/1.0) + (2.5/1.9)} = 1.0$$

$$S(\text{centrifuge discharge}) = \frac{80 + 20}{(80/1.0) + (20/1.9)} = 1.1$$

$$S(\text{filter cake}) = \frac{60 + 40}{(60/1.0) + (40/1.9)} = 1.2$$

2. Substituting these values into Eq. (13.3) with $W_s = 10{,}000$ lb/day gives

$$V(\text{thickened sludge}) = \frac{10{,}000}{(2.5/100)8.34 \times 1.0} = 48{,}000 \text{ gpd}$$

$$= \frac{4540 \text{ kg}}{(2.5/100)1000 \text{ kg/m}^3 \times 1.0} = 182 \text{ m}^3\text{/d}$$

$$V(\text{centrifuge discharge}) = \frac{10{,}000}{(20/100)8.34 \times 1.1} = 5400 \text{ gpd } (20.6 \text{ m}^3\text{/d})$$

$$V(\text{filter cake}) = \frac{10{,}000}{(40/100)8.34 \times 1.2} = 2500 \text{ gpd } (9.5 \text{ m}^3\text{/d})$$

Example 13.2

Estimate the quantity of sludge produced by a trickling-filter plant treating 1.0 mgd of domestic wastewater. Assume a suspended-solids concentration of 220 mg/l in the raw wastewater, a solids content in the sludge equivalent to 90% removal, and a sludge of 5.0% concentration withdrawn from the settling tanks.

Solution:

$$\text{solids in the sludge} = 1.0 \times 220 \times 8.34 \times 0.90 = 1650 \text{ lb/day}$$

$$\text{volume of sludge [using Eq. (13.3)]} = \frac{1650}{0.05 \times 62.4} = 530 \text{ ft}^3\text{/day}$$

13.2 CHARACTERISTICS AND QUANTITIES OF WASTEWATER SLUDGES

The purpose of primary sedimentation and secondary aeration is to remove waste organics from solution and concentrate them in a much smaller volume to facilitate dewatering and disposal. The concentration of organic matter in wastewater is approximately 200 mg/l (0.02%), while that in a typical raw-waste sludge is about 40,000 mg/l (4%). Based on these approximate values, treatment of 1.0 mil gal of wastewater produces about 4000 gal of sludge. This raw, odorous, and putrescible residue must be further processed and reduced in volume for land disposal or incineration. Common methods include gravity or mechanical thickening, biological digestion, and mechanical dewatering after chemical conditioning.

The quantity and nature of sludge generated relates to character of the raw wastewater and processing units employed. Daily sludge production may fluctuate over a wide range, depending on size of municipality, contribution of industrial wastes, and other factors. Both maximum and average daily sludge volumes are considered in designing facilities. A limited quantity of solids can be stored temporarily in clarifiers and aeration tanks to provide short-term equalization of peak loads. Mechanical thickening and dewatering units may be sized to handle sludge quantities as high as double the daily average. Other processes, such as conventional anaerobic digestion, have substantial equalizing capacity and are designed on the basis of maximum average monthly loading. Often, selection of conservative design parameters and liberal estimates of sludge yield take into account anticipated quantity variations. In this manner, designers disguise the fact that the maximum sludge yield is being considered in sizing unit processes. For example, in designing a trickling-filter plant, the dry solids may be calculated assuming 0.24 lb/capita/day when the actual amount realized in the treatment process is closer to 0.15 lb/capita/day. Furthermore, the required digester volume may be computed using a conservative figure of 5 ft^3/population equivalent.

Primary sludge is a gray-colored, greasy, odorous slurry of settleable solids, accounting for 50%–60% of the suspended solids applied, and tank skimmings. Scum is usually less than 1% of the settled sludge volume. Primary precipitates can be dewatered readily after chemical conditioning because of their fibrous and coarse nature. Typical solids concentrations in raw primary sludge from settling municipal wastewater are 4%–6%. Overpumping during sludge withdrawal can result in a thinner sludge by drawing in wastewater from above the settled solids, and biological activity in warm wastewater can result in gas production, decreasing solids concentration in the settled sludge. The portion of volatile solids varies from 60% to 80%.

Trickling-filter humus from secondary clarification is dark brown in color, flocculent, and relatively inoffensive when fresh. The suspended particles are biological growths washed from the filter media. Although they exhibit good settleability, the precipitate does not compact to a high density. For this reason and because sloughing is irregular, underflow from the final clarifier containing filter humus is returned to the wet well for mixing with the inflowing raw wastewater. Thus humus is settled with raw organics in the primary clarifier. The combined sludge has a solids content of 4%–5%, which is slightly thinner than primary residue with raw organics only.

Waste-activated sludge is a dark-brown, flocculent suspension of active microbial masses inoffensive when fresh, but it turns septic rapidly because of biological activity. Mixed-liquor solids settle slowly, forming a rather bulky sludge of high water content. The thickness of return activated sludge is 0.5%–1.5% suspended solids, depending on the rate of recirculation pumping, with a volatile fraction of 0.7–0.8. Excess activated sludge in most processes is wasted from the return sludge line. A high water content, resistance to gravity thickening, and the presence of active microbial floc make this residue difficult to handle. Routing of waste-activated sludge to the wet well for settling with raw wastewater is not recommended. Carbon dioxide, hydrogen sulfide, and odorous organic compounds are liberated from the settlings in the primary basin as a result of anaerobic decomposition, and the solids concentration is rarely greater than 4%. Waste-activated sludge can be thickened effectively by gravity belt thickening and dissolved-air flotation with chemical addition to ensure high solids capture in the concentration process.

Anaerobically digested sludge is a thick slurry of dark-colored particles and entrained gases, principally carbon dioxide and methane. When well digested, it dewaters rapidly on sand-drying beds, releasing an inoffensive odor resembling that of garden loam. Substantial additions of chemicals are needed to coagulate a digested sludge prior to mechanical dewatering, owing to the finely divided nature of the solids. The dry residue is 30%–60% volatile, and the solids content of digested liquid sludge ranges from 3% to 12%, depending on the mode of digester operation.

Aerobically digested sludge is a dark-brown, flocculent, relatively inert waste produced by long-term aeration of sludge. The suspension is bulky and difficult to gravity-thicken, thus creating problems of ultimate disposal. Since decanting clear supernatant can be difficult, the primary functions of an aerobic digester are stabilization of organics and temporary storage of waste sludge. The solids concentration in thickened, aerobically digested sludge is generally in the range 1.0%–2.0% as determined by digester design and operation. The thickness of aerobically digested sludge can be less than that of the influent, since approximately 50% of the volatile solids are converted to gaseous end products. Stabilized sludge is often disposed of by spreading on land for its fertilizer value, particularly at small treatment plants. For these reasons, aerobic digestion is limited to treatment of waste-activated sludge from aeration plants without primary clarifiers.

Mechanically dewatered sludges vary in characteristics based on the type of sludge, chemical conditioning, and unit process employed. The density of dewatered cakes ranges from 15% to 40%. The thinner cake is similar to a wet mud, while the latter is a chunky solid. The method of ultimate disposal and economics dictate the degree of moisture reduction necessary.

Biosolids refer to liquid and dewatered sludges that have been suitably processed for land application as agricultural fertilizer and soil conditioner. To control the concentration of heavy metals and other toxins in the sludge, industrial wastewaters are pretreated prior to discharge to the sewer collection system. The common sludge treatment is anaerobic or aerobic digestion to reduce pathogenic microorganisms and to stabilize the organic matter. Quality standards for application of biosolids are established by Environmental Protection Agency regulations [1].

Waste solids production in primary and secondary processing can be estimated using the following formulas:

$$W_s = W_{s_p} + W_{s_s} \tag{13.4}$$

where

W_s = total dry solids, lb/day

W_{s_p} = raw primary solids, lb/day

W_{s_s} = secondary biological solids, lb/day

$$W_{s_p} = f \times \text{SS} \times Q \times 8.34 \tag{13.5}$$

where

W_{s_p} = primary solids, lb of dry weight/day

f = fraction of suspended solids removed in primary settling

SS = suspended solids in unsettled wastewater, mg/l

Q = daily wastewater flow, mgd

8.34 = conversion factor, lb/mil gal per mg/l

$$W_{s_s} = k \times \text{BOD} \times Q \times 8.34 \tag{13.6}$$

where

W_{s_s} = biological sludge solids, lb of dry weight/day

k = fraction of applied BOD that appears as excess biological growth in waste-activated sludge or filter humus, assuming about 30 mg/l of BOD and suspended solids remaining in the secondary effluent

BOD = concentration in applied wastewater, mg/l

Q = daily wastewater flow, mgd

The first expression states that the total weight of dry solids produced equals the sum of the primary plus secondary residues. Settleable matter removed in primary clarification can be considered a function of the suspended-solids concentration [Eq. (13.5)]. For typical municipal wastewaters, the value for f is between 0.4 and 0.6. The settleable fraction of suspended solids in a fresh domestic wastewater is about 0.5, but septic conditions and industrial waste contributions are likely to decrease the portion of settlings in a wastewater. For example, many food-processing discharges are high in colloidal matter and exhibit BOD/suspended-solids ratios of 2:1 or greater. Thus, a combined wastewater may exhibit an f value that is considerably less than the average 0.5 for domestic waste.

Organic matter entering secondary biological treatment is colloidal in nature and best represented by its BOD value. Most is synthesized into flocculent biological growths that entrain nonbiodegradable material. Therefore, excess activated-sludge solids from aeration and humus from biological filtration can be estimated by Eq. (13.6), which relates residue production to BOD load. The coefficient k is a function of process food/microorganism ratio and biodegradable (volatile) fraction of the matter in suspension. For trickling-filter humus, k is assumed to be in the range 0.3–0.5, with the lower value for light BOD loadings and the larger number applicable to high-rate filters and rotating biological contactors. The k for secondary activated-sludge processes can be estimated using Figure 13.1 by entering the diagram along the ordinate with a known food/microorganism ratio.

Excess solids production for activated-sludge processes treating unsettled wastewater can be estimated using Eq. (13.7) (based on influent BOD), without considering suspended solids input, by increasing the calculated quantity by 100%. Thus, for aeration systems without primary clarifiers, the k factor is the value determined from Figure 13.1 multiplied by 2.0.

$$W_{as} = 2.0k \times \text{BOD} \times Q \times 8.34 \tag{13.7}$$

where

W_{as} = total dry solids from activated-sludge processing without primary sedimentation, lb/day of dry weight

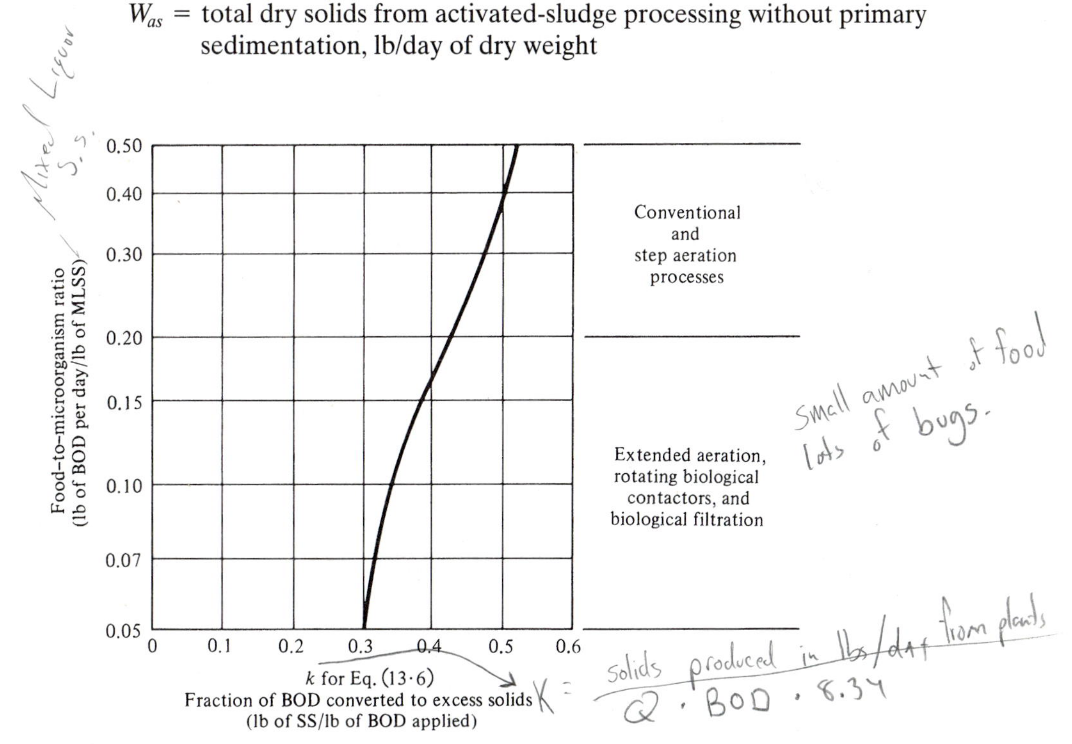

FIGURE 13.1 Hypothetical relationship between the food-to-microorganism ratio and the coefficient k in Eqs. (13.6) and (13.7).

The design of a sludge-handling system is based on the volume of wet sludge as well as dry solids content. Once the dry weight of residue has been estimated, the volume of sludge is calculated by applying Eq. (13.3).

The foregoing formulations are reasonable for sludge quantities from processing domestic wastewater at average daily design flow. Real sludge yields may differ considerably from anticipated values when treating a municipal discharge containing substantial contributions of industrial wastes and when loading or operational conditions create unanticipated peak sludge volumes.

Peak loads must be assessed for each treatment plant design based on local conditions such as seasonal industrial discharges, anticipated trends in per capita solids contribution, and the type of unit operations employed in wastewater processing. A rule-of-thumb approach assumes that the maximum weekly dry solids yield will be approximately 25% greater than the yearly mean. Variations in daily sludge volumes may be considerably greater, owing to changes in moisture content. For example, if the concentration of settlings drawn from primary clarifiers shifts from 6% to 4%, the total quantity of wet sludge increases 50% for the same amount of dry solids. These deviations are normally taken into account by selecting conservative design criteria in sizing biological digesters. Where mechanical dewatering is practiced, increased quantities of sludge can be handled by applying higher chemical dosages and by extending the time of operation each day. Perhaps the most difficult parameter to predict is the solids concentration of waste-activated sludge. Bulking can easily reduce the solids content by one-half, say, from 15,000 mg/l to 7500 mg/l, thus doubling the volume. Processing of this larger quantity must be evaluated by the designer. One guideline is to size mechanical thickeners at 200% of the estimated volume during normal operation in order to ensure consolidation of the diluted slurry. Provisions are required for the addition of coagulants to aid in concentrating waste-activated sludge.

Example 13.3

One million gallons of municipal wastewater with a BOD of 260 mg/l and suspended solids of 220 mg/l are processed in a primary plus secondary activated-sludge plant. Estimate the quantities and solids contents of the primary, waste-activated, and mixed sludges. Assume the following: 60% SS removal and 40% BOD reduction in clarification, a water content of 94.0% in the raw sludge, and an operating F/M ratio of 1:3 in the aeration basin; the solids concentration of 15,000 mg/l in the waste-activated sludge is increased to 45,000 mg/l by gravity belt thickening. If the sludges are blended, calculate the consolidated sludge volume.

Solution: The primary sludge solids and volume based on Eqs. (13.5) and (13.3) are, respectively,

$$W_{s_p} = 0.60 \times 220 \times 1.0 \times 8.34 = 1100 \text{ lb}$$

$$V = 1100/[(100 - 94)/100]8.34 = 2200 \text{ gal}$$

To determine k for Eq. (13.6), enter Figure 13.1 with F/M = 0.33, and read 0.48 lb of SS produced/lb of BOD applied.

The thickened waste-activated sludge, based on Eq. (13.6), is

$$W_{s_s} = 0.48 \times 260 \times 0.60 \times 1.0 \times 8.34 = 620 \text{ lb}$$

$$V = \frac{620}{0.045 \times 8.34} = 1650 \text{ gal}$$

The blended sludge volume, solids content, and solids concentration are, respectively,

$$V = 2200 + 1650 = 3850 \text{ gal}$$

$$W_s = 1100 + 620 = 1720 \text{ lb}$$

$$s = \frac{1720 \times 100}{3850 \times 8.34} = 5.4\%$$

Example 13.4

The rational formulas for waste solids production [Eqs. (13.4)–(13.6)] can be used to predict future sludge yields, provided the characteristics of the wastewater are not expected to change significantly.

Listed in columns 1–5 of Table 13.1 are average monthly operating data for a conventional activated-sludge plant treating a municipal wastewater containing discharges from several industries. The average suspended solids removal in primary sedimentation is 53% ($f = 0.53$) and the BOD removal is 22%, leaving an average of 490 mg/l in the settled wastewater. The current solids concentration in the combined waste sludge varies from 4.0% to 4.5%. The sludge is dewatered by belt filter presses 5 days per week using a sludge holding tank for storing weekend production.

TABLE 13.1 Operating Data and Calculated Values of Solids Production for Example 13.4

	Measured Values				Calculated Values	
Month	Flow (mgd)	Suspended Solids in Unsettled Wastewater (mg/l)	BOD in Settled Wastewater (mg/l)	Sludge Solids Produced from Wastewater (mg/l)	Sludge Solids Production from Eq. (13.8) (mg/l)	Percentage Difference of Columns 5 and 6 (%)
Feb.	2.3	480	480	340	410	−17
Mar.	2.5	620	460	530	480	10
Apr.	2.5	640	500	660	500	32
May	2.3	780	600	460	610	−24
June	2.5	740	400	580	520	11
July	2.4	870	550	450	640	−30
Aug.	2.4	600	430	450	460	−2
Sept.	2.5	460	480	520	400	30
Mean	2.4	650	490	500	500	

Develop equations to calculate the average annual weekday sludge solids production and the average weekday solids production during the maximum month. Calculate these values and sludge volumes at 4.0% solids concentration for a future raw wastewater flow of 3.0 mgd with 600 mg/l suspended solids and 580 mg/l BOD.

Solution: The average solids production expressed in milligrams per liter of wastewater processed is

$$W_s = W_{s_p} + W_{s_s} = f \times \text{SS} + k \times \text{BOD}$$

We substitute the mean measured values from Table 13.1 and solve for k:

$$500 \text{ mg/l} = 0.53 \times 650 + k \times 490$$

$$k = 0.32$$

The equation for average annual weekday sludge solids production in milligrams per liter of wastewater processed is then

$$W_s = 0.53 \times \text{SS} + 0.32 \times \text{BOD} \tag{13.8}$$

Using this equation, the sludge solids production is calculated for each month and listed in column 6 of Table 13.1. The percentage difference between the measured and calculated solids production values is computed and listed in column 7 by subtracting the numbers in column 6 from those in column 5 and dividing the resultants by the numbers in column 6 and multiplying by 100 to yield percentages.

Based on Eq. (13.8), the equation for average annual weekday sludge solids production in pounds based on a 5-day work week is

$$W_{s_d} = (7/5)(Q)(8.34)(0.53 \times \text{SS} + 0.32 \times \text{BOD})$$

The results in column 7 of Table 13.1 show that the average weekday solids production during the maximum month is 30% greater than the average; therefore,

$$W_{s_m} = 1.3 W_{s_d}$$

For a future raw-wastewater flow of 3.0 mgd with 600 mg/l suspended solids and 580 mg/l of BOD (settled BOD of $0.78 \times 580 = 450$ mg/l), the average annual values are

$$W_{s_d} = (7/5)(3.0)(8.34)(0.53 \times 600 + 0.32 \times 450) = 16{,}000 \text{ lb}$$

$$V_{s_d} = \frac{16{,}000}{0.04 \times 8.34} = 48{,}000 \text{ gal}$$

The average weekday values during the maximum month are $W_{s_m} = 20{,}000$ lb and $V_{s_m} = 61{,}000$ gal.

13.3 CHARACTERISTICS AND QUANTITIES OF WATER-PROCESSING SLUDGES

The characteristics and quantities of water treatment plant residuals vary greatly depending on the water source and kinds of treatment processes; the most common plants are coagulation–filtration systems to remove turbidity and pathogens and precipitation softening to reduce hardness [2]. Water treatment residuals are derived from sedimentation and filtration of chemically conditioned water. Surface supplies yield wastes containing colloidal matter removed from the raw water and chemical flocs, while groundwater-processing precipitates are mineral with little or no organic material. Sludges vary widely in composition, depending on character of the water source and chemicals added during treatment. A typical method of handling a turbid river supply includes presedimentation for reduction of settleable solids, lime softening for hard water, alum coagulation, and filtration for removal of colloids, plus the addition of activated carbon for taste and odor control, and chlorination for disinfection. The presedimentation deposit is silt plus detritus; settling-basin sludge is a mixture of inerts, organics, and chemical precipitates, including metal hydroxides; and filter backwash water contains floc from agglomerated colloids and unspent coagulant hydroxides. Lake and reservoir waters are often dosed with alum and flocculation aids, plus activated carbon. Settlings during the summer may include significant quantities of algae. Precipitates from lime–soda ash softening are predominantly calcium carbonate and magnesium hydroxide with traces of other minerals, such as oxides of iron and manganese.

Sludge storage capacity and the time intervals between withdrawals are governed by the installation design, the type of water processing, and operations management. Clarifiers equipped with mechanical scrapers discharge sludge either continuously at a low rate or intermittently, often daily. Settled sludge is allowed to accumulate and consolidate in plain rectangular or hopper-bottomed basins. These tanks are cleaned at time intervals varying from a few weeks to several months by draining and removing the compacted sludge. Backwashing of filters produces a high flow of dilute wastewater for a few minutes, usually once a day for each filter. Obviously, any system for handling water treatment wastes must consider temporary storage and thickening of wash water.

Alum-coagulation sludge is dramatically influenced by the gelatinous nature of the aluminum hydroxides formed in the reaction with raw-water alkalinity. Particles entrained in the floc and other coagulation precipitates do not suppress the jellylike consistency that makes an alum slurry difficult to dewater. Coagulation settlings and backwash water normally can be gravity thickened to a liquid with 2%–3%, although polymers may be needed to achieve this consolidation. Studies have shown that centrifugal dewatering can concentrate this waste to a truckable semisolid with 10%–15% with a consistency similar to a soft wet clay. Pressure filtration will produce a 30%–40% cake with the consistency of stiff clay that breaks easily. Complete dehydration by drying or freezing results in a granular material that does not revert to its original gelatinous form if again mixed with water. Enmeshed water of hydration, not water of suspension, causes the original jelly consistency. Iron coagulants yield slightly denser sludges that are somewhat easier to handle.

Surface water wastes are highly variable, owing to changes in raw-water quality. High turbidities during spring runoff and periods of high rainfall result in a decreased percentage of aluminum hydroxide solids. The result is a precipitate that settles better and is easier to dewater. Water temperature changes affect algal growth in surface supplies, the rate of chemical reactions in treatment, and filterability of the sludge. In designing a waste-handling system, changes in raw-water quality and accompanying variations in sludge characteristics must be investigated by long-term studies of daily and seasonal records.

Sludge production from surface water treatment can be estimated from chemical additions and raw-water characteristics. Based on empirical data, 1 mg of commercial alum applied as a coagulant produces 0.44 mg of aluminum precipitate, which is nearly double the 0.26 theoretical amount [Eq. (11.34)]. The observed relationship between turbidity of the raw water, expressed in nephelometric turbidity units (NTU), and weight of impurities removed varies from 0.5 to 2.0 mg/NTU, with an average value of 0.74. The following equation for estimating dry sludge solids production from alum coagulation is based on these data:

$$\text{Total sludge solids (lb/mil gal)} = 8.34(0.44 \times \text{alum dosage} + 0.74 \times \text{turbidity}) \tag{13.9}$$

The majority of sludge solids settle by gravity in the sedimentation basins; the remainder are removed by subsequent filtration. The nonsettleable portion depends on the kinds of impurities in the water and chemicals applied in coagulation. The sludge withdrawn from sedimentation basins is 1%–2%, and the solids content of filter wash water is less than 0.05%. Gravity thickening of settled sludge and backwash water from alum coagulation can be thickened in a clarifier–thickener to 2%–6%.

Coagulation–softening sludges result from processing hard, turbid surface waters, such as those found in Midwestern rivers. A typical treatment plant flow arrangement is presedimentation followed by two-stage or split treatment; lime softening and coagulation with alum or iron salts; and filtration. Solids concentration in settled sludges varies with turbidity in the raw water, ratio of calcium to magnesium in the softening precipitate, type and dosage of metal coagulant, and filter aids used. In general, filter wash water gravity thickens to about 4%, alum–lime sludges have densities of up to 10%, and lime–iron precipitates range between 10% and 20%, with a consistency of a viscous liquid. The quantity of sludge produced is difficult to predict because the chemical treatment varies with hardness and turbidity of the river water. In-plant modifications to help even out fluctuations in sludge yield may include applying polymers in coagulation, closer pH control to reduce the amount of magnesium hydroxide produced in softening, and providing flexibility to thicken sludges separately or combined.

Lime–soda ash softening sludges produced in treating groundwaters contain calcium carbonate and to a lesser extent magnesium hydroxide. Aluminum hydroxides and other coagulant aids may be present if added in water processing. The quantity of $Mg(OH)_2$ depends on magnesium hardness in the raw water and the softening process employed. In general, dry solids are 85%–95% $CaCO_3$. The residue is stable, dense, inert, and relatively pure, since groundwater does not contain colloidal inorganic or organic matter. Calcium carbonate compacts readily, while magnesium hydroxide, like aluminum hydroxide, is gelatinous and does not consolidate as well nor dewater as easily. Slurry wasted from flocculator–clarifiers (upflow units) has a solids content in the range 2%–5%. Stoichiometrically, from Eq. (13.10), 3.6 lb of calcium carbonate is precipitated for each pound of lime applied. However, in actual practice, the dry solids yield from softening is closer to 2.6 lb/lb of lime applied, owing to incomplete chemical reaction, impurities in commercial-grade lime, and precipitation of variable amounts of magnesium.

$$CaO + Ca(HCO_3)_2 = 2\ CaCO_3 + H_2O \tag{13.10}$$

Iron and manganese oxides removed from groundwater by aeration and chemical oxidation are flocculent particles with poor settleability. The amount of sludge produced in the removal of these metals without simultaneous precipitation softening is relatively small. The majority of hydrated ferric and manganic oxides pass through sedimentation tanks, are trapped in the filters, and appear in the dilute backwash water.

Filter wash water is a relatively large volume of wastewater with a low solids concentration of 100–400 mg/l. The exact amount of water used in backwashing is a function of the type of filter system, cleansing technique, and quality and source of the raw water being treated. Generally, 2%–3% of the water processed in a plant is used for filter washing. The fraction of total waste solids removed by filtration depends on efficiency of the coagulation and sedimentation stages, type of treatment system, and characteristics of the raw water. The amount may be a substantial portion, say, 30% of the dry solids resulting from treatment.

Example 13.5

A reservoir water supply with a turbidity of 10 units in the summer is treated by applying an alum dosage of 30 mg/l. For each million gallons of water processed, estimate the total solids production, volume of settled sludge, and quantity of filter wash water. Assume a settled sludge concentration of 1.5% solids, a filter solids loading of 26 lb/mil gal, and a backwash water with 400 mg/l of suspended solids; also assume that Eq. (13.9) is applicable. Compute the composite sludge volume after the two wastes are gravity thickened to 3.0% solids.

Solution: By Eq. (13.9),

$$\text{Total sludge solids} = 8.34(0.44 \times 30 + 0.74 \times 10) = 172 \text{ lb/mil gal}$$

$$\text{Sedimentation basin solids} = \text{total sludge solids} - \text{solids to filters}$$

$$= 172 - 26 = 146 \text{ lb/mil gal}$$

Applying Eq. (13.3) yields

$$V \text{ of settled sludge} = \frac{146}{0.015 \times 8.34} = 1170 \text{ gal}$$

$$V \text{ of wash water} = \frac{26}{(400/1{,}000{,}000)8.34} = 7800 \text{ gal}$$

$$V \text{ of thickened sludge} = \frac{172}{0.030 \times 8.34} = 690 \text{ gal}$$

Example 13.6

The lime–soda ash softening process described in Example 11.5 requires a lime dosage of 2.7 meq/l CaO and a soda ash addition of 1.2 meq/l of Na_2CO_3. Based on the appropriate chemical reactions, calculate the calcium carbonate residue produced in the softening of 10^6 m^3 of water, assuming the practical limit of $CaCO_3$ precipitation is 0.60 meq/1 (30 mg/l).

Solution: Based on Eq. (11.50), the 0.4 meq/l of CO_2 is precipitated by addition of 0.4 meq/l of lime to form 0.4 meq/l of $CaCO_3$:

$$Ca(OH)_2 + CO_2 = CaCO_3\downarrow + H_2O$$

From Eq. (11.45), the 2.3 meq/l of $Ca(HCO_3)_2$ reacts with 2.3 meq/l of lime to form 4.6 meq/l of $CaCO_3$:

$$Ca(HCO_3)_2 + Ca(OH)_2 = 2CaCO_3\downarrow + 2H_2O$$

Finally, 1.2 meq/l of soda ash precipitates 1.2 meq/l of $CaCO_3$ by the reaction of Eq. (11.49a):

$$CaSO_4 + Na_2CO_3 = CaCO_3\downarrow + Na_2SO_4$$

The residue is theoretically equal to the stoichiometric quantities of $CaCO_3$ formed minus the practical limit of treatment (solubility):

$$0.4 + 4.6 + 1.2 - 0.6 = 5.6 \text{ meq/l}$$

$$5.6 \times 50 = 280 \text{ mg/l of } CaCO_3$$

and

$$280 \text{ g/m}^3 \times 10^6 \text{ m}^3 \times 10^{-3} \text{ kg/g} = 280{,}000 \text{ kg}/10^6 \text{ m}^3$$

ARRANGEMENT OF UNIT PROCESSES IN SLUDGE DISPOSAL

Many processes for sludge handling are applied to both wastewater sludges and water treatment plant residues. Individual unit operations are discussed in the latter sections of this chapter, with comments relative to their method of operation and application. Under this heading, selection and arrangement of units are outlined to illustrate how they integrate with each other. Referring to these discussions while studying the individual operations will help relate them to complete sludge-disposal schemes.

13.4 SELECTION OF PROCESSES FOR WASTEWATER SLUDGES

Techniques selected for processing waste sludges are a function of the type, size, and location of the wastewater plant, the unit operations employed in treatment, and the method of ultimate solids disposal [3]. The system adopted must be able to accept the primary and secondary sludges produced and economically convert them to a residue that is environmentally acceptable for disposal.

TABLE 13.2 Processes for Storage, Treatment, and Disposal of Wastewater Sludges

Process
Storage prior to processing
In the primary clarifiers
Separate holding tanks
Sludge lagoons
Thickening prior to dewatering or digestion
Gravity settling in tanks
Gravity belt thickening
Dissolved-air flotation
Centrifugation
Conditioning prior to dewatering
Stabilization by anaerobic digestion
Stabilization by aerobic digestion
Chemical coagulation
Heat treatment or wet oxidation
Mechanical dewatering
Belt filter pressure filtration
Plate-and-frame pressure filtration
Centrifugation
Composting
Air drying of digested sludge
Sand drying beds
Shallow lagoons
Disposal of liquid or dewatered digested sludge
Spreading as biosolids on agricultural land
Application on dedicated surface disposal site
Disposal of dewatered raw or digested solids
Codisposal in municipal solid-waste landfill
Burial in dedicated sludge landfill
Incineration
Production of bagged fertilizer and soil conditioner

Methods for storage, treatment, and disposition are listed in Table 13.2. Settled raw solids may be stored in the bottom of primary clarifiers during the day, or perhaps over a weekend, and then pumped directly to a processing unit. In small plants, sludge is transferred to anaerobic digesters either once or twice a day. Belt filter presses and centrifuges require a steady flow of sludge during the operating period, which may extend from 4 to 24 hr/day. Separate holding tanks can be used to receive and blend primary and secondary sludges. Primary sludge may be pumped intermittently using an automatic time-clock control, while secondary sludge wasting may be either continuous or periodic. Aerated holding tanks for accumulating and biologically stabilizing waste from aeration plants are called *aerobic digesters*.

The economics of chemical conditioning and biological treatment are directly related to the sludge density. As a general rule, the solids content of settled waste must be at least 4% for feasible handling. Primary precipitates and mixtures of primary and secondary settlings are amenable to thickening by sedimentation; therefore, gravity tank thickeners are used to increase the concentration of sludges withdrawn from either clarifiers or holding tanks. Because of the flocculent nature of waste-activated sludge, separate thickening is performed by either dissolved-air flotation or gravity belt thickening. The float or thickened sludge is then blended with primary waste and directed to holding tanks or the next dewatering or treatment step.

The purpose of anaerobic digestion is to convert bulky, odorous, and putrescible raw sludge to a well-digested material that can be rapidly dewatered without emission of noxious odors. In addition to stabilizing and gasifying the organic matter, the volume of residue is significantly reduced by withdrawal of supernatant from digesters to thicken the sludge. Aerobic digestion is almost exclusively used to treat excess sludge from plants without primary clarifiers. Although it stabilizes the organic matter, solids thickening and dewatering are troublesome, owing to the bulky nature of overaerated sludge. Chemical coagulation with polymers or other coagulants is required for mechanical dewatering of both raw and digested wastes.

Heat treatment and wet oxidation are less commonly applied than other methods of sludge conditioning prior to dewatering because of higher costs of operation and maintenance. This conditioning sterilizes, deodorizes, and prepares the waste for mechanical dewatering without addition of chemicals. Processes using heat treatment apply steam to heat the reactor vessel to about 300°F under a pressure of 150 psi or greater. Sludge from the reactor is discharged to a decant tank, from which the underflow is withdrawn for dewatering. The supernatant contains high concentrations of water-soluble organic compounds and must therefore be returned to the treatment plant for processing. This is a serious drawback in handling some sludges, since it can result in cycling of solids through the heating process back to the treatment plant, which extracts them and returns them again to the heat treatment. Wet oxidation can be achieved under high pressure at elevated temperatures. Liquid sludge with compressed air is fed into a pressure vessel, where the organic matter is stabilized. Inert solids are separated from the effluent by dewatering in lagoons or by mechanical means.

Belt filter presses in small and medium-sized plants and centrifuges in large plants are the most common methods of dewatering chemically conditioned sludges. Belt presses are effective in dewatering a wide range of waste concentrations from raw primary to thin aerobically digested sludges. In operating performance, the main advantages are low energy consumption, high cake density, and clarity of filtrate. Centrifuges dewater digested sludges to a high cake density. Economy in centrifugation results from high-capacity machines that require much less space than an equivalent capacity of belt filter presses. Polymers are applied to sludges for chemical conditioning prior to dewatering by both belt filter presses and solid-bowl decanter centrifuges. Plate-and-frame filter presses are rarely used to dewater wastewater sludges because of the high installation cost. Their operation is a batch process with a cycle of filling with sludge, pressure dewatering, discharging of cake, and washing of filter cloths. The sludge is chemically conditioned with lime and ferric chloride, which are chemicals more difficult to store and feed than polymers. For very small treatment plants, any kind of mechanical dewatering may be uneconomical because of the minimum available sizes of dewatering units. The smallest belt filter press can dewater the sludge production from a population of 5000. The minimum size of centrifuges restricts their application to even larger treatment plants serving a population greater than 100,000.

The oldest technique for drying digested sludge is on open sand beds contained by short concrete walls. Well-digested slurry is drained or pumped onto the surface of the bed to a depth of 8–12 in. Moisture leaves by evaporation and seepage; the latter is collected in underdrain piping for return to the plant influent. In current design, the beds have vertical draw-off pipes to decant supernatant after the solids settle and paved surfaces that gently slope to a narrow strip of sand bed with an underdrain located along the center line of the bed. This construction allows cleaning with a small front-end loader, since manual cleaning with hand shovels is no longer economical. Shallow lagoons can also be used for air drying of digested sludge. This permits the use of front-end loaders or buckets on draglines, but operations can be disrupted by inclement weather.

Liquid and dewatered digested sludges (biosolids) are commonly spread on and tilled into agricultural land as a fertilizer and soil conditioner. Farmland application is often originated by the operator of a small plant to avoid codisposal in a municipal solid-waste landfill. Currently, many large plants digest and dewater sludges for application on agricultural land, with a sludge management firm acting as the intermediary to sell and haul the biosolids to applicators and to ensure that the quality meets the requirements of the Environmental Protection Agency. Where agricultural land is not available, the sludge can be applied on a dedicated surface disposal site.

Dewatered solids, both raw and digested, can be disposed of in municipal solid-waste or dedicated-sludge landfills. Dried anaerobic cake and incinerator ash are relatively inoffensive, but cake composed of raw solids putrefies and may contain pathogens. The latter type must be covered every day to prevent nuisance and a health hazard. Incineration is much more costly than land disposal, yet in highly urbanized areas it is often the only feasible alternative. Rather than burning, digested sludge may be dried and bagged as soil conditioner.

The following process flow diagrams graphically illustrate the selection and arrangement of common sludge-processing layouts. Figure 13.2 is typical for communities with a population of less than 10,000. Raw solids and filter humus are settled and stored in the primary clarifiers. Once or twice a day, sludge and scum are pumped to

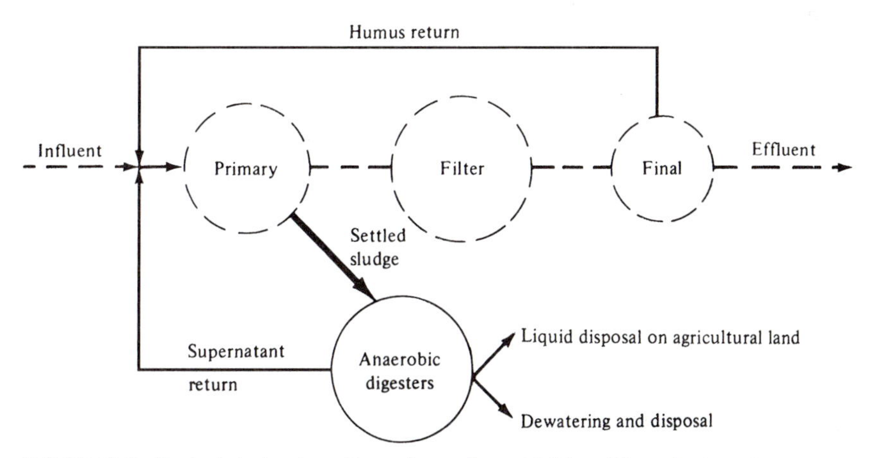

FIGURE 13.2 Typical sludge-handling scheme for a trickling-filter plant serving a community of fewer than 10,000 people.

anaerobic digesters, and supernatant is withdrawn and returned to the treatment plant influent. Stabilized and thickened sludge accumulates in the digesters for withdrawal when weather conditions permit disposal. These biosolids can be spread as liquid digested sludge on agricultural land or dried on sand beds or in lagoons followed by hauling to a landfill for burial.

The optimum sludge-process scheme for an activated-sludge plant, shown in Figure 13.3(a), thickens the waste-activated sludge separately prior to blending with the primary sludge. (Activated-sludge plants that return waste-activated sludge to the plant influent commonly experience serious difficulties resulting from anaerobiosis and thinning of settled sludge in the primary clarifier.) The waste-activated sludge is concentrated independently by gravity belt thickening or dissolved-air flotation, processes that give reliable and effective results. Water of separation is returned for reprocessing, while the thickened sludge is pumped along with a primary sludge to a mixed blending tank. If the sludge is anaerobically digested, the dewatered biosolids are likely to be applied to land as a fertilizer and soil conditioner. If the blended raw

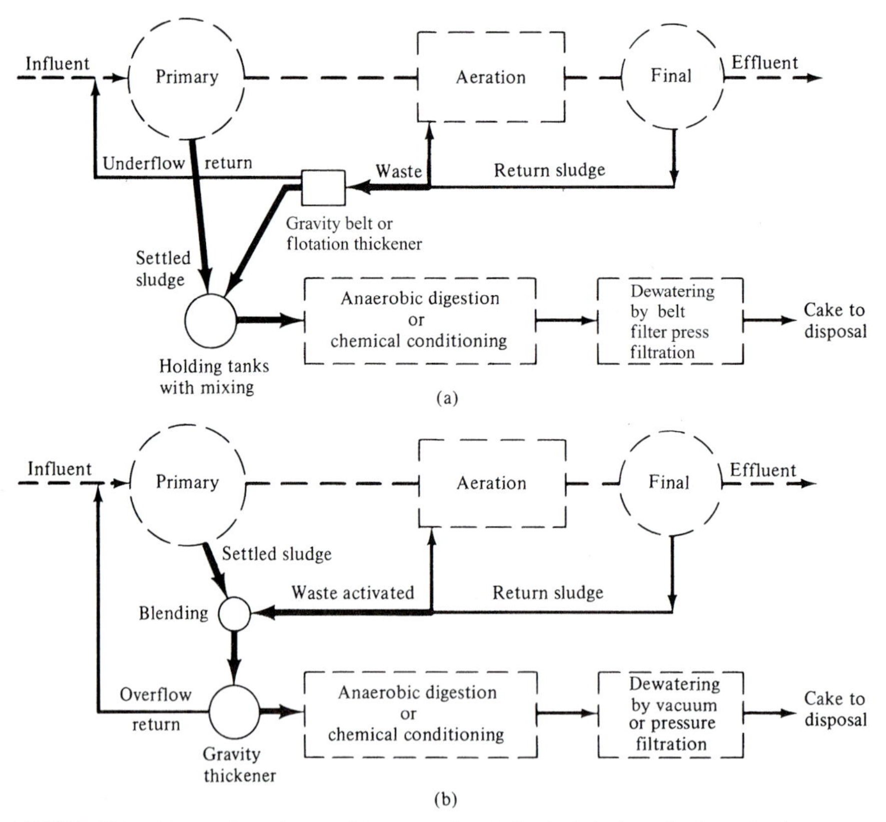

FIGURE 13.3 Alternative schemes for processing activated-sludge plant wastes by thickening in advance of conditioning and dewatering. (a) Separate thickening of waste-activated sludge before mixing with primary sludge. (b) Gravity thickening of combined raw primary and waste-activated sludges.

sludge is dewatered without prior digestion, the blending tank is sized so that sludge can accumulate when not being dewatered, and the installed mixing impellers are adequate to ensure a homogeneous feed for mechanical dewatering. Unless the retention time of the blended sludge is relatively short by continuous removal, the tank may release foul odors. The active microbial floc in the waste-activated sludge decomposes the raw organic matter in the primary sludge under anaerobic conditions, releasing hydrogen sulfide and other reduced volatile organic compounds. Dewatered raw solids are disposed of by burial in a landfill or incineration.

An alternative arrangement, shown in Figure 13.3(b), blends waste-activated and raw sludges in a gravity thickening tank. Consolidation of this waste mixture by gravity settling may yield only marginal results because of carryover of flocculent solids, thus providing poor solids capture. Assuming good operation and chemical additions, performance at the allowable solids loading can recover (capture) up to 90% of the solids. The thickened underflow is likely to be only in the range of 4%–6% solids (Section 13.6). Supernatant from the thickener is returned to the plant inlet and underflow sludge is pumped for processing and disposal.

Aerobic processing of wastewater without primary sedimentation often utilizes aerobic digestion for stabilization (Fig. 13.4). Decanting chambers for return of supernatant, installed in the digesters of package plants, rarely provide efficient solids separation. However, tanks with separate controls for aeration and decanting can be effective in concentrating aerobically digested solids. The most common method for eliminating the liquid stabilized sludge is by spreading it on farmland.

Designing a sludge-processing system requires a thorough understanding of the characteristics of the waste being produced and the most feasible method for solids disposal. Selection of the latter is dictated by local conditions and practices. Intermediate steps of thickening, treatment, and dewatering must be integrated so that each relates to the prior operation and prepares the residue for subsequent handling. A scheme should have flexibility to allow alternative modes of operation, since actual conditions may differ from those assumed at the time of design. Too often, built-in rigidity, by limiting piping and pumping facilities, does not permit plant personnel to vary operations to meet changing conditions. The practice of sanitary design requires both knowledge and foresight to consider all available options.

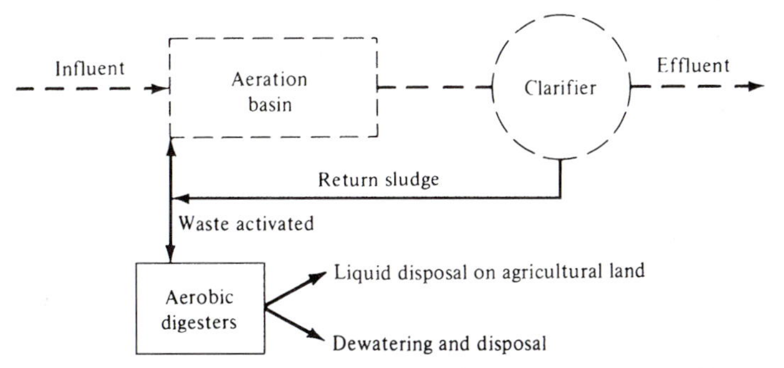

FIGURE 13.4 Common disposal methods for waste-activated sludge from small treatment plants without settling prior to aeration.

The following list illustrates typical sludge-processing and disposal problems that have confronted plant operators:

1. Returning excess activated sludge to the head of the treatment plant with no alternative for separate thickening or disposal. The result can be upset primary clarifiers, reduced efficiency, thin sludge, and higher chemical costs for conditioning prior to mechanical dewatering.
2. Employing gravity thickeners to concentrate combined primary and waste-activated sludges [Fig. 13.3(b)] or waste-activated sludge alone. Besides limited thickening of the sludges, the thickener overflow returns to the head of the plant a substantial quantity of suspended solids that are then recycled through the system.
3. Returning poor-quality digester supernatant to the head of a treatment plant increases the BOD load and recycling of suspended solids.
4. Excessive concentrations of heavy metals in biologically stabilized sludges that limit their application as biosolids on agricultural land.
5. Being forced to incinerate sludge when an alternate method would be much less expensive if facilities were available.
6. Providing only sand drying beds for dewatering anaerobic or aerobically digested sludges at small plants with no provision for spreading on adjacent grassland or farmland. Plant operators usually modify the piping and buy a tank wagon or truck for liquid disposal because of the labor involved in scooping up dried cake from the beds.

13.5 SELECTION OF PROCESSES FOR WATER TREATMENT SLUDGES

The pollution of surface waters and groundwaters from the discharge of residuals from water treatment is controlled by state regulations under authorization from the Environmental Protection Agency. In general, clarified waters like settled backwash water from filters and overflow from solids-separation processes can be discharged to flowing waters, provided that in-stream water quality standards are not violated. After separation of supernatant, the sludge is often disposed of by land application. Water treatment plants may also discharge to a sanitary sewer or directly to a wastewater plant, provided that the wastewaters comply with pretreatment requirements for industrial wastewater.

Water treatment processes can be modified to change the characteristics and reduce the quantities of residuals. Polymers as coagulant aids lower the required dosages of alum and auxiliary chemicals. The result is less sludge that is easier to dewater because of the reduced content of hydroxide precipitate. Polymers also enhance presedimentation of turbid river waters, thus controlling carryover of solids to subsequent chemical coagulation. The addition of specially manufactured clays can be used to aid flocculation of relatively clear surface supplies by producing a denser floc that settles more rapidly. Coagulant aids with alum should be considered as a cost-effective technique for modifying sludge-handling processes at surface water plants. Groundwater softening plants can also change their mode of operation to lessen the volume of

TABLE 13.3 Processes for Storage, Treatment, and Disposing of Water Treatment Sludges

Storage prior to processing
Sedimentation basins
Separate holding tanks
Flocculator–clarifier basins
Thickening prior to dewaterng
Gravity settling
Chemical conditioning prior to dewatering
Polymer application
Lime addition to alum sludges
Mechanical dewatering
Centrifugation
Pressure filtration
Air drying
Shallow lagoons
Sand drying beds
Disposal of dewatered solids
Codisposal in municipal solid-waste landfill
Burial in dedicated landfill
Disposal of liquid and dewatered sludge
Spread on agricultural land
Application on dedicataed surface disposal site

waste sludge. Emphasis is placed on preventing magnesium hydroxide precipitation because it inhibits dewaterability and the feasibility of processing sludge for recovery of lime. Also, hardness reduction can be limited to produce less solid material while still supplying a moderately soft water acceptable to the general public.

The common processes for storage, treatment, and disposal of water treatment sludges are listed in Table 13.3. Each waterworks is unique so that local conditions and existing facilities tend to dictate techniques applied in processing and disposing of waste solids [4].

Modern clarifiers equipped with mechanical scrapers discharge sludge at regular intervals, usually daily. Separate holding tanks can be installed to accumulate this slurry prior to dewatering. Filter backwash can be stored in clarifier–flocculators that serve as both temporary holding tanks and wash-water settling basins. Equalization and settling are generally the only prerequisites if sludges are discharged to a sewer for processing at the municipal wastewater treatment plant.

A typical system for thickening coagulation waste and filter wash water is shown schematically in Figure 13.5. The two primary sources of waste are sludge from the clarifier, following chemical coagulation, and wash water from backwashing filters. The latter is discharged to a clarifier holding tank for gravity separation of the suspended solids and flow equalization. Settled solids consolidate to a sludge volume less than one-tenth of the wash-water volume. Supernatant is withdrawn slowly and recycled to the plant inlet. After a sufficient portion has been drained, the holding tank is able to receive the next backwash surge. Settled sludges from both the wash-water tanks and in-line clarifiers are given second-stage consolidation in a clarifier–thickener. Polymer

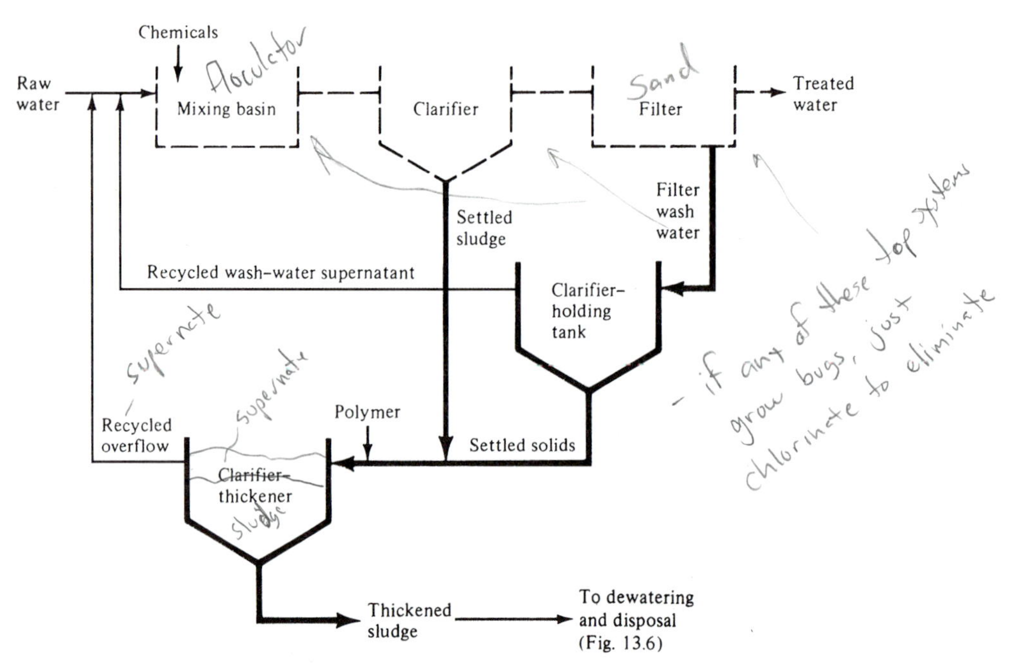

FIGURE 13.5 Thickening coagulation waste and filter wash water in preliminary handling of water treatment plant sludges.

is normally added to enhance solids capture, overflow is recycled, and thickened sludge withdrawn for further dewatering and disposal.

Recycled wash-water supernatant and recycled overflow without treatment can return microorganisms, organic precursors, and disinfection by-products to the influent raw water. Figure 13.5 satisfies the EPA Filter Backwash Recycling Rule that requires treatment of recycled flows and their point of return ahead of coagulation (i.e., before the rapid mix/coagulant process preceding clarification and/or filtration). The primary objective of this regulation is to ensure the removal of 99% of *Cryptosporidium* oocysts in treatment of surface water. This rule applies to all recycled flows including filter backwash, supernatant from sludge thickening, and wastewater from sludge dewatering.

Mechanical dewatering of chemical sludges can be accomplished by centrifugation or plate-and-frame pressure filtration and in some instances by belt filter pressure filtration (Fig. 13.6). A major advantage of a solid-bowl decanter centrifuge is operational flexibility. Machine variables, such as speed of rotation, allow a range of moisture content in the discharged solids varying from a dry cake to a thickened slurry. Feed rates, sludge solids content, and chemical conditioning are also variables that influence performance. Polymers or other coagulants are normally applied with the slurry feed to enhance solids capture. Lime sludges compact readily, producing a cake with 50% or greater solids from a feed of 5%–10%. Alum sludges do not dewater as readily and discharge with a toothpaste consistency suitable for either further processing or transporting by truck to a disposal site. Gravity-thickened sludge containing about one-half aluminum hydroxide slurry can be concentrated to 10%–15% solids, while one-quarter hydrate slurry can be dewatered to 20% or greater. Removal of solids by centrifugation

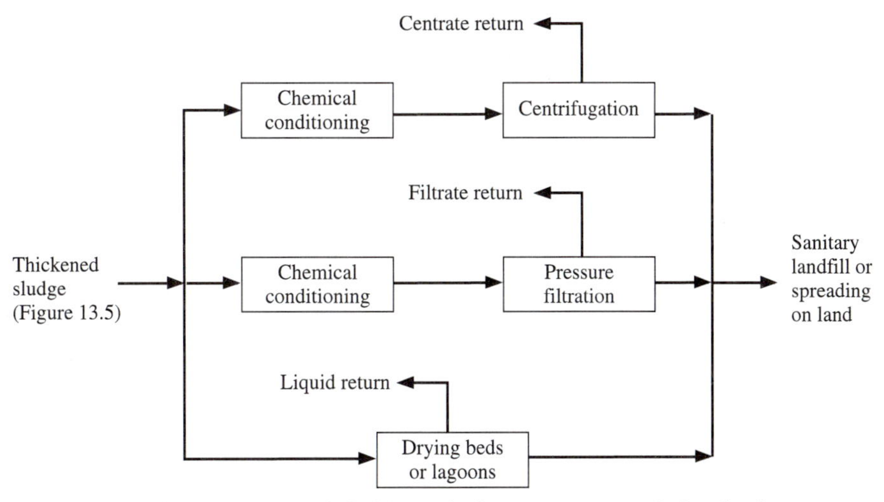

FIGURE 13.6 Alternative methods for disposal of water treatment sludges by dewatering, drying, or chemical recovery of thickened chemical coagulation wastes.

varies over a broad range, depending on operating situations and chemical conditioning; in general, solids recovery and density of cake are related to polymer dosage.

A plate-and-frame filter press is particularly advantageous for dewatering alum sludges if a high solids concentration in the filter cake is desired. Aluminum hydroxide wastes are often conditioned with lime to improve their filterability. The filter medium is precoated with either diatomaceous earth or fly ash before applying sludge solids. Precoat protects against blinding of the filter cloth by fines and ensures easy cake discharge without sticking. With proper chemical conditioning, alum sludges can be pressed to a solids content of about 30%–40%, which can be handled as a chunky solid rather than the paste consistency associated with a 15% density. Belt filter presses can dewater alum sludges to 15%–20% and lime sludges to 50% or greater with polymer conditioning.

Lagooning is a common method for dewatering, thickening, and temporary storage of waste sludge where suitable land area is available [4]. The diked pond area needed relates to character of the sludge, climate, design features such as underdrains and decanters, and method of operation. Clarified overflow may be returned to the treatment plant, particularly if filter wash water is directed to the lagoons without prior thickening. Sludges from lime softening consolidate to about 50% solids after drying by evaporation and can be removed by a scraper or dragline and hauled to land burial. Alum sludge dewaters and dries more slowly to a density of only 10%–15%. Although the surface may dry to a hard crust, the underlying sludge turns to a viscous liquid upon agitation. This slurry must be removed, usually by dragline, and spread on the banks to air dry prior to hauling. Freezing enhances the dewatering of alum sludge by breaking down its gelatinous character. Neither lime nor alum sludges made a good, stable landfill. Air drying at small water plants can be done on sand beds with tile underdrains. Repeated sludge applications over a period of several months can be made in depths up to several feet. Dewatering action is by drainage and air drying, although operation may include decanting supernatant. Dried cake is removed by either hand shoveling or mechanical means.

GRAVITY THICKENING

Gravity thickening is the simplest and least expensive process for consolidating waste sludges. Thickeners in wastewater treatment are employed most successfully in consolidating primary sludge separately or in combination with trickling-filter humus. When raw primary and waste-activated sludges are blended and concentrated, results are often marginal because of poor solids capture. Water treatment wastes from both sedimentation and filter backwashing can be compacted effectively by gravity separation.

13.6 GRAVITY SLUDGE THICKENERS IN WASTEWATER TREATMENT

The tank of a gravity thickener resembles a circular clarifier except that the depth–diameter ratio is greater and the hoppered bottom has a steeper slope. The three settling zones are the clear supernatant on top, feed zone characterized by hindered settling, and compression near the bottom where consolidation occurs. Figure 13.7 is a schematic diagram with the main components labeled. Influent sludge is applied continuously. However, if this is not practical, sludge application can be intermittent at frequent intervals. The circular inlet baffle is partly submerged but not so deep as to disturb the sludge blanket. The discharge weir is peripheral for maximum length. Although not always installed, a skimmer can be used to push scum over the weir with the overflow or down a pipe for separate collection. Thickened sludge is scraped to a central outlet for continuous or intermittent discharge. Withdrawal is often governed by the operation of the feed pump to the next processing step, e.g., anaerobic digestion. Attached to the scraper arms are picket fences consisting of vertical rods or pickets that are slowly pulled along. By gentle agitation of the settled solids, consolidation is increased by dislodging gas bubbles and preventing the bridging of solids. A properly designed picket fence is considered essential in thickening of organic waste [5].

The principal design criterion is solids loading expressed in units of pounds of solids applied per square foot of bottom area per day ($lb/ft^2/day$ or $kg/m^2 \cdot d$). Typical loading values and thickened-sludge concentrations based on operational experience

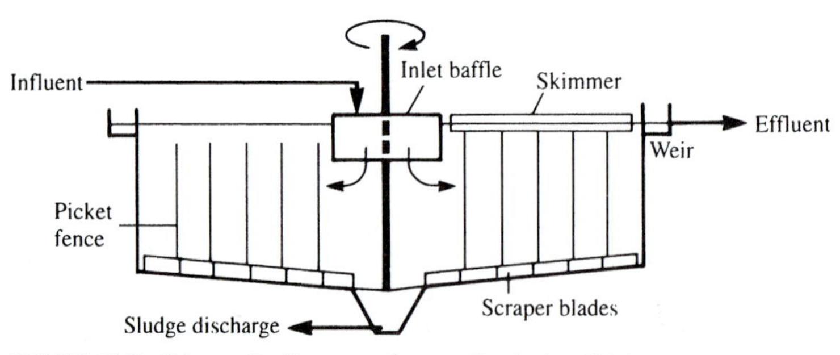

FIGURE 13.7 Schematic diagram of a gravity sludge thickener.

TABLE 13.4 Gravity Thickener Design Loadings and Underflow Concentrations for Wastewater Sludges

Type of Sludge	Average Solids Loading ($lb/ft^2/day$)[a]	Underflow Concentration (% solids)
Primary	20	8–10
Primary plus filter humus	10	6–8
Primary plus activated sludge	8	4–6

[a] $1.0\ lb/ft^2/day = 4.88\ kg/m^2 \cdot d$

are listed in Table 13.4. These data assume good operation and chemical additions, such as chlorine, if necessary to inhibit biological activity. Solids recovery in a properly functioning unit is 90%–95%, with perhaps the exception of a unit handling primary plus waste activated where it is difficult to achieve this degree of solids capture. Most continuous-flow thickeners are designed with a side-water depth of approximately 10 ft to provide an adequate clear-water zone, sludge-blanket depth, and space for temporary storage of consolidated waste. Sludge-blanket depths (feed plus compaction zones) should be 3 ft or greater to ensure maximum compaction, using a suggested solids retention time of 24 hr. This is estimated by dividing the volume of the sludge blanket by the daily sludge withdrawal; values vary from 0.5 to 2 days, depending on operation. Overflow rates should be 400–900 gpd/ft^2 (16–37 $m^3/m^2 \cdot d$) and are defined by the quantity of sludge plus supplementary dilution water applied. Dilution water blended with sludge feed increases the overflow rate to carry out fine solids in the supernatant to enhance thickening and reduce the emission of foul odors from anaerobiosis.

Gravity thickeners are normally sized to handle the maximum seasonal or monthly sludge yield anticipated. Peak daily sludge production often requires storage in the thickener or other sludge-processing units. Low liquid overflow rates result in malodors from septicity of the thickener contents. A common remedy is to feed dilution water to the thickener along with the sludge to increase hydraulic loading. An alternative is to apply chlorine to reduce bacterial activity. The design of pumps and piping should be sufficiently flexible to allow regulation of the quantity of dilution water and have the capacity to transport viscous, thickened sludges.

Example 13.7

The daily quantity of primary sludge from a trickling-filter plant contains 1130 lb of solids at a concentration of 4.5%. Size a gravity thickener based on a solids loading of 10 $lb/ft^2/day$. Calculate the daily volumes of applied and thickened sludges, assuming an underflow of 8.0% and 95% solids capture. What is the flow of dilution water required to attain an overflow rate of 400 gpd/ft^2? If the blanket of consolidated sludge in the tank has a depth of 3.0 ft, estimate the solids retention time.

Solution:

$$\text{tank area required} = \frac{1130}{10} = 113 \text{ ft}^2$$

$$\text{diameter} = \left(\frac{113 \times 4}{\pi}\right)^{1/2} = 12.0 \text{ ft}$$

Use a depth of 10.0 ft.

$$\text{volume of applied sludge} = \frac{1130}{0.045 \times 62.4} = 402 \text{ ft}^3/\text{day}$$

$$= 3010 \text{ gpd}$$

$$\text{overflow rate of applied sludge} = \frac{3010}{113} = 27 \text{ gpd/ft}^2$$

$$\text{supplemental dilution flow to attain } 400 \text{ gpd/ft}^2 = (400 - 27)113 = 42{,}000 \text{ gpd}$$

$$\text{volume of thickened sludge} = \frac{1130 \times 0.95}{(8.0/100)62.4} = 215 \text{ ft}^3/\text{day}$$

$$\text{solids retention time} = \frac{3 \times 113 \times 24}{215} = 38 \text{ hr}$$

13.7 GRAVITY SLUDGE THICKENERS IN WATER TREATMENT

The performance of thickeners processing water treatment plant wastes varies with the character of the water being treated and the chemicals applied. Alum sludges from surface water coagulation settle to a density in the range of 2%–6% solids. Coagulation–softening mixtures from the treatment of turbid river waters gravity thicken approximately as follows: alum–lime sludge, 4%–10%; iron–lime settlings, 10%–20%; alum–lime filter wash water, about 4%; and iron–lime backwash, up to 8%. The density achieved in gravity thickening relates to the calcium–magnesium ratio in the solids, quantity of alum, nature of impurities removed from the raw water, and other factors. Calcium carbonate residue from groundwater softening consolidates to 15%–25% solids. In most cases, special studies have to be conducted at a particular water works to determine settleability of solids in waste sludges and wash water. Flocculation aids are used to improve clarification in most cases.

Relatively dense chemical slurries are thickened in tanks similar to the one shown in Figure 13.7. Thin sludges and backwash waters may be concentrated in clarifier–thickeners that have an inlet well equipped with mixing paddles, where the feed can be flocculated with polymers or other coagulants. Holding tanks are used to dampen hydraulic surges of filter wash water. These units can be plain tanks with mixers or clarifiers equipped for removing settled solids and decanting clear supernatant.

Evaluating the performance of a thickener often involves mass balance calculations. Overflow plus underflow solids equals influent solids. Also, the sum of overflow and underflow volumes is equal to the quantity of applied sludge and supplementary dilution water. These values can be calculated using Eq. (13.3), as illustrated in Example 13.8.

Example 13.8

An alum–lime slurry with 4.0% solids content is gravity thickened to 20% with a removal efficiency of 95%. Calculate the quantity of underflow per 1.0 m^3 of slurry applied and the concentration of solids in the overflow. Assume a specific gravity of 2.5 for the dry solids.

Solution:

$$\text{solids applied} = 1.0\ \text{m}^3 \times 1000\ \text{kg/m}^3 \times 0.04 = 40\ \text{kg}$$

$$\text{underflow solids} = 0.95 \times 40 = 38\ \text{kg}$$

The specific gravity of underflow using Eq. (13.2) is

$$S = \frac{80 + 20}{(80/1.0) + (20/2.5)} = 1.14$$

$$\text{volume of underflow [Eq. (13.3)]} = \frac{38}{0.20 \times 1000 \times 1.14} = 0.17\ \text{m}^3$$

$$\text{volume of overflow} = 1.0 - 0.17 = 0.83\ \text{m}^3$$

From Eq. (13.3), the concentration of solids in the overflow is

$$s = \frac{0.05 \times 40 \times 100}{0.83 \times 1000 \times 1.0} = 0.24\% = 2400\ \text{mg/l}$$

GRAVITY BELT THICKENING

The stand-alone gravity belt thickener is an outgrowth of the development of the belt filter press (Section 13.20), which has a gravity drainage zone for initial thickening prior to the pressure dewatering zone. After flocculation of a sludge, conveyance on a continuous porous belt supported horizontally on an open framework allows the separated water to drain freely through the belt. In addition to mechanical development, success of gravity belt thickening is attributable to the availability of improved polymers for flocculation of sludge solids to release bound water. The common application is separate thickening of waste-activated sludge prior to blending with primary sludge for anaerobic digestion.

13.8 DESCRIPTION OF A GRAVITY BELT THICKENER

The feed sludge is conditioned by applying a polymer solution through an injection ring ahead of a variable orifice venturi mixer that discharges into the bottom of a feed retention tank. Adequate conditioning is necessary to ensure flocculation by agglomerating the solids to reduce their affinity to water so that drainage can occur. After flowing up through the retention tank, to allow formation of larger floc and free water, the flocculated sludge is uniformly applied across the belt through a chute.

As illustrated in Figure 13.8, the porous belt moves horizontally across the drainage zone to the drive roller where the thickened sludge slurry is discharged. It then passes underneath where the belt is washed, maintained in alignment by a steering roller, and returns to the top by passing over the belt tensioning roller. In the drainage zone, the belt is supported by bars that wipe the underside of the belt, and sludge is prevented from flowing off the sides of the belt by retainers and rubber seals. To open channels in the sludge layer being conveyed along the belt, plastic drainage elements are supported just above the belt surface. The filtrate is collected in a pan under the drainage zone and discharged below the machine. The belt is fabricated of monofilament polyester with the mesh design, porosity, and tensile properties based on the kind of sludge being thickened. The effective dewatering length (horizontal travel distance of the belt) varies with manufacturer from 12 to 14 ft (3.8 to 4.3 m). The effective widths of belts are 1.0, 1.5, 2.0, and 3.0 m.

The thickened sludge, usually in the range of 5%–8% solids, is a loose, wet slurry subject to splashing; therefore, the scraper blade and discharge hopper are surrounded by a shield. The rubber-covered drive roller that provides traction to pull the belt is powered by an electric motor with a drive train for variable speed up to 100 fpm. Since woven monofilament polyester fabrics have visible clear openings, solids can penetrate the pores of the fabric, requiring washing with a high-pressure water spray. The spray containment housing has a wash tube with jet nozzles and an internal brush with a handwheel for cleaning nozzles during operation. The steering assembly monitors the position of the belt and shifts the steering roller to maintain belt alignment. A sensing unit in contact with the edge of the belt signals the roller positioning unit, which can be either a hydraulic or a pneumatic system. Variable belt tension is established and controlled by a roller held in a yoke tensioned by either hydraulic or pneumatic cylinders.

13.9 LAYOUT OF A GRAVITY BELT THICKENER SYSTEM

A complete separate system of auxiliary equipment, as shown in Figure 13.9, is recommended for each gravity belt thickener. The sludge feed must be applied at a uniform rate by a plunger pump, progressing cavity pump, or at higher feed rates by a centrifugal pump. A variable-speed drive is necessary with a maximum pumping capacity greater than the design hydraulic loading. The polymer preparation and feed equipment is designed with flexibility to allow for variations in feed rate and use of different kinds of polymers. Dry flake or microbead polymer has greater than 95% active solids and is commonly prepared for feed at a concentration of 0.2% total solids; emulsion forms are 25%–50% active solids and prepared at 0.5%; and liquid forms are 2%–8% active solids prepared at 2.0%. For accuracy in polymer addition, the metering pump is

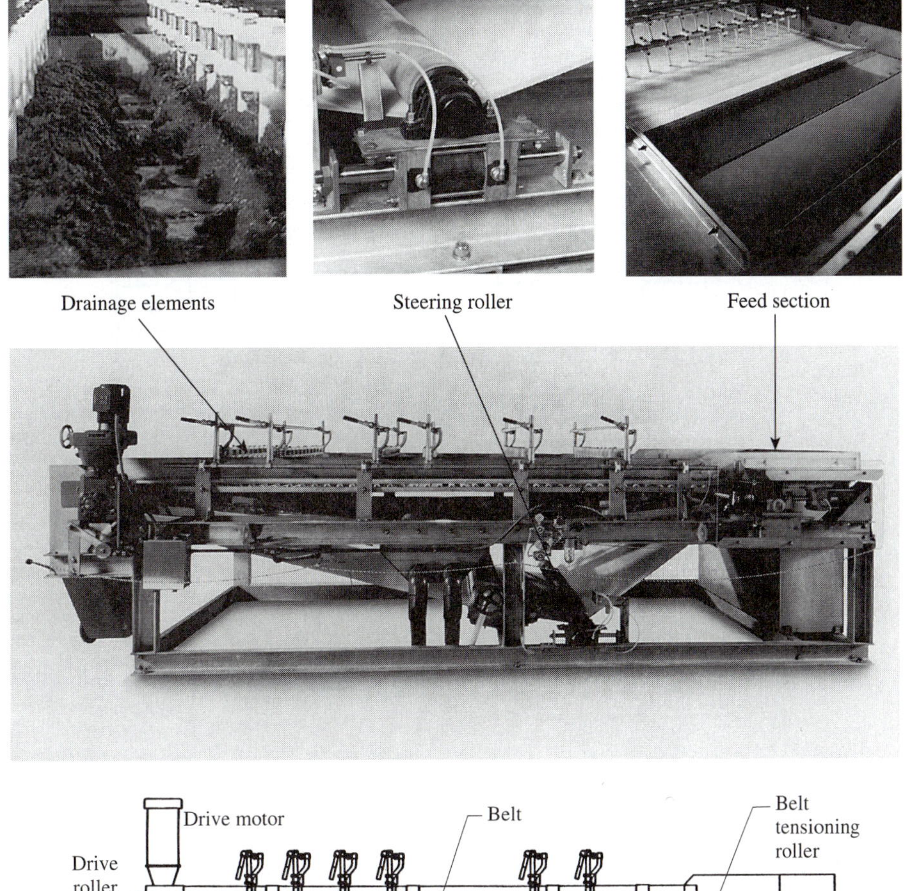

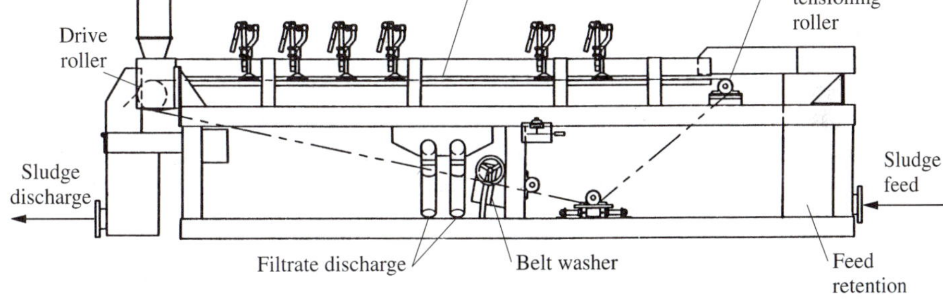

FIGURE 13.8 Gravity belt thickener. Sludge feed after polymer injection overflows a retention tank onto the porous belt for gravity drainage. Elements positioned close to the belt open channels for free drainage and to decelerate the agglomerated sludge solids. The thickened sludge slurry is released from the belt by a discharge blade. *Source:* Courtesy of Komline-Sanderson, Peapack, NJ.

positive displacement with a speed regulator. The sludge mixed with polymer is applied to the belt after upward flow through the feed retention tank.

Shown at the top of Figure 13.9 are the air compressor, or hydraulic unit, to control the belt tension cylinder and steering mechanism and the wash-water booster pump. The supply of wash water is 15–20 gpm/m $(1\text{–}1.3 \text{ l/s} \cdot \text{m})$ at a pressure of 80–100 psi. For

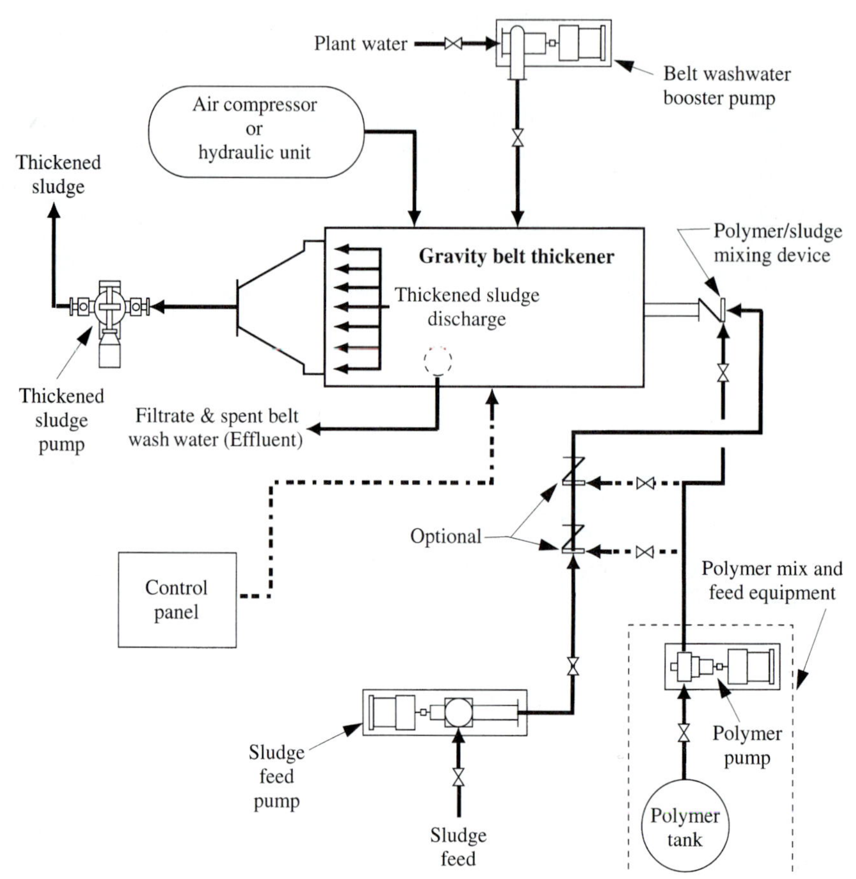

FIGURE 13.9 Schematic layout showing the major components of a gravity thickener system. *Source:* Courtesy of Komline-Sanderson, Peapack, NJ.

cleaning the belt, either clear plant effluent or potable water can be used as a supply. The electric control panel provides for automatic, sequenced start-up and shut-down, with monitoring at each step in the cycle to provide prewetting of the belt prior to sludge application and discharge of sludge slurry and belt washing upon shut-down. Alarm conditions shut down the system in case of serious belt misalignment, low air or hydraulic pressure, low wash-water pressure, or broken belt or belt drive. All auxiliary equipment for each press is controlled from one panel. For return to the plant inlet, the filtrate and belt wash water are collected in a drain under each machine. The thickened sludge is pumped for further processing.

13.10 SIZING OF GRAVITY BELT THICKENERS

The main operating parameters of a gravity belt thickener are hydraulic loading, solids loading, and polymer dosage. Hydraulic loading is expressed in gallons per minute per meter of belt width (cubic meters per meter per hour). Solids loading is expressed as the pounds of total dry solids feed per meter per hour (kilograms per meter per hour). Polymer dosage is calculated as pounds applied per ton of total dry solids in the sludge feed (kilograms per tonne). The key performance parameter is concentration of dry solids by weight in the thickened sludge expressed as a percentage.

Thickened sludge is usually in the range of 5%–8% solids when the feed is conditioned with sufficient polymer and solids capture is 95% or more. Underdosing produces unstable floc formation and poor drainage, resulting in reduced solids concentration, lower solids capture, and the necessity to reduce feed rate. Overdosing, though giving the appearance of improved operation by better drainage and higher solids concentration, can, after an extended period of operation, prevent proper drainage by blinding the porous belt as the excess polymer forms a skin on the surface of the belt. Maintaining optimum polymer feed rate is important for cost-effective operation.

Drainage belts are available in various porosities affecting drainage rate and solids capture. Design selection is based on documented experience at other plants thickening a sludge with similar characteristics. Similarly, hydraulic loading and speed of the belt are based on the kind and concentration of waste solids and dewaterability after polymer conditioning. To account for seasonal and operational variations in the volume of sludge produced, a conservative capacity is normally selected in design. The maximum weekly production of dry solids may be greater than 25% of the annual average, and the water content may increase as much as 50% to double the sludge volume. The maximum hydraulic capacity of a gravity belt thickener is usually 50%–80% greater than the design rating to allow increased loading of conditioned sludge at increased belt speed. Conversely, decreasing capacity can be achieved by reducing hydraulic loading, polymer feed rate, and belt speed. The minimum capacity is normally 25% of design rating. Since the horizontal travel distance between sludge application and discharge is fixed, belt speed affects the time allowed for drainage. At higher belt speed, the thickened sludge solids concentration is likely to decrease, while at slower speed, the solids content is likely to increase.

Design of a thickening facility considers the size of the treatment plant and desired flexibility of operation. The selection of belt filters depends on the quantity of sludge production, design feed rate, and operating time. Gravity thickeners are manufactured with active belt widths of 1.0, 1.5, 2.0, and 3.0 m. Unless adequate storage is available in sludge treatment units, at least two machines are installed so one can operate while the other is out of service for maintenance. In a small plant, the operating time may be 7 hr/day, whereas in a larger plant, operating time may be for a period of 15 or 23 hr/day. These schedules allow 1 hr for start-up and shut-down. Approximate operating parameters for gravity belt thickening of two kinds of sludge are listed in Table 13.5. Because published operational data are limited, these listed values should not be relied on for design without verification. For a machine of a particular manufacturer, operating experience at other installations thickening a sludge of similar characteristics is essential. In general, belt filter capacity should be prudently selected for new installations to account for the probable inaccuracy in projecting performance and peak sludge volume.

TABLE 13.5 Approximate Results from Gravity Belt Thickening of Polymer Flocculated Wastewater Sludges

Type of Sludge	Feed Solids (%)	Hydraulic Loading (gpm/m)[a]	Solids Loading (lb/m/hr)[b]	Thickened Sludge Solids (%)	Active Polymer Dosage (lb/ton)[c]
Waste activated	0.5–2	150–250	600–2000	5–8	4–8
Anaerobically digested primary plus waste activated	2.5–5	100–150	2000–3000	6–10	5–10

[a]1.0 gpm/m = 0.225 $m^3/m \cdot h$
[b]1.0 lb/m/hr = 0.454 $kg/m \cdot h$
[c]1.0 lb/ton = 0.500 kg/tonne

Example 13.9

As originally constructed, a conventional activated-sludge plant was piped to return waste-activated sludge to the plant inlet or blend it with primary sludge. Returning the waste sludge to the inlet upset the primary clarifiers, resulting in floating solids, foul odors, and thinner sludges. After blending the waste-activated and primary sludges, the mixture was less than 3% solids, which was too low for good operation of anaerobic digestion. The proposed solution is to modify the sludge processing system by installing gravity belt thickeners to concentrate the solids in the waste-activated sludge and a blending tank to mix this thickened sludge with the primary sludge prior to digestion, as diagrammed in Figure. 13.3(a). If the waste-activated sludge were thickened to 5%–8%, the mixed sludge would be in the range of 6%–7%.

Size two gravity belt thickeners to increase the solids content of the waste-activated sludge to 5%–8% at the plant design load of 15.0 mgd. During the maximum month, the average plant wastewater load is 10.9 mgd. Normal operation of the activated-sludge process produces an average of 72,000 gpd with a solids concentration of 1.12% (11,200 mg/l). Poor operation, which may periodically last for one or two weeks, results in an average of 102,000 gpd with 0.81% solids.

Solution: Assuming similar operating conditions at design load, the volume of waste-activated sludge production would be

$$\text{Volume with normal operation} = (15.0/10.9)72{,}000$$
$$= 99{,}100 \text{ gpd}$$
$$\text{Volume with poor operation} = (15.0/10.9)102{,}000$$
$$= 140{,}000 \text{ gpd}$$

Design operation of the gravity belt thickeners is to be 7 days per week for a maximum of 7 hr/day. During normal operation, assume a conservative hydraulic loading of 150 gpm/m from Table 13.5 to calculate the required belt width and solids loading.

$$\text{Belt width required at 150 gpm/m} = \frac{99{,}100}{150 \times 7 \times 60} = 1.57 \text{ m}$$

$$\text{Solids loading at 150 gpm/m} = 150 \times 8.34 \times 0.0112 \times 60$$
$$= 840 \text{ lb/m/hr}$$

During poor operation, calculate the required belt widths and solids loadings for both the lower hydraulic loading limit of 150 gpm/m and the upper limit of 250 gpm/m from Table 13.5.

$$\text{Belt width required at 150 gpm/m} = \frac{140{,}000}{150 \times 7 \times 60} = 2.22 \text{ m}$$

$$\text{Solids loading at 150 gpm/m} = 150 \times 8.34 \times 0.0081 \times 60$$
$$= 608 \text{ lb/m/hr}$$

$$\text{Belt width required at 250 gpm/m} = \frac{140{,}000}{250 \times 7 \times 60} = 1.33 \text{ m}$$

$$\text{Solids loading at 250 gpm/m} = 250 \times 8.34 \times 0.0081 \times 60$$
$$= 1010 \text{ lb/m/hr}$$

Installation of two 1.5-m gravity belt thickeners satisfies both the hydraulic and solids loading ranges given in Table 13.5 during the maximum month with one unit in operation.

During normal activated-sludge process operation,

$$\text{Hydraulic loading} = \frac{99{,}100}{1.5 \times 7 \times 60} = 157 \text{ gpm/m}$$

$$\text{Solids loading} = 157 \times 8.34 \times 0.0112 \times 60 = 880 \text{ lb/m/hr}$$

During poor activated-sludge process operation,

$$\text{Hydraulic loading} = \frac{140{,}000}{1.5 \times 7 \times 60} = 222 \text{ gpm/m}$$

$$\text{Solids loading} = 222 \times 8.34 \times 0.0081 \times 60 = 900 \text{ lb/m/hr}$$

For the few weeks of poor operation, both thickeners could be operated for 7 hr or less using the combined belt width of 3.0 m to lower the hydraulic loading to 150 gpm/m. Alternately, if one thickener were out of service, the other one could be operated for 10.4 hr at 150 gpm/m.

FLOTATION THICKENING

Air flotation is applicable in concentrating waste-activated sludge and pretreatment of industrial wastes to separate grease or fine particulate matter. Fine bubbles to buoy up particles may be generated by air dispersed through a porous medium, by air drawn from the liquid under vacuum, or by air forced into solution under elevated pressure followed by pressure release. The latter, called *dissolved-air flotation*, is the process used to thicken waste-activated sludge.

13.11 DESCRIPTION OF DISSOLVED-AIR FLOTATION

The major components of a typical flotation system are sludge pumps, chemical feed equipment to apply polymers, an air compressor, a control panel, and a flotation unit. Figure 13.10 is a schematic diagram of a dissolved-air system. Influent enters near the tank bottom and exits from the base at the opposite end. Float is continuously swept from the liquid surface and discharged over the end wall of the tank. Effluent is recycled at a rate of 30%–150% of the influent flow through an air dissolution tank to the feed inlet. In this manner, compressed air at 60–80 psi is dissolved in the return flow. After pressure release, minute bubbles with a diameter about 80 μm pop out of solution. They attach to solid particles and become enmeshed in sludge flocs, floating them to the surface. The sludge blanket, varying from 8 to 24 in. thick, is skimmed from the surface. Flotation aids are introduced in a mixing chamber at the tank inlet.

The operating variables for flotation thickening are air pressure, recycle ratio, detention time, air/solids ratio, solids and hydraulic loading rates, and application of chemical aids. The operating air pressure in the dissolution tank influences the size of bubbles released. If too large, they do not attach readily to sludge particles, while too fine a dispersion breaks up fragile floc. Generally, a bubble size less than 100 μm is best; however, the only practical way to establish the proper rise rate is by conducting experiments at various air pressures.

Recycle ratio is interrelated with feed solids concentration, detention time, and air/solids ratio. Detention time in the flotation zone is not critical, providing that particles rise rapidly enough and the horizontal velocity does not scour the bottom of the sludge blanket. An air/solids ratio of 0.01–0.03 lb of air/lb of solids is sufficient to achieve acceptable thickening of waste-activated sludge. Optimum recycle ratio must be determined by on-site studies.

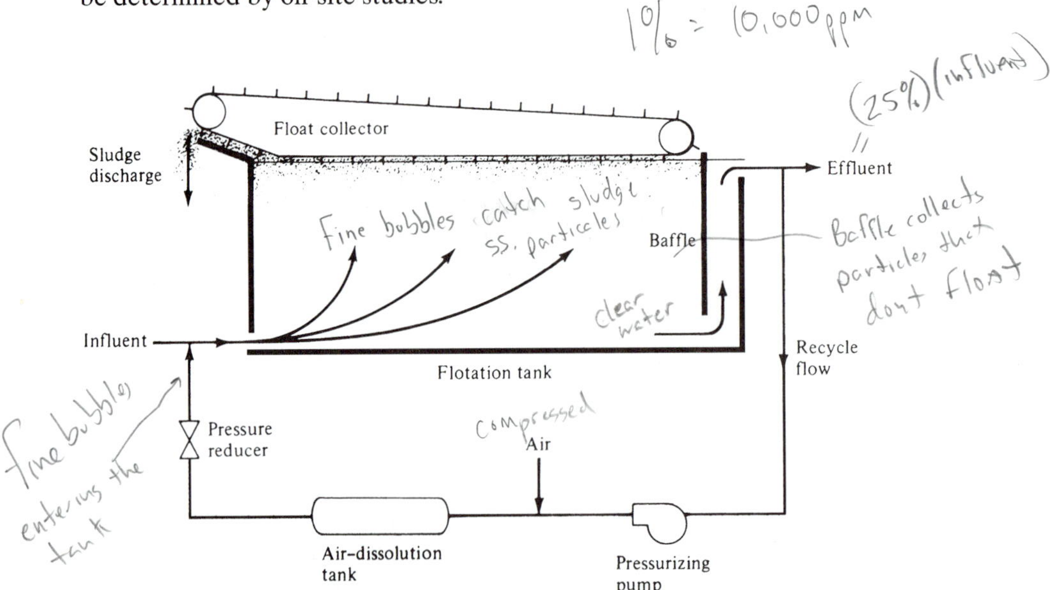

FIGURE 13.10 Schematic diagram of a dissolved-air flotation system.

Operating data from plant-scale units indicate that solids loadings of 2–4 lb/ft^2/hr, with hydraulic flows of about 1 gpm/ft^2, can produce floats of 4%–8% solids. Without polymer addition, solids capture is 70%–90%. However, removal efficiency increases to a mean of 97%, with an active polymer dosage of approximately 10 lb/ton of dry suspended solids. This is the reason most wastewater installations use polymers as flotation aids.

13.12 DESIGN OF DISSOLVED-AIR FLOTATION UNITS

Wherever possible, laboratory and pilot-scale tests are recommended to help determine specific design criteria for a given waste. Notwithstanding, the suggested design criteria for flotation thickening of typical waste-activated sludges are listed in Table 13.6. A conservative solids design loading is 2 lb/ft^2/hr (10 kg/m$^2 \cdot$ h) with the use of flotation aids. From actual operating data, at least 3 lb/ft^2/hr can be expected, and most thickeners have a built-in capacity for 4–5 lb/ft^2/hr loadings. While a 4% minimum float concentration is specified for design purposes, 5%–6% solids can normally be expected. Flotation without polymers generally results in a concentration that is about 1 percentage point less than with chemical aids. Removal efficiency varies from 90% to 98% with polymer addition. The maximum hydraulic loading for design is set at 0.8 gpm/ft^2 (0.54 l/m$^2 \cdot$ s); this is equivalent to applying a waste with a solids concentration of 5000 mg/l at a loading of 2 lb/ft^2/hr. Lower solids levels or higher hydraulic loadings result in lower removal efficiencies and/or float densities.

The typical design values recommended in Table 13.6 apply to anticipated average sludge production. This procedure provides a significant safety factor and permits flexibility in operations. Peak solids loads at municipal treatment plants can usually be accommodated, since these conservative design criteria allow a maximum loading of nearly 100% greater than the average without a serious drop in performance. Perhaps

TABLE 13.6 Design Parameters for Dissolved-Air Flotation of Waste-Activated Sludge with Addition of Polymer Flotation Aids

Parameter	Typical Design Value	Anticipated Results
Solids loading (lb/ft^2/hr)[a]	2	3–5
Float concentration (%)	4	5–6
Removal efficiency	90–95	97
Polymer addition (lb/ton of dry solids)	10	5–10
Air/solids ratio (lb of air/lb of solids)	0.02	
Effluent recycle ratio (% of influent)	40–70	
Hydraulic loading (gpm/ft^2)	0.8 maximum	

[a]1.0 lb/ft^2/hr = 4.88 kg/m$^2 \cdot$ h

the most critical condition is during a period of sludge bulking when the waste mixed liquor is more difficult to thicken and maximum hydraulic loading is applied to the flotation unit.

Sizing of flotation units for an existing plant can be calculated from available data on sludge quantities, characteristics, and solids concentrations. For new plant design, raw wastewater is often assumed to contain 0.24 lb of dry solids/capita/day. A portion of these solids is removed in primary settling, and a conservative estimate for secondary activated-sludge production is 0.12 lb/capita/day. The actual amount is likely to be closer to one-half of this value, because of biological decomposition. Solids yield in an activated-sludge process without primary settling may be safely assumed to bc 0.20 lb/capita/day for domestic wastewater. If the waste sludge from such a system is aerobically digested, the concentration of solids is reduced by about 35%.

Operating hours of a flotation unit depend on size of plant and the working schedule. Although a unit does not require continuous operator attention, periodic checks of a system are scheduled. Generally, a 48-hr week is adequate for plants with capacities of less than 2 mgd. For systems of 2–5 mgd, two shifts 5 days per week establishes an operating period of 80 hr/week. Treatment plants handling more than 20 mgd have operators on duty continuously, and thickening units are run on a schedule appropriate for sludge dewatering and disposal.

Example 13.10

A dissolved-air flotation thickener is being sized to process waste-activated sludge based on the design criteria given in Table 13.6. The average waste flow is 33,600 gpd at 15,000 mg/l (1.5%) suspended solids, and the maximum daily quantity contains 50% more solids at a reduced concentration of 10,000 mg/l. What is the pcak daily hydraulic loading that can be processed? Base all computations on a 14-hr/day operating schedule.

Solution: The flotation tank surface area required for the average daily flow at a design loading of 2.0 lb/ft^2/hr for a 14-hr/day schedule is

$$\text{area} = \frac{33{,}600 \times 0.015 \times 8.34}{2.0 \times 14} = \frac{4200}{28} = 150 \text{ ft}^2$$

Check the solids loading and overflow rate at maximum daily sludge production:

$$\text{maximum solids loading} = \frac{1.5 \times 4200}{150 \times 14} = 3.0 \text{ lb/ft}^2\text{/hr} \quad \text{(OK)}$$

$$\text{maximum sludge volume} = \frac{1.5 \times 4200}{0.01 \times 8.34} = 75{,}500 \text{ gpd}$$

$$\text{maximum hydraulic loading} = \frac{75{,}500}{150 \times 14 \times 60} = 0.60 \text{ gpm/ft}^2 \quad \text{(OK)}$$

$$\text{peak hydraulic loading based on } 0.80 \text{ gpm/ft}^2 = 0.80 \times 150 \times 14 \times 60$$
$$= 100{,}000 \text{ gpd}$$

BIOLOGICAL SLUDGE DIGESTION

Biological digestion of sludge from wastewater treatment is widely practiced to stabilize the organic matter prior to ultimate disposal. Anaerobic digestion is used in plants employing primary clarification followed by either trickling-filter or activated-sludge secondary treatment. Aerobic digestion stabilizes waste-activated sludge from aeration plants without primary settling tanks. The fundamental differences between aerobic and anaerobic digestion are illustrated in Figures 12.2 and 12.3. The end product of aerobic digestion is cellular protoplasm, and growth is limited by depletion of the available carbon source. The end products of anaerobic metabolism are methane, unused organics, and a relatively small amount of cellular protoplasm. Growth is limited by a lack of hydrogen acceptors. Anaerobic digestion is basically a destructive process, although complete degradation of the organic matter under anaerobic conditions is not possible.

13.13 ANAEROBIC SLUDGE DIGESTION

Anaerobic digestion consists of two distinct stages that occur simultaneously in digesting sludge (Fig. 12.13). The first consists of hydrolysis of the high-molecular-weight organic compounds and conversion to organic acids by acid-forming bacteria [Eq. (12.9)]. The second stage is gasification of the organic acids to methane and carbon dioxide by the acid-splitting, methane-forming bacteria [Eq. (12.10)].

Methane bacteria are strict anaerobes and very sensitive to conditions of their environment. The optimum temperature and pH range for maximum growth rate are limited. Methane bacteria can be adversely affected by excess concentrations of oxidized compounds, volatile acids, soluble salts, and metal cations and also show a rather extreme substrate specificity. Each species is restricted to the use of only a few compounds, mainly alcohols and organic acids, whereas the normal energy sources, such as carbohydrates and amino acids, are not attacked. An enrichment culture developed on a feed of acetic or butyric acid cannot decompose propionic acid. The sensitivity exhibited by methane bacteria in the second stage of anaerobic digestion, coupled with the rugged nature of the acid-forming bacteria in the first stage, creates a biological system where the population dynamics are easily upset. Any shift in environment adverse to the population of methane bacteria causes a buildup of organic acids, which in turn further reduces the metabolism of acid-splitting methane formers.

Pending failure of the anaerobic digestion process is evidenced by a decrease in gas production, a lowering in the percentage of methane gas produced, an increase in the volatile acids concentration, and eventually a drop in pH when the accumulated volatile acids exceed the buffering capacity created by the ammonium bicarbonate in solution. Therefore, the operation of a digester can be monitored by any of the following methods: plotting the daily gas production per unit raw sludge fed, the percentage of carbon dioxide in the digestion gases, or the concentration of volatile acids in the digesting sludge. A reduction in gas production, an increase in carbon dioxide percentage, and a rise in volatile acids concentration all indicate reduced activity of the acid-splitting methane-forming bacteria. Digester failure may be caused by any of the following: a significant increase in organic loading, a sharp decrease in digesting sludge

TABLE 13.7 General Conditions for Sludge Digestion

Temperature	
Optimum	98°F (37°C)
General range of operation	85°–95°F (29°–35°C)
pH	
Optimum	7.0–7.1
General limits	6.7–7.4
Gas production	
Per pound of volatile solids added	8–12 ft^3 (230–340 l)
Per pound of volatile solids destroyed	16–18 ft^3 (450–510 l)
Gas composition	
Methane	65%–69%
Carbon dioxide	31%–35%
Hydrogen sulfide	Trace
Volatile acids concentration as acetic acid	
Normal operation	200–800 mg/l
Maximum	Approx. 2000 mg/l
Alkalinity concentrations as $CaCO_3$	
Normal operation	2000–3500 mg/l
Minimum solids retention times	
Single-stage digestion	25 d
High-rate digestion	15 d
Volatile solids reduction	
Single-stage digestion	50%–70%
High-rate digestion	50%

volume (i.e., when digested sludge is withdrawn), a sudden increase in operating temperature, or the accumulation of a toxic or inhibiting substance.

The Environmental Protection Agency specifies processing requirements for application of digested sludge as biosolids on agricultural land [1]. Anaerobic digestion reduces pathogens so that public health is not threatened, provided site restrictions minimize the potential for human and animal contact until after natural die-off reduces any remaining pathogens. The criteria for adequate digestion are a solids retention time (mean cell residence time) and temperature between 15 days at 35°C (95°F) and 60 days at 20°C (68°F). Stabilization for vector reduction is at least 38% reduction in volatile solids. Of the 10 toxic metals controlled, cadmium is of greatest concern since it is suspected of being taken up by plants to enter the human food chain.

General conditions for mesophilic sludge digestion are given in Table 13.7.

13.14 SINGLE-STAGE FLOATING-COVER DIGESTERS

The cross section of a floating-cover digestion tank is shown in Figure 13.11. Raw sludge is pumped into the digester through pipes terminating either near the center of the tank or in the gas dome. Pumping sludge into the dome helps to break up the scum layer that forms on the surface.

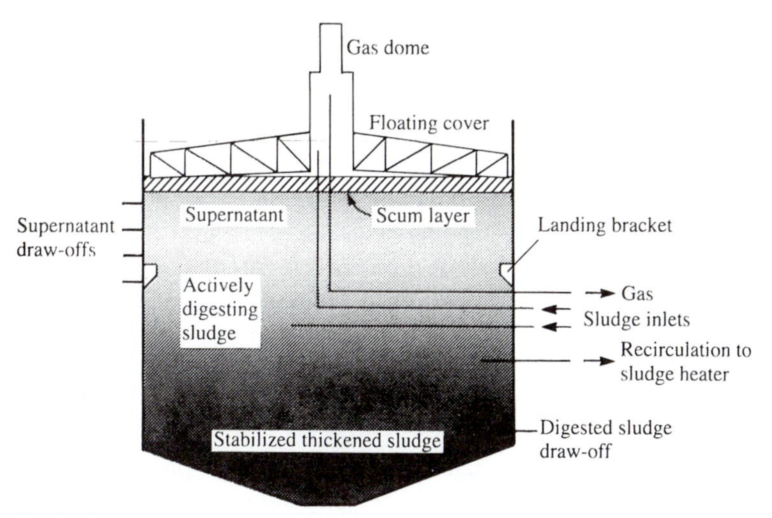

FIGURE 13.11 Cross-sectional diagram of a floating-cover anaerobic digester.

Digested sludge is withdrawn from the tank bottom. The contents are heated in the zone of digesting sludge by pumping them through an external heater and returning the heated slurry through the inlet lines. The tank contents stratify with a scum layer on top and digested thickened sludge on the bottom. The middle zones consist of a layer of supernatant (water of separation) underlain by the zone of actively digesting sludge. Supernatant is drawn from the digester through any one of a series of pipes extending out of the tank wall. Digestion gas from the gas dome is burned as fuel in the external heater or wasted to a gas burner.

The weight of the cover is supported by sludge, and the liquid forced up between the tank wall and the side of the cover provides a gas seal. Gas rises out of the digesting sludge, moves along the ceiling of the cover, and collects in the gas dome. The cover can float on the surface of the sludge between the landing brackets and the height of the overflow pipe. Rollers around the circumference of the cover keep it from binding against the tank wall.

Three functions of a single-stage floating-cover digester are (1) anaerobic digestion of the volatile solids, (2) gravity thickening, and (3) storage of the digested sludge. A floating-cover feature of the tank provides for a storage volume equal to approximately one-third that of the tank. The unmixed operation of the tank permits gravity thickening of sludge solids and withdrawal of the separated supernatant. Anaerobic digestion of the sludge solids is promoted by maintaining near optimum temperature and stirring the digesting sludge through the recirculation of heated sludge. However, the rate of biological activity is inhibited by the lack of mixing; on the other hand, good mixing would prevent supernatant formation. Therefore, in single-tank operation, the biological process is compromised to allow both digestion and thickening to occur in the same tank.

In the operation of an unmixed digester, raw sludge is pumped to the digester from the bottom of the settling tanks once or twice a day. Supernatant is withdrawn daily and returned to the influent of the treatment plant. It is normally returned by gravity flow to the wet well during periods of low raw-wastewater flow, or, in the case of an activated-sludge plant, it may be pumped to the head end of the aeration basin. Because of the floating cover, supernatant does not have to be drawn off simultaneously with the pumping of raw sludge into the digester.

Digested sludge is stored in the tank and withdrawn periodically for disposal. Spreading of liquid sludge from smaller plants on grassland or cropland is common practice in agricultural regions. In larger plants, it is mechanically dewatered and used as a fertilizer and soil conditioner or hauled to land burial. In either case, weather often dictates the schedule for digested sludge disposal. In northern climates, the cover is lowered as close as possible to the corbels (landing brackets) in the fall of the year to provide maximum volume for winter sludge storage.

13.15 HIGH-RATE (COMPLETELY MIXED) DIGESTERS

The biological process of anaerobic digestion is significantly improved by complete mixing of the digesting sludge, either mechanically or by use of compressed digestion gases. Mechanical mixing is normally accomplished by an impeller suspended from the cover of the digester [Fig. 13.12(a)]. Three common methods of gas mixing are the injection of

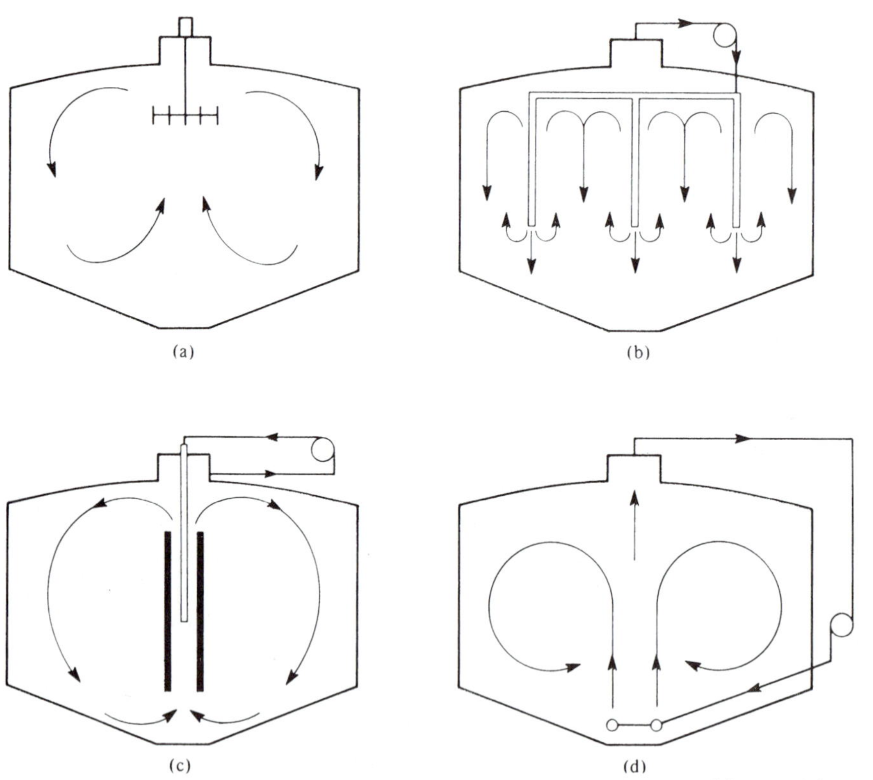

FIGURE 13.12 High-rate digester-mixing systems. (a) Mechanical mixing. (b) Gas mixing using a series of gas discharge pipes. (c) Gas mixing using a central draft tube. (d) Gas mixing using diffusers mounted on the tank bottom.

compressed gas through a series of small-diameter pipes hanging from the cover into the digesting sludge [Fig. 13.12(b)]; the use of a draft tube in the center of the tank, with compressed gas injected into the tube to lift recirculating sludge from the bottom and spill it out on top [Fig. 13.12(c)]; and supplying compressed gas to a number of diffusers mounted in the center at the bottom of the tank [Fig. 13.12(d)].

A completely mixed digester may be either a fixed- or a floating-cover tank. Digesting sludge is displaced when raw sludge is pumped into a fixed-cover digester. By use of a floating cover, tank volume is available for the storage of digesting sludge, and withdrawals do not have to coincide with the introduction of raw sludge.

The homogeneous nature of the digesting sludge in a high-rate digester does not permit formation of supernatant. Therefore, thickening cannot be performed in a completely mixed digester. High-rate digestion systems normally consist of two tanks operated in series (Fig. 13.13). The first stage is a completely mixed, heated, floating-, or fixed-cover digester fed as continuously as possible, whose function is anaerobic digestion of the volatile solids. The second stage may be heated or unheated, and it accomplishes gravity thickening and storage of the digested sludge. Two-stage systems may consist of two similar floating-cover tanks with provisions for mixing in one tank.

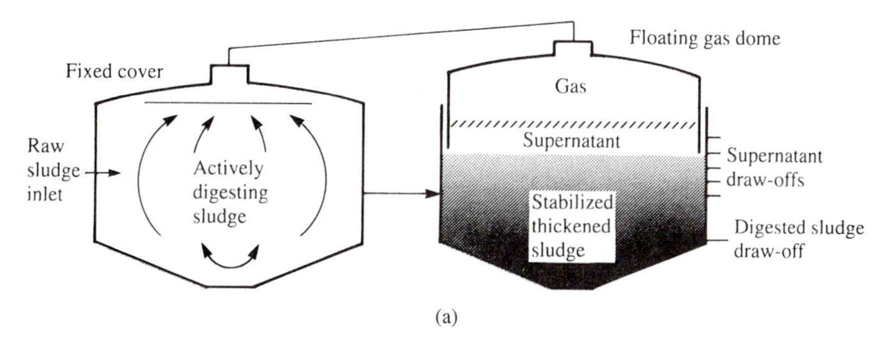

(a)

(b)

FIGURE 13.13 Sketch of a two-stage anaerobic digestion system. (a) The first stage is a completely mixed high-rate digester with a fixed cover. The second stage is a thickening and storage tank covered with a dome for collecting and storing gas. (b) Photograph of two-stage digesters. *Source*: Lincoln, NE.

13.16 VOLATILE SOLIDS LOADINGS AND DIGESTER CAPACITY

Typical ranges of loadings and detention times employed in the design and operation of heated anaerobic digestion tanks treating domestic waste sludge are listed in Table 13.8. Values given for volatile solids loading and digester capacity for conventional, single-stage digesters are based on the total sludge volume available in the tank (i.e., the volume with the floating cover fully raised). Figures given for high-rate digestion apply only to the volume needed for the first-stage tank. There are no established design standards for the tank capacity required in second-stage thickening and supernatant separation.

The loading applied to a digester is expressed in terms of pounds of volatile solids applied per day per cubic foot of digester capacity. Detention time is the volume of the tank divided by the daily raw-sludge pumpage. Digester capacity in Table 13.8 is given in terms of cubic feet of tank volume provided per design population equivalent of the treatment plant.

The *Recommended Standards of the Great Lakes–Upper Mississippi River Board of State Public Health & Environmental Managers* [6] recommends a maximum of 0.08 lb/ft^3/day of VS (1.3 kg/m$^3 \cdot$ d) for high-rate digestion and a maximum of 0.04 lb (0.6 kg) of VS loading for single-stage operation. These loadings assume that the raw sludge is derived from domestic wastewater, the digestion temperature is in the range of 85°–95°F (29°–35°C), volatile solids reduction is 40%–50%, and the digested sludge is removed frequently from the digester.

The capacity required for a single-stage floating-cover digester can be determined by the formula

$$V = \frac{V_1 + V_2}{2} \times T_1 + V_2 \times T_2 \tag{13.11}$$

where

V = total digester capacity, ft^3 (m^3)

V_1 = volume of average daily raw-sludge feed, ft^3/day (m^3/d)

TABLE 13.8 Loadings and Detention Times for Heated Anaerobic Digesters

	Conventional Single-Stage (Unmixed)	First-Stage High-Rate (Completely Mixed)
Loading (lb/ft^3/day of VS)[a]	0.02–0.05	0.1–0.2
Detention time (days)	30–90	15
Capacity of digester (ft^3/population equivalent)[b]		
Primary only	2–1	0.4–0.6
Primary and secondary	4–6	0.7–1.5
Volatile solids reduction (%)	50–70	50

[a]1.0 lb/ft^3/day = 16.0 kg/m$^3 \cdot$ d
[b]1.0 ft^3 = 0.0283 m^3

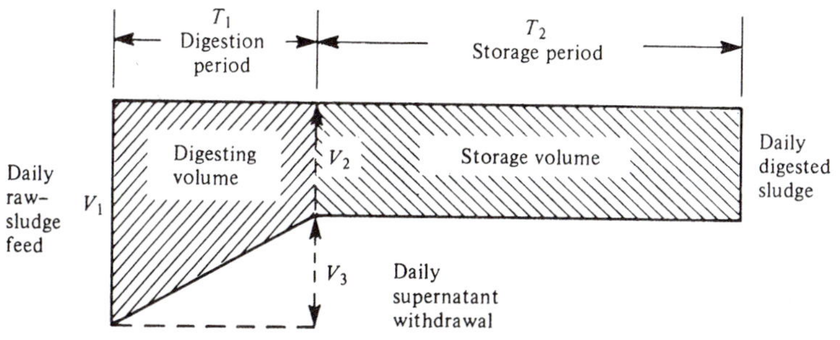

FIGURE 13.14 Pictorial presentation of Eq. (13.11).

V_2 = volume of daily digested sludge accumulation in tank, ft^3/day (m^3/d)

T_1 = period required for digestion, days [approximately 25 days at a temperature of 85°–95°F (29°–35°C)]

T_2 = period of digested sludge storage, days (normally 30–120 days)

Figure 13.14 is a pictorial representation of Eq. (13.11).

Predicting daily volumes of raw sludge produced and the digested sludge accumulated as required in Eq. (13.11) is often difficult. Therefore, the capacity of conventional digesters frequently is based on empirical values relating digester capacity to the equivalent population design of the plant (Table 13.8). Values of 5 and 6 ft^3/capita (0.14 and 0.17 m^3/capita) are frequently used for high-rate trickling-filter plants and activated-sludge plants, respectively.

The minimum detention time for satisfactory high-rate digestion at 95°F (35°C) is approximately 15 days. In general, this limiting period depends on the minimum time required to digest the grease component of raw sludge. Also, too little detention time results in depletion of the methane-bacteria populations, since they are washed out of the digester. The maximum volatile solids loadings, at a 15-day detention time, vary from 0.1 to 0.2 lb/ft^3/day (1.6–3.2 kg/$m^3 \cdot$ d) for adequate volatile solids destruction and gas production. For larger treatment plants with uniform loading conditions, design values of a 15-day minimum detention time and maximum 0.15 lb/ft^3/day of VS loading appear to be satisfactory. Digesters for small treatment plants with wider variations in daily sludge production should be planned using more conservative loading rates.

The capacities required for a high-rate digestion system can be determined by the following equations:

$$V_I = V_1 \times T \qquad (13.12)$$

where

V_I = digester capacity required for first-stage high-rate, ft^3 (m^3)

V_1 = volume of average daily raw sludge feed, ft^3/day (m^3/d)

T = period required for digestion, days

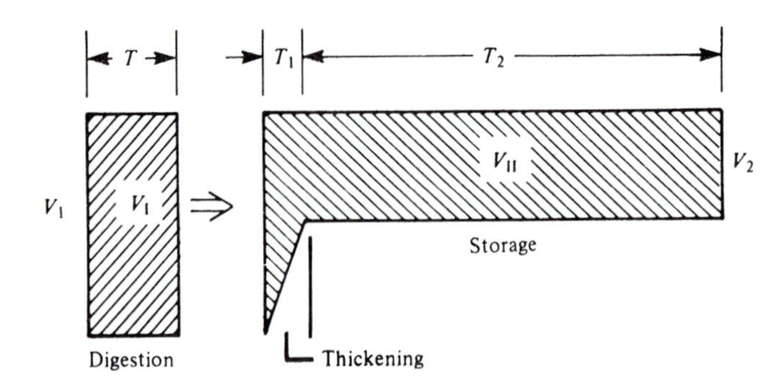

FIGURE 13.15 Diagrams for Eqs. (13.12) and (13.13).

and

$$V_{II} = \frac{V_1 + V_2}{2} \times T_1 + V_2 \times T_2 \tag{13.13}$$

where

V_{II} = digester capacity required for second-stage digested sludge thickening and storage, ft^3 (m^3)

V_1 = volume of digested sludge feed = volume of average daily raw sludge, ft^3/day (m^3/d)

V_2 = volume of daily digested sludge accumulation in tank, ft^3/day (m^3/d)

T_1 = period required for thickening, days

T_2 = period of digested sludge storage, days

Figure 13.15 explains Eqs. (13.12) and (13.13).

Example 13.11

A high-rate trickling-filter plant treats a domestic wastewater flow of 0.48 mgd. Characteristics of the wastewater are identical to those in Table 12.1. Determine the digester capacities required for a single-stage floating-cover digestion system. Digested sludge is to be dried on sand beds, and the longest anticipated storage period required is 90 days.

Solution:

$$\text{Equivalent population} = 4000 \text{ at } 0.24 \text{ lb per capita}$$

Assume the following:

$$\text{water content in raw sludge} = 96\%$$
$$\text{volatile solids in raw sludge solids} = 70\%$$
$$\text{water content of digested sludge} = 94\%$$
$$\text{volatile solids reduction} = 50\%$$

The volume of raw sludge, from Eq. (13.3), is

$$V = \frac{4000 \times 0.24}{[(100 - 96)/100]62.4} = 385 \text{ ft}^3/\text{day}$$

The volume of digested sludge is

$$V = \frac{0.30(4000 \times 0.24) + 0.70 \times 0.50(4000 \times 0.24)}{[(100 - 94)/100]62.4} = 167 \text{ ft}^3/\text{day}$$

Substituting into Eq. (13.11) yields

$$V = [(385 + 167)/2]25 + 167 \times 90 = 21{,}900 \text{ ft}^3$$

Check the volatile solids loading:

$$\frac{0.70 \times 4000 \times 0.24}{21{,}900} = 0.031 \text{ lb/ft}^3\text{/day of VS}$$

Verify the digester capacity per capita:

$$\frac{21{,}900}{4000} = 5.5 \text{ ft}^3\text{/population equivalent}$$

(This value is in the range of the empirical design figure of 4–6 ft^3/population equivalent given in Section 13.16.)

Example 13.12

A high-rate digester operating at a minimum temperature of 30°C with a volume of 2800 m^3 is fed 180,000 l/d of raw sludge with 7400 kg of solids that are 70% volatile. Calculate the concentration of solids in the raw sludge, in the digested sludge assuming 50% volatile solids destruction, volatile solids loading, and detention time. Compare the loading and detention times to those given in Table 13.8.

Solution:

$$\text{Raw sludge solids} = \frac{7400 \times 100}{180{,}000} = 4.1\%$$

$$\text{Digested sludge solids} = \frac{(0.30 \times 7400 + 0.50 \times 0.70 \times 7400)100}{180{,}000} = 2.7\%$$

$$\text{Volatile solids loading} = \frac{0.70 \times 7400}{2800} = 1.85 \text{ kg/m}^3 \cdot \text{d}$$

$$\text{Detention time} = \frac{2800}{180} = 15.6 \text{ days}$$

Values in Table 13.8 are 1.6–3.2 $kg/m^3 \cdot d$ and 15 d.

13.17 AEROBIC SLUDGE DIGESTION

The function of aerobic digestion is to stabilize waste sludge solids by long-term aeration, thereby reducing the BOD and destroying volatile solids. The most common application of aerobic digestion is in handling waste activated sludge. Customary methods for disposal of the digested sludge are spreading on farmland, lagooning, and drying on sand beds.

The Environmental Protection Agency specifies the degree of aerobic digestion and quality of the biosolids for application on agricultural land [1]. For pathogen reduction, the solids retention time and temperature must be between 40 days at 20°C (68°F) and 60 days at 15°C (59°F). (This range of 800–900 degree-days appears to be excessive for aerobic digestion of waste-activated sludge, without the addition of raw primary sludge, since the sludge age in the aeration tank is usually already 20 days or more prior to transfer to the aerobic digester.) For vector attraction reduction, the volatile solids reduction during sludge treatment must be at least 38%. If this is not achieved, the digested sludge can be tested in a bench-scale batch digester to determine if the stability is adequate. The additional volatile solids reduction at 20°C is to be less than 15%. The concentrations of 10 toxic metals in the solids are limited for land application.

Aerobic digestion is accomplished in one or more tanks mixed by diffused aeration. Since dilute solids suspensions have a low rate of oxygen demand, the need for effective mixing rather than microbial metabolism usually governs the air supply required. The volume of air supplied for aerobic digestion is normally in the range of 15–30 cfm/1000 ft^3 of digester.

Design criteria vary with the type of activated-sludge system, BOD loading, and the means provided for ultimate disposal of the digested sludge. Small activated-sludge plants (e.g., package and factory-built, field-erected plants) without primary sedimentation are generally provided with 2–3 ft^3 (57–85 l) of aerobic digester volume per design population equivalent of the plant. This provides a conservative loading in the range of 0.01–0.02 lb/ft^3/d (0.16–0.32 kg/$m^3 \cdot$ d) of volatile solids loading. For stabilizing waste-activated sludge with a suspended-solids concentration of 1%–2%, the volatile solids loading should be limited to 0.04–0.08 lb/ft^3/d (0.64–1.28 kg/$m^3 \cdot$ d), and the aeration period should be 200–300 degree-days, computed by multiplying the digesting temperature in degrees Celsius times the sludge age. This equates to a minimum aeration period of 10 days at 20°C or 20 days at 10°C. Volatile solids and BOD reductions at these loadings are in the range of 30%–50%, and the digested sludge can be disposed of without causing odors or other nuisance conditions.

Long-term aeration of waste-activated sludge creates a bulking material that resists gravity thickening. The solids concentration of aerobically digested sludge is usually in the range of 1.0%–2.0%. The maximum concentration in a well-operated system is not likely to exceed 2.5%. This poor settleability frequently creates problems in disposing of the large volume of sludge produced. Thickening or dewatering by mechanical methods is often too expensive for incorporation in small treatment plants. Therefore, plant design should consider storage of aerobically digested sludge for hauling to land disposal.

An aerobic digester is operated as a semibatch process with continuous feed and intermittent supernatant and digested sludge withdrawals. The contents of the digester are continuously aerated during filling and for a specified period after the tank is full. Aeration is then discontinued, allowing the stabilized solids to settle. Supernatant is decanted and returned to the head of the plant, and a portion of the gravity-thickened sludge removed for disposal. In practice, aeration and settlement may be a daily cycle with feed applied early in the day and clarified water decanted later in the day. Digested solids are withdrawn when the sludge in the tank does not gravity thicken to provide a supernatant with adequate clarity.

Example 13.13

Refer to the data noted in Example 12.11. Calculate the cubic meters of aerobic digester volume provided per design population equivalent, and estimate the volatile solids loading on the aerobic digester.

Solution: The data from Example 12.11 are

$$\text{volume of aerobic digester} = 153 \text{ m}^3$$

$$\text{design population of plant} = 2000$$

Therefore,

$$\text{volume provided} = \frac{153{,}000}{2000} = 76 \frac{\text{liters of digester volume}}{\text{design population equivalent}}$$

$$\text{BOD load on plant} = 182 \text{ kg/d}$$

Assuming a BOD loading of 0.20 g of BOD/g of MLSS applied to the aeration tank, the estimated excess sludge produced per day from Figure 13.1 is 0.42 g of SS/g of BOD load. For the digester volume of 153 m^3 and assuming 70% of the SS as volatile, the estimated volatile solids loading applied to the aerobic digester is $2.0(182 \times 0.70 \times 0.42)/153 = 0.70 \text{ kg/m}^3 \cdot \text{d}$.

Example 13.14

The wastewater treatment plant serving a resort community in a warm climate is an extended aeration system without primary clarifiers. Processing of the waste-activated sludge is by gravity thickening in basins without mechanical scrapers. After settling for about 12 hr, supernatant is drawn off and returned to the plant inlet and thickened sludge discharged to open sand drying beds. This method of sludge processing has proven to be unsatisfactory for several reasons: high temperature of the waste-activated sludge with biological activity reduces settleability, resulting in significant overflow of solids in the supernatant; thickened sludge of only 1.5%–1.8% total solids; emission of foul odors from the drying beds; and excessive drying time. Thus, for several years, the majority of waste sludge has had to be hauled away as a liquid for disposal.

The wastewater flows currently exceed plant capacity during most of the tourist season, while during the off-season when the weather is cool the flows decrease by more than one-half. The proposed plant expansion to meet the high-season flows is to double the aeration capacity, construct aerobic digesters to stabilize and thicken the waste-activated sludge, and install belt presses to dewater the digested sludge. Laboratory and operations reports from the tourist season were studied for the past five years.

From this study, the following data were established as values during the maximum month to be used for design of the expanded plant:

$$\text{Design flow} = 22{,}500\ \text{m}^3/\text{d}$$

$$\text{Design BOD load} = 10{,}100\ \text{kg/d}\ (450\ \text{mg/l})$$

$$\text{Total volume of aeration tanks} = 26{,}800\ \text{m}^3$$

$$\text{Operating MLSS in aeration tanks} = 4000\ \text{mg/l}$$

$$\text{Wastewater temperature} = 20^\circ\ \text{to}\ 25^\circ\text{C}$$

$$\text{Effluent BOD} = <15\ \text{mg/l}$$

$$\text{Effluent suspended solids} = 15\ \text{mg/l}$$

$$\text{Waste-activated sludge volume} = 552\ \text{m}^3/\text{d}$$

$$\text{Waste-activated sludge solids} = 6900\ \text{kg/d}\ (1.25\%)$$

$$\text{Waste-activated sludge temperature} = 20^\circ\text{C}$$

Perform preliminary calculations to determine the number and size of aerobic digesters required and discuss the general method of operation.

Solution: An aerobic digestion tank should be deep enough for gravity settling of solids with aeration off so that the underflow of sludge is thickened for dewatering and the supernatant is clear enough to return to the plant inlet. Minimum thickening requires zone settling with a compression depth of 1–2 m. Zone settling is represented by clear supernatant over a hindered settling layer underneath. Compression of settled solids occurs near the bottom of the tank, as illustrated in Figure 10.21. Adequate depth is also necessary to mount nonclog air diffusers for complete mixing and good oxygen transfer. For draw-and-fill batch operation, several identical tanks are necessary to allow aeration between raw sludge additions to stabilize the solids before withdrawal.

Four tanks are proposed to operate on a 4-day cycle. After adding raw sludge, the tank is aerated continuously for 3 days. On the fourth day, aeration is turned off and the solids are allowed to settle for several hours before withdrawal of digested sludge and draw-off of supernatant. Upon completion, waste-activated sludge is withdrawn from the wastewater aeration system to refill the digester.

Design criteria for the digesters are

$$\text{Minimum aeration period for solids} = 240\ \text{degree-days}$$

$$\text{Maximum volatile solids loading} = 1.28\ \text{kg/m}^3 \cdot \text{d}$$

$$\text{Total solids reduction during digestion} = 30\%$$

$$\text{Solids concentration in settled digested sludge} = 2.50\%$$

Consider 4 tanks each 15 m square, side-water depth of 4.0 m plus 0.5 freeboard, and a hopper bottom with slopes toward a draw-off of approximately 1:3. The liquid volume of each tank is 1100 m^3 with 200 m^3 of this in the hopper.

$$\text{Volatile solids loading} = \frac{0.70 \times 6900}{4 \times 1100} = 1.10 \text{ kg/m}^3 \cdot \text{d} \quad \text{(OK)}$$

$$\text{Liquid detention time} = \frac{4 \times 1100}{552} = 8.0 \text{ d}$$

To increase solids stabilization, the sludge age is increased by retaining the maximum solids concentration in the tank. After aeration, the raw solids are reduced by 30%; therefore,

$$\text{Digested solids remaining} = 0.70 \times 6900 = 4830 \text{ kg}$$

$$\text{Volume of sludge withdrawn} = \frac{4830}{0.025 \times 1000} = 193 \text{ m}^3$$

Resulting from the withdrawal of this volume of sludge,

$$\text{Liquid level in the tank lowers} = \frac{193}{15 \times 15} = 0.86 \text{ m}$$

To add 552 m^3 of raw sludge,

$$\text{Supernatant draw-off required} = 552 - 193 = 359 \text{ m}^3$$

$$\text{Liquid level in the tank lowers} = \frac{359}{15 \times 15} = 1.60 \text{ m}$$

$$\text{Liquid remaining in the tank} = 1100 - 193 - 359$$
$$= 548 \text{ m}^3$$

$$\text{Liquid level above the hopper} = 4.0 - 0.86 - 1.60$$
$$= 1.54$$

For ideal solids retention, the upper boundary of zone settling with a 2.5% solids concentration would be at the top of liquid level, but this would be an unrealistic expectation. Instead, assume the top of the zone settling is 0.5 m below the liquid surface, then

Volume of settled sludge in tank

$$= (15 \times 15)(1.54 - 0.5) + 200(\text{in hopper}) = 434 \text{ m}^3$$

$$\text{Dry weight of solids in tank} = 434 \times 1000 \times 0.025$$
$$= 10{,}900 \text{ kg}$$

Total solids in each tank before sludge withdrawal is this amount plus the digested solids resulting from the addition of raw waste solids after withdrawal of sludge and supernatant.

$$\text{Total solids} = 10{,}900 + 0.70 \times 6900 = 15{,}700 \text{ kg}$$

$$\text{Sludge age} = \frac{\text{solids in 4 tanks}}{\text{solids withdrawn}} = \frac{15{,}700 \times 4}{0.70 \times 6900} = 13.0 \text{ d}$$

$$\text{Aeration period of solids} = 13.0 \times 20$$
$$= 260 \text{ degree-days} \quad \text{(OK)}$$

13.18 OPEN-AIR DRYING BEDS

Historically, small communities have dewatered digested sludge on open beds because of their simplicity, rather than operating more complex mechanical systems. The disadvantages include poor drying during damp weather, potential odor problems, large land area required, and labor for removing the dried cake. In construction of outmoded drying beds, the entire surface area was a level layer of coarse sand supported on a bed of graded gravel. Tile or perforated pipe underdrains were spaced about 20 ft apart in the bottom gravel layer to collect and return drainage to the treatment plant influent. Cleaning dried sludge cake from the beds was a laborious job. The cake had to be lifted from the surface of the sand by hand shoveling and taken off in wheelbarrows for loading in a truck. Attempts to use mechanical equipment resulted in excessive loss of sand and disturbance of the gravel underdrain.

In modern design, drying beds are constructed to permit the use of tractors with front-end loaders to scrape up the sludge cake and dump it into a truck. The main features as illustrated in Figure 13.16 are (1) watertight walls extending 18–24 in. above the surface of the bed; (2) an end opening in the wall, sealed by inserting planks, for

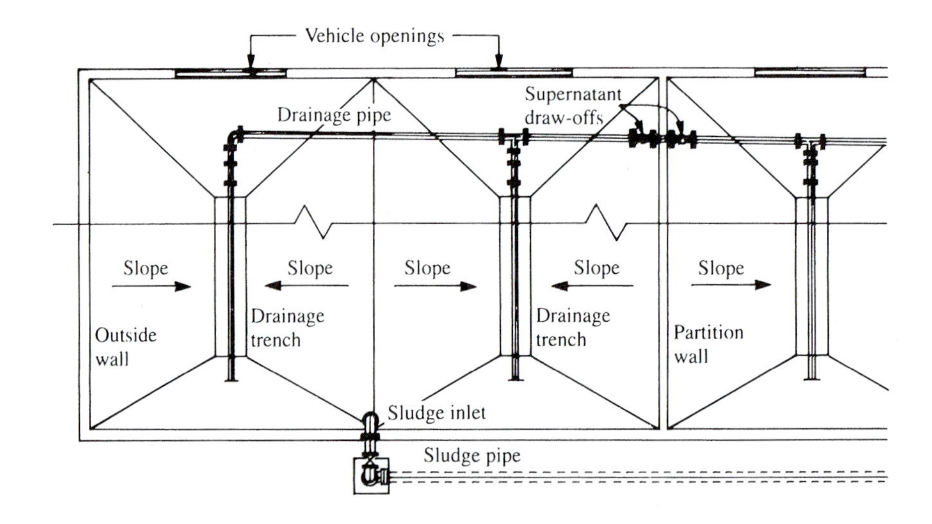

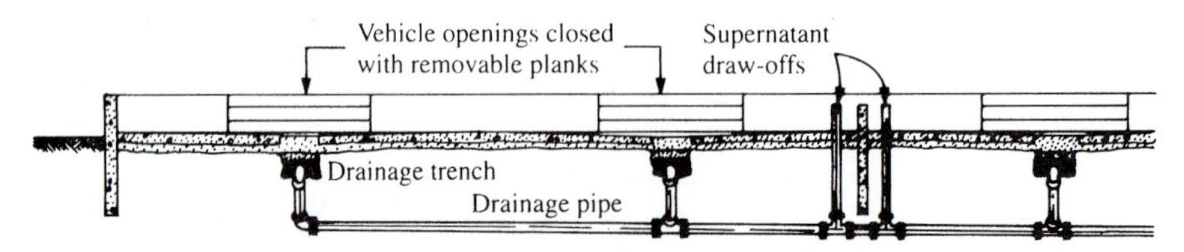

FIGURE 13.16 Open-air beds for drying digested sludge with paved surfaces to allow mechanical removal of dried cake. Water is separated by decanting supernatant, draining to trenches, and evaporation. *Source*: Courtesy of HDR Engineering, Inc.

entrance of a front-end loader; (3) centrally located drainage trenches filled with a coarse sand bed supported on a gravel filter with a perforated pipe underdrain; (4) paved areas on both sides of the trenches with a 2%–3% slope for gravity drainage; and (5) a sludge inlet at one end and supernatant draw-off at the opposite end. For this illustrated bed, the width is 40 ft and the length 100 ft. The operating procedure is to apply digested sludge to a depth of 12 in, or more, draw off supernatant after the solids settle, and allow the sludge to dry. A well-digested sludge forms a cake 3–5 in. thick that is thoroughly dry, black in color, and has cracks resulting from horizontal shrinkage. Paved drying beds have been constructed with more limited drainage than shown in Figure 13.16; however, climatic conditions must be very favorable since the major water loss is by evaporation.

Rational design for sludge beds is difficult, owing to the multitude of variables that affect drying rate. These include climate and atmospheric conditions, such as temperature, rainfall, humidity, and wind velocity; sludge characteristics, including degree of stabilization, grease content, and solids concentration; depth and frequency of sludge application; and condition of the sand stratum and drainage piping. The bed area furnished for desiccating anaerobically digested sludge is from 1 to 2 ft^2/BOD design population equivalent of the treatment plant. Solids loadings average about 20 $lb/ft^2/yr$ (100 $kg/m^2 \cdot y$) in northern states, while unit loading may be as high as 40 $lb/ft^2/yr$ in southern climates. Drying time ranges from several days to weeks, depending on drainability of the sludge and suitable weather conditions for evaporation. Dewatering may be improved and exposure time shortened by chemical conditioning, such as addition of a polymer.

Air drying of digested sludge may be practiced in shallow lagoons where permitted by soil and weather conditions. Water removal is by evaporation, and the groundwater table must remain below the bottom of the lagoon to prevent contamination by seepage. Sludge is normally applied to a depth of about 2 ft and residue removed by a front-end loader after an extended period of consolidation. Because of long holding times, odor problems are more likely to occur. Design data and operational techniques are defined by local experience.

13.19 COMPOSTING

The objectives of sludge composting are to biologically stabilize putrescible organics, destroy pathogenic organisms, and reduce the volume of waste [7]. The optimum moisture content for a compost mixture is 50%–60%; less than 40% may limit the rate of decomposition, while over 60% is too wet to stack in piles. Volatile solids reduction during composting is similar to biological digestion, averaging about 50%. The compost product is a moist, friable humus with a water content less than 40%. For most efficient stabilization and pasteurization, the temperature in the compost piles should rise to 130°–150°F (55°–65°C) but not above 176°F (80°C). Moisture content, aeration rates, size and shape of pile, and climatic conditions affect composting temperature. The finished compost, although too low in nutrients to be classified as fertilizer, is an excellent soil conditioner. When mixed with soil, one advantage of the added humus content is increased capacity for retention of water.

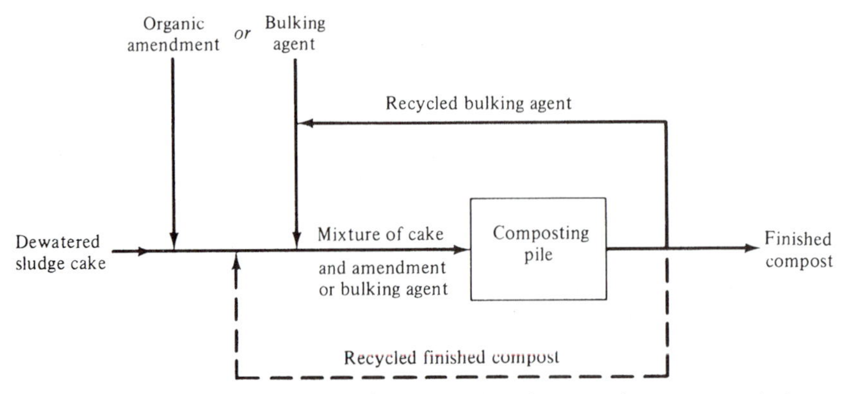

FIGURE 13.17 Generalized diagram for composting dewatered wastewater sludges.

The main products of biological metabolism in aerobic composting are carbon dioxide, water, and heat. Anaerobic composting produces intermediate organics, such as organic acids, and gases including carbon dioxide and methane. Since anaerobic decomposition has a higher odor potential and releases less heat, most systems are designed for aerobic composting. Nevertheless, all forms of composting have the potential for problems such as odors and dust.

Dewatered sludge cake, usually with a moisture content in the range of 70%–85%, is too wet to maintain adequate porosity for aeration. If it is not mixed with another substance, a pile of sludge cake tends to slump and compact to a dense mass with a wet, anaerobic interior and a dried exterior crust. Figure 13.17 is a generalized diagram for composting organic sludges. Dewatered cake is mixed with either an organic amendment (e.g., dried manure, straw, or sawdust) or a recoverable bulking agent (e.g., wood chips) to reduce the unit weight and increase air voids. Finished compost may also be recycled and added to the wet cake. Although composting can be performed in an enclosed reactor, the common processes use outdoor piles either exposed or sheltered under a roofed structure. Compost may be placed in either windrows agitated by periodic turning for remixing and aeration or static piles with forced aeration. The choice between these two processes is based on several factors, including climate, environmental considerations, the availability of a bulking agent, and economics.

In the windrow system, mixed compost material is arranged in long parallel rows. These windrows are turned at regular intervals by mobile equipment to restructure the compost. The piles may be triangular or trapezoidal in shape and may vary in height and width, as determined by the equipment used for turning and the characteristics of the composting material. The height of windrows is usually 4–8 ft and the width 8–12 ft.

Windrow composting is used in agricultural regions where manure from confined feeding of cattle is available for an amendment. The manure is aged and dried by stacking in the feedlot. The wastewater sludge is unstable (raw) filter cake collected immediately after mechanical dewatering with polymer conditioning. Combined in approximately equal portions, the wet cake and dried manure are mixed using a modified manure spreader with the back beaters reversed so that the compost is deposited on the ground in a row rather than being thrown upward by the back beaters for widespread distribution.

A large machine straddling the rows, equipped with an auger-type agitator between the outboard wheels, forms the shaped windrows; periodic turning is performed by the same machine. With weekly turning, stabilization requires 4–6 weeks in good weather. In northern climates, the windrows may freeze on the outside and be covered with snow for several weeks, preventing turning and slowing the rate of decomposition. The finished compost is stored and applied at appropriate times on grassland and cropland.

In the aerated static-pile process, oxygen is supplied by mechanically drawing air through the pile. Porosity is maintained by wood chips or similar recyclable bulking agent, which also reduces the initial water content by absorption. The ratio of sludge to wood chips on a volumetric basis is in the range of 1:2 to 1:3. After preparing a base of wood chips over perforated aeration piping, the mixture is placed on a pile 8–10 ft high. This is layered with finished compost to form a cover. Air is then drawn through the pile for a period of about 3 weeks by a blower operating intermittently to prevent excessive cooling. Exhaust air is vented and deodorized through a pile of finished compost. After stabilization, the mixture is cured and dried for several weeks either in the original pile or after moving to a stockpile. The wood chips are separated from the compost by vibrating screens for reuse.

PRESSURE FILTRATION

Sludges can be dewatered by pressure filtration using either a belt filter press or a plate-and-frame filter press. The belt filter press consists of two continuous porous belts that pass over a series of rollers to squeeze water out of the sludge layer compressed between the belts. A filter press consists of a series of recessed plates with cloth filters and intervening frames held together to form enclosed filter chambers. Sludge pumped under high pressure into these chambers forces water out through the cloth filters, filling the chamber with dewatered cake. At the end of the feed and pressure cycles, the plates are separated to remove the sludge cake. This type of pressure filter is noted for producing a dry cake.

13.20 DESCRIPTION OF BELT FILTER PRESS DEWATERING

The two-belt filter press shown in Figure 13.18 illustrates the basic operational steps. Before wet sludge is distributed on the top of the upper belt, it is conditioned with polymer to aggregate the solids. Initial dewatering takes place in the gravity drainage zone where the belt is supported on closely spaced small-diameter bars to allow separated water to drain freely through the belt into a collection pan. Some presses use an open framework or grid to support the belt. Small adjustable plastic drainage elements (cones, plows, or vanes, depending on manufacturer) are supported just above the belt surface to open channels in the sludge to aid the release of free water. Depending on the characteristics of the applied sludge, up to one-half of the water is removed in the gravity zone; thus, the solids content is nearly doubled and the sludge volume halved. After dropping onto the lower belt, the sludge is gradually compressed between the two belts as they come together in the low-pressure, cake-forming zone. This wedge zone terminates with the two belts wrapping over the first of a series of rollers. Some

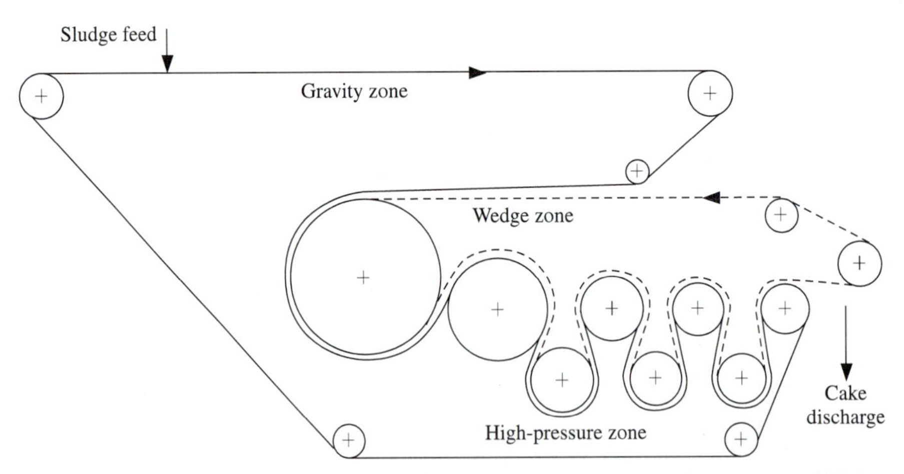

FIGURE 13.18 Two-belt filter press with a gravity drainage zone, wedge zone, and high-pressure zone. *Source:* Courtesy of Komline-Sanderson, Peapack, NJ.

machines have uniform diameter rollers, while on others the subsequent rollers decrease in diameter to gradually increase pressure on the cake. As the belts pass over these rollers, the confined sludge layer is subjected to both compression and shearing action caused by the outer belt being a greater distance from the center of the roller than the inner belt. Depending on the manufacturer, the rollers may be perforated stainless steel cylinders or plain carbon steel with a coating for protection against corrosion. The belt tension, alignment, and drive rollers have a rubber coating to increase frictional resistance and prevent slippage. The cake is scraped from the belts by doctor blades held against the belts.

Belts are made from several fabrics of synthetic fibers. Monofilament polyester woven fabrics with visible clear openings are used in dewatering wastewater sludges. As a result, solids pressed tightly on the surface can penetrate the pores of the fabric and belts require washing with a high-pressure water spray. To confine and collect the wash water, the nozzles are housed in enclosures through which the belts pass. These wash boxes are located on both the upper and lower belts after cake discharge. Belt tensioning is performed by pressing one roller against each belt using hydraulic or pneumatic cylinders connected to the roller shaft. A tension that is too low results in belt slippage, while excess pressure can extrude sludge from between the belts. Each belt is kept centered on its rollers, using a steering assembly consisting of a device that senses the track of the belt and signals adjustment by an alignment roller. By actuating a hydraulic cylinder attached to one end of the shaft, this roller swivels, causing the track of the belt to move along the axis of the roller.

The option of a separate belt for the gravity drainage zone allows a longer time for water separation through a more porous belt than those used for the pressure zone. Nevertheless, many two-belt presses are used in dewatering sludges where an independent gravity zone is not of additional benefit. In a three-belt press as diagrammed in Figure 13.19, the independent upper belt conveys wet sludge in the gravity zone and drops it on the lower belt of a two-belt press that compresses the sludge through the wedge zone and the high-pressure zone.

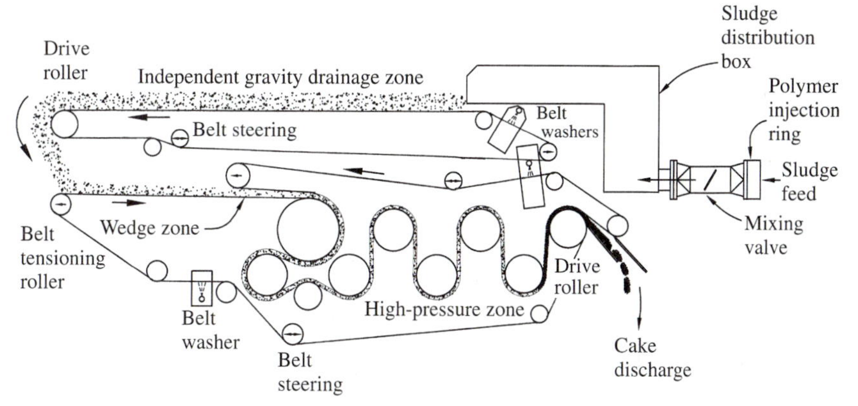

FIGURE 13.19 Three-belt filter press with one belt for the independent gravity zone for drainage of the flocculated sludge followed by two belts in the high-pressure zone for dewatering sludge confined between the belts. *Source:* Courtesy of Komline-Sanderson, Peapack, NJ.

The layout of a belt filter press system is similar to that of a gravity belt thickener, shown in Figure 13.9 (Section 13.9), with one major exception. The wet cake from dewatering sludge at solids concentrations of 15%–25% is more like a semisolid than the liquid slurry of a thickened sludge of 5%–8% solids. Therefore, the cake is transported away from the belt press by a belt conveyor, screw conveyor, or cake pumps.

13.21 APPLICATION OF BELT FILTER DEWATERING

The most significant variables that affect dewatering performance of a belt filter press are the sludge characteristics, polymer conditioning, sludge feed rate, belt tension, and belt speed. The characteristics of greatest importance in wastewater sludges are the solids concentration, the nature of the solids, and prior biological or chemical conditioning. A press is limited to a hydraulic capacity essentially independent of solids concentration less than about 4%. For thin sludges, the recommended maximum loading is 50 gpm (11.4 m^3/h) per meter of belt width. At solids contents greater than about 6%, the capacity of a press is restricted by solids loading. The nature of the solids influences both polymer flocculation and mechanical dewatering. Fibrous solids, commonly associated with primary clarifier settlings, are much easier to dewater than the fine, bulky biological solids wasted from secondary activated-sludge processing.

A polymer is applied to flocculate a sludge, forming aggregates of the particles to allow easier release of the water. The dosage of polymer applied per mass of solids dewatered depends essentially on the sludge characteristics and machine loading, which is in turn directly related to belt speed (detention time in the gravity dewatering zone). Besides inadequate flocculation, the maximum sludge feed rate can be governed by poor solids recovery, too wet a cake, or excessive polymer dosage. If sludge loading is too high, the detention time in the gravity drainage section cannot provide sufficient water release. The result is the extrusion of fine solids through the fabric and from between the belts at the edges. Excessive belt tension can also cause extrusion of solids. In actual plant operation, the acceptable loading is based on the economy of operation, the two major costs being the polymer consumption and hours of operation. A logical sequence for adjusting processing variables at a given sludge loading rate is to select a polymer dosage, adjust the belt speed, set the belt tension, and then readjust these three settings to achieve the desired cake dryness and solids recovery with the minimum polymer dosage.

The main performance parameters of a belt filter press are the hydraulic and solids loading rates, polymer dosage, solids recovery, cake dryness, wash-water consumption, and wastewater discharge. Hydraulic loading is expressed in gallons per minute of sludge feed per meter of belt width (cubic meters per meter per hour). Solids loading is expressed as the pounds of total dry solids feed per meter per hour (kilograms per meter per hour). The polymer dosage is calculated as the pounds applied per ton of total dry solids in the sludge feed (kilograms per tonne). Although the fraction of solids recovery is the quantity of dry solids in the cake divided by the dry solids in the feed sludge, it is often calculated based on the suspended solids in the wastewater (filtrate plus wash water) as follows:

$$\text{solids recovery} = \frac{\begin{pmatrix}\text{total solids}\\ \text{in feed sludge}\end{pmatrix} - \begin{pmatrix}\text{suspended solids}\\ \text{in wastewater}\end{pmatrix}}{\text{total solids in feed sludge}} \qquad (13.14)$$

Cake dryness is expressed as the percentage of dry solids by weight in the cake. For easy comparison with hydraulic sludge loading, wash-water consumption and wastewater discharge are usually expressed in units of gallons per minute per meter of belt width (cubic meters per meter per hour). Example 13.15 illustrates the calculation of these parameters.

Example 13.15

A belt filter press with an effective belt width of 2.0 m is used to dewater an anaerobically digested sludge. The machine settings during operation are a sludge feed rate of 18.2 m^3/h (80 gpm), polymer dosage of 1.8 m^3/h (8.0 gpm) containing 0.20% powdered polymer by weight, belt speed of 6.1 m/min, belt tension of 4.7 kN/m of roller, and wash-water application of 15.4 m^3/h (68 gpm) at 550 kN/m^2. Based on laboratory analyses, total solids in the feed sludge equal 3.5%, total solids in the cake are 32%, wastewater from belt washing contains 2600 mg/l suspended solids, and filtrate production measures 17.7 m^3/h (78 gpm) with a suspended solids concentration of 550 mg/l. From these data calculate the hydraulic loading rate, solids loading rate, polymer dosage, and solids recovery. Comment on the production water usage and wastewater generated relative to the hydraulic sludge feed.

Solution:

$$\text{hydraulic loading rate} = \frac{18.2}{2} = 9.1\ \text{m}^3/\text{m}\cdot\text{h}\ (40\ \text{gpm/m})$$

$$\text{solids loading rate} = \frac{18.2\ \text{m}^3/\text{h} \times 1000\ \text{kg/m}^3 \times 0.035}{2\ \text{m}}$$

$$= 320\ \text{kg/m}\cdot\text{h}\ (700\ \text{lb/m/hr})$$

$$\text{polymer dosage} = \frac{1.8\ \text{m}^3/\text{h} \times 0.002 \times 1000\ \text{kg/m}^3 \times 1000\ \text{kg/t}}{320\ \text{kg/m}\cdot\text{h} \times 2.0\ \text{m}}$$

$$= 5.7\ \text{kg/t}\ (11.4\ \text{lb/ton})$$

wastewater suspended solids

$$= \text{wash-water solids} + \text{filtrate solids}$$

$$= \frac{15.4\ \text{m}^3/\text{h} \times 2600\ \text{g/m}^3 + 17.7\ \text{m}^3/\text{h} \times 500\ \text{g/m}^3}{2\ \text{m} \times 1000\ \text{g/kg}}$$

$$= 24\ \text{kg/m}\cdot\text{h}\ (54\ \text{lb/m/hr})$$

(Note that approximately 80% of the waste solids are in the wash water.)

$$\text{solids recovery} = \left(\frac{320 - 24}{320}\right)100 = 93\%$$

Wash-water consumption equals 7.7 $m^3/m \cdot h$ and the polymer feed is 0.9 $m^3/m \cdot h$ for a total of 8.6 $m^3/m \cdot h$, and hence the process water added very nearly equals the 9.1 $m^3/m \cdot h$ sludge feed. Wastewater production is 16.5 $m^3/m \cdot h$, composed of

7.7 $m^3/m \cdot h$ wash water and 8.8 $m^3/m \cdot h$ filtrate from the sludge and polymer solution water. This equals 1.8 times the sludge feed rate of 9.1 $m^3/m \cdot h$.

13.22 SIZING OF BELT FILTER PRESSES

Belt widths of presses range from 0.5 to 3.0 m, with the most common sizes between 1.0 and 2.5 m. Some manufacturers supply only 1.0- and 2.0-m machines, while others build 1.5- and 2.5-m units. The selection during design of a sludge dewatering facility depends on such factors as the size of the plant, the desired flexibility of operations, anticipated conditions of dewatering, and economics. Typical results from filter pressing of wastewater sludges are listed in Table 13.9. The solids loading rates relate to both the feed solids concentration and hydraulic loading. For example, in the first line, a 4% feed at 50 gpm/m yields a solids loading of 1000 lb/m/hr. Also, the cake solids percentage decreases and the polymer dosage increases with greater dilution of the sludges and for those containing waste-activated sludge. Since performance of filter pressing depends on the character of the sludge, sizing of presses for new treatment plants without an existing sludge to test must be based on operating experience at other installations. While such data are more reliable than the values given in Table 13.9, new installations should be conservatively sized to account for the probable inaccuracy in projecting performance.

Design of a belt filter press installation at an existing facility can be reliably done by conducting field testing using a narrow-belt machine enclosed in a mobile trailer.

TABLE 13.9 Approximate Results from Belt Filter Press Dewatering of Polymer Flocculated Wastewater Sludges

Type of Sludge	Feed Solids (%)	Hydraulic Loading (gpm/m)[a]	Solids Loading (lb/m/hr)[b]	Cake Solids (%)	Active Polymer Dosage (lb/ton)[c]
Anaerobically digested primary only	4–6	40–60	1000–1600	20–28	4–8
Anaerobically digested primary plus waste activated	2–4	40–60	500–1000	16–22	6–12
Aerobically digested without primary	1–3	30–45	200–500	12–16	8–14
Raw primary and waste activated	3–6	40–50	800–1200	18–24	4–10
Thickened waste activated	3–5	40–50	800–1000	14–16	6–10
Extended aeration waste activated	1–3	30–45	200–500	12–16	8–14

[a] 1.0 gpm/m = 0.225 $m^3/m \cdot h$
[b] 1.0 lb/m/hr = 0.454 $kg/m \cdot h$
[c] 1.0 lb/ton = 0.500 kg/tonne

Most manufacturers use a full-scale 0.5- or 1.0-m press that is representative of their larger machines. During the preliminary design phase, a rented trailer unit can be used to determine the dewaterability of the sludge and to establish testing criteria for the performance specifications. After sizing and design of the press facility, selection of the press manufacturer can be based on both competitive bidding and qualification testing using trailer units, either individually, with the lower bidder's press first, or as a group of several proprietary machines operating in parallel. This procedure reduces the risk in design by demonstrating that the selected manufacturer's press can achieve the results required by the performance specifications. This testing does not replace acceptance testing after construction, when the installed presses are evaluated to ensure compliance with the specifications.

Example 13.16

An existing wastewater treatment plant is going to install belt filter presses to dewater anaerobically digested sludge prior to stockpiling and spreading on agricultural land. The current system of lagooning and disposal of liquid sludge is expensive and has created environmental concerns. Past sludge production records were studied to determine the following values for design. The average annual sludge quantity equals 80,000 gpd, with an average solids concentration of 5.0%. The design quantity during the peak month is 130,000 gpd at 4.0% solids, which contain 30% more dry solids than the annual average. The digesters have sufficient capacity to equalize the variations in raw-sludge feed during the peak month.

Field testing using a trailer-mounted press resulted in the following performance data. When the sludge feed had a solids concentration of 4.0%, the allowable hydraulic loading was 50 gpm/m (solids loading of 1000 lb/m/hr), producing a cake with a solids content of 22% with a polymer dosage of 9.6 lb/ton. At 5.0% sludge feed, acceptable operation was a hydraulic loading of 45 gpm/m (solids loading of 1100 lb/m/hr), producing a 24% cake with a polymer dosage of 8.1 lb/ton. Solids recoveries for all tests were between 94% and 96%. The polymer solution was 0.20% dry powder by weight, and wash-water usage was 32 gpm/m.

The design operating schedule is a maximum of 12 hr/day during the peak month. A minimum of two machines is desired so that sludge can be continuously dewatered if one unit is out of service. Based on these criteria, size the belt filter presses and determine the operating times under design conditions. For the average annual sludge quantity, calculate the polymer usage and weight of cake produced and estimate the wastewater generated.

Solution: The belt width required based on peak month operation is

$$\frac{130{,}000 \text{ gal}}{50 \text{ gpm/m} \times 60 \text{ min/hr} \times 12 \text{ hr}} = 3.6 \text{ m}$$

or

$$\frac{130{,}000 \text{ gal} \times 8.34 \times 0.040}{1000 \text{ lb/m/hr} \times 12 \text{ hr}} = 3.6 \text{ m}$$

Specify two belt filter presses with effective belt widths of 2.0 m. For these units, the operating time at the peak monthly load is:

$$\frac{130{,}000 \text{ gal}}{2 \times 2.0 \text{ m} \times 50 \text{ gpm/m} \times 60 \text{ min/hr}} = 10.8 \text{ hr/day}$$

with both presses operating; with one unit operating, the time required is 21.6 hr/day. The operating time at the average annual load is

$$\frac{80{,}000}{2 \times 2.0 \times 45 \times 60} = 7.4 \text{ hr/day}$$

with both presses operating. Polymer usage at the average daily annual load is

$$\frac{80{,}000 \text{ gal} \times 8.34 \times 0.05 \times 8.1 \text{ lb/ton}}{2000 \text{ lb/ton}} = 140 \text{ lb/day}$$

The weight of cake produced is

$$\frac{80{,}000 \times 8.34 \times 0.05}{0.24 \times 2000} = 70 \text{ tons/day}$$

The flow of wastewater generated equals the wash water plus filtrate. The filtrate can be estimated by subtracting the theoretical volume of sludge cake from the sludge feed plus polymer solution.

$$\text{sludge feed} = 45 \text{ gpm/m}$$

polymer solution feed

$$= \frac{1100 \text{ lb/m/hr} \times 8.1 \text{ lb/ton}}{60 \text{ min/hr} \times 2000 \text{ lb/ton} \times 0.002 \text{ lb/lb} \times 8.34 \text{ lb/gal}} = 4 \text{ gpm/m}$$

theoretical flow of sludge cake

$$= \frac{1100 \text{ lb/m/hr}}{60 \text{ min/hr} \times 0.24 \text{ lb/lb} \times 8.34 \text{ lb/gal} \times 1.05} = 9 \text{ gpm/m}$$

$$\text{filtrate} = 45 - 4 - 9 = 32 \text{ gpm/m}$$

$$\text{wash water} = 32 \text{ gpm/m}$$

Therefore, the wastewater flow is $32 + 32 = 64$ gpm/m. The quantity of wastewater at the average daily annual load is

$$64 \text{ gpm/m} \times 4.0 \times 60 \text{ min/hr} \times 7.4 \text{ hr} = 110{,}000 \text{ gpd}$$

13.23 DESCRIPTION OF FILTER PRESS DEWATERING

The two types of plate-and-frame filter presses are the fixed-volume press and the variable-volume diaphragm press. Removal of dewatered cake from a fixed-volume press is done by manually separating the press frames and loosening the layers of cake from the recessed plates with a wooden paddle if they do not drop by force of gravity. The diaphragm press is designed for automatic operation. After opening, the cakes are forcefully discharged and the filter cloths automatically washed before the press closes for another cycle.

Compared to belt presses, filter presses are more expensive, have higher operating costs, and are substantially larger machines for the same sludge processing capacity. Dewatering of wastewater sludge requires lime and ferric chloride conditioning; polymer flocculation is not suitable. High cake dryness is the principal advantage of pressure filtration with cake solids content greater than 35% and up to 40%–50% possible.

A pressure filter consists of depressed plates held vertically in a frame for proper alignment and pressed together by a hydraulic cylinder (Fig. 13.20). Each plate is constructed with a drainage surface on the depressed portion of the face. Filter cloths are

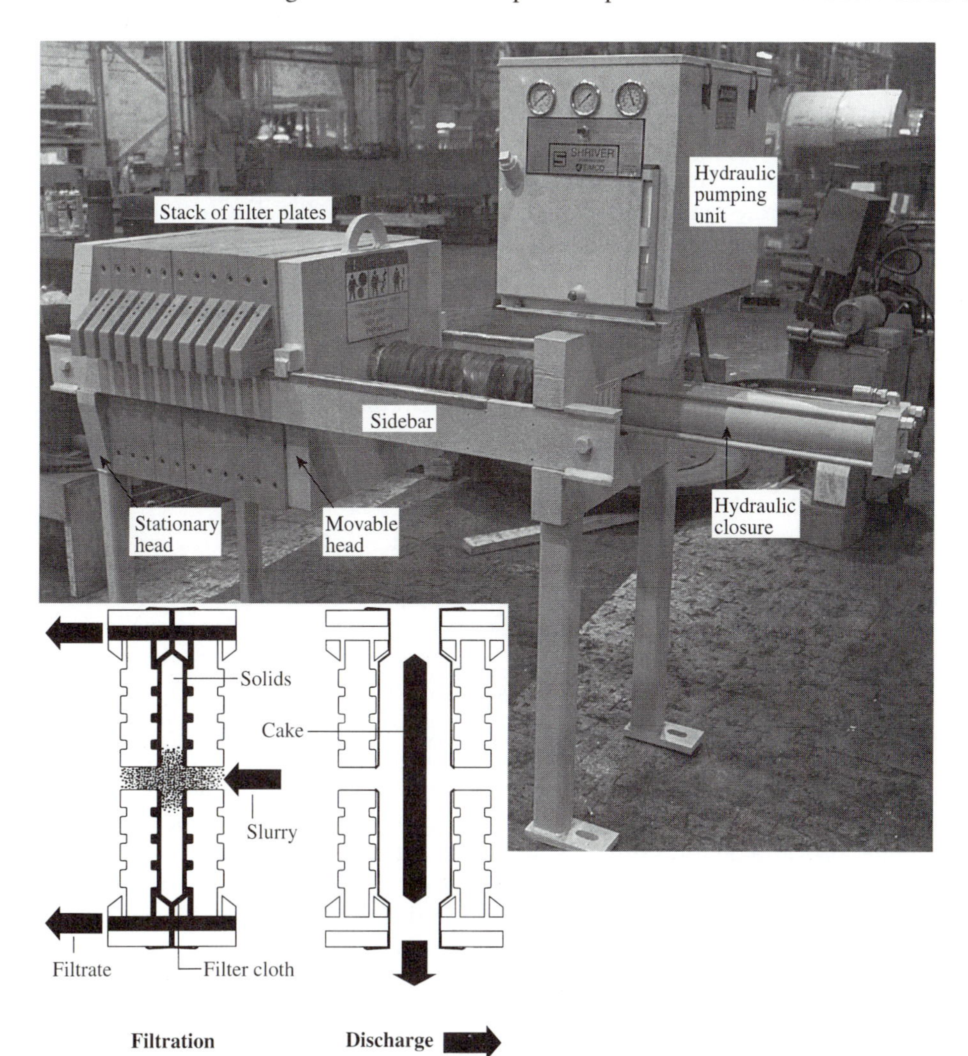

FIGURE 13.20 Pressure filter for dewatering waste slurries from water and wastewater processing. The two major components of the filter press are the skeleton frame and plate stack. With the plates clamped together, as shown on the left, waste slurry is pumped under pressure into the cavities between the plates and filtrate passes through the filter cloths to discharge as the cavities fill with solids. The cake is released, as shown on the right, by opening the plates. *Source*: Courtesy of EIMCO Process Equipment.

caulked onto the plate and peripheral gaskets seal the frames when the press is closed. Influent and filtrate ports are formed by openings that extend through the press. Sludge is pumped under pressure into the chambers between the plates of the assembly, and water passing through the media drains to the filtrate outlets. Solids retained form cakes between the cloth surfaces and ultimately fill the chambers. High pressure consolidates the cakes by applying air to the sludge inlet at about 225 psi (1550 kN/m^2). After this filter cycle, which requires from 2 to 3 hr, compressed air blows the feed sludge remaining in the influent ports back to a holding tank. The filter plates are separated and dewatered cakes drop out of the chambers into a hopper equipped with a conveyor mechanism. Cake release is assisted by introducing compressed air behind the filter cloths and by manually prodding with a paddle if necessary.

Chemical conditioning improves sludge filterability by flocculating fine particles so that the cake remains reasonably porous, allowing passage of water under high pressure. Dosages for conditioning wastewater sludges, expressed as percentages of dry solids in the feed sludge, are commonly 10%–20% CaO and 5%–8% $FeCl_3$. Precoating the media with diatomaceous earth or fly ash helps to protect against blinding and ensures easy separation of the cake for discharge. Filter aid for the precoat is placed by feeding a water suspension through the filter before applying sludge. In some cases, the aid may be added to the conditioned sludge mixture to improve porosity of the solids as they collect. Solids capture in pressure filtration is very high, commonly measuring 98%–99%. The organic sludge solids content in a typical cake is 35%. If the application of conditioning chemicals were 20%, the cake would have a total solids concentration including chemicals of 40%.

13.24 APPLICATION OF PRESSURE FILTRATION

Water treatment plant wastes are suited to pressure filtration since they are often difficult to dewater, particularly alum sludges and softening precipitates containing magnesium hydroxide. Figure 13.21 shows a schematic flow diagram for a pressure filtration process. Gravity-thickened alum wastes are conditioned by the addition of lime slurry. A precoat of diatomaceous earth or fly ash is applied prior to each cycle, and conditioned sludge is then fed continuously to the pressure filter until filtrate ceases and the cake is consolidated under high pressure. A power pack holds the chambers closed during filtration and transports the movable head for opening and closing. An equalization tank provides uniform pressure across the filter chambers as the cycle begins. Prior to cake discharge, excess sludge in the inlet ports of the filter is removed by air pressure to a core separation tank. Filtrate is measured through a weir tank and recycled to the inlet of the water treatment plant. Cake is transported by truck to a disposal site.

Alum sludges are conditioned using lime and/or fly ash. Lime dosage is in the range of 10%–15% of the sludge solids. Ash from an incinerator, or fly ash from a power plant, is applied at a much higher dosage, approximately 100% of dry sludge solids. Polymers may also be added to aid coagulation. Fly ash and diatomaceous earth are used for precoating; the latter requires about 5 lb/100 ft^2 of filter area. Under normal operation, cake density is 40%–50% solids and has a dense, dry, textured appearance.

Wastewater sludges are amenable to dewatering by pressure filtration after conditioning with ferric chloride and lime or fly ash. Minimum ferric chloride and lime

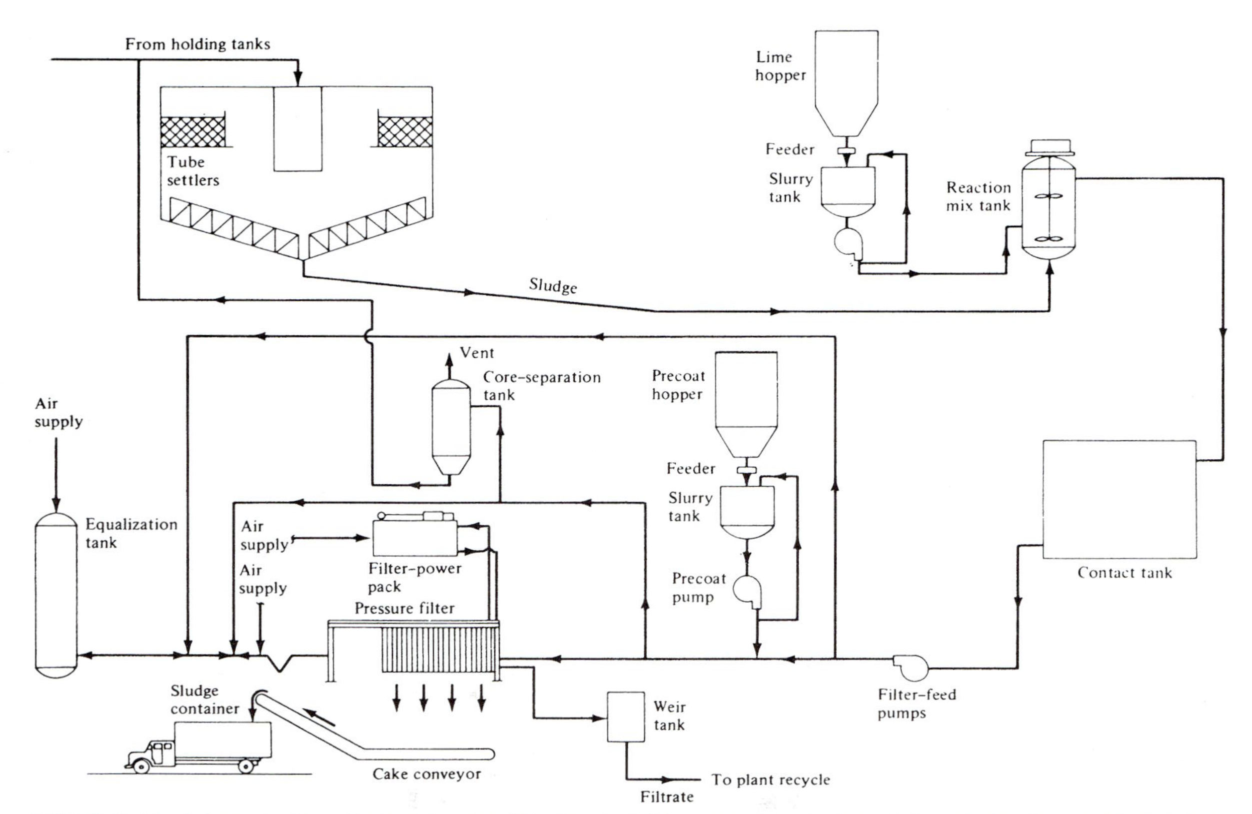

FIGURE 13.21 Schematic flow diagram for conditioning and filter-press dewatering of alum sludge from surface water treatment. *Source:* G. P. Westerhoff and M. P. Daly, "Water-Treatment-Plant Waste Disposal," *J. Am. Water Works Assoc.* 66, no. 7 (1974): 443. Copyright 1974 by the American Water Works Association, Inc.

dosages are 5% and 10%, respectively, for a waste concentration of 5% solids or greater. Conditioning with fly ash requires 100%–150% additions. Either diatomaceous earth or fly ash is used for precoating the filter media.

CENTRIFUGATION

Solid-bowl decanter centrifuges are used to dewater waste sludges from both wastewater and water treatment plants. Dewatering of wastewater sludge by centrifugation is an economic choice in large plants (usually greater than 40 mgd) where costs of machines and a facility building, operating costs, and disposal of drier cake are less than for an equivalent belt filter press system. For water treatment plant wastes, the advantages are thickening sludges from which pore water is difficult to separate, e.g., alum coagulation residues, and for which high solids concentration is desirable, e.g., lime-softening precipitates prior to recalcining.

13.25 DESCRIPTION OF CENTRIFUGATION

The basic operating principles of a solid-bowl decanter centrifuge are illustrated in Figure 13.22. A centrifuge is like a clarifier with the base wrapped around a center line so that its rotation generates gravitational force of 1000–4000 times the force of gravity. The greater the rotational speed, the more rapidly the solids in the sludge are spun out against the rotating bowl wall. While the separated water forms a concentric inner layer and overflows the adjustable plate dam, the settled solids are compacted and moved onto a conical drainage area for further dewatering. The scroll (helical screw conveyor) pushing the solids operates at a higher rotational speed than the bowl.

The solid-bowl decanter centrifuge is the best kind of centrifuge to provide centrate clarity and cake dryness for a wide variety of granular, fibrous, flocculent, and gelatinous solids in sanitary sludges. From the cutaway view in Figure 13.23, the construction of an actual centrifuge can be seen, including feed input, conveyor, bowl, and adjustable plate dam. Feed slurry enters at the center and is spun against the bowl wall. Settled solids are moved by the conveyor to one end of the bowl and out of the liquid for drainage before discharge, while clarified effluent discharges at the other end over a dam plate. This system is best suited for separating solids that compact to a firm cake and can be conveyed easily out of the water pool. If solids compact poorly, moving a soft cake causes redispersion, resulting in poor clarification and a wet concentrate. Flocculent solids are made scrollable by chemical conditioning of the sludge.

A major advantage of scroll dewatering is operational flexibility. Machine variables include pool volume, bowl speed, and conveyor speed. The depth of liquid in the bowl and the pool volume can be controlled by an adjustable plate dam. Pool volume adjustment varies the liquid retention time and changes the drainage deck surface area in the solids-discharge section. The bowl speed affects gravimetric forces on the settling particles, and conveyor rotation controls the solids retention time. The driest cake results when the bowl speed is increased, the pool depth is the minimum allowed, and the differential speed between the bowl and conveyor is the maximum possible. Flexibility of operation allows a range of densities in the solids discharge varying from a dry

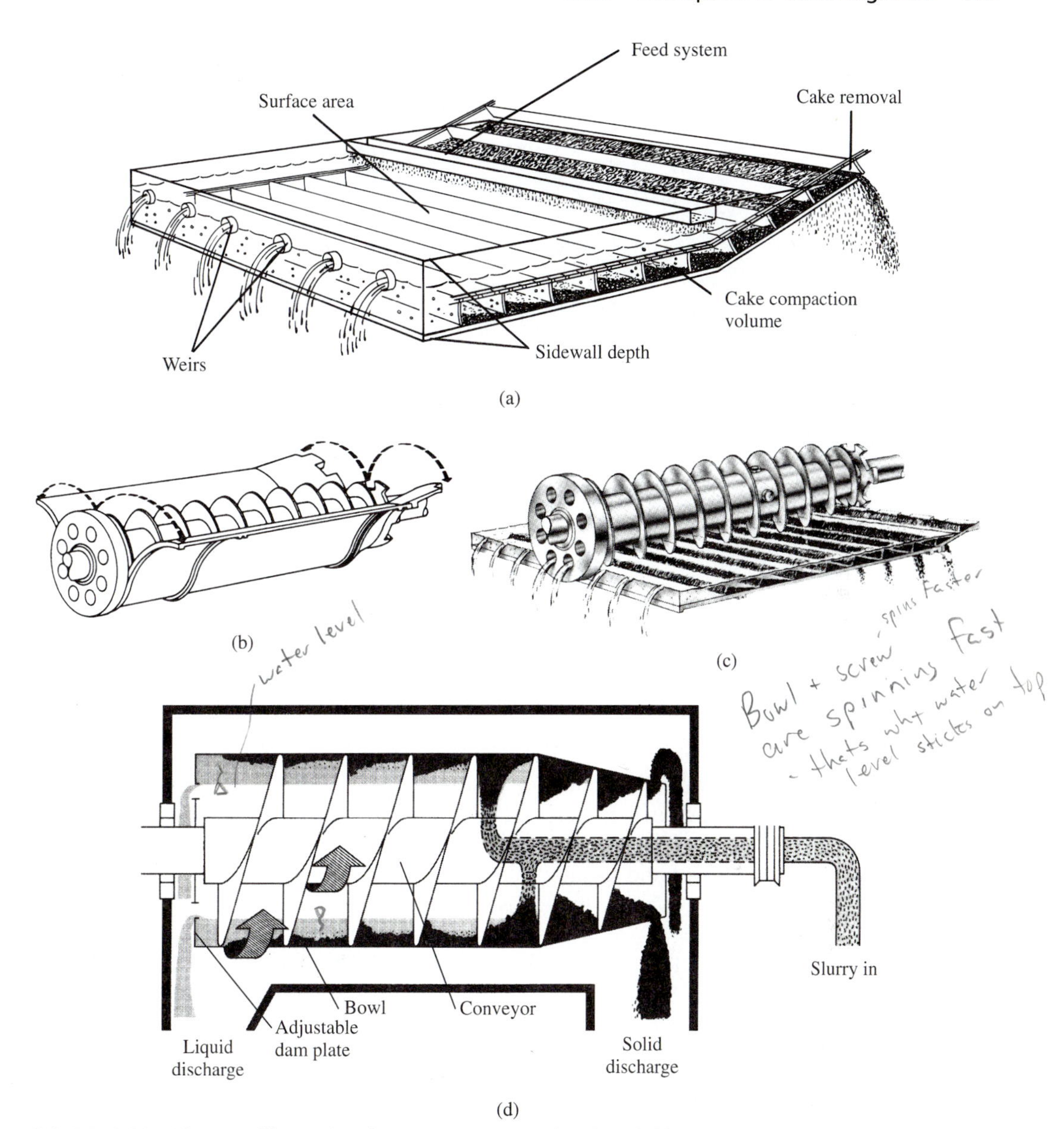

FIGURE 13.22 Diagrams illustrating the operating principle of a solid-bowl decanter centrifuge. (a) The bowl represents a clarifier with defined surface area and retention time with overflow weirs. (b) The bottom of the clarifier is wrapped around a centerline to form a bowl that rotates to increase the gravitational force for sedimentation. (c) The liquid flows through the long narrow channel formed by the helical screw conveyor against the bowl and out over the weirs. (d) As the liquid discharge flows out over the weir (adjustable dam plate), the settled solids are moved by the conveyor out of the liquid onto a conical drainage area for dewatering prior to discharge. *Source*: Courtesy of Alfa Laval Sharples, Alfa Laval Separation Inc.

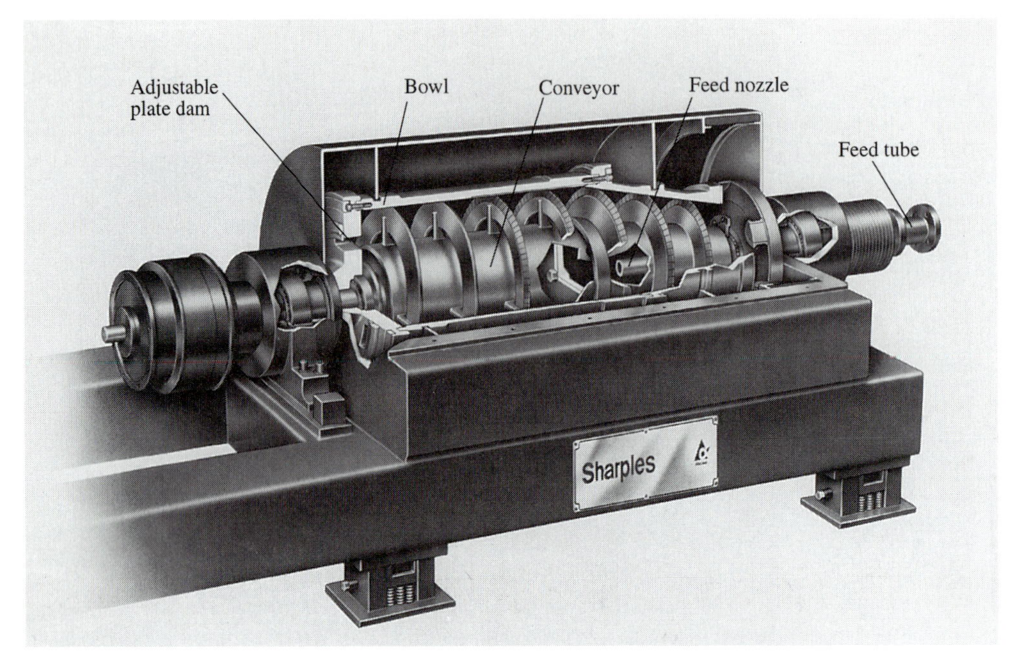

FIGURE 13.23 Solid-bowl decanter centrifuge. *Source:* Courtesy of Alfa Laval Sharples, Alfa Laval Separation Inc.

cake to a thickened liquid slurry. Feed rates, solids content, and prior chemical conditioning can also be varied to influence performance.

The performance of centrifuge dewatering for given feed and machine-operating conditions depends on the dosage of polymers and other chemical coagulants. Suspended-solids removal and usually cake dryness increase with greater chemical additions, while carryover of solids in the centrate decreases. There is, however, a saturation point at which flocculent dosage does not significantly improve centrate clarity. Optimum chemical conditioning without overdosing can be determined most reliably by full-scale or pilot-plant tests. For some wastes, centrate recycling can improve overall suspended-solids removal, but for others it may cause upset, owing to an accumulation of fine particles.

13.26 APPLICATIONS OF CENTRIFUGATION

The characteristics of a wastewater sludge to be dewatered determine the centrifugation capacity, chemical conditioning, cake dryness, and solids recovery. The ratio of primary to waste-activated solids in a sludge has a significant effect. For illustration, consider the following typical performance data in dewatering different kinds of sludges by using an adequate polymer dosage for a minimum of 90% solids recovery: raw primary, cake solids of 28%–34% and polymer dosage of 2–4 lb/ton of dry solids; raw primary plus waste-activated sludge, cake solids of 18%–25% and polymer dosage of 6–14 lb/ton; and raw waste-activated sludge, cake solids of 14%–18% and polymer dosage of 12–20 lb/ton [3]. Usually, the cake solids can be increased a few percentage points by applying an excessive polymer dosage. Machine loadings have notable influence on performance. Underloading allows reduced polymer dosage and produces a drier cake. As the loading increases toward the capacity of the centrifuge, the required polymer dosage increases and the cake becomes wetter. Therefore, to be cost effective, dewatering must take into account machine loading relative to hours of operation per day and the operation of standby centrifuge capacity.

The number of centrifuges and operating hours are designated in design. An installation requires at least one standby machine. More may be required depending on availability of maintenance service and alternative methods of disposal. For example, wastewater plants under 20 mgd should have two full-sized machines (one + one standby); between 20–100 mgd, a minimum of three machines (two + one); and over 250 mgd, a minimum of six machines (four + two) [3]. The designer's performance specification designates the sludge characteristics (raw or digested, solids concentrations, ratio of primary to waste activated, and temperature) and performance requirements (sludge flow rates, minimum cake solids, minimum solids recovery, and polymer dosage). The design engineer does not normally select the actual model and size of the centrifuges unless on-site testing has been conducted. Field tests of two or more machines are run concurrently so that all are dewatering sludge with the same characteristics. The testing range of flow and solids loadings should be adequate to fully evaluate plant operations. The test machines must represent the full-scale centrifuges that will be provided by the manufacturer for installation.

Water treatment plant wastes are also amenable to centrifuge dewatering. For alum sludges from surface water treatment, centrifugation performance must be verified by testing at each plant, since sludge characteristics vary considerably. In general, aluminum hydroxide slurries from coagulation settling and gravity-thickened backwash waters can be concentrated to a truckable pasty sludge of about 20% solids. The removal efficiency in a scroll centrifuge ranges from 50% to 95%, based on operating conditions and polymer dosage, and the centrate is correspondingly turbid or clear. Lime-softening precipitates compact more readily than alum floc. A gravity-thickened sludge of 15%–25% solids can be dewatered to a solidified cake of 65%. Solids recovery is often 85%–90% with polymer flocculation.

Example 13.17

A conventional activated-sludge plant mixes primary and thickened waste-activated sludges prior to high-rate anaerobic digestion. Currently, the digested sludge is stored and thickened in sludge lagoons. Using a dredge, the bottom layer of sludge with a solids concentration of about 8% is pumped into tank trucks and hauled to a dedicated land disposal site. This method of disposal has resulted in complaints about truck traffic, foul odors, potential environmental problems, and high cost of hauling. Because of these factors and the steady increase in raw wastewater flow into the plant, installation of a biosolids dewatering facility using centrifugation has been proposed. An environmental solid-waste management company would be hired under contract to haul and sell the biosolids to local farmers for fertilizer and soil conditioner, with a portion of the profit applied to treatment plant expenses.

Table 13.10 lists the current and design wastewater flows, BOD and suspended-solids loads, waste solids produced based on existing records, and projected digested-solids production for 15 years in the future when the plant is expected to reach design

TABLE 13.10 Current and Design Wastewater Flows, BOD and Suspended-Solids Loads, and Projected Anaerobically Digested Sludge Production for Example 13.16

Parameter	Current		Design	
	Annual Average	Maximum Month	Annual Average	Maximum Month
Wastewater flow (mgd)	150	190	220	260
BOD load (lb/d)	260,000	310,000	450,000	670,000
Suspended-solids load (lb/d)	210,000	260,000	400,000	450,000
Primary solids (lb/d)	140,000	—	—	—
Waste-activated solids (lb/d)	100,000	—	—	—
Total solids (lb/d)	240,000	—	—	—
Digested-solids production (lb/d)	170,000	—	—	—
Digested-solids concentration (%)	2–3	—	—	—
Digested-sludge volume at 2.0% (gal/d)	1,020,000	—	—	—
Projected digested-solids production (lb/d)	—	—	250,000	330,000
Projected digested-solids concentration (%)			2.0	2.0
Projected digested-sludge volume at 2.0% (gal/d)			1,500,000	2,000,000

flow. Outline the primary considerations for design of a biosolids dewatering facility using centrifugation.

Solution:

Equalization of Digested Sludge Feed

Because of existing structures on the plant site, the biosolids dewatering building, loadout structure, and truck parking area must be sited remote from the anaerobic digesters. Therefore, the digested sludge must be pumped to near the dewatering facility and stored in feed tanks. Based on design maximum volume of digested sludge (Table 13.10), four 250,000-gal storage tanks are needed for a half-day retention time. This storage time with mixing at a 30-min turnover rate is to maintain uniform conditions to moderate changes in solids concentration, since even a minor change in concentration can reduce consistency of the cake discharged from the centrifuges. Screening of the feed sludge between the storage tanks and centrifuges should be considered.

Size and Number of Centrifuges

Selecting the size of centrifuges is based on estimated costs, operation, and service experience. Mid-size machines are chosen, rather than large, because they have been in service for a much longer operating record. Although capacities vary somewhat with manufacturers, typical recommended design values of flow rate and solids loading for mid-size centrifuges are as follows: flow rate of 200–300 gpm (up to 450 gpm maximum) and solids loading of 2400–5500 lb/hr (up to 8000 lb/hr maximum). The centrifuges are to be operated continuously, since the plant is staffed 24 hr/d for 7 d/wk and significant cost savings result from continuous operation. The sludge cake hoppers are to provide 2 days' storage so that biosolids do not have to be trucked away on the weekend.

The following calculations determine the required number of operating centrifuges based on data given in Table 13.10, assuming a median flow rate of 250 gpm to each centrifuge.

$$\text{At current annual average: } \frac{1{,}020{,}000}{250 \times 1440} = 2.8 \text{ centrifuges}$$

$$\text{Solids loading with 3 centrifuges} = \frac{170{,}000}{3 \times 24} = 2400 \text{ lb/hr}$$

$$\text{At design annual average: } \frac{1{,}500{,}000}{250 \times 1440} = 4.2 \text{ centrifuges}$$

$$\text{Solids loading with 4 centrifuges} = \frac{250{,}000}{4 \times 24} = 2600 \text{ lb/hr}$$

$$\text{At design maximum month: } \frac{2{,}000{,}000}{250 \times 1440} = 5.6 \text{ centrifuges}$$

$$\text{Solids loading with 6 centrifuges} = \frac{330{,}000}{6 \times 24} = 2300 \text{ lb/hr}$$

Two additional machines are to be added, one for standby and one for maintenance. Therefore, at design capacity of the treatment plant, the recommended number of midsize centrifuges is 8.

Polymer Feed System

Polymers are available in dry-powder or granular form and in concentrated liquid as emulsion, solution, and mannich. (Refer to Section 11.10.) Figure 13.24 is a day tank system for feeding either dry or liquid polymers. The components to prepare a diluted aged polymer solution from dry forms are: bulk storage for a 20–30-day supply, volumetric feeder calibrated to deliver the required amount per batch, mixed aging tank to improve activation, and transfer pump to day tank. Liquid polymers require little or no aging time and can be pumped through a mixing unit for dilution to a day tank. Since activated polymers are unstable after 2–3 days, the diluted polymer solution should be used daily. Consideration should be given to installing only one polymer feed system with space provided for future installation of the other. Polymer solution is fed to each centrifuge by a positive displacement pump.

The expected performance in dewatering the dilute anaerobically digested sludge is as follows: active polymer dosage of 13 lb/ton with an operating range of 9–17 lb/ton, cake solids of 28% with an operating range of 25%–32%, and solids capture of 95% with an operating range of 95%–98%.

Cake Conveyance and Loadout

Cake can be conveyed by belt conveyors, augers, and cake pumps. The belt conveyor is a continuous revolving belt supported by a framework of rollers and is most effective in long straight runs. The auger is a rotating screw that pushes solids along the bottom of a U-shaped trough that can be covered to contain spillage and odors. The cake pump

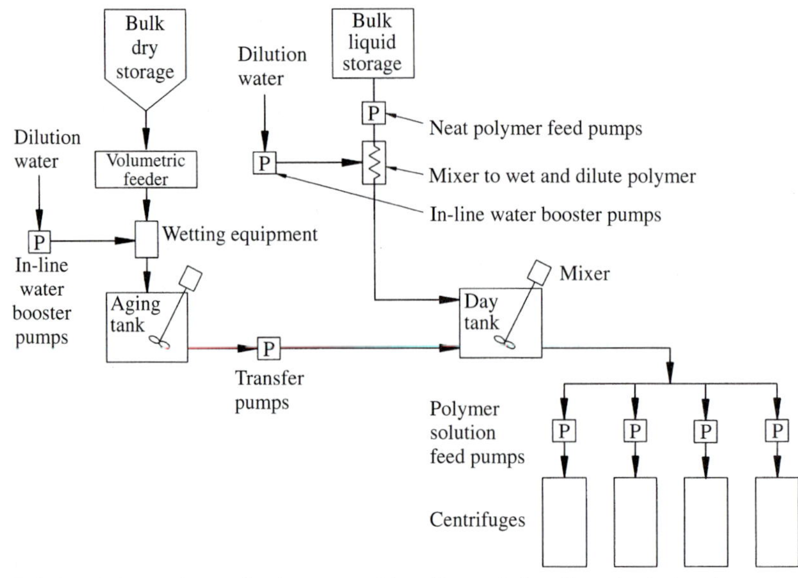

FIGURE 13.24 Polymer feed system with a day tank for diluted solution from either dry or liquid forms of polymer. *Source:* Courtesy of HDR Engineering, Inc.

is a hydraulically driven high-pressure positive-displacement twin-cylinder reciprocating piston pump capable of pushing semisolid cake through a pipeline. Thus, the solids are enclosed, preventing spillage and emission of odors.

The cake conveyance system recommended is a combination of augers and cake pumps. Figure 13.25 is a schematic plan view showing the installation locations of 8 centrifuges with additional spaces for future installation of 4 more machines. An inclined auger receives the cake discharge from each centrifuge and conveys it to one of two horizontal augers. Each inclined auger is sized for 40 gpm, which is the maximum volume of cake discharge of 5500 lb/hr of dry solids at a solids concentration of

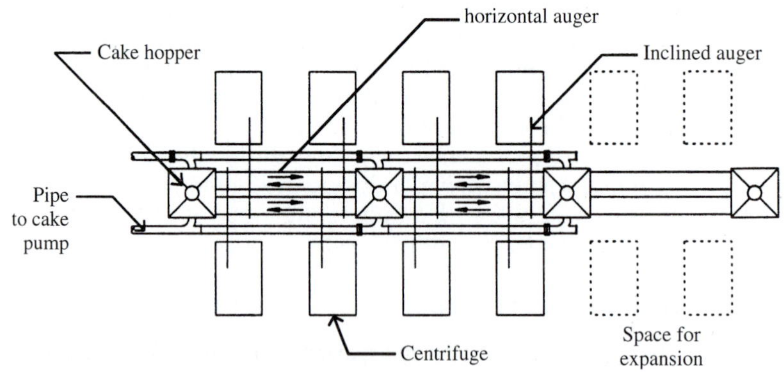

FIGURE 13.25 Plan view of eight installed centrifuges showing locations of machines, inclined augers, horizontal augers, cake hoppers, and piping to cake pumps. *Source:* Courtesy of HDR Engineering, Inc.

25%. Although normal operation is for each horizontal conveyor to transport cake discharge from only 2 centrifuges, the design capacity is 160 gpm for receiving cake from 4 centrifuges. The cake pumps are located at each cake hopper for a total of 6 pumps, each with a pumping rate of 120–160 gpm with discharge pressure up to 600–800 psi for a maximum discharge pipe length of 200–250 ft. The pipes terminate at the top of the loadout hoppers 50 ft above ground level.

The loadout structure is to have 2 truck bays each with 2 hoppers, for a total 2-day cake storage capacity of 40,000 ft^3 weighing 2,600,000 lb. The hopper configuration is cylindrical, with slide gates and openings with splash protection. Loadout operations are to be designed for an 8-hr shift. The truck loading time is assumed to be 10–15 min. With a maximum truck load of 45,000 lb and 2 bays in operation, 64–96 loads can be hauled away in an 8-hr period.

CYCLING OF WASTE SOLIDS IN TREATMENT PLANTS

Centrate from dewatering sludge is returned to the head of a treatment plant, contributing suspended solids and BOD to the raw-wastewater influent. In addition to centrate from centrifuges, other recycled flows from sludge processes are supernatant from anaerobic or aerobic digesters, flow from gravity thickeners, and filtrate from belt presses and pressure filters. Consequently, poor solids capture in sludge thickening and dewatering contribute to the load on the plant and cycling of solids within the system. Being colloidal in nature, many of the finer solids pass through primary sedimentation for capture in biological aeration and return in the waste secondary sludge for thickening and dewatering again. Cycling solids can lead to overloading and upset of all treatment processes. However, their presence is normally first noticed by a thinner primary sludge and an increase in the oxygen demand of biological aeration.

13.27 SUSPENDED-SOLIDS REMOVAL EFFICIENCY

A relatively easy method of estimating solids capture by a sludge-thickening or dewatering unit is to measure solids concentrations in the process flows. The relationship is developed as follows. The solids mass balance is given in Eq. (13.15), and the liquid volumetric balance in Eq. (13.16):

$$M_S = M_R + M_C \tag{13.15}$$

$$Q_S = Q_R + Q_C \tag{13.16}$$

where

M_S, Q_S = mass of solids and quantity of flow of feed sludge

M_R, Q_R = mass of solids and quantity of return flow

M_C, Q_C = mass of solids and quantity of flow of thickened sludge or cake

Without introducing significant error, the specific gravity of all flows can be assumed to be 1.00. Hence, the mass of solids M in a process flow stream is the rate of flow Q times the solids concentration S.

$$M = Q \times S \tag{13.17}$$

Combining these equations results in the following relationships:

$$\frac{M_C}{M_S} = \frac{S_C(S_S - S_R)}{S_S(S_C - S_R)} \tag{13.18}$$

where

M_C/M_S = fraction of solids removal (solids capture)

S_C = solids concentration in thickened sludge or cake

S_S = solids concentration in feed sludge

S_R = solids concentration in return flow

$$\frac{Q_R}{Q_S} = \frac{S_C - S_S}{S_C - S_R} \tag{13.19}$$

where

Q_R/Q_S = fraction of feed sludge appearing as return flow

The solids concentrations S can be either total solids (residue upon evaporation) or suspended solids (nonfilterable residue). Testing for suspended solids by the standard laboratory technique of filtration through a glass filter is feasible for dilute suspensions where dissolved solids are a major portion of the total solids. In contrast, wastes with high solids contents are difficult to test accurately by laboratory filtration. Therefore, suspended-solids analysis of a sludge sample is performed by a total-solids test that is corrected by subtracting an estimated dissolved-solids concentration. The procedure does not create significant error since the filterable solid content usually amounts to less than 5% of the total solids in the sludge samples.

Example 13.18

A scroll centrifuge dewaters an alum–lime sludge containing 8.0% solids at a feed rate of 20 gpm. The cake produced has a solids concentration of 55% and the centrate contains 9000 mg/l. Calculate the solids removal efficiency and the centrate flow.

Solution: From Eq. (13.18),

$$\text{solids removal} = \frac{M_C}{M_S} = \frac{55(8.0 - 0.9)}{8.0(55 - 0.9)} = 0.90\ (90\%)$$

Substituting into Eq. (13.19) yields

$$Q_R = 20 \text{ gpm} \times \frac{55 - 8.0}{55 - 0.9} = 17.4 \text{ gpm}$$

Example 13.19

The operation of a vacuum filter dewatering wastewater sludge was analyzed by sampling and testing the feed sludge and cake for total-solids content and the filtrate for both total-solids and suspended-solids concentrations. The presence of conditioning chemicals was ignored, since the polymer addition amounted to only 3% of the solids content in the sludge. Based on the following laboratory results, estimate the suspended-solids capture and the fraction of flow of feed sludge appearing as filtrate.

	Total Solids (mg/l)	**Suspended Solids (mg/l)**
Sludge	36,600	
Cake	158,000	
Filtrate	10,800	8700

Solution: Calculate the dissolved solids in the filtrate:

$$DS = TS - SS = 10{,}800 - 8700 = 2100 \text{ mg/l}$$

Applying this value to the sludge and cake, one sees that suspended-solids concentrations for all three flows are

$$S_C = 158{,}000 - 2100 = 156{,}000 \text{ mg/l}$$

$$S_S = 36{,}600 - 2100 = 34{,}500 \text{ mg/l}$$

$$S_R = 8700 \text{ mg/l}$$

Substituting into Eqs. (13.18) and (13.19) yields

$$\text{solids removal} = \frac{M_C}{M_S} = \frac{156{,}000 \times (34{,}500 - 8700)}{34{,}500 \times (156{,}000 - 8700)} = 0.79\ (79\%)$$

$$\frac{\text{filtrate flow}}{\text{sludge feed}} = \frac{Q_R}{Q_S} = \frac{156{,}000 - 34{,}500}{156{,}000 - 8700} = 0.82\ (82\%)$$

The recommendation is to replace the vacuum filters with belt filter presses with the next treatment plant renovation.

FINAL DISPOSAL OR USE

Table 13.11 summarizes the estimated final disposition of waste solids from municipal wastewater treatment plants. Spreading on agricultural land as a fertilizer and soil conditioner is the fastest growing method of use, currently amounting to approximately two-thirds of land-applied biosolids. Codisposal with municipal solid waste accounts

TABLE 13.11 Estimated Final Disposal or Use of Waste Sludge Based on Survey Data of the Majority of Municipal Wastewater Treatment Plants.[a]

	Treatment Plants		Quantity of Sludge	
Method of Ultimate Disposal or Use	Number	Percentage (%)	Dry Solids (1000 tons)	Percentage (%)
Land application on agricultural land, forest, and reclamation sites	4700	49	1600	35
Codisposal in municipal solid waste sites	3000	32	1600	35
Surface disposal by burial or injection at dedicated sites	1400	15	600	13
Incineration in furnaces and other incinerators	400	4	800	17
Ocean disposal	0	Ocean Dumping Ban Act of 1988		

[a] Adopted from Ref. 1.

for the elimination of about one-third of waste sludge. Nevertheless, because of the difficulty in burying wet sludge and high cost of preparing an environmentally safe landfill, treatment plants are often encouraged to process waste sludge to produce biosolids for land application. Where land is economically available, often near the plant, a site may be dedicated solely for the purpose of sludge disposal by burial, impoundment in lagoons, or subsurface injection.

Land application, codisposal, and surface disposal are discussed separately in subsequent sections.

Incineration involves drying sludge cake to evaporate the water and burning for complete oxidation of the volatile solids at high temperature to deodorize exhaust gases. Natural gas or fuel oil is frequently needed for complete combustion. The fuel value of sludge solids is usually unable to sustain combustion because of heat absorbed for evaporation of water and losses in stack gases and radiation through furnace walls [4]. Incineration is rarely used where land disposal or use is possible; hence, most sludge burning is done in large metropolitan areas. The survey data in Table 13.11 estimates that 4% of the treatment plants dispose of 17% of waste sludge by incineration.

Water treatment plant waste solids are commonly disposed of by codisposal in municipal solid waste sites and burial in monofills at dedicated sites. As a result of increasing regulatory and environmental constraints, surface application on agricultural land, dedicated sites, and for land reclamation is increasing as a disposal option. Comprehensive quantitative data on final disposal or use of water treatment plant wastes are not available.

13.28 LAND APPLICATION

The EPA Standards for the Use and Disposal of Sludge define the quality parameters for biosolids applied on land as a soil conditioner and fertilizer. The key concerns are selected metals, vector attraction, and human pathogens. Sludge samples are tested for the following 10 metals: arsenic, cadmium, chromium, copper, lead, mercury, molybdenum, nickel, selenium, and zinc. For each metal, numerical limits are established for the maximum concentration expressed in milligrams of metal per kilogram of biosolids and annual loading rate in kilograms of metal applied per hectare of land. Vector attraction limits and pathogen reductions are applied to two categories of biosolids, Class A and Class B.

Class A biosolids can be applied in land areas open to the public and sold in bag or container form. To reduce vector attraction and pathogen levels, disinfection treatment for production of Class A is usually by thermal treatment defined by time and temperature or high pH-high temperature alkaline treatment. The levels of pathogens cannot exceed one plaque-forming unit (PFU) of enteric virus and cannot exceed one viable helminth ovum in 4 g of total-sludge biosolids (dry weight), and the level of bacteria must be less than 3 most probable number (MPN) of *Salmonella* spp in 4 g of total-sludge biosolids (dry weight) or less than 1000 MPN of fecal coliforms per 1 g of total-sludge biosolids (dry weight) [8].

Class B biosolids can be applied on agricultural land for fodder, fiber, and seed crops and selected food crops with different site restrictions imposed for harvesting, animal grazing, and public contact based on the particular crop. The common sludge treatment processes for Class B are biological digestion for a minimum volatile-solids reduction at a defined temperature and mean cell residence time for anaerobic digestion (Section 13.13) and aerobic digestion (Section 13.17). The pathogen requirement for application of digested sludge on agricultural land is a fecal coliform density in sludge solids not to exceed 2 million CFU (colony forming units) or MPN (most probable number) per gram of total-sludge biosolids (dry weight) [8]. The monitoring procedure for Class B requires seven treated sludge samples be collected and the average geometric mean fecal coliform density of these sample be less than 2 million CFU or MPN per gram of total-sludge biosolids (dry weight).

Biosolids contain plant macronutrients of nitrogen and phosphorus and micronutrients such as boron, copper, iron, manganese, molybdenum, and zinc. The organic matter is a valuable soil conditioner, making clayey soils more friable and increasing the water-holding capacity of sandy soils.

Approximately two-thirds of biosolids are applied on cropland to raise corn, soybeans, cotton, small grains, and forages like alfalfa and clover. The typical rate of application is 5 ton/ac/yr (10 tonne/ha · y); thus, a large land area is needed to spread biosolids from a large plant. Also, application scheduling must be compatible with planting, harvesting, and weather conditions. Since the farmland is usually privately owned, conditions of application and storage of biosolids on farm sites are included in an agreement between the farmers and the municipality or contract hauler.

Spreading on forest land accounts for only a few percent of biosolids applied to land. The major constraints are the cost of long-haul distances and, in the forest, limited-access roads and uneven terrain. Yet, newly established plantations and recently cleared land prior to planting may be advantageous sites for nearby treatment plants.

Reclamation sites, usually owned by mining firms or governmental agencies, account for about one-tenth of biosolids application. Renovation of unsightly and often useless land areas has a positive environmental impact. Biosolids improve the poor soils by providing nutrients, pH buffering, water retention, and organic matter.

Approximately one-fifth of biosolids are sold or given away in a bag or other container for application on public parks, highway median strips, golf courses, and private lawns and gardens. The limitations on toxic metals and pathogens are stricter than for agricultural use because of potential human contact and lack of direct control in application. For necessary reduction of pathogens, the biosolids are often thermally treated for a specific time–temperature regime, e.g., heat drying to reduce moisture to 10% or lower at a temperature exceeding 80°C. The bag or container must be labeled giving the nitrogen concentration and allowable application rate information so the user does not exceed the annual pollutant-loading rate for metal contaminants.

Agricultural Land Application

The main environmental and health concerns with regard to application of biosolids on agricultural land are the degree of stabilization of the organic matter, reduction of pathogenic organisms, and presence of toxic chemicals. EPA regulations have been promulgated to specify reduction of pathogens, limitation of toxic chemicals, application of nitrogen to no more than the agronomic rate, and management practices to protect the environment [1,2].

The variety and numbers of bacteria, viruses, and parasitic organisms pathogenic to humans and animals found in wastewater relate to the state of health of the contributing community. Although treatment processes reduce their numbers, often considerably, the effluent and sludge still contain the species present in the wastewater. The certainty of pathogens being present in raw sludge is the reason for spreading only stabilized sludges on agricultural land. The effectiveness of reducing pathogenic populations during sludge stabilization is a subject of considerable controversy. In general, anaerobic or aerobic digestion of sludge is effective in reducing the numbers of viruses and bacteria, but not in reducing roundworm and tapeworm ova or other resistant parasites. Being the most fragile microorganisms, bacteria and viruses are inactivated by sunlight, drying, and competition in the soil environment. In contrast, parasitic ova are more resistant and may persist in soils or on vegetation for several weeks or months. Even though the risk of infecting livestock is not great, farmers are usually advised to allow 30 days or more after sludge application before grazing animals or harvesting a fodder crop. Regarding human health, despite the possibility of communicable disease transmission, the lack of epidemiological evidence suggests that the current practice of sludge disposal to land is safe [8].

Anaerobic and aerobic digestion are the common processes for reducing the organic content in sludges. Besides eliminating unaesthetic conditions, particularly offensive odors, stabilization reduces the attraction of vectors like flies and rodents. Although disease transmission in an agricultural setting is unlikely, vectors can be a nuisance and do have the potential of carrying pathogens from sludge. Digestion processes also substantially reduce the numbers of enteric bacteria and viruses. The adverse environments and passage of time during storage and spreading on the land contribute significantly to continued die-off of pathogens.

The cultivation of fodder, fiber, and seed crops is unrestricted when digested sludge is applied on farmland. In contrast, food crops with harvested parts that contact the soil–biosolids mixture and with harvested parts underground cannot be harvested for more than a year after application of biosolids. Animals should be prevented from grazing for 30 days after application. For land with high potential for public exposure, such as a park, public access is restricted for 1 yr after applying biosolids. For private farmland, public access is restricted for 30 days [1].

Haul distance, climate, and availability of storage facilities are key factors in land application of sludge. At large plants, mechanical dewatering is used to reduce the sludge mass and cost of hauling. The cake can be distributed by a spreader with backbeaters that throw the solids outward from the rear of a wagon or truck-mounted box. Disk cultivators are used to incorporate the sludge solids into the surface soils. Biologically stabilized cake can be stockpiled at farm sites between planting seasons and during winter months. Storage sites are prepared to prevent pollution of air, groundwater, and surface waters. At small plants, liquid sludge is often transported in a tank truck or tractor-drawn wagon for spreading on fields. Storage can be provided at the plant in digesters or holding tanks. Also, drying beds can be used for dewatering and storage of sludge when spreading is not feasible.

Liquid biosolids can be applied by a vehicle equipped with a rear splash plate for surface spreading or by chisel plows for subsurface incorporation. The flexibility of vehicular hauling allows application at a variety of locations, often privately owned farmlands. Spraying from fixed or portable irrigation nozzles can be practiced where odor and insects are not problems. Subsurface injection is the most environmentally acceptable method, since the sludge is incorporated directly into the soil, eliminating exposure to the atmosphere. A tractor pulls an injection unit that is supplied with liquid biosolids through a trailing hose. Several injectors are on spring-loaded chisel plow shanks with wide cultivator sweeps. As a sweep passes through the soil, a cavity is formed approximately 6 in deep for injecting the sludge. Sludge is supplied to the injection unit directly from a remote holding tank through underground piping and a flexible delivery hose. Another means of incorporating sludge, which is less effective in covering the liquid, is disk plowing with hoses discharging sludge ahead of each disk. Valves and sludge pumps located near the holding tank are operated by radio control from the tractor.

The environmental concerns of chemicals in sludge spread on agricultural land are surface water and groundwater pollution and contamination of soil and crops with toxic substances. Laboratory analyses of a sludge normally include solids content; nitrate, ammonia, and organic nitrogen; soluble and organic phosphate; potassium; and metals of arsenic, cadmium, chromium, copper, lead, mercury, molybdenum, nickel, selenium, and zinc. These 10 toxic metals are regulated as maximum allowable concentrations in biosolids applied to agricultural land [1]. Cadmium is the heavy metal of greatest concern to human health, since it can be taken up by plants to enter the human food chain. The movement of cadmium to groundwater is very unlikely to occur at the pH values of greater than 6.0 commonly associated with agricultural soils. The primary chronic health effect of excessive dietary intake is kidney damage.

The cadmium content of agricultural soils ranges from near zero to several hundred milligrams per kilogram of soil contaminated by industrial wastes. In addition to sludge, many phosphate fertilizers contain cadmium. Predicting plant uptake from soils with accumulated cadmium is difficult because of interactive controlling conditions. Soil pH is an important chemical factor, but the kind of plant species is just as important. Within plants, the highest amounts of cadmium absorption occur in the fibrous roots, followed by the leaves; the lowest concentrations are in fruits, seeds, and storage organs. Potential high-risk food plants for humans are leafy vegetables (e.g., lettuce and spinach). Fodder crops and cereal grains appear to present the least risk. Davis and Coker [9] have written a comprehensive literature review and discussion of the distribution of cadmium, plant uptake and hazards associated with sludge used in agriculture.

Example 13.20

A municipal wastewater contains 200 mg/l of BOD, 220 mg/l of suspended solids, 35 mg/l of nitrogen, and 0.016 mg/l of cadmium. Wastewater processing is primary sedimentation and secondary activated sludge, with the waste sludge anaerobically digested prior to spreading on agricultural land. The plant effluent is discharged to a river. Based on operational data, 25% of the influent cadmium appears in the 110 mg of digested sludge solids produced per liter of wastewater processed. The fertilizer value of the dried sludge solids is 4.0% available nitrogen.

1. Calculate the concentration of cadmium in the plant effluent in milligrams per liter. If the concentration limit of cadmium in the river water is 0.0013 mg/l to protect aquatic life, what dilution ratio of plant effluent to cadmium-free river water is needed?
2. Calculate the concentration of cadmium in the dry biosolids in units of milligrams per kilogram. Compare this value to the maximum allowable concentration of 85 mg/kg. The digested sludge is being applied to land cultivated for growing soybeans with a recommended agronomic nitrogen application rate of 220 kg/ha. Calculate the maximum allowable biosolids application rate for nitrogen uptake in tonnes per hectare.
3. Ferric chloride (waste pickle liquor) added to the wastewater for increasing phosphorus removal in treatment also increases the cadmium removal to 70%. What dilution ratio of plant effluent to cadmium-free river water is needed? Calculate the concentration of cadmium in the dry biosolids.

Solution:

1. The cadmium in effluent amounts to $0.75 \times 0.016 = 0.012$ mg/l. Assume the dilution ratio is X (i.e., when the effluent flow is 1, the river flow is X). Then

$$X \times 0 + 1 \times 0.012 = (1 + X) \times 0.0013$$

$$X = \text{flow of river/flow of wastewater discharge} = 8.2$$

2. The concentration of cadmium in sludge solids equals

$$\frac{0.25 \times 0.016 \text{ mg}}{110 \text{ mg} \times 10^{-6} \text{ kg/mg}} = 36 \text{ mg/kg } (<85 \text{ mg/kg})$$

The nitrogen content of the sludge solids equals

$$\frac{110 \text{ mg/l} \times 0.04 \times 1000 \text{ kg/t}}{110 \text{ mg/l}} = 40 \text{ kg/t}$$

The maximum allowable biosolids application rate equals

$$\frac{220}{40} = 5.5 \text{ t/ha}$$

3. The cadmium in effluent is $0.30 \times 0.016 = 0.0048$ mg/l. Therefore, the dilution ratio required is 2.7. The concentration in the sludge is 100 mg/kg, which exceeds the maximum allowable concentration of 85 mg/kg. The maximum allowable sludge application rate based on nitrogen is 5.5 t/ha.

These computations illustrate that treatment for removal of phosphate by chemical precipitation also increases the removal of cadmium. Therefore, although the effluent quality improves, the contamination of the sludge is greater.

13.29 CODISPOSAL IN A MUNICIPAL SOLID-WASTE LANDFILL

Sludge cake disposed of with household refuse can be either dewatered raw or digested sludge [10,11]. The estimated percentage of sludge buried in municipal landfills for final disposal is 35% (Table 13.11).

The methods of codisposal can be to mix the wet sludge cake with either solid waste or soil. In a sludge–solid-waste operation, sludge is deposited on top of a layer of solid waste at the working face of the landfill and mixed as thoroughly as possible. Using bulldozers and landfill compactors, each mixed layer is then spread, compacted, and covered with a layer of soil. If the sludge is too wet, it is difficult to confine on the working face and equipment can slip and become stuck. Application of sludge should not exceed the absorptive capacity of the refuse. If difficulties persist, the sludge must be dewatered to a higher solids content or the quantity of sludge accepted by the landfill reduced. The recommended design bulking ratios (weight of refuse to weight of wet sludge) for various sludge solids contents are as follows: 6 tons of refuse for 1 wet ton of sludge at 10%–17% solids content, 5 for 1 at 17%–20%, and 4 for 1 at 20% or greater [10]. Water treatment plant sludges are buried in codisposal landfills in a similar manner [2].

In a sludge–soil mixture operation, the sludge is mixed with soil and placed as intermediate layers between layers of refuse and as a final cover. Although this method is not strictly landfilling, it is a viable option with the advantage of removing sludge from the working face where difficulties with equipment occur. A final cover of sludge–soil mixture promotes growth of vegetation over the fill area to minimize wind and water erosion, without the need for fertilizer.

Codisposal regulations for municipal solid-waste landfills define wastewater sludges and water treatment plant residuals as nonhazardous wastes. If toxicity is suspected, testing for controlled toxic substances is advisable. A key restriction in codisposal applies to liquids to reduce the amount of landfill leachate. Using conventional dewatering processes, the solids contents of wastewater and water wastes are usually adequate.

13.30 SURFACE LAND DISPOSAL

Surface disposal is the placement of sludges on land at high application rates for final disposal rather than as a beneficial soil amendment [10]. Of the approximately 13% of wastewater sludge (Table 13.11) disposed of in surface sites, about one-half is placed in dedicated sites, one-quarter in monofills, and one-quarter in others. Water treatment plant residuals are usually buried in monofills [2].

Dedicated sites are where liquid wastewater sludge is injected below the surface using a chisel plow injector unit or spread on the surface by tank truck or sprayed by an irrigation system. Because of the high rate of disposal, dedicated sites do not qualify as land application even though some are designated as beneficial use sites to grow a vegetative cover.

Monofills are burial sites where raw or digested wastewater sludges or water treatment residuals as semisolids are spread and covered with soil. On sites with deep groundwater or bedrock, trenches can be excavated with a layer of stratum or soil between the deposited sludge and the top of the groundwater or bedrock. The sludge mass is usually dumped from the transport trucks directly into the trenches and on-site equipment is used for trench excavation and covering. The other method of monofill construction is an area fill, with sludge placed on the ground surface and the fill built as a mound with a soil cover. Area fills are better for sites with shallow groundwater or bedrock.

Other surface disposal methods are piles and impoundments. Sludge piles are uncovered mounds of stabilized cake used for storage prior to final disposal. Surface impoundments are above-ground or below-ground ponds for storage of liquid sludges, such as anaerobically digested wastewater sludge and water treatment plant clarifier and backwash waters. In order to maintain a constant liquid depth, an overflow pipe drains supernatant back to the treatment plant. Seepage is controlled by either a liner or a leachate system. Settled solids accumulate until the impoundment is either dredged or covered and closed.

PROBLEMS

13.1 The settled sludge from coagulation of a surface water is 1.5% solids, of which 30% are volatile. Compute the specific gravity of the dry solids and specific gravity of the wet sludge. Assume specific gravity values of 2.5 for the fixed matter and 1.0 for the volatile. What is the volume of waste in cubic meters per 1000 kg of dry solids?

13.2 A primary wastewater sludge contains 6.0% dry solids that are 65% volatile. Calculate the specific gravities of the solid matter and the wet sludge. If this residue is thickened to a cake of 22.0% solids, what is the specific gravity of the moist cake?

13.3 Compute the volume of a waste sludge with 96% water content containing 1000 lb of dry solids. If the moisture content is reduced to 92%, what is the sludge volume?

13.4 An activated-sludge wastewater plant with primary clarification treats 10 mgd of wastewater with a BOD of 240 mg/l and suspended solids of 200 mg/l. (a) Calculate the daily primary and waste-activated sludge yields in pounds of dry solids and gallons, assuming the following: 60% SS removal and 35% BOD reduction in primary settling; a primary sludge concentration of 6.0% solids; an operating food/microorganism ratio of 1:3 in the aeration basin; and a solids concentration of 15,000 mg/l (70% volatile) in the waste-activated sludge. (b) What would be the solids content of the sludge mixture if the two waste volumes were blended together? (c) If the waste activated is thickened separately to 4.5% solids before blending, what would be the combined sludge volume and solids content?

13.5 Estimate the excess solids production of an extended aeration process treating unsettled wastewater with a BOD of 200 mg/l and suspended solids concentration of 240 mg/l, assuming $k = 0.35$. Express the answer in milligrams per liter of wastewater treated.

13.6 A municipal wastewater with 220 mg/l of BOD and 260 mg/l of suspended solids is processed in a two-stage trickling-filter plant. Calculate the following per 1.0 m^3 of wastewater treated: (a) the dry solids production in primary and secondary treatment, assuming 50% SS removal and 35% BOD reduction in primary settling and a k value of 0.40 applicable for the trickling-filter secondary; (b) the daily sludge volume, both primary and secondary solids, assuming 5.0% solids content. (A specific gravity of 1.0 can be assumed for the wet sludge.)

13.7 What is the estimated waste-activated sludge generated from a conventional aeration process treating 29,000 m^3/d with 173 mg/l of BOD after primary settling? The operating F/M is 0.24 g BOD/day per gram MLSS. Assume a suspended-solids concentration of 9800 mg/l in the waste sludge.

13.8 A river water treatment plant coagulates raw water having a turbidity of 12 units with an alum dosage of 40 mg/l. (a) Estimate the total sludge solids production in lb/mil gal of water processed. (b) Compute the volumes of waste from the settling basins and filter backwash water, assuming that 70% of the total solids are removed in sedimentation and 30% in filtration, a settled sludge solids concentration of 1.2%, and a solids content of 600 mg/l in the filter wash water. (c) Calculate the composite sludge volume after the two wastes are gravity thickened to 3.5% solids.

13.9 A water plant treats a surface water with a mean turbidity of 18 NTU by applying an average of 19.8 mg/l of alum. From plant records, the sludge yield from the sedimentation basins averages 1400 m^3 per Mm^3 of water treated with a solids content of 1.4%. The filter wash water, at a volume of 3.2% of the water treated, contains an average of 160 mg/l of solids. Determine the total dry solids produced per 1.0 Mm^3 of water treated and compare this value with the total sludge solids calculated by using Eq. (13.9).

13.10 Example 11.4 describes softening a water by excess-lime treatment in a two-stage system. Based on the appropriate chemical reactions, both precipitation softening and recarbonation removal of excess lime, calculate the total residue produced per million gallons of water processed.

13.11 Example 11.6 describes softening a water by split treatment in a two-stage system. Based on the appropriate chemical reactions, calculate the precipitate produced per million gallons of water processed.

13.12 The superintendent of a water treatment plant requests your assistance in calculating the dissolved mineral solids removed in the lime softening of a well water and the sludge solids produced. He wants to present these data to the city council to emphasize the improvement in water quality resulting from treatment. The plant processes 0.45 mgd by the addition of 300 mg/l of hydrated lime, which is 74% CaO by weight, followed by recarbonation. The lime treatment also removes iron and manganese from the water. Analyses of the well water and treated water are listed below. (a) Tabulate the ionic concentrations, calculate the milliequivalents per liter, and sketch milliequivalent-per-liter bar graphs. (Refer to Example 11.1.) (b) Calculate the dissolved minerals (calcium, magnesium, iron, manganese, and bicarbonate) removed in pounds per day for the 0.45 mil gal of water treated. (c) Calculate the dry sludge solids produced per day and the weight of the wet sludge, assuming a solids concentration of 5%.

Well Water				Treated Water			
(All values are given in milligrams per liter, except pH.)							
Ca	94	HCO_3	390	Ca	21	HCO_3	107
Mg	24	SO_4	73	Mg	14	SO_4	64
Na	27	Cl	2	Na	23	Cl	6
Fe	0.7	F	0.3	Fe	0	F	0.2
Mn	0.4			Mn	0		
Total hardness = 332				Total hardness = 112			
pH = 7.2				pH = 7.7			

13.13 The purpose of this problem is to view conventional wastewater treatment as a thickening process where pollutants removed from solution are concentrated in a small volume convenient for ultimate disposal. Figure 13.3(a) is a schematic diagram of conventional wastewater treatment. Starting with 120 gpcd of average domestic wastewater with characteristics as listed in Table 12.1, show in step-by-step calculations how 120 gal of wastewater is reduced to one-half pint of filter cake with 20% solids. Assume 35% solids reduction in anaerobic digestion. List your assumptions. (1 gal = 8 pints.)

13.14 The proposed sludge-processing scheme for a conventional step-aeration activated-sludge plant is as follows: return of waste-activated sludge to the head of the plant, withdrawal of the combined sludge (raw and waste activated) from the primary clarifier, concentration in a gravity thickener, application of plant effluent to the thickener for dilution water, return of the thickener overflow to the plant inlet, and mechanical dewatering of the thickened underflow. Briefly comment on the operating problems you would anticipate with this system. What type of sludge handling and thickening would you recommend to replace the proposed scheme?

13.15 As originally constructed, a conventional activated-sludge plant cannot thicken the waste-activated sludge before mixing with the primary sludge. The combined sludge has an average solids concentration of 2.5%, which is too low for good operation of anaerobic digestion. The proposed solution is to modify the sludge-processing system by installing gravity belt thickeners to concentrate solids in the waste-activated sludge before mixing with the primary sludge, which has a solids concentration averaging 3.5%. If the waste-activated sludge were thickened to 6%, the mixed sludge would have a solids concentration of 4.0% to 4.5%, which is considered the minimum for satisfactory anaerobic digestion. Determine the size of two gravity belt thickeners to increase solids content of the waste-activated sludge to the 6% at the plant design load of 57,000 m^3/d of wastewater. Currently, during the maximum month, the average plant wastewater load is 41,000 m^3/d. Normal operation of the activated-sludge process produces an average of 270 m^3/d, with a solids concentration of 1.12% (11,200 mg/l). Poor operation, which may periodically last for one or two weeks, results in an average of 380 m^3/d with 0.81% solids.

13.16 Outline the alternative methods for processing and disposal of alum-coagulation wastes from treatment of a surface water. If two lagoons are used for storage and dewatering the waste slurry, what is your recommended method for operation of the lagoons? If the sludge in the lagoon is a viscous liquid at the time a lagoon must be emptied, how can the sludge be dewatered for disposal in a landfill?

13.17 A gravity thickener handles 33,000 gpd of wastewater sludge, increasing the solids content from 3.0% to 7.0% with 90% solids recovery. Calculate the quantity of thickened sludge.

13.18 A waste sludge flow of 40 m^3/day is gravity thickened in a circular tank with a diameter of 6.8 m. The solids concentration is increased from 4.5% to 7.5% with 85% suspended-solids capture. Calculate the solids loading in $kg/m^2 \cdot d$ and the quantity of thickened sludge in m^3/d.

13.19 Size a gravity thickener based on 10 lb/ft^2/day for a waste sludge flow of 25,000 gpd with 5.0% solids. Assume a side-water depth of 10.0 ft. After installation, operation at design flow yields an underflow of 8.0% with 90% solids removal. What is the flow of dilution water needed to maintain an overflow rate of 400 gpd/ft^2? If the consolidated sludge blanket in the tank is 4.0 ft thick, compute the estimated solids retention time in the thickener.

13.20 An activated-sludge treatment plant without primary sedimentation processing a warm wastewater has 2 aeration basins with a total volume of 6730 m^3 and 2 clarifiers with scraper mechanisms to push the settled solids to a central hopper. Each aeration basin is prismoidal in shape, with the length twice the width and completely mixed by pedestal-mounted mechanical aerators. The waste-activated sludge is discharged to prismoidal thickening and holding tanks with a maximum liquid depth of 3.9 m and designed for fill-and-draw operation. Before withdrawing supernatant, the suspended solids thicken by hindered settling, and at least some compression, for several hours. The underflow of thickened sludge is discharged to open-air sand drying beds for dewatering and the supernatant returned to the plant inlet. The loading on the plant is 13,000 m^3/d with 260 g/m^3 of BOD, and the effluent contains 7 mg/l of BOD and 13 mg/l of suspended solids. The operating MLSS is 2300 mg/l, and the mixed liquor is normally 20°C or greater. The waste-activated sludge averages 340 m^3/d with 0.80% solids. The designer assumed the waste sludge would thicken to 3.0% solids and drying time on the sand beds would be 32 d. In actual operation, the performance of the sludge processing system was a failure. The solids concentration in the thickened sludge averaged 1.8%, which is 67% greater in volume than at 3.0%. Furthermore, the wet sludge required two to three times longer than 32 d to dry sufficiently for removal from the sand beds. Since the cake remained moist on the bottom and did not shrink sufficiently to open cracks, the organic solids adhered to the sand grains, sealing the pores on the surface of the bed. Foul odors were emitted during drying. As a result, most of the settled sludge is hauled away in tank trucks for disposal in a dedicated landfill, which is a costly operation.

a. How would you classify this activated-sludge process? Keep in mind the high temperature of the wastewater. (Refer to Sections 12.19 and 12.21.)

b. Calculate the sludge age and multiply it by 20°C. Compare this to the aeration period of 200–300 degree-days given in Section 13.17. What does this value indicate?

c. Could the solids concentration of 0.80% in the waste-activated sludge have been improved by installing different clarifier mechanisms? (Refer to Section 10.14.) Could the return sludge rate in operation of the activated-sludge process influence the solids concentration in the waste sludge? (Refer to Section 12.19.)

d. Why doesn't the waste-activated sludge thicken to 3.0% by plain sedimentation? (Refer to Section 13.2.)

e. Why is the thickened waste-activated sludge difficult to dewater on sand drying beds?

f. What sludge processing and disposal system would you recommend for this treatment plant?

13.21 Settled sludge and filter wash water from water treatment are thickened in clarifier–thickeners prior to mechanical dewatering. The settled sludge volume is

1150 gpd with 1.0% solids, and the backwash is 9800 gpd containing 500 mg/l of suspended solids. Calculate the daily quantity of thickened sludge, assuming a concentration of 3.0% solids.

13.22 Lagoon disposal of alum sludge is being considered for a water treatment plant processing 10,000 m^3/d. The proposed site can accommodate four lagoon cells, each with a surface area of 800 m^2 and depth of 3.0 m. Sludge from the sedimentation basins averages 15 m^3/d and contains 1.4% solids; the quantity of filter wash water is 400 m^3/d and contains 160 mg/l of suspended solids. The wash water is to be gravity thickened to greater than 1% with polymer addition before discharging to the lagoons. The plan of operation is to discharge from the sedimentation basins and thickeners to fill one cell at a time, withdrawing supernatant and allowing consolidation and air drying for at least one year after filling before cleaning. This aging period is expected to enhance dewatering by subjecting the sludge to freeze–thaw cycles during the winter months. Based on field studies, this procedure will shrink the sludge layer to a depth of less than one meter with an average solids content of 10%. Is this proposed plan likely to be successful?

13.23 After expansion of an activated-sludge plant, the production of waste-activated sludge during the maximum month is 500 m^3/d with 0.80% solids. The old existing flotation thickeners are to be replaced by gravity belt thickeners. Determine the size of 2 identical units to increase the solids content to a minimum of 5%. Assume that operation is to be 7 d/wk for a maximum of 7 hr/d; design hydraulic loading of 34 $m^3/m \cdot h$ and maximum solids loading of 270 $kg/m \cdot h$. If the polymer solution is 0.20% dry powder by weight, what is the solution feed rate for a polymer dosage of 3.0 kg/tonne?

13.24 Waste-activated sludge from a biological aeration system is 293,000 gpd with 1.0% solids consisting of 70% volatile solids. The sludge is thickened to 4% with a 98% solids capture by gravity belt thickening. The thickened sludge is digested, without withdrawal of supernatant, in high-rate anaerobic digestion (Section 13.15) resulting in a 50% volatile solids reduction. The sludge is then dewatered by belt filter pressing (Section 13.20) to 18% solids concentration with a 95% capture of solids. Calculate the solids concentration and flow at each step in the process, including the wastewater flow recycled from gravity belt thickening and filtrate from the pressure dewatering.

13.25 A flotation thickener processes 250 m^3 of waste-activated sludge in a 16-hr operating period. The solids content is increased from 10,000 mg/l to 40,000 mg/l with 92% solids capture. Calculate the quantity of float in cubic meters per 16-hr period. If the solids loading is 12 $kg/m^2 \cdot hr$, what is the hydraulic loading in $m^3/m^2 \cdot d$?

13.26 Waste-activated sludge processed by dissolved-air flotation is concentrated from 9800 mg/l to 4.7% with 95% suspended-solids capture. During a 24-hr operating period, 50,000 gal of sludge was applied with a polymer dosage of 32 mg/l. The thickener surface area is 75 ft^2. Calculate the solids loading, volume of float produced in 24 hr, and polymer addition in lb/ton of dry solids.

13.27 A single-stage anaerobic digester has a capacity of 13,800 ft^3, of which 10,600 ft^3 is below the landing brackets. The average raw-waste sludge solids fed to the digesters are 580 lb of solids/day. (a) Calculate the digester loading in lb of volatile solids fed/ft^3 of capacity below the landing brackets/day. Assume that 70% of the solids are volatile. (b) Determine the digester capacity required based on Eq. (13.11), using the following data:

$$\text{average daily raw-sludge solids} = 580 \text{ lb}$$

$$\text{raw-sludge moisture content} = 96\%$$

$$\text{digestion period} = 30 \text{ days}$$

solids reduction during digestion = 45%

digested-sludge moisture content = 94%

digested-sludge storage required = 90 days

13.28 Calculate the digester capacity in cubic meters required for conventional single-stage anaerobic digestion based on the following parameters:

daily raw-sludge solids production = 630 kg

volatile solids in raw sludge = 70%

moisture content of raw sludge = 95%

digestion period = 30 days

volatile solids reduction during digestion = 50%

moisture content of digested sludge = 93%

storage volume required for digested sludge = 90 days

13.29 The design wastewater loading for a proposed two-stage trickling-filter plant is 1200 m^3/d containing 260 mg/l BOD and 310 mg/l suspended solids. (a) Calculate the required capacity of a conventional digester based on a raw-sludge concentration of 4.5%, digested-sludge solids concentration of 8.0%, total solids reduction during digestion of 40%, an f value of 0.50 for Eq. (13.5), a k value of 0.40 for Eq. (13.6), and a storage period for digested sludge of 90 days. (b) Compute the digester capacity per population equivalent load on the treatment plant.

13.30 The average daily quantity of thickened raw waste sludge produced in a municipal treatment plant is 15,000 gal containing 10,000 lb of solids. The solids are 70% volatile. (a) Calculate the percentage of moisture in the thickened sludge. (b) Determine the volume required for a first-stage high-rate digester based on the following criteria: a maximum loading of 100 lb of volatile solids per 1000 ft^3/day and a minimum detention time of 15 days. (c) If the volatile solids reduction in the completely mixed digester is 60%, what is the percentage of moisture in the digested sludge?

13.31 The waste sludge production from a trickling-filter plant is 12.5 m^3/d containing 620 kg of solids (70% volatile). The proposed anaerobic digester design to stabilize this waste consists of two floating-cover heated tanks, each with a volume of 480 m^3 when the covers are fully raised and a volume of 310 m^3 with the covers resting on the landing brackets. (a) Calculate the digester volume per equivalent population with the covers fully raised. Assume 100 g of solids production per capita. (b) Calculate the digester loading in terms of kg volatile solids fed/m^3 of tank volume below the landing brackets/day. (c) Calculate the digested-sludge storage time available between lowered and raised cover positions. Assume a volatile solids reduction of 60% during digestion and a digested-sludge moisture content of 93%.

13.32 A domestic wastewater plant using rotating biological contactors for secondary treatment (Fig. 12.30) has a two-stage anaerobic digestion system (Fig. 13.13). The first stage is a heated, completely mixed fixed-cover digester with a liquid volume of 15,500 ft^3. The second stage is an unheated and unmixed digester with a liquid volume of 15,500 ft^3 equipped with a gas-dome cover and supernatant draw-offs. Alum is added to the RBC effluent ahead of the final clarifiers to precipitate phosphate. (The plant was retrofitted with alum feeders and new final clarifiers after original construction to meet a revised effluent standard for phosphorus.) The average wastewater flow is 0.63 mgd. Influent characteristics are 166 mg/l of BOD, 128 mg/l of suspended solids, and 7.1 mg/l of inorganic phosphorus. Effluent characteristics are 7 mg/l of BOD, 16 mg/l of suspended solids, and 1.3 mg/l of inorganic phosphorus. Alum addition is 405 lb/day. The quantity of raw sludge produced averages 2400 gpd with 4.4% solids, which are 67% volatile. The quantity of digested sludge accumulated in the second-stage tank averages 1500 gpd with 4.3% solids, which are 45% volatile. The total gas production from digestion is 4200 ft^3/day. (a) Calculate the volatile solids loading and liquid detention time of the first-stage digester. How do these values compare with those in Table 13.8? (b) Calculate the digestion gas production per pound of volatile solids added and per pound of volatile solids destroyed. Compare these values with those in Table 13.6. (c) Estimate the total dry solids sludge yield as the sum of raw primary solids [Eq. (13.5)], biological sludge solids [Eq. (13.6)], and alum precipitate as aluminum phosphate [Eq. (14.5)]. How does this compare with actual sludge solids yield?

13.33 The two-stage anaerobic digesters at a trickling-filter plant are a fixed-cover tank for first-stage high-rate sludge stabilization with a liquid volume of 42,000 ft^3 and a floating-cover second-stage storage and thickening tank with a volume of 70,000 ft^3 between the lowered and fully raised cover positions. The quantity of raw sludge applied is 2800 ft^3/day containing 5.0% solids that are 70% volatile. (a) If 60% of the volatile solids in the raw sludge is reduced by digestion, calculate the solids concentration in the digested sludge leaving the first-stage digester. (b) Estimate the daily methane production in first-stage digestion. (c) If the digested, thickened sludge contains 7.0% solids, compute the maximum number of days of digested-sludge storage available. (d) Assume the first-stage digester has been operated at a digesting sludge temperature of 35°C for several weeks. How would the gas production change if the operating temperature were decreased slowly to 25°C? How would gas production be affected by a slow increase in operating temperature from 35° to 45°C?

13.34 A two-stage stone-media trickling-filter plant was evaluated for the town of Nancy in Problem 12.32. Determine the anaerobic digester capacity for stabilizing the sludge from this plant in two identical single-stage floating-cover tanks, and the required area for open-air drying beds for a northern state in the Great Lakes region. Using Eq. (13.11), calculate the volume required based on the following: (a) The design quantity of raw sludge solids during the maximum month is the amount calculated by Eq. (13.4) using Eq. (12.37) and Eq. (12.38). (b) A raw-sludge solids content of 4.0%, digested-sludge solids content of 6.0%, total solids reduction of 35%, and required storage period of 120 days. How does the calculated volume compare with the suggested capacity of 5 ft^3/PE? (c) Calculate the required area of open-air drying beds.

13.35 A wastewater sludge is stabilized and thickened in anaerobic digestion. The daily raw-sludge feed is 100,000 lb containing 5500 lb of dry solids. Forty percent of the matter applied is converted to gases during digestion, and the digested residue is increased to 10% solids by supernatant withdrawal. Calculate the volumetric reduction of wet sludge achieved in the digestion process.

13.36 The raw sludge pumped to an anaerobic digester contains 5.0% solids that are 65% volatile. The digested sludge withdrawn from the tank has 6.5% solids that are 40% volatile. Based on these data, estimate by calculations (a) the percentage of volatile solids converted to gas during digestion and (b) the quantity of supernatant withdrawn for every 1000 gal of raw sludge fed to the digester.

13.37 The solutions to Problem 12.43 include flow diagrams for trickling-filter plants to treat a wastewater flow of 2.0 mgd. Expand one of these diagrams to include anaerobic digestion of the waste sludge and pressure dewatering of the digested sludge. Show the arrangement for two single-stage digesters, belt filter presses, sludge and supernatant return piping, and utilization of the digestion gas. List numerical design guidelines for sizing the digesters and belt filter presses.

13.38 The solution to Problem 12.53 is a flow diagram for a treatment plant with primary sedimentation and secondary step aeration. Expand this diagram to include anaerobic digestion of the waste sludge and dewatering of the digested sludge by belt-filter presses. Show the arrangement for handling both the primary sludge and waste-activated sludge, a two-stage digestion system, belt filter presses, sludge piping, and piping for return of digester supernatant and press wastewater. List numerical design guidelines for sizing the digesters, waste sludge thickeners, and belt presses.

13.39 A new two-stage anaerobic digestion system is being sized for an activated-sludge plant with primary sedimentation. Based on an average raw wastewater flow, during the maximum month, of 50,000 m^3/d containing 280 mg/l of suspended solids and 240 mg/l of BOD, primary waste solids are calculated to be 7000 kg/d and waste-activated sludge solids are calculated to be 3900 kg/d for a total of 10,900 kg/d with a volatile solids content of 70%. Both sludges are to be thickened so that the solids concentration will be 5.0% for a volume of 218 m^3/d. For a solids retention time in the first-stage tank of 20 d, calculate the volatile solids loading. If digestion reduces the volatile solids by 50%, what is the calculated concentration of solids in the digested sludge flowing into the second-stage tank? If 50 m^3/d of reasonably clear supernatant can be withdrawn from the second tank, what is the estimated concentration of digested solids in the thickened sludge withdrawn from the second tank.

13.40 Aerobic digesters with a total capacity of 50,000 ft^3 stabilize waste-activated sludge from an extended-aeration treatment plant without primary clarification. The average daily sludge flow pumped to the digestion tanks is 32,000 gal with 1.5% solids, of which 65% are volatile. The estimated solids production per capita is 0.17 lb of VS/person/day. Calculate the sludge detention time in the digesters, VS loading lb/ft^3/day and volume provided per capita.

13.41 The oxidation-ditch treatment plant described in Problem 12.50 has two aerobic digesters each with a liquid volume of 8000 ft^3 and 200 cfm aeration capacity. The digesters are operated as a semibatch process. Each morning the air is turned off, and the digesting sludge solids are allowed to thicken. After several hours of quiescent settling, supernatant is withdrawn and returned to the plant inlet so waste sludge can be pumped into the digesters. Once a week, settled sludge is withdrawn from the hopper bottom of the digesters and hauled away for disposal by spreading on agricultural land. In order to maintain the MLSS concentration at 2500 mg/l in the oxidation ditches, an average of 2000 ft^3/day of sludge containing 7600 mg/l of suspended solids (65% volatile) is pumped to the digesters, after withdrawal of supernatant. The average suspended-solids concentration in the digesting sludge is 11,000 mg/l, and the lowest temperatures measured in the aerating sludge are 8–10°C. The withdrawal of digested gravity-thickened sludge averages 700 ft^3/day containing 14,000 mg/l of suspended solids (55% volatile). Calculate the volatile solids loading and air supply and compare the results with the recommended values given in Section 13.17. Calculate the sludge aeration period in degree-days. (The sludge age can be calculated by dividing the product of the volume of the digesters and the suspended-solids concentration in the digesting sludge by the average daily sludge withdrawal times its suspended-solids concentration.) How does the calculated aeration period in degree-days compare to the recommended value for adequate sludge stabilization?

13.42 A waste-activated sludge with a total solids concentration of 12,500 mg/l and volatile solids content of 8800 mg/l is applied to an aerobic digester with a detention time of 25 days. The volatile solids destruction is 50% during digestion. Calculate the volatile solids loading in units of $kg/m^3 \cdot d$ and the solids content of the digested sludge, assuming no supernatant is withdrawn. If the solids content at the bottom of the digester can be increased to 1.5% by quiescent settling, what percentage of the waste sludge can be decanted as supernatant?

13.43 The activated-sludge treatment plant without primary sedimentation described in Problem 13.20 is recommended for expansion to process a wastewater flow of 19,500 m^3/d by adding one more aeration basin and clarifier. Using a waste activated sludge of 510 m^3/d with 0.80% solids, determine the dimensions of 3 aerobic digesters using the same design data (except for waste-activated sludge volume and solids concentration) and criteria listed in Example 13.14.

13.44 The extended aeration system for the town of Nancy is described in Problem 12.52. Determine the aerobic digestion capacity for stabilizing the sludge from this plant in two diffused-air digesters equipped with supernatant draw-offs. Although sludge dewatering on open-air drying beds is being considered, the feasibility of dewatering by belt filter presses is also under consideration because of the long cold winter. The aerating sludge is expected to be as low as 8°C, and drying beds probably could not be cleaned for up to 3 months. (a) Calculate the digester volume required based on the following: (1) The daily waste-solids yield for design can be calculated from average flow and BOD, the waste sludge has a total solids content of 1.0%, and the solids are 70% volatile. (2) The volatile solids loading of the digester should not exceed 0.04 $lb/ft^3/day$, and the aeration period should not be less than 200 degree-days, computed by multiplying the estimated sludge age by 8°C. (3) A thickened digested sludge with a solids concentration of 1.5%. (4) Volatile solids reduction during digestion of 50%. (5) An average operating suspended-solids concentration in the digesters of 1.0%. (b) For the drying beds, assume a maximum allowable solids loading of 10 $lb/ft^2/yr$. (c) For sizing belt filter presses, assume a maximum hydraulic sludge loading of 30 gpm per meter of belt width (Table 13.9) and a maximum operating period of 35 hr per week, 7 hr/day for 5 days. Two presses are required for reliability by the design-reviewing authority.

13.45 The design wastewater flow for a small community is 400 m^3/d, containing 200 mg/l BOD and 240 mg/l suspended solids. The proposed treatment is extended aeration without primary sedimentation. After aerobic digestion, the waste sludge can be taken directly from the aerobic digesters for spreading on agricultural land or pumped into a large asphalt-lined storage basin for winter storage. (a) Calculate the required capacity for the aerobic digesters assuming the F/M loading on the aeration tank is 0.15 g BOD/d per gram MLSS, the waste-activated sludge solids content is 1.5% (70% volatile), and the criteria for sizing the aerobic digesters are a maximum VS loading of 0.60 $kg/m^3 \cdot d$ and a minimum digestion period of 15 days. (b) How large must the storage basin be to hold a 90-day volume of digested sludge, assuming that the total solids reduction during aerobic digestion is 25% of the total solids in the applied sludge and the solids concentration in the storage basin can be increased to 2.0% by withdrawing supernatant?

13.46 What are the reasons for mixing either an organic amendment or a recoverable bulking agent with dewatered sludge cake before composting?

13.47 A treatment plant produces an average of 75,000 gpd of primary sludge with a solids concentration of 6.0% and 230,000 gpd of waste-activated sludge having a water content of 99.0%. (a) If the waste-activated sludge is thickened to 4.0% solids by flotation and then combined with the primary sludge prior to dewatering, what is the quantity and solids content of the blended sludge? (Assume 100% solids capture in flotation thickening.) (b) If belt filter presses dewater the sludge to 23% solids, how many 8-ton truckloads of cake must be hauled to land burial each day? (Assume 100% solids capture in filtration and negligible weight added by chemical conditioning.)

13.48 The smallest belt filter press manufactured for sludge dewatering has an effective belt width of 1 m. Estimate the equivalent population that can be served by this unit assuming 0.20 lb of dry solids/capita/day of sludge production, a solids loading of 100 lb/m/hr, and a 7-hr operating period each day.

13.49 A belt filter press with an effective belt width of 1.5 m dewaters anaerobically digested sludge at a sludge feed rate of 70 gpm. The polymer dosage is 6.0 gpm containing 0.20% polymer by weight, and the wash-water usage is 50 gpm. Based on laboratory analyses, total solids in the feed sludge are 4.0%, total solids in the cake are 35%, and suspended solids in the wastewater (filtrate, polymer feed, and wash water) are 1800 mg/l. Calculate the hydraulic feed rate, solids loading rate, polymer dosage, and solids recovery. Compare these values with those listed in Table 13.9.

13.50 A belt filter press with an effective belt width of 2.0 m dewaters 100 gpm of anaerobically digested sludge with a solids content of 6.5%. The polymer dosage is 6.4 gpm, containing 0.20% powdered polymer by weight. The wash-water consumption for belt cleaning is 60 gpm. The cake solids content is 30% and the suspended-solids concentration in the wash water measures 1800 mg/l. Calculate the hydraulic loading rate, solids loading rate, and polymer dosage and estimate the solids recovery.

13.51 A belt filter press housed in a mobile trailer was used to test the dewaterability of an anaerobically digested sludge. The effective width of the belt was 0.50 m. The operating data were a sludge feed rate of 4.5 m^3/hr, polymer dosage of 0.45 m^3/hr containing 0.20% powdered polymer by weight, and wash-water flow of 1.07 l/s. Based on laboratory analyses, total solids in the feed sludge equaled 3.5%, total solids in the cake equaled 32%, wastewater from belt washing contained 2600 mg/l suspended solids, and the filtrate was 1.20 l/s with 500 mg/l of suspended solids. Calculate the hydraulic loading, solids loading, polymer dosage, and solids recovery.

13.52 Recommend the size of belt filter presses and schematic layout for dewatering the aerobically digested sludge in Problem 13.43.

13.53 At an existing treatment plant, hauling away liquid digested sludge for spreading on farmland is being replaced by belt filter presses to dewater anaerobically digested sludge prior to stockpiling and spreading on agricultural land. The design sludge production, which is the average quantity during the maximum month of the year plus 25% for future plant expansion, equals 90,000 gpd with an average solids concentration of 5.0%. Using a trailer-mounted press, the performance data for dewatering the 5.0% sludge were a hydraulic loading of 45 gpm/m, cake solids content of 24%, polymer dosage of 4.0 gpm/m with a concentration of 0.20% powdered polymer, wash-water usage of 32 gpm/m, and wastewater production of 64 gpm/m containing 2200 mg/l of suspended solids. Calculate the following: solids loading, polymer dosage, wash-water usage and wastewater production per 1000 gal of sludge dewatered, solids recovery, and daily cake production. For a design operating schedule of 12 hr/day with two presses, recommend the size of the presses for installation.

13.54 An alum sludge is dewatered by pressure filtration. A daily volume of 40 m^3 of slurry is pressed from 2.0% solids concentration to a cake of 40% solids, after conditioning with 10% lime. Calculate the volume of filtrate and weight of the cake produced per day. (The 40% concentration of cake solids includes the alum precipitate and the lime addition.)

13.55 A diaphragm plate-and-frame filter press is used to dewater alum sludge from a water treatment plant. Each chamber of the press has a filtering area of 36 ft^2. During normal operation, the total cycle time (feed, compression, discharge, and wash) is 20 min. The volume of alum sludge applied per chamber per cycle is 300 gal, and the lime dosage as CaO is 10% of the alum solids. The alum sludge feed averages 3.6% solids, and the sludge cake is 38.5% solids, including the lime. Solids capture can be assumed to be 100%, since the filtrate contained less than 200 mg/l of suspended solids. Calculate the cake production per gallon of alum sludge in pounds, assuming a specific gravity of 1.0, and the filter yield in pounds per square foot per hour based on applied alum solids.

13.56 The solution to Problem 11.20 is a process flow diagram for a surface water treatment plant, and the solution to Problem 11.30 is a process flow diagram for a water-softening plant. Expand each diagram to include a scheme to process the sludge from sedimentation and the wash water from filtration. The wastes must be thickened and dewatered for hauling to land disposal, since sufficient land area at the plant sites is not available for lagoons or drying beds. (Refer to Sections 13.5 and 13.3.)

13.57 Refer to Example 13.17. Why were centrifuges selected for dewatering the sludge from this plant rather than belt filter presses? Why is equalization of feed sludge important in centrifuge dewatering? What were the selected methods of cake conveyance?

13.58 A solid-bowl centrifuge dewaters an anaerobically digested blend of primary and waste-activated sludges at a flow rate of 8.0 l/s with a polymer addition of 10 kg per 1000 kg of dry solids. The sludge feed contains 2.4% solids, the cake averages 19%, and the suspended-solids concentration in the centrate is 3200 mg/l. Calculate the solids recovery and centrate flow rate.

13.59 A lime precipitate from a water-softening plant is concentrated in a scroll centrifuge from 10% to 60%. The suspended-solids capture is 90%. Calculate the suspended-solids concentration in the centrate and the volumetric reduction of the waste slurry as a percentage of the feed.

13.60 **a.** Compare the required processing and quality for biosolids sold in bulk for agricultural land application to biosolids sold in bags for application on private lawns and gardens.

b. Compare the required processing and quality for biosolids sold in bulk for agricultural land application to dewatered sludge buried by codisposal in a municipal solid-waste landfill.

13.61 Anaerobically digested sludge dewatered by belt filter press is tested for fecal coliform density by the CFU (colony-forming units) method as specified by EPA for application of biosolids on agricultural land (Section 13.28). The results of seven sludge samples collected and tested on consecutive days expressed in number of fecal coliforms per gram of total solids (dry weight) are as follows: 140,000; 320,000; 130,000; 330,000; 300,000; 220,000; and 305,000. Calculate the average density based on geometric mean for these data. (Geometric mean of a number is the logarithm to base 10.) Does this sludge meet the pathogen reduction for Class B biosolids?

REFERENCES

[1] "Process Design Manual, Land Application of Sewage Sludge and Domestic Septage," U.S. Environmental Protection Agency, Office of Research and Development, EPA-625/R-95/001 (Cincinnati, OH: September 1995).

[2] *Management of Water Treatment Plant Residuals*, AWWA Technology Transfer Handbook (Denver, CO: Am. Water Works Assoc., 1996).

[3] "Design Manual, Dewatering Municipal Wastewater Sludges," U.S. Environmental Protection Agency, Office of Research and Development, EPA/625/1-87/014 (September 1987).

[4] *Handbook of Public Water Systems*, 2nd ed., Chapter 23, "Residuals Management," HDR Engineering, Inc. (New York: John Wiley & Sons, 2001).

[5] G. Hoyland, A. Dee, and M. Day, "Optimum Design of Sewage Sludge Consolidation Tanks," *J. Inst. Water Env. Manage.* no. 3, (1989): 505–516.

[6] *Recommended Standards for Wastewater Facilities, Great Lakes–Upper Mississippi River Board of State Public Health & Environmental Managers* (Albany, NY: Health Research Inc., Health Education Services Division, 1978).

[7] "Composting of Municipal Wastewater Sludges," U.S. Environmental Protection Agency, Technology Transfer, EPA 625/4-85-014 (August 1985).

[8] "Control of Pathogens and Vector Attraction in Sewage Sludge," U.S. Environmental Protection Agency, Office of Research and Development, EPA/625/R-92/013, http://www.epa.gov/ORD/NRMRL (October 1999).

[9] R. D. Davis and E. G. Coker, *Cadmium in Agriculture With Special Reference to the Utilization of Sewage Sludge on Land*, Water Research Centre, Stevenage Laboratory, England, TR 139 (June 1980).

[10] "Process Design Manual, Surface Disposal of Sewage Sludge and Domestic Septage," U.S. Environmental Protection Agency, Office of Research and Development, EPA 625/R-95/002 (Washington, DC: September 1995).

[11] "Process Design Manual, Municipal Sludge Landfills," U.S. Environmental Protection Agency, Office of Solid Waste, EPA 625/1-78-010 SW-705 (October 1978).

CHAPTER 14

Advanced Wastewater Treatment Processes and Water Reuse

OBJECTIVES

The purpose of this chapter is to:

- Acquaint the reader with the quality standards and uses for reclaimed water
- Discuss the characteristics of specific contaminants and treatment processes for removal
- Present existing systems for agricultural and urban landscape irrigation
- Discuss indirect reuse to augment groundwater supplies

Advanced wastewater treatment refers to methods and processes that remove more contaminants from wastewater than are usually taken out by conventional techniques. The term *tertiary treatment* is often used as a synonym for advanced waste treatment, but the two are not precisely the same. Tertiary suggests an additional step applied only after conventional primary and secondary wastewater processing; for example, granular-media filtration followed by chlorination with an extended contact time for removal of pathogens. Advanced treatment actually means any process or system that is used after conventional treatment, or to modify or replace one or more steps, to remove refractory contaminants. Several of these pollutants not taken out by secondary biological methods can adversely affect aquatic life in streams, accelerate eutrophication of lakes, hinder use of surface waters for municipal supplies, and restrict reuse of wastewater for irrigation, groundwater recharge, or other beneficial applications.

The term *water reclamation* implies that the combination of conventional and advanced treatment processes employed returns the wastewater to nearly original quality (i.e., reclaims the water). Reclamation to drinking water quality is required for infiltration or injection into an aquifer used or classified as a potable water source. Nevertheless, reclaiming wastewater for direct reuse as drinking water is not permitted and even when diluted with natural water is not considered safe. If reclaimed water were consumed as drinking water, the long-term health effects would not be known.

Water reuse is the planned direct use of reclaimed wastewater for beneficial applications. Wastewater treatment requirements consider protection of human health, potential environmental impacts, and public access. The most common reuse is restricted agricultural irrigation of fodder, fiber, and seed crops requiring conventional treatment and disinfection. In regions with limited water supplies, restricted urban irrigation is used to water highway medians and golf courses. Unrestricted irrigation, after tertiary filtration and disinfection, is applied to open access areas, such as parks and playgrounds. Groundwater recharge by surface spreading requires advanced wastewater treatment based on specific site conditions. For groundwater recharge by well injection, water reclamation is required to achieve drinking water quality.

LIMITATIONS OF SECONDARY TREATMENT

Contamination of municipal water results from human excreta, food preparation wastes, and a wide variety of organic and inorganic industrial wastes. Conventional treatment employs physical–biological processes, and possibly chlorination, to reduce biochemical oxygen demand, suspended solids, and pathogens. Water quality deterioration from municipal use results in the following approximate increases of pollutants after ordinary secondary treatment: 300 mg/l of total solids, 200 mg/l of volatile solids, 30 mg/l of suspended solids, 30 mg/l of BOD, 26 mg/l of nitrogen, and 5 mg/l of phosphorus. Treated wastewater can also contain traces of organic chemicals, heavy metals, excreted pathogens, and other contaminants.

14.1 EFFLUENT STANDARDS

The maximum acceptable level of organic matter in a wastewater effluent after biological treatment is defined in terms of BOD and suspended-solids concentrations. For each of these parameters, the arithmetic mean of daily measurements in a period of 30 consecutive days is limited to 30 mg/l, and that in a period of 7 consecutive days is limited to 45 mg/l. Conventional wastewater processing can include chemical disinfection, usually chlorination. Where public health is a concern, the effluent standard is often a maximum geometric mean concentration of fecal coliform bacteria in a period of 30 consecutive days of 200 per 100 ml and in a period of 7 consecutive days of 400 per 100 ml. In some cases, these standards are not adequate to protect the receiving water body to ensure adequate quality for uses involving human contact, that is, recreation and water supply. Advanced wastewater processes can further reduce BOD and suspended-solids concentrations, increase removals of pathogens, precipitate phosphorus, oxidize ammonia nitrogen, and diminish the content of trace organics and heavy metals.

Phosphorus is the key nutrient for growth of algae and aquatic plants associated with eutrophication. Where effluents are discharged to lakes and estuaries, either directly or via rivers, standards have been adopted by most states to limit phosphorus input. A commonly established maximum effluent concentration is 1.0 mg/l or less. This criterion requires a reduction of 85% or greater in domestic wastewater treatment.

Ammonia nitrogen, the most common form of nitrogen in an effluent from biological treatment, is toxic to fish at relatively low concentrations and can exert a significant oxygen demand. The maximum allowable ammonia nitrogen concentration to protect warm-water fish is about 5 mg/l at a pH of 8; for cold-water fish, the limit is about 1 mg/l. With regard to oxygen demand, 4.6 lb of oxygen are used theoretically to bacterially nitrify 1.0 lb of ammonia nitrogen. For disposal in a receiving stream with limited dilutional flow, oxidation of a major portion of the ammonia is sometimes needed along with more stringent BOD and suspended-solids removals. For example, effluent limits may be established at 10 mg/l of BOD, 10 mg/l of SS, and 5 mg/l of NH_3—N. Partial denitrification is occasionally specified where the receiving watercourse is used for a public water supply. The limit for nitrate concentration in drinking water is 10 mg/l NO_3—N.

Wastewater effluent can be chlorinated in an attempt to reduce the number of pathogens. Usually the chlorine is applied at a dosage up to 15 mg/l with a minimum contact time of 30 min. Even though fecal coliforms in conventional wastewater treatment plus chlorination are removed to a level of 200 per 100 ml, a 99.99% + reduction, the effluent may not be free of excreted pathogens. Helminth eggs are unharmed by chlorination and protozoal cysts are extremely resistant, so they must be physically removed, and viruses are more resistant to chlorination than fecal bacteria. Furthermore, the presence of suspended solids inhibits disinfection by harboring bacteria and viruses within biological floc, shielding them from the action of the chlorine.

The ability of gravity sedimentation to remove suspended solids following biological aeration is a major limitation of conventional treatment. Plain granular-media filtration can remove suspended solids to a concentration of about 10 mg/l to further reduce BOD. With chemical coagulation, the suspended solids can be reduced to less than 5 mg/l, and in a well-operating coagulation and filtration system an effluent turbidity of less than 10 NTU can be achieved. A pathogen-free effluent is produced by tertiary coagulation and filtration, with or without sedimentation, and chlorination with an extended contact time. Helminth eggs, protozoal cysts, and suspended solids that interfere with the disinfecting action of chlorine are removed in filtration. Chlorination of the clear filtered effluent with a high chlorine residual-contact time inactivates enteroviruses and kills pathogenic bacteria. The microbiological effluent standard for reuse of tertiary effluent in unrestricted irrigation and groundwater recharge is commonly 2.2 fecal coliforms per 100 ml. Sometimes allowable limits are also specified for enteric viruses and helminth eggs, such as *Ascaris*.

Effluent chlorination results in the formation of chloramines from the reaction between free chlorine and ammonia. Chloramines are toxic to fish at very low levels. The recommended maximum concentrations of residual chlorine are 0.01 mg/l for warm-water fish and 0.002 mg/l for cold-water fish. In advanced wastewater treatment, dechlorination is performed by the addition of sulfur dioxide to the effluent.

Many organic and inorganic chemicals are refractory to conventional treatment, and local conditions often dictate discharge limits for specific compounds. Some cause aesthetic problems, e.g., foam or color in the receiving watercourse or taste and odor in downstream water supplies. Toxic organic compounds or metal ions are injurious to aquatic life and, in sufficiently high concentrations, prevent safe reuse of the receiving water as a public supply source. The main problems are detecting, identifying, and establishing discharge limits for toxic substances. The traditional approach has been to compare instream concentrations of known pollutants with surface water quality standards. However, sometimes the toxic substances cannot be identified by chemical testing, standards may not exist, and the impact of toxicity may depend on interacting factors such as pH, alkalinity, and hardness. As a result, biomonitoring has been developed for toxicity assessment.

Effluent biological toxicity testing to evaluate acute and chronic toxicities is performed by short- and long-term tests on aquatic organisms like minnows, crustaceans, and invertebrates. The tests are conducted on full-strength wastewater and/or several dilutions, using water from the receiving stream to determine the no-observed-effect-level (NOEL) values for acute and chronic exposures. Based on these data and the dilution provided by the receiving stream, exposure criteria can be calculated and discharge permit limits established. The toxic substances and their sources are determined by additional bioassays in combination with chemical testing on samples from the wastewater collection system, particularly industrial discharges.

14.2 FLOW EQUALIZATION

Wastewater flows have a diurnal variation ranging from less than one-half to more than 200% of the average flow rate. In addition, daily volumes are increased by inflow and infiltration into the sewer collection system during wet weather. The strength of a municipal waste also has a pronounced diurnal variation resulting from nonuniform discharge of domestic and industrial wastes. Treatment plants traditionally have been sized to handle these deviations without significant loss of efficiency. For conventional primary and secondary treatment, designing units to accommodate peak hourly flow is more economical than installing flow equalization basins to eliminate the diurnal influent pattern. Industrial wastewaters entering a municipal system can cause excessively large flows and peak organic loads, so it is better to install facilities at the industrial site for flow smoothing prior to discharge.

Many advanced wastewater treatment operations, such as filtration and chemical clarification, are adversely affected by flow variation and sudden changes in solids loading. Maintaining a relatively uniform influent allows improved chemical feed control and process reliability. Costs saved by installing smaller units for chemical precipitation and filtration, together with reduced operating expenses, may more than compensate for the added costs of flow equalization facilities.

Equalization basins to level diurnal flow variations may be designed for either side-line or in-line operation. A side-line basin may receive, temporarily store, and return either raw wastewater after screening and degritting or clarified wastewater after primary sedimentation. A raw-wastewater basin requires aeration mixing to reduce odors and maintain organic solids in suspension. The tank is often equipped with a sludge scraper to prevent consolidation of settled solids. If the side-line basin follows primary sedimentation, aeration mixing and sludge scrapers may not be necessary. In a warm climate, the basin may be a paved pond that is flushed clean with chlorinated effluent after each use by an automatic sprinkling system. All the wastewater passes through an in-line basin. If placed between grit removal and primary sedimentation, mechanical or diffused aeration is required to keep solids in suspension and to prevent septicity. Holding tanks may also be placed at other locations in the treatment scheme. For example, a basin serving as a pump suction pit can be located just ahead of filters to dampen hydraulic surges without providing complete flow equalization.

Basin volume required for flow equalization is determined from mass diagrams based on average diurnal flow patterns. The capacity needed is usually equivalent to 10%–20% of the average daily dry-weather flow. Extra volume should be provided below the low water level, since both mechanical and diffused aeration systems must have a minimum depth to maintain mixing. Example 14.1 includes sample calculations for determining equalization basin volume.

Example 14.1

Determine the basin volume needed to equalize the diurnal wastewater flow pattern diagrammed in Figure 12.16(b).

Solution: Hourly flow rates measured from Figure 12.16(b) and calculated cumulative volumes are listed in Table 14.1; the mass diagram of wastewater flow from Table 14.1 is drawn in Figure 14.1. The slope of the line connecting the origin and final point on the mass curve equals the average 24-hr rate of 18.2 mgd. To find the required volume for equalization, construct lines parallel to the average flow rate and tangent to the mass curve at the high and low points. The vertical distance between these two parallels is the required basin capacity; in this case, the value is 2.5 mil gal or 13.7% of the daily influent of 18.2 mil gal.

TABLE 14.1 Wastewater Flows From Figure 12.16(b)

Time	Flow Rate (mgd)	Cumulative Volume (mil gal)
Midnight	—	0
2 a.m.	13.1	1.1
4	10.7	2.0
6	9.7	2.8
8	14.8	4.0
10	22.5	5.9
Noon	26.6	8.1
2 p.m.	25.6	10.3
4	23.5	12.2
6	21.8	14.0
8	18.9	15.6
10	16.7	17.0
Midnight	14.4	18.2
Average	18.2	

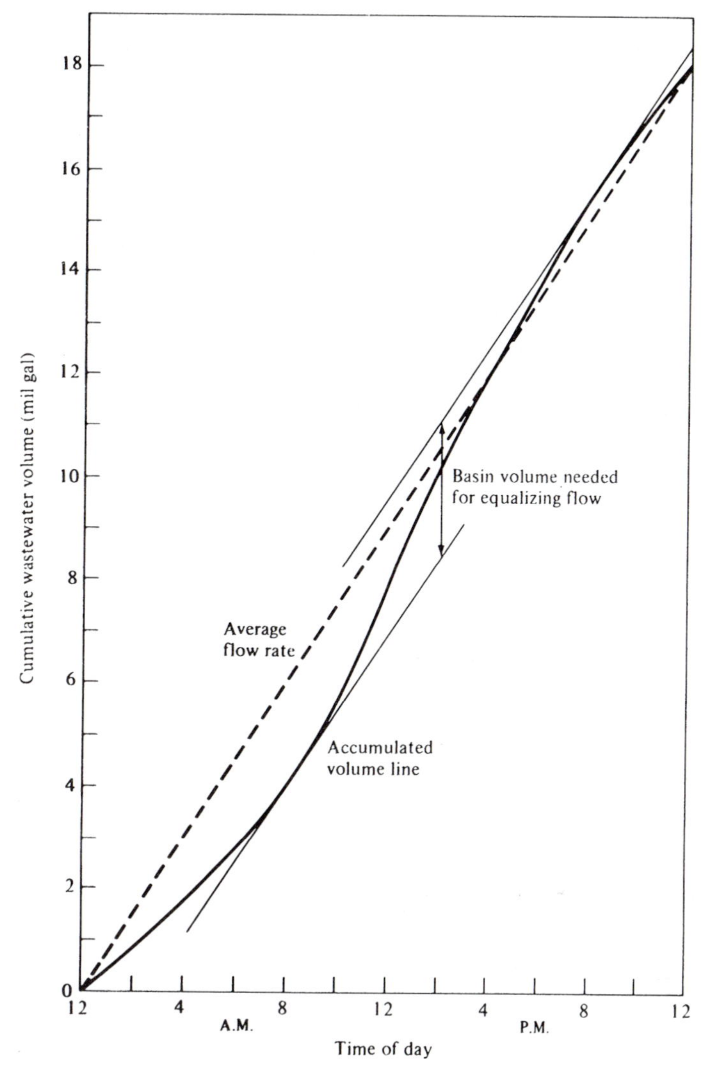

FIGURE 14.1 Mass diagram of wastewater flow, from Table 14.1 for Example 14.1, to determine the basin volume needed to equalize flow.

SELECTION OF ADVANCED WASTEWATER TREATMENT PROCESSES

Table 14.2 lists popular advanced treatment methods. Upgrading treatment to increase BOD and suspended-solids removal is usually accomplished by tertiary operations, such as filtration. Emphasis is also being placed on unit operations for nutrient removal, particularly precipitation of phosphates by either separate chemical coagulation or biological–chemical aeration and by nitrogen extraction by air stripping or biological nitrification–denitrification.

TABLE 14.2 Selected Advanced Wastewater Treatment Processes

Process
Suspended-solids removal
Plain filtration through granular media
Direct filtration after chemical coagulation
Organic chemical removal
Adsorption on granular activated carbon
Phosphorus removal
Biological–chemical precipitation and clarification
Chemical coagulation and clarification
Irrigation of cropland
Nitrogen removal
Biological nitrification–denitrification
Irrigation of cropland
Pathogen removal
Granular-media filtration after chemical coagulation
Disinfection by chlorine or ozone
Heavy metals removal
Lime precipitation

14.3 SELECTING AND COMBINING UNIT PROCESSES

Major factors affecting unit process selection are influent wastewater characteristics, effluent quality required, reliability, sludge handling, process compatibility, and costs. The degrees of treatment needed at present and in the future are the primary considerations, because neither raw-wastewater characteristics nor effluent quality specifications can be safely considered permanent. When considering alternative schemes, the best choice for the original plant allows for expansion and change to meet future needs. Maximum flexibility for modifying operations to improve performance is of major importance.

The reliability of a process is directly related to the experience gained from operating full-scale systems. Pilot-plant studies can provide satisfactory information on liquid processing, but often they do not adequately evaluate sludge problems or compatibility with previous and subsequent unit operations. Problems involved in going from pilot to full scale are not easily recognized; thus, for maximum reliability, plant designers should take advantage of the practice developed from existing full-scale installations.

Sludge handling and disposal dictate to a considerable extent the selection of processes that are most feasible for separating contaminants from wastewater. A unit operation, even though successful in extracting pollutants from water, can be unacceptable if the waste sludge produced is difficult and costly to dewater; hence, sludge disposal must always be considered an integral part of any treatment technique. The elimination of residue by spreading on land, burial, or incineration necessitates concentration of waste slurries by biological or chemical stabilization and mechanical dewatering methods. Advanced treatment processes require greater study because they often include chemical precipitates as well as organic solids. A critical question is whether organic and inorganic wastes should be kept separate or combined for thickening and disposal. The answer often determines whether chemical treatment is applied as a tertiary step or combined with conventional operations of primary sedimentation and biological aeration.

Compatibility of unit processes applied in the overall treatment scheme is essential. Optimum pH is important and can influence the sequence of unit operations. For instance, lime precipitation is performed at high pH, while disinfection by chlorine is most effective at neutral or low pH values. Possible adverse effects of recycled waste streams from individual unit processes to the plant inlet, or intermediate locations, must be considered in the overall treatment scheme. Segregation or blending of centrate, filtrate, backwash water, and other return flows and their point of return must be evaluated in design.

Finally, capital and operating costs influence process selection and often dictate design decisions. Common factors include electric power consumption; application of chemicals; methods of sludge disposal; and separation or combination of biological and chemical treatment.

SUSPENDED-SOLIDS REMOVAL

Removal of suspended solids from the effluent of a conventional treatment plant may be in order to reduce the organic content or to pretreat the wastewater for subsequent processing. Effective disinfection requires removal of suspended solids that can harbor and protect pathogenic bacteria and viruses from the oxidizing action by chlorine or ozone. Carbon adsorption columns are preceded by filtration to prevent fouling of the granular activated carbon medium.

14.4 GRANULAR-MEDIA FILTRATION

Design criteria for wastewater filters cannot be derived directly from experience in potable water systems. Waterworks filters are generally operated at constant rates under relatively steady suspended-solids loading. Unless equalization is provided, a wastewater plant must handle a varying rate of flow with peak hydraulic and solids loadings occurring simultaneously. Particulate matter found in typical wastewaters is less predictable and much more "sticky" than water-plant solids, thus making filter backwashing more difficult. Also, the nature of suspended matter is not consistent and varies with the preceding treatment processes. Microbial flocs are the dominant suspended solids following secondary biological treatment, while carryover from biological–chemical and physical–chemical methods contains a significant quantity of coagulant residue.

Dual-media anthracite–sand filters with downward flow by either gravity or pressure dominate in American practice. This design permits production of a high-quality effluent with reasonable filter runs between bed cleanings. Typically, the layer of coarse coal with a specific gravity of 1.35–1.75 is 12–18 in. (0.3–0.5 m) deep, and the underlying sand layer with a specific gravity of 2.65 is 8–12 in. (0.2–0.25 m) [1]. After backwashing, the media are arranged with the coarse lighter coal on top and the finer heavier sand on the bottom. The actual distribution and degree of intermixing depend on both relative particle-size gradations and specific gravities of the media. A filter bed with coarser coal over finer sand allows both surface straining and "in-depth" filtration, reducing premature surface plugging and delaying solids breakthrough. Another kind of filter is a deep-bed single-medium filter of sand or anthracite ranging in size from 1 to 3 mm with bed depths of 3–6 ft (1–2 m).

Cleaning solids from granular media requires air scouring in addition to fluidizing the bed by upflow of wash water. Fluidization alone does not provide the necessary abrasive action among media grains. Common sequences are air scouring by vigorous aeration up through the underdrain followed by backwash or concurrent air-and-water scouring with simultaneous subfluidization before initiating the full flow of backwash water. The source of backwash water is effluent from the tertiary filters. If the filtrate is disinfected, the chlorine contact tank serves as the source of supply. Otherwise, a storage tank is needed with sufficient capacity to clean the filters during their peak usage. Dirty wash water is collected in an equalizing tank and returned to the plant influent at a nearly constant rate for treatment. The quantity of backwash water returned is 2%–5% of the filtered water.

Design considerations for granular-media filters [1, 2, 3] include the filter configuration; types, size, gradation, and depths of filter media; method of flow control; backwash requirements; filtration rate; terminal head loss; quantity and characteristics of the water applied; and desired effluent quality. Figure 14.2 is a schematic diagram of a biological treatment plant with tertiary filtration. Normally, at least two and usually four filter cells are provided for operational flexibility. The bed area should be sufficient to allow peak design flows with one unit out of service for backwashing or repair.

Gravity filters in wastewater processing are usually designed with flow control by either influent flow splitting or declining-rate filtration. These methods are described in Section 10.18.

Pressure filters (Fig. 10.37) are normally operated at a constant rate of flow, although some units use a constant applied pressure without effluent control, resulting in declining-rate filtration. Two advantages of enclosed filters are that high pressure can be used to overcome a greater head loss and negative head conditions are not created in the filter.

The filtration rate, terminal head loss, length of filter run, and solids loading are interrelated parameters. Production from gravity-flow beds is normally 3–6 gpm/ft^2 (2.0–4.1 l/m$^2 \cdot$ s), and terminal head loss is 8–10 ft (2.5–3.0 m) for gravity filters and 20–30 ft when using pressure filters. The length of filter runs should be a minimum of 6 hr to avoid excessive use of wash water and shorter than 40 hr to reduce bacterial decomposition of organics trapped in the media. These design limits provide an average run length of about 24 hr. The highest solids loading on tertiary

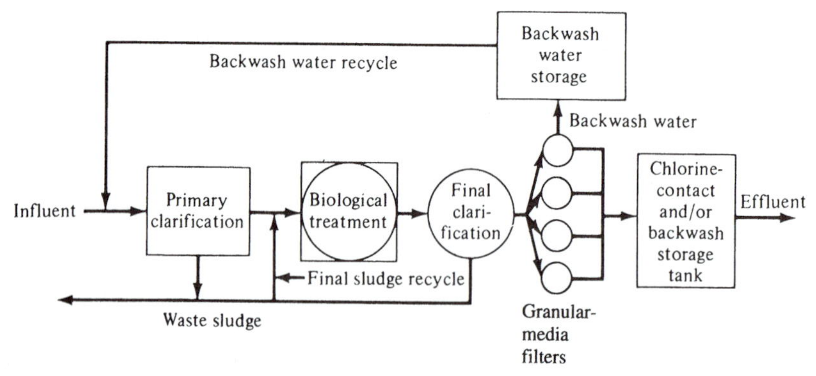

FIGURE 14.2 Typical layout of a biological treatment plant with tertiary granular-media filters. *Source: Wastewater Filtration,* Environmental Protection Agency, Technology Transfer (July 1974): 6.

filters coincides with maximum hydraulic flow, since clarification following secondary treatment is least efficient at peak overflow. Even in well-operated plants, the suspended-solids content during high flows can range from 30–50 mg/l (15–25 turbidity units). Therefore, the critical design condition is often based on the maximum 4-hr flow rate at the worst expected suspended-solids concentration.

Experience indicates that filtration following secondary biological treatment can reduce suspended solids to a level of 5–10 mg/l, so the expected performance of a well-designed and properly operated system is an effluent with SS and BOD concentrations of less than 10 mg/l.

Example 14.2

Compute the area of granular-media filters required for suspended-solids removal from a trickling-filter plant effluent. The average daily flow is 18.2 mgd (68,900 m^3/d), and the maximum wet-weather flow for a 4-hr period is 31.0 mgd (117,000 m^3/d). Data from a pilot-plant study are plotted in Figure 14.3. Filters are dual-media, gravity-flow beds, downtime for backwashing is 30 min, and water usage equals 150 gal/ft^2 (6.1 m^3/m^2). Assume a nominal filtration rate of 3 gpm/ft^2 (54 $m^3/m^2 \cdot d$). Check the peak rate with one filter cell out of service.

Solution: For the average daily wastewater flow and nominal filtration rate of 3.0 gpm/ft^2,

$$\text{filter area} = \frac{18.2 \text{ mgd} \times 694 \text{ gpm/mgd}}{3.0 \text{ gpm/ft}^2} = 4200 \text{ ft}^2$$

Use four filters, each 1050 ft^2 (97.5 m^2).

From Figure 14.3, the length of filter run is 21 hr for 3.0 gpm/ft^2 and 30 mg/l of SS.

$$\text{filter cycle time} = \text{run length} + \text{backwash time}$$

$$= 21.0 + 0.5 = 21.5 \text{ hr}$$

$$\text{volume of filtrate/cycle} = 21 \times 3.0 \times 60 = 3800 \text{ gal/ft}^2$$

$$\text{backwash volume/cycle} = 150 \text{ gal/ft}^2$$

$$\text{backwash as a percentage of filtrate} = \frac{150}{3800} = 4.0\%$$

$$\left(\begin{array}{c}\text{needed filtration rate to account for} \\ \text{downtime and return of backwash} \\ \text{water}\end{array}\right) = 3.0 \times 1.04 \times \frac{21.5}{21}$$

$$= 3.2 \text{ gpm/ft}^2 \text{ } (57 \text{ m}^3/\text{m}^2 \cdot \text{d})$$

The quantity of suspended solids removed assuming 30 mg/l applied and 5 mg/l remaining in the effluent is

$$(30 - 5) \times 18.2 \times 8.34 = 3800 \text{ lb of dry SS/day } (1700 \text{ kg/d})$$

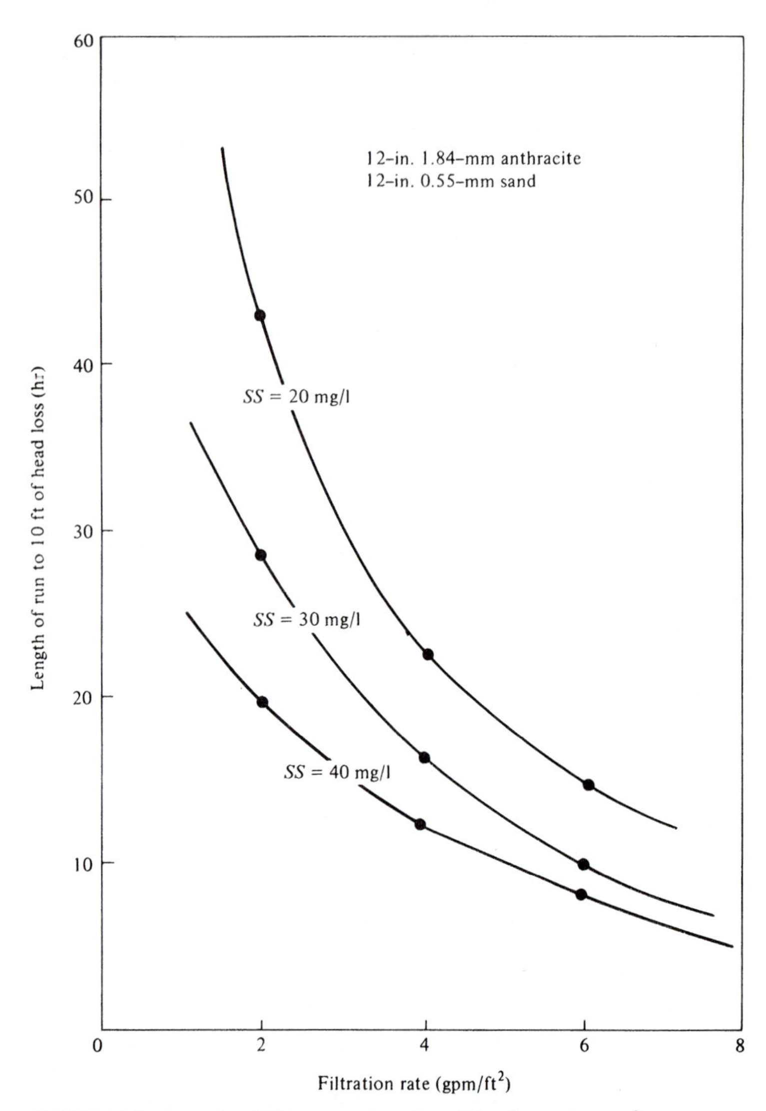

FIGURE 14.3 Length of filter run at various filtration rates and suspended-solids concentrations for a bed of anthracite coal and sand media, and an applied wastewater of trickling-filter plant effluent. *Source:* E. R. Baumann and J. Y. C. Huang, "Granular Filters for Tertiary Wastewater Treatment," *J. Water Poll. Control Fed.* 46, no. 8 (August 1974).

For the maximum 4-hr flow of 31.0 mgd, estimate the required filtration rate with three filter cells operating and the fourth out of service if the return backwash water is 10% of the wastewater flow.

$$\text{filtration rate} = \frac{31.0 \times 1.1 \times 694}{3 \times 1050} = 7.5 \text{ gpm/ft}^2 \ (130 \text{ m}^3/\text{m}^2 \cdot \text{d})$$

The length of filter run from Figure 14.3 at 40 mg/l is 6 hr, which should be sufficient to allow operation when one cell is out of operation for cleaning or emergency repair.

14.5 DIRECT FILTRATION WITH CHEMICAL COAGULATION

In Southern California, inland wastewater treatment plants discharge to flood control channels or river beds during periods of little or no dilutional flow. Impoundments in these watercourses are open to the public for recreation, including swimming. They also serve as recharge basins to replenish groundwater withdrawn by municipal wells. As a result of these activities in water reuse, effluent standards are needed to protect public health, with emphasis on virus removal.

The State of California Health Department established both quality and treatment criteria for wastewaters discharged into recreational impoundments. The quality criterion was a 7-day median coliform concentration less than or equal to 2.2 per 100 ml. Because of the difficulties in monitoring for the presence of viruses, no specific limit on virus concentration was proposed. The specified tertiary treatment included coagulation, sedimentation, filtration, and disinfection. Alternative tertiary treatment could be substituted for this criterion, provided the same degree of virus removal is achieved by the proposed processing. The prospect of being able to apply advanced wastewater treatment other than conventional water treatment resulted in researching less expensive alternatives.

The Pomona Virus Study [4] investigated virus-removal efficiency of several tertiary treatment systems on a pilot-plant scale. Flow diagrams for two of the systems are given in Figure 14.4. The first scheme, serving as the baseline for alternative processing, includes coagulation with 150 mg/l of alum, sedimentation, dual-media filtration at 5 gpm/ft^2, and chlorine contact for 2 hr with minimum residuals of 5 and 10 mg/l. Figure 14.4(b) is direct filtration (no flocculation or sedimentation) with a low alum dosage of 5 mg/l. The rate of filtration was 5 gpm/ft^2 and the time and dosage of chlorine was 2-hr contact with residuals of 5 and 10 mg/l, which are the same as the baseline system. Other processes tested, but not discussed here, were two-stage granular activated carbon filtration with intermediate chlorination and ozonation in place of chlorination for disinfection processing of a nitrified wastewater effluent. The major objectives of this study were to rank the treatment systems based on their efficiency and reliability of virus removal and to estimate treatment costs for those systems that performed equivalent to the baseline system.

Naturally occurring viruses were present in relatively low concentrations in the wastewater and were rarely detected in the tertiary effluents. Therefore, seeding experiments had to be conducted with sufficient virus concentrations to produce measurable effluent virus counts. With polio I as a seed, viruses were added to the influent in the concentration range of 10^5–10^6 PFU/gal.

The key conclusions of the study were that the majority of virus inactivation occurred during disinfection, and the main function of the preceding treatment was to remove substances that interfere with effluent disinfection. Also, the magnitude of the chlorine residual directly affected the effluent virus concentration. The systems equivalent in efficiency and reliability of virus removal were the baseline system [Fig. 14.4(a)]

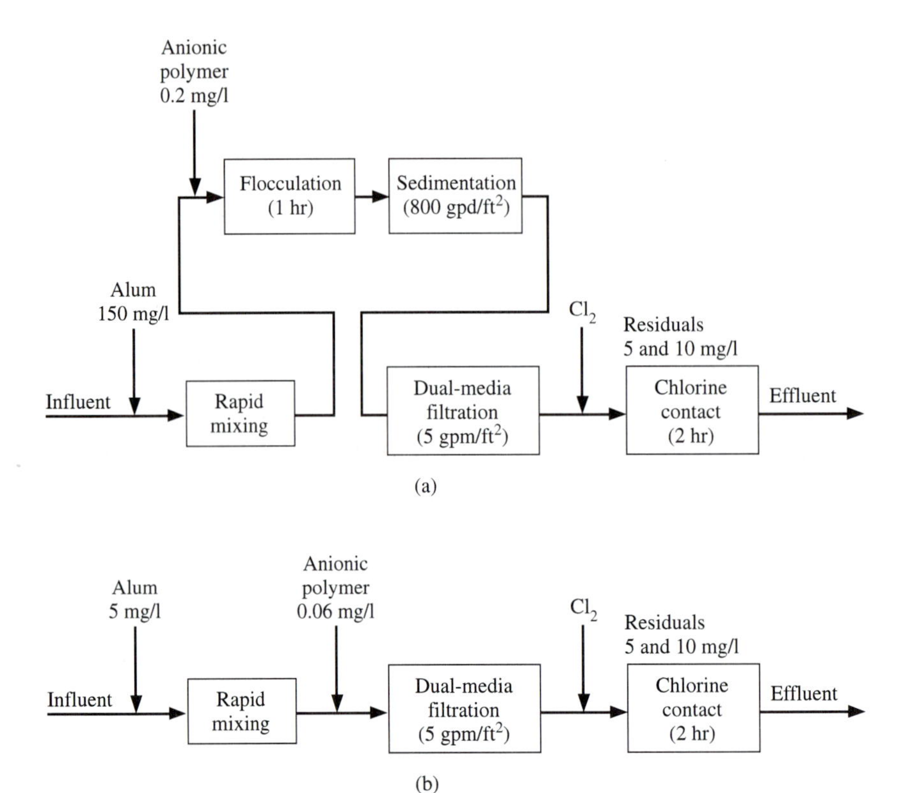

FIGURE 14.4 Two of the tertiary treatment systems investigated for virus-removal efficiency in the Pomona, CA, pilot-plant studies. (a) Coagulation, sedimentation, filtration, and disinfection. (b) Coagulation, filtration, and disinfection.

with 5 mg/l of combined chlorine residual, direct filtration [Fig. 14.4(b)] with 10 mg/l of chlorine residual, two-stage carbon adsorption with a 5-mg/l residual, the baseline system with a 10-mg/l ozone dosage, and direct filtration with a 50-mg/l ozone dosage. The least-cost system that performed equivalent to the baseline system was direct filtration and disinfection at the higher chlorine residual of 10 mg/l.

As a result of the Pomona Virus Study, tertiary treatment has been implemented at 10 plants of the Sanitation Districts of Los Angeles County (La Canada, Long Beach, Los Coyotes, Pomona, San Jose Creek, Whittier Narrows, Valencia, Saugus, Lancaster, and Palmdale) treating approximately 190 mgd [5]. The water reclamation at all of these plants is primary clarification and conventional activated-sludge treatment, followed by direct tertiary filtration, with prior chemical coagulation using alum and anionic polymer, and plug-flow chlorination with an extended contact time [Fig. 14.4(b)]. The effluent quality is microbiologically and chemically safe for unrestricted reuse, except for direct drinking water. The requirements for final effluent turbidity of 2 nephelometric turbidity units (NTU) has been met over 99.8% of the time and the total coliform 7-day median of 2.2 per 100 ml has been met over 99% of the time. Chlorine dosages applied to filtered effluent for the various plants range from 5.3 mg/l to 7.7 mg/l to maintain the minimum required residual of 3 mg/l leaving the chlorination tanks. The mean $C \cdot t$ values for chlorine disinfection at the plants average 310 (mg/l) · min, and the median $C \cdot t$ values average 290 (mg/l) · min. Effluent monitoring for 20 years (1979–1999) has demonstrated that these tertiary plants produce essentially virus-free reclaimed water. Of the over 1000 samples of reclaimed water tested, only one sample was positive for an enteric virus (*Coxsackie B*).

The concentrations of heavy metals, nitrate nitrogen, and chlorination by-products meet drinking water standards. The excellent chemical quality is controlled by both limiting the industrial wastewaters entering the sewer system and efficient treatment plant operations.

Distribution of the 190 mgd of reclaimed water produced is 60% for groundwater recharge, 16% landscape irrigation, 9% environmental enhancement, 7% agriculture irrigation, and 8% industrial process supply [6]. Since the water reclamation plants are located along existing rivers, creeks, or flood-control channels, groundwater recharge is convenient for indirect reuse by infiltration to groundwater and subsequent withdrawal as a drinking water supply from wells. These waterways can also convey the effluent by gravity to existing off-stream recharge basins for percolation into the aquifer. Urban landscape irrigation is the second largest use of recycled water for parks, schools, roadway greenbelt areas, golf courses, and other landscaping sites.

CARBON ADSORPTION

Activated carbon can remove soluble organic chemicals at low concentrations from water (Section 11.36). Both granular activated carbon and powdered activated carbon are used in advanced wastewater treatment processes.

14.6 GRANULAR-CARBON COLUMNS

Organic chemicals not extracted by conventional biological treatment can be mostly removed by adsorption and biodegradation on granular activated carbon (GAC). The large surface area of activated carbon assimilates organics while microbial degradation reopens pores in the granules. Because of this biological contribution, toxic substances in the applied wastewater can reduce the removal capacity. Some readily biodegradable substances are difficult to adsorb on carbon, making it difficult to predict the quality of effluent for a given wastewater. The ability to remove soluble organics should be demonstrated by pilot-plant tests as well as experience from existing full-scale plants.

The configuration and operation of a carbon contactor, in addition to the nature of waste organics and character of the granular carbon, influence the effectiveness of adsorption. Contact times are generally less than 30 min, since longer periods do not substantially enhance removals but do produce a favorable environment for generating hydrogen sulfide. Anaerobiosis can be controlled by decreased contact time, more frequent backwashing, oxygen addition, or prechlorination.

The two alternative carbon-contacting systems are downward passage through the bed, either under pressure or gravity flow, and upflow through a packed or expanded column. GAC in a downflow unit adsorbs organics and filters out suspended materials. However, this dual-purpose approach can result in an unsatisfactory effluent owing to losses of efficiency in both filtration and adsorption. Downflow beds are periodically backwashed to remove accumulated solids, and therefore usually have an underdrain system and water piping similar to granular-media filters. Pretreatment by granular-media filtration is common to remove contaminants that interfere with carbon adsorption. For regeneration, the GAC is removed from the bed, processed, and returned.

Packed-bed upflow GAC columns are countercurrent with water entering the bottom of the column through a screen manifold and overflowing through outlet screens at the top. Fresh carbon slurry is added by gravity, and the spent packing is withdrawn from the bottom. With the column in service, replacement can be performed continuously at a slow rate, or intermittently by replacing 5%–10% of the column contents at a time. About 10% of the contactor volume is void space at the top of the tank. This permits upward flow to be increased for bed expansion and flushing particulate matter to waste to reduce head loss through the bed. The upflow-to-waste cycle can also be used to clean the medium of excess carbon fines, if necessary. Countercurrent operation allows carbon near the inlet to become fully saturated with impurities prior to withdrawal for regeneration.

Expanded-bed upflow contactors can be either pressurized or constructed with open tops. Tank volume includes a space of about 50% for bed enlargement during operation. Expanded-bed units have the advantage of being able to treat wastewaters relatively high in suspended solids without the excessive head losses that occur in a downflow unit.

Exhausted activated carbon must be regenerated and reused to make adsorption economically feasible. Restoration is accomplished by heating in a multiple-hearth furnace with a low-oxygen steam atmosphere. Adsorbed organics are volatilized and released in gaseous form at a temperature of about 1700°F. With proper control, granular carbon can be restored to near-virgin adsorptive capacity while limiting weight loss. Thermal regeneration involves drying, baking (pyrolysis of adsorbates), and activating by oxidation of the remaining residue. The time required is about 30 min—15 min for drying, 5 min for gasifying the volatiles, and 10 min for reactivation. Furnace controls include the temperature, rate of steam feed, and rotational speed of the rabble arms. The latter two determine the carbon depth on the hearths and residence time in the furnace. Temperature is the most critical factor because insufficient heat will not volatilize the organic matter, while too high a temperature burns the carbon.

14.7 ACTIVATED-SLUDGE TREATMENT WITH POWDERED ACTIVATED CARBON

The addition of powdered activated carbon (PAC) to the biological mixed liquor of an activated-sludge process can significantly improve removal of soluble organic chemicals. The mechanism for removal in the biological activated carbon (BAC) process is a combination of biodegradation and carbon adsorption [7]. This system, under the trade name PACT® (U.S. Filter/Zimpro), has been used primarily in treatment of industrial wastewaters to remove a wide range of volatile and synthetic chemicals. In water reclamation of municipal wastewater, the application is for removal of organic chemicals limited by drinking water standards and to enhance nitrification. (Refer to Section 14.22 for application of the PACT system at the Fred Hervey Water Reclamation Plant.)

The flow scheme is the same as a conventional activated-sludge system. The raw wastewater, usually after settling, flows through an aeration tank to a final clarifier. The overflow is the effluent and underflow the return sludge, a portion of which is wasted. Fresh PAC is added to the wastewater entering the aeration tank to make up for the losses in the waste sludge and effluent. To aid in agglomeration of the fine PAC, a polymer is often added to the mixed liquor entering the final clarifier. Depending on the chemical characteristics of the wastewater being processed, the operating sludge age (carbon-microorganism solids retention) ranges from 5 to 30 days, the fresh carbon dose ranges from 20 to 1000 mg/l, and the mixed-liquor concentration is 10,000–20,000 mg/l. Maintaining a large quantity of PAC adsorbent in the process improves removal of soluble organic chemicals, minimizes air stripping of volatile organic chemicals, and provides stability against shock and variable organic loads. Nitrification may be enhanced, apparently due to adsorption of compounds inhibitory to nitrifying bacteria.

Wet air regeneration of PAC consists of the following process sequence. After thickening of the carbon-microorganism sludge to 6%–10% solids, it is pumped with the addition of pressurized air at a minimum of 700 psi through heat exchangers to raise the temperature to greater than about 430°F. Under this high pressure and temperature, the sludge enters a reactor where wet-air oxidation reduces organic chemicals and suspended solids to carbon dioxide, water, some low-molecular-weight organics, and ash. A small percentage (about 5%) of the PAC is lost by oxidation. After pressure reduction and cooling, a portion of the slurry is wasted to control buildup of ash in the BAC process. The slurry may also be processed in an effort to reduce the ash content of the regenerate product.

PHOSPHORUS REMOVAL

Most phosphorus entering surface waters is from human-generated wastes and land runoff. Contributions from nonpoint sources in surface drainage vary from 0 to 15 lb of phosphorus/acre/year, depending on land use, agricultural practice, fertilizer additions, topography, soil conservation practices, and other factors. Domestic waste contains approximately 2 lb (0.9 kg) of phosphorus/capita/year.

The most common forms of phosphorus are organic phosphorus, orthophosphates ($H_2PO_4^-$, HPO_4^{2-}, PO_4^{3-}), and polyphosphates. Typical polyphosphates are sodium hexametaphosphate, $Na_3(PO_3)_6$; sodium tripolyphosphate, $Na_5P_3O_{10}$; and tetrasodium pyrophosphate, $Na_4P_2O_7$. All polyphosphates gradually hydrolyze in aqueous solution and revert to the ortho form. Domestic wastewater contains approximately 7 mg/l of total phosphorus, of which about 80% is soluble.

Reactions involving phosphorus are

$$PO_4 + NH_3 + CO_2 \xrightarrow{\text{sunlight}} \text{green plants} \tag{14.1}$$

$$\text{organic P} \xrightarrow[\text{decomposition}]{\text{bacterial}} PO_4 \tag{14.2}$$

$$\text{polyphosphates} \xrightarrow[\text{in water}]{\text{hydrolysis}} PO_4 \tag{14.3}$$

$$PO_4 + \text{multivalent metal ions} \xrightarrow[\text{coagulant}]{\text{excess}} \text{insoluble precipitates} \tag{14.4}$$

The reaction in Eq. (14.1) is photosynthesis. Equations (14.2) and (14.3) are decomposition and hydrolysis reactions in which complex phosphates are converted to the stable orthophosphate form. Equation (14.4) is the precipitation of orthophosphate by chemical coagulation. Substantial excess concentrations of hydrolyzing coagulants (aluminum or iron) or lime are required for effective phosphate precipitation.

14.8 BIOLOGICAL PHOSPHORUS REMOVAL

Primary sedimentation in conventional treatment settles only a small percentage of the phosphorus in wastewater, since the majority is in solution. Secondary biological processing involves removal of soluble phosphate taken up by the microbial floc. The amount synthesized into growth is related primarily to the concentration of phosphates in the wastewater relative to the BOD content. Treating waste with a high BOD/P ratio eliminates a large percentage of the phosphorus, whereas processing with phosphorus in excess of biological needs results in lower removal efficiency. Domestic wastewater has a surplus of phosphorus relative to the quantities of nitrogen and carbon necessary for synthesis. In general, the amount of P embodied in the biological floc of a conventional activated-sludge process is equal to about 1% of the BOD applied. Anticipated removal in treatment of a typical wastewater with 200 mg/l of BOD is 2 mg/l of P, a reduction from 7 mg/l to 5 mg/l or 30% phosphorus reduction.

The method of processing and disposal of sludge withdrawn from primary and secondary settling tanks is an important consideration in nutrient removal. The only phosphorus considered extracted, which does not end up in surface waters, is the amount in solids disposed of or hauled away from the treatment plant site. An extended aeration system operating without sludge wasting extracts no phosphorus. Dewatering of raw-waste sludge followed by land burial of solids results in maximum phosphorus removal. Conventional sludge stabilization by anaerobic or aerobic digestion returns phosphorus to the influent of the treatment plant a supernatant liquid containing nutrients.

Remedial action for phosphorus pollution is the treatment of wastewaters that discharge directly into lakes, and into rivers and streams that flow into lakes. States have adopted standards to control phosphorus discharges. Effluent limits range from 0.1 to 2.0 mg/l as P, with many standards established at 1.0 mg/l. To protect the lakes and surface waters of the Great Lakes and Chesapeake Bay drainage basins, phosphorus removal has been implemented at many wastewater treatment plants [8]. Of 526 plants in the Chesapeake Bay drainage basin (Maryland, Pennsylvania, and Virginia), 99 are removing phosphorus.

14.9 BIOLOGICAL–CHEMICAL PHOSPHORUS REMOVAL

Chemical precipitation, using aluminum and iron coagulants or lime, is effective in phosphate removal. Three alternatives employed with conventional biological treatment are illustrated in Figure 14.5. The process most frequently employed is that in Figure 14.5(a), chemical precipitation with biological treatment.

Chemical Precipitation and Biological Treatment

Coagulation with biological aeration is used both to upgrade existing plants and in new design. For this process [Fig. 14.5(a)], chemicals are added to the activated-sludge tank or to the effluent of the aeration basin before final settling. In a trickling-filter or rotating biological contactor system, the coagulant is usually applied only to the effluent of the biological process. Proper mixing at the point of addition and a few minutes of flocculation prior to clarification are essential for maximum effectiveness. The location for best floc formation and subsequent settling is determined experimentally in the field by varying the position of chemical application and monitoring settleability of the solids.

Both alum and ferric chloride are used in combined chemical–biological flocculation, with lime and polymers occasionally applied as coagulation aids. The theoretical chemical reaction between alum and phosphate is

$$Al_2(SO_4)_3 \cdot 14.3\,H_2O + 2\,PO_4^{3-} = 2\,AlPO_4\downarrow + 3\,SO_4^{2-} + 14.3\,H_2O \quad (14.5)$$

The molar ratio of aluminum to phosphorus in this equation is 1 : 1, which is equivalent to a weight ratio of 0.87 : 1.00. Since alum contains 9.0% Al, 9.7 lb of coagulant is theoretically required to precipitate 1.0 lb of P. The actual coagulation reaction in wastewater is only partially understood and more complex than Eq. (14.5) because of secondary reactions with colloidal solids and alkalinity [Eq. (11.34)].

Alum demand is also a function of the degree of phosphorus removal, as diagrammed in Figure 14.6 [9]. The data are from operation of a bench-scale activated-sludge unit consisting of an aeration chamber and separate settling tank with a sludge return line. The wastewaters applied were settled domestic and a glucose–glutamic acid medium, both with a BOD of 150–190 mg/l and average phosphorus level of 10 mg/l. Without alum added to the aeration basin, phosphorus removal averaged 23% (Fig. 14.6). Chemical–biological processing improved phosphate extraction with increased Al to P dosages (aluminum applied to total influent phosphorus), indicating a drop in coagulation efficiency as the amount of phosphorus remaining decreases. The aluminum-to-phosphorus ratios for 90% P removal from the municipal and synthetic wastes were 1.5:1 and 2.0:1 Al to P, respectively, equivalent to 170 and 220 mg/l of alum. In full-scale activated-sludge and trickling-filter plants, alum applications vary from 50 to 200 mg/l for 80%–95% phosphorus removal [8]. Diffused aeration

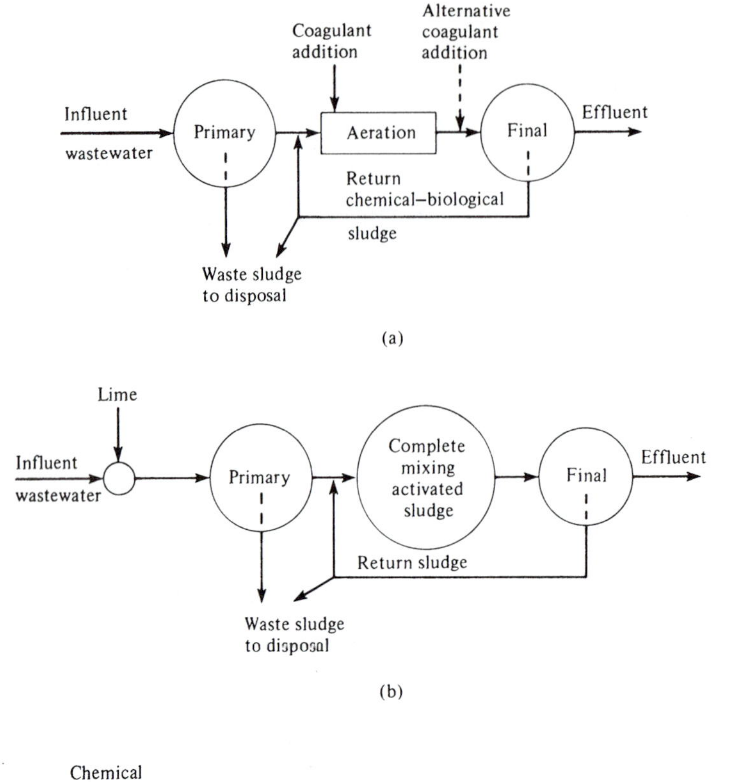

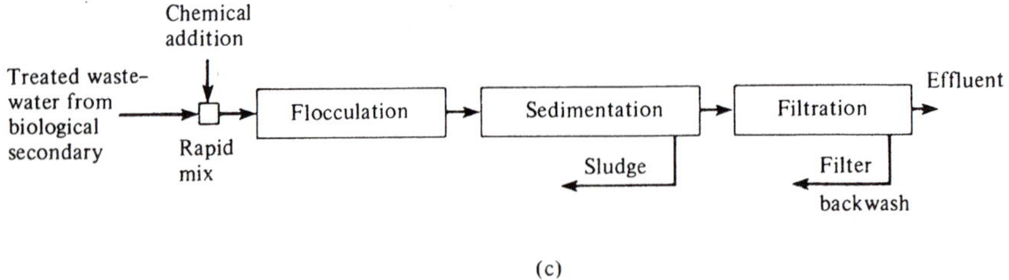

FIGURE 14.5 Phosphorus removal schemes. (a) Chemical precipitation with biological treatment. (b) Lime precipitation in primary sedimentation followed by secondary completely mixed activated sludge. (c) Tertiary treatment by chemical precipitation.

provides more efficient flocculation and, for the same degree of phosphorus removal, uses less alum than trickling-filter installations.

A large addition of aluminum coagulant has a marked influenced on the biota of activated sludge. Anderson and Hammer [9] reported that free-swimming and stalked protozoans are adversely affected to the extent that higher life forms are virtually absent with alum dosages in excess of 150 mg/l. Under these conditions it appears that chemical flocculation replaces the role of protozoans in clarifying settled effluent. BOD and suspended-solids removals are enhanced by coagulant addition to aeration tanks and trickling filters and generally result in effluent concentrations in the range of 10 to 20 mg/l.

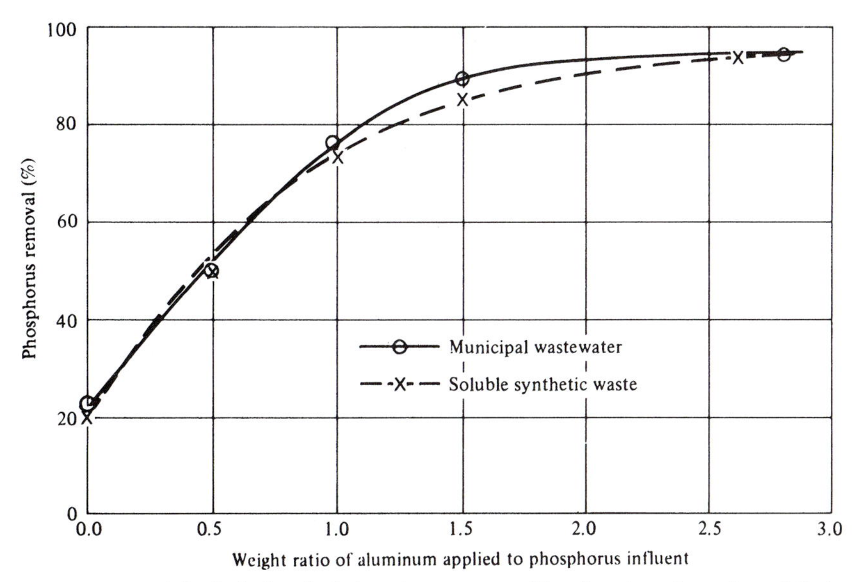

FIGURE 14.6 Biological–chemical phosphorus removal in a bench-scale, activated-sludge unit with alum added to the aeration chamber. The substrates applied were settled municipal wastewater and a synthetic medium of glucose and glutamic acid, with a phosphorus content of approximately 10 mg/l. *Source:* D. T. Anderson and M. J. Hammer, "Effects of Alum Addition on Activated Sludge Biota," *Water and Sewage Works* 120, no. 1 (January 1973).

Ferric chloride can also be applied with biological aeration to complex the phosphate ion. The hypothetical chemical reaction is

$$FeCl_3 + PO_4^{3-} = FePO_4\downarrow + 3\ Cl^- \tag{14.6}$$

The molar and weight ratios of Fe to P are 1:1 and 1.8:1, respectively. Theoretically, 5.2 lb of $FeCl_3$ is needed to precipitate 1 lb of P since ferric chloride is 34% iron. The actual dosage is usually greater than this equation predicts for 85%–95% phosphate removal, and lime is commonly applied to maintain optimum pH and aid coagulation. The ferric iron reacts with both natural alkalinity and lime to precipitate as ferric hydroxide [Eqs. (11.40) and (11.41)].

The least expensive and most common source of iron coagulants is waste pickle liquor from the steel industry [10]. Pickle liquor is variable in composition, depending on the metal treatment process. Ferrous sulfate resulting from pickling with sulfuric acid and ferrous chloride from pickling with hydrochloric acid are the two common waste liquors from metal finishing. Waste pickle liquors have an iron content of 5%–10% and free acid in the range of 0.5%–15%. The addition of lime or sodium hydroxide is necessary for neutralization and pH adjustment.

Aluminum and iron coagulants can be mixed with raw wastewaters to precipitate phosphates in primary clarification. This point of chemical addition, however, is less common than secondary chemical–biological processing. The advantage of first-stage chemical settling is increased suspended-solids removal, thus reducing the organic load to the biological secondary. The major disadvantage is higher coagulant usage for the same degree of phosphorus removal. In many plants, chemical feeders permit coagulation at either location to allow more flexibility in operation.

Characteristics of waste chemical–biological sludge are influenced by the coagulant dosage, the nature of the wastewater solids, the system design, and operating conditions. The quantity of chemical precipitate produced can be estimated using the theoretical equations for a given dosage, and the organic residue calculated as described in Section 13.2. Addition of metal salts improves settleability of microbial floc resulting in a denser waste slurry. Chemical residues do not hamper sludge thickening and stabilization by any conventional methods; in fact, mechanical dewatering is improved. Studies indicate that coagulation precipitates do not release phosphate back into solution during biological digestion [11]. Of course, the greater inert content of inorganic–organic sludge will influence ultimate disposal by incineration.

Lime Precipitation of Raw Wastewater

Lime applied prior to primary clarification [Fig. 14.5(b)] precipitates phosphates and hardness cations along with organic matter. Reaction with alkalinity [Eq. (14.7)] consumes most of the lime and produces calcium carbonate residue that aids in settled suspended solids. Calcium ion also combines with orthophosphate in an alkaline solution to form gelatinous calcium hydroxyapatite [Eq. (14.8)]. Treating domestic wastewater requires a dosage of 100–200 mg/l as calcium hydroxide to remove 80% of the phosphate. The actual amount applied depends primarily on phosphorus concentration and hardness of the wastewater.

$$Ca(HCO_3)_2 + Ca(OH)_2 = 2\,CaCO_3\downarrow + 2\,H_2O \tag{14.7}$$

$$5\,Ca^{2+} + 4\,OH^- + 3\,HPO_4^{\,2-} = Ca_5(OH)(PO_4)_3\downarrow + 3\,H_2O \tag{14.8}$$

The system sequence of lime precipitation followed by activated-sludge treatment, rather than reversing their order, uses less lime when an effluent of low phosphorus concentration is desired. The principal reason is that a biological system can readily extract low concentrations of phosphorus that would require excessive lime addition to precipitate. Secondary treatment by complete mixing of activated sludge is not adversely affected by chemical pretreatment. With proper control exercised, microbial production of carbon dioxide in the activated-sludge unit is sufficient to maintain a pH near neutral in the aeration compartment. A chemical–biological system of phosphate extraction can remove 90%–95% of the total phosphorus from a domestic wastewater containing lime dosages generally less than 150 mg/l as calcium hydroxide.

Use of excess lime in chemical treatment has two potential problems: scale formation on tanks, pipes, and other equipment, and disposal of the large quantity of lime sludge produced. Only operation of a full-scale installation will reveal the significance of these possible problems. The quantity of sludge produced is about 1.5–2 times that obtained by conventional treatment.

Tertiary Treatment by Chemical Precipitation

This process, diagrammed in Figure 14.5(c), is biological secondary treatment followed by chemical treatment with a flow diagram similar to that used in processing surface water supplies. The mixing and sedimentation system can consist of either separate rapid mix, flocculation, and sedimentation units in series or a flocculator–clarifier with these three operations in a single-compartmented tank. Filters are usually multimedia beds operated by pressure or gravity flow. Possible chemical additives are lime, alum, ferric chloride, and ferric sulfate, with polymers as flocculation aids. A major design consideration of tertiary treatment is processing and disposal of settled sludge and filter backwash water.

14.10 TRACING PHOSPHORUS THROUGH TREATMENT PROCESSES

Tracing phosphorus through a hypothetical treatment plant assists in understanding the mechanisms of phosphate removal. In order to do this, the phosphorus content of primary and biological sludges must be estimated. The phosphorus in dry primary-sludge solids from settling domestic wastewater is in the range of 0.4%–1.3% with a typical value of 0.9%, which is used in subsequent calculations. The phosphorus content of waste-activated sludge is more variable; nevertheless, recorded values are often in the 1.5%–2.5% range. In the following calculations, biological sludge solids are assumed to contain 2.0% phosphorus on a dry-weight basis.

Biological Phosphorus Removal

Consider a conventional activated-sludge plant with primary clarification. Assume an influent with the characteristics of an average domestic wastewater as listed in Table 12.1. A flow diagram for the plant is shown in Figure 14.7 with the influent concentrations of BOD, suspended solids (SS), insoluble organic phosphorus (oP), soluble inorganic phosphorus (iP), and total phosphorus (tP). Sedimentation reduces the wastewater BOD to 130 mg/l (35% removal), SS to 120 mg/l (50% removal), and tP to 5.9 mg/l (16% removal). From the sedimentation of 1 liter of wastewater, the quantity of sludge solids is 120 mg of SS containing $0.009 \times 120 = 1.1$ mg of oP.

Next, consider biological phosphorus removal by the activated-sludge process. Based on the method presented in Section 13.2, the waste biological solids produced per liter of wastewater treated using $k = 0.5$ (Fig. 13.1) and an applied BOD of 130 mg/l are $0.5 \times 130 = 65$ mg of SS. Assuming a phosphorus content of 2.0%, the oP in the waste sludge is $0.02 \times 65 = 1.3$ mg.

The tP in the plant effluent equals the influent (tP $= 7.0$ mg/l) minus the oP removals in the primary and waste-activated sludges (1.1 and 1.3 mg/l, respectively) for a remainder of 4.6 mg/l. The oP concentration in the effluent is 2.0% of the 30 mg/l of SS for 0.6 mg/l, which leaves an iP of 4.0 mg/l. Overall phosphorus reduction, assuming none of the phosphorus in the sludge is recycled to the plant in return flows, is from 7.0 to 4.6 mg/l, for a removal efficiency of 34%.

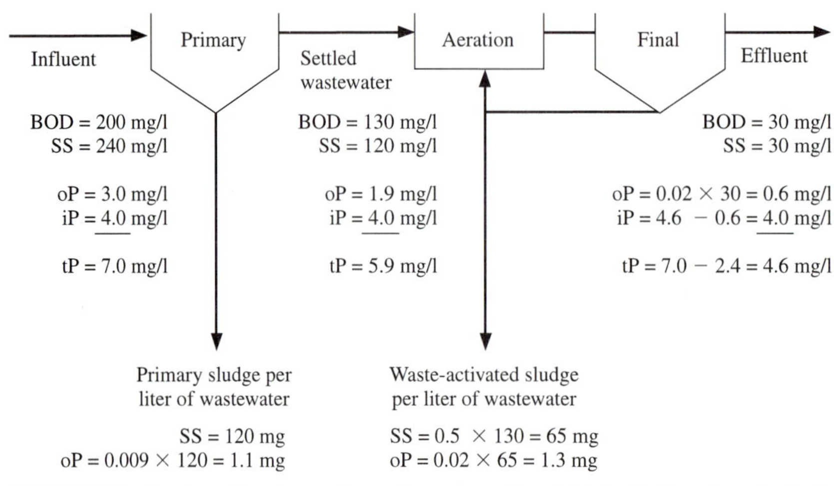

FIGURE 14.7 Tracing phosphorus through a conventional biological treatment plant.

Biological–Chemical Phosphorus Removal

Figure 14.8 is a flow diagram for a hypothetical biological–chemical treatment plant using alum addition in the activated-sludge process to precipitate the inorganic phosphate. (Phosphorus removal in primary sedimentation is the same as in Figure 14.7.) The required alum dosage was established by gradually increasing the amount applied while monitoring the total phosphorus in the plant effluent. For this hypothetical plant, the minimum alum dosage to reduce the effluent phosphorus concentration from 4.6 mg/l without coagulation to the maximum allowable tP of 1.0 mg/l is assumed to be 80 mg/l alum (molecular weight, 600). Addition of this coagulant also enhances the SS removal, reducing the effluent SS from 30 to 15 mg/l. The calculated oP in the effluent is therefore $0.02 \times 15 = 0.3$ mg/l, and the iP concentration is $1.0 - 0.3 = 0.7$ mg/l.

Now consider phosphorus removal by the biological–chemical process. The waste biological solids include 65 mg of SS resulting from the applied BOD (Fig. 14.7) plus the 15 mg/l from improved SS removal, for a total of 80 mg. The calculated oP in this organic sludge is $0.02 \times 80 = 1.6$ mg. The quantity of phosphorus precipitated by the alum as $AlPO_4$ [Eq. (14.5)] is the influent tP (7.0 mg/l) minus the effluent tP (1.0 mg/l) and the oP removal in the organic sludge solids ($1.1 + 1.6$ mg) for a remainder of 3.3 mg. The overall phosphorus removal efficiency is 86%. For an alum dosage of 80 mg/l and phosphorus feed concentration of 5.9 mg/l, the weight ratio of aluminum to phosphorus applied to the activated-sludge process is

$$\frac{\text{Al}}{\text{P}} = \frac{80(2 \times 27/600)}{5.9} = \frac{80 \times 0.09}{5.9} = \frac{1.2}{1.0}$$

Example 14.3 illustrates the procedure for calculating the theoretical sludge production for this biological–chemical treatment plant.

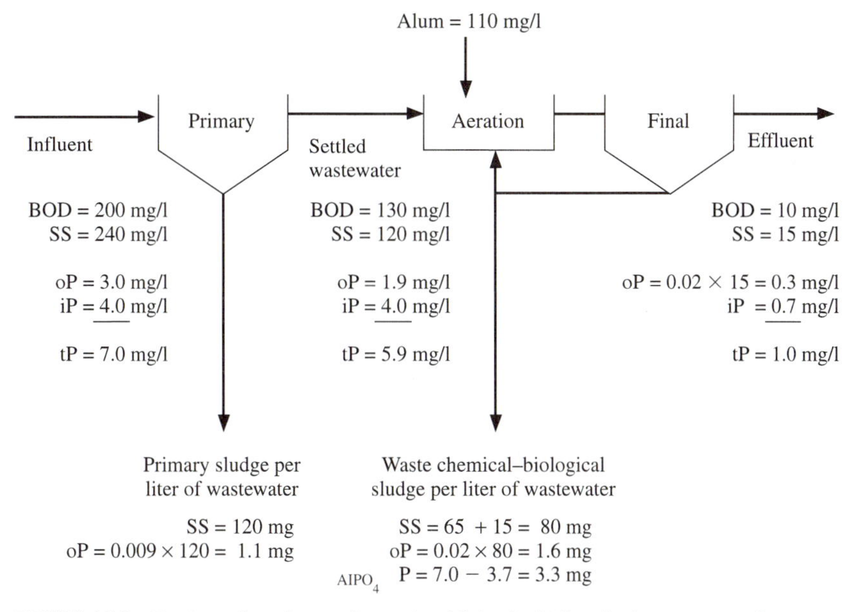

FIGURE 14.8 Tracing phosphorus through a biological–chemical treatment plant.

Example 14.3

Calculate the theoretical sludge solids production for the hypothetical biological–chemical plant diagrammed in Figure 14.8. Assume the influent wastewater characteristics listed in Figure 14.8, primary removals of 50% SS and 35% BOD, an operating F/M in the activated-sludge process of 0.40, an alum dosage of 80 mg/l, 15 mg/l of effluent SS, and an effluent phosphorus concentration of 1.0 mg/l.

Solution: The following calculations express sludge solids production in terms of milligrams per liter of wastewater treated:

$$\text{SS removal in primary sedimentation} = 0.50 \times 240 = 120 \text{ mg}$$

$$\text{SS removal in the activated-sludge process} = k \times \text{BOD} \quad [\text{Eq. (13.6)}]$$

From Figure 13.1, $k = 0.50$ for an F/M of 0.40, and therefore

$$\text{SS removal} = 0.50 \times 0.65(200) = 65 \text{ mg}$$

Since the k from Figure 13.1 assumes an effluent SS of 30 mg/l and the actual effluent concentration is 15 mg/l,

$$\text{total SS removal} = 65 + (30 - 15) = 80 \text{ mg}$$

From Eq. (14.5),

$$AlPO_4 \text{ precipitate} = \frac{(\text{P precipitated})(\text{MW of } AlPO_4)}{(\text{MW of P})}$$

Based on the 6.3 mg of phosphorus precipitate calculated in Figure 14.8,

$$AlPO_4 \text{ precipitate} = \frac{(3.3)(122)}{31} = 13 \text{ mg}$$

From Eq. (14.5),

$$\text{unused alum} = \text{alum dosage} - \frac{(\text{P precipitated})\ (\text{MW of alum})}{(2 \times \text{MW of P})}$$

$$= 80 - \frac{3.3 \times 600}{2 \times 31} = 48 \text{ mg/l}$$

The alum not used in phosphorus precipitation reacts with the natural alkalinity to precipitate as $Al(OH)_3$. From Eq. (11.34) in Section 11.9,

$$Al(OH)_3 \text{ precipitate} = \frac{(\text{unused alum})[2 \times \text{MW of } Al(OH)_3]}{(\text{MW of alum})}$$

$$= \frac{(48)(2 \times 78)}{600} = 12 \text{ mg}$$

The total sludge solids production equals the sum of the primary SS removal, activated-sludge SS removal, effluent SS reduction below 30 mg/l, $AlPO_4$ precipitate, and $Al(OH)_3$ precipitate. For this hypothetical plant, solids production per liter of wastewater treated equals 120 + 65 + 15 + 13 + 12 = 225 mg.

NITROGEN REMOVAL

Most nitrogen found in surface waters is derived from land drainage (3–24 lb of N/acre/yr) and dilution of wastewater effluents. Feces, urine, and food-processing discharges are the primary sources of nitrogen in domestic waste with a per capita contribution in the range of 8–12 lb of N/yr. About 60% is in the form of ammonia and 40% bound in organic matter. Conventional primary and secondary processing extracts approximately 25% of the total nitrogen, leaving most of the remainder as ammonia in the effluent.

The nitrogen forms of interest are organic, inorganic, and gaseous nitrogen. Bacterial decomposition releases ammonia by deamination of nitrogenous organic compounds [Eq. (14.9)], and continued aerobic oxidation results in nitrification [Eq. (14.10)]. Equation (14.11) is biochemical denitrification that occurs with heterotrophic metabolism in an anaerobic or anoxic environment. These three reactions in sequence define the biological nitrification–denitrification process. Water-soluble inorganic nitrogens (NH_3, NO_2^-, NO_3^-) serve as plant nutrients in photosynthesis [Eq. (14.12)].

$$\text{Organic N} \xrightarrow[\text{decomposition}]{\text{bacterial}} NH_4^+ \tag{14.9}$$

$$NH_4^+ + O_2 \xrightarrow[\text{bacteria}]{\text{nitrifying}} NO_2^- \tag{14.10a}$$

$$NO_2^- + O_2 \xrightarrow[\text{bacteria}]{\text{nitrifying}} NO_3^- \tag{14.10b}$$

$$NO_3^- \xrightarrow[\text{denitrification}]{\text{bacterial}} N_2\uparrow \tag{14.11}$$

$$\text{Inorganic N} + CO_2 \xrightarrow{\text{sunlight}} \text{green plants} \tag{14.12}$$

14.11 TRACING NITROGEN THROUGH TREATMENT PROCESSES

Tracing nitrogen through a hypothetical treatment plant assists in understanding the transformations of the various forms of nitrogen. For this purpose, nitrogen contents must be assumed for the raw wastewater and waste sludges. Wastewater characteristics are assumed to be those of an average domestic wastewater as listed in Table 12.1. The nitrogen content of dry sludge solids in primary sludge is in the range of 2%–4%, that in waste-activated sludge is 2%–6%, and that in anaerobically digested solids is 2%–6%. For the subsequent calculations, the values are assumed to be 4.0% for primary solids, 6.0% for activated sludge, and 4.0% for digested solids.

Consider a conventional activated-sludge plant with primary clarification. Figure 14.9 is a flow diagram listing the concentrations of BOD, suspended solids (SS), insoluble organic nitrogen (org—N), soluble ammonia nitrogen (NH_3—N), soluble nitrate nitrogen (NO_3—N), and total nitrogen (tN). Sedimentation reduces the wastewater BOD to 130 mg/l (35% removal), SS to 120 mg/l (50% removal), and tN to 30 mg/l (14% removal). From the sedimentation of 1 liter of wastewater, the quantity of sludge solids is 120 mg of SS containing $0.04 \times 120 = 5$ mg/l of org—N.

Next consider biological nitrogen removal by the activated-sludge secondary. In the method presented in Section 13.2, the waste biological solids produced per liter of wastewater treated, assuming $k = 0.5$ (Fig. 13.1) and an applied BOD of 130 mg/l, is $0.5 \times 130 = 65$ mg/l of SS. The nitrogen content at 6.0% is org—N $= 0.06 \times 65 = 4$ mg/l. The org—N in the effluent is 6.0% of the SS, which is $0.06 \times 30 = 2$ mg/l. During biological metabolism in the activated-sludge process, nitrogen in the waste organic matter is released to solution in the form of ammonia (deamination). Therefore, the remainder of the 26 mg/l of tN in the effluent is 24 mg/l of NH_3—N.

Aeration in the activated-sludge process can induce nitrification, converting a portion of the NH_3—N to NO_3—N. The occurrence of nitrification, and the degree to which it proceeds, depends on the environmental and operating conditions, including temperature, dissolved-oxygen concentration, and sludge age. In Figure 14.9, the diagram with substantial nitrification assumes that the effluent inorganic nitrogen concentrations are 5 mg/l of NH_3—N and 15 mg/l of NO_3—N. When the wastewater in a final clarifier becomes anaerobic, microorganisms can use the oxygen in the nitrate

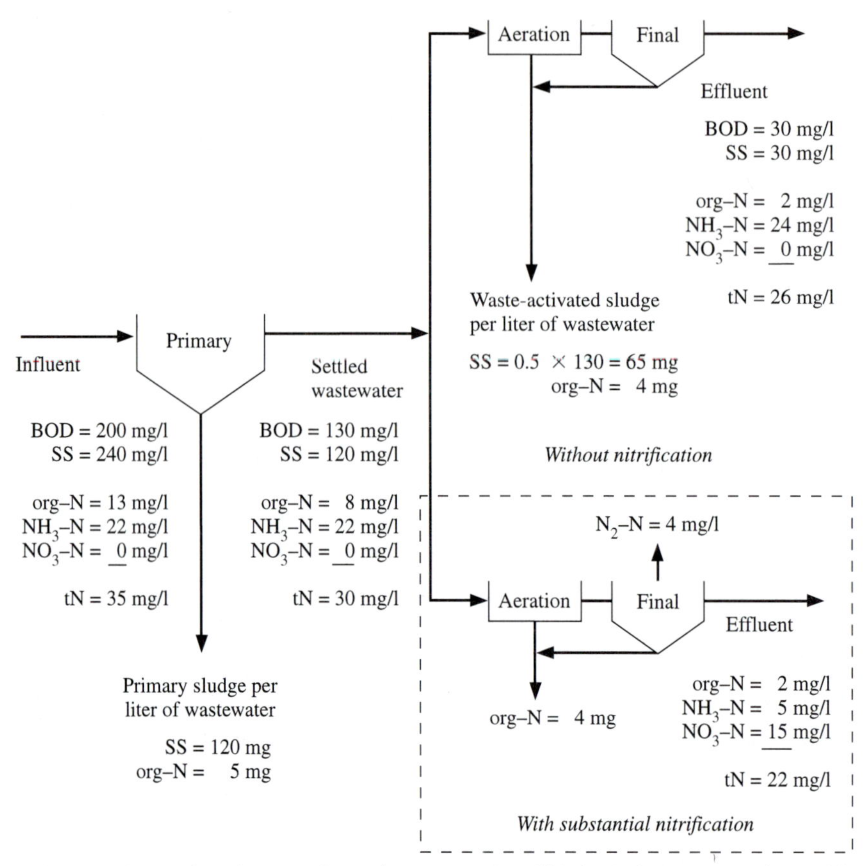

FIGURE 14.9 Tracing nitrogen through a conventional biological treatment plant without nitrification and with substantial nitrification in activated-sludge aeration.

for respiration, releasing nitrogen gas. In this example, denitrification is assumed to release 5 mg/l of N_2—N, reducing the effluent tN to 22 mg/l.

Figure 14.9 is based on the assumption that none of the nitrogen withdrawn in the waste sludges is recycled to the plant in return flows. While this is possible by mechanical dewatering of the raw sludges, biological stabilization by either anaerobic or aerobic digestion releases nitrogen to supernatant returned to the plant influent. Anaerobic digestion is likely to return 40%–50% of the org—N in the sludge solids as ammonia, 4 mg/l of the 9 mg/l removed in the wastewater processing. Depending on operation, aerobic digestion can result in nitrogen loss through nitrification–denitrification.

Nitrogen removal is affected by several factors: The forms and concentrations in the raw wastewater, synthesis in aerobic treatment, nitrification–denitrification, and methods of sludge processing. From the scheme in Figure 14.9, without nitrification and assuming a return of 4 mg from anaerobic digestion, the effluent tN would be 30 mg/l for a removal of $5/35 = 14\%$. With nitrification and no nitrogen return from sludge processing, the removal would be $(35 - 22)/35 = 37\%$.

14.12 BIOLOGICAL NITRIFICATION

Nitrification does not remove ammonia but converts it to the nitrate form, thereby eliminating problems of toxicity to fish and reducing the nitrogen oxygen demand (NOD) of the effluent. Ammonia oxidation to nitrate is a diphasic process performed by autotrophic bacteria with nitrite as an intermediate product [Eq. (14.10)]. These aerobic reactions yield energy for metabolic functions such as synthesis of carbon dioxide into new cell growth. Conversion of ammonia to nitrite is the rate-limiting step that controls the overall reaction; therefore, nitrite concentrations normally do not build up to significant levels. The rate of nitrification in wastewater, being essentially linear, is a function of time and is independent of ammonia nitrogen concentration (zero-order kinetics):

$$\begin{aligned} NH_4^- + 1.5\,O_2 &\xrightarrow{Nitrosomonas} NO_2^- + 2\,H^+ + H_2O + \text{energy} \\ NO_2^- + 0.5\,O_2 &\xrightarrow{Nitrobacter} NO_3^- + \text{energy} \end{aligned} \qquad (14.13)$$

Temperature, pH, and dissolved-oxygen concentration are important parameters in nitrification kinetics [12]. The nitrification rate decreases by about one-half for every 10°–12°C temperature drop above 10°C and then decreases more rapidly in cold wastewater, such that lowering the temperature from 10° to 5°C halves the rate of ammonia oxidation. Equation (14.14) is a suggested expression of the effect of temperature on the maximum growth rate of *Nitrosomonas* over a range of 5°–30°C (41°–86°F) [12].

$$\hat{\mu}_N = 0.47e^{0.098(T-15)} \qquad (14.14)$$

where

$\hat{\mu}_N$ = maximum specific growth rate of *Nitrosomonas*, d^{-1}

e = base of Napierian logarithms, 2.718

T = temperature, °C

Based on temperature alone, a winter aeration period would have to be several times longer than in the summer; however, this seasonal effect can be largely overcome by increasing mixed-liquor suspended solids (MLSS) and adjusting pH to a more favorable level.

The optimum pH for nitrification is approximately 8.0, with 90% of the maximum occurring at 7.5 and 8.5, and less than 50% of the optimum below 6.4 and above 9.6. The percent of maximum rate of nitrification versus pH of the mixed liquor in the aeration tank is graphed in Figure 14.10 with data from a variety of operating systems plotted as circles [12].

No detectable inhibition of nitrification occurs at dissolved-oxygen levels above 1.0 mg/l. Nevertheless, a minimum dissolved-oxygen level of 2.0 mg/l is recommended in practice to prevent reduced nitrification during the passage of peak ammonia concentrations through the aeration tank.

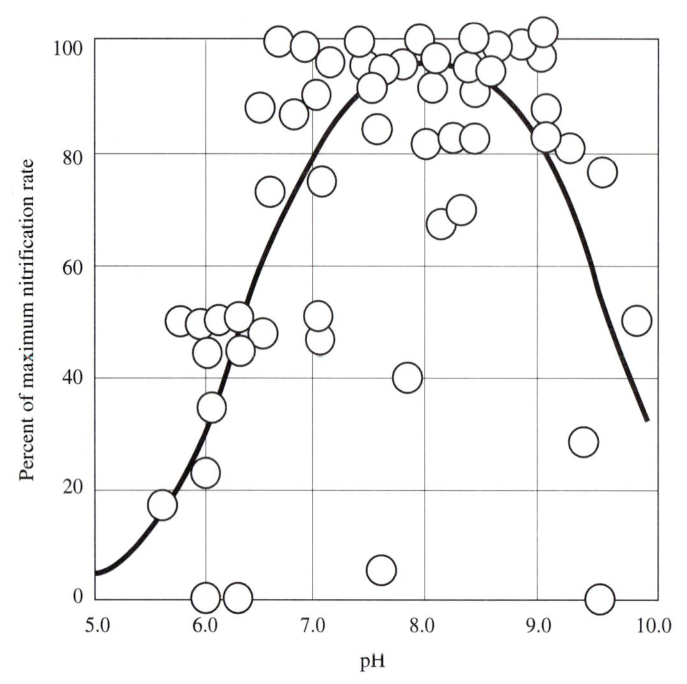

FIGURE 14.10 Rate of nitrification relative to maximum rate versus pH of the mixed liquor. *Source: Manual for Nitrogen Control,* Environmental Protection Agency, EPA/625/R-93/010 (September 1993): 93.

Sludge age and temperature are interrelated factors in establishing and maintaining healthy nitrifier populations essential to efficient ammonia oxidation. In continuous-flow aeration systems, a long sludge age is required to prevent excessive loss of viable bacteria (i.e., the growth rate must be rapid enough to replace microbes lost through sludge wasting and washout in the plant effluent). The supply of organic matter controls the growth of heterotrophic organisms, while the quantity of ammonia applied governs synthesis of nitrifiers. Increased sludge wasting, as a result of organic loading, reduces the sludge retention time and removes both heterotrophs and nitrifiers from the system. On a substrate of domestic waste having a BOD/total nitrogen ratio of approximately 200 mg/l to 35 mg/l, growth rates of nitrifying bacteria are substantially lower than those of decomposers. Therefore, in activated-sludge processes under normal operating conditions, nitrification is limited because of loss of the autotrophic populations. For performance reliability, current design for biological nitrification in cool wastewater requires two-step treatment. The first stage reduces BOD without oxidation of the ammonia–nitrogen to produce an effluent having a lower BOD/ammonia ratio—about 40 mg/l to 25 mg/l. Applying this flow to a second-stage nitrification unit provides an adequate growth potential for nitrifiers relative to heterotrophs, since the system can be operated at an increased sludge age.

The relative reproduction rates of heterotrophic and nitrifying bacteria are also influenced to a measurable extent by temperature. In southern climates, nitrification may be possible in a single-stage extended aeration unit treating domestic waste if pH and sludge wasting are carefully controlled to compensate for the relatively high BOD/ammonia N feed. However, winter wastewater temperature in northern states is often 10°–15°C, requiring a two-stage system with the best combination of aeration tank capacity, MLSS concentration, and pH control. Operation is possible at a less favorable pH level and lower mixed-liquor solids at a warm temperature provided the first stage is properly controlled. If nitrification is allowed to occur in carbonaceous aeration, the reduced ammonia supply to the secondary leads to starving the nitrifier populations. The sludge age, rate of recirculation, and air supply can be adjusted to minimize ammonia oxidation in the first stage; nevertheless, provision for chlorinating the effluent of the aeration tank prior to clarification is recommended in design. The addition of 2–8 mg/l of chlorine is effective in inhibiting nitrifying bacteria and can also help control sludge bulking caused by denitrification in the carbonaceous-phase clarifier.

Nitrification in Activated-Sludge Systems

The following mathematical equations apply to nitrification in completely mixed activated-sludge systems based on the relationship by Monod [Eq. (12.25), Section 12.6]. The *Nitrosomonas* growth is limited by the concentration of ammonium ion, while *Nitrobacter* growth is limited by the concentration of nitrite. Since nitrite does not accumulate, the rate-limiting step is modeled after *Nitrosomonas* as follows:

$$\mu_N = \hat{\mu}_N \frac{N}{K_N + N} \tag{14.15}$$

where

μ_N = specific growth rate of *Nitrosomonas*, d^{-1}

$\hat{\mu}_N$ = maximum specific growth rate of *Nitrosomonas* (at a growth-limiting concentration of NH_3—N at or above saturation), d^{-1}

N = concentration of growth-limiting NH_3—N, mg/l

K_N = saturation constant (equal to the limiting NH_3—N concentration at one-half the maximum growth rate), mg/l

Although reported K_N values vary, the low value of the saturation constant and reported range of values imply that selecting a value of 1.0 mg/l NH_3—N should be acceptable for design purposes [12]. This equation assumes that the specific growth rate is not limited by the dissolved-oxygen concentration or pH of the mixed liquor in the aeration tank. The specific growth rate μ_N must be adjusted for the pH of the mixed liquor based on Figure 14.10.

At steady state, the solids leaving the aeration system equal new solids production. Therefore, the growth rate and solids retention time of the nitrifiers are related by the following equation:

$$\theta_c = \frac{1}{\mu_N - b_N} = \frac{1}{\mu'_N} \tag{14.16}$$

where

θ_c = mean cell residence time, d

μ_N = specific growth rate of *Nitrosomonas*, d^{-1}

b_N = endogenous decay coefficient for *Nitrosomonas*, d^{-1}

μ'_N = net specific growth rate of *Nitrosomonas*, d^{-1}

With nitrifying bacteria, b_N is considered to be negligible (i.e., $b_N = 0$), thus the μ_N equals μ'_N. Mean cell residence time is the sludge age calculated using MLVSS in the aeration tank and VSS in the effluent and waste sludge [Eq. (12.60), Section 12.19].

The calculated mean cell residence time is a theoretical value based on steady-state conditions at minimum solids retention time with ideal conditions for growth of nitrifying bacteria. Peaking factors are applied to reflect anticipated field conditions. Compensating for peak flow conditions is important because of potential short-circuiting in the aeration tank and loss of suspended solids in the clarifier overflow. If the influent is not equalized, the calculated θ_c should be increased by a peaking factor of 1.2–1.6, depending on the peak hourly flow and aeration period. An extended aeration period of 12–24 hr can absorb a peak diurnal flow better than a conventional aeration period of 5–7 hr. In addition, if average annual wastewater flow is used to calculate θ_c, the value should be further increased to represent the average during the peak month (Section 12.10). The commonly assumed peaking factors for maximum-month design for nitrification are 1.5 for flow and 1.2 for total nitrogen during peak diurnal flow for a total multiplier of 1.8 [12].

Design engineers may apply a safety factor to θ_c after adjustment with peaking factors to compensate for uncertainty in performance, such as unknown variations in temperature, ammonia concentration, dissolved-oxygen concentration, pH, and inhibiting substances. The safety factor may be 1.5 or greater. The selection depends primarily on reliability of the data used in design from laboratory tests and plant operational records.

Example 14.4

An activated-sludge plant with primary clarification has two completely mixed aeration tanks with a total volume of 1.50 mil gal. The wastewater characteristics are similar to those listed in Figure 14.9 without nitrification. When partial nitrification does occur and results in floating solids on the final clarifiers from denitrification, the air supply is reduced to inhibit the nitrification. Because of potential toxicity to aquatic life in the receiving stream during low flow, the department of natural resources is initiating a study to determine the feasibility of nitrification in the activated-sludge process. Substantial nitrification as characterized in Figure 14.9 would be satisfactory since the remaining 5 mg/l of ammonia nitrogen would be sufficiently diluted in the mixing zone of the stream even at the one-in-ten-year low flow. Unfortunately, full-scale nitrification cannot be evaluated because the capacity of existing diffused aeration was designed for only BOD oxidation and the plant is already at 70% of the 5.0-mgd design capacity. Installation of additional aeration capacity is too costly to conduct full-scale nitrification tests. The plant design capacity is 5.0 mgd based on the average flow during the maximum month. By extrapolating existing performance from plant records, the data for nitrification calculations are given below for the critical summer month with maximum wastewater flow and the critical winter month with minimum temperature of the mixed liquor.

Parameter	Summer	Winter
Influent wastewater flow, mgd	5.0	4.2
Aeration period, hr	7.2	8.6
Influent wastewater characteristics as given in Fig. 14.9		
Effluent characteristics with substantial nitrification, Fig. 14.9		
Effluent volatile suspended solids, mg/l	20	20
Activated-sludge operation		
MLSS, mg/l	2000	2000
MLVSS, mg/l	1400	1400
Minimum dissolved oxygen, mg/l	1.0	1.0
Minimum pH	6.8	6.8
Average temperature, °C	17	10
Waste-activated sludge		
Quantity, mgd	0.026	0.031
Solids concentration, mg/l	8200	9500

Solution:

Calculations of Mean Cell Residence Times in Activated-Sludge Operations Using Eq. (12.60) Based on MLVSS and VSS

$$\text{Summer:} \quad \theta_c = \frac{1400 \times 1.50}{20 \times 5.0 + 8200 \times 0.026} = 6.7 \text{ d}$$

$$\text{Winter:} \quad \theta_c = \frac{1400 \times 1.50}{20 \times 4.2 + 9500 \times 0.031} = 5.6 \text{ d}$$

Calculations of Mean Cell Residence Times for Nitrification

Using Eq. (14.14),

$$\text{Summer:} \quad \hat{\mu}_N = 0.47e^{0.098(17-15)} = 0.572 \text{ d}^{-1}$$

$$\text{Winter:} \quad \hat{\mu}_N = 0.47e^{0.098(10-15)} = 0.288 \text{ d}^{-1}$$

Using Eq. (14.15),

$$\text{Summer:} \quad \mu_N = 0.572[5.0/(1.0 + 5.0)] = 0.477 \text{ d}^{-1}$$

$$\text{Winter:} \quad \mu_N = 0.288[5.0/(1.0 + 5.0)] = 0.240 \text{ d}^{-1}$$

Correcting for a pH of 6.8 using Figure 14.10,

$$\text{Summer:} \quad \mu_N = 0.477 \times 0.71 = 0.339 \text{ d}^{-1}$$

$$\text{Winter:} \quad \mu_N = 0.240 \times 0.71 = 0.170 \text{ d}^{-1}$$

Using Eq. (14.16),

$$\text{Summer:}\quad \theta_c = 1/0.339 = 2.95 \text{ d with steady-state flow}$$

$$\text{Winter:}\quad \theta_c = 1/0.170 = 5.88 \text{ d with steady-state flow}$$

The assumed peaking factor to compensate for diurnal flow is assumed to be 1.4; therefore,

$$\text{Summer:}\quad \theta_c = 1.4 \times 2.95 = 4.1 \text{ d with diurnal flow}$$

$$\text{Winter:}\quad \theta_c = 1.4 \times 5.88 = 8.2 \text{ d with diurnal flow}$$

Conclusion

Comparing θ_c values required for nitrification with θ_c based on operation, nitrification could be achieved in the summer (6.7 d $>$ 4.1 d) but not in the winter (5.6 d $<$ 8.2 d) because of the cooler mixed-liquor temperature. Although an increase in MLVSS to a maximum of 2100 mg/l would increase the mean cell residence time to 8.3 d, this is still barely sufficient and may result in increased suspended solids in the effluent above the 30 mg/l limit. Also, the calculated values for nitrification include a peaking factor to compensate for diurnal flow variation, but no safety factor was applied for uncertainty in performance at design flow. The recommendation, if nitrification is necessary, is to consider construction of a second-stage aeration system for nitrification.

Nitrification by Second-Stage Suspended-Growth Aeration

Theoretical mathematical equations for nitrification by second-stage suspended-growth systems have been developed [12]. Nevertheless, kinetics models for plug-flow reactors (rather than complete mixing) have not been substantiated sufficiently for reliable use in design. Engineers rely on empirical data from existing systems with similar environments to design new systems and records of performance to expand existing systems. The following discussion is based on empirical data collected from laboratory and field studies in the early development of design criteria for nitrification [13]. The optimum aeration tank is either a long, narrow tank with diffused aeration or a shorter tank divided into a series of at least three compartments for diffused or mechanical aeration equipment, with intervening ports (Fig. 14.11). This tank configuration simulates plug flow compatible with the zero-order kinetics of ammonia oxidation.

Biological nitrification destroys alkalinity, which can result in a drop of pH when processing wastewaters of moderate hardness, or where alum precipitation has been used for phosphate removal in the preceding activated-sludge phase. Theoretically, 7.2 lb of alkalinity is destroyed per pound of ammonia nitrogen oxidized to nitrate, as follows:

$$2\,NH_4HCO_3 + 4\,O_2 + Ca(HCO_3)_2 = Ca(NO_3)_2 + 4\,CO_2 + 6\,H_2O \qquad (14.17)$$

Whether pH should be controlled by chemical addition depends on the rate of nitrification desired, as limited by other environmental conditions. For example, in operation at low temperature, lime can be applied to maintain oxidation efficiency for the aeration tank capacity available. New plant design should provide for installation of chemical feeders and instrumentation for monitoring pH in the aeration basin.

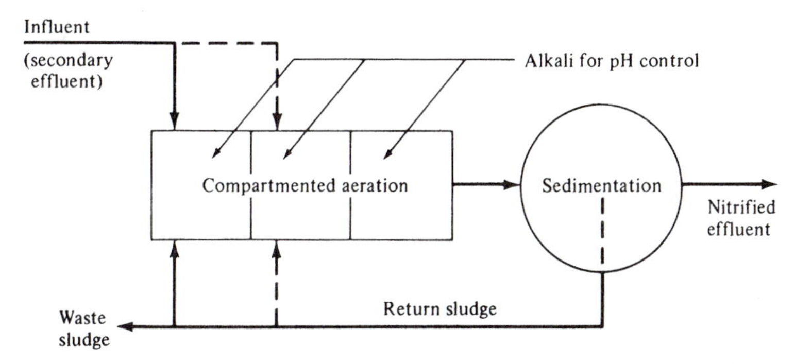

FIGURE 14.11 Flow diagram for nitrification by diffused or mechanical aeration of wastewater following conventional biological treatment.

The recommended design mixed-liquor concentration for a nitrification process receiving normal secondary effluent is in the range 1500–2000 mg/l of volatile suspended solids. Mean cell residence time must be longer than in a carbonaceous activated-sludge process, greater than 7 days.

Loading on nitrification basins is expressed in units of lb ammonia N/1000 ft^3/day of aeration tank volume. Figure 14.12 shows recommended loadings for various temperatures and volatile mixed-liquor concentrations (MLVSS) at optimum pH, based on studies at Marlboro, Massachusetts. Corrections for permissible loadings at pH values other than optimum can be taken from Figure 14.10. The value selected for design loading should include a factor to allow for reasonable peaking of the influent ammonia content. A commonly adopted peak loading is 1.2–1.6 times the average daily nitrogen load under low-temperature conditions.

Stoichiometrically, nitrification of 1.0 lb of ammonia N in the form of ammonium bicarbonate requires 4.6 lb of oxygen [Eq. (14.17)]; however, additional oxygen allowance must be made for carbonaceous BOD carried over to the nitrification stage. A dissolved oxygen concentration of 3.0 mg/l in the mixed liquor is suggested under average loading conditions with a lower concentration permitted during peak loads, but not below 1.0 mg/l.

Suggested design criteria for final clarifiers following nitrification are an overflow rate of 400–500 gpd/ft^2 (16–20 $m^3/m^2 \cdot d$) based on average daily discharge, with a maximum permissible value of 1000 gpd/ft^2 (41 $m^3/m^2 \cdot d$) at peak hourly flow, and a side-water depth of at least 10 ft (3.0 m). Because of the relatively slow settling velocities of nitrifying sludges, more than two clarifiers are desirable to ensure satisfactory operation when one tank is out of service for maintenance. Hydraulic-type collector arms are recommended for rapid sludge return because denitrification can occur in settled sludge, creating problems of floating solids. When present, the float should be collected by skimmers and returned to the aeration basins. A desirable capacity for sludge recirculation pumps is 100% of the raw wastewater influent, although normal operation will probably be at a rate of only 50% return. Accumulation of biological floc is limited by the low organic loading and slow growth of nitrifying bacteria. Consequently, the volume of excess sludge produced is small, with a reasonable estimate being less than 1% of the quantity of wastewater processed.

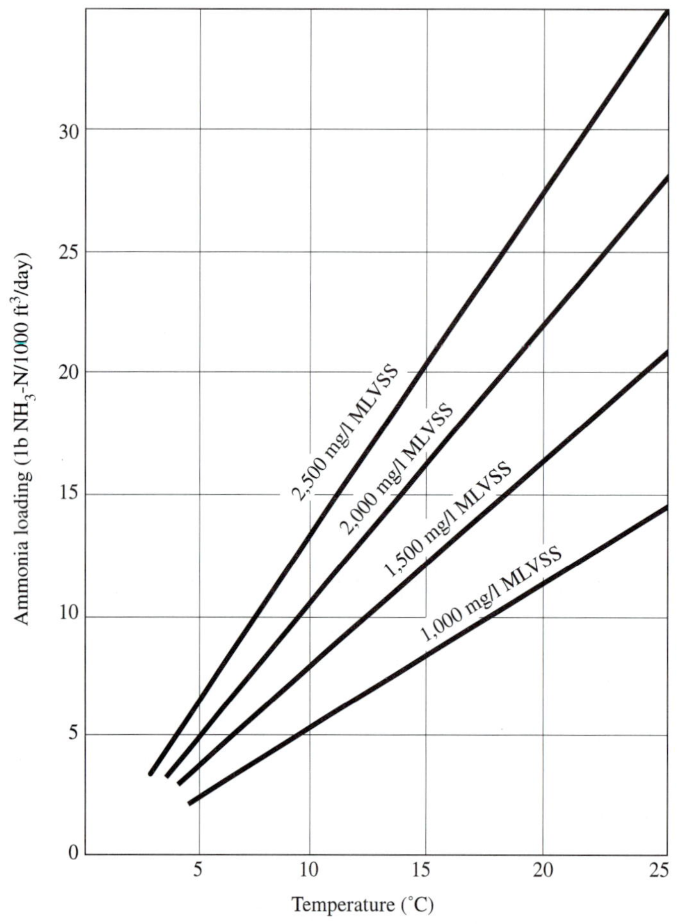

FIGURE 14.12 Permissible nitrification-tank loadings at an optimum pH of 8.4. *Source: Nitrification and Denitrification Facilities,* Environmental Protection Agency, Technology Transfer (August 1973): 22.

Example 14.5

Calculate the aeration basin volume for suspended-growth nitrification following conventional secondary treatment. The wastewater characteristics are the following:

$$\text{average daily design flow} = 10 \text{ mgd}$$

$$\text{ammonia nitrogen} = 20 \text{ mg/l}$$

$$\text{BOD} = 40 \text{ mg/l}$$

$$\text{minimum operating temperature} = 10°\text{C}$$

$$\text{operating pH} = 7.2$$

$$\text{design volatile MLSS} = 1500 \text{ mg/l}$$

Solution:

$$\text{average ammonia load} = 10 \times 20 \times 8.34 = 1670 \text{ lb/day of N}$$

$$\text{maximum ammonia load} = 1.5 \times 1670 = 2500 \text{ lb/day of N}$$

Permissible nitrification tank loading from Figure 14.12 for a temperature of 10°C and 1500 mg/l of MLVSS is 8.1 lb of NH_3—N/1000 ft^3/day. Correcting this to a pH of 7.2 using Figure 14.10, the allowable loading is $8.1 \times 0.88 = 7.1$ lb/1000 ft^3/day.

$$\text{aeration basin volume} = 2500\frac{1000}{7.1} = 350{,}000 \text{ ft}^3$$

$$\text{resulting aeration period} = \frac{350{,}000 \times 7.48 \times 24}{10{,}000{,}000} = 6.3 \text{ hr}$$

$$\text{BOD load on aeration basin} = 10 \times 40 \times 8.34 = 2300 \text{ lb/day}$$

The oxygen uptake using 4.6 lb of O_2/lb of NH_3—N and 1.0 lb of O_2/lb of BOD is $2500 \times 4.6 + 2300 \times 1.0 = 13{,}800$ lb/day.

14.13 BIOLOGICAL DENITRIFICATION

Nitrite and nitrate are bacterially reduced to gaseous nitrogen by a variety of facultative heterotrophs in an anoxic (lacking oxygen) environment. An organic carbon source, such as acetic acid, ethanol, methanol, or organic matter, is needed to act as a hydrogen donor (oxygen acceptor) and to supply carbon for biological synthesis. Certain autotrophic bacteria are also capable of denitrification by oxidizing an inorganic compound for energy and using carbon dioxide for synthesis. Although denitrification is considered an anoxic process because it occurs in the absence of dissolved oxygen, strict anaerobiosis characterized by hydrogen sulfide and methane production is not necessary.

Methanol is the preferred carbon source because it is the least expensive synthetic compound available that can be applied without leaving a residual BOD in the process effluent—but this does not imply that methanol treatment is cheap. Introduction of methanol first reduces the dissolved oxygen present by Eq. (14.18); then biological reduction of nitrate and nitrite occurs [Eqs. (14.19 and (14.20)].

$$3\,O_2 + 2\,CH_3OH = 2\,CO_2\uparrow + 4\,H_2O \tag{14.18}$$

$$6\,NO_3^- + 5\,CH_3OH = 3\,N_2\uparrow + 5\,CO_2\uparrow + 7\,H_2O + 6\,OH^- \tag{14.19}$$

$$2\,NO_2^- + CH_3OH = N_2\uparrow + CO_2\uparrow + H_2O + 2\,OH^- \tag{14.20}$$

From these reactions, the amount of methanol required as a hydrogen donor for complete denitrification is

$$CH_3OH = 0.7\,DO + 1.1\,NO_2\text{—N} + 2.0\,NO_3\text{—N} \tag{14.21}$$

where

$$CH_3OH = \text{methanol, mg/l}$$
$$DO = \text{dissolved oxygen, mg/l}$$
$$NO_2—N = \text{nitrite nitrogen, mg/l}$$
$$NO_3—N = \text{nitrate nitrogen, mg/l}$$

Approximately 30% excess methanol feed is needed for synthesis; hence chemical consumption to satisfy both energy and synthesis can be estimated from the relationship

$$CH_3OH = 0.9\,DO + 1.5\,NO_2—N + 2.5\,NO_3—N \tag{14.22}$$

The optimum pH falls in the same range as for most heterotrophic bacteria (between 6.5 and 7.5), with the rate decreasing to about 80% of maximum when the pH is lowered to 6.1 or raised to 7.9. Nitrified wastewaters, which tend to be basic, are naturally controlled from excessively high pH by carbon dioxide generated in a denitrification unit; thus there appears to be no need for addition of chemicals to control pH in actual systems. The effect of temperature on the rate of denitrification is sketched in Figure 14.13 [14].

Denitrification by Suspended-Growth Systems

The kinetics of biological denitrification using a chemical carbon source have been developed based on the Monod model. Presentation of the mathematical equations are in the *Manual for Nitrogen Control* [12]. As few installations perform denitrification using a chemical like methanol because of high cost, limited operational data are available to calibrate the models. In engineering practice, design of denitrification processes is founded on empiricism, and this discussion is based on studies conducted on biological denitrification processes [14].

The process studied most extensively consists of a completely mixed tank followed by a clarifier for sludge separation and return (Fig. 14.14). Although a single, mixed chamber is common in laboratory studies, plug flow minimizes short-circuiting and becomes more suitable for the relatively short detention period of 2–3 hr required. Underwater stirrers, comparable to those used in waterworks flocculation tanks, or impeller mixers stir the contents sufficiently to keep microbial floc in suspension without producing undue aeration. A power supply in the range of 1 hp for each 2000–4000 ft^3 of tank volume appears to be adequate. Whether tanks should be covered to minimize absorption of oxygen is a matter of conjecture, but certainly airtight covers should be avoided.

Denitrification reactions form carbon dioxide and nitrogen gas bubbles that inhibit gravity settling by adhering to the biological floc. Supersaturation of the mixed liquor with gases can be relieved by short-term aeration of 20–60 min in an open channel or tank between the denitrification basin and the final clarifier. The settleability of sludge solids following this air stripping appears to be similar to that of other biological sludges. Recommended clarifier depths and overflow rates are the same as those suggested for final settling tanks following the nitrification process. Basins should be equipped with

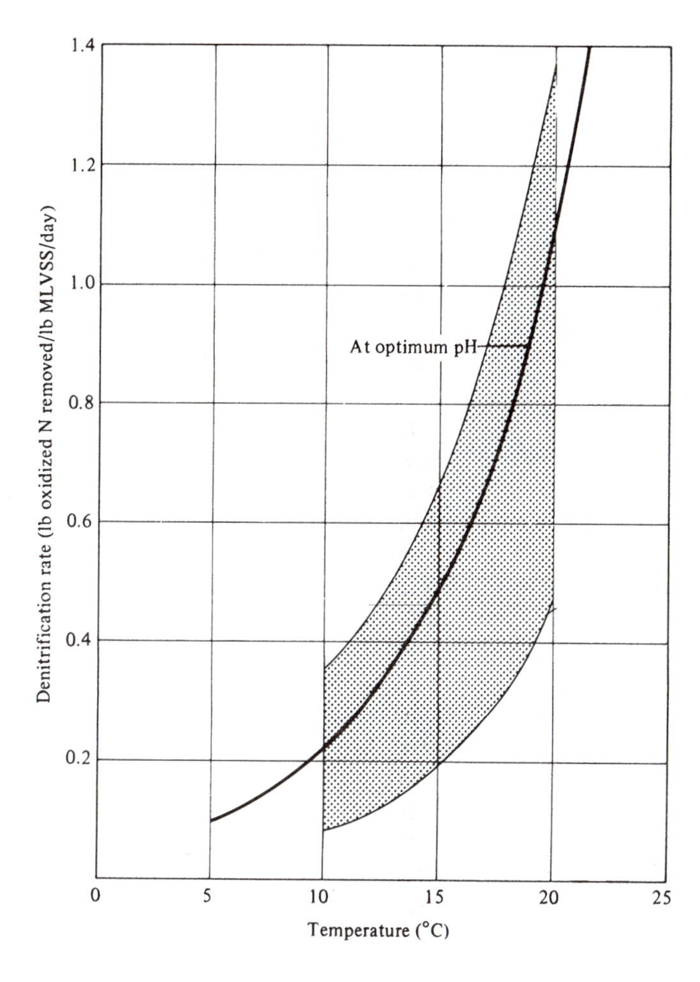

FIGURE 14.13 Effect of temperature on the rate of denitrification. *Source: Nitrification and Denitrification Facilities,* Environmental Protection Agency, Technology Transfer (August 1973): 28.

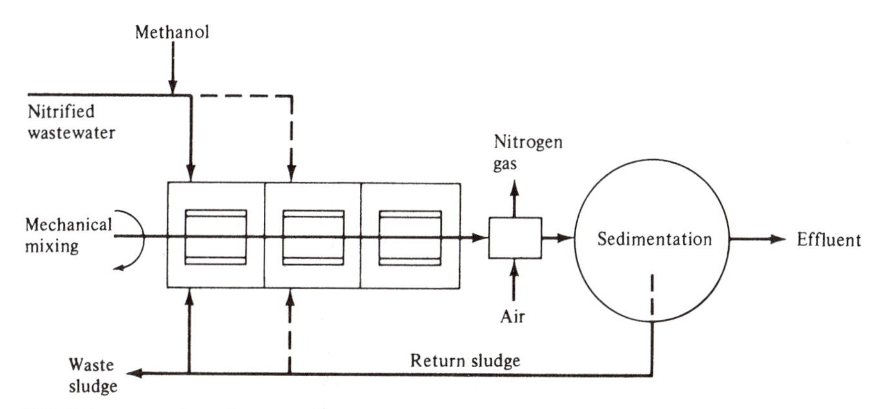

FIGURE 14.14 Flow diagram for a completely mixed, compartmented denitrification basin and clarifier.

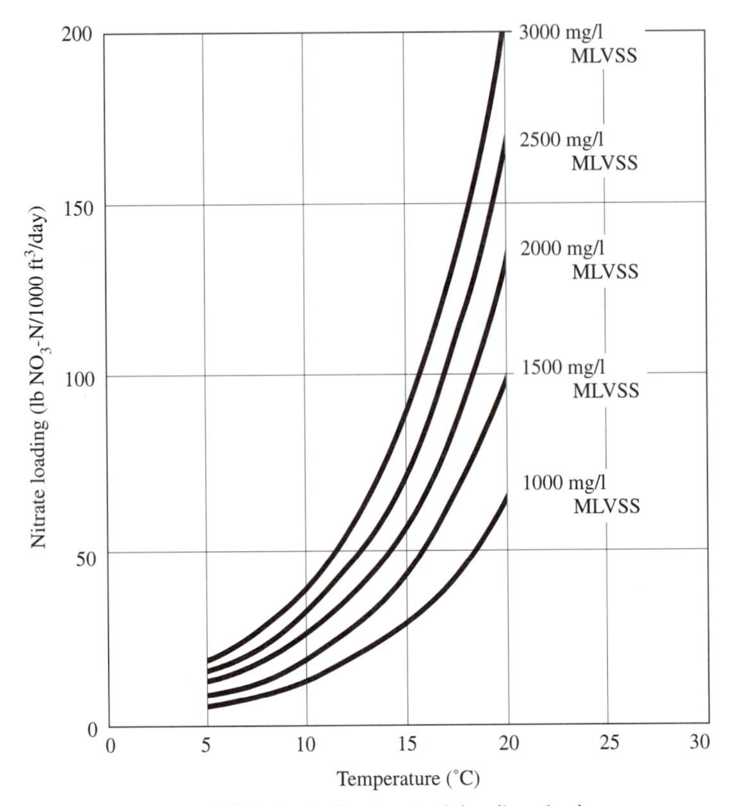

FIGURE 14.15 Permissible denitrification tank loadings in the pH range 6.5–7.5. *Source: Nitrification and Denitrification Facilities,* Environmental Protection Agency, Technology Transfer (August 1973): 30.

rapid-sludge-return collector arms and skimming devices. A sludge recirculation capacity equal to the average wastewater flow is recommended, and scum may be returned to the denitrification tank or routed to disposal. Withdrawal of excess microbial solids, to keep the biological system in balance, ranges from 0.2 to 0.3 lb/lb of methanol applied.

The volumetric capacity needed for denitrification can be estimated using Figure 14.15, which is based on pilot-plant studies. Denitrifying sludges after degasification have good settling characteristics, allowing design mixed-liquor solids of 2000–3000 mg/l that are approximately 65% volatile.

Nitrification–Denitrification

A two-stage system composed of the units illustrated in Figures 14.11 and 14.14, following secondary biological treatment, can achieve about 90% inorganic nitrogen reduction and 80%–85% total nitrogen removal under normal operating conditions. The biological cultures performing ammonia oxidation are more sensitive to heavy metals and toxic chemicals than conventional activated sludge. Therefore, industrial wastewaters discharged to municipal sewers must be carefully monitored and necessary controls established to ensure that the nitrifying microorganisms are not inhibited.

Example 14.6

Based on the following data, calculate the volume needed for suspended-growth denitrification.

$$\begin{aligned} \text{average daily design flow} &= 10 \text{ mgd} \\ \text{nitrate nitrogen} &= 20 \text{ mg/l of N} \\ \text{dissolved oxygen} &= 8 \text{ mg/l} \\ \text{minimum operating temperature} &= 8°\text{C} \\ \text{operating pH} &= 7.8 \\ \text{design volatile MLSS} &= 2000 \text{ mg/l} \end{aligned}$$

Solution:

$$\text{average nitrate load} = 10 \times 20 \times 8.34 = 1670 \text{ lb/day of N}$$

$$\text{maximum nitrate load} = 1.5 \times 1670 = 2500 \text{ lb/day of N}$$

Permissible denitrification tank loading from Figure 14.15 for a temperature of 8°C and 2000 mg/l is 20 lb of NO_3—N/1000 ft^3/day. Correcting this to a pH of 7.8, the allowable loading is $20 \times 0.9 = 18$ lb/1000 ft^3/day.

$$\text{denitrification basin volume} = 2500\frac{1000}{18} = 140{,}000 \text{ ft}^3$$

$$\text{resulting detention time} = \frac{140{,}000 \times 7.48 \times 24}{10{,}000{,}000} = 2.5 \text{ hr}$$

The average methanol dosage for 8 mg/l of DO and 20 mg/l of NO_3—N, from Eq. (14.22), is

$$CH_3OH = 0.9 \times 8 + 2.5 \times 20 = 57 \text{ mg/l}$$

14.14 SINGLE-SLUDGE BIOLOGICAL NITRIFICATION–DENITRIFICATION

Single-sludge systems for nitrogen removal combine BOD removal, ammonia oxidation, and nitrate reduction within the same activated-sludge process, followed by clarification [12]. Rather than supplying methanol as a carbon source, unoxidized organic matter in the wastewater is used as an oxygen acceptor for nitrate reduction. Although single-stage processes have varying flow patterns, all rely on aerobic zones for nitrification and anoxic zones (zones lacking in dissolved oxygen) for denitrification.

The single anoxic zone shown in Figure 14.16 blends recirculation with raw wastewater for denitrification in the anoxic zone and nitrification in the aerobic zone.

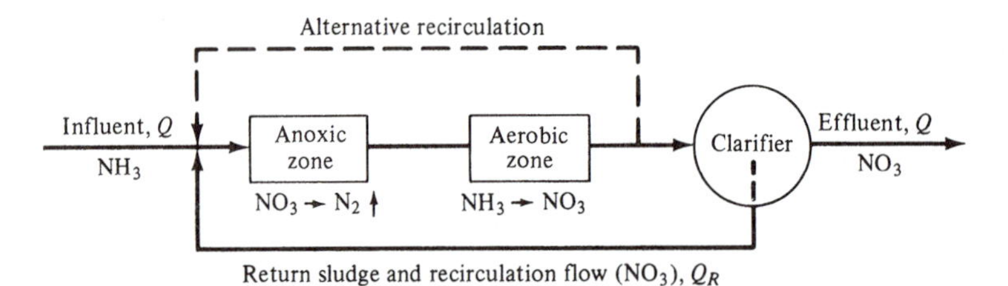

FIGURE 14.16 Biological nitrification–denitrification by recirculation of return sludge for blending with raw wastewater before flow through an anoxic zone and subsequent aerobic zone.

Both zones reduce the BOD content of the wastewater. The anoxic zone provides the oxygen in nitrate for biological respiration, releasing nitrogen gas. In the aerobic zone, the BOD is reduced by microbial uptake of dissolved oxygen, and the majority of the nitrogen is converted to nitrate. Since only that nitrate returned to the process influent is denitrified, the degree of nitrogen removal is limited by the rate of recirculation and the nitrate respiration achievable in the anoxic zone. Ideally, with high recirculation and efficient denitrification in the anoxic zone, nitrogen removal could be 70%–80% in processing domestic wastewater. In reality, the removal is more likely to be about 50% for operational reasons. For instance, the recirculation flow cannot reduce excessively the hydraulic detention time in the nitrification–denitrification process. The relative concentrations of BOD and nitrogen also influence the efficiency of nitrogen removal since the oxygen demand of the wastewater controls the rate and extent of microbial nitrate respiration. A reduction of 1 mg/l of NO_3—N theoretically satisfies 3.4 mg/l of BOD, which is the oxygen/nitrogen ratio in nitrate.

Nitrification–Denitrification in a Plug-Flow Process

Nitrification–denitrification by modifying a plug-flow activated-sludge process was studied by the Water Research Centre, Great Britain [15]. The laboratory units, illustrated schematically in Figure 14.17, had a series of four completely mixed reactors to simulate plug flow in an aeration tank and to allow isolation of an anoxic zone. The influent was a settled municipal wastewater with an average BOD of 240 mg/l and total nitrogen of approximately 75 mg/l. Operating parameters are listed on the diagrams. The nitrogen concentrations are expressed as percentages of the total influent nitrogen.

Figure 14.17(a) is the flow scheme and nitrogen data for operation as a normal activated-sludge process. A mass balance based on measurements of nitrogen in the influent, effluent, and waste sludge showed an unaccountable loss of 14%. This was assumed to result from nitrate respiration in the final clarifier, resulting in denitrification. Total nitrogen removal was 33% (100 to 67), with 93% of the effluent nitrogen in the form of nitrate.

Figure 14.17(b) is the flow scheme and nitrogen data for operation as an activated-sludge process with denitrification. The first reactor was converted to an anoxic zone by replacing the air supply with a stirrer to blend the wastewater influent with the return activated sludge. The total nitrogen removed averaged 59% (100 to 41), and the nitrate content in the effluent was 40% less than during normal operation ($[(62 - 37)/62] \times 100 = 40\%$).

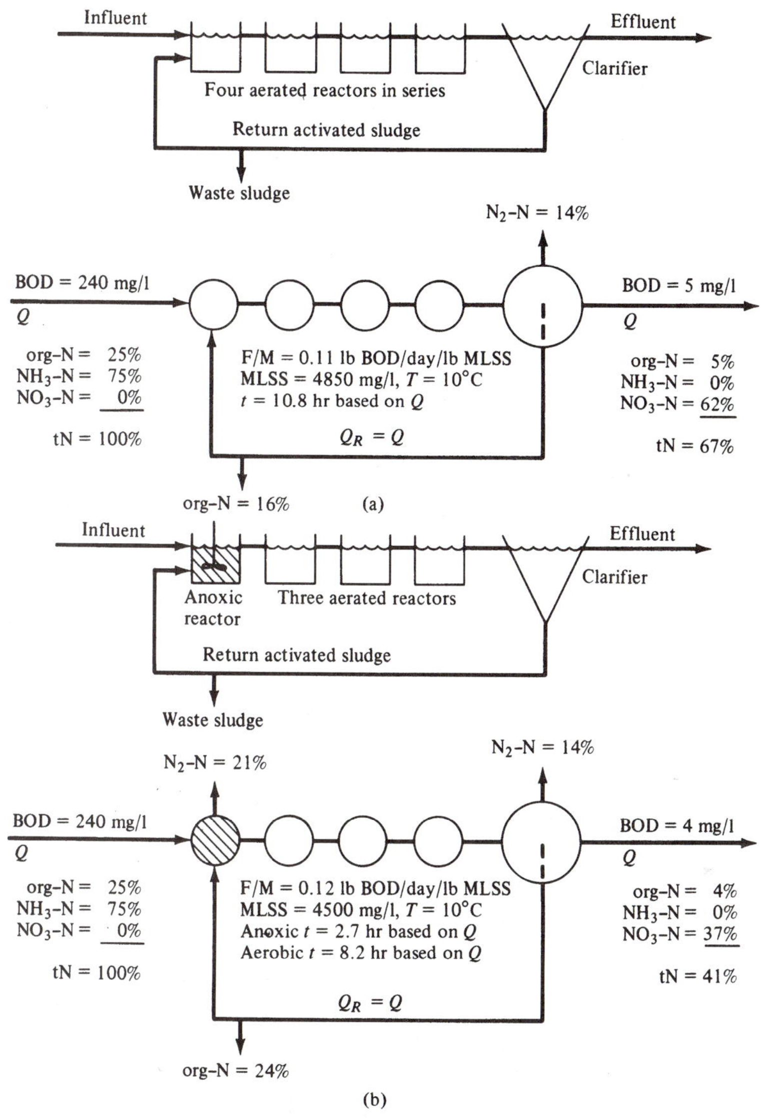

FIGURE 14.17 Flow schemes and average nitrogen amounts as percentages of influent nitrogen from laboratory testing of (a) normal activated-sludge process with nitrification and (b) modified activated-sludge process with nitrification–denitrification [15].

Following these laboratory studies, a full-scale research project was undertaken at the Rye Meads wastewater plant, Thames Water Authority, Great Britain [15,16]. The plug-flow aeration tanks were modified to provide anoxic zones for denitrification. The arrangements of anoxic zones and wastewater feed are diagrammed in Figure 14.18; each aeration tank received a settled wastewater flow of approximately 9000 m^3/d. Unit 1 was modified to create a mechanically stirred (unaerated) anoxic zone in the first one-fourth

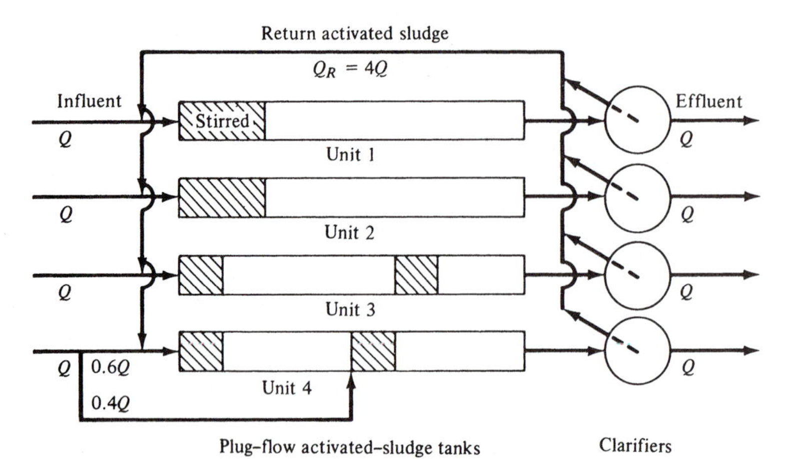

FIGURE 14.18 Modified activated-sludge arrangements tested for nitrification–denitrification efficiency at the Rye Meads wastewater plant, Thames Water Authority, Great Britain. Shaded areas: anoxic zones; unshaded rectangles: aerobic zones.

of the tank; this was the same process tested in the laboratory (Fig. 14.17). Unit 2 was the same arrangement except that mixing in the anoxic zone was obtained by reduced aeration rather than by mechanical stirring. Two-thirds of the air diffusers had been removed from this section of the tank. Nitrate reductions in these units relative to normal activated-sludge processing were both in the range of 47%–49%.

Units 3 and 4, depicted in Figure 14.18, were placed into operation with two anoxic zones, each one-eighth of the tank length, separated by an aerobic zone. The anoxic zones were mixed by reduced aeration with three-quarters of the diffusers removed. As in previous schemes, the first anoxic zone was to reduce nitrate in the return activated sludge. The purpose of the second anoxic zone was to increase denitrification efficiency by nitrate respiration of the influent ammonia oxidized in the immediately preceding aerobic zone. Unit 3, when operated with all of the wastewater entering at the inlet of the tank, did not achieve any greater denitrification than did units 1 and 2. This was attributed to difficulty in achieving an anoxic condition in the second zone. In the operation of unit 4, anoxia was created in the second zone by applying 40% of the influent wastewater directly to the head of this zone. Under these operating conditions, the nitrate removal was 54% greater than normal activated-sludge treatment, which is approximately 6% greater than the removals in units 1 and 2. Denitrification was expected to be greater than the observed 54%. Based on subsequent laboratory testing, the process with dual anoxic zones—which were mechanically stirred—achieved up to 70% nitrate reduction using a similar mode of operation. Apparently, mixing the anoxic zone with diffused air limited nitrate respiration.

After the field studies at Rye Meads, the Water Research Centre, Great Britain, conducted a laboratory investigation to evaluate the use of dual anoxic zones in a modified activated-sludge process [16]. As shown in Figure 14.19, the laboratory system consisted of a series of eight equal-sized compartments followed by a clarifier. Underflow from the final clarifier was returned to the first reactor, and the first and fifth compartments were

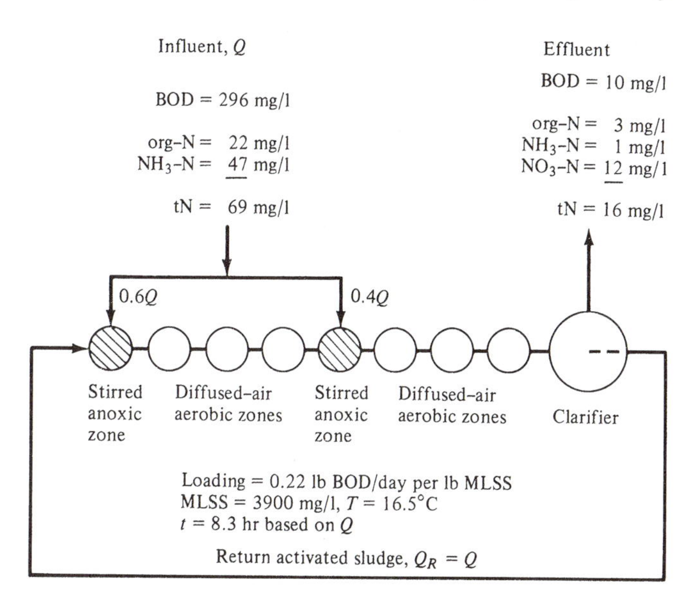

FIGURE 14.19 Flow scheme of the laboratory-scale modified activated-sludge system tested for nitrification–denitrification efficiency using two anoxic zones. The division of flow between anoxic zones and nitrogen concentrations in the effluent is given for optimum denitrification based on the listed operating conditions.

stirred anoxic reactors receiving influent wastewater. A second laboratory system with eight aerated compartments (no anoxic zones) was operated as a control unit to determine the effluent nitrate concentrations processed by a normal activated-sludge process. In a series of test runs the proportions of wastewater applied to the fifth compartment were 0%, 30%, 35%, 40%, and 50% of the influent. At zero flow to the second anoxic zone, only the nitrate in the return activated sludge was denitrified for a reduction of 41% of the oxidized nitrogen in the effluent. Anoxia in the second zone was inadequate to produce significant denitrification. At a 30% feed rate, the percentage of nitrate removal increased to 52%. When 35%–40% of the influent was applied to the second anoxic zone, the degree of denitrification increased to 68%–70%, since part of the nitrate produced from the wastewater fed to the first anoxic zone was removed in the second. The operating conditions and wastewater data for a 40% portion applied to the second zone are given in Figure 14.19. (The normal activated-sludge effluent from the control unit contained 40 mg/l of NO_3—N as compared to 12 mg/l from the modified process.) At a 50% feed rate to the second zone, the performance deteriorated to 40% nitrate reduction since insufficient BOD was applied to the first anoxic zone to satisfy the nitrate respiration demand.

Two other modes of operation were tested using this laboratory apparatus. First, with a single initial anoxic zone consisting of compartments 1 and 2, the sludge recycle was doubled, so $Q_R = 2Q$. The measured nitrate removal was 66% because more effluent nitrate was being returned for denitrification. The second test was conducted with a 60%–40% division of influent between the first and second anoxic zones and a sludge recycle of $2Q$. Compared with the normally operated process, the nitrate reduction was 77% with an effluent nitrate level of 9 mg/l; however, this was really a false value because the effluent contained 4 mg/l of NH_3—N, which is an increase of 3 mg/l from previous tests. Thus, the nitrate reduction relative to the other tests should be considered to be approximately 70%. The additional ammonia nitrogen appeared in the effluent since the detention times in the aerobic zones were too short to achieve complete nitrification.

Nitrification–Denitrification in a Closed-Loop Process

Closed-loop activated-sludge processes are commonly referred to as the oxidation ditch process (Fig. 12.36) or the *Carrousel®* process (Fig. 12.37), which are described in Section 12.21. These extended-aeration processes are usually installed in small and medium-sized plants because of reliable performance, ease of operation, and minimal maintenance.

Most closed-loop activated-sludge plants are designed for an aeration period of 20–30 hr and sludge age in excess of 20 days with actual operation at 25–35 days. If the wastewater temperature is warm, greater than 10°C, nitrification is likely to exceed 90%. In fact, significant nitrification cannot be prevented if aeration maintains an adequate dissolved-oxygen concentration in the mixed liquor. The rate of nitrification can be reduced if the mixed liquor temperature drops below about 10°C from the cooling effect of mechanical-aerator operation in cold air.

Significant biological denitrification is possible by proper operational control of an oxidation ditch or *Carrousel®* process, as illustrated in Figure 14.20. By controlling the rate of aeration, the aerobic zone along the channel can be shortened to create an anoxic zone before the mixed liquor recirculates back to the aerator. A dissolved-oxygen sensor located downstream from the aerator is used to monitor the concentration and control the aerator. Therefore, as the mechanical aerator propels the mixed liquor around the channel, the first section is aerobic enough for nitrification and the return section sufficiently anoxic for denitrification as the activated sludge takes up the oxygen bound in nitrate to satisfy wastewater BOD. Total nitrogen removal in the range of 50%–70% is possible.

Biological nitrification–denitrification can be improved by adding anoxic tanks with mechanical stirrers to the oxidation process, as illustrated in Figure 14.21. Return activated sludge from the final clarifiers and recycled mixed liquor are blended with the raw wastewater influent. BOD in the organic matter takes up oxygen from nitrate recirculating in the mixed liquors to release nitrogen gas. Nitrogen removals up to 90% are possible with proper environmental conditions of temperature, pH, and dissolved oxygen, operation with a long sludge age, and wastewater with a high enough BOD-to-nitrogen ratio that the oxygen demand available is sufficient to take up the oxygen in the oxidized nitrogen.

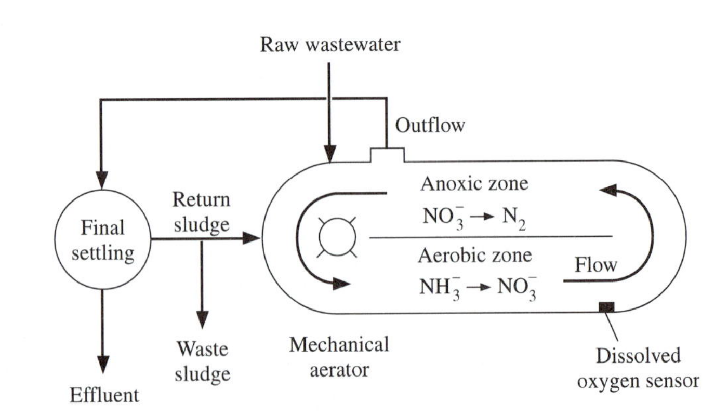

FIGURE 14.20 Schematic diagram of a *Carrousel®* system operating as an extended-aeration process with concurrent nitrification and denitrification.

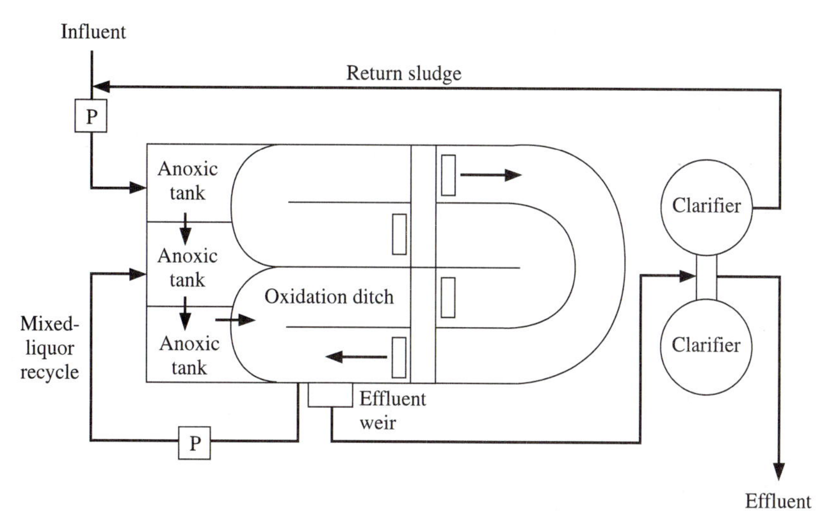

FIGURE 14.21 Biological nitrification–denitrification in a closed-loop reactor (oxidation ditch) with three anoxic tanks to blend raw wastewater with return sludge and recycled mixed liquor from the aeration tank for reduction of nitrate to nitrogen gas. *Source:* Courtesy of Lakeside Equipment Corporation.

WATER REUSE

The major reuse of processed wastewater is for agricultural irrigation, amounting to about 60% of the total. The second largest reuse, accounting for approximately 30%, is for industrial cooling and process waters. Since the public is generally not exposed to these reused waters, high-quality biological processing with or without chemical disinfection is often considered satisfactory treatment. The remainder of water reuse, less than 10% of the total, is for urban landscape irrigation and groundwater recharge through surface percolation. For these applications, control of industrial wastewaters entering the sewer system and tertiary treatment for a higher degree of organic matter removal and disinfection for virus inactivation are usually the prescribed pretreatment. Dilution with surface water and dispersion of seepage in groundwater are relied on for reducing the concentrations of refractory contaminants.

A small fraction of groundwater recharge is performed by direct injection through wells to recharge groundwater aquifers or coastal areas as a barrier against underground intrusion of saltwater from the ocean. Because injected water seeps toward potable groundwater supplies, the wastewater is reclaimed to meet drinking water standards. This is the highest degree of water reclamation required in the United States, since direct potable reuse is not recommended.

14.15 WATER QUALITY AND REUSE APPLICATIONS

Protection of public health is the primary concern in establishing water-quality standards for reuse. Important environmental considerations are protection of groundwater, soils, and crops. In establishing standards, consideration must be given to the economics of integrating wastewater treatment with the reuse application. For example, a properly designed agricultural irrigation system restricted to fodder, fiber, and seed crops can reuse conventionally treated wastewater without health or environmental risks. In contrast, surface infiltration of wastewater, whether planned or unintentional, can result in significant deterioration of groundwater quality unless the groundwater in storage provides abundant dilution or the infiltration water is treated to drinking water quality.

Guidelines for Water Reuse, Environmental Protection Agency

Many states have regulations specifying water quality requirements, treatment processes, or both for defined reuse applications. Though beneficial reuse encourages conservation of water resources, the complementary intent is often to provide a disposal alternative to discharge in surface waters or in the absence of suitable surface waters for disposal.

The guidelines for water reuse in Table 14.3 recommend treatment, reclaimed water quality, and reclaimed water monitoring for municipal wastewaters [17]. Since water reclamation and reuse standards are the responsibility of state agencies, the guidelines are based on practices in the states and are not standards proposed by the Environmental Protection Agency. The *Manual of Guidelines for Water Reuse* assembles data to assist agencies in development of reuse programs and appropriate regulations. The overriding consideration is to develop a reuse system to ensure that reclaimed water quality is appropriate for the intended use.

Agricultural reuse to irrigate nonfood crops, commercially processed food crops, and surface irrigation of orchards and vineyards requires biological treatment with effluent chlorination for disinfection. The reclaimed water quality is based on effluent standards (Section 9.1) and monitoring requirements of the National Pollutant Discharge Elimination System (NPDES). Nonfood crops are fodder, fiber, and seed crops, and commercially processed food crops undergo sufficient chemical and/or physical processing to destroy pathogens prior to sale.

Irrigation of food crops not commercially processed, including those eaten raw, requires biological treatment and extensive disinfection. The reclaimed water quality specified is to ensure high clarity, less than or equal to 2 NTU, and removal of pathogens, with no detectable fecal coliforms. To achieve this quality, the common treatment is tertiary filtration with chemical coagulation, followed by chlorination using an extended contact time (Section 14.5). In agricultural regions where water reuse is practiced, reclaimed water is commonly applied only to nonfood crops and groundwater or surface water to food crops because of the high cost of tertiary processing.

The suggested water quality for urban reuse in Table 14.3 is biological treatment, filtration, and disinfection with quality limits of 2 NTU and no detectable fecal coliforms. Though this is the common quality for reuse in unrestricted public areas such as parks, school yards, and residential lawns, state regulations usually allow secondary effluent with disinfection to be used for irrigation of restricted sites with controlled access such as golf courses, cemeteries, and roadway medians. Often the time of watering is limited to night or early morning. Although some states allow a fecal coliform limit of 200 per 100 ml, most specify 23 per 100 ml, which requires more efficient disinfection.

TABLE 14.3 Suggested Guidelines for Water Reuse Based on Water Reclamation and Reuse Practices in the United States [17]

Types of Reuse	Treatment	Reclaimed Water Quality	Reclaimed Water Monitoring
Agricultural Reuse—Nonfood Crops Pasture for milking animals; fodder, fiber, and seed crops	• Secondary • Disinfection	• pH = 6–9 • ≤30 mg/l BOD • ≤30 mg/l SS • ≤200 fecal coli/100 ml • 1 mg/l Cl_2 residual (min.)	• pH—weekly • BOD—weekly • SS—daily • Coliform—daily • Cl_2 residual—continuous
Agricultural Reuse—Food Crops Commercially Processed; Surface Irrigation of Orchards and Vineyards	• Secondary • Disinfection	• pH = 6–9 • ≤30 mg/l BOD • ≤30 mg/l SS • ≤200 fecal coli/100 ml • 1 mg/l Cl_2 residual (min.)	• pH—weekly • BOD—weekly • SS—daily • Coliform—daily • Cl_2 residual—continuous
Restricted Access Area Irrigation Sod farms, silviculture sites, and other areas where public access is prohibited, restricted, or infrequent	• Secondary • Disinfection	• pH = 6–9 • ≤30 mg/l BOD • ≤30 mg/l SS • ≤200 fecal coli/100 ml • 1 mg/l Cl_2 residual (min.)	• pH—weekly • BOD—weekly • SS—daily • Coliform—daily • Cl_2 residual—continuous
Agricultural Reuse Food Crops Not Commercially Processed Surface or spray irrigation of any food crop, including crops eaten raw	• Secondary • Filtration • Disinfection	• pH = 6–9 • ≤10 mg/l BOD • ≤2 NTU • No detectable fecal coli/100 ml • 1 mg/l Cl_2 residual (min.)	• pH—weekly • BOD—weekly • Turbidity—continuous • Coliform—daily • Cl_2 residual—continuous
Urban Reuse All types of landscape irrigation (e.g., golf courses, parks, cemeteries)—also vehicle washing toilet flushing, use in fire protection system and commercial air conditioners, and other uses with similar access or exposure to the water	• Secondary • Filtration • Disinfection	• pH = 6–9 • ≤10 mg/l BOD • ≤2 NTU • No detectable fecal coli /100 ml • 1 mg/l Cl_2 residual (min.)	• pH—weekly • BOD—weekly • Turbidity—continuous • Coliform—daily • Cl_2 residual—continuous
Indirect Potable Reuse Groundwater recharge by spreading into potable aquifers	• Site specific • Secondary and disinfection (min.). May also need filtration and/or advanced wastewater treatment	• Site specific • Meet drinking water standards after percolation through vadose zone	Includes, but not limited to, the following: • pH—daily • Coliform—daily • Cl_2 residual—continuous • Drinking water standards—quarterly • Other—depends on constituent
Groundwater recharge by injection into potable aquifers	• Secondary • Filtration • Disinfection • Advanced wastewater treatment	Includes, but not limited to, the following: • pH = 6.5–8.5 • ≤2 NTU • No detectable fecal coli/100 ml • 1 mg/l Cl_2 residual (min.) • Meet drinking water standards	Includes, but not limited to, the following: • pH—daily • Turbidity—continuous • Coliform—daily • Cl_2 residual—continuous • Drinking water standards— quarterly

Source: Manual of Guidelines for Water Reuse, EPA/625/R-92/004.

Indirect potable reuse is intentional or unintended groundwater recharge by seepage of treated wastewater spread or impounded on the ground surface. The recommended treatment is specific for each site, based on soils, percolation rate, thickness of unsaturated soil profile (vadose zone), natural groundwater quality, and dilution. In general, the reclaimed water should meet drinking water standards and contain no measurable levels of pathogens after percolation through the vadose zone.

Direct groundwater recharge is injection of reclaimed water into potable aquifers. The recommended processing is secondary treatment, filtration, disinfection, and advanced unit processes such as chemical precipitation, carbon adsorption, and reverse osmosis. The most inclusive water quality criteria are to meet drinking water standards before injection and retention underground for an extended period of time (e.g., at least one year) prior to withdrawal from wells.

Water Recycling Criteria, State of California

Title 22 Code of Regulations [18] specifies both treatment processes and quality standards for recycled water applications in irrigation and for impoundments. In this edition of the regulations, the term *water reclamation* is statutorily changed to *water recycling*.

Disinfected tertiary recycled water can be applied for landscape irrigation with unrestricted access (i.e., parks, playgrounds, schoolyards, residential landscaping, and golf courses) and irrigation of food crops (i.e., all edible root crops, where the recycled water comes in contact with the edible portion of the crop). After high-quality biological treatment, preferably by activated-sludge processing, the influent to tertiary granular-media filtration is not to exceed 5 nephelometric turbidity units (NTU) for more than 15 min and never exceed 10 NTU. After coagulation [commonly with alum and polymer, Fig. 14.4(b)], the rate of filtration is not to exceed 5 gpm/ft^2 in mono-, dual or mixed-media gravity, upflow or pressure filters, or the rate is not to exceed 2 gpm/ft^2 in traveling bridge automatic backwash filters. The turbidity in the filtered water is limited to a maximum average of 2 NTU within a 24-hr period, 5 NTU for more than 5% of the time in a 24 hr period, and 10 NTU any time. If membrane filtration (microfiltration, ultrafiltration, nanofiltration, or reverse osmosis) is used instead of granular-media filtration, the turbidity of the filtered water is not to exceed 0.2 NTU for more than 5% of the time within a 24-hr period and 0.5 NTU at any time. Filtered effluent turbidity shall be measured for any filtration system using a continuous turbidity meter and recorder.

The tertiary filtered water has to be disinfected by either (1) chlorination in a plug-flow contact tank providing a $C \cdot t$ value of not less than 450 (mg/l) $\cdot$ min (concentration of chlorine residual in milligrams per liter multiplied by modal contact time in minutes measured at the same point) and a modal contact time of at least 90 min based on peak dry weather design flow; or (2) a disinfection process in combination with a filtration process that has been demonstrated to inactivate and/or remove 99.999% of plaque-forming units of F-specific bacteriophage MS2 or polio viruses. The median concentration of total coliform bacteria in the disinfected filtered effluent is not to exceed a most probable number (MPN) of 2.2 per 100 ml based on the results of the last 7 days for which analyses have been completed, and the number is not to exceed a MPN of 23 per 100 ml in more than one sample in any 30-day period. No sample shall exceed a MPN of 240 per 100 ml. Samples shall be analyzed at least once daily. [Definitions of terms: "Modal contact time" is the time elapsed between introduction of a tracer into the influent of the chlorination tank and the time when the highest concentration of the tracer is observed in the effluent. "Peak dry weather design flow" is the arithmetic mean of the maximum peak flow rates sustained over some period of time (e.g., 3 hr) during the maximum 24-hr dry weather of little or no rain. "F-specific bacteriophage MS2" means a strain of specific type that infects coliform bacteria and is traceable to the American Type Culture Collection and is grown on growths of *Escherichia coli*.]

Disinfected tertiary recycled water can be stored in an unrestricted recreational impoundment with no limitations imposed on body-contact water recreational activities. Total coliform bacteria concentrations measured at a point between the disinfection process and the point to entry to the impoundment shall comply with the limits of disinfected filtered effluent as given above.

Disinfected secondary-2.2 recycled water is high-quality biologically treated secondary effluent; chemically disinfected with an extended contact time so that the median concentration of total coliform bacteria does not exceed a MPN of 2.2 per 100 ml based on results of the last 7 days and does not exceed a MPN of 23 per 100 ml in more than one sample in any 30-day period. This quality of recycled water can be used (1) for the surface irrigation of food crops where the edible portion is produced above ground and not contacted by recycled water and (2) as the source of supply for restricted recreational impoundments.

Disinfected secondary-23 recycled water is secondary effluent with chemical disinfection so that total coliform bacteria do not exceed a 7-day mean MPN of 23 per 100 ml and do not exceed a MPN of 240 per 100 ml in any 30-day period. This quality of recycled water can be used (1) for surface irrigation where public access is restricted and where water can be applied only at night (e.g., freeway landscaping, golf courses, nursery stock and sod farms, and pasture for animals) and (2) as the source of supply for landscape impoundments storing water for aesthetic enjoyment without decorative fountains.

Undisinfected secondary recycled water can be used for surface irrigation of orchards and vineyards where recycled water does not come in contact with the edible portion of the crop; fodder and fiber crops and pasture for animals not producing milk for human consumption; seed crops not eaten by humans; and food crops that must undergo commercial pathogen-destroying processing before being consumed by humans.

Regulations for the Reuse of Wastewater, State of Arizona

The regulations established by Arizona [19] are listed in Table 14.4. The allowable limits for various reuses are specified without reference to any required treatment processes, and the water quality for several water reuse applications differs from the California standards. For irrigation of fodder, fiber, and seed crops, the fecal coliform allowable limit is 1000 per 100 ml as the geometric mean of a minimum of five samples and 4000 per 100 ml as the maximum of a single sample.

The Arizona standards for landscape irrigation are less restrictive than the California standards. Landscape irrigation with restricted access has a fecal coliform allowable limit of 200 per 100 ml and a maximum of 1000 per 100 ml, compared with the California total coliform standard of 23 per 100 ml and 240 per 100 ml, respectively. Landscape irrigation with open access has a fecal coliform allowable limit of 25 per 100 ml and a maximum of 75 per

TABLE 14.4 Allowable Permit Limits for Specific Reuses of Reclaimed Water, State of Arizona [19]

	A	B	C	D	E	F	G	H	I	J
						Landscaped Areas				
Parameter	Orchards	Fiber, Seed, & Forage	Pastures	Livestock Watering	Processed Food	Restricted Access	Open Access	Food Consumed Raw	Incidental Human Contact	Full Body Contact
pH	4.5–9	4.5–9	4.5–9	6.5–9	4.5–9	4.5–9	4.5–9	4.5–9	6.5–9	6.5–9
Fecal coliform (CFU/100 ml)[a]										
geometric mean (5 sample minimum)	1000	1000	1000	1000	1000	200	25	2.2	1000	200
single sample not to exceed	4000	4000	4000	4000	2500	1000	75	25	4000	800
Turbidity (NTU)[b]	—	—	—	—	—	—	5	1	5	1
Enteric virus[c]	—	—	—	—	—	—	125 per 40 l	1 per 40 l	125 per 40 l	1 per 40 l
Entamoeba histolytica	—	—	—	—	—	—	—	None detectable	—	None detectable
Giardia lamblia	—	—	—	—	—	—	—	None detectable	—	None detectable
Ascaris lumbricoides	—	—	—	—	—	—	None detectable	None detectable	None detectable	None detectable
Common large tapeworm	—	—	None detectable	None detectable	—	—	—	—	—	—

[a] CFU = colony-forming units.
[b] NTU = nephelometric turbidity units.
[c] Expressed as PFU, plaque-forming units; MPN, most probable numbers; or immunofluorescent foci per liter. "None detectable" means no pathogenic microorganisms observed during examination.

100 ml, compared with the California total coliform standard of 2.2 per 100 ml and 25 per 100 ml, respectively. Nevertheless, the coliform and turbidity standards for irrigation of food consumed uncooked are similar: Arizona specifies a fecal coliform geometric mean of 2.2 per 100 ml and a maximum of 25 per 100 ml, which are numerically the same as the California total coliform limits, and an allowable turbidity of 1 NTU, which is similar to the California turbidity standard of less than 2 units.

For higher-quality reuse applications, allowable limits are also specified for enteric viruses, protozoa, and helminths. The operators of reuse systems are not required to monitor routinely for these pathogens; however, if the Arizona Department of environmental Control requires corrective action to improve the microbiological quality of the reclaimed water, monitoring for pathogens will be necessary for verification.

Both of these reuse regulations specify reliability in the design to ensure that the plant can meet the designated water quality standards, an operational plan for safety of public health, and a monitoring program for quality control. Reliability in performance requires provision for safe disposal or storage of wastewater in case of emergency, installation of multiple treatment units for flexibility in operational control and shutdown for maintenance, standby chemical feeders and chlorinators, alarm devices installed to warn of process malfunction, alternative electric power supply, and adequate funding for operation and maintenance. Protection of public health includes isolation of restricted agricultural irrigation areas by buffer zones and fencing, signs identifying the water supply as reclaimed water, isolation and identification of reclaimed water piping and appurtenances from any potable water supply, landscape irrigation scheduling, and other management practices to reduce the contact of people with reclaimed water. Management of a reuse system includes scheduled laboratory testing of reclaimed water quality and recording of plant operations and maintenance.

International Microbial Guidelines for Agricultural Irrigation

The wastewater reclamation regulations established in the United States have been replicated or used as the basis for similar regulations in many countries. Unfortunately, the construction and operation requirements of complex wastewater treatment systems dictated by these restrictive quality standards have inhibited beneficial water reuse in developing areas of the world. As a result, wastewaters are often not officially approved for agricultural irrigation pending some unresolved solution in order to achieve compliance with standards that cannot be attained within local resources. This results in the practice of totally uncontrolled, unsafe irrigation of salad and other food crops with inadequately treated wastewater.

A group of environmental experts and epidemiologists meeting in Engelberg, Switzerland, under the auspices of several international organizations, formulated new tentative microbiological guidelines for treated wastewater reuse in agricultural irrigation [20]. As shown in Table 14.5, the guidelines require effective wastewater treatment to remove helminths to a level of one egg per liter, since helminthic disease transmission has been identified as the top-priority health problem in developing countries. For unrestricted irrigation, the guideline of 1000 fecal coliforms per 100 ml is recommended. In conventional wastewater treatment, tertiary coagulation and filtration are needed to ensure removal of helminth eggs, but in stabilization ponds, removal of eggs can be

TABLE 14.5 Tentative Microbiological Quality Guidelines for Treated Wastewater Reuse in Agricultural Irrigation,[a] *The Engelberg Report* [20]

Reuse Process	Intestinal Nematodes[b] (eggs/liter)	Fecal Coliforms (no./100 ml)
Restricted irrigation[c]		
Irrigation of trees, industrial crops, fodder crops, fruit trees,[d] and pasture[e]	Less than 1	Not applicable[c]
Unrestricted irrigation		
Irrigation of edible crops, sports fields, and public parks[f]	Less than 1	Less than 1000[g]

[a] In specific cases, local epidemiologic, sociocultural, and hydrogeologic factors should be taken into account, and these guidelines modified accordingly.
[b] *Ascaris, Trichuris*, and hookworms.
[c] A minimum degree of treatment equivalent to at least a 1-day anaerobic pond followed by a 5-day facultative pond or its equivalent is required in all cases.
[d] Irrigation should cease 2 weeks before fruit is picked, and no fruit should be picked off the ground.
[e] Irrigation should cease 2 weeks before animals are allowed to graze.
[f] Local epidemiologic factors may require a more stringent standard for public lawns, especially hotel lawns in tourist areas.
[g] When edible crops are always consumed well cooked, this recommendation may be less stringent.

achieved by simply impounding the water for a sufficiently long time to allow settlement of the eggs. Other pathogens are reduced by natural die-off. Therefore, with the availability of adequate land area and tolerance for a reduced aesthetic quality of the reclaimed water, a low-cost, easy-to-operate stabilization pond system in a warm climate can provide reclaimed water for irrigation.

Investigations of the effectiveness of stabilization ponds to remove helminth eggs and protozoal cysts showed 100% removal in all cases where the total retention time in multicelled ponds was greater than 20 days. Hookworm larvae may survive in aerobic ponds with an overall retention time of less than 10 days but not if the retention time is greater than 20 days [21]. Inactivation of enteric viruses is rapid in warm waters and is in the range of 1 to 2 log units per 5 days retention in ponds at a temperature greater than 25°C. Thus, a pond system with an overall retention time of 20 days in a warm climate would be expected to achieve a reduction of excreted viruses of 4 to 6 log units (99.99% to 99.9999%) [21]. Fecal bacteria are also significantly reduced in stabilization ponds with warm water and long retention times. A summary of reported bacterial removal efficiencies in multicelled ponds with retention time greater than 25 days reduced fecal coliforms to 100 per 100 ml or less [22]. (Raw domestic wastewater contains approximately 3,000,000 per 100 ml, so the reduction was between 99.99% and 99.999%.)

14.16 AGRICULTURAL IRRIGATION

Reuse of wastewater for agricultural irrigation is often considered land treatment and/or disposal. Regardless of the designated primary objective, irrigation systems are being used extensively in regions where available sites have suitable soil conditions and groundwater hydrology and a climate favorable to grow grasses, crops, or trees. Land irrigation is also used for advanced wastewater treatment to recycle nutrients to land instead of polluting surface waters. Reuse of reclaimed water has a strong ecological appeal; however, the requirements for large tracts of land and reclaimed water storage are major disadvantages in humid climates, northern states, and metropolitan areas.

Agricultural irrigation has the following features [23]: Water loading rate of 0.5–4 in./wk (13–100 mm/wk); annual application of 2–20 ft/yr (0.6–6 m/yr); field area for 1 mgd (3785 m^3/d) flow of 56–560 acres (23–230 ha); minimum depth to groundwater of 5 ft (1.5 m); moderately permeable soils with good productivity when irrigated; loss of wastewater by evapotranspiration and percolation; and climatic restrictions with storage needed for cold weather and to contain runoff from irrigation and precipitation.

Water distribution is by fixed or moving sprinkling systems, or surface spreading. Fixed nozzles are attached to risers from either surface or buried pipe networks, the most popular moving sprinkling system is a center-pivot spray boom that rotates around a central tower with the distribution piping suspended between wheel supports riding on circumferential tracks. On flat land, having less than 1% slope, surface irrigation is possible by the ridge-and-furrow method. Water applied to furrows, spaced about 3 ft apart, flows down slope by gravity and seeps into the ground. Border-strip irrigation, the second method for surface spreading, uses parallel soil ridges constructed in the direction of slope. Water introduced between the ridges at the upper end flows down the 20–100-ft-wide strips several hundred feet long.

Restricted Irrigation

Restricted irrigation refers to the use of low-quality reclaimed water in specific areas where only fodder, fiber, and seed crops are grown, such as alfalfa, cotton, and wheat. Public access is controlled by fencing and warning signs around the perimeter. If watering is by spray irrigation, buffer zones are established along the boundaries of the site to prevent aerosols from drifting into adjacent public access areas. Thus, the only health risk is to the agricultural workers.

The objectives of wastewater treatment for restricted agricultural irrigation are (1) maximum removal of helminth eggs and protozoal cysts, (2) effective removal of pathogenic bacteria, (3) reduction of enteric viruses, and (4) substantial removal of organic matter to clarify the water and eliminate offensive odors. This treatment can be accomplished best by conventional biological processing followed by detention in storage reservoirs or by completely mixed aerated lagoons followed by stabilization ponds in series prior to detention in storage reservoirs. Storage is required to equalize the demand for irrigation water with the supply of reclaimed water. The crop selections, growing seasons, soil conditions, and types of irrigation systems affect the volume of storage needed. In warm semiarid or arid regions, reservoirs have water depths of 5 to 10 m with a storage volume equivalent to about 90 days of reclaimed water production. Thus, the detention time of the reclaimed water in the reservoirs varies from 2 to several months. In cold humid regions, the required storage volume may be significantly greater to accumulate wastewater flow for 4 or more months when irrigation is not possible. Besides equalizing storage, reservoirs provide supplementary treatment for removal of helminth eggs, protozoal cysts, pathogenic bacteria, enteric viruses, and organic matter. The potential unaesthetic conditions that can occur in storage are the growth of algae, which increase the turbidity and suspended solids, and the potential generation of odors if prior biological treatment is not adequate.

The management of an agricultural irrigation system is an interdisciplinary function of agronomy and environmental engineering. The engineering aspect is to ensure adequate operation and maintenance of the biological treatment system to produce a reclaimed water suitable for storage in the reservoirs and to monitor water quality by conducting routine tests, such as biochemical oxygen demand, suspended solids, and coliform bacteria, and occasional tests for helminth eggs and enteric viruses. The agronomic aspect involves selection of crops, fertilizer and pesticide applications, and scheduling of water flow through the reservoirs to the fields. The cultivation of only approved fodder, fiber, and seed crops must be monitored and strictly enforced; food crops for human consumption, particularly those eaten uncooked, cannot be irrigated with the low-quality reclaimed water designated for restricted irrigation.

Unrestricted Irrigation

Unrestricted irrigation refers to the application of high-quality reclaimed water for irrigation of food crops for human consumption, even those eaten uncooked. Public access to the irrigation site is not controlled, but persons must be warned not to use the water for drinking or domestic purposes. Signs reading "Irrigated with Reclaimed Wastewater," or a similar warning, should be posted at the boundaries of the site, and faucets discharging reclaimed water should be posted with signs reading "Reclaimed Water, Do Not Drink," or a similar warning, or be secured to prevent public use. To preclude inadvertent connection as a potable water source, the irrigation piping in areas accessible to the public should be color-coded, buried with colored tape, or otherwise suitably marked to indicate nonpotable water.

The wastewater for unrestricted irrigation must be treated for the removal of helminth eggs and protozoal cysts, elimination of pathogenic bacteria, and inactivation of enteric viruses. Of greatest importance is the removal of helminth eggs, since they are extremely persistent, surviving in harsh environmental conditions, and because of their latency period are transmitted primarily through salad crops eaten uncooked. Recommended microbiological standards published for reclaimed water applied to unrestricted irrigation of food crops are extremely restrictive, with the quality near that of drinking water (Tables 14.3 and 14.4). An extensive study in California using reclaimed water of this quality to irrigate food crops eaten uncooked demonstrated no health or environmental risks [24]. To achieve this high quality, the required wastewater processing is conventional biological treatment, chemical coagulation, granular-media filtration, and disinfection by chlorination. The coliform standard in Table 14.5 is applicable in developing countries only with warm climates and abundant sunshine where stabilization ponds are effective for both wastewater treatment and disinfection.

Chemical Quality

The chemical characteristics of irrigation water are important for public health and agronomy. Reuse standards applied to edible crops include heavy metals and organic compounds detrimental to the health of consumers. The agronomic standards include salinity, sodium absorption ratio, and specific ion toxicity of sodium, chloride, boron, and

TABLE 14.6 Guidelines for Interpretation of Water Quality for Irrigation

Potential Irrigation Problem	Units	Degree of Restriction on Use		
		None	Slight to Moderate	Severe
Salinity (affects crop water availability)[a]				
EC_w	dS/m	<0.7	0.7–3.0	>3.0
TDS	mg/l	<450	450–2000	>2000
Infiltration (affects infiltration rate of water into the soil. Evaluation using EC_w and SAR together)[b]				
SAR = 0–3 and EC_w =		>0.7	0.7–0.2	<0.2
= 3–6 =		>1.2	1.2–0.3	<0.3
= 6–12 =		>1.9	1.9–0.5	<0.5
= 12–20 =		>2.9	2.9–1.3	<1.3
= 20–40 =		>5.0	5.0–2.9	<2.9
Specific ion toxicity (affects sensitive crops)				
Sodium (Na)[b]				
surface irrigation	SAR	<3	3–9	>9
sprinkler irrigation	me/l	<3	<3	
Chloride (Cl)[c]				
surface irrigation	me/l	<4	4–10	>10
sprinkler irrigation	me/l	<3	>3	
Boron (B)	mg/l	<0.7	0.7–3.0	>3.0

[a] EC_w means electrical conductivity, a measure of the water salinity, reported in deciSiemens per meter at 25°C (dSm) or in units millimhos per centimeter (mmho/cm). Both are equivalent. TDS means total dissolved solids, reported in milligrams per liter (mg/l).
[b] SAR means sodium adsoprtion ratio, and is sometimes reported by the symbol RNa. At a given SAR, infiltration rate increases as water salinity increases. Evaluate the potential infiltration problem by SAR as modified by EC_w.
[c] For surface irrigation, most tree crops and woody plants are sensitive to sodium and chloride; use the values shown. Most annual crops are not sensitive. With overhead sprinkler irrigation and low humidity (<30%), sodium and chloride may be absorbed through the leaves of sensitive crops.
Source: Adapted from the Food and Agricultural Organization of the United Nations (FAO) (1985) [25].

trace elements affect sensitive crops. Normally, sanitary wastewater from domestic and commercial sources does not have chemical contaminants in excess of the allowable limits for agricultural cultivation. Table 14.6 gives the guidelines for interpretation of water quality for irrigation related to salinity from the Food and Agricultural Organization of the United Nations [25]. The nitrogen content of irrigation water is restricted where percolation to groundwater is possible or where the crops are sensitive to high nitrate. Trace elements, including several heavy metals, are toxic to selected plants. Of the 20 elements listed by FAO [25], only a few are likely to be present in treated municipal wastewater in concentrations exceeding the recommended guidelines for irrigation water. An example of a possible trace element in wastewater is cadmium, with a guideline limit of 0.01 mg/l for long-term use [17]. This conservative value is recommended due to its potential for accumulation in plants and soils in amounts that may be harmful to humans. Also, cadmium is toxic to beans, beets, and turnips at concentrations as low as 0.1 mg/l in nutrient solutions.

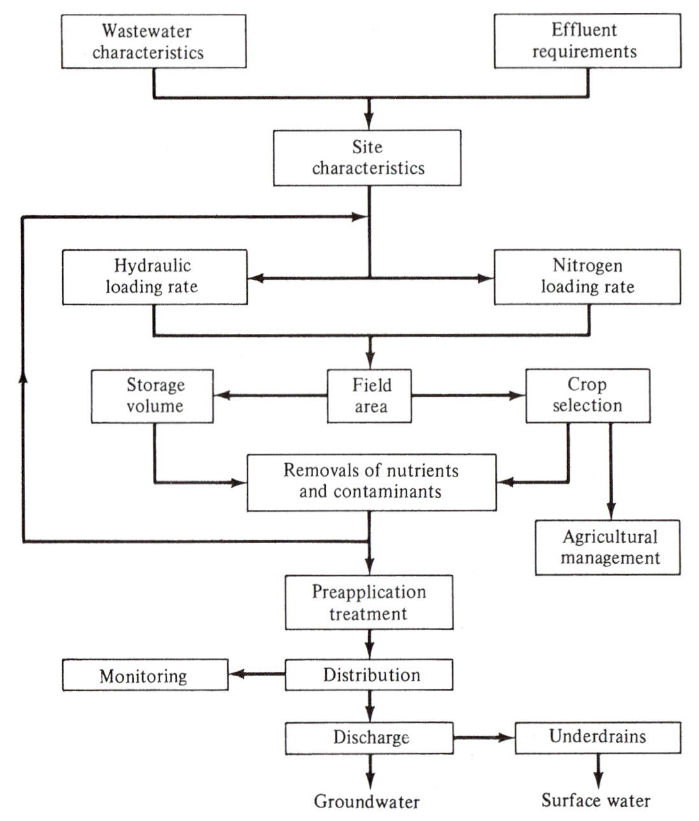

FIGURE 14.22 Iterative design process for an irrigation system.
Source: Process Design Manual for Land Treatment of Municipal Wastewater, U.S. Environmental Protection Agency, EPA/625/1-77-008 (October 1977): 5-5.

Design Process for Land Treatment

The recommended design for an irrigation system is the iterative process diagrammed in Figure 14.22 [23]. The initial step in design is to define the characteristics of the wastewater, effluent quality, and the site. Regulatory limits on effluent quality are established to protect both the groundwater and surface water. Investigations of the site include climate, geology, soils, plant cover, and topography. The iterative procedure for determining the field area needed for irrigation involves the interdependence of the hydraulic loading rate, nitrogen loading rate, water storage volume, and crop selection. System monitoring, the method of wastewater distribution, discharge control, and agricultural management are final considerations.

The water balance for a land application system can be calculated by the relationship

$$\begin{aligned} &\text{Precipitation} + \text{wastewater loading} \\ &\qquad = \text{evapotranspiration} + \text{percolation} + \text{runoff} \end{aligned} \tag{14.23}$$

Runoff is zero for most irrigation and infiltration–percolation installations, while precipitation and evapotranspiration are of little significance compared to wastewater loading and percolation for rapid infiltration systems.

The storage required can be calculated using monthly, weekly, or daily water-balance determinations. If the sum of evapotranspiration and percolation is less than precipitation plus available effluent, the balance must be stored. Conversely, when losses exceed the quantity available, water can be drawn from storage to supplement irrigation. These relationships are summarized in the following statement:

$$\begin{pmatrix}\text{Precipitation } + \\ \text{effluent available}\end{pmatrix} \pm \begin{pmatrix}\text{change in} \\ \text{storage}\end{pmatrix} = \begin{pmatrix}\text{evapotranspiration} \\ + \text{ percolation}\end{pmatrix} \tag{14.24}$$

The nitrogen balance for an irrigation system can be calculated from the relationship

$$\begin{aligned}\text{Applied nitrogen} = \text{crop uptake} &+ \text{loss by denitrification} \\ &+ \text{loss by percolation}\end{aligned} \tag{14.25}$$

The greatest nitrogen uptake is by perennial grasses, often in the range of 200–500 lb/acre/yr (220–560 kg/ha · y). Field crops (e.g., corn, soybeans, and cotton) assimilate considerably less nitrogen in one growing season, 70–170 lb/acre/yr (80–190 kg/ha · y). Loss by denitrification, commonly assumed to be 15%–25% of the applied nitrogen, is difficult to predict accurately. Based on the maximum contaminant level for drinking water, the allowable nitrogen concentration in percolate is usually taken as 10 mg/l.

Example 14.7

These data are from sample calculations by Pound and Crites [26]. Compute the storage requirements using monthly water balances for a 1-mgd irrigation system based on the following: (1) Design evapotranspiration and precipitation data, listed in columns 2 and 5 of Table 14.7, are with average monthly distributions for the wettest year in 25. (2) A perennial grass is grown and irrigated year-round. (3) Runoff is contained and reapplied. (4) The design year begins in October with the storage reservoir empty. (5) Nitrogen is the limiting factor for a land area of 120 field acres to balance the quantity applied with the amount of nitrogen in harvested grass. (6) The design allowable percolation rate is 10 in./month from March through November and 5 in./month for the remaining months (column 3, Table 14.7).

Solution: The effluent available per month is

$$\frac{10 \text{ mil gal/day} \times 30.4 \text{ days/month} \times 36.8 \text{ acre} \cdot \text{in./mil gal}}{120 \text{ acres}} = 9.3 \text{ in.}$$

Water losses (column 4 of Table 14.7) are found by summing evapotranspiration and percolation. The effluent applied (column 6) is the difference between water losses

TABLE 14.7 Water Balance and Storage Calculations for Example 14.7 (all values given in inches)

(1)	(2)	(3)	(4)	(5)	(6)	(7)	(8)	(9)	(10)
Month	Evapo-transpiration	Percolation	Water Losses: (2) + (3)	Precipitation	Effluent Applied	Effluent Available	Total Water Available: (5) + (7)	Change in Storage: (8) − (4)	Total Storage
Oct.	2.3	10.0	12.3	1.6	10.7	9.3	10.9	−1.4	0
Nov.	1.0	10.0	11.0	2.4	8.6	9.3	11.7	0.7	0.7
Dec.	0.5	5.0	5.5	2.7	2.8	9.3	12.0	6.5	7.2
Jan.	0.2	5.0	5.2	3.0	2.2	9.3	12.3	7.1	14.3
Feb.	0.3	5.0	5.3	2.8	2.5	9.3	12.1	6.8	21.1
March	1.1	10.0	11.1	3.4	7.7	9.3	12.7	1.6	22.7
April	3.0	10.0	13.0	3.0	10.0	9.3	12.3	−0.7	22.0
May	3.5	10.0	13.5	2.1	11.4	9.3	11.4	−2.1	19.9
June	4.8	10.0	14.8	1.0	13.8	9.3	10.3	−4.5	15.4
July	6.0	10.0	16.0	0.5	15.5	9.3	9.8	−6.2	9.2
Aug.	5.7	10.0	15.7	1.1	14.6	9.3	10.4	−5.3	3.9
Sept.	3.9	10.0	13.9	2.0	11.9	9.3	11.3	−2.6	1.3
Total	32.3	105.0	137.3	25.6	111.7	111.8	137.2		

and precipitation. The total water available (column 8) equals the sum of the effluent available and precipitation.

The monthly change in storage (column 9) from Eq. (14.24) is the difference between the total water available (column 8) and water losses (column 4). The total accumulated storage (column 10), computed by summing the monthly changes, reaches a maximum of 22.7 in. in March.

$$\text{basin capacity equivalent to this value} = \frac{22.7 \text{ in.} \times 120 \text{ acre}}{12 \text{ in./ft}}$$

$$= 227 \text{ acre} \cdot \text{ft} = 74 \text{ mil gal}$$

The wastewater effluent applied is 111.7 in. (9.3 ft) on an annual basis; the maximum rate, in July, being about 3.5 in./wk.

Example 14.8

Calculate the crop uptake of nitrogen required for the irrigation design outlined in Example 14.7. Base calculations on an allowable concentration of nitrogen in the percolate of 10 mg/l and an assumed denitrification loss of 20%. The average nitrogen content of the applied wastewater is 25 mg/l with a hydraulic loading of 1.0 mgd on 120 acres (112 in./yr). From Table 14.7, the percolation is 105 in./yr.

Solution:

$$\text{applied nitrogen} = \frac{1.0 \text{ mil gal/day} \times 365 \text{ days/yr} \times 25 \text{ mg/l} \times 8.34}{120 \text{ acres}}$$

$$= 634 \text{ lb/acre/yr}$$

$$\text{loss by denitrification} = 0.20 \times 634 = 127 \text{ lb/acre/yr}$$

$$\text{loss by percolation} = \frac{105 \text{ in./yr} \times 10 \text{ mg/l} \times 8.34}{36.8 \text{ acre} \cdot \text{in./mil gal}} = 238 \text{ lb/acre/yr}$$

From Eq. (14.25),

$$\text{crop uptake required} = 634 - 127 - 238 = 269 \text{ lb/acre/yr}$$

14.17 AGRICULTURAL IRRIGATION REUSE, TALLAHASSEE, FLORIDA

The city of Tallahassee, with a population of 130,000, is located in the panhandle of northwest Florida. The climate there is temperate, having rainy summers and mild rainy winters that produce an average annual rainfall of 60 in./yr (1500 mm/yr). With no industrial wastewaters and an abundant groundwater source, the municipal wastewater is essentially a low-strength domestic wastewater. Initially, biologically treated wastewater was discharged to surface waters, causing environmental degradation resulting in eutrophication. In 1980, the Southeast Farm with center-pivot irrigators was established for agricultural reuse to recycle all of the treated wastewater from two treatment plants. Agricultural reuse was the most economical and environmental method of disposal.

The principal process at the treatment plants is extended aeration followed by effluent chlorination. The waste-activated sludge is thickened by dissolved-air flotation, with the separated underflow returned to the plant inlet and dense sludge stabilized by anaerobic digestion. The digested sludge, averaging 3.2% suspended solids, is thickened in a screw press yielding 14% biosolids for a soil conditioner and fertilizer. As an alternate, liquid-digested sludge can be applied by spreader trucks. Both dewatered and liquid biosolids are spread on a 700-ac site of forest and grassland southwest of the city, which is not part of the irrigation operation at the Southeast Farm.

The primary plant is the Thomas P. Smith Wastewater Treatment Facility with a design flow of 27.5 mgd (104,000 m^3/d) and an actual average daily flow of 14 mgd (53,000 m^3/d) ranging between 12–17 mgd. The smaller plant is the Lake Bradford Road Wastewater Treatment Plant, with a design flow of 4.5 mgd (17,000 m^3/d) and an actual average flow of 2.9 mgd (11,000 m^3/d) ranging between 1.8–3.4 mgd. Table 14.8 lists typical average yearly influent and effluent wastewater characteristics for these plants. The Florida Department of Environmental Protection sets effluent limits for these plants at a carbonaceous BOD of 20 mg/l, total suspended solids of 20 mg/l, and fecal coliform of 200 per 100 ml. For the Thomas P. Smith plant, operating at approximately one-half design flow, the high quality of the effluent can be attributed to the low organic loading (low food/microorganism ratio) and temperate climate, resulting in a warm wastewater temperature. After pumping to the Southwest Farm, the total nitrogen concentration in the reuse water is 8–10 mg/l, probably lower than the effluent as a result of denitrification.

TABLE 14.8 Typical Average Yearly Wastewater Flows and Influent and Effluent Wastewater Characteristics of Two Tallahassee Treatment Plants

	Thomas P. Smith		Lake Bradford Road	
	Influent	Effluent	Influent	Effluent
Average flow (mgd)	14.0		4.5	
Carbonaceous BOD (mg/l)	102	5	165	8
BOD (mg/l)	129	11	201	15
Total suspended solids (mg/l)	170	9	269	16
Total phosphorus (mg/l)	4.7	1.7	5.4	2.8
Total Kjeldahl nitrogen[a] (mg/l)	29.8	6.2	35.7	22.6
Nitrate nitrogen (mg/l)		5.6		2.5
Total nitrogen (mg/l)		12		25
Fecal coliform (number/100 ml)		4		18

[a] Organic and ammonia nitrogens.

The effluents from the treatment plants are discharged into a common holding pond with a capacity of 12 mil gal near the Thomas P. Smith plant. In addition to storage, the primary purpose is flow equalization for pumping by three constant-speed pumps to the Farm ponds through a 36-in.-diameter pipeline, 8.5 miles in length. Figure 14.23(a) is an aerial view of the four ponds at the Southeast Farm with a maximum capacity of approximately 140 mil gal. The only method of discharge is through the irrigation system; no surface discharge is possible. Even though all of the ponds are interconnected to receive reuse water, the three constructed in parallel are for emergency storage. Under normal operation, all inflow enters the pond on the right through an inlet at the far end. Pond water is drawn out by irrigation pumps through a traveling self-cleaning screen with a bar spacing of 6 mm, and then discharged through a large inline self-cleaning basket strainer with clear openings of 1.0 mm^2. This is necessary to prevent plugging of the irrigation nozzles. Solids from both the bar screen and strainer are drained on gravity screen filters and discharged to a dumpster. The drainage is returned to the holding pond.

Figure 14.23(b) is an aerial view of a typical center-pivot circle of 134 ac (54 ha) with a diameter of 2720 ft (800 m) sprinkling grass on one half and corn on the other half. Center-pivot irrigators can be operated in a full circle (at an adjustable rotational speed) or over only one-half of the circle, back-and-forth reversing directions. This option allows better management of watering different crops and environmental control of nutrients, particularly nitrogen. Bermuda grass has excellent nutrient uptake capacity at even a high rate of watering. The uptake of nutrients by corn or soybeans varies with maturity of the crop and application of water. At harvest time, corn can be allowed to dry while still applying water on the grass.

A center-pivot irrigator is a series of spans connected by flexible joints at the support towers as diagrammed in Figure 14.24(a). To sprinkle water over a circular area, each tower has two large wheels with rubber tires so that the irrigator can rotate around the pivot end. The last tower regulates movement of the entire irrigator. The other towers are controlled as the end tower moves by use of microswitches and mechanical linkage to maintain alignment. The curvilinear pipes over the top of the spans convey the water. The pipeline is supported by braces, truss angles, and rods for structural stability. For application of water, spray nozzles are mounted on the crest of the pipeline because loss of the water by evaporation is not of concern [Fig. 14.24(c)]. (In the case of groundwater applied in a hot, arid climate, hose drops with nozzles hang down from the pipeline to distribute water close to the plants and reduce evaporation loss.) Water is pumped into an irrigator at a fixed design pressure to provide the design flow for proper distribution of water through the nozzles. Since the flow of water is constant, the only way to reduce the quantity of water applied per day is to periodically turn off the supply being pumped to the irrigator. Operation of irrigators in the field can be controlled by computer in the field office or from control panels at the pivot towers. The telemetry program monitors operational parameters plus alarms warning of system failures.

Reuse water must be screened prior to pumping to irrigators to reduce the incidence of nozzle plugging by suspended solids, algae, insect parts, and debris. The guideline for determining the allowable size of screen opening is that the particle size passing

FIGURE 14.23 Aerial views of Southeast Farm Wastewater Reuse Facility for Tallahassee, FL. (a) The holding pond on the right receives treated wastewater providing flow equalization between water supply and irrigation demand. (b) Center-pivot irrigation circle with a radius of 1360 ft (400 m) sprinkling grass on one half and corn on the other half, for a total of 134 ac (54 ha). *Source:* Courtesy of the city of Tallahassee.

FIGURE 14.24 Center pivot irrigation. (a) The irrigator is a series of spans from pivot to overhang with a curved pipeline on top supported by a truss for stability. (b) A center pivot sprinkling bermuda grass on the Southeast Farm. (c) An irrigator tower with large rubber tires and sprinkler nozzles mounted on the curved overhead pipeline. In the background are bundles of hay harvested from the field.

through the screen should not be greater than one-fifth (1/5) the area of the nozzle opening. For the smallest-diameter nozzle opening of 3.18 mm (0.125 in.), the area is 7.92 mm^2. One-fifth of this area is 1.58 mm^2, which is equivalent to a particle with a diameter of 1.42 mm. Therefore, a strainer or screen with a clear opening of 1.0 mm by 1.0 mm is the best selection, which is that of the inline basket strainer installed in the discharge pipe of the irrigation pumps at the Southeast Farm.

Center-pivot irrigation is the only suitable method of water application on highly permeable soils like the fine sandy soil at the farm site and, for this large a project, it is the only economical method of sprinkling. The largest irrigators at the Southeast Farm have nine 140-ft (42.7-m) spans for a total of 1260 ft, with an end gun that extends the irrigated distance another 100 ft for a total circle with a radius of 1360 ft (400 m) and watered area of 134 ac (54 ha) [Fig. 14.23(b)].

Bermuda grass, which is the most successful crop, is used as cattle feed either by herds of cows grazing both summer and winter in the fields or as harvested hay. To a lesser extent, ryegrass is grown for fodder. For maintenance of pivots and harvesting of crops, the irrigation circles are fenced with access roads between the fence lines. Fences that divide the irrigation circles into halves have spring-loaded gates to allow the pivot towers to pass through from one side of the field to the other side during rotation. During summer months, as many as 2000 cows are on the farm. Young stock arrive in the fall of the year and nearly double in weight before being shipped to market; pasture grass is supplemented with dry rations as necessary. Cattle shipped to market must be held in a feedlot for 30 days prior to slaughter.

Corn and soybeans are grown, and after harvest are stored in large silos and sold for revenue. For these crops, 0.25–0.50 in. of water is applied per irrigation pass to satisfy evapotranspiration of the plants without flushing the nutrients in the water past the root zone. Providing adequate nutrients for plant growth while reclaiming the applied reuse water requires careful irrigation control. Limiting the rate of water application for the first 8–10 weeks of growth also enhances root growth of the plants. The highly successful farming system derives from flexibility in management including crop selection, planting, harvesting, irrigation, nutrient scheduling, cattle feeding, and marketing. New crops are tested for suitability; for example, canola was planted for several years.

Arrangement of the 16 irrigation circles, holding ponds, farm buildings, silos, and groundwater monitoring wells is shown in Figure 14.25. Currently, 2000 ac of the 4000-ac farm site is under the 16 irrigators. Of the remaining 2000 ac, approximately 400 ac can be converted for irrigation with the remainder designated as tree areas and wetlands. The permitted water application with 16 irrigators is an average continuous application of 3.1 in. (79 mm) per week throughout the year, which is equivalent to the permitted flow of 23 mgd (87,000 m^3/d) of effluent from the treatment plants. Currently, the actual application rate is a yearly average of 2.3 in. (58 mm) per week.

The soil profile under the farm site is a fine sand layer up to 40 ft in thickness, with a vadose zone of 10–20 ft, overlying the limestone Floridan Aquifer, which is a source of groundwater for drinking water supplies. (The vadose zone is the unsaturated soil profile above the water table.) The rate of water application in irrigation greatly exceeds evapotranspiration of the crops since the primary purpose of the Tallahassee water reuse is environmentally safe disposal of wastewater rather than crop yield. Consequently, the

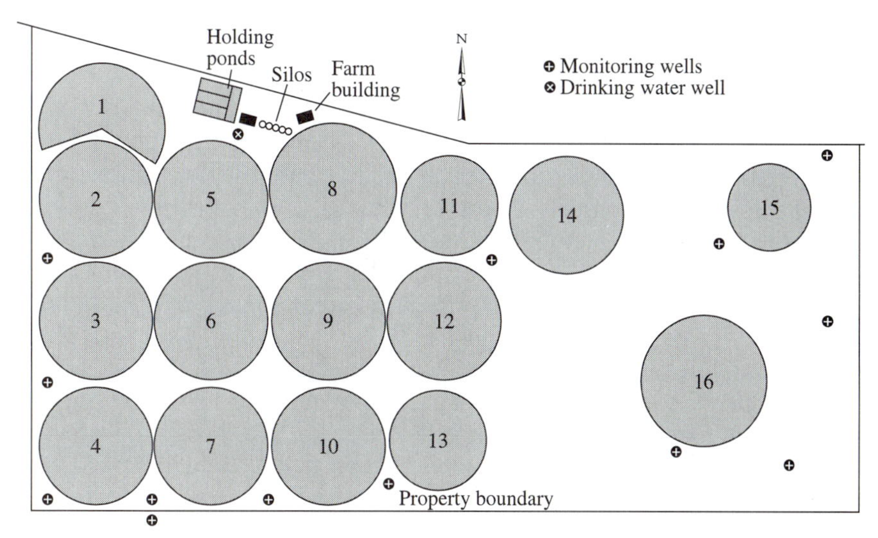

FIGURE 14.25 Layout of the 4000-acre Southeast Farm showing 16 irrigation circles, holding ponds, farm buildings, silos, and locations of monitoring wells. *Source:* Courtesy of the city of Tallahassee.

underlying limestone aquifer is recharged by percolation of irrigation water. Fine sand is effective in removing suspended solids, organic matter, microorganisms, and phosphorus [27]. The "living filter" provided by deep-rooted bermuda grass and ryegrass, along with other harvested crops, is effective in removal of nitrogen and other plant nutrients [27]. In addition, slow seepage through the vadose zone can result in nitrogen loss by nitrification and denitrification, and laminar groundwater flow through the saturated fine sand completes reclamation by adsorption.

Groundwater monitoring wells are located throughout the farm site at varying depths in the saturated sand layer inside the irrigated areas and at depths up to 120 ft upstream and downstream of the groundwater movement, which flows from the northeast to southwest toward the Gulf of Mexico. Slightly elevated nitrate–nitrogen levels may be recognized in samples of groundwater withdrawn from the interior monitoring wells; however, by the time the groundwater flow reaches the southern boundary of the farm, the concentrations at 40 ft are greatly diminished or altogether nonexistent. Analyses of deeper monitoring wells upstream and downstream from the farm site can be used to identify any changes in quality related to groundwater recharge under the farm site.

The Florida Environmental Protection Agency requires comprehensive testing and reporting of contaminants in groundwater withdrawn from monitoring wells. The monitoring report for each 3-month period requires analyses of groundwater from all monitoring wells for the following: arsenic, cadmium, chloride, chromium, fecal coliform, filterable residue, lead, nitrate, pH, and sulfate. Groundwater from the underlying aquifer is analyzed in accordance with drinking water regulations for the following: inorganic, volatile organic, and synthetic organic chemicals; radionuclides; disinfection by-products; and secondary standards. The results of all analyses on record are less than the maximum contaminant levels established for drinking water.

14.18 CITRUS IRRIGATION AND GROUNDWATER RECHARGE, ORANGE COUNTY AND CITY OF ORLANDO, FLORIDA

The primary reasons for reuse of reclaimed waters by Orange County and the city of Orlando are the need for effluent disposal, conservation of the groundwater source, citrus irrigation, recharge of groundwater by infiltration basins, and urban landscape irrigation. The Water Conserv II process flow diagram is illustrated in Figure 14.26. Reclaimed water from the county's South Regional Water Reclamation Facility and the city's McLeod Road Water Reclamation Facility are connected by a 21-mi transmission pipeline to the Water Conserv II Distribution Center in West Orange County. Forty-three miles of distribution piping serve 76 agricultural areas, with 11,500 ac and five infiltration sites that contain 58 rapid infiltration basins (RIBs) on 2000 ac. "Water Conserv II has become one of the largest reuse projects in Florida, now supplying reclaimed water to 60 citrus growers, nine landscape and foliage nurseries, three tree farms, the Orange County National Golf Center, and 17.9 mgd (68,000 m^3/d) to the RIB system" [28].

The Orange County South Regional Water Reclamation Facility, shown in the aerial view of Figure 14.27, has a design flow of 30.5 mgd (115,000 m^3/d), current average flow of 22.5 mgd (85,200 m^3/d), and peak flow of 32 mgd (121,000 m^3/d). The raw wastewater is composed of approximately 55% domestic, 25% commercial, and 20% industrial waters. To prohibit toxic wastewaters, regulations are established for pretreatment at industrial sites to remove toxic substances prior to discharge to the sewer system. The two wastewater aeration processes, without primary sedimentation, are the original *Carrousel*® activated-sludge plant and the newer and larger diffused-air activated-sludge plant. The diffused-air process has two physically separated stages, with the first operated as an anoxic zone for denitrification of return activated sludge mixed with the raw wastewater and the second stage for nitrification. Nitrification–denitrification is operated by a progammable logic controller to allow automated adjustment of aeration-tank air valves and blower inlet valves. Typical process data are as follows: At a raw wastewater flow of 13.5 mgd and return activated-sludge flow of 10.8 mgd, the influent has a Kjeldahl nitrogen concentration of 39 mg/l and the effluent after final clarification has a Kjeldahl nitrogen concentration of 1.1 mg/l and nitrate nitrogen concentration of 7.0 mg/l. The influent carbonaceous BOD is 215 mg/l and settled effluent 4.3 mg/l [29]. Following secondary clarification, both activated-sludge aeration plants have peak storage for flow equalization prior to tertiary treatment by chemical coagulation using alum and polymer, granular-media filtration, and plug-flow chlorination with an extended contact time. The two ground-level storage tanks are for holding effluent prior to pumping to the Water Conserv II Distribution Center and other local users of reclaimed water. For solids processing, the waste-activated sludge is thickened by dissolved-air flotation and applied to anaerobic digesters. The stabilized sludge is dewatered by belt filter presses and biosolids applied on land.

The quality of the reclaimed water is microbiologically and chemically safe for unrestricted reuse (public access), except as a direct drinking water source. For urban irrigation, reuse requires a water free of all potential human pathogens, including enteric viruses. To prevent clogging of microsprinklers in citrus irrigation, the limit established for total suspended solids is 5 mg/l and the distributed water is chlorinated to 0.5 mg/l to

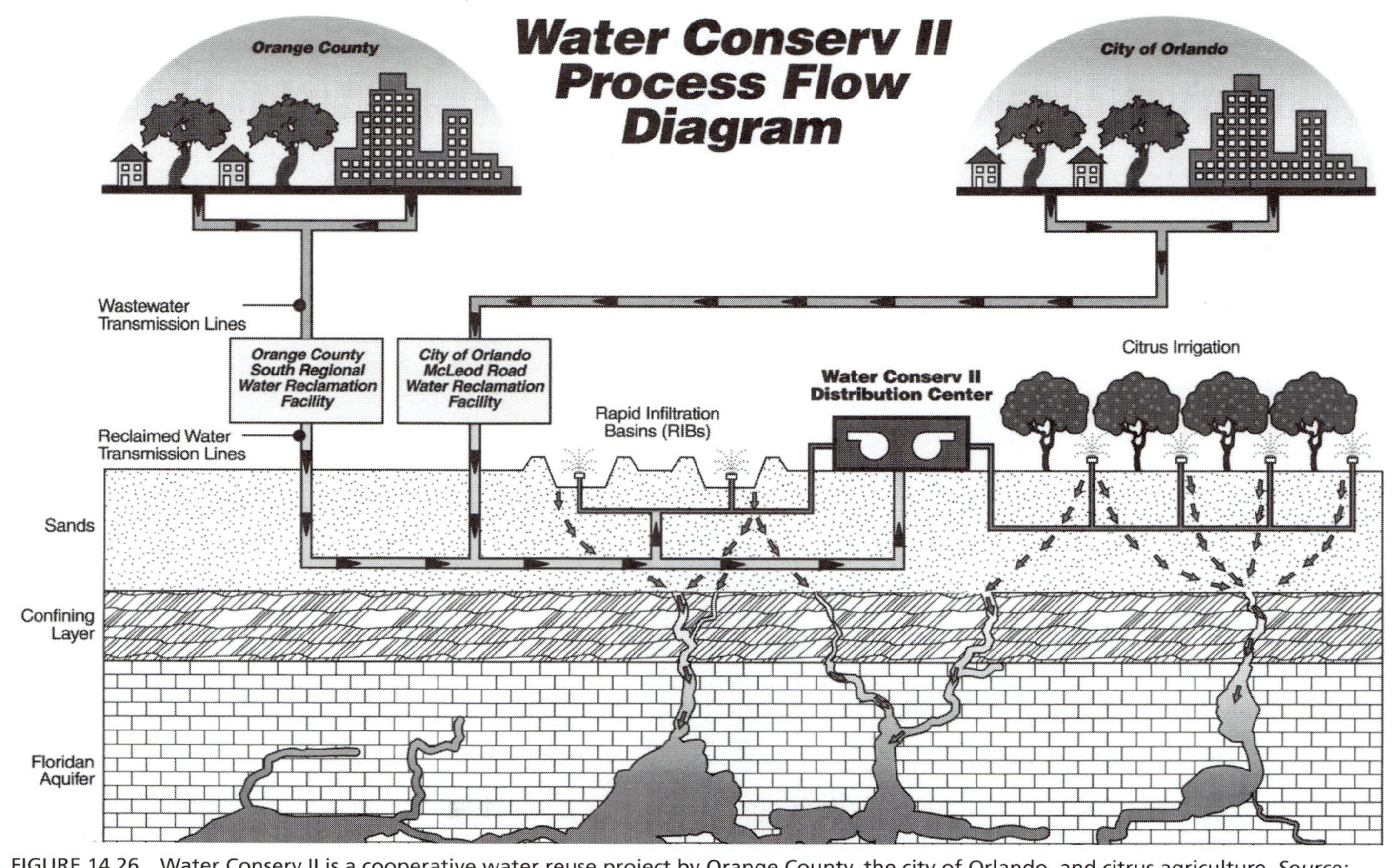

FIGURE 14.26 Water Conserv II is a cooperative water reuse project by Orange County, the city of Orlando, and citrus agriculture. *Source:* Courtesy of Orange County, FL.

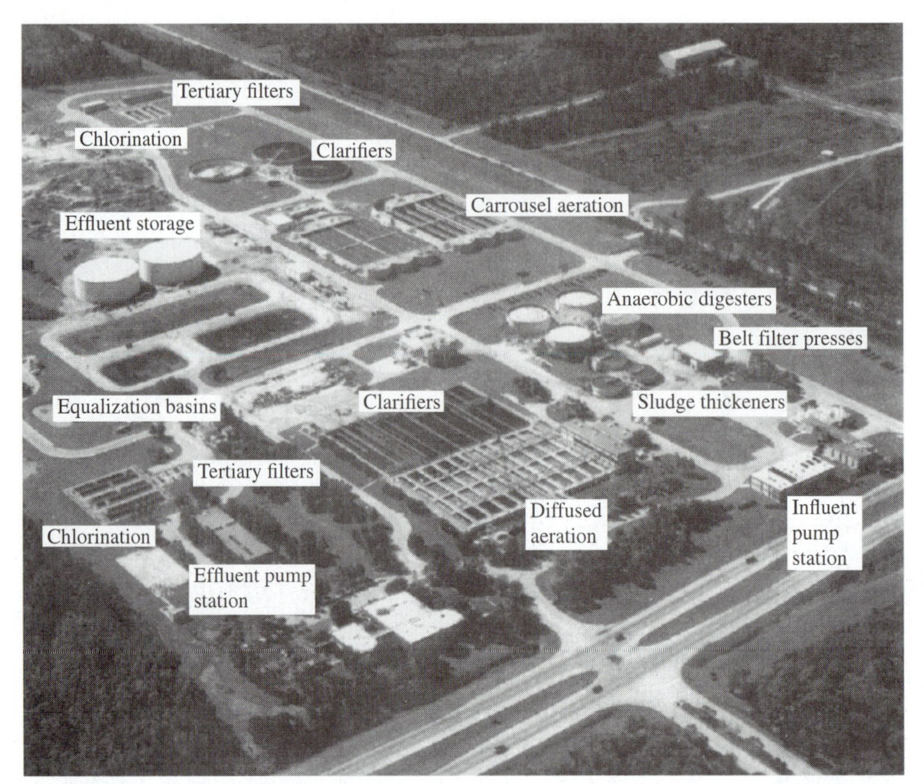

FIGURE 14.27 Orange County South Regional Water Reclamation Facility with tertiary treatment to produce a reclaimed water microbiologically and chemically safe for unrestricted reuse. *Source:* Courtesy of Orange County, FL.

prevent bacterial growth. In addition, the quality limits for citrus irrigation water includes 27 chemical limits (arsenic through zinc) plus BOD, COD, conductivity, and pH. Reclaimed water applied to rapid infiltration basins cannot exceed 10 mg/l of nitrate nitrogen and must be free of toxic subtances. Samples of recharged groundwater withdrawn from monitoring wells located in infiltration sites must meet drinking water standards.

The Water Conserv II Distribution Center is about 20 miles west of the city of Orlando surrounded by citrus groves as shown in Figure 14.28(a). The center started operation in 1986, involving a few citrus growers and the rapid infiltration system. The irrigation project is permitted for 44 mgd (167,000 m^3/d), designed for 50 mgd with peak flows to 75 mgd. Daily flows transmitted to the distribution center average 30 mgd (114,000 m^3/d). Water supplied to the center is pumped into four storage tanks for flow equalization, each with a capacity of 5 mil gal. If the tanks are full, the excess water flows into an overflow standpipe that discharges to nearby rapid infiltration basins (RIBs). The distribution pump station has a total of 8 vertical turbine pumps with a total capacity of 76,000 gpm. Sixty percent of the reclaimed water is piped to agricultural and commercial users and 40% is distributed to RIB sites.

The project currently delivers water to 74 customer turnouts, which include irrigation of 4500 ac (1800 ha) of citrus and nurseries. The distribution network is

FIGURE 14.28 Citrus irrigation. (a) Water Conserv II Distribution Center. (b) Citrus trees under irrigation. (c) Rapid infiltration basins for groundwater recharge. *Source:* Courtesy of Orange County, FL.

approximately 49 miles of pipeline, ranging in size from 6-in. in diameter to 54-in. in diameter, and buried electric and instrumentation cables. The typical piping for a citrus grove is a buried water main up to 12 in. in diameter or less, with buried submains 6 in. or less and risers for each row of trees. From each riser, a 1-in. plastic tube is laid above ground along a tree line with one microsprinkler under each tree [Fig. 14.28(b)].

Twenty-five supplemental water wells with a total production of 80 mgd (303,000 m^3/d) are strategically located on the distribution network to provide water for citrus freeze protection and drought protection when reclaimed water supply is inadequate. The normal daily irrigation flow averages 18 mgd (68,000 m^3/d), while freeze protection demand currently exceeds 100,000 gpm (380 m^3/min), or 150 mgd (568,000 m^3/d), which is more than eight times the normal flow rate. In Orange County, devastating freezes occurred in 1983, 1985, and 1989, putting many citrus growers out of business.

Five rapid infiltration basin (RIB) sites with a total of 58 RIBs and an infiltration capacity of 17.9 mgd (67,800 m^3/d) receive the reclaimed water not needed by agricultural and commercial users [Fig. 14.27(c)]. The primary purpose of rapid infiltration is to recharge the groundwater while disposing of surplus reclaimed effluent from the treatment plants. Approximately 10 miles of pipeline, ranging in size from 6 in. to 24 in. in diameter, and 10 miles of communication and 480-volt electric cables provide communication and electricity to the RIBs. Each RIB site and individual RIB is computer-controlled from the operations building at the distribution center. The computerized management system provides the ability to forecast the impact on groundwater storage of loading the RIBs at precribed rates and duration. Each of the 58 RIBs has 1 to 5 cells, totaling 124, for a total bottom percolation area of 166 ac (67.2 ha). The total area of RIB sites is 2000 ac (809 ha). The location of a RIB site is selected based on permeability of the natural sand profile and topography. In warm weather, the sand is disked to inhibit weed growth and maintain percolation.

14.19 URBAN REUSE

The predominant urban reuse is landscape irrigation, with a variety of lesser nonpotable commercial applications.

Restricted landscape irrigation can be used in areas where public access is limited and/or where the water application is controlled to prevent direct contact with people. Examples are areas landscaped for beautification only, highway median strips, and golf courses. Where transient human activities take place in the area, the water is applied only during night hours without airborne drift or surface runoff into public areas. The vegetation is allowed to dry, and excess water is allowed to soak into the ground before start of use. Warning signs should be prominently displayed, reading "Irrigation with Reclaimed Wastewater," or a similar warning, and faucets should be posted with signs reading "Reclaimed Water, Do Not Drink," or a similar warning, or be secured to prevent public use. The irrigation piping should be color-coded, buried with colored tape, or otherwise suitably marked to indicate nonpotable water.

Unrestricted landscape irrigation with a high-quality reclaimed water can be used on public parks and playgrounds, private and public lawns and gardens, and other places where people might contact the reclaimed water. This quality of water can be used to spray-irrigate areas where airborne drift or surface runoff into public areas may occur; however, the timing of watering should minimize public contact. Warning signs, secure faucets, and marking of piping as nonpotable water are required.

The allowable limits for reclaimed water quality for reuse in restricted and unrestricted landscape irrigation are given for California and Arizona in Section 14.15. For restricted irrigation, high-quality biological treatment and chlorination can meet the coliform standards. The water applied for unrestricted irrigation of parks, playgrounds, lawns, and gardens must be free of pathogens; thus, the standards are very restrictive, requiring tertiary filtration and disinfection.

14.20 URBAN REUSE, ST. PETERSBURG, FLORIDA

The city of St. Petersburg is located on a peninsula between Tampa Bay and the Gulf of Mexico. Since only insignificant potable groundwater is available, the municipal supply is imported from well fields in adjacent counties up to 60 miles away. Located 30 miles to the north, the water treatment facility collects the groundwater from well fields, processes it by lime–soda ash softening, and pumps it to the city. With population growth in the region, competition increased for the available groundwater resources. Concurrently, the Florida legislature in 1972 enacted a bill requiring advanced wastewater treatment for discharge to Tampa Bay with effluent limits of 5 mg/l BOD, 5 mg/l suspended solids, 3 mg/l total nitrogen, and 1 mg/l phosphorus. In order to both conserve drinking water and avoid the extremely high cost of wastewater-reclamation processing, the city chose to upgrade its biological treatment plants to allow unrestricted urban reuse and deep-well injection for zero discharge. The tertiary treatment added is effluent granular-media filtration and chlorination with an extended contact time for effective pathogen removal. The reclaimed water system began operation in 1978, and effluent discharge to Tampa Bay ceased in 1987.

The city has four tertiary plants with a total capacity of 68.4 mgd. Figure 14.29 is a picture of the southwest 20-mgd plant showing the major unit processes. After screening and grit removal, unsettled wastewater is treated in two long rectangular

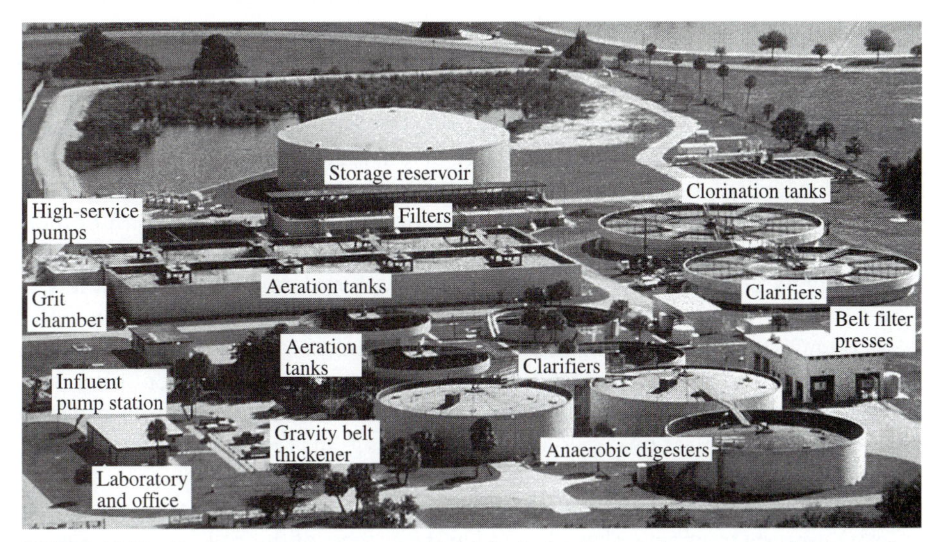

FIGURE 14.29 Tertiary treatment plant with biological activated-sludge processing, granular-media filtration, and chlorine disinfection for unrestricted urban reuse in St. Petersburg, FL. *Source:* Courtesy of the city of St. Petersburg.

tanks each with four platform-mounted mechanical aerators in series and two circular aeration tanks. Because of the warm wastewater temperature of 20°–25°C, the activated-sludge processes can be operated at a BOD loading of up to 50 lb/1000 ft^3/day. The flocculator–clarifiers have both radial and circumferential weirs for efficient settling of biological solids. Following clarification, the wastewater is filtered through four dual-media coal–sand beds at a nominal rate of 2.5 gpd/ft^2, with alum applied as needed for coagulation. Filtered water is chlorinated in long narrow contact tanks with a detention time of 40 min based on average daily flow and then pumped into a 5-mil-gal storage tank. Filter backwash water is discharged to a retention basin and returned at a uniform rate to the plant inlet. After thickening on a gravity belt, the waste-activated sludge is anaerobically digested, dewatered by belt filter presses, and applied on agricultural land.

The reclaimed water is pumped from storage into the distribution pipe network. During times of wet weather resulting in low reuse, the excess reclaimed water is pumped to injection wells that discharge into a saltwater aquifer 900 ft below ground. Reject water that does not meet reuse standards is also injected for disposal.

The quality criteria of reclaimed water for unrestricted urban reuse established by the state of Florida are: 20 mg/l of carbonaceous BOD as an annual average; a maximum of 5 mg/l of suspended solids in any single sample; minimum chlorine residual of 1 mg/l after 15 min contact time at peak hourly flow; and fecal coliform concentrations below detection in 75% of the samples tested during a 30-day period, with no sample greater than 25 per 100 ml. The monitoring requirement for the suspended-solids limit is daily sampling of the effluent prior to disinfection and for fecal coliforms is one sample per day of the disinfected effluent. All four of the tertiary plants meet the standards imposed by these limitations. Effluent suspended solids are in the range of 1–3 mg/l and fecal coliforms are rarely detected. Rather than the minimum chlorine residual of 1 mg/l required, the operating standard is a minimum of 4.0 mg/l at the outlet of the contact tank.

An unexpected quality problem encountered shortly after initiating residential irrigation was damage to ornamental plants and trees. The results of a research study revealed that high chloride levels affect certain ornamental plants. However, if the concentration is below 600 mg/l, only a few plant species show significant adverse effect. Much of the reported plant damage, presupposing the quality of reclaimed water, was actually attributable to severe drops in air temperature and other climatic changes. All varieties of turf grasses commonly used in central Florida display excellent appearance and growth when irrigated with reclaimed water with moderate chloride concentrations. Nevertheless, application rates in excess of 1.5 in./wk increases weed infestations and incidence of fungal diseases. Excessive chloride concentrations have been observed in the wastewater entering only one treatment plant serving two beach communities. During drought conditions, saltwater infiltration can increase the chloride concentration to 700 mg/l, resulting in the necessary rejection of the reclaimed water for reuse.

The reclaimed water system illustrated in Figure 14.30 shows the arterial pipe network, tertiary treatment plants located in the corners of the city, and locations of irrigated parks, school yards, and golf courses. The arterial mains are ductile iron 12–36 in. in diameter and secondary mains are a light-purple-color PVC plastic 4–12 in. in diameter for a

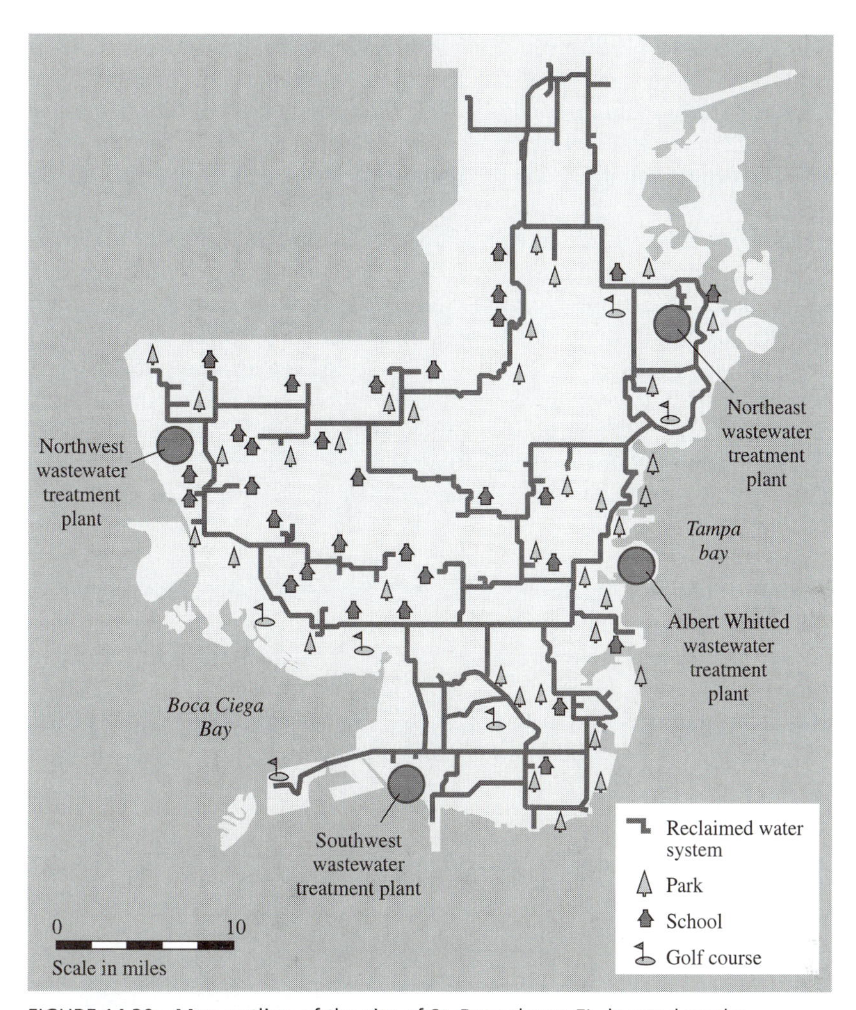

FIGURE 14.30 Map outline of the city of St. Petersburg, FL, located on the peninsula with Tampa Bay on the east and Boca Ciega Bay on the west. Reclaimed water system showing four tertiary treatment plants, arterial pipe network, and locations of irrigated parks, schoolyards, and golf courses. *Source:* Courtesy of the city of St. Petersburg.

total of approximately 300 miles of pipe. Distribution pressures are maintained in the range of 50–85 psi, which allows the hydrants to provide secondary fire service. In addition to irrigation of recreational areas and that provided to commercial customers, more than 95% of the 9000 connections are for residential irrigation. Backflow prevention devices are installed in the potable water service connections to residential customers to isolate the public drinking water supply from potential cross connections that may be installed illegally by homeowners.

During the dry season of greatest usage, normally March–June, the typical residential lawn requires about 1.5 in./wk or 30,000 gal/month, while the average wastewater discharge per household is 6000 gal/month. In other words, five residential sewer connections are required to produce enough for one reclaimed water connection. Therefore, only 20% of the 80,000 potential residential customers are able to receive reclaimed water. Because of the seasonal variation in demand, approximately 50% of the annually available reclaimed water is injected into the deep wells for disposal.

Economic benefits of the reclaimed water system are the reduced cost of tertiary wastewater treatment compared to the water reclamation required for surface discharge, substantial reduction in potable water consumption, and beneficial fertilization of lawns and gardens. Perhaps the greatest of these has been the delay in expansion of the drinking water system, including well fields, treatment works, transmission pipelines, and distribution mains. St. Petersburg has increased in population and commercial activity since 1976 without significant increase in potable water consumption because of reclaimed water usage, as diagrammed in Figure 14.31. Prior to the reuse system, the average potable water usage was 435 gpd per residence; after the availability of reclaimed water the usage decreased to 220 gpd.

The city has established a fee schedule to offset the costs of installation of distribution piping and operation and maintenance. Service to a residential area requires at least 50% approval by homeowners. Depending on lot size, the fee assessed against each property is \$1200–\$1500, payable over a period of 3–10 yr. Then each resident requesting reclaimed water service pays a one-time connection fee of \$300. Currently, for an unmetered service, the usage fee is \$11 per month for one acre of property or less and \$6 for each additional half-acre and, for a metered service, \$0.30 per 1000 gal, with a minimum of \$11 per month.

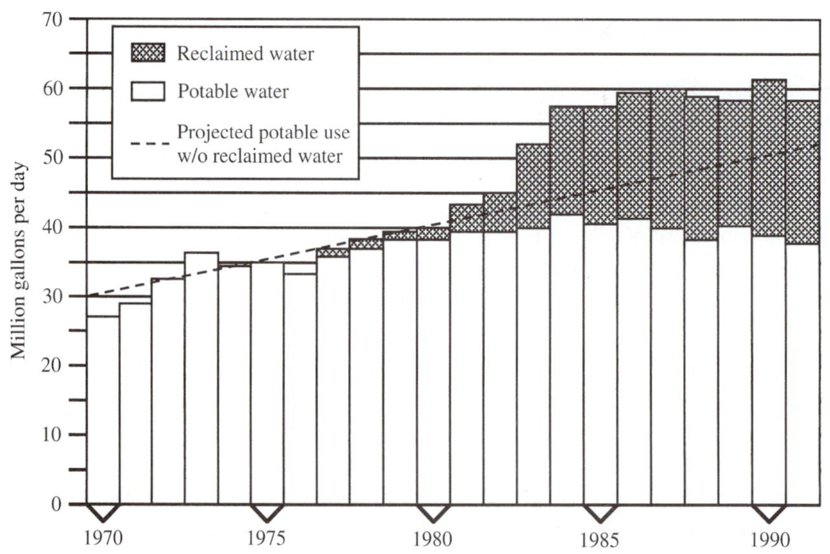

FIGURE 14.31 Estimated potable water conservation achieved through urban reuse by the city of St. Petersburg. *Source:* "Manual of Guidelines for Water Reuse," Environmental Protection Agency, EPA 625/R-92/004 (September 1992).

14.21 INDIRECT REUSE TO AUGMENT DRINKING WATER SUPPLY

Groundwater recharge from surface infiltration of wastewater may be intentional, for aquifer recharge, or inadvertent, resulting from disposal of wastewater. Deterioration of groundwater quality because of infiltration depends on the characteristics of the recharge water, removal of contaminants by the soils, and dilution by the groundwater in storage. Before undertaking aquifer recharge with the intention of recovering the reclaimed water as a drinking water supply, a demonstration project is recommended to evaluate the feasibility of storage and controlled subsurface movement of the recharged water, improvement of recharge water quality by filtration through the soil profile, and treatment of the reclaimed water necessary to maintain drinking water quality in the recovered groundwater withdrawn from wells. Numerous monitoring wells are necessary for tracking water movement and sampling for water quality testing.

The minimum quality of reclaimed water for surface infiltration, from basins or river channels, is equivalent to the quality for unrestricted landscape irrigation with processing by biological treatment, chemical coagulation, granular-media filtration, and chlorine disinfection. Where the soils do not provide significant removal of contaminants, and dilution of recharged water in the aquifer is limited, more extensive treatment is necessary to ensure that the groundwater remains in compliance with drinking water standards. Injection of reclaimed water into an aquifer through recharge wells requires that the water meet drinking water standards. Water reclamation is required to remove microorganisms, heavy metals, organic chemicals, and, if necessary, inorganic salts. The sequence of biological, chemical, and physical processes often includes activated sludge, lime precipitation, denitrification, granular-media filtration, carbon adsorption, and disinfection (Section 14.22) or filter screening, microfiltration, reverse osmosis, and disinfection (Section 14.24).

The rules for *Aquifer Water Quality Standards, State of Arizona* [30] state the following for wastewater discharges:

> A discharge shall not cause a pollutant to be present in an aquifer classified for [protected use as drinking water] in a concentration which endangers human health. [Nor] ... which impairs existing or reasonably foreseeable uses of water in an aquifer.

Dry riverbeds and ephemeral streambeds in Arizona are often permeable and the groundwater has little dilution capacity for infiltration of wastewater. Therefore, all aquifers are classified as protected for drinking water use, except for those reclassified as nondrinking water. The most likely reasons for reclassification are existing contamination, either from wastewater or natural origins, or that the aquifer is not used as a drinking water source and contamination cannot be prevented.

The degree of treatment of wastewaters discharged where groundwater is potable and susceptible to contamination is optimal treatment, which has been demonstrated to be reliable, disregarding the economic limitations of costs of construction and operation. Requiring this highest possible degree of treatment promotes effluent reuse as an alternate to discharge. The significant cost savings by eliminating the need for a complex water reclamation plant is a major economic benefit. In Arizona, water is a scarce resource and augmentation by water reuse is incorporated in water policy.

The 10 wastewater reclamation plants of the Sanitation Districts of Los Angeles County have recharged groundwater for indirect use by infiltration in dry or ephemeral watercourses and recreational impoundments for the past two decades. The tertiary treatment process produces an effluent that meets drinking water standards for heavy metals, organic chemicals, radionuclides, and microorganisms (Section 14.5). Replenishment of the groundwater basin is the most efficient, cost-effective means of storage for recycled water along with natural runoff and imported surface water for subsequent withdrawal from wells as a drinking water supply. The California *Draft Regulations for Groundwater Recharge Reuse* propose including the nonregulated contaminants of total organic carbon (TOC) and total nitrogen [31]. TOC, which is the carbon that is bound in a variety of organic compounds in recycled water, would become a quality parameter for groundwater replenishment by both surface spreading and direct injection. Allowable nitrogen concentration in the recycled water would be based on total nitrogen to assure the recharged groundwater will not exceed the maximum contaminant levels for inorganic nitrogen. The draft regulations also require the installation of monitoring wells to test the water in the mound formed by the recharge water for TOC and nitrogen compounds.

Epidemiological studies have been conducted by Orange and Los Angeles counties in the Montebello Forebay area of Los Angeles County to examine health risks associated with exposure to reclaimed water recycled to replenish the groundwater supply since 1962 [32]. When the five-year study was initiated in 1978, the reclaimed water supplied about 16% of the total percolation flow after blending with local storm water and river water prior to infiltration. The second study evaluated health outcomes from 1987 to 1991. "Overall, neither study observed consistently higher rates of either general or specific mortality or morbidity in the populations who lived in areas receiving higher percentages of reclaimed water" [32]. "Although the overall analyses do not suggest an association between adverse outcomes and reclaimed water in the drinking water supply, the public health significance of no or low rates of incidence (of chronic diseases), especially for specific outcomes, should always be interpreted in the light of the statistical power of the study to detect an elevated risk" [32].

Every five years, the U.S. Geological Survey compiles and disseminates estimates of water use in the United States [33]. In addition to water withdrawals, public-supply deliveries, and consumptive use, treated wastewater releases are compiled for publicly owned (municipal) treatment plants. In Table 14.9, wastewater releases under the subheading "return flow" are the quantities of treated wastewaters returned to surface waters. Under the heading "reclaimed water," the releases are for beneficial uses including agricultural irrigation, urban reuse, and groundwater recharge for indirect reuse to augment drinking water supplies. Reclaimed water quantities are tabulated for the 9 states with the largest releases. Seven other states released 1 to 10 mgd, and the remaining 34 states had no recorded reclaimed water releases. About 16,400 publicly owned plants released 41,000 mgd of treated wastewater in 1995. In addition, over 2% (983 mgd) of the treated wastewater that was released was reclaimed for beneficial uses. Arizona, California, and Florida reported the largest uses of reclaimed water, each exceeding 200 mgd. Arizona has a semiarid climate and because of limited water resources strongly encourages water reuse (Section 14.15). Arizona's reuse is evidenced by the large portion of wastewater releases that are reclaimed (see Table 14.9).

TABLE 14.9 Treated Wastewater Flows Returned to Streams or Other Surface Water Bodies and Treated Wastewater Reclaimed for Beneficial Uses

		Wastewater Releases	
State	Number of Municipal Treatment Plants	Return Flow (mgd)	Reclaimed Water (mgd)
Arizona	150	359	209
California	1049	3250	216
Colorado	393	422	11
Florida	387	1540	271
Maryland	161	422	70
Nevada	68	179	24
South Carolina	274	404	22
Texas	1308	2180	96
Utah	50	236	39

Source: Adopted from W. B. Solley, R. R. Pierce, and H. A. Perlman, *Estimated Use of Water in the United States in 1995, Wastewater Release: U.S. Geological Survey Circular 1200* (Denver, CO: U.S. Geological Survey, 1998).

In southern California, indirect reuse of recycled water to augment groundwater supplies is being expanded because of the finite supply of imported water and the depletion of groundwater in storage. Orange County Water District is expanding its groundwater replenishment system (Section 14.24). In Florida, reuse of wastewater is an important issue in water resources. Wastewater is commonly given tertiary treatment for agricultural irrigation (Sections 14.17 and 14.18) and for urban reuse (Section 14.20). In addition to the beneficial reuse in irrigation, tertiary treatment is less costly than the advanced wastewater treatment required for discharge to environmentally sensitive surface waters. Indirect reuse to recharge groundwater aquifers for public supplies is being evaluated for regions of Florida where demand of the expanding population is exceeding the groundwater supply.

Direct Potable Reuse

None of the water reuse regulations recommend reclaiming wastewater directly as drinking water [17]. The highest level of reclamation is for groundwater recharge after advanced treatment in an attempt to comply with drinking water standards. Although a surface water supply may be polluted, usually only a small percentage of the raw water is actually derived from treated wastewater. Thus, the practice of indirect potable reuse through polluted water sources is only marginally related to direct water reuse. The revision of drinking water regulations, currently being undertaken nationally and internationally, is resulting in much more restrictive maximum contaminant levels for more chemicals, and for many of the trace organic compounds the proposed maximum contaminant level goal is zero. Although monitoring and advanced treatment technologies are available with respect to pathogens, inorganic chemicals, and radionuclides, the large number and low concentrations of organic chemicals pose uncertainties. Most have not been identified or quantified. The health effects over a lifetime of consumption of these trace organic chemicals are unknown. Therefore, direct reuse of reclaimed wastewater for a domestic water supply, even when diluted with a natural water, is not considered safe. Until the effectiveness of fail-safe advanced treatment can be clearly demonstrated, the risk of wastewater contaminants being present in the reclaimed water is too great to justify domestic reuse, particularly when natural water sources are available.

14.22 FRED HERVEY WATER RECLAMATION PLANT, EL PASO, TEXAS

The reclaimed water from the Fred Hervey Water Reclamation Plant recharges groundwater in the Hueco Bolson aquifer through injection wells. The plant has been in operation since 1985. Limited amounts of reclaimed water are also provided for cooling water to an electric generating station and irrigation water for a golf course. Nevertheless, the primary purpose of the Fred Hervey Plant is to augment the municipal water supply for El Paso.

The flow diagram in Figure 14.32 shows the major processes used for water reclamation. After preliminary screening and grit removal, conventional primary sedimentation clarifies the raw wastewater. The sludge is stabilized by anaerobic digestion and dried on open beds prior to hauling to a landfill. Storage ponds are used for flow equalization. During the daytime, the settled wastewater flow rates greater than the average are diverted and, during periods of lower flow, the stored wastewater is pumped back to maintain a constant flow through the plant. The storage ponds, which were originally stabilization ponds for raw wastewater treatment, can store 300 mil gal to allow complete shut-down of the plant for major repair, if necessary.

The biological activated carbon process, BAC, (Section 14.7), and the subsequent biological denitrification process (Section 14.13) are linked to the same carbon wet-air regeneration system. The current hydraulic loading is 7.6 mgd, 76% of the 10-mgd design capacity. The BAC process is activated sludge with coarse-bubble diffused aeration. The addition of powdered activated carbon (PAC) is for the purposes of adsorbing toxic chemicals, enhancing the removal of organic matter, and nitrification. The concentration of ammonia leaving the aeration tank is less than 1 mg/l and the BOD is reduced to 5 mg/l or less. Based on total suspended solids, the solids retention time is 18 days, ranging 10–31 days, and the mixed-liquor suspended solids (MLSS) averages 13,500 mg/l, including an ash content averaging 5300 mg/l (40%). The biological denitrification process uses methanol as a soluble carbon source to accept oxygen in reducing nitrate to nitrogen gas. The solids retention time is 15 days, ranging 7–26 days, and the MLSS averages 2000 mg/l, with an ash content averaging 400 mg/l (20%). By carrying the PAC through both processes, the contact time is extended to effectively adsorb toxic chemicals.

Waste BAC sludge is processed by wet-air regeneration first by thickening to an average solids content of 6.8% with supernatant recycled to the plant inlet. After heating the sludge in a heat exchanger to 475°F with a system pressure of 820 psi, it enters a reactor at a flow rate of 60 gpm for thermal regeneration of the PAC in a low-oxygen environment. Carbon loss in regeneration is about 5%. When the sludge flow rate decreases to 30 gpm, to keep the process thermally self-sustaining the process is shut down to descale the heat exchangers. Gravity separation of the ash from the regenerated

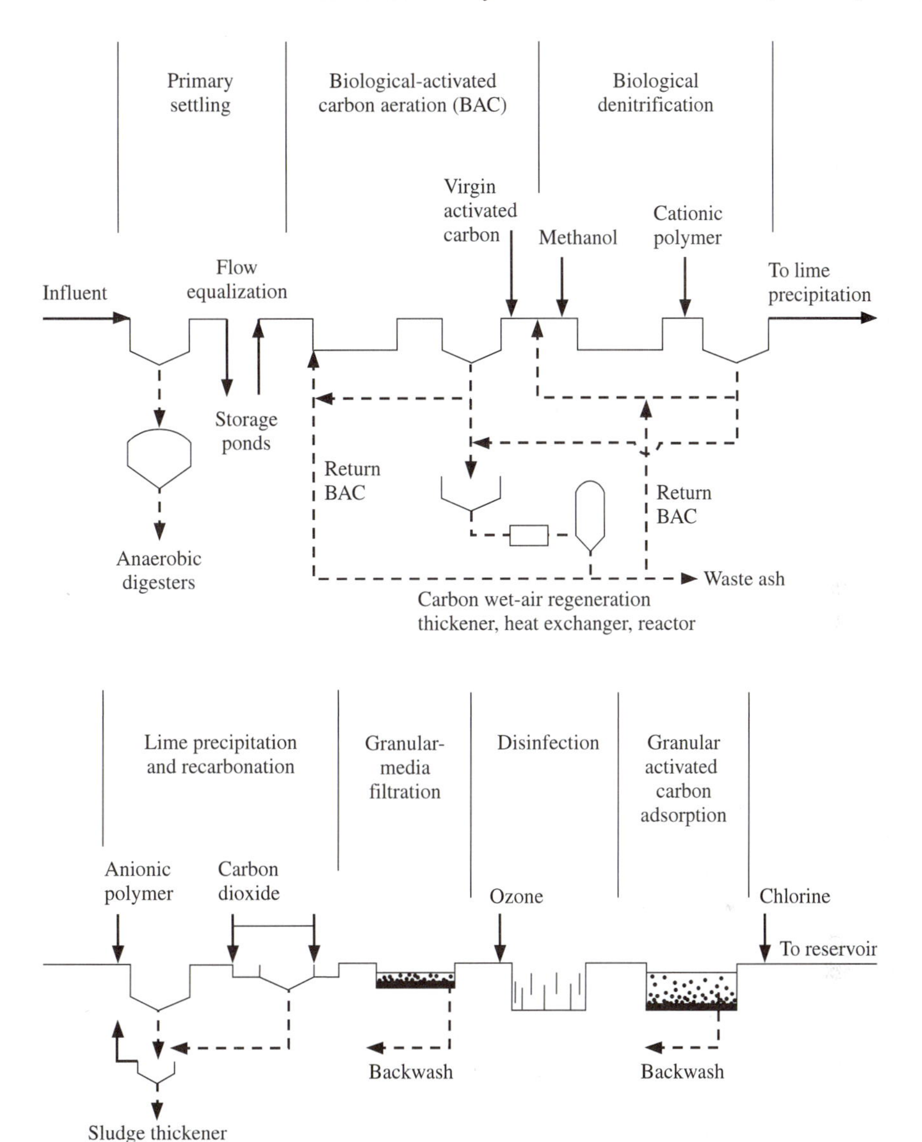

FIGURE 14.32 Flow diagram showing major processes at the Fred Hervey Water Reclamation Plant, El Paso, TX.

slurry is not effective, since the specific gravities of ash and regenerated PAC are similar [34]. Therefore, a portion of the slurry containing regenerated carbon is wasted to drying beds for disposal. The desired concentration of PAC in the BAC and denitrification processes is maintained by recycling regenerated-carbon slurry and addition of virgin PAC. The overall performance of the BAC system has proved satisfactory in reliable removal of toxic chemicals, organic matter, and nitrification–denitrification.

Lime precipitation at pH 11 in flocculator–clarifiers reduces phosphorus to less than 1 mg/l and removes a variety of contaminants. Concentrations of heavy metals are significantly reduced along with removal of organic substances as measured by COD and turbidity reductions. A secondary benefit of high lime treatment is disinfection, with significant reductions of coliforms and viruses. Recarbonation is performed in a two-stage process with intermediate sedimentation to reduce calcium content of the stabilized water. Waste lime slurry is thickened with the supernatant returned to the plant inlet, and the underflow is discharged to drying beds. Subsequent granular-media filtration at a nominal rate of 2.5 gpm/ft^2 is done to remove nonsettleable solids. Backwash water is recycled for treatment.

Ozonation through porous diffusers in a compartmented ozone contactor with a 23-min detention time at 10 mgd is for disinfection. Final filtration through a bed of granular activated carbon is for removal of any remaining trace amounts of adsorbable contaminants and final reduction of turbidity to 0.1–0.2 NTU. Chlorine is applied to the influent and effluent of the storage reservoir as necessary to prevent bacterial growth before injection of the reclaimed water.

An aerial view of the plant in Figure 14.33 shows the relative positions of the major treatment processes. The raw wastewater enters at the bottom left and the reclaimed water is stored in the underground reservoir at the top right. The anaerobic digesters are on the lower left; the wet-air regeneration system for the BAC sludge is on the right; the granular-media filters and granular activated carbon filters are further up to the right; and the lime storage silos are located below the storage reservoir. Notice the dark color of the biological activated carbon aeration tanks, clarifiers, and

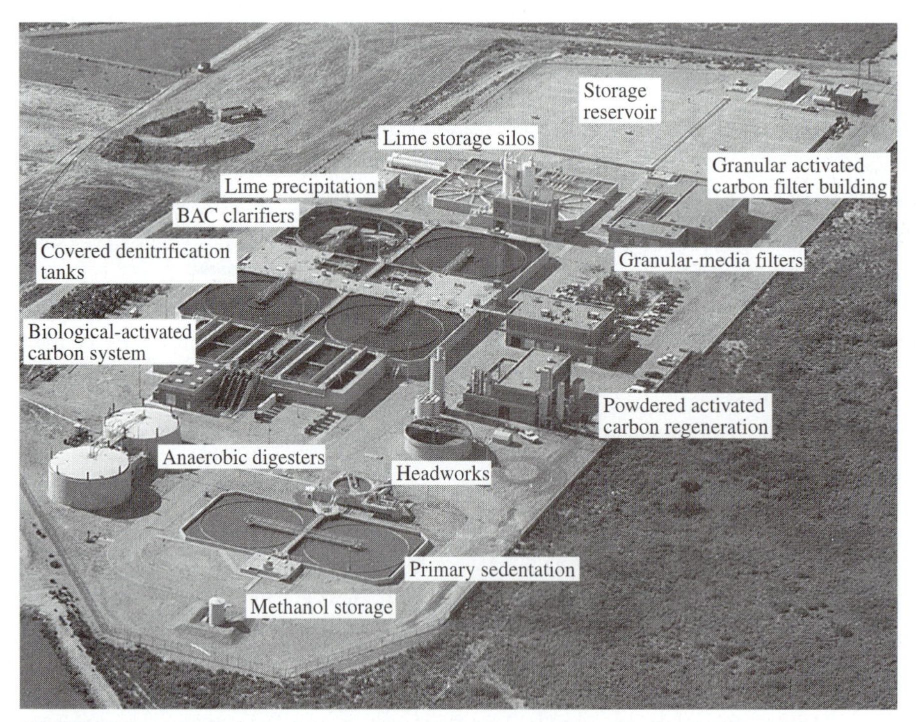

FIGURE 14.33 Aerial view of the Fred Hervey Water Reclamation Plant, El Paso, TX.
Source: Courtesy of the city of El Paso.

denitrification clarifiers in comparison to the lighter color of the primary tanks. Also, note the white color of the lime flocculator–clarifiers.

The plant, with a design capacity of 10 mgd, is currently processing an average of 6.5 mgd. At this hydraulic loading, the average process chemical dosages are: virgin powdered activated carbon 25 mg/l, methanol 60 mg/l, cationic polymer 0.4 mg/l, lime as CaO 170 mg/l, anionic polymer 0.6 mg/l, carbon dioxide 120 mg/l, ozone 5 mg/l at 1.5% ozone, granular activated carbon replaced once in 1988, and chlorine 4 mg/l.

The reclaimed water meets current drinking water standards as verified by scheduled testing for toxins most likely to be present, including heavy metals and organic chemicals. To assure compliance with the Safe Drinking Water Act, reclaimed water from the Fred Hervey Plant is analyzed for 12 regulated metals every two weeks; 5 regulated pesticides every three months; nitrate nitrogen, turbidity, and chlorine residual every day; and total coliforms, heterotrophic plate count, *Pseudomonas*, and coliphage three times a week. Biological testing for chronic toxicity is performed by bioassays on *Ceriodaphnia dubia* (cladocera) and larval *Pimephales promelas* (fathead minnow) every year. Selected typical concentrations for water quality parameters compared to government standards (given in parentheses) are: chloride 200 mg/l (<300 mg/l), fluoride 0.8 mg/l (<2 mg/l), nitrate-N 3 mg/l (<10 mg/l), sulfate 80 mg/l (<300 mg/l), barium 0.01 mg/l (<1.0 mg/l), iron <0.02 (<0.30 mg/l), and total dissolved solids 600 mg/l (<1000 mg/l).

The Fred Hervey Water Reclamation Plant has successfully produced an effluent that meets the standards established for injection into a water supply aquifer. In the first 17 years of operation, approximately 34 billion gal of reclaimed water has been injected, resulting in a rising rather than declining water table in the region around the injection wells.

14.23 DIRECT INJECTION FOR POTABLE SUPPLY, EL PASO, TEXAS

The reason for presenting, in some detail, the history of the Hueco Bolson Recharge Project is to inform the reader of the complexity of indirect reuse of reclaimed water by aquifer injection to augment a drinking water supply. Defining the hydrogeology of the aquifer was a difficult task, extending over a two-year period. Extensive water-quality testing was required to assure that the demonstration project after 6 years of operation was producing a drinking water safe for human consumption and the ambient groundwater in the aquifer was not being degraded. Finally, operation of the water supply system required water-quality analyses of reclaimed water, analyses of recharged groundwater, and production drinking water.

The city of El Paso has a population of 606,000 and is located in the Chihuahuan Desert at an elevation of 3700 ft above sea level. Located on the U.S.–Mexico border, El Paso is a fast-growing metropolitan area. The climate is sunny and semiarid, with hot summers and mild winters. The yearly average air temperature is 63°F and the average yearly rainfall is 8 in. with minimal rainfall during several months. The estimated pan evaporation is 81 in. per year.

The Hueco Bolson, as shown in cross section in Figure 14.34(a) is a sequence of sedimentary layers that fill the basin lying between two mountain ranges. (The word

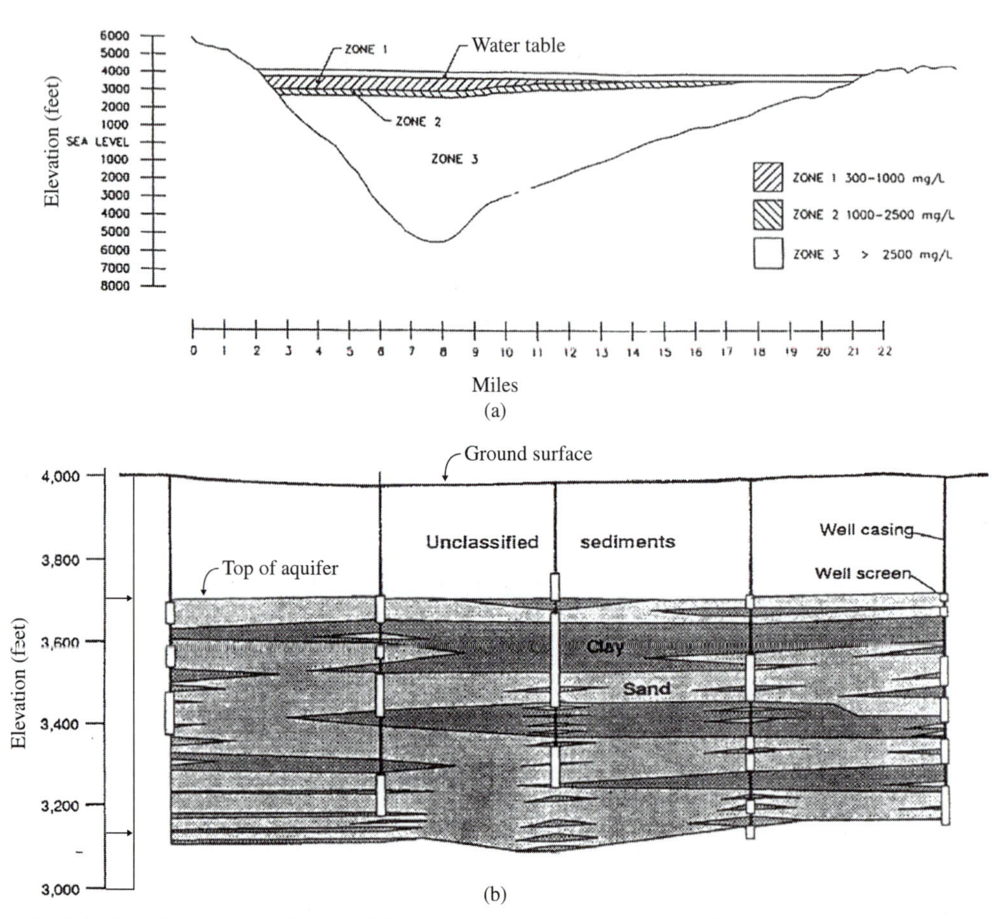

FIGURE 14.34 Hueco Bolson basin. (a) Cross section showing the zones of increasing total dissolved solids in groundwater with depth. The scale on the left is depth in feet relative to sea level, with the maximum depth of alluvium at 9000 ft. The horizontal scale is in miles. (b) A typical cross section of the freshwater zone composed of sand and clay strata. The five injection wells have gravel-packed multiple screens to discharge reclaimed water into the granular strata. *Source:* [Reference 35].

bolson pertains to a sediment-filled, intermontane basin surrounded by mountains.) The basin with a maximum depth of 9000 ft is composed of outwash alluvium underlain by lacustrine deposits with the groundwater increasing in salinity with depth. The groundwater in the upper portion of the saturated section, Zone 1, has a total dissolved solids concentration is 300–1000 mg/l. (The secondary standard for aesthetics of drinking water is 500 mg/l.) Beneath this "freshwater," Zone 2 contains 1000–2500 mg/l and Zone 3 is greater than 2500 mg/l.

Figure 14.34(b) is a cross section of the Hueco Bolson aquifer 13 miles wide where the freshwater zone is approximately 900 ft in depth. The five injection wells shown in this cross section are 600 to 800 ft deep and are screened to discharge reclaimed water by gravity flow into the granular strata. The freshwater zone is very susceptible to

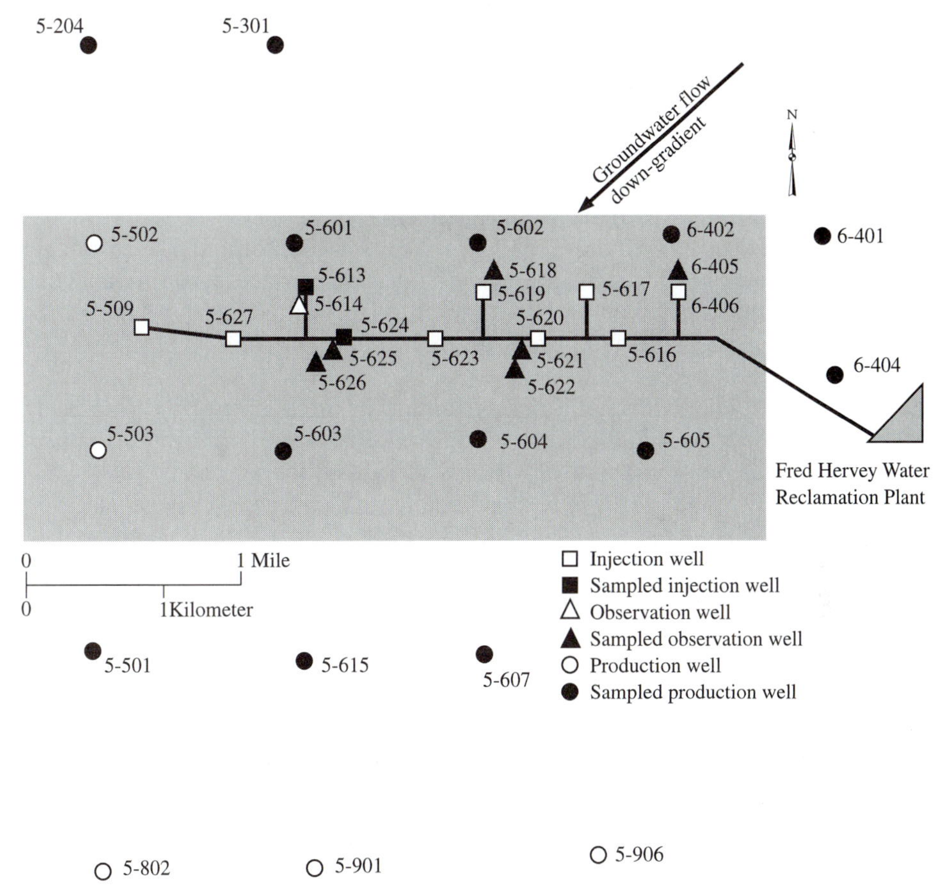

FIGURE 14.35 Hueco Bolson Recharge Project area for U.S. Geological Survey studies on the hydrogeology and selected water-quality aspects of the aquifer. *Source:* [Reference 35].

contamination by surface infiltration because permeable sand and gravel strata, the unclassified sediments, overlie much of the recharge area.

In 1991, the U.S. Geological Survey initiated studies on the hydrogeology and selected water-quality aspects of the Hueco Bolson aquifer at the recharge site [35]. About 8 mil gal of reclaimed water had been injected into the aquifer through the 10 injection wells between May 1985 and March 1991. Figure 14.35 diagrams the project area with the locations of injection, observation, and production wells. Strata thickness and lithology of aquifer sediments were classified using borehole-geophysical data from 16-in. normal resistivity, spontaneous potential, and natural gamma radiation logs. A 15 ohm-m or larger response on the 16-in. normal log indicated a sand unit. A lesser response indicated clay strata.

The 10 injection wells are located about three-quarters of a mile (1.2 km) up-gradient and one-third of a mile down-gradient from the production wells. The design injection rate of each well is 700 gpm by gravity flow under a pressure head of 16 to 20 ft in the water tank at the end of the pipeline. The location in the basin of the recharge project was selected to be near the treatment plant, and the anticipated minimum residence time of reclaimed water within the aquifer was 2 years.

The production wells have continuous gravel-packed screens of varying lengths extending below the water table in the freshwater zone. The drinking water withdrawn from the production wells is treated by chlorination prior to distribution. Hueco Bolson provides approximately 65% of El Paso's water supply. (The remainder is chlorinated groundwater from the Mesilla Bolson.)

The five observation wells are 6-in. gravel-packed slotted pipe extending through the freshwater zone and sealed near the ground surface. Groundwater samples for analysis were collected at selected depth intervals to match up with the screened intervals in the nearest injection well. Hydraulic continuity among injection, observation, and production wells existed.

The groundwater monitoring program for the hydrogeology included sampling selected injection, production, and observation wells. The samples collected were analyzed for inorganic and organic chemicals and tracer constituents. In order to explore mixing between injected and ambient waters, chemical analyses were used to develop mixing diagrams using the numerical technique of end-member mixing analysis (EMMA). Among the chemical constituents in the reclaimed water, the three that were successful for tracing groundwater flow were chloride, nitrate, and del-values of oxygen-18.

Average linear groundwater velocities near the zone of highest hydraulic flow were about 1.3 ft per day. Using this velocity, the calculated time of flow to production wells closest to the injection wells is 3.5 years. Based on EMMA results, the residence time of injected water indicated a significantly longer residence time of 6 years.

The Bureau of Reclamation, under the High Plains Demonstration Program Act, evaluated the impacts to groundwater quality resulting from the Hueco Bolson Groundwater Recharge Demonstration Project [36]. Until the time of this study, which started in January 1989, the indirect reuse of reclaimed water from the Fred Hervey Plant was considered a demonstration. This study was completed in June 1990.

The quality of the reclaimed water from the Fred Hervey Plant is the most important determinant of the impact on groundwater quality in the Hueco Bolson. After injection, filtration through the aquifer involves a number of potential changes, such as dilution of contaminants by mixing in the natural groundwater, adsorption of contaminants onto sediments, chemical interactions, decay of unstable substances, physical filtration of particulate matter, and die-off of microorganisms. Analyses of samples of the mixed waters can provide data on the alterations in the natural groundwater. The aquifer water quality was monitored on a regular basis from injection, monitoring, and production wells throughout the study period. As the final evaluation of water quality, the drinking water withdrawn from the aquifer was monitored by the El Paso Water Utilities prior to distribution and in the pipe network in compliance with the Safe Drinking Water Act.

At the Fred Hervey Plant, records of sample analyses for quality and operational control were assessed since inception in 1985. Monitoring of reclaimed water quality includes regulated metals; volatile organic compounds; regulated pesticides; total Kjeldahl nitrogen and nitrate nitrogen; chloride, fluoride, phosphate, and sulfate ions; cyanide; silicon; 5 microbiological tests; and biological testing for chronic toxicity. The established industrial waste pretreatment program was judged, and records of the 5 industries reviewed. None have discharged significant wastewater to the municipal sewer system. During the 6-year period of records, mercury and selenium were both reported as being above their maximum contaminant levels once. With regard to pathogen removal, the reclaimed water met drinking water standards for total coliform and enteric viruses. Based on the biological testing for chronic toxicity, the reclaimed water was not toxic to aquatic organisms.

During the 18-month study period, reclaimed water was tested for all of the 129 priority toxic pollutants (from acenaphthene to zinc) prohibited from discharge by the Clean Water Act [37]. Only mercury exceeded the maximum contaminant level once, and this was never repeated. To evaluate the removal of pollutants by the reclamation treatment processes, 5 consecutive days of 24-composite samples were collected at 7 locations along the sequence of treatment processes. Analyses were conducted for 207 pollutants, and of these 11 were detected. The overall removal efficiency in reclamation treatment was very high, with only strontium and vanadium resistant to removal; however, the concentrations did not exceed recommended health-based standards.

About 8 billion gallons of reclaimed water were injected into the aquifer from May 1985 to March 1991, and about 10 billion gallons were withdrawn from the 6 adjacent production wells. Based on the U.S. Geological Survey tracer studies, the maximum movement of the injected water was 3200 ft south and 1800 ft north from the line of injection wells. By 1990, reclaimed water constituted a significant portion of groundwater in wells less than 1000 ft from injection wells. Production wells at greater distances showed lower values. The chemical analysis of groundwater in production wells revealed little or no impact of injected water on water quality. The only suspected exceptions were boron and trihalomethanes.

Boron from perborate bleach additives in the effluent from treatment plant was typically 300 μg/l in contrast to about 100 μg/l in the freshwater aquifer. The EPA lifetime advisory for health is 600 μg/l. Nevertheless, all boron levels in production wells were well within the health limit. Boron can adsorb onto clays in the aquifer sediments.

The presence and transport of trihalomethanes (THMs), which are chlorination by-products, in groundwater is an important issue in reuse as drinking water. Concern for the Fred Hervey Plant effluent is not justified because the total THM concentration is not likely to exceed 20 μg/l. This is because the formation potential is limited by the low concentration of total organic carbon in the treated wastewater. Furthermore, where observed in the groundwater, the concentration of total THMs decreased significantly with time as the injected water moved through the aquifer.

Groundwater quantity in the Hueco Bolson aquifer is of concern in the future. The original consideration of augmenting the groundwater supply was to extend the life of the aquifer for one year for every 10 years that the full capacity of the treatment plant injected reclaimed water. Unfortunately, in recent years, El Paso has pumped 24 billion gallons per year while the natural recharge of freshwater is 1.6 billion gallons per year and the injection of reclaimed water is 2.9 billion gallons per year, for a total of 4.5 billion gallons per year [36]. Future problems include interference with production wells and salinity encroachment from the groundwater of high salinity in the deeper zones of the basin.

14.24 WATER FACTORY 21 AND GROUNDWATER REPLENISHMENT SYSTEM, ORANGE COUNTY, CALIFORNIA

The Orange County Water District (OCWD) is responsible for managing the groundwater reserves that support about 500 wells within the district. Increased water withdrawal in this semiarid coastal plain, with rainfall of 13–15 in. annually, caused overdraft of the freshwater basin. To prevent seawater from the Pacific Ocean from mixing with the inland freshwater, an advanced wastewater treatment facility, known as Water Factory 21 (WF21), was placed in service in 1976. The recycled water is injected into coastal aquifers to establish an intrusion barrier.

The evolution of WF21 presented in this section exemplifies the progress of advancing technologies of water reclamation for groundwater replenishment. The existing WF21 is being demolished and replaced by a new facility named the Groundwater Replenishment System, planned for completion by 2006. The original water reclamation processes are being replaced by new state-of-the-art technologies. The authors decided to continue reporting on water reclamation by the Orange County Sanitation District for the following reasons: (1) We want to keep our readers cognizant of the evolution of WF21 rather than abandoning an innovative system. (2) Presenting the processes in the Groundwater Replenishment System allows preliminary discussions of new technologies. (3) The public information officer for Orange County Water District provided the authors with information and process schematics for publication.

Initial Water Factory 21 and Coastal Barrier

Water Factory 21 reclaims water from sanitary wastewater after activated-sludge treatment and equalization of flow [38]. The design capacity is 15 mgd (57,000 m^3/d) at a uniform flow rate. The final process of demineralization by reverse osmosis has a design flow rate of only 5 mgd, since removal of dissolved solids from one-third of the water is adequate to provide a blended effluent of suitable quality. As diagrammed in Figure 14.36, the processing consists of lime precipitation with sludge recalcination, recarbonation, multimedia filtration, granular activated-carbon adsorption with carbon regeneration, disinfection by chlorination, and reduction of dissolved solids by reverse osmosis.

Lime treatment is performed in separate rapid mixing, flocculation, and sedimentation tanks at a controlled pH of 11.0 by application of 350 to 400 mg/l of CaO with a small amount of anionic polymer to aid coagulation. The primary objectives in lime precipitation are extraction of heavy metals and reduction of organic solids. The concentrations of heavy metals are significantly reduced. Occasionally, cadmium and chromium have exceeded the maximum contaminant levels for drinking water, but only during periods of high influent concentrations. Suspended solids reduction results in an average COD removal of 50 percent and turbidity reduction of over 90%. Another benefit of high-lime treatment is disinfection resulting in very high coliform reductions and virus removals greater than 98 percent.

The sludge from lime precipitation of the wastewater and calcium carbonate from first-stage recarbonation are concentrated in a gravity thickener to 8%–15% solids. After dewatering by centrifugation, approximately 50% of the solids are recalcined in a multiple-hearth furnace for recovery of lime for reuse. By wasting one-half of the precipitate, the buildup of impurities is controlled in the recalcined lime to prevent recycling of contaminants to the wastewater. Storage is provided in the lime-processing building for

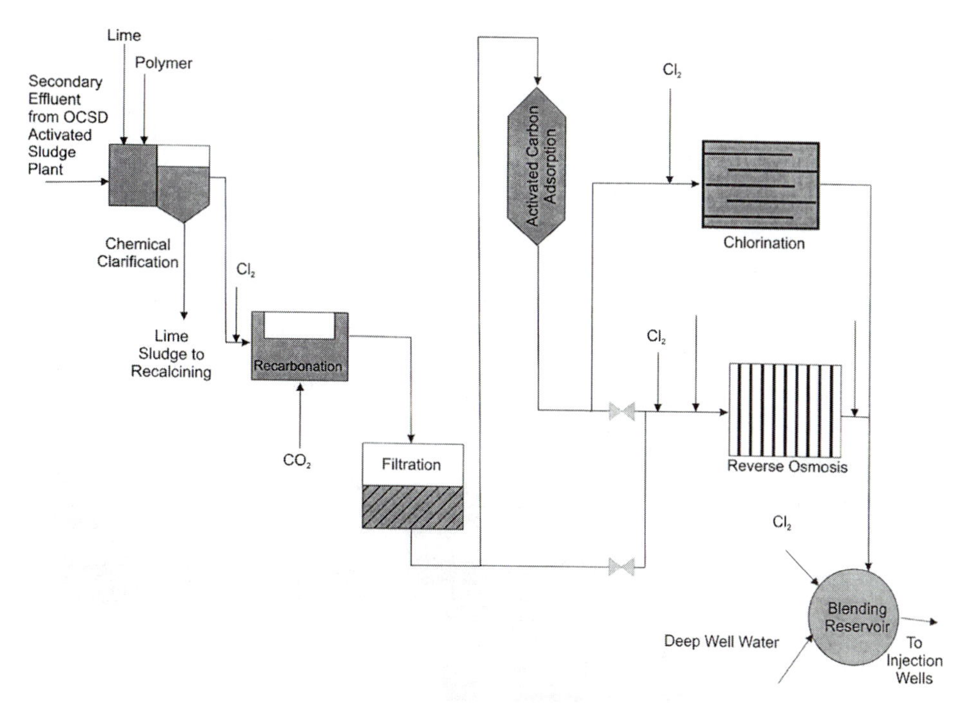

FIGURE 14.36 Flow diagram showing the complex series of physical-chemical processes of the initial Water Factory 21, which was in operation from 1976 to 1991. *Source:* Courtesy of the Orange County Water District, CA.

both recalcined lime and unused lime. Carbon dioxide from the recalcination process is used for two-stage recarbonation. The stabilized wastewater is then filtered through gravity granular-media beds or suspended solids removal prior to activated-carbon adsorption.

Air stripping of ammonia in large countercurrent-induced towers (not shown in Fig. 14.36) was abandoned in the late 1980s because the towers were no longer needed for nitrogen removal and because of excessive maintenance.

The granular activated-carbon columns are operated in a downflow direction at an empty-bed contact time of approximately 30 min. In upflow operation, as originally designed, carbon fines smaller than 25 μm were carried out of the columns and accumulated on the membranes in the reverse-osmosis modules. Efficiency of COD removal by carbon adsorption is generally in the range of 50%–60% by the adsorption of a wide variety of organic compounds. Chlorinated compounds with one or two carbons, however, were not readily adsorbed. Some heavy metals, such as chromium and copper, are partially removed.

One-third of the effluent from carbon columns is processed by reverse osmosis to remove dissolved inorganic solids. Typical operation is 90% reduction of dissolved solids to a level of about 100 mg/l with a recovery of approximately 85% of the feed water. In addition, the passage of the water through membranes also removes high-molecular-weight organic substances. The influent is treated with sodium hexametaphosphate as a scale inhibitor and chlorine to control biological growth on the membranes. Cartridge filters with replaceable polypropylene elements with an effective size of 25 μm are used to remove particulate matter. After pressurizing by feed pumps to 550 psi, sulfuric acid is injected to adjust the pH to 5.5 to minimize membrane hydrolysis and scale formation, primarily calcium carbonate and calcium sulfate.

Each reverse osmosis module consists of six spiral-wound cellulose acetate elements enclosed in a 23-ft-long, 8-in.-diameter fiberglass cylinder. The water passes through 3 stages consisting of 42 modules. First, the feed water flows in parallel through 24 first-stage modules; the concentrate (brine) from these units is passed through 12 second-stage modules; and the concentrate from the second pass flows through the 6 remaining modules. The reject brine, which is approximately 15% of the feed water, is discharged to an outfall sewer for ocean disposal. The product water is air stripped in a packed tower to remove carbon dioxide, which resulted from prior acidic pH adjustment.

Loss of productivity is a major problem in the reverse-osmosis treatment of wastewater. The membranes must be periodically cleaned to restore the rate of water passage. The inhibiting layer that forms on the membrane surface consists primarily of bacteria that attach to the surface of the cellulose acetate. Common cleaning solutions contain detergents and enzymes. After cleaning, a flushing solution is used to remove the cleaning chemicals; if the process is to be shut down for several days, the flushing solution is used to displace wastewater from the modules. When productivity of reverse osmosis is too low, high-purity deep well water is added to dilute the processed water prior to injection.

The initial Water Factory 21 reliably produced a treated water with contaminant concentrations less than the maximum levels dictated by drinking water standards. Therefore, the reclaimed water was considered satisfactory for underground injection to prevent saltwater intrusion by forming a water mound blocking infiltration of seawater from the Pacific Ocean. The injected water was required to meet drinking water standards since it can flow through the aquifer toward inland wells used for public water supplies. Heavy metals are removed, viruses inactivated, and trace organic compounds reduced to extremely low levels. The safety of reusing reclaimed wastewater for groundwater recharge was increased by dilution with natural groundwater and filtration as the water moved through the aquifer.

The area of saltwater intrusion, known as the Talbert Gap, is the area between Newport Beach and Huntington Beach. Talbert Gap is the mouth of an alluvial fan formed millions of years ago by the Santa Ana River; it has since been buried along the coast by a depth of 100 ft or more of clay. Figure 14.37 shows well locations in the Talbert Gap and the injection well construction options. The injected water forms a downhill gradient to the ocean forming a hydraulic barrier. For more than 25 years, the barrier system has prevented seawater from entering freshwater aquifers. On average, 50% of the injected recycled water flows inland to augment the general groundwater supply for Orange County.

Interim Water Factory 21

Demolition of portions of WF21 started in 2002 in preparation of constructing the new reclamation plant for the Groundwater Reclamation System using the new technologies of microfiltration as pretreatment for reverse osmosis and ultraviolet (UV) radiation for effluent disinfection. Figure 14.38 is the interim process flow scheme to maintain an adequate supply of reclaimed water, along with supplementary potable water and deep well, for injection to prevent seawater intrusion. The UV radiation system replacing chlorination for disinfection is similar to the system described in Section 11.26. The UV radiation

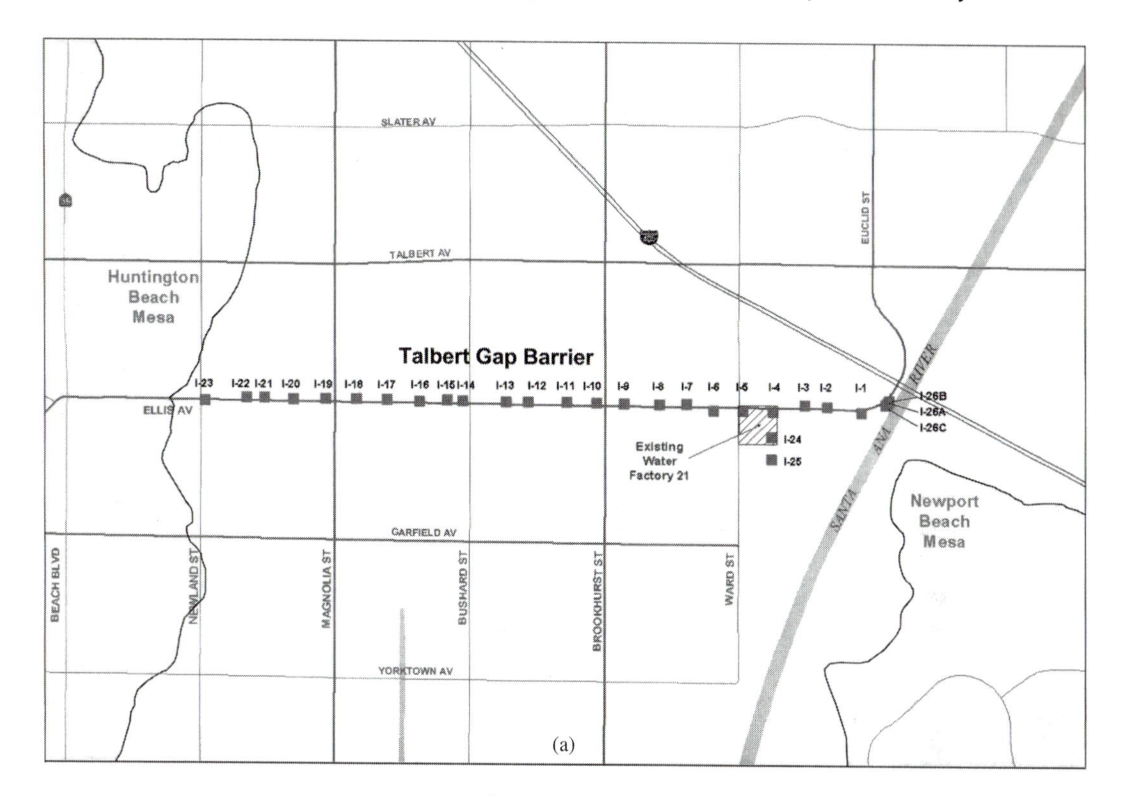

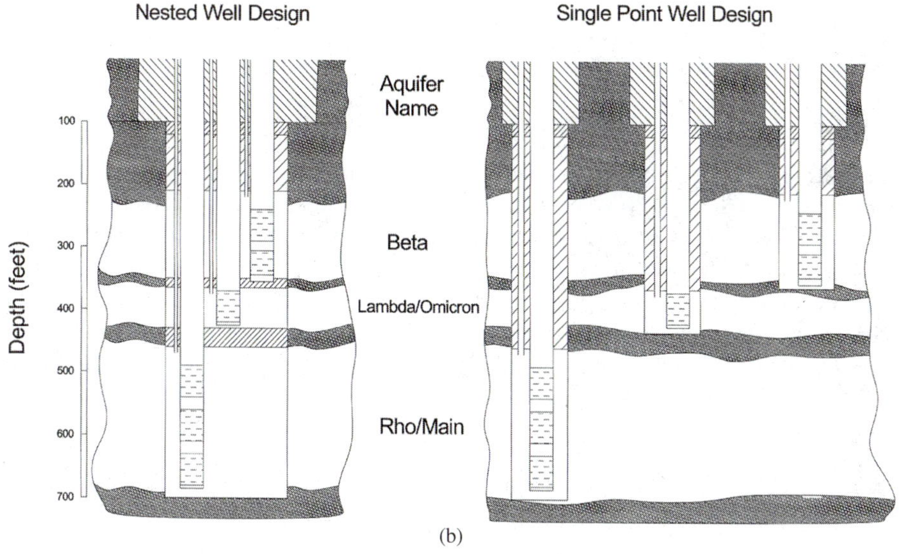

FIGURE 14.37 Talbert Gap barrier. (a) Twenty-eight multicasing injection wells placed approximately every 600 ft (182 m) span the 2.5-mi gap. (b) Barrier well construction options with wells screened in the three named aquifers ranging in depth below ground surface from 200 to 700 ft. *Source:* Courtesy of the Orange County Water District, CA.

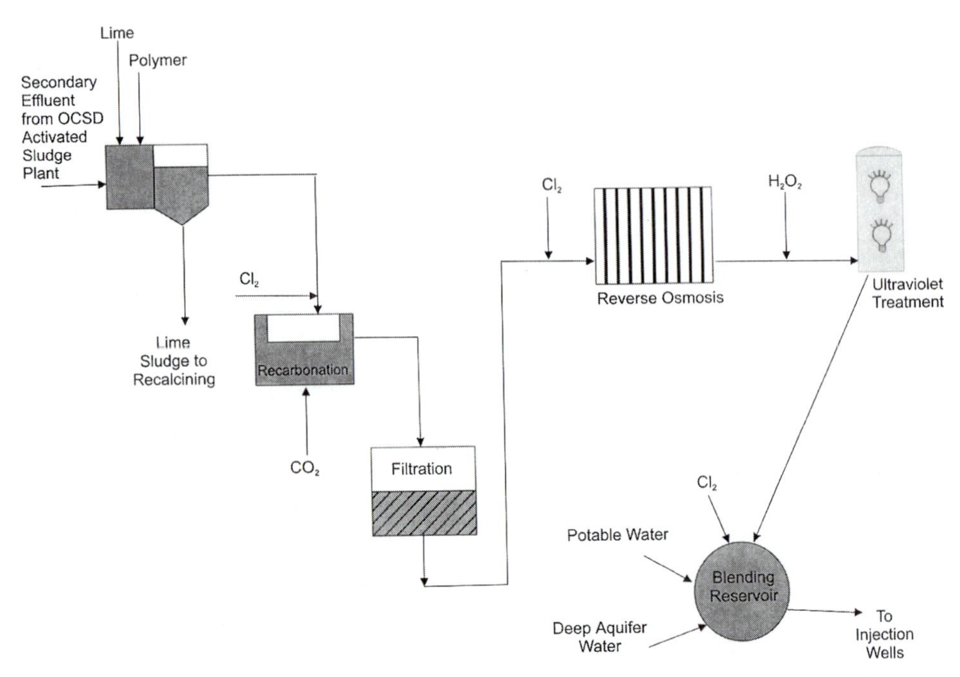

FIGURE 14.38 Interim process schematic of Water Factory 21 with ultraviolet radiation for disinfection of reclaimed water in operation in 2002. *Source:* Courtesy of the Orange County Water District, CA.

is expected to inactivate 99.99% (4 log) of the enteric viruses in the water, which in series following reverse osmosis will increase the overall inactivation to 6-log removal.

Groundwater Replenishment System

Orange County's population currently at 2.5 million is forecast to increase by more than 20% in the next 20 years. Groundwater supplies about 75% of the water demand with the remaining 25% being imported aqueduct water. Since an increase in imported water is unlikely, groundwater usage in this semiarid region is expected to increase beyond natural recharge from rainfall. Diversifying the sources of water supply will lessen the future dependence on imported water from the Colorado River and Northern California. In addition to maintaining a safe, reliable supply of drinking water, the increasing production of wastewater will require an additional Pacific Ocean outfall pipe unless the wastewater is reclaimed for recycling. The Groundwater Replenishment System (GRS), a joint project of the Orange County Sanitation District (OCSD) and the Orange County Water District (OCWD), is now under construction.

The process schematic for the GRS, as shown in Figure 14.39, replaces lime precipitation, filtration, and activated carbon adsorption with microfiltration. The filtrate is is expected to have less particulate matter and lower microorganism concentrations than the physical–chemical treated wastewater of WF21 to reduce the problem of fouling of the reverse osmosis membranes. Pretreatment for microfiltration includes self-cleaning filter screens with 1-mm openings to remove fine suspended solids and the addition of sodium hypochlorite for disinfection. The break tank provides a storage

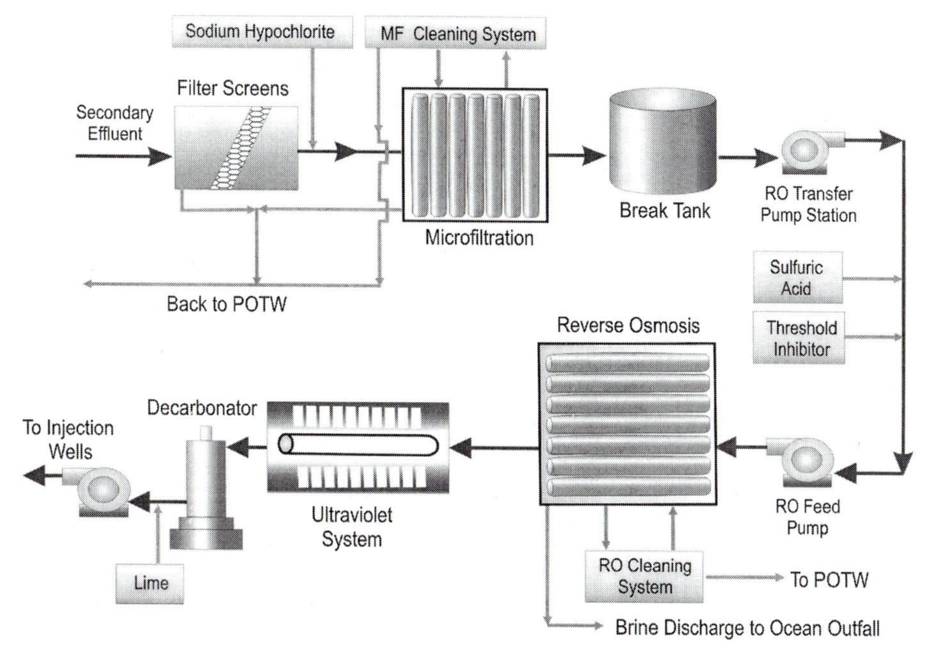

FIGURE 14.39 Process schematic for the Groundwater Replenishment System with a capacity of 100 mgd to replace Water Factory 21, which is planned for completion by 2006. *Source:* Courtesy of the Orange County Water District, CA.

volume to interrupt the direct flow between microfiltration and reverse osmosis. The microfiltration system reduces the space needed for pretreatment compared to the previous physical–chemical system, requires less labor-intensive maintenance, and the possibility of producing an overall recovery of up to 90% of the feed water.

Reverse osmosis is the primary barrier in the water renovation process by producing a product water low in dissolved salts and total organic carbon. Polyamide composite membranes have been successfully demonstrated at WF21. With microfiltration pretreatment, the recovery rate in reverse osmosis is expected to be 85%. Following ultraviolet radiation, the product water is stripped to remove carbon dioxide, and lime is added for stabilization. (Reverse osmosis is described in Section 11.34.)

Figure 14.40 illustrates the microfiltration process being constructed for the GRS [39]. Memcor® installations consist of multiple cells, with each cell self-contained with peripheral equipment including filtrate suction pumps, air blowers, chemical system for cleaning membranes, and a control system. As shown in Figure 14.40(b), the cell is a tank containing a large number of modules filled with hundreds of thousands of individual hollow fibers that filter the water from outside-in. This transverse flow of water from the cell to the inside of the membrane fibers is drawn by vacuum through suction pumps into the filtered water tank. Membrane modules are grouped together in manifolds [Fig. 14.40(c)] that are submerged under the water level in the cell and attached to the filtrate (clean water) manifold.

Combined air-and-water backwash reduces the quantity of the wash water required for flushing the modules resulting in a higher percentage of filtrate recovery, which is a major advantage of outside-in microfiltration. At backwash initiation, the

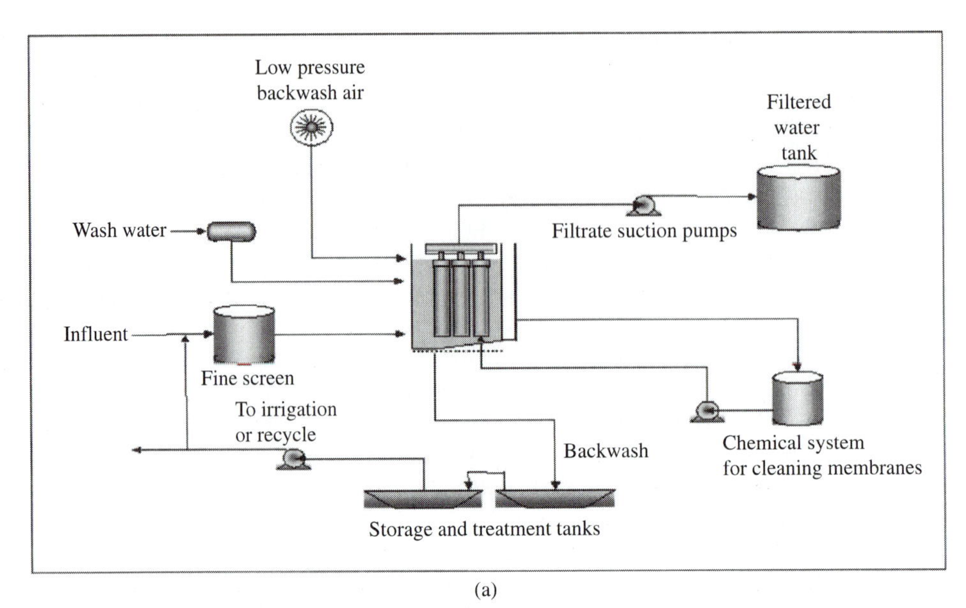

(a)

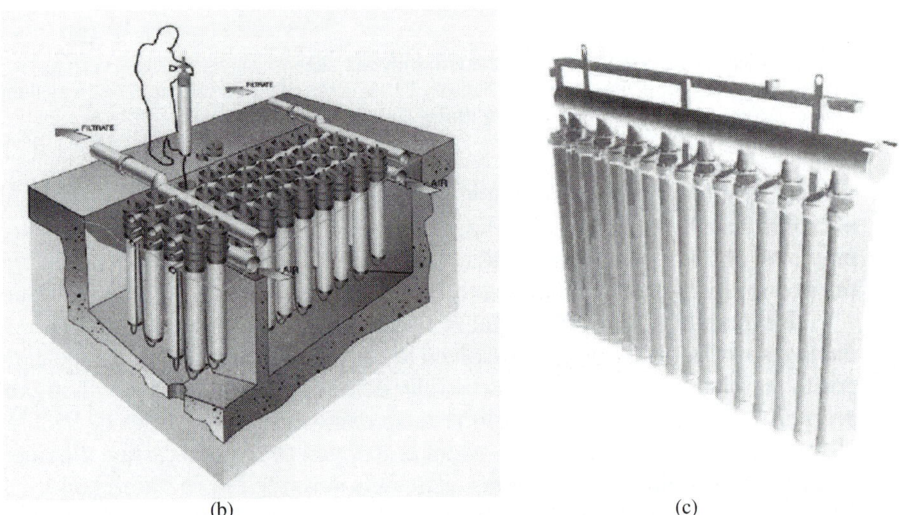

(b) (c)

FIGURE 14.40 Memcor® continuous microfiltration-submerged (CMF-S) process for installation in the Groundwater Replenishment System to replace Water Factory 21. (a) Cell containing membrane modules and peripheral equipment including influent screen, filtrate suction pumps, blowers for air scour, wash-water tank, chemical membrane cleaning system, and backwash processing. (b) Modules in a cell are grouped together in several manifolded assembles. (c) A manifold with suspended microfiltration modules. *Source:* Courtesy of US Filter/Memcor.

influent valve to the cell is closed and filtration is continued until the water level within the cell is lowered to slightly above the top of the modules. Then, air is introduced at the base of the modules to loosen the dirt and debris from the surface of the membrane fibers. To facilitate scouring inside the modules, air is introduced within the base of the modules through an air connection in the bottom manifold. As air scour continues outside of the fibers, clean filtered water is fed back inside the fibers to assist in flushing the debris from the outside of the membrane fibers. Upon completion of the air-and-water backwash, the cell is drained to remove the dirty water and debris and refilled with feed water. The entire backwash sequence takes less than 3 min.

Membrane fouling is the primary cause of reduced productivity in microfiltration. Chemical cleaning is required for restoration commonly using a caustic chemical mixed with surfactants, or dilute citric acid, or sodium hypochlorite. The control system continuously measures and displays the transmembrane pressure of a cell to indicate the need for cleaning. The control system can also perform integrity testing of the membrane barrier, cartridge construction, O-ring seals, and potential cross connections. If a cell fails the integrity test, maintenance personnel can locate the leakage by introducing air into the microfiltration system and observing the module with the leak by release of air bubbles. The control system also monitors operation of the system, e.g., the flow of filtrate based on automatic or manual production command.

Figure 14.41 shows the proposed replenishment system for the groundwater basin in Orange County. A 13-mile pipeline in the Santa Ana River right-of-way is to convey reclaimed water to the Anaheim spreading ponds. The recharge water will mix with the natural groundwater to augment well water withdrawn for drinking water supplies.

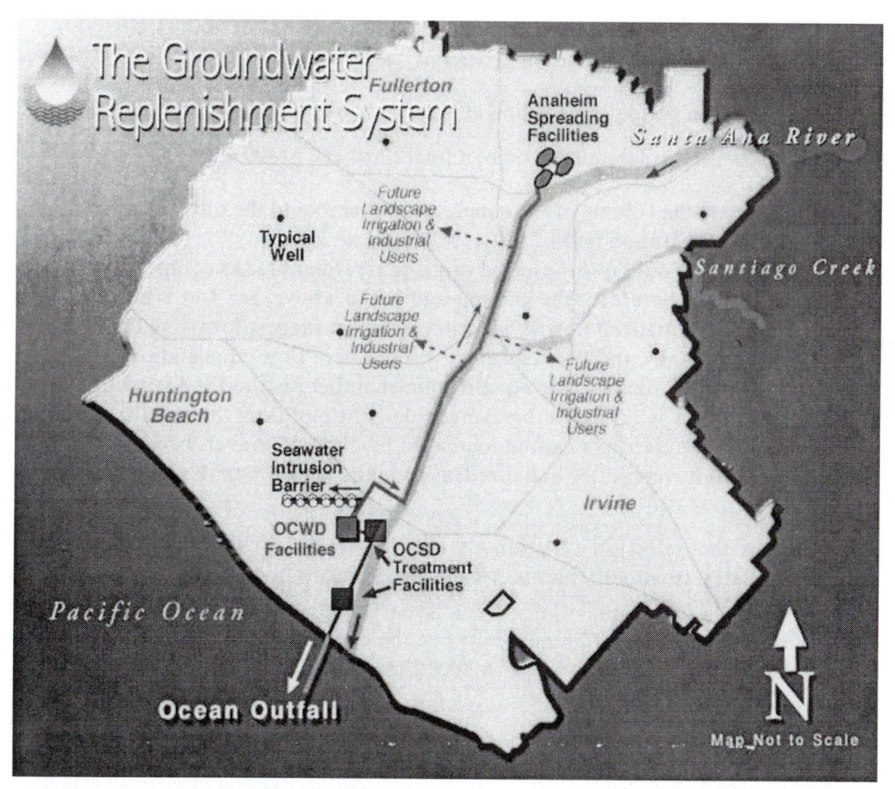

FIGURE 14.41 Groundwater replenishment system showing the proposed route of the pipeline from the Orange County Sanitation District (OCSD) treatment facilities and the Orange County Water District (OCWD) facilities to Anaheim recharge basins, a joint project of the OCSD and the OCWD. *Source:* http://www.gwrsystem.com/pages/gwrs_maps.htm.

Future recycling along the pipeline will be for landscape irrigation and industrial users. Reclaimed water from the new reclamation plant will continue to be injected as a seawater intrusion barrier.

PROBLEMS

14.1 List the common characteristics of a wastewater discharged from conventional biological processing. Which contaminants are most likely to cause pollution if the effluent is discharged to a stream with limited dilutional flow? Which constituents adversely affect a lake or reservoir receiving the wastewater discharge?

14.2 What are the potential benefits of flow equalization in municipal wastewater processing?

14.3 Compute the basin volume required to equalize the diurnal wastewater flow pattern diagrammed in Figure 12.16(a). Follow the procedure described in Example 14.1.

14.4 The feasibility of an in-line equalization chamber is being considered to reduce diurnal variations in flow through a conventional activated-sludge plant. The design wastewater flow is 10 mgd. The treatment units are sized based on the following parameters:

$$\text{equalization chamber volume} = 15\% \text{ of the design flow}$$

$$\text{overflow rate and depth of primary clarifiers} = 600 \text{ gpd/ft}^2,\ 8 \text{ ft}$$

$$\text{aeration period and loading of aeration tanks} = 5 \text{ hr},\ 40 \text{ lb BOD/1000 ft}^3\text{/day}$$

$$\text{overflow rate and detention time of final clarifiers} = 800 \text{ gpd/ft}^2,\ 2.2 \text{ hr}$$

a. Compare the volume of the equalization chamber to the total liquid volume of the clarifiers and aeration tanks.

b. The typical design values used in sizing treatment tanks compensate for diurnal flow variation; therefore, the parameters listed above are too conservative for design based on equalized flow. What criteria do you suggest for sizing the primary clarifiers, aeration tanks, and final clarifiers with influent flow equalization? If these values are used in tank sizing, is the equalization chamber justified? (Assuming the peak hourly rate of flow is 1.5 times the average daily rate of flow, the equalized design overflow rate of the clarifiers can be increased by 50%. However, because the BOD loading rate is not completely equalized, assume the aeration tank volume can be reduced by only 20%.)

14.5 Discuss the selection of methods for disposal of dewatered solids from advanced wastewater treatment plants. (Refer to Section 13.4, Table 13.2, and Section 13.5, Table 13.3.)

14.6 What are the basic differences between the characteristics of water and wastewater that influence the design of granular-media filters?

14.7 The solution to Problem 12.53(a) is a flow diagram for a treatment plant with primary sedimentation and secondary step aeration. Expand this diagram to include direct tertiary granular-media filtration and effluent chlorination. Show the arrangement of filters and backwash system. Recommend the kind of filter media, underdrain system, method of backwashing, and flow control. (Refer to Sections 14.4 and 10.16–10.18.) List numerical design guidelines for sizing the processing units.

14.8 Assume that flow equalization is provided prior to filtration in Example 14.2, thus allowing a nominal application rate of 5.0 gpm/ft^2 for design. Compute the filter area required, cycle time, backwash as a percentage of filtrate, and rate of filtration with three of four cells operating.

14.9 Calculate the area of granular-media filters needed for suspended-solids removal from a wastewater effluent with an average daily flow of 1500 m^3/d and peak 4-hr discharge of 3300 m^3/d. The nominal filtration rate should exceed neither 180 m^3/m^2 · d with all beds operating nor 350 m^3/m^2 · d at maximum discharge with only three of four beds in service.

14.10 Sketch a schematic flow diagram of the least-cost system for virus removal tested in the Pomona Virus Study. Label the treatment units and the chemicals applied.

14.11 What effluent quality, in terms of BOD, SS, turbidity, chlorine residual, dissolved solids, nitrogen, and coliforms, would you anticipate for treatment including (a) primary sedimentation, conventional activated sludge, dual-media filtration, chlorination, and dechlorination? (b) What effluent quality would be expected if 5 mg/l of alum and anionic polymer were applied before filtration?

14.12 What kinds of pollutants are removed by granular carbon columns?

14.13 The following data correlated suspended-solids concentrations with fecal coliform MPN indexes in effluents from wastewater treatment plants with different degrees of chlorination. Plot the data in a log–log scale. For each of the fecal coliform values, list the minimal treatment processes required to achieve that fecal coliform limit.

Suspended Solids (mg/l)	**Fecal Coliform (MPN index/100 ml)**
3	2.2
10	23
21	100
30	200

14.14 Study the application of the biological activated carbon (BAC) process at the Fred Hervey Water Reclamation Plant described in Section 14.22. Why was this process selected? What contaminants are removed by this process? How is the powdered activated carbon (PAC) in the waste-activated sludge regenerated? Why must new PAC be continuously added to the BAC process?

14.15 For Eq. (14.5), verify by appropriate calculations the molar ratio for Al to P of 1 to 1 and the weight ratio of commercial alum to phosphorus of 9.7 to 1.0.

14.16 A laboratory activated-sludge system was used to evaluate chemical–biological phosphorus removal by applying alum to the aeration tank. The laboratory apparatus was a diffused-air complete-mixing aeration tank, separate gravity clarifier, and airlift return sludge pump. Twelve liters of settled municipal wastewater were applied daily at a constant rate to the aeration tank, which had a volume of 3.6 l. The aeration tank MLSS concentration was held near 2000 mg/l by wasting 200–250 ml of mixed liquor each day. The temperature was 22°–24°C, pH 7.3–7.7, and sludge volume index varied from 90–130 ml/g. The alum solution feed to the aeration tank had a strength of 10.0 mg of commercial alum per milliliter. The following data were collected at various alum feed rates after arriving at steady-state conditions.

Wastewater Feed (l/d)	Alum Applied (ml/d)	Influent			
		BOD (mg/l)	SS (mg/l)	P (mg/l)	Alk (mg/l)
12.0	0	150	94	10.3	350
12.0	70	158	104	10.9	350
12.0	140	169	114	10.6	340
12.0	210	173	135	10.4	360
12.0	390	173	123	9.3	320

Wastewater Feed (l/d)	Alum Applied (ml/d)	Effluent			
		BOD (mg/l)	SS (mg/l)	P (mg/l)	Alk (mg/l)
12.0	0	6	7	7.9	230
12.0	70	6	8	5.4	200
12.0	140	5	11	2.4	160
12.0	210	10	10	1.0	140
12.0	390	8	17	0.5	80

1. Calculate the phosphorus removal and the weight ratio of alum applied to total phosphorus in the influent wastewater for each run. Plot a graph of percentage of phosphorus removal versus the weight ratio of alum to phosphorus.
2. Calculate the average BOD and SS (suspended solids) removal efficiencies. Suggest reasons why the effluent BOD and SS values are lower and resulting efficiencies higher than are normally achieved by a full-scale treatment plant.
3. Why did the concentration of alkalinity in the effluent decrease with increasing alum dosage?

14.17 A domestic wastewater, with 200 mg/l of BOD, 240 mg/l of SS, and 10 mg/l of P, is treated by extended aeration without primary sedimentation at a loading of 12.5 lb of BOD/1000 ft^3/day and an aeration period of 24 hr. Alum dosing for phosphate precipitation is graphed in Figure 14.6. What is the alum addition in milligrams per liter required for 80% phosphorus removal? Estimate the biological–chemical solids produced per million gallons of wastewater processed using the technique presented in Section 13.2 for biological waste and Eqs. (14.5) and (11.34) to approximate the amount of chemical precipitate. Compute the volume of sludge in gal/mil gal of wastewater treated, assuming a solids content of 1.5%.

14.18 Alum at a dosage of 100 mg/l is added to a conventional secondary aeration process for 90% removal of total phosphorus. The settled wastewater entering the aeration tank has the following characteristics: BOD = 130 mg/l, SS = 120 mg/l, and organic phosphorus (oP) = 1.9 mg/l plus inorganic phosphorus (iP) = 4.0 mg/l for a total phosphorus (tP) = 5.9 mg/l. The effluent has the following characteristics: BOD = 10 mg/l, SS = 10 mg/l, and tP = 0.5 mg/l. The aeration process is operated at an F/M ratio of 0.30 lb of SS/lb of BOD applied. Calculate the weight ratio of aluminum to phosphorus applied to the aeration process. Determine the biological–chemical solids in the waste-activated sludge in milligrams per liter of wastewater treated. (For help, review the procedures under Section 14.9 and Example 14.3.)

14.19 The characteristics of a settled wastewater (primary effluent) are as follows:

calcium = 34 mg/l	chloride = 43 mg/l
magnesium = 12 mg/l	BOD = 130 mg/l
sodium = 64 mg/l	suspended solids = 120 mg/l
potassium = 12 mg/l	volatile SS = 100 mg/l
alkalinity = 185 mg/l	nitrogen = 30 mg/l
sulfate = 44 mg/l	phosphorus = 7 mg/l

This wastewater is processed in a diffused-air activated-sludge system at an F/M loading of 1.0 g of BOD applied per day for each 3.0 g of MLSS in the aeration basin. The final clarifiers are appropriately sized.

1. During conventional operation, what are the estimated effluent BOD, SS, and P concentrations in milligrams per liter? Calculate the average waste organic solids production in milligrams per liter of wastewater treated. (For oP removal in waste-activated sludge, refer to Fig. 14.7.)
2. Alum is then added to the aeration tanks at a dosage equivalent to an Al/P ratio (aluminum added to phosphorus in the applied wastewater) of 2.0:1.0. With this chemical addition, the effluent BOD, SS, and P decrease to 10, 15, and 0.5 mg/l, respectively. No aluminum appears in the effluent. Calculate the average amounts of chemical precipitates and waste organic solids produced per liter of wastewater processed. How do these reactions affect the ionic character of the wastewater? (For help, refer to Fig. 14.8 and Example 14.3.)
3. Assume that granular-media filtration is added as a tertiary step following the biological–chemical phosphorus removal. What are the estimated BOD, SS, and P concentrations in the effluent? Estimate the additional waste solids removed by tertiary filtration.

14.20 Problem 13.32 describes the operation of an RBC plant that adds alum to the biodisk effluent to precipitate phosphate in the final clarifiers. Assume that the total phosphorus in the raw unsettled wastewater is 10.1 mg/l, consisting of 3.0 mg/l organic phosphorus and 7.1 mg/l inorganic phosphorus.

1. Trace phosphorus through the plant from influent to effluent as illustrated in Figure 14.7, and write the appropriate chemical and biological reactions resulting in phosphorus removal.
2. Calculate the alum dosage in milligrams per liter as alum and as the weight ratio of A1/P. How do these values compare with values given in Section 14.9 under the heading "Chemical Precipitation and Biological Treatment"?

14.21 A conventional activated-sludge plant treats a domestic wastewater with 220 mg/l of suspended solids, 180 mg/l of BOD, and 6.4 mg/l of total phosphorus. After primary settling, the characteristics are 110 mg/l of suspended solids, 120 mg/l of BOD, and 5.4 mg/l of total phosphorus. Based on a pilot-plant study, a dosage of 116 mg/l of alum is required to reduce the effluent total phosphorus to 0.5 mg/l with the aeration tank operating at an F/M of 0.20 g BOD/day/g MLSS. The effluent suspended-solids concentration averages 10 mg/l. Calculate the estimated production of sludge solids per cubic meter of wastewater treated. If the primary sludge has a solids content of 5.0% and the waste sludge from the chemical–biological aeration system has a concentration of 1.5%, calculate the liters of combined sludge produced per cubic meter of wastewater treated.

14.22 A domestic wastewater with 220 mg/l of suspended solids, 180 mg/l of BOD, and 6.4 mg/l of phosphorus is treated by lime precipitation in primary settling followed by conventional activated-sludge processing. A dosage of 170 mg/l of CaO removes 80% of the influent phosphorus. Of the remaining 1.3 mg/l of phosphorus, 0.8 mg/l is removed by biological uptake during aeration and 0.5 mg/l appears in the plant effluent. Lime precipitation in the primary also removes 80% of the suspended solids and 60% of the BOD. The plant effluent suspended-solids concentration averages 20 mg/l. Calculate the estimated production of sludge solids per cubic meter of wastewater treated based on the following data: The chemical precipitate produced is 2.0 mg of calcium hydroxyapatite and calcium carbonate per 1.0 mg of CaO applied. The aeration process operates at an F/M of 0.2 g BOD/day/g MLSS. How does this sludge production compare to the sludge solids produced by the biological–chemical treatment in Problem 14.21?

14.23 Advanced treatment of the wastewater from a small city consists of aeration tanks without primary sedimentation with the addition of alum for phosphorus precipitation, final clarifiers with sludge recirculation back to the aeration tanks, tertiary granular-media filters with a backwash storage tank that drains back to the head of the plant, effluent chlorine-contact chambers for disinfection, aerobic digesters with supernatant return to the head of the plant, and a belt filter press for dewatering digested sludge withdrawn from the digesters with filtrate returned to the head of the plant. The influent wastewater is 2000 m^3/d with 200 mg/l BOD, 240 mg/l SS, and 10.0 mg/l P. The effluent characteristics are 3 mg/l BOD, 3 mg/l SS, and 0.5 mg/l P. The fecal coliform count is less than 200 per 100 ml. The activated-sludge process is operated at $F/M = 0.15$, and the alum addition is 190 mg/l. Aerobic digestion reduces the organic content of the waste-activated sludge by 25% and thickens the digested sludge to 2.0% organic–chemical solids. The dewatered filter cake has 18% solids.

1. Draw a flow diagram of the treatment system.
2. Estimate the weight of filter cake produced per day by calculating the following: the organic solids produced by aeration, organic solids removed by tertiary filtration, organic solids remaining after digestion, chemical precipitates, and weight of total solids at 18% concentration.
3. What is the principal form of chlorine residual formed in chlorination of the wastewater? What chemical can be used for dechlorination?

14.24 Phosphorus is precipitated during biological–chemical treatment by the addition of waste pickle liquor (ferrous sulfate) to the aeration tanks of an activated-sludge process. The influent wastewater flow is 17.7 mgd containing 210 mg/l BOD, 190 mg/l SS, and 7.1 mg/l P. The effluent characteristics are 20 mg/l BOD, 26 mg/l SS, 1.0 mg/l P, and 1.4 mg/l Fe. The waste pickle liquor feed is 2300 gpd with an iron content of 8.8% by weight. The dry weight of sludge solids produced per day is 28,000 lb.

1. Calculate the weight ratios of iron applied to influent phosphorus and iron precipitated to phosphorus removed.
2. Compute the percentage of iron in the dry sludge solids.

14.25 Chemical coagulation and sedimentation of a raw municipal wastewater using 60 mg/l of $FeCl_3$ and 50 mg/l of CaO precipitates 230 mg/l of suspended solids and 8 mg/l of phosphorus. What is the weight ratio of iron applied to phosphorus removed? Calculate the weight of organic–chemical sludge solids per million gallons of wastewater treated using the method in Section 13.2 for organic matter and Eqs. (14.6) and (14.7) for chemical precipitates. If the sludge has an 8.0% solids concentration, compute the volume of waste in gal/mil gal of wastewater processed.

14.26 Why is it impractical to perform biological nitrification concurrently with BOD reduction in an activated-sludge aeration tank operating at the wastewater temperatures common in most regions of the United States?

14.27 Convert the concentration data for phosphorus given in milligrams in Figure 14.7 to percentages with 7 mg/l equivalent to 100%. Convert the concentration data for nitrogen given in milligrams in Figure 14.9 to percentages with 35 mg/l equivalent to 100%. How do these percentages compare with percentages that would be expected for BOD removal?

14.28 How would anaerobic digestion of the waste sludge affect the nitrogen data given in Figure 14.9?

14.29 What is the reduction in rate of biological nitrification

1. if the wastewater temperature decreases from 20° to 10°C and
2. if the operating pH is 7.0 rather than 8.0? How can reduced temperature and pH be compensated for in operating a suspended-growth nitrification system?

14.30 An oxidation ditch plant, which is a completely mixed process, has an aeration volume of 14,300 m^3. The average raw wastewater flow is 13,000 m^3/d with a BOD of 315 mg/l and total nitrogen concentration of 50 mg/l. Based on plant records, the operating conditions are MLVSS = 1200 mg/l, pH = 7.0, temperature = 18°C, and waste sludge solids = 1400 kg/d of volatile solids. Calculate the aeration period, volumetric BOD loading, and sludge age. Using evaluation computations as in Example 14.4, determine if nitrogen oxidation to an ammonia concentration of 2.0 mg/l (essentially complete nitrification) is likely to occur based on the operating conditions.

14.31 Refer to Figure 14.9. Calculate the theoretical oxygen uptake in aeration for BOD demand and 100% nitrification of ammonia nitrogen, assuming supernatant from anaerobic digestion is returned to the plant inlet. If the influent alkalinity is 320 mg/l, what is the alkalinity remaining in the effluent? [For BOD demand, look in Section 12.25. For NOD, see Eq. (14.17).]

14.32 An activated-sludge treatment plant without primary sedimentation has 2 aeration basins and 2 clarifiers. The waste-activated sludge is stabilized and thickened in aerobic digesters, dewatered by belt filter presses, and hauled to land disposal. Each aeration basin is prismoidal in shape and the length is twice the width, with 2 pedestal-mounted mechanical aerators. Since the plant is currently over design load, a proposed expansion is to add another identical aeration basin and clarifier. The original design loadings were 10,000 m^3/d with 300 $g/m^3 \cdot d$ of BOD (200 mg/l); the current maximum-month loading is 13,000 m^3/d with a BOD concentration of 260 mg/l. Although the BOD and suspended solids in the effluent are below the specified maximums of 20 mg/l and 20 mg/l, the state regulatory agency has established a new effluent quality standard of a maximum of 5 mg/l of ammonia nitrogen during the maximum month upon completion of the plant expansion. Even though the wastewater temperature is usually 20°C or greater throughout the year, the original designer did not think that extensive nitrification would occur. Nevertheless, partial nitrification is occurring as the total nitrogen (ammonia plus organic nitrogen) in the effluent is reduced to 21 mg/l.

The volume of the aeration basins and oxygen-transfer capacity of the mechanical aerators are as follows: Each basin is 3365 m^3, for a total liquid volume of 6730 m^3 for the 2 basins. Each aerator is 55 kW, with an oxygen transfer capacity of 84 kg/h in clean water at 20°C and a barometric pressure of 760 mm of mercury, for a total of 336 kg/h for the 4 installed aerators. The barometric pressure at plant elevation is 730 mm of mercury.

The average flow during the maximum month is 13,000 m^3/d, with high peak and low hourly flow variations of 200% maximum and 10% minimum. The typical mixed-liquor temperature is 19°C during the maximum month, pH = 7.0, and mixed-liquor concentrations are MLSS = 2300 mg/l and MLVSS = 1600 mg/l. During the hours of maximum diurnal flow, the dissolved oxygen decreases to 0.4 mg/l near the aerators and to zero in the corners of the aeration basins. The average waste-activated sludge is 340 m^3/d with 0.80% solids, which are 70% volatile. The average wastewater parameters during the maximum month based on 24-hr composite samples are as follows:

	Influent	**Effluent**
BOD	260	7
Suspended solids	180	13
Volatile suspended solids	130	9
Total nitrogen	56	21
Organic nitrogen	22	1
Ammonia nitrogen	34	20

You have been asked to evaluate the existing operation and make preliminary recommendations regarding the plant expansion to ensure that the effluent meets the standards of 20 mg/l BOD, 20 mg/l suspended solids, and 5 mg/l of ammonia nitrogen.

1. Calculate the aeration period and volumetric BOD loading at the current plant loadings. Compare these to the design values.
2. Using current loading and performance parameters, calculate the sludge age based on volatile suspended-solids values [Eq. (12.60)]. Using Eqs. (14.14)–(14.16) and Figure 14.10, calculate the theoretical mean cell residence time for an effluent value of N of 5 mg/l NH_3—N and peaking factor of 1.5 to compensate for diurnal flow variation. Also calculate the mean cell residence for N of 20 mg/l NH_3—N.
3. Are the low dissolved-oxygen concentrations during the period of maximum loading likely to inhibit nitrification? Calculate the total oxygen demand, both BOD and NOD, for nitrification at current loading and performance. BOD demand is equal to 1.4 times the applied BOD. To determine NOD, calculate the following: total nitrogen, which is the applied nitrogen in kilograms per day; effluent organic nitrogen; organic nitrogen in the waste-activated sludge, which is 6% of the waste solids; and effluent ammonia nitrogen. The nitrified nitrogen is then the applied tN minus the sum of org—N in effluent plus org—N in waste sludge plus effluent NH_3—N. NOD equals nitrified N times 4.6. Total oxygen demand is BOD plus NOD.
4. Calculate the rate of oxygen transfer in kilograms per hour for one mechanical aerator, using Eq. (12.102). Assume $\beta = 0.8$, $\alpha = 0.9$, $C_t = 1.0$ mg/l, and C_s based on mixed-liquor temperature corrected for barometric pressure. (Refer to Table A.10 and the equation given in the footnote.) Estimate total oxygen transfer by multiplying the calculated R by four aerators and 24 hr of operation. How does this value compare to the total oxygen demand?
5. Since the effluent quality is satisfactory other than the degree of nitrification, should design of the expanded plant consider a design loading of 6500 m^3/d per aeration tank

for the maximum month rather than the original design loading of 5000 m^3/d? This could be an economical way of increasing the capacity of the expanded plant. For evaluation, calculate the total oxygen demand with an effluent ammonia nitrogen concentration of 5.0 mg/l. Compare this value to the total oxygen transfer with larger installed aerators each with a transfer capacity of 150 kg/h in clean water at 20°C and 730 mm Hg and a C_s of 2.0 mg/l (rather than 1.0 mg/l).

6. Suggest field studies of the existing plant to evaluate the possible cause of limited nitrification either by short-circuiting of ammonia during the period of maximum diurnal flow or reduced nitrification caused by low dissolved oxygen.

14.33 Calculate the theoretical design mean cell residence time for nitrification in a completely mixed activated-sludge process based on the following: mixed-liquor temperature of 15°C and pH of 6.8; growth-limiting ammonia concentration of 2.0 mg/l; and the commonly assumed peaking factors given in the text for maximum-month flow and peak diurnal flow.

14.34 Calculate the aeration volume for second-stage suspended-growth nitrification of a daily design flow equal to 1500 m^3/d containing 25 mg/l of ammonia nitrogen. Assume a minimum operating temperature of 16°C, operating pH 7.4, peaking factor of 1.5, and design MLVSS of 1500 mg/l. If flow equalization is provided for the plant influent, what basin volume would you recommend?

14.35 For the town of Nancy, the solution to Problem 12.30 gives design data, the solution to Problem 12.52 sizes an extended aeration system, and the solution to Problem 13.44 sizes aerobic digesters and belt filter presses. Assume this plant has been constructed and is in operation. Proposed new effluent standards specified for this plant are a maximum total phosphorus concentration of 2.0 mg/l and a maximum ammonia nitrogen concentration of 5.0 mg/l. The city council has employed you to study the implications of these new effluent limitations on the operation and modification of the treatment plant. You are to submit a preliminary report within 12 months.

1. Outline a plan to evaluate performance of the existing treatment system by wastewater sampling, laboratory testing, and modifying or changing the operation of unit processes of the treatment system if necessary. These data are also to be used for future design if the new effluent standards are implemented.
2. Propose a treatment system with processing schemes for phosphorus removal and nitrification. Based on assumed data, calculate preliminary sizes of treatment units and estimate chemical usage.
3. Propose full-scale and/or pilot-plant studies to evaluate modified or additional treatment processes.

14.36 1. Why is methanol, or some other carbon source, needed in biological denitrification?

2. Calculate the methanol demand for denitrifying 1 m^3 of nitrified effluent with a nitrate nitrogen concentration of 25 mg/l and dissolved oxygen concentration of 8 mg/l. Express answers in grams of methanol and liters of 95% methanol with a specific gravity of 0.82.

14.37 Tertiary nitrification–denitrification processing is being considered for a 30-mgd secondary-treatment plant. The wastewater characteristics selected for nitrification process design are 30 mg/l of NH_3—N with a peaking factor of 1.5, 30 mg/l of BOD, a temperature of 18°C, pH 6.8, and 1500 mg/l of MLVSS. Using these parameters, compute the aeration basin volume required, recommended clarifier surface area, and estimated oxygen utilization. Assume the following characteristics for subsequent denitrification design: 27 mg/l of NO_3—N, 5 mg/l of DO, a temperature of 16°C, pH 7.0, and 1500 mg/l of MLVSS. Calculate the denitrification basin volume, recommended clarifier surface area, and methanol dosage.

14.38 The methanol dosage for denitrification at the Fred Hervey Water Reclamation Plant (Section 14.22) is 60 mg/l. Does this dosage appear to be reasonable? The settled raw wastewater contains 26–30 mg/l of organic plus ammonia nitrogen, and the effluent from denitrification contains an average of 1 mg/l of ammonia nitrogen, 2–4 mg/l of nitrate nitrogen, and negligible organic nitrogen.

14.39 The overall nitrogen removal in Figure 14.19 is $(69 - 16)100/69 = 77\%$, and in Figure 14.17(b) the removal is $100 - 41 = 59\%$. Why is the flow scheme in Figure 14.19 more effective in nitrogen removal than the one in Figure 14.17(b)?

14.40 How many milligrams of oxygen are in 1.0 mg of nitrate nitrogen? Using this value, estimate the milligrams per liter of BOD satisfied in the anoxic zone of the scheme in Figure 14.17(b). What are the advantages of recirculation of nitrified wastewater to the anoxic zone ahead of the aerobic nitrification zone?

14.41 An oxidation ditch system as diagrammed in Figure 14.21 can be used for nitrification–denitrification of domestic wastewater. The influent has a BOD of 178 mg/l and total nitrogen of 27 mg/l, and the effluent has a BOD of 8 mg/l and 4 mg/l inorganic nitrogen, 3 mg/l nitrate, and 1 mg/l ammonia. The aeration period averages 24 hr and, being in a warm climate, the temperature of the mixed liquor is rarely below 16°C and is usually 18°C or greater. Calculate the dissolved-oxygen uptake for nitrification per liter of wastewater. [Based on Eq. (14.16), 1.0 mg of ammonia N requires 4.6 mg of oxygen.] Calculate the quantity of BOD satisfied by denitrification per liter. (One mg of nitrate contains 3.4 mg of oxygen.)

14.42 The Citizens Committee on Environmental Issues has proposed to the state environmental protection agency the adoption of an effluent standard for all the wastewater treatment plants in the state to include a maximum allowable inorganic nitrogen limit of 10 mg/l, which is based on the maximum contaminant level (MCL) for drinking water of 10 mg/l of nitrate nitrogen. The committee's main arguments are that treatment plants are not properly designed to remove nitrogen and that nitrate nitrogen in surface waters is an environmental risk and health hazard. The state agency has data showing that no surface waters used as drinking water sources exceed 10 mg/l of nitrate nitrogen, and analyses of groundwaters near flowing waters downstream from wastewater discharges show little or no nitrate contamination. Nevertheless, these data have not deterred the committee from insisting on reducing the concentration of contaminants listed in the Safe Drinking Water Act to less than their MCLs in wastewater discharges. You have been asked to present a general assessment of nitrogen removal by conventional wastewater treatment and the feasibility of available technology to retrofit existing plants and design new plants to meet an effluent limitation of 10 mg/l of inorganic nitrogen. (Assume other speakers will address the topics of economic feasibility and other sources of nitrogen contamination, such as agricultural land drainage to surface waters and percolation of nitrate fertilizers to groundwaters under cropland.)

14.43 Compare the reclaimed water quality required by California and Arizona for irrigation of food crops consumed where the recycled (reclaimed) water comes in contact with the edible portion of the crop.

14.44 Define the concept of reliability in wastewater processing and discuss some ways to increase reliability in tertiary treatment.

14.45 A city in a semiarid region plans to reuse wastewater for agricultural irrigation. The soils at the proposed irrigation site are composed of poorly drained alkaline loams underlain by dense, clayey strata. The groundwater is unsuitable for potable use because of naturally occurring high salt and nitrate contents. The proposed crops include alfalfa, barley, corn, cotton, grain sorghums, and grasses. What kind of wastewater treatment is necessary, and what quality standards must be met for this reuse application? (Assume the wastewater is primarily from domestic water use and the concentrations of heavy metals, boron, and toxic chemicals are negligible.) What controls are necessary for operation of the irrigation site?

14.46 Using Eq. (14.23), estimate the land area in acres required for irrigation of 1.0 mgd based on precipitation of 20 in./yr, evapotranspiration of 40 in./yr, percolation equal to 10 in./month, and zero runoff.

14.47 Using Eq. (14.23), estimate the land area in acres for one month's (30 days) irrigation for a flow of 1.6 mgd. Base calculations on a precipitation of 4.2 in./month, evapotranspora-tion of 5.7 in./month, and percolation of 12 in./month. Nitrogen crop uptake is 400 lb/ac/yr. Runoff is pumped back to the storage reservoir and returned for irrigation. Check the nitrogen loading, assuming a wastewater nitrate concentration of 27 mg/l, 20% denitrification in the soil, and a maximum of 10 mg/l allowed in the water percolating to groundwater.

14.48 The total field area at an agricultural irrigation site in a semiarid region is 4800 acres. The proposed total surface area of the storage reservoirs is 300 acres. By water-balance calculations, determine the storage requirement based on the data listed below. The total water available each month is calculated in acre-inches by summing the effluent available plus precipitation minus evaporation loss from the storage reservoirs. The monthly irrigation requirements listed in the table were determined by estimating the consumptive use by the crops most likely to be grown at the site, taking into account optimum economic farming operation, crop rotation, and planting and harvesting times. Assume that the storage reservoirs are empty on October 1.

Month	Effluent Available (acre-in.)	Precipitation (in.)	Evaporation (in.)	Irrigation Requirement (acre-in.)
Oct.	20,100	0.3	4.6	0
Nov.	19,400	0.5	2.3	0
Dec.	19,800	1.0	1.0	0
Jan.	19,000	1.2	1.1	0
Feb.	16,600	1.1	1.6	0
Mar.	17,800	1.1	4.0	22,000
Apr.	17,900	0.7	5.0	22,000
May	19,400	0.2	7.4	36,500
June	20,400	0.1	8.9	36,500
July	22,300	0	9.5	25,500
Aug.	22,300	0	8.6	44,000
Sept.	20,700	0.1	6.2	33,000
	235,700	6.3	60.2	219,500

14.49 What quality standards and regulations apply to irrigation of a golf course (tee boxes, fairways, and greens), excluding the landscaped area around the clubhouse? The play area for a golf course is considered to have restricted access if fenced along the boundary with public property.

14.50 List the sequential steps, or processes, that the raw wastewater passes through for reuse in agricultural irrigation at Tallahassee, Florida, from the sewer collection system to recharge of the groundwater.

14.51 Describe a center-pivot irrigator and its operation. Why is center-pivot irrigation the only suitable method of water application at the Tallahassee farm site, as compared to surface spreading? How is the groundwater monitored to determine if the percolation of irrigation water is contaminating the natural groundwater?

14.52 List the sequential steps, or processes, in citrus irrigation and groundwater recharge in Orange County and the city of Orlando, Florida, from the raw wastewater to reuse.

14.53 In the reuse of reclaimed water in Orange County and the city of Orlando, Florida, what analyses are conducted to monitor water quality?

14.54 Name three examples of restricted urban landscape irrigation and three examples of unrestricted irrigation sites. The following questions apply to urban reuse in St. Petersburg, Florida. (1) What are the two major economical benefits for the city by practicing urban reuse of reclaimed water? (2) What is the primary purpose (objective) in tertiary processing for urban reuse? (3) Why can only a limited percentage of the residential customers in the city be connected to the reclaimed water system?

14.55 What is the primary limiting problem related to the biological activated carbon (BAC) process? (Refer to Sections 14.7 and 14.22.)

14.56 The effluent from the Fred Hervey Water Reclamation Plant is chlorinated and pumped into a large storage reservoir (Fig. 14.33). Describe the route of the water from the reservoir to the water distribution system in the city of El Paso.

14.57 The city of El Paso is a fast growing metropolitan area located on the international border with Mexico. The climate is semiarid with an average yearly rainfall of 8 in. The upper freshwater zone below the water table in Hueco Bolson (Zone 1) averages 900 ft in depth with dissolved solids concentrations in the range of 300 to 1000 mg/l. Production wells have continuous gravel-packed screens to varying depths extending below the water table in this freshwater zone. Beneath this freshwater zone, Zone 2 has 1000–2000 mg/l and the deeper Zone 3 has greater than 2500 mg/l. [Refer to Fig. 14.34(a).] In recent years, production wells have pumped more than the natural recharge of runoff from rainfall plus the injection of reclaimed water. What is the likely future problem with Hueco Bolson as a water source?

14.58 In wastewater processing at the initial Water Factory 21, what was the purpose for each of the following processes in water reclamation: lime precipitation, recarbonation, granular-media filtration, and activated-carbon adsorption? What replaced these processes in the new Groundwater Replenishment System?

14.59 For each of the following reuse systems, provide one or more reasons for selecting reclaimed water reuse (recycling) for wastewater disposal. List the primary reason first.

El Paso, TX
Orange County and city of Orlando, FL
Orange County Water District, CA
St. Petersburg, FL
Tallahassee, FL
Water Factory 21 (original facility)

REFERENCES

[1] "Tertiary Filtration of Wastewaters," by the Task Committee on Design of Wastewater Filtration Facilities, *Proc. Am. Soc. Civil Engrs., Env. Eng. Div.* 112 (EE6) (December 1986): 1008–1025.

[2] "Wastewater Filtration," EPA Technology Transfer Seminar Publication, Environmental Protection Agency (July 1974).

[3] E. R. Baumann and J. Y. C. Huang, "Granular Filters for Tertiary Wastewater Treatment," *J. Water Poll. Control Fed.* 46 no.8 (August 1974): 1958–1972.

[4] *Pomona Virus Study*, Sanitation Districts of Los Angeles County (Sacramento, CA: California State Water Resources Control Board, 1977).

[5] "Wastewater Treatment Plants," Sanitation Districts of Los Angeles County, http://www.LACSD.ORG/waswater/wrp/wrp2.htm.

[6] "Water Reuse Summary for Fiscal Year 1995–96," Sanitation Districts of Los Angeles County, http://www.LACSD.ORG/WEBREUSE/refy9596.htm.

[7] Z. Ziaojian, W. Zhansheng, and G. Xiasheng, "Simple Combination of Biodegradation and Carbon Adsorption—The Mechanism of the Biological Activated Carbon Process," *Water Research*, 25 no.2 (February 1991): 165–172.

[8] "Retrofitting POTWs for Phosphorus Removal in the Chesapeake Bay Drainage Area," U.S. Environmental Protection Agency, Technology Transfer, EPA/625/6-87/017 (September 1987).

[9] D. T. Anderson and M. J. Hammer, "Effects of Alum Addition on Activated Sludge Biota," *Water and Sewage Works* 120, no.1 (January 1973): 63–67.

[10] R. D. Leary, L. A. Ernest, R. M. Manthe, and M. Johnson, "Phosphorus Removal Using Waste Pickle Liquor," *J. Water Poll. Control Fed.* 46, no.2 (February 1974): 284–300.

[11] J. C. O'Shaughnessy, J. B. Nesbitt, D. A. Long, and R. R. Kountz, "Digestion and Dewatering of Phosphorus-Enriched Sludges," *J. Water Poll. Control Fed.* 46, no.8 (August 1974): 1914–1926.

[12] "Manual for Nitrogen Control," Office of Water, Office of Research and Development, Environmental Protection Agency, EPA/625/R-93/010 (September 1993).

[13] "Process Design Manual for Nitrogen Control," U.S. Environmental Protection Agency, Technology Transfer (October 1975).

[14] "Nitrification and Denitrification Facilities," EPA Technology Transfer Seminar Publication, U.S. Environmental Protection Agency (August 1973).

[15] P. F. Cooper, E. A. Drew, D. A. Bailey, and E. V. Thomas, "Recent Advances in Sewage Effluent Denitrification: Part I," Water *Pollut. Control* 76, no.3 (1977): 287–300.

[16] P. F. Cooper, B. Collinson, and M. K. Green, "Recent Advances in Sewage Effluent Denitrification: Part II," *Water Pollut. Control* 76, no.4 (1977): 389–401.

[17] "Manual of Guidelines for Water Reuse," Office of Water, Office of Research and Development, Environmental Protection Agency and U.S. Agency for International Development, EPA/625/R-92/004 (September 1992).

[18] *California Health Laws Related to Recycled Water, Title 22 Code of Regulations, Division 4, Environmental Health*, California State Department of Health Services (Sacramento, CA: June 2001).

[19] *Environmental Quality, Title 18, Chapter 9, Water Pollution Control, Article 7, Regulations for the Reuse of Wastewater*, Department of Environmental Quality, State of Arizona (Phoenix, AZ: 1985).

[20] *Health Aspects of Wastewater and Excreta Use of Agriculture and Aquaculture*, The Engelberg Report (Dubendorf, Switzerland: International Reference Centre for Wastes Disposal, 1985).

[21] R. G. Feachem, D. J. Bradley, H. Garelick, and D. D. Mara, *Sanitation and Disease, Health Aspects of Excreta and Wastewater Management; World Bank Studies in Water Supply and Sanitation 3* (Chichester: Wiley, 1983).

[22] C. R. Bartone, "Development of Health Guidelines for Water Reuse in Agriculture: Management and Institutional Aspects," in *Implementing Water Reuse* (Denver: AWWA Research Foundation, 1987), 489–504.

[23] "Process Design Manual for Land Treatment of Municipal Wastewater," U.S. Environmental Protection Agency, EPA/625/1-77-008; U.S. Army Corps of Engineers, COE EM 1110-1-501 (October 1977).

[24] R. Cort, B. Sheikh, R. Jaques, and R. Cooper, "Safety, Feasibility, and Cost of Reuse of Wastewater Irrigation of Raw-Eaten Vegetables," in *Implementing Water Reuse* (Denver: AWWA Research Foundation, 1987): 445–474.

[25] R. S. Ayers and D. W. Westcot, *Water Quality for Agriculture 8*, FAO Irrigation and Drainage Paper No. 29, Rev. 1 (Rome: Food and Agriculture Organization of the United Nations, 1985).

[26] C. E. Pound and R. W. Crites, *Wastewater Treatment and Reuse by Land Application*, Vols. I and II, Office of Research and Development, Environmental Protection Agency (August 1973).

[27] M. N. Allhands, S. A. Allick, A. R. Overman, W. G. Leseman, and W. Vilak, "Municipal Water Reuse at Tallahassee, Florida," *Am. Society of Agricultural Engineers, Soil and Water Div.* 38, no.2 (November 1994): 411–418.

[28] P. Cross, G. Delneky, and T. Lothrop, "Worth Its Weight in Oranges," *Water Environment & Technology, Water Environment Federation*, 12, no.1 (January 2000): 26–30.

[29] D. S. Sloan, R. A. Pelletier, and T. L. Lothrop, "Not by Design," *Water Environment & Technology, Water Environment Federation*, 13, no.12 (December 2001): 62–65.

[30] "Arizona Standards, Aquifer Water Quality Standards, Rules," Department of Environmental Quality, State of Arizona (Phoenix, AZ, 1990).

[31] *California Health Law Related to Recyled Water, Title 22, Code of Regulations, Division 4, Environmental Health*. Chapter 3, "Recycling Criteria, Groundwater Recharge Reuse, DRAFT Regulations" (Sacramento, CA: April 23, 2001).

[32] *Issues in Potable Reuse: The Viability of Augmenting Drinking Water Supplies With Reclaimed Water*, National Research Council (Washington, DC: National Academy Press, 1998).

[33] W. B. Solley, R. R. Pierce, and H. A. Perlman, *Estimated Use of Water in the United States in 1995: U.S. Geological Survey Circular 1200, Federal Center* (Box 25386, Denver, CO: 1998), http://water.usgs.gov/watuse/.

[34] D. B. Knorr, J. Hernandez, and W. M. Copa, "Wastewater Treatment and Groundwater Recharge: A Learning Experience at El Paso, Texas," *Proc. Am. Water Works Assoc. Conference* (Denver, CO, 1987).

[35] P. M. Buszka, R. D. Brock, and R. P. Hooper, *Hydrogeology and Selected Water-Quality Aspects of the Hueco Bolson Aquifer at the Heuco Bolson Recharge Project, El Paso, Texas*, U.S. Geological Survey, Water-Resources Investigations Report 94-4092 (Austin, TX: U.S. Geological Survey, 1994).

[36] *Hueco Bolson Ground Water Recharge Demonstration Project, El Paso, Texas, Part II Water Quality Analysis*, for inclusion in the Bureau of Reclamation's final report to Congress on the Hueco Bolson Ground Water Recharge Project under the High Plains States Ground Water Demonstration Act of 1984 (Dallas, TX: U.S. Environmental Protection Agency, Region 6, January 1995).

[37] *Clean Water Act*, 25th Anniversary Edition Alexandria, VA: Water Environment Federation, (1997).

[38] "Wastewater Contaminants Removal for Groundwater Recharge at Water Factory 21," U.S. Environmental Research Laboratory, EPA 600/2-80/144 (August 1980).

[39] "US Filter Memcor® CMF-S Microfiltration Technology Data," and "Case Study Water Factory 21, Orange County Water District, CA, Memcor® CMF-S Demonstration" (Ames, IA: US Filter/Memcor, 2001).

Appendix

TABLE A.1 Weights and Measures

Length					
cm	m	in.	ft	yd	mi
1 cm = 1	0.01	0.3937008	0.03280840	0.01093613	6.213712×10^{-6}
1 m = 100	1	39.37008	3.280840	1.093613	6.213712×10^{-4}
1 in. = 2.54	0.0254	1	0.08333333 . . .	0.02777777 . . .	1.578283×10^{-5}
1 ft = 30.48	0.3048	12	1	0.3333333 . . .	$1.893939\ldots \times 10^{-4}$
1 yd = 91.44	0.9144	36	3	1	$5.681818\ldots \times 10^{-4}$
1 mi = 1.609344×10^{5}	1.609344×10^{3}	6.336×10^{4}	5280	1760	1

Area					
cm^2	m^2	$in.^2$	ft^2	acre	mi^2
1 cm^2 = 1	10^{-4}	0.155003	1.076391×10^{-3}	2.5×10^{-8}	3.861022×10^{-11}
1 m^2 = 10^{4}	1	1550.003	10.76391	2.5×10^{-4}	3.861022×10^{-7}
1 $in.^2$ = 6.4516	6.4516×10^{-4}	1	6.944444×10^{-3}	1.59×10^{-7}	2.490977×10^{-10}
1 ft^2 = 929.0304	0.09290304	144	1	2.3×10^{-5}	3.587007×10^{-8}
1 acre = 40.47×10^{6}	4047	6.27×10^{6}	43,560	1	1.56×10^{-3}
1 mi^2 = 2.589988×10^{10}	2.589988×10^{6}	4.014490×10^{9}	2.78784×10	640	1

(*continued*)

TABLE A.1 (Continued)

Volume					
ml	l	in.^{3}	ft^3	gal	acre · ft
1 cm^3 = 1	10^{-3}	0.06102374	3.531467×10^{-5}	2.641721×10^{-4}	8.1×10^{-10}
1 l = 1000	1	61.02374	0.03531467	0.2641721	8.1×10^{-7}
1 in.^{3} = 16.38706	0.01638706	1	5.787037×10^{-4}	4.329004×10^{-3}	1.33×10^{-8}
1 ft^3 = 28,316.85	28.31685	1728	1	7.480520	2.3×10^{-5}
1 gal (U.S.) = 3875.412	3.785412	231	0.1336806	1	3.07×10^{-6}
1 acre · ft = 1.23×10^9	1.23×10^6	75.3×10^6	43,560	325,851	1

Mass				
g	kg	oz	lb	ton
1 g = 1	10^{-3}	0.03527396	2.204623×10^{-3}	1.102311×10^{-6}
1 kg = 1000	1	35.27396	2.204623	1.102311×10^{-3}
1 oz (avdp) = 28.34952	0.02834952	1	0.0625	5×10^{-4}
1 lb (avdp) = 453.5924	0.4535924	16	1	0.0005
1 ton = 907,184.7	907.1847	32,000	2000	1

Système International d'Unités (SI)

The SI metric system is based on meter-kilogram-second units. The principal units applicable to water and wastewater engineering are listed in Tables A.2 and A.3. Derived SI units are consistent, since the conversion factor among various units is unity (e.g., 1 joule = 1 newton × 1 meter). Prefixes given in Table A.4 may be added to write large or small numbers and thus avoid the use of exponential values of 10. Groups of three digits, on either side of the decimal point, are separated by spaces. Tables A.5 and A.6 list common conversion factors from customary units to SI metric. [Reference: *Units of Expression for Wastewater Management*, Manual of Practice No. 6 (Washington, DC: Water Pollution Control Federation, 1982).]

TABLE A.2 Basic SI Units

Quantity	Unit	Symbol
Length	meter	m
Mass	kilogram	kg
Time	second	s
Thermodynamic temperature	Kelvin	K
Molecular weight	mole	mol
Plane angle	radian	rad

TABLE A.3 Derived SI Units

Quantity	Unit	Symbol	Formula
Energy	joule	J	$N \cdot m$
Force	newton	N	$kg \cdot m/s^2$
Power	watt	W	J/s
Pressure	pascal	Pa	N/m^2

TABLE A.4 Multiples and Submultiples of SI Units

Multiplier		Prefix	Symbol
1 000 000	$= 10^6$	mega	M
1 000	$= 10^3$	kilo	k
0.001	$= 10^{-3}$	milli	m
0.000 001	$= 10^{-6}$	micro	μ

TABLE A.5 Conversion Factors from English Units to SI Metric Units

Customary Units			Metric Units	
	Symbol	Multiplier	Symbol	
Description	Multiply ...	By ...	To Obtain ...	Reciprocal
Acre	acre	0.404 7	ha	2.471
British thermal unit	Btu	1.055	kJ	0.947 0
British thermal units per cubic foot	Btu/ft^3	37.30	J/l	0.026 81
British thermal units per pound	Btu/lb	2.328	kJ/kg	0.429 5
British thermal units per square foot per hour	$Btu/ft^2/hr$	3.158	$J/m^2 \cdot s$	0.316 7
Cubic foot	ft^3	0.028 32	m^3	35.31
Cubic foot	ft^3	28.32	l	0.035 31
Cubic feet per minute	cfm	0.471 9	l/s	2.119
Cubic feet per minute per thousand cubic feet	cfm/1000 ft^3	0.016 67	$l/m^3 \cdot s$	60.00
Cubic feet per second	cfs	0.028 32	m^3/s	35.31
Cubic feet per second per acre	cfs/acre	0.069 98	$m^3/s \cdot ha$	14.29
Cubic inch	$in.^3$	0.016 39	l	61.01
Cubic yard	yd^3	0.764 6	m^3	1.308
Fathom	f	1.839	m	0.546 7
Foot	ft	0.304 8	m	3.281
Feet per hour	ft/hr	0.084 67	mm/s	11.81
Feet per minute	fpm	0.005 08	m/s	196.8
Foot-pound	ft · lb	1.356	J	0.737 5
Gallon, U.S.	gal	3.785	l	0.264 2
Gallons per acre	gal/acre	0.009 35	m^3/ha	106.9
Gallons per day per linear foot	gpd/lin ft	0.012 42	$m^3/m \cdot d$	80.53
Gallons per day per square foot	gpd/ft^2	0.040 74	$m^3/m^2 \cdot d$	24.54
Gallons per minute	gpm	0.063 08	l/s	15.85
Grain	gr	0.064 80	g	15.43
Grains per gallon	gr/gal	17.12	mg/l	0.058 41
Horsepower	hp	0.745 7	kW	1.341
Horsepower-hour	hp · hr	2.684	MJ	0.372 5
Inch	in.	25.4	mm	0.039 37
Knot	knot	1.852	km/h	0.540 0
Knot	knot	0.514 4	m/s	1.944
Mile	mi	1.609	km	0.621 5
Miles per hour	mph	1.609	km/h	0.621 5
Million gallons	mil gal	3 785.0	m^3	0.000 264 2
Million gallons per day	mgd	43.81	l/s	0.022 82
Million gallons per day	mgd	0.043 81	m^3/s	22.82
Ounce	oz	28.35	g	0.035 27

TABLE A.5 (Continued)

Customary Units		Multiplier	Metric Units	
	Symbol		Symbol	
Description	Multiply ...	By ...	To Obtain ...	Reciprocal
Pound (force)	lbf	4.448	N	0.224 8
Pound (mass)	lb	0.453 6	kg	2.205
Pounds per acre	lb/ha	1.121	kg/ha	0.892 1
Pounds per cubic foot	pcf	16.02	kg/m^3	0.062 42
Pounds per foot	lb/ft	1.488	kg/m	0.672 0
Pounds per horsepower-hour	lb/hp · hr	0.169 0	mg/J	5.918
Pounds per square foot	lb/ft^2	0.047 88	kN/m^2	20.89
Pounds per square inch	psi	6.895	kN/m^2	0.145 0
Pounds per thousand cubic feet per day	$lb/1000\ ft^3/day$	0.016 02	$kg/m^3 \cdot d$	62.43
Square foot	ft^2	0.092 90	m^2	10.76
Square inch	$in.^2$	645.2	mm^2	0.001 550
Square mile	mi^2	2.590	km^2	0.386 1
Square yard	yd^2	0.836 1	m^2	1.196
Ton, short	ton	0.907 2	t	1.102
Yard	yd	0.914 4	m	1.094

Acceleration of gravity $g = 32.174\ ft/s^2 = 9.806\ 65\ m/s^2$.

TABLE A.6 Selected English–Metric Conversion Factors

	English Unit	Multiplier	Metric Unit
Mass	lb	0.453 6	kg
Length	ft	0.304 8	m
Area	ft^2	0.092 90	m^2
	acre	4 047	m^2
	acre	0.404 7	ha
Volume	gal	0.003 785	m^3
	gal	3.785	l
	ft^3	0.028 32	m^3
	ft^3	28.32	l
Velocity	fpm	0.005 08	m/s
Flow	mgd	3 785	m^3/d
	gpm	5.450	m^3/d
	gpm	0.063 09	l/s
	cfs	0.028 32	m^3/s
BOD loading	lb/1000 ft^3/day	16.02	$g/m^3 \cdot d$
	lb/acre/day	1.121	$kg/ha \cdot d$
Solids loading	lb/ft^2/day	4.883	$kg/m^2 \cdot d$
	lb/ft^3/day	16.02	$kg/m^3 \cdot d$
Hydraulic loading	gpd/ft^2	0.040 75	$m^3/m^2 \cdot d$
	gpm/ft^2	0.679 0	$l/m^2 \cdot s$
Concentration	lb/mil gal	0.119 8	mg/l

	Metric Unit	Multiplier	English Unit
Mass	kg	2.205	lb
Length	m	3.281	ft
Area	m^2	10.76	ft^2
	m^2	0.000 247	acre
	ha	2.471	acre
Volume	m^3	264.2	gal
	m^3	35.31	ft^3
	l	0.264 2	gal
	l	0.035 31	ft^3
Velocity	m/s	196.8	fpm
Flow	m^3/d	0.000 264 2	mgd
	m^3/d	0.183 5	gpm
	m^3/s	35.31	cfs
	l/s	15.85	gpm
BOD loading	$g/m^3 \cdot d$	0.062 43	lb/1000 ft^3/day
	$kg/ha \cdot d$	0.892 1	lb/acre/day
Solids loading	$kg/m^2 \cdot d$	0.204 8	lb/ft^2/day
	$kg/m^3 \cdot d$	0.062 42	lb/ft^3/day
Hydraulic loading	$m^3/m^2 \cdot d$	24.54	gal/ft^2/day
	$l/m^2 \cdot s$	1.473	gpm/ft^2
Concentration	mg/l	8.345	lb/mil gal

TABLE A.7 Relative Atomic Weights

Name	Symbol	Atomic Number	Atomic Weight	Name	Symbol	Atomic Number	Atomic Weight
Actinium	Ac	89	—	Mercury	Hg	80	200.59
Aluminum	Al	13	26.9815	Molybdenum	Mo	42	95.94
Americium	Am	95	—	Neodymium	Md	60	144.24
Antimony	Sb	51	121.75	Neon	Ne	10	20.183
Argon	Ar	18	39.948	Neptunium	Np	93	—
Arsenic	As	33	74.9216	Nickel	Ni	28	58.71
Astatine	At	85	—	Niobium	Nb	41	92.906
Barium	Ba	56	137.34	Nitrogen	N	7	14.0067
Berkelium	Bk	97	—	Nobelium	No	102	—
Beryllium	Be	4	9.0122	Osmium	Os	76	190.2
Bismuth	Bi	83	208.980	Oxygen	O	8	15.9994
Boron	B	5	10.811	Palladiium	Pd	46	106.4
Bromine	Br	35	79.904	Phosphorus	P	15	30.9738
Cadmium	Cd	48	112.40	Platinum	Pt	78	195.09
Calcium	Ca	20	40.08	Plutonium	Pu	94	—
Californium	Cf	98	—	Polonium	Po	84	—
Carbon	C	6	12.01115	Potassium	K	19	39.102
Cerium	Ce	58	140.12	Praseodymium	Pr	59	140.907
Cesium	Cs	55	132.905	Promethiuim	Pm	61	—
Chlorine	Cl	17	35.453	Protactinium	Pa	91	—
Chromium	Cr	24	51.996	Radium	Ra	88	—
Cobalt	Co	27	58.9332	Radon	Rn	86	—
Copper	Cu	29	63.546	Rhenium	Re	75	186.2
Curium	Cm	96	—	Rhodium	Rh	45	102.905
Dysprosium	Dy	66	162.50	Rubidium	Rb	37	85.47
Einsteinium	Es	99	—	Ruthenium	Ru	44	101.07
Erbium	Er	68	167.26	Samarium	Sm	62	150.35
Europium	Eu	63	151.96	Scandium	Sc	21	44.956
Fermium	Fm	100	—	Selenium	Se	34	78.96
Fluorine	F	9	18.9984	Silicon	Si	14	28.086
Francium	Fr	87	—	Silver	Ag	47	107.868
Gadolinium	Gd	64	157.25	Sodium	Na	11	22.9898
Gallium	Ga	31	69.72	Strontium	Sr	38	87.62
Germanium	Ge	32	72.59	Sulfur	S	16	32.064
Gold	Au	79	196.967	Tantalum	Ta	73	189.948
Hafnium	Hf	72	178.49	Technetium	Tc	43	—
Helium	He	2	4.0026	Tellurium	Te	52	127.60
Holmium	Ho	67	164.930	Terbiuim	Tb	65	158.924
Hydrogen	H	1	1.00797	Thallium	Tl	81	204.37
Indium	In	49	114.82	Thorium	Th	90	232.038
Iodine	I	53	126.9044	Thulium	Tm	69	168.934
Iridium	Ir	77	192.2	Tin	Sn	50	118.69
Iron	Fe	26	55.847	Titanium	Ti	22	47.90
Krypton	Kr	36	83.80	Tungsten	W	74	183.85
Lanthanum	La	57	138.91	Uranium	U	92	238.03
Lead	Pb	82	207.19	Vanadium	V	23	50.942
Lithium	Li	3	6.939	Xenon	Xe	54	131.30
Lutetium	Lu	71	174.97	Ytterbium	Yb	70	173.04
Magnesium	Mg	12	24.312	Yttrium	Y	39	88.905
Manganese	Mn	25	54.9380	Zinc	Zn	30	65.37
Mendelevium	Md	101	—	Zirconium	Zr	40	91.22

Source: "Report of the International Commission on Atomic Weights—1961," *J. Am. Chem. Soc.* 84 (1962): 4192.

TABLE A.8 Properties of Water in English Units

Temperature (°F)	Specific Weight γ (lb/ft^3)	Mass Density ρ (lb · sec^2/ft^4)	Absolute Viscosity μ (× 10^{-5} lb · sec/ft^2)	Kinematic Viscosity ν (× 10^{-5} ft^2/sec)	Vapor Pressure p_v(psi)
32	62.42	1.940	3.746	1.931	0.09
40	62.43	1.938	3.229	1.664	0.12
50	62.41	1.936	2.735	1.410	0.18
60	62.37	1.934	2.359	1.217	0.26
70	62.30	1.931	2.050	1.059	0.36
80	62.22	1.927	1.799	0.930	0.51
90	62.11	1.923	1.595	0.826	0.70
100	62.00	1.918	1.424	0.739	0.95
110	61.86	1.913	1.284	0.667	1.24
120	61.71	1.908	1.168	0.609	1.69
130	61.55	1.902	1.069	0.558	2.22
140	61.38	1.896	0.981	0.514	2.89
150	61.20	1.890	0.905	0.476	3.72
160	61.00	1.896	0.838	0.442	4.74
170	60.80	1.890	0.780	0.413	5.99
180	60.58	1.883	0.726	0.385	7.51
190	60.36	1.876	0.678	0.362	9.34
200	60.12	1.868	0.637	0.341	11.52
212	59.83	1.860	0.593	0.319	14.70

TABLE A.9 Properties of Water in SI Metric Units

Temperature (°C)	Specific Weight γ (kN/m^3)	Mass Density ρ (kg/m^3)	Absolute Viscosity μ (× 10^{-3} kg/m · s)[a]	Kinematic Viscosity ν (× 10^{-6} m^2/s)	Vapor Pressure p_v(kPa)
0	9.805	999.8	1.781	1.785	0.61
5	9.807	1000.0	1.518	1.518	0.87
10	9.804	999.7	1.307	1.306	1.23
15	9.798	999.1	1.139	1.139	1.70
20	9.789	998.2	1.002	1.003	2.34
25	9.777	997.0	0.890	0.893	3.17
30	9.764	995.7	0.798	0.800	4.24
40	9.730	992.2	0.653	0.658	7.38
50	9.689	988.0	0.547	0.553	12.33
60	9.642	983.2	0.466	0.474	19.92
70	9.589	997.8	0.404	0.413	31.16
80	9.530	971.8	0.354	0.364	47.34
90	9.466	965.3	0.315	0.326	70.10
100	9.399	958.4	0.282	0.294	101.33

[a] N · s/m^2

TABLE A.10 Saturation Values of Dissolved Oxygen in Water Exposed to Water-Saturated Air Containing 20.90% Oxygen Under a Pressure of 760 mm Hg[a]

Temperature (°C)	Chloride Concentration in Water (mg/l): 0	5000	10,000	Difference per 100 mg Chloride	Temperature (°C)	Vapor Pressure (mm)
	Dissolved Oxygen (mg/l)					
0	14.6	13.8	13.0	0.017	0	5
1	14.2	13.4	12.6	0.016	1	5
2	13.8	13.1	12.3	0.015	2	5
3	13.5	12.7	12.0	0.015	3	6
4	13.1	12.4	11.7	0.014	4	6
5	12.8	12.1	11.4	0.014	5	7
6	12.5	11.8	11.1	0.014	6	7
7	12.2	11.5	10.9	0.013	7	8
8	11.9	11.2	10.6	0.013	8	8
9	11.6	11.0	10.4	0.012	9	9
10	11.3	10.7	10.1	0.012	10	9
11	11.1	10.5	9.9	0.011	11	10
12	10.8	10.3	9.7	0.011	12	11
13	10.6	10.1	9.5	0.011	13	11
14	10.4	9.9	9.3	0.010	14	12
15	10.2	9.7	9.1	0.010	15	13
16	10.0	9.5	9.0	0.010	16	14
17	9.7	9.3	8.8	0.010	17	15
18	9.5	9.1	8.6	0.009	18	16
19	9.4	8.9	8.5	0.009	19	17
20	9.2	8.7	8.3	0.009	20	18
21	9.0	8.6	8.1	0.009	21	19
22	8.8	8.4	8.0	0.008	22	20
23	8.7	8.3	7.9	0.008	23	21
24	8.5	8.1	7.7	0.008	24	22
25	8.4	8.0	7.6	0.008	25	24
26	8.2	7.8	7.4	0.008	26	25
27	8.1	7.7	7.3	0.008	27	27
28	7.9	7.5	7.1	0.008	28	28
29	7.8	7.4	7.0	0.008	29	30
30	7.6	7.3	6.9	0.008	30	32

[a] Saturation at barometric pressures other than 760 mm (29.92 in.), C'_σ is related to the corresponding tabulated values C_s by the equation

$$C'_\sigma = C_s \frac{P - p}{760 - p}$$

where

C'_σ = solubility at barometric pressure P and given temperature, mg/l
C_s = saturation at given temperature from table, mg/l
P = barometric pressure, mm
p = pressure of saturated water vapor at temperature of the water selected from table, mm

Index

A

D